ANNUAL REVIEW OF
BIOCHEMISTRY

EDITORIAL COMMITTEE (1998)

ANNUAL REVIEW OF BIOCHEMISTRY

VOLUME 67, 1998

CHARLES C. RICHARDSON, *Editor*
Harvard Medical School

JOHN N. ABELSON, *Associate Editor*
California Institute of Technology

CHRISTIAN R. H. RAETZ, *Associate Editor*
Duke University Medical Center

JEREMY W. THORNER, *Associate Editor*
University of California, Berkeley

http://www.AnnualReviews.org   science@annurev.org   650-493-4400

ANNUAL REVIEWS   4139 EL CAMINO WAY   P.O. BOX 10139   PALO ALTO, CALIFORNIA 94303-0139

ANNUAL REVIEWS
Palo Alto, California, USA

International Standard Serial Number: 0066-4154
International Standard Book Number: 0-8243-0867-0
Library of Congress Catalog Card Number: 32-25093

Annual Review and publication titles are registered trademarks of Annual Reviews.

⊗ The paper used in this publication meets the minimum requirements of American National Standard for Information Sciences—Permanence of Paper for Printed Library Materials. ANSI Z39.48-1992.

Annual Reviews and the Editors of its publications assume no responsibility for the statements expressed by the contributors to this *Annual Review*.

TYPESET BY TECHBOOKS, FAIRFAX, VA
PRINTED AND BOUND IN THE UNITED STATES OF AMERICA

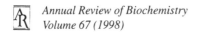 *Annual Review of Biochemistry*
Volume 67 (1998)

CONTENTS

SOME RELATED ARTICLES IN OTHER *ANNUAL REVIEWS*

From the *Annual Review of Biophysics and Biomolecular Structure*, Volume 27, 1998:

Structure, Dynamics, and Function of Chromatin in Vitro, Johnathan Widom

Crystallographic Structures of the Hammerhead Ribozyme: Relationship to Ribozyme Folding and Catalysis, Joseph E. Wedekind, David R. McKay

Minor Groove-Binding Architectural Proteins: Structure, Function, and DNA Recognition, Carole A. Bewley, Angela M. Gronenborn, and G. Marius Clore

RNA Recognition by RNP Proteins During RNA Processing, Gabriele Varani and Kiyoshi Nagai

The Three-Dimensional Structure of the Ribosome and Its Components, Peter B. Moore

DNA Nanotechnology: Novel DNA Constructions, Nadrian C. Seeman

Inhibitors of HIV-1 Protease: A Major Success of Structure-Assisted Drug Design, Alexander Wlodawer and Jiri Vondrasek

From the *Annual Review of Cell and Developmental Biology*, Volume 13, 1997:

Genetics of Transcriptional Regulation in Yeast: Connections to the RNA Polymerase II CTD, Marian Carlson

The Design Plan of Kinesin Motors, Ronald D. Vale, Robert J. Fletterick

Bacterial Cell Division, David Bramhill

Implications of Atomic-Resolution Structures for Cell Adhesion, Daniel J. Leahy

Adipocyte Differentiation and Leptin Expression, Cheng-Shine Hwang, Thomas M. Loftus, Susanne Mandrup, and M. Daniel Lane

Cyclin-Dependent Kinases: Engines, Clocks, and Microprocessors, David O. Morgan

Initiation of DNA Replication in Eukaryotic Cells, Anindya Dutta and Stephen P. Bell

Plant Cell Morphogenesis: Plasma Membrane Interactions with the Cytoskeleton and Cell Wall, John E. Fowler and Ralph S. Quatrano

From the *Annual Review of Genetics*, Volume 31, 1997:

Viral Transactivating Proteins, Jane Flint and Thomas Shenk

Understanding Gene and Allele Function with Two-Hybrid Methods, Roger Brent and Russell L. Finley, Jr.

Genetics of Prions, Stanley B. Prusiner and Michael R. Scott

(*continued*)

Annu. Rev. Biochem. 1998. 67:xiii–xxxii

AN ACCIDENTAL BIOCHEMIST

Edwin G. Krebs

Departments of Pharmacology and Biochemistry, University of Washington, Seattle, Washington 98195

KEY WORDS: phosphorylase, protein phosphorylation, glyceraldehyde 3-phosphate
              dehydrogenase, cyclic AMP-dependent protein kinase, MAP kinase
              kinase, growth factors

CONTENTS

INTRODUCTION

One of the topics usually discussed in a prefatory chapter of the *Annual Review of Biochemistry* is how the author decided to become a biochemist. This section is preceded by a short family history, accounts of pertinent childhood events, and aspects of schooling that are deemed important. Specific teachers and others who were influential in the life of the writer are identified and their roles discussed. Eventually the chapter turns to the scientific fields in which the author worked. The writer's own scientific contributions are then discussed

xiii

in an appropriately modest vein. I follow this same pattern more or less, but I am struck by the fact that in contrast to writing a scientific paper, for which one has a comprehensive set of written notes and descriptions of experiments to draw upon, here one relies largely on memory and secondhand accounts of what occurred. Senior scientists, who are usually the ones asked to write prefatory chapters, often have poor memories, or they may be prone to present events in a manner that puts them in a favorable light. This last problem is particularly troublesome when career choices are discussed, because a natural tendency is to make one's decisions appear to be based on a logical process rather than the result of chance events. I have tried to tell it the way it was.

FAMILY BACKGROUND AND EARLY SCHOOLING

I was born on June 6, 1918, in Lansing, Iowa, the third of four children of Louise Helen (Stegeman) Krebs and William Carl Krebs, whose families had settled in Wisconsin after immigrating to the United States from Germany in the latter part of the nineteenth century. My grandfather Stegeman was a harness maker and my grandfather Krebs a house painter. My father, a minister, received his training at Moravian Seminary in Bethlehem, Pennsylvania, but after a few years in various Moravian churches in southern Wisconsin, he switched to the Presbyterian church and the family moved to Lansing, Iowa. My mother attended Milwaukee Normal School and taught school for several years prior to her marriage. In 1919 the family moved from Lansing to Newton, Illinois, where we stayed until 1925, at which time we moved to Greenville, Illinois. I went through grade school and one year of high school in Greenville, which I always think of as my hometown; I still like to visit there from time to time when something takes me to St. Louis, which is about 50 miles away. Greenville is a small college town located in a pleasant region of southern Illinois. Its schools were and probably still are very good, and it had a free Carnegie library, which I visited many times. For reasons I don't recall, I was fascinated by the Civil War and remember reading virtually every book that they had on this subject. I don't believe, however, that I had any particular interest in science, and the closest that I came to participating in anything of a scientific nature was building a shortwave radio receiver and becoming a licensed ham radio operator (W9SIX). My interest in ham radio stemmed from a desire to be able to communicate with a friend, also a ham, who had moved to Chicago. I dabbled in chemistry to the extent of learning how to make gunpowder, the ingredients of which were available from the local drugstore—even to ten-year-olds. If I wanted the gunpowder to burn with a red flame, I added some strontium nitrate obtained from my brother's Gilbert chemistry set. As a kid I loved to fish and swim in local streams and go on hikes in the country. Although I joined the

Boy Scouts and attended meetings for a year or more, I dropped out because belonging to this organization wasn't necessary to gain access to the out-of-doors. In addition, the Scouts started holding their meetings in churches, which took some of the fun out of it for me.

I have very positive feelings about my elementary school years. Not that I wasn't like all the other kids and hated to see school start in the fall and rejoiced when it was over in the spring, but most of the time, I found it very satisfying. Both of my parents, especially my mother, encouraged all of us to do as well as possible in school, and it was made clear that school comes first. My receiving a good report card was rewarded with a quarter unless the good grades were offset by comments such as "whispers too much" and "doesn't always pay attention." I liked my teachers and can still readily recall their names. The happy feelings that I had about my elementary education probably set the stage for my spending my entire life in schools of one sort or another.

URBANA HIGH SCHOOL AND THE UNIVERSITY OF ILLINOIS

In 1933, when I was 15, my father died. This unexpected event affected me strongly. Although I had absolutely no desire to follow in my father's footsteps and become a minister, I had great affection for him, and I especially admired his outside pursuits. Dad was an excellent carpenter, painted in oils (landscapes), and loved to garden. My mother was devastated by his death, but she soon began to pick up the pieces and started planning for the future. Uppermost in her mind was completion of the education of her children. It was decided that the family should move to Urbana, Illinois, where my brothers were already students at the University of Illinois (U of I). One brother had just finished undergraduate work in chemical engineering but wanted to go on to graduate school. The other was majoring in chemistry and biology. My younger sister was ready to enter grade school, and I was about to start my second year of high school. Living in Urbana was ideal for us because the main cost of higher education at that time was the living expense; tuition was only $35 a semester.

There were striking differences between Urbana High School and the high school I had been attending in Greenville. Part of this was due to the relative sizes of the two schools, but another difference stemmed from the fact that Urbana High was strongly influenced by the presence of the U of I. A large fraction of my high school class were headed for college, a situation that affected the general atmosphere of the school. The faculty were excellent and were closely attuned to the immediate requirements we would be facing in college. Apart from course work, I plunged into a variety of extracurricular activities including the band (I played the tuba), the school paper, plays, and various clubs, including the ham radio club. It may seem peculiar, but I did not belong to the science club.

My father's death marked the point at which I began to think seriously about what sort of a career I should follow. Preoccupation with this problem was to continue for many years. This was the 1930s, the time of the Great Depression, and young people were worried about what they could do just to make a living. My own family's financial situation was precarious, which did not help my feelings of insecurity. Although I would hardly compare our status with that of the destitute Irish family in *Angela's Ashes*, we certainly could not have been considered well off. Widows' pensions from the Presbyterian church were strictly borderline, and the only kind of work I could find was a few lawn jobs. My early attempts at door-to-door selling were remarkably unsuccessful, perhaps because my standard opening was, "You wouldn't want to buy any magazines—or books—or Christmas cards—or whatever—would you?" I remember how the family rejoiced when my older brother was awarded a $600 per year teaching assistantship after his first year in graduate school. It is perhaps no wonder that I entered college with the idea that whatever course of study I followed, it had to be one that would lead to some definite job. For better or for worse, I did not think of college as a place to take courses that would simply enrich my life; instead I felt that it should also serve as a trade school. The trouble was that I couldn't decide what trade I should follow. In 1937, I was greatly impressed when my brother got what was to me a fantastic job with Standard Oil of New Jersey immediately after he received his PhD in chemical engineering. My other brother, who at this point was not fully decided as to what he should do, was also leaning toward chemistry.

I took myself far too seriously at this stage in life and failed to grasp the flexibility that exists in schooling. My experience as a salesman had assured me that I wasn't cut out for the business world, and, indeed, I even wondered whether business, as we knew it, would survive the Depression. My political leanings were fairly far to the left, but I was not a joiner and didn't belong to any organizations that were later to be investigated by Senator McCarthy. Among possible careers that I found myself considering were law and medicine. These professions appeared to be more people oriented than a career in science, although the latter was appealing because it seemed solid and dependable. A dull lawyer, speaking at a vocational guidance conference, made law sound so boring that I soon dropped that idea. Medicine and science, then, remained the best possibilities, and the courses that I took during my first two years of college—mathematics, various sciences, foreign language—were suitable for either of these fields. Although students were forced to declare majors eventually, I discovered something called individual curriculum at the U of I. Under this plan, which was available to students whose grades were good enough, I found that I could take whatever I needed to prepare myself either for medical school or for graduate work in chemistry. The added flexibility that the individual curriculum offered allowed me to take professional series chemistry, physics, and math courses and still have time for a sufficient number of biology courses to qualify for medical school. What I lost out on was that I took very few cultural courses, something that I have always regretted.

Of considerable importance in my senior year at the U of I was that I enrolled for undergraduate thesis research in organic chemistry and had as my mentor the late Harold Snyder. I thoroughly enjoyed this new type of laboratory experience. In addition, I worked as a dishwasher on an NYA job (a New Deal program) under Charles Price, who was studying the role of organic peroxides in the initiation and termination of polymerization reactions. Price indicated that it would be more fun for me if I simply did the dishes in my spare time and during scheduled hours helped him with his research. This experience was wonderful, and as it later turned out, it resulted in my being included as a contributor on several of his papers. Although my role in this research was not great, I remember being excited by seeing my name in print as an author (1). For the most part, these research experiences took place after I was already committed to attending medical school, but they were clearly important with respect to choices that I was to make later on. In addition to Harold Snyder and Charles Price, another member of the organic chemistry faculty at Illinois who influenced me greatly was Carl S (Speed) Marvel, who was a superb teacher and, in addition, showed a special interest in students. Marvel was instrumental in arranging for teaching assistantships at various universities for those who planned to go on for PhDs, and he was kind enough to reserve one of these for me in the event that I decided to go to graduate school instead of medical school. However, the latter was my first choice, assuming that I could get financial help. If not, then I would become an organic chemist. As it turned out, I was awarded a scholarship at Washington University and from that point on started thinking of myself as a future physician.

MEDICAL SCHOOL AND RESIDENCY TRAINING

While still a senior at the U of I, I visited St. Louis and met Philip A Shafer, dean of Washington University School of Medicine and also chairman of the Department of Biological Chemistry. It was very unusual for a PhD bio-chemist to serve as the dean of a medical school, but as I was soon to learn, Washington University was a highly research-oriented institution; perhaps it was not unique in this respect, but it came as a happy surprise to me. I had opted for medical school rather than graduate school, naively thinking that I had now made a firm decision to become a practicing physician. Dean Shafer, however, made it clear that in addition to training students for medical practice, Washington University was dedicated to training students for research in the various fields of medicine. We discussed research versus practice, but I never felt that he was trying to pressure me in any way. Nonetheless, I went back to Urbana to finish my senior year with the realization that I might still have a few decisions to make in life.

The first year of medical school involved intensive course training in basic human biology. As an extracurricular activity, I worked on a research project under the direction of Dean Shafer, but he was so heavily occupied with other duties that I saw comparatively little of him. I did, however, become acquainted with members of the Department of Pharmacology, which was headed by Carl Cori, and I was invited to attend their weekly seminar series. I met Gerty Cori, as well as Arda Green, who worked closely with her. It was from Arda that I first heard about the enzyme phosphorylase, and I recall her showing me one of the first crystalline preparations of that enzyme (2). Others whom I knew in Pharmacology were Sidney Colowick, long associated with the Coris, and Earl Sutherland, who was a year ahead of me in medical school and served as a student assistant in the department; we often played tennis together.

Although these contacts with the laboratory rekindled my interest in research, any serious thoughts that I might have had about research were abruptly terminated on December 7, 1941, with the start of the US involvement in World War II. Essentially all of the medical students joined the Army or Navy as reserve officers, which exempted us from active military duty until we had finished our medical training. I opted for the Naval Reserve and plunged into preparation for my future role as a naval medical officer. During this period I missed out on one trimester of the regular curriculum, because the school asked me to help teach biochemistry. Later, we found out from the Navy that the school had no authority to let me do this and that I must graduate with the rest of my class. I was never required to make up the work that I had missed. The result of this irregularity was that I missed the surgical service in which students learned how to put in sutures and sew up wounds. This proved to be a little embarrassing some years later when I had independent duty as a medical officer aboard ship.

Under wartime rules, medical school graduates were required to take a minimum of 9 months of internship and a maximum of 18 months of assistant residency after graduation before going on active duty. I opted for 9 months of internship and 9 months of assistant residency, which I took in internal medicine at Barnes Hospital in St. Louis. In addition to teaching practical aspects of patient care, these services were excellent from an academic standpoint: Each patient was considered to be unique and was viewed almost like a research problem. My most important professors during this period were Dr. WB Wood, chairman of the Department of Medicine, and Dr. Carl Moore, chief of the Hematology

Division; both were excellent teachers and wonderful human beings. My earlier contact with the Department of Pharmacology in the medical school enabled me to utilize the department's newly acquired Tiselius apparatus, so I was able to obtain free boundary electrophoretic patterns of blood sera from several of my patients. This type of analysis was a first at Barnes Hospital and won me quite a few brownie points from my supervisors. During this period, I carried out a short study on the effects of malnutrition on serum protein patterns, and the study was eventually published (3). As a result of my internship and assistant residency in internal medicine I obtained the training that was needed to take care of one category of patients, but I still had had essentially no experience in surgery. This deficiency did not appear to be a source of concern to the US Navy.

A major event in my life that occurred during this period was my marriage to Virginia (Deedy) Frech. She was a student in the Washington University School of Nursing and I was working at Barnes Hospital when we met. Deedy and I have three children and five grandchildren, and we are currently a few years past our fiftieth wedding anniversary. She has always been extremely supportive of me, not only from a personal standpoint but also with respect to my career. The latter form of support became apparent almost immediately and has continued to the present day. Without it, I have no idea where I might have strayed.

ACTIVE DUTY IN THE NAVY

I reported for active duty shortly before the end of World War II and was then obliged to serve for approximately a year after the war was over. My duties were light, but I had an opportunity to see other parts of the world, and I also learned how to play poker. A memorable part of this period was the time that I spent as the sole medical officer on the USS *Sirona*, an AKA, in the Pacific. Here my lack of training in surgery finally caught up with me, and I had to depend on my chief pharmacist mate to teach me how to sew up lacerations. Interestingly, one of my patients, who had suffered a very severe cut on his face that required at least 20 sutures, wrote to me 46 years later on the occasion of my name appearing in his local newspaper. He identified himself and expressed his gratitude for the fine job that I had done in taking care of him. I suspect that he had derived satisfaction over the years from the fact that he carried such obvious evidence of a wound received during the war.

POSTDOCTORAL TRAINING IN BIOCHEMISTRY

After being released from the Navy, I returned to St. Louis to continue my training in internal medicine. At this point, it was my intent to become an academic internist. However, I found that I would have had to wait at least two years before I could be accommodated in the residency program because a system had been established to give priority to those whose training had been interrupted early in the war. Dr. Wood advised me that this would be an ideal time for me to obtain more experience in biochemistry, i.e. as an adjunct to my career in medicine. Carl Cori, who was now chairman of the Department of Biological Chemistry, agreed to take me, and in fall 1946 I took up my duties as a postdoctoral fellow.

My first research problem under the Coris was to carry out phase rule solubility studies on rabbit skeletal muscle glyceraldehyde 3-phosphate dehydrogenase or triose phosphate dehydrogenase (TPD), which they had just crystallized, in order to determine whether their preparation was pure. This approach was subsequently extended to skeletal muscle phosphorylase. Neither protein behaved the way a pure substance was supposed to, but the study nonetheless taught me how to prepare and handle proteins. Another project that I undertook was to compare the properties of muscle and yeast TPD, the latter having been purified earlier by Warburg. In this study I was joined by Victor Najjar, who became a very good friend and with whom I thoroughly enjoyed working. We made antibodies to the yeast enzyme and found that they would inhibit its activity but that they did not react with the corresponding muscle enzyme (4). As a final project in the Coris' laboratory, I studied a biphasic effect that protamine had on muscle phosphorylase b and phosphorylase a activities measured in the presence and absence of AMP. The Km of phosphorylase b for AMP was markedly reduced by protamine, and in its presence, IMP would also activate the enzyme. At this time (1947) the concepts of allosterism and conformational changes in proteins were unknown, and at first the Coris hesitated to let me publish these purely "phenomenological" findings.

During the years that I spent in the Coris' laboratory (1946–1948), I was in good company. In addition to the Coris, faculty members in the department were Sidney Velick, John Taylor, Mildred Cohn, Earl Sutherland, Barbara Illingworth, and David Brown. The postdoctoral fellows and visiting scientists included, in addition to Victor Najjar, Arthur Kornberg, Christian de Duve, Joe Lampen, Rollo Park, Henry Sable, Michael Krahl, Thadeus Baranowski, Shlomo Hestrin, and Milton Stein. Our major instruments were a Beckman DU spectrophotometer, a Klett colorimeter, two Beckman pH meters, a Sorval tabletop centrifuge, and the slower-speed International centrifuges. A Warburg apparatus was still available, if needed. Chromatography was in its infancy. As mentioned above, there was also a Tiselius apparatus, which was very useful. I don't recall any use being made of radioactive isotopes, although a mass spectrometer was being built by Mildred Cohn in anticipation of work using heavy isotopes.

I can't say enough about the benefit that I derived from my training with the Coris, yet it is difficult for me to pinpoint exactly what about their laboratory made it such a good place. Both Carl and Gerty set very high standards with respect to the quality of the research that was going on, and nothing was ever accepted as having been established until all possible evidence was brought to bear on the problem. Everyone in the laboratory worked very hard, but nobody thought of work as drudgery, because it led to new results, and new results were exciting. Gerty's highest accolade for a fellow scientist was that he or she was a good worker. We were given great freedom with respect to what we chose to work on, except that problems were examined as to whether they were appropriate for the laboratory. We were encouraged to pursue problems that were new and original; the Coris had no use for anyone who jumped on the bandwagon and did "me too" research. It was absolutely essential that we always give credit to other scientists for their contributions. Access to either Carl or Gerty was always available if one really had something to say or show to them, but one didn't interrupt them just to engage in idle chatter. I remember the time when I first went to see Carl Cori about carrying out postdoctoral work just after getting out of the Navy. Not having seen him for two or more years, I was engaging in the usual pleasantries that I considered appropriate for the occasion, when he asked somewhat abruptly, "Are you here just to pass the time of day, or do you want to talk business?" But Carl mellowed considerably as time went by.

Toward the end of my time in the Coris' laboratory I started giving serious thought to what I would do after my two years were over. The most likely course for me to follow was to return to residency training in internal medicine at Barnes Hospital. To my surprise, however, I found that I now had new opportunities based on my training in biochemistry, as brief as it had been. For example, the chairman of a department of pathology wanted an MD with biochemical training to join his department as an assistant professor; the person selected could then learn pathology "on the job." A research institute in Cleveland wanted a physician with biochemical training to carry out research on hypertension. Finally, I learned that even departments of biochemistry considered two years of postdoctoral work, combined with a medical degree, sufficient training for a faculty appointment. (The job market at that time was obviously not as competitive as it is today.) A friend of mine in the Department of Chemistry at the University of Washington (U of W), Arthur G Anderson, informed me that a new medical school was being established in Seattle and that he would be happy to recommend me for a position to Earl Norris, the acting chairman of the new Department of Biochemistry that had been created.

While I was in the Navy, my ship had visited Seattle, and although it was January, the sky was blue, the weather absolutely delightful, and the mountains brilliant: Nature had conspired to make Seattle look like the garden spot of the universe. Hearing that a job might be there, I couldn't help but be interested. To make a long story short, I visited, was offered a job, and accepted. A little rationalization soon made me sure that becoming an assistant professor of biochemistry was the logical thing for me to do. Although I hated to give up internal medicine, I would also have hated to give up biochemistry, which I was enjoying. To be on the safe side, however, I took the requisite examinations necessary for me to obtain a license to practice medicine in Washington should I be unsuccessful as a biochemist. As it turned out, I had taken care of my last patient while I was in the Navy. The license was nice to have, however, because I could write my own prescriptions for family use and it made me feel secure.

THE UNIVERSITY OF WASHINGTON, 1948–1968

The Department of Biochemistry in the School of Medicine

Becoming a faculty member in a well-established biochemistry department would in itself have been a big step, but my joining an undeveloped department in a newly formed medical school might have turned out to be a disaster; fortunately it wasn't. By 1950, Hans Neurath had come to Seattle as the first permanent chairman of the department and things got under way. Hans (Dr. Neurath, at first, although he soon let us know that his first name was Hans) was an excellent chairman and was unusually effective in that position. He was appropriately aggressive, when the occasion warranted it, but was open and fair in his dealing with people. More important than his administrative skills, however, was the example that he set in research and other academic aspects of the department. Even before Hans arrived, several of us had obtained research grants and had our programs under way. I well remember the occasion in 1949 when Donald Hanahan, one of the other new biochemists at the U of W, came by my office and asked whether I had heard about National Institutes of Health (NIH) research grants. He indicated that all one had to do to get one of them was to write a letter explaining what he or she proposed to work on. So I took an afternoon off, composed a letter, and in a few months I had my grant!

Research on Glyceraldehyde 3-Phosphate Dehydrogenase

Glyceraldehyde 3-phosphate dehydrogenase (TPD), with which I was familiar, became the focus of my first research at the U of W. I determined that an unknown crystalline protein, "Yeast Protein No. 2," which had been isolated at the Rockefeller Institute, was, in reality, TPD, which I had been working on all along. Gale Rafter, my first graduate student, was the one who made the identification. I recall his coming into my office and saying, "The crystals of Protein No. 2 look just like those of TPD, the enzyme that you are studying." In one project I carried out on TPD, it was determined that if glucose fermentation, catalyzed by fresh yeast extract as a source of glycolytic enzymes, was inhibited using antibody to the dehydrogenase, one could then reestablish glycolysis by adding the rabbit skeletal muscle dehydrogenase. Based on studies utilizing free boundary electrophoresis, multiple forms of TPD were determined to be present in yeast; surprisingly, this early demonstration of what later became known as isoenzymes attracted very little attention. Finally, I discovered a slow structural modification reaction of NADH that was catalyzed of TPD. This conversion of NADH to "NADH-X," the name I used for the product, was a hydration reaction. DPNH-X became my major area of research because I felt that it might have some role in oxidative phosphorylation. This problem was eventually discontinued, however, so that I would have more time available for work on glycogen phosphorylase. I was sorry to leave the problem because we had just learned that NADH-X could be converted back to NADH in the presence of ATP and a yeast enzyme (5). In the DPNH-X work, I was assisted by my first two technicians, Doris Cleveland and Jane Crosby; my second graduate student, Sterling Chaykins; and two postdoctoral associates, Dr. Josephine Junge and Dr. Josephine Meinhart.

Conversion of Phosphorylase b to Phosphorylase a

In 1953, Edmond H Fischer joined the Department of Biochemistry at the U of W as our seventh faculty member. The others, in addition to Hans Neurath and me, were Donald Hanahan, Walter Dandliker, Frank Huennekens, and Philip Wilcox. Ed Fischer had worked on potato glycogen phosphorylase before coming to Seattle, and I had worked on the rabbit muscle enzyme, as mentioned earlier. We discussed the differences between plant and animal phosphorylases and the possible existence of a prosthetic group that might be present in form a of animal phosphorylase. Based on the work of the Coris, the removal of a prosthetic group by the PR (prosthetic group–removing) enzyme was believed to account for the conversion of phosphorylase a to phosphorylase b, the latter form requiring AMP for activity. It was believed that the prosthetic group was some form of AMP. As a result of our discussions, we decided to initiate a joint project centered on the role of AMP in the activation of phosphorylase b. Thus was born a collaboration that has lasted for about 50 years!

The first step in our joint research was to isolate phosphorylase a according to the published procedure of Green & Cori (6). To our surprise, and particularly surprising to me because I had isolated phosphorylase a a number of times while I was with the Coris, we could not obtain any phosphorylase a. Activity tests showed that all of the phosphorylase that we could extract from muscle was in the b form. Insofar as we could tell, the only step of our procedure that deviated from the Green & Cori procedure was that we centrifuged crude rabbit muscle extract instead of clarifying it by filtering through coarse paper. We had elected to alter this step, because the centrifuge equipment that we had in 1953 was much better than that available 10 years earlier, when Green & Cori had developed their isolation procedure. When we carried out the preparation exactly as they had described it, we obtained phosphorylase *a*! It was apparent that phosphorylase *b* was being converted to phosphorylase *a* in vitro as a result of a reaction triggered by filtration. We found that this reaction did not occur if muscle extracts were aged prior to filtration, which suggested that ATP was required for the conversion. We then realized that we were probably dealing with a kinase reaction in which an enzyme, phosphorylase kinase, was catalyzing the phosphorylation of phosphorylase *b* to yield phosphorylase *a* (7). The role of the filtration step, carried out with unwashed filter paper, turned out to be that it provided Ca^{2+} ions needed to obtain the kinase in an active form. Eventually it was determined that calcium ions brought about the activation of the kinase as a result of two different mechanisms: first, a mechanism that required a protein factor, the kinase activity factor (KAF), and second, a direct requirement of phosphorylase kinase for calcium ions (8). KAF was eventually shown (9) to be a Ca^{2+}-dependent protease (calpain). The conversion of phosphorylase *a* to phosphorylase *b* was found to be the result of a phosphatase reaction, i.e. the PR enzyme was in fact a protein phosphatase. Our inability to isolate phosphorylase in its *a* form, which is what led us to the finding that the *b* to *a* reaction is due to protein phosphorylation, was a classic example of how a failure to obtain an expected result is often something to rejoice about.

The demonstration that the interconversion reactions of glycogen phosphorylase are enzyme-catalyzed phosphorylation-dephosphorylation reactions constituted the first demonstration of protein phosphorylation as a dynamic process involved in the regulation of protein function. In addition to the work that Ed Fischer and I performed with the skeletal muscle enzyme, a parallel study was being carried out independently by Earl Sutherland, together with his coworkers Walter Wosilait and Theodore Rall, on liver phosphorylase. These investigators showed that liver phosphorylase is activated as a result of a protein phosphorylation reaction and is inactivated by a protein phosphatase. Another very important early finding with respect to the phosphorylation and dephosphorylation of proteins was the very early study of Burnett & Kennedy (10) on the phosphorylation and dephosphorylation of casein by a protein kinase present in their tissue preparation.

The Cyclic AMP-Dependent Protein Kinase

In their work on the interconversion reactions of liver phosphorylase, Sutherland and his coworkers discovered cyclic AMP (cAMP) as a heat-stable factor that serves as an intracellular second messenger for epinephrine and glucagon in promoting the phosphorylation and activation of phosphorylase. This effect could have occurred as a result of either inhibition of phosphorylase phosphatase or activation of phosphorylase kinase. Ed Fischer and I determined that the cAMP effect could also be observed in skeletal muscle extracts and that the target in this case was phosphorylase kinase, which was shown to exist in a highly active phosphorylated form and a less active nonphosphorylated form. cAMP accelerated the activation of phosphorylase kinase in a reaction that required ATP (11); a key coworker in establishing this point was my graduate student, Donald A Graves. (Many years later Don's son, Lee, was to join my laboratory as a postdoctoral fellow.)

It took us several years to establish that the effect of cAMP on the phosphorylation and activation of phosphorylase kinase was due to a separate protein kinase, the cAMP-dependent protein kinase (PKA), present as a contaminant in partially purified phosphorylase kinase preparations. The problem was complicated because phosphorylase kinase also undergoes an autophosphorylation reaction that causes its activation. This difficulty and others were overcome by Donal A Walsh and John P Perkins, fellows in my laboratory during the mid- to late 1960s (12). PKA was found to catalyze the phosphorylation of proteins other than phosphorylase kinase; hence we proposed the general name rather than calling the new enzyme phosphorylase kinase kinase. The discovery of PKA established the existence of the first protein kinase cascade in which one kinase phosphorylates and activates a second kinase. Key technical assistance for work on PKA was provided by Carol Ann Leavitt. Paul Greengard was the first to show that PKA is widely distributed, and he emphasized the broad nature of its functions (13).

The discovery of PKA in 1968 stimulated work on the protein phosphorylation-dephosphorylation process in general. Prior to this time essentially all of the known protein phosphorylation reactions had involved enzymes important with respect to glycogen metabolism, and the idea prevailed that this sort of regulatory mechanism might be limited to that area. However, because it was well known that cyclic AMP is involved in many diverse processes, the existence of a protein kinase that served to mediate its functions suggested a broad role for protein phosphorylation. Around this same time (late 1960s), Lester Reed and his group showed that pyruvate dehydrogenase, an enzyme clearly outside the sphere of glycogen metabolism, is regulated by protein phosphorylation. In addition, Tom Langan had just demonstrated that histones undergo phosphorylation. In any event, by the beginning of the 1970s the field of protein phosphorylation was ready to take off.

THE UNIVERSITY OF CALIFORNIA, DAVIS, 1968–1977

In 1968 I moved to the University of California, Davis, to become chairman of the Department of Biological Chemistry in a new medical school that was being established. For several years I had been interested in taking a chairmanship, the motivation for this being an interest that I had in teaching biochemistry for medical students, and it was my feeling that only by chairing a department could I give full rein to whatever ideas I might have in this regard. The attractiveness of Davis stemmed partly from the fact that it was a new school, not too different in this respect from what the University of Washington had been in 1948, and partly because the person in charge of The Sciences Basic to Medicine at Davis was a former colleague, Loren Carlson, whom I had known and respected when he was in the Department of Physiology at the University of Washington. The dean of the new school was CJ Tupper, whom I met and had confidence in. Another factor was that some of the problems that would ordinarily beset one in setting up a department in a new school were softened by the fact that Paul Stumpf, chairman of the Department of Biochemistry in the College of Agriculture at Davis, was doing whatever he could to help get the new medical school department started. Furthermore, an active biochemistry graduate program already existed on the Davis campus, which enjoys a multidisciplinary graduate group system. Not the least of other attractive factors was the enthusiasm of my postdoctoral fellow, Donal Walsh, who accompanied me to Davis and helped me in setting up the new department.

My research program was obviously affected by the move to California. One negative factor was the loss of my direct contacts with Ed Fischer. Although by 1968 our joint research project had expanded appreciably, and each of us now had individual areas of concentration, we nonetheless held joint research conferences and still worked together on many problems up to the time of my move to Davis. My contacts with other colleagues at the U of W were also lost. The major problem, however, was that I found that being the chairman of a department (especially in a new school) carries with it an awesome array of duties— interesting and enjoyable, perhaps, but not closely related to research. In addition, before I moved to Davis, we had just established the existence of PKA, and there were now a thousand leads to exploit with respect to that enzyme. In addition, much more work remaining to be done on phosphorylase kinase.

It is doubtful that I would have been able to keep my head above water in research if I had not had such an exceptionally talented set of students, postdoctoral fellows, and visiting scientists in the laboratory at that time. Also, even though Donal Walsh was now busy setting up an independent research program, he still found time to work on problems related to PKA and phosphorylase kinase. The postdoctoral fellows in the lab during the Davis years included Erwin Reimann, Jackie Corbin, Margaret Brostrom, Charles Brostrom, James Stull, Paul England, Joseph Beavo, Taro Hayakawa, Tung-shiuh Huang, Peter Bechtel, William Dills, Bruce Kemp, Ramji Khandelwal, Jackie Vandenheede, Franz Hofmann, Yutaka Shizuta, James Maller, David Glass, and Lloyd Waxman. My graduate students were Tom Soderling, David Bylund, Ronald

Alexander, and James Feramisco. Visiting scientists included Donald Graves, Felix Hunkeler, and Jerry Wang.

During the Davis period it was established that the hormone-sensitive lipase is a target of PKA. Evidence was obtained that PKA is made up of catalytic (C) and regulatory (R) subunits and can be dissociated by cAMP. Improved methods for purifying the holoenzyme and its subunits were devised, and the reversibility of PKA-catalyzed protein phosphorylation reactions was studied. The concentration of PKA subunits in various tissues and the amino acid sequence of phosphorylation sites in PKA and phosphorylase kinase were obtained. In other studies, the subunit composition of phosphorylation kinase was determined, and its phosphorylation of troponin T was examined. A new method for purifying muscle glycogen synthase was developed, and more extensive sequence data for its phosphorylation sites were obtained. A major new area of research that was initiated was the use of synthetic peptides for studying protein kinase specificity and inhibition. Finally, a completely new area of research brought to the laboratory by Jim Maller was the microinjection of protein kinase subunits and related proteins into *Xenopus laeris* oocytes to determine the role of protein phosphorylation during maturation.

A SECOND HITCH AT THE UNIVERSITY OF WASHINGTON, 1977 TO PRESENT

In winter 1976, I was spending a sabbatical quarter in Ed Fischer's laboratory in Seattle when I was approached with a plan for me to return to the U of W as chairman of the Department of Pharmacology and an investigator of the Howard Hughes Medical Institute. The opportunity was appealing for a variety of reasons, and after negotiations with the dean, Robert Van Citters, I accepted the offer. This time my decision to move was based on the attractiveness of a second opportunity to build up a department and because there were research advantages to moving back to the U of W. Happily, several people from my laboratory in Davis accompanied me to Seattle, which was obviously important from a research standpoint. In particular, I want to mention Edwina Beckman and Floyd Kennedy, who moved with me and provided superb help for a number of years with physical instrumentation and enzyme preparations, respectively.

My second departmental chairmanship was less demanding than the first, in part because I was more experienced and had become a firm believer in the delegation of administrative duties, which has obvious advantages. Also, the U of W School of Medicine was well established. Finally, the year was 1977, instead of 1968, and students no longer automatically assumed that anyone in a position of authority must be a bad person. One disadvantage, but possibly an advantage, was that I had never been identified as a pharmacologist and could thus think of the discipline differently than some. I felt that pharmacology should simply be considered a basic biology department that emphasized those aspects of cellular function that are of particular importance in the action of drugs. One factor that was very good for the department was the generous attitude of the Hughes Institute in the remodeling of space, i.e. well beyond my own immediate needs. In addition, as a result of its support, which for me lasted until the end of 1990, my own laboratory was able to pursue a broader array of projects than heretofore, even though essentially all of my efforts still dealt with aspects of protein phosphorylation. Fortunately, when the institute terminated my research support during my seventy-second year, I was able to reestablish funding from the NIH and other agencies, which has allowed me to continue my work.

Studies on the Primary Structure of Protein Kinases

A fruitful collaboration involving the primary structure of protein kinases was established between my laboratory and that of Kenneth Walsh and Koiti Titani of the Department of Biochemistry. Before my arrival, this group, together with Ed Fischer, had determined the amino acid sequence of the catalytic subunit of PKA. We now went on to obtain the sequences of Types I and II regulatory subunits of this enzyme. In other projects, we sequenced the cGMP-dependent protein kinase, the α, α', and β subunits of CK2, and skeletal muscle myosin light chain kinase (MLCK). A number of coworkers were involved in this effort, but I would single out Koji Takio as being particularly important. The major contribution of my immediate laboratory to the sequencing project was that of providing proteins for this work. However, our association with the total effort prompted our carrying out a variety of structure-function studies that we might not have undertaken otherwise. These included the affinity labeling of nucleotide triphosphate and effector binding sites in PKA, the cGMP-dependent protein kinase, and other protein kinases; phosphorylation-dephosphorylation studies on phosphorylation sites in PKA subunits; the effect of partial proteolysis on glycogen synthase activity; and identification of the Ca^{2+}-calmodulin binding site of MLCK (14). Donald Blumenthal and Arthur Edelman were the primary investigators in connection with the latter study. In collaboration with Michael James of the University of Edmonton, we tried valiantly to obtain suitable crystals of MLCK for him to use in crystallographic work but were unsuccessful. John Scott determined the complete amino acid sequence of PKI (15) and mapped its inhibitory site.

Protein Tyrosine Phosphorylation and the Mechanism of Action of Growth Factors

Three major findings relating to the phosphorylation-dephosphorylation of proteins occurred in the late 1970s. These were (a) the discovery by Ray Erickson and coworkers that the transforming protein encoded by the Rous sarcoma virus (p60^{src}) is a protein kinase; (b) the finding of Tony Hunter that proteins can be phosphorylated on tyrosine as well as serine/threonine residues, which quickly suggested that p60^{src} is a protein tyrosine kinase (PTK); and (c) the finding of Stanley Cohen and his collaborators that the epidermal growth factor (EGF) receptor is a PTK, which was followed by Ronald Kahn's determination that the insulin receptor is also a PTK. The lure of the new field that was opened by these findings was great, and my laboratory soon became involved in this area. Projects carried out in the early 1980s included the development of a tyrosine-containing synthetic peptide substrate for PTKs by John Casnellie, a comparison by Linda Pike and the late Elizabeth Kuenzel of the EGF and insulin receptors as PTKs, and a determination that LSTRA (a T-cell lymphoma cell line) cells contain elevated levels of a PTK structurally related to p60^{src}. One of my graduate students, Jamie Marth, in a collaborative study with Roger Perlmutter, cloned this enzyme, which was named p56^{lck}.

A project that soon dominated our efforts in the laboratory was to determine a mechanism that could explain the coupling of protein tyrosine phosphorylation to protein serine/threonine phosphorylation (16, 17). Evidence for such a coupling had existed since the early 1960s, when Larner and his group showed that the state of phosphorylation of glycogen synthase on serine residues is affected by insulin. These and other observations made it seem likely that many cellular responses to growth factors whose receptors are PTKs are mediated by changes in serine/threonine phosphorylation. Our own studies in this area, which were carried out primarily by Byron Gallis, Christine Chan, Katherine Meyer, James Weiel, Natalie Ahn, and Rony Seger, eventually helped to establish what is now known as the mitogen-activated protein (MAP) kinase cascade. Our approach was to work upstream from the effect of growth factors on the phosphorylation of an S6 peptide kinase. In this research, Natalie Ahn was the first investigator to demonstrate the existence of a MAP kinase kinase (18), which was then cloned and characterized by Rony Seger (19).

Ongoing Research

An area of research that has been a consistent focus of attention in the laboratory for a number of years is the enzyme casein kinase 2, recently renamed CK2, which is a ubiquitous protein serine/threonine kinase having more than 100 different substrates (20). Despite this aspect, as well as the fact that CK2 was one of the first protein kinases to be discovered, the enzyme is poorly understood with respect to its function(s) and regulation. CK2 was included as one of the protein kinases selected for total sequencing, which was carried out in Ken Walsh's laboratory (see above), and it was later cloned. We and others are currently identifying binding proteins for CK2 and are also studying the effect of its overexpression in cells. Recent work in the laboratory on CK2 has been carried out by Dongxia Li and Grazyna Dobrowolska; earlier studies involved James Sommercorn, David Litchfield, and Fred Lozeman.

In collaboration with Karin Bornfeldt, Russell Ross, and Elaine Raines, the laboratory is investigating the role of several different protein kinase cascades in the proliferation and mobility of smooth-muscle cells (21). Of great interest has been the cross talk that occurs between the MAP kinase cascade and other signaling pathways. The major investigator from my group who has been involved in this research is Lee Graves.

We have extended work on the MAP kinase cascade to include an investigation of "downstream" events, especially the regulation and function of glycogen synthase kinase 3 (GSK3), which can be inactivated by the cascade. An interesting finding that has come out of this work is that GSK3 can phosphorylate the insulin receptor substrate-1 (IRS-1) and cause it to become an inhibitor of the insulin receptor. This research is being carried out by Hagit Eldar-Finkelman.

Finally, a relatively new project in the laboratory centers on programmed cell death, or apoptosis (a fitting topic for someone nearing the end of his career). As might be expected, we are looking at the role of protein phosphorylation in this process (22) and are exploring what appears to be an intermingling of reversible protein phosphorylation reactions and irreversible proteolytic events. It has long been known that in vitro many protein kinases can be activated by partial proteolysis, but this type of regulation has seldom been viewed as being of physiological relevance. Now, however, we have evidence that in apoptotic cascades some of the steps involve proteolysis and some protein phosphorylation-dephosphorylation. These studies are being carried out by Jonathan Graves and Nancy Chamberlain from my group in collaboration with Edward Clark's laboratory.

EPILOGUE

In a textbook of biochemistry published 50 years ago (23), coincident in time with the period in which I was deciding to become a biochemist, proteins were classified as simple proteins, conjugated proteins, and derived proteins. Conjugated proteins differed from simple proteins in that they consisted of proteins that were combined with some nonprotein substances "in a manner that confers new and characteristic properties on the complex formed." One of the subclasses of conjugated proteins was phosphoproteins. These were described as "compounds of the protein molecule with phosphoric acid, e.g. casein from milk, ovovitillin from egg yolk, and other proteins associated with the feeding of the young." I think it is now safe to conclude that there is more to it than that. Nonetheless, the authors were prophetic in their suggestion that the phosphorylation of proteins might confer new and characteristic properties on the complexes formed. The phosphorylation of proteins is now considered the major means by which protein functions are regulated. Although there are other types of reversible covalent modification of proteins, none of them approaches phosphorylation in its frequency of occurrence. Those of us who started working on reversible protein phosphorylation in the 1950s had no inkling that this process would turn out to be as important as it is now known to be. The challenge that now faces investigators is that of determining how the vast protein phosphorylation-dephosphorylation network is integrated and controlled.

In concluding this retrospective article, I want to return to the theme that I addressed at the beginning, i.e. that of choosing a profession. I became a biochemist by a circuitous route and was never completely sure that I was making the right decisions along the way. Even after having taken a faculty position in a department of biochemistry, I occasionally asked myself whether becoming a biochemist was the proper choice for me. Now, however, after 50 years in the field, I am becoming convinced that this was the right thing to do. If it weren't, why would I hesitate to give up now?

ACKNOWLEDGMENTS

I want to thank the many secretaries and others who have helped me with various administrative and editorial tasks over the years, and I especially want to express my gratitude to Sally Hopkins, Evelyn Mercier, and Christina Boyd, and the late Frances Worthington, who have been my personal secretaries since around 1965. Their help has been enormous, not only for me but for all of those who have worked in my laboratory. In this connection I am grateful to the Howard Hughes Medical Institute, which has made it possible for Chris to continue working with me for the past seven years in my capacity an investigator emeritus.

> **Visit the *Annual Reviews* home page at**
> **http://www.AnnualReviews.org.**

Literature Cited

1. Price CC, Chapin EC, Goldman A, Krebs EG, Shafer HM. 1941. *J. Am. Chem. Soc.* 63:1857–61
2. Green AA, Cori GT, Cori CF. 1942. *J. Biol. Chem.* 142:447–48
3. Krebs EG. 1946. *J. Lab. Clin. Med.* 31:85–89
4. Krebs EG, Najjar VA. 1948. *J. Exp. Med.* 88:569–77
5. Meinhart JO, Chaykin S, Krebs EG. 1956. *J. Biol. Chem.* 220:821–29
6. Green AA, Cori GT. 1943. *J. Biol. Chem.* 151:21–29
7. Fischer EH, Krebs EG. 1955. *J. Biol. Chem.* 216:121–32
8. Meyer WL, Fischer EH, Krebs EG. 1964. *Biochemistry* 3:1033–39
9. Huston RB, Krebs EG. 1968. *Biochemistry* 7:2116–22
10. Burnett G, Kennedy EP. 1954. *J. Biol. Chem.* 211:969–80
11. Krebs EG, Graves DJ, Fischer EH. 1959.

J. Biol. Chem. 234:2867–73

12. Walsh DA, Perkins JP, Krebs EG. 1968. *J. Biol. Chem.* 243:3763–65

13. Miyamoto E, Kuo JF, Greengard P. 1969. *J. Biol. Chem.* 244:6395–402

14. Blumenthal DK, Takio K, Edelman AM, Charbonneau H, Titani K, et al. 1985. *Proc. Natl. Acad. Sci. USA* 82:3187–91

15. Scott JD, Fischer EH, Takio K, Demaille JG, Krebs EG. 1985. *Proc. Natl. Acad. Sci. USA* 82:5732–36

16. Pike LJ, Bowen-Pope DF, Ross R, Krebs EG. 1983. *J. Biol. Chem.* 258:9383–90

17. Krebs EG. 1986. *J. Anim. Sci.* 63(Suppl. 2):39–47

18. Ahn NG, Seger R, Bratlien RL, Diltz CD, Tonks NK, Krebs EG. 1991. *J. Biol. Chem.* 266:4220–27

19. Seger R, Seger D, Lozeman FJ, Ahn NG, Graves LM, et al. 1992. *J. Biol. Chem.* 267:25628–31

20. Pinna LA, Meggio F. 1997. *Prog. Cell Cycle Res.* 3:1–21

21. Bornfeldt KE, Campbell JS, Koyama H, Argast GM, Leslie CC, et al. 1997. *J. Clin. Invest.* 100:875–85

22. Graves JD, Draves KE, Craxton A, Saklatvala J, Krebs EG, Clark EA. 1996. *Proc. Natl. Acad. Sci. USA* 93:13814–18

23. Hawk PB, Oser BL, Summerson WN. 1947. *Practical Physiological Chemistry.* 12th ed. Toronto Blackiston. 323 pp.

Annu. Rev. Biochem. 1998. 67:1–25

HIV-1: Fifteen Proteins and an RNA

Alan D. Frankel
Department of Biochemistry and Biophysics, University of California, San Francisco,
San Francisco, California 94143-0448; e-mail: frankel@cgl.ucsf.edu

John A. T. Young
Department of Microbiology and Molecular Genetics, Harvard Medical School,
Boston, Massachusetts 02115; jatyoung@warren.med.harvard.edu

KEY WORDS: human immunodeficiency virus, retrovirus, AIDS, viral proteins, viral RNA

ABSTRACT

Human immunodeficiency virus type 1 is a complex retrovirus encoding 15 distinct proteins. Substantial progress has been made toward understanding the function of each protein, and three-dimensional structures of many components, including portions of the RNA genome, have been determined. This review describes the function of each component in the context of the viral life cycle: the Gag and Env structural proteins MA (matrix), CA (capsid), NC (nucleocapsid), p6, SU (surface), and TM (transmembrane); the Pol enzymes PR (protease), RT (reverse transcriptase), and IN (integrase); the gene regulatory proteins Tat and Rev; and the accessory proteins Nef, Vif, Vpr, and Vpu. The review highlights recent biochemical and structural studies that help clarify the mechanisms of viral assembly, infection, and replication.

CONTENTS

1

0066-4154/98/0701-0001$08.00

INTRODUCTION

Human immunodeficiency virus type 1 (HIV-1) has been the subject of intense investigation for 15 years, and a great deal has been learned about how the retrovirus infects cells, replicates, and causes disease. In the past few years, substantial progress has been made toward understanding the detailed biochemical function of each viral component, and in many cases structures have been determined. In this review, we attempt to integrate structural and biochemical information into a view of the HIV-1 particle as a whole, emphasizing key interactions among viral and cellular components during the viral replication cycle. Given this broad scope, we can barely scratch the surface of the relevant literature and have thus chosen a limited number of reviews and recent references that should help guide the reader to more detailed aspects of the viral components. We apologize to all investigators in the field for the arbitrary selection of references, and the reader should recognize that many classic papers have not been cited.

HIV-1 GENES AND THE VIRUS LIFE CYCLE

The HIV-1 genome encodes nine open reading frames (Figure 1). Three of these encode the Gag, Pol, and Env polyproteins, which are subsequently proteolyzed into individual proteins common to all retroviruses. The four Gag proteins, MA (matrix), CA (capsid), NC (nucleocapsid), and p6, and the two Env proteins, SU (surface or gp120) and TM (transmembrane or gp41), are structural components that make up the core of the virion and outer membrane envelope. The three Pol proteins, PR (protease), RT (reverse transcriptase), and IN (integrase), provide essential enzymatic functions and are also encapsulated within the particle. HIV-1 encodes six additional proteins, often called accessory proteins, three of which (Vif, Vpr, and Nef) are found in the viral particle. Two other accessory proteins, Tat and Rev, provide essential gene regulatory functions, and the last protein, Vpu, indirectly assists in assembly of the virion. The retroviral genome is encoded by an ~9-kb RNA, and two genomic-length RNA molecules are also packaged in the particle. Thus, in simplistic terms, HIV-1 may be considered

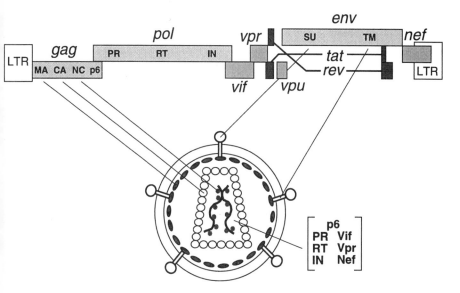

Figure 1   Organization of the HIV-1 genome and virion.

as a molecular entity consisting of 15 proteins and one RNA. We first describe how each component fits into the viral life cycle, and then we review selected structural and biochemical studies to highlight important functional aspects of each protein.

In our view, the HIV-1 replication cycle may be broken into 15 discrete steps, as depicted in Figure 2. We begin the cycle with the viral genome integrated into a host chromosome, and we describe the order of events that lead to expression of the viral gene products, production of virus particles, infection of a new cell, and reintegration of the viral genome. In step 1, viral transcripts are expressed from the promoter located in the 5′ long terminal repeat (LTR), with Tat greatly enhancing the rate of transcription. In step 2, a set of spliced and genomic-length RNAs are transported from the nucleus to the cytoplasm, where they can be translated or packaged. This step is regulated by Rev. In step 3, viral mRNAs are translated in the cytoplasm, and the Gag and Gag-Pol polyproteins become localized to the cell membrane. The Env mRNA is translated at the endoplasmic reticulum (ER). In step 4, the core particle is assembled from the Gag and Gag-Pol polyproteins (later processed to MA, CA, NC, p6, PR, RT, and IN), Vif, Vpr, Nef, and the genomic RNA, and an immature virion begins to bud from the cell surface. To provide SU and TM proteins for the outer membrane coat during budding, the Env polyprotein must first be released from complexes with CD4 (the cell surface HIV-1 receptor), which is coexpressed with Env in the ER. Vpu assists this process by promoting CD4 degradation, as shown in step 5. Env is then transported to the cell surface (step 6), where again it must be prevented from binding CD4. Nef promotes endocytosis and degradation of surface CD4 (step 7). As the particle buds and is released from the cell surface coated with SU and TM (step 8), the virion undergoes a morphologic change known as maturation (step 9). This step involves proteolytic processing of the Gag and Gag-Pol polyproteins by PR and a less well defined function of Vif. The mature virion is then ready to infect the next cell, which is targeted by interactions between SU and CD4 and CC or CXC chemokine coreceptors (step 10). Following binding, TM undergoes a conformational change that promotes virus-cell membrane fusion, thereby allowing entry of the core into the cell (step 11). The virion core is then uncoated to expose a viral nucleoprotein complex, which contains MA, RT, IN, Vpr, and RNA (step 12). This complex is transported to the nucleus (step 13), where the genomic RNA is reverse transcribed by RT into a partially duplex linear DNA (step 14). IN then catalyzes integration of the viral DNA into a host chromosome and the DNA is repaired (step 15), thereby completing the viral replication cycle. We now describe our current understanding of the biochemistry and structure of each viral component.

VIRAL COMPONENTS

Tat

The HIV-1 promoter is located in the 5' LTR and contains a number of regulatory elements important for RNA polymerase II transcription. Sites for several cellular transcription factors are located upstream of the start site, including sites for NF-κB, Sp1, and TBP (1). These cellular factors help control the rate of transcription initiation from the integrated provirus (step 1, Figure 2), and their abundance in different cell types or at different times likely determines whether a provirus is quiescent or actively replicating. Despite the importance of these factors, transcription complexes initiated at the HIV-1 promoter are rather inefficient at elongation and require the viral protein Tat to enhance the processivity of transcribing polymerases. Under some conditions, Tat may also enhance the rate of transcription initiation. Tat increases production of viral mRNAs $\sim$100-fold and consequently is essential for viral replication. It is not yet clear which features of the HIV-1 promoter cause initiating transcription complexes to be poorly processive, but experiments in which the TATA box and downstream sequences have been interchanged with different promoters suggest an important role for these regions (1). In the absence of Tat, polymerases generally do not transcribe beyond a few hundred nucleotides, though they do not appear to terminate at specific sites. It is not yet clear how Tat causes transcribing polymerases

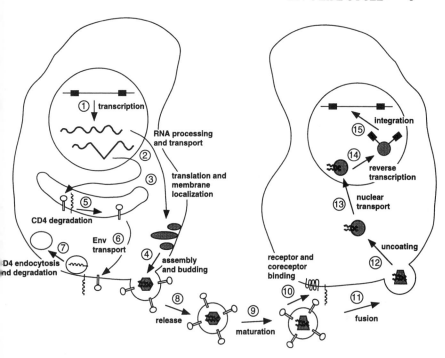

Figure 2   HIV-1 replication cycle. Steps 1–15 are described in the text.

to become sufficiently processive to completely transcribe the ~9-kb viral genome, but recent experiments suggest that Tat may assemble into transcription complexes and recruit or activate factors that phosphorylate the RNA polymerase II C-terminal domain (CTD), including the general transcription factor TFIIH and other novel kinases (2–6, 6a). These findings support a model in which Tat enhances phosphorylation of the CTD, a process known to occur as RNA polymerase II converts from an initiating to an elongating enzyme.

   Unlike typical transcriptional activators, Tat binds not to a DNA site but rather to an RNA hairpin known as TAR (trans-activating response element), located at the 5′ end of the nascent viral transcripts. An arginine-rich domain of Tat helps mediate binding to a three-nucleotide bulge region of TAR, with one arginine residue being primarily responsible for recognition. NMR studies of TAR complexed to arginine (7, 8) show a base-specific contact between the arginine side chain and a guanine in the RNA major groove (Figure 3A). The complex is stabilized by additional contacts to the phosphate backbone and by formation of a U-A:U base triple between a bulge nucleotide and a base pair above the bulge (Figure 3A). NMR studies of the full-length 86–amino acid Tat protein have suggested that a hydrophobic core region of about 10 amino acids adopts a defined structure but that the rest of the molecule, including the arginine-rich RNA-binding domain, is relatively disordered (9). It seems likely that Tat requires interactions with cellular proteins in addition to TAR to adopt a stable structure. Aside from proteins of the transcription apparatus, another protein is needed to bind to the loop of the TAR hairpin, apparently helping to stabilize the Tat-TAR interaction (10). Functional data suggest that the loop-binding protein is encoded by human chromosome 12. Several candidates have been identified, but none have yet been definitively shown to be essential for Tat activity.

RNA

The transcript produced from the viral promoter is ~9 kb long and may be thought of as a large macromolecular component of the virion containing structured subdomains throughout its length. Beginning at the 5' end, several essential regions have been defined [nucleotide numbers (nts) are approximate and vary among HIV-1 isolates]:

1. The TAR hairpin (nts 1–55) is the Tat-binding site.

2. The primer-binding site (nts 182–199) is important for initiating reverse transcription by annealing to a cellular tRNA^{Lys}.

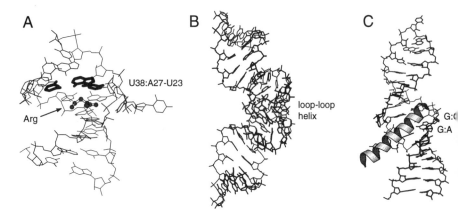

Figure 3 Structures of HIV-1 proteins and RNA. (*A*) TAR-arginine complex (7), (*B*) kissing-loop dimerization hairpins (A Mujeeb, T Parslow, T James, personal communication) (*C*) Rev peptide- RRE IIB complex (22), (*D*) MA trimer (24), (*E*) CA C-terminal dimerization domain (37), (*F*) CA N-terminal domain dimer (39), (*G*) NC (53), (*H*) Nef (73), (*I*) PR-inhibitor complex (133), (*J*) TM$_{core}$ trimer (105), (*K*) RT-nevirapine complex (120), (*L*) IN N-terminal domain dimer (129), (*M*) IN catalytic domain dimer (130), (*N*) IN DNA-binding domain dimer (132). Figures were prepared using MOLSCRIPT (134).

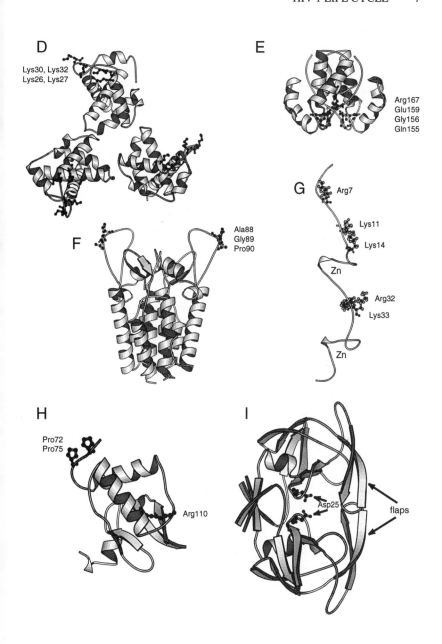

Figure 3   (*Continued*)

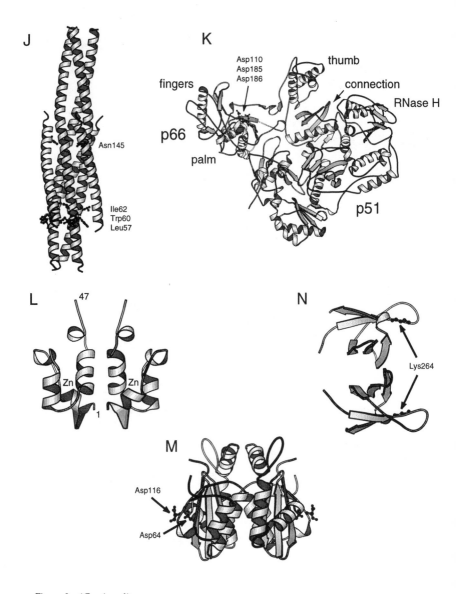

Figure 3   (Continued)

3. The packaging signal or ψ (nts 240–350) binds NC and is critical for incorporation of genomic RNA into the virion (11).

4. The dimerization site includes a "kissing loop" hairpin (nts 248–271) that facilitates incorporation of two genomic RNAs into the virion (11–13).

5. The major splice donor site (nt 290) is used to generate all subgenomic spliced mRNAs.

6. The Gag-Pol frameshifting region (nts 1631–1673) comprises a heptanucleotide slippery sequence and RNA hairpin that promote -1 ribosomal frameshifting, thereby translating a fused Gag-Pol polyprotein at a frequency of $\sim$5–10% (14).

7. The Rev response element (RRE) (nts 7362–7596) is the Rev-binding site.

8. Splice acceptor sites are present at several downstream regions of the RNA and allow production of a relatively large number of spliced products (the two major sites are at nts 5358 and 7971).

9. The polyadenylation signal (nts 9205–9210) is used to generate the 3' end.

The structures of several important RNA elements have been solved by NMR. A kissing-loop complex, which helps mediate dimerization of the genomic RNA, contains two identical RNA hairpins (one from each genomic strand) with complementary loops that form a six–base pair helix. The loop-loop helix lies perpendicular to the helical stems of the hairpins, introducing a slight bend that is stabilized by the stacking of unpaired adenosines (Figure 3*B*) (A Mujeeb, T Parslow, and T James, personal communication). The junctions cause unpairing of the adjacent stems, which may facilitate a transition to a more stable duplex structure. As described in other sections, the structures of TAR-arginine and RRE-Rev peptide complexes have also been determined, and the structure of a ψ-NC complex is expected in the near future.

Rev

When viral mRNAs are first produced, most are doubly spliced and encode the Tat, Rev, and Nef proteins. Later, when other viral components are needed to assemble infectious virions, singly spliced and unspliced transcripts are transported to the cytoplasm (step 3, Figure 2), where they are translated and where genomic RNAs are packaged. Rev is important in this switch because it overcomes the default pathway in which mRNAs are spliced prior to nuclear export and functions by binding to the RRE site located in the *env* coding region. Whether Rev directly enhances the export of unspliced mRNAs or inhibits splicing has been unclear, but recent studies lend strong support to a role in export (14a). Microinjection experiments in *Xenopus* oocytes have shown that Rev is required to export unspliced RNAs that contain an RRE (15). Rev contains a leucine-rich nuclear export signal (NES) that allows it to shuttle between the nucleus and cytoplasm (16) and that interacts with a nucleoporin-like protein (hRip/Rab) located at the nuclear pore (17, 18, 18a). The interaction with hRip/Rab may be bridged by CRM1, a nuclear export receptor that is important for Rev export (18b). Thus, Rev binding to the RRE is believed to target the attached mRNA to the nuclear export machinery. There is evidence that entry into the splicing pathway may also be important for Rev function because mutating 5′ splice sites on RRE-containing mRNAs eliminates Rev activity but compensatory mutations in U1 snRNA, which binds at 5′ splice sites, can restore activity (19). Furthermore, Rev can directly inhibit splicing by preventing entry of additional snRNPs during the later stages of spliceosome assembly (20). Possible relationships between the splicing and transport pathways and the precise mechanism of Rev function remain to be clarified.

The RRE contains several hairpins and binds several Rev monomers, nucleated by the interaction of a single monomer with a high-affinity site, hairpin IIB (21). Oligomeric binding is important for Rev function, presumably because it increases the concentration of NES sites on a single mRNA. Binding is mediated by an arginine-rich domain that forms an α-helix and specifically recognizes an internal loop in the IIB stem. The structure of a Rev peptide complexed to IIB has been determined by NMR (Figure 3*C*) as well as complexed to an in vitro selected RNA (22, 23). The internal loop contains G:G and G:A base pairs that widen an otherwise narrow major groove. The widened groove allows amino acids on the Rev α-helix to recognize specific features of the site, primarily through hydrogen bonds between three arginines and specific bases and phosphates and between Asn40 and the G:A pair. The structures of the intact 116-residue Rev protein and oligomeric complexes remain to be determined.

MA

MA is the N-terminal component of the Gag polyprotein and is important for targeting Gag and Gag-Pol precursor polyproteins to the plasma membrane prior to viral assembly (step 3, Figure 2). In the mature viral particle, the 132-residue MA protein lines the inner surface of the virion membrane (Figure 1). Two discrete features of MA are involved in membrane targeting: an N-terminal myristate group and basic residues located within the first 50 amino acids. The crystal structure of residues 1 through 104 (24) shows five α-helices capped by a three-stranded mixed β-sheet, with three monomers arranged like a triskelion (Figure 3*D*). The trimeric form is presumed to be biologically relevant because mutation of residues involved in trimerization (residues 42–77) abolishes viral assembly and because basic residues important for membrane localization (lysines 26, 27, 30, 32) are arranged on the putative membrane-binding surface of the trimer (Figure 3D). The MA structure suggests an obvious model for membrane binding that involves the insertion of three myristate groups into the lipid bilayer located directly above the trimer and interactions between basic residues of the membrane-binding surface and phospholipid head groups. However, the N-terminal basic region is not strictly required for the formation of virus particles because noninfectious virus particles that lack MA can be produced if a myristate group is placed directly upstream of CA (25). In addition to targeting Gag and Gag-Pol to the membrane, MA also appears to help incorporate Env glycoproteins with long cytoplasmic tails into viral particles (26, 27). Indeed, the array of threefold symmetric holes located between matrix trimers appears to be large enough to accommodate the long cytoplasmic tails of full-length Env (24, 28).

In addition to its function in viral assembly, MA facilitates infection of nondividing cell types, principally macrophages. Its precise role in viral entry is controversial. Some studies have shown that a subset of phosphorylated MA proteins are associated with viral preintegration complexes and that MA contains a nuclear localization signal (NLS) that interacts with Rch1, a member of the karyopherin-α family, to facilitate rapid nuclear transport (29–32) (step 13, Figure 2). Phosphorylation of Tyr131 was shown to mediate association with IN, thereby linking MA to the preintegration complex (33, 34). Other studies, however, have found no evidence for an MA NLS and suggest that phosphorylation of Tyr131 is not important for macrophage infection (26, 31, 35, 36). Instead, mutation of the putative MA NLS in a macrophage-tropic HIV-1 isolate decreased infectivity in both nondividing and dividing cells and resulted in delayed proteolytic processing of the Gag polyprotein, presumably because the mutations affect association of MA with the membrane (36). Additional studies are needed to clarify the role of MA in infection of nondividing cells.

CA

CA is the second component of the Gag polyprotein and forms the core of the virus particle, with $\sim$2000 molecules per virion (Figure 1). The C-terminal domain (residues 152–231) functions primarily in assembly (step 4, Figure 2) and is important for CA dimerization and Gag oligomerization (37). Although mutations in the N-terminal domain (residues 1–151) do not prevent assembly or budding, the domain is important for infectivity, apparently by participating in viral uncoating (step 12, Figure 2) through its association with a putative cellular chaperone, cyclophilin A (CypA) (38).

Structures of the C-terminal domain, N-terminal domain, and N-terminal domain complexed to CypA have been solved by crystallography and NMR

(37, 39–41). The C-terminal domain is composed of an extended strand followed by four α-helices, with an extensive dimer interface (37) (Figure 3E). The major homology region (MHR), a 20–amino acid sequence that is one of the most highly conserved within all retroviral Gag proteins, adopts a compact fold in which the four most conserved residues (Gln155, Gly156, Glu159, and Arg167) form a stabilizing hydrogen-bonding network (Figure 3E). Additional hydrophobic residues from the MHR contribute to the hydrophobic core. The MHR is essential for particle assembly and may have a role in incorporation of Gag-Pol precursors through interactions with Gag (42), though not all mutants show this phenotype (43). Biochemical experiments also suggest a possible role for the MHR in membrane affinity, perhaps reflecting exposure of hydrophobic residues (44). The structures of two dimeric forms of the N-terminal domain, one complexed to an antibody fragment and the other complexed to CypA, show the same monomeric CA structure but different subunit interfaces (39, 40). The CA-CA interfaces observed in the CypA complex are blocked in the antibody complex, but given that the C-terminal domain is largely responsible for dimerization (the N-terminal domain is monomeric at mM concentrations), it still remains to be determined whether the observed N-terminal domain interfaces represent true subunit interactions. The most extensive dimer interface from the CypA complex is shown in Figure 3F. The CA subunits are also seen to arrange in strips within the crystal, consistent with a plausible packing arrangement in the virion core (39). Residues from an extended region of CA interact with CypA, with Ala88, Gly89, and Pro90 buried in the CypA active site groove (Figure 3F). A short spacer peptide located between CA and NC—p2—may also influence CypA incorporation into the virion (45).

NC

NC is the third component of the Gag polyprotein and coats the genomic RNA inside the virion core (Figure 1). The primary function of NC is to bind specifically to the packaging signal and deliver full-length viral RNAs into the assembling virion (step 4, Figure 2). The packaging signal, ψ, is not completely defined but is probably composed of three RNA hairpins located around the major splice donor site (11, 13), the first of which contains the kissing loop involved in RNA dimerization. Studies with a chimeric Gag containing NC from HIV-1 and the remainder of Gag from Moloney murine leukemia virus (Mo-MLV) demonstrate that genomic HIV-1 RNA is preferentially packaged but that additional downstream sequences, which result in packaging of spliced RNAs, may contribute (46). NC is a basic protein that also binds single-stranded nucleic acids nonspecifically, leading to coating of the genomic RNA that presumably protects it from nucleases and compacts it within the core. Nonspecific binding also provides chaperone-like functions that enhance other nucleic acid–dependent steps in the life cycle; for example, by promoting annealing of the tRNA primer, melting of RNA secondary structures, or DNA strand exchange reactions during reverse transcription (47–49) or by stimulating integration (50).

NC is 55 residues long and contains two zinc finger domains (of the CCHC type) flanked by basic amino acids. The specific NC-ψ interaction requires intact fingers as well as several basic amino acids (51, 52). The residues that make specific versus nonspecific contacts are not yet well defined. The structure of NC has been determined by NMR (53, 54) and shows two well-ordered zinc domains with a relatively flexible linker in the absence of RNA (Figure 3G). Basic residues shown to be particularly important for in vitro RNA binding (Arg7, Arg32, and Lys33) and viral replication (Lys11 and Lys14) are indicated, though mutation of Arg32 or Lys33 seems to have little effect on RNA packaging in vivo (52). Disulfide-substituted benzamide compounds specifically remove zinc from the NC domains and inhibit viral replication (55), providing additional evidence of the importance of these structures.

p6

p6 comprises the C-terminal 51 amino acids of Gag and is important for incorporation of Vpr during viral assembly (step 4, Figure 2). Residues 32–39 and three hydrophobic residues within a highly conserved sequence motif (Leu41-X42-Ser43-Leu44-Phe45-Gly46) are important for Vpr binding (56–58). In Vpr, a predicted α-helical structure located near its N-terminus contains amino acids responsible for p6 binding (59). p6 also helps mediate efficient particle release (step 8, Figure 2), and a region of four amino acids (Pro7-Thr8-Ala9-Pro10) has been implicated in this function (60).

Vpu

Newly synthesized Env glycoproteins (gp160), which are later cleaved into SU (gp120) and TM (gp41), are sometimes held in the endoplasmic reticulum through interactions with newly synthesized CD4 molecules. Vpu promotes degradation of CD4 in these complexes, thus allowing Env transport to the cell surface for assembly into viral particles (steps 5 and 6, Figure 2). Vpu is an 81-residue oligomeric integral membrane protein with an N-terminal 24-residue hydrophobic membrane-spanning domain and a C-terminal cytoplasmic tail (59, 61). Amino acids important for receptor binding and degradation have been mapped to the C-terminal region of Vpu and to putative α-helices in the cytoplasmic tail of CD4 (62). Coimmunoprecipitation experiments have shown that Vpu associates with wild-type CD4 or with recombinant proteins containing the CD4 cytoplasmic tail, but it is not yet known if the interaction is direct or indirect (63). These complexes are probably relevant to CD4 degradation because there is a direct correlation between the extent of Vpu association and their relative levels of degradation (63).

The effect of Vpu on CD4 degradation appears to be regulated by posttranslational modification. Vpu is phosphorylated on Ser52 and Ser56 by a casein kinase-2-related protein, and mutation of these positions decreases the levels of CD4 degradation (59). The mechanism of degradation is not clear but may involve the cytoplasmic proteasome, because Vpu-mediated degradation can be blocked by proteasome inhibitors such as lactacystin (64). Vpu can also downregulate cell surface expression of MHC class I proteins, which may protect infected cells from recognition and killing by cytotoxic T lymphocytes (65).

In addition to its role in CD4 degradation, Vpu can also stimulate virion release, and it has been proposed to be an ion channel (61). In Vpu mutant viruses, significantly increased numbers of particles either remain associated with the cell surface or are localized to intracellular membranes (59, 61). In contrast to Vpu-mediated CD4 degradation, its effect on particle release requires the hydrophobic N-terminal domain and is not influenced by serine phosphorylation (59, 61). The mechanism appears to be relatively nonspecific in that Vpu can also promote the release of heterologous retroviral particles (59, 61).

Nef

Nef is a 206–amino acid, N-terminally myristoylated protein that, like Vpu, reduces the levels of cellular CD4. Nef facilitates the routing of CD4 from the cell surface and golgi apparatus to lysosomes, resulting in receptor degradation and preventing inappropriate interactions with Env, as for Vpu (step 7, Figure 2) (66). A dileucine-based sorting signal located in the cytoplasmic tail of CD4 is essential for Nef-mediated downregulation and is presumed to interact with Nef (66). Nef has been proposed to serve as a direct bridge between CD4 and the cellular endocytic machinery by interacting with β-COP and adaptins, which link proteins in the golgi and plasma membrane to clathrin-coated pits (66). By downregulating CD4, Nef may enhance Env incorporation into virions, promote particle release, and possibly affect CD4$^+$ T-cell signaling pathways (66). As with Vpu, Nef can also downregulate expression of MHC class I molecules, which may help protect infected cells from killing by cytotoxic T cells (67).

Nef mutant viruses also exhibit decreased rates of viral DNA synthesis following infection (68). This defect can be overcome if Nef is supplied in *trans* in virus-producing cells but not in target cells, suggesting possible roles in virus assembly, maturation, or entry. Such roles are consistent with the observation that ~70 Nef molecules are incorporated per virion; these virion-associated proteins are cleaved by PR at residue 57 to generate a soluble C-terminal fragment (68). The mechanism of Nef incorporation has not been defined but is probably relatively nonspecific, because Nef can also be incorporated into Mo-MLV particles (69).

Nef contains a consensus SH3 domain binding sequence (PXXP) that mediates binding to several Src-family proteins (e.g. Src, Lyn, Hck, Lck, Fyn), thereby regulating their tyrosine kinase activities (70–72). These interactions appear to be important for enhancing viral infectivity but not for downregulating CD4 (72). It is not yet clear which SH3-containing proteins are relevant for Nef function. The crystal structure of a Nef-SH3 complex (73) shows that the PXXP motif is in a left-handed polyproline type II helix and interacts directly with the SH3 domain (Figure 3H). Two residues that define the motif, Pro72 and Pro75, are important for enhancing viral replication and pack against hydrophobic residues of the SH3 domain (73). The central core of Nef comprises two antiparallel α-helices packed against a layer of four antiparallel β strands (Figure 3H) (73, 74). A hydrophobic crevice, which is presumably a ligand-binding site, is located between the two helices and is close to Arg110. Arg110 has been defined as an important residue for association with NAK, a Nef-associated serine/threonine kinase related to a p21 kinase (PAK) (75). PAKs are known to bind the p21 Rho-like GTP-binding proteins Rac-1 and CDC42hs, suggesting possible mechanisms by which Nef can interfere with both endocytosis and T-cell signaling (76). However, mutations that disrupt the Nef-NAK complex do not affect Nef-mediated CD4 downregulation (72). Nef has also been reported to bind other cellular proteins, including p53, MAP kinase, and TEase-II (70, 72), but the significance of these interactions remains to be determined.

PR

As the core virion is assembled to include the Gag and Gag-Pol polyproteins, the Vif, Vpr, and Nef proteins, and the genomic RNA, and as the membrane coat containing SU and TM surrounds the particle, the virus buds from the membrane surface and is released (steps 4 and 8, Figure 2). The immature particles formed are noninfectious. The Gag and Gag-Pol polyproteins must be cleaved by PR, and conformational rearrangements must occur within the particle, to produce mature infectious viruses (step 9, Figure 2). Some of these "maturation" events may occur simultaneously with assembly and budding (77); the precise timing is not clear. PR cleaves at several polyprotein sites to produce the final MA, CA, NC, and p6 proteins from Gag and PR, RT, and IN proteins from Pol. The final stoichiometries are determined largely by the amount of Gag-Pol produced by ribosomal frameshifting and incorporated into the virion ($\sim$5–10% of Gag). Because assembly and maturation must be highly coordinated, factors that influence PR activity can have dramatic effects on virus production. PR functions as a dimer and is part of Pol, so PR activity initially depends on the concentration of Gag-Pol and the rate of autoprocessing, which may be influenced by adjacent p6 sequences (78). Cleavage efficiencies can vary substantially among sites, thereby influencing the order of appearance of different processed proteins (79). The p2 spacer peptide located between CA and NC may also help control relative cleavage rates and infectivity (80, 81), and processing of NC and p6 may be further influenced by RNA binding to NC (82). Overexpression of PR can lead to aberrant rates of processing and decreased infectivity (83).

PR has been a prime target for drug design, and crystal structures of many PR-inhibitor complexes have been solved, including peptidomimetic and nonpeptide inhibitors (84). The enzyme active site is formed at the dimer interface, with each 99-residue monomer contributing a catalytically essential aspartic acid (Asp25) (Figure 3*I*). The active site resembles that of other aspartyl proteases and contains a conserved triad sequence, Asp-Thr-Gly. The PR dimer contains flexible flaps (Figure 3*I*) that close down on the active site upon substrate binding. Amino acid side chains surrounding the cleavage site bind within hydrophobic pockets of PR, helping to explain some of the rate differences observed between different sites. Several PR inhibitors are in wide clinical use, and mutants resistant to multiple inhibitors have been observed (85, 86). Resistance mutations are located both within the inhibitor binding pocket and at distant sites, and some mutants show increased catalytic activities (87). An alternative approach to inhibitor design involves the use of inactive subunits that act as dominant negative inhibitors (88).

Vif

Vif is a 192-residue protein that is important for the production of highly infectious mature virions. Vif mutant viruses show markedly reduced levels of viral DNA synthesis and produce highly unstable replication intermediates (59, 89), suggesting that Vif functions before or during DNA synthesis. It is intriguing that Vif mutants show defects in infectivity only when produced in certain cell types, designated nonpermissive or semipermissive, but not when produced in permissive cells. It is possible that permissive cells produce a factor or factors that compensate for a lack of Vif or that expression of Vif in permissive cells blocks an inhibitor of viral infectivity (59). Vif activity may be regulated by posttranslational modification because mutation of one of three serine phosphorylation sites (Ser144) causes a defect in viral infectivity (90).

Compared with mature wild-type virions, Vif mutant viruses have similar protein and RNA contents but grossly altered core structures, suggesting that Vif may play a role in viral assembly and/or maturation (59) (steps 4 and 9, Figure 2). Consistent with this role, the infectivity defect can be complemented by supplying Vif in *trans* in virus-producing cells but not in target, nonpermissive cells (59), as also seen with Nef. It has been estimated that 7 to 100 molecules of Vif are packaged into the virion (91–94), suggesting that Vif may function directly within the particle. Incorporation of Vif is probably nonspecific because there is no apparent requirement for any viral protein or RNA and, like Nef, Vif can be incorporated into Mo-MLV particles (91).

SU

Viral entry is initiated by the binding of the SU glycoprotein, located on the viral membrane surface (Figure 1), to specific cell surface receptors (step 10, Figure 2). The major receptor for HIV-1 is CD4, an immunoglobulin (Ig)-like protein expressed on the surface of a subset of T cells and primary macrophages. The 515-residue SU protein binds CD4 with high affinity (K_d ~4 nM), and amino acids important for binding have been mapped primarily to four separate conserved regions of SU and to the C'-C'' ridge of CD4, which protrudes from the first Ig-like extracellular domain (95). Structural details of the interactions are not yet known.

The SU-CD4 interaction is not sufficient for HIV-1 entry. Instead, a group of chemokine receptors (a family of seven transmembrane G-coupled proteins) that mobilize intracellular calcium and induce leukocyte chemotaxis serve as essential viral coreceptors (96). There are two major classes of HIV-1: those that are macrophage (M)-tropic and non–syncytium inducing (NSI) and those that are T-cell (T)-tropic and syncytium inducing (SI). CXCR4/fusin was the first coreceptor identified; it permits entry of T-tropic but not M-tropic viruses. CCR5 is a major coreceptor for M-tropic but not T-tropic viruses. Other molecules, including CCR3, CCR2b, Bonzo/STRL33, and BOB/GPR15, serve as coreceptors for some HIV-1 isolates (96). The physiological ligands for CXCR4, CCR5, and CCR3 (SDF-1, RANTES/MIP-1α/MIP-1β, and eotaxin, respectively) are each able to inhibit viral entry by competing with the cognate coreceptor (97–100). Some ligand derivatives have been described that block infection without activating chemokine signaling pathways and may represent a novel class of HIV-1 therapeutics (100).

Binding of CD4 to SU appears to cause structural changes in Env that facilitate coreceptor binding and subsequent viral entry (100). The variable V3 loop of SU is an important determinant of viral tropism. It becomes exposed upon CD4 binding and presumably interacts with the cognate coreceptor (100). However, the V3 loop is probably not the sole determinant of coreceptor specificity, because HIV-1 isolates that use the same coreceptor can have highly variable V3 sequences (99, 101, 102). Determinants for virus specificity are located in each of the extracellular regions of the coreceptors, and the signaling functions of these receptors apparently are not important for viral infection (100).

TM

The primary function of TM, a 345–amino acid protein located in the viral membrane (Figure 1), is to mediate fusion between the viral and cellular membranes following receptor binding (step 11, Figure 2). An N-terminal hydrophobic glycine-rich "fusion" peptide has been predicted to initiate fusion, and a transmembrane region is important both for fusion and for anchoring Env in the viral membrane (103). Two crystal structures of the core region (TM_{core}) have been reported (104, 105). In the larger of the two structures (shown in Figure 3J), residues 30–79 and 113–154 of TM were fused to a 31-residue trimeric coiled-coil from GCN4 in place of the N-terminal fusion peptide (105). TM_{core} lacks residues 80–112 of TM.

TM_{core} forms a trimer containing a central parallel α-helical coiled-coil (residues 1–77) and an outer antiparallel α-helical layer (residues 117–154) (105). The structure of TM_{core} probably does not represent the native TM structure but rather a structure formed during the fusion reaction, as suggested by the following. First, mutations at the interface between the outer and central helical layers (including Ile62, Figure 3J) specifically block membrane fusion. Second, TM_{core} is extremely thermostable, a feature predicted of the fusion-active protein and not the native protein (104, 105). Third, the structures of TM_{core} and a low pH fusion-active form of the influenza virus HA_2 protein are strikingly similar (104, 105). Fourth, the estimated distance between the C-terminus of TM_{core} and the viral lipid bilayer cannot be spanned by the 18 C-terminal extracellular residues missing from the structure. However, the distance is consistent with a conformation in which the fusion peptides and transmembrane regions are located at the same end of the central rod structure when viral and cell membranes are brought together (Figure 3J) (105). The structure helps to explain how two peptides known to inhibit fusion may act. A peptide from the C-terminus may bind to the central trimer, disrupting the structure of the N-terminal region, whereas a peptide from the N-terminus may either compete with folding of the central trimer or bind to the C-terminal region and prevent association with the central core (104, 105). Emerging rules for the design of coiled-coils may aid in the development of new fusion inhibitors.

Vpr

Following fusion and entry, the virus is "uncoated" in the cytoplasm (a poorly defined process; see step 12, Figure 2) and nucleoprotein complexes are rapidly transported to the host cell nucleus, mediated by the 96–amino acid Vpr protein (59) (step 13, Figure 2). The components of the transported complexes are not completely defined but certainly include RT, IN, and MA (106) and probably the genomic RNA and partially reverse-transcribed DNA. Vpr is especially important for nuclear localization in nondividing cells, such as macrophages, because it contains an NLS that directs transport even in the absence of mitotic nuclear envelope breakdown (59). Vpr does not contain a canonical karyophilic NLS but instead contains two important putative N-terminal amphipathic α-helices (107). This unusual NLS localizes Vpr to the nuclear pores rather than to the interior of the nucleus and does not use an importin-dependent pathway (32, 107). Vpr is incorporated into viral particles through an interaction with p6 and may later become associated with the nucleoprotein complexes through an interaction with the C-terminal region of MA (108).

In addition to its nuclear uptake function, Vpr can also induce G2 cell cycle arrest prior to nuclear envelope breakdown and chromosome condensation, and sustained expression can reportedly kill T cells by apoptosis (107). Vpr acts before dephosphorylation of the $p34^{cdc2}$ cyclin-dependent kinase by CDC25, which is required to initiate mitosis (107). Although G2 arrest occurs with Vpr proteins from different primate lentiviruses, it is not known how the activity contributes to viral replication (107). Amino acids important for G2 arrest are located in the C-terminal region of Vpr, and cellular proteins have been identified that bind Vpr, including the 65-kDa regulatory subunit of protein phosphatase 2A (PP2A), a serine/threonine phosphatase that regulates the transition from G2 to mitosis (107). In addition to roles in nuclear localization and cell cycle arrest, Vpr can also influence mutation rates during viral DNA synthesis (109) and has been proposed to form an ion channel (61).

RT

Before the viral genome can be integrated into the host chromosome, it must first be reverse transcribed into duplex DNA (step 14, Figure 2). RT catalyzes both RNA-dependent and DNA-dependent DNA polymerization reactions and contains an RNase H domain that cleaves the RNA portion of RNA-DNA hybrids generated during the reaction. Reverse transcription initiates from the 3' end of a $tRNA_3^{Lys}$ primer annealed to the primer binding site near the 5' end of the genomic RNA. RT can use other tRNAs if complementary binding sites are provided, but reverse transcription is most efficient with $tRNA_3^{Lys}$ (110). $tRNA_3^{Lys}$ is incorporated into virions during assembly and is often extended by several nucleotides inside the particle (110, 111). The remainder of the reaction probably occurs after uncoating in the cytoplasm. The kinetic properties of RT during the initiation and elongation phases of the reaction are quite different, becoming highly processive during elongation, and posttranscriptional modifications of $tRNA_3^{Lys}$ enhance the formation of initiation complexes (112). These kinetic transitions are reminiscent of those observed in transcription complexes, with $tRNA_3^{Lys}$ performing a role analogous to σ factor (112). Following tRNA-primed initiation, reverse transcription involves two DNA strand transfer reactions that are catalyzed by RT and are important for priming the synthesis of both minus and plus strands (see references 113 and 114 for details of the mechanism).

RT has also been a major target for drug design, and crystal structures of unliganded RT, an RT-DNA complex, and RT-inhibitor complexes have been solved (115–120). RT is a heterodimer containing a 560-residue subunit (p66) and a 440-residue subunit (p51) both derived from the Pol polyprotein. Each subunit contains a polymerase domain composed of four subdomains called fingers, palm, thumb, and connection, and p66 contains an additional RNase H domain (Figure 3K). Even though their amino acid sequences are identical, the polymerase subdomains are arranged differently in the two subunits, with p66 forming a large active-site cleft and p51 forming an inactive closed structure (121). The p66 polymerase active site contains a catalytic triad (Asp110, Asp185, and Asp186) conserved in many polymerases (Figure 3K), and the 3'OH group of the primer strand in an RT-DNA complex is positioned close to the active site for nucleophilic attack on the incoming nucleoside triphosphate (118). The DNA in this complex has primer and template strands clamped between the palm, thumb, and fingers subdomains of p66 and is bent. Portions of the DNA near the active site adopt an A-form geometry expected of RNA-DNA hybrids or RNA duplexes bound during reverse transcription. The positioning of the DNA near the RNase H domain does not explain how RNA-DNA hybrids are cleaved. Cross-linking experiments suggest that the 5' part of the tRNA$_3^{Lys}$ primer may contact both the RT dimer interface and a C-terminal region of p66 during initiation (122).

Two classes of RT inhibitors are in clinical use: nucleoside analogs such as AZT and ddI that are presumed to bind to the polymerase active site and non-nucleoside inhibitors such as nevirapine. The structures of several non-nucleoside inhibitor-RT complexes show a common hydrophobic binding site near to, but distinct from, the polymerase active site that rearranges to fit the particular drug and lock RT into an inactive conformation (119, 120). Mutations that confer resistance to nucleoside or non-nucleoside inhibitors map to different parts of RT, including regions in and around the active site and DNA-binding cleft, suggesting that some mutations directly alter the drug-binding site while others have more indirect effects (120, 123). Structures of the unliganded RT show substantial variability in the positioning of the p66 thumb subdomain (115–117), indicating that large-scale conformational rearrangements occur upon nucleic acid or drug binding. Such conformational changes may be important during reverse transcription; for example, to allow translocation of RT along the nucleic acid or to correctly position the RNase H and polymerase active sites.

IN

Following reverse transcription, IN catalyzes a series of reactions to integrate the viral genome into a host chromosome (step 15, Figure 2). In the first step, IN removes two 3' nucleotides from each strand of the linear viral DNA, leaving overhanging CA$_{OH}$ ends (113). The CA dinucleotide is found at the ends of many retrotransposons, and mutation of these nucleotides substantially reduces the efficiency of 3'-end processing. In the second step, the processed 3' ends are covalently joined to the 5' ends of the target DNA. In the third step, which probably involves additional cellular enzymes, unpaired nucleotides at the viral 5' ends are removed and the ends are joined to the target site 3' ends, generating an integrated provirus flanked by five base-pair direct repeats of the target site DNA. The viral substrate used for integration is a linear DNA molecule containing a complete minus strand and a discontinuous plus strand, which is presumably completed by cellular enzymes following integration (124). In vitro systems using viral preintegration complexes, or purified IN with short oligonucleotides, have helped define important nucleotides near the viral DNA ends, important features of the target DNA, and critical amino acids of IN (113). The enzymatic mechanism involves two sequential transesterification reactions and requires no exogenous energy source, but an appropriate metal cofactor (either Mn^{2+} or Mg^{2+}) is needed (113).

Integration can occur at many target sites within the genome. In vitro studies have indicated a preference for kinked or distorted DNA, such as that found in nucleosomes, but it is not yet clear how these target sites relate to those used in vivo (125). It has been suggested that interactions with other DNA-binding proteins might target IN to specific sites, and a yeast two-hybrid screen has identified a human Snf5-related protein, Ini1, as a possible partner (126). Ini1 influences the efficiency of integration, but its effect on target site selection in vivo is unknown. Another cellular factor, HMG I(Y), is associated with preintegration complexes and plays a crucial role in integration (127).

IN is active as an oligomer, probably a tetramer (128), and the 288-residue monomer can be divided into three domains whose structures have been determined. The N-terminal domain (residues 1–55) contains a zinc-binding site (coordinated by two histidines and two cysteines) and forms a dimer with a largely hydrophobic interface, as shown by NMR (Figure 3L) (129). Each monomer contains a helix-turn-helix structure very similar to those found in DNA-binding proteins and exists in two closely related conformational states. The catalytic domain (residues 50–212) contains a D,D(35),E motif. This motif is conserved among integrases, is crucial for the processing and joining reactions, and is proposed to bind the active site metal ion (128). The isolated catalytic domain cannot perform processing or joining reactions but can perform an apparent reverse reaction, termed disintegration, indicating that it contains the catalytic site for polynucleotidyl transfer. The crystal structure of the catalytic domain shows a dimeric structure, with each monomer containing a five-stranded β-sheet and six α-helices similar to other polynucleotidyl transfer enzymes (Figure 3M) (130). Asp64 and Asp116 of the D,D(35),E motif are clearly seen in the structure, but Glu152 is located on a disordered loop. The two active sites in the dimer are too far apart to permit five base pair staggered cleavage of the target DNA, suggesting either that a very large conformational change occurs during catalysis or, more likely, that IN functions as a tetramer or other oligomeric form during some steps of the reaction (128, 129). The C-terminal domain (residues 220–270) has nonspecific DNA-binding activity and forms a dimer of parallel monomers, as shown by NMR (Figure 3N). The structure of each monomer consists of a five-stranded β barrel strikingly similar to a SH3 domain, with a saddle-shaped groove that might accommodate double-stranded DNA and containing Lys264, an important DNA-binding residue (Figure 3N) (131, 132). The relative orientations of the three IN domains remain to be established (128, 129).

CONCLUDING REMARKS

The past decade has seen remarkable progress in elucidating the structures and functions of many of the HIV-1 proteins. Achieving such a sophisticated level of understanding of this complex retrovirus in such a short period of time is a true testament to the collaborative efforts of the scientific community. With this basic framework in hand, it should now be possible to probe each viral component in greater detail and to focus attention on the remaining gaps in our knowledge of HIV-1 biology, including issues pertaining to virus-host interactions and pathogenesis. At the molecular level, the structures of several viral proteins and RNAs and the domain arrangements in proteins such as CA and IN remain to be solved, and interaction surfaces between viral factors and other viral or cellular partners remain to be mapped. Many fundamental questions are unanswered: How do the Gag and Gag-Pol proteins, RNA, SU, and TM interact to form the virus particle? How do Vif, Nef, cyclophilin A, Vpr, and the Gag proteins contribute to viral uncoating, nuclear transport, and other early steps of replication? How are the activities of the viral enzymes PR, RT, and IN regulated at the appropriate steps in the replication cycle? What cellular factors are required for the functions of Tat, Rev, and other viral proteins? How do viral factors take advantage of existing cellular mechanisms, including transcription, RNA processing, protein synthesis and degradation, protein and RNA transport, and membrane trafficking? It is hoped that answers to these and other questions will lead to the discovery of new classes of effective HIV-1 therapeutics.

ACKNOWLEDGMENTS

We thank our many colleagues who have contributed structure coordinates to the Brookhaven Protein Database, and we thank Marius Clore, Andrea Dessen, Theresa Gamble, Angela Gronenborn, Tom James, Peter Kim, John Kuriyan, Anwer Mujeeb, Tris Parslow, Michael Summers, Wes Sundquist, and Don Wiley for contributing coordinates of recent structures. We also thank Ron Fouchier and Michael Malim for sharing unpublished results and our many colleagues who have made substantial intellectual contributions to our thoughts on HIV, particularly members of our laboratories and Mark Feinberg and Raul Andino. Work in our laboratories has been supported by NIH grants AI29135, GM47478, and GM39589 (ADF) and CA62000 and CA70810 (JATY).

> Visit the *Annual Reviews home page* at
> http://www.AnnualReviews.org.

Literature Cited

1. Jones KA, Peterlin BM. 1994. *Annu. Rev. Biochem.* 63:717–43
2. Zhou QA, Sharp PA. 1996. *Science* 274:605–10
3. Parada CA, Roeder RG. 1996. *Nature* 384:375–78
4. Yang XZ, Herrmann CH, Rice AP. 1996. *J. Virol.* 70:4576–84
5. Garcia-Martinez LF, Mavankal G, Neveu JM, Lane WS, Ivanov D, Gaynor RB. 1997. *EMBO J.* 16:2836–50
6. Cujec TP, Cho H, Maldonado E, Meyer J, Reinberg D, Peterlin BM. 1997. *Mol. Cell. Biol.* 17:1817–23
6a. Jones KA. 1997. *Genes Dev.* 11:2593–99
7. Puglisi JD, Tan RY, Calnan BJ, Frankel AD, Williamson JR. 1992. *Science* 257:76–80
8. Aboul-ela F, Karn J, Varani G. 1995. *J. Mol. Biol.* 253:313–32
9. Bayer P, Kraft M, Ejchart A, Westendorp M, Frank R, Rosch P. 1995. *J. Mol. Biol.* 247:529–35
10. Alonso A, Cujec TP, Peterlin BM. 1994. *J. Virol.* 68:6505–13
11. Clever JL, Parslow TG. 1997. *J. Virol.* 71:3407–14
12. Paillart JC, Skripkin E, Ehresmann B, Ehresmann C, Marquet R. 1996. *Proc. Natl. Acad. Sci. USA* 93:5572–77
13. Laughrea M, Jette L, Mak J, Kleiman L, Liang C, Wainberg MA. 1997. *J. Virol.* 71:3397–406
14. Cassan M, Delaunay N, Vaquero C, Rousset JP. 1994. *J. Virol.* 68:1501–8
14a. Hope TJ. 1997. *Chem. Biol.* 4:335–44
15. Fischer U, Huber J, Boelens WC, Mattaj IW, Luhrmann R. 1995. *Cell* 82:475–83
16. Meyer BE, Malim MH. 1994. *Genes Dev.* 8:1538–47
17. Fritz CC, Zapp ML, Green MR. 1995. *Nature* 376:530–33
18. Bogerd HP, Fridell RA, Madore S, Cullen BR. 1995. *Cell* 82:485–94
18a. Stutz F, Neville M, Rosbash M. 1995. *Cell* 82:495–506
18b. Ullman KS, Powers MA, Forbes DJ. 1997. *Cell* 90:967–70
19. Lu XB, Heimer J, Rekosh D, Hammarskjold ML. 1990. *Proc. Natl. Acad. Sci. USA* 87:7598–602
20. Kjems J, Sharp PA. 1993. *J. Virol.* 67:4769–76
21. Zemmel RW, Kelley AC, Karn J, Butler PJ. 1996. *J. Mol. Biol.* 258:763–77
22. Battiste JL, Mao H, Rao NS, Tan R, Muhandiram DR, et al. 1996. *Science* 273:1547–51
23. Ye XM, Gorin A, Ellington AD, Patel DJ. 1996. *Nat. Struct. Biol.* 3:1026–33
24. Hill CP, Worthylake D, Bancroft DP, Christensen AM, Sundquist WI. 1996. *Proc. Natl. Acad. Sci. USA* 93:3099–104
25. Lee PP, Linial ML. 1994. *J. Virol.* 68:6644–54
26. Freed EO, Englund G, Martin MA. 1995. *J. Virol.* 69:3949–54
27. Mammano F, Kondo E, Sodroski J, Bukovsky A, Gottlinger HG. 1995. *J. Virol.* 69:3824–30
28. Massiah MA, Worthylake D, Christensen AM, Sundquist WI, Hill CP, Summers MF. 1996. *Protein Sci.* 5:2391–98
29. Gallay P, Swingler S, Song JP, Bushman F, Trono D. 1995. *Cell* 83:569–76
30. Bukrinsky MI, Haggerty S, Dempsey MP, Sharova N, Adzhubel A, et al. 1993. *Nature* 365:666–69
31. Bukrinskaya AG, Ghorpade A, Heinzinger NK, Smithgall TE, Lewis RE, Steverson M. 1996. *Proc. Natl. Acad. Sci. USA* 93:367–71
32. Gallay P, Stitt V, Mundy C, Oettinger M, Trono D. 1996. *J. Virol.* 70:1027–32
33. Gallay P, Swingler S, Aiken C, Trono D. 1995. *Cell* 80:379–88
34. Trono D, Gallay P. 1997. *Cell* 88:173–74
35. Freed EO, Englund G, Maldarelli F, Martin MA. 1997. *Cell* 88:171–73
36. Fouchier RAM, Meyer BE, Simon JHM, Fischer U, Malim MH. 1997. *EMBO J.* 16:4531–39

37. Gamble TR, Yoo S, Vajdos FF, Von Schwedler UK, Worthylake DK, et al. 1997. *Science.* In press
38. Luban J. 1996. *Cell* 87:1157–59
39. Gamble TR, Vajdos FF, Yoo S, Worthylake DK, Houseweart M, et al. 1996. *Cell* 87:1285–94
40. Momany C, Kovari LC, Prongay AJ, Keller W, Gitti RK, et al. 1996. *Nat. Struct. Biol.* 3:763–70
41. Gitti RK, Lee BM, Walker J, Summers MF, Yoo S, Sundquist WI. 1996. *Science* 273:231–35
42. Srinivasakumar N, Hammarskjold ML, Rekosh D. 1995. *J. Virol.* 69:6106–14
43. Mammano F, Ohagen A, Hoglund S, Gottlinger HG. 1994. *J. Virol.* 68:4927–36
44. Ebbets-Reed D, Scarlata S, Carter CA. 1996. *Biochemistry* 35:14268–75
45. Dorfman T, Gottlinger HG. 1996. *J. Virol.* 70:5751–57
46. Berkowitz RD, Ohagen A, Hoglund S, Goff SP. 1995. *J. Virol.* 69:6445–56
47. Huang Y, Khorchid A, Wang J, Parniak MA, Darlix JL, et al. 1997. *J. Virol.* 71:4378–84
48. Guo JH, Henderson LE, Bess J, Kane B, Levin JG. 1997. *J. Virol.* 71:5178–88
49. Cameron CE, Ghosh M, Le Grice SF, Benkovic SJ. 1997. *Proc. Natl. Acad. Sci. USA* 94:6700–5
50. Carteau S, Batson SC, Poljak L, Mouscadet J-F, deRocquigny H, et al. 1997. *J. Virol.* 71:6225–29
51. Schmalzbauer E, Strack B, Dannull J, Guehmann S, Moelling K. 1996. *J. Virol.* 70:771–77
52. Poon DT, Wu J, Aldovini A. 1996. *J. Virol.* 70:6607–16
53. Summers MF, Henderson LE, Chance MR, Bess JW Jr, South TL, et al. 1992. *Protein Sci.* 1:563–74
54. Morellet N, Jullian N, De Rocquigny H, Maigret B, Darlix JL, Roques BP. 1992. *EMBO J.* 11:3059–65
55. Rice WG, Supko JG, Malspeis L, Buckheit RW Jr, Clanton D, et al. 1995. *Science* 270:1194–97
56. Kondo E, Gottlinger HG. 1996. *J. Virol.* 70:159–64
57. Lu YL, Bennett RP, Wills JW, Gorelick R, Ratner L. 1995. *J. Virol.* 69:6873–79
58. Checroune F, Yao XJ, Gottlinger HG, Bergeron D, Cohen EA. 1995. *J. AIDS Hum. Retrovirol.* 10:1–7
59. Cohen EA, Subbramanian RA, Gottlinger HG. 1996. *Curr. Top. Microbiol. Immunol.* 214:219–35
60. Huang M, Orenstein JM, Martin MA, Freed EO. 1995. *J. Virol.* 69:6810–18
61. Lamb RA, Pinto LH. 1997. *Virology* 229:1–11
62. Tiganos E, Yao XJ, Friborg J, Daniel N, Cohen EA. 1997. *J. Virol.* 71:4452–60
63. Bour S, Schubert U, Strebel K. 1995. *J. Virol.* 69:1510–20
64. Fujita K, Omura S, Silver J. 1997. *J. Gen. Virol.* 78:619–25
65. Kerkau T, Bacik I, Bennink JR, Yewdell JW, Hunig T, et al. 1997. *J. Exp. Med.* 185:1295–305
66. Mangasarian A, Trono D. 1997. *Res. Virol.* 148:30–33
67. Le Gall S, Heard JM, Schwartz O. 1997. *Res. Virol.* 148:43–47
68. Guatelli JC. 1997. *Res. Virol.* 148:34–37
69. Bukovsky AA, Dorfman T, Weimann A, Gottlinger HG. 1997. *J. Virol.* 71:1013–18
70. Greenway A, McPhee D. 1997. *Res. Virol.* 148:58–64
71. Moarefi I, LaFevre-Bernt M, Sicheri F, Huse M, Lee CH, et al. 1997. *Nature* 385:650–53
72. Benichou S, Liu LX, Erdtmann L, Selig L, Benarous R. 1997. *Res. Virol.* 148:71–73
73. Lee CH, Saksela K, Mirza UA, Chait BT, Kuriyan J. 1996. *Cell* 85:931–42
74. Grzesiek S, Bax A, Hu J-S, Kaufman J, Palmer I, et al. 1997. *Protein Sci.* 6:1248–63
75. Sawai ET, Cheng-Mayer C, Luciw PA. 1997. *Res. Virol.* 148:47–52
76. Cullen BR. 1996. *Curr. Biol.* 6:1557–59
77. Kaplan AH, Manchester M, Swanstrom R. 1994. *J. Virol.* 68:6782–86
78. Zybarth G, Carter C. 1995. *J. Virol.* 69:3878–84
79. Dunn BM, Gustchina A, Wlodawer A, Kay J. 1994. *Methods Enzymol.* 241:254–78
80. Pettit SC, Moody MD, Wehbie RS, Kaplan AH, Nantermet PV, et al. 1994. *J. Virol.* 68:8017–27
81. Krausslich HG, Facke M, Heuser AM, Konvalinka J, Zentgraf H. 1995. *J. Virol.* 69:3407–19
82. Sheng N, Pettit SC, Tritch RJ, Ozturk DH, Rayner MM, et al. 1997. *J. Virol.* 71:5723–32
83. Luukkonen BG, Fenyo EM, Schwartz S. 1995. *Virology* 206:854–65
84. Wlodawer A, Erickson JW. 1993. *Annu. Rev. Biochem.* 62:543–85
85. Condra JH, Schleif WA, Blahy OM, Gabryelski LJ, Graham DJ, et al. 1995. *Nature* 374:569–71
86. Ridky T, Leis J. 1995. *J. Biol. Chem.* 270:29621–23

87. Schock HB, Garsky VM, Kuo LC. 1996. *J. Biol. Chem.* 271:31957–63
88. McPhee F, Good AC, Kuntz ID, Craik CS. 1996. *Proc. Natl. Acad. Sci. USA* 93:11466–81
89. Simon JH, Malim MH. 1996. *J. Virol.* 70:5297–305
90. Yang X, Goncalves J, Gabuzda D. 1996. *J. Biol. Chem.* 271:10121–29
91. Camaur D, Trono D. 1996. *J. Virol.* 70:6106–11
92. Fouchier RA, Simon JH, Jaffe AB, Malim MH. 1996. *J. Virol.* 70:8263–69
93. Karczewski MK, Strebel K. 1996. *J. Virol.* 70:494–507
94. Liu HM, Wu XY, Newman M, Shaw GM, Hahn BH, Kappes JC. 1995. *J. Virol.* 69:7630–38
95. Luciw PA. 1996. In *Fundamental Virology,* ed. BM Fields, DM Knipe, PM Howley, pp. 845–916. Philadelphia: Lippincott-Raven. 3rd ed.
96. Clapham PR, Weiss RA. 1997. *Nature* 388:230–31
97. Bleul CC, Farzan M, Choe H, Parolin C, Clark-Lewis I, et al. 1996. *Nature* 382:829–33
98. Oberlin E, Amara A, Bachelerie F, Bessia C, Virelizier JL, et al. 1996. *Nature* 382:833–35
99. Choe H, Farzan M, Sun Y, Sullivan N, Rollins B, et al. 1996. *Cell* 85:1135–48
100. Clapham PR. 1997. *Trends Cell Biol.* 7:264–68
101. Cocchi F, DeVico AL, Garzino-Demo A, Cara A, Gallo RC, Lusso P. 1996. *Nat. Med.* 2:1244–47
102. Oravecz T, Pall M, Norcross MA. 1996. *J. Immunol.* 157:1329–32
103. Hernandez LD, Hoffman LR, Wolfsberg TG, White JM. 1996. *Annu. Rev. Cell Dev. Biol.* 12:627–61
104. Chan DC, Fass D, Berger JM, Kim PS. 1997. *Cell* 89:263–73
105. Weissenhorn W, Dessen A, Harrison SC, Skehel JJ, Wiley DC. 1997. *Nature* 387:426–30
106. Miller MD, Farnet CM, Bushman FD. 1997. *J. Virol.* 71:5382–90
107. Emerman M. 1996. *Curr. Biol.* 6:1096–1103
108. Sato A, Yoshimoto J, Isaka Y, Miki S, Suyama A, et al. 1996. *Virology* 220:208–12
109. Mansky LM. 1996. *Virology* 222:391–400
110. Oude Essink BB, Das AT, Berkhout B. 1996. *J. Mol. Biol.* 264:243–54
111. Huang Y, Wang J, Shalom A, Li Z, Khorchid A, et al. 1997. *J. Virol.* 71:726–28
112. Lanchy JM, Ehresmann C, Le Grice SF, Ehresmann B, Marquet R. 1996. *EMBO J.* 15:7178–87
113. Katz RA, Skalka AM. 1994. *Annu. Rev. Biochem.* 63:133–73
114. Peliska JA, Benkovic SJ. 1992. *Science* 258:1112–18
115. Rodgers DW, Gamblin SJ, Harris BA, Ray S, Culp JS, et al. 1995. *Proc. Natl. Acad. Sci. USA* 92:1222–26
116. Esnouf R, Ren J, Ross C, Jones Y, Stammers D, Stuart D. 1995. *Nat. Struct. Biol.* 2:303–8
117. Hsiou Y, Ding J, Das K, Clark AD Jr, Hughes SH, Arnold E. 1996. *Structure* 4:853–60
118. Jacobo-Molina A, Ding JP, Nanni RG, Clark AD Jr, Lu XD, et al. 1993. *Proc. Natl. Acad. Sci. USA* 90:6320–24
119. Kohlstaedt LA, Wang J, Friedman JM, Rice PA, Steitz TA. 1992. *Science* 256:1783–90
120. Ren JS, Esnouf R, Garman E, Somers D, Ross C, et al. 1995. *Nat. Struct. Biol.* 2:293–302
121. Wang J, Smerdon SJ, Jager J, Kohlstaedt LA, Rice PA, et al. 1994. *Proc. Natl. Acad. Sci. USA* 91:7242–46
122. Mishima Y, Steitz JA. 1995. *EMBO J.* 14:2679–87
123. Tantillo C, Ding J, Jacobo-Molina A, Nanni RG, Boyer PL, et al. 1994. *J. Mol. Biol.* 243:369–87
124. Miller MD, Wang B, Bushman FD. 1995. *J. Virol.* 69:3938–44
125. Miller MD, Bor YC, Bushman F. 1995. *Curr. Biol.* 5:1047–56
126. Miller MD, Bushman FD. 1995. *Curr. Biol.* 5:368–70
127. Farnet CM, Bushman FD. 1997. *Cell* 88:483–92
128. Rice P, Craigie R, Davies DR. 1996. *Curr. Opin. Struct. Biol.* 6:76–83
129. Cai M, Zheng R, Caffrey M, Craigie R, Clore GM, Gronenborn AM. 1997. *Nat. Struct. Biol.* 4:567–77
130. Dyda F, Hickman AB, Jenkins TM, Engelman A, Craigie R, Davies DR. 1994. *Science* 266:1981–86
131. Eijkelenboom AP, Lutzke RA, Boelens R, Plasterk RH, Kaptein R, Hard K. 1995. *Nat. Struct. Biol.* 2:807–10
132. Lodi PJ, Ernst JA, Kuszewski J, Hickman AB, Engelman A, et al. 1995. *Biochemistry* 34:9826–33
133. Fitzgerald PMD, McKeever BM, VanMiddlesworth JF, Springer JP, Heimbach JC, et al. 1990. *J. Biol. Chem.* 264:14209–19
134. Kraulis P. 1991. *J. Appl. Crystallogr.* 24:924–50

Annu. Rev. Biochem. 1998. 67:27–48

SPHINGOLIPID FUNCTIONS IN *SACCHAROMYCES CEREVISIAE*: Comparison to Mammals

Robert C. Dickson

Department of Biochemistry, University of Kentucky College of Medicine and the Lucille P. Markey Cancer Center, Lexington, Kentucky 40536; e-mail: bobd@pop.uky.edu

KEY WORDS: heat shock, ceramide, signal transduction, stress response, lipid second messengers

ABSTRACT

Many roles for sphingolipids have been identified in mammals. Available data suggest that sphingolipids and their intermediates also have diverse roles in *Saccharomyces cerevisiae*. These roles include signal transduction during the heat stress response, regulation of calcium homeostasis or components in calcium-mediated signaling pathways, regulation of the cell cycle, and functions as components in trafficking of secretory vesicles from the endoplasmic reticulum to the Golgi apparatus and as the lipid moiety in many glycosylphosphatidylinositol-anchored proteins. *S. cerevisiae* is likely to be the first organism in which all genes involved in sphingolipid metabolism are identified. This information will provide an unprecedented opportunity to determine, for the first time in any organism, how sphingolipid synthesis is regulated. Through the use of both genetic and biochemical techniques, the identification of the complete array of processes regulated by sphingolipid signals is likely to be possible, as is the quantification of the physiological contribution of each.

CONTENTS

INTRODUCTION

Sphingolipids, particularly sphingomyelin and glycosphingolipids, have been well established as essential components of mammalian cells, where they are predominantly found on the outer leaflet of the plasma membrane (1–3). Sphingolipids are composed of a long chain sphingoid base, generally 18 carbons long (e.g. sphinganine, Figure 1), with the 2-amino group amide-linked to a fatty acid and with various polar constituents added to the C_1 OH group (4). Once thought of as primarily structural components of membranes, our view of sphingolipids has expanded so that they are now also appreciated as diverse and dynamic regulators of a rapidly growing number of cellular functions. This review focuses on the functions of sphingolipids in the unicellular baker's yeast *Saccharomyces cerevisiae*, an important model organism for deciphering the mechanisms of eukaryotic cellular processes and their modes of regulation. The role of sphingolipids as second messengers for regulating intracellular signal transduction pathways is emphasized and compared with multicellular eukaryotes, primarily mammals.

Interest in sphingolipids as second messengers for regulating intracellular signal transduction pathways was spurred by the demonstration that sphingosine, the primary long chain base component of mammalian sphingolipids, is a potent inhibitor of protein kinase C (5). Interest was also spurred by the subsequent suggestion of a sphingomyelin cycle (6) in which agonists [for example, tumor necrosis factor α (7, 8)] bind to extracellular membrane receptors, triggering the hydrolysis of sphingomyelin to yield the second messenger ceramide. An increasing body of evidence has since shown that ceramide, produced in response to such diverse agents and insults as ultraviolet light, Fas ligands, heat shock, DNA damage, chemotherapeutic agents, and others, activates intracellular signal transduction pathways. These pathways regulate many cellular processes, ranging from cell cycle control and apoptosis to senescence (9–11), immune responses (12, 13), and cell-cell interactions (1).

In addition to ceramide, other sphingolipids are important mammalian intracellular signaling molecules; these include sphingosine and sphingosine-1-phosphate (10, 14, 15), sphingosylphosphorylcholine (15, 16), and di- and tri-N-methylsphingosine (1). As discussed below, some of these molecules have also been implicated as signaling molecules in *S. cerevisiae*, and others are likely to be implicated in the near future.

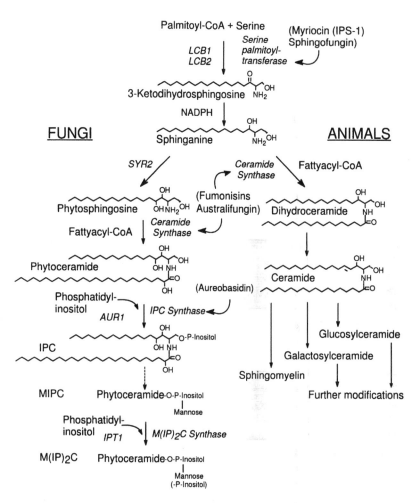

Figure 1   Sphingolipid biosynthetic pathway of fungi and animals. Known pathway intermediates, substrates, and enzymes (italics) are indicated for the synthesis of *S. cerevisiae* sphingolipids. Similarities and differences between fungal and mammalian sphingolipid biosynthesis are indicated. Genes necessary for specific steps in yeast sphingolipid synthesis are shown in italics, and these along with other genes involved in sphingolipid metabolism are discussed in the text and listed in Table 1. In the fungal pathway it is unclear whether 4-hydroxylation occurs on the free base (sphinganine and phytosphingosine) or on ceramide. Inhibitors are shown in parentheses.

S. cerevisiae, because of the recent determination of its complete genomic sequence (17) combined with the facility of gene manipulation in both haploid and diploid cells, is likely to be the first eukaryote for which all genes involved in sphingolipid metabolism will be identified. This knowledge will provide a unique opportunity to uncover new functions for sphingolipids and to understand the mechanisms for regulating sphingolipid synthesis—something that is not known in any organism. It should also reveal how sphingolipid metabolism is integrated with global cellular metabolism under changing conditions, many of which are stressful and potentially life threatening.

SACCHAROMYCES CEREVISIAE SPHINGOLIPIDS

Biosynthetic Pathway

The early steps in mammalian and *S. cerevisiae* sphingolipid synthesis are similar, but they produce structurally and chemically different types of long chain bases and ceramides. These differences may have no functional significance and simply represent species differences, but more likely, the differences may be used in species-specific ways to control unique signaling pathways by a novel mechanism or mechanisms. Our understanding of signaling mechanisms is too superficial to distinguish between these possibilities.

The proposed pathway for sphingolipid synthesis in *S. cerevisiae* is outlined in Figure 1 (18). None of the enzymes in this critical lipid biosynthetic pathway have been purified, and the relative order of several steps remains to be determined. The steps in the pathway up to the formation of sphinganine (also called dihydrosphingosine) appear to be the same in yeast and animals (Figure 1). The predominant ceramide found in aerobically grown *S. cerevisiae* is composed of an αOH-C_{26}-fatty acid amide-linked to phytosphingosine. It is unclear whether the 4-hydroxylation of sphinganine, a reaction requiring the *SYR2* gene (Figure 1 and Table 1), occurs at the level of free sphinganine, as shown in Figure 1, or at the level of ceramide or both. Likewise the substrate or substrates for α-hydroxylation of the fatty acid are unknown.

Mammals typically add fatty acids with shorter alkyl chains, C_{16} to C_{24}, to sphinganine to yield *N*-acylsphinganine (dihydroceramide), which is rapidly reduced to give ceramide. A 4,5 *trans*-double bond is introduced in the sphinganine moiety to form a ceramide containing sphingosine, not PHS, as its long chain base (14, 19). Thus, animals synthesize ceramide, whereas fungi and plants synthesize phytoceramide. The addition of one or two OH groups to the fatty acid portion of ceramide is common in yeast but rare in mammals (18). There are exceptions to these generalizations. For example, some mammals make small amounts of phytosphingosine (1), and the ceramides in skin have unusual structures and some contain very long chain fatty acids (20).

Table 1   Genes for *S. cerevisiae* sphingolipid metabolism

Gene	Function	Reference
AUR1	Synthesis of IPC, possible IPC synthase or a subunit of the enzyme	(27)
BST1	Breakdown of long chain base phosphates, possible long chain base phosphate lyase	(28, 28a)
CSG1	Required for mannosylation of sphingolipids	(138)
CSG2	Required for mannosylation of sphingolipids	(96)
ELO2	Synthesis of C_{24}-fatty acids, possible component of fatty acid elongation system	(132)
ELO3	Conversion of C_{24}- to C_{26}-fatty acids, possible component of fatty acid elongation system	(132)
IPT1	Synthesis of $M(IP)_2C$, possible $M(IP)_2C$ synthase	(133, 134)
LCB1	Synthesis of long chain bases, possible subunit of serine palmitoyltransferase	(135)
LCB2/ SCS1	Synthesis of long chain bases, possible catalytic subunit of serine palmitoyltransferase	(54, 96)
LCB3/ YSR2	Dephosphorylation of long chain base phosphates, possible long chain base-1-phosphate phosphatase	(29, 30)
SYR2	Hydroxylation of sphinganine or sphinganine-containing dihydroceramides at the C4 position to yield phytoceramide	(136, 137)

All fungi and plants that have been examined (18) add inositol phosphate to phytoceramide to form inositol-phosphorylceramide (IPC) (Figure 1). Animals do not do this; instead they transfer phosphocholine from phosphatidylcholine to ceramide to form the major sphingolipid sphingomyelin. Animals also add glucose or galactose to ceramide, and these additions are generally further decorated by addition of sugars and sometimes sulfates to yield a large number of complex glycosphingolipids (Figure 1) (21). In contrast, *S. cerevisiae* synthesizes only three types of complex sphingolipids: IPC, mannose-inositol-P-ceramide (MIPC), and mannose-(inositol-P)$_2$-ceramide [$M(IP)_2C$] (Figure 1).

Inhibitors of specific steps in yeast and mammalian sphingolipid biosynthesis have been identified and have proven useful in understanding sphingolipid functions. Myriocin (ISP-1) (22) and sphingofungins (23) inhibit serine palmitoyltransferase. Their use is limited, however, because they are not available commercially. Fumonisins (24) have proven very useful for studying sphingolipid functions in mammals, but they are less useful in yeast because *S. cerevisiae* cells are naturally quite resistant to them (25). A more useful drug for blocking

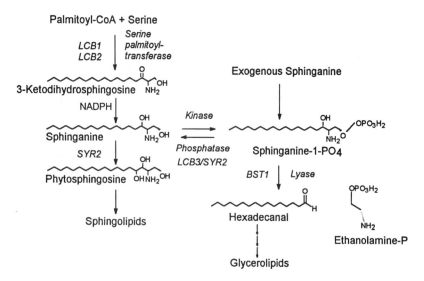

Figure 2 Pathways for sphinganine-1-PO$_4$ formation and breakdown in *S. cerevisiae*. Sphinganine-1-PO$_4$ can be synthesized de novo from palmitoyl-CoA and serine or directly by phosphorylation of exogenous sphinganine. It is not known whether yeast can also synthesize phytosphingosine-1-PO$_4$ from phytosphingosine.

ceramide synthesis in *S. cerevisiae* cells is australifungin (26). Unfortunately, it also is not available commercially. Recent results show that the antifungal drug aureobasidin A inhibits IPC synthase in *S. cerevisiae* cells [IC$_{50}$ = 0.2 nM, (27)]. No specific inhibitors of the other steps in yeast sphingolipid metabolism have been identified. Inhibitors for other steps in mammalian glycosphingolipid synthesis have been reviewed (21).

Sphingoid long chain bases, particularly sphingosine, are modified in animals to give a variety of effector molecules. Few comparable studies have been done in *S. cerevisiae*, but recent data indicate that sphinganine-1-PO$_4$ (Figure 2) is likely to be a signaling molecule in yeast. A multicopy plasmid that gives resistance to sphingosine was identified and shown to encode a gene (*BST1*; Figure 2 and Table 1) necessary for sphinganine-1-PO$_4$ lyase activity (28, 28a). A *bst1*-deletion mutant displays several phenotypes, including accumulation of sphinganine-1-PO$_4$, slow growth in the exponential phase, and increased growth in the stationary phase, suggesting that sphinganine-1-PO$_4$ is somehow involved in growth regulation. Further analysis of sphinganine-1-PO$_4$ metabolism (Figure 2) in *S. cerevisiae* should be very fruitful because its concentration can be manipulated by mutation of the lyase gene [*BST1* (28a)], the phosphatase gene [*LCB3/YSR2* (29, 30)], or the kinase gene, whose identification should be completed soon (31).

Analysis of a mutant defective in the sphinganine-1-PO$_4$ phosphatase gene (30) shows that exogenous long chain bases are not directly used as substrates for ceramide formation, as has been assumed, but are first phosphorylated and then dephosphorylated before being incorporated into sphingolipids or glycerophospholipids (Figure 2). This unexpected finding suggests that there may be different pools of long chain bases in yeast. Perhaps such pools are important for signaling events, but this possibility awaits experimental confirmation.

Cellular Location and Abundance

The three complex sphingolipids in *Saccharomyces cerevisiae* are located primarily in the plasma membrane (32, 33), where they account for 7–8% of the total mass of the membrane (30% of the plasma membrane phospholipids) (32). M(IP)$_2$C, the terminal sphingolipid, accounts for about 75% of the mass of total sphingolipids; the remainder are IPC and MIPC (34). Sphingomyelin and glycosphingolipids are highly concentrated in the outer leaflet of the plasma membrane of mammals. Lesser amounts are found in several other cellular membranes (see 35 and references therein). Whether yeast sphingolipids are concentrated in the inner leaflet or in the outer leaflet of the plasma membrane is not known.

Sphingolipid synthesis begins in the endoplasmic reticulum (ER) in both *S. cerevisiae* and mammals. In yeast, some phytoceramide is converted to IPC before transport to the Golgi apparatus, where the remaining polar head groups (Figure 1) are added (36). In mammals, the unmodified ceramide is transported from the ER by secretory vesicles to the Golgi apparatus, where it is converted to complex sphingolipids (37, 38).

Although primarily thought of as constituents of the plasma membrane, sphingolipids are found in other locations. Fractionated yeast cells showed complex sphingolipids, especially IPC, in the Golgi as well as the vacuole (the yeast equivalent of lysosomes). Trace amounts of sphingolipids were detected in mitochondria but none was detected in lipid particles; nuclei were not examined (33). Sphingomyelin has been detected in mammalian nuclear membranes (39), as has the ganglioside GM1 (40), but their function remains a mystery.

Recent studies in *S. cerevisiae* have shown that C$_{26}$-fatty acids are present in the nucleus. Whether they are free or incorporated into sphingolipids, glycerolipids, or some novel lipid species (41) is unknown, but most likely they are in sphingolipids. These C$_{26}$-fatty acids have been hypothesized to play a role in stabilizing the sharp curvature in the membrane that occurs at nuclear pore complexes.

Regulation of Sphingolipid Synthesis

The general regulatory schema in yeasts and mammals for controlling glycerolipid synthesis (42–45) and cholesterol biosynthesis (46, 47) are known, and the molecular mechanism or mechanisms for some parts of the regulatory circuits are understood in depth. In contrast, little is known about the regulation of de novo ceramide synthesis or synthesis of complex sphingolipids in any organism.

Based on a variety of correlative experiments in mammalian cells, it has been suggested that the first unique and committed step in sphingolipid biosynthesis, catalyzed by serine palmitoyltransferase (Figure 1), is likely to be regulated and to be important for controlling the rate of de novo ceramide synthesis and the rate of complex sphingolipid synthesis (19). One of the most direct pieces of evidence for this hypothesis is the demonstration that exogenous long chain bases down-regulate serine palmitoyltransferase transferase activity in cultured primary rat neurons (48). Further experimentation, particularly at the mechanistic level, is needed to verify that the rate of de novo ceramide synthesis is regulated at the serine palmitoyltransferase transferase step. In addition, this step is unlikely to be the only regulated step in the biosynthetic pathway based upon what is known about other lipid biosynthetic pathways.

An unexpected hint about the regulation of sphingolipid synthesis in *S. cerevisiae* was uncovered during studies of the Sec14 protein, which is the yeast phosphatidylinositol transfer protein necessary for secretory vesicle formation from the Golgi apparatus (49). Mutations in several genes including *SAC1* can rescue or bypass the Sec14 protein requirement (see 50 for original references). Kearns et al (50) recently demonstrated that the rate of flux through the sphingolipid biosynthetic pathway was six times higher than normal in a *sac1* bypass strain, with a commensurate increase in the formation of the terminal sphingolipid $M(IP)_2C$. The Sac1 protein appears to increase the rate of conversion of MIPC to $M(IP)_2C$, a step catalyzed by the Ipt1 protein (Table 1), so as to generate diacylglycerol, postulated to be necessary for vesicle budding from the Golgi (50). This is the first indication that the Sac1 protein regulates inositol phospholipid metabolism. What needs to be determined now is whether the wild-type Sac1 protein regulates sphingolipid synthesis and, if so, whether it interacts directly with the Ipt1 protein (Figure 1) or influences its activity indirectly.

FUNCTIONAL ROLES FOR SPHINGOLIPIDS IN *S. CEREVISIAE*

Stress Responses

Prokaryotes and eukaryotes have evolved multiple and complex mechanisms for defending themselves against stressful insults. A large body of evidence supports the hypothesis that many diverse stresses generate ceramide, which acts as a second messenger to activate intracellular signal transduction pathways.

Depending on the stress and the cell type, these pathways can produce cell cycle arrest, apoptosis, or senescence (1, 9–13).

Hannun (9) has suggested that ceramide may function in a broader sense than simply as a second messenger and act as a component of a biostat that measures and initiates responses to cellular stresses. This suggestion is made for two reasons. First, unlike many second messengers, such as adenosine $3',5'$-monophosphate (cAMP), ceramide is an essential intermediate in sphingolipid synthesis. Second, changes in its concentration often occur over long time spans, on the order of hours, rather than over short time spans, on the order of seconds or minutes, as seen with cAMP and many other second messengers (9). Evidence discussed below demonstrates that the phytoceramide level in *S. cerevisiae* cells increases to a stable plateau during heat stress, as expected for a biostat, and that this increase may regulate thermoprotection processes.

The first indication of an essential role for sphingolipids in stress resistance in *S. cerevisiae* came from studies of mutant strains (termed SLC for sphingolipid compensatory) that can grow without making sphingolipids (51). Although sphingolipids are normally essential for the viability of *S. cerevisiae* cells (52), this requirement can be bypassed by the *SLC1-1* suppressor mutation (53). Besides the suppressor mutation, SLC strains contain a deletion of the *LCB1* gene or the *LCB2* gene that encodes subunits of serine palmitoyltransferase (54), the initial enzyme in sphingolipid synthesis (Figure 1). Strains mutated in either *LCB* gene are auxotrophic for a long chain base (51, 52, 55). This requirement for a long chain base and for sphingolipids is bypassed or suppressed by mutation of the *SLC1* gene, believed to encode a fattyacyltransferase (53). The suppressor gene enables cells to produce a novel set of glycerophospholipids that mimic sphingolipid structures (56) and probably some of their functions. The novel lipids are phosphatidylinositol (PI), mannosyl-PI, and inositol-P-(mannosyl-PI), all containing a C_{26}-fatty acid in the *sn*-2-position of the glycerol moiety. Normally the C_{26}-fatty acid is found not in diacylglycerol phospholipids but only in phytoceramide, IPC, MIPC, $M(IP)_2C$ (Figure 1), and some lipid anchors, as described below.

SLC strains are unable to grow at pH 4, at 37°C, or in the presence of 0.75-M sodium chloride when they lack sphingolipids. When SLC cells contain sphingolipids, they grow under these stress conditions as well as wild-type yeast cells (57). These results imply that sphingolipids are necessary for a normal stress response. In addition, SLC cells begin to die at the start of the diauxic shift and are unable to enter the stationary phase, suggesting that sphingolipids are necessary for this long-term survival phase of growth (M Nagiec, M Skrzypek, R Lester, P Tillman & R Dickson, unpublished data).

HEAT STRESS   Because of the impaired stress responses of SLC cells lacking sphingolipids and the role of ceramide signaling in mammalian stress responses, stresses seem likely to induce an increase in the level of ceramide and other sphingolipids or their precursors in yeast. Examination of ceramide and long chain base intermediates following a shift from a nonstressful (23–25°C) to a stressful (37–39°C) temperature demonstrated 2- to 3-fold transient increases in the concentration of C_{18}-sphinganine and C_{18}-PHS, peaking between 5 and 10 min after the temperature shift. It showed a more than 100-fold transient increase in C_{20}-sphinganine and C_{20}-PHS, peaking 10 min after the stress. In addition, there was a 3-fold increase in total ceramides that reached a stable plateau after 20 min (58, 59; G Jenkins & Y Hannun, personal communication). Increases in ceramide, the only intermediate analyzed, were not observed during osmotic or low pH stress. These data are the first to suggest that one or more sphingolipid intermediates are signaling molecules specific for heat stress in yeast. If these are signaling molecules, what processes are they regulating?

S. cerevisiae cells have multiple responses to heat stress. One universal response is termed the heat shock response. It refers to induced synthesis of the heat shock proteins following exposure to an elevated but not extreme temperature (60). For example, switching S. cerevisiae cells from 25 to 37°C induces heat shock proteins. Heat shock proteins are primarily but not exclusively chaperones that facilitate refolding of proteins damaged by heat stress (61, 62). Recent experiments have shown that heat shock proteins are induced in SLC cells lacking sphingolipids, as well as in SLC cells containing sphingolipids and in wild-type cells. This finding indicates that sphingolipids are not necessary for signaling induction of the heat shock proteins, at least not the major heat shock proteins (59).

Another thermoprotective mechanism in S. cerevisiae is accumulation of the disaccharide trehalose (α-D-glucopyranosyl-α-D-glucopyranoside), which occurs rapidly following a shift from 25 to 37°C (63, 64, and references therein). Trehalose is known to protect proteins and biological membranes from heat denaturation (65–67). Mutant strains deleted for *tps1* or *tps2*, encoding subunits of trehalose synthase, have reduced thermotolerance, indicating the importance of trehalose in thermotolerance (63). Examination of SLC cells shows that they fail to accumulate trehalose when switched from 25 to 37°C in the absence of sphingolipid synthesis. Sphingolipids thus play a role in the induction of trehalose accumulation (59).

A common technique in the field of signal transduction is to determine if the effect of a biological stimulus, such as heat shock, can be mimicked by treating cells with a suspected signaling molecule or a precursor of a signaling molecule. Ceramide signaling in mammalian cells has been studied extensively through the use of more water-soluble, cell-permeable ceramides such as *N*-acetylsphingosine (C_2-ceramide) to attempt to mimic the effect of biological ligands (8, 68, 69). Natural ceramides cannot be used because they are highly insoluble in water.

C_2-ceramide is unlikely to be a good mimic in *S. cerevisiae* because, as shown in Figure 1, it is structurally distant and less hydrophobic than natural yeast ceramides. Thus, to attempt to mimic the effect of heat treatment, cells grown at 25°C were treated with sphinganine or other long chain bases. Such treatments mimicked the effect of heat and induced trehalose accumulation (59).

The way in which heat induces trehalose accumulation is not well characterized in *S. cerevisiae*, but the mechanism or mechanisms must induce de novo trehalose synthesis and block turnover by the hydrolytic enzyme trehalase (63). Trehalose accumulation is transient during stresses up to about 40°C and stably accumulates only with stresses over 40°C (64 and references therein). Transcription of one trehalose biosynthetic gene, *TPS2*, is heat inducible (70–72). Sphinganine treatment induces *TPS2* transcription 12-fold compared with a 9-fold induction by heat (59). The effect of sphinganine was specific: Transcription of another heat-inducible gene, *HSP104* [encoding a heat shock protein (73)], was induced only 2-fold by sphinganine, whereas it was induced 10-fold by heat. The ability of sphinganine to mimic the effect of a 40°C heat treatment is consistent with the hypothesis that sphinganine, PHS, and phytoceramide, or an uncharacterized derivative or derivatives of these compounds, produced during heat stress, activate a signal transduction pathway or pathways leading to induction of *TPS2* transcription (Figure 3).

Many stresses in *S. cerevisiae* including heat activate gene transcription through a promoter element termed *STRE* (stress response element) (74–76), which contains an CCCCT or AGGGG base sequence. Four *STRE*s are present in the *TPS2* promoter (72), and these may mediate the sphinganine-responsiveness of *TPS2* expression (59). Heat and other stresses activate an unknown signaling pathway or pathways connecting to two transcription activators, Msn2 and Msn4, which bind *STRE*s (77, 78). Sphinganine signaling is likely to regulate Msn2 and Msn4 activity, but this possibility awaits experimental conformation. Whether mammals have Msn2 and Msn4 homologues is not known, but if they do, it will be of interest to determine if they respond to sphingolipid signals.

GENERATION OF SPHINGOLIPID SIGNALS   The increase in ceramide produced during mammalian stress responses is generated in most cases by sphingomyelinase action on membrane-bound sphingomyelin (9, 10). How ceramide is generated during heat stress in mammals is not known (79, 80). Two lines of evidence indicate that the increase in phytoceramide caused by heat stress in *S. cerevisiae* arises by de novo synthesis, not by breakdown of preformed sphingolipids (58, 59). First, heat treatment does not induce accumulation of sphingolipid polar head groups, therefore indicating an absence of sphingolipid breakdown. Second, treatment of cells with australifungin (Figure 1) blocks ceramide accumulation following heat shock, as expected when the increase is due to de novo ceramide synthesis. Thus, *S. cerevisiae* appears to represent a

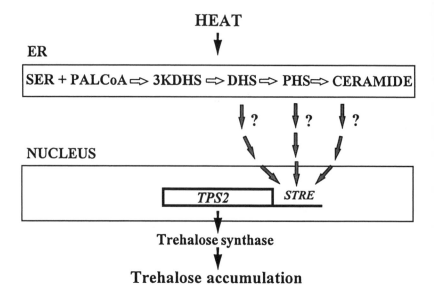

Figure 3   Possible roles for sphingolipid intermediates in the *S. cerevisiae* heat stress response. This diagram attempts to explain why heat shock (switching cells from 25 to 37–39°C) causes transient increases in the sphingolipid intermediates sphinganine and PHS and stable increases in ceramides. It also attempts to show why addition of sphinganine to the culture medium mimics the effect of heat shock and induces transcription of the *TPS2* gene and accumulation of trehalose, a known thermoprotectant. The sphingolipid intermediates sphinganine and PHS can have chain lengths of C_{18} or C_{20} and give rise to C_{18} or C_{20}-ceramides, all of which have the potential to be signaling molecules (indicated by question marks). The *STRE* promoter element is known to bind the transcription activator proteins Msn2 and Msn4 during heat stress (77, 78), but it is not known if sphingolipid signals use these proteins to activate *TPS2* transcription. Abbreviations are SER, serine; PALCoA, palmitoylCo-A; 3KDHS, 3-ketodihydrosphingosine; DHS, sphinganine; PHS, phytosphingosine; and CERAMIDE, phytoceramide.

new paradigm for the generation of sphingolipid signals. There are indications that the mechanism for generating increased ceramide in mammals occurs under some circumstances through de novo synthesis rather than through breakdown of sphingomyelin (81, 81a), although this possibility needs further study (82). Other undefined mechanisms for generating increases in ceramide have been observed (6).

Although there is no indication in vivo of rapid sphingolipid turnover in *S. cerevisiae*, a phospholipase C-like activity capable of hydrolyzing yeast sphingolipids has been detected in membrane preparations (R Lester, personal communication). A sphingomyelinase-like activity has been detected in *S. cerevisiae*. However, this lipase was not able to hydrolyze authentic yeast sphingolipids to yield phytoceramide (83). The physiological function and identity of this lipase are unknown.

An important area for future research is the molecular mechanism or mechanisms for increasing the level of phytoceramide threefold to fivefold by de novo synthesis following heat stress in *S. cerevisiae*; it could enhance our understanding of ceramide signaling in mammals. Likewise, future research should reveal how the sphingolipid signal or signals are generated in response to heat. This information may help to reveal how ceramide is generated during mammalian heat stress (79, 80). Finally, it needs to be determined which sphingolipid intermediate or intermediates, C_{18} or C_{20}-PHS, C_{18} or C_{20}-sphinganine, phytoceramide or some unrecognized derivative, activate the signaling pathway leading to *STRE*s and how the signal is recognized and transduced to the nucleus (Figure 3).

Components of the signaling pathway or pathways upstream of the Msn2 and Msn4 proteins that connect to heat stress are unknown (Figure 3). The identification of these components is a major mystery waiting to be solved. Sphingolipid signals such as sphinganine, PHS, and phytoceramide would be expected to be generated by de novo synthesis in the ER, where they might interact with a receptor and activate a signaling pathway terminating in the nucleus with the Msn2 and Msn4 proteins. An example of a signaling pathway emanating in the ER and terminating in the nucleus is the unfolded protein response pathway (84, 85). This pathway is composed of the transmembrane protein kinase Ire1, present in the ER, whose C terminus is thought to project into the nucleus and interact with the transcription coactivator Gcn5. The Gcn5 protein then activates transcription of genes including *KAR2*, the yeast homologue of mammalian BiP disulfide isomerase, necessary for folding of proteins in the ER (86). Whether sphingolipid signals use a similar or different type of signaling pathway awaits determination.

LOW pH STRESS AND PLASMA MEMBRANE FUNCTIONS   The previously mentioned inability of SLC cells to grow at pH 4.1 when they lack sphingolipids (57) is a strong indication that sphingolipids perform one or more essential functions during low pH stress. A clue to the function or functions of sphingolipids at low pH comes from the observation that *S. cerevisiae* mutants defective in the proton ATPase (H^+-ATPase) encoded by *PMA1* (87) also fail to grow at low pH.

The Pma1 protein, which is related to the mammalian P-type Na^+, K^+-ATPases, regulates intracellular pH and generates the proton gradient necessary for nutrient transport into *S. cerevisiae* cells (88). Acidification or addition of glucose to the culture medium causes a drop in intracellular pH. This drop activates the H^+-ATPase, thereby pumping protons out of the cell and restoring intracellular pH to its optimal range of 6.8–7.3 (88, 89). Extracellular pHs below

4 reduce the K_m of the ATPase twofold, but the mechanism or mechanisms for this mode of regulation are unknown. Addition of glucose to the culture medium reduces the K_m twofold, increases the V_{max} threefold, increases the pH optimum from 5.6 to 6–7, and reduces the K_i for vanadate. These data suggest that acid and glucose regulate the activity of the H^+-ATPase by different mechanisms. Glucose activation of the ATPase is probably due to phosphorylation (90). Current evidence indicates that two protein kinases regulate the activity of the Pma1 protein in response to glucose. A casein kinase I, encoded by *YCK1* and *YCK2*, down-regulates the activity of the H^+-ATPase, and a second but unknown protein kinase (or kinases) up-regulate the enzyme. The net effect of glucose is to increase ATPase activity (91). The protein phosphatases responsible for removing the phosphates and the pathway from glucose to the ATPases are unknown. Results from drug inhibition studies suggest that the unknown glucose-activated kinase or kinases may involve protein kinase C and one or more Ca^{2+}/calmodulin-dependent protein kinases that are activated by a phosphatidylinositol-type signaling pathway (92). Although detailed mechanisms are lacking, these data indicate that one or more signal transduction pathways regulate Pma1 activity. The failure of SLC cells to grow at low pH when sphingolipids are absent may indicate a second messenger function for sphingolipids in regulating one or more of the signaling pathways that regulate Pma1 activity. This appears to be an interesting and fertile area for future studies.

Alternatively, or in addition, sphingolipids may help to fulfill the known lipid requirement of Pma1 (89). In vitro measurements showed that the ATPase activity of Pma1 could be satisfied by *S. cerevisiae* sphingolipids, as well as by other yeast lipids (93).

Calcium Homeostasis

Sphingosine and sphingosine-1-phosphate cause Ca^{2+} to be released from intracellular stores in a variety of mammalian cell types (94), and the accumulating data indicate that sphingolipid mediators are important regulators of Ca^{2+} homeostasis.

A possible relationship between *S. cerevisiae* sphingolipids and Ca^{2+}-regulated processes or Ca^{2+} homeostasis has been suggested from studies of the *CSG2* gene (95). Strains mutated in *csg2* do not grow in the presence of high concentrations of Ca^{2+} (100 mM) and make only one type of sphingolipid, an uncharacterized species of IPC (95). Second-site mutations, termed *SCS* (for suppressor of calcium-sensitive growth), which suppress the calcium-sensitive growth phenotype, were isolated and shown to define seven complementation groups, *SCS1–7* (96). The *scs* mutations partially restore synthesis of some sphingolipids, but none of the strains have a wild-type sphingolipid profile. One complementation group, *SCS1* (96), is identical to the *LCB2* gene (54), necessary for the first step in sphingolipid synthesis (Figure 1 and Table 1). Based on these and other results, it has been suggested that sphingolipid metabolism in *S. cerevisiae* is regulated by Ca^{2+} or is required for Ca^{2+} homeostasis or both (96).

Another role for sphingolipids in Ca^{2+} signaling is suggested from studies of the cytosolic Nth1 trehalase (97) that degrades trehalose. The enzyme is inhibited by an unidentified protein capable of binding to calmodulin (98). Because heat stress causes sphingolipid intermediates to increase and trehalose to accumulate, and because cells treated with sphingoid long chain bases accumulate trehalose (59), it is possible that during heat stress a sphingolipid intermediate signals calmodulin, causing it to down-regulate neutral trehalase activity so that trehalose accumulates.

Regulation of Growth and the Cell Cycle

Several long chain base derivatives and ceramide are thought to modulate signaling pathways that regulate the cell cycle. Ceramide is typically generated from sphingomyelin in response to stresses including heat, ionizing radiation, chemotherapeutic agents, and serum deprivation, to name a few (9–12). Such stimuli signal cells to arrest growth at the G_0/G_1 boundary, which allows the stress-induced damage to be repaired or, if the damage cannot be repaired, signals cells to undergo apoptosis (programmed cell death). A ceramide-activated protein kinase (99) and a ceramide-activated protein phosphatase (9) have been implicated in transmission of ceramide signals during some of these stresses, but the detailed mechanism or mechanisms of the signaling pathways remain unknown.

Long chain bases inhibit growth and are cytotoxic to mammalian cells under some conditions (14), but under other conditions they stimulate division of several mammalian cell types (94). These seemingly contradictory effects are not completely understood, but most likely they reflect the way cells metabolize long chain bases and ceramide. If cells primarily convert the long chain base to ceramide, then growth is arrested and apoptosis may ensue. If, however, the long chain base, particularly sphingosine, is phosphorylated on the 1-OH group, it induces a mitogenic response. The mechanism of this response is not fully understood, but possible mechanisms have been presented (94).

Growth of *S. cerevisiae* cells is inhibited by C_2-ceramide, but only when cells are maintained at a low density, less than about 5×10^4/ml (100, 101). However, growth inhibition is not always observed (83), and there is no clear explanation for these variable results. *S. cerevisiae* cells have a ceramide-activated protein phosphatase (100) that is a member of the PP2A family of phosphatases, as is the mammalian ceramide-activated protein phosphatase (102). The yeast enzyme contains a catalytic subunit encoded by *SIT4* and regulatory subunits encoded by *CDC55* and *TPD3* (101). It remains unknown how this phosphatase blocks cells in the G_1 phase of the cell division cycle in response to C_2-ceramide treatment. It also remains to be determined if *S. cerevisiae* contains a ceramide-activated protein kinase. Finally, it seems likely that mammalian and yeast ceramide-activated protein phosphatases are homologues, but mammalian protein or proteins await identification.

Vesicle Transport of GPI-Anchored Proteins in S. cerevisiae Requires Sphingolipids

Higher eukaryotes and *S. cerevisiae* transport membrane proteins from organelle to organelle by vesicle transport (103, 104). Horvath et al (105) first suggested that ceramide (phytoceramide) synthesis was necessary for vesicle transport of glycosylphosphatidylinositol (GPI)-anchored but not unanchored proteins in *S. cerevisiae*. This suggestion followed from the observation that myriocin, an inhibitor of serine palmitoyltransferase activity (Figure 1), causes the 105-kDa precursor form of the Gas1 protein to accumulate in the ER. Precursor accumulation indicates failure of the Gas1 protein to reach the Golgi, where it would be further glycosylated to yield a 125-kDa mature species.

Other support for the idea that phytoceramide is necessary for vesicle transport of GPI-anchored proteins comes from studies of suppressor genes that when present in multiple copies partially compensate for the lack of sphingolipids in SLC cells and enhance survival at low pH (pH 4.1) (106). One such gene is *CWP2* (cell wall protein 2), which encodes a major mannoprotein of the yeast cell wall (107). Most yeast cell wall proteins, including Cwp2, are thought to be GPI-anchored on their way from the ER to the plasma membrane (108, 109). One hypothesis to explain how multiple copies of the *CWP2* gene enhance survival of SLC cells lacking sphingolipids at low pH is that sphingolipids are necessary for a normal rate of transport of GPI-anchored proteins, including Cwp2, from the ER to the Golgi apparatus. In the absence of sphingolipids, less Cwp2p would reach the cell wall, and the wall would be less able to defend against the stress of a high proton concentration. Several lines of evidence support this hypothesis (106). These results, along with those of Horvath et al (105), make a strong case that a sphingolipid, perhaps phytoceramide or IPC, plays a vital role in vesicle transport of GPI-anchored proteins in *S. cerevisiae*.

What needs to be determined now is the step or steps that require sphingolipids in the maturation of GPI-anchored proteins. One step may be vesicle budding from the ER or docking to the Golgi apparatus. Because phytoceramide is made de novo in the ER (36), its concentration is likely to be highest there. This high concentration may thus serve to label the ER membrane and thereby play a key role in vesicle formation. The Gas1 protein is known to be transported from the ER to the Golgi in vesicles containing a COPII protein coat (110). Because the COPII budding process can be reproduced in vitro through the use of membranes and either purified coat proteins or a cytosolic extract (111–113), it may be possible to begin identifying which step and protein(s) in vesicle transport require sphingolipids.

An alternative hypothesis, the clustering hypothesis (105), is based on studies of animal cells showing clustering of GPI-anchored proteins with glycosphingolipids and cholesterol to form microdomains that originate in the *trans* Golgi network and move to specific plasma membrane targets by vesicle transport (114–116). Thus, in *S. cerevisiae* clustering of sphingolipids (phytoceramide and IPC) and GPI-anchored proteins may occur in the ER, with the clusters playing a role in formation of or partitioning into COPII transport vesicles. Microdomain formation in the ER of *S. cerevisiae* may also be promoted by the presence of ergosterol, the primary membrane sterol in *S. cerevisiae*, but this possibility has not been examined experimentally.

The requirement of sphingolipids for transport of GPI-anchored proteins from the ER to Golgi has not been directly examined in higher eukaryotes, so detergent-insoluble glycosphingolipid-enriched complexes containing GPI-anchored proteins may form in the ER to facilitate vesicle transport to the Golgi. However, one argument against this possibility is the difference in the site of synthesis of complex sphingolipids in yeast versus higher eukaryotes. Ceramide is the only sphingolipid thought to be present in high quantity in the ER of higher eukaryotes (35). It is transported to the Golgi, where the complex sphingolipids, sphingomyelin, and glycosphingolipids are synthesized (35). Examination of temperature-sensitive *S. cerevisiae* mutants blocked in the protein secretory pathway suggest that besides ceramide, IPC is present in the ER. If IPC had a propensity to form microdomains with GPI-anchored proteins, then such domains might enhance vesicle transport of GPI-anchored proteins in yeast (36).

One other potentially significant difference between *S. cerevisiae* and mammalian ceramides and sphingolipids is the presence in all yeast ceramides and sphingolipids of a very long chain fatty acid, primarily C_{26} (18). The C_{26}-fatty acid also appears in the *sn-2* position of the glycerol backbone of some GPI-anchored proteins in *S. cerevisiae*. If the C_{26}-fatty acid had a tendency to form microdomains, perhaps with ergosterol, then *S. cerevisiae* cells would form such domains in their ER and mammals would not.

As described in the next section, the diacylglycerol moiety of many GPI anchors is replaced in *S. cerevisiae* by a ceramide moiety after transport from the ER, most probably in the Golgi apparatus (117–119). If such remodeling is necessary for transport or for covalent attachment of Cwp2p to the extracellular glycan layer, then in cells lacking sphingolipids there would be reduced transport or attachment and a weakened cell wall. These explanations cannot be ruled out because it is not possible to isolate and analyze the GPI anchor on intermediates in Cwp2 protein maturation. However, this possibility does not explain the reduced rate of Gas1 transport in the absence of sphingolipids because the GPI anchor on Gas1 does not contain ceramide (120).

Phytoceramide Replaces Diacylglycerol in Many GPI Anchors

Many proteins at the surface of eukaryotic cells are attached at their C terminus to the plasma membrane by means of a GPI anchor that has a core carbohydrate structure consisting of protein-ethanolamine-PO_4-6Manα1-2Manα1-6Manα1-4GlcNH$_2$-*myo*-inositol-PO_4-lipid (121). GPI-anchored proteins include receptors, adhesion molecules, lymphoid antigens and antigens of unknown function, hydrolytic enzymes, the scrapie prion protein, and the variant surface glycoproteins of some parasitic protozoans (122, 123).

GPI-anchor synthesis is essential in *S. cerevisiae* because mutants blocked in synthesis are not viable (124–127). One essential yeast gene, *SPT14*, is necessary for the first step in GPI-anchor synthesis (125, 126). *SPT14* is a homologue of the human *PIG-A* gene, in which somatic mutations cause paroxysmal nocturnal hemoglobinuria, a hemolytic disease (123).

Many of the dozen or so identified GPI-anchored *S. cerevisiae* proteins either are anchored to the plasma membrane and project outside the cell or are cleaved from the lipid anchor and covalently attached to the cell wall, where such proteins constitute a major wall component (109, 128, 129). Synthesis of the GPI moiety and attachment to the C terminus of a protein occur in the ER, followed by vesicular transport to the plasma membrane via the Golgi (123).

Initial studies of the lipid component of GPI anchors in *S. cerevisiae* (117) showed the presence of both base-sensitive, GPI-containing lipid anchors and base-resistant, phytoceramide-containing lipid anchors. The GPI-anchored proteins were synthesized in the ER, and then the diacylglycerol component of GPI was replaced by phytoceramide, probably in the Golgi (117, 119, 130). Further analysis has shown that the diacylglycerol moiety in most GPI anchors is remodeled and replaced by phytoceramide to give what should be called ceramidephosphoinositol anchors. Some proteins such as Gas1 retain their GPI anchors (120). Strangely, the GPI anchors on the Gas1 protein and probably some other proteins have very long chain fatty acids (C_{26}) in the *sn-2* position (118). Such long chain fatty acids are found primarily in phytoceramide (18). Future research should reveal how and why GPI anchors are remodeled with phytoceramide and C_{26}-fatty acids and why some proteins do not have remodeled GPI anchors.

SUMMARY AND FUTURE DIRECTIONS

Knowledge of sphingolipids in *S. cerevisiae* is entering an exciting period of unparalleled expansion. Soon all steps in *S. cerevisiae* sphingolipid biosynthesis should be identified and their order established. *S. cerevisiae* is likely to be the first organism for which all genes and proteins directly involved in sphingolipid metabolism will be identified. This information will offer an unprecedented opportunity to study how the synthesis of these genes and proteins is regulated, and such information should be a great aid to understanding tissue-specific regulatory mechanisms in multicellular eukaryotes. Cloned genes will also provide new tools for the challenging task of purifying enzymes (none have been purified) and determining their biochemical mechanisms.

Recent progress in identifying putative sphingolipid signaling pathways in *S. cerevisiae* opens the way to more detailed mechanistic studies and the possibility of quantifying the role of sphingolipid signals in a variety of physiological processes, something that is still almost impossible to do in multicellular eukaryotes. The generation of sphingolipid signaling molecules during heat stress appears to be due to de novo synthesis of intermediates rather than to breakdown of mature sphingolipids present in the plasma membrane, as is the case in most mammalian stress responses examined. Thus, *S. cerevisiae* may be useful for studying this new paradigm in the generation of sphingolipid signals. Through the use of DNA array technology (131) it will be possible to examine global effects of sphingolipid signals on gene expression. This information should provide new and unique insights into the physiological role of sphingolipid signaling in a way that was not possible previously. This information will provide a starting point for the far more complex task of understanding how the signals are sent from the signal recognition machinery to the nucleus and to other cellular targets that await discovery.

ACKNOWLEDGMENTS

The critical comments and suggestions of Drs. Robert L Lester and Charles J Waechter, which helped to clarify and improve this review, are gratefully acknowledged. The preparation of this review and work in the author's laboratory were supported by NIH grant GM 41302.

Visit the *Annual Reviews home page* at
http://www.AnnualReviews.org.

Literature Cited

1. Hakomori S. 1996. *Cancer Res.* 56:5309–18
2. Schnaar RL, Mahoney JA, Swank-Hill P, Tiemeyer M, Needham LK. 1994. *Prog. Brain Res.* 101:185–97
3. Karlsson K-A. 1989. *Annu. Rev. Biochem.* 58:309–50
4. Hakomori S. 1981. *Annu. Rev. Biochem.* 50:733–64
5. Hannun YA, Loomis CR, Merrill AH Jr, Bell RM. 1986. *J. Biol. Chem.* 261:12604–9
6. Hannun YA, Bell RM. 1993. *Adv. Lipid. Res.* 25:27–41
7. Kim M-Y, Linardic C, Obeid LM, Hannun YA. 1991. *J. Biol. Chem.* 266:484–89
8. Mathias S, Dressler KA, Kolesnick RN. 1991. *J. Biol. Chem.* 88:10009–13

9. Hannun YA. 1996. *Science* 274:1855–59
10. Spiegel S, Foster D, Kolesnick RN. 1996. *Curr. Opin. Cell Biol.* 8:159–67
11. Kyriakis JM, Avruch J. 1996. *J. Biol. Chem.* 271:24313–16
12. Ballou LR, Laulederkind SJF, Rosloniec EF, Raghow R. 1996. *Biochim. Biophys. Acta Lipids. Lipid. Metab.* 1301:273–87
13. Adam D, Adam-Klages S, Kronke M. 1995. *J. Inflamm.* 47:61–66
14. Merrill AH Jr, Liotta DC, Riley RE. 1996. In *Lipid Second Messengers*, ed. RM Bell, 8:205–37. New York: Plenum
15. zu Heringdorf DM, van Koppen CJ, Jakobs KH. 1997. *FEBS Lett.* 410:34–38
16. Seufferlein T, Rozengurt E. 1995. *J. Biol. Chem.* 270:24343–51
17. Goffeau A, Barrell BG, Bussey H, Davis RW, Dujon B, et al. 1996. *Science* 274:546–63
18. Lester RL, Dickson RC. 1993. *Adv. Lipid. Res.* 26:253–72
19. Merrill AH Jr, Jones DD. 1990. *Biochim. Biophys. Acta* 1044:1–12
20. Downing DT. 1992. *J. Lipid Res.* 33:301–13
21. Merrill AH Jr, Sweeley CC. 1996. In *Biochemistry of Lipids, Lipoproteins and Membranes*, ed. DE Vance, JE Vance, pp. 309–39. New York: Elsevier Science
22. Miyake Y, Kozutsumi Y, Nakamura S, Fujita T, Kawasaki T. 1995. *Biochem. Biophys. Res. Commun.* 211:396–403
23. Zweerink MM, Edison AM, Wells GB, Pinto W, Lester RL. 1992. *J. Biol. Chem.* 267:25032–38
24. Wang E, Norred WP, Bacon CW, Riley RT, Merrill AH Jr. 1991. *J. Biol. Chem.* 266:14486–90
25. Wu W-I, McDonough VM, Nickels JT Jr, Ko J, Fischl AS, et al. 1995. *J. Biol. Chem.* 270:13171–78
26. Mandala SM, Thornton RA, Frommer BR, Curotto JE, Rozdilsky W, et al. 1995. *J. Antibiot.* 48:349–56
27. Nagiec MM, Nagiec EE, Baltisberger JA, Wells GB, Lester RL, et al. 1997. *J. Biol. Chem.* 272:9809–17
28. Saba JD, Nara F, Gottlieb D, Bielawska A, Garrett S, et al. 1997. *The BST1 gene of* Saccharomyces cerevisiae *is the sphingosine-1-phosphate lyase, and its deletion identifies phytosphingosine-1-phosphate as a regulator of proliferation.* Presented at Int. Union Biochem. Mol. Biol. Satellite Symp., Napa Valley, CA (Abstr.)
28a. Soba JD, Nora F, Brelawska A, Garrett S, Hannun YA. 1997. 272:26087–90
29. Qie LX, Nagiec MM, Baltisberger JA, Lester RL, Dickson RC. 1997. *J. Biol. Chem.* 272:16110–17
30. Mao CG, Wadleigh M, Jenkins GM, Hannun YA, Obeid LM. 1997. *J. Biol. Chem.* 272:28690–94
31. Lanterman MM, Saba JD. 1997. *Characterization of sphingosine kinase activity in* S. cerevisiae *and isolation of sphingosine kinase deficient mutants.* Presented at Int. Union Biochem. Mol. Biol. Satellite Symp., Napa Valley, CA (Abstr.)
32. Patton JL, Lester RL. 1991. *J. Bacteriol.* 173:3101–8
33. Hechtberger P, Zinser E, Saf R, Hummel K, Paltauf F, et al. 1994. *Eur. J. Biochem.* 225:641–49
34. Smith SW, Lester RL. 1974. *J. Biol. Chem.* 249:3395–3405
35. van Echten G, Sandhoff K. 1993. *J. Biol. Chem.* 268:5341–44
36. Puoti A, Desponds C, Conzelmann A. 1991. *J. Cell Biol.* 113:515–25
37. Rosenwald AG, Pagano RE. 1993. *Adv. Lipid. Res.* 26:101–18
38. Sandhoff K, van Echten G. 1993. *Adv. Lipid. Res.* 26:119–42
39. Allan D, Raval PJ. 1987. *Biochim. Biophys. Acta* 897:355–63
40. Wu G, Lu ZH, Ledeen RW. 1995. *J. Neurochem.* 65:1419–22
41. Schneiter R, Hitomi M, Ivessa AS, Fasch EV, Kohlwein SD, et al. 1996. *Mol. Cell. Biol.* 16:7161–72
42. Carman GM, Zeimetz GM. 1996. *J. Biol. Chem.* 271:13293–96
43. Jackowski S. 1996. *J. Biol. Chem.* 271:20219–22
44. Greenberg ML, Lopes JM. 1996. *Microbiol. Rev.* 60:1–20
45. Paltauf F, Kohlwein SD, Henry SA. 1992. In *The Molecular and Cellular Biology of the Yeast* Saccharomyces, ed. EW Jones, JR Pringle, JR Broach, 2:415–500. Cold Spring Harbor, NY: Cold Spring Harbor Lab. Press. 1st ed.
46. Goldstein JL, Brown MS. 1990. *Nature* 343:425–30
47. Parks LW, Casey WM. 1995. *Annu. Rev. Microbiol.* 49:95–116
48. Mandon EC, van Echten G, Birk R, Schmidt RR, Sandhoff K. 1991. *Eur. J. Biochem.* 198:667–74
49. Cleves AE, McGee TP, Bankaitis VA. 1991. *Trends Cell Biol.* 1:30–34
50. Kearns BG, McGee TP, Mayinger P, Gedvilaite A, Phillips SE, et al. 1997. *Nature* 387:101–5
51. Dickson RC, Wells GB, Schmidt A, Lester RL. 1990. *Mol. Cell. Biol.* 10:2176–81

52. Wells GB, Lester RL. 1983. *J. Biol. Chem.* 258:10200–3
53. Nagiec MM, Wells GB, Lester RL, Dickson RC. 1993. *J. Biol. Chem.* 268:22156–63
54. Nagiec MM, Baltisberger JA, Wells GB, Lester RL, Dickson RC. 1994. *Proc. Natl. Acad. Sci. USA* 91:7899–7902
55. Pinto WJ, Srinivasan B, Shepherd S, Schmidt A, Dickson RC, et al. 1992. *J. Bacteriol.* 174:2565–74
56. Lester RL, Wells GB, Oxford G, Dickson RC. 1993. *J. Biol. Chem.* 268:845–56
57. Patton JL, Srinivasan B, Dickson RC, Lester RL. 1992. *J. Bacteriol.* 174:7180–84
58. Wells GB, Dickson RC, Lester RL. 1996. *Heat shock increases ceramide levels in Saccharomyces cerevisiae.* Presented at ASBMB Satellite Meet. "Lipid Second Messengers," Am. Soc. Biochem. Mol. Biol., New Orleans, LA (Abstr. L31)
59. Dickson RC, Nagiec EE, Skrzypek M, Tillman P, Wells GB, et al. 1997. *J. Biol. Chem.* 272:47 In press
60. Parsell DA, Lindquist S. 1994. In *The Biology of Heat Shock Proteins and Molecular Chaperones,* ed. RI Morimoto, A Tissieres, C Georgopoulos, pp. 457–94. Cold Spring Harbor, NY: Cold Spring Harbor Lab. Press
61. Craig EA, Weismann JS. 1994. *Cell* 78:365–72
62. Johnson JL, Craig EA. 1997. *Cell* 90:201–4
63. De Virgilio C, Hottiger T, Dominguez J, Boller T, Wiemken A. 1994. *Eur. J. Biochem.* 219:179–86
64. Parrou JL, Teste MA, Francois J. 1997. *Microbiology* 143:1891–1900
65. Crowe JH, Hoekstra FA, Crowe LM. 1992. *Annu. Rev. Physiol.* 54:579–99
66. Iwahashi H, Obuchi K, Fujii S, Komatsu Y. 1995. *Cell. Mol. Biol.* 41:763–69
67. Colaco C, Kampinga J, Roser B. 1995. *Science* 268:788
68. Schutze S, Potthoff K, Machleidt T, Berkovic D, Wiegmann K, et al. 1992. *Cell* 71:765–76
69. Obeid LM, Linardic CM, Karolak LA, Hannun YA. 1993. *Science* 259:1769–71
70. De Virgilio C, Burckert N, Bell W, Jeno P, Boller T, et al. 1993. *Eur. J. Biochem.* 212:315–23
71. Sur IP, Lobo Z, Maitra PK. 1994. *Yeast* 10:199–209
72. Gounalaki N, Thireos G. 1994. *EMBO J.* 13:4036–41
73. Lindquist S, Kim G. 1996. *Proc. Natl. Acad. Sci. USA* 93:5301–6
74. Marchler G, Schuller C, Adam G, Ruis H. 1993. *EMBO J.* 12:1997–2003
75. Kobayashi N, McEntee K. 1993. *Mol. Cell. Biol.* 13:248–56
76. Ruis H, Schuller C. 1995. *BioEssays* 17:959–65
77. Martinez-Pastor MT, Marchler G, Schuller C, Marchler-Bauer A, Ruis H, et al. 1996. *EMBO J.* 15:2227–35
78. Schmitt AP, McEntee K. 1996. *Proc. Natl. Acad. Sci. USA* 93:5777–82
79. Chang Y, Abe A, Shayman JA. 1995. *Proc. Natl. Acad. Sci. USA* 92:12275–79
80. Verheij M, Bose R, Lin XH, Yao B, Jarvis WD, et al. 1996. *Nature* 380:75–79
81. Bose R, Verheij M, Haimovitz-Friedman A, Scotto K, Fuks Z, et al. 1995. *Cell* 82:405–14
81a. Balsinde J, Balboa MA, Dennis EA. 1997. *J. Biol. Chem.* 272:20373–77
82. Jaffrezou JP, Levade T, Bettaieb A, Andrieu N, Bezombes C, et al. 1996. *EMBO J.* 15:2417–24
83. Ella KM, Qi C, Dolan JW, Thompson RP, Meier KE. 1997. *Arch. Biochem. Biophys.* 340:101–10
84. Cox JS, Shamu CE, Walter P. 1993. *Cell* 73:1197–1206
85. Mori K, Ma W, Gething MJ, Sambrook J. 1993. *Cell* 74:743–56
86. Welihinda AA, Tirasophon W, Green SR, Kaufman RJ 1997. *Proc. Natl. Acad. Sci. USA* 94:4289–94
87. McCusker JH, Perlin DS, Haber JE. 1987. *Mol. Cell. Biol.* 7:4082–88
88. Serrano R. 1993. *FEBS Lett.* 325:108–11
89. Serrano R. 1983. *FEBS Lett.* 156:11–14
90. Chang A, Slayman CW. 1991. *J. Cell Biol.* 115:289–95
91. Estrada E, Agostinis P, Vandenheede JR, Goris J, Merlevede W, et al. 1996. *J. Biol. Chem.* 271:32064–72
92. Brandao RL, de Magalhaes-Rocha NM, Alijo R, Ramos J, Thevelein JM. 1994. *Biochim. Biophys. Acta* 1223:117–24
93. Patton JL, Lester RL. 1992. *Arch. Biochem. Biophys.* 292:70–76
94. Spiegel S, Merrill AH Jr. 1996. *FASEB J.* 10:1388–97
95. Beeler T, Gable K, Zhao C, Dunn T. 1994. *J. Biol. Chem.* 269:7279–84
96. Zhao C, Beeler T, Dunn T. 1994. *J. Biol. Chem.* 269:21480–88
97. Nwaka S, Kopp M, Holzer H. 1995. *J. Biol. Chem.* 270:10193–98
98. DeMesquita JF, Paschoalin VMF, Panek AD. 1997. *Biochim. Biophys. Acta Gen. Subj.* 1334:233–39
99. Zhang YH, Yao B, Delikat S, Bayoumy S, Lin XH, et al. 1997. *Cell* 89:63–72

100. Fishbein JD, Dobrowsky RT, Bielawska A, Garrett S, Hannun YA. 1993. *J. Biol. Chem.* 268:9255–61
101. Nickels JT, Broach JR. 1996. *Genes Dev.* 10:382–94
102. Dobrowsky RT, Kamibayashi C, Mumby MC, Hannun YA. 1993. *J. Biol. Chem.* 268:15523–30
103. Schekman R. 1996. *Cell* 87:593–95
104. Rothman JE, Wieland FT. 1996. *Science* 272:227–34
105. Horvath A, Suetterlin C, Manning-Krieg U, Movva NR, Riezman H. 1994. *EMBO J.* 13:3687–95
106. Skrzypek M, Lester RL, Dickson RC. 1997. *J. Bacteriol.* 179:1513–20
107. van der Vaart JM, Caro LHP, Chapman JW, Klis FM, Verrips CT. 1995. *J. Bacteriol.* 177:3104–10
108. Lu C-F, Kurjan J, Lipke PN. 1994. *Mol. Cell. Biol.* 14:4825–33
109. de Nobel H, Lipke PN. 1994. *Trends Cell Biol.* 4:42–45
110. Doering TL, Schekman R. 1996. *EMBO J.* 15:182–91
111. Wuestehube LJ, Schekman R. 1992. *Methods Enzymol.* 219:124–36
112. Barlowe C, Orci L, Yeung T, Hosobuchi M, Hamamoto S, et al. 1994. *Cell* 77:895–907
113. Bednarek SY, Ravazzola M, Hosobuchi M, Amherdt M, Perrelet A, et al. 1995. *Cell* 83:1183–96
114. Simons K, Van Meer G. 1988. *Biochemistry* 27:6197–6202
115. Brown DA. 1992. *Trends Cell Biol.* 2: 338–43
116. Simons K, Ikonen E. 1997. *Nature* 387: 569–72
117. Conzelmann A, Puoti A, Lester RL, Desponds C. 1992. *EMBO J.* 11:457–66
118. Sipos G, Reggiori F, Vionnet C, Conzelmann A. 1997. *EMBO J.* 16:3494–3505
119. Reggiori F, Canivenc-Gansel E, Conzelmann A. 1997. *EMBO J.* 16:3506–18
120. Fankhauser C, Homans SW, Thomas-Oates JE, McConville MJ, Desponds C, et al. 1993. *J. Biol. Chem.* 268:26365–74
121. McConville MJ, Ferguson MA. 1993. *Biochem. J.* 294:305–24
122. Englund PT. 1993. *Annu. Rev. Biochem.* 62:121–38
123. Takeda J, Kinoshita T. 1995. *Trends Biochem. Sci.* 20:367–71
124. Leidich SD, Drapp DA, Orlean P. 1994. *J. Biol. Chem.* 269:10193–96
125. Vossen JH, Horvath A, Fassler J, Riezman H. 1995. *EMBO J.* 14:1637–45
126. Vossen JH, Ram AFJ, Klis FM. 1995. *Biochim. Biophys. Acta* 1243:549–51
127. Benghezal M, Lipke PN, Conzelmann A. 1995. *J. Cell Biol.* 130:1333–44
128. Lu C-F, Montijn RC, Brown JL, Klis F, Kurjan J, et al. 1995. *J. Cell Biol.* 128: S333–40
129. Vosson JH, Muller WH, Lipke PN, Klis FM. 1997. *J. Bacteriol.* 179:2202–9
130. Sipos G, Puoti A, Conzelmann A. 1994. *EMBO J.* 13:2789–96
131. Schena M, Shalon D, Davis RW, Brown PO. 1995. *Science* 270:467–70
132. Oh C-S, Toke DA, Mandala S, Martin CE. 1997. *J. Biol. Chem.* 272:17376–84
133. Dickson RC, Nagiec EE, Wells GB, Nagiec MM, Lester RL. 1997. *J. Biol. Chem.* 272:29620–25
134. Leber A, Fischer P, Schneiter R, Kohlwein SD, Daum G. 1997. *FEBS Lett.* 411:211–14
135. Buede R, Rinker-Schaffer C, Pinto WJ, Lester RL, Dickson RC. 1991. *J. Bacteriol.* 173:4325–32
136. Grilley MM, Stock SD, Dickson RC, Lester RL, Takemoto JY. 1997. *J. Biol. Chem.* In press
137. Heak D, Gable K, Beeler T, Dunn T. 1997. *J. Biol. Chem.* 272:29704–10
138. Beeler TJ, Fu D, Rivera J, Monaghan E, Gable K, Dunn TM. 1997. *Mol. Gen. Genet.* 2SS:570–79

Annu. Rev. Biochem. 1998. 67:49–69

TRANSPORTERS OF NUCLEOTIDE SUGARS, ATP, AND NUCLEOTIDE SULFATE IN THE ENDOPLASMIC RETICULUM AND GOLGI APPARATUS

Carlos B. Hirschberg, Phillips W. Robbins, and Claudia Abeijon

Department of Molecular and Cell Biology, Boston University Goldman School of
Dental Medicine, Boston, Massachusetts 02118-2392

KEY WORDS: membrane proteins, antiporters, glycosylation, sulfation, phosphorylation, lipids,
proteoglycans

ABSTRACT

The lumens of the endoplasmic reticulum and Golgi apparatus are the subcellular sites where glycosylation, sulfation, and phosphorylation of secretory and membrane-bound proteins, proteoglycans, and lipids occur. Nucleotide sugars, nucleotide sulfate, and ATP are substrates for these reactions. ATP is also used as an energy source in the lumen of the endoplasmic reticulum during protein folding and degradation. The above nucleotide derivatives and ATP must first be translocated across the membrane of the endoplasmic reticulum and/or Golgi apparatus before they can serve as substrates in the above lumenal reactions. Translocation of the above solutes is mediated for highly specific transporters, which are antiporters with the corresponding nucleoside monophosphates as shown by biochemical and genetic approaches. Mutants in mammals, yeast, and protozoa showed that a defect in a specific translocator activity results in selective impairments of the above posttranslational modifications, including loss of virulence of pathogenic protozoa. Several of these transporters have been purified and cloned. Experiments with yeast and mammalian cells demonstrate that these transporters play a regulatory role in the above reactions. Future studies will address the structure of the above proteins, how they are targeted to different organelles,

49

their potential as drug targets, their role during development, and the possible
occurrence of specific diseases.

CONTENTS

PERSPECTIVES AND SUMMARY

Secreted and membrane proteins and lipids are often glycosylated, sulfated, and
phosphorylated in the lumen of the endoplasmic reticulum and/or the Golgi apparatus. The substrates for these posttranslational modifications are nucleotide
sugars, nucleotide sulfate, and ATP. The latter is also part of energy-requiring
reactions in the lumen of the endoplasmic reticulum. All these nucleotide substrates must first be translocated from their respective sites of synthesis within
the cell, most often the cytosol, into the lumen of the above organelles by specific
transporters. In general, they are antiporters with the corresponding nucleoside monophosphate. They have been detected in mammals, yeast, protozoa,
and plants and thus appear to have been conserved throughout evolution. Several of the transporters of the Golgi apparatus membrane have been purified and
cloned; they are very hydrophobic, multitransmembrane-spanning proteins that
appear to regulate posttranslational modifications in the Golgi lumen. Several
mutants in Golgi transport activities result in distinct phenotypes, including loss
of virulence of pathogenic protozoa. Future studies will address the structure
of these proteins, how they are targeted to different organelles, their role during development, their potential as drug targets, and the occurrence of specific
diseases that may result from deficiencies in these transporters.

The Requirement for Transporters of ATP, Nucleotide Sugars, and Nucleotide Sulfate in the Membrane of the Endoplasmic Reticulum and Golgi Apparatus

The pioneering studies of Palade (1) showed that secreted proteins and those in transit to the cell surface and lysosomes become compartmentalized in the lumen of the endoplasmic reticulum, from which they travel via vesicles (2) through the Golgi apparatus toward their final destination. Nascent polypeptides are translocated into the lumen of the endoplasmic reticulum (3, 4), where many undergo initial N-glycosylation (5). Although most of the sugars required for these reactions do not enter the lumen as nucleotide sugars but as derivatives of dolichol (6), UDP-glucose and UDP-N-acetylglucosamine, like ATP, can be substrates in the lumen of the endoplasmic reticulum (7–9). ATP is used for energy-requiring reactions and is a substrate in protein phosphorylation (10–13). From the endoplasmic reticulum, these proteins reach the Golgi apparatus, where terminal N-glycosylation, O-glycosylation, phosphorylation, and sulfation occur (14, 15). Many glycosphingolipids also undergo glycosylation and sulfation reactions in the lumen of the Golgi apparatus. Nucleotide derivatives are synthesized to a large extent in the cytosol (16), not in the lumen of the endoplasmic reticulum or Golgi apparatus. The exceptions are ATP, which is synthesized mostly in mitochondria, and CMP-sialic acid, which is synthesized in the nucleus (16, 17). Therefore a mechanism must exist for these charged solutes to enter the lumen of these organelles and serve as substrates in the above posttranslational reactions.

Transport Characteristics of Nucleotide Sugars, Nucleotide Sulfate, and ATP into Vesicles from the Endoplasmic Reticulum and Golgi Apparatus

The following transport characteristics of nucleotide sugars, nucleotide sulfate, and ATP were determined through examination of vesicles from the above organelles; the vesicles were sealed and of in vivo membrane topological orientation. Transport was measured through the use of filtration (18, 19) and through centrifugation assays (20, 21), yielding comparable results for apparent Kms and Vmax.

1. Transport of the above solutes is organellar specific, with some solutes entering the lumen of the Golgi and others entering the lumen of the endoplasmic reticulum (Table 1; Figures 1 and 2). A few are transported into the lumen of both organelles but with different velocities (Table 1). Transport of nucleotide sugars and nucleotide sulfate has been detected most often solely into organelles for which the corresponding glycosyl- or sulfo- transferases have been detected. Occasionally, however, transport into the endoplasmic

Table 1 Organellar specificity of mammalian transporters of nucleotide sugars, ATP, and nucleotide sulfate

Nucleotide derivative	ER	Golgi	References
CMP-sialic acid	—	••••	22–33
GDP-fucose	—	••••	28, 34
UDP-galactose	—	••••	18, 35–45
PAPS	—	••••	25, 26, 29, 31, 46–51
GDP-mannose	—	••••	43, 52–54
UDP-N-acetylglucosamine	••	••••	22, 25, 26, 54–59
UDP-N-acetylgalactosamine	••	••••	26, 60–62
UDP-xylose	••	••••	42, 63–65
ATP	•••	••••	66–70
UDP-glucuronic acid	••••	••••	63, 64, 71–74
UDP-glucose	••••	•	26, 75, 76

Note: The number of black dots is proportional to the rates of transport/mg protein. GDP-mannose is transported only into the lumen of the Golgi apparatus of yeast and protozoa.

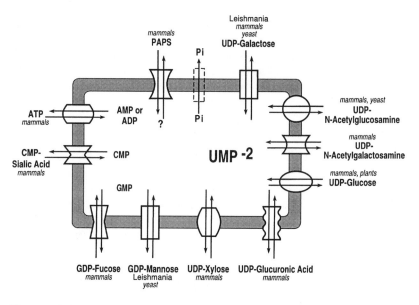

Figure 1 Golgi membrane nucleotide sugar, ATP, and PAPS transporters have been detected in mammals, yeast, protozoa, and plants. The transporters are antiporters with the corresponding nucleoside monophosphate. Exceptions are for PAPS, where the antiporter molecule is not known, and for ATP, where it is AMP, ADP, or both. The existence of an inorganic phosphate transporter is hypothesized.

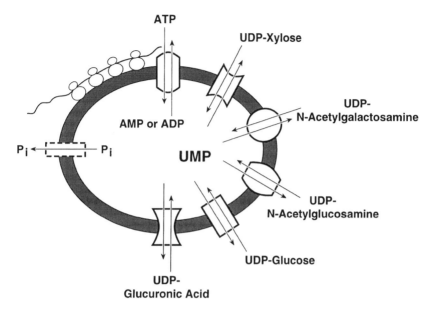

Figure 2 Nucleotide sugar and ATP transporters in the endoplastic reticulum have been detected in mammals and yeast. These are antiporters with the corresponding nucleoside monophosphate except for ATP, where AMP or ADP or both are active. The existence of an inorganic phosphate transporter is hypothesized.

reticulum has been detected without a previous characterization of a corresponding transferase.

Recent studies have also shown that transport of some nucleotide derivatives into the above subcellular organelles may not be the same in different eukaryotes. Thus, GDP-mannose is transported into the lumen of the Golgi apparatus of yeast and *Leishmania* but not into that of mammals. Figures 1 and 2 show transport activities detected to date in the membranes of the Golgi apparatus and endoplasmic reticulum of eukaryotes.

2. The entire molecule of nucleotide sugar, nucleotide sulfate, and ATP is transported into the organelle lumen in a temperature-dependent and saturable manner. The apparent Kms are generally between 1 and 10 micromolar.

3. Transport of solutes is competitively inhibited by the corresponding nucleoside mono- and diphosphates; it is not inhibited by monosaccharides, sulfate, or phosphate.

4. When sealed vesicles derived from the endoplasmic reticulum and Golgi apparatus, with their orientation as in vivo, are treated with proteases while vesicle latency is not compromised, transport of solutes is greatly reduced.

Therefore transport is mediated by proteins with a domain facing the cytosolic side of the membrane.

5. Solutes become concentrated in the lumen of vesicles up to 50-fold relative to their concentration in the reaction medium; this phenomenon does not require ATP and is neither inhibited nor stimulated by ionophores.

The following sections summarize experiments demonstrating that (a) Golgi membrane transporters are required for transport into the lumen of the above solutes in vivo and for subsequent posttranslational modifications, (b) transporters are antiporters, and (c) they regulate the amounts of substrates available in the lumen of organelles and thereby play an important role in posttranslational modifications.

Phenotype of Mutants Defective in Transport of Nucleotide Sugars into the Golgi Lumen

Mutants in Golgi transport of nucleotide sugars in mammals, yeast, and *Leishmania* have demonstrated that this function is necessary for subsequent addition of the corresponding sugars to proteins and lipids in vivo. These mutants were isolated because of their resistance to plant lectins. CHO Lec8 (77, 78), CHO clone 13 (78), and MDCK II-RCA^R (37) belong to the same genetic complementation group. They have a 70–90% deficiency of galactose in their glycoproteins, glycosphingolipids, and selective proteoglycans. Glycoproteins and glycosphingolipids are also deficient in sialic acid because this sugar is most often attached to galactose. Transport of UDP-galactose into Golgi vesicles from these cells was 90% defective compared to vesicles from wild-type cells, whereas transport of other nucleotide sugars, including uridine derivatives and CMP-sialic acid, was normal (25). Biochemical studies had previously shown that the resulting phenotype was not due to an impairment in the biosynthesis of nucleotide sugars, glycosyltransferases, or macromolecular and lipid acceptors (37, 77–79).

CHO Lec2 (75, 76) and CHO clone 1021 (77) cells are other mutants that belong to the same genetic complementation group and have a 70–90% deficiency of sialic acid in their glycoproteins and gangliosides. Transport of CMP-sialic acid into the Golgi vesicles of these cells was 90% defective compared to vesicles from wild-type cells, whereas transport of other nucleotide sugars was normal (25). These cells do not have a defect in biosynthesis of CMP-sialic acid, asialoglycoproteins, and asialoglycosphingolipids, and they have normal levels of sialyltransferases.

A mutant of the yeast *Kluyveromyces lactis* was found to be defective in transport of UDP-GlcNAc into the Golgi lumen (80). Previous biochemical studies had shown that it lacked terminal N-acetylglucosamine on its cell surface mannan chains and that this defect was not a consequence of impaired biosynthesis of the nucleotide sugar, N-acetylglucosaminyltransferase or endogenous acceptors (81). The defect was specific, because transport of GDP-mannose was the same as transport into vesicles from wild-type cells (80).

A mutant of *Schizosaccharomyces pombe*, defective in transport of UDP-galactose into the Golgi lumen, has been characterized recently (45). The mutant, which lacks galactose in its outer oligosaccharide chains, appears to have the ability to synthesize nucleotide sugars, endogenous mannan acceptors, and galactosyltransferase at normal rates.

A mutant of *Leishmania donovani* defective in Golgi transport of GDP-mannose was described recently (43) as defective in the biosynthesis of cell surface lipophosphoglycan chains that are required for infectivity of this flagellate, thus rendering the mutant avirulent. The transport defect appears to be specific because transport of UDP-galactose into Golgi vesicles from the mutant was essentially the same as for wild type.

These studies demonstrate that transport of nucleotide sugars into the Golgi lumen is mediated by highly specific transporter proteins and is necessary for subsequent glycosylation of proteins, lipids, and proteoglycans in vivo. As shown later, because the above mutants have a distinct phenotype, they have been used to clone the corresponding Golgi transporters. They also provide an experimental system for testing substrate specificity of putative nucleotide sugar transporters appearing in available data banks.

Transporters of Nucleotide Sugars and ATP in the Golgi Membrane Are Antiporters

Earlier studies had shown that during transport of nucleotide sugars, ATP, and PAPS into Golgi vesicles of mammals and yeast, these solutes are concentrated between 50- and 100-fold in the lumen relative to the incubation medium, suggesting active transport (82). Because ATP and ionophores have no effect on transport rates or final nucleotide sugar steady-state levels, it was hypothesized that the energy required for lumenal concentration of the solutes must be provided, at least in part, by exit of solutes down a concentration gradient from the lumen into the cytosol (33). This was first demonstrated by incubating Golgi vesicles of rat liver with GDP-fucose radiolabeled with tritium in the guanosine (33). When the same vesicles were then incubated with GDP-fucose radiolabeled with C14 in the fucose, exit of tritiated GMP was detected and appeared to be stoichiometric with entry of GDP-[^{14}C] fucose. Exit of the nucleoside monophosphate was obligatory with entry of the nucleotide sugar because GDP-[^{14}C] mannose, which does not enter the above vesicles, did not effect tritiated GMP exit. Similar studies with other nucleotide sugars suggested that entry of nucleotide sugars and ATP into the Golgi lumen is coupled to exit of the corresponding nucleoside monophosphate from the Golgi lumen.

The latter nucleoside derivatives were postulated to arise from a Golgi nucleoside diphosphatase that is known to occur in the lumen of the Golgi apparatus (18, 35, 83, 84) and whose substrates (nucleoside diphosphates) would originate from transfer of the corresponding sugars, phosphate, or sulfate to appropriate macromolecular acceptors.

In an important contribution supporting the antiporter hypothesis, Waldman & Rudnick concluded that UDP-GlcNAc transport across the Golgi membrane occurs via an electroneutral exchange for dianionic UMP (19). They determined that preloading Golgi vesicles with UMP or UDP-GlcNAc markedly stimulated the accumulation of this latter nucleotide sugar compared to vesicles loaded with buffer, ATP, uridine, or uracyl. The rate of exchange between UDP-GlcNAc and UMP was faster when vesicles were preloaded at pH 7.5, where UMP is dianionic, versus pH 5.45, where it is monoanionic. (UDP-GlcNAc is dianionic at both pHs.) Finally, exit of UDP-GlcNAc from the lumen was faster when the external UMP was dianionic. The Golgi membrane should also have a transporter for inorganic phosphate to remove this anion from the lumen into the cytosol; no direct evidence for such activity has been demonstrated. Alternatively, it has been suggested that in mammary glands, phosphate is secreted together with protein and lactose by exocytosis of Golgi vesicles (85, 86).

Independent evidence that nucleotide sugars enter the Golgi lumen via antiporters has been obtained with proteoliposomes containing Golgi membrane proteins. When these vesicles are prepared in the presence of lumenal CMP (31) or UMP (41), the rate of entry of CMP-sialic acid, UDP-xylose, UDP-glucuronic acid, and UDP-galactose is stimulated two- to threefold compared to vesicles prepared in the presence of buffer alone. The antiporter involved in the transport of PAPS is unclear because no stimulation of PAPS entry into proteoliposomes was observed when these were preloaded with 3'5' ADP, 3' AMP, or 5' AMP (50). A summary of the mammalian Golgi antiporter mechanism is shown in Figure 3.

A combination of biochemical and genetic studies with yeast strongly support the above antiporter model. Golgi vesicles from S. cerevisiae are postulated to contain a lumenal GDPase whose role would be to generate GMP following entry of GDP-mannose and transfer of mannose to endogenous mannan acceptors. It was also postulated that exit of GMP would be required for additional entry of GDP-mannose (9). Such a Golgi lumenal GDPase was detected, characterized, and purified (87), and its gene was cloned (88). It encodes a type II membrane protein. Upon disruption of this gene, Golgi mannosylation of O- and N-linked proteins as well as mannosyl inositol sphingolipids in vivo was greatly reduced (88). In vitro studies showed a fivefold reduction in entry of GDP-mannose into Golgi vesicles (53). These results strongly support the

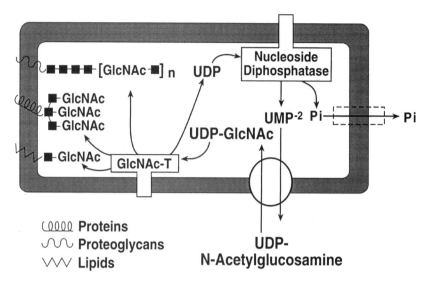

Figure 3   Mechanism of mammalian Golgi membrane antiporters. UDP-N-acetylglucosamine is transported into the lumen of the Golgi apparatus, where N-acetylglucosamine is transferred to proteins, proteoglycans, and lipids in reactions catalyzed by specific N-acetylglucosaminyltransferases. UDP, the other reaction product, is converted by a lumenal nucleoside diphosphatase to UMP^{-2} and inorganic phosphate. The latter is postulated to exit the Golgi lumen via a specific transporter.

existence of a GDP-mannose/GMP antiporter in *S. cerevisiae* Golgi membranes. A summary of the Golgi antiporter mechanism in yeast is shown in Figure 4.

Reconstitution of Nucleotide Sugar, ATP, and Nucleotide Sulfate Transport into Proteoliposomes

Reconstitution of transporters into proteoliposomes serves several important objectives: (*a*) to assay their purification; (*b*) to study mechanisms of transport, including their arrangement within the membrane; (*c*) to establish functions of putative transporter proteins following expression of homologs; and (*d*) to understand regulation of the transporter proteins. Different approaches have been used to reconstitute Golgi transporters and transporters of the endoplasmic reticulum (see below). The former were reconstituted by solubilizing Golgi membrane integral proteins in buffers containing Triton X-100, removing most of the detergent with Extracti-gel D, and intercalating these proteins into unilamellar phosphatidylcholine liposomes following repeated freeze-thaw cycles. The following Golgi membrane transporters have been reconstituted into proteoliposomes: CMP-sialic acid (31), UDP-galactose (41), UDP-glucuronic acid (41), ATP (62), UDP-GalNAc (62), and PAPS (31). For reconstitution of

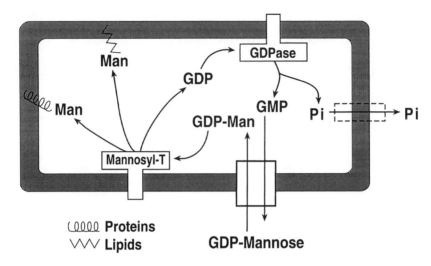

Figure 4 Mechanism of yeast Golgi membrane antiporters. GDP-mannose is transported into the lumen of the Golgi apparatus, where mannose is transferred to proteins and lipids in reactions catalyzed by specific mannosyltransferases. GDP, the other reaction product, is converted by a lumenal guanosine diphosphatase to GMP and inorganic phosphate. The latter is postulated to exit the Golgi lumen via a specific transporter.

Golgi UDP-xylose transport, liposomes containing Golgi lipids were required (41).

This reconstituted system retained its solute transport specificity, saturability, temperature dependence, solute inhibition, and membrane protein specificity. This system has been used for the purification of Golgi membrane nucleotide sugar (62), ATP (62), and PAPS (50) transporters. Proteoliposomes were also used to demonstrate that the Kms determined previously for transport of nucleotide sugars, PAPS, and ATP into Golgi vesicles, where transport of these solutes is coupled to transfer of sugars, sulfate, and phosphate to endogenous lipids and macromolecules, are those of the transporters and not those of glycosyl- and sulfo- transferases or kinases. These studies also suggest that the transporters and not the corresponding transferases or kinases are rate limiting in posttranslational modifications; thus the transporters are possible candidates for regulation of posttranslational modifications (38, 88, 89). Finally, as mentioned in the previous section, proteoliposomes provided independent evidence for nucleotide sugar and ATP Golgi transporters being antiporters.

Reconstitution of the ATP transporter in rat liver endoplasmic reticulum (90) required a different protocol from that used for the Golgi apparatus transporters. Apparently, the former cannot withstand the repeated freeze-thaw cycles used to intercalate the protein into phosphatidylcholine unilamellar liposomes. Following solubilization of ER membrane proteins with octylglucoside and partial purification on a DEAE-Sephacel column, proteins were incorporated into phosphatidylcholine liposomes by dilution. The proteoliposomes, recovered after high-speed centrifugation, showed the same specificity and affinity for ATP transport (and not for other nucleotides and nucleotide derivatives) as ER-derived vesicles (90).

Structures of Golgi Nucleotide Sugar and Nucleotide Sulfate Transporters

Very recently several Golgi nucleotide sugar transporters from mammals (42, 42a, 91), yeast (45, 54), and protozoa (43a) have been cloned by phenotypic correction of the corresponding transport mutants. These transporters appear to be very hydrophobic, multitransmembrane-spanning proteins (Figure 5). The K. lactis UDP-GlcNAc (54) and murine CMP-sialic acid (91) transporters contain leucine-zipper motifs, whereas the others do not. Because the PAPS transporter appears to be a homodimer in situ (50), it is possible that leucine-zipper motifs participate in their oligomerization. Other transporters may also be homodimers in situ.

The recent functional expression of the mammalian murine CMP-sialic acid transporter in S. cerevisiae (32) has shown that (a) the transporter cloned by phenotypic correction probably does not require accessory proteins and (b) the CMP-sialic acid transporter, even though it has 40% amino acid sequence identity with the human UDP-galactose transporter, is highly specific for CMP-sialic acid. This fact had previously been inferred from the characterization of mutants defective in CMP-sialic acid (25) and UDP-galactose transport (26). These transporters have N-linked glycosylation motifs [Asn-X-Ser(Threo)] but do not appear to be glycosylated, probably because the potential glycosylation sites are close to putative transmembrane regions (54, 91).

Transport of ATP and Nucleotide Sugars into the Endoplasmic Reticulum

As shown in Figure 2, several nucleotide sugars and ATP are transported into the lumen of the endoplasmic reticulum from rat liver and yeast. Whereas some of these transport activities make nucleotide sugars and ATP available in the Golgi lumen for known reactions, in some instances the function of the transport activities in the endoplasmic reticulum is not clear. (The possibility of contamination of a membrane preparation with other intracellular membranes must obviously be rigorously excluded.) This question concerning transport function relates, for example, to the case for UDP-xylose and UDP-GalNAc, where a role in the lumen of the endoplasmic reticulum has not yet been described. The

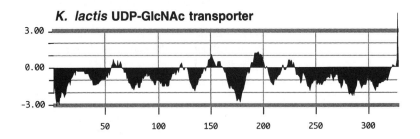

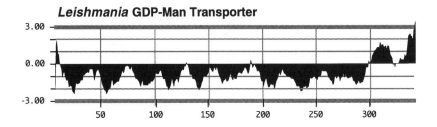

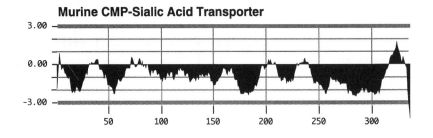

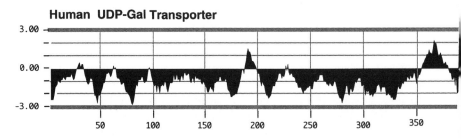

Figure 5   Kyte-Doolittle hydrophilicity plots of the Golgi membrane nucleotide sugar transporters (42, 54, 91, and 43a, respectively). Window size: 18 amino acids.

corresponding glycosyltransferases for N-glycosylation appear to be solely in the Golgi membrane (60, 63). These transport activities in the endoplasmic reticulum may reflect incomplete sorting of the corresponding Golgi proteins during membrane biogenesis. Or they may show that the endoplasmic reticulum transporters are primarily for other nucleotide sugars and have a lower but detectable activity for the above nucleotide sugars. Thus the endoplasmic reticulum transporter for UDP-glucose (75) may also mediate transport of UDP-xylose and UDP-GalNAc.

The transporter for ATP is thought to be required for making the nucleotide triphosphate available in the lumen for energy-requiring reactions (11). These reactions include dissociation of complexes among chaperones and correctly folded and assembled proteins, disulfide bond formation, and protein polymerization (10–13). ATP is also the substrate in the lumenal phosphorylation of proteins such as Bip (10) and protein disulfide isomerase (13), although the role of this phosphorylation is not clear. Degradation of misfolded or over-expressed proteins, a process that requires ATP, has been postulated to occur within the lumen of the ER (92, 93). The accumulation of oligosaccharides with a single reducing GlcNAc has been attributed to a hypothetical ER endo-β-acetylglucosaminidase with release of oligosaccharides in the ER (94). An attractive new alternative view of deglycosylation and proteosomal degradation of nascent or newly formed glycoproteins derives from the recent work of the Ploegh and Inoue groups (95, 96). The former group of authors showed that in the presence of the human cytomegalovirus gene product US11, newly synthesized MHC class 1 heavy chains are rapidly translocated from the ER to the cytosol, where they are deglycosylated and degraded. The deglycosylation leaves both N-acetylglucosamine residues attached to the reducing end of the polymannose oligosaccharide, which suggests that the soluble (presumably cytosolic) N-glycanase isolated and characterized by the Inoue group plays a role in the deglycosylation that precedes proteosomal digestion. If this cytosolic pathway serves as the major route for ER glycoprotein degradation, only a small amount of ATP may be required for degradation within the ER. Consistent with this view is the recent work of Römisch & Ali (97), which showed that ATP-dependent cytosolic factors that lead to glycopeptide export from the ER are interchangeable between yeast and mammalian cells. This finding clearly suggests an important, conserved metabolic system.

UDP-glucose is required for reglucosylation of proteins during their folding (7), whereas UDP-GlcNAc might be required for the synthesis of lumenally facing O- N-acetylglucosamine-linked proteins (9). A transporter for UDP-glucuronic acid in the endoplasmic reticulum membrane has been described (63, 71–74); its possible function has received attention from investigators studying the subcellular site and topography of glucuronidation of xenobiotics and bile acids. Considerable discussion has focused on whether the active site of glucuronosyltransferase (98–101) is lumenal or cytosolic. Proponents of the latter arrangement have cast doubt on the criteria for protein-mediated transport of UDP-glucuronic acid into the ER lumen, including the rates of entry of nucleotide sugars (101). In light of the studies that have been undertaken on nucleotide sugar transporters in the Golgi membrane and a recent report that transport of UDP-glucuronic acid is a prerequisite for glucuronosylation (74), such transport measurements undoubtedly are of physiological relevance (63, 71–74). However, mutants in these activities have not been described, and the putative proteins have not yet been purified.

Of the endoplasmic reticulum nucleotide and nucleotide-derivative transporters, only transporter for ATP has been reconstituted into proteoliposomes (68, 90). The ATP transporter from *S. cerevisiae* has been purified and cloned (69), whereas that from rat liver has been reconstituted, partially purified (90), and identified through use of a photoaffinity label as a 56-kDa protein (70). The ATP transporter from the endoplasmic reticulum of *S. cerevisiae* was characterized as a 68-kDa protein by combining fractionation studies with reconstitution of proteoliposomes. These reconstitution experiments may well have measured transport, although simple binding of ATP to the isolated protein has not been ruled out. The authors also did not report measurements of transport by native ER vesicles in their studies (69). Cloning of the gene that encodes the isolated protein resulted in it being identified as Sac1p (69), a suppressor of a mutation in the ACT 1 gene, which is the single essential gene that encodes actin in this yeast (102). The phenotype of the null mutant is very pleotropic, including major disruptions of cytosolic events (102). In addition, hydrophobicity plots reveal a protein that is mainly hydrophilic with one or perhaps two hydrophobic domains. By contrast, transporters associated with antigen-processing molecules (TAP) across the endoplasmic reticulum, for example, appear to be very hydrophobic multitransmembrane domain proteins (103), analogous to Golgi membrane nucleotide sugar transporters. This characteristic raises a question concerning whether the putative yeast endoplasmic reticulum ATP transporter is, indeed, a transporter or a contaminant of the gel from which the protein was excised for amino acid sequencing (69). Future studies should shed further light on this important issue.

The cystic fibrosis transmembrane conductance regulator (CFTR) has been associated recently with an adenosine 3′ phosphate 5′ phosphosulfate channel in the endoplasmic reticulum (104). Because the concentrations of PAPS in these studies were between 10 and 100 mM and the apparent Km for PAPS transport into Golgi vesicles is approximately 1 μM (31, 50), the physiological significance of these studies is not clear. In addition, during the experiments,

PAPS is being degraded, raising further questions concerning the molecules actually responsible for the observed currents.

FUTURE DIRECTIONS

Mechanisms of Transport of the Endoplasmic Reticulum and Golgi Membrane Transporters

An opportunity to understand the molecular basis of transport is offered by the recent molecular cloning of several nucleotide sugar transporters from mammals, yeast, and protozoa and the heterologous expression of the mammalian CMP-sialic acid transporter in *S. cerevisiae* (32). First one must determine how many transmembrane segments each transporter has, whether the transmembrane segments are helices and/or beta sheets, and relative orientations of each within the membrane. Then one needs to determine which residues are essential for the transport cycle and which are part of substrate specificity domains. Different uridine-containing nucleotide sugar transporters, though highly substrate specific, are likely to have certain common recognition features that allow all of these transporters to use UMP as an antiporter.

Membrane-spanning domains were once believed to contain exclusively alpha-helical secondary structures. Current results indicate that beta strands may also be present in membrane-spanning segments such as those of the nicotinic acetylcholine receptor (105). The periodicity of photolabeling with ^{125}I-TID [3-trifluoromethyl-3-(m-^{125}I]iodophenyl)diazirine] may indicate whether a transmembrane segment is exposed to the lipid environment and whether it has alpha-helical secondary structure. Fourier-transformed infrared spectroscopy may also give insight into the secondary structure of membrane proteins, and measurements of linear dichroism may be used to investigate the orientation of secondary-structure elements with respect to the membrane (106).

Tertiary structure can be determined by cysteine-scanning mutagenesis. Starting with the cysteineless version of a transporter, cysteine residues are introduced one at a time, at each position along the protein. Then, the single-cysteine variants can be probed with hydrophilic, membrane-impermeant-SH-reactive agents such as p-chloromercuribenzosulfonate or N-ethyl maleimide to determine their accessibility from each side of the membrane in the presence and absence of a substrate. This approach was used to show that the seventh transmembrane segment of the glucose-6-phosphate/phosphate antiporter of *E. coli* has a central portion of the helix that is exposed to both surfaces, forming part of the translocation pathway through this transporter (107).

Relative positions between transmembrane helices can be documented by spin-labeling of the single-cysteine variants and by thiol cross-linking experiments in which only two cysteines are present at specific positions in the molecule. Spatial relationships between residues in different helices can be analyzed also by engineering pairs of divalent metal binding sites (bis-histidine residues) at different positions. Binding of Mn^{++} by pairs of substituted histidine residues establishes not only the proximity of the residues but also their accessibility to a hydrophilic probe.

Data derived with these and other techniques allowed Kaback (108) to propose a model for coupling of lactose and H^+ translocation by the $E.$ $coli$ lactose permease. Two separate pathways are proposed: one for the driving substrate (H^+) and another for the driven substrate (lactose). Each has gates, binding sites, and a cycle for alternating affinities and accessibilities. Only 4 of the 417 residues in the molecule are essential, and only 6 of the 12 transmembrane helices are central to the mechanism that transduces a proton-electrochemical gradient into a lactose-concentration gradient.

Time-resolved, multiwavelength Laue X-ray diffraction was used to explore the solvent exposure and protonation of individual atoms of different residues in the photoactive yellow protein photoreceptor of $E.$ $halophilum$ and to clearly determine a gated mechanism (109).

With the ongoing availability of genome databases, many putative homologs to the characterized transporters have appeared. Are these homologs functional or solely structural? Great caution must be exercised in interpreting the significance of such homologies, particularly in view of the lack of functional homology in cases where there is relatively significant structural homology. Examples are between the mammalian CMP-sialic acid and UDP-galactose transporters and between the yeast UDP-GlcNAc and the yet-uncharacterized homolog in $S.$ $cerevisiae$ that does not appear to transport the same nucleotide sugar (110). In this context, the transporters are required either to be expressed in a heterologous system, as was recently described for mammals and yeast (32), or to be reconstituted into liposomes.

Regulation of Nucleotide Sugar, ATP, and Nucleotide Sulfate Transport into the Lumen of the Endoplasmic Reticulum and Golgi Apparatus: Possible Correlations with Disease

Recent studies in mammals and yeast have shown that Golgi membrane transporters of nucleotide sugars appear to regulate specific glycosylation reactions in the Golgi lumen (38, 88, 89). In an MDCK cell mutant (38) where availability of UDP-galactose in the Golgi lumen is limited, proteoglycans such as chondroitin sulfate and heparan sulfate are galactosylated normally while keratan sulfate and galactose-containing proteins and lipids are synthesized in reduced amounts (38). We hypothesize that this differential macromolecular galactosylation is the result of transport of UDP-galactose being limiting. Lumenal reactions with low Kms for this nucleotide sugar would thus be favored over reactions with higher Kms.

Another interesting question is whether the supply of nucleotide derivatives into the lumen of the Golgi apparatus and endoplasmic reticulum can be increased by the number of transporter molecules. Transport of nucleotide sugars, nucleotide sulfate, and ATP into the lumen of the endoplasmic reticulum and Golgi apparatus is competitively inhibited by the corresponding nucleoside mono- and diphosphates (29). Do these solutes play a role under physiological conditions, thereby affecting the amounts of nucleotide sugars, nucleotide sulfate, and ATP that can be translocated into the lumen of the above organelles?

Recent studies have shown that several derivatives of purines affect glycosylation of proteins and lipids in vivo, presumably as a result of inhibition of nucleotide sugar transport into the Golgi apparatus (111–114). This characteristic has been postulated as a potential target in the design of drugs that might inhibit specific glycosylation of virus glycoproteins and possibly the infectivity and pathogenesis of these viruses (111, 112). AZT, or 3′-azido-3′-deoxythymidine, and 3′ azido thymidine inhibit protein and lipid glycosylation in vivo in addition to inhibiting nucleic acid biosynthesis. This effect could lead to development of novel therapeutic strategies, which prevent viral replication and do not affect endogenous glycosylation pathways (113, 114). The lack of virulence of the *L. donovani* mutant defective in Golgi transport of GDP-mannose also makes this transporter a possible drug target (43).

Do diseases exist that involve defects in nucleotide sugar, nucleotide sulfate, and ATP transport into the lumen of the endoplasmic reticulum and Golgi apparatus? We believe that such diseases, if they exist, either are lethal or, as a result of "leaky mutations," have very pleiotropic effects, because defects in N-glycosylation, O-glycosylation, biosynthesis of glycosphingolipids, GPI-anchoring of proteins, and proteoglycans could all be affected. Gene disruption of N-acetylglucosaminyltransferase I, an enzyme that uses UDP-N-acetylglucosamine as substrate, is already known to be lethal (121), and further genetic evidence suggests that each nucleotide derivative has only one Golgi transporter gene. Recently, an isoform of the human UDP-galactose transporter (probably an alternative splicing transcript of the corresponding gene) has been characterized (41).

The description of diseases in transport of sugars and amino acids across lysosomal membranes raises the possibility that diseases caused by lack of Golgi transporters may also occur (115, 116); the ongoing cloning and chromosomal localization of such proteins may lead to identification of such diseases. To what extent the absence of a given transporter for a certain nucleotide derivative can be compensated by other transporters mediating transport of a given solute depends on the intrinsic specificity of these proteins. Answers to some of these questions are likely to be obtained in the near future, following the heterologous expression of some of these transporters in systems in which such transport activities do not normally occur.

Targeting Signals for Transporters of Nucleotide Sugars, ATP, and Nucleotide Sulfate to the Endoplasmic Reticulum and Golgi Apparatus

Elucidation of the structural features that result in transporters for nucleotide sugars, nucleotide sulfate, and ATP being localized to the endoplasmic reticulum and the Golgi apparatus, and not to other organelles, is of major importance. We predict that transporters for solutes that can enter both the endoplasmic reticulum and Golgi membrane, such as UDP-N-acetylglucosamine ATP and UDP-glucuronic acid, will have different structures and organellar targeting signals. Are the Golgi transporters polarized in a manner analogous to the glycosyltransferases in this organelle? Is this putative polarization cell dependent, as has been found for glycosyltransferases (117)? Are the transporters in a complex with enzymes that use nucleotide sugars as substrates? Do the transporters from the Golgi apparatus recycle into the endoplasmic reticulum (118–120)? Golgi transporters do not appear to have a Golgi-ER retrieval signal (KKXX) (118–120), but the S. cerevisiae homolog of the K. lactis Golgi UDP-GlcNAc transporter does (54, 110).

CONCLUSIONS

Work of the past decade has firmly established a biochemical understanding of the membrane-bound antiporters responsible for transport of nucleotide sugars, nucleotide sulfate, and ATP into the lumen of the endoplasmic reticulum and Golgi apparatus. Future studies will explore the intricate mechanisms of transport and the biological consequences of mutagenic changes in and overexpression of transporters. Finally, heterologous expression of transporters will provide information concerning the mechanism of targeting of transporters to specific membrane systems, and it will be important for biotechnology-based efforts to reconstruct mammalian glycosylation systems in yeast and other organisms.

ACKNOWLEDGMENTS

We thank Karen Welch and Annette Stratton for excellent secretarial assistance and David Raden for help with computing. Work in the authors' laboratory was supported by grants from the National Institutes of Health, GM30365, GM34396, and GM45188.

Literature Cited

1. Palade G. 1975. *Science* 189:347–58
2. Pryer NK, Wuestehube LJ, Schekman R. 1992. *Annu. Rev. Biochem.* 61:471–516
3. Katz FN, Rothman JE, Lingappa VR, Blobel G, Lodish HF. 1977. *Proc. Natl. Acad. Sci. USA* 74:3278–82
4. Lingappa VR, Lingappa JR, Prassad R, Ebner K, Blobel G. 1978. *Proc. Natl. Acad. Sci. USA* 75:2338–42
5. Kornfeld R, Kornfeld S. 1985. *Annu. Rev. Biochem.* 54:631–34
6. Abeijon C, Hirschberg CB. 1992. *Trends Biochem. Sci.* 17:31–36
7. Sousa MC, Ferrero-Garcia MA, Parodi AJ. 1992. *Biochemistry* 31:97–105
8. Hebert DN, Foelmer B, Helenius A. 1995. *Cell* 81:425–33
9. Abeijon C, Hirschberg CB. 1988. *Proc. Natl. Acad. Sci. USA* 85:1010–14
10. Quemeneur E, Guthapfel R, Gueguen P. 1994. *J. Biol. Chem.* 269:5485–88
11. Pfeffer SR, Rothman JE. 1987. *Annu. Rev. Biochem.* 56:835–51
12. Munro S, Pelham HRB. 1986. *Cell* 46:291–300
13. Braakman I, Helenius J, Helenius A. 1992. *Nature* 356:260–62
14. Neutra M, Leblond CP. 1966. *J. Cell Biol.* 30:137–50
15. Herscovics A. 1969. *Biochem. J.* 112:709–19
16. Coates SW, Gurney T, Sommers LW, Yeh M, Hirschberg CB. 1980. *J. Biol. Chem.* 255:9225–29
17. Muenster A, Eckhardt M, Potvin B, Muehlenhoff M, Stanley P, et al. 1997. *Glycoconjugate J.* 14:S134
18. Kuhn NJ, White A. 1976. *Biochem. J.* 154:243–44
19. Waldman BC, Rudnick G. 1990. *Biochemistry* 29:44–52
20. Perez M, Hirschberg CB. 1987. *Methods Enzymol.* 138:709–15
21. Capasso JM, Keenan TW, Abeijon C, Hirschberg CB. 1989. *J. Biol. Chem.* 264:5233–40
22. Fleischer B. 1983. *J. Histochem. Cytochem.* 32:1033–40
23. Carey DJ, Sommers LW, Hirschberg CB. 1980. *Cell* 19:597–605
24. Creek KE, Morre DJ. 1981. *Biochim. Biophys. Acta* 643:292–305
25. Deutscher SL, Nuwayhid N, Stanley P,

Briles EIB, Hirschberg CB. 1984. *Cell* 39:295–99
26. Deutscher SL, Hirschberg CB. 1986. *J. Biol. Chem.* 261:96–100
27. Hanover JA, Lennarz WJ. 1982. *J. Biol. Chem.* 257:2787–94
28. Sommers LW, Hirschberg CB. 1982. *J. Biol. Chem.* 10811–17
29. Capasso JM, Hirschberg CB. 1984. *Biochim. Biophys. Acta* 777:133–39
30. Lepers A, Shaw L, Schneckenburger P, Cacan R, Verbert A, Schauer R. 1990. *Eur. J. Biochem.* 193:715–23
31. Milla ME, Hirschberg CB. 1989. *Proc. Natl. Acad. Sci. USA* 86:1786–90
32. Berninsone P, Eckhardt M, Gerardy-Schahn R, Hirschberg CB. 1997. *J. Biol. Chem.* 272:12616–19
33. Hayes BK, Varki A. 1993. *J. Biol. Chem.* 268:16155–69
34. Capasso JM, Hirschberg CB. 1984. *Proc. Natl. Acad. Sci. USA* 81:7051–55
35. Brandan E, Fleischer B. 1982. *Biochemistry* 21:4640–45
36. Brandan E, Fleischer B. 1981. *Fed. Proc.* 40:681 (Abstr.)
37. Brandli AW, Hansson GC, Rodriguez-Boulan E, Simons K. 1988. *J. Biol. Chem.* 263:16283–90
38. Toma L, Pinhal MAS, Dietrich CP, Nader HB, Hirschberg CB. 1996. *J. Biol. Chem.* 271:3897–901
39. Yusuf HKM, Pohlentz G, Sandhoff K. 1983. *Proc. Natl. Acad. Sci. USA* 80:7075–79
40. Barthelson R, Roth S. 1985. *Biochem. J.* 225:67–75
41. Milla ME, Clairmont CA, Hirschberg CB. 1992. *J. Biol. Chem.* 267:103–7
42. Miura N, Ishida N, Hoshino M, Yamauchi M, Hara T, et al. 1996. *J. Biochem.* 120:236–41
42a. Ishida N, Miura N, Yoshioka S, Kawakita M. 1996. *J. Biochem.* 120:1074–78
43. Ma D, Russell DG, Beverley SM, Turco SJ. 1997. *J. Biol. Chem.* 272:3799–805
43a. Descoteaux A, Luo Y, Turco SJ, Beverley SM. 1995. *Science* 269:1869–72
44. Etchison JR, Freeze HH. 1996. *Glycobiology* 6:177–89
45. Tabuchi M, Tanaka N, Iwahara S, Takegawa K. 1997. *Biochem. Biophys. Res. Commun.* 232:121–25

46. Schwarz JK, Capasso JM, Hirschberg CB. 1984. *J. Biol. Chem.* 259:3554–59
47. Habuchi O, Conrad HE. 1985. *J. Biol. Chem.* 260:13102–8
48. Capasso JM, Hirschberg CB. 1984. *J. Biol. Chem.* 259:4263–66
49. Zaruba ME, Schwartz NB, Tennekoon GI. 1988. *Biochem. Biophys. Res. Commun.* 155:1271–77
50. Mandon E, Milla ME, Kempner E, Hirschberg CB. 1994. *Proc. Natl. Acad. Sci. USA* 91:10707–11
51. Ozeran JD, Westley J, Schwartz NB. 1996. *Biochemistry* 35:3685–94
52. Abeijon C, Orlean P, Robbins PW, Hirschberg CB. 1989. *Proc. Natl. Acad. Sci. USA* 86:6935–39
53. Berninsone P, Miret JJ, Hirschberg CB. 1994. *J. Biol. Chem.* 269:207–11
54. Abeijon C, Robbins PW, Hirschberg CB. 1996. *Proc. Natl. Acad. Sci. USA* 93:5963–68
55. Traynor AJ, Hall ET, Walker G, Miller WH, Melancon P, Kuchta RD. 1996. *J. Med. Chem.* 39:2894–99
56. Bossuyt X, Blanckaert N. 1995. *Eur. J. Biochem.* 223:981–88
57. Bossuyt X, Blanckaert N. 1995. *Biochem. J.* 305:321–28
58. Perez M, Hirschberg CB. 1985. *J. Biol. Chem.* 260:4671–78
59. Hayes BK, Freeze HH, Varki A. 1993. *J. Biol. Chem.* 268:16139–54
60. Abeijon C, Hirschberg CB. 1987. *J. Biol. Chem.* 262:4153–59
61. Hayes BK, Varki A. 1993. *J. Biol. Chem.* 268:16170–78
62. Puglielli L, Mandon E, Hirschberg CB. 1997. *Glyconjugate J.* 14:S8 (Abstr.)
63. Nuwayhid N, Glaser JH, Johnson JC, Conrad HE, Hauser SC, Hirschberg CB. 1986. *J. Biol. Chem.* 261:12936–41
64. Kearns AE, Vertel BM, Schwartz NB. 1993. *J. Biol. Chem.* 268:11097–104
65. Vertel BM, Walters LM, Flay N, Kearns AE, Schwartz NB. 1993. *J. Biol. Chem.* 268:11105–12
66. Capasso JM, Keenan TW, Abeijon C, Hirschberg CB. 1989. *J. Biol. Chem.* 264:5233–40
67. Clairmont CA, DeMaio A, Hirschberg CB. 1992. *J. Biol. Chem.* 267:3983–90
68. Mayinger P, Meyer DI. 1993. *EMBO J.* 12:659–66
69. Mayinger P, Bankaitis VA, Meyer DI. 1995. *J. Cell Biol.* 131:1377–86
70. Kim SH, Shin SJ, Park JS. 1996. *Biochemistry* 35:5418–25
71. Hauser SC, Ziurys JC, Gollan JL. 1988. *Biochim. Biophys. Acta* 967:149–57
72. Berg CL, Radominska A, Lester R, Gollan JL. 1995. *Gastroenterology* 108:183–92
73. Bossuyt X, Blanckaert N. 1994. *Biochem. J.* 302:261–69
74. Bossuyt X, Blanckaert N. 1997. *Biochem. J.* 323:645–48
75. Perez M, Hirschberg CB. 1986. *J. Biol. Chem.* 260:6822–30
76. Vanstapel F, Blanckaert N. 1988. *J. Clin. Invest.* 82:1113–22
77. Stanley P. 1980. *Am. Chem. Soc. Symp. Ser. B.* 128:213–21
78. Stanley P. 1985. *Mol. Cell. Biol.* 5:923–29
79. Briles EB, Li E, Kornfeld S. 1977. *J. Biol. Chem.* 252:1107–16
80. Abeijon C, Mandon EC, Robbins PW, Hirschberg CB. 1996. *J. Biol. Chem.* 271:8851–54
81. Douglas RH, Ballou CE. 1982. *Biochemistry* 21:1561–70
82. Hirschberg CB, Snider MD. 1987. *Annu. Rev. Biochem.* 56:63–87
83. Kuhn NJ, White A. 1977. *Biochem. J.* 168:423–33
84. Novikoff AB, Goldfischer S. 1961. *Proc. Natl. Acad. Sci. USA* 47:802–10
85. Neville MC, Peaker M. 1979. *J. Physiol.* 290:59–67
86. Shillingford JM, Calvert DT, Beechey RB, Shennan DB. 1996. *Exp. Physiol.* 81:273–84
87. Yanagisawa K, Resnick D, Abeijon C, Robbins PW, Hirschberg CB. 1990. *J. Biol. Chem.* 265:19351–55
88. Abeijon C, Yanagisawa K, Mandon EC, Hausler A, Moremen K, et al. 1993. *J. Cell Biol.* 122:307–23
89. Rijcken WRP, Overdijk B, Van den Eijnden DH, Ferwerda W. 1995. *Biochem. J.* 305:865–70
90. Guillen E, Hirschberg CB. 1995. *Biochemistry* 34:5472–76
91. Eckhardt M, Muehlenhoff M, Bethe A, Gerardy-Schahn R. 1996. *Proc. Natl. Acad. Sci. USA* 93:7572–76
92. Cacan R, Villers C, Belard M, Kaiden A, Krag S, Verbert A. 1992. *Glycobiology* 2:127–36
93. Stafford FJ, Bonifacino JS. 1991. *J. Cell Biol.* 115:1225–36
94. Villers C, Cacan R, Mir A-M, Labiau O, Verbert A. 1994. *Biochem. J.* 298:135–42
95. Wiertz EJHJ, Jones TR, Sun L, Bogyo M, Geuze HJ, Ploegh HL. 1996. *Cell* 84:769–79
96. Suzuki T, Kitajima K, Inoue S, Inoue Y. 1995. *Glycoconjugate J.* 12:183–93
97. Römisch K, Ali BRS. 1997. *Proc. Natl. Acad. Sci. USA* 94:6730–34
98. Shepard SRP, Baird SJ, Hallinan T,

Burchell B. 1989. *Biochem. J.* 259:617–20

99. Banhegyi G, Garzo T, Fulceri R, Benedetti A, Mandl J. 1993. *FEBS Lett.* 328:149–52

100. Banhegyi G, Csala M, Mandl J, Burchell A, Burchell B, et al. 1996. *Biochem. J. Lett.* 320:343–44

101. Zakim D, Dannenberg A. 1992. *Biochem. Pharm.* 43:1385–93

102. Shortle D, Haber JE, Botstein D. 1982. *Science* 217:371–73

103. Momburg F, Neefjes JJ, Hammerling GJ. 1994. *Curr. Opinion Immunol.* 6:32–37

104. Pasyk EA, Foskett KJ. 1997. *J. Biol. Chem.* 272:7746–51

105. Hucho F, Gorne-Tschelnokow U, Strecker A. 1994. *Trends Biochem. Sci.* 19:383–87

106. Gorne-Tschelnokow U, Strecker A, Kaduk C, Naumann D, Hucho F. 1994. *EMBO J.* 13:338–41

107. Yan RT, Maloney PC. 1995. *Proc. Natl Acad. Sci. USA* 92:5973–76

108. Kaback HR. 1997. *Proc. Natl. Acad. Sci. USA* 94:5539–43

109. Genick UK, Borgstahl GE, Ng K, Ren Z, Pradervand C, et al. 1997. *Science* 225:1471–75

110. Abeijon C, Chen L. 1997. *Glycoconjugate J.* 14:S128 (Abstr.)

111. Olofsson S, Milla M, Hirschberg C, De-Clercq E, Datema R. 1988. *Virology* 166:440–50

112. Olofsson S, Sjoblom I, Hellstrand K, Shugar D, Clairmont C, Hirschberg C. 1993. *Arch. Virol.* 128:241–56

113. Hall ET, Yan JP, Melancon P, Kuchta RD. 1994. *J. Biol. Chem.* 269:14355–58

114. Yan JP, Ilsley DD, Frohlick S, Street R, Hall ET, et al. 1995. *J. Biol. Chem.* 270:22836–41

115. Gahl WA, Basham N, Tietz F, Bernardini I, Schulman JD. 1982. *Science* 217:1263–64

116. Mancini GMS, deJonge HR, Galjaard H, Verheijen FM. 1989. *J. Biol. Chem.* 264:15247–54

117. Colley KJ. 1997. *Glycobiology* 7:1–13

118. Jackson MR, Nilsson T, Peterson PA. 1990. *EMBO J.* 9:3153–62

119. Townsley FM, Pelham HRB. 1994. *Eur. J. Cell Biol.* 64:211–16

120. Schröder S, Schimmöeller F, Singer-Krueger B, Riezman H. 1995. *J. Cell. Biol.* 131:895–912

121. Ioffe E, Stanley P. 1994. *Proc. Natl. Acad. Sci. USA* 91:728–32

Annu. Rev. Biochem. 1998. 67:71–98

RIBONUCLEOTIDE REDUCTASES

A. Jordan
Department of Genetics and Microbiology, Faculty of Sciences, Autonomous
University of Barcelona, Bellaterra, 08193 Barcelona, Spain

P. Reichard
Department of Biochemistry 1, Medical Nobel Institute, MBB, Karolinska Institutet,
17177 Stockholm, Sweden

KEY WORDS: deoxyribonucleotide synthesis, protein free radicals, adenosylcobalamin,
iron-sulfur cluster, allosteric regulation, evolution

ABSTRACT

Ribonucleotide reductases provide the building blocks for DNA replication in all
living cells. Three different classes of enzymes use protein free radicals to activate
the substrate. Aerobic class I enzymes generate a tyrosyl radical with an iron-
oxygen center and dioxygen, class II enzymes employ adenosylcobalamin, and the
anaerobic class III enzymes generate a glycyl radical from S-adenosylmethionine
and an iron-sulfur cluster. The X-ray structure of the class I *Escherichia coli*
enzyme, including forms that bind substrate and allosteric effectors, confirms
previous models of catalytic and allosteric mechanisms. This structure suggests
considerable mobility of the protein during catalysis and, together with experiments involving site-directed mutants, suggests a mechanism for radical transfer
from one subunit to the other. Despite large differences between the classes, common catalytic and allosteric mechanisms, as well as retention of critical residues
in the protein sequence, suggest a similar tertiary structure and a common origin
during evolution. One puzzling aspect is that some organisms contain the genes
for several different reductases.

CONTENTS

71

INTRODUCTION

The history of ribonucleotide reductases is full of surprises. The first surprise was their existence. The discovery of the first ribonucleotide reductase in *Escherichia coli* (1) was met with considerable skepticism since organic chemistry then did not know of a reaction whereby a carbon-bound OH-group could be replaced directly by hydrogen. The second surprise was that the enzyme did not contain vitamin B_{12} (2). A few years after the discovery of the *E. coli* enzyme a second reductase was discovered in *Lactobacillus leichmannii* (3). This enzyme required adenosylcobalamin, which led researchers to believe that ribonucleotide reductases were B_{12} enzymes and that their malfunction was the cause of the defective DNA synthesis in pernicious anemia, a disease that could be cured by administration of B_{12}.

When the *E. coli* reductase was shown not to require B_{12} this was the first sign that more than one class of ribonucleotide reductases exists. The *E. coli* reductase became the prototype of class I enzymes, to which all eukaryotic reductases belong, whereas the *L. leichmannii* enzyme became the prototype for a large group of microbial class II reductases that depend on adenosylcobalamin. Since the human enzyme lacks vitamin B_{12}, a different explanation, involving folate metabolism, is required to explain the symptoms of the megaloblastic anemia (4).

Originally it looked as if each of the four ribonucleotides required a specific reductase. During purification of the enzyme, with CDP reduction as an assay, the activity toward purine ribonucleotides disappeared. It soon became clear that only a single enzyme was involved, with allosteric effects directing its substrate specificity (5). Then came the biggest surprise of all: An organic free radical (6), identified as tyrosyl-122 (7), forms part of the polypeptide structure of one of the subunits of the *E. coli* enzyme and is required for activity. The radical is extremely stable and survived the two weeks it then took to purify the enzyme. This highly unexpected result had no precedent.

Ribonucleotide reductase was the first protein free radical to be discovered, followed by many others that contain tyrosyl or other amino acid radicals (8).

From then on, radical chemistry became the leading theme in mechanistic studies that resulted in the proposal of a detailed mechanism involving the transfer of the radical function from the enzyme to the substrate (9). The substitution of OH by H, so difficult to envisage in 1960, could be explained.

The discovery of a second ribonucleotide reductase in anaerobically growing *E. coli* (10) came as no surprise, as the generation of the tyrosyl radical of class I enzymes requires oxygen for its formation. The anaerobic enzyme is also a protein radical, with the unpaired electron located on a glycyl residue at the COOH-terminal region of the polypeptide chain (11). It contains an iron-sulfur cluster in a separate subunit and uses this cluster and *S*-adenosylmethionine to generate the glycyl radical (12). The *E. coli* anaerobic reductase is the prototype for class III reductases, whose genes have been found in other anaerobically growing eubacteria, in phage T4, and in archaebacteria (13–15).

A big surprise was, however, the discovery in enterobacteria of an additional class I enzyme that has an unknown physiological role (16, 17). The importance of this discovery also lies in the subsequent realization that active class I reductases of many other bacteria resemble the new reductase in structure and properties (18–23). Therefore, two subgroups of class I reductases exist: One (class Ia) has the original *E. coli* enzyme as the prototype and also includes the eukaryotic enzymes; the second (class Ib) has the new enzyme as its prototype and is present exclusively in bacteria. Some bacteria contain genes for both class I and class II reductases (24). The explanation of this tantalizing recent discovery is awaited with great interest.

A major advance has been the elucidation of the structure of the *E. coli* class Ia enzyme (25–28), including the binding of substrate and allosteric effectors (29). The work not only substantiated earlier hypotheses about allosteric effects and the reaction mechanism, but also laid the foundation for an understanding of the radical transfer within the protein. From this work one can deduce that during catalysis reductases undergo an amazing diversity of conformational changes that are involved in the reduction of the catalytic cysteines and open up and adapt the catalytic site for the appropriate substrate. Conformational changes are also a prerequisite for the channeling of the radical between the two subunits and participate in the final expulsion of the deoxyribonucleotide.

The multiplicity of reductases may seem bewildering. How can a fundamental metabolic reaction that probably was a prerequisite for the appearance of DNA during evolution appear in so many different variations today? At first sight the members of the three different classes show little resemblance. Their amino acid sequences are different, and they use different mechanisms to generate their protein radicals. However, they catalyze the same radical chemistry at the nucleotide level, and all employ amino acid free radicals. A second common denominator is the way in which they regulate their substrate specificity by allosteric effects. This suggests that they share a common tertiary structure, at least in part, in spite of large differences in primary structure.

We describe here our present understanding of how the members of the three classes have solved specific aspects of ribonucleotide reduction. We propose that the considerable differences between the classes reflects the appearance of oxygen during evolution, which drastically changed the conditions for the radical-based reaction mechanism. Despite these differences we can distinguish a pattern that suggests a common evolutionary origin for the classes. Several excellent reviews of ribonucleotide reduction have been published during recent years, often stressing special aspects such as radical formation in class I and class II enzymes (30), metal centers (31, 32), enzyme regulation (33), evolution (34), or mammalian reductases (35).

THREE CLASSES: AN OVERVIEW

Ribonucleotide reductases can be grouped into three major classes (34, 36) based on the mechanisms they use for radical generation and on their structural differences (Table 1). Class I can be subdivided further into two subclasses, Ia and Ib, based mainly on differences in allosteric regulation but also on involvement of auxiliary proteins (18).

All ribonucleotide reductases, irrespective of class, use radical chemistry for ribonucleotide reduction (30, 31, 34). Class I enzymes consist of two

Table 1 Overview of the characteristics of the ribonucleotide reductase classes[a]

	Class Ia	Class Ib	Class II	Class III
Oxygen dependence	Aerobic	Aerobic	Aerobic/anaerobic	Anaerobic
Structure	$\alpha_2\beta_2$	$\alpha_2\beta_2$	$\alpha(\alpha_2)$	$\alpha_2\beta_2$
Genes	nrdAB	nrdEF	—[b]	nrdDG
Radical	Tyr...Cys	Tyr...Cys	AdB12...Cys	AdoMet...Gly...Cys?
Metal site	Fe-O-Fe	Fe-O-Fe	Co	Fe-S
		Mn-O-Mn		
Substrate	NDP	NDP	NDP/NTP	NTP
Reductant	Thioredoxin	NrdH-redoxin	Thioredoxin	Formate
	Glutaredoxin	Glutaredoxin		
Allosteric sites/ polypeptide chain	2	1	1	2
dATP inhibition	Yes	No	No	Yes
Occurrence	Eukaryotes		Archaebacteria	Archaebacteria
	Eubacteria	Eubacteria	Eubacteria	Eubacteria
	Bacteriophages		Bacteriophages	Bacteriophages
	Viruses			
Prototype	E. coli	S. typhimurium	L. leishmannii	E. coli
	Mouse	C. ammoniagenes		

[a]Class I has been divided in two subclasses (Ia and Ib) as they present some interesting differential characteristics.
[b]The gene encoding the class II prototype from L. leishmannii has never been named. On the other hand, the gene encoding the ribonucleotide reductase from T. maritima is named nrdJ (24), and this name can be extended to other typical class II enzymes.

homodimeric proteins, R1 (α_2), coded by the *nrdA* gene, and R2 (β_2), coded by *nrdB* (37). The large α chain harbors the catalytic site and binding sites for allosteric effectors. The small β chain contains an oxygen-linked diferric center and, in its active form, a stable tyrosyl free radical. Class II enzymes have a simpler structure (α or sometimes α_2) (38, 39). Class III enzymes are $\alpha_2\beta_2$ heterotetramers, similar to class I (12). The α polypeptide, coded by *nrdD*, contains a stable oxygen-sensitive glycyl radical (11). β_2, coded by *nrdG*, contains an iron-sulfur cluster that together with S-adenosylmethionine can generate the glycyl radical (12).

The large (80–100 kDa) α polypeptide chains represent the "business end" of the enzymes, where catalysis and allosteric interactions occur. Depending on the allosteric configuration, one of the four common ribonucleotides binds to the catalytic site (34, 36). Ribonucleoside diphosphates are the substrates for class I enzymes, whereas the two class III enzymes investigated so far use triphosphates (10, 14). The prototype class II *L. leishmannii* enzyme also uses triphosphates (38), but other, later discovered members of this class use diphosphates (24, 39–41).

The enzyme reaction requires an external electron donor. These are either small proteins with redox-active thiols (thioredoxins or glutaredoxins) for class I and class II enzymes (42, 43), or formate for class III enzymes (44). Class I, class II, and probably class III enzymes contain a redox-active cysteine pair for the direct reduction of the ribose moiety (45–47). An additional cysteine, capable of forming a thiol radical, is also present (48). Within each class the enzymes show clear amino acid sequence homologies, whereas the prototypes for the three classes are not homologous. This observation originally led to the conclusion that the three classes differed completely in their primary sequence. The elucidation of additional class II sequences (24, 40, 41) is at present changing this conclusion by demonstrating considerable homologies between class II and the α chain of class I. Some of the new class II proteins (40, 41) show at their N-terminal ends some homology with that of the class III α chain, thus linking the primary structures of three classes.

The small β subunits differ in mass (43 kDa for class I; 17.5 kDa for class III) and participate in the reaction through different mechanisms (12, 37). They can generate the stable amino acid free radical. For this reason, class I β contains an oxygen-linked diferric center, whereas class III β_2 contains an iron-sulfur cluster. The amino acid radical in class I enzymes is located on a tyrosine residue of β itself, whereas the glycyl radical of class III is located on α. Class II enzymes lack the small subunit and contain no stable radical. In these enzymes adenosylcobalamin generates a transient radical during catalysis (48).

One important difference among the three classes is the relation of their free radicals to dioxygen. For class I enzymes, oxygen is a component of the system that generates the tyrosyl radical (37). These enzymes are therefore limited to aerobic conditions. In contrast, the glycyl radical of class III is rapidly destroyed by oxygen, with cleavage of the polypeptide chain at the location of the radical (49). Class III enzymes are therefore strictly anaerobic. Radical formation from adenosylcobalamin of class II enzymes does not require oxygen nor is the radical sensitive to oxygen; therefore, class II enzymes function in both aerobic and anaerobic organisms.

One unique feature of all ribonucleotide reductases concerns their substrate specificity. To satisfy the requirements of DNA replication, four separate deoxyribonucleotides must be produced, and in each case this is done by a single reductase. All reductases are equipped with a special allosteric site (substrate specificity site); binding of a specific deoxyribonucleoside triphosphate then adapts the catalytic site for the reduction of a specific substrate (50). In addition, binding of dATP to a second allosteric site (activity site) present in many class I enzymes inhibits the enzyme's activity.

Class I enzymes have been found in aerobic eubacteria and in eukaryotes but so far not in archaebacteria. Class II enzymes occur in both aerobic and anaerobic bacteria from both domains. With one (51), or possibly two (52), exceptions they have not been found in eukaryotes. Class III enzymes are used by facultative anaerobic eubacteria during anaerobic growth (10, 13, 14) and by anaerobic archaebacteria (15).

Some Enterobacteriaceae have the coding potential for two class I enzymes (16). The two reductases show only modest sequence homology and differ in certain functional aspects (16, 17). We have therefore proposed a subdivision of class I into Ia and Ib, with the original *E. coli* reductase being the prototype for Ia and the new enzyme from *Salmonella typhimurium* the prototype for Ib (18). In Enterobacteriaceae and related species the Ia enzyme is physiologically active and the Ib enzyme, when present, has no obvious function. In the microbial world this is an exception, since the active reductase in other bacteria belongs to subgroup Ib. This also applies to the enzyme from *Corynebacterium ammoniagenes* that has been reported to depend on manganese (23, 53). The only class Ib reductase that has been demonstrated to contain iron is the *S. typhimurium* enzyme (17), and it is unknown whether the majority of Ib enzymes contain iron or manganese.

A distinguishing feature of all known Ib enzymes is that they lack approximately 50 amino-terminal residues of the Ia enzyme and are not inhibited by dATP (54). This observation led to the suggestion that the allosteric activity site is located in this area, a theory that recent crystallographic work has confirmed (29). A second characteristic feature is that Ib operons code for an additional auxiliary protein (NrdI) of unknown function and often also for a specific external electron donor (NrdH) (55, 56).

STRUCTURE

Class I

High-resolution structures of several forms of *E. coli* R1 (27, 28) and R2 proteins (25, 26, 57, 58), a plausible model for the R1:R2 holoenzyme (27), and the structure of the mouse R2 protein (59) are now available. Also, the structures of complexes between R1 and allosteric effectors and substrate have been solved and now provide a solid foundation for models concerning chemical mechanisms and allosteric behavior of class I reductases (29). Unless stated otherwise the following discussion concerns the *E. coli* enzyme.

R1 PROTEIN   The large homodimer (2 × 85.5 kDa) mediates both catalysis and allosteric interactions. Figure 1*A* shows the high-resolution structure of the dimer with substrate and effectors in place. Each monomer consists of three domains: one mainly helical domain comprising the 220 N-terminal residues; a second large (480 residues), novel ten-stranded β/α barrel; and a small $\alpha\beta\alpha\alpha\beta$ domain comprising 70 residues. The barrel is composed of two halves connected in an antiparallel way, each containing five parallel strands and four connecting helices (27, 28).

The active site (Figure 1*B*) was identified from a GDP-R1 complex (29). It is located deep inside each protomer in the center of a cleft between the N-terminal and barrel domains. C225 and C462, which form the redox-active cysteine pair involved in the reduction of the ribose (46, 47), are located on one of the two connections between the halves. Redox-active cysteines, capable of forming a disulfide bond, are usually separated by two or four amino acid residues (60). Here, the distance is over 200 residues, but the cysteines are brought into proximity by the unique barrel structure.

A third cysteine (C439), responsible for the transient thiol radical required for catalysis (48), is located on the tip of a finger-like loop that inserts into the center of the barrel. During reduction of the disulfide bridge, C462 moves approximately 6 Å deeper into the interior of the protein, permitting substrate binding. The disulfide bridge of the oxidized form sterically hinders access to the catalytic site. The diphosphate moiety of the substrate is bound deepest in the catalytic cleft, and the ribose ring is placed between the redox active C225/C462 pair on one side and the thiol radical–providing C439 on the other.

A second redox pair (C754/C759) at the C-terminal end receives the electrons from the external hydrogen donor (46–48). Its mobility permits it to swing in and transfer electrons to the catalytic site.

The structures of the two allosteric sites were established from complexes between R1 and effectors (29). The two effectors used in the crystallographic studies were the ATP analog AMPPNP (activity site) and dTTP (specificity site).

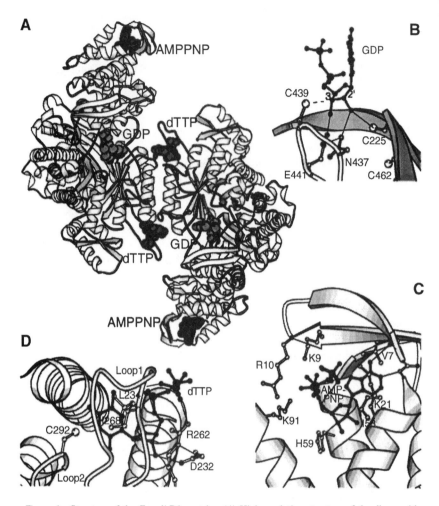

Figure 1 Structure of the *E. coli* R1 protein. (*A*) High-resolution structure of the dimer with substrate (GDP) and effectors (AMPPNP and dTTP) in place. (*B*) Detail of the active site. (*C*) Activity site containing a molecule of the ATP analog AMPPNP. (*D*) Specificity site occupied by the effector molecule dTTP.

AMPPNP was found attached in a cleft at the tip of the N terminus of R1 with the base bound in a hydrophobic pocket with main chain hydrogen bonds and the phosphates bound to basic residues at the entrance (Figure 1*C*). In the model for R1:R2 interaction (see below), the activity site on R1 is close to the C terminus of R2 and may influence the relative positions of R1 and R2 and thereby the radical transfer from the tyrosyl radical of protein R2 to C439 of protein R1 during catalysis.

The specificity site is at the interface between the two subunits of R1 (Figure 1D). dTTP is bound across the subunit interface with the phosphates to the N terminus of a helix from one subunit and thymine close to the C termini of two helices from the other. The closest distance to the substrate binding site is about 15 Å (Figure 1A), which is too far to allow direct interaction between effector and substrate. The effector stabilizes three loops that are highly flexible in the nonliganded protein, one of which may mediate the interaction with the substrate binding site.

R2 PROTEIN   The elucidation of the R2 structure in 1990 represented a milestone (25). Not only did it initiate our understanding of the structure of class I reductases, but it also shed light on the structure and function of dinuclear iron centers in general and the question of free radical generation and transmission (26). One important finding was that Y122, responsible for the stable free radical of the enzyme (7), is buried deep inside the protein, in a highly hydrophobic environment and close to the iron center. The tyrosyl radical cannot participate directly in the radical chemistry taking place at the catalytic site of R1 but must therefore generate a second, transient radical. A plausible model for such a radical transfer between R2 and R1 has been suggested from experiments with site-directed mutants (see below).

Several excellent reviews (31, 32, 37) have discussed the R2 structure, including that of its μ-oxo bridged dinuclear iron center. Briefly, the two μ-oxo-linked irons of the oxidized center are coordinated by four carboxylates and two histidines. On reduction the iron center is still coordinated by the same carboxylate-dominated ligand sphere, but several of the carboxylate ligands have undergone major conformational changes, with loss of the oxo bridge and of two bound water molecules, and a decrease of the coordination number from six to four. The iron center has opened up and become accessible to dioxygen, a prerequisite for radical generation on the neighboring Y122. Also the structure of the iron-free apoR2 (57) has been solved to high resolution and found to be very similar to those of redR2 (58) and metR2 (27).

The structure of mouse R2 closely resembles that of *E. coli*, with a similar coordination of the iron center (59). However, the radical/diferric site is less shielded and communicates via a narrow hydrophobic channel with the solvent, which explains its greater susceptibility to radical scavengers and metal chelators. The resemblance of the two structures suggests a similar structure for other class I proteins, including class Ib proteins, which usually have retained the important invariable amino acids of Ia (16), with two notable exceptions: R2 from *Mycoplasma* species (21, 22) and RNR4 from *Saccharomyces cerevisiae* (61). In both cases the DNA sequences suggest that important iron ligands are missing.

THE HOLOENZYME   Crystals of the R1:R2 complex suitable for structure determination have as yet not been obtained. However, a plausible model for the holoenzyme was proposed, mainly from considerations of symmetry, complementarity, and conserved amino acid residues (27). The C terminus of R2 interacts with a region close to the C terminus of the R1 protein with some species-specificity. This model forms the basis for the design of peptidomimetic drugs that specifically inhibit the activity of the herpes reductase but not the mammalian enzyme (62).

Class II and Class III Enzymes

Crystallization and structure determination of representatives of classes II and III has not yet been achieved. The tertiary structure of class II is probably in part closely related to that of the R1 proteins. This conclusion is based on the relative positions of the five catalytically active cysteines in the primary structures of the two classes (63), the high sequence similarities around the hydrogen-abstracting cysteine, as well as the similarity between the allosteric effects that regulate their substrate specificity (34, 36). Also the recently determined primary structures of several class II enzymes show considerable homology to R1 proteins, with retention of residues critically involved in effector binding (24).

Our knowledge of the structure of the anaerobic E. coli reductase, the prototype of class III, is in its infancy. Binding between the large α_2 and the small β_2 proteins is very strong (12). The α_2 protein contains active and allosteric sites and thus corresponds functionally to R1 of class I (64). The location of putative active cysteines has not been established. In contrast to class I enzymes, α_2 and not β_2 carries the stable free radical. In class III enzymes, β_2 is probably required only for the generation of the glycyl radical. For this reason, β_2 contains an Fe-S center, which after chemical reduction has the characteristics of a [4Fe-4S] cluster, with a stoichiometry of one cluster per β_2 dimer (12, 65). This stoichiometry suggests that the cluster is bound by two monomers. Before reduction, β_2 contains a small amount of a 3Fe-4S cluster, probably a degradation product. The nature of the major form of the oxidized cluster remains to be established.

CATALYTIC MECHANISM

External Electron Donors

Class I and class II enzymes receive the electrons required for the reduction of ribose from small proteins (thioredoxins and glutaredoxins) with redox-active thiols (42, 43). Ultimately, the reducing power comes from NADPH, which reduces the cysteines of the redoxins either via a specific thioredoxin reductase or via glutathione and glutathione reductase. Additional glutaredoxins (66) and thioredoxins (67, 68) have been found that can substitute for the classical ones. An interesting newcomer is the NrdH-redoxin (18, 56), which has thioredoxin-like properties but a glutaredoxin-like amino acid sequence. It is a specific reductant of the NrdEF system. After each catalytic cycle of a class I or II enzyme, a redoxin delivers electrons to a specific pair of cysteines and these, in turn, re-reduce by transthiolation the two cysteines at the catalytic site directly involved in the reduction of C-2' (46, 47). A fifth invariable cysteine is responsible for thiol radical formation (48).

Class III enzymes use formate instead of a redoxin as an external reductant (44). With ^{3}H-formate as reductant, tritium is recovered in water rather than in the deoxyribonucleotide, demonstrating that the hydrogen of formate equilibrates with an enzyme-bound intermediate that exchanges protons with water. Cysteine-thiols are good candidates for this reaction. During catalysis in D_2O, deuterium replaced the OH group at C-2' with retention of the original configuration (69), identical to earlier results with class I (70) and class II (71) enzymes. A very small amount of deuterium was also recovered in the 3'-position demonstrating that to some extent this hydrogen also exchanges with the protons of water during the reaction (69).

A similar labilization of the 3'-H was demonstrated earlier for class I and class II enzymes (72). These results suggest that in spite of the differences in the external reductant, class III enzymes also use redox-active cysteines and a cysteine-thiol radical for catalysis. Experiments now under way involving site-directed mutagenesis of selected cysteines should localize the catalytically important residues.

Radical Involvement During Catalysis

The reaction sequence depicted in Figure 2 shows a generic radical mechanism for class I and class II enzymes, with details taken from the *E. coli* class Ia enzyme. The general mechanism was proposed by Stubbe (9, 47), largely from experiments with mechanism-based inhibitors (73–75) and from the demonstration that the reaction involved breakage of the C-H bond at carbon 3' (72). Strong support for the proposed mechanism comes from experiments with site-directed mutants (76–78) and from the recently elucidated structure of the enzyme (27–29). Results from experiments with mechanism-based inhibitors (78a) suggest that a similar mechanism also is valid for class III reductases.

The reaction is initiated by binding of the substrate into the active site of the reduced enzyme. In the *E. coli* enzyme this leads to a transfer of the radical function from Y122 of the R2 protein to C439 of the R1 protein (see below), generating a thiol radical. A thiol radical is also formed in class II enzymes with the aid of adenosylcobalamin (48). The radical initiates the reduction of the ribonucleotide by abstracting the 3'-hydrogen atom, thereby generating a

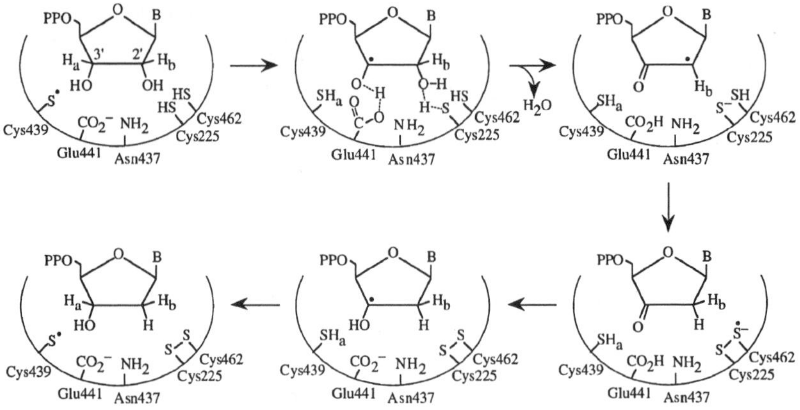

Figure 2   The reaction mechanism of ribonucleotide reductase (adapted from Stubbe & coworkers (9, 77).

substrate radical. Radical formation facilitates the leaving of the protonated OH-group at C-2′. A substrate cation radical is generated that subsequently is reduced by the redox-active cysteine pair C225 and C462. Finally the hydrogen atom stored at C439 is returned to C-3′ with regeneration of the thiol radical at C439. E441 and N437 stabilize the interaction between enzyme and substrate by hydrogen bonding to the oxygens at C-3′ and C-2′, respectively.

Generation of the Thiol Radical

In class I enzymes the thiol radical is formed transiently during each catalytic cycle by radical transfer from the stable tyrosyl radical. A specific pathway leading from Y-122 of R2 to C439 of R1 is probably involved. In it participate the iron center; the R2 residues D237, W48, and Y356; and the R1 residues Y731 and Y730. Evidence for this pathway comes from structural considerations (25–29) and site-directed mutants (79–82). The long-range electron transfer may be aided by concomitant switching of protons in hydrogen bonds (31). The specific channel protects the unpaired electron from unwanted side reactions.

Class II enzymes contain no stable radical in the resting enzyme. The thiol radical is formed from adenosylcobalamin during catalysis (48). Radical generation from adenosylcobalamin is usually considered to occur by homolytic cleavage of the carbon-cobalt bond, leading to the formation of a 5′-deoxy-adenosyl radical. Recent evidence suggests, however, that the *L. leishmannii* class II reductase generates its thiol radical by a concerted pathway leading to formation of 5′-deoxyadenosine and cob(II)alamin, with the thiol radical very close to cob(II)alamin (48). These experiments provide the first direct evidence for formation of a thiol radical. Since adenosylcobalamin functions catalytically, class II enzymes—in contrast to class I enzymes—do not require regeneration of their thiol radical by adenosylcobalamin for each catalytic cycle.

For class III enzymes, indirect evidence points to the possible participation of a transient thiol radical in the reaction (69). The stable glycyl radical then stores the unpaired electron and does not itself catalyze the abstraction of the 3'-hydrogen atom. This is the most appealing general picture, but more evidence is required on this point.

Generation of the Stable Radicals

The tyrosyl radical of the class I R2 protein is located close to a diferric iron center (25, 26). The generation of the tyrosyl radical is intimately linked to the center (reviewed in 30–32). Either the iron-free apoR2 or metR2, the form that lacks the radical but has retained an intact diferric center, can be the starting point for radical formation. ApoR2 requires addition only of ferrous iron and oxygen. In metR2 the diferric center must first be reduced in situ, and the tyrosyl radical is formed subsequently by oxygen. In both cases, reduced R2, containing a diferrous center, is an intermediate. On addition of oxygen a diferric peroxide is formed, followed by other transient species whose structures are under discussion.

In *E. coli* an enzyme system exists that generates the tyrosyl radical from metR2 (83, 84). The main component is a flavin reductase that reduces external riboflavin, FMN, or FAD. In the presence of ferrous iron, or an iron-rich protein, the reduced flavin reduces the ferric iron of metR2 in situ to give reduced R2 in which the tyrosyl radical can be generated by oxygen. Regeneration of metR2 becomes important when R2 has been inactivated by radical scavengers, for example, by treatment with hydroxyurea.

Mouse R2 contains a rather unstable diferric iron center that must be regenerated continuously with ferrous iron (35). Also, in this case, a mammalian flavin reductase may be required for the reduction of iron before its introduction into apoR2.

Much less is known about the generation of the glycyl radical in the α_2 protein of class III reductases. The reaction is catalyzed by the β_2 protein and requires a reducing enzyme system (NADPH, flavodoxin, and flavodoxin reductase) and S-adenosylmethionine (12). First, the iron-sulfur cluster of β_2 is reduced either by the flavodoxin system or chemically by dithionite or deazaflavin and light (65). Chemical reduction gives an electron paramagnetic resonance (EPR) spectrum typical for a reduced 4Fe-4S cluster, but not enzymatic reduction. After both types of reduction, β_2 binds S-adenosylmethionine tightly and generates the glycyl radical on α_2. In this reaction, S-adenosylmethionine is cleaved reductively, and methionine is formed. It is not clear if the reduced 4Fe-4S cluster is the species involved in the physiological reaction.

Other enzymes that employ S-adenosylmethionine, together with a reduced iron center for the generation of an amino acid free radical, are pyruvate formate lyase (85–88) and lysine 2,3-amino mutase (89, 90). In all cases, formation of a deoxyadenosyl radical appears likely. For the anaerobic reductase this implies the following steps for radical generation (12, 65): 1) the iron cluster of β_2 is reduced by reduced flavodoxin, 2) S-adenosylmethionine is bound to the reduced cluster, and 3) the nucleoside is cleaved reductively and forms a deoxyadenosyl radical that generates a glycyl radical on residue G682 of the α protein.

ALLOSTERIC REGULATION

In general, allosteric regulation rapidly adapts an enzyme to changing requirements for its product by binding of effectors that increase or decrease its activity. Most class Ia reductases are regulated in this way by binding either ATP (activating) or dATP (inhibitory) to the activity site of protein R1 (50). However, reductases are also endowed with an additional and unique allosteric mechanism that regulates their substrate specificity and ensures that the enzyme produces equal amounts of each dNTP for DNA synthesis. Disturbances in pool sizes lead to genetic damage and in severe cases to cell death (reviewed in 91). Such disasters are prevented by binding of end products (dATP, dGTP, and dTTP) to the specificity site of the reductase (50). The allosteric sites communicate with the substrate-binding site and also affect each other. A detailed model for class Ia enzymes involving various effectors and substrates was developed in 1979 (92). It has survived well over the years.

The discovery of new classes of ribonucleotide reductases raises the question of their allosteric regulation. Figure 3 illustrates effector binding to the allosteric sites from all three classes. Class Ib (54) and class II (R Eliasson, E Pontis, A Jordan & P Reichard, unpublished information) enzymes harbor only specificity sites, and these enzymes are not inhibited by dATP. Allosteric effects of thermophilic class II reductases from thermophilic bacteria are more pronounced at high temperature even though binding of effectors occurs in the cold room. Apparently the increased flexibility of the proteins at high temperature permits better communication between effector and catalytic sites.

The anaerobic $E.$ $coli$ reductase (class III) resembles the aerobic class Ia enzyme in that it has two separate allosteric sites (93) (Figure 3). Both sites affect substrate specificity and overall activity, with the same physiological end result as for class Ia enzymes.

Allosteric regulation of the enzymes from different classes thus shows similarities and differences. The similarity between the effects on substrate specificity is striking. In all cases, binding of (d)ATP to the specificity site induces

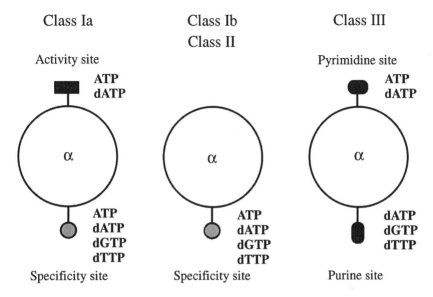

Figure 3 Models for the allosteric regulation of the ribonucleotide reductase classes. The model shows binding of effectors to one monomer of the large protein of the classes Ia, Ib, II, and III.

reduction of CDP and UDP, dGTP induces reduction of ADP, and dTTP induces reduction of GDP. Considering that these effects must depend on subtle conformational variations at the catalytic site and that the primary structures of the enzymes differ, the identity of effects is probably of evolutionary significance.

An obvious difference concerns the effect of dATP that functions as general inhibitor only for class Ia and class III enzymes that have two allosteric sites. Eukaryotic class Ia enzymes enjoy the most sophisticated regulation. The difference in phosphorylation levels between substrates and effectors facilitates regulation, and the presence of an activity site can turn off de novo synthesis on short notice.

Crystallographic studies of enzyme-nucleotide complexes (29) have confirmed and extended the above model. Only structures of the *E. coli* enzyme are known so far, but extrapolation to other reductases is beginning to provide an explanation for the structural basis for some of the variations in the allosteric behavior of different reductases.

Thus, with a few exceptions, the amino acid residues involved in effector binding in the *E. coli* enzyme at both the activity and the specificity sites are highly conserved in eukaryotic R1 proteins, suggesting a similar organization of both allosteric sites. The reductases of herpes viruses are one exception (94). They lack most of the important residues (95) and, as a result, are not regulated at all.

Other exceptions among class Ia are the enzymes from T4 bacteriophage (96) and *Trypanosoma brucei* (97) that are not inhibited by dATP. The T4 enzyme apparently lacks a dATP binding activity site. Its structure retains many of the residues of the *E. coli* enzyme required for binding but has exchanged the *E. coli* H59 for glutamine. This position is clearly crucial since in the mouse R1 protein a mutation resulted in the loss of inhibition by dATP (98). The trypanosoma R1 protein has retained almost all of the residues of the mammalian R1 proteins required for effector binding and consequently has two separate allosteric sites per monomer that bind dATP (97). In this case we must assume that binding of dATP does not transmit a message to the catalytic site that results in enzyme inhibition.

What about class Ib and class II enzymes that are not inhibited by dATP? For Ib the answer is simple: Compared with class Ia they lack the approximately 50 N-terminal amino acids of importance for the activity site (54). At the specificity site these enzymes—as well as the T4 enzyme—have retained crucial residues involved in binding and show the same allosteric regulation as the *E. coli* reductase. Also many class II reductases lack the specific N-terminal region that provides the activity site (24). As for the specificity site, most of the critical binding residues are present in the primary structure of class II enzymes. In the *E. coli* enzyme these residues are located at the first part of the 10-stranded β/α barrel (29), suggesting an identical structure for class II enzymes.

GENE ORGANIZATION AND REGULATION OF ENZYME SYNTHESIS

Ribonucleotide reduction occupies a central role in the regulation of the pool sizes of the four dNTPs required for DNA synthesis, even though deoxynucleoside kinases and nucleotidases also are important (reviewed in 99). The small dNTP pools suffice in mammalian cells for only a few minutes of DNA replication and must therefore be renewed continuously during S-phase. But also cells that are not in S-phase require dNTPs for DNA repair and mitochondrial DNA replication. They also synthesize dNTPs de novo, albeit at a much slower rate (100). To satisfy the changing demands for dNTPs the synthesis of ribonucleotide reductases is tightly adapted to the cell cycle. Regulation of ribonucleotide reduction by class Ia enzymes has been reviewed in depth elsewhere (33). Our overview of regulation includes the meager knowledge we have concerning enzymes belonging to other classes.

Enterobacteria

THE NrdAB AND NrdEF ENZYMES FROM ENTEROBACTERIA   The *nrdA* and *nrdB* genes that code for the R1 and R2 proteins, respectively, constitute a tightly regulated transcriptional unit that does not include the gene for either thioredoxin or glutaredoxin, the putative external electron donors (101, 102). Its transcription is coupled to the cell cycle and is increased by DNA damage (103). In general, transcription is increased when DNA replication is inhibited or when conditional DNA mutants are shifted to nonpermissive conditions (104). DnaA and Fis proteins bind to specific promoter sites and activate transcription but do not couple transcription to the cell cycle (105, 106). Instead, a *cis*-acting ATP-rich region located downstream of the Fis footprint area and preceding an inverted repeat is important for coupling *nrdAB* expression to the cell cycle and also for increasing transcription after inhibition of DNA synthesis (107). We still lack an integrated view of the regulation of this operon.

The recently discovered *nrdE* and *nrdF* genes code for the R1E and R2F proteins of the class Ib reductase (16–18). In *S. typhimurium* and *E. coli* they form an operon where the promoter is followed by four genes (55). In front of *nrdEF* are *nrdH*, coding for a 9-kDa protein acting as a specific electron donor in the R1E/R2F reaction, and *nrdI*, coding for a 15-kDa stimulatory protein whose specific function is not known (56). The promoter is induced by the reductase inhibitor hydroxyurea but not by inhibitors of DNA replication or by DNA damage. The chromosomal operon is poorly transcribed under all growth conditions tested so far and cannot fully complement conditional *nrdAB* mutants. Complementation is, however, achieved when a second *nrdEF* is introduced into the chromosome or when the genes are present on a plasmid (55). The physiological function of *nrdEF* in enterobacteria remains unclear.

In some other bacteria the genes are organized *nrdHIEF*, as in enterobacteria, with (18, 23) or without (20) an *nrdH* gene. *Mycobacterium tuberculosis* contains the *nrdHIE* cluster and two copies of the *nrdF* gene in separate locations on the chromosome (19, 108). In *Mycoplasma* species the organization is *nrdFIE* (21, 22). *Streptococcus pyogenes* contains two differently organized clusters, *nrdHEF* and *nrdFIE*.

THE NrdDG SYSTEM IN *E. COLI*   The regulation of the *nrdD* and *nrdG* genes poses interesting and as yet poorly resolved questions concerning anaerobic metabolism. In *E. coli*, synthesis of NrdD and NrdG is induced by microaerophilia and is accompanied by the turnoff of the aerobic *nrdAB* system. The *nrdDG* operon is probably regulated by the Arc/Fnr system, and an Fnr binding site has indeed been localized in the promoter region (109). Cells lacking functional *nrdD* or *nrdG* genes do not grow during strict anaerobiosis (110). The two genes are not required under microaerophilic conditions, however. In their absence the mutants overproduce the R1/R2 system to compensate for the insufficient oxygen activation of R2.

Eukaryotes

THE MOUSE R1 AND R2 PROTEINS   The genes for mouse R1 and R2 proteins are regulated separately and are located on separate chromosomes, *nrdA* (R1) on chromosome 7, *nrdB* (R2) on chromosome 12 (reviewed in 35). Reductase activity is very low during G_0/G_1 and high in S phase (111–113). Analyses of the separate subunits showed that R1 activity remains essentially constant, whereas R2 fluctuates during the cycle (114–116). In spite of this, mRNA levels fluctuate in parallel and are negligible during G_0/G_1, with a large increase on entrance into S (117). The observed fluctuations in enzyme activity are caused by differences in half-lives of the two proteins: 3 h for R2 and more than 20 h for R1. A complex pattern of regulation affecting transcription, mRNA stability, and protein phosphorylation is responsible for these effects.

Protein R1   Transcription of the TATA-less promoter is controlled by three specific protein-DNA interactions (118). One protein is S-phase specific; the other two exist during the whole cycle. Two additional regulatory regions, one at the transcription start and the other downstream, are also important (119). mRNA stability is affected by binding of specific proteins to the 3'-untranslated region and is controlled by the protein kinase C pathway (120).

Protein R2   Promoter activity during G_1 depends on four DNA-protein interaction regions. The region closest to the transcription start binds the NF-Y transcription factor and is required for basal transcription; the other three regions located farther upstream are required for proliferation-specific induction (121). Early during G_1, mRNA synthesis is initiated but blocked in the first intron by a unique cell cycle–dependent transcriptional block (122). During S-phase the polymerase is able to read through the block and transcribe the whole gene. R2-mRNA stability is regulated similarly to R1-mRNA by specific RNA-protein interactions in the 3'-untranslated region (123). Protein R2 (but not R1) can be phosphorylated by cyclin-dependent kinases (124). Phosphorylation may be related to the degradation of R2 during G_2. Finally, a recent report claimed that aberrant expression of R2 altered the malignant potential of cells (125).

UV irradiation, but not other DNA-damaging agents, activates the R1 and R2 promoters in quiescent cells, permitting the synthesis of dNTPs required for nucleotide excision repair. The mechanism for this induction differs from that for proliferation-specific induction (126).

RIBONUCLEOTIDE REDUCTASES IN YEAST   *S. cerevisiae* contains four genes involved in ribonucleotide reduction: *RNR1* and *RNR3* encode R1 proteins, *RNR2* encodes an R2 protein (127, 128), and *RNR4* encodes an R2-like protein (61). *RNR1* and *RNR2* are essential for viability. *RNR3* is not, but when present on a high-copy number plasmid, complements mutations in *RNR1*. This is reminiscent of the NrdEF system in *E. coli*. *RNR4* is also required for normal growth but the coded protein by itself has no R2 activity. It may be required for the formation of a functional holoenzyme complex (61).

Expression of the *RNR1* and *RNR2* coded proteins is cell cycle–controlled and maximal at the G_1-S phase transition. Extensive studies concerning the effects of various kinds of DNA damage gave the highest maximal inducibility factor for *RNR3*, although mutations did not confer a higher sensitivity to DNA damage (127, 128).

RIBONUCLEOTIDE REDUCTASES OF VIRUSES   Infection of *E. coli* with phage T4 induces the formation of phage-specific reductases. Aerobiosis induces a class I enzyme (129), and anaerobiosis induces a class III enzyme (14, 130). Both are highly homologous to the host enzymes. The phage genes possess an important distinguishing feature: *nrdB* (131) and *nrdD* (14) contain introns.

The large herpes viruses contain information for their own R1 and R2 proteins as well as for other enzymes required for DNA synthesis. This makes it possible for them to replicate in quiescent host cells that are deficient in these enzymes. The viral reductases present some unusual characteristics. The complete lack of allosteric regulation was mentioned previously (94). The R1 proteins of both HSV-1 and HSV-2 contain at their N termini a transmembrane helical segment followed by a Ser/Thr protein kinase (132, 133).

DISTRIBUTION AND EVOLUTION OF RIBONUCLEOTIDE REDUCTASES

Ribonucleotide reduction is an ancient reaction. It was probably a prerequisite for the appearance of DNA during the evolution of life. The existence today of three different classes of enzymes may therefore seem surprising. Originally, no or little homology was found between the prototypes of the three classes, and one could therefore argue that each class was invented separately during evolution (134), even though similarities in reaction mechanisms and allosteric regulation were in favor of a common ancestral origin for the large proteins (34, 36).

With the elucidation of an increasing number of primary sequences of class II and class III enzymes, it is now almost certain that the large proteins of the three classes arose by evolution from a common ancestor. In particular the sequences of class I and class II enzymes contain identical, strategically placed amino acids, involved in catalysis and allosteric regulation. Earlier negative results were due to the choice of the class II *L. leishmannii* enzyme as the object of comparison (134). This enzyme has a sequence atypical of class II reductases (135). Homologies with α proteins from class III reductases are less apparent and can at present be recognized only for the N-terminal portions of some enzymes. In contrast, the β proteins of class I and class III reductases differ both in structure and function and must have evolved separately.

Table 2 Occurrence of ribonucleotide reductase classes in nature[a]

Groups and species	Class	References/Source
Archaebacteria		
Euryarchaeota		
Methanococcus jannaschii[b]	III	15
Methanobacterium thermoautotrophicum	III	137[e]
Halobacteria[c]	II	40
Archaeglobus fulgidus[b]	II	f
Thermoplasma acidophilum	II	40
Pyrococcus furiosus	II	41
Eubacteria		
Thermotogales		
Thermotoga maritima	II	24
Deinococci and relatives		
Deinococcus radiodurans[b]	Ib II	24
Thermus X-1[c]	II	138
Green non-sulfur bacteria		
Chloroflexus aurantiacus[c]	II	24
Cyanobacteria[c, d]	II	51
Synechocystis sp.[b]	Ia	139
Low G+C gram-positive bacteria		
Lactococcus lactis	Ib III	18[g]
Streptococcus pyogenes[b]	Ib III	h
Mycoplasma genitalium[b]	Ib	21
Mycoplasma pneumoniae[b]	Ib	22
Bacillus subtilis	Ib	20
Bacillus megaterium[c]	II	140
Lactobacillus leishmannii	II	135
High G+C gram-positive bacteria		
Mycobacterium tuberculosis[b]	Ib II	19, 108[f]
Corynebacterium ammoniagenes	Ib	23, 53
Corynebacterium nephridii	II	141[i]
Spirochaetes		
Treponema pallidum[b]	Ia	f
Proteobacteria		
Delta and epsilon subdivision		
Helicobacter pylori[b]	Ia	142
Alpha subdivision		
Rhizobium sp.[c]	II	143
Beta subdivision		
Neisseria meningitidis[b]	Ia	f
Neisseria gonorrhoeae[b]	Ia	j
Alcaligenes eutrophus	I III	k
Gamma subdivision		
Haemophilus influenzae[b]	Ia III	13
Vibrio cholerae[b]	Ia III	f
Escherichia coli[b]	Ia Ib III	55, 101, 109, 144, 145
Salmonella typhimurium	Ia Ib III	16, 102[l]

(Continued)

Table 2   (*Continued*)

Groups and species	Class	References/Source
Eukarya[d]	Ia	146
Euglena gracilis[c]	II	147
Viruses		
Bacteriophages		
T4	Ia III	14, 148
Mycobacteriophage L5[b]	II	149
Animal viruses[d]	Ia	146

[a]Additional data has been published by Harder (150). Here we limit ourselves largely to species with known sequences and also include crucial data from important phylogenetic groups.
[b]Complete genome sequence known. Consequently the number of reductases per genome is known.
[c]Sequence not known, data from enzyme work.
[d]Data from several species.
[e]E Torrents, A Jordan, J Nolling, R Thauer, I Gilbert & P Reichard, unpublished.
[f]The Institute for Genomic Research, personal communication.
[g]G Buist, personal communication.
[h]The Streptococcal Genome Sequencing Project, personal communication.
[i]M Karlsson, personal communication.
[j]The Gonococcal Genome Sequencing Project, personal communication.
[k]A Siedow & B Friedrich, personal communication.
[l]E Torrents, A Jordan, I Gilbert, unpublished.

We propose the following scenario for the appearance of the three classes during evolution. A hypothetical "ur" reductase probably existed before oxygen was put into the atmosphere by primitive photosynthetic oxygenic cells. Therefore, it could not require oxygen for its function, which excludes a class I–type enzyme as a close relative. Instead, we suggest a class III–type protein. These enzymes generate the free radical required for catalysis with an iron-sulfur cluster. Iron in combination with sulfur has been suggested as a very early player during the chemical evolution of life (136). The cluster is located in the β_2 protein of the reductase (12, 65), which is closely related to the activase of pyruvate formate lyase (85–88), another ancient enzyme.

An additional argument arises from the hydrogen donors: Class III reductases—but not the other classes—use formate (44), one of the simplest organic reductants that might have been available on primitive earth. When oxygen began to appear in the atmosphere, class III enzymes could no longer function outside anaerobic environments, which we believe was the incitement for evolution of class II enzymes, functioning under both anaerobic and aerobic conditions. Finally, when oxygen metabolism became more sophisticated, with the iron center activating dioxygen, an oxidative radical–generating mechanism was developed and class I reductases evolved.

Table 2 gives the present day distribution of reductases from different classes in nature. Figure 4 shows a phylogenetic tree of the large subunits with established primary sequences. In Figure 4 the proteins are clearly grouped according to the earlier discussed classes, and relationships within the classes are also discernible.

Class Ia exists as two major and well-delineated groups in eukaryotes and in a limited number of eubacteria, most of which are of recent origin. The animal herpes viruses form an additional, smaller group. Class Ib exists in many major groups of eubacteria, from the deeply branched *Deinococcus* genus to enterobacteria. The enzyme from *C. ammoniagenes*, whose NrdF protein contains manganese, also belongs to this class (23). It is most closely related to the protein from *M. tuberculosis* (Figure 4). Class II is spread widely among

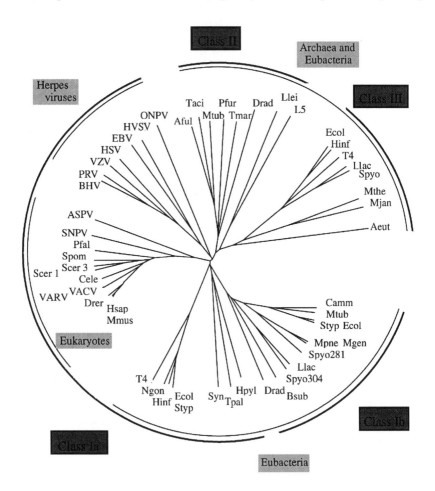

both eubacteria and archaebacteria. Class III is found in both domains, although it is more restricted. The widespread occurrence of the three classes suggests that they existed before the evolution of the kingdoms of the Archaebacteria, Eubacteria, and Eukaryotes, in agreement with the belief that oxygen appeared on earth before that event.

Aside from the limitations posed by oxygen, we can see no discernible rule for class selection by bacteria (Table 2). Among eubacteria, closely related species may have either a class I or a class II enzyme. As an example, most Cyanobacteria contain active class II proteins (51), as shown by enzyme analyses, but the genome of *Synechocystis* sp. contains only the gene for a class I enzyme (139). In other cases a bacterium may have the coding potential for both a class I and a class II reductase, with apparently only one of them being expressed during normal growth. In *M. tuberculosis*, enzymological work first described an active class Ib system (19), whereas genome sequencing subsequently showed the additional existence of a class II gene. The opposite occurred with *Deinococcus radiodurans* (24). These results resemble the situation in enterobacteria, where the *nrdAB* and *nrdEF* systems occur together (16). Finally, the genome of *Streptococcus pyogenes* contains two separate class Ib operons, one similar to that of some *Mycoplasma* species (21, 22), both with respect to amino acid sequence and gene organization, and the other more

---

Figure 4 Unrooted phylogenetic tree of deduced amino acid sequences of the large polypeptide chain of the three classes. All sequences were obtained from the Swiss-Prot database or from the reference or source stated in Table 2, corresponding to the following species: *Alcaligenes eutrophus* (Aeut), *Archaeglobus fulgidus* (Aful), African swine fever virus (ASFV), Bovine herpes virus (BHV), *Bacillus subtilis* (Bsub), *Corynebacterium ammoniagenes* (Camm), *Caenorhabditis elegans* (Cele), *Deinococcus radiodurans* (Drad), *Danio rerio* (Drer), Epstein-Barr virus (EBV), *Escherichia coli* (Ecol), *Haemophilus influenzae* (Hinf), *Helicobacter pylori* (Hpyl), *Homo sapiens* (Hsap), Herpes Simplex virus type-1 (HSV), Herpesvirus Saimiri (HVSA), mycobacteriophage L5 (L5), *Lactococcus lactis* (Llac), *Lactobacillus leishmannii* (Llei), *Mycoplasma genitalium* (Mgen), *Methanococcus jannaschii* (Mjan), *Mus musculus* (Mmus), *Mycoplasma pneumoniae* (Mpne), *Methanobacterium thermoautotrophicum* (Mthe), *Mycobacterium tuberculosis* (Mtub), *Neisseria gonorrhoeae* (Ngon), *Orgya pseudotsugata* nuclear polyhedrosis virus (ONPV), *Plasmodium falciparum* (Pfal), *Pyrococcus furiosus* (Pfur), Pseudorabies virus (PRV), *Saccharomyces cerevisiae* (Scer), *Spodoptera exigua* nuclear polyhedrosis virus (SNPV), *Schizosaccharomyces pombe* (Spom), *Streptococcus pyogenes* (Spyo), *Salmonella typhimurium* (Styp), *Synechocystis* sp. (Syn), bacteriophage T4 (T4), *Thermoplasma acidophilum* (Taci), *Thermotoga maritima* (Tmar), *Treponema pallidum* (Tpal), Vaccinia virus (VACV), Variola virus (VARV), and Varicella Zoster virus (VZV). The *outer circle* delimits the extension of each class, and the *inner circle* refers to its occurrence within the three domains of life. The CLUSTAL W 1.7 program (151) was used for multiple sequence alignment and to compute the phylogenetic tree.

related to that of *L. lactis* (18). In all these cases, except in enterobacteria, the functionality of both genes remains to be established. If both genes code for active enzymes, the mechanisms by which the second gene can be switched on remains to be determined.

Class II and class III reductases have not been found in the same organism. In eubacteria, class III exists together with class I in facultative anaerobes (10, 13, 18). Among strict anaerobes only methanobacteria have been found to contain a class III enzyme (15). All other anaerobic bacteria investigated so far contain a class II reductase. Again, there is no obvious reason for the choice of class. In fact, methanobacteria produce large amounts of cobalamines and would be able to make a class II reductase.

The highest degree of homology between members of all classes is found in the 50 to 100 N-terminal amino acids of the large proteins with the motif VXKRDG (positions 7 to 12 in *E. coli*) as an often recurring theme. The N-terminal part of the reductases should therefore represent an ancient section of the molecule. It is involved in effector binding in class Ia (29) and probably also class III. So far, this section has been found in all class III proteins (13, 15, 109), except T4 phage. Class Ib enzymes lack this sequence (54), whereas some class II enzymes contain it (40, 41).

We speculate that this motif originally was part of one of two allosteric sites in an ancient class III–type reductase that regulated both substrate specificity and activity (see Figure 3). During the evolution of class II and class I reductases, the second site took over all specificity functions, and the first site was either lost or modified. Modification resulted in either loss of allosteric effects (class II) or evolution into an activity site (class Ia).

CONCLUSION AND PERSPECTIVES

For many years the research on ribonucleotide reduction has provided a spectrum of fascinating results with ramifications into other areas of biochemistry: protein-free radicals, adenosylcobalamin, iron-proteins, active thiols, and allosteric mechanisms, to name a few. The evolutionary implications of the existence of several distantly related forms of the enzyme also continue to challenge the imagination.

The future is invented, not predicted (152). The recently emerging X-ray structures of class I ribonucleotide reductases have unveiled a highly flexible protein with a built-in ability to change and adapt its structure to the requirements of a sophisticated catalytic mechanism and complicated allosteric regulation. We eagerly await the structures for class II and class III enzymes. In particular the intricacies of the latter enzymes are largely unexplored. An understanding of the mechanism of glycyl radical formation should provide insight into more general questions concerning the function of iron-sulfur clusters as well as deoxyadenosyl radicals.

A different puzzle for future research arises from the appearance of an increasing number of bacterial species that have more than one ribonucleotide reductase. It started with the discovery of a functional enzyme related to but different from the usual class I enzyme in enterobacteria and, via ongoing genome projects, has developed into findings of genes coding for class I and class II reductases in the same bacterium. Yeast presents a related picture. Only one of the enzymes appears to be active during normal growth. What does the other do?

Finally, why has more work not been done on medical applications of inhibitors of ribonucleotide reduction? The feasibility of such an approach is illustrated by the work of the Biomega group, which demonstrated a strong inhibition of the herpes enzyme by peptidomimetic drugs that compete specifically with the binding of the viral R2 to R1, resulting in inhibition of viral growth (62). The beauty of this result lies in its specificity, with no inhibition of the host enzyme. A similar approach is feasible in other instances when the amino acid sequence of the C terminus of R2 differs between the parasite and host enzymes. *M. tuberculosis* might present such a case.

ACKNOWLEDGMENTS

We thank Drs. Vera Bianchi and Hans Eklund for constructive criticism of the original version of this review and Mathias Eriksson for providing Figure 1. We acknowledge the Institute for Genomic Research and the Streptococcal and Gonococcal Genome Sequencing Projects for making sequence data available prior to publication. AJ was supported by a fellowship from Axel Wenner-Grens Stiftelse, and PR by a grant from the Swedish Medical Research Council.

Literature Cited

1. Reichard P, Baldesten A, Rutberg L. 1961. *J. Biol. Chem.* 236:1150–57
2. Holmgren A, Reichard P, Thelander L. 1965. *Proc. Natl. Acad. Sci. USA* 54:830–36
3. Blakley RL, Barker HA. 1964. *Biochem. Biophys. Res. Commun.* 10:391–97
4. Haurani FL. 1973. *Science* 182:78–79
5. Larsson A, Reichard P. 1966. *Biochem. Biophys. Acta* 113:407–8
6. Ehrenberg A, Reichard P. 1972. *J. Biol. Chem.* 247:3485–88
7. Sjöberg B-M, Reichard P, Gräslund A,

Ehrenberg A. 1978. *J. Biol. Chem.* 253:6863–65
8. Stubbe J. 1989. *Annu. Rev. Biochem.* 58:257–85
9. Stubbe J. 1990. *Adv. Enzymol. Relat. Areas Mol. Biol.* 63:349–419
10. Fontecave M, Eliasson R, Reichard P. 1989. *Proc. Natl. Acad. Sci. USA* 86:2147–51
11. Sun XY, Ollagnier S, Schmidt PP, Atta M, Mulliez E, et al. 1996. *J. Biol. Chem.* 271:6827–31
12. Ollagnier S, Mulliez E, Gaillard J, Elias-

son R, Fontecave M, Reichard P. 1996. *J. Biol. Chem.* 271:9410–16
13. Fleischmann RD, Adams MD, White O, Clayton RA, Kirkness EF, et al. 1995. *Science* 269:496–512
14. Young P, Öhman M, Xu MQ, Shub DA, Sjöberg B-M. 1994. *J. Biol. Chem.* 269:20229–32
15. Bult CJ, White O, Olsen GJ, Zhou L, Fleischmann RD, et al. 1996. *Science* 273:1058–73
16. Jordan A, Gibert I, Barbé J. 1994. *J. Bacteriol.* 176:3420–27
17. Jordan A, Pontis E, Atta M, Krook M, Gibert I, et al. 1994. *Proc. Natl. Acad. Sci. USA* 91:12892–96
18. Jordan A, Pontis E, Åslund F, Hellman U, Gibert I, Reichard P. 1996. *J. Biol. Chem.* 271:8779–85
19. Yang FD, Lu GZ, Rubin H. 1994. *J. Bacteriol.* 176:6738–43
20. Scotti C, Valbuzzi A, Perego M, Galizzi A, Albertini AM. 1996. *Microbiology* 142:2995–3004
21. Fraser CM, Gocayne JD, White O, Adams MD, Clayton RA, et al. 1995. *Science* 270:397–403
22. Himmelreich R, Hilbert H, Plagens H, Pirkl E, Li B-C, Herrmann R. 1996. *Nucleic Acids Res.* 24:4420–49
23. Fieschi F, Torrents E, Toulokhonova L, Jordan A, Hellman U, et al. 1997. *J. Biol. Chem.* In press
24. Jordan A, Torrents E, Jeanthon C, Eliasson R, Hellman U, et al. 1998. *Proc. Natl. Acad. Sci. USA* 94:13487–92
25. Nordlund P, Sjöberg B-M, Eklund H. 1990. *Nature* 345:593–98
26. Nordlund P, Eklund H. 1993. *J. Mol. Biol.* 232:123–64
27. Uhlin U, Eklund H. 1994. *Nature* 370:533–39
28. Uhlin U, Eklund H. 1996. *J. Mol. Biol.* 262:358–69
29. Eriksson M, Uhlin U, Ramaswamy S, Ekberg M, Regnström K, et al. 1997. *Structure* 5:1077–92
30. Stubbe J, van der Donk WA. 1995. *Chem. Biol.* 2:793–801
31. Sjöberg B-M. 1997. *Struct. Bond.* 88:139–73
32. Gräslund A, Sahlin M. 1996. *Annu. Rev. Biophys. Biomol. Struct.* 25:259–86
33. Greenberg GR, Hilfinger JM. 1996. *Progr. Nucleic Acid Res. Mol. Biol.* 53:345–95
34. Reichard P. 1997. *Trends Biochem. Sci.* 22:81–85
35. Thelander L, Gräslund A. 1994. In *Metal Ions in Biological Systems*, ed. H Sigel, A Sigel, pp. 109–29. New York: Dekker

36. Reichard P. 1993. *Science* 260:8383–86
37. Fontecave M, Nodlund P, Eklund H, Reichard P. 1992. *Adv. Enzymol. Relat. Areas Mol. Biol.* 65:147–83
38. Panagou D, Orr MD, Dunstone JR, Blakley RL. 1972. *Biochemistry* 11:2378–88
39. Tsai PK, Hogenkamp HPC. 1980. *J. Biol. Chem.* 255:1273–78
40. Tauer A, Benner SA. 1997. *Proc. Natl. Acad. Sci. USA* 94:53–58
41. Riera J, Robb FT, Weiss R, Fontecave M. 1997. *Proc. Natl. Acad. Sci. USA* 94:475–78
42. Holmgren A, Björnstedt M. 1995. *Methods Enzymol.* 252:199–208
43. Holmgren A, Åslund F. 1995. *Methods Enzymol.* 252:2183–92
44. Mulliez E, Ollagnier S, Fontecave M, Eliasson R, Reichard P. 1995. *Proc. Natl. Acad. Sci. USA* 92:8759–62
45. Thelander L. 1974. *J. Biol. Chem.* 249: 4858–62
46. Åberg A, Hahne S, Karlsson M, Larsson A, Ormö M, et al. 1989. *J. Biol. Chem.* 264:12249–52
47. Mao SS, Holler TP, Yu GX, Bollinger JM, Booker S, et al. 1992. *Biochemistry* 31:9733–43
48. Licht S, Gerfen GJ, Stubbe JA. 1996. *Science* 271:477–81
49. King D, Reichard P. 1995. *Biochem. Biophys. Res. Commun.* 206:731–35
50. Brown NC, Reichard P. 1969. *J. Mol. Biol.* 46:39–55
51. Gleason FK, Wood JM. 1976. *Science* 92:1343–44
52. Stuzenberger F. 1973. *J. Gen. Microbiol.* 81:501–3
53. Willing A, Follman H, Auling G. 1988. *Eur. J. Biochem.* 170:603–11
54. Eliasson R, Pontis E, Jordan A, Reichard P. 1996. *J. Biol. Chem.* 271:26582–87
55. Jordan A, Aragall E, Gibert I, Barbe J. 1996. *Mol. Microbiol.* 19:777–90
56. Jordan A, Åslund F, Pontis E, Reichard P, Holmgren A. 1997. *J. Biol. Chem.* 272:18044–50
57. Åberg A, Nordlund P, Eklund H. 1993. *Nature* 361:276–78
58. Logan D, Su XD, Åberg A, Rengström K, Hajdu J, et al. 1996. *Structure* 4:1053–64
59. Kauppi B, Nielsen BB, Ramaswamy S, Larsen IK, Thelander M, et al. 1996. *J. Mol. Biol.* 262:706–20
60. Holmgren A. 1980. In *Dehydrogenases Requiring Nicotinamide Coenzymes*, ed. J Jeffery, pp. 149–80. Basel: Birkhäuser
61. Wang PJ, Chabes A, Casagrande R, Tian XC, Thelander L, Huffaker TC. 1997. *Mol. Cell. Biol.* 17:6114–121

62. Liuzzi M, Déziel R, Moss N, Beaulieu P, Bonneau A-M, et al. 1994. *Nature* 372:695–98
63. Booker S, Licht S, Broderick J, Stubbe J. 1994. *Biochemistry* 33:12676–85
64. Reichard P. 1993. *J. Biol. Chem.* 268: 8383–86
65. Ollagnier S, Mulliez E, Schmidt PP, Eliasson R, Gaillard J, et al. 1997. *J. Biol. Chem.* 272:24201–23
66. Åslund F, Ehn B, Miranda-Vizuete A, Pueyo C, Holmgren A. 1994. *Proc. Natl. Acad. Sci. USA* 91:9813–17
67. Spyrou G, Enmark E, Miranda-Vizuete A, Gustafsson J-Å. 1997. *J. Biol. Chem.* 272:2936–41
68. Miranda-Vizuete A, Damdimopulos A, Gustafsson J-Å, Spyrou G. 1997. *J. Biol. Chem.* In press
69. Eliasson R, Reichard P, Mulliez E, Ollagnier S, Fontecave M, et al. 1995. *Biochem. Biophys. Res. Commun.* 214:28–35
70. Durham LJ, Larsson A, Reichard P. 1967. *Eur. J. Biochem.* 1:92–95
71. Batterham TJ, Ghambeer RK, Blakley RL, Brownson C. 1967. *Biochemistry* 6:1203–8
72. Stubbe J, Ator M, Krenitzky T. 1983. *J. Biol. Chem.* 258:1625–30
73. Thelander L, Larsson B, Hobbs J, Eckstein F. 1976. *J. Biol. Chem.* 251:1398–1405
74. Ator M, Stubbe J. 1985. *Biochemistry* 24:7214–21
75. Harris G, Ashley GW, Robins MJ, Tolman RL, Stubbe J. 1987. *Biochemistry* 26:1895–1902
76. Mao SS, Holler TP, Bollinger JM, Yu GX, Johnston MI, Stubbe J. 1992. *Biochemistry* 31:9744–51
77. Mao SS, Yu GX, Chalfoun D, Stubbe J. 1992. *Biochemistry* 31:9752–59
78. Persson AL, Eriksson M, Katherle B, Pötsch S, Sahlin M, Sjöberg B-M. 1997. *J. Biol. Chem.* In press
78a. Eliasson R, Pontis E, Eckstein F, Reichard P. 1994. *J. Biol. Chem.* 269:26116–20
79. Climent I, Sjöberg B-M, Huang CY. 1992. *Biochemistry* 31:4801–7
80. Ekberg M, Sahlin M, Eriksson M, Sjöberg B-M. 1996. *J. Biol. Chem.* 271:20655–59
81. Persson BO, Karlsson M, Climent I, Ling JS, Sanders-Loehr J, et al. 1996. *J. Biol. Inorg. Chem.* 1:247–56
82. Rova U, Goodtzova K, Ingemarson R, Behravan G, Gräslund A, Thelander L. 1995. *Biochemistry* 34:4267–75
83. Fontecave M, Eliasson R, Reichard P. 1989. *J. Biol. Chem.* 264:9164–70

84. Coves J, Delon B, Climent I, Sjöberg B-M, Fontecave M. 1995. *Eur. J. Biochem.* 233:357–63
85. Frey M, Rothe M, Wagner AFV, Knappe J. 1994. *J. Biol. Chem.* 269:12432–37
86. Knappe J, Wagner AFV. 1995. *Methods Enzymol.* 258:343–62
87. Wong KK, Murray BW, Lewisch SA, Baxter MK, Ridky TW, et al. 1993. *Biochemistry* 32:14102–10
88. Broderick JB, Duderstadt RE, Fernandez DC, Wojtuszewski K, Henshaw TF, Johnson MK. 1997. *J. Am. Chem. Soc.* 119:7396–97
89. Frey PA, Reed GH. 1993. *Adv. Enzymol. Relat. Areas Mol. Biol.* 66:1–39
90. Reed GH, Bollinger MD. 1995. *Methods Enzymol.* 258:362–77
91. Kunz BA, Kohalmi SE, Kunkel TA, Mathews CK, McIntosh EM, Reidy JA. 1994. *Mutat. Res.* 318:1–64
92. Thelander L, Reichard P. 1979. *Annu. Rev. Biochem.* 48:133–58
93. Eliasson R, Pontis E, Sun XY, Reichard P. 1994. *J. Biol. Chem.* 269:26052–57
94. Averett DR, Furman PA, Spector T. 1984. *J. Virol.* 52:981–83
95. Nikas I, McLauchlan J, Davison AJ, Taylor WR, Clements JB. 1986. *Proteins* 1:376–84
96. Berglund O. 1972. *J. Biol. Chem.* 247: 7276–81
97. Hofer A, Schmidt PP, Gräslund A, Thelander L. 1997. *Proc. Natl. Acad. Sci. USA* 94:6959–64
98. Caras IW, Martin DW Jr. 1988. *Mol. Cell. Biol.* 8:2698–704
99. Reichard P. 1988. *Annu. Rev. Biochem.* 57:349–74
100. Bianchi V, Borella S, Rampazzo C, Ferraro P, Calderazzo F, et al. 1997. *J. Biol. Chem.* 272:16118–24
101. Carlson J, Fuchs JA, Messing J. 1984. *Proc. Natl. Acad. Sci. USA* 81:4292–97
102. Jordan A, Gibert I, Barbé J. 1995. *Gene* 167:75–79
103. Sun L, Fuchs JA. 1992. *Mol. Biol. Cell* 3:1095–105
104. Fuchs JA. 1977. *J. Bacteriol.* 130:957–59
105. Augustin LB, Jacobson BA, Fuchs JA. 1994. *J. Bacteriol.* 176:378–87
106. Sun L, Fuchs JA. 1994. *J. Bacteriol.* 176:4617–26
107. Sun L, Jacobson BA, Dien BS, Srienc F, Fuchs JA. 1994. *J. Bacteriol.* 176:2415–26
108. Yang F, Curran SC, Li L-S, Avarbock D, Graf JD, et al. 1997. *J. Bacteriol.* 179:6408–15
109. Sun XY, Harder J, Krook M, Jörnvall H, Sjöberg B-M, Reichard P. 1992. *Proc.*

Natl. Acad. Sci. USA 90:577–81
110. Garriga X, Eliasson R, Torrents E, Jordan A, Barbé J, et al. 1996. *Biochem. Biophys. Res. Commun.* 229:189–92
111. Turner MK, Abrams R, Lieberman I. 1968. *J. Biol. Chem.* 243:3725–28
112. Elford HL, Freese M, Passamani E, Morris HP. 1970. *J. Biol. Chem.* 245:5228–33
113. Nordenskjöld BA, Skoog L, Brown NC, Reichard P. 1970. *J. Biol. Chem.* 245:5360–68
114. Eriksson S, Gräslund A, Skog S, Thelander L, Tribukait B. 1984. *J. Biol. Chem.* 259:11695–700
115. Engström Y, Eriksson S, Jildevik I, Skog S, Thelander L, Tribukait B. 1985. *J. Biol. Chem.* 260:9114–16
116. Mann GJ, Musgrove EA, Fox RM, Thelander L. 1988. *Cancer Res.* 48:5151–56
117. Björklund S, Skog S, Tribukait B, Thelander L. 1990. *Biochemistry* 29:5452–58
118. Björklund S, Hjortsberg K, Johansson E, Thelander L. 1993. *Proc. Natl. Acad. Sci. USA* 90:11322–26
119. Johansson E, Skogman E, Thelander L. 1995. *J. Biol. Chem.* 270:23698–704
120. Chen FY, Amara FM, Wright JA. 1994. *Nucleic Acids Res.* 22:4796–97
121. Filatov D, Thelander L. 1995. *J. Biol. Chem.* 270:25239–43
122. Björklund S, Skogman E, Thelander L. 1992. *EMBO J.* 11:4953–59
123. Amara FM, Sun J, Wright JA. 1996. *J. Biol. Chem.* 271:22126–31
124. Chan AK, Litchfield DW, Wright JA. 1993. *Biochemistry* 32:12835–40
125. Fan H, Villegas C, Wright JA. 1996. *Proc. Natl. Acad. Sci. USA* 93:14036–40
126. Filatov D, Björklund S, Johansson E, Thelander L. 1996. *J. Biol. Chem.* 271:23698–704
127. Elledge SJ, Zhou Z, Allen JB. 1992. *Trends Biochem. Sci.* 17:119–23
128. Elledge SJ, Zhou Z, Allen JB, Navas TA. 1993. *BioEssays* 15:333–39
129. Berglund O, Karlström O, Reichard P. 1969. *Proc. Natl. Acad. Sci. USA* 62:829–35
130. Young P, Öhman M, Sjöberg B-M. 1994. *J. Biol. Chem.* 269:27815–18

131. Sjöberg B-M, Hahne S, Mathews CZ, Mathews CK, Rand KN, Gait MJ. 1986. *EMBO J.* 5:2031–36
132. Chung TD, Wymer JP, Smith CC, Kulka M, Aurelian L. 1989. *J. Virol.* 63:3389–98
133. Cooper J, Conner J, Clements JB. 1995. *J. Virol.* 69:4979–85
134. Benner SA, Ellington AD, Tauer A. 1989. *Proc. Natl. Acad. Sci. USA* 86:7054–58
135. Booker S, Stubbe J. 1993. *Proc. Natl. Acad. Sci. USA* 90:8352–56
136. Wächtershäuser G. 1990. *Proc. Natl. Acad. Sci. USA* 87:200–4
137. Sze IS-Y, McFarlan SC, Spormann A, Hogenkamp HPC. 1992. *Biochem. Biophys. Res. Commun.* 184:1101–7
138. Sando GN, Hogenkamp HPC. 1973. *Biochemistry* 12:3316–22
139. Kaneko T, Sato S, Kotani H, Tanaka A, Asamizu E, et al. 1996. *DNA Res.* 3:109–36
140. Yau S, Wachsman JT. 1973. *Mol. Cell. Biochem.* 1:101–5
141. Tsai PK, Hogenkamp HPC. 1980. *J. Biol. Chem.* 255:1273–78
142. Tomb JF, White O, Kerlavage AR, Clayton RA, Sutton GG, et al. 1997. *Nature* 388:539–47
143. Cowles JR, Evans HJ, Russell SA. 1969. *J. Bacteriol.* 97:1460–65
144. Nilsson O, Åberg A, Lundqvist T, Sjöberg B-M. 1988. *Nucleic Acids Res.* 16:4174
145. Blattner FR, Plunkett G III, Bloch CA, Perna NT, Burland V, et al. 1997. *Science* 227:1453–74
146. Sjöberg B-M. 1995. *Nucleic Acids Mol. Biol.* 9:192–221
147. Hamilton FD. 1974. *J. Biol. Chem.* 249:4428–34
148. Tseng MJ, Hilfinger JM, Walsh A, Greenberg GR. 1988. *J. Biol. Chem.* 263:16242–51
149. Hatfull GF, Sarkis GJ. 1993. *Mol. Microbiol.* 7:395–405
150. Harder J. 1993. *FEMS Microbiol. Rev.* 12:273–92
151. Thompson JD, Higgins DG, Gibson TJ. 1994. *Nucleic Acids Res.* 22:4673–80
152. Kornberg A. 1997. *Trends Biochem. Sci.* 22:282–83

Annu. Rev. Biochem. 1998. 67:99–134

MODIFIED OLIGONUCLEOTIDES:
Synthesis and Strategy for Users

Sandeep Verma and Fritz Eckstein

Max-Planck-Institut für Experimentelle Medizin, Hermann-Rein Strasse 3, D-37075
Göttingen, Germany

KEY WORDS: oligonucleotide analogs, backbone modifications, base modifications, sugar
modifications, reporter groups

ABSTRACT

Synthetic oligonucleotide analogs have greatly aided our understanding of several biochemical processes. Efficient solid-phase and enzyme-assisted synthetic methods and the availability of modified base analogs have added to the utility of such oligonucleotides. In this review, we discuss the applications of synthetic oligonucleotides that contain backbone, base, and sugar modifications to investigate the mechanism and stereochemical aspects of biochemical reactions. We also discuss interference mapping of nucleic acid–protein interactions; spectroscopic analysis of biochemical reactions and nucleic acid structures; and nucleic acid cross-linking studies.

The automation of oligonucleotide synthesis, the development of versatile phosphoramidite reagents, and efficient scale-up have expanded the application of modified oligonucleotides to diverse areas of fundamental and applied biological research. Numerous reports have covered oligonucleotides for which modifications have been made of the phosphodiester backbone, of the purine and pyrimidine heterocyclic bases, and of the sugar moiety; these modifications serve as structural and mechanistic probes. In this chapter, we review the range, scope, and practical utility of such chemically modified oligonucleotides. Because of space limitations, we discuss only those oligonucleotides that contain phosphate and phosphate analogs as internucleotidic linkages.

CONTENTS

0066-4154/98/0701-0099$08.00

SYNTHETIC METHODS

Solid-Phase Chemical Synthesis

The ease of customizing reaction cycles in automated, solid-phase DNA synthesizers has allowed for efficient and site-specific introduction of chemical modifications in oligodeoxynucleotides. In the past decade, significant progress has also been made toward developing novel synthetic and deprotection strategies for oligoribonucleotide synthesis. In this section we outline the solid-phase synthetic process. Details of other synthetic methods are discussed in subsequent sections.

For the synthesis of oligodeoxynucleotide, automated solid-phase synthesizers make use of 5'-protected deoxynucleosides anchored to a controlled pore glass support at the 3'-end via a chemically cleavable linker. A series of programmed reactions sequentially couple deoxynucleoside phosphoramidites to the support-bound nucleoside. This approach circumvents the need to purify synthetic intermediates at each step of the reaction cycle, thereby saving enormous amounts of time. Finally, base hydrolysis is used to cleave the linker from the anchored nucleoside to yield a fully deprotected oligomer (1, 2). A detailed description of the synthetic process and related chemistry is beyond the scope of this chapter and is described elsewhere (3). Similarly, oligoribonucleotide synthesis is also performed on solid supports through the use of appropriately protected phosphoramidites. Continued development of 2'-OH protecting groups, the use of efficient activators, and purification methods have greatly improved RNA synthesis (4, 5). Despite these advances, RNA synthesis still remains less efficient, necessitating further improvement in the synthetic methodology (6).

Commercial availability of phosphoramidites of suitably protected natural and nonnatural nucleosides, such as 2′-substituted-2′deoxy- and 2′-O-methyl nucleosides, terminal nonradioactive reporter groups, and functionalized alkyl tethers for postsynthetic modifications, have greatly added to the convenience of solid-phase synthesis. The phosphoramidite approach, developed for the synthesis of oligonucleotides containing a phosphodiester backbone, can also be used to synthesize phosphorothioates by replacing the oxidation step with sulfurization (7). The solid-phase synthesis of other modified backbones, such as phosphorodithioates (8) or methylphosphonates (9), however, requires special nucleoside amidites.

The key features of automated solid-phase oligonucleotide synthesis include ease of operation, efficiency, and reproducibility. Despite these advantages, this method is difficult to use for the synthesis of long oligonucleotides (100 or more nucleotides) in high yields. In such cases, it is more convenient to use the enzymatic ligation approach, in tandem with chemical synthesis, to give an oligomer of desired length.

Enzymatic Ligation

The enzymatic approach to ligate short oligonucleotides offers an attractive route for the synthesis of long deoxynucleic and ribonucleic acids. These reactions are catalyzed by DNA or RNA ligases that promote intermolecular ligation of the 5′ and 3′ termini of oligonucleotides through the formation of a phosphodiester bond.

In the case of DNA ligase, two oligodeoxynucleotides, one bearing a 5′-phosphoryl donor group and another with a free 3′-hydroxyl acceptor, constitute the substrate requirement. It is essential that the phosphorylated and the hydroxyl termini-bearing oligonucleotides be present either on homo-(DNA:DNA) or hetero-duplexes (DNA:RNA) (10, 11). Oligonucleotides bearing terminal fluorescent and chemiluminescent labels can also act as substrates for DNA ligase. Such modified oligomers have been successfully employed to study cystic fibrosis gene mutations in hybridization-ligation assays (12, 13).

T4 RNA ligase catalyzes the formation of a phosphodiester bond between a 5′-phosphoryl donor and a free 3′-hydroxyl acceptor group of two single-stranded oligoribonucleotides. It accepts a variety of oligonucleotide substrates for ligation. With the use of this enzyme, nucleoside analogs can be site-specifically introduced in oligoribonucleotides at the ligation junction (10). The use of this enzyme makes possible the introduction of modifications in oligoribonucleotides that are longer than those accessible by chemical synthesis alone. This methodology has been used to introduce nonstandard nucleosides in oligoribonucleotides (14, 15), for fluorescent labeling of tRNA and oligodeoxy-nucleotides (16), for DNA strand ligation within a duplex (17), and for the synthesis of mRNA and tRNA segments (18, 19). In general, the use of chemical synthesis in conjunction with enzymatic ligation provides a more efficient route to the synthesis of long oligoribonucleotides compared to the use of solid-phase chemical synthesis alone.

Two novel approaches for the ligation of oligoribonucleotides have been described. In the first method, two oligoribonucleotide fragments are annealed to a complementary oligodeoxynucleotide (cDNA) bridge and then ligated through use of T4 DNA ligase (20). With this ligation strategy, modified RNA substrates are synthesized, allowing study of the importance of the 2'-hydroxyl group in nuclear pre-mRNA splicing. It was demonstrated that T4 DNA ligase efficiently incorporated 2'-deoxy or 2'-O-methyl nucleotides immediately adjacent to the splice site of pre-mRNA substrates, with a high degree of acceptor specificity. In another method, two oligoribonucleotide fragments are annealed to a partially complementary oligodeoxynucleotide splint and then ligated through use of T4 RNA ligase (21). The DNA splint used in the latter method was designed to lack base-pairing nucleotides in the vicinity of the ligation site, thus providing single-stranded 5'-phosphoryl and 3'-hydroxyl termini for enzymatic ligation.

Enzymatic Incorporation of Nonnatural Nucleotides

The enzymatic synthesis of oligonucleotides containing modified bases, sugars, or phosphate groups is a viable alternative to chemical synthesis. Enzymatic incorporation of modified nucleoside triphosphates, with T7 or a similar phage RNA polymerase, is particularly well suited for template-directed transcription reactions, provided that they are good substrates for the polymerase. This requirement limits the number of modified nucleotides that can be used in transcription reactions; thus only a few modifications can be enzymatically introduced into oligoribonucleotides. For oligodeoxynucleotides, these modifications must be made by chain extension of a primer annealed to a template, through use of a DNA polymerase such as the Klenow fragment of *Escherichia coli* DNA polymerase I.

Among internucleotidic linkages, phosphorothioate is the most common enzymatically introduced modification, as all of the four nucleoside α-thiotriphosphates are good substrates for DNA and RNA polymerases (Figure 1; 22). In phosphorothioates, substitution of one of the nonbridging oxygen atoms with sulfur introduces chirality at the phosphorus center. Only the Sp-diastereomer of nucleoside α-thiotriphosphates is a good substrate for the polymerase. The enzymatic incorporation proceeds with inversion of configuration at phosphorus, resulting in the formation of the Rp internucleotidic linkage. Therefore,

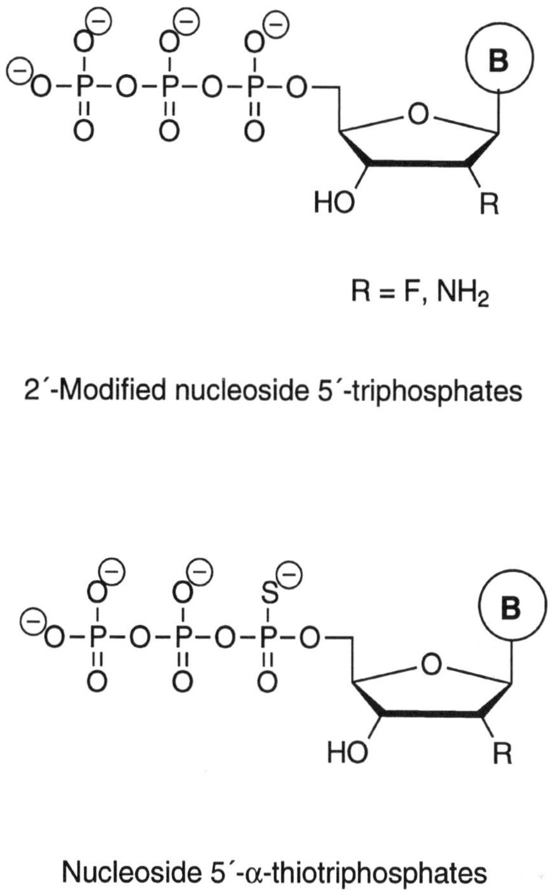

R = F, NH$_2$

2'-Modified nucleoside 5'-triphosphates

Nucleoside 5'-α-thiotriphosphates

Figure 1 Modified nucleotides for transcription.

enzymatic synthesis is limited to the incorporation of Rp-phosphorothioate linkage (22). This methodology has been applied to the synthesis of all Rp-phosphorothioate transcripts (23) and oligodeoxynucleotides through use of either a modified T7 DNA polymerase (24) or Taq DNA polymerase (25). The applications of phosphorothioate modification are discussed later in this chapter.

2'-Deoxynucleoside triphosphates were the first sugar-modified analogs to be examined as substrates for T7 RNA polymerase (26). However, in their presence, transcription was aborted 30% of the time. Recently, the use of Mn^{2+} instead of Mg^{2+}, in carefully controlled ratios with respect to the ribonucleotide triphosphates, was shown to permit more efficient incorporation of these analogs

(27). This report also describes the incorporation of 2'-O-methyl nucleotides, which are not incorporated under normal transcription conditions. Efficient incorporation of deoxynucleoside triphosphates can also be achieved by a mutant T7 RNA polymerase, without the requirement for special transcription conditions (28). 2'-Amino-2'-deoxynucleoside triphosphates are good substrates for the wild-type enzyme. As an example, synthesis of luciferase transcripts 2500 nt in length has been achieved (Figure 1; 29). However, corresponding 2'-fluoro-2'-deoxy analogs are less readily incorporated (29). The mutant T7 RNA polymerase improves the incorporation of the 2'-fluoro derivatives (30).

T7 RNA polymerase normally initiates transcription with a guanosine triphosphate, but it can also accept other guanosine derivatives such as guanosine 5'-monophosphorothioate (31), guanosine 5'-β-thiotriphosphate (32), or guanosine 5'-γ-thiotriphosphate (33–35) for initiation. Applications of terminal phosphorothioates are discussed in later sections.

Postsynthetic Modification of Oligonucleotides

Conjugation of oligonucleotides to other biological macromolecules, to nonradioactive reporter groups, to spin labels, and to cross-linking agents has necessitated the development of postsynthetic modification procedures. The chemical strategies for oligonucleotide conjugation have been reviewed (36).

The modification of either the 3'- or the 5'-terminus is a convenient method for equipping an oligonucleotide with a reactive aminoalkyl or a mercaptoalkyl group. Introduction of these functionalities in an oligomer can be achieved either through use of appropriate commercial phosphoramidite reagents or by solution-phase chemistry. In the case of solution-phase modification, a diamino compound such as ethylenediamine or cystamine can be reacted with the 5'-phosphate group in the presence of a water-soluble carbodiimide. This reaction results in the formation of an oligonucleotide containing a reactive functional group at the 5'-terminal. The introduction of a 5'-terminal phosphorothioate group that can be used as a nucleophile is easily achieved through use of ATPγS and polynucleotide kinase (33). Alternatively, a 3'-terminal phosphorothioate can be introduced through use of 2'-deoxycytidine-5'-phosphate 3'-phosphorothioate (pdCpS) and RNA ligase (16).

Such oligonucleotide conjugates have been used extensively for a number of applications, which include cellular delivery of antisense oligonucleotides, synthesis of artificial nucleases, and hybridization probes for biological detection. Examples include conjugation of oligonucleotides to viral fusogenic peptides (37, 38), staphylococcal nuclease and other enzymes (39, 40), neoglycopeptides (41), phospholipids (42), thiolated polysaccharides (43), vitamin E (44), lipophilic groups such as cholesterol (45–47), and nonradioactive detection labels (48–50).

INTERNUCLEOTIDE LINKAGE MODIFICATION

Phosphorothioates

The exchange of one nonbridging oxygen atom with a sulfur atom in the phosphorothioate modification is a conservative change, as the negative charge of the phosphate group is retained and the size of the sulfur atom is only slightly larger than the oxygen atom. However, this exchange makes the phosphorus center chiral; therefore, two diastereomers of the phosphorothioate internucleotide linkage exist (Figure 2). The chirality of the phosphorus center permits the determination of stereospecificity and the stereochemical course of reactions occurring at phosphorus. In addition, sulfur is a "soft" atom that coordinates preferentially with "soft" metal ions, whereas oxygen is a "hard" atom that coordinates to "hard" metal ions. Therefore, use of phosphorothioates also permits the identification of metal-ion binding sites in the biological systems. The phosphorothioate internucleotidic linkage is considerably more stable than the phosphodiester bond toward nucleases, and this feature makes it useful for cell culture and in vivo studies. The literature on nucleoside phosphorothioates up to 1984 has been reviewed by Eckstein (22).

Oligonucleotide phosphorothioates can be synthesized either by chemical synthesis, with use of a sulfurizing reagent in the oxidation step, or by enzymatic incorporation. Chemical synthesis produces a mixture of diastereomers, which in some cases can be separated, while the enzymatic incorporation invariably results in the formation of the Rp-diastereomeric internucleotidic linkage (7, 22, 23, 51).

ANALYSIS OF THE STEREOCHEMICAL COURSE OF ENZYMATIC REACTIONS   Site-specific incorporation of phosphorothioates has been used to study the stereochemical course of many nucleotidyl-transfer reactions (22). These studies have been extended to investigate reaction mechanisms involving restriction enzymes such as *Eco*RI (52) and *Eco*RV (53), DNA integration (54), and site-specific DNA recombination (55). All of these reactions proceed with an inversion of configuration as a result of a single nucleophilic substitution.

The chirality of the P-S bond in phosphorothioates can be used to identify oxygen atoms responsible for or involved in the interaction of specific phosphate groups with proteins. Such studies have been conducted extensively by the incorporation of Rp and Sp linkages at several positions in the recognition sequences for *Eco*RI (56) and *Eco*RV (57) restriction endonucleases. The latter study also identified several phosphate groups, besides those at the cleavage site, as important for catalysis.

The stereochemical course of ribozyme-catalyzed reactions such as RNA cleavage by the group I intron (58, 59), pre-mRNA splicing (60, 61), and group II

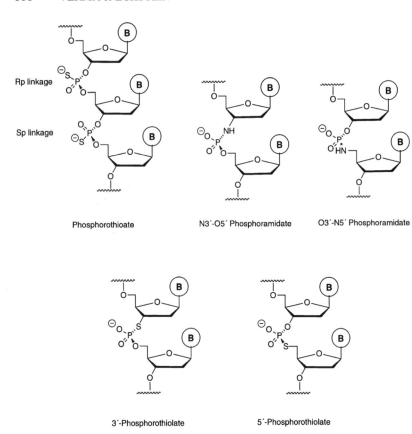

Figure 2 Modified oligonucleotide internucleotidic linkages.

intron self-splicing (62, 63) have been studied with the use of phosphorothioates as mechanistic probes. All of these reactions proceed with an inversion of configuration of phosphorus present at the cleavage site. A comparison of the stereochemical requirements for the phosphorothioates in the individual steps allows for the establishment of similarities for different splicing systems. This analysis emphasizes a close relationship between pre-mRNA splicing and the group II splicing reaction (63, 64). Inversion of configuration at the phosphorus center has also been reported for the hammerhead ribozyme-mediated catalysis (65, 66), indicating an in-line mechanism that is not obvious from the X-ray structures (67, 68). In another study, phosphorothioate substitution in one of the two strands involved in bacteriophage lambda recombination showed a strong bias in the choice of strands to initiate crossing-over (69).

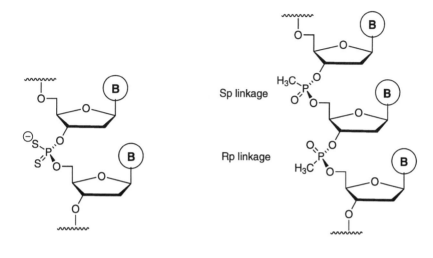

Phosphorodithioate

Methylphosphonate

Figure 2   (*Continued*)

A thorough and critical discussion of how to treat rate differences observed between phosphate and phosphorothioate substrates in enzyme-catalyzed reactions has been outlined (70). From a purely chemical viewpoint, reactivity of the phosphorothioate group should be only marginally lower than that of the phosphate diester. One study involving mechanistic investigation of the polymerase activity of Klenow DNA polymerase (71) and another study investigating the interaction of GTPγS with transducin-α (72) have pinpointed the source of the rate difference. It was suggested that steric clashes in the transition state because of the larger van der Waals radius of sulfur, rather than the electronegativity difference, result in slowing the reaction rate for the phosphorothioates.

PHOSPHOROTHIOATE INTERFERENCE   Phosphorothioate interference studies have been widely employed to identify the role of specific phosphate groups in biochemical reactions. These studies exploit the change in binding specificity when sulfur is substituted for oxygen. Thus, any observed interference by phosphorothioates should identify specific phosphate groups important for recognition. For example, binding of a metal ion to a specific phosphate group might be altered if it is replaced by a phosphorothioate, making it easier to identify the exact position of the phosphate group involved in metal ion binding. Although the reasons for phosphorothioate interference may not always be readily understood, it is still possible to assess the mechanistic contribution of specific phosphate groups.

For interference studies with oligoribonucleotides, phosphorothioates are generally incorporated randomly by transcription. The enzymatic incorporation results in the formation of an Rp phosphorothioate linkage, and from an observed interference, the mechanistic role of the pro-Rp oxygen atom of the replaced phosphate group can be identified. The success of this method depends on the actively interfering species being separable from the inactive ones, so that the precise positions of phosphorothioate incorporation can be determined. The latter can be accomplished by iodine footprinting (73, 74). A comparison of the cleavage patterns allows one to identify positions of interference with the reaction or interaction being studied.

A study using phosphorothioate interference has identified 44 phosphate groups important for Mg^{2+} ion binding in the group I intron reaction (75, 76). However, for most of these positions the cleavage reaction could be restored or rescued with Mn^{2+} ions. This metal ion coordinates to both oxygen and sulfur atoms, whereas Mg^{2+} ion coordinates only to oxygen atoms (77, 78). Thus, the restoration of catalytic activity indicates that the pro-R oxygen atoms are coordinated to the metal ion in the splicing reaction.

The phosphorothioate interference analysis and footprinting with iodine can also be used as a probe for the identification of phosphate groups involved in RNA-protein interaction. This methodology was first applied to study the affinity of an oligoribonucleotide to the R17 coat protein (79). This method was extended to probe the interaction of E. coli tRNA[Ser] with its cognate aminoacyl synthetase. The results suggested that the contact with the tRNA variable loop might be located on the characteristic long extended arm of the synthetase (75). With one exception, all of these contacts were later confirmed by the X-ray crystal structure analysis of the related *Thermus thermophilus* complex (80). Results with the yeast tRNA[Asp] system were also consistent with the interaction of the identity elements of the tRNA with the synthetase protein (81). Phosphorothioate footprinting has also been used to investigate the interaction between tRNA[Phe] transcripts from *T. thermophilus* and its synthetase. In this case, iodine footprinting revealed a complex and novel pattern of tRNA[Phe]-synthetase interaction (82).

The phosphorothioate protection strategy has also been used to study mito-chondrial group I intron RNA splicing, which involves interaction with tyrosyl-tRNA synthetase (83). These results have aided the construction of a model that suggests that the recognition of the synthetase occurs by a tRNA-like structure in the group I intron (84). Similarly, tRNA interaction with the ribosome A and P sites was also investigated to map the essential "identity elements" involved in ribosome-tRNA recognition (85). Four phosphate groups in the RNase P RNA were identified as important for an intramolecular pre-tRNA cleavage reaction (86). Besides the 4 positions identified in this study, 12 additional phosphate positions were found to be important for the binding of the pre-tRNA (87). Phosphorothioate substitution at the cleavage site of the pre-tRNA resulted in a 1000-fold reduction of the cleavage rate for both diastereomers, but rate inhibition of the Rp diastereomer could be largely restored by Cd^{2+}, a soft metal ion (88). In yet another study, phosphorothioate substitutions of the *pro*-Rp oxygens in a pre-tRNA transcript revealed that direct coordination of Mg^{2+} ion to the *pro*-Rp oxygen of the scissile bond is required for RNase P catalysis (89).

Recently, a "nucleotide analog interference mapping" approach has been applied to identify the functional groups essential for group I intron activity (90). Modified group I intron constructs were prepared through the substitution of inosine α-thiotriphosphate for GTP in the transcription reactions. Iodine-mediated phosphorothioate cleavage (73) was used to identify the positions of essential guanosine residues. This assay successfully mapped almost every exocyclic amino group of guanosine residues essential for the catalytic activity of the group I intron ribozyme.

Rp-phosphorothioate interference has also identified three phosphate groups, besides the group at the cleavage site, as important for catalysis by the hammerhead ribozyme (91). The importance of one of these sites, where a metal ion is coordinated to the *pro*-R oxygen of the A9 phosphate, has been confirmed by X-ray structural analyses (67, 68). Knöll and coworkers have also used synthetic hammerhead ribozymes, containing a mixture of phosphorothioate diastereomers, for mechanistic investigations (92). They identified interference by the Sp isomer at A6 and U16.1. A metal-ion rescue could be observed with the Rp isomer at A9, which complements the observation of metal ion coordination to this site as determined by the X-ray structure. In a more recent X-ray structure, a Mg^{2+} ion has been found coordinated to the *pro*-R oxygen of the phosphate group at the cleavage site (93). These results are consistent with the cleavage of the Rp phosphorothioate in the presence of Mn^{2+} ion and that of the Sp-diastereomer in the presence of Mg^{2+} ion (66, 94). The phosphate groups required for the self-cleavage reaction by the extended hammerhead ribozyme, present in satellite 2 transcripts, has been studied by phosphorothioate interference analysis (95). This approach has also been used to map the phosphate oxygens critical for the hepatitis delta virus (HDV) ribozyme self-cleavage reaction. A set of phosphate groups and their pro-Rp oxygen atoms were identified as being important for the ribozyme activity. In particular, positions 0, 1, and 21 were deduced as the most critical for the ribozyme catalysis (96).

As is the case with other chemical cleavage methods, caution should be exercised when interpreting phosphorothioate-iodine footprinting results obtained with RNA-RNA and RNA-protein complexes. An observed protection might not be due entirely to direct nucleic acid–macromolecule interaction but may be due partly to induced conformational changes in the nucleic acid, which can bury the phosphorothioate groups accessible to iodine cleavage. Conversely, phosphorothioates can also become hyperreactive by a similar mechanism.

NUCLEASE STABILITY   Phosphorothioate modification increases nuclease stability of oligonucleotides. This fact was first demonstrated in cell culture experiments with interferon-inducing polynucleotides (97). This property is particularly important when oligonucleotides are employed for in vivo therapeutic purposes such as the antisense oligonucleotide-mediated inhibition of gene expression. The applications of antisense phosphorothioate oligonucleotides were reviewed recently (98).

Resistance of phosphorothioate-modified oligonucleotides to exonuclease III has allowed for the delineation of the borders of the thymine dimer removal by the human nucleotide excision nuclease by filling the resulting gap through polymerization with the four dNTPαS nucleotides (99). Surprisingly, analysis of the excision-repair mechanism with UV-sensitive rodent cell lines, through use of a phosphorothioate patch assay, shows that such cells have normal incision, excision, and repair steps (100). In the case of the Klenow fragment, 3′-exonuclease hydrolysis differs more than 100-fold for a phosphorothioate substitution when compared to the phosphate group (101). Similarly, introduction of a single 3′-phosphorothioate also inhibits degradation of oligonucleotide primers by the 3′-exonuclease activity of the thermostable DNA polymerases *Vent* and *Pfu* (102), an important consideration for the use of such primers in polymerase chain reactions.

Nuclease resistance of phosphorothioate internucleotide linkages, particularly to the restriction enzyme *Nci* I, is the basis of an oligonucleotide-directed, site-specific mutagenesis procedure (103, 104). Interestingly, to confer complete nuclease resistance, the position 5′ to the cleavage site also has to be a phosphorothioate, a phenomenon very similar to that observed for the enzymes *Ban* II and *Eco*RV (105).

The isothermal single-strand displacement amplification (SDA) method, originally developed for the detection of DNA from pathogens of the *Mycobacterium* genus, is based on the resistance of phosphorothioates to the restriction endonuclease *Hinc* II (106). This methodology has been extended to mimic the rolling-circle replication of single-stranded phages (107).

PHOSPHOROTHIOATE DIESTERS AS NUCLEOPHILES   The sulfhydryl anion of internucleotidic phosphorothioates can act as an efficient nucleophile for the site-specific attachment of reporter groups, such as photoaffinity agents (108) and fluorescent dyes (109), to oligodeoxynucleotides via alkylation reactions. The resulting triesters are very stable below pH 7 but are rapidly hydrolyzed under alkaline conditions. However, the rate of hydrolysis is remarkably slow for double-stranded oligonucleotides (109). Such triester formation is not possible for phosphorothioate-substituted oligoribonucleotides because the 2'-hydroxyl group readily reacts to form a cyclic 2',3'-phosphate by displacing the reporter group via an intramolecular reaction. A similar chemical approach is the basis for the determination of the position of phosphorothioate groups in oligodeoxyribonucleotides by using iodoethanol (73) or iodine for oligoribonucleotides (74).

MERCURY ION INTERACTION   As mentioned before, phosphorothioate monoesters interact strongly with mercury (II), and this property has been used to purify modified oligonucleotides by mercury-affinity techniques. In contrast, phosphorothioate diesters, which are present in the internucleotidic linkage, interact poorly with mercury (II), as evident by the lack of retention of thiolated tRNA on mercury (II)-polyacrylamide gels (33). Surprisingly, a recent study demonstrated that a mercury (II) complex could be prepared using a phosphorothioate-containing oligonucleotide and $\gamma\delta$-resolvase. This complex was used further for X-ray structural analysis (110). A similar structural study has been reported for a phosphorothioate oligodeoxynucleotide and yeast RAP1 protein complex (111). However, this method might not be general because nonspecific interactions with protein sulfhydryl groups may cause interference.

TERMINAL PHOSPHOROTHIOATES FOR COUPLING REACTIONS   Oligonucleotides with a phosphorothioate at the termini can be prepared by chemical synthesis (112, 113), by enzymatic thiophosphorylation at the 5' end with ATPγS (33, 114), by attachment of pdCpS with RNA ligase to the 3' end of oligoribonucleotides (16), or by initiating transcription with guanosine derivatives such as GMPS (31), GTPβS (32), and GTPγS (33, 35). Terminal phosphorothioate groups in oligonucleotides have been used for a variety of purposes, such as (a) the separation of oligonucleotides or transcripts on mercury gels or columns (32, 33, 35), (b) reaction with haloacetyl derivatives of fluorescent dyes (16) and bromoacetyl agarose for the affinity purification of DNA-binding proteins (115), (c) reaction with photoaffinity agents (31), (d) reaction with intercalators (112), (e) the template-directed coupling of two oligonucleotides (113, 116), and (f) cross-linking a 5'-phosphorothioate oligonucleotide to its complementary sequence by reacting it with trans-platinum(II)diamine dichloride (114).

NH- AND S-BRIDGING PHOSPHATE INTERNUCLEOTIDIC LINKAGES   Another series of oligonucleotides have been prepared through the replacement of either the 3'- or the 5'-bridging oxygen of the phosphodiester internucleotidic linkage with sulfur or nitrogen. Oligodeoxynucleotides containing N3'-O5'phosphoramidate or N5'-O3'phosphoramidate linkages where either the 3'- or the 5'-oxygen is respectively exchanged with nitrogen possess interesting biophysical and biochemical properties (Figure 2). The amidate linkage in N3'-P5' (117) or N5'-P3' phosphoramidates (118, 119) can be selectively cleaved by mild acidic treatment. However, N5'-O3' (120) and N3'-O5' phosphoramidate oligodeoxynucleotides are resistant to various nucleases and have high binding affinity for RNA (121). These properties make them suitable for antisense studies even though they do not invoke the RNaseH cleavage mechanism (122).

Oligodeoxynucleotides with 3'-S- or 5'-S-phosphorothiolate internucleotide linkage, where the bridging oxygen at position 3'- or 5'- is replaced by sulfur, have also generated considerable interest (Figure 2). These linkages are susceptible to Ag^+, Hg^{2+} and iodine-mediated chemical cleavage (123, 124). However, the 3'-S-phosphorothiolate linkage is stable to the action of restriction endonuclease EcoRV (124) and is refractory to the corresponding methylase (125). Interestingly, topoisomerase I and lambda integrase protein can cleave a 5'-S-phosphorothiolate linkage in an oligodeoxynucleotide by forming a covalent enzyme intermediate with a 5'-thio-nucleoside leaving group. In contrast, the reverse reaction with a 5'-thio-nucleoside as the attacking nucleophile, to accomplish topoisomerase-mediated ligation, is inhibited (126). This observation is not surprising because sulfur is a poor nucleophile at phosphorus and requires a good leaving group to form the P-S bond (127).

Oligonucleotides with such linkages have also been used to probe the mechanism of ribozyme catalysis. In the group I intron ribozyme-catalyzed transesterification reaction, cleavage of a substrate containing a 3'-S-phosphorothiolate internucleotidic substitution was about 1000-fold slower in the presence of Mg^{2+} ion when compared to an unmodified phosphodiester substrate (128). However, in the presence of soft metal ions such as Mn^{2+} and Zn^{2+}, the cleavage activity was restored. This observation indicates a possible role of metal ions in stabilizing the negative charge on 3'-oxygen or sulfur in the transition state. Recently, the presence of a second metal ion involved in group I ribozyme catalysis was demonstrated through the use of 3'-(thioinosylyl)-(3'→5')-uridine (IspU) dinucleotide as a substrate (129). In a reaction mimicking the exon ligation step, the IspU substrate required a second, thiophilic metal ion for optimum activity. Based on these observations, it was proposed that a second metal ion that activates the 3'-hydroxyl group of guanosine is involved in the first step of splicing.

The mechanism of the hammerhead ribozyme–catalyzed reaction has also been probed with 5'-S-phosphorothiolates. One group has used this modification at the cleavage site in a DNA substrate (130). Because there was no appreciable thio effect from changing the metal ion from Mg^{2+} to Mn^{2+}, the authors concluded that there is no metal ion coordination with the 5'-oxygen of the leaving group in the transition state. Contradicting arguments were presented by another group, which used an all-RNA substrate for similar experiments (131). The RNA 5'-S-phosphorothiolate substrate was cleaved almost 100 times faster than the unmodified substrate in the presence of low concentrations of Mg^{2+} ions. It was reasoned that attack at the 2'-OH is rate-limiting for the modified substrate, whereas the departure of the leaving group is rate-limiting for the unmodified one. Rate differences observed for the hydrolysis of 3'- and 5'-S-phosphorothiolate susbtrates with different metal ions should be interpreted with caution because the metal ion-dependent hydrolysis of simple dinucleotides is not well understood (132).

Phosphorodithioates

In the phosphorodithioate linkage, both of the nonbridging oxygens of the phosphodiester backbone are replaced by sulfur atoms (Figure 2; 133–135). Such modified biopolymers are isostructural and isopolar to the natural nucleic acid backbone, and unlike phosphorothioates, they retain the achiral nature of the phosphorus atom.

Stability studies with phosphorodithioate oligonucleotides have demonstrated resistance to a variety of nucleases. The effect was most prominent when the dithioate modification was introduced at alternating internucleotide positions (136). However, introduction of phosphorodithioate linkages in an oligonucleotide reduces its binding to a complementary target, induces secondary structures in single-stranded oligomers (137), and enhances binding of modified oligonucleotides to proteins (138). Despite these potential drawbacks, phosphorodithioate-modified antisense oligonucleotides have been shown to interfere with the expression of erbB-2 mRNA associated with breast cancer (139), to inhibit HIV-1 reverse transcriptase activity (140), and to induce B-cell proliferation and differentiation (141).

Methylphosphonates

The methylphosphonate internucleotidic linkage is nonionic in contrast to the polyanionic natural phosphodiester backbone (Figure 2). Solid-phase synthesis of methylphosphonate oligomers requires nucleoside methylphosphonamidites, and following the synthesis, the oligomers require special deprotection chemistry (9). Akin to the phosphorothioate linkage, replacing one nonbridging oxygen with the methyl group introduces chirality at the phosphorus center. Efforts have been made to devise new stereospecific methods for the synthesis of Rp-methylphosphonate linkage to improve the binding properties of the modified oligomers (142, 143). Also determined was a high-resolution NMR structure of an alternating, chirally pure, Rp-methylphosphonate chimeric oligodeoxynucleotide hybridized to an RNA target (144).

Methylphosphonate oligomers have been shown to hybridize sequence-specifically to complementary nucleic acid targets. Examples for biological effects include antisense inhibition of herpes simplex virus replication (145), inhibition of human collagenase IV mRNA expression (146), and triplex-directed inhibition of chloramphenicol acetyltransferase mRNA expression in cell cultures (147).

In a study on deformations caused by DNA bending, several methylphosphonate linkages were introduced to create a neutral patch in the anionic phosphodiester backbone. Electrophoretic experiments revealed that the modified DNA bends toward the neutral surface (148, 149). Electrophoretic gel retardation analysis with Oct-2 POU and Oct-2 POU$_{HD}$ proteins and 14-mer duplex targets containing diastereomeric methylphosphonate linkages was used to map phosphate groups essential for protein–nucleic acid interaction (150). This interference method relies on the fact that the introduction of methylphosphonate linkages preserve B-DNA conformation while abolishing electrostatic contacts (151). Methylphosphonate linkages were also incorporated in a four-way DNA junction. The results support the notion that charge-charge interactions influence the structure of the Holliday junction (152).

BASE MODIFICATIONS

The heterocyclic ring of purine and pyrimidine bases provides hydrogen-bonding functional groups in nucleic acids. Carefully designed base analogs, when introduced into oligonucleotides, can provide information on the importance of specific functional groups in natural bases. In interpreting results obtained with base analogs, even a subtle change in the analog can have a dramatic effect because of the change in size, electronic distribution, nucleoside sugar conformation, tautomeric structure, or functional group pKa values. Representative structures of several modified bases are presented in Figure 3, and their use is described in subsequent sections. Base analogs possessing unusual spectroscopic properties or a potential for cross-linking are discussed in appropriate sections.

Purine Modifications

The imine nitrogen at position 7 of purine nucleosides is located in the major groove of B-form DNA and can act as a potential hydrogen bond acceptor site in the recognition processes. For example, this nitrogen atom is used as a hydogen bond acceptor in Hoogsteen base pairing. 7-Deazapurine nucleosides, which lack the N-7 nitrogen atom, are therefore valuable probes for studying the role of N-7 nitrogens in recognition. These analogs can easily be incorporated in oligonucleotides, either by enzymatic means (153) or by automated solid-phase synthesis (154, 155). Because of altered base stacking, oligonucleotides containing 7-deazapurine analogs destabilize DNA duplex slightly, and the effect is sequence dependent (156).

The interaction of the *trp* repressor with the *trp EDCBA* operon was studied through the use of 7-deazaadenine and 7-deazaguanine to validate the importance of four purine N-7 nitrogens implicated in recognition (157). Other studies also have used these analogs to investigate protein-DNA (158, 159) and restriction enzyme–DNA interactions (125, 160, 161).

The role of purine N-7 nitrogens in the folding of G-rich telomere sequences was investigated through the use of 7-deazaadenine and 7-deazaguanine substitutions (162). G-rich oligonucleotide sequences possess a high propensity to form tetraplexes through hydrogen bonding between the N-7 nitrogen of one guanine and the N-2 hydrogen of another. The effect of 7-deaza substitutions on tetraplex formation was studied by gel electrophoretic mobility shift experiments and guanine- and adenine-specific chemical footprinting reactions. Based on these experiments, the role of guanine N-7 nitrogen in stabilizing tetraplexes by specific hydrogen bond formation was confirmed.

Other examples that have exploited 7-deazapurine analogs are studies of DNA-DNA contacts in homologous recombination (163), the role of tertiary interactions and self-cleavage of hepatitis delta virus ribozyme (164), and tertiary structure mapping of the retinoblastoma gene (165).

The N-3 nitrogen of purines is located in the minor groove of DNA and is not directly involved in base pairing through hydrogen bond formation. However, 3-deazaadenosine substitution in the hammerhead ribozyme considerably reduces its catalytic efficiency, which indicates an important role for the N-3 nitrogen atom in the reaction (166). Similarly, when 3-deazaadenosine was substituted in the recognition sequence of *Eco*RV methylase, the enzymatic reaction was inhibited, thus implying a role for N-3 nitrogen in protein-DNA recognition (167).

Another purine analog, N-1-methylguanosine, when substituted in the core region of the hammerhead ribozyme, completely inhibits its catalytic activity (168). This analog has also been incorporated into a tRNA^{Asp}, where it abolished mischarging (169).

2-Aminopurine is often used for fluorescence measurements, which are discussed further in a subsequent section. Other analogs such as purine, isoguanine, 2,6-diaminopurine, 6-thioguanine, and hypoxanthine have been used as probes for the identification of purine nucleosides essential for protein-DNA interaction (167, 170–172) and RNA catalysis (166, 173–175) and to estimate the contribution of the hydrophobic interaction in duplex stability (176). A more

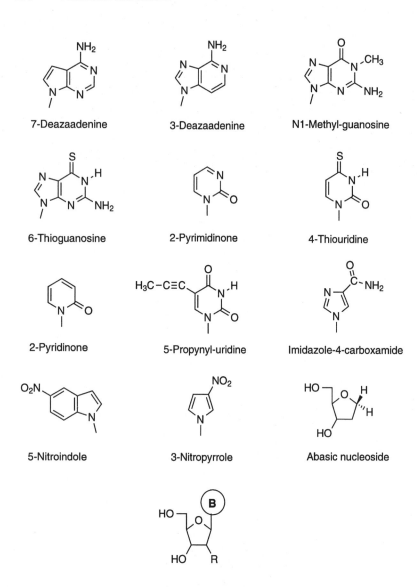

7-Deazaadenine

3-Deazaadenine

N1-Methyl-guanosine

6-Thioguanosine

2-Pyrimidinone

4-Thiouridine

2-Pyridinone

5-Propynyl-uridine

Imidazole-4-carboxamide

5-Nitroindole

3-Nitropyrrole

Abasic nucleoside

R : OCH₃ , 2′-O-Methylnucleoside
SH , 2′-Thionucleoside
OCH₂CH=CH₂ , 2′-O-Allylnucleoside

Figure 3   Base and sugar modifications.

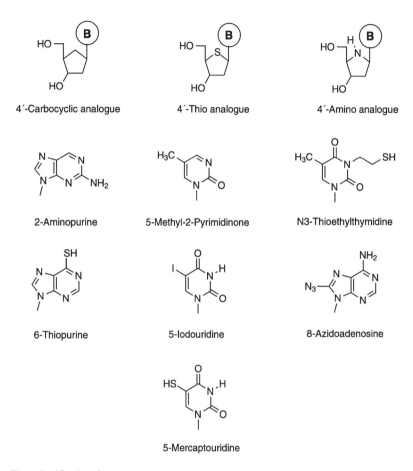

4´-Carbocyclic analogue   4´-Thio analogue   4´-Amino analogue

2-Aminopurine   5-Methyl-2-Pyrimidinone   N3-Thioethylthymidine

6-Thiopurine   5-Iodouridine   8-Azidoadenosine

5-Mercaptouridine

Figure 3 (*Continued*)

detailed account of purine and pyrimidine nucleoside analogs and their utility in delineating biochemical mechanisms can be found elsewhere (177, 178).

Pyrimidine Modifications

2-Pyrimidinone and 4-thiouridine ribonucleosides have been used to assess the participation of conserved pyrimidine bases in making hydrogen-bonding contacts in the catalytic core of hammerhead ribozyme (179). This study provided evidence for a magnesium ion-binding site in the catalytically active ribozyme. 2- and 4-pyridinone ribonucleosides have also been used to study the structure-function relationship of the hammerhead ribozyme (180).

The role of conserved deoxycytidine residues in template recognition by bacteriophage T7 primase was studied with oligonucleotides containing 2-pyrimidinone and 5-methyldeoxycytidine bases (181). The importance of the hydrogen-bonding framework for protein-DNA recognition in conjunction with other nucleoside substitutions was demonstrated.

Oligonucleotides containing a propynyl substituent at the C-5 position of pyrimidines possess high binding affinity toward complementary RNA targets and display a gene-specific antisense effect (182, 183). When such modified oligonucleotides additionally contain phosphorothioates, they can be used to target low and high copy messages with equal efficiency (184). Remarkably, a sequence-specific inhibition of gene expression has been observed with a C-5 propyne antisense phosphorothioate as short as a heptanucleotide (185).

Universal Bases

Natural nucleobases display exquisite selectivity in recognizing complementary bases. A universal base can be defined as an analog that can substitute for any of the four natural bases in oligonucleotides without significantly impairing the duplex stability. In general, universal base analogs use aromatic ring stacking, instead of specific hydrogen bonds, to stabilize a duplex. However, universal recognition by imidazole-4-carboxamide nucleoside has been attributed to specific hydrogen-bonding contacts (186).

The efficiency of oligonucleotide primers containing multiple substitutions of 5-nitroindole (187, 188) and 3-nitropyrrole (189, 190) has been studied and compared in DNA sequencing and in polymerase chain reactions. These analogs can be used at primer sites corresponding to degenerate base positions and specifically where the sequence data is incomplete. The use of imidazole-4-carboxamide nucleoside in oligonucleotide templates results in nucleotide misincorporation during PCR amplification, and it has been suggested that this property can be further used to generate mutant gene libraries (186).

Depending on the length of the oligonucleotides, a true mismatch in a DNA hybrid is often difficult to detect. A recent study used 3-nitropyrrole-containing oligonucleotide probes to analyze either a PCR-amplified or an allele-specific PCR-amplified polymorphic region of the HLA-DRB locus by artificial mismatch hybridization (191). The presence of 3-nitropyrrole in a hybridization probe created an artificial mismatch, and it provided an excellent discrimination between perfectly matched and mismatched duplexes when compared to conventional hybridization methods.

Abasic Nucleosides

Deletion of the purine or pyrimidine bases from the deoxyribose or ribose sugar moiety gives rise to abasic nucleosides. These analogs preserve the sugar hydroxyl groups; therefore, any biochemical effect observed from their use results directly from the loss of the heterocyclic ring. A deoxy-abasic nucleoside was used to replace both dA and T in a dA/T base pair to study the binding of *Eco*RV DNA methyltransferase to its recognition sequence (171). Such a substitution is expected to disrupt hydrogen bonding and was therefore used to assess the role of base pairing to protein-binding affinity.

Substitution of abasic ribonucleosides in the hammerhead ribozyme significantly impairs its catalytic activity (192). However, it was possible to restore the activity of four ribozymes by a simple addition of the missing base. This finding indicated that the ribozyme structure can be engineered to create a binding site for small ligands. Other studies using ribo-abasic residue include the recognition of branch-point adenosine in the Group II introns (175); the identification of essential base residues in loop B of the hairpin ribozyme (193); the importance of the stem-loop II; and the importance of conserved U residues in the catalytic core of hammerhead ribozyme (194, 195).

SUGAR MODIFICATION

The 2′-deoxyfuranose ring in oligodeoxynucleotides can adopt C_3'-endo and C_3'-exo as two major, interchangeable conformations, whereas the C_3'-endo sugar ring conformation is maintained in oligoribonucleotides (196). The electronegativity of the 2′-substituent has a profound effect on the sugar conformation, and as a result of such substitution, the population of C_3'-endo conformer increases with an increase in the electronegativity. Several sugar modifications, primarily at the 2′-position, have been described, and there has been renewed interest in oligoribonucleotides containing such modifications to generate nuclease-resistant RNA aptamers and ribozymes.

Sugar-modified analogs have been used extensively to study ribozymes and RNA aptamers to probe the importance of specific 2′-hydroxyl groups in catalysis and to provide resistance against nucleases. The 2′-hydroxyl modifications introduced to probe hammerhead ribozyme catalysis include deoxy (197, 198); fluoro, amino (199, 200); methoxy, allyloxy (201); and 2′-*C*-allyl substituents (202). 2′-Deoxy- or 2′-*O*-methyl sugar modifications have been introduced to identify the positions of 2′-hydroxyl groups necessary for the hairpin ribozyme activity (203). Interactions involving 2′-hydroxyl groups, and the importance of having a ribose moiety at the cleavage site in *Tetrahymena* ribozyme reaction, have been studied through the use of 2′-deoxy- and 2′-fluoro-substituted nucleosides (204–207). 2′-Deoxy substitution has also been used to investigate the role of active-site 2′-hydroxyl groups in group II intron molecular recognition and catalysis (208). Recently, incorporation of 2′-deoxy-2′-mercaptocytidine into an oligonucleotide, via suitably protected phosphoramidite, has been reported (209). However, enzymatic incorporation of 2′-mercapto group-containing modified nucleotides remains to be established.

2'-Fluoro- and 2'-difluoro-2'-deoxyuridine–containing oligonucleotides have been used to investigate the mechanism of human G/T glycosylase (210). The modified oligonucleotides resisted the excision repair activity of the enzyme and resulted in the formation of a tight-binding DNA-glycosylase complex. Based on these observations, a rationale for preferential repair of deaminated CpG sites was presented.

When a variety of selection protocols were employed, high-affinity nucleic acid aptamers were isolated. These aptamers can recognize a wide range of molecular targets such as antibiotics, amino acids, nucleoside triphosphates, cofactors, organic dyes, porphyrins, and proteins (211). Improved chemical and enzymatic stability is essential for the development and use of RNA-based aptamers in therapeutics and diagnostics. The substitution of 2'-fluoro- and 2'-aminopyrimidines in RNA imparts resistance to nucleases (200). Moreover, these modifications can be introduced easily during aptamer selection because the corresponding nucleotide triphosphates are good substrates for T7 RNA polymerase (29, 30). When these analogs were used, nuclease-resistant aptamers were isolated against the basic fibroblast growth factor (212), human thyroid-stimulating hormone (213), a monoclonal antibody that recognizes the main immunogenic region of the human ACh receptor (214), and the keratinocyte growth factor (215).

2'-O-methyl-modified, biotinylated oligoribonucleotides have been used as antisense probes for mapping U2 snRNP-pre-mRNA interactions and for studying structure and function of U4/U6 snRNPs (216, 217). Modified nucleosides such as 2'-deoxy and 2'-O-methyl have also been placed at either the 3'- or the 5'-splice site of a nuclear pre-mRNA to aid in understanding chemical steps involved in the spliceosomal assembly (20). In vitro studies revealed that the presence of the 2'-hydroxyl group is more important at the 3'-splice site for the second step in the splicing reaction, which results in ligation of two exons with a concomitant release of an intron.

Posttranscriptional nuclear processing of most eukaryotic mRNAs involves 3'-end polyadenylation, which is suggested to confer stability to the transcript. A highly conserved hexameric-sequence AAUAAA, present in the transcript, acts as a polyadenylation signal by binding a specificity factor protein. 2'-O-methyl ribose-modified nucleosides were chemically substituted at all nucleotide positions in synthetic RNA substrates, including the signal sequence, to aid in the study of the polyadenylation reaction in vitro (218). A single 2'-O-methyl substitution, within the AAUAAA sequence, inhibited the polyadenylation and binding of the specificity factor. Oligoribonucleotides with substitutions at any position, except the 3'-terminal adenosine, were efficiently polyadenylated.

Another sugar modification involves replacement of the O-4' furanose ring oxygen with a methylene group to give carbocyclic analogs. Such sugar-modified analogs of thymidine and 5-methyl-2'-deoxycytidine have been incorporated in oligodeoxynucleotides to allow study of their effect on triplex formation (219). The mechanism of action of the MutY protein of *E. coli*, which deletes mismatched deoxyadenosine residues from a DNA substrate, has been studied through use of the carbocyclic analog of 2'-deoxyadenosine (220). The proposed mechanism of excision involves an enzyme-catalyzed attack of water on the C-1' carbon to displace the base from the duplex, followed by the opening of the furanose ring to give an acyclic structure. In accordance with the proposed mechanism, it was found that DNA substrates containing carbocyclic dA analog, designed to prevent the ring opening step, were not excised by the MutY protein. Ribocarbocyclic analogs of cytidine and adenosine have been chemically incorporated in the hammerhead ribozyme to confer RNase resistance to these ribozyme constructs (221).

The O-4' ring oxygen can also be replaced by sulfur to result in 4'-thiofuranose modification, and oligonucleotides containing such modification have been used to probe protein-DNA interactions such as in the *Eco*RV restriction endonuclease and methylase systems (125, 222). In one study, modified deoxynucleosides possessing an inverted C4' configuration or capable of forming C4' adducts were incorporated in DNA templates to investigate structural the requirement of the Klenow fragment of *E. coli* DNA pol I (223). The polymerase halts transiently at inversions while adduct-forming modifications completely inhibit the DNA synthesis.

The furanose ring oxygen has also been replaced with a pyrrolidine ring. This modified base has been incorporated in a 25-mer to study mechanism-based inhibition of an *E. coli* DNA repair enzyme, 3-methyladenine DNA glycosylase II (224).

NONRADIOACTIVE LABELS OF OLIGONUCLEOTIDES

Fluorescent Labels

BASE ANALOGS   2-Aminopurine (2AP), a fluorescent nucleoside analog, has been used extensively to detect changes in oligonucleotide conformation (225). It can substitute for adenosine in base pairing with thymidine without distorting the double helix. The absorption and excitation maximum for 2AP is at 330 nm and has an emission at 380 nm. The quantum yield of 2AP fluorescence, when substituted in oligonucleotides, depends on the degree of base stacking (226). Therefore, any change in its fluorescence is a sensitive indicator of structural perturbations in the modified oligonucleotide and provides an insight into the dynamics of such processes. Temperature-dependent conformational changes in oligonucleotides (227) and the dynamics of mismatched base pairs in oligodeoxynucleotides have been followed in the monitoring of change in fluorescence intensity of 2AP (228). Several reports have described the use of 2AP-containing oligonucleotides to study base-pairing interactions in DNA polymerase $3'$-$5'$-exonuclease proofreading (229–231), kinetics of nucleotide incorporation with the Klenow fragment and T4 DNA polymerase, (232, 233), and the helicase-catalyzed unwinding of duplex DNA (234). In one study, Mg^{2+} ion-induced conformational perturbations in the hammerhead ribozyme tertiary structure were followed by the monitoring of change in fluorescence of 2AP. The data were used to determine the affinity constants for Mg^{2+} ion binding (235).

A fluorescent pteridine analog, 3-methyl-8-(2-deoxy-β-D-ribofuranosyl) isoxanthopterin, has been introduced as a guanosine mimic for mechanistic studies (236). The fluoroscence of this analog, when incorporated into an oligodeoxynucleotide identical to the U5 terminal sequence of HIV-1, increases with $3'$-processing reaction of integrase. This analog extends the repertoire of fluorescent nucleosides, even though its base-pairing properties are not well documented. 5-Methyl-2-pyrimidinone has been reported as an alternative to 2AP, but it has not found wider application, presumably because of the lack of base-pairing capabilities (237).

Fluorescent reporter groups have also been attached to the 5-position of deoxyuridine, via aminoalkyl linkers, to follow oligonucleotide-protein interactions. One such study with template/primer oligonucleotides, where a fluorescent dansyl probe was appended at various nucleotide positions from the $3'$-end of the primer, indicated that the Klenow polymerase remains in contact with the template/primer up to six nucleotides from the $3'$-end (238). Similarly, contacts between a 42-mer oligonucleotide and the *E. coli* regulatory protein TyrR were also identified through the use of such nucleoside analogs (239).

A fluorescent derivative has also been attached to 5-sulfhydryl-deoxyuridine triphosphates for polymerization into a template/primer for the determination of dissociation constants of DNA and the dNTPs for the Klenow fragment of *E. coli* DNA polymerase I (240).

NONBASE ATTACHMENT   Fluorescent dyes have been attached to oligodeoxynucleotides by the reaction of internucleotidic phosphorothioates with 5-iodoacetamido-eosin and -fluorescein (241). Distances calculated from fluorescence resonance energy transfer (FRET) studies that used the fluorescently tagged oligonucleotides were in reasonable agreement with B-DNA helical structure.

A variety of reporter groups have been attached sequence-specifically to oligodeoxynucleotides by the oxidation of an internucleotidic H-phosphonate in presence of cystamine to form a phosphoramidate linkage (242). Reduction of cystamine results in a free sulfhydryl group that can act as a nucleophile to react with thiol-specific reporter groups. However, this modification results in a pair of diastereomers that might have to be separated for an accurate interpretation of the spectroscopic data.

Fluorescent dyes have also been attached to the sugar of dideoxynucleoside triphosphates to act as chain terminators in fluorescence-based DNA sequencing (243). In one study, two dyes were attached to the sequencing primer, with a common donor at the 5'-end, and to four different acceptors at a thymidine within the primer, which facilitate DNA sequencing with the use of FRET measurements (244). Recently, guanosine-specific fluorescence quenching of 3'-fluorescein-labeled oligoribonucleotides was used to determine rate constants for substrate binding, cleavage, and dissociation of the hairpin ribozyme (245). There is considerable literature on fluorescent dyes attached at the termini of oligonucleotides, particularly for FRET studies; this literature has been reviewed elsewhere (225).

Electron Spin Resonance Probes

Electron paramagnetic resonance (EPR) is an elegant method of monitoring global tumbling and internal motion of oligonucleotides. A nitroxide spin-label probe has been attached to the 5-position of deoxyuridine, via a mono- or diacetylene tether, to form a complementary base pair with an adenosine (246). A quinolonyl nitroxide analog has base paired with 2-aminopurine nucleoside (247). This derivative seems to be particularly attractive because the motion of the spin label can be directly correlated to that of the base. A nitroxide spin label has also been introduced at the phosphate backbone, via a phosphoramidate sulfhydryl tether, but EPR measurements were not reported (242).

Resonance Raman Spectroscopy

The UV-absorbance maximum at 330–340 nm of 4-thiothymidine facilitates assignment of the bands in the resonance Raman spectrum to a specific nucleotide. Raman spectroscopy was performed on a complex of EcoRV endonuclease with an oligodeoxynucleotide containing 4-thiothymidine, which represents the enzyme recognition sequence (248). The observations were found to be consistent with the distortion of the DNA as seen in the X-ray structure. Such distortion was not observed with a nonsubstrate oligonucleotide. These results document the utility of this analog for its use in Raman spectroscopic studies.

NUCLEIC ACID CROSS-LINKING

Chemically modified oligonucleotides are extremely useful in fixing conformations by introducing cross-links in DNA, RNA, or nucleic acid–protein complexes. Site-specific cross-linking permits the determination of the proximity of nucleotides or amino acids because the cross-linking would not be possible at distances greater than the length of the spacer. Such reagents are usually attached to either the base or the sugar moiety, and in most cases, the sulfhydryl group is used for introducing cross-links.

Chemical Cross-Links

A versatile strategy for the cross-linking of DNA and RNA oligomers results from the recently described convertible nucleoside approach (249–251). In this method, uridine and inosine derivatives possessing good leaving groups at the 4 or 6 position, respectively, are incorporated site-specifically in oligonucleotides. Subsequent reaction of the convertible nucleoside, at the oligonucleotide level, with a symmetrical ω, ω-dithiobis(alkylamine), results in the formation of cytidine and adenosine derivatives bearing an alkyl disulfide tether. Reduction of the disulfide linkage results in the formation of a free sulfhydryl group, which can be reoxidized to form a cross-link. For example, this approach has been applied to form adenine-adenine cross-links at 5'-AT-3' sequences in the major groove and guanine-guanine cross-links at 5'-GC-3' sequences in the minor groove of oligodeoxynucleotides. Such an interstrand cross-link mimics a bent DNA structure and binds to a high mobility group protein, thereby supporting the view that this protein recognizes bent DNA (252). An interstrand disulfide bridge has been utilized to prevent DNA strand separation in the DNA-(cytosine-5) methyltransferase-catalyzed reaction (253). It was suggested that reduction in the rate of methylation of the cross-linked oligonucleotide was due to the extrusion of the cytosine to an extrahelical position in the normal substrate oligonucleotide. Oligoribonucleotides can also be cross-linked in this manner, as exemplified by the formation of an intramolecular disulfide in an RNA ministem loop (254).

In another approach, a mercaptoalkyl group is attached to the N-3 of a pyrimidine base, such as in N-3-thioethylthymidine (255). When incorporated at the 3'-end and 5'-end of two oligonucleotides, the sulfhydryl group can be oxidized, thus linking two oligonucleotides through the formation of disulfide loops. The cross-link imparts greater thermal stability and a unique conformation. Mercaptoethyl groups can also be attached to C5 of pyrimidines and oxidized to the disulfides, as described for the cross-linking of triple helices and tRNA (256, 257). Interstrand disulfide cross-linking can also be obtained through the use of 6-mercaptopurine and 4-thiothymidine bases in an oligodeoxynucleotide without the employment of tether (258).

Cross-linked duplexes have been prepared to allow the study of the mechanism of DNA polymerases through the introduction of an aziridinyl group at the 4-position of a cytidine or at position 6 of guanosine, which can react with the 4-amino group of a mismatched cytidine in the complementary strand to link two cytidines or a 2,6 diaminopurine and cytidine, respectively, by an ethylene bridge (259, 260). In some systems, however, cross-linking via base modification is undesirable. Therefore, attachment of the cross-linking reagent via the sugar moiety is preferred. This approach has been used to discriminate between two models of the hammerhead ribozyme, one based on X-ray crystallographic analysis and the other based on FRET results (261). In this method, the nucleotides to be linked were replaced by the corresponding 2'-amino nucleotides, which were then reacted with 2-pyridyl 3-isothiocyanatobenzyl disulfide. After reduction, an intrastrand disulfide was reintroduced by oxidation, and the catalytic activity of cross-linked ribozymes was determined. Recently, more reactive aliphatic isocyanates have been used for the cross-linking of the hairpin ribozyme as part of an effort to build a first 3-D model (262; DJ Earnshaw, B Masquida, S Müller, STh Sigurdsson, F Eckstein, et al, manuscript in preparation). This method has been further extended to introduce interstrand disulfide cross-links, by a sulfhydryl exchange reaction, in a group I ribozyme (264). The cross-linked ribozymes were catalytically competent, and this observation reveals interesting dynamic properties of large catalytic RNAs.

Alternatively, 2'-hydroxyl groups have also been derivatized with ethylthio groups to introduce intra- and interhelical disulfide cross-links in tRNA^{Phe} (256).

Photoaffinity Cross-Linking

BASE CROSS-LINKS   Natural nucleosides can be photoactivated to form new bonds under favorable distance conditions. A good example is the formation of a cross-link between a cytidine and a uridine residue, which was observed upon irradiation in yeast tRNA^{Phe} and was found to be consistent with the X-ray structure (265). Another example is described for the hairpin ribozyme, in which a uridine and a guanosine in a loop were cross-linked in high yields (266).

Usually, the cross-linking yields obtained with the natural nucleotides are low. Moreover, the wavelength required for activation is around 260 nm, which is not specific for a particular nucleotide. Thus, analogs that are more reactive and absorb at a higher wavelength are usually preferred for photoactivated cross-linking. 4-Thiothymidine and -uridine, 5-bromo- and -iodouridine, and 6-thioguanosine meet this requirement and are also able to form Watson-Crick base pairs. The most frequently used analogs are 4-thiouridine and 4-thiothymidine, which absorb at 331 nm and can be irradiated between 300 and 400 nm for activation (267). The first intrastrand RNA site-specific cross-link with 4-thiouridine was achieved upon irradiation of yeast tRNA^{Val}, which contains 4-thiouridine as a natural constituent (268). In one study, 4-thiouridine was incorporated by enzymatic methods in oligoribonucleotides. Depending on its position, cross-links were observed to ribosomal proteins and 16S RNA (269).

The chemical synthesis of oligodeoxy- and oligoribonucleotides containing 4-thiopyrimidine and 6-thiodeoxyguanosine requires special protection of the sulfhydryl group (267, 270). The *Eco*RV restriction endonuclease and modification methylase have been investigated with oligonucleotides containing these analogs (267). These analogs could be cross-linked to the enzymes with higher yields for 4-thiothymidine compared to 6-thiodeoxyguanosine. Unfortunately, oligonucleotides without the recognition sequences were also cross-linked as a result of nonspecific binding.

4-Thiouridine has been incorporated at specific positions in an adenovirus pre-mRNA, through the Moore and Sharp ligation approach (20), to identify contacts to snRNAs during splicing reaction (271). Cross-links were observed with a loop sequence in U5 snRNA and with an invariant sequence of U6. Also investigated through use of a similar strategy is strand-specific cross-linking between 4-thiouridine-modified human tRNA[Lys] with HIV-1 reverse transcriptase subunits p66 and p51 (272). 4-Thiouridine was used to cross-link small nuclear ribonucleoprotein particles U11 and U6atac to the 5′-splice site of an AT-AC intron found in metazoan genes (273).

6-Thioinosine and 4-thiouridine have been used as photoaffinity probes to study hammerhead and hairpin ribozyme conformations (274). Multiple cross-links were found in both of the cases, and the results were difficult to explain on the basis of models. It was concluded that these ribozymes can adopt multiple conformations, making it difficult to identify catalytically competent conformations by this approach alone.

5-Bromo- and 5-iodouridine, which can be activated at 308 and 325 nm, respectively, are also convenient photoaffinity labels, as exemplified by the cross-linking of an RNA transcript and an oligodeoxynucleotide containing these modified nucleosides to the bacteriophage R17 coat protein (275, 276). Protein photodamage was considerably reduced with the iodo derivative. A telomeric DNA containing 5-bromodeoxyuridine has been cross-linked to the *Oxytricha* telomeric protein (277). Two of the three cross-links were traced to the tyrosine residues in the hydrophobic region, suggesting that intercalation plays a major role in protein-telomeric DNA binding.

The 5-position of uridine and deoxyuridine can be conveniently functionalized, without hydrogen bonding interaction being affected, to attach various nonradioactive probes. In one of the first reports, biotin was covalently attached at the 5-position of thymidine or uridine triphosphate for incorporation into DNA or RNA, respectively, for a subsequent streptavidin-affinity isolation (278). Many biotin phosphoramidites are commercially available for solid-phase chemical coupling to oligonucleotides. A similar approach uses incorporation of probes such as digoxigenin (DIG) or fluorescein instead of biotin. For photo-cross-linking, azido-nitrobenzoate has been coupled to the 5′-position, via an amino linker, and the nucleotide has been incorporated chemically into oligonucleotides to allow study of the contact with the Klenow fragment (279), T4 DNA polymerase (280), and *E. coli* DNA polymerase III (281).

In one study, multiple 5-methyleneaminouridine triphosphate residues were introduced in oligoribonucleotides by T7 RNA polymerase transcription of synthetic DNA templates (282). The free amino group in modified uridine was postsynthetically derivatized, at the RNA level, with an azirinyl group-containing reagent. The photo-cross-linking results with the azirinyl-uridine derivative and 6-thioguanosine were used to study the binding of radiolabeled mRNA analogs to *E. coli* ribosomes. Precise cross-link positions were determined by a combination of ribonuclease H and T_1 digestion and by primer extension analysis.

8-Azidoadenosine and -guanosine are frequently used for the photoaffinity labeling of enzymes (283, 284). Due to incompatibility of the azido group with phosphoramidite-based chemical synthesis, these analogs are not suitable for automated synthesis. However, enzymatic methods have been used to incorporate 8-azidoadenosine into the acceptor stem of tRNAs to probe its interaction with ribosomes (285). 5-Azido-uridine and -deoxyuridine triphosphates, with maximum absorption at around 300 nm, are also suitable for photoactivated cross-linking and are substrates for RNA and DNA polymerases. However, affinity labeling studies have not been reported with these analogs (286).

5-Sulfhydryluridine triphosphate has been enzymatically incorporated into RNA, followed by subsequent derivatization with azidophenacyl bromide for cross-linking to several RNA polymerases (287).

TERMINAL CROSS-LINK   Posttranscriptional reaction of a 5'-terminal guanosine phosphorothioate is a versatile method to attach a photoactivatable azidophenacyl group to the transcipts. Upon irradiation, intra- and intermolecular cross-links in RNase P RNA were obtained in the absence or presence of pre-tRNA (288). In another study, RNA containing a 5'-terminal phosphorothioate was derivatized with an azidophenacyl group to determine the tertiary structure of the ribonuclease P ribozyme-substrate cross-linked complex (289). A similar cross-linking study was also used to identify "neighbors" in the group I intron ribozyme (290).

METAL DERIVATIVES   A solid-phase synthesis of an oligodeoxynucleotide that was site-specifically platinated, employing a cisplatin-derivatized deoxyguanosine 3'-H-phosphonate monomer, has been described (291). In another study, 2'-O-methyl oligoribonucleotides were postsynthetically modified with transplatin to give a 1,3-intrastrand cross-linked oligomer (292). When hybridized with complementary RNA strands, intrastrand platinated oligomers trigger the formation of specific interstrand cross-links. The utility of transplatin-modified oligoribonucleotides was demonstrated through the introduction of cross-links in Ha-ras mRNA and through concomitant inhibition of cell proliferation.

CONCLUSIONS

Chemically modified oligonucleotides have been immensely helpful in understanding mechanistic and stereochemical aspects of numerous biochemical reactions and processes. The imagination and ability of chemists will ensure that tailor-made nucleoside analogs will be synthesized to answer specific questions in the future. There is, however, a real challenge to widen the arsenal of nucleotide analogs that can be enzymatically incorporated into nucleic acids. Novel strategies, such as mutant polymerases, will have to be devised to achieve this goal.

ACKNOWLEDGMENT

We thank Paul A. Heaton for valuable comments on the manuscript.

> **Visit the *Annual Reviews home page* at
> http://www.AnnualReviews.org.**

Literature Cited

1. Caruthers MH, Beaton G, Wu JV, Wiesler W. 1992. *Methods Enzymol.* 211:3–20
2. Beaucage SL. 1993. *Methods Mol. Biol.* 20:33–61
3. Brown T, Brown DJS. 1991. In *Oligonucleotides and Analogues. A Practical Approach,* ed. F Eckstein, pp. 1–24. Oxford: IRL
4. Wincott F, Direnzo A, Grimm S, Tracz D, Workman C, et al. 1995. *Nucleic Acids Res.* 23:2677–84
5. Davis RH. 1995. *Curr. Opin. Biotechnol.* 6:213–17
6. Gait MJ, Pritchard C, Slim G. 1991. See Ref. 3, pp. 25–48
7. Zon G, Stec WJ. 1991. See Ref. 3, pp. 87–108
8. Wiesler WT, Marshall WS, Caruthers MH. 1993. *Methods Mol. Biol.* 20:191–206
9. Miller PS, Cushman CD, Levis JT. 1991. See Ref. 3, pp. 137–54
10. Maunders MJ. 1993. *Methods Mol. Biol.* 16:213–30
11. Yu YT, Steitz JA. 1997. *RNA* 3:807–10
12. Eggerding FA, Iovannisci DM, Brinson E, Grossman P, Winn-Deen ES. 1995. *Hum. Mutat.* 5:153–65
13. Martinelli RA, Arruda JC, Dwivedi P. 1996. *Clin. Chem.* 42:14–18
14. Barrio JR, Barrio MCG, Leonard NJ, England TE, Uhlenbeck OC. 1978. *Biochemistry* 17:2077–81
15. Voegel JJ, Benner SA. 1996. *Helv. Chim. Acta* 79:1881–98
16. Cosstick R, McLaughlin LW, Eckstein F. 1984. *Nucleic Acids Res.* 12:1791–810
17. Kinoshita Y, Nishigaki K, Husimi Y. 1996. *Chem. Lett.* 9:797–98
18. Breitschopf K, Gross HJ. 1996. *Nucleic Acids Res.* 24:405–10
19. Ohtsuki T, Kawai G, Watanabe Y, Kita K, Nishikawa K, Watanabe K. 1996. *Nucleic Acids Res.* 24:662–67
20. Moore MJ, Sharp PA. 1992. *Science* 256:992–97
21. Bain JD, Switzer C. 1992. *Nucleic Acids Res.* 20:4372
22. Eckstein F. 1985. *Annu. Rev. Biochem.* 54:367–402

23. Griffiths AD, Potter BVL, Eperon IC. 1987. *Nucleic Acids Res.* 15:4145–62
24. Hacia JG, Wold BJ, Dervan PB. 1994. *Biochemistry* 33:5367–69
25. Tang JY, Roskey A, Li Y, Agrawal S. 1995. *Nucleosides Nucleotides* 14:985–90
26. Milligan JF, Uhlenbeck OC. 1989. *Methods Enzymol.* 180:51–62
27. Conrad E, Hanne A, Gaur RK, Krupp G. 1995. *Nucleic Acids Res.* 23:1845–53
28. Sousa R, Padilla R. 1995. *EMBO J.* 14:4609–21
29. Aurup H, Williams DM, Eckstein F. 1992. *Biochemistry* 31:9636–41
30. Huang Y, Eckstein F, Padilla R, Sousa R. 1997. *Biochemistry* 36:8231–42
31. Burgin AB, Pace NR. 1990. *EMBO J.* 9:4111–18
32. Zhang-Keck Z-y, Eckstein F, Washington LD, Stallcup MR. 1988. *J. Biol. Chem.* 263:9550–56
33. Igloi G. 1988. *Biochemistry* 27:3842–49
34. Logsdon N, Lee CGL, Harper JW. 1992. *Anal. Biochem.* 205:36–41
35. Vaish NK, Heaton PA, Eckstein F. 1997. *Biochemistry* 36:6495–501
36. Goodchild J. 1990. *Bioconjug. Chem.* 1:165–87
37. Bongartz JP, Aubertin AM, Milhaud PG, Lebleu B. 1994. *Nucleic Acids Res.* 22:4681–88
38. Soukchareun S, Tregear GW, Haralambidis J. 1995. *Bioconjug. Chem.* 6:43–53
39. Corey DR, Munoz-Medellin D, Huang A. 1995. *Bioconjug. Chem.* 6:93–100
40. Ruth JL. 1994. *Methods Mol. Biol.* 26:167–85
41. Hangeland JJ, Levis JT, Lee YC, T'so POP. 1995. *Bioconjug. Chem.* 6:695–701
42. Shea RG, Marsters JC, Bischofberger N. 1990. *Nucleic Acids Res.* 18:3777–83
43. Buijsman RC, Kuijpers WHA, Basten JEM, Kuylyeheskiely E, Vandermarel GA, et al. 1996. *Chemistry* 2:1572–77
44. Will DW, Brown T. 1992. *Tetrahedron Lett.* 33:2729–32
45. Letsinger RL, Zhang G, Sun DK, Ikeuchi T, Sarin PS. 1989. *Proc. Natl. Acad. Sci. USA* 86:6553–56
46. Zelphatti O, Wagner E, Leserman L. 1994. *Antiviral Res.* 25:13–25
47. Henderson GB, Stein CA. 1995. *Nucleic Acids Res.* 23:3726–31
48. Hagmar P, Bailey M, Tong G, Haralambidis J, Sawyer WH, Davidson BE. 1995. *Biochim. Biophys. Acta Gen. Subj.* 1244:259–68
49. Soukup GA, Cerny RL, Maher LJ. 1995. *Bioconjug. Chem.* 6:135–38
50. Holtke HJ, Ankenbauer W, Muhlegger K, Rein R, Sagner G, et al. 1995. *Cell. Mol. Biol.* 41:883–905
51. Stec WJ, Grajkowski A, Kobylanska A, Karwowski B, Koziolkiewicz M, et al. 1995. *J. Am. Chem. Soc.* 117:12019–29
52. Connolly BA, Potter BVL, Eckstein F, Pingoud A, Grotjahn L. 1984. *Biochemistry* 23:3443–53
53. Grasby JA, Connolly BA. 1992. *Biochemistry* 31:7855–61
54. Engelman A, Mizuuchi K, Craigie R. 1991. *Cell* 67:1211–21
55. Mizuuchi K, Adzuma K. 1991. *Cell* 66:129–40
56. Kurpiewski MR, Koziolkiewicz M, Wilk A, Stec WJ, Jen-Jacobson L. 1996. *Biochemistry* 35:8846–54
57. Thorogood H, Grasby JA, Connolly BA. 1996. *J. Biol. Chem.* 271:8855–62
58. McSwiggen JA, Cech TR. 1989. *Science* 244:679–83
59. Rajagopal J, Doudna JA, Szostak JW. 1989. *Science* 244:692–94
60. Moore MJ, Sharp PA. 1993. *Nature* 365:364–68
61. Maschhoff KL, Padgett RA. 1993. *Nucleic Acids Res.* 21:5456–62
62. Chanfreau G, Jacquier A. 1994. *Science* 266:1383–87
63. Padgett RA, Podar M, Boulanger SC, Perlman PS. 1994. *Science* 266:1685–88
64. Sharp PA. 1994. *Cell* 77:805–15
65. van Tol H, Buzayan JM, Feldstein PA, Eckstein F, Bruenig G. 1990. *Nucleic Acids Res.* 18:1971–75
66. Slim G, Gait MJ. 1991. *Nucleic Acids Res.* 19:1183–88
67. Pley HW, Flaherty KM, McKay DB. 1994. *Nature* 372:68–74
68. Scott WG, Finch JT, Klug A. 1995. *Cell* 81:991–1002
69. Kitts PA, Nash HA. 1988. *J. Mol. Biol.* 204:95–107
70. Herschlag D, Piccirilli JA, Cech TR. 1991. *Biochemistry* 30:4844–54
71. Polesky AH, Dahlberg ME, Benkovic SJ, Grindley NDF, Joyce CM. 1992. *J. Biol. Chem.* 267:8417–28
72. Noel JP, Hamm HE, Sigler PB. 1993. *Nature* 366:654–63
73. Gish G, Eckstein F. 1988. *Science* 240:1520–22
74. Schatz D, Leberman R, Eckstein F. 1991. *Proc. Natl. Acad. Sci. USA* 88:6132–36
75. Waring RB. 1989. *Nucleic Acids Res.* 17:10281–93
76. Christian EL, Yarus M. 1993. *Biochemistry* 32:4475–80
77. Jaffe EK, Cohn M. 1979. *J. Biol. Chem.* 254:10839–45

78. Pecoraro VL, Hermes JD, Cleland WW. 1984. *Biochemistry* 23:5262–71
79. Milligan JF, Uhlenbeck OC. 1989. *Biochemistry* 28:2849–55
80. Biou V, Yaremchuk A, Tukalo M, Cusack S. 1994. *Science* 263:1404–10
81. Rudinger J, Puglisi JD, Pütz J, Schatz D, Eckstein F, et al. 1992. *Proc. Natl. Acad. Sci. USA* 89:5882–86
82. Kreutzer R, Kern D, Giege R, Rudinger J. 1995. *Nucleic Acids Res.* 23:4598–4602
83. Caprara MG, Mohr G, Lambowitz AM. 1996. *J. Mol. Biol.* 257:512–31
84. Caprara MG, Lehnert V, Lambowitz AM, Westhof E. 1996. *Cell* 87:1135–45
85. Dabrowski M, Spahn CMT, Nierhaus KH. 1995. *EMBO J.* 14:4872–82
86. Harris ME, Pace NR. 1995. *RNA* 1:210–18
87. Hardt WD, Warnecke JM, Erdmann VA, Hartmann RK. 1995. *EMBO J.* 14:2935–44
88. Warnecke JM, Fürste JP, Hardt WD, Erdmann VA, Hartmann RK. 1996. *Proc. Natl. Acad. Sci. USA* 93:8924–28
89. Chen Y, Li XQ, Gegenheimer P. 1997. *Biochemistry* 36:2425–38
90. Strobel SA, Shetty K. 1997. *Proc. Natl. Acad. Sci. USA* 94:2903–8
91. Ruffner DE, Uhlenbeck OC. 1990. *Nucleic Acids Res.* 18:6025–29
92. Knöll R, Bald R, Fürste JP. 1997. *RNA* 3:132–40
93. Scott WG, Murray JB, Arnold JRP, Stoddard BL, Klug A. 1996. *Science* 274:2065–69
94. Dahm SC, Uhlenbeck OC. 1991. *Biochemistry* 30:9464–69
95. Mitrasinovic O, Epstein LM. 1997. *Nucleic Acids Res.* 25:2189–96
96. Jeoung Y-H, Kumar PKR, Suh Y-A, Taira K, Nishikawa S. 1994. *Nucleic Acids Res.* 22:3722–27
97. De Clercq E, Eckstein F, Sternbach H, Merigan TC. 1970. *Virology* 42:421–28
98. Agrawal S. 1996. *Trends Biotechnol.* 14:376–87
99. Huang JC, Svoboda DL, Reardon JT, Sancar A. 1992. *Proc. Natl. Acad. Sci. USA* 89:3664–68
100. Reardon JT, Thompson LH, Sancar A. 1997. *Nucleic Acids Res.* 25:1015–21
101. Gupta AP, Benkovic PA, Benkovic SJ. 1984. *Nucleic Acids Res.* 12:5897–911
102. Skerra A. 1992. *Nucleic Acids Res.* 20:3551–54
103. Nakamaye K, Eckstein F. 1986. *Nucleic Acids Res.* 14:9679–98
104. Sayers JR, Schmidt W, Eckstein F. 1988. *Nucleic Acids Res.* 16:791–801
105. Olsen DB, Kotzorek G, Eckstein F. 1990. *Biochemistry* 29:9546–51
106. Walker GT, Nadeau JG, Spears PA, Schram JL, Nycz CM, Shank DD. 1994. *Nucleic Acids Res.* 22:2670–77
107. Walter NG, Strunk G. 1994. *Proc. Natl. Acad. Sci. USA* 91:7937–41
108. Musier-Forsyth K, Schimmel P. 1994. *Biochemistry* 33:773–79
109. Fidanza JA, Ozaki H, McLaughlin LW. 1992. *J. Am. Chem. Soc.* 114:5509–17
110. Yang W, Steitz TA. 1995. *Cell* 82:193–207
111. König P, Giraldo R, Chapman L, Rhodes D. 1996. *Cell* 85:125–36
112. Francois J-C, Saison-Behmoaras T, Barbier C, Chassignol M, Thuong NT, Helene C. 1989. *Proc. Natl. Acad. Sci. USA* 86:9702–6
113. Gryaznov SM, Letsinger RL. 1993. *Nucleic Acids Res.* 21:1403–8
114. Chu BCF, Orgel LE. 1990. *DNA Cell Biol.* 9:71–76
115. Kang S-H, Xu X, Heidenreich O, Gryaznov S, Nerenberg M. 1995. *Nucleic Acids Res.* 23:2344–45
116. Gryaznov SM, Letsinger RL. 1993. *J. Am. Chem. Soc.* 115:3808–9
117. Mag M, Schmidt R, Engels JW. 1992. *Tetrahedron Lett.* 33:7319–22
118. Bannwarth W. 1988. *Helv. Chim. Acta* 71:1517–27
119. Mag M, Engels JW. 1989. *Nucleic Acids Res.* 17:5973–88
120. Letsinger RL, Mungall WS. 1970. *J. Org. Chem.* 35:3800–3
121. Chen J-K, Schultz RG, Lloyd DH, Gryaznov SM. 1995. *Nucleic Acids Res.* 23:2661–68
122. Gryaznov S, Skorski T, Cucco C, Nieborowska-Skorska M, Chiu CY, et al. 1996. *Nucleic Acids Res.* 24:1508–14
123. Mag M, Lüking S, Engels JW. 1991. *Nucleic Acids Res.* 19:1437–41
124. Vyle JS, Connolly BA, Kemp D, Cosstick R. 1992. *Biochemistry* 31:3012–18
125. Szczelkun MD, Connolly BA. 1995. *Biochemistry* 34:10724–33
126. Burgin AB, Huizenga BN, Nash HA. 1995. *Nucleic Acids Res.* 23:2973–79
127. Dantzman CL, Kiessling LL. 1996. *J. Am. Chem. Soc.* 118:11715–19
128. Piccirilli JA, Vyle JS, Caruthers MH, Cech TR. 1993. *Nature* 361:85–88
129. Weinstein LB, Jones BCNM, Cosstick R, Cech TR. 1997. *Nature* 388:805–8
130. Kuimelis RG, McLaughlin LW. 1996. *J. Am. Chem. Soc.* 35:5308–17
131. Zhou D-M, Usman N, Wincott FE, Matulic-Adamic J, Orita M, et al. 1996. *J. Am. Chem. Soc.* 118:5862–66

132. Thomson JB, Patel BK, Jimenez V, Eckart K, Eckstein F. 1996. *J. Org. Chem.* 61: 6273–81

133. Beaton G, Dellinger D, Marshall WS, Caruthers MH. 1991. See Ref. 3, pp. 109–35

134. Wiesler WT, Caruthers MH. 1996. *J. Org. Chem.* 61:4272–81

135. Greef CH, Seeberger PH, Caruthers MH, Beaton G, Bankaitisdavis D. 1996. *Tetrahedron Lett.* 37:4451–54

136. Cummins L, Graff D, Beaton G, Marshall WS, Caruthers MH. 1996. *Biochemistry* 35:8734–41

137. Piotto ME, Granger JN, Cho Y, Gorenstein DG. 1990. *J. Am. Chem. Soc.* 112: 8632–34

138. Tonkinson JL, Guvakova M, Khaled Z, Lee J, Yakubov L, et al. 1994. *Antisense Res. Dev.* 4:269–74

139. Vaughn JP, Stekler J, Demirdji S, Mills JK, Caruthers MH, et al. 1996. *Nucleic Acids Res.* 24:4558–64

140. Marshall WS, Caruthers MH. 1993. *Science* 259:1564–70

141. Krieg AM, Matson S, Fisher E. 1996. *Antisense Nucleic Acid Drug Dev.* 6:133–39

142. Pirrung MC, Chidambaram N. 1996. *J. Org. Chem.* 61:1540–42

143. Reynolds MA, Hogrefe RI, Jaeger JA, Schwartz DA, Riley TA, et al. 1996. *Nucleic Acids Res.* 24:4584–91

144. Mujeeb A, Reynolds MA, James TL. 1997. *Biochemistry* 36:2371–79

145. Kean JM, Kipp SA, Miller PS, Kulka M, Aurelian L. 1995. *Biochemistry* 34: 14617–20

146. Delong RK, Miller PS. 1996. *Antisense Nucleic Acid Drug Dev.* 6:273–80

147. Reynolds MA, Arnold LJ, Almazan MT, Beck TA, Hogrefe RI, et al. 1994. *Proc. Natl. Acad. Sci. USA* 91:12433–37

148. Strauss JK, Maher LJ. 1994. *Science* 266:1829–34

149. Strauss JK, Prakash TP, Roberts C, Switzer L, Maher LJ. 1996. *Chem. Biol.* 3:671–78

150. Botfield MC, Weiss MA. 1994. *Biochemistry* 33:2349–55

151. Hausheer FH, Singh UC, Palmer TC, Saxe JD. 1990. *J. Am. Chem. Soc.* 112:9468–74

152. Duckett DR, Murchie AIH, Lilley DMJ. 1990. *EMBO J.* 9:583–90

153. Seela F, Tran-Thi QH, Franzen D. 1982. *Biochemistry* 21:4338–43

154. Seela F, Driller H. 1986. *Nucleic Acids Res.* 14:2319–32

155. Seela F, Kehne A. 1985. *Biochemistry* 24:7556–61

156. Seela F, Grein T. 1992. *Nucleic Acids Res.* 20:2297–2306

157. Smith SA, Rajur SB, McLaughlin LW. 1994. *Nat. Struct. Biol.* 1:18–22

158. Zhang XL, Gottlieb PA. 1993. *Biochemistry* 32:11374–84

159. Duggan LJ, Hill TM, Wu S, Garrison K, Zhang XL, Gottlieb PA. 1995. *J. Biol. Chem.* 270:28049–54

160. Lesser DR, Kurpiewski MR, Jen-Jacobson L. 1990. *Science* 250:776–86

161. Aiken CR, Gumport RI. 1991. *Methods Enzymol.* 208:433–57

162. Murchie AIH, Lilley DMJ. 1994. *EMBO J.* 13:993–1001

163. Kim MG, Zhurkin VB, Jernigan RL, Camerini-Otero RD. 1995. *J. Mol. Biol.* 247:874–89

164. Wieczorek A, Dinter-Gottlieb G, Gottlieb PA. 1994. *Bioorg. Med. Chem. Lett.* 4:987–94

165. Murchie AIH, Lilley DMJ. 1992. *Nucleic Acids Res.* 20:49–53

166. Bevers S, Xiang GB, McLaughlin LW. 1996. *Biochemistry* 35:6483–90

167. Newman PC, Nwosu VU, Williams DM, Cosstick R, Seela F, Connolly BA. 1990. *Biochemistry* 29:9891–901

168. Limauro S, Benseler F, McLaughlin LW. 1994. *Bioorg. Med. Chem. Lett.* 4:2189–92

169. Pütz J, Florentz C, Benseler F, Giege R. 1994. *Struct. Biol.* 1:580–82

170. Adams CJ, Murray JB, Farrow FA, Arnold JRP, Stockley PG. 1995. *Tetrahedron Lett.* 36:5421–24

171. Min CH, Cushing TD, Verdine GL. 1996. *J. Am. Chem. Soc.* 118:6116–20

172. Cal S, Connolly BA. 1997. *J. Biol. Chem.* 272:490–96

173. Tuschl T, Ng MMP, Pieken W, Benseler F, Eckstein F. 1993. *Biochemistry* 32: 11658–68

174. Grasby JA, Mersmann K, Singh M, Gait M. 1995. *Biochemistry* 34:4068–76

175. Liu Q, Green JB, Khodadadi A, Haeberli P, Beigelman L, Pyle AM. 1997. *J. Mol. Biol.* 267:163–71

176. Schweitzer BA, Kool ET. 1994. *J. Org. Chem.* 59:7238–42

177. McKay DB. 1996. *RNA* 2:395–403

178. McLaughlin LW, Wilson M, Ha SB. 1998. *Comp. Nat. Prod. Chem.* 7:In press

179. Murray JB, Adams CJ, Arnold JRP, Stockley PG. 1995. *Biochem. J.* 311:487–94

180. Burgin AB, Gonzalez C, Matulic-Adamic J, Karpeisky AM, Usman N, et al. 1996. *Biochemistry* 35:14090–97

181. Mendelman LV, Kuimelis RG, McLaughlin LW, Richardson CC. 1995. *Biochemistry* 34:10187–93

182. Wagner RW, Matteucci MD, Lewis JG, Gutierrez AJ, Moulds C, Froehler BC. 1993. *Science* 260:1510–13
183. Moulds C, Lewis JG, Froehler BC, Grant D, Huang T, et al. 1995. *Biochemistry* 34: 5044–53
184. Flanagan WM, Kothavale A, Wagner RW. 1996. *Nucleic Acids Res.* 24:2936–41
185. Wagner RW, Matteucci MD, Grant D, Huang T, Froehler BC. 1996. *Nat. Biotechnol.* 14:840–44
186. Sala M, Pezo V, Pochet S, Wain-Hobson S. 1996. *Nucleic Acids Res.* 24:3302–6
187. Loakes D, Brown DM. 1994. *Nucleic Acids Res.* 22:4039–43
188. Loakes D, Brown DM, Linde S, Hill F. 1995. *Nucleic Acids Res.* 23:2361–66
189. Nichols R, Andrews PC, Zhang P, Bergstrom DE. 1994. *Nature* 369:492–93
190. Bergstrom DE, Zhang P, Toma PH, Andrews PC, Nichols R. 1995. *J. Am. Chem. Soc.* 117:1201–9
191. Guo Z, Liu QH, Smith LM. 1997. *Nat. Biotechnol.* 15:331–35
192. Perrachi A, Beigelman L, Usman N, Herschlag D. 1996. *Proc. Natl. Acad. Sci. USA* 93:11522–27
193. Schmidt S, Beigelman L, Karpeisky A, Usman N, Sorensen US, Gait MJ. 1996. *Nucleic Acids Res.* 24:573–81
194. Beigelman L, Karpeisky A, Usman N. 1994. *Bioorg. Med. Chem. Lett.* 4:1715–20
195. Beigelman L, Karpeisky A, Matulic-Adamic J, Gonzalez C, Usman N. 1995. *Nucleosides Nucleotides* 14:907–10
196. Saenger W. 1983. *Principles of Nucleic Acid Structure.* New York: Springer-Verlag
197. Bratty J, Chartrand P, Ferbeyre G, Cedergren R. 1993. *Biochim. Biophys. Acta* 1216:345–59
198. Yang J-H, Usman N, Chartrand P, Cedergren R. 1992. *Biochemistry* 31:5005–9
199. Olsen DB, Benseler F, Aurup H, Pieken WA, Eckstein F. 1991. *Biochemistry* 30: 9735–41
200. Pieken WA, Olsen DB, Benseler F, Aurup H, Eckstein F. 1991. *Science* 253:314–17
201. Paolella G, Sproat BS, Lamond AI. 1992. *EMBO J.* 11:1913–19
202. Jarvis TC, Wincott FE, Alby LJ, Mc-Swiggen JA, Beigelman L, et al. 1996. *J. Biol. Chem.* 29107–12
203. Chowrira BM, Berzal-Herranz A, Keller CF, Burke JM. 1993. *J. Biol. Chem.* 268: 19458–62
204. Pyle AM, Cech TR. 1991. *Nature* 350: 628–31
205. Pyle AM, Murphy FL, Cech TR. 1992. *Nature* 358:123–28

206. Herschlag D, Eckstein F, Cech TR. 1993. *Biochemistry* 32:8299–311
207. Herschlag D, Eckstein F, Cech TR. 1993. *Biochemistry* 32:8312–21
208. Abramovitz DL, Friedman RA, Pyle AM. 1996. *Science* 271:1410–13
209. Hamm ML, Piccirilli JA. 1997. *J. Org. Chem.* 62:3415–20
210. Schärer OD, Kawate T, Gallinari P, Jiricny J, Verdine GL. 1997. *Proc. Natl. Acad. Sci. USA* 94:4878–83
211. Osborne SE, Ellington AD. 1997. *Chem. Rev.* 97:349–70
212. Jellinek D, Green LS, Bell C, Lynott CK, Gill N, et al. 1995. *Biochemistry* 34: 11363–72
213. Lin Y, Nieuwlandt D, Magallanez A, Feistner B, Jayasena SD. 1996. *Nucleic Acids Res.* 24:3407–14
214. Lee S-W, Sullenger BA. 1997. *Nat. Biotechnol.* 15:41–45
215. Pagraitis NC, Bell C, Chang Y-F, Jennings S, Fitzwater T, et al. 1997. *Nat. Biotechnol.* 15:68–73
216. Barabino SM, Sproat BS, Ryder U, Blencowe BJ, Lamond AI. 1989. *EMBO J.* 8:4171–78
217. Blencowe BJ, Sproat BS, Ryder U, Barabino S, Lamond AI. 1989. *Cell* 59:531–39
218. Bardwell VJ, Wickens M, Bienroth S, Keller W, Sproat BS, Lamond AI. 1991. *Cell* 65:125–33
219. Froehler BC, Ricca DJ. 1992. *J. Am. Chem. Soc.* 114:8320–22
220. Bulychev NV, Varaprasad CV, Dorman G, Miller JH, Eisenberg M, et al. 1996. *Biochemistry* 35:13147–56
221. Burlina F, Favre A, Fourrey J-L, Thomas M. 1996. *Chem. Commun.,* pp. 1623–24
222. Hancox EL, Connolly BA, Walker RT. 1993. *Nucleic Acids Res.* 21:3485–91
223. Hess MT, Schwitter U, Petretta M, Giese B, Naegeli H. 1997. *Biochemistry* 36:2332–37
224. Schärer OD, Ortholand J-Y, Ganesan A, Ezaz-Nikpay K, Verdine GL. 1995. *J. Am. Chem. Soc.* 11:6623–24
225. Millar DP. 1996. *Curr. Opin. Struct. Biol.* 6:322–26
226. Ward DC, Reich E, Stryer L. 1969. *J. Biol. Chem.* 244:1228–37
227. Xu D, Evans KO, Nordlund TM. 1994. *Biochemistry* 33:9592–99
228. Guest CR, Hochstrasser RA, Sowers LC, Millar DP. 1991. *Biochemistry* 30:3271–79
229. Hochstrasser RA, Carver TE, Sowers LC, Millar DP. 1994. *Biochemistry* 33:11971–79

230. Bloom LB, Otto MR, Eritja R, Reha-Krantz LJ, Goodman MF, Beechem JM. 1994. *Biochemistry* 33:7576–86
231. Marquez LA, Reha-Krantz LJ. 1996. *J. Biol. Chem.* 271:28903–11
232. Bloom LB, Otto MR, Beechem JM, Goodman MF. 1993. *Biochemistry* 32: 11247–58
233. Frey MW, Sowers LC, Millar DP, Benkovic SJ. 1995. *Biochemistry* 34:9185–92
234. Raney KD, Sowers LC, Millar DP, Benkovic SJ. 1994. *Proc. Natl. Acad. Sci. USA* 91:6644–48
235. Menger M, Tuschl T, Eckstein F, Porschke D. 1996. *Biochemistry* 35:14710–16
236. Hawkins ME, Pfleiderer W, Mazumder A, Pommier YG, Balis FM. 1995. *Nucleic Acids Res.* 23:2872–80
237. Wu PG, Nordlund TM, Gildea B, McLaughlin LW. 1990. *Biochemistry* 29: 6508–14
238. Guest CR, Hochstrasser RA, Dupuy CG, Allen DJ, Benkovic SJ, Millar DP. 1991. *Biochemistry* 30:8759–70
239. Bailey M, Hagmar P, Millar DP, Davidson BE, Tong G, et al. 1995. *Biochemistry* 34:15802–12
240. Allen DJ, Darke PL, Benkovic SJ. 1989. *Biochemistry* 28:4601–7
241. Ozaki H, McLaughlin LW. 1992. *Nucleic Acids Res.* 20:5205–14
242. Fidanza JA, McLaughlin LW. 1992. *J. Org. Chem.* 57:2340–46
243. Lee LG, Connell CR, Woo SL, Cheng RD, McArdle BF, et al. 1992. *Nucleic Acids Res.* 20:2471–83
244. Ju JY, Glazer AN, Mathies RA. 1996. *Nucleic Acids Res.* 24:1144–48
245. Walter NG, Burke JM. 1997. *RNA* 3:392–404
246. Hustedt EJ, Kirchner JJ, Spalstenstein A, Hopkins PB, Robinson BH. 1995. *Biochemistry* 34:4369–75
247. Miller TR, Alley SC, Reese AW, Solomon MS, McCallister WV, et al. 1995. *J. Am. Chem. Soc.* 117:9377–78
248. Thorogood H, Waters TR, Parker AW, Wharton CW, Connolly BA. 1996. *Biochemistry* 35:8723–33
249. Ferentz AE, Keating TA, Verdine GL. 1993. *J. Am. Chem. Soc.* 115:9006–14
250. Ferentz AE, Verdine GL. 1994. *Nucleic Acids Mol. Biol.* 8:14–40
251. Allerson CR, Chen SN, Verdine GL. 1997. *J. Am. Chem. Soc.* 119:7423–33
252. Wolfe SA, Ferentz AE, Grantcharova V, Churchill MEA, Verdine GL. 1995. *Chem. Biol.* 2:213–21
253. Erlanson DA, Chen L, Verdine GL. 1993. *J. Am. Chem. Soc.* 115:12583–84
254. Allerson CR, Verdine GL. 1995. *Chem. Biol.* 2:667–75
255. Osborne SE, Völker J, Stevens SY, Breslauer KJ, Glick GD. 1996. *J. Am. Chem. Soc.* 118:11993–12003
256. Goodwin JT, Osborne SE, Scholle EJ, Glick GD. 1996. *J. Am. Chem. Soc.* 118: 5207–15
257. Osborne SE, Cain RJ, Glick GD. 1997. *J. Am. Chem. Soc.* 119:1171–82
258. Milton J, Connolly BA, Nikiforov TT, Cosstick R. 1993. *J. Chem. Soc. Chem. Commun.*, pp. 779–80
259. Cowart M, Gibson KJ, Allen DJ, Benkovic SJ. 1989. *Biochemistry* 28:1975–83
260. Cowart M, Benkovic SJ. 1991. *Biochemistry* 30:788–96
261. Sigurdsson STh, Tuschl T, Eckstein F. 1995. *RNA* 1:575–83
262. Sigurdsson STh, Eckstein F. 1996. *Nucleic Acids. Res.* 24:3129–33
263. Deleted in proof
264. Cohen SB, Cech TR. 1997. *J. Am. Chem. Soc.* 119:6259–68
265. Behlen LS, Sampson JR, Uhlenbeck OC. 1992. *Nucleic Acids Res.* 20:4055–59
266. Burke JM, Butcher SE, Sargueil B. 1996. *Nucleic Acids Mol. Biol.* 10:129–43
267. Nikiforov TT, Connolly BA. 1992. *Nucleic Acids Res.* 20:1209–14
268. Favre A, Michelson AM, Yaniv M. 1971. *J. Mol. Biol.* 58:367–79
269. Dontsova O, Kopylov A, Brimacombe R. 1991. *EMBO J.* 10:2613–20
270. McGregor A, Rao MV, Duckworth G, Stockley PG, Connolly BA. 1996. *Nucleic Acids Res.* 24:3173–80
271. Sontheimer EJ, Steitz JA. 1993. *Science* 262:1989–96
272. Mishima Y, Steitz JA. 1995. *EMBO J.* 14:2679–87
273. Yu Y-T, Steitz JA. 1997. *Proc. Natl. Acad. Sci. USA* 94:6030–35
274. Favre A, Fourrey J-L. 1995. *Acc. Chem. Res.* 28:375–82
275. Gott JM, Willis MC, Koch TH, Uhlenbeck OC. 1991. *Biochemistry* 30:6290–95
276. Willis MC, Hicke BJ, Uhlenbeck OC, Cech TR, Koch TH. 1993. *Science* 262: 1255–57
277. Hicke BJ, Willis MC, Koch TH, Cech TR. 1994. *Biochemistry* 33:3364–73
278. Langer PR, Waldrop AA, Ward DC. 1981. *Proc. Natl. Acad. Sci. USA* 78:6633–37
279. Catalano CE, Allen DJ, Benkovic SJ. 1990. *Biochemistry* 29:3612–21
280. Capson TL, Benkovic SJ, Nossal NG. 1991. *Cell* 65:249–58

281. Reems JA, Wood S, McHenry CS. 1995. *J. Biol. Chem.* 270:5606–13
282. Sergiev PV, Lavrik IN, Wlasoff VA, Dokudovskaya SS, Dontsova OA, et al. 1997. *RNA* 3:464–75
283. Jayaram B, Haley BE. 1994. *J. Biol. Chem.* 269:3233–42
284. Chavan A, Nemoto Y, Narumiya S, Kozaki S, Haley BE. 1992. *J. Biol. Chem.* 267:14866–70
285. Wower J, Hixson SS, Zimmermann RA. 1989. *Proc. Natl. Acad. Sci. USA* 86:5232–36
286. Evans RK, Haley BE. 1987. *Biochemistry* 26:269–76
287. He BK, Riggs DL, Hanna MM. 1995. *Nucleic Acids Res.* 23:1231–38
288. Harris ME, Nolan JM, Malhotra A, Brown JW, Harvey SC, Pace NR. 1994. *EMBO J.* 13:3953–63
289. Harris ME, Kazantsev AV, Chen J-L, Pace NR. 1997. *RNA* 3:561–76
290. Wang J-F, Downs WD, Cech TR. 1993. *Science* 260:504–8
291. Manchanda R, Dunham SU, Lippard SJ. 1996. *J. Am. Chem. Soc.* 118:5144–45
292. Boudvillain M, Guerin M, Dalbies R, Saison-Behmoaras T, Leng M. 1997. *Biochemistry* 36:2925–31

Annu. Rev. Biochem. 1998. 67:135–52

THE MOLECULAR CONTROL OF CIRCADIAN BEHAVIORAL RHYTHMS AND THEIR ENTRAINMENT IN *DROSOPHILA*

Michael W. Young
National Science Foundation Science and Technology Center for Biological Timing, and Laboratory of Genetics, The Rockefeller University, New York, New York 10021; e-mail:young@rockvax.rockefeller.edu

KEY WORDS: *period, timeless, doubletime*, protein interaction, molecular clocks

ABSTRACT

Molecular and genetic characterizations of circadian rhythms in *Drosophila* indicate that function of an intracellular pacemaker requires the activities of proteins encoded by three genes: *period* (*per*), *timeless* (*tim*), and *doubletime* (*dbt*). RNA from two of these genes, *per* and *tim*, is expressed with a circadian rhythm. Heterodimerization of PER and TIM proteins allows nuclear localization and suppression of further RNA synthesis by a PER/TIM complex. These protein interactions promote cyclical gene expression because heterodimers are observed only at high concentrations of *per* and *tim* RNA, separating intervals of RNA accumulation from times of PER/TIM complex activity. Light resets these molecular cycles by eliminating TIM. The product of *dbt* also regulates accumulation of *per* and *tim* RNA, and it may influence action of the PER/TIM complex. The recent discovery of PER homologues in mice and humans suggests that a related mechanism controls mammalian circadian behavioral rhythms.

CONTENTS

Introduction

Most forms of life show prominent adaptations to daily cycles of light and dark and have developed endogenous, temperature-compensated, circadian (from the Latin "about a day") clocks. Complex behaviors such as the wake/sleep cycle have come under the control of these clocks, and these physiological clocks allow temporal ordering of gene and protein expression throughout the day and night (1–3). Most circadian behavioral and molecular rhythms have become dependent on the activity of these biological clocks: Eliminating clock function by tissue or gene ablation produces behavioral and molecular arrhythmicities that cannot be reversed by provision of environmental cycles (4–7). Thus, cellular pacemakers use environmental cycles to establish the phase of a biological oscillation, which in turn regulates the behavioral and physiological response. Molecular and phylogenetic evidence suggests that intracellular mechanisms for circadian timekeeping arose before the divergence of the prokaryota and eukaryota (8). Mammals, insects, and fungi construct circadian oscillators, at least in part, from related proteins (discussed below), which suggests a common origin for eukaryotic clocks.

Circadian rhythms have certain shared properties, regardless of the organism studied. The rhythms persist with a species-specific period, usually 22–25 h, in constant darkness; this shows their endogenous origin. The phase of the rhythm can be reset (entrained) by pulses of daylight, and the period of the rhythm shows little tendency to vary with changes in temperature. Molecular mechanisms underlying these biological oscillators are being studied in humans (9, 9a), hamsters (10), mice (9, 9a, 11), *Drosophila* (12–14), *Neurospora* (15), *Arabidopsis* (16), and *Cyanobacteria* (17). For each of these systems, genetic screening for clock mutations has been directly or indirectly responsible for most progress. In this review I describe the organization of the *Drosophila* clock, for which there is now a fairly detailed understanding of the molecular origins of intracellular circadian oscillations and their entrainment by light.

Genetic Screens for Clock Mutants

Early screens in *Drosophila* were the first to prove that single gene mutations could affect circadian rhythms and alter fundamental properties of these oscillations such as period length. Konopka & Benzer (18), in a limited screen of the X chromosome, isolated three alleles of a single gene. The gene was named *period*, with allele *per⁰* giving arrhythmicity; *per^L*, long-period rhythms of

28 h; and per^S, short-period 19-h rhythms (18). Only recently have new genes been discovered that are also required for *Drosophila's* circadian rhythms. Mutations of the *timeless* (*tim*) locus, on chromosome 2, have produced arrhythmic, short-period, and long-period alleles (7, 19; A Rothenfluh-Hilfiker & MW Young, unpublished data), and a suppressor of the per^L mutation (20). Mutations of *doubletime* (*dbt*), chromosome 3, have provided arrhythmic, long-period and short-period alleles (J Price, B Kloss & MW Young, unpublished data). The most important feature of this collection of genes is that they all produce proteins that interact to regulate the progression and timing of a single intracellular circadian oscillator (19, 21–25; J Price, B Kloss & MW Young, unpublished data). Thus, *Drosophila* has likely adopted a single biochemical strategy for producing all circadian behavioral and physiological rhythms. I review the properties of each of these genes and their interactions in the following sections.

Molecular Characterization of per

The *per* locus was isolated independently by two groups. Chromosomal rearrangement breakpoints affecting *per* (26, 27) identified the gene in a chromosomal walk (28). Subsequently the gene was recovered by chromosomal microdissection (29). Behavioral rhythms were also restored through the transfer of cloned wild-type DNA to per^0 *Drosophila* (30, 31). Behavioral studies of such transgenic flies, and genetic studies of *per* aneuploids, demonstrated that the period of *Drosophila* circadian rhythms is sensitive to *per* dosage, with lower doses of *per* increasing period length (27, 30, 32, 33).

Sequence analysis of wild-type and mutant alleles showed that per^0 was null and that per^L and per^S were derived by single amino acid substitutions (33, 34). The gene encodes a protein of ~1200 amino acids (35–37) that is found predominantly in cell nuclei (38–40).

per has been cloned from representatives of four insect orders: *Diptera, Lepidoptera, Hymenoptera,* and *Blattodea* (41–43). Sequence comparisons indicate five discrete regions of high homology that are interspersed with sequences that are poorly conserved (41, 42). A high degree of sequence conservation is seen at the PER N-terminus (~75 aa), including a sequence forming the PER nuclear localization signal (NLS) (Figure 1; 19, 25, 42). Two more regions of high homology are referred to as PAS and CLD (together ~300 aa; Figure 1). Both domains mediate protein-protein interactions that are described in detail below. Downstream of CLD is found the short-mutable domain (~70 aa). Amino acid substitutions and deletions in a portion of this interval usually generate short-period circadian rhythms, and the original per^S mutation maps to this conserved interval (33, 34, 44). It has been suggested that this region regulates activity or stability of the protein (44). A fifth region

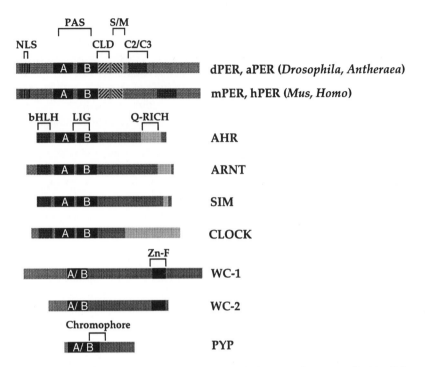

Figure 1 PAS family proteins include both products of clock genes and genes encoding blue-light phototransducers and photoreceptors. PER proteins encoded by *Drosophila*, mouse, and human genes can be related by five common sequence blocks: NLS, PAS, CLD, short-mutable domain (S/M), and C2/C3. NLS, nuclear localization signal; CLD, cytoplasmic localization domain; S/M, short-mutable domain. Arylhydrocarbon receptor (AHR) and AHR nuclear transporter (ARNT) heterodimerize to regulate transcription in response to xenobiotic factors. LIG stands for ligand-binding sequence of AHR. SIM, single minded, is required for aspects of neurectoderm development in *Drosophila*. Clock is required for circadian rhythmicity in the mouse. Q-RICH stands for possible transcriptional activator sequences, and bHLH, predicted DNA binding sequences in AHR, ARNT, SIM, and Clock. Zn-F stands for zinc finger linked to DNA binding of WC1 and WC2. *Neurospora* circadian rhythms and blue-light phototransduction depend on WC-1 and WC-2. PYP (bacterial photoactive yellow protein) is shown with the region of chromophore association.

of conserved sequence, C2/C3 (~130 aa), has not been functionally defined (41, 42). The level of conservation observed in these insect comparisons is sufficient for inter-order function, because a *per* transgene from the moth *A. pernyi* rescues rhythmicity in *per*[0] *Drosophila* (45).

per homologues were recently isolated from mouse and human DNA libraries (9, 9a). Homologies among the *Drosophila*, mouse, and human PER proteins are extensive: They involve all five conserved domains of the protein defined by structural and functional studies in insects (9, 9a; Figure 1). The mouse gene is expressed in the suprachiasmatic nucleus (SCN), which controls behavioral rhythmicity in rodents (cf. 46). As in *Drosophila* (reviewed below), the mouse gene is expressed in clock tissues with a circadian rhythm (9, 9a). Thus, mammalian circadian pacemakers are probably organized after the fashion of the *Drosophila* clock.

PER (insect, mouse, and human) is linked to a family of proteins by inclusion of the ~260-aa protein interaction domain PAS (47; Figure 1). Although PER has not been shown to bind DNA or to contain a DNA binding protein motif, most PAS proteins include an established DNA binding sequence and are proven, or likely, transcription factors (Figure 1). The role of PAS was first revealed by studies of dioxin receptor assembly and activation and by in vitro studies of PAS protein associations. The two subunits of the dioxin receptor, AHR and ARNT, associate at least in part through their PAS domains in response to ligand binding through the PAS element of AHR (48–50). Dimerization of AHR and ARNT permits DNA binding and activation of target genes such as cytochrome P450 (51–53). The domain promotes heterodimerization of different PAS-containing proteins in vitro, which suggests that PER may interact with an unidentified PAS-containing protein (54). Although no PAS-containing partner has been identified, PER's PAS domain is an element of this protein's physical interaction with TIM (21, 25; described below), indicating a role for PAS in heterotypic as well as homotypic protein associations.

Recently PAS has been discovered in circadian clock proteins other than PERIOD in three noninsect species. In *Neurospora*, two proteins *white collar-1* (*wc-1*) and *white collar-2* (*wc-2*), are required for blue-light phototransduction (55–57). Both genes are also required for proper expression of the *Neurospora* clock gene *frequency* (*frq*), and loss of *wc-1*, *wc-2*, or *frq* results in arrhythmicity (55). Cloning of *wc-1* and *wc-2* revealed that both are PAS-containing (Figure 1) zinc-finger proteins that may dimerize to regulate transcription of *frq* and several previously identified light-responsive genes (56, 57). Both proteins bind DNA associated with regulated target genes (56, 57).

In the mouse, a gene referred to as *Clock* is required for circadian rhythmicity. Homozygotes produce long-period rhythms that grade into arrhythmicity following transfer from light/dark cycles to constant darkness (58). Cloning of *Clock* revealed a PAS-domain–containing protein (11, 58a; Figure 1). Like most PAS-family proteins, CLOCK carries a putative DNA binding sequence (bHLH), which suggests function in mouse timekeeping as a transcription factor.

Photoactive yellow protein (PYP) and certain Algal phytochromes have recently been added to the PAS family (Figure 1). Both of these proteins function as photoreceptors in conjunction with an associated chromophore (59). Inclusion of PYP, a phytochrome, wc-1, and wc-2 in the PAS family raises the possibility that PER and CLOCK arose from a group of proteins originally dedicated to blue-light photoreception and phototransduction (55, 59).

Patterns of per RNA and Protein Expression

Studies of genetically mosaic *Drosophila* and of transgenic flies indicate that circadian behavioral rhythms are controlled by 20–30 neurons of the central brain (60–62). Some of the cells, referred to as lateral neurons (LN) (60), express the transcription factor GLASS, which is required for development of all known *Drosophila* photoreceptors (62–65). When the *glass* promoter is used to direct *per* expression to only these central brain cells in transgenic per^0 flies, behavioral rhythms are restored, even if the eyes and ocelli are removed (62). In a light/dark cycle, the phase of the rhythm produced by these eyeless, ocelliless flies is reset, which suggests that *glass*-expressing brain cells can mediate both rhythmicity and entrainment (62).

per RNA and proteins are expressed with a circadian rhythm. Mutations of *per* alter these molecular oscillations in a fashion that corresponds to their effects on behavioral rhythmicity. per^0 mutants eliminate *per* RNA cycling. per^L and per^S respectively lengthen and shorten the period of the RNA and protein cycles (5, 66). These rhythms have been documented in the eye and brain, including the LNs, and in tissues outside the nervous system of the fly (5, 39, 67–71). Oscillations in some of these tissues appear to be autonomously sustained. For example, circadian cycles of *per* RNA expression are observed in Malpighian tubules even when dissected from the carcasses of decapitated flies that have been maintained in culture for several days (70, 71). RNA and protein cycles are produced with a constant-phase relationship in all of these tissues. Highest levels of *per* RNA are observed about two hours after lights off in a 12-h/12-h light/dark cycle, and PER protein is subsequently detected in nuclei with a delay in peak accumulation of 4–6 h. The Malpighian rhythms can also be entrained following decapitation, suggesting involvement of a novel photoreceptor.

In addition to cycles of protein accumulation, cycles of PER phosphorylation have been observed (66). Phosphorylation appears to involve several sites on PER, and progressive, phosphorylation-dependent increases in mobility are seen until PER is degraded near dawn (66, 72).

Cycles of *per* RNA and PER protein expression have been reported in the silk-moth *Antheraea pernyi* (73, 74). In the eyes of the moth, *per* RNA and proteins cycle in a fashion that is indistinguishable from the *Drosophila* brain and eye rhythms (73). Yet in the moth brain, cytoplasmic rather than nuclear cycles of PER protein have been immunocytochemically observed. Biochemical studies of PER protein cycles in the moth are not available, and it is not known whether the immunocytochemically detected cycling is autonomously generated in any of the moth tissues studied. Cells controlling behavioral rhythms also have not been identified in the moth. In some exceptional *Drosophila* tissues (e.g. ovaries), PER proteins accumulate cytoplasmically, as in *Antheraea* (38, 68), but no *per* RNA or protein rhythms are observed in these *Drosophila* tissues (68). As mentioned above, RNA from the mouse *period* gene is expressed with a circadian cycle in an established pacemaker tissue, the SCN (9, 9a). Thus, in both *Drosophila* and mice, cycling *per* expression can be linked to cells known to control circadian behavioral rhythms.

Although studies of *per* alone were unable to uncover a mechanism for generating clocklike cycles of gene expression, several observations suggested that PER proteins somehow inhibit *per* transcription:

1. Constitutively high levels of *per* transcription are found in *per*[0] mutants (75, 76).

2. *per* transcription declines as PER protein accumulates, and it does not rise again until PER proteins decay (66).

3. Constitutive overexpression of a *per* transgene that is limited to the eye blocks cycling of *per* transcription in that tissue but not in the brain (77).

4. Transient suppression of PER function leads to increased accumulation of *per* RNA (14, 23).

Cycles of clock gene expression are also associated with the circadian oscillator of *Neurospora* (6, 15). The clock gene *frequency* (*frq*) is required for the circadian regulation of conidiation (78), and *frq* RNA and proteins cycle with a circadian period (6, 79). Overexpression of FRQ protein suppresses *frq* transcription, and transitory reductions in the FRQ protein shift the phase of the conidiation rhythm and the molecular rhythms of *frq* RNA and protein accumulation (6, 15).

timeless *Promotes Cycles of* per *RNA and Protein Accumulation*

Null mutations of the second chromosome-linked clock gene *timeless* (*tim*[0]) eliminate circadian behavioral rhythms, stop cycling of *per* RNA, and block nuclear accumulation of PER proteins (7, 19). *tim*[0] mutations also suppress accumulation of PER and lead to constitutive hypophosphorylation of residual PER proteins (72). *tim*[L] mutants produce long-period behavioral rhythms and correspondingly lengthen the period of *per* RNA and protein rhythms (A Rothenfluh-Hilfiker & MW Young, unpublished data).

timeless encodes a large (~1400 aa) previously undescribed protein (80) that is well conserved in other *Drosophila* species (80a, 80b). *tim*[01], the most completely characterized arrhythmic mutation, contains an intragenic deletion of 64 bp that blocks TIM protein production (80–82).

tim RNA and proteins cycle with a circadian rhythm (22, 24, 81, 82). The apparent molecular size of the protein also cycles as for PER (24, 82) because of rhythmic TIM phosphorylation (24). The phase of the *tim* RNA rhythm is similar to that of *per* (22), and TIM proteins accumulate in nuclei of lateral neurons and photoreceptors of the eyes and ocelli with kinetics that correspond to those of PER protein (81, 82). tim^0 and per^0 mutations block molecular cycles of *tim* RNA production (22). The per^S mutation shortens the period of *tim* RNA and protein cycles (22), whereas per^L and tim^L mutations lengthen the periods of these cycles (22; A Rothenfluh-Hilfiker & MW Young, unpublished data). Thus, molecular rhythms of *per* and *tim* are interdependent, with mutations at either locus eliciting corresponding changes in the cycles produced by both loci.

tim regulates PER's subcellular localization by encoding a protein that must heterodimerize with PER to permit movement to the nucleus. Coexpression of PER and TIM in cultured *Drosophila* cells (S2 cells) results in physical association and nuclear localization of both proteins, but expression of PER in the absence of TIM leads to cytoplasmic accumulation in S2 cells (25; see also further description below). PER also physically interacts with TIM in yeast (21), in vitro (21, 25), and in cells of the *Drosophila* head (23, 24). Just as TIM is required for PER nuclear localization, PER must be present for nuclear localization of TIM, because TIM accumulates cytoplasmically in per^0 mutants (82) and when expressed without PER in S2 cells (25). Although PER accumulation is suppressed in tim^0 mutants, high levels of TIM accumulate in the cytoplasm of per^0 flies (82). The defect in PER accumulation in tim^0 flies suggests that heterodimerization influences the stability of cytoplasmic PER proteins, and studies of truncated, PER–β-galactosidase fusion proteins have located sequences on PER that are likely to confer TIM-dependent accumulation (19, 76, 83). Because both *tim* and *per* RNAs accumulate with a circadian rhythm, the timing and rate of formation of PER/TIM dimers should be influenced by concentrations of *per* and *tim* RNA, by the affinity of PER for TIM, and by rates of decay of the PER monomer (14, 22).

The specificity of PER's physical interaction with TIM has been demonstrated in yeast. Gekakis et al (21) searched for proteins expressed in the *Drosophila* head that would associate with PER in a yeast two-hybrid assay. Of 20 million transformants, each expressing a PER bait and a prey protein from a *Drosophila* head cDNA library, 48 clones signaled a significant protein-protein interaction. When these clones were subsequently probed with *tim* cDNA, 16 clones, by far the largest subset, were found to carry a *tim* prey (21).

Studies of the PER/TIM interaction in yeast also support the expectation that rates of physical association of the two proteins can set the period of the circadian rhythm. The mutation *per*[L] lengthens the period of molecular and behavioral rhythms to ∼28 h when *Drosophila* are maintained at 25°C, and the period is further increased through the elevation of temperature, with ∼30-h rhythms observed at 29°C (84). By comparing the times of accumulation of PER proteins in lateral neurons of wild-type and *per*[L] *Drosophila*, Curtin et al (85) showed that nuclear localization was delayed by the *per*[L] mutation. The delays varied in *per*[L] so that they corresponded to the changes in period length produced by the mutant at different temperatures. The *per*[L] mutation is due to a single amino acid substitution in PER's PAS domain (33). Because PAS mediates PER's physical interaction with TIM (see below), Gekakis et al (21) compared wild-type and mutant protein interactions at varying temperatures in yeast. Higher temperatures reduced the affinity of TIM for PER[L], suggesting that the temperature-sensitive delays in nuclear localization were due to depressed association of TIM and the PER[L] protein in vivo (21). These observations, and the finding that lowering PER dosage lengthens the period (27, 30, 32, 33), show that rates of PER/TIM association regulate the duration of a part of the circadian cycle.

Physical Association of PER and TIM Regulates Activity of Cytoplasmic Localization Domains

Sequences on PER and TIM responsible for their physical association have been mapped in vitro and by sequencing TIM fragments derived from yeast two-hybrid interactions with PER baits (21, 25; Figure 2). Two binding sites on PER involve PAS and an adjacent sequence referred to as CLD (cytoplasmic localization domain). An N-terminal region referred to as PAS-A interacts with a region of TIM that includes its nuclear localization signal. CLD interacts with a second region of TIM downstream from the TIM NLS (Figure 2; 25).

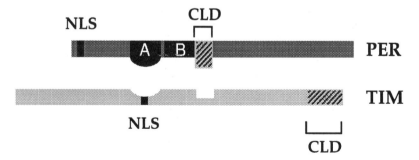

Figure 2 Binding sites for PER/TIM heterodimerization regulate cytoplasmic localization domains (CLDs). TIM binds two regions of PER, one of which, CLD, promotes cytoplasmic localization of PER monomers. A TIM CLD, of unrelated sequence, is also suppressed by formation of the PER/TIM complex.

These patterns of PER/TIM binding are pertinent to studies of PER and TIM sequences regulating subcellular localization (25). As first reported by Vosshall et al (19), PER contains a sequence that promotes cytoplasmic accumulation, and the activity of this sequence element is dominant to the PER NLS in monomeric proteins. Saez & Young (25) identified a comparable cytoplasmic localization domain (CLD) for TIM. The PER and TIM CLDs function in pacemaker cells of the fly head, but their activities have been studied most completely in cultured *Drosophila* cells. In S2 cells, full-length monomeric PER or TIM proteins accumulate in the cytoplasm, while certain truncations permit nuclear localization (25). The PER CLD was mapped to an ~60 amino acid region corresponding to the binding site for TIM that is adjacent to PAS-B (Figure 2). Thus, for PER a TIM binding site produces cytoplasmic localization of monomeric PER proteins. The TIM CLD was similarly mapped to ~160 C-terminal amino acids (25). Because coexpression of full-length PER and TIM proteins in S2 cells induced nuclear accumulation of both proteins, physical association of PER and TIM must inhibit the function of both the PER and TIM CLDs. Thus, the activities of the PER and TIM CLDs confine assembly of PER/TIM complexes to the cytoplasm.

Additional insights into CLD regulation have come from localization studies of reporter proteins carrying the PER and TIM nuclear localization signals coupled to a single CLD provided by either TIM or PER. In both cases localization of the reporter was cytoplasmic in S2 cells (25; L Saez & MW Young, unpublished data). The dominant functions of the PER and TIM CLDs, and their continued function in chimeric proteins, suggest that a cytoplasmic factor or factors may interact with monomeric PER and TIM proteins to inhibit nuclear localization. This regulation of subcellular localization may also influence rates of PER/TIM complex accumulation if monomeric PER and TIM proteins must compete with cytoplasmic factors for CLD binding.

Progression of a Self-Sustaining, Intracellular Oscillator

Sehgal et al (22) proposed a model to explain generation of self-sustaining circadian rhythms through the connected activities of PER and TIM. As indicated in Figure 3, transcription of *per* and *tim* is initiated 3-6 h after subjective dawn (time in constant darkness corresponding to lights on during entrainment) because of absence of nuclear PER and TIM proteins. Although *per* and *tim* RNA levels rise during the subjective day, unstable PER monomers fail to accumulate as a result of insufficient levels of TIM. Near subjective dusk, pools of *per* and *tim* RNA are sufficiently large to promote heterodimerization of PER and TIM. PER/TIM binding suppresses action of CLDs on each protein, allowing nuclear translocation. Nuclear PER/TIM complexes directly or indirectly suppress further *per* and *tim* transcription. As *per* and *tim* RNA levels fall, a threshold for PER/TIM heterodimerization is again reached, eliminating further nuclear

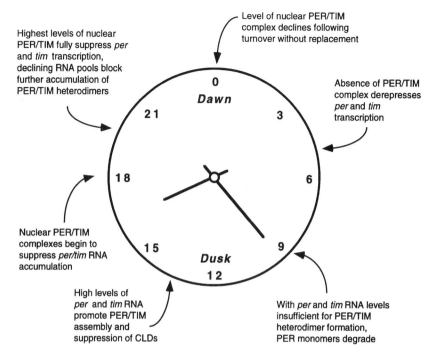

Figure 3 Model of the PER/TIM oscillator. Progression of molecular cycles is predicted for a self-sustained oscillator in constant darkness (after 22).

translocation and leaving a pool of nuclear PER/TIM complex that was amassed in the early subjective evening. Nuclear PER/TIM complexes will continue to suppress transcription until late subjective night, when the heterodimers are seen to turn over without replacement. As indicated above, period length can be significantly altered experimentally by changing the affinity of PER for TIM (21) or by reducing the concentration of PER (27, 30, 32, 33). Thus, self-sustained oscillations are promoted because times of RNA synthesis are separated from times of nuclear protein accumulation and function by an interval of PER/TIM heterodimerization. In the absence of a mechanism generating such temporal separations, autoregulation of *per* and *tim* transcription by PER and TIM proteins should not produce clocklike cycles of accumulation (22).

Light Entrains the Clock by Rapidly Eliminating TIM

Exposure of wild-type flies to constant daylight produces arrhythmicity (84, 86) and suppresses accumulation of the PER protein as in *tim*[01] mutants (67, 72). Because the low levels of PER found in *tim*[01] flies were not further suppressed through the exposure of the mutants to constant light, light's effects on wild-type flies might be mediated by TIM (72).

Several studies have now confirmed that light rapidly lowers the level of TIM protein (24, 81, 82). For example, a 10-min pulse of daylight is sufficient to eliminate immunocytochemical staining of TIM in the eyes and central brain (81, 82). The response of TIM-expressing brain cells (LNs) does not appear to require the eyes, as the protein is eliminated from these cells by exposing eyeless flies to the same light pulses (K Wager-Smith & MW Young, unpublished data). These effects of light on TIM do not require clock function, as TIM is rapidly lost when *per⁰ Drosophila* are exposed to pulses of light (24, 82). Such clock-independent responses to light suggested that TIM's light sensitivity is responsible for entrainment of the phase of the circadian rhythm to the phase of the environmental cycle. TIM's light sensitivity could also explain realignment of the behavioral rhythm to a light/dark cycle with an altered phase, such as would be encountered on crossing time zones.

Drosophila maintained in constant darkness continue to show rhythmic behavior with a phase dictated by their prior experience in a light/dark cycle. However, pulses of daylight will reset the phase of these "free-running" rhythms. The response of a fly to light pulses provided at different times of day is reproducible and can be represented by a phase-response curve (PRC) (Figure 4, *top*). As can be seen for the *Drosophila* PRC, light pulses given between subjective dusk and midnight (CT 12 to CT 18) lead to phase delays in the rhythm whereas advances are produced by comparable pulses between subjective midnight and dawn (CT 18 to CT 24; Figure 4). Thus the response to the same stimulus differs in magnitude and direction, depending on the time of day.

Immunoblot analysis has shown that phase advances and delays also occur in the molecular rhythms of TIM protein in response to resetting light pulses (82). Light pulses delivered in the advance zone of the behavioral cycle cause premature loss of TIM protein. TIM accumulates on the following subjective day with an advanced phase. In contrast, light pulses in the delay zone cause a loss of TIM that is promptly followed by its reaccumulation. TIM subsequently decays, but with a new phase that is established by the secondary accumulation. Myers et al (82) proposed that advances are differentiated from delays by the different *tim* RNA titers found in the early versus late subjective night. Because *tim* and *per* RNA levels are highest during the first part of the subjective night (Figure 4, *bottom*), TIM proteins eliminated by a light pulse at this time are replaced by new translation in the current cycle. This would not be the case for light-induced loss of TIM near the end of the subjective night, as *tim* and *per* RNA levels are low. Thus, light late at night causes TIM to decay prematurely, and it advances the following day's cycle of *per* and *tim* RNA and protein accumulation (Figure 4; 82).

Three observations suggest that TIM protein changes mediate entrainment of behavioral rhythmicity through the above mechanism.

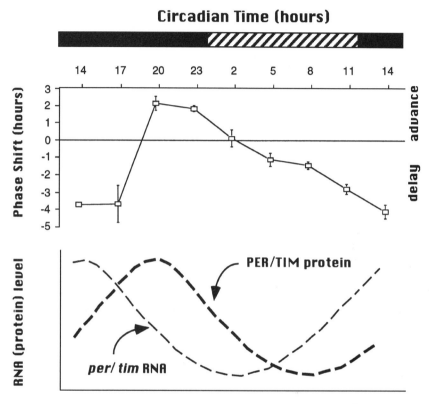

Figure 4   Time-specific levels of *tim* and *per* RNA may determine whether light pulses advance or delay the phase of the behavioral rhythm. *Top*: phase response curve (PRC) for 10-min pulses of daylight delivered at different times of subjective day or night. Pulses in early subjective night (CT 12 to CT 18) delay the behavioral rhythm by up to ~4 h. Comparable light pulses in late subjective night (CT 18 to CT 24) can advance the rhythm by ~2 h (after 82). *Bottom*: PER and TIM protein accumulation follows *per* and *tim* RNA accumulation by several hours in constant darkness. Behavioral delays occur when RNA levels are high at the time of the light pulse. TIM that is degraded by light is replaced by continued translation, further delaying the decline in TIM protein levels. Behavioral advances occur for light pulses given when RNA levels are low. Replacement of TIM requires new transcription. Premature loss of TIM without replacement in the current molecular cycle allows advanced *per* and *tim* RNA accumulation in the following cycle.

1. TIM levels drop when arrhythmic per^0 flies are exposed to light but tim RNA levels are unaffected (22, 24, 82). In rhythmic flies (per^+ or per^S), light-dependent loss of TIM is followed by increased transcription of per and tim (14, 23, 87). Thus, acute effects of light are on TIM protein and not tim RNA.

2. As indicated above, TIM's light sensitivity is clock-independent because it occurs in per^0 flies. Therefore changes in the accumulation and diminution profiles of the protein should dictate changes in the behavioral rhythm and not vice versa.

3. The new phase of the TIM protein rhythm is apparent within hours of the administered light pulse and thus accompanies the first evidence of a behavioral change in the flies (82, 88).

doubletime, *a Third Essential Component of the* Drosophila *Oscillator*

Further genetic screening might be expected to identify additional components of the fly clock. These could include genes controlling TIM's light sensitivity, PER's instability in the absence of TIM, and functions controlling PER and TIM phosphorylation. There is also evidence for posttranscriptional regulation of per and tim RNA accumulation. For example, per RNA coding sequences promote cycling of per mRNA in the absence of the per promoter (61). They also promote cycling when fused to a novel, constitutively active promoter (62). Sequences within the transcribed region of per delay the timing of per RNA accumulation. These findings suggest effects on rates of per transcription or RNA stability (76).

Another gap in our understanding involves the biochemical activity of the PER/TIM complex. Accumulation of the complex results in suppression of per and tim transcription, but there is no evidence that PER, TIM, or the complex directly interacts with DNA or a mediating transcription factor or factors. There is some evidence that effects on transcription are indirect: Phase-advancing or -delaying light pulses that transiently eliminate TIM give rise to brief increases in per and tim RNA pools that appear 4–5 h after administration of the pulse (14, 23).

A third clock gene, *doubletime* (*dbt*), encodes a factor required for behavioral rhythmicity and progression of the PER/TIM cycle. Three classes of mutant alleles have been recovered at *dbt*. The mutation originally defining the locus, dbt^S, produces flies with short-period, 18-h rhythms when homozygous (J Price & MW Young, unpublished data). Thus, dbt^S produces a more extreme period-shortening phenotype than per^S. Gene dosage studies suggest that increasing the level of wild-type *dbt* function shortens the period (J Price, B Kloss & MW Young, unpublished data). A long-period mutant has also been isolated, dbt^L (∼27.5 hours) (A Rothenfluh-Hilfiker & MW Young, unpublished data). Both dbt^S and dbt^L are semidominant. $dbt^S/+$ heterozygotes produce ∼21-h rhythms, whereas $dbt^L/+$ flies have ∼25-h periods (J Price, A Rothenfluh-Hilfiker & MW Young, unpublished data). The third class of *dbt* mutations, dbt^0, are null. However, unlike per^0 and tim^0 mutations, dbt^0 alleles confer recessive lethality, establishing a double role for the gene. Although adult behavior cannot be assessed, molecular studies of developing dbt^0 *Drosophila* indicate that they are arrhythmic (B Kloss, J Blau & MW Young, unpublished data).

dbt mutations affect *per* and *tim* RNA cycling. For example, dbt^S shortens rhythms of *tim* RNA expression to 18 h (B Kloss and MW Young, unpublished data). The effects of dbt^0 mutations are of the most interest, as these mutations block production of *dbt* RNA and *tim* RNA. In addition to these effects on RNA levels, PER proteins accumulate abnormally in dbt^0 *Drosophila* (B Kloss, J Blau, J Price & MW Young, unpublished data).

The *dbt* locus has been cloned (B Kloss, J Price & MW Young, unpublished data), and *dbt* RNA localization studies in situ have shown that *per*, *tim*, and *dbt* are expressed in the same cells in the *Drosophila* head (J Blau & MW Young, unpublished data). Therefore *dbt*, *per*, and *tim* appear to represent three essential components of a single intracellular oscillator.

How Does Cycling of an Intracellular PER/TIM Oscillator Regulate Behavior?

Tissue transplantation studies in silkmoths (89, 90) and *Drosophila* (91) have shown that cells composing the clock for the moth and fly can control rhythmic behavior through action of a diffusible substance. Handler & Konopka (91) transplanted brains from dissected heads of per^S adults into abdomens of adult per^0 recipients. Recipient flies showed transformed locomotor activity rhythms with periods expected for the donor strain (19 h). per^0 recipients that received brain transplants from per^0 donors continued to show behavioral arrhythmicity (91).

A diffusible signal or signals similarly couple cells of the hamster suprachiasmatic nucleus to control of circadian locomotor rhythms. Silver et al (46) transplanted encapsulated SCN tissues from wild-type donors into SCN-ablated, *tau*/+ recipient hamsters. The *tau* mutation shortens the locomotor activity rhythm so that *tau*/+ hamsters produce ∼22- rather than 24-h rhythms (10). Transplant recipients were recovered that displayed donor (24-h) rather than recipient rhythms. Because the transplanted cells were encased in semipermeable membranes, it was concluded that cell contact was not required for establishment of the behavioral rhythm (46). The suprachiasmatic nucleus is composed of cells that autonomously produce circadian oscillations in culture (92), which suggests that diffusible signals controlling behavior may be products of SCN pacemaker cells themselves.

These and other factors that must integrate cellular responses to circadian molecular oscillators are likely to be products of clock-controlled genes (CCGs) (2). Clock-controlled genes produce transcripts that accumulate with a circadian rhythm. But these genes can be distinguished as targets of the circadian pacemaker rather than cycling clock components because CCG loss fails to block pacemaker cycling. Some of the earliest descriptions of CCGs emerged from studies of plants (cf. 1, 93) and fungi (2), and known or likely CCGs are now well characterized in *Drosophila* (3, 94), *Cyanobacteria* (17), and cold- and warm-blooded vertebrates (cf. 95–100).

CCGs that cycle with stable phase differences can be identified in the same organism (cf. 2, 3, 94). This regulation of CCGs indicates that a cell can evaluate progression of the molecular oscillator over much of its cycle. In *Drosophila* this suggests hierarchical control by the PER/TIM oscillator. It is as yet unknown how the pattern of expression of any CCG is determined, but possibly the same regulatory activity of the PER/TIM complex that mediates cycling transcription of *per* and *tim* will be found to time expression of *Drosophila*'s CCGs.

Visit the *Annual Reviews* home page at
http://www.AnnualReviews.org.

Literature Cited

1. Nagy F, Kay SA, Chua N-H. 1988. *Genes Dev.* 2:376–82
2. Loros JJ, Denome SA, Dunlap JC. 1989. *Science* 243:385–88
3. Van Gelder R, Krasnow M. 1996. *EMBO J.* 15:1625–31
4. Takahashi JS, Zatz M. 1982. *Science* 217: 1104–11
5. Hardin PE, Hall JC, Rosbash M. 1990. *Nature* 343:536–40
6. Aronson BD, Johnson KA, Loros JJ, Dunlap JC. 1994. *Science* 263:1578–84
7. Sehgal A, Price JL, Man B, Young MW. 1994. *Science* 263:1603–6
8. Young MW. 1998. In *Handbook of Behavioral Neurobiology*. New York: Plenum. In press
9. Tei H, Okamura H, Shigeyoshi Y, Fukuhara C, Ozawa R, et al. 1997. *Nature* 389:512–16
9a. Sun ZS, Albrecht U, Zuchenko O, Bailey J, Eichele G, Lee CC. 1997. *Cell* 90:1003–11
10. Ralph MR, Menaker M. 1988. *Science* 241:1225–27
11. King DP, Zhao YL, Sangoram AM, Wilsbacher LD, Tanaka M, et al. 1997. *Cell* 89:641–53
12. Rosbash M, Allada R, Dembinska M, Guo WQ, Le M, et al. 1996. *Cold Spring Harbor Symp. Quant. Biol.* 61:265–78
13. Sehgal A, Ousley A, Hunter-Ensor M. 1996. *Mol. Cell. Neurosci.* 7:165–72
14. Young MW, Wager-Smith K, Vosshall LB, Saez L, Myers MP. 1996. *Cold Spring Harbor Symp. Quant. Biol.* 61:279–84
15. Dunlap JC. 1996. *Annu. Rev. Genet.* 30:579–601
16. Millar AJ, Carre IA, Strayer CA, Chua N-H, Kay S. 1995. *Science* 267:1161–63
17. Liu Y, Tsinoremas NF, Johnson CH, Lebedeva NV, Golden SS, et al. 1995. *Genes Dev.* 9:1469–78
18. Konopka RJ, Benzer S. 1971. *Proc. Natl. Acad. Sci. USA* 68:2112–16
19. Vosshall LB, Price JL, Sehgal A, Saez L, Young MW. 1994. *Science* 263:1606–9

20. Rutila JE, Zeng H, Le M, Curtin KD, Hall JC, Rosbash M. 1996. *Neuron* 17:921–29
21. Gekakis N, Saez L, Delahaye-Brown A-M, Myers MP, Sehgal A, et al. 1995. *Science* 270:811–15
22. Sehgal A, Rothenfluh-Hilfiker A, Hunter-Ensor M, Chen YF, Myers MP, Young MW. 1995. *Science* 270:808–10
23. Lee CG, Parikh V, Itsukaichi T, Bae K, Edery I. 1996. *Science* 271:1740–44
24. Zeng H, Qian Z, Myers MP, Rosbash M. 1996. *Nature* 380:129–35
25. Saez L, Young MW. 1996. *Neuron* 17:911–20
26. Young MW, Judd B. 1978. *Genetics* 88:723–42
27. Smith RF, Konopka RJ. 1982. *Mol. Gen. Genet.* 185:30–36
28. Bargiello TA, Young MW. 1984. *Proc. Natl. Acad. Sci. USA* 81:2142–46
29. Reddy P, Zehring WA, Wheeler DA, Pirrota V, Hadfield C, et al. 1984. *Cell* 38:701–10
30. Bargiello TA, Jackson FR, Young MW. 1984. *Nature* 312:752–54
31. Zehring WA, Wheeler DA, Reddy P, Konopka RJ, Kyriacou CP, et al. 1984. *Cell* 39:369–76
32. Cote GC, Brody S. 1986. *J. Theor. Biol.* 121:487–503
33. Baylies MK, Bargiello TA, Jackson FR, Young MW. 1987. *Nature* 326:390–92
34. Yu Q, Jacquier AC, Colot HV, Citri Y, Hamblen M, et al. 1987. *Proc. Natl. Acad. Sci. USA* 84:784–88
35. Jackson FR, Bargiello TA, Yun S-H, Young MW. 1986. *Nature* 320:185–88
36. Citri YV, Colot HV, Jacquier AC, Yu Q, Hall JC, et al. 1987. *Nature* 326:42–47
37. Baylies MK, Weiner L, Vosshall LB, Saez L, Young MW. 1993. In *Molecular Genetics of Biological Rhythms*, pp. 123–53. New York: Marcel Dekker
38. Saez L, Young MW. 1988. *Mol. Cell. Biol.* 8:5378–85
39. Siwicki KK, Eastman C, Petersen G, Rosbash M, Hall JC. 1988. *Neuron* 1:141–50
40. Liu X, Zweibel LJ, Hinton D, Benzer S, Hall JC, Rosbash M. 1992. *J. Neurosci.* 12:2735–44
41. Colot HV, Hall JC, Rosbash M. 1988. *EMBO J.* 7:3929–37
42. Reppert SM, Tsai T, Roca AL, Sauman I. 1994. *Neuron* 13:1167–76
43. Toma DP, Robinson GE. 1996. Presented at Meet. Soc. Res. Biol. Rhythms, 5th, Jacksonville, Fla.
44. Baylies MK, Sehgal A, Young MW. 1992. *Neuron* 9:575–81
45. Levine JD, Sauman I, Imbalzano M, Reppert SM, Jackson FR. 1995. *Neuron* 15:147–57
46. Silver R, LeSauter J, Tresco PA, Lehman MN. 1996. *Nature* 382:810–13
47. Crews ST, Thomas JB, Goodman CS. 1988. *Cell* 52:143–51
48. Hoffman EC, Reyes J, Chu FF, Sander F, Conley LH, et al. 1991. *Science* 252:954–58
49. Burbach KM, Poland A, Bradfield CA. 1992. *Proc. Natl. Acad. Sci. USA* 89:8185–89
50. Reyes H, Reisz-Porszasa S, Hankinson O. 1992. *Science* 256:1193–95
51. Gonzalez FJ, Tukey RH, Nebert DW. 1984. *Mol. Pharmacol.* 26:117–21
52. Denison MS, Fisher JM, Whitlock JP Jr. 1988. *Proc. Natl. Acad. Sci. USA* 85:2528–32
53. Hapgood J, Cuthill S, Denis M, Poellinger L, Gustafsson JA. 1989. *Proc. Natl. Acad. Sci. USA* 86:60–64
54. Huang ZJ, Edery I, Rosbash M. 1993. *Nature* 364:259–62
55. Crosthwaite SK, Dunlap JC, Loros JJ. 1997. *Science* 276:763–69
56. Ballario P, Vittorioso P, Magrelli A, Talora C, Cabibbo A, Macino G. 1996. *EMBO J.* 15:1650–57
57. Linden H, Macino G. 1997. *EMBO J.* 16:98–109
58. Vitaterna MH, King DP, Chang A-M, Kornhauser JM, Lowrey PL, et al. 1994. *Science* 264:719–25
58a. Antoch MP, Song EJ, Chang A-M, Vitaterna MH, Zhao YL, et al. 1997. *Cell* 89:655–010167
59. Kay SA. 1997. *Science* 276:753–54
60. Ewer J, Frisch B, Hamblen-Coyle MJ, Rosbash M, Hall JC. 1992. *J. Neurosci.* 12:3321–49
61. Frisch B, Hardin PE, Hamblen-Coyle MJ, Rosbash M, Hall JC. 1994. *Neuron* 12:555–70
62. Vosshall LB, Young MW. 1995. *Neuron* 15:345–60
63. Moses K, Ellis MC, Rubin GM. 1989. *Nature* 340:531–36
64. Moses K, Rubin GM. 1991. *Genes Dev.* 5:583–93
65. Ellis MC, O'Neill EM, Rubin GM. 1993. *Development* 119:855–65
66. Edery I, Zweibel LJ, Dembinska ME, Rosbash M. 1994. *Proc. Natl. Acad. Sci. USA* 91:2260–64
67. Zerr DM, Hall JC, Rosbash M, Siwicki KK. 1990. *J. Neurosci.* 10:2749–62
68. Hardin PE. 1994. *Mol. Cell. Biol.* 14:7211–18

69. Emery IF, Noveral JM, Jamison CF, Siwicki KK. 1997. *Proc. Natl. Acad. Sci. USA* 94:4092–96
70. Giebultowicz JM, Hege D. 1997. *Nature* 386:664
71. Hege DM, Stanewsky R, Hall JC, Giebultowicz JM. 1997. *J. Biol. Rhythms* 12: 300–8
72. Price JL, Dembinska ME, Young MW, Rosbash M. 1995. *EMBO J.* 14:4044–47
73. Sauman I, Reppert SM. 1996. *Neuron* 17:889–900
74. Sauman I, Tsai T, Roca AL, Reppert SM. 1996. *Neuron* 17:901–9
75. Brandes C, Plautz JD, Stanewsky R, Jamison CF, Straume M, et al. 1996. *Neuron* 16:687–92
76. Stanewsky R, Jamison CF, Plautz JD, Kay SA, Hall JC. 1997. *EMBO J.* 16:5006–18
77. Zeng H, Hardin PE, Rosbash M. 1994. *EMBO J.* 13:3590–98
78. Feldman JF, Hoyle MN. 1973. *Genetics* 75:605–13
79. Garceau NY, Liu Y, Loros JJ, Dunlap JC. 1997. *Cell* 89:469–76
80. Myers MP, Wager-Smith K, Wesley CS, Young MW, Sehgal A. 1995. *Science* 270:805–8
80a. Myers MP, Rothenfluh-Hilfiker A, Chang M, Young MW. 1997. *Nucleic Acids Res.* 25:4710–14
80b. Ousley A, Zafarullah K, Chen Y, Emerson M, Hickman L, Sehgal A. 1998. *Genetics.* In press
81. Hunter-Ensor M, Ousley A, Sehgal A. 1996. *Cell* 84:677–85
82. Myers MP, Wager-Smith K, Rothenfluh-Hilfiker A, Young MW. 1996. *Science* 271:1736–40
83. Dembinska ME, Stanewsky R, Hall JC, Rosbash M. 1997. *J. Biol. Rhythms* 12: 157–72
84. Konopka RJ, Pittendrigh CS, Orr D. 1989. *J. Neurogen.* 6:1–10
85. Curtin KD, Huang ZJ, Rosbash M. 1995. *Neuron* 14:365–72
86. Pittendrigh CS. 1966. *Z. Pflanzenphysiol.* 54:275–307
87. Marrus SB, Zeng H, Rosbash M. 1996. *EMBO J.* 15:6877–86
88. Pittendrigh CS. 1981. In *Handbook of Behavioral Neurobiology*, pp. 95–124. New York: Plenum
89. Truman JW. 1972. *J. Comp. Physiol.* 81: 99–114
90. Truman JW. 1974. *J. Comp. Physiol.* 95: 281–96
91. Handler AM, Konopka RJ. 1979. *Nature* 279:236–38
92. Welsh DK, Logothetis DE, Meister M, Reppert, SM. 1995. *Neuron* 14:697–706
93. Giuliano G, Hoffman NE, Ko K, Scolnik PA, Cashmore AR. 1988. *EMBO J.* 7:3635–42
94. Van Gelder R, Bae H, Palazzolo M, Krasnow M. 1995. *Curr. Biol.* 5:1424–36
95. Korenbrot JI, Fernald RD. 1989. *Nature* 337:454–57
96. Pierce ME, Sheshberadaran H, Zhang Z, Fox LE, Applebury ML, Takahashi JS. 1993. *Neuron* 10:579–84
97. Borjigin J, Wang MM, Snyder SH. 1995. *Nature* 378:783–85
98. Coon SL, Roseboom PH, Baler R, Weller JL, Namboodiri MAA, et al. 1995. *Science* 270:1681–83
99. Foulkes NS, Duval G, Sassone-Corsi P. 1996. *Nature* 381:83–85
100. Green CB, Besharse JC. 1996. *Proc. Natl. Acad. Sci. USA* 93:14884–88

Annu. Rev. Biochem. 1998. 67:153–80

RIBONUCLEASE P: Unity and Diversity in a tRNA Processing Ribozyme

Daniel N. Frank and Norman R. Pace
Departments of Plant and Microbial Biology and Molecular and Cellular Biology,
University of California, Berkeley, California 94720-3102;
e-mail: dfrank@nature.berkeley.edu, nrpace@nature.berkeley.edu

KEY WORDS: RNA enzyme, RNA structure, metalloenzyme, phylogenetic-comparative analysis

ABSTRACT

Ribonuclease P (RNase P) is the endoribonuclease that generates the mature $5'$-ends of tRNA by removal of the $5'$-leader elements of precursor-tRNAs. This enzyme has been characterized from representatives of all three domains of life (Archaea, Bacteria, and Eucarya) (1) as well as from mitochondria and chloroplasts. The cellular and mitochondrial RNase Ps are ribonucleoproteins, whereas the most extensively studied chloroplast RNase P (from spinach) is composed solely of protein. Remarkably, the RNA subunit of bacterial RNase P is catalytically active in vitro in the absence of the protein subunit (2). Although RNA-only activity has not been demonstrated for the archaeal, eucaryal, or mitochondrial RNAs, comparative sequence analysis has established that these RNAs are homologous (of common ancestry) to bacterial RNA. RNase P holoenzymes vary greatly in organizational complexity across the phylogenetic domains, primarily because of differences in the RNase P protein subunits: Mitochondrial, archaeal, and eucaryal holoenzymes contain larger, and perhaps more numerous, protein subunits than do the bacterial holoenzymes. However, that the nonbacterial RNase P RNAs retain significant structural similarity to their catalytically active bacterial counterparts indicates that the RNA remains the catalytic center of the enzyme.

CONTENTS

153

0066-4154/98/0701-0153$08.00

INTRODUCTION: PROPERTIES OF RNASE P

Despite their relatively small sizes, transfer-RNAs are subjected to a complex set of posttranscriptional modifications during their biosynthesis (3, 4). Almost every species of tRNA is transcribed as a precursor with both 5' and 3' terminal extensions. tRNA genes also often contain introns. These extraneous elements must all be accurately excised to produce functional tRNAs. Maturation of tRNA also often entails the addition of the 3' terminal CCA sequence, which is required for aminoacylation (some bacterial tRNA genes, such as those of *Escherichia coli*, encode CCA). Finally, immature tRNA transcripts undergo extensive base modification; tRNAs are the most highly modified class of functional RNAs. Thus, the synthesis of mature tRNA requires the interplay of a complex suite of enzymatic factors. In this review we focus on the characterization of one participant in the tRNA processing pathway: ribonuclease P (RNase P), the enzyme that forms the mature 5' end of tRNA (Figure 1). The defining characteristic of RNase P (RNase P; E.C. 3.1.26.5) is specific cleavage of precursor-tRNA (pre-tRNA) molecules by phosphodiester hydrolysis. Cleavage is endonucleolytic, and in all cases examined, the products of pre-tRNA cleavage retain 3' OH and 5' PO_4 groups. RNase P activity absolutely requires divalent metal ions (preferably Mg^{2+}), which probably fill both structural and functional roles in the catalytic cycle (5, 6).

RNase P activities have been identified, and in several cases highly purified, from organisms representing all three domains of life (Archaea, Bacteria, and Eucarya) (1) as well as from mitochondria and chloroplasts. Where possible, genetic analyses of RNase P have demonstrated an essential cellular function of the enzyme (7–10). All cells and organelles in which pre-tRNAs are synthesized are expected to possess RNase P activity.

Cellular RNase Ps are unique: They are complex holoenzymes composed of essential RNA and protein subunits. Despite the many shared characteristics

A

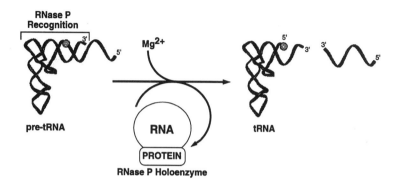

B

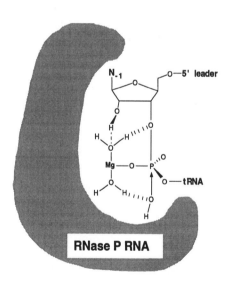

Figure 1 RNase P cleaves precursor tRNAs (pre-tRNAs) to generate the mature 5'-ends of tRNA. *A.* The scissile bond of pre-tRNA is indicated by a *sphere*. All known examples of RNase P, except those of the chloroplast, are composed of protein and RNA subunits; in the bacteria, the RNA subunit alone is catalytically active. RNase P recognizes its substrates primarily through interactions with the coaxially stacked acceptor-stem/T-hairpin domain of mature tRNA. *B.* Schematic diagram of possible transition-state structure of RNase P. Evidence for the direct coordination of a Mg^{2+} ion to the *pro*-Rp oxygen of the scissile bond has been provided by References 84 and 90.

of RNase Ps, the structural and functional organization of diverse species of RNase P differs significantly across phylogenetic domains. In the case of bacterial RNase P, the RNA component of the complex is catalytically active in the absence of the protein subunit; it is a ribozyme, or RNA-enzyme (2). Ribozyme activity has not yet been demonstrated in nonbacterial RNase P RNAs, although the RNAs of these organisms are essential for catalytic activity (Table 1). The archaeal and eucaryal holoenzymes are enriched in protein, relative to the bacterial enzyme, which suggests that the protein subunit(s) of these holoenzymes assume more complex functions than do their bacterial counterparts. The bacterial RNase P thus would seem to be of an entirely different character than the other species of RNase P. Indeed, the diversity of holoenzyme structures that exists between phylogenetic domains has begged the question of whether the known RNase Ps truly are all homologous (of common ancestry). Recent comparative analyses of the RNA subunits of RNase P have established, however, that all known RNase P RNAs contain a similar core structure, which in bacteria is the seat of catalytic activity. It is thus likely that nonbacterial versions of RNase P, even though inactive without protein, also are fundamentally RNA-based catalysts. Understanding how bacterial RNase P RNAs function therefore sheds light on the functions of nonbacterial RNase P RNAs.

In contrast to the RNA subunits, the protein components of RNase P vary greatly in structure across phylogenetic boundaries. In this review, we address both the unity of RNA structure and diversity of holoenzyme composition that exists in cellular and organellar RNase Ps. The capacity to engineer, synthesize, and assay ribozymes in vitro has led to an explosion of information regarding the structure and function of bacterial RNase P RNAs, particularly those of *E. coli, Bacillus subtilis*, and *Thermus thermophilus*. Much of this work has been reviewed extensively (11–14). Our understanding of archaeal, eucaryal, and organellar enzymes is less developed, but significant progress has been made. Information about the bacterial RNase P therefore will be used as a comparative framework for interpreting the archaeal, eucaryal, and organellar RNase P systems.

RNASE P RNA STRUCTURE

Obtaining Sequences

Much of the structural information concerning RNase P RNA has come from comparing the sequences from different organisms. Genes encoding the RNA subunits of RNase P have been identified by two basic approaches: (*a*) genetic screens for factors required for pre-tRNA processing (8, 15, 16) and (*b*) hybridization of RNAs extracted from highly purified RNase P to genomic DNA

Table 1 Diverse types of ribonuclease P holoenzymes

Organism/Plastid	Buoyant density	Micrococcal nuclease sensitive	Protein subunits	RNA subunits	Active ribozyme	Additional substrates	References
Bacteria							
Escherichia coli	1.71 g/ml[a]	+	14 kDa (*rnaA*)	377 nt (*rnpB*)	+	4.5S, 10Sa RNAs	2, 7, 8, 16, 20, 116, 118, 153, 154
Bacillus subtilis	1.7[a]	+	14 kDa	401	+		2, 19, 117, 119
Archaea							
Sulfolobus acidocaldarius	1.27[b]	−	+	315 nt	−	None known	17, 125
Haloferax volcanii	1.61[b]	+	+	435 nt	−	None known	33, 124
Eukaryotes							
Schizosaccharomyces pombe	1.40[b]	+	100 kDa	285	−	None known	21, 100, 129, 155
Saccharomyces cerevisiae	n.d.	+	100 kDa (POP1p) 23 kDa (POP3p) 31 kDa (POP4p)	369 (*RPR1*)	−	None known	9, 22, 130–132
Xenopus laevis	1.34[b]	+	+	320	−	None known	24
Homo sapiens	1.28[b]	+	115 kDa (hPOP1p) 38 kDa (Rpp38) 30 kDa (Rpp30)	340	−	None known	18, 128 133–136, 156
Rattus rattus	1.36[b]	−	+	+	−	None known	126
Dictyostelium discoidium	1.23[b]	+	+	+	−	None known	127
Mitochondria							
S. cerevisiae	1.28[b]	+	105 kDa (RPM2)	490 (RPM1)	−	None known	15, 42, 138, 140
H. sapiens (HeLa)	n.d.	+	+	+	−	None known	157
Chloroplast							
Spinach	1.28[a]	−	+	−	n.a.	None known	38, 39

[a]Measured in CsCl.
[b]Measured in Cs_2SO_4.
n.d. not determined.
n.a. not applicable.

libraries (17–24). Once identified, RNase P RNA sequences can be used to isolate homologous RNAs from other organisms by either heterologous probing under conditions for low-stringency hybridization (25, 26), inspection of genomic sequence databases (27–29), or polymerase chain reaction (PCR) amplification (30, 31). RNase P RNA sequences have been obtained from organisms and organelles spanning most of phylogeny. These sequences are compiled in the RNase P database maintained by J Brown (North Carolina State University; URL: http://www.mbio.ncsu.edu/RNaseP/home.html) (32). More than 130 bacterial RNase P RNA sequences, representing many of the major phylogenetic groups of bacteria, have been reported. Full or partial genes for RNase P RNAs also have been isolated from many archaeal species, including representatives of the two main phylogenetic branches of Archaea: Crenarchaeota and Euryarchaeota (e.g. 17, 33, 34). In contrast, all known eucaryal RNase P RNA sequences are from either yeasts (e.g. 21, 26) or vertebrates (e.g. 24, 31); only a limited degree of eucaryal phylogenetic diversity is represented by these sequences. Whereas all mitochondrial RNase P holoenzymes that have been examined contain essential RNA subunits (Table 1), only a few of the corresponding sequences have been reported, primarily from fungi (e.g. 15, 29, 35, 36). Only one RNase P RNA gene has been identified in a photosynthetic organelle: the cyanelle of the protist *Cyanophora paradoxa* (37). No RNase P RNAs have been isolated from the chloroplasts of higher plants, and, intriguingly, at least one chloroplast RNase P probably functions without an RNA subunit (see below; 38, 39).

Identification of a putative RNase P RNA must be verified by some means. Because the RNA subunits of bacterial RNase P are catalytically active in the absence of protein, verification that a novel clone encodes an RNase P RNA can be accomplished simply by assaying in vitro–synthesized transcripts of the gene for pre-tRNA cleavage activity (19, 40). In the absence of ribozyme activity, as is the case with all nonbacterial RNase P RNAs, conclusive evidence that a given RNA is a bona fide subunit of RNase P can be more difficult to obtain. While copurification of an RNA with RNase P activity may be encouraging, it does not constitute definitive proof that the RNA is a component of the enzyme. In only a few instances [i.e. the nuclear and mitochondrial RNase P RNA genes of *Saccharomyces cerevisiae* (9, 15, 22, 41–45)] is there ironclad experimental evidence (usually a combination of genetic and biochemical data) linking a putative RNase P RNA gene to RNase P function.

Even in the absence of functional data, it is possible to identify RNase P RNA genes based on their similarity to known RNase P RNAs. However, nonbacterial RNase P RNAs can differ so greatly at the sequence level that it is often difficult to align and compare RNA sequences across even relatively shallow phylogenetic boundaries. In such cases, alignment of sequences and assessment of homology between putative and known RNase P RNAs requires consideration of the higher-order structures of these RNAs.

Secondary Structure: Defining the Universal Core Structure

Functional RNAs fold into compact, three-dimensional structures of defined shape. The basic unit of RNA folding—the secondary structure—we define as two or more contiguous Watson-Crick base pairs aligned in an antiparallel helix. The determination of RNase P RNA secondary structure has relied primarily on the phylogenetic-comparative approach, which remains the most successful means of determining the higher-order structures of large RNAs (46, 47). In this method, structure is inferred by identifying covariations, or concerted changes, in the nucleotide sequences of homologous RNAs (i.e. those sharing common ancestry and function). In the most straightforward case, Watson-Crick base pairing within a potential RNA helix is maintained despite changes in the primary sequences of the RNAs. Thus, by aligning and comparing homologous RNA sequences it is possible to test, reject, and reformulate hypotheses of RNA structure. As a rule of thumb, a structure is considered proven when at least two instances of covariation are identified for any particular pairing.

The bacterial RNase P RNA secondary structure is now well proven (30). Representative structures of two of the most disparate bacterial RNase P RNAs, from *E. coli* and *B. subtilis*, are shown in Figure 2. Although the details of the two structures differ, both contain a core structure that is common to all bacterial RNase P RNAs. This core structure, summarized in Figure 2, constitutes only about half the length of the typical native RNA (30, 48). The strict phylogenetic conservation of structures and sequences present in this hypothetical, consensus RNA implies their functional importance; the phylogenetic-minimum structure likely approximates the minimum functional structure in vivo (below).

The secondary structures of archaeal RNase P RNAs (Figure 3; 34) are similar to those of bacterial RNAs. This similarity extends to the sequence level, where several elements (e.g. J3/4, J2/4, J18/2, J5/15) are conserved between the two phylogenetic domains. Overall, the high degree of sequence and structural similarity shared by archaeal and bacterial RNase P RNAs clearly indicates that these RNAs are homologues. Considering their similarities, it is perplexing that bacterial, but not archaeal, RNAs are catalytically active independently of protein.

Eucaryal RNase P RNAs display much greater variability in sequence and content of structural elements than do their bacterial and archaeal counterparts, so it has been difficult to align and compare sequences from even such closely related groups as yeasts and vertebrates. Nevertheless, incremental additions to the eucaryal sequence alignment, as well as the development of bacterial and archaeal phylogenetic-consensus structures, have permitted an extensive

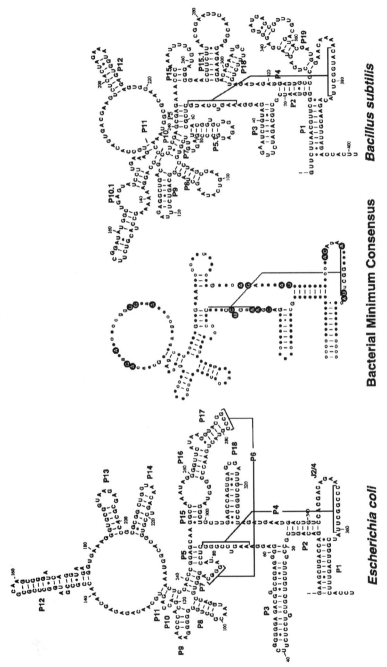

Figure 2 Secondary structures of representative bacterial and phylogenetic minimum-consensus RNase P RNAs. P: paired (i.e. helical) regions, which are enumerated according to their occurrence with respect to the 5'-end of the RNA (114). J: joining region, which connects helices (e.g. J3/4). Helices P4 and P6, which are disrupted in this representation, are connected by *line segments*. The structural elements common to all bacterial RNase P RNAs are shown in the phylogenetic minimum-consensus structure. Nucleotides that are invariant in bacterial RNase P RNAs are represented by *letters*. *White letters* represent nucleotides that are not conserved superimposed on *filled circles* denote nucleotides that are universally conserved in cellular RNAs (51). *Circles* represent nucleotides that are not conserved

Escherichia coli   **Bacterial Minimum Consensus**   *Bacillus subtilis*

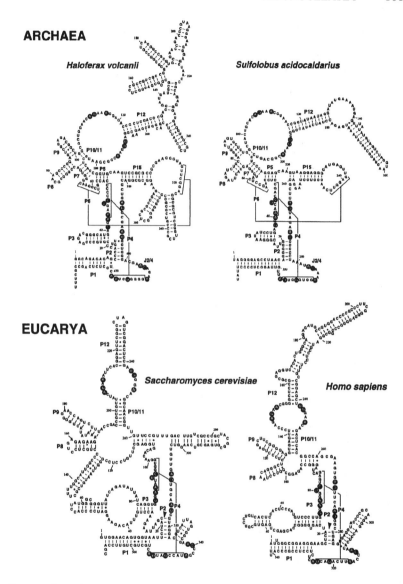

Figure 3 Secondary structures of representative archaeal and eucaryal RNase P RNAs. P: paired (i.e. helical) regions, which are enumerated according to their occurrence with respect to the 5'-end of the RNA (114). J: joining region, which connects helices (e.g. J3/4). Only those helices believed to be homologous to helices of the bacterial RNase P are labeled. Helices P4 and P6, which are disrupted in this representation, are connected by *line segments*. *White letters* superimposed on *filled circles* denote nucleotides that are universally conserved in cellular RNAs (51). Eucaryal structures are adapted from Chen & Pace (51), and archaeal structures were obtained from the RNase P database (32).

delineation of the structural features common to all eucaryal RNase P RNAs (Figure 3; 49–51). Comparison of eucaryal RNase P RNA secondary structures with those of bacteria and archaea reveals a core structure that is conserved across all three domains of life (Figures 2 and 3). Homologues of helices P1, P2, P3, P4, and P10/11 are identifiable in all instances of cellular RNase P RNA, although these structures can differ significantly in sequence. In some cases, such as helix P3, the eucaryal structure is elaborated considerably relative to the archaeal and bacterial structures; this finding suggests a more complex functional role for the eucaryal helix P3. Eucaryal RNase P RNAs probably share other homologous structural elements (e.g. P5/7, P8, and P9) with archaeal and bacterial RNAs. However, the eucaryal secondary structure is not sufficiently resolved at this time to conclusively make these comparisons.

Only 21 of about 200 nucleotides in the universal core of RNase P are conserved in nucleotide identity across phylogenetic groups (*circled* nucleotides in Figures 2 and 3; 44, 51). Most of these universally conserved nucleotides are clustered in three regions: (*a*) J3/4-P4-J2/4, (*b*) J15/2, and (*c*) J11/12-J12/13. Identification of these nucleotides is the key to establishing the homology of the conserved secondary-structural elements of cellular RNase P RNAs (e.g. helix P10/P11), which are typically located a set distance from universally conserved nucleotides.

Eleven of the 21 invariant nucleotides in RNase P RNA occur within or adjacent to helix P4. This is unusual because elements of secondary structure are rarely as well conserved in terms of primary sequence as are single-stranded nucleotides. Based solely on the comparative data, then, it is predictable that these nucleotides play critical roles in RNase P structure or function. Both in vitro selection and site-directed mutagenesis experiments have confirmed that at least a subset of the invariant nucleotides found in bacterial RNase P RNA contribute to catalytic activity (MAT Rubio & NR Pace, unpublished data; 52, 53). Indeed, optimal catalytic activity demands the native residue at each conserved position within the J3/4-P4-J2/4 region; few, if any, alternative solutions to the structure formed in this region are functionally viable (52).

Because mitochondria evolved from bacterial endosymbionts (54), mitochondrial RNase P RNAs are expected to be structurally similar to their bacterial counterparts. Indeed, a recently identified gene from the mitochondrial genome of the protozoan *Reclinomonas americana* encodes a putative RNase P RNA with striking similarity to the bacterial consensus structure (29). In contrast, the RNase P RNAs encoded by the mitochondria of fungi are extremely rich in adenosine and uridine residues (>85%), complicating the alignment of these sequences and the derivation of plausible secondary structures (15, 35). Nevertheless, two short, relatively guanosine- and cytidine-rich elements in these RNAs can be aligned with bacterial and nuclear RNase P RNAs (35). Significantly, these elements constitute the 5′ and 3′ halves of the highly conserved helix P4; this structure has thus been identified in all known RNase P RNAs. Mutations that disrupt the 5' half of mitochondrial helix P4 cause an accumulation of 5'-immature pre-tRNA, which indicates that this structure is of critical importance for mitochondrial RNase P activity (43), as it is for bacterial RNase P activity.

The important theme that emerges from analysis of the consensus structures of eucaryal, archaeal, and bacterial RNase P RNAs is that each RNA consists of a phylogenetically conserved core of helices and joining nucleotides that is embellished with species- or group-specific structural modifications. Sequence length variation between species is usually manifest in one of two ways: (a) by large, complex extensions to core helices or (b) by the presence or absence of entire helices. Whereas phylogenetically variable structures are, in general, dispensable for catalytic activity, they often serve to stabilize global folding of ribozymes (see below).

Toward the Tertiary Structure

The secondary structure of RNase P RNA provides considerable information about its tertiary structure because paired regions are expected to form helices with A-form geometry (intrahelical bulges can cause some deviations). However, the relative orientations of RNA helices in space cannot be determined from secondary structure. X-ray diffraction and nuclear magnetic resonance (NMR) spectroscopy can provide such information, but these techniques have not yet been applied successfully to RNase P RNA. Lower-resolution spatial information has been obtained from other methodologies, however. Site-specific cross-linking, for instance, has been used to orient the helices of bacterial RNase P RNA with respect both to each other and to the pre-tRNA substrate (55–59). In this approach, photoactivatable cross-linking reagents are attached specifically to the 5' or 3' ends of ribozyme or substrate RNAs, and cross-linking is performed by exposing ribozyme-substrate complexes to UV irradiation under native conditions. Because reverse transcriptase stalls at chemically modified nucleotides, sites of cross-linking can be mapped by reverse transcription of either enzyme or substrate RNAs and by separation of radiolabeled cDNA products on denaturing polyacrylamide gels. In many instances, the 5' and 3' ends of an RNA can be circularly permuted to internal sites, thus allowing the convenient positioning of cross-linking reagents at virtually any residue within the RNA.

Phylogenetic-comparative analysis also can contribute information about RNA tertiary structure by the detection, through covariations, of long-range interactions between bases (either Watson-Crick base pairs or more complex interactions) that serve to juxtapose helices (60). Tertiary interactions tend to be both more complex and more highly conserved than Watson-Crick base pairs within helices. Consequently, much larger sequence data sets must be analyzed to identify possible tertiary contacts (61). The isolation of numerous genes one at a time is a tedious process. One means of rapidly acquiring sequences for comparative analysis has been through the isolation of RNase P RNA genes from complex, natural microbial communities. Most natural environments are characterized by extensive microbial diversity, from which hundreds of species of a particular gene can be rapidly isolated by PCR amplification, provided that sufficiently conserved primer-binding sites occur in the gene. Brown et al (30) used this approach to isolate more than 50 unique bacterial RNase P RNA genes, which increased substantially the number of known sequences in the bacterial RNase P RNA database. Comparative analysis of this expanded set of data revealed several potential sites of tertiary contacts. For example, sequence covariation provides evidence for tertiary contacts between the conserved tetraloops of helices P14 and P18 and base pairs within helix P8 (Figure 4); cross-linking studies are consistent with this inference. The docking of GNRA-tetraloops with helices has been proposed on the basis of mutational analyses (62, 63) and observed in the crystal structures of the hammerhead (64, 65) and group I ribozymes (66). Extrapolation of this structural data to RNase P suggests that the P14 and P18 helices form a coaxially stacked helix that is anchored to the stem of P8 through tetraloop-helix interactions (30). Additional long-range interactions in bacterial RNase P RNAs have been proposed on similar comparative grounds (67–69).

The combined data sets from the cross-linking and comparative approaches provide a library of information concerning the spatial organization of the RNase P RNA. The distance constraints indicated by these studies have been used to model the structure of the ribozyme-substrate complexes of both *E. coli* and *B. subtilis* RNase P RNAs (57, 59, 70, 71). Of particular interest in these models is the positioning of the scissile bond of pre-tRNA, because any region of the enzyme that is closely juxtaposed with this bond is potentially a component of the enzyme's active site. In the models of Harris et al (57, 59) and Chen et al (for the *B. subtilis* version, 71a), the scissile bond is surrounded by a core of phylogenetically well-conserved RNase P structures and abutted against helix P4 (Figure 5). As described below, evidence that helix P4 forms a portion of the enzyme's active-site structure is also provided by both modification interference (72) and in vitro selection (52) experiments.

RNASE P RNA FUNCTION

Kinetic and Thermodynamic Studies of the Ribozyme Reaction

The bacterial RNase P cleavage reaction consists of three basic steps: (*a*) binding of the pre-tRNA substrate, (*b*) cleavage of the scissile phosphodiester bond,

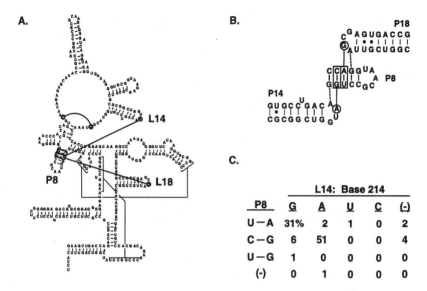

Figure 4 Phylogenetically conserved tertiary pairings within bacterial RNase P RNAs. P: pairing region (i.e. a helix). L: Loop. (*A*) Nucleotides that participate in long-distance interactions are connected by *line segments*. The secondary structure is that of *Escherichia coli* RNase P RNA (30, 67). (*B*) Details of proposed GNRA-tetraloop interactions with helix P8. Purine residues within the loops of both P14 and P18 are proposed to form base-triple interactions with adjacent base pairs within helix P8. (*solid lines*) Interactions predicted from comparative analysis; (*dashed lines*) Interactions observed in other ribozymes. See text for details. (*C*) Analysis of covariations between nucleotide 214 of L14 and the base pair in P8 with which it potentially interacts. Percentile values indicate the frequency of a given nucleotide's occurrence at position 214, given a particular base pair in P8. (–): Absence of the helix (for L14) or base pairing (for P8). Adapted from Brown et al (30).

and (*c*) dissociation of 5' leader and mature-tRNA products (Equation 1).

$$E + S \underset{k_{-1}}{\overset{k_1}{\rightleftharpoons}} E \cdot S \xrightarrow[180–360 \text{ min}^{-1}]{k_2} E \cdot P \underset{k_{-3}}{\overset{k_3}{\rightleftharpoons}} E + P. \qquad 1.$$

The rate constants of the various steps of the reaction pathway have been measured for the ribozymes of *E. coli*, *B. subtilis*, and *T. thermophilus* by a combination of pre-steady-state and steady-state analyses (e.g. 73–77). Under steady-state conditions and in the presence of excess substrate, k_{cat} is determined by the rate of product release (k_3) (75, 77, 78). The actual rate of substrate cleavage therefore must be measured by single-turnover reactions (i.e. [E] ≫ [S]). Under such conditions, substrate cleavage occurs with a first-order rate constant (k_2) in the range of 180–360 min^{-1} for the *E. coli* (74) and *B. subtilis* (75) enzymes, respectively. In comparison, the k_{cat} of the multiple-turnover

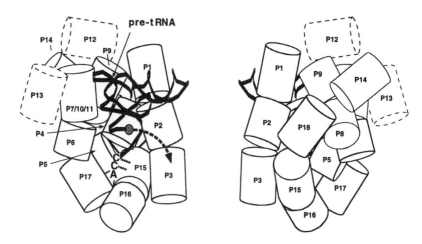

Figure 5 Model of *Escherichia coli* RNase P RNA tertiary structure bound to pre-tRNA. The model is derived from molecular modeling of phylogenetic and photoaffinity cross-linking data (see text for details). RNase P RNA helices are drawn as cylinders with diameters proportional to that of A-form RNA helices. The scissile phosphate bond is depicted by a *sphere* in close proximity to P4. The 5′-leader element is depicted as a *dashed arrow*. Interactions between the pre-tRNA 3′-CCA and residues within the J15/16 internal bulge are denoted by *dashed lines*. The two enzyme-substrate structures are related by a rotation of 180°. Figure courtesy of Dr. ME Harris (Case Western Reserve).

reaction is approximately 100-fold slower. At least in the case of the *E. coli* RNA, k_2 increases linearly with pH in the range 5–8 but plateaus above pH 8, which suggests that hydroxide ion acts as the nucleophile in the cleavage reaction and that the active site forms a saturable binding site (or generating site) for the nucleophile (74). Finally, the rates of substrate and product association are an order of magnitude slower than predicted by simple diffusion, suggesting that significant conformational barriers must be overcome during the binding step of the reaction (74, 75).

The affinity of RNase P RNA for tRNA has been assayed by a variety of methods, including gel mobility shift (75, 79, 80), gel-exclusion chromatography (75), Pb^{2+}-hydrolysis (76), and site-specific photoaffinity cross-linking (55, 73). The latter three methods are of particular utility because substrate-binding affinity is assayed under reaction conditions. The cross-linking assay has an added benefit in that the patterns of cross-links that are obtained can be used to assess whether the enzyme-substrate complex is in the native conformation. Under optimal salt conditions (e.g. 100 mM $MgCl_2$, 800 mM NH_4Cl), *B. subtilis* RNase P RNA binds mature tRNA with a K_d of approximately 3 nM; the 5′-leader sequence binds at least two orders of magnitude less well and rapidly dissociates from the enzyme following catalysis (75). Ribozyme and substrate RNAs bind noncooperatively and with a 1:1 stoichiometry (75, 76, 79).

Role of Divalent Metals

Although the RNase P reaction absolutely requires divalent metal ions (preferably Mg^{2+}) for catalysis, tRNA can bind to the RNase P RNAs of *B. subtilis, E. coli*, and *Chromatium vinosum* in the absence of metal ions, as assayed by photoaffinity cross-linking of tRNA to the ribozyme (73). An important finding was that the patterns of the cross-linked species observed in the experiments were identical in the presence and absence of divalent ions. This indicates that the structure surrounding the scissile bond (the attachment site for the cross-linking reagent) was not grossly affected by the presence or absence of Mg^{2+}. Furthermore, the results of kinetic burst experiments indicate that pre-tRNA productively binds the *C. vinosum* ribozyme in the absence of Mg^{2+} (73). Although Mg^{2+} is not necessary for substrate binding, the efficiency at which cross-linked species are obtained is notably improved (ca 7- to 10-fold) by the addition of Mg^{2+} in the case of the *B. subtilis* RNase P RNA. Using different methodologies, Beebe et al (81) also ascertained that the affinity of *B. subtilis* RNase P RNA for tRNA exhibits a strong dependence on Mg^{2+} concentration (73). Furthermore, Pan (82) and Zarrinkar et al (83) reported that the folding of *B. subtilis* RNase P RNA (as monitored by resistance to hydroxyl-radical attack or oligonucleotide hybridization, respectively) is cooperatively dependent on Mg^{2+} concentration and that at least three Mg^{2+} ions contribute to folding.

In contrast to their contribution to ribozyme folding and substrate binding, divalent metal ions, preferably Mg^{2+}, are required absolutely for catalysis. Maximal catalytic activity requires multiple metal ions (Hill coefficient $= 3.2$ for *E. coli* RNase P RNA; 74, 84). Mn^{2+} can substitute for Mg^{2+} with only a slight decrease in catalytic efficiency, whereas Ca^{2+} results in a 10^4-fold reduction in k_2 (74). No other metals are known to stimulate RNase P activity (85, 86). The high degree of specificity that RNase P exhibits for catalytic metal ions suggests that the ribozyme-substrate complex forms a highly ordered metal-binding pocket that constitutes a portion of the active site. Given the ability of divalent metal ions to lower the pK_a of bound water molecules, it is likely that at least one of the metal ions that are required for catalysis serves to direct an attacking nucleophilic hydroxide ion into the scissile phosphodiester bond.

A number of possible coordination sites for catalytic metal ions within both enzyme and substrate RNAs have been proposed. Several 2'-OH groups, of both the enzyme and substrate, are known to influence reaction kinetics (74, 87, 88). For instance, deoxy replacement of the 2'-OH group of the -1 residue of pre-tRNA (i.e. the residue immediately adjacent to the cleaved phosphodiester bond) confers a 3400-fold decrease in k_2 (6). Because the cooperative dependence on Mg^{2+} concentration of the RNase P reaction is diminished by this deoxy substitution (Hill coefficient is decreased from 3.2 to 2.0), it is possible that removal of the 2'-OH group of residue -1 results in the loss of one of the catalytic Mg^{2+} ions (74). Additionally, according to mutational and kinetic studies, the 3'-CCA of the tRNA, which is close to the cleavage site in the tRNA tertiary structure, contributes to the binding of divalent cations that participate in catalysis (BK Oh, DN Frank & NR Pace, manuscript in preparation).

Phosphate oxygens that bind divalent metals important for catalysis can be detected by replacing the oxygen atoms of these groups with sulfur, which coordinates Mg^{2+} several orders of magnitude less well than does oxygen. In contrast, more thiophilic metals, such as Mn^{2+} and Cd^{2+}, are relatively insensitive to sulfur substitution (89). Thiophosphate substitutions that inhibit activity in the presence of Mg^{2+}, but not thiophilic metals, are therefore candidate sites for metal coordination. Thio substitution of either the *pro*-Rp or *pro*-Sp nonbridging oxygens at the cleavage site in pre-tRNA severely inhibits catalysis (i.e. >2000-fold decreased k_2) in a Mg^{2+}-dependent reaction (84, 90). In the presence of Cd^{2+} or Mn^{2+}, however, the effect of *pro*-Rp substitution is ameliorated, whereas *pro*-Sp substitution remains deleterious, thus suggesting that the *pro*-Rp oxygen directly interacts with a divalent metal ion during catalysis (Figure 1*B*; 84, 90); the role of the *pro*-Sp oxygen in catalysis is less clear.

Phosphate oxygens in the ribozyme that are important for substrate binding (91) and catalysis (72) have been identified by phosphorothioate modification-interference techniques. In this type of experiment, modified nucleotides are randomly incorporated into an RNA, during in vitro transcription or solid-phase synthesis. The population of modified RNAs is allowed to react, and then functional and nonfunctional molecules are separated from one another. Positions of nucleotide modification are then mapped by some means (in the case of phosphorothioate modifications, by iodine cleavage of end-radiolabeled material followed by gel electrophoresis). Any modification that inhibits the activity of the RNA is expected to be enriched in the inactive pool of molecules and correspondingly depleted in the active pool.

To map phosphate oxygens involved in binding catalytically critical metals, Harris & Pace (72) used a self-cleaving pre-tRNA–RNase P RNA conjugate (Figure 6; 92) that allows discrimination of functional and nonfunctional molecules on the basis of catalytic activity rather than substrate binding. Substitution with sulfur at any of four nonbridging oxygens was found to inhibit catalysis (Figure 6); one of these sites (A67) was rescued by reaction in Mn^{2+}, consistent with a direct role for this oxygen in coordinating metal ions (72). All four phosphorothioate-sensitive sites cluster at the ends of the universally conserved helix P4 (Figure 6), which sits adjacent to the cleavage site of pre-tRNA

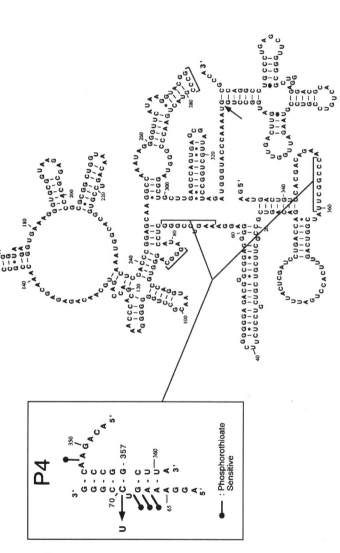

Figure 6 Secondary structure of the self-cleaving pre-tRNA–RNase P RNA conjugate, PT332 (92). The site of substrate cleavage is denoted by an *arrow*. The structure of helix P4 is shown in the *inset box*. Sites at which phosphorothioate substitution inhibits the self-cleavage activity of PT332 RNA are marked by *lines* connected to *circles* (72). These sites may coordinate Mg²⁺ ions involved in catalysis. Also shown is a C-to-U transition that is sufficient to alter the metal specificity of RNase P (see text; DN Frank & NR Pace, 92a).

in the enzyme-substrate complex (57, 59). It is thus likely that helix P4 coordinates metal ions that participate in the catalytic reaction.

The specific structure of the metal-binding pocket in RNase P is expected to be dictated by interactions among the base moieties of the RNA chain. As a means of identifying the residues that determine metal-binding specificity, in vitro RNA selection experiments have been used to evolve RNase P RNA variants with an altered preference for divalent metal ions (92a). In addition to showing improved catalytic activity in Ca^{2+}, the final selected population of RNAs exhibited markedly reduced activity in Mg^{2+}, which indicates that the relative preference of RNase P for divalent metals was changed (ca 100-fold shift in preference). The activation of Ca^{2+}-dependent activity and the diminution of Mg^{2+}-dependent activity were the result of a single nucleotide change: a cytosine-to-uracil mutation at position 70 (Figure 6). This mutation converts a naturally invariant C-G base pair within helix P4 to a U-G pairing (the pairing partner of U70 must be a G for enhanced activity in Ca^{2+}). The finding that a single nucleotide change within helix P4 can alter the specificity of RNase P RNA for catalytic divalent metal ions provides further, independent evidence that helix P4 constitutes a portion of the ribozyme's active site.

Substrate Recognition

All tRNAs in a particular cell or organelle are processed by the same RNase P, which therefore must be capable of recognizing features common to all pre-tRNA species. Our current understanding is that the primary determinants for recognition by RNase P are arranged by the tertiary structure of the mature domain of tRNA. Several lines of evidence indicate that the 5'-leader sequence is not a significant determinant of the recognition process. First, early sequence comparisons revealed no structure or sequence element common to 5'-leader sequences (93–99). Second, substrates with artificial 5'-leaders, some as small as a single nucleotide, are processed efficiently (74, 100, 101). Finally, mature tRNA is a strong competitive inhibitor of the RNA-alone reaction (K_i of mature tRNA is similar to K_m of pre-tRNA), which indicates that the 5'-leader sequence does not contribute substantially to the energetics of substrate binding (73, 75, 78). Thus, mature tRNA must embody the determinants that are recognized by RNase P RNA.

Studies of small model substrates have demonstrated that an RNA mini-helix consisting solely of T and acceptor stems (and including a 5' leader) is cleaved by RNase P, albeit with increased K_m (ca 100-fold) (102–104). The loss of binding energy associated with the interaction of RNase P with the mini-helix could be due either to the loss of recognition elements in the D and anticodon arms or to a loosening of T-loop structure, which is normally fixed in space by interactions with the D loop. The results of chemical modification-protection experiments have corroborated the deletion analyses and further delineated the contacts made between enzyme and substrate RNAs. RNase P is seen to occupy a patch on the substrate that overlaps the acceptor and T stems (87, 105, 106). In addition to the structure formed by the acceptor and T stems, the 3'-CCA sequence found in all tRNAs is an important sequence-specific feature recognized by bacterial RNase P (58, 76, 85, 102, 107–111). *E. coli* strains deficient in nucleotidyl transferase, the enzyme required for 3'-CCA addition (*cca⁻* strains), accumulate 5'-immature pre-tRNA, suggesting that the RNase P holoenzyme requires the 3'-CCA for optimal activity in vivo (107). In vitro, an intact 3'-CCA improves the efficiency of 5'-leader removal by bacterial RNase P RNA but is not absolutely required for activity. In the RNA-alone reaction, the structure of the 3'-CCA influences both substrate binding (i.e. K_m effects; 58, 76, 111) and catalysis (i.e. k_2 effects; BK Oh, DN Frank & NR Pace, manuscript in preparation). Both cross-linking and mutational analyses have located the binding site for the 3'-CCA in the J15/J16 bulges of bacterial RNase P RNAs, although the details of this interaction remain to be elucidated (58, 111, 112). It is surprising, considering the in vivo data, that the protein subunit of the bacterial holoenzyme strongly attenuates the deleterious effects of 3'-CCA deletions on RNase P activity in vitro (58, 109). Whether the activities of nonbacterial RNase P holoenzymes are affected by the structure of the 3'-CCA has not been elucidated. However, the RNA subunits of eucaryal RNase P apparently lack the structure that is homologous to the 3'-CCA–binding domain of bacterial RNase P. This suggests that 3'-CCA is not recognized by eucaryal RNase P or that other RNA structures (e.g. helix P3) have assumed the 3'-CCA–binding function (51).

Properties of Minimum-Consensus RNase P RNAs

The articulation of a phylogenetically defined minimum-consensus RNA structure (Figure 2) postulates that phylogenetically conserved sequences and structures present in the model are sufficient for RNA-mediated catalysis. This hypothesis has been tested in the case of the current version of the bacterial minimum-consensus structure by modifying the RNase P RNA gene of *Mycoplasma fermentans* to conform to the minimum-consensus RNA structure (Figure 3; 48). Rather than simply deleting portions of the *M. fermentans* gene, phylogenetically variable structures were replaced with the homologous, minimal regions of other RNase P RNA genes. In this way, structural distortions that might be engendered by simple deletion are minimized. For instance, helix P18 of *M. fermentans* was replaced with the single cytidine residue found in the homologous position of the *Chlorobium tepidum* RNA. The minimum-consensus RNA is active in pre-tRNA cleavage assays, albeit with reduced catalytic efficiency relative to the full-length *M. fermentans* RNA (k_{cat}/K_m reduced ~600-fold). Several lines of evidence indicate that the inefficiency of this minimum

RNA is due to a global destabilization of enzyme structure: (a) It requires higher concentrations of monovalent and divalent salts, which normally stabilize structure; (b) its temperature optimum is substantially lower than that of the native ribozyme; and (c) its affinity for substrate is reduced. Other deletions of phylogenetically variable domains within RNase P similarly disrupt enzyme structure (113–115). Some of the variable helices are known to dock into distal sites on the ribozyme (30). It is thus hypothesized that these structural elements function as struts that add stability to the conserved core structure of the enzyme (48).

CHARACTERIZATION OF RNASE P PROTEINS

Although the low cellular abundance of RNase P holoenzymes has hindered extensive purification, enzymes have been at least partially purified from a number of organisms and organelles. The biochemical characteristics of a sample of these holoenzymes are summarized in Table 1. The results of such basic assays as equilibrium-density centrifugation and micrococcal nuclease (MN) digestion have revealed that RNase P holoenzymes vary greatly in overall composition. Because the RNA subunits of RNase P do not vary appreciably in size throughout phylogeny, the protein subunits must be the source of diversity in holoenzyme composition.

Bacterial Holoenzymes

In general, bacterial RNase P holoenzymes have densities in the range of 1.7 g/ml in CsCl, which indicates that the RNA subunit contributes most of the mass of the ribonucleoprotein (116, 117). Indeed, bacterial RNase P proteins are uniformly small (ca 14 kDa) relative to their corresponding RNA subunits (>100 kDa). The gene encoding the RNase P protein of *E. coli, rnpA*, was identified through a genetic screen for factors involved in pre-tRNA processing (7) and subsequently cloned and sequenced (118). Bacterial homologues of *rnpA* have been identified by sequence similarity from numerous species (119, 120).

   The bacterial RNase P protein is required for activity both in vivo and in vitro under physiological conditions, but its function is dispensable at higher ionic strengths (e.g. 1 M NH_4Cl); thus, the protein is not essential for catalysis (2, 40, 86). Elevated ionic strength improves substrate binding (i.e. decreases K_m) in the RNA-alone cleavage reaction. Because the function of the bacterial RNase P protein can largely be replaced by high salt, one likely role of the highly basic protein is to shield unfavorable electrostatic repulsion that develops when the polyanionic ribozyme and substrate RNAs interact (78). The bacterial RNase P protein subunit may also broaden the specificity of the enzyme, thus allowing the holoenzyme to productively interact with a wider spectrum of substrates (e.g. pre-4.5S and pre-10Sa RNAs) than does the ribozyme [2, 121,

122; see Gopalan et al (123) for a more detailed review of the function of bacterial RNase P proteins].

Archaeal Holoenzymes

Archaeal examples of the RNase P holoenzyme have been isolated from only two species: *Sulfolobus acidocaldarius* and *Haloferax volcanii*. The biochemical properties of the two holoenzymes differ substantially, suggesting rather different organizations to the enzymes. The activity of the *H. volcanii* enzyme, for instance, is reportedly sensitive to treatment with MN, as are bacterial RNase P holoenzymes (124). Similarly, the *H. volcanii* holoenzyme exhibits a buoyant density of 1.61 g/ml in Cs_2SO_4 (*E. coli* RNase P is 1.55 g/ml in Cs_2SO_4), suggestive of a ribonucleoprotein complex consisting primarily of RNA (124). The RNase P of *S. acidocaldarius*, in contrast, is both insensitive to MN and has a buoyant density of 1.27 g/ml in Cs_2SO_4 (125), indicating a considerable protein content. Presumably, the protein protects the RNA from MN digestion. (The holoenzyme could also be modified by low-density molecules such as lipids or carbohydrates; see Reference 17.) Although the holoenzyme of *S. acidocaldarius* seems structurally more complex than that of *H. volcanii*, the holoenzyme of *S. acidocaldarius* is expected to resemble more closely the ancestral state and the complexity of eucaryal RNase P holoenzymes (below). Additional archaeal holoenzymes need to be characterized to provide generality.

Eucaryal Holoenzymes

Eucaryal nuclear RNase P holoenzymes have been partially purified from a number of organisms, primarily yeasts and vertebrates. The buoyant densities of eucaryal holoenzymes (1.28–1.4 g/ml in Cs_2SO_4) are less than those of the bacterial enzymes, though still intermediate between the densities of protein and RNA, and the exclusion volumes of holoenzymes in size chromatography are larger in eucarya than in bacteria (Table 1; 18, 24, 100, 126, 127). These data suggest a ribonucleoprotein particle in which protein in the eucaryal holoenzyme contributes a higher percentage of the total mass of the complex than it does in the bacterial holoenzyme. Despite the greater mass of proteins, the eucaryal RNA subunits are, in general, susceptible to inactivation by ribonuclease treatment, indicating that the RNAs are essential for catalytic activity (24, 100, 126–128).

Several candidate RNase P proteins have been identified in eucaryal organisms through the use of a variety of biochemical and genetic methods (Table 1). For example, the RNase P holoenzyme of *Schizosaccharomyces pombe* has been purified to apparent homogeneity (129). Both 100-kDa protein and 285-nucleotide (nt) RNA species copurify with RNase P activity throughout the purification procedure, which includes a tRNA-affinity purification step, and are presumably components of RNase P (21, 129). Intriguingly, the 100-kDa protein can be cross-linked to 4-thiouridine–modified pre-tRNA but not to mature tRNA, suggesting a role for this protein in discriminating between substrate and product RNAs (129). However, final proof that these factors are indeed constituents of RNase P awaits reconstitution of active enzyme from purified components. A gene for the 100-kDa protein has not been identified.

Genetic screens for rRNA-processing mutants fortuitously have identified at least three genes (*POP1, POP3*, and *POP4*) that potentially encode protein subunits of *S. cerevisiae* RNase P (130–132). Genetic depletion of any of the three proteins results in the in vivo accumulation of precursors of both 5.8S rRNA and tRNA. Furthermore, both RNase P RNA and the RNA associated with a ribosomal RNA-processing nuclease, RNase MRP, are specifically coimmunoprecipitated with antibodies directed against epitope-tagged POP1p, POP3p, or POP4p (130–132). These findings clearly implicate the *POP* gene products in the function of RNase P, but it is not known whether these proteins play direct or indirect roles (e.g. in the biosynthesis of the holoenzyme) in this activity. That all the POP proteins associate with both MRP and RNase P RNAs may indicate that other proteins, specific for RNase P, remain to be discovered.

The *POP1* gene of *S. cerevisiae* (yPOP1) encodes an open reading frame (ORF) with a predicted mass of approximately 100 kDa, similar in size to the protein isolated from *S. pombe* (130); it remains to be seen whether the *S. pombe* and *S. cerevisiae* proteins are homologues. Both human and *C. elegans* homologues of the *POP1* gene (*hPOP1* and *cPOP1*, respectively) were recently discovered by database searches, despite their lack of strong similarity to *yPOP1*. At the amino-acid level, only three short sequences are conserved between the yeast and metazoan *POP1* peptides (133). Despite the differences in *POP1* sequence, antibodies raised against hPOP1p recognize a 115-kDa protein in HeLa cell extracts and can immunoprecipitate RNase P activity along with human RNase P RNA (133). Thus, hPOP1p, like its yeast counterpart, is likely to be a component of RNase P.

Partial characterization of the human RNase P has revealed several potential subunits of the holoenzyme. Autoimmune sera that can both deplete RNase P activity from extracts and immunoprecipitate RNase P RNA also recognize ca 38-kDa and 30-kDa polypeptides in highly purified preparations of human RNase P (134–136). The genes encoding the 38-kDa (*RPP38*) and 30-kDa (*RPP30*) proteins have been isolated; both differ in sequence from hPOP1p (136). Furthermore, a 40-kDa protein can be cross-linked to human RNase P RNA and immunoprecipitated with anti-Th sera (135, 137); its relationship to the product of *RPP38* is not known. Several other proteins copurify with the human RNase P holoenzyme but are not recognized by anti-Th sera. Whether the association of these proteins with RNase P activity is fortuitous, or indicates a functional role for these proteins in the holoenzyme, has yet to be determined (136).

Mitochondrial Holoenzymes

Mitochondrial RNase P activities have been characterized to various extents from several species, including yeasts, vertebrates, plants, and trypanosomes (Table 1). By far the most extensively characterized mitochondrial RNase P is that of *S. cerevisiae*. The holoenzyme from yeast mitochondria has a buoyant density similar to that of protein (1.28 g/ml in Cs_2SO_4) (138), but it clearly contains an essential RNA component, as shown by genetic analysis (15, 41) and ribonuclease sensitivity (Table 1; 42). A 105-kDa protein encoded by the nuclear gene *RPM2* is enriched in highly purified preparations of yeast mitochondrial RNase P (139). Haploid strains carrying a disruption of *RPM2* are viable but are unable to respire, indicating that the gene is essential for mitochondrial function. The mitochondria of these deletion-carrying strains accumulate pre-tRNAs with immature 5′-ends, as would be expected for disruption of a locus encoding a component of RNase P (139). Finally, antibodies raised against the product of the *RPM2* gene specifically coimmunoprecipitate the mitochondrial RNase P RNA (140). Although the mitochondrial RPM2 and nuclear POP1 proteins are of similar mass, they do not share significant sequence similarity.

Chloroplast RNase P

Functional characterization of chloroplast RNase P holoenzymes has been limited primarily to activities isolated from tobacco or spinach chloroplasts. In the latter system, RNase P–enriched fractions exhibit buoyant densities of 1.28 g/ml in CsCl, suggesting that most of the mass of the enzyme is composed of protein rather than RNA (38). Initial reports concluded that the RNase P activity of spinach chloroplasts could be inactivated by MN (141). However, more detailed subsequent analysis demonstrated that spinach chloroplast RNase P is not susceptible to MN inactivation (38). The earlier results were probably due to a substrate-masking phenomenon, in which inactivated MN binds to pre-tRNA and thereby occludes the substrate from cleavage by the enzyme (38, 142). The proteinlike buoyant density and micrococcal nuclease insensitivity of chloroplast RNase P suggest that this enzyme does not require an RNA component. Indeed, spinach chloroplast RNase P has been purified extensively without the concomitant enrichment of an RNA species (38, 143).

In addition to lacking an RNA subunit, spinach chloroplast RNase P differs significantly in mechanism from bacterial RNase P, as indicated by susceptibility to thio-phosphate–modified substrates. Placement of a *pro*-Rp phosphorothioate at the scissile bond of pre-tRNA substantially reduces the activity of bacterial RNase P (84, 90) but has little effect on the activity of spinach chloroplast RNase P (39, 143; BC Thomas & P Gegenheimer, manuscript in preparation). In bacterial RNase P, this nonbridging phosphate oxygen is believed to coordinate a Mg^{2+} ion that is an integral part of the enzyme's active site (Figure 1*B*; 84, 90). The absence of a thio effect on the activity of chloroplast RNase P suggests that the active-site structure of this enzyme differs substantially from that of bacterial RNase P and that cleavage occurs by a mechanism different from that used by other types of RNase P (143). The evolutionary lineage leading to chloroplasts stemmed from cyanobacteria, which possess RNA-based enzyme (144). Thus, chloroplasts seem to have discarded the RNA in favor of an all-protein enzyme.

CONCLUSIONS AND FUTURE DIRECTIONS

RNase P activity, that of *E. coli*, was first reported 25 years ago (145). The intervening years have witnessed the characterization of RNase P from all domains of organisms (Bacteria, Eucarya, Archaea) and organelles. Aside from its importance in pre-tRNA processing, RNase P has garnered much attention as one of the first examples of RNA-mediated catalysts. Along with group I self-splicing introns and hairpin and hammerhead ribozymes, the RNase P ribozyme has become one of the better-understood examples of these novel RNA enzymes. Substantial progress has been made toward a description of the three-dimensional folding of bacterial RNase P RNAs. Although robust, the current models of RNase P structure are derived from molecular-modeling studies based on comparative and cross-linking data (e.g. 57, 59, 70) and thus lack the resolution that can be achieved by NMR and/or X-ray diffraction analyses. The recent crystallographic success with the hammerhead ribozyme (65, 146) and the P4/P6 domain of group I intron (66) lends hope that an atomic-resolution structure of RNase P RNA will be achievable.

Another important avenue of research concerns the protein subunits of the archaeal and eucaryal RNase Ps. Because of the apparent inactivity of the RNA subunits of these types of RNase P, elucidation of how the RNA and protein subunits collaborate to form an active enzyme remains an intriguing question. A solution to this problem will first necessitate a description of the components of archaeal and eucaryal holoenzymes that are sufficient to mediate catalysis; this, in turn, demands an efficient system for reconstituting RNase P activity from defined subunits.

Endeavors to localize the subcellular compartment(s) in which eucaryal RNase P functions are now beginning to provide information. In *Xenopus* oocytes, 5'-leader elements are removed prior to transport of tRNA transcripts from the nucleus to the cytoplasm (147, 148), indicating that RNase P is resident in the nucleus. However, antisense oligonucleotides directed against human RNase P RNA, and detected by fluorescent in situ hybridization, stain cytoplasmic and nucleolar compartments of HeLa cells (149, 150). Whether the cytoplasmic staining is artifactual or reveals a previously undescribed cytoplasmic function for RNase P awaits further study. The nucleolar localization of RNase P is intriguing in light of reports that the enzyme may participate in processing pre-rRNAs (151, 152). As described above, RNase P shares an epitope with the nucleolar-localized, pre-rRNA processing enzyme, RNase MRP. Lee et al (150) recently reported that RNase P and RNase MRP are associated in HeLa cell extracts. They proposed that the enzymes form a supramolecular structure in the nucleolus that functions in the processing of pre-tRNA and pre-rRNA. Given the homologous RNAs and shared protein subunits of the enzymes, a detailed comparison of RNase P and RNase MRP structures and functions will provide an interesting glimpse of how ribonucleoproteins evolve.

We close by restating a question about RNase P that has been asked many times: Why has an RNA subunit remained the heart of the enzyme? The shared history of all RNase P RNAs indicates that this enzyme first arose before the evolutionary radiation that resulted in Bacteria, Archaea, and Eucarya. RNase P RNA is certainly of ancient origin. Its evolutionary longevity implies a function that, aside from the spinach chloroplast RNase P, has not been superseded by protein. Many other RNAs of the putative RNA world have faded away. Why has RNase P RNA remained?

ACKNOWLEDGMENTS

We thank our colleagues for their contributions to the material reviewed here; we regret that space constraints did not allow us to cite all of their work. We thank Drs. Michael Harris and Brian Thomas for providing figures as well as for helpful criticisms of the manuscript. This work was supported by grants from the National Institutes of Health and the Department of Energy (to NRP) and a fellowship from the American Cancer Society (to DNF).

> Visit the *Annual Reviews home page* at
> http://www.AnnualReviews.org.

Literature Cited

1. Woese CR, Kandler O, Wheelis ML. 1990. *Proc. Natl. Acad. Sci. USA* 87: 4576–79
2. Guerrier-Takada C, Gardiner K, Marsh T, Pace N, Altman S. 1983. *Cell* 35:849–57
3. Deutscher MP. 1984. *CRC Crit. Rev. Biochem.* 17:45–71
4. Söll D, RajBhandary UL, eds. 1995.

tRNA: Structure, Biosynthesis, and Function, p. 572. Washington, DC: ASM Press
5. Yarus M. 1993. *FASEB* 7:31–39
6. Smith D. 1995. In *The Biological Chemistry of Magnesium*, ed. JA Cowan, pp. 111–36. New York: VCH
7. Schedl P, Primakoff P. 1973. *Proc. Natl. Acad. Sci. USA* 70:2091–95

8. Sakano H, Yamada S, Ikemura T, Shimura Y, Ozeki H. 1974. *Nucleic Acids Res.* 1:355–71
9. Lee JY, Rohlman CE, Molony LA, Engelke DR. 1991. *Mol. Cell Biol.* 11:721–30
10. Cherayil B, Krupp G, Schuchert P, Char S, Söll D. 1987. *Gene* 60:157–61
11. Altman S, Kirsebom L, Talbot S. 1993. *FASEB J.* 7:7–14
12. Pace NR, Brown JW. 1995. *J. Bacteriol.* 177:1919–28
13. Kirsebom LA. 1995. *Mol. Microbiol.* 17:411–20
14. Nolan JM, Pace NR. 1996. In *Nucleic Acids and Molecular Biology*, ed. F Eckstein, DMJ Lilley, pp. 109–28. Berlin: Springer-Verlag
15. Miller DL, Martin NC. 1983. *Cell* 34:911–17
16. Sakamoto H, Kimura N, Nagawa F, Shimura Y. 1983. *Nucleic Acids Res.* 11:8237–51
17. LaGrandeur TE, Darr SC, Haas ES, Pace NR. 1993. *J. Bacteriol.* 175:5043–48
18. Bartkiewicz M, Gold H, Altman S. 1989. *Genes Dev.* 3:488–99
19. Reich C, Gardiner KJ, Olsen GJ, Pace B, Marsh TL, Pace NR. 1986. *J. Biol. Chem.* 261:7888–93
20. Reed RE, Baer MF, Guerrier-Takada C, Donis-Keller H, Altman S. 1982. *Cell* 30:627–36
21. Krupp G, Cherayil B, Frendewey D, Nishikawa S, Söll D. 1986. *EMBO J.* 5:1697–703
22. Lee JY, Engelke DR. 1989. *Mol. Cell. Biol.* 9:2536–43
23. Eder PS, Srinivasan A, Fishman MC, Altman S. 1996. *J. Biol. Chem.* 271:21031–36
24. Doria M, Carrara G, Calandra P, Tocchini-Valentini GP. 1991. *Nucleic Acids Res.* 19:315–20
25. Brown JW, Haas ES, James BD, Hunt DA, Pace NR. 1991. *J. Bacteriol.* 173:3855–63
26. Tranguch AJ, Engelke DR. 1993. *J. Biol. Chem.* 268:14045–55
27. Bult CJ, White O, Olsen GJ, Zhou LX, Fleischmann RD, et al. 1996. *Science* 27:1058–73
28. Shareck F, Biely P, Morosoli R, Kluepfel D. 1995. *Gene* 153:105–9
29. Lang BF, Burger G, O'Kelly CJ, Cedergren R, Golding GB, et al. 1997. *Nature* 387:493–97
30. Brown JW, Nolan JM, Haas ES, Rubio MAT, Major F, Pace NR. 1996. *Proc. Natl. Acad. Sci. USA* 93:3001–6
31. Altman S, Wesolowski D, Puranam RS. 1993. *Genomics* 18:418–22
32. Brown JW. 1997. *Nucleic Acids Res.* 25:263–64
33. Nieuwlandt DT, Haas ES, Daniels CJ. 1991. *J. Biol. Chem.* 266:5689–95
34. Haas ES, Armbruster DW, Vucson BM, Daniels CJ, Brown JW. 1996. *Nucleic Acids Res.* 24:1252–59
35. Shu HH, Wise CA, Clark-Walker GD, Martin NC. 1991. *Mol. Cell Biol.* 11:1662–67
36. Ragnini A, Grisanti P, Rinaldi T, Frontali L, Palleschi C. 1991. *Curr. Genet.* 19:169–74
37. Baum M, Cordier A, Schön A. 1996. *J. Mol. Biol.* 257:43–52
38. Wang MJ, Davis NW, Gegenheimer P. 1988. *EMBO J.* 7:1567–74
39. Gegenheimer P. 1996. *Mol. Biol. Rep.* 22:147–50
40. Guerrier-Takada C, Altman S. 1984. *Science* 223:285–86
41. Martin NC, Underbrink-Lyon K. 1981. *Proc. Natl. Acad. Sci. USA* 78:4743–47
42. Hollingsworth MJ, Martin NC. 1986. *Mol. Cell. Biol.* 6:1058–64
43. Sulo P, Groom KR, Wise C, Steffen M, Martin N. 1995. *Nucleic Acids Res.* 23:856–60
44. Pagan-Ramos E, Lee Y, Engelke DR. 1996. *RNA* 2:441–51
45. Pagan-Ramos E, Tranguch AJ, Kindelberger DW, Engelke DR. 1994. *Nucleic Acids Res.* 22:200–7
46. Woese CR, Pace NR. 1993. In *The RNA World*, ed. RF Gesteland, JF Atkins, pp. 91–117. Cold Spring Harbor, NY: Cold Spring Harbor Lab. Press
47. Gutell RR. 1993. *Curr. Opin. Struct. Biol.* 3:313–22
48. Siegel RW, Banta AB, Haas ES, Brown JW, Pace NR. 1996. *RNA* 2:452–62
49. Forster AC, Altman S. 1990. *Cell* 62:407–9
50. Pagan-Ramos E, Lee Y, Engelke DR. 1996. *RNA* 2:1100–9
51. Chen J-L, Pace NR. 1997. *RNA* 3:557–60
52. Frank DN, Ellington AE, Pace NR. 1996. *RNA* 2:1179–88
53. Warfe W-D, Hartmann RK. 1996. *J. Mol. Biol.* 259:422–33
54. Woese CR. 1987. *Microbiol. Rev.* 51:221–71
55. Burgin AB, Pace NR. 1990. *EMBO J.* 9:4111–18
56. Nolan JM, Burke DH, Pace NR. 1993. *Science* 261:762–65
57. Harris ME, Nolan JM, Malhotra A, Brown JW, Harvey SC, Pace NR. 1994. *EMBO J.* 13:3953–63

58. Oh B-K, Pace NR. 1994. *Nucleic Acids Res.* 22:4087–94
59. Harris ME, Kazantsev AV, Chen J-L, Pace NR. 1997. *RNA* 3:561–76
60. Gautheret D, Damberger SH, Gutell RR. 1995. *J. Mol. Biol.* 248:27–43
61. Gutell RR, Power A, Hertz GZ, Putz EJ, Stormo GD. 1992. *Nucleic Acids Res.* 20:5785–95
62. Jaeger L, Michel F, Westhof E. 1994. *J. Mol. Biol.* 236:1271–76
63. Costa M, Michel F. 1995. *EMBO J.* 14:1276–85
64. Pley HW, Flaherty KM, McKay DB. 1994. *Nature* 372:111–13
65. Scott WG, Finch JT, Klug A. 1995. *Cell* 81:991–1002
66. Cate JH, Gooding AR, Podell E, Zhou K, Golden BL, et al. 1996. *Science* 273: 1678–85
67. Tallsjö A, Svard SG, Kufel J, Kirsebom LA. 1993. *Nucleic Acids Res.* 21:3927–33
68. Tanner MA, Cech TR. 1995. *RNA* 1:349–50
69. Massire C, Jaeger L, Westhof E. 1997. *RNA* 3:553–56
70. Westhof E, Altman S. 1994. *Proc. Natl. Acad. Sci. USA* 91:5133–37
71. Westhof E, Wesolowski D, Altman S. 1996. *J. Mol. Biol.* 258:600–13
71a. Chen J-LC, Nolan JM, Harris ME, Pace NR. 1998. *EMBO J.* 17:1515–25
72. Harris ME, Pace NR. 1995. *RNA* 1:210–18
73. Smith D, Burgin AB, Haas ES, Pace NR. 1992. *J. Biol. Chem.* 267:2429–36
74. Smith D, Pace NR. 1993. *Biochemistry* 32:5273–81
75. Beebe JA, Fierke CA. 1994. *Biochemistry* 33:10294–304
76. Hardt WD, Schlegl J, Erdmann VA, Hartmann RK. 1995. *J. Mol. Biol.* 247:161–72
77. Tallsjö A, Kirsebom LA. 1993. *Nucleic Acids Res.* 21:51–59
78. Reich C, Olsen GJ, Pace B, Pace NR. 1988. *Science* 239:178–81
79. Hardt WD, Schlegl J, Erdmann VA, Hartmann RK. 1993. *Nucleic Acids Res.* 21:3521–27
80. Guerrier-Takada C, Altman S. 1993. *Biochemistry* 32:7152–61
81. Beebe JA, Kurz JC, Fierke CA. 1996. *Biochemistry* 36:10493–505
82. Pan T. 1995. *Biochemistry* 34:902–9
83. Zarrinkar PP, Wang J, Williamson JR. 1996. *RNA* 2:564–73
84. Warnecke JM, Furste JP, Hardt W, Erdmann VA, Hartmann RK. 1996. *Proc. Natl. Acad. Sci. USA* 93:8924–28
85. Surratt CK, Carter BJ, Payne RC, Hecht SM. 1990. *J. Biol. Chem.* 265:22513–19
86. Gardiner KJ, Marsh TL, Pace NR. 1985. *J. Biol. Chem.* 260:5415–19
87. Perreault JP, Altman S. 1992. *J. Mol. Biol.* 226:399–409
88. Kleineidam RG, Pitulle C, Sproat B, Krupp G. 1993. *Nucleic Acids Res.* 21: 1097–1101
89. Eckstein F. 1985. *Annu. Rev. Biochem.* 54:367–402
90. Chen Y, Li XQ, Gegenheimer P. 1997. *Biochemistry* 36:2425–38
91. Hardt WD, Warnecke JM, Erdmann VA, Hartmann RK. 1995. *EMBO J.* 14:2935–44
92. Frank DN, Harris ME, Pace NR. 1994. *Biochemistry* 33:10800–8
92a. Frank DN, Pace NR. 1997. *Proc. Natl. Acad. Sci. USA.* 94:14355–60
93. Chang S, Carbon J. 1975. *J. Biol. Chem.* 250:5542–55
94. Guthrie C. 1975. *J. Mol. Biol.* 95:529–47
95. Barrell BG, Seidman JG, Guthrie C, McClain WH. 1974. *Proc. Natl. Acad. Sci. USA* 71:413–16
96. Engelke DR, Gegenheimer P, Ableson J. 1985. *J. Biol. Chem.* 260:1271–79
97. Goodman HM, Olson MV, Hall BD. 1977. *Proc. Natl. Acad. Sci. USA* 74:5453–57
98. Schmidt O, Mao J, Ogden R, Beckmann J, Sakano H, et al. 1980. *Nature* 287:750–52
99. Mao J, Schmidt O, Söll D. 1980. *Cell* 21:509–16
100. Kline L, Nishikawa S, Söll D. 1981. *J. Biol. Chem.* 256:5058–63
101. Surratt CK, Lesnikowski Z, Schifman AL, Schmidt FJ, Hecht SM. 1990. *J. Biol. Chem.* 265:22506–12
102. McClain WH, Guerrier-Takada C, Altman S. 1987. *Science* 238:527–30
103. Schlegl J, Hardt WD, Erdmann VA, Hartmann RK. 1994. *EMBO J.* 13:4863–69
104. Hardt WD, Schlegl J, Erdmann VA, Hartmann RK. 1993. *Biochemistry* 32:13046–53
105. Kahle D, Wehmeyer U, Krupp G. 1990. *EMBO J.* 9:1929–37
106. Thurlow DL, Shilowski D, Marsh TL. 1991. *Nucleic Acids Res.* 19:85–91
107. Seidman JG, McClain WH. 1975. *Proc. Natl. Acad. Sci. USA* 72:1491–95
108. Schmidt FJ, Seidman JG, Bock RM. 1976. *J. Biol. Chem.* 251:2440–45
109. Guerrier-Takada C, McClain WH, Altman S. 1984. *Cell* 38:219–24
110. Green CJ, Vold BS. 1988. *J. Biol. Chem.* 263:652–57
111. Kirsebom LA, Svard SG. 1994. *EMBO J.* 13:4870–76

112. Easterwood TR, Harvey SC. 1997. *RNA* 3:577–85
113. Darr SC, Zito K, Smith D, Pace NR. 1992. *Biochemistry* 31:328–33
114. Haas ES, Brown JW, Pitulle C, Pace NR. 1994. *Proc. Natl. Acad. Sci. USA* 91:2527–31
115. Green CJ, Rivera-Leon R, Vold BS. 1996. *Nucleic Acids Res.* 24:1497–1503
116. Stark BC, Kole R, Bowman EJ, Altman S. 1978. *Proc. Natl. Acad. Sci. USA* 75:3717–21
117. Gardiner K, Pace NR. 1980. *J. Biol. Chem.* 255:7507–9
118. Hansen FG, Hansen EB, Atlung T. 1985. *Gene* 38:85–93
119. Ogasawara N, Moriya S, von Meyerburg K, Hansen FG, Yoshikawa H. 1985. *EMBO J.* 4:3345–50
120. Brown JW, Pace NR. 1992. *Nucleic Acids Res.* 20:1451–56
121. Peck-Miller KA, Altman S. 1991. *J. Mol. Biol.* 221:1–5
122. Liu F, Altman S. 1994. *Cell* 77:1093–1100
123. Gopalan V, Talbot SJ, Altman S. 1994. In *RNA-Protein Interactions*, ed. K Nagai, IW Mattaj, pp. 103–26. Oxford: IRL
124. Lawrence N, Wesolowski D, Gold H, Bartkiewicz M, Guerrier-Takada C, et al. 1987. *Cold Spring Harbor Symp. Quant. Biol.* 52:233–38
125. Darr SC, Pace B, Pace NR. 1990. *J. Biol. Chem.* 265:12927–32
126. Jayanthi GP, Van TG. 1992. *Arch. Biochem. Biophys.* 296:264–70
127. Stathopoulos C, Kalpaxis DL, Drainas D. 1995. *Eur. J. Biochem.* 228:976–80
128. Altman S, Gold HA, Bartkiewicz M. 1988. In *Structure and Function of Major and Minor Small Nuclear Ribonucleoprotein Particles*, ed. M Birnstiel, pp. 183–95. New York: Springer-Verlag
129. Zimmerly S, Drainas D, Sylvers LA, Söll D. 1993. *Eur. J. Biochem.* 217:501–7
130. Lygerou Z, Mitchell P, Petfalski E, Seraphin B, Tollervey D. 1994. *Genes Dev.* 8:1423–33
131. Dichtl B, Tollervey D. 1997. *EMBO J.* 16:417–29
132. Chu S, Zengel JM, Lindahl L. 1997. *RNA* 3:382–91
133. Lygerou Z, Pluk H, van Venrooij WJ, Seraphin B. 1996. *EMBO J.* 15:5936–48
134. Gold HA, Craft J, Hardin JA, Bartkiewicz M, Altman S. 1988. *Proc. Natl. Acad. Sci. USA* 85:5483–87

135. Yuan Y, Tan E, Reddy R. 1991. *Mol. Cell Biol.* 11:5266–74
136. Eder PS, Kekuda R, Stolc V, Altman S. 1997. *Proc. Natl. Acad. Sci. USA* 94:1101–6
137. Liu MH, Yuan Y, Reddy R. 1994. *Mol. Cell. Biochem.* 130:75–82
138. Morales MJ, Wise CA, Hollingsworth MJ, Martin NC. 1989. *Nucleic Acids Res.* 17:6865–82
139. Morales MJ, Dang YL, Lou YC, Sulo P, Martin NC. 1992. *Proc. Natl. Acad. Sci. USA* 89:9875–79
140. Dang YL, Martin NC. 1993. *J. Biol. Chem.* 268:19791–96
141. Yamaguchi-Shinozaki K, Shinozaki K, Sugiura M. 1987. *FEBS Lett.* 215:132–36
142. Wang MJ, Gegenheimer P. 1990. *Nucleic Acids Res.* 18:6625–31
143. Thomas BC. 1996. *Structural and mechanistic differences between a protein and an RNA-containing RNase P.* PhD thesis. Univ. Kansas, Lawrence. 142 pp.
144. Banta AB, Haas ES, Brown JW, Pace NR. 1992. *Nucleic Acids Res.* 20:911
145. Robertson HD, Altman S, Smith JD. 1972. *J. Biol. Chem.* 247:5243–51
146. Pley HW, Flaherty KM, McKay DB. 1994. *Nature* 372:68–74
147. Melton DA, Cortese R. 1979. *Cell* 18:1165–72
148. Melton DA, DeRobertis EM, Cortese R. 1980. *Nature* 284:143–48
149. Matera AG, Frey MR, Margelot K, Wolin SL. 1995. *J. Cell. Biol.* 129:1181–93
150. Lee B, Matera AG, Ward DC, Craft J. 1996. *Proc. Natl. Acad. Sci. USA* 93: 11471–76
151. Lygerou Z, Allmang C, Tollervey D, Seraphin B. 1996. *Science* 272:268–70
152. Chamberlain JR, Pagan-Ramos E, Kindelberger DW, Engelke DR. 1996. *Nucleic Acids Res.* 24:3158–66
153. Bothwell ALM, Garber RL, Altman S. 1976. *J. Biol. Chem.* 251:7709–16
154. Komine Y, Kitabatake M, Yokogawa T, Nishikawa K, Inokuchi H. 1994. *Proc. Natl. Acad. Sci. USA* 91:9223–27
155. Zimmerly S, Gamulin V, Burkard U, Söll D. 1990. *FEBS Lett.* 271:189–93
156. Mamula MJ, Baer M, Craft J, Altman S. 1989. *Proc. Natl. Acad. Sci. USA* 86: 8717–21
157. Doerson C, Guerrier-Takada C, Altman S, Attardi G. 1985. *J. Biol. Chem.* 260:5942–49

Annu. Rev. Biochem. 1998. 67:181–98

BASE FLIPPING

Richard J. Roberts
New England Biolabs, Beverly, Massachusetts 01915; e-mail: roberts@neb.com

Xiaodong Cheng
Department of Biochemistry, Emory University School of Medicine, Atlanta, Georgia 30322; e-mail: xcheng@emory.edu

KEY WORDS: DNA repair enzymes, DNA methyltransferases, DNA-modifying enzymes, evolution

ABSTRACT
Base flipping is the phenomenon whereby a base in normal B-DNA is swung completely out of the helix into an extrahelical position. It was discovered in 1994 when the first co-crystal structure was reported for a cytosine-5 DNA methyltransferase binding to DNA. Since then it has been shown to occur in many systems where enzymes need access to a DNA base to perform chemistry on it. Many DNA glycosylases that remove abnormal bases from DNA use this mechanism. This review describes systems known to use base flipping as well as many systems where it is likely to occur but has not yet been rigorously demonstrated. The mechanism and evolution of base flipping are also discussed.

CONTENTS

INTRODUCTION

The binding of proteins to DNA is crucial to life. Whereas some proteins bind to control transcription and replication, others both bind to DNA and catalyze chemical reactions on the DNA. The latter proteins include nucleases, glycosylases, DNA methyltransferases, and various enzymes such as integrases and recombinases, which rearrange DNA segments. For proteins such as transcription factors and repressors, whose function rests with DNA binding, some common structural features have been identified, such as zinc fingers and helix-turn-helix motifs (1). The binding to DNA by these proteins frequently involves small deformations in the usual B helix, although in some cases extreme distortions can be found, such as the saddle structure on which the TATA-binding protein sits (2, 3) or the bends induced by integration host factor (4). Some nucleases, such as R.*Eco*RI (5) and R.*Eco*RV (6), which cleave the phosphodiester backbone, also induce conformational changes in the DNA helix, but for the most part these changes are not dramatic. This is probably because the target phosphodiester bonds lie on the outside of the helix and thus are readily accessible to the enzyme. For proteins that need to interact with the bases rather than the phosphodiester backbone and then perform chemistry on those bases, accessibility is less clear. How might such proteins gain access to the interior of the helix? Although it seems possible that the bases might become accessible by distortion of the helix through bending and kinking, it is by no means simple to envision. Obviously, a structure is needed for such a protein actually binding to DNA.

In 1993 a structure was obtained for the cytosine-5 DNA methyltransferase, M.*Hha*I, an enzyme that needed access to its target cytosine base to perform the chemistry of methylation within the aromatic ring (7, 8). The crystal contained both the methyltransferase and its cofactor, *S*-adenosyl-L-methionine (AdoMet), but no DNA. From the structure, and the accumulated biochemical knowledge about the enzyme, some clear predictions could be made about where the DNA would be located. However, the structure gave little hint of the rather dramatic conformational change in DNA that would take place upon its binding. That change was revealed in 1994, when a ternary structure was reported for a complex between M.*Hha*I, its DNA substrate, and the reaction product, *S*-adenosyl-L-homocysteine (9). Surprisingly, the enzyme did not distort the DNA in some crude fashion by bending or kinking, but rather the target cytosine had swung completely out of the helix and into the active-site pocket of the enzyme (Figure 1, C-1, color section at end of volume). The base had undergone a conformational shift of 180°, and the first example of base flipping had been discovered. In retrospect, this shift makes sense, because it probably provides the simplest method for a protein to access the interior of a DNA helix.

Because of the overall similarity at the primary sequence level of all cytosine-5 DNA methyltransferases (10), it seemed reasonable that base flipping was not just an isolated quirk unique to M.*HhaI* but would be found in all members of this set of methyltransferases. It was thus reassuring when a structure for a second methyltransferase, M.*HaeIII*, appeared (11). It showed essentially the same phenomenon, albeit with some differences that could be attributed to differences in the recognition sequences. The big question, though, was whether base flipping would show up in other situations. In our original paper we proposed that other classes of DNA methyltransferases and some DNA glycosylases might also use base flipping to gain access to the bases within the helix (9). Both of these predictions proved accurate, although in the case of T4 endonuclease V (12), the base flipping is not quite as we had envisioned (see below). Many other examples have now appeared, either as co-crystal structures of DNA-protein complexes or from structures of the native enzymes with or without cofactors. In the latter cases, it has often proved possible to infer that base flipping may be involved because modeling of normal B-DNA into these structures fails to bring the target base close to the active site—a situation that is remedied if the base is flipped.

It has been argued that base flipping is an ancient process that may even have preceded the use of DNA as the genetic material (13). If that is the case, then one might expect to find other examples of base flipping among RNA enzymes and in other proteins that interact with DNA. One might even imagine that base flipping could nucleate the opening of a helix, as would be required during the initiation of replication or transcription. In this review we summarize the progress that has been made in studies of base flipping since its discovery in 1994 and discuss, somewhat speculatively, its mechanism and evolution. We present the range of processes where base flipping has been proven to occur, further cases where it is postulated to occur, and a few examples where it might be worth looking for its involvement.

KNOWN BASE-FLIPPING SYSTEMS

Base flipping has been directly observed in four systems (Table 1): in the co-crystal structures for the cytosine-5 DNA methyltransferases M.*HhaI* (9) and M.*HaeIII* (11), in a catalytically compromised form of T4 endonuclease V (12), and in human uracil DNA glycosylase (14).

Table 1   Known base-flipping systems

Specific protein	Catalytic reaction	Reference
*Hha*I DNA methyltransferase	Forms G-5mC-GC in DNA	9
*Hae*III DNA methyltransferase	Forms GG-5mC-C in DNA	11
T4 endonuclease V	Removes pyrimidine dimers from DNA	12
Human uracil DNA glycosylase	Removes uracil from DNA	14

*Hha*I DNA Methyltransferase

*Hha*I DNA methyltransferase (M.*Hha*I), a 327–amino acid residue protein, methylates the internal cytosine of its recognition sequence 5′-GCGC-3′/3′-CGCG-5′ (15, 16). Despite its palindromic recognition sequence, the protein functions as a monomer, methylating only one strand at a time. The methyl donor AdoMet is required for enzymatic activity and is converted to S-adenosyl-L-homocysteine (AdoHcy) during methylation.

The structure of M.*Hha*I has been characterized extensively by X-ray crystallography in complexes with various forms of its DNA substrate (9, 17–19). The protein has two lobes: a conserved AdoMet-dependent catalytic domain and a DNA-recognition domain (7, 8, 20, 21). Nine ternary structures of M.*Hha*I-DNA-cofactor have been determined, with either AdoMet or AdoHcy. Each structure contains a 13-mer oligonucleotide duplex with a different combination of modifications at the position of the target cytosine within the recognition sequence in the two DNA strands. Table 2 lists the oligonucleotides used in the M.*Hha*I ternary complexes and the nomenclatures used.

5-FLUORO-2′-DEOXYCYTIDINE, 2′-DEOXYCYTIDINE, AND 5-METHYL-2′-DEOXYCYTIDINE   The first three structures listed in Table 2 contain identical self-complementary DNA sequences. The complex F13 contains 5-fluoro-2′-deoxycytidine (5FC or F) and was crystallized in the presence of AdoMet, which resulted in the formation of a covalent linkage between the enzyme and DNA and generation of 5-methyl-5-fluoro-2′-deoxydihydrocytidine and AdoHcy (9). The N13 structure contains unmodified deoxycytidine, whereas the M13 contains 5-methyl-2′-deoxycytidine (5mC or M) (17). The structures are rather similar, with one of the target nucleotides flipped out of the DNA helix and fitting snugly into the active site of the enzyme (Figure 1, C-1, color section).

The 5mC residue in the fully methylated oligonucleotide in M13, the reaction product, is flipped out of the DNA helix in the same manner as with the C in N13 and 5FC in F13. This is surprising but consistent with biochemical data, which suggest that the binding specificity for M.*Hha*I is asymmetric 5′-GXGC-3′/3′-CGCG-5′ and determined by the nucleotides neighboring the target (X)

Table 2 DNA sequences used in M.*Hha*I ternary structures

DNAᵃ	Xᵇ	Cofactor	Name	Resolution (Å)	Reference
5'-TGATA*GXGC*TATC -3'	F	AdoMet	F13	2.8	9
3'- CTAT*CGX*GATAGT-5'	C	AdoHcy	N13	2.7	17
	M	AdoHcy	M13	2.7	17
5'-TGTCA*GXGC*ATGG -3'	C	AdoHcy	HM13	2.7	18
3'- CAGT*CGMG*TACCT-5'	S	AdoMet	S13	2.05	19
	Z	AdoHcy	Z13	2.5	c
5'-TGTCA*GXGC*ATGG -3'	AP	AdoHcy	AP13	2.4	d
3'- CAGT*CGCG*TACCT-5'					
5'-TGTCA*GCGC*ATGG -3'	A	AdoHcy	GA13	2.8	d
3'- CAGT*CGXG*TACCT-5'	U	AdoHcy	GU13	2.7	d

ᵃThe recognition sequence is in italics, and the flipped nucleotide is in bold and underlined.
ᵇA = adenine, C = cytosine, U = uracil, M = 5-methyl-2'-deoxycytidine, F = 5-fluoro-2'-deoxycytidine, S = 4'-thio-2'-deoxycytidine, Z = 5,6-dihydro-5-azacytidine, AP = abasic sugar.
ᶜJK Christman, R Sheikhnejad, V Marquez, C Marasco, J Sufrin, M O'Gara, X Cheng, unpublished data.
ᵈM O'Gara, JR Horton, RJ Roberts, X Cheng, unpublished data.

nucleotide (22). In other words, the methyltransferase does not depend on the flippable base for its binding specificity.

The HM13 structure (the fourth in Table 1) contained a hemimethylated DNA substrate with a nonpalindromic sequence (18). The two strands of DNA are clearly nonequivalent, because only the unmethylated target C flips out of the DNA helix, whereas the 5mC on the complementary strand remains stacked in the DNA helix. This seems to be an important aspect of DNA methylation; that is, the ability of DNA methyltransferases to distinguish DNA substrates with methyl groups on one strand (hemimethylated DNA) from those that carry no methyl groups. However, many of the bacterial type II enzymes such as M.*Hha*I are equally active on unmethylated and hemimethylated DNA, with an increased affinity for asymmetrically methylated DNA. Consequently they function as both maintenance and de novo methyltransferases. Both the N4 (NH₂) group and the methyl group of 5mC on the complementary strand interact with the protein (18); these interactions may enable the de novo methyltransferase to recognize both unmodified cytosine and methylated cytosine.

4'-THIO-2'-DEOXYCYTIDINE AND DIHYDRO-5-AZACYTIDINE   That M.*Hha*I does not show much binding specificity for the flippable base may reflect a need to leave that base unencumbered by recognition contacts. This provides an opportunity to probe the structural and chemical interactions involved in sequence-specific recognition and catalysis using nucleotide analogs incorporated into synthetic oligonucleotides in the position of the target nucleotide of M.*Hha*I. So far, two nucleotide analogs—a 4′-thionucleoside and 5,6-dihydro-5-azacytosine —have been used in our studies.

When 4′-thio-2′-deoxycytidine is incorporated as the target cytosine in the recognition sequence for M.*Hha*I, binding to the 4′-thio–modified DNA is almost identical to that of the unmodified DNA under equilibrium conditions (19). In contrast, the methyl transfer was strongly inhibited in solution. Surprisingly, the flipped 4′-thio-2′-deoxycytidine in the S13 crystal structure was partially methylated (19). These results show that 4′-thio-2′-deoxycytidine does not disrupt DNA recognition, binding, or base flipping by M.*Hha*I. Instead they suggest that it interferes with a step in the methylation reaction after flipping but prior to methyl transfer.

5,6-Dihydro-5-azacytosine (DHAC) riboside was originally synthesized to obtain a hydrolytically stable replacement for the antileukemic drug 5-azacytosine riboside (23, 24). DHAC contains a cytosine-like ring lacking aromatic character with an sp^3-hybridized carbon (CH_2 group) at position 6 and an NH group at position 5, resembling the transition state of a dihydrocytosine intermediate in the reaction of cytosine-5 DNA methyltransferases (25). The Z13 structure, containing DHAC as the target (Table 2), showed that DHAC is also flipped out of the DNA helix similarly to C, 5mC, and 5FC, but with no covalent bond formed between the sulfur atom of nucleophile Cys81 and the pyrimidine C6 carbon (JK Christman, R Sheikhnejad, V Marquez, C Marasco, J Sufrin, M O'Gara & X Cheng, unpublished data). This result indicates that the DHAC-containing DNA is sufficient to produce strong inhibition of the DNA methyltransferase and that the DHAC moiety occupies the active site of M.*Hha*I as a transition-state mimic.

MISMATCHED BASES (G:A, G:U, AND G:AP)   Following the discovery of base flipping by M.*Hha*I, the effects of replacing the target cytosine by mismatched bases, including adenine, guanine, thymine, and uracil, were investigated (22, 26). By electrophoretic mobility shift analysis, M.*Hha*I and M.*Hpa*II were found to bind even more tightly to such mismatched substrates; the highest affinity was for a gap formed by removal of the target nucleotide and both phosphodiester linkages (22). Furthermore, the uracil can be enzymatically methylated and converted to thymine at low efficiencies (22, 26), and the binding of these methyltransferases at the G:U mismatch prevented its repair by uracil DNA glycosylase in vitro (26).

So far, three well-refined ternary structures of M.*Hha*I complexed with AdoHcy and a nonpalindromic oligonucleotide containing a G:A, G:U, or G:AP (AP = apurinic/apyrimidinic = abasic) mismatch at the target base pair, respectively, have been determined (M O'Gara, JR Horton, RJ Roberts & X Cheng, unpublished data). The mismatched adenine, uracil, and abasic site are flipped out and located in the enzyme's active site, respectively. It seems likely that this active-site pocket is nonspecific for binding but specific for methylation. In light of the nonspecific binding pocket, the DNA methyltransferase may be related more to repair enzymes such as 3-methyladenine DNA glycosylase II and endonuclease III, which have broad substrate specificity. On the other hand, the methylation reaction is specific in that catalysis occurs only when the flipped base is cytosine or uracil (at low efficiency), more closely resembling uracil DNA glycosylase.

*Hae*III DNA Methyltransferase

*Hae*III DNA methyltransferase (M.*Hae*III), which also belongs to the monomeric type II bacterial methyltransferases, acts at its substrate site in palindromic DNA and modifies the recognition sequence in two independent methyl transfer events—5'-GG<u>C</u>C-3'/3'-C<u>C</u>GG-5'—in double-stranded DNA. It methylates either of the underlined cytosines. In the structure of M.*Hae*III bound to hemimethylated DNA, the substrate cytosine is flipped out from the DNA helix (Figure 2, C-1, color section), as observed for M.*Hha*I. M.*Hae*III differs from M.*Hha*I in that the flipping out of the cytosine is accompanied by rearrangement of the remaining recognition bases in their pairing (11).

Between M.*Hha*I and M.*Hae*III, the protein-base contacts in the recognized sequence are expected to differ because of their different specificity. Indeed, the folding of the corresponding DNA-recognition domains is different (11). However, these protein-base contacts share two important features (27). For both methyltransferase-DNA complexes, six phosphates on the methylated strand contact the protein: three on the 5' side and three on the 3' side of the target cytosine, regardless of the position of the target nucleotide within the recognition sequence (second of four for M.*Hha*I, 5'-G<u>C</u>GC-3', and third of four for M.*Hae*III, 5'-GG<u>C</u>C-3'). From the protein side, a conserved arginine is responsible for the recognition of the guanine 5' to the target cytosine. These shared recognition patterns may be common to other 5mC methyltransferases recognizing a guanine 5' to the target base (27, 28).

Human Uracil DNA Glycosylase

Human uracil DNA glycosylase (UDG) is responsible for the removal of uracil residues within either single- or double-stranded DNA. The crystal structures of the human (29, 30) and herpes simplex viral (31, 32) forms of the enzyme have been solved. They revealed an extraordinarily specific binding pocket that excludes normal DNA bases and uracil within RNA from specific binding. Clearly these interactions could not form if the uracil was positioned inside a B-DNA helix, so it seemed likely that base flipping could be involved. The involvement of base flipping has now been confirmed with the description of a co-crystal structure for human UDG complexed with a uracil-containing double-stranded DNA (Figure 3, C-2, color section) (14). This structure, which was solved using a double mutant of UDG (Leu272Arg and Asp145Asn) showed that the uracil, the deoxyribose, and the 5′ phosphate are rotated 180° from their starting structure within DNA, with Arg272 inserted into the position previously occupied by uracil. Even though the catalytic efficiency of this double mutant was severely reduced, the N-Cl′ glycosidic bond had been broken (Figure 3, C-2, color section).

The DNA conformation essentially maintained the B form, except for the flipped nucleotide. The conformation of the flipped-out nucleotide is very similar to that observed in the M.HhaI-DNA complex, and the protein-DNA interactions are concentrated along the sugar-phosphate backbone of the strand containing the flipped-out uracil (14). Thus, the four structurally characterized base-flipping proteins, M.HhaI, M.HaeIII, human UDG, and T4 endonuclease V (see below), have all shown that the DNA phosphate-protein contacts are made mainly on one particular strand (reviewed in 27).

It was suggested that UDG uses a push-and-pull mechanism of base flipping to ultimately achieve specific binding and catalysis (14). The push (protein invasion) is contributed by Leu272: It displaces the target base through hydrophobic interactions with DNA. The pull (trapping the flipped nucleotide) is derived by satisfying the specific binding pocket with uracil.

T4 Endonuclease V

T4 endonuclease V is a DNA glycosylase/AP lyase that can initiate repair of cis-syn cyclobutane pyrimidine dimers in DNA by cleaving the glycosydic bond of the 5′ pyrimidine and then cleaving the phosphodiester backbone (33). Unlike M.HhaI and UDG, T4 endonuclease V kinks the dimer-containing DNA, at an angle of approximately 60° (Figure 4, C-3, color section) (12). In contrast to UDG, endonuclease V does not flip out the damaged bases into a binding pocket. Instead, it moves the nucleotide opposite the 5′ pyrimidine of the dimer into a binding pocket on the surface of the enzyme (Figure 4, C-3, color section). In the co-crystal structure, a flipped adenine lies in proximity of and sandwiched between two layers of protein atoms, both of which are arranged parallel to the base plane. Vassylyev et al (12) suggested that the adenine was stabilized by van der Waals interactions. Interestingly, the pocket into which the adenine flips does not provide specific contacts that allow unambiguous recognition of the base. This lack of specific recognition has been seen biochemically (34; unpublished data cited in 12) and shows that the glycosylase activity is unaffected by the nature of the base opposite the 5′-lesion.

However, the key feature associated with endonuclease V's base flipping is that the hole in DNA, which is created by movement of the base, is filled by the enzyme inserting its active-site residues into that hole. Thus, through the change in the structure of the DNA, the enzyme is correctly positioned to carry out a nucleophilic attack on Cl′ of the 5′ pyrimidine of the dimer.

PROBABLE BASE-FLIPPING SYSTEMS

Several crystal structures have appeared for proteins that interact with DNA. In these structures, it seems improbable that proper interaction could occur if the DNA maintained its normal B conformation. However, in these cases modeling with DNA containing a flipped base gives a much more reasonable chance of interaction and suggests that base flipping is used in these systems. Several recent reviews have appeared that discuss flipping in the context of enzymes involved in the repair of DNA lesions (35–38). These systems are described individually below and summarized in Table 3. Although definitive proof of base flipping for those enzymes listed in Table 3 awaits successful co-crystallization studies, the locations of concave active sites within deep clefts in these enzymes strongly suggest that a simple base flipping in the DNA takes place rather than a complicated refolding of the protein.

The Amino-Methyltransferases M.TaqI and M.PvuII

One reason to flip out a DNA nucleotide is to subject it to chemical modification within a catalytic pocket (13, 27), as noted above for the cytosine-5 DNA methyltransferases (Figures 1 and 2, C-1, color section). Another class of DNA methyltransferases, the amino-methyltransferases, methylate the exocyclic amino groups of adenine (N6) or cytosine (N4). None of these amino-methyltransferases have been structurally characterized in complex with DNA,

Table 3 Probable base-flipping systems

Specific protein	Catalytic reaction	Reference
TaqI DNA methyltransferase	Forms TCG-6mA in DNA	39
PvuII DNA methyltransferase	Forms CTG-4mC-AG in DNA	40
Escherichia coli photolyase	Converts pyrimidine dimers to TT	41
E. coli endonuclease III	Removes pyrimidine radiolysis products from DNA	42
3-Methyladenine DNA glycosylase	Removes 3-methyladenine from DNA	43, 44
E. coli exonuclease III	Cleaves 5′ to Ap sites	45
T4 β-glucosyltransferase	Transfers glucose residues to T4 DNA	46
E. coli Ada O6-methylguanine DNA methyltransferase	Removes Me from O6-Me-guanosine	47
E. coli mismatch-specific uracil DNA glycosylase	Removes thymine from DNA	a

[a]TE Barrett, R Savva, LH Pearl, unpublished data.

but M.*Taq*I (adenine-N6 specific) (Figure 5, C-4, color section) and M.*Pvu*II (cytosine-N4 specific) (Figure 6, C-5, color section) have a catalytic-domain structure with a concave active-site pocket very similar to that of M.*Hha*I (39, 40, 48–50). When these structures are modeled with normal B-DNA [Figures 5 (*top*) and 6 (*bottom*); see color section], the AdoMet is far away from the target base, but by flipping this base out of the helix the two reactants are brought into close proximity [Figures 5 (*bottom*), and 6 (*bottom*); see color section]. Thus, both N^6A-methyltransferases and N^4C-methyltransferases almost certainly flip out their target nucleotides.

In the case of M.*Eco*RI, which forms GA(^{6me}A)TTC, biochemical evidence in favor of base flipping has been obtained through a fluorescence-based assay to detect conformational alterations in DNA induced by protein binding (51). This method, which uses 2-aminopurine to replace the adenine in the substrate DNA, may have general applications to probe potential base flipping in other systems. Thus it has recently been shown that when M.*Hha*I and M.*Taq*I bind to oligonucleotides containing 2-aminopurine as the target base within their recognition sequences, the fluorescence intensity is greatly enhanced, consistent with base flipping (B Holz, S Klimasauskas, S Serva & E Weinhold, unpublished data). Biochemical evidence in favor of base flipping has also been developed for another N^6A-methyltransferase, M.*Eco*RV (52). Here it has been shown that enhanced binding takes place when the target adenine, the first A in the recognition sequence GATATC, is replaced by a modified base that weakens base pairing. These results parallel those found for the known base-flipping cytosine-5 methyltransferase, M.*Hha*I (22). In all of these studies it is believed that destabilization of the base pair leads to an increased rate of base flipping and the concomitant appearance of enhanced binding.

E. coli *DNA Photolyase*

In addition to DNA repair enzymes such as UDG that function within the base excision repair pathway, DNA photolyase—an enzyme that acts through a direct reversal pathway—also appears to incorporate nucleotide flipping in its mechanism of action (41). *E. coli* photolyase uses a blue light–harvesting chromophore 5,10 methenyl-tetrahydrofolylpolyglutamate (MTHF) to absorb a photon and transfer the excitation energy to a catalytic chromophore, flavin adenine dinucleotide (FAD) (53). The enzyme further transfers an electron from FADH to the cyclobutane pyrimidine dimer to catalyze its fission to yield the two original pyrimidines (53).

Examination of the solvent-accessible surface of the photolyase revealed that the FAD cofactor is accessible to the dimer, only by way of a cavity in the enzyme (Figure 7, C-6, color section). The dimensions of the putative binding-site cavity are sufficient to bind the bases of the dimer but to exclude the intradimer phosphate. The cavity is lined with polar residues on one side and hydrophobic residues on the other. Park et al (41) point out that the asymmetric polarity of the cavity matches well with the asymmetric polarity of the pyrimidine dimer, in which the cyclobutane ring is hydrophobic and the opposite edges of the thymine bases have nitrogens and oxygens capable of forming hydrogen bonds.

E. coli *Endonuclease III*

The X-ray crystal structure of endonuclease III showed that this enzyme possesses an elongated bilobal structure with a deep cleft separating two distinct domains linked by two loops (Figure 8, C-6, color section) (42, 54). The two domains are α-helical in structure, with one containing a 4Fe-4S cluster and the other containing a helix-hairpin-helix motif.

The cleft separating the two domains can accommodate B-form DNA; thus both domains have been suggested to be involved in DNA binding. Endonuclease III substrate binding was investigated by soaking the enzyme crystals in thymine glycol, a known inhibitor of glycosylase activity (54). The thymine glycol binding site was located within a water-filled pocket in the cleft. Thus, the damaged base is suggested to be flipped extrahelical and inserted into this pocket for catalysis to occur, in a manner similar to that of M.*Hha*I and UDG (Figure 8, C-6, color section).

3-Methyladenine DNA Glycosylase (AlkA)

AlkA is an *N*-alkylpurine DNA glycosylase that can efficiently remove 3-methyladenine, 7-methylguanine, 3-methylguanine, $O2$-methylcytosine, and $O2$-methylthymine from double-stranded DNAs. The crystal structure was published by Yamagata et al (43) and Labahn et al (44). These studies revealed that AlkA is composed of three approximately equal-sized domains. Domain 1 has the topology and shape of the TATA-binding protein (2, 3, 55), whereas domains 2 and 3 have the same topology and fold as endonuclease III (see above). Domain 1 appears to serve as a platform for the other two domains. The interface of domains 2 and 3 creates the base-binding and catalytic site, and the flexibility of domain 3 permits the binding of various substrates.

Examination of the surface topography of AlkA revealed a cleft at the junction of domains 2 and 3, which is hydrophobic in nature and is dominated by aromatic amino acid side chains (Figure 9, C-7, color section) (43). The hydrophobic cleft also contains several charged residues, including the catalytically essential Asp. However, to position the methylated base correctly within the active site, it was suggested (44) that AlkA uses the π electron-rich aromatic rings to interact with the electron-deficient, positively charged alkylated base, which is flipped out of the DNA helix. Thus, broad substrate specificity can likely be achieved through these strong π donor-acceptor interactions.

E. coli *Exonuclease III*

The only hydrolytic AP endonuclease that has its structure solved to date is *E. coli* exonuclease III (45). The overall structure was shown to be similar to that of DNase I (56), even though there is less than 20% sequence similarity. It consists of two six-stranded β sheets, flanked by four α-helices forming a four-layered $\alpha\beta$-sandwich motif. A ternary complex of exonuclease III with Mn^{2+} and dCMP has also been reported (45). This complex showed the dCMP bound at one end of the $\alpha\beta$ sandwich, a region of the enzyme that contains a high concentration of basic residues.

It was suggested that DNA-containing AP sites may be significantly distorted such that the abasic sugar may protrude into the active site to allow the chemistry to occur (45). As discussed above, M.*Hha*I is also able to flip an abasic sugar. Alternatively, Mol et al (45) suggest that the nucleotide opposite the AP site may be flipped into a pocket on the surface of the protein. If this flipping does occur, then exonuclease III will be an example similar to T4 endonuclease V.

T4 β-Glucosyltransferase

Bacteriophage T4 uses 5-hydroxymethylcytosine in place of cytosine when polymerizing its DNA. Following DNA synthesis, most (if not all) of these hydroxymethylcytosines in duplex DNA are further modified by the addition of glucose. This reaction is catalyzed by the β-glucosyltransferase, which transfers the glucose from uridine diphosphoglucose to the hydroxymethyl groups in 5-hydroxymethylcytosine. The structure of the enzyme was solved in both the presence and absence of the glucose donor, uridine diphosphoglucose (46). The structure comprises two domains of similar topology (Figure 10, C-7, color section). The two domains are separated by a cleft that contains a possible site for duplex DNA binding. The catalytic site can be inferred from the position of the UDP-glucose, which is deeply bound in a pocket at the bottom of the cleft. When normal B-DNA is modeled into the structure, the glucose donor and the target base are widely separated. However, as the authors suggest, base flipping of the 5-hydroxymethylcytosine would nicely juxtapose the two interacting moieties (46).

E. coli *Ada O6-Methylguanine DNA Methyltransferase*

O6-methylguanine DNA methyltransferase, otherwise known as the ada repair enzyme, is a suicidal DNA-repair protein that removes the dangerous methyl group from the promutagenic lesion O6-methylguanine and transfers it onto a cysteine in the protein. The resulting self-methylation of the active-site cysteine renders the protein inactive. A structure has been reported for the protein without DNA, but as noted (47), when B-DNA is modeled into the structure, a target base would be more than 20 Å away from the active center! Thus a significant conformational change would be required to bring the O6-methylguanine into close proximity with the buried catalytic cysteine that is the acceptor for the methyl group (Figure 11, C-8, color section). Although Moore et al (47) proposed a model in which the protein undergoes a drastic conformational distortion to bind DNA, it seems more likely that flipping the O6-methylguanine out of the DNA helix would solve the problem of accessibility.

E. coli *Mismatch-Specific Uracil DNA Glycosylase*

Mismatch-specific uracil DNA glycosylase (MUG) removes uracil by a glycolytic mechanism from mismatches that contain uracil or thymine opposite guanine, which would arise from deamination of cytosine or 5-methylcytosine. The enzyme was originally discovered and characterized on the basis of its amino acid similarity to human thymine glycosylase (57). Recently a crystal structure has been obtained for *E. coli* MUG binding to DNA (TE Barrett, R Savva & LH Pearl, unpublished data). The structure shows great similarity to uracil DNA glycosylase, especially around the active site. In the crystal, the uracil base has been hydrolyzed and an abasic site is left. Interestingly, an arginine residue is intercalated on the opposite strand, and this may be partly responsible for the double-stranded DNA specificity of this enzyme.

T7 ATP-Dependent DNA Ligase and T7 DNA Polymerase

Finally, some proteins may mimic base flipping without actually flipping anything. The T7 ATP-dependent DNA ligase has a structure that broadly resembles that of DNA methyltransferases (58). Subramanya et al (58) have proposed that the reaction intermediate, which has AMP covalently bonded to the target nick in the DNA, is formally analogous to the DNA methyltransferase intermediates: The AMP is flipped out, even though no bases are missing from the target DNA.

A recent crystal structure for T7 DNA polymerase in complex with a primer template, a nucleoside triphosphate, and its processivity factor (thioredoxin) has been obtained (S Doublie, S Tabor, A Long, CC Richardson & T Ellenberger, unpublished data). In this complex the template strand is sharply kinked such that the base immediately 5' of the template base paired to the incoming nucleotide is flipped out of the active site. This conformation exposes the template base for interactions with the polymerase and its nucleotide substrate.

MECHANISM

We have little hard data about the mechanism of base flipping. Two theories are debated. In the first, it is suggested that base flipping is an active process in which the base is pushed out of the helix by some appropriate residues on the protein. Once flipped, it is pulled into the active site of the enzyme, where it remains trapped during the reaction. A three-step pathway has been proposed for the DNA methyltransferases in which they first recognize the target site and increase the interstrand phosphate-phosphate distance nearby. They then initiate base flipping by protein invasion of the DNA and finally trap the flipped DNA structure (27). An alternative possibility is that during the normal breathing of DNA, the bases naturally spend some time in the completely flipped-out position, and it is this transient conformation in DNA that is recognized and caught by the protein. Although some biophysical measurements of DNA motion have been made and lifetimes assigned to the flipped-out state, all such measurements rely on assumptions about the states being modeled (59). For instance, there is no direct experimental evidence showing that the lifetime assigned to the flipped-out state is correct.

Most recently, nuclear magnetic resonance (NMR) techniques have been applied to study the interaction of M.*Hha*I with DNA containing 5-fluorocytosine in solution (S Klimasauskas, T Szyperski, S Serva & K Wuthrich, unpublished data). Surprisingly, the dynamics of base-pair opening appear unaffected by binding M.*Hha*I alone. Only when the cofactor, or a cofactor analog, is added can a clear signal be detected that is attributable to the flipped state of the target base. Furthermore, these NMR studies suggest that at least three protein-DNA complexes are involved: (*a*) an initial binding complex that forms between the DNA and the protein in which the DNA maintains its normal stacked B conformation; (*b*) a complex, or series of complexes, in which the target base is flipped from the helix but is not yet locked into its final flipped position observed crystallographically; and (*c*) a final, catalytically competent complex in which the base is locked into the active site. This last complex would correspond to the complex seen in the crystal structure (9). Based on the NMR evidence, the initial complex and the flipped-out complex are major equilibrium species in solution, and all three types of complexes appear to be in rapid equilibrium until the cofactor binds and shifts the equilibrium to favor the final complex. This initial flipping is suggested to be an enzyme-assisted process even in the absence of cofactor. Unfortunately, there is no crystallographic evidence about the details of initial recognition. This evidence is needed to help unravel what has become a complicated reaction pathway. In addition, T4 endonuclease V has also been studied biochemically; evidence has been obtained for an initial binding complex in which the adenine residue is not flipped (A McCullough, ML Dodson, OD Scharer & RS Lloyd, unpublished data).

What is clear from the recent structure between M.*Hha*I and an abasic site (M O'Gara, JR Horton, RJ Roberts & X Cheng, unpublished data) is that the enzyme does not require a flipped-out cytosine residue as part of its initial recognition mechanism. Similarly, T4 endonuclease V can flip bases other than its normal substrate adenine (34; also cited in 12). Both observations suggest that what is flipped is not the base, but rather the deoxyribose and/or its flanking phosphates. Many of the other enzymes that use base flipping, such as uracil DNA glycosylase, operate much more efficiently as enzymes than does M.*Hha*I. In this case it seems unlikely that UDG simply waits for the natural flip of the uracil residue.

THE ORIGINS OF BASE FLIPPING

It has been proposed that base flipping is an ancient process that may have arisen even before the use of DNA as genetic material (13). The argument is based on two lines of reasoning: Base flipping is used by many enzymes involved in DNA repair, and the DNA methyltransferase M.*Hha*I can bind tightly to mismatches in DNA. In both cases abnormal base pairing is recognized with dramatic consequences. A possible connection between DNA methylation and repair has been suggested by Smith (60), who proposed that the human methyltransferase is a mismatch and DNA damage-recognition protein. Perhaps the DNA methyltransferases have evolved from DNA mismatch binding proteins by combining, into a single molecule, protein domains able to recognize mismatches in DNA, to perform sequence-specific recognition, to bind AdoMet, and to carry out the methylation reaction. In this scenario, the key original module is a mismatch recognition system that could accomplish base flipping.

Imagine an early stage in evolution when DNA first assumed its role as the genetic material. Undoubtedly, the DNA polymerase of that time was less faithful than are current DNA polymerases, and incorrect bases would be inserted with alarming frequency. In addition, spontaneous deamination of cytosine to uracil could also lead to loss of fidelity, as would other chemical modifications of the bases that might disrupt pairing. In short, a repair system would be essential if this early DNA were to be competent to serve as the genetic material. How might this be accomplished?

A first step must be recognition that a mismatch was present, followed by removal of the incorrect nucleotide and its replacement, most likely accomplished by the DNA polymerase having a second chance to copy faithfully. Because it is inconceivable that a full-blown repair process would have arisen immediately in a single enzyme, the repair process would require a series of steps, each involving a separate catalytic molecule. Base flipping seems tailor-made for the first step. One could imagine a small molecule, such as an RNA or peptide, being able to recognize and bind a mismatch and, in the process, flipping the mismatched base out of the DNA helix. Once the base is out of the helix, it would then be simple for a second molecule such as an endonuclease to cleave the phosphodiester chain and essentially create a new primer for the DNA polymerase. Experimental evidence shows that at least one small molecule, a metallo-porphyrin, can cause base flipping during intercalation into a B-DNA helix (61). The real strength of this proposal is that it allows the separation of mismatch recognition from the further steps of repair. Such stepwise accumulation of functions in a single molecule is undoubtedly how modern-day proteins evolved.

If base flipping was indeed used for mismatch detection in the early DNA polymers, one might reasonably ask how mismatch correction was dealt with in the earlier RNA world. Presumably, exactly the same scenario could be painted during the early course of RNA replication. Although there is no evidence for base flipping by enzymes involved in RNA metabolism, there is also a dearth of crystal structures for proteins involved in such processes.

FUTURE PROSPECTS

The theme common to all examples of base flipping is the requirement for enzymes to perform chemistry on the bases normally embedded in a B-DNA helix. As more proteins that perform chemistry on DNA are examined, further examples of base flipping are likely to be discovered. Of particular interest will be studies of the various mismatch repair systems, such as the methyl-directed and the short patch-repair systems in *E. coli*, which might also be expected to use base flipping.

Could there be other examples of base flipping within the realm of nucleic acid metabolism that have so far been elusive? Perhaps! If base flipping truly was an early evolutionary discovery, then one might anticipate its occurrence in other processes where it might prove advantageous. One such process is the local unwinding in DNA that is needed during the initiation of transcription or replication. We do not know how this is achieved. Examination of any of the DNA structures that contain flipped bases immediately suggests that base flipping would provide an easy answer. With one base flipped out of helix and the concomitant loss of both base pairing and stacking interactions, it would be simple to disrupt an adjacent base pair and so begin the unzipping of the helix. This is not the only mechanism that could be envisioned to initiate strand separation, but it should be considered a possibility.

Within the area of RNA metabolism, all enzymes that perform chemistry on RNA must also be considered candidates to employ base flipping. The enzymes that modify tRNA or rRNA within base-paired regions, either by methylation or by introducing more complex side chains, might use this mechanism. Another likely candidate is double-stranded RNA adenosine deaminase. This enzyme converts adenosine residues in RNA into inosine. It has been implicated in

RNA editing and is suggested to resemble the DNA methyltransferases at the sequence level (62).

In summary, the phenomenon of base flipping is appearing in a number of systems following its initial discovery in the DNA methyltransferases, and many additional enzyme systems must be considered candidates to use this novel mechanism. Unfortunately, detailed mechanistic information is lacking, and it remains to be proven whether base flipping is an active process in which the protein pushes the base out of the helix or a passive one in which the protein binds to a transiently flipped base. Most authors, including ourselves, favor the active process, but further work is needed to settle the issue. Particularly intriguing is the recent finding that when M.*HhaI* binds to an abasic site, it flips the deoxyribose ring and its flanking phosphates into the same conformation that is adopted during base flipping on its normal substrate. Thus if the process is active, one might expect that the push will take place not on the base, but rather on the sugar-phosphate backbone. Perhaps we should have termed the phenomenon the less evocative "DNA backbone rotation."

ACKNOWLEDGMENTS

We thank Drs S Kumar, S Pradhan, L Robow, and X Zhang for critical reading of the manuscript; Dr. S Kumar and C Lin for help with the preparation of figures; Drs T Barrett, P Freemont, C Mol, L Pearl, W Saenger, G Schluckebier, J Tainer, M Thayer, and Y Yamagata for providing coordinates; and T Ellenberger, S Klimasauskas, and E Weinhold for providing preprints. We also wish to thank most warmly the members of our laboratories whose hard work was responsible for much of the methyltransferase story. Work in our laboratories is supported by grants from the National Institutes of Health (GM46127 to RJR; GM/OD52117 and GM49245 to XC).

Visit the *Annual Reviews home page* at
http://www.AnnualReviews.org.

Literature Cited

1. Pabo CO, Sauer RT. 1992. *Annu. Rev. Biochem.* 61:1053–95
2. Kim YC, Geiger JH, Hahn S, Sigler PB. 1993. *Nature* 365:512–20
3. Kim JL, Nikolov DB, Burley SK. 1993. *Nature* 365:520–27
4. Travers A. 1997. *Curr. Biol.* 7:252–54
5. Kim YC, Grable JC, Love R, Greene PJ, Rosenberg JM. 1990. *Science* 249:1307–9
6. Winkler FK, Banner DW, Oefner C, Tsernoglou D, Brown RS, et al. 1993. *EMBO J.* 12:1781–95
7. Cheng XD, Kumar S, Posfai J, Pflugrath JW, Roberts RJ. 1993. *Cell* 74:299–307
8. Cheng X, Kumar S, Klimasauskas S, Roberts RJ. 1993. *Cold Spring Harbor Symp. Quant. Biol.* 58:331–38
9. Klimasauskas S, Kumar S, Roberts RJ, Cheng XD. 1994. *Cell* 76:357–69
10. Kumar S, Cheng XD, Klimasauskas S, Mi S, Posfai J, et al. 1994. *Nucleic Acids Res.* 22:1–10
11. Reinisch KM, Chen L, Verdine GL, Lipscomb WN. 1995. *Cell* 82:143–53

12. Vassylyev DG, Kashiwagi T, Mikami Y, Ariyoshi M, Iwai S, et al. 1995. *Structure* 4:1381–85

13. Roberts RJ. 1995. *Cell* 82:9–12

14. Slupphaug G, Mol CD, Kavli B, Arvai AS, Krokan HE, Tainer JA. 1996. *Nature* 384:87–92

15. Roberts RJ, Myers PA, Morrison A, Murray K. 1976. *J. Mol. Biol.* 103:199–208

16. Mann MB, Smith HO. 1979. In *Proc. Conf. Transmethylation*, ed. E Usdin, RT Borchardt, CR Creveling, pp. 483–92. New York: Elsevier

17. O'Gara M, Klimasauskas S, Roberts RJ, Cheng XD. 1996. *J. Mol. Biol.* 261:634–45

18. O'Gara M, Roberts RJ, Cheng XD. 1996. *J. Mol. Biol.* 263:597–606

19. Kumar S, Horton JR, Jones GD, Walker RT, Roberts RJ, Cheng XD. 1997. *Nucleic Acids Res.* 25:2773–83

20. Cheng XD. 1995. *Curr. Opin. Struct. Biol.* 5:4–10

21. Cheng XD. 1995. *Annu. Rev. Biophys. Biomol. Struct.* 24:293–318

22. Klimasauskas S, Roberts RJ. 1995. *Nucleic Acids Res.* 23:1388–95

23. Beisler JA. 1978. *J. Med. Chem.* 21:204–8

24. Creusot F, Acs G, Christman JK. 1982. *J. Biol. Chem.* 257:2041–48

25. Wu JC, Santi DV. 1987. *J. Biol. Chem.* 262:4778–86

26. Yang AS, Shen J-C, Zingg J-M, Mi S, Jones PA. 1995. *Nucleic Acids Res.* 23:1380–87

27. Cheng XD, Blumenthal RM. 1996. *Structure* 4:639–45

28. Lange C, Wild C, Trautner TA. 1996. *EMBO J.* 15:1443–50

29. Mol CD, Arvai AS, Sanderson RJ, Slupphaug G, Kavli B, et al. 1995. *Cell* 82:701–8

30. Mol CD, Arvai AS, Slupphaug G, Kavli B, Alseth I, et al. 1995. *Cell* 80:869–78

31. Savva R, McAuley-Hecht K, Brown T, Pearl L. 1995. *Nature* 373:487–93

32. Savva R, Pearl LH. 1995. *Nat. Struct. Biol.* 373:752–57

33. Dodson ML, Michaels ML, Lloyd RS. 1994. *J. Biol. Chem.* 266:17631–39

34. McCullough AK, Scharer O, Verdine GL, Lloyd RS. 1996. 271:32147–52

35. Vassylyev DG, Morikawa K. 1996. *Structure* 4:1381–85

36. Vassylyev DG, Morikawa K. 1997. *Curr. Opin. Struct. Biol.* 7:103–9

37. Verdine GL, Bruner SD. 1997. *Chem. Biol.* 4:329–34

38. Cunningham RP. 1997. *Mutat. Res.* 383:189–96

39. Labahn J, Granzin J, Schluckebier G, Robinson DP, Jack WE, et al. 1994. *Proc. Natl. Acad. Sci. USA* 91:10957–61

40. Gong W, O'Gara M, Blumenthal RM, Cheng XD. 1997. *Nucleic Acids Res.* 25:2702–15

41. Park H-W, Kim S-T, Sancar A, Deisenhofer J. 1995. *Science* 268:1866–72

42. Thayer MM, Ahern H, Xing DX, Cunningham RP, Tainer JA. 1995. *EMBO J.* 14:4108–20

43. Yamagata Y, Kato M, Odawara K, Tokuno Y, Nakashima Y, et al. 1996. *Cell* 86:311–19

44. Labahn J, Scharer OD, Long A, Ezaz-Nikpay K, Verdine GL, Ellenberger TE. 1996. *Cell* 86:321–29

45. Mol CD, Kuo C-F, Thayer MM, Cunningham RP, Tainer JA. 1995. *Nature* 374:381–86

46. Vrielink A, Ruger W, Driessen HPC, Freemont PS. 1994. *EMBO J.* 13:3413–22

47. Moore MH, Gulbis JM, Dodson EJ, Demple B, Moody PCE. 1994. *EMBO J.* 13:1495–501

48. Schluckebier G, Labahn J, Granzin J, Schildkraut I, Saenger W. 1995. *Gene* 157:131–34

49. Schluckebier G, O'Gara M, Saenger W, Cheng XD. 1995. *J. Mol. Biol.* 247:16–20

50. Malone T, Blumenthal RM, Cheng XD. 1995. *J. Mol. Biol.* 253:618–32

51. Allan BW, Reich NO. 1996. *Biochemistry* 35:14757–62

52. Cal S, Connolly BA. 1997. *J. Biol. Chem.* 272:490–96

53. Sancar A. 1994. *Biochemistry* 33:2–9

54. Kuo CF, McRee DE, Fisher CL, O'Hadley SF, Cunningham RP, Tainer JA. 1992. *Science* 258:434–40

55. Nikolov DB, Hu SH, Lin J, Gasch A, Hoffmann A, et al. 1992. *Nature* 360:40–46

56. Oefner C, Suck D. 1986. *J. Mol. Biol.* 192:605–32

57. Gallinari P, Jiricny J. 1996. *Nature* 383:735–38

58. Subramanya HS, Doherty AJ, Ashford SR, Wigley DB. 1996. *Cell* 85:607–15

59. Guest CR, Hochstrasser RA, Sowers LC, Millar DP. 1991. *Biochemistry* 30:3271–79

60. Smith SS. 1994. *Prog. Nucleic Acid Res. Mol. Biol.* 49:65–111

61. Lipscomb LA, Zhou FX, Presnell SR, Woo RJ, Peek ME, et al. 1996. *Biochemistry* 35:2818–23

62. Hough RF, Bass BL. 1997. *RNA* 3:356–70

Annu. Rev. Biochem. 1998. 67:199–225

THE CAVEOLAE MEMBRANE SYSTEM

Richard G. W. Anderson

KEY WORDS: potocytosis, cell signaling, GPI-anchored protein, cholesterol

ABSTRACT

The cell biology of caveolae is a rapidly growing area of biomedical research. Caveolae are known primarily for their ability to transport molecules across endothelial cells, but modern cellular techniques have dramatically extended our view of caveolae. They form a unique endocytic and exocytic compartment at the surface of most cells and are capable of importing molecules and delivering them to specific locations within the cell, exporting molecules to extracellular space, and compartmentalizing a variety of signaling activities. They are not simply an endocytic device with a peculiar membrane shape but constitute an entire membrane system with multiple functions essential for the cell. Specific diseases attack this system: Pathogens have been identified that use it as a means of gaining entrance to the cell. Trying to understand the full range of functions of caveolae challenges our basic instincts about the cell.

CONTENTS

0066-4154/98/0701-0199$08.00

INTRODUCTION

Yamada proposed the name caveolae intracellularis (little caves) to define "a small pocket, vesicle, cave or recess communicating with the outside of the cell" in gallbladder epithelial cells (1). Although he did not attribute any special shape to these invaginations, nor did he distinguish coated from non-coated varieties, the name later became synonymous with "flask-shaped" or "omega-shaped" membrane owing to the prominence of such membrane profiles in endothelial and smooth-muscle cells. In fact, two years earlier Palade had described morphologically similar invaginations in endothelial cells (2). He later named them plasmalemmal vesicles (3) because they appeared to shuttle molecules across these cell.

Over the ensuing years, many studies supported Palade's hypothesis that caveolae were endocytic structures involved in the transcellular movement of molecules across endothelial cells (4). Unfortunately, little was learned during this time about how they might function in other cell types. The modern era of caveolae research was ushered in by two discoveries: (a) receptor-mediated uptake of folate by caveolae (5) and (b) caveolin, the first marker protein for caveolae (6). The former provided a general model for how caveolae might function in diverse cell types, whereas the latter was the critical tool needed to purify this membrane domain for analysis. Purified caveolae not only yielded information about their chemical composition but also resulted in the unexpected finding that caveolae are rich in a variety of cell-signaling molecules. Current research is focused on caveolae as a membrane system responsible for compartmentalizing signal transduction, thereby facilitating the integration of nutritional, mechanical, and humoral information at the cell surface.

DEFINING A CAVEOLA

The shift in research from the use of morphological tools to the use of biochemical tools has brought a changing perspective of caveolae. The original intent of the word caveolae was to describe membrane invaginations at the cell surface (Figure 1A), but membranes with the classic morphologic features of

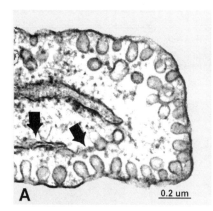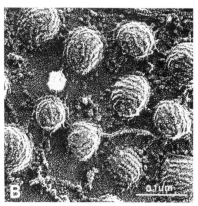

Figure 1 (*A*) Thin-section electron microscopy (EM) image and (*B*) rapid-freeze deep-etch image of fibroblast caveolae. *Arrows* point to endoplasmic reticulum near invaginated caveolae.

caveolae are not found in all cells. Purification methods using the caveolae marker protein caveolin established new criteria for identifying this membrane. These included (*a*) resistance to solubilization by Triton X-100 at 4°C (7); (*b*) a light buoyant density (8); and (*c*) richness in glycosphingolipids (GSLs), cholesterol, and lipid-anchored membrane proteins (Table 1). Membrane fractions with these properties, however, can be obtained from virtually all cells, even those not expressing detectable caveolin. The evidence presented in this review emphasizes that all cells have plasma membrane domains with the biochemical features of caveolae but only a subset of these membranes display the flask-shaped morphology. Caveolae assume a variety of shapes, including flat, vesicular, and tubular. They can be either open at the cell surface or closed off to form a unique endocytic/exocytic compartment. The use of the word caveolae in this review, therefore, is not restricted to membranes with a particular shape. Rather it is meant to encompass a membrane system with specific functions essential for normal cell behavior.

MOLECULAR COMPOSITION OF CAVEOLAE

The structure, function, and molecular composition of caveolae are dependent on the phase properties of a unique set of membrane lipids. Resident molecules freely move in and out of caveolae many times during their lifetime.

Caveolae Coat

Caveolae in endothelial cells (9) and fibroblasts (6) have a striated coat. Rapid-freeze deep-etch images show that the coat decorates membranes with different

Table 1 Partial list of molecules enriched in caveolae

Class of molecules	Name of molecule	Biochemical localization	Morphological localization	References
Lipid	Ganglioside	√	√	8, 143, 159, 160
	Sphingomyelin	√	—	43, 82, 191
	Ceramide	√	—	43, 82, 191
	Diacylglycerol (DAG)	√	—	191
	Cholesterol	√	√	19, 43, 216
Acylated protein	Heterotrimeric G_α and G_β	√	√	7, 8, 31, 35
	Src, Fyn, Hck, Lck	√	—	7, 53, 168, 169
	E-NOS	√	√	90, 91
	CD-36	√	—	32
	Caveolin	√	√	6, 7
Glycosylphospha-tidylinositol (GPI)-anchored protein	Folate receptor	√	√	58, 63
	Thy 1	√	√	32, 76
	Alkaline phosphatase	√	√	70, 72, 191
	Prion	√	√	76, 206, 207, 210
	Urokinase Rec	√	√	65, 217
	Multiple GPI proteins	√	—	31
	5′-Nucleotidase	√	√	203, 218–220
	CD14	√	—	221
Prenylated protein	Rap1A	√	√	31, 222
	Ras	√	—	36, 179
Membrane receptor	Platelet-derived growth factor (PDGF)	√	√	79, 223
	Insulin	—	√	224, 225
	Epidermal growth factor (EGF)	√	—	8, 179
	Receptor for advanced glycation end product (RAGE)	√	—	32
	Cholecystokinin (CCK) receptor	√	√	150
	m2 acetylcholine	√		181
	Tissue factor	√	—	226, 227
	β Adrenergic	—	√	100
	Bradykinin	√	—	83
	Endothelin	√	—	84
	SR-B1	√	√	121
Signal transducer	PKC_α	√	√	32, 99, 144
	SHC	√	—	79
	SOS	√	—	179
	GRB_2	√	—	179

(*Continued*)

Table 2 (*Continued*)

Class of molecules	Name of molecule	Biochemical localization	Morphological localization	References
	Mitogen-activated protein (MAP) kinase	√	√	32, 35, 79, 172
	Adenylyl cyclase	√	√	182–185
	SYP	√	—	79
	PI3 kinase	√	—	79
	Raf1	√	—	94, 179
	Calmodulin	√	—	90
	Phosphoinositides	√		192, 193
	Polyphosphoinositide phosphatase	√	—	193
	Engrailed	√	—	228
Membrane transporter	Porin	√	—	32
	IP$_3$ receptor	√	√	33, 190
	Ca^{+2} ATPase	√	√	33, 189, 229
	Aquaporin-1	√	—	135
	H$^+$ ATPase	√	—	230
Structural molecules	Annexin II	√	—	32, 231
	Ezerin	√	—	32
	Myosin	√	—	32
	VAMP	√	—	158
	NSF	√	—	158
	MAL	√	—	119, 232
	Actin	√	√	8, 31, 32
Miscellaneous	Atrial natriuretic Peptide	—	√	233
	Flotillin	√	—	234

amounts of curvature (Figure 1*B*), suggesting that it may control the shape of the membrane. The coat is composed of integral membrane proteins, one of which is caveolin (6). At least four caveolin gene products are present in mammals—caveolin-1α and -1β, -2, and -3 (10–14)—and possibly two in *Caenorhabditis elegans* (15). Each contains a 33-amino-acid hydrophobic domain that is thought to anchor the protein in the membrane, leaving the amino and carboxyl portions free in the cytoplasm (16). Caveolin-1 and -3 have cysteine residues at positions 134, 144, and 157 that in caveolin-1 are acylated (17). The expression of caveolin-1 in cells is correlated with the appearance of invaginated caveolae (18, 19) as well as the presence of the striated coat material (20). Although caveolin-1 is able to form homotypic oligomers both in vitro (21–23) and in vivo (24), it probably does not have a mechanical function in shaping the membrane because invaginated caveolae sometimes lack the molecule (25, 26). Sequestration of membrane cholesterol with drugs such as filipin (6) or depletion of intracellular cholesterol (27) causes the coat to disassemble. At the same time, invaginated caveolae disappear. Because caveolin-1 appears to be a cholesterol-binding protein (28, 29) and cholesterol stabilizes caveolin-1 oligomers (23), sterol and caveolin-1 must work together to form the coat.

Purification

Caveolin-1 is the marker protein used to isolate caveolae by cell fractionation. Six methods have been reported for purifying caveolae from either tissues (30–35) or tissue culture cells (7, 8, 36). The methods fall into four categories: (a) flotation of detergent-insoluble membrane on sucrose gradients (7), (b) flotation of sonicated plasma membranes on OptiPrep gradients (8), (c) differential centrifugation of tissue homogenates (31), and (d) recovery, either by centrifugation (33) or immunoadsorption (34), from endothelial cell plasma membranes purified by adsorption to cationized silica. The caveolae obtained by these procedures are not strictly comparable, largely because no morphologic standard exists by which to judge purity. A coatlike material is visible in some preparations (Figure 2), but it generally is hard to recognize. Caveolin-1 is not the ideal molecule for assessing purity because the concentration in caveolae is variable (see below). Finally, the physical aids used during purification (e.g. cationized silica, Triton X-100, sonication, high pH carbonate, and immunoadsorption)

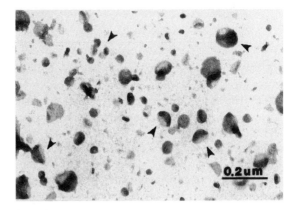

Figure 2   Whole-mount electron microscopy (EM) image of isolated caveolae prepared by the method of Chang et al (31). *Arrows* point to cup-shaped caveolae having a visible coat covering a portion of the membrane.

can alter the molecular composition of caveolae, making the comparison of results from different laboratories difficult. Detergents, for example, solubilize resident proteins (31), yielding extracted preparations of caveolae. The lack of a standard purification method, combined with the potential for contamination and extraction artifacts, obligates researchers to verify whenever possible all fractionation results by independent methods.

The resistance of core lipids to detergents at 4°C is a consistent property of caveolae (7). Simons & van Meer (37, 38) proposed that hydrogen bonding between GSLs caused the formation of GSL-rich membrane domains in the Golgi apparatus. Earlier reports had indicated that GSL-rich membranes (39–42) were insoluble in Triton X-100 at 4°C. Sucrose density centrifugation subsequently showed that GSL, cholesterol, and sphingomyelin (SPH), along with glycosylphosphatidylinositol-anchored membrane proteins (GPI protein), form a detergent-insoluble membrane domain with a light buoyant density (43). These light fractions are rich in caveolin-1 (7). Chang et al (31) directly demonstrated that caveolae are detergent insoluble by showing that partially purified caveolae, obtained without detergents, are resistant to Triton X-100. Caveolae in situ also are not extracted by Triton X-100 (44).

Because not all resident caveolae proteins are detergent insoluble, another useful identifier for caveolae is their light buoyant density relative to the bulk plasma membrane. They float on both velocity (8) and equilibrium (7) gradients independently of whether they have been treated with Triton X-100. Light membrane fractions with the same vesicular morphology and marker proteins are obtained regardless of whether cells express detectable caveolin-1 (8, 35).

Dynamics of Caveolae Molecules

Biochemical and morphologic techniques have identified a number of molecules that appear to be concentrated in caveolae relative to the surrounding membrane (Table 1). Approximately 35% have been localized by both morphological and biochemical methods (see check marks). The list does not take into consideration the purification method used to identify the molecule. Taken as a whole, the list indicates that caveolae have a specific lipid composition and are enriched in lipid-modified proteins. They are also rich in receptors and signal-transducing molecules.

GSL, SPH, and cholesterol, which form the lipid core of caveolae, govern the phase properties of the lipids in this domain. The detergent insolubility of caveolae at low temperature (45) is a characteristic of the liquid-ordered phase (referred to as the β phase) (46, 47). The detergent-insoluble properties of caveolae have been reconstituted using liposomes composed of cholesterol and sphingomyelin (48, 49). In general, the lateral and rotational mobilities of phospholipids are high in β phase membranes (50), whereas in sphingomyelin/cholesterol-rich membranes, the fluidity appears to be lower than that of the bulk cell membrane (48).

The GSL/SPH/cholesterol lipid core plays an important role in attracting lipid-modified membrane proteins to caveolae (Table 1). Proteins modified with either GPI or fatty acids are found to be enriched in caveolae fractions obtained by most methods of purification (7, 8). Mutations that abolish either the GPI-anchor addition (51, 52) or fatty acylation (53, 54) shift the protein to other fractions, which suggests that the lipid moiety is required for targeting to caveolae. These two different covalent modifications, therefore, are responsible for targeting proteins with a wide range of biochemical activities to opposing surfaces of the same membrane domain. Because the acyl chains on these proteins intercalate in the lipid bilayer, they probably collect in caveolae as a result of a slowed lateral mobility upon encountering the β phase lipids (55). Perturbing the β phase with cholesterol-sequestering drugs such as filipin (56) disperse GPI proteins in the plane of the membrane (57, 58). Protein-protein and protein-lipid (59, 60) interactions within caveolae influence how long the molecules remain at this site. The phase properties of the core lipids therefore play a major role in generating the complex molecular environment found in caveolae.

GPI proteins are dynamically associated with caveolae: They spontaneously insert into membranes of living cells (61). When the GPI-anchored complement inhibitor CD59 is inserted into the promonocyte cell U937, it is initially dispersed in the membrane but becomes clustered after a brief time (62). The clustered molecules are active in cell signaling, whereas unclustered CD59 is inactive. The GPI-anchored folate receptors are also mobile in the membrane. Ordinarily about 60–70% of the receptors in the monkey kidney cell MA104 are recovered in caveolae fractions (58), and indirect immunogold shows ~80% in discrete clusters (63). Surprisingly, incubation of these cells in the presence of a monoclonal anti-receptor immunoglobulin (Ig)G shifts receptors to the noncaveolae fraction (58), and by direct immunofluorescence with the same antibody, they appear dispersed in the membrane (64). These and other experiments (55) show that GPI proteins are not static on the cell surface. They are constantly moving, potentially accessing many different membrane compartments during their lifetime. In some cases, protein ligands for GPI proteins shift the protein from caveolae to other compartments, where they may become tethered by a resident protein (65–68). A relatively fast lateral mobility combined with a natural attraction for β phase lipids allows GPI proteins to shuttle information among different membrane compartments.

A persistent problem in caveolae research has been the conflicting reports on the native distribution of GPI proteins. Electron microscopy (EM) histochemistry clearly shows that GPI alkaline phosphatase (AP) is clustered both on the surface and within invaginated caveolae (69–71). Immunocytochemistry, by contrast, shows AP either clustered or diffusely distributed, depending on (*a*) the primary antibody used (72, 73), (*b*) whether a second antibody (or protein A) is applied (74), and (*c*) the fixation conditions (56, 64, 74). Other GPI proteins, as well as GSL and SPH, appear diffuse on the surface of cells after direct antibody labeling but clustered when indirect labeling methods are used (75). The clustered GPI proteins visualized after antibody additions are not randomly distributed. Instead they are always nearby (63) or inside (74, 76) invaginated caveolae or tightly co-localized with caveolin-1 (64, 75). In contrast to morphology, both detergent-dependent (7) and independent (8, 31) purification methods find GPI proteins concentrated in caveolae fractions. Only when membranes are pretreated—with antibodies against the protein (58), with cationized silica (77), or with pH 9.5 Tris buffer (36)—do they appear in noncaveolae fractions. Obviously these molecules cannot be clustered and diffuse at the same time! The most likely explanations for these discrepant results is that GPI proteins remain mobile in the plane of the membrane after weak aldehyde fixation and that their distribution is easily altered by physical agents such as antibodies. The majority of the morphologic and functional data supports the conclusion that GPI proteins tend to cluster and associate with caveolae.

Caveolin-1 interacts with GPI proteins as well as several other proteins and lipids enriched in caveolae. Immunoprecipitates of caveolin-1 from cells exposed to insulin (78) or platelet-derived growth factor (PDGF) (79) contain different sets of tyrosine-phosphorylated proteins. Anticaveolin IgG precipitates can also contain endothelial nitric oxide synthase (eNOS) (80), Ras (36, 81), P75(Ntr) (82), bradykinin receptors (83), and both endothelin and endothelin receptor subtype A (84). Immunoprecipitates of alpha integrin (85), dystrophin (86), and the GPI urokinase receptor (87) contain caveolin-1, whereas specific caveolin-1 peptides bind heterotrimeric $G\alpha$ (88), Ras (81), and Src (81) in vitro. A photoreactive derivative of GM1 ganglioside binds caveolin-1 after it is inserted into cells (89). Despite the proposal that caveolin-1 is a scaffolding protein (21), there is still no direct experimental evidence that any of these interactions with caveolin-1 are required for targeting to caveolae. To the contrary, both eNOS (90, 91) and nonreceptor tyrosine kinases (53, 54, 92) lacking fatty acids are not concentrated in caveolin-rich caveolae. GM1 gangliosides (35), GPI proteins (35, 76, 93), heterotrimeric G proteins (93), and nonreceptor tyrosine kinases (93) are concentrated in caveolae fractions that do not contain detectable caveolin-1. Furthermore, addition of the K-Ras consensus sequence for prenylation to Raf-1 targets the kinase to caveolae (94), which suggests that farnesylation plays a role in targeting Ras to caveolae.

Reconciling the Biochemical and Morphological Caveola

Not all membrane domains with the biochemical and physical characteristics outlined above have a flask shape. Synaptic plasma membranes, which have numerous membrane invaginations of unspecified origin (95), yield fractions with the properties of caveolae (35). Many cells have tubular invaginations lined by flask-shaped out-pocketings (76) or terminating in clusters of typical flask-shaped membrane. Similar structures appear to form into T tubules during skeletal muscle cell differentiation (96). A tubular intracellular compartment (71) rich in detergent-insoluble GPI proteins (97) is prominent in neutrophils. GPI proteins also co-localize with tubular invaginations in placental epithelial cells (76). Clustered GPI proteins in the neuronal cell line N2A are associated with membrane invaginations that have variable morphology (76) but apparently lack a striated coat (98).

Thus, the shape of a caveola in thin-section EM is variable, not all flask-shaped membranes have detectable caveolin-1, and a striated coat is not always visible. The common features of this domain are a light buoyant density, a GSL/SPH/cholesterol lipid core in the liquid-order phase, a high concentration of lipid-anchored proteins and signaling molecules, and a discrete size. The only viable conclusion is that patches of membrane with these properties are dynamic domains that assume different shapes depending on their activity in the cell. Flat caveolae that contain caveolin-1 and have a striated coat, for example, become flask shaped during internalization (99). Modifiers such as flat, invaginated, tubular, and vesicular should be used to indicate more precisely which caveolae shape is being considered. In cases where the native shape is unclear, the term caveolae-like is a suitable identifier.

BIOGENESIS AND MAINTENANCE OF CAVEOLAE

The building of caveolae is a multistep process that involves both an initial assembly step and a mechanism for actively maintaining the structure so that it can function properly at the cell surface (Figure 3).

Initial Assembly

The detergent-insoluble, GSL/SPH/cholesterol lipid core of the caveola membrane forms in the transitional region of the Golgi apparatus (Figure 3A) (24, 43). GPI proteins and caveolin-1 (24), arriving from the endoplasmic reticulum (ER) after synthesis, are then incorporated to complete the initial assembly step. Anticaveolin-1 immunoprecipitation and chemical cross-linking experiments indicate that other proteins associate with caveolin-rich membrane at this point (24). Caveolae are shipped to the cell surface embedded in the membrane of exocytic vesicles (Figure 3A) (100). There is no direct evidence that they bud off and migrate as independent vesicles, although this is a formal possibility.

BIOGENESIS AND MAINTENANCE OF CAVEOLAE

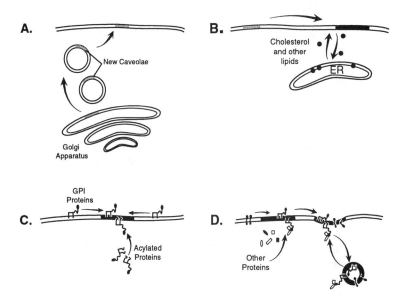

Figure 3 Biogenesis and maintenance of caveolae. (*A*) Caveolae biogenesis begins with the formation of glycosphingolipid (GSL)/sphingomyelin (SPH)/cholesterol-rich domains in the Golgi apparatus. Here the phase transition takes place in which the detergent-insoluble properties of this membrane are created. This is also where caveolin-1, glycosylphosphatidylinositol (GPI)-anchored proteins, and other proteins initially join the membrane. (*B*) New caveolae are shipped to the cell surface, where the lipid shuttle (*solid circle*) begins transporting cholesterol and other lipids from the endoplasmic reticulum (ER). (*C*) The lipid shuttle maintains the liquid-order phase of the caveolae core lipids, which is essential for concentrating GPI and acylated proteins migrating in and out of the domain. (*D*) Protein-protein interactions between lipid-anchored proteins in caveolae and nearby peripheral and transmembrane proteins bring additional molecules to the domain. Once the assembly process is completed, caveolae internalize molecules and deliver them to specific locations in the cell.

A natural outcome of the assembly process is the sorting of specific proteins and lipids away from bulk membrane components. Extensive work in polarized epithelial cells has documented how sorting by GSL/SPH/cholesterol domains—sometimes referred to as rafts (101) or DIGS (102)—contributes to the overall membrane polarity of the cell (37, 38).

The biogenesis of caveolae points out a major difference between this membrane domain and those coated with peripheral proteins such as clathrin. Caveolae are assembled in the Golgi apparatus and then shipped to other locations while clathrin-coated membrane is assembled de novo at sites of vesicle formation (103). Therefore, caveolae-like domains may exist in all membranes that traffic to and from the cell surface (104). They behave as coherent patches of membrane (47) immersed in the lipid bilayer, like icebergs floating in a sea (101, 105, 106). Surprisingly, the core lipids do not melt into the surrounding bilayer. In part, this is because the lipid composition of the domain is continuously maintained (Figure 3B,C).

Maintenance

Both cholesterol and SPH contribute to the lipid β phase (107). Cholesterol, however, is constantly fluxing out of the cell (108). Several immediate consequences result when caveolae cholesterol levels get too low. GPI proteins no longer cluster properly in caveolae (27). The striated coat disassembles (6), and the number of invaginated caveolae declines (27). Eventually internalization by caveolae ceases (27). Pharmacologic agents that block cholesterol transport to the cell surface have exactly these effects (19), which suggests that cholesterol is continuously transported to caveolae. A novel transport system has been identified that appears to be necessary for maintaining the proper level of cholesterol, and maybe other lipids, in caveolae.

Cholesterol moves bidirectionally between the ER and the plasma membrane (109). Transport of newly synthesized cholesterol to the cell surface is rapid (110, 111) and occurs in a light membrane fraction that does not appear to pass through the Golgi apparatus (110, 112–115). Newly synthesized cholesterol that has accumulated in the ER at 14°C is rapidly transferred to caveolae upon shifting the temperature to 37°C (19). This process suggests that the light membrane fraction is related to caveolae. The arrival of new cholesterol in caveolae is followed by the immediate movement of the sterol to noncaveolae membrane and possibly out of the cell (19). Caveolae have also been identified as intermediates in the cellular efflux of both newly synthesized cholesterol and cholesterol delivered to the cell by low-density lipoprotein (LDL) (116, 117). Therefore, cholesterol, and possibly other lipids, are constantly flowing through caveolae.

Bidirectional ER-to-caveolae transport of cholesterol appears to involve caveolin-1. The caveolae fraction from cells expressing caveolin-1 has a cholesterol-to-protein ratio that is four- to fivefold higher than that of noncaveolae fractions (19, 25). Selective oxidation of caveolae cholesterol with cholesterol oxidase causes caveolin-1 to move from caveolae to the ER and eventually accumulate in the Golgi apparatus (25). After the enzyme is removed, caveolin-1 reappears in caveolae at the same time that the cholesterol levels return to normal. Caveolin-1 also leaves the surface and accumulates in internal membranes when cells are exposed to progesterone (19), a condition that depletes caveolae cholesterol. Finally, transfection experiments show that caveolae fractions become enriched in cholesterol when caveolin-1 is expressed. Expression is

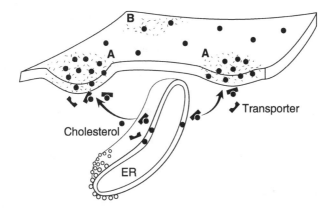

Figure 4 Endoplasmic reticulum (ER)-caveolae lipid shuttle. Cells have both high-cholesterol (*A*) and low-cholesterol (*B*) caveolae. The cholesterol level is maintained by a lipid transporter system that in some cells uses caveolin-1 as the cholesterol carrier. The molecular details of this system remain to be elucidated. Most likely the transporter can work in both directions and therefore is able to transport cholesterol from the plasma membrane to the ER. It may also transport lipids other than cholesterol.

accompanied by a four- to fivefold increase in the rate of cholesterol transport to the surface (19).

Caveolin-1 binds cholesterol (28), preferentially incorporates into cholesterol-containing membranes in vitro (28, 29), and moves between caveolae and internal membranes. High cholesterol levels in cells also cause an increase in caveolin-1 mRNA (118). Therefore, caveolin-1 appears to be part of an intracellular lipid transport system capable of moving sterols between ER and caveolae (Figure 4). The proteolipid MAL/VIP17, which is found in caveolae fractions (119, 120), may also be part of this system. The ER in all cells is near the plasma membrane, sometimes entwined with invaginated and vesicular caveolae (Figure 1*A*). Thus these molecules can easily move between the two membranes. Cholesterol acceptor/donor molecules such as high-density lipoprotein (HDL) influence the net direction of cholesterol movement. Recently the high-affinity HDL receptor SR-B1 was localized to caveolae (121), placing it in a position to facilitate the transfer of cholesterol between HDL or LDL and caveolin-1.

Caveolae contain different levels of cholesterol (and may be other essential lipids), depending on the activity of the lipid shuttle (*A*, Figure 4). High-cholesterol domains perform the full range of caveolae functions such as maintaining clustered proteins and lipids (Figure 3*C*) or internalizing molecules (Figure 3*D*). Low-cholesterol domains (*B*, Figure 4) perform only a subset of these duties. Cells containing low levels of caveolin-1 have reduced cholesterol transport, so formation of invaginated caveolae is suppressed (122). Conversely, cells expressing high levels of caveolin-1 like adipocytes (123) transport more cholesterol and as a consequence have greater numbers of invaginated caveolae (124).

INTERNALIZATION BY CAVEOLAE

Caveolae are involved in endocytosis. The underlying mechanism of invagination, budding, and vesicle trafficking differs significantly from the coated pit pathway.

Transcellular Transport

Space precludes a critical evaluation of the extensive literature on transcellular transport in endothelial cells. Both morphologic (4, 125, 126) and biochemical (127, 128) evidence supports the view that caveolae are the source of vesicles that move between the two surfaces of the cell. The process is inhibited by N-ethylmaleimide (NEM) (129, 130) and filipin (127) and may require the hydrolysis of GTP (128). Vesicles appear to move directly without merging with an intermediate compartment (125, 126, 131). Serial section and dye penetration show that many caveolae vesicle profiles in endothelial cells are actually open to the cell surface (132). Endothelial cell caveolae can fuse to form transcellular channels that allow the passage of small molecules across the cell (126). The vesiculo-vacuolar organelle (VVO) in the cytoplasm of the endothelial cells lining tumor microvessels may also be channels formed by the fusion of multiple swollen caveolae (133). Topical application of vascular endothelial growth factor (VEGF) rapidly induces the swelling and fusion of caveolae (134), which indicates that the formation of caveolae channels is under hormonal control. The swelling might depend on the water channel aquaporin-1 (135). Therefore, the formation of tubule-shaped caveolae appears to be part of a regulated transendothelial transport pathway. Tubular caveolae in muscle cells, neutrophils, and placental epithelia may have a similar origin.

Potocytosis

The GPI-anchored folate receptor provided the first biochemical clues that caveolae could mediate the uptake of molecules and ions in a variety of cells. The cardinal features of this pathway were discovered in MA104 cells (5). These cells express a limited number of clustered receptors that often appear to be associated with invaginated caveolae and are not detected in the coated pit pathway (63). Folate receptors internalize bound 5-methyltetrahydrofolate (136) while maintaining a constant ratio of internal and external receptors. Internalized 5-methyltetrahydrofolate dissociates from its receptor in response to an acidic environment (136, 137) and diffuses directly into the cytoplasm of the cell (138). Accumulation of folate in the cytoplasm plateaus even though the receptor continues to internalize (137). A minor population of folate receptors appear in the coated vesicle pathway of cells expressing high numbers of receptors (139). But when chimeric receptors are targeted specifically to coated pits, folate delivery is inefficient and unregulated (52). The process was named potocytosis (5) to emphasize the special ability of caveolae to concentrate and move molecules or ions into the cell.

Potocytosis was confirmed and extended by the discovery that caveolae mediate the delivery of molecules to the ER. The first molecule found to travel this route was caveolin-1 (25, 26). Viral pathogens also appear to reach the ER by caveolae (140–142). Membrane-bound simian virus 40 (SV40) becomes trapped in tight-fitting membrane invaginations that have a light buoyant density (141), are detergent insoluble (141), and contain caveolin-1 (141, 142). From monopinocytic vesicles (142), viruses next appear in smooth, tubular membrane extensions of the ER. A variety of molecules and ions may be delivered to the ER during potocytosis. These include lipids such as cholesterol, Ca^{2+} (see Table 1) as well as other ions, and ligands bound to receptors in caveolae. Opportunistic molecules such as cholera toxin (143) possibly reach the ER by this pathway too.

Complementary studies on the internalization of alkaline phosphatase (74), the folate receptor (99, 144), and cholera toxin (74) indicate that membrane recycling occurs during potocytosis. The mechanism of recycling appears to depend on the activity of protein kinase $C\alpha$ (PKCα) and a serine/threonine phosphatase activity present in purified caveolae (144). During folate uptake, a constant pool of internal and external receptors is maintained. Cells depleted of PKCα no longer internalize receptors; instead, the internal pool returns to the cell surface (99). By contrast, the phosphatase inhibitor okadaic acid causes a decline in the number of invaginated caveolae and the intracellular accumulation of cholera toxin–positive vesicle and tubular caveolae profiles (74). More of these profiles appear when coated pit uptake is blocked. Okadaic acid also causes alkaline phosphatase to leave the surface, consistent with a block in the recycling of internalized enzyme. The effects of okadaic acid are prevented by staurosporine, an inhibitor of PKCα. The internalization of GPI CD59 by lymphocytes (145) and SV40 by fibroblasts (141) also appears to depend on PKCα.

If these two enzymes control opposing limbs of a caveolae recycling pathway, then inactivation of PKCα would inhibit sequestration by caveolae, and inhibiting the phosphatase(s) would cause the accumulation of caveolae vesicles. Several studies indicate that the internalization cycle can be regulated at these two sites. Histamine binding to H_1 receptors transiently inhibits the internalization of clustered folate receptors (144) by inactivating PKCα. Exposure of MA104 cells to indomethacin prevents both internalization and externalization of folate receptor by an unknown mechanism (146). Tubular caveolae in unstimulated neutrophils sequester three GPI proteins [alkaline phosphatase, decay accelerating factor (DAF), and CD16] away from the surface, but all three rapidly reappear when cells are stimulated with chemotactic peptide (71, 147–149). Up to 20% of the cholecystokinin receptors are sequestered by caveolae after ligand binding, but when the coated pit pathway is blocked, nearly all the receptors become sequestered (150). Ligand binding also stimulates caveolae sequestration of bradykinin receptors (83) and possibly endothelin (151). Therefore, caveolae vesicles may be used regularly as compartments for storage, processing, and rapid deployment of molecules to the surface (152).

Besides a requirement for cholesterol (27) and possibly caveolin-1, little is known about how caveolae invaginate and bud from membranes (Figure 3D). Based on sensitivity to various treatments, however, they use a mechanism that is different from the one used by coated pits. Inhibitors of caveolae vesicle formation include cholesterol-binding drugs (127, 141, 153), cytochalasin D (74, 145), and in some cases PKC inhibitors (141, 154). Clathrin-coated vesicle formation is selectively (74, 150) blocked by hypertonic treatment (155) and K$^+$ depletion (156). Both processes are sensitive to NEM (129, 157). Even though several proteins implicated in vesicle trafficking have been localized to caveolae fractions (158), it is not clear whether tubular or vesicular caveolae ever fuse with endosomes from coated pits. Cholera toxin, for example, can reach endosomes (74, 159, 160), but quantitative studies (160) indicate it is only a subfraction of the total surface-bound toxin and could easily have arrived by rapidly recycling coated pits. In fact, neither labeled toxin (74) nor cholecystokinin (150) are usually found in typical endosomes when the coated pit pathway is blocked.

Caveolae and clathrin-coated pits are specialized to internalize different types of molecules. Therefore, potocytosis and receptor-mediated endocytosis (157) are parallel, but not redundant, endocytic pathways. Molecules internalized by potocytosis follow one of four distinct intracellular routes (Figure 5). From the cell surface they travel to the cytoplasm (Figure 5A), the endoplasmic reticulum (Figure 5B), the opposite cell surface (Figure 5C), or a caveolae-derived tubular/vesicular compartment (Figure 5D). Rarely do receptors or ligands appear in an intermediate compartment during ligand delivery. That is, the carrier vesicles retain many of the morphological and biochemical properties of caveolae. As a consequence, vesicular compartments generated during internalization can transform into exocytic vesicles that carry molecules back to the surface. This type of compartmentalization most likely has a variety of special uses in the cell (152).

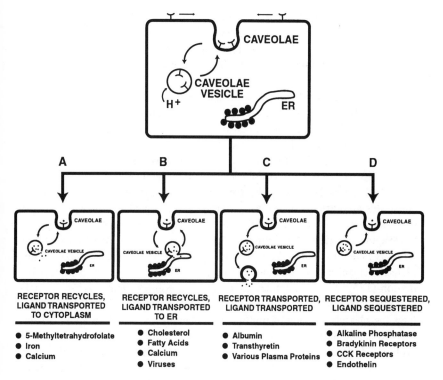

Figure 5 Multiple pathways of potocytosis. Molecules and receptors internalized by caveolae can have one of four fates: (*A*) The ligand is delivered to the cytoplasm while the receptor recycles back to the surface. (*B*) The ligand is delivered to the ER while the receptor recycles back to the surface. (*C*) The ligand is transported across the cell, and the receptor recycles. (*D*) Both the ligand and the receptor remain in a vesicular caveolae compartment. Examples of molecules that follow these routes are listed.

SIGNAL TRANSDUCTION FROM CAVEOLAE

The original potocytosis model (5) predicted that caveolae were involved in cell signaling at the plasma membrane (152, 161). No one was prepared, however, for the variety of signaling molecules that have now been found concentrated in this domain.

Tyrosine Kinases

Receptor and nonreceptor tyrosine kinases (PTK) are reliable markers for caveolae (Table 1). With one exception (34), all the purification methods find PTKs to be substantially enriched in caveolae isolated from a variety of cells and tissues. Moreover, the immunoprecipitation of multiple GPI proteins coprecipitates

PTKs (162). Immunoblotting, enzymatic activity, and immunocytochemistry all indicate that this is a major location for PTKs. For example, by immunofluorescence nearly all of the PDGF receptor in quiescent fibroblasts colocalizes with caveolin-1, and the majority of the receptor is recovered in isolated caveolae (79).

The first indication that PTKs were in caveolae can be traced to the paradoxical finding that antibodies directed against GPI proteins stimulated PTK activity (162–164) and PTK-dependent cell functions (164–166). A complex of GPI proteins was then found to be associated with nonreceptor PTKs (167–169). The complexes exhibited several properties of caveolae, including detergent insolubility, richness in glycolipids (170), a light buoyant density (171), and a size of ~100 nm. These studies were also the first to show that multiple kinase substrates are enriched in caveolae-like membranes and that phosphorylation of these substrates can occur in vitro. PTK activity has now been localized to invaginated caveolae by immunocytochemistry (172). A major substrate for PTKs is caveolin-1 (173–175). Tyrosine phosphorylation of caveolin-1 is stimulated by insulin (78), oxidants (176), sulfonylurea (177), and cell transformation (173, 178). A peptide sequence in caveolin-1 (amino acids 82–101) that interacts with c-Src possibly modulates PTK activity (81). PDGF stimulates the phosphorylation of multiple caveolae proteins, both in vivo (79) and in vitro (172), and its receptor is linked to a preassembled mitogen-activated protein (MAP) kinase module (172). Both PDGF (79) and epidermal growth factor (EGF) (179) stimulate the recruitment to caveolae of multiple signal-transducing molecules as well as the migration of the respective receptor out of caveolae.

GTP-Binding Proteins

A long-standing issue in signal transduction is whether receptors, G proteins, and effectors are organized or randomly distributed at the cell surface (180). A combination of cell fractionation and immunocytochemistry has now documented that all three classes of molecules are enriched in caveolae (Table 1). G proteins, for example, are found in most caveolae preparations. Receptors for endothelin (84), cholecystokinin (150), m2 acetylcholine (181), and bradykinin (83) are dynamically associated with these membranes. Receptors appear to be functionally connected to effectors in caveolae. Histochemistry has localized isoproterenol-stimulated adenylyl cyclase activity to membrane invaginations resembling caveolae (182–184). Isolated caveolae fractions contain a significant proportion of the total adenylyl cyclase activity (185). Isoproterenol-stimulated cyclic AMP formation cofractionates with the enzyme. Caveolae-like fractions from *Dictyostelium* are highly enriched in chemoattractant receptors as well as in adenylyl cyclase and phosphodiesterase

(186). Finally, bradykinin stimulates the recruitment of $G_{\alpha q}$ and $G_{\alpha i}$ to caveolae (83). Several studies suggest that caveolin-1 has a role in recruiting G proteins to caveolae as well as in modulating their activity (12, 36, 81, 88). The functional organization of the cyclase in S49 cells, however, is not dependent on caveolin-1 (185).

Calcium

A model of excitation-contraction coupling mediated by caveolae was proposed in 1974 (187). Since then, considerable evidence has accumulated that caveolae are sites of calcium storage and entry. For example, pyroantimonate precipitates of calcium are present in caveolae of relaxed smooth-muscle cells (188). Stimulation of contraction generates a diffuse distribution of the precipitate in the myoplasm, consistent with a movement of calcium into the cell. A rich collection of morphologic observations documents smooth-muscle–cell caveolae interacting with smooth ER, just as sarcoplasmic reticulum interacts with T tubules in skeletal muscle. Indeed, caveolae play a direct role in the biogenesis of the T-tubule system (96). Ca^{2+} ATPase (33, 189), IP3 receptors (33, 190), and calmodulin (90)—key molecular components of calcium transport—have all been localized to caveolae. These findings suggest a role for ER-caveolae interactions during calcium signaling.

Lipid Signals

Some of the lipids and lipid-anchored proteins incorporated into caveolae in the Golgi apparatus (see Biogenesis and Maintenance of Caveolae, above) are important sources of signaling intermediates. Sphingomyelin, phosphatidylinositol 4,5, bisphosphate, and GPI proteins/lipids are substrates for enzymes that release ceramide (82, 191), inositol trisphosphate (IP3) (192, 193), and inositolphosphoglycans (IPG) (194, 195), respectively. Each is produced in caveolae after a specific stimuli. Ceramide increases after Il-1β (191) or neurotrophin (82) stimulation. IP3 is released after exposure to either bradykinin or EGF (193), and IPG forms in response to insulin (196). These responses appear to be specific because neither ceramide nor IP3 is generated in noncaveolae fractions, and the IPG released on the extracellular side of the membrane is internalized, presumably by caveolae. All three molecules elicit characteristic cellular responses. These are just three examples of what must be a general mechanism whereby lipids sorted to caveolae become the source of critical signaling intermediates.

Compartmentalized Signaling

Caveolae compartmentalize enzymatic reactions at the surface that are important for signaling. Immunocytochemistry (90, 197), cell fractionation (90, 91, 197–199), and immunoprecipitation (200) show that the majority of cell surface endothelial nitric oxide synthase (eNOS) is located in caveolae. This finding suggests that caveolae are the site of nitric oxide (NO) production. A unique tubular compartment in neutrophils with the biochemical properties of caveolae (71) produces superoxide (O_2^-) in response to chemotactic peptides (201), and NAD(P)H oxidase has been localized by histochemistry to invaginated caveolae (202). GPI 5'-nucleotidase targeted to caveolae may convert extracellular 5'-AMP to adenosine where it locally activates receptors (203). Finally, caveolae are likely to be the site where α_7 integrin is ribosylated by a GPI ADP-ribosyltransferase (204). Integrins have recently been found to interact with caveolin-1 (85), and integrin function is regulated by the urokinase receptor (87).

Signal Integration

With so many different signaling molecules in one location, caveolae are the logical place to look for signal integration. Integration refers to the feedback interplay between two or more signaling processes that result in a reciprocal modulation of the interacting pathways. The stimulation of GPI proteins in endothelial cells is used to illustrate the concept. GPI proteins can activate PTKs and generate a Ca^{2+} influx (62, 164). PTKs phosphorylate eNOS, thereby inhibiting the enzyme and promoting its interaction with caveolin-1 (200). But the released Ca^{2+} will bind calmodulin, which activates eNOS. Any NO produced will stimulate the MAP kinase pathway through Ras (205), in synergy with PTKs (172). All the underlined components are in endothelial cell caveolae (Table 1), allowing the cross talk between pathways to occur at one site on the plasma membrane. Furthermore, the ability of caveolae to sequester molecules provides an opportunity for locally produced or imported molecules to modulate these signaling events.

CAVEOLAE AND HUMAN DISEASE

A number of human diseases appear to involve the caveolae membrane system. The system is the target of several pathogens and becomes altered during cell transformation.

Prion Diseases

Prions are a class of proteins that cause fatal encephalopathies in humans and other animals. The posttranslational conversion of cellular prion (Pr^C) to the scrapie isoform (Pr^{Sc}) is the mechanism of transmission. GPI Pr^C has been localized to invaginated caveolae in both fibroblasts and neuronal N2A cells (76) and fractionates with caveolae (35, 206–208). A caveolae localization appears to be necessary for conversion of Pr^C to Pr^{Sc} because replacement of the GPI anchor of Pr^C with a coated pit targeting sequence prevents conversion (209). Lowering cellular cholesterol, which disperses GPI proteins in the membrane (27), also inhibits conversion (210). Accumulation of Pr^{Sc} may impair many different caveolae functions.

Pathogens

Caveolae appear to be the cellular entrance point for pathogens as well as molecules produced by pathogens. The internalization of SV40 by caveolae has already been discussed (see Potocytosis, above). *Campylobacter jejuni* (211) may enter cells by caveolae too. A portion of the surface-bound cholera toxin is internalized by caveolae (74), but it is not known whether the entering A subunit reaches the cytoplasm by this route. Finally, because GPI proteins are able to spontaneously insert and cluster in caveolae (62), they may be the target for GPI proteins shed by parasites such as *Plasmodium, Trypanosoma,* and *Leishmania* (212, 213). Purified GPI proteins from these organisms simultaneously activate macrophage PKC and $p59^{hck}$ (213).

Cancer

Invaginated caveolae are substantially reduced in many types of transformed cells (214). The loss of these invaginations is correlated with the tyrosine phosphorylation of caveolin-1 (173, 214) and its loss from the cell (214). Although caveolin-1 was originally discovered as a PTK substrate in v-Src transformed cells (173), other oncogenic viruses have the same effect (214). Caveolin-1 may be a tumor suppresser because expression of the cDNA in transformed cells reverses anchorage-independent growth in soft agar (215). Expression of caveolin-1 could be essential for normal signal transduction from caveolae. Alterations in the permeability of tumor blood vessels resulting from VVOs (133) suggest that caveolae also have an indirect role in tumor formation.

Cardiovascular Disease

Caveolae are abundant in most parenchymal cells of the cardiovascular system. Having key roles in calcium metabolism, cell signaling, blood clotting, and cholesterol transport, caveolae are vulnerable sites in these cells. Caveolae are sensitive to oxidized cholesterol and contain receptors that bind HDL, LDL, and oxidized lipoproteins (121). This raises the possibility of a direct link between damage by oxysterols, inappropriate activation of multiple signaling pathways in caveolae, and cell proliferation during atherogenesis.

CONCLUSION

The rapid growth in caveolae research has brought with it a changing view of this membrane domain. Caveolae constitute a membrane system equal in complexity to any cellular compartment or organelle. Specific diseases have been identified that can attack this system, making it important to learn more about its normal biology. From another perspective, caveolae are an important research tool. They clearly contain a variety of signal-transducing molecules that interact in characteristic patterns after cell stimulation. The ease of caveolae isolation makes it possible to study at the molecular level how the natural organization of these molecules imparts cell function. Compartmentalized signal transduction is a growing area of research, and caveolae promise to provide many new insights.

ACKNOWLEDGMENTS

I would like to thank my colleagues Drs. Fred Grinnell, George Bloom, and Michael White for their many helpful discussions and suggestions during the preparation of this manuscript. I would also like to thank Dr. John Heuser for contributing the rapid-freeze deep-etch images of caveolae. I am indebted to Ms. Stephanie Baldock for her diligent editorial assistance. Finally, I would like to thank the past and present members of my laboratory who have made so many contributions to our understanding of caveolae biology. Their spirit and dedication are a continuing source of inspiration.

> Visit the *Annual Reviews home page* at
> http://www.AnnualReviews.org.

Literature Cited

1. Yamada E. 1955. *J. Biophys. Biochem. Cytol.* 1:445–58
2. Palade GE. 1953. *J. Appl. Phys.* 24:1424
3. Bruns RR, Palade GE. 1968. *J. Cell Biol.* 37:244–76
4. Simionescu N. 1983. *Physiol. Rev.* 63: 1536–60
5. Anderson RGW, Kamen BA, Rothberg KG, Lacey SW. 1992. *Science* 255:410–11
6. Rothberg KG, Heuser JE, Donzell WC, Ying YS, Glenney JR, Anderson RG. 1992. *Cell* 68:673–82
7. Sargiacomo M, Sudol M, Tang ZL, Lisanti MP. 1993. *J. Cell Biol.* 122:789–807
8. Smart EJ, Ying YS, Mineo C, Anderson RGW. 1995. *Proc. Natl. Acad. Sci. USA* 92:10104–8
9. Peters KR, Carley WW, Palade GE. 1985. *J. Cell Biol.* 101:2233–38
10. Way M, Parton RG. 1996. *FEBS Lett.* 378:108–12
11. Tang ZL, Scherer PE, Okamoto T, Song K, Chu C, et al. 1996. *J. Biol. Chem.* 271: 2255–61
12. Scherer PE, Okamoto T, Chun MY, Nishimoto I, Lodish HF, Lisanti MP. 1996. *Proc. Natl. Acad. Sci. USA* 93:131–35
13. Glenney JR Jr. 1992. *FEBS Lett.* 314:45–48
14. Kurzchalia TV, Dupree P, Parton RG, Kellner R, Virta H, et al. 1992. *J. Cell Biol.* 118:1003–14
15. Tang ZL, Okamoto T, Boontrakulpoontawee P, Katada T, Otsuka AJ, Lisanti MP. 1997. *J. Biol. Chem.* 272: 2437–45
16. Kurzchalia TV, Dupree P, Monier S. 1994. *FEBS Lett.* 346:88–91
17. Dietzen DJ, Hastings WR, Lublin DM. 1995. *J. Biol. Chem.* 270:6838–42
18. Fra AM, Williamson E, Simons K, Parton RG. 1995. *Proc. Natl. Acad. Sci. USA* 92:8655–59
19. Smart EJ, Ying YS, Donzell WC, Anderson RGW. 1996. *J. Biol. Chem.* 271: 29427–35

20. Chung KN, Elwood PC, Heuser JE. 1996. *Mol. Biol. Cell* 7:276a
21. Sargiacomo M, Scherer PE, Tang ZL, Kubler E, Song KS, et al. 1995. *Proc. Natl. Acad. Sci. USA* 92:9407–11
22. Monier S, Parton RG, Vogel F, Behlke J, Henske A, Kurzchalia TV. 1995. *Mol. Biol. Cell* 6:911–27
23. Monier S, Dietzen DJ, Hastings WR, Lublin DM, Kurzchalia TV. 1996. *FEBS Lett.* 388:143–49
24. Lisanti MP, Tang ZL, Sargiacomo M. 1993. *J. Cell Biol.* 123:595–604
25. Smart EJ, Ying YS, Conrad PA, Anderson RG. 1994. *J. Cell Biol.* 127:1185–97
26. Conrad PA, Smart EJ, Ying YS, Anderson RG, Bloom GS. 1995. *J. Cell Biol.* 131:1421–33
27. Chang WJ, Rothberg KG, Kamen BA, Anderson RG. 1992. *J. Cell Biol.* 118:63–69
28. Murata M, Peranen J, Schreiner R, Wieland F, Kurzchalia TV, Simons K. 1995. *Proc. Natl. Acad. Sci. USA* 92:10339–43
29. Li SW, Song KS, Lisanti MP. 1996. *J. Biol. Chem.* 271:568–73
30. Jacobson BS, Schnitzer JE, McCaffery M, Palade GE. 1992. *Eur. J. Cell Biol.* 58:296–306
31. Chang WJ, Ying YS, Rothberg KG, Hooper NM, Turner AJ, et al. 1994. *J. Cell Biol.* 126:127–38
32. Lisanti MP, Scherer PE, Vidugiriene J, Tang ZL, Hermanowski-Vosatka A, et al. 1994. *J. Cell Biol.* 126:111–26
33. Schnitzer JE, Oh P, Jacobson BS, Dvorak AM. 1995. *Proc. Natl. Acad. Sci. USA* 92:1759–63
34. Stan RV, Roberts WG, Predescu D, Ihida K, Saucan L, et al. 1997. *Mol. Biol. Cell* 8:595–605
35. Wu CB, Butz S, Ying YS, Anderson RGW. 1997. *J. Biol. Chem.* 272:3554–59
36. Song KS, Li SW, Okamoto T, Quilliam LA, Sargiacomo M, Lisanti MP. 1996. *J. Biol. Chem.* 271:9690–97
37. Simons K, van Meer G. 1988. *Biochemistry* 27:6197–202
38. van Meer G, Simons K. 1988. *J. Cell Biochem.* 36:51–58
39. Hagman J, Fishman PH. 1982. *Biochim. Biophys. Acta* 720:181–87
40. Yu J, Fischman DA, Steck TL. 1973. *J. Supramol. Struct.* 3:233–48
41. Davies AA, Wigglesworth NM, Allan D, Owens RJ, Crumpton MJ. 1984. *Biochem. J.* 219:301–8
42. Mescher MF, Apgar JR. 1985. *Adv. Exp. Med. Biol.* 184:387–400
43. Brown DA, Rose JK. 1992. *Cell* 68:533–44
44. Moldovan NI, Heltianu C, Simionescu N, Simionescu M. 1995. *Exp. Cell Res.* 219:309–13
45. Ribeiro AA, Dennis EA. 1973. *Biochim. Biophys. Acta* 332:26–35
46. Sankaram MB, Thompson TE. 1990. *Biochemistry* 29:10670–75
47. Sankaram MB, Thompson TE. 1991. *Proc. Natl. Acad. Sci. USA* 88:8686–90
48. Schroeder R, London E, Brown D. 1994. *Proc. Natl. Acad. Sci. USA* 91:12130–34
49. Ahmed SN, Brown DA, London E. 1997. *Biochemistry* 36:10944–53
50. Vist MR, Davis JH. 1990. *Biochemistry* 29:451–64
51. Keller GA, Siegel MW, Caras IW. 1992. *EMBO J.* 11:863–74
52. Ritter TE, Fajardo O, Matsue H, Anderson RGW, Lacey SW. 1995. *Proc. Natl. Acad. Sci. USA* 92:3824–28
53. Robbins SM, Quintrell NA, Bishop JM. 1995. *Mol. Cell. Biol.* 15:3507–15
54. Shenoy-Scaria AM, Dietzen DJ, Kwong J, Link DC, Lublin DM. 1994. *J. Cell Biol.* 126:353–63
55. Hannan LA, Lisanti MP, Rodriguez-Boulan E, Edidin M. 1993. *J. Cell Biol.* 120:353–58
56. Cerneus DP, Ueffing E, Posthuma G, Strous GJ, van der Ende A. 1993. *J. Biol. Chem.* 268:3150–55
57. Rothberg KG, Ying YS, Kamen BA, Anderson RG. 1990. *J. Cell Biol.* 111:2931–38
58. Smart EJ, Mineo C, Anderson RGW. 1996. *J. Cell Biol.* 134:1169–77
59. Zhang F, Crise B, Su B, Hou Y, Rose JK, Bothwell A, Jacobson K. 1991. *J. Cell Biol.* 115:75–84
60. Fiedler K, Parton RG, Kellner R, Etzold T, Simons K. 1994. *EMBO J.* 13:1729–40
61. Ilangumaran S, Robinson PJ, Hoessli DC. 1996. *Trends Cell Biol.* 6:163–67
62. van den Berg CW, Cinek T, Hallett MB, Horejsi V, Morgan BP. 1995. *J. Cell Biol.* 131:669–77
63. Rothberg KG, Ying Y-S, Kolhouse JF, Kamen BA, Anderson RGW. 1990. *J. Cell Biol.* 110:637–49
64. Mayor S, Rothberg KG, Maxfield FR. 1994. *Science* 264:1948–51
65. Stahl A, Mueller BM. 1995. *J. Cell Biol.* 129:335–44
66. Conese M, Nykjaer A, Petersen CM, Cremona O, Pardi R, et al. 1995. *J. Cell Biol.* 131:1609–22
67. Gliemann J, Nykjaer A, Petersen CM, Jorgensen KE, Nielsen M, et al. 1994. *Ann. NY Acad. Sci.* 737:20–38

68. Andreasen PA, Sottrup-Jensen L, Kjoller L, Nykjaer A, Moestrup SK, et al. 1994. *FEBS Lett.* 338:239–45
69. Ide C, Saito T. 1980. *J. Neurocytol.* 9:207–18
70. Latker CH, Shinowara NL, Miller JC, Rapoport SI. 1987. *J. Comp. Neurol.* 264:291–302
71. Kobayashi T, Robinson JM. 1991. *J. Cell Biol.* 113:743–56
72. Jemmerson R, Klier FG, Fishman WH. 1985. *J. Histochem. Cytochem.* 33:1227–34
73. Jemmerson R, Agree M. 1987. *J. Histochem. Cytochem.* 35:1277–84
74. Parton RG, Joggerst B, Simons K. 1994. *J. Cell Biol.* 127:1199–215
75. Fujimoto T. 1996. *J. Histochem. Cytochem.* 44:929–41
76. Ying Y-S, Anderson RGW, Rothberg KG. 1992. *Cold Spring Harbor Symp. Quant. Biol.* 57:593–604
77. Schnitzer JE, McIntosh DP, Dvorak AM, Liu J, Oh P. 1995. *Science* 269:1435–39
78. Mastick CC, Brady MJ, Saltiel AR. 1995. *J. Cell Biol.* 129:1523–31
79. Liu PS, Ying YS, Ko YG, Anderson RG. 1996. *J. Biol. Chem.* 271:10299–303
80. Feron O, Belhassen L, Kobzik L, Smith TW, Kelly RA, Michel T. 1996. *J. Biol. Chem.* 271:22810–14
81. Li SW, Couet J, Lisanti MP. 1996. *J. Biol. Chem.* 271:29182–90
82. Bilderback TR, Grigsby RJ, Dobrowsky RT. 1997. *J. Biol. Chem.* 272:10922–27
83. de Weerd WFC, Leeb-Lundberg LMF. 1997. *J. Biol. Chem.* 272:17858–66
84. Chun MY, Liyanage UK, Lisanti MP, Lodish HF. 1994. *Proc. Natl. Acad. Sci. USA* 91:11728–32
85. Wary KK, Mainiero F, Isakoff SJ, Marcantonio EE, Giancotti FG. 1996. *Cell* 87:733–43
86. Song KS, Scherer PE, Tang ZL, Okamoto T, Li SW, et al. 1996. *J. Biol. Chem.* 271:15160–65
87. Wei Y, Lukashev M, Simon DI, Bodary SC, Rosenberg S, et al. 1996. *Science* 273:1551–55
88. Li SW, Okamoto T, Chun MY, Sargiacomo M, Casanova JE, et al. 1995. *J. Biol. Chem.* 270:15693–701
89. Fra AM, Masserini M, Palestini P, Sonnino S, Simons K. 1995. *FEBS Lett.* 375:11–14
90. Shaul PW, Smart EJ, Robinson LJ, German Z, Yuhanna IS, et al. 1996. *J. Biol. Chem.* 271:6518–22
91. Garcia-Cardena G, Oh P, Liu JW, Schnitzer JE, Sessa WC. 1996. *Proc. Natl. Acad. Sci. USA* 93:6448–53
92. Zlatkine P, Mehul B, Magee AI. 1997. *J. Cell Sci.* 110:673–79
93. Gorodinsky A, Harris DA. 1995. *J. Cell Biol.* 129:619–27
94. Mineo C, Anderson RGW, White MA. 1997. *J. Biol. Chem.* 272:10345–48
95. Zamora AJ, Garosi M, Ramirez VD. 1984. *Neuroscience* 13:105–17
96. Parton RG, Way M, Zorzi N, Stang E. 1997. *J. Cell Biol.* 136:137–54
97. Cain TJ, Liu YJ, Takizawa T, Robinson JM. 1995. *Biochim. Biophys. Acta* 1235:69–78
98. Shyng SL, Heuser JE, Harris DA. 1994. *J. Cell Biol.* 125:1239–50
99. Smart EJ, Foster DC, Ying YS, Kamen BA, Anderson RG. 1994. *J. Cell Biol.* 124:307–13
100. Dupree P, Parton RG, Raposo G, Kurzchalia TV, Simons K. 1993. *EMBO J.* 12:1597–605
101. Simons K, Ikonen E. 1997. *Nature* 387:569–72
102. Parton RG, Simons K. 1995. *Science* 269:1398–99
103. Schmid S. 1997. *Annu. Rev. Biochem.* 66:511–548
104. Harder T, Kellner R, Parton RG, Gruenberg J. 1997. *Mol. Biol. Cell* 8:533–45
105. Karnovsky MJ, Kleinfeld AM, Hoover RL, Klausner RD. 1982. *J. Cell Biol.* 94:1–6
106. Thompson TE, Tillack TW. 1985. *Annu. Rev. Biophys. Biophys. Chem.* 14:361–86
107. Hanada K, Nishijima M, Akamatsu Y, Pagano RE. 1995. *J. Biol. Chem.* 270:6254–60
108. Oram JF, Yokoyama S. 1996. *J. Lipid Res.* 37:2473–91
109. Lange Y, Steck TL. 1994. *J. Biol. Chem.* 269:29371–74
110. Urbani L, Simoni RD. 1990. *J. Biol. Chem.* 265:1919–23
111. DeGrella RF, Simoni RD. 1982. *J. Biol. Chem.* 257:14256–62
112. Lange Y. 1994. *J. Biol. Chem.* 269:3411–14
113. Lange Y, Matthies HJG. 1984. *J. Biol. Chem.* 259:14624–30
114. Lange Y, Steck TL. 1986. *J. Biol. Chem.* 260:15592–97
115. Kaplan MR, Simoni RD. 1985. *J. Cell Biol.* 101:446–53
116. Fielding PE, Fielding CJ. 1996. *Biochemistry* 35:14932–38
117. Fielding PE, Fielding CJ. 1995. *Biochemistry* 34:14288–92
118. Fielding CJ, Bist A, Fielding PE. 1997. *Proc. Natl. Acad. Sci. USA* 94:3753–58

119. Millan J, Puertollano R, Fan L, Rancano C, Alonso MA. 1997. *Biochem. J.* 321:247–52
120. Zacchetti D, Peranen J, Murata M, Fiedler K, Simons K. 1995. *FEBS Lett.* 377:465–69
121. Babitt J, Trigatti B, Rigotti A, Smart EJ, Anderson RGW, et al. 1997. *J. Biol. Chem.* 272:13242–49
122. Fra AM, Williamson E, Simons K, Parton RG. 1994. *J. Biol. Chem.* 269:30745–48
123. Scherer PE, Lisanti MP, Baldini G, Sargiacomo M, Mastick CC, Lodish HF. 1994. *J. Cell Biol.* 127:1233–43
124. Fan JY, Carpentier JL, van Obberghen E, Grunfeld C, Gorden P, Orci L. 1983. *J. Cell Sci.* 61:219–30
125. Simionescu N, Simionescu M, Palade GE. 1973. *J. Cell Biol.* 57:424–52
126. Simionescu N, Siminoescu M, Palade GE. 1975. *J. Cell Biol.* 64:586–607
127. Schnitzer JE, Oh P, Pinney E, Allard J. 1994. *J. Cell Biol.* 127:1217–32
128. Schnitzer JE, Oh P, McIntosh DP. 1996. *Science* 274:239–42
129. Predescu D, Horvat R, Predescu S, Palade GE. 1994. *Proc. Natl. Acad. Sci. USA* 91: 3014–18
130. Schnitzer JE, Allard J, Oh P. 1995. *Am. J. Physiol.* 268:H48–55
131. Ghitescu L, Bendayan M. 1992. *J. Cell Biol.* 117:745–55
132. Severs NJ. 1988. *J. Cell Sci.* 90:341–48
133. Dvorak AM, Kohn S, Morgan ES, Fox P, Nagy JA, Dvorak HF. 1996. *J. Leukocyte Biol.* 59:100–15
134. Roberts WG, Palade GE. 1997. *Can. Res.* 57:765–72
135. Schnitzer JE, Oh P. 1996. *Am. J. Physiol.* 39:H416–22
136. Kamen BA, Wang MT, Streckfuss AJ, Peryea X, Anderson RGW. 1988. *J. Biol. Chem.* 263:13602–9
137. Kamen BA, Johnson CA, Wang MT, Anderson RGW. 1989. *J. Clin. Invest.* 84:1379–86
138. Kamen BA, Smith AK, Anderson RGW. 1991. *J. Clin. Invest.* 87:1442–49
139. Rijnboutt S, Jansen G, Posthuma G, Hynes JB, Schornagel JH, Strous GJ. 1996. *J. Cell Biol.* 132:35–47
140. Kartenbeck J, Stukenbrok H, Helenius A. 1989. *J. Cell Biol.* 109:2721–29
141. Anderson HA, Chen YZ, Norkin LC. 1996. *Mol. Biol. Cell* 7:1825–34
142. Stang E, Kartenbeck J, Parton RG. 1997. *Mol. Biol. Cell* 8:47–57
143. Parton RG. 1994. *J. Histochem. Cytochem.* 42:155–66
144. Smart EJ, Ying YS, Anderson RG. 1995. *J. Cell Biol.* 131:929–38

145. Deckert M, Ticchioni M, Bernard A. 1996. *J. Cell Biol.* 133:791–99
146. Smart EJ, Estes K, Anderson RGW. 1995. *Cold Spring Harbor Symp. Quant. Biol.* 60:243–48
147. Borregaard N, Miller LJ, Springer TA. 1987. *Science* 237:1204–6
148. Berger M, Medof ME. 1987. *J. Clin. Invest.* 79:214–20
149. Tosi MF, Zakem H. 1992. *J. Clin. Invest.* 90:462–70
150. Roettger BF, Rentsch RU, Pinon D, Holicky E, Hadac E, et al. 1995. *J. Cell Biol.* 128:1029–41
151. Chun MY, Lin HY, Henis YI, Lodish HF. 1995. *J. Biol. Chem.* 270:10855–60
152. Anderson RGW. 1993. *Proc. Natl. Acad. Sci. USA* 90:10909–13
153. Kiss AL, Geuze HJ. 1997. *Eur. J. Cell Biol.* 73:19–27
154. Deckert M, Ticchioni M, Mari B, Mary D, Bernard A. 1995. *Eur. J. Immunol.* 25:1815–22
155. Heuser JE, Anderson RGW. 1988. *J. Cell Biol.* 108:389–400
156. Larkin JM, Brown MS, Goldstein JL, Anderson RGW. 1983. *Cell* 33:273–85
157. Goldstein JL, Brown MS, Anderson RGW, Russell DW, Schneider WJ. 1985. *Annu. Rev. Cell Biol.* 1:1–39
158. Schnitzer JE, Liu J, Oh P. 1995. *J. Biol. Chem.* 270:14399–404
159. Montesano R, Roth J, Robert A, Orci L. 1982. *Nature* 296:651–53
160. Tran D, Carpentier JL, Sawano F, Gorden P, Orci L. 1987. *Proc. Natl. Acad. Sci. USA* 84:7957–61
161. Lisanti MP, Tang ZL, Scherer PE, Kubler E, Koleske AJ, Sargiacomo M. 1995. *Mol. Membr. Biol.* 12:121–24
162. Stefanová I, Horejsí V, Ansotegui IJ, Knapp W, Stockinger H. 1991. *Science* 254:1016–19
163. Thomas PM, Samelson LE. 1992. *J. Biol. Chem.* 267:12317–22
164. Morgan BP, van den Berg CW, Davies EV, Hallett MB, Horejsi V. 1993. *Eur. J. Immunol.* 23:2841–50
165. Shenoy-Scaria AM, Kwong J, Fujita T, Olszowy MW, Shaw AS, Lublin DM. 1992. *J. Immunol.* 149:3535–41
166. Brown D. 1993. *Curr. Opin. Immunol.* 5: 349–54
167. Cinek T, Horejsí V. 1992. *J. Immunol.* 149:2262–70
168. Bohuslav J, Cinek T, Horejsí V. 1993. *Eur. J. Immunol.* 23:825–31
169. Dráberová L, Dráber P. 1993. *Proc. Natl. Acad. Sci. USA* 90:3611–15

170. Kniep B, Cinek T, Angelisova P, Horejsí V. 1994. *Biochem. Biophys. Res. Commun.* 203:1069–75
171. Cerny J, Stockinger H, Horejsí V. 1996. *Eur. J. Immunol.* 26:2335–43
172. Liu P, Ying Y-S, Anderson RGW. 1997. *Proc. Natl. Acad. Sci. USA* 94:13666–70
173. Glenney JR. 1989. *J. Biol. Chem.* 264: 20163–66
174. Glenney JR. 1986. *Proc. Natl. Acad. Sci. USA* 83:4258–62
175. Glenney JR Jr, Zokas L. 1989. *J. Cell Biol.* 108:2401–8
176. Vepa S, Scribner WM, Natarajan V. 1997. *Free Radic. Biol. Med.* 22:25–35
177. Muller G, Geisen K. 1996. *Horm. Metab. Res.* 28:469–87
178. Li SW, Seitz R, Lisanti MP. 1996. *J. Biol. Chem.* 271:3863–68
179. Mineo C, James GL, Smart EJ, Anderson RGW. 1996. *J. Biol. Chem.* 271:11930–35
180. Neubig RR. 1994. *FASEB J.* 8:939–46
181. Feron O, Smith TW, Michel T, Kelly RA. 1997. *J. Biol. Chem.* 272:17744–48
182. Wagner RC, Kreiner P, Barrnett RJ, Bitensky MW. 1972. *Proc. Natl. Acad. Sci. USA* 69:3175–79
183. Slezak J, Geller SA. 1984. *J. Histochem. Cytochem.* 32:105–13
184. Rechardt L, Hervonen H. 1985. *Histochemie* 82:501–5
185. Huang C, Hepler JR, Chen LT, Gilman AF, Anderson RGW, Mumby SM. 1997. *Mol. Biol. Cell* 8:2365–78
186. Xiao Z, Devreotes PN. 1997. *Mol. Biol. Cell* 8:855–69
187. Popescu LM. 1974. *Stud. Biophys.* 44: 141–53
188. Sugi H, Suzuki S, Daimon T. 1982. *Can. J. Physiol. Pharmacol.* 60:576–87
189. Fujimoto T. 1993. *J. Cell Biol.* 120:1147–57
190. Fujimoto T, Nakade S, Miyawaki A, Mikoshiba K, Ogawa K. 1992. *J. Cell Biol.* 119:1507–13
191. Liu PS, Anderson RG. 1995. *J. Biol. Chem.* 270:27179–85
192. Hope HR, Pike LJ. 1996. *Mol. Biol. Cell* 7:843–51
193. Pike LJ, Casey L. 1996. *J. Biol. Chem.* 271:26453–56
194. Clemente R, Jones DR, Ochoa P, Romero G, Mato JM, Varela-Nieto I. 1995. *Cell Signal* 7:411–21
195. Stralfors P. 1997. *BioEssays* 19:327–35
196. Parpal S, Gustavsson J, Stralfors P. 1995. *J. Cell Biol.* 131:125–35
197. Liu JW, Garcia-Cardena G, Sessa WC. 1996. *Biochemistry* 35:13277–81

198. Michel JB, Michel T. 1997. *FEBS Lett.* 405:356–62
199. Sase K, Michel T. 1997. *Trends Cardiovas. Med.* 7:28–37
200. Garcia-Cardena G, Fan R, Stern DF, Liu JW, Sessa WC. 1996. *J. Biol. Chem.* 271: 27237–40
201. Robinson JM, Badwey JA. 1995. *Histochem. Cell Biol.* 103:163–80
202. Miyazaki T, Inoue Y, Takano K. 1995. *Acta Histochem. Cytochem.* 28:365–70
203. Strohmeier GR, Lencer WI, Patapoff TW, Thompson LF, et al. 1997. *J. Clin. Invest.* 99:2588–601
204. Zolkiewska A, Moss J. 1993. *J. Biol. Chem.* 268:25273–76
205. Lander HM, Jacovina AT, Davis RJ, Tauras JM. 1996. *J. Biol. Chem.* 271: 19705–9
206. Harmey JH, Doyle D, Brown V, Rogers MS. 1995. *Biochem. Biophys. Res. Commun.* 210:753–59
207. Vey M, Pilkuhn S, Wille H, Nixon R, Dearmond SJ, et al. 1996. *Proc. Natl. Acad. Sci. USA* 93:14945–49
208. Naslavsky N, Stein R, Yanai A, Friedlander G, Taraboulos A. 1997. *J. Biol. Chem.* 272:6324–31
209. Kaneko K, Vey M, Scott M, Pilkuhn S, Cohen FE, Prusiner SB. 1997. *Proc. Natl. Acad. Sci. USA* 94:2333–38
210. Taraboulos A, Scott M, Semenov A, Avrahami D, Laszlo L, et al. 1995. *J. Cell Biol.* 129:121–32
211. Wooldridge KG, Williams PH, Ketley JM. 1996. *Microb. Pathog.* 21:299–305
212. Tachado SD, Gerold P, McConville MJ, Baldwin T, Quilici D, et al. 1996. *J. Immunol.* 156:1897–907
213. Tachado SD, Gerold P, Schwarz R, Novakovic S, McConville M, Schofield L. 1997. *Proc. Natl. Acad. Sci. USA* 94: 4022–27
214. Koleske AJ, Baltimore D, Lisanti MP. 1995. *Proc. Natl. Acad. Sci. USA* 92: 1381–85
215. Engelman JA, Wykoff CC, Yasuhara S, Song KS, Okamoto T, Lisanti MP. 1997. *J. Biol. Chem.* 272:16374–81
216. Simionescu N, Lupu F, Simionescu M. 1983. *J. Cell Biol.* 97:1592–600
217. Okada SS, Tomaszewski JE, Barnathan ES. 1995. *Exp. Cell Res.* 217:180–87
218. Andersson Forsman C, Gustafsson LE. 1985. *J. Neurocytol.* 14:551–62
219. Kittel A, Bacsy E. 1994. *Cell Biol. Int.* 18:875–79
220. Parkin ET, Turner AJ, Hooper NM. 1996. *Biochem. J.* 319:887–96
221. Wang PY, Kitchens RL, Munford RS. 1996. *J. Inflamm.* 47:126–37

222. Lisanti MP, Scherer PE, Vidugiriene J, Tang ZL, Hermanowski-Vosatka A, et al. 1994. *J. Cell Biol.* 126:111–26

223. Liu J, Oh P, Horner T, Rogers RA, Schnitzer JE. 1997. *J. Biol. Chem.* 272: 7211–22

224. Goldberg RI, Smith RM, Jarett L. 1987. *J. Cell Physiol.* 133:203–12

225. Smith RM, Jarett L. 1988. *Lab. Invest.* 58:613–29

226. Sevinsky JR, Rao LVM, Ruf W. 1996. *J. Cell Biol.* 133:293–304

227. Mulder AB, Smit JW, Bom VJJ, Blom NR, Halie MR, Vandermeer J. 1996. *Blood* 88:3667–70

228. Joliot A, Trembleau A, Raposo G, Calvet S, Volovitch M, Prochiantz A. 1997. *Development* 124:1865–75

229. Tachibana T, Nawa T. 1992. *Arch. Histol. Cytol.* 55:375–79

230. Mineo C, Anderson RGW. 1996. *Exp. Cell Res.* 224:237–42

231. Harder T, Gerke V. 1994. *Biochim. Biophys. Acta* 1223:375–82

232. Millan J, Puertollano R, Fan L, Alonso MA. 1997. *Biochem. Biophys. Res. Commun.* 233:707–12

233. Page E, Upshaw-Earley J, Goings GE. 1994. *Circ. Res.* 75:949–54

234. Bickel PE, Scherer PE, Schnitzer JE, Oh P, Lisanti MP, Lodish HF. 1997. *J. Biol. Chem.* 272:13793–802

Annu. Rev. Biochem. 1998. 67:227–64

HOW CELLS RESPOND
TO INTERFERONS

George R. Stark,[1] *Ian M. Kerr,*[2] *Bryan R. G. Williams,*[1]
Robert H. Silverman,[1] *and Robert D. Schreiber*[3]

[1]The Lerner Research Institute, The Cleveland Clinic Foundation, Cleveland, Ohio 44195; [2]Imperial Cancer Research Fund, Lincoln's Inn Fields, London WC2A 3PX, United Kingdom; [3]Department of Pathology, Washington University School of Medicine, St. Louis, Missouri 63110-1093; e-mail: starkg@cesmtp.ccf.org, Kerr@icrf.icnet.uk, williab@cesmtp.ccf.org, silverr@cesmtp.ccf.org, schreiber@immunology.wustl.edu

KEY WORDS: JAKs, STATs, signaling, antiviral, antigrowth, immunity

ABSTRACT

Interferons play key roles in mediating antiviral and antigrowth responses and in modulating immune response. The main signaling pathways are rapid and direct. They involve tyrosine phosphorylation and activation of signal transducers and activators of transcription factors by Janus tyrosine kinases at the cell membrane, followed by release of signal transducers and activators of transcription and their migration to the nucleus, where they induce the expression of the many gene products that determine the responses. Ancillary pathways are also activated by the interferons, but their effects on cell physiology are less clear. The Janus kinases and signal transducers and activators of transcription, and many of the interferon-induced proteins, play important alternative roles in cells, raising interesting questions as to how the responses to the interferons intersect with more general aspects of cellular physiology and how the specificity of cytokine responses is maintained.

CONTENTS

227

INTRODUCTION

Type I (predominantly α and β) and type II (γ) interferons (IFNs) signal through distinct but related pathways. Enormous progress has been made in recent years in understanding how cells respond to IFNs, especially in uncovering the pathways that mediate inducible gene expression. We now know that these pathways involve (a) specific type I and type II receptors, which bind to the Janus kinases (JAKs), and (b) the signal transducers and activators of transcription (STATs), which in turn propagate the signals. Moreover, JAKs and STATs, discovered through investigations of IFN signaling, also are involved in many different cytokine- and growth factor–mediated pathways. We know of four mammalian JAKs and seven STATs. Several recent reviews describe signaling by IFNs in relation to other cytokines and growth factors (1–5) and more general aspects of JAK-STAT function and the family relationships (6–10).

After activation by JAKs through phosphorylation of a specific tyrosine residue, STATs form homo- or heterodimers through mutual phosphotyrosine-Src homology region 2 (SH2) interactions. STAT dimers bind to gamma-activated sequence (GAS) elements, which drive the expression of nearby target genes. Different GAS elements prefer different STAT dimers, helping to establish specificity. Both STAT1-2 heterodimers and STAT1 homodimers bind to p48, a member of the interferon regulatory factor (IRF) family. The resulting trimers—called IFN-stimulated gene factor 3 (ISGF3) in the case of the STAT1-2 heterodimer—bind to IFN-stimulated regulatory elements (ISREs) that are distinct from the GAS elements. ISREs drive the expression of most IFNα/β-regulated genes and a few IFNγ-regulated genes. This review describes the signaling pathways used to turn the IFN responses on and off and the functions of the induced proteins in mediating the major cellular responses to IFN. Many of the proteins involved in both signaling and responses have important alternative functions, which are also reviewed.

SIGNALING PATHWAYS

Interferon γ

The proximal events of IFNγ signaling require the obligatory participation of five distinct proteins: type I integral membrane proteins IFNGR1 and IFNGR2 (the subunits of the IFNγ receptor) and JAK1, JAK2, and STAT1 (2, 11). Recent work has revealed that this signaling pathway is necessary, though not always sufficient, for induction of most if not all IFNγ-dependent biological responses in vitro and in vivo. IFNγ receptors are expressed on nearly all cell types, with the possible exception of mature erythrocytes, and display strict species specificity in their ability to bind IFNγ (12). Functionally active IFNγ receptors consist of at least two species-matched polypeptide chains (Figure 1). IFNGR1 (previously the α chain or CD119w), a 90-kDa polypeptide encoded by genes on human chromosome 6 and murine chromosome 10, plays important roles in mediating ligand binding, ligand trafficking through the cell, and signal transduction (11, 12). IFNGR2 (previously the β chain or accessory factor-1),

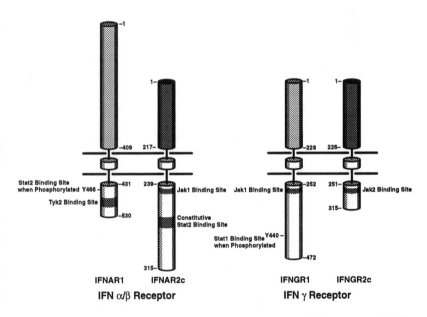

Figure 1   Schematic diagram of the human interferon (IFN) α/β and IFNγ receptors. (*left*) The IFNAR1 and IFNAR2c subunits of the IFNα/β receptor. (*right*) The IFNGR1 and IFNGR2 subunits of the IFNγ receptor. The positions of amino acid residues are shown inside each subunit, and functionally important intracellular domains are also identified. STAT: signal transducer and activator of transcription; JAK: Janus kinase.

a 62-kDa polypeptide encoded by a gene on human chromosome 21 and murine chromosome 16, plays only a minor role in ligand binding but is required for signaling (11, 13, 14).

Three sets of experiments have implicated JAKs and STATs in mediating IFNγ-dependent cellular responses. First, isolation and complementation of mutant human cell lines have revealed that JAK1 and JAK2 become selectively activated in IFNγ-treated cells and are required for the ligand-dependent activation of IFNγ-inducible target genes (1). Second, through biochemical approaches, STAT1—a novel latent cytosolic transcription factor—was isolated and shown to undergo rapid tyrosine phosphorylation and activation in IFNγ-treated cells (1, 2). Third, structure-function analyses of the intracellular domains of the two IFNγ receptor subunits identified constitutive, specific binding sites for JAK1 and JAK2. Moreover, IFNγ induced the formation of a specific phosphotyrosine binding site on the receptor for STAT1, thereby providing the mechanism linking the activated receptor to its signal transduction apparatus (11).

Based on these and other observations, a relatively complete model of IFNγ signaling has been formulated (Figure 2). In unstimulated cells, the IFNγ receptor subunits do not preassociate with one another strongly (15), but their intracellular domains associate specifically with JAK1 and JAK2 (15–18). JAK1 binds to IFNGR1 through a 4-residue sequence ($_{266}$LPKS$_{269}$) in the membrane-proximal region of the IFNGR1 intracellular domain. JAK2 binds to a 12-residue, proline-rich Box 1–like sequence ($_{263}$PPSIPLQIEEYL$_{274}$) in the membrane-proximal region of the intracellular domain of IFNGR2.

Functionally active IFNγ is a homodimer that binds to two IFNGR1 subunits, thereby generating binding sites for two IFNGR2 subunits (15, 19–22). Within the resulting symmetrical signaling complex, the intracellular domains of the receptor subunits are brought into close proximity, together with the inactive JAKs that they carry. JAK1 and JAK2 are then sequentially activated by auto- and transphosphorylation. Activation of JAK2 occurs first and is needed for the subsequent activation of JAK1, which has a structural as well as enzymatic role (23). Work with chimeric JAK1 proteins and receptors has shown that the

→

Figure 2   Signaling through the interferon (IFN)γ receptor. The details of this model are described in the text. In unstimulated cells, IFNGR1 associates with Janus kinase (JAK)1, and IFNGR2 associates with JAK2. IFNγ induces oligomerization of the IFNγ receptor subunits, which leads to the transphosphorylation and activation of JAK1 and JAK2. The activated JAKs then phosphorylate Y440 of IFNGR1, creating a docking site for signal transducer and activator of transcription (STAT) 1. While bound to the receptor, STAT1 is phosphorylated on Y701 and is released from the receptor, forming a homodimer that translocates to the nucleus.

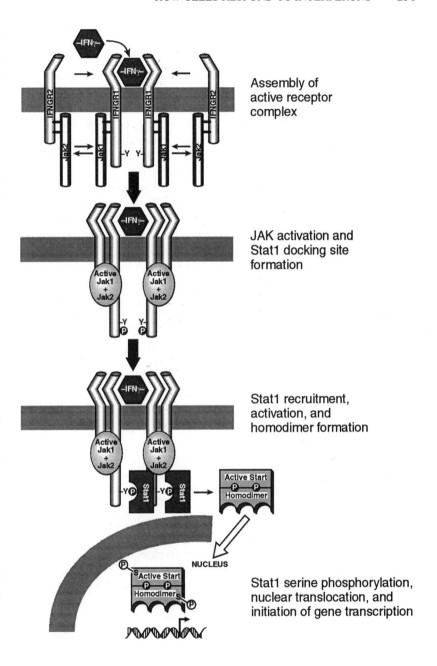

Assembly of active receptor complex

JAK activation and Stat1 docking site formation

Stat1 recruitment, activation, and homodimer formation

Stat1 serine phosphorylation, nuclear translocation, and initiation of gene transcription

specificity of JAKs lies in their capacity to associate with particular cytokine receptor subunits rather than because of a high degree of substrate specificity (24, 25).

Once activated, the receptor-associated JAKs phosphorylate a functionally critical, tyrosine-containing five-residue sequence ($_{440}$YDKPH$_{444}$) near the C terminus of IFNGR1, thereby forming paired ligand-induced docking sites for STAT1 (26–28). Two latent STAT1 proteins then bind to these sites because the SH2 domain of each recognizes the tyrosine-phosphorylated YDKPH sequence (28, 29). The receptor-associated STAT1 proteins are thus phosphorylated by the receptor-bound kinases at tyrosine 701, near the C terminus (30–32). The phosphorylated STAT1 proteins dissociate from the receptor and form a reciprocal homodimer, which translocates to the nucleus by a mechanism dependent on the GTPase activity of Ran/TC4 (33). The active STAT1 homodimers bind to specific GAS elements of IFNγ-inducible genes (1, 2) and stimulate their transcription. The transcriptional activity of STAT1 homodimers is enhanced at some point in the activation cascade by serine phosphorylation (at position 727) by an enzyme with MAP kinase–like specificity (34, 35).

Thus, the biological responses of cells to IFNγ result from ligand-dependent, affinity-driven assembly of a multimolecular signal transduction complex that derives at least some of its specificity from selective recruitment of only one member of the STAT family to its ligand-induced, tyrosine phosphate–docking site on the receptor. Importantly, subsequent work in other labs has shown that other members of the STAT family are recruited to their respective cytokine receptors by similar ligand-induced mechanisms. As a result, the IFNγ signaling model is now an accepted paradigm that explains an important mechanism of how cytokine receptors are coupled to their specific STAT signaling systems.

Processes that negatively regulate IFNγ signaling are only now being defined. In certain cells, such as T cells, IFNγ can induce desensitization by down-regulating the expression of the IFNGR2 mRNA and protein (36, 37). However, whether this mode of desensitization occurs in other cell types remains unclear. Dephosphorylation of the activated IFNGR1 subunit occurs rapidly following stimulation with IFNγ (26–27). However, because no data suggest that the IFNγ receptor associates with a particular phosphatase, dephosphorylation of the receptor may result from the action of general cellular phosphatases. More work in this area is needed. IFNγ (as well as several other cytokines) can induce the expression of a family of proteins termed SOCS/JAB/SSI, which bind to and inhibit activated JAKs (38–40). This work reveals that cytokines can desensitize cells in either a homologous or a heterologous manner by inducing proteins that block JAK activity. Although some of these proteins have been shown to inhibit IFNγ-induced biological responses when overexpressed in cells, little information is available to define their enzyme or cytokine specificities. Nevertheless, further investigation of those novel proteins is likely to produce new insights into how JAK-STAT pathways are regulated.

The physiological relevance of IFNγ signaling through the JAK-STAT pathway and the basis for signaling specificity has been established unequivocally through the generation and characterization of mice with a targeted disruption of the STAT1 gene (41, 42). STAT1-null mice show normal tissue and organ development, produce normal numbers and distributions of immune cell populations, and are able to reproduce. However, cells from these mice are incapable of manifesting any biologic responses to either IFNγ or IFNα, and the mice display severe defects in the ability to resist microbial and viral infections. In contrast, STAT1-null mice do not display abnormalities in responses induced by a variety of other cytokines [such as growth hormone, epidermal growth factor (EGF), and interleukin (IL)-10] that have been shown to activate STAT3 and STAT1 in vitro. Taken together, these results show that under physiological conditions, the development of biological responses induced by IFNγ (and in many cases also by IFNα/β) requires the participation of STAT1. These results further suggest that the use of STAT1 in signaling pathways under physiological conditions is restricted largely to the IFN systems. Thus, the specificity of IFNγ signaling is due predominantly to two temporally and topographically distinct processes involving STAT1. The first process is the recruitment of STAT1 to a specific docking site formed on the activated receptor at the membrane. In the second process, activated STAT1 dimers, once they arrive in the nucleus, activate a distinct set of cytokine-inducible genes.

The ability of STAT1 to activate gene expression may also be modulated by its interaction with other transcription factors. For example, IFNγ-dependent induction of the 9-27 gene is mediated by the interaction of a STAT1 homodimer-p48 complex with an ISRE rather than with a GAS element (43, 44). In addition, induction of the ICAM-1 gene by IFNγ depends on the interaction of STAT1 and the transcription factor Sp1, which occurs when both proteins are bound to DNA (45). Thus, cell type–specific gene induction by IFNγ may be explained, at least in part, by the ability of additional cell-specific positive and negative factors to modulate the actions of STAT1 (46, 47).

Interferons α and β: The Common Pathways

The main pathway of response to IFNα/β requires two receptor subunits, two JAKs, two STATs, and the IRF-family transcription factor p48 (Figure 3). The IFNα/β signaling pathways are understood at least as well as any other, but we are still at a relatively early stage, capable of drawing blobs to illustrate the major interactions but ignorant of the fine mechanistic detail. It will require analysis of the three-dimensional structures of the major components, individually and

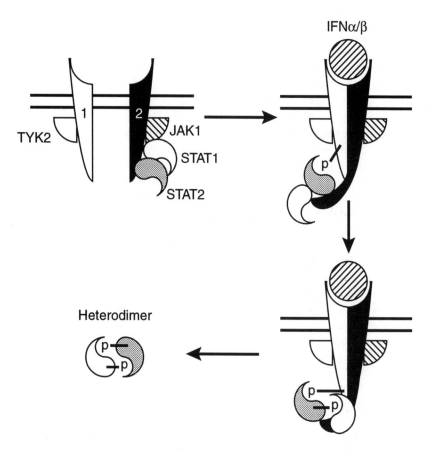

Figure 3 A model for the ordered formation of signal transducer and activator of transcription (STAT) 1 and STAT2 heterodimers at the interferon (IFN)α/β receptor. In the unliganded receptor, IFNAR1 associates with Tyk2, and IFNAR2 associates with Janus kinase (JAK) 1, STAT1, and STAT2. The binding of STAT1 to IFNAR2 depends on STAT2 but not vice versa. The IFN-mediated association of IFNAR1 and -2 facilitates the cross-phosphorylation and activation of Tyk2 and JAK1, which in turn phosphorylate Y466 of IFNAR1, creating a docking site for the SH2 domain of STAT2. This new interaction positions STAT2 for phosphorylation on Y690, thus creating a docking site for the SH2 domain of STAT1, positioning it for phosphorylation on Y701. Release of the STAT1-2 heterodimer from the receptor follows.

complexed with one another, and then manipulation of those structures to reveal the detailed interactions that are crucial for function.

The overall plan of IFNα/β signaling (Figure 3) involves five major steps: (a) IFN-driven dimerization of the receptor outside the cell leads to (b) initiation of a tyrosine phosphorylation cascade inside the cell, resulting in (c) dimerization of the phosphorylated STATs, activating them for (d) transport into the nucleus, where they (e) bind to specific DNA sequences and stimulate transcription. Current understanding of this initial part of the response is greater than of the full response, which additionally involves the suppression of IFN-stimulated genes (ISGs) in the absence of IFN and down-regulation of the initial response in the continued presence of IFN. In addition to this main pathway, IFNα/β activates several other pathways. Although the biochemical evidence for additional signaling is persuasive, unfortunately we still have little knowledge of the physiological roles. It is also clear that different IFNα/β subtypes can stimulate distinct and different ancillary responses. As discussed below, the mechanism probably involves a novel pathway in addition to the one shown in Figure 3.

The receptor has two major subunits (Figure 1): IFNAR1 (the a subunit in the older literature) and IFNAR2c (the β_L subunit). IFNAR2a is a soluble form of the extracellular domain of the IFNAR2 subunit (48), and IFNAR2b (also called the β_S subunit) is an alternatively spliced variant with a short cytoplasmic domain (48) that, when overexpressed, can have dominant negative activity (49). Only IFNAR2c restores IFNα/β signaling to a mutant cell line in which the IFNAR2 gene has been inactivated (50). In contrast to the situation for the IFNγ receptor, neither IFNAR1 nor IFNAR2 alone binds to IFNα/β with the high affinity of the two-subunit combination (51, 52). After IFNα/β is bound, the cascade begins with the phosphorylation of Tyk2, which is preassociated with IFNAR1 (53, 54). JAK1, bound to IFNAR2c, can phosphorylate and activate Tyk2 (55), which can then cross-phosphorylate JAK1 to activate it further. Tyk2 also plays a structural role because the amount of IFNAR1 is low in Tyk2-null cells (56). The domains required for this role are distinct from those required to transduce the signal (57). Activated JAK1 and Tyk2 are almost certainly responsible for the sequential phosphorylation of Y466 of IFNAR1 (58), Y690 of STAT2, and Y701 of STAT1.

As shown in Figure 3, both STAT1 and STAT2 preassociate with IFNAR2c in untreated cells (59). STAT2 binds in the absence of STAT1, but STAT1 binds well only to the IFNAR2c-STAT2 complex (59). STAT1 and STAT2 also seem to associate with each other in the cytosol of untreated cells (60), but the physiological significance of this interaction is unclear. When Y466 of IFNAR1 is phosphorylated, the SH2 domain of STAT2 binds to it (61), followed by the phosphorylation of both STATs and dissociation of the phosphorylated heterodimer from the receptor.

Experiments in which the SH2 domains of STAT1 and STAT2 have been interchanged reveal that the specificity of the IFNγ receptor for STAT1 requires the SH2 domain (29) because a variant STAT1 carrying the SH2 domain of STAT2 does not function. However, in the case of the IFNα/β receptor, a very different result is obtained. STAT2 works equally well with either its own SH domain or that of STAT1 (29, 59), revealing that different domains of STAT2 are more important in establishing specific interactions with the receptor. The N-terminal third of STAT2 has been identified as the dominant subregion in determining specificity (59).

In addition to the seven major components discussed above, there is evidence that the tyrosine phosphatase SHP-2 may also be required for signaling. This enzyme preassociates with IFNAR1 and is phosphorylated in response to IFNα/β (62). In transient cotransfection experiments, a dominant negative form of SHP-2 inhibits the IFNα/β-induced expression of a reporter gene (62). Experiments showing that SHP-2 is required for IFNα/β-dependent activation of endogenous genes in untransfected cells would help to complete this interesting story.

How activated STATs reach the nuclei of IFN-treated cells is not yet clear. ISGF3, the major transcription factor formed in response to IFNα/β (Figure 4), is required to drive the expression of most ISGs via their ISREs, as shown by the specific defects in p48-null human cells (63, 64). Interestingly, p48-null mice also show severe defects in the induction by viruses of the IFNα/β genes themselves, consistent with the binding of ISGF3 to virus-inducible elements within the IFNα/β promoters (65). STAT1-2 heterodimers and STAT1 homodimers form in response to IFNα/β independently of p48, and each can drive the expression of a minority of ISGs, such as the IRF1 gene, through GAS elements (66, 67). The relative amounts of STAT1-2 heterodimer and ISGF3, or of STAT1 homodimer and its complex with p48 (64), will obviously depend on the levels of p48, which can vary widely among different cell types. Because the STAT1 homodimer that forms in response to IFNα/β does not drive the expression of IFNγ-responsive genes that contain GAS elements, it stands to reason that an additional response to IFNγ is required and that a secondary modification of STAT1 homodimers in response to IFNγ may be involved (66). A prime candidate is the phosphorylation of serine 727 of STAT1 (34).

Initial analysis of the interaction of ISGF3 with the 6-16 and 9-27 ISREs showed that the protected region is about 35 nucleotides long (68, 69). The most meaningful contacts between the ISRE and ISGF3 involve STAT1, with p48 playing a less important role and STAT2 serving to provide a potent transactivation domain (70, 71). The region between residues 400 and 500 of STAT1 provide binding-site specificity (72), and the region between residues 150 and 250 is involved in contacting the C-terminal portion of p48 (73). The STAT

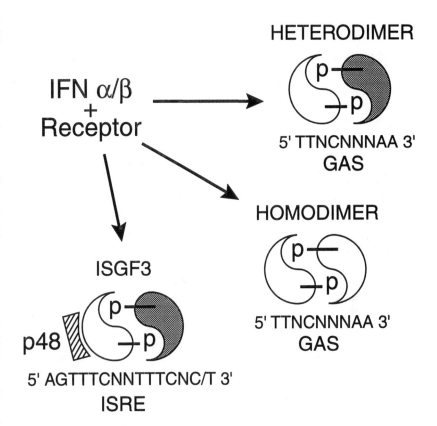

Figure 4 Transcription factors formed at the liganded interferon (IFN)α/β receptor and their DNA recognition elements. Signal transducer and activator of transcription (STAT) 1 and STAT2 heterodimers and STAT1 homodimers bind to identical gamma-activated sequence (GAS) elements, whereas interferon-stimulated gene factor 3 (ISGF3) binds to interferon-stimulated regulatory elements (ISREs).

dimers that are formed in response to many different ligands, including the IFNs, bind to GAS elements whose sequences determine the specificity of the interactions (74). Little current information is available on the regions of STAT dimers that are in contact with the DNA of GAS elements.

Xu et al (75) and Vinkemeier et al (76) have found that the N-terminal domain of STATs 1 and 4 are required for the respective pairs of homodimers to cooperate in binding to tandem GAS sites, which are found in the promoters of some genes that are induced in response to activation of these STATs, for example, the IFNγ gene (75) and the *mig* gene (77). Other genes, for example,

6–16 (78), also have tandem ISREs, and it may be that the cooperative binding of two ISGF3 moieties is required for their optimum expression.

Recent publications have established important connections between STATs 1 and 2 and the CREB-binding protein (CBP)/p300 transcription factors. Zhang et al (79) showed that the N-terminal region of STAT1 interacts with the cyclic-AMP response element binding protein (CREB)–binding domain of CBP/p300 and that the C-terminal region of STAT1 interacts with the domain of CBP/p300, which also binds to the adenovirus protein E1A. Furthermore, both unphosphorylated STAT1 monomers and the phosphorylated STAT1 dimers formed in response to IFNγ are competent to bind to CBP/p300. In transient expression assays, cotransfection of CBP/p300 potentiated and E1A inhibited the activation of a GAS-driven reporter in response to IFNγ. Similarly, Horvai et al (80) showed that the STAT1 and AP1/ets factors that are activated by *Ras*-dependent signaling compete for the limiting amounts of CBP/p300 that each requires for activity. Impressively, microinjection of antibodies directed against CBP/p300 blocks transcriptional responses to IFNγ. Furthermore, Rutherford et al (81) found that the Ets-1 protein of mice binds to an ISRE and may negatively regulate activation by ISGF3. It remains to be seen whether CBP/p300 is required for the transcriptional response to STAT1 homodimers formed in response to IFNα/β, but Bhattacharya et al (82) have shown that CBP/p300 also binds to the C-terminal region of STAT2 and that blockade of this interaction by the adenovirus E1A protein inhibits ISRE-mediated responses to IFNα/β. These fascinating studies provide the first indication of how STATs may interact with the transcriptional machinery.

The mitogen-activated protein kinase (MAPK) cascade is activated by IFNα/β, and the effect of this activation on signaling has been explored. David et al (35) found that ERK2 (the 42-kDa MAPK) binds to a glutathione S-transferase (GST) fusion protein containing the membrane-proximal 50 residues of the cytoplasmic domain of IFNAR1 but not to the full-length cytoplasmic domain of ~100 residues. However, there was association between ERK2 and full-length IFNAR1 in vivo. Treatment of cells with IFNβ induced the tyrosine phosphorylation and activation of ERK2 and caused it to associate with STAT1, as judged by coprecipitation. Furthermore, expression of a dominant negative form of MAPK inhibited IFNβ-induced transcription in a transient cotransfection assay employing an ISRE-driven reporter. It is tempting to connect these observations with those of Wen et al (34), who showed that serine 727 of STAT1, which lies in a MAPK consensus site, is phosphorylated in response to IFNγ and that this phosphorylation increases the response of a GAS-driven promoter to IFNγ. Unfortunately, the connection is not clear. It has not yet been shown that STAT1 is phosphorylated on serine in response to IFNα/β. Wen et al (34) have argued that such phosphorylation is unlikely to be important for the activation of ISRE-driven genes because STAT1β, an alternatively spliced form of STAT1 lacking serine 727, can form ISGF3 and drive the expression of such genes, albeit not as well as STAT1α (83). This is presumably because STAT2 provides a potent transactivation domain. STAT2 does not contain a serine residue in a MAPK consensus site and is not known to be phosphorylated in response to IFN. Thus, the basis of the cross talk between the IFNα/β and MAPK pathways requires further clarification. In more recent work, Stancato et al (84) showed that Raf1, which lies between Ras and ERK2 in the MAPK cascade, is activated by IFNβ in a manner that does not require Ras but does require JAK1. Furthermore, Raf1 activated by IFNβ can be coprecipitated with either JAK1 or Tyk2.

IFNα/β treatment causes phosphorylation and activation of cytosolic phospholipase A$_2$ (CPLA$_2$), which requires JAK1. Furthermore, JAK1 and CPLA$_2$ can be coprecipitated (85). Inhibitors of CPLA$_2$ inhibit IFNα/β-induced expression of ISRE-driven genes, but not of a GAS-driven gene, which implies that CPLA$_2$ is somehow required for the formation of ISGF3 but not of STAT1 homodimer. The basis for this interesting effect remains to be discovered.

The insulin receptor substrate-1 (IRS1) is phosphorylated on tyrosine residues in response to IFNα/β and, in this state, plays its usual role in bringing phosphatidylinositol 3'-kinase to receptors for activation by tyrosine phosphorylation, by engaging the SH2 domains of the p85 regulatory subunit (86). Burfoot et al (87) found that this activation of IRS1 depends on both JAK1 and Tyk2, thus requiring full function of the IFNα/β receptor. However, the physiological significance of this activation is unclear, because cells lacking both IRS1 and IRS2 show little difference from cells that express these proteins in cell growth inhibition in response to IFNα/β or -γ (87). More experiments need to be done to test the requirement for phosphatidylinositol 3'-kinase in a variety of IFN responses.

Interferon β: A Subtype-Specific Pathway

Humans have at least 12 functional IFNαs, a single antigenically distinct IFNβ, and a related IFNω (88), whereas other species have multiple IFNβ subtypes. The human IFNαs are synthesized predominantly by a subset of lymphocytes, and IFNβ is made by fibroblasts. Work with different IFNα subtypes, produced by recombinant DNA technology and by purification of natural leukocyte IFNs, has revealed substantial differences in their specific antiviral activities, in the ratios of antiviral to antiproliferative activities, and in a number of additional functions (89–92). A priori, it might be expected that different cell types might respond differentially to the different IFNs, but to date there is no clear evidence of this (for example, see Reference 91). The various IFNα/βs appear to interact with the same receptor and to have antiviral, antiproliferative, and immunomodulatory activities in a number of cell types. The functional significance of the multiple species, and how functional differences are mediated through apparently identical receptors, remain intriguing questions in this area of research. Interesting differences are emerging, both with respect to IFN-receptor interactions and the induced mRNAs. Mutant cells in the U1 complementation group, lacking Tyk2, are completely defective in response to a purified mixture of natural IFNαs or to recombinant IFNα1 or -α2; nevertheless, they retain partial responses to IFNβ (56) and IFNα8 (91). How the residual IFNβ response is mediated is not yet known, but, importantly, it is not seen in mutant cell lines lacking JAK1, STAT1, or STAT2. It is likely, therefore, to be mediated through JAK-STAT pathways but without an absolute requirement for Tyk2 in the receptor complex. Consistent with this model, IFNβ engages the receptor in a distinct fashion. The groups of Revel and Colamonici first noted the rapid, transient tyrosine phosphorylation of a receptor-associated 100-kDa protein in response to IFNβ but not to IFNα (93, 94). This protein has recently been identified as the IFNAR2c subunit of the receptor (95, 96). Importantly, IFNAR2c is phosphorylated apparently equivalently in response to IFNα or -β, the difference lying in its ability to coprecipitate with IFNAR1 from the complex with IFNβ but not IFNα (95, 96). It remains to be established whether this intriguing finding reflects a tighter association of the two subunits in the β versus the α complex, which might in turn reflect structural differences that, if transmitted through the membrane to the intracytoplasmic domains, mediate a differential response. Importantly, βR1, a gene transcribed preferentially in response to IFNβ, has been discovered (97). The isolation and characterization of the corresponding promoter should, through the identification of known and novel motifs, provide evidence for the involvement of known signaling pathways and experimental handles to investigate unknown ones.

Modulation of IFN Responses

Proteins of the IRF family, such as IRF2 (98), ICSBP (99), and ICSAT (100), bind to ISREs and negatively regulate expression of the associated genes. These repressors may help to prevent the expression of ISGs in the absence of IFN, to down-regulate the induced response, or both. Treatment of cells with IFN in the presence of protein synthesis inhibitors prolongs ISG transcription (101, 102), indicating that some IFN-induced proteins may help to shut off the response. These may be repressors or other types of inhibitors (see below). A mutant cell line with IFN-independent constitutive expression of ISGs has been isolated. There is little or no defect in shutting off the response to IFNα/β in these cells, suggesting that these two aspects of negative regulation can be distinguished (DW Leaman, A Salvekar, R Patel, GC Sen & GR Stark, unpublished data).

The amount of active STAT1 can be reduced by dephosphorylation (103, 104). Proteasome-mediated degradation may also have a role (105), though this aspect is controversial (104). Inhibition of phosphatases by potent agents such as per-oxyvanadate stabilizes the ligand-induced phosphorylation of STATs (106, 107) and can also lead, more slowly, to the ligand-independent accumulation of phosphorylated STATs (107). The inhibited phosphatases may operate on phosphorylated STATs in the nucleus or on phosphorylated JAKs or receptor subunits at the plasma membrane (103, 106). Phosphatases with SH2 domains are especially good candidates for the latter function, and SHP-1 has been implicated in this function in hematopoietic cells (108, 109). Other phosphatases would have to assume this role in most other cell types, where SHP-1 is not expressed. Phorbol esters, which inhibit signaling in response to IFNα/β, can do so by activating one or more tyrosine phosphatases that selectively dephosphorylate Tyk2 but not JAK1 or IFNAR1 (110). Decreased availability of p48 may also play a role (111). The ISGF3-mediated response to IFNα/β is initiated by the rapid formation of ISGF3, but at least for some genes, it is likely to be sustained by IRF1 (69). In human fibroblasts, the level of ISGF3 declines over the course of a few hours and returns to a basal level after 4 h. However, at this time, the transcription rate of the 6-16 gene is still at a maximum, coincident with the maximum in the IFN-induced expression of IRF1 (69). The eventual return of 6-16 transcription to a basal level, in about 8 h, corresponds to the decline of IRF1. By studying mice null for expression of p48, IRF1, or both, Kimura et al (112) have shown that p48 and IRF1 do not have redundant functions but instead complement each other in the responses to both IFNα/β and IFNγ.

FUNCTIONS INDUCED BY INTERFERONS

Antiviral Activities

The ability of IFNs to confer an antiviral state on cells is their defining activity as well as the fundamental property that allowed their discovery (113). IFNs are essential for the survival of higher vertebrates because they provide an early line of defense against viral infections—hours to days before immune responses. This vital role has been demonstrated by the exquisite sensitivity to virus infections of mice lacking both IFNα/β and γ receptors (114). Multiple, redundant pathways have evolved to combat different types of viruses and the various compensatory defense mechanisms that different viruses have evolved (see below). Any stage in virus replication appears to be fair game for inhibition by IFNs (115), including entry and/or uncoating [simian virus 40 (SV40), retroviruses], transcription [influenza virus, vesicular stomatitis virus (VSV)], RNA stability (picornaviruses), initiation of translation (reoviruses, adenovirus, vaccinia), maturation, and assembly and release (retroviruses, VSV).

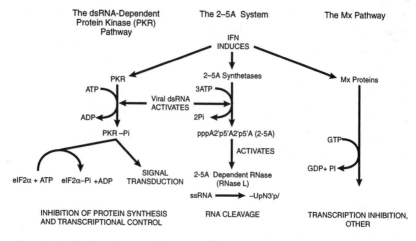

Figure 5   Antiviral mechanisms of interferon (IFN) action.

PKR   The best-characterized IFN-induced antiviral pathways utilize the dsRNA-dependent protein kinase (PKR), the 2-5A system, and the Mx proteins (Figure 5). PKR is a serine-threonine kinase with multiple functions in control of transcription and translation (116, 117). PKR is normally inactive, but on binding to dsRNA, it undergoes autophosphorylation and subsequent dsRNA-independent phosphorylation of substrates. Two conserved dsRNA-binding motifs are present in the N-terminal regulatory half of PKR. The first mediates dsRNA-binding activity and includes residues critical for binding to dsRNA, highly conserved within a large family of dsRNA-binding proteins. No RNA sequence specificity is required for dsRNA to bind to PKR. The resulting conformational change in the enzyme probably unmasks its C-terminal catalytic domain (118). The antiviral effect of PKR is due to its phosphorylation of the alpha subunit of initiation factor eIF2. This phosphorylation results in the formation of an inactive complex that involves eIF2-GDP and the recycling factor eIF2B, resulting in rapid inhibition of translation. Apoptosis may also play a role in the antiviral effect of PKR (see below). Overexpression of PKR leads to the suppression of encephalomyocarditis virus (EMCV) replication in cultured cells (119). In addition, a dominant negative PKR mutant or an antisense PKR cDNA construct suppresses the anti-EMCV effect of IFNα and IFNγ in promonocytic U-937 cells (120). Poly(I):poly(C) or IFNγ treatment extends the survival of wild-type but not PKR-null mice after infection with EMCV (121). In contrast, redundancy in the antiviral pathways is apparent because IFNα extends to the same extent the survival of wild-type and PKR-null mice after EMCV infection.

THE 2-5A SYSTEM   This system is a multienzyme pathway (Figure 5) in which IFN-inducible 2-5A synthetases are stimulated by dsRNAs, often of viral origin, to produce a series of short, $2',5'$-oligoadenylates (2-5A) that activate the 2-5A–dependent RNase L (122, 123). Activation of this pathway leads to extensive cleavage of single-stranded RNA (124–126). The 2-5A synthetases (40, 46, 67, 69, 71, and 100 kDa) are encoded by multiple genes and reside in different parts of the cell (127–130). 2-5A binds to inactive, monomeric RNase L, inducing the formation of the homodimeric, active enzyme (131–133). The activation of RNase L is reversible (134). Its N-terminal half is a repressor that contains a repeated P-loop motif and nine ankyrin repeats, both involved in 2-5A binding. The C-terminal half contains a region of protein kinase homology, a cysteine-rich domain, and the ribonuclease domain (135, 136). The isolated C-terminal half of RNase L cleaves RNA in the absence of 2-5A (137). There are striking similarities between RNase L and IRE1p, a yeast endoribonuclease that functions in Hac1 mRNA splicing in the unfolded protein response (138). An intriguing possibility is that RNase L might exist as a member of a family of regulated nucleases with diverse functions in different organisms.

The functions of the 2-5A system have been explored through genetic manipulation of RNase L. Cells expressing a dominant negative derivative were defective in expressing the anti-EMCV and antiproliferative activities of IFNα/β, whereas overexpression of RNase L blocked vaccinia virus and HIV-1 replication (136, 139; RK Maitra & RH Silverman, unpublished data). RNase L–null mice are deficient in both the anti-EMCV effect of IFNα and in several apoptotic pathways (140). Although IFNα treatment extended the survival of both wild-type and RNase L–null mice after EMCV infection, the RNase L–null mice died several days earlier, lending support to cell culture studies linking the 2-5A system to the anti-EMCV effect of IFN (123). The ability of RNase L to be activated by small molecules opens up possibilities for drug design and development. In one such example, RNase L was recruited by 2-5A–antisense oligonucleotides to cleave respiratory syncytial virus M2 RNA selectively, thus blocking viral replication in human tracheal epithelial cells (141, 142). A mammalian 2-5A system has also been cloned in transgenic tobacco plants, resulting in resistance to several different viruses (143, 144).

THE MX PROTEINS   Mx proteins are IFN-inducible, high-abundance 70- to 80-kDa GTPases in the dynamin superfamily (145, 146). Mx proteins and dynamins self-assemble into horseshoe- and ring-shaped helices and other helical structures (147–149). Human MxA forms tight oligomeric complexes in cell-free systems and in intact cells (150, 151). The Mx proteins interfere with viral replication, impairing the growth of influenza and other negative-strand RNA viruses at the level of viral transcription and at other steps. The murine nuclear protein Mx1 suppresses the growth of influenza, Thogoto, and tick-born Dhori viruses, and the human cytoplasmic protein MxA inhibits the growth of influenza, VSV, measles, Thogoto, bunya, phlebo, hanta, and human parainfluenza 3, but not Dhori viruses (152–157). Mutant forms of Mx proteins lacking the ability to bind or hydrolyze GTP fail to suppress viral replication. However, the binding, not the hydrolysis, of GTP is required to inhibit VSV transcription by MxA in vitro (158, 159). Mx proteins are believed to interfere with the trafficking or activity of viral polymerases (160). Furthermore, MxA specifically binds to Thogoto-virus ribonucleoprotein complex (O Haller, personal communication). Murine Mx1 inhibits the primary transcription of influenza virus, whereas human MxA acts in the cytoplasm to inhibit a later step in the viral life cycle (161). Although PKR and Mx genes are induced preferentially by type I IFNs, 2-5A-synthetase and RNase L are induced by both types (116, 135, 162, 163). Also, the induction by IFNγ (but not IFNα/β) of nitric oxide synthase in mouse macrophages inhibits the growth of ectromelia, vaccinia, and HSV-1 viruses (164). Therefore, different antiviral pathways may be induced in different cell types, depending on the type of IFN involved.

Many IFN-induced proteins are poorly characterized, and some of these are very likely to possess antiviral activity. For instance, expression of the IFN-inducible 9-27 protein led to a partial inhibition of VSV replication (165). Clearly, the enormous selective pressures imposed by viruses have resulted in a rich and diverse set of antiviral pathways.

VIRAL INHIBITION OF THE IFN RESPONSE   Hardly surprising, viruses fight back, not only against host defenses in general (166–169) but also against the IFN systems in particular, both through novel mechanisms and by subverting host systems through the synthesis of novel proteins and proteins that mimic and thus interfere with host proteins (e.g. the IFN receptors; see References 169, 170). There is evidence for the inhibition of the 2-5A–dependent RNase L in response to EMC infection (171) and for a cellular protein inhibitor of RNase L (172), but the most extensively studied examples involve the inhibition of PKR. At least four different mechanisms are used, including inhibitory viral RNA, inhibitory viral or cellular proteins, and proteolytic cleavage. Best studied is the adenovirus virus-associated (VA) RNA, which binds to but does not activate PKR (173). An important fact is that mutant viruses lacking VA RNA are more sensitive to IFN-mediated inhibition (174). Epstein-Barr virus-encoded small nonpolyadenylated RNAs (EBER) may perform a similar function for Epstein-Barr (EB) virus (175), although an EBER-negative strain shows no obviously enhanced sensitivity to IFNs in vitro (176). Examples of proteins that sequester the viral dsRNA activators of PKR are the reovirus sigma 3 capsid protein (177) and the vaccinia virus E3L protein (178). The HIV transcriptional transactivator (TAT) (179, 180) and hepatitis C virus NS5A (181) proteins appear to inhibit by interacting with PKR directly. In response to EMC infection, proteolytic cleavage of PKR in poliovirus-infected cells and sequestration of the enzyme occur (182, 183). Particularly interesting are the cellular protein systems that inhibit PKR in response to influenza virus infection. p58(IPK), a cellular protein inhibitor of PKR, is inhibited by I-p58(IPK), which is apparently inactivated in response to infection. I-P58(IPK) has recently been identified as the molecular chaperone hsp40; this identification revealed that the influenza virus regulates PKR activity by recruiting a cellular stress protein (184). Both cells and viruses have developed elegant mechanisms to control PKR, which shows the importance of this enzyme in controlling cellular functions (see below) and virus replication. As a variant on this theme, SV40 can restore efficient translation in cells, despite the elevated levels of phosphorylated eIF2α that result from activating PKR, because the translational rescue mediated by the SV40 large T antigen occurs downstream of the phosphorylation of eIF2α (185).

Cell death in response to virus infection may be mediated by apoptosis as well as necrosis (186). Interestingly, cells from mice lacking the 2-5A–dependent RNase L or PKR show defects in apoptosis (140, 187), consistent with a possible role for these enzymes in virus-induced, IFN-mediated cell death. Poxviruses produce CrmA, and the Kaposi's sarcoma herpes virus produces FLIPs (inhibitors of the apoptotic ICE and FLICE proteases, respectively), presumably to suppress host-cell suicide and inflammatory responses (188, 189). Other viral anti-apoptotic genes resemble the mammalian bcl-2 gene, which suppresses apoptosis (186, 190).

The adenovirus E1A and human papilloma virus E6 and E7 proteins inhibit the production and action of IFN at the level of transcription (66, 191–194). For E1A, the effect on the transcription of ISGs is mediated, in part at least, through a reduction in functional p48 (192, 193) and probably also by sequestering p300/CBP, required for transcriptional activation through STAT2 (82). A similar inhibition at a transcriptional level of IFN production and action is mediated by the Kaposi's sarcoma herpes virus through the production of an inhibitory mimic of IRF (169). In additional strategies, the poxviruses produce soluble IFNα and IFNγ receptors (116), and the EB virus generates an IL-10 analog. Interestingly, the type I IFN receptor mimic produced by the vaccinia virus shows a wide species specificity, consistent with the broad host range of the virus (195). The IL-10 analog probably performs a dual function for the virus, inhibiting the production of IFNγ and activating the B lymphocytes necessary for virus replication (196).

The IFN systems are subject to cellular control during development and differentiation and are subject to inhibition by viruses. The multiple mechanisms involved emphasize the importance of these systems to both cells and viruses. Conversely, the IFN systems are not, of course, the only host defense systems against which viruses retaliate. Indeed, to know which host defenses are important, viruses must be investigated.

Inhibition of Cell Growth

IFNs inhibit cell growth and control apoptosis, activities that affect the suppression of cancer and infection. Genes have been identified that are important for the apoptotic, but not the growth inhibitory, effects of IFNγ (197). Therefore, these two activities of IFNs are considered separate but related topics.

Different cells in culture exhibit varying degrees of sensitivity to the antiproliferative activity of IFNs. In some cases, growth arrest may be due to differentiation, particularly when IFNs are used in combination with other agents such as retinoids (198, 199). Specific IFN-induced gene products have not been linked directly to antiproliferative activity. However, IFNα has been shown to target specific components of the cell-cycle control apparatus, including c-myc, pRB, cyclin D3, and cdc25A (200–203). Lymphoblastoid Daudi cells are exquisitely sensitive to the antiproliferative effects of IFNα, which lead to a rapid shutdown of c-myc transcription, possibly through a decrease in the activity of the transcription factor E2F (202). Cells expressing a transdominant mutant of PKR fail to suppress c-myc in response to IFN, although the phosphorylation of pRB is suppressed (204). PKR may play a subtle role in cell growth regulation. Early-passage embryo fibroblasts (MEFs) established from PKR-null mice achieve saturation densities similar to those of wild-type MEFs, whereas PKR-null cells consistently achieve higher saturation densities beyond five passages. The doubling times of wild-type and PKR-null cells do not differ appreciably between early and late passages, however (S Der & BRG Williams, unpublished data). This phenotype is similar to the increased saturation densities described for MEFs derived from p53-null or p21/WAF1-null mice (205) and may result from increased resistance to apoptosis, induced by growth-factor deprivation in the absence of PKR.

The phosphorylation of pRB by IFN is suppressed by the inhibition of cdk4 and cdk6. This inhibition is achieved through the suppression of cyclin D3 and by preventing the activation of cdk2-cyclinA and cdk-cyclinE, thereby inhibiting the phosphatase cdc25A (203). This mechanism of growth suppression is distinct from that of other growth-inhibitory cytokines such as TGFb because it does not appear to involve the induction of cdk inhibitors such as p21, p27, and p57^{kip}. Cell-type differences clearly complicate any mechanistic understanding of the antiproliferative effects of IFN. For example, in contrast to the Daudi cells discussed above, the induction of p21-cdk2 has been correlated with growth inhibition of the prostate cancer cell line DU145 by IFNα (206).

Control of Apoptosis

IFNs are essential for host responses to viruses and some other microbial pathogens, events that often culminate in apoptosis. However, development in the mouse proceeds normally in the absence of functional IFNα/β and IFNγ receptors (114). IFNs have either pro- or anti-apoptotic activities, depending on factors such as the state of cell differentiation. For instance, IFNγ either induced or inhibited the apoptosis of murine pre-B cells or B chronic lymphocytic leukemia cells, respectively (207–209). Similarly, IFNγ promoted either proliferation or apoptosis in malignant human T cells, depending on the presence or absence of serum and the levels of the IFNγ receptor (210). The involvement of IFNs in apoptosis is interwoven with the roles of other modulators of apoptosis and the enzymes they regulate. For example, dsRNA produced during viral infections, and lipopolysaccharide (LPS), an endotoxin of gramnegative bacteria, are potent inducers of apoptosis. Interestingly, either dsRNA or LPS induces the synthesis of IFNs. The induction of IFNγ by LPS requires the activity of caspase-1 (ICE) to process the IFNγ-inducing factor IGIF or IL-18 (211, 212). Similarly, dsRNA induces the production of IFNα/β and is a pro-apoptotic agent (187).

Investigations into molecular pathways mediating IFN-induced apoptosis have focused either on the antiviral enzymes (see below) or on identifying other proteins in the pathway. Several genes have been cloned that, when downregulated, suppress the growth inhibitory or apoptotic activities of IFNγ in HeLa cells. Five novel genes for death-associated proteins, DAP-1 to -5, and two other proteins (thioredoxin and the protease cathepsin D) play a role in these processes (197, 213–218). Down-regulation of DAP-1 (a proline-rich protein), DAP-2 (DAP kinase), DAP-3, or cathepsin D with antisense RNAs blocked the apoptotic activity but not the cytostatic effect of IFNγ. DAP-2, a calmodulin-dependent protein kinase with a consensus death domain, localizes to the cytoskeleton, and its expression is lost frequently in human tumors (218, 219). DAP-2 is a tumor suppressor gene that couples the control of apoptosis to metastasis (219a). DAP-5 was identified from a partial cDNA encoding a dominant negative protein related to the protein-synthesis initiation factor eIF4G (217). It will be interesting to determine how all of these proteins interact with each other and with other intracellular apoptotic factors.

Effects of IFNs on the Immune System

The immunomodulatory actions of IFNs have been studied extensively, but because of space limitation, they cannot be discussed in detail here. Several reviews on IFNα/β and IFNγ biology have been published recently, and the reader is referred to these for more details (12, 115, 220, 221). Here we identify major recent advances in understanding the roles of IFNs in promoting immune responses, and we provide examples of how the actions of IFNα/β and IFNγ diverge. IFNs are known to profoundly affect nearly all phases of innate and adaptive immune responses. Within the IFN family, IFNγ plays the predominant immunomodulatory role. It is produced by a restricted set of immune cells (T cells and natural killer cells) in response to immune and/or inflammatory stimuli and functions to stimulate the development and actions of immune effector cells. The immunomodulatory actions of IFNα/β are more restricted: They are directed largely at promoting responses that provide the host with adaptive immune response mechanisms to resist viral infection.

IFN, ANTIGEN PROCESSING AND PRESENTATION, AND DEVELOPMENT OF CD8$^+$ T-CELL RESPONSES   One unarguable role of IFNs in promoting protective immune responses is their ability to regulate the expression of proteins encoded in the major histocompatibility complex (MHC). All IFN family members share the ability to enhance the expression of MHC class I proteins and thereby to promote the development of CD8$^+$ T-cell responses (221). This expression is known to be driven by IRF1, the transcription factor predominantly responsible for activating MHC class I gene transcription (222, 223). Cells from mice with targeted mutations in either the IFNγ or IFNα/β receptor systems, STAT1, PKR, or IRF1 fail to up-regulate MHC class I proteins on their surface in response to stimulation by the appropriate IFN. In contrast, IFNγ is uniquely capable of inducing the expression of MHC class II proteins on cells, thereby promoting enhanced CD4$^+$ T-cell responses (221, 224). This response depends on a distinct transactivating factor, CIITA. Cells from human patients with the rare abnormality bare-lymphocyte syndrome, characterized by the absence of CIITA, fail to express MHC class II proteins either constitutively or following exposure to IFNγ. IFNγ induces MHC class II protein expression in a wide variety of different cell types, such as mononuclear phagocytes, endothelial cells, and epithelial cells, but it inhibits IL-4–dependent class II expression on B cells (225). The molecular basis for this discordant effect is unknown.

IFNs also play an important role in antigen processing by regulating the expression of many proteins required to generate antigenic peptides. IFNγ modifies the activity of proteasomes by modulating the expression of both enzymatic and nonenzymatic components (221, 226). The proteasome is a multisubunit enzyme complex that is responsible for the generation of all peptides that bind to MHC class I proteins. In unstimulated cells, it contains three enzymatic subunits: x, y, and z. However, following treatment of cells with IFNγ, transcription of the x, y, and z genes decreases, and transcription of three additional genes encoding different enzymatic proteasome subunits, LMP2, LMP7, and MECL1, increases. This leads to the formation of different proteasomes containing these subunits and possessing a different substrate specificity, thereby altering the types of peptides produced and eventually presented to the immune system. IFNγ also induces the expression of a nonenzymatic proteasome subunit, PA28 (also known as the 11S regulator), which binds to proteasome enzyme components and alters their specificity (227, 228). Finally, IFNγ increases the expression of TAP1 and TAP2, which transfer peptides generated by the proteasome in the cytoplasm into the endoplasmic reticulum, where they bind to nascent MHC class I chains (229, 230). Thus, IFNs enhance immunogenicity by increasing the quantity and repertoire of peptides displayed in association with MHC class I proteins.

IFNγ AND DEVELOPMENT OF THE CD4$^+$ HELPER T-CELL PHENOTYPE   Activated human and murine CD4$^+$ T cells can differentiate into two polarized subsets, defined by the cytokines they produce when stimulated (231). In mice, T helper 1 (Th1) cells have the selective ability to synthesize IFNγ, lymphotoxin (LT), and IL-2 and to promote cell-mediated immunity and delayed type hypersensitivity (DTH) responses. In contrast, murine Th2 cells selectively produce IL-4, IL-5, IL-6, and IL-10 and thereby facilitate antibody production and the development of humoral immune responses. IFNγ has an important effect on Th1 cell development. In vitro, antibody-mediated neutralization of IFNγ greatly reduces the development of Th1 cells and augments the development of Th2 cells (232). Similar effects are seen in mice lacking the ability to respond to IFNγ, i.e. STAT1-null mice. However, administration of exogenous IFNγ, in vitro or in vivo, does not drive a Th1 response. Thus, IFNγ is necessary but not sufficient for Th1 development.

IFNγ plays a dual role in this process. First, it facilitates Th1 production by enhancing the synthesis of IL-12 in antigen-presenting cells (233–235). IL-12 is the proximal effector that drives developing CD4$^+$ T cells to become Th1 cells (232, 236). In addition, IFNγ maintains expression of the β2 subunit of the IL-12 receptor on developing CD4$^+$ T cells, thereby preserving their capacity to respond to IL-12 (237). Second, IFNγ blocks the development of Th2 cells by inhibiting the production of IL-4, which is required for Th2 formation (238), and by preventing Th2-cell proliferation (239). Th1 cells are not affected in this manner because they become insensitive to IFNγ as a result of IFNγ-dependent down-regulation of the expression of IFNGR2 (36, 37).

IFN, MACROPHAGE ACTIVATION, AND CELLULAR IMMUNITY   Macrophages function as a key effector cell population in innate and adaptive immune responses. To carry out these functions, they must first become activated, a process involving a reversible series of biochemical and functional alterations that provide them with enhanced cytocidal activities (240). Through the use of neutralizing IFNγ-specific monoclonal antibodies and gene-targeted mice, it has been possible to establish unequivocally the predominant role played by IFNγ in generating activated macrophages, both in vitro and in vivo (241–243). Importantly, the macrophage-activating activity of IFNγ is not provided by IFNα/β. Supporting data come from studies demonstrating that IFNγ-unresponsive mice or humans (i.e. IFNγ-null, IFNγ receptor–null, or STAT1-null mice or patients with inactivating mutations in the IFNGR1 gene) are highly susceptible to infection with a variety of microbial pathogens such as *Listeria monocytogenes, Toxoplasma gondii, Leishmania major*, and several different strains of *Mycobacteria* (41, 242–245). Increased susceptibility to infection occurs in IFNγ-unresponsive hosts despite their ability to maintain an unaltered capacity to produce and respond to IFNα/β.

Activated macrophages use a variety of IFNγ-induced mechanisms to kill microbial targets. Two of the most important involve the production of reactive oxygen and reactive nitrogen intermediates. Reactive oxygen intermediates are generated as a result of the IFNγ-induced assembly of NADPH oxidase, formed as a result of the induced translocation of two cytosolic enzyme subunits to the plasma membrane, where they combine with a membrane-associated electron transport chain component, cytochrome $b558$ (246). This enzyme effects a one-electron transfer to oxygen, producing superoxide anion, which, in turn, is used to generate additional toxic oxygen compounds such as hydrogen peroxide, hydroxyl radicals, and singlet oxygen. Reactive nitrogen intermediates, particularly nitric oxide (NO), are generated in murine macrophages as a result of the IFNγ-dependent transcription of the gene encoding the inducible form of nitric oxide synthase (iNOS), which catalyzes the formation of large amounts of NO (247). NO is thought to kill target cells by one of two mechanisms. First, it can form an iron-nitrosyl complex with the Fe-S groups of aconitase, complex I and complex II, thereby inactivating the mitochondrial electron transport chain. Alternatively, NO can react with superoxide anion to form peroxynitrite, which decays rapidly to form highly toxic hydroxyl radicals. Although iNOS is induced in murine macrophages in an IFNγ-dependent manner, it is not induced in human mononuclear phagocytes exposed to the same stimuli. The molecular basis for this difference has not yet been defined.

IFN AND HUMORAL IMMUNITY   IFNs play complex and sometimes conflicting roles in regulating humoral immunity. Most analyses have attempted to define the influence of IFNγ in the process, although more recent observations suggest that IFNα/β may also induce many of the same biological effects. IFNs exert their effects either indirectly (as described above), by regulating the development of specific T helper cell subsets, or directly at the level of B cells. In the latter case, IFNs are predominantly responsible for regulating three specialized B-cell functions: development and proliferation, immunoglobulin (Ig) secretion, and Ig heavy-chain switching.

The best-characterized action of IFNs directed toward B cells is their ability to influence Ig heavy-chain switching. Ig class switching is significant because the different Ig isotypes promote distinct effector functions in the host. By favoring the production of certain Ig isotypes while inhibiting the production of others, IFNs can facilitate interactions between the humoral and cellular effector limbs of the immune response and increase the host defense against certain bacteria and viruses. In vitro, IFNγ is able to direct immunoglobulin class switching from IgM to the IgG2a subtype in LPS-stimulated murine B cells (248) and to IgG2a and IgG3 in murine B cells that have been stimulated with activated T cells (249). Moreover, IFNγ blocks IL-4–induced Ig class switching in murine B cells from IgM to IgG1 or IgE (250). The validity of these observations has been tested stringently by injecting mice with polyclonal anti-IgD serum, a polyclonal activator of B cells. These mice produced large quantities of IgG1 and IgE. However, when IFNγ was administered prior to anti-IgD treatment, the mice produced high levels of IgG2a and decreased levels of IgG1. Thus, IFNγ is clearly an important regulator of Ig class switching in vivo.

A role for type I IFNs in this process has also been identified (251). Of particular importance are experiments using mice that lack receptors for IFNγ, IFNα/β, or both (114). The mice were infected with lymphocytic choriomeningitis virus (LCMV), and the profiles of the LCMV-specific antibodies generated were determined. Comparable levels of LCMV-specific IgG2a antibodies were observed in the sera of normal mice and of mice unresponsive to either IFNγ or IFNα/β. In contrast, IgG2a antibodies were not produced in mice lacking responses to both types of IFN. These results demonstrate that if induced during the immune response, IFNs α/β can indeed function in a manner redundant to IFNγ in effecting Ig class switching.

ADDITIONAL FUNCTIONS OF PROTEINS INVOLVED IN IFN RESPONSES

JAKs

JAKs can auto- and transphosphorylate, and it is reasonable to assume that they phosphorylate the receptors and the proteins recruited to them, foremost among which are STATs. The interaction of SH2 domains with receptor subunit phosphotyrosine motifs clearly plays a major role in recruitment. But it is increasingly unlikely that this is the whole story, and recruitment directly by JAKs is an interesting alternative. For example, it appears possible that signaling by growth hormone can be achieved with a receptor entirely lacking phosphotyrosine motifs (252). Fujitani et al (253) have presented evidence for the recruitment of STAT 5 through JAK2, and a number of additional JAK-signaling component interactions have been reported (see below). More generally, JAK1 and JAK2 are present in the cell nucleus as well as in the cytoplasm and at membranes (254; A Ziemiecki, personal communication). Initial results with a dominant negative derivative of JAK1 raise the possibility of a constitutive requirement for JAKs early in zebra fish development (255). JAK3-null mice show no obvious defect in early development (256, 257). JAK1-null mice, however, are runted, fail to nurse, and die perinatally. They also appear to have a sensory neuron defect, which includes a failure of explanted dorsal root ganglion neurons to survive when cultured in the presence of IL-6, leukemia inhibitory factor (LIF), ciliary neurotrophic factor (CNTF), oncostatin M (OSM), or cardiotrophin 1 (CT-1), all neurotrophic factors that signal to the JAKs through the IL-6 receptor family (S Rodig & R Schreiber, unpublished data). JAK2-null mice die early in embryogenesis, consistent with a failure of hematopoeisis (E Parganas & JN Ihle, personal communication). Of course, it remains to be established that any of the defects in the knockout mice reflect a requirement for additional JAK functions. That said, it has become increasingly clear that just as JAKs may not be the only mediators of STAT activation, STATs are not the only targets of activated JAKs. Evidence for this comes from both protein association data and functional experiments. For example, growth hormone, IL-11, and OSM all promote the association of JAK2 with Shc and Grb2 (258–260). Work with a JAK2-null cell line has established that the phosphorylation of Shc in response to growth hormone depends on JAK2 (261), consistent with a requirement for cotransfected JAK2 to achieve MAP kinase activation in response to activation of a transfected growth hormone receptor (262). Raf1 associates with JAK2 when coexpressed in the baculovirus system and in erythropoietin- or IFN-treated cells (262). Therefore, JAK-dependent MAP kinase activation by different cytokines or growth factors may occur through recruitment to the JAKs as well as through the well-established pathway involving receptor tyrosine motifs (262). Additional proteins reported to interact with JAKs include SHP1 and SHP2, Vav, Fyn, Btk, Tec, and c-Abl (108, 109, 263–268). For c-Abl, constitutive JAK activation correlated with transformation and was lost on inactivation of a temperature-sensitive c-Abl protein (267). Early suggestive evidence for an additional role for JAKs came from the demonstration that JAK1 is activated in response to EGF but is not required for STAT activation (269). The function of activated JAK1 in this response remains to be established. Similarly, the work of many groups on truncated cytokine receptors has established a requirement for the juxtamembrane Box1 and Box2 motifs and for JAK (but not STAT) activation to stimulate a mitogenic response through pathways yet to be defined (270, 271). IRS1 has been shown to coimmunoprecipitate with

JAK2, and the ability of the growth hormone receptor to transduce the signal for IRS1 depends on the same region of the receptor required for JAK2 binding (272). More recently, work with JAK-null cell lines has established that the activations of IRS1 and phosphatidylinositol 3′ kinase by IL-4, OSM, and IFNs are JAK dependent (87, 273). For IFNs, work with JAK1-null cells has established that JAK1 is required for the activation of cytosolic phospholipase A2 by IFNα (85). Data obtained with a kinase-negative derivative of JAK1 and mutant receptors have raised the possibility that additional JAK-dependent pathways may be required in the antiviral responses to IFNγ and IFNα/β (23, 274). Also, in the most detailed study of the activation of the MAP kinase pathway by IFNα, Larner et al have concluded that JAK1 is essential for the activation of Raf1 and the ERK/MAP kinases (84). Finally, Sugamura et al have implicated JAKs 2 and 3 in activating the signal transducing adaptor molecule, which is involved in both c-myc induction and cell growth in response to IL-2 and GM-CSF (275).

STATs

Evidence is accumulating that STATs play an important role distinct from their well-known function as inducible transcription factors. Three distinct observations reveal an important role for STAT1 in the constitutive expression of certain genes. The expression of IRF1 is low in STAT1-null U3A cells and becomes significant when STAT1 is expressed from a transgene in U3A-R cells (83). The caspase family members ICE, Cpp32, and Ich-1 are expressed at levels 10- to 15-fold lower in U3A cells than in U3A-R or wild-type cells, leading to substantial defects in response to pro-apoptotic signals (275a). Expression of both LMP2 and LMP7 is almost completely absent in U3A cells and is restored in U3A-R cells (Chatterjee-Kishore et al, unpublished data). The defects in caspase expression were corrected when U3A cells were complemented with the Y701F mutant of STAT1, ruling out the possibility that a STAT-STAT dimer stabilized by SH2-phosphotyrosine interactions can be responsible (275a). A strong conclusion is that STAT1 is required for the constitutive expression of some genes, either alone as monomers or, more likely, in combination with transcription-factor partners still to be identified. It remains to be seen whether other STATs have similar functions. As noted above (see also Reference 75), STAT dimers interact with each other and with several different transcription factors, primarily through the N-terminal domains, and STAT2 also uses its N-terminal domain to bind to the IFNAR2c subunit of the receptor (59). We thus imagine that the N-terminal domain of STAT1 may mediate its binding to transcription-factor partners required for constitutive gene expression.

A recent report (49) reveals a scaffolding role for STAT3, which uses its SH2 domain to bind to the tyrosine-phosphorylated cytoplasmic tail of IFNAR1 in the activated IFNα/β receptor. STAT3 also binds to phosphatidylinositol 3'-kinase, thus bringing it to the receptor. This binding is followed by phosphorylation of phosphatidylinositol 3'-kinase on tyrosine. The functional consequences of this activation by IFNα/β of an additional signaling pathway remain to be elucidated. That at least some of the STATs can serve alternative functions alerts us to the possibility that STAT-null mice may exhibit phenotypes that do not result solely from the lack of STAT activation in response to cytokines or growth factors.

PKR

The activity of PKR in regulating translation is supplemented by its role as a signal-transducing kinase in pathways activated by dsRNA, LPS, and different cytokines (117, 276). In human and mouse cells, activation of PKR by dsRNA leads to activation of NFκB through PKR-mediated phosphorylation of IκB (121, 277–279), and recombinant PKR can activate NFκB and induce DNA-binding activity in cell lysates (277). It is likely that PKR regulates an IκB kinase, and two recent publications have now identified such an enzyme, capable of phosphorylating IκB on the two serine residues appropriate for in vivo function (280, 281).

PKR also plays a role in signal transduction by IFNα, IFNγ, dsRNA, TNFα, LPS, and platelet-derived growth factor (PDGF), revealed largely through experiments with MEFs derived from PKR-null mice. IFNs, dsRNA, TNFα, and LPS all fail to activate the DNA-binding activity of IRF1 in PKR-null MEFs, resulting in a selective defect in the induction of genes dependent on IRF1 (or NFκB) (121, 187, 279). Several genes important in mounting different aspects of host resistance to infection can now be classified as wholly or partially dependent on PKR, including genes involved in antigen presentation (class I MHC), chemotaxis (the chemokines IP-10, myg, JE, and Rantes), antimicrobial activity (iNOS), and apoptosis (FAS). The induction of the cell adhesion molecules VCAM and E-selectin by dsRNA is also mediated through a PKR-dependent pathway (282; S Bandyopadhyay & BRG Williams, unpublished data). Induction of the immunoglobulin κ gene by LPS or IFNγ is mediated by PKR, probably through activation of IRF1 (283).

The mechanisms of activation of PKR by cytokines require further investigation. It is not known whether activation occurs through JAK-dependent pathways or through other signals generated by receptor engagement (for example, Ca^{2+}). An interaction of PKR with STAT1 has been reported (284) but does not appear to be functional, because STAT1-dependent activities are unaffected in PKR-null cells (279). In contrast, the induction of c-fos and c-myc expression by PDGF can be blocked by inhibitors of PKR or by an antisense oligonucleotide against PKR mRNA (285). In accord with this, the

PDGF-induced binding of STAT3 to the GAS element of the c-fos promoter is defective in extracts from PKR-null MEFs compared with extracts from wild-type cells, although the response of STAT3 to other stimuli remains unaffected (A Deb & BRG Williams, unpublished data).

In addition to mediating an important antiviral activity of PKR, the phosphorylation of eIF2α is involved in antiproliferative activities because of this kinase. The most direct evidence comes from studies of the expression of PKR in *Saccharomyces cerevisiae*, where inducible expression of wild-type but not kinase-inactive PKR results in inhibition of growth, which can be reversed by the coexpression of a mutant yeast eIF2α that is not phosphorylated by PKR (286, 287). The induction of tumor formation by mutant PKR proteins (288–290) could be due to failure to appropriately regulate eIF2 or to interactions with other cellular proteins involved in cell growth control.

INVOLVEMENT OF PKR IN APOPTOSIS   The mechanisms and signaling mediators that regulate virus-induced apoptosis are not well understood, but it has long been recognized that a combination of IFN and dsRNA is cytotoxic. Because PKR inhibits the growth of yeast and mammalian cells, it is an attractive candidate for involvement in the apoptosis mediated by dsRNA. In support of this idea, overexpression of PKR induces apoptosis through a mechanism dependent on Bcl2 and ICE (291, 292). Normal levels of PKR are required to mediate an apoptotic response to different stimuli, including dsRNA. For example, reduction of PKR levels by antisense oligonucleotides in promonocytic U937 cells inhibits the apoptosis induced by TNFα (293). MEFs derived from PKR-null mice resist apoptotic cell death in response to dsRNA, TNFα, or LPS through a mechanism linked to a defect in activating the DNA-binding activity of IRF1 (187). These results reveal an unexpected role for PKR in mediating stress-induced apoptosis through regulation of IRF1 activity.

Apoptosis is also important in T-cell development. Although thymocytes of wild-type mice express relatively high levels of PKR, the size of the thymus and the ratios of peripheral T-cell subsets are normal in PKR-null mice (S Kadererit & BRG Williams, unpublished data). Therefore, PKR expression in thymocytes is not essential for apoptosis associated with negative and positive selection. Fas mRNA expression is strongly induced in wild-type cells by dsRNA, LPS, and IFNγ, but with the exception of IFNγ, the induction is much reduced in MEFs derived from PKR-null mice (187). Death signals transduced by the Fas receptor depend on the presentation of the ligand Fas L and are largely restricted to a few cell types, such as activated cytotoxic T cells. However, the induction of Fas on wild-type MEFs by dsRNA results in sensitization of the cells to killing by an anti-Fas antibody that stimulates the Fas receptor. MEFs derived from PKR-null mice remain insensitive to killing by this antibody when treated with dsRNA (S Der & BRG Williams, unpublished data). A role for Fas in virus-induced apoptosis has been suggested for influenza virus (294, 295), and it is likely that PKR is required, although experiments to prove this point remain to be carried out.

RNase L and 2-5A Synthetase

The possible wider role of the 2-5A system in cell metabolism extends beyond the antiviral activity of IFNs. The 2-5A system has been implicated in the action of IFNs. Although RNase L is not essential for normal mouse development (140), the 2-5A system has long been suspected to be involved in RNA decay during cell death. The regression of chick oviducts upon estrogen withdrawal and of rat mammary glands after lactation was correlated with the induction of 2-5A synthetase or with 2-5A per se (296–298). RNase L–null mice have enlarged thymus glands as a result of a defect in apoptosis (140). Thymocytes from these mice were resistant to inducers of apoptosis anti-CD3, anti-fas, staurosporine, and TNFα plus actinomycin D, whereas RNase L–null fibroblasts were resistant to staurosporine or the combination of IFNα and 2-5A. Expression of a dominant negative derivative of RNase L also suppressed apoptosis in cultured cells (J Castelli, BA Hassel, J Paranjape, A Maran, RH Silverman & R Youle, unpublished data). These findings suggest that the control of RNA stability by RNase L plays a role in apoptosis.

An intriguing but unresolved question is whether the 2-5A synthetases do something other than synthesize activators of RNase L. These enzymes differ in structure, intracellular location, activation profiles, and lengths of the 2-5A oligomers produced (127, 130, 299–302). The 2-5A synthetases are versatile enzymes that not only produce 2-5A but also transfer AMP residues in 2′, 5′-linkage to a variety of molecules that terminate in an adenosine residue, such as A5′p35′A, A5′p45′A, NAD, ADP-ribose, and tRNA (303–307). Also, the final nucleotide added by 2-5A synthetase can be something other than AMP (302, 304). Recently, 2-5A synthetases have been used to make pppG2′p5′G by using GTP as a (relatively poor) substrate (302). 2′, 5′-Oligoadenylates with structures different from 2-5A have been observed in cells and tissues of higher vertebrates (298), and some virus-induced alternative 2′, 5′-oligoadenylates can function as inhibitors of RNase L (308). In summary, suggestions of a wider role for the 2-5A synthetases are tantalizing but remain largely unexplored.

IRFs

The IRF family of DNA-binding transcription factors, including IRF1, IRF2, IRF3, ISGF3γ (p48), ICSBP, and ICSAT/PiP/LSIRF, has been implicated in IFN production, cell growth regulation, and induction of gene expression by IFN (100, 309, 310). Experiments using mice null for IRF family members have recently complemented studies of transfected cell lines and have also provided a link to PKR. IRF1 is essential for mouse *gbp* gene induction by IFNγ (311), and PKR is a signal transducer in this pathway (279). In the absence of PKR, IRF1 DNA binding activity induced in response to IFNγ (or LPS, TNFα, or dsRNA) is deficient. However, certain phenotypes and cellular responses of PKR- and IRF1-null mice are distinct, suggesting both shared and nonoverlapping pathways. For example, IRF1-null mice exhibit reduced levels of CD8+ T cells resulting from a failure of IFNγ to appropriately up-regulate the LMP1 and TAP2 genes, essential for class I MHC function (312–314). Recently it has been shown that IRF1 is required for a TH1 response in vivo (315, 316). PKR-null mice have a normal complement of CD8+ cells in the periphery but exhibit exaggerated contact hypersensitivity, possibly because they fail to induce fas-dependent apoptosis appropriately (121, 187; S Kadereit, R Fairchild & BRG Williams, unpublished data). Both IRF1 and PKR appear to be essential for induction of the inducible nitric oxide synthase gene by IFNγ (317), but unlike IRF1, PKR is not involved in the induction of cell-cycle arrest in response to DNA damage (318; S Der, M Zamanian-Daryouch & BRG Williams, unpublished data). The phenotype of ICSBP-null mice (a lymphoid-specific member of the IRF1 family) is enhanced susceptibility to virus infection as a result of a deficiency in IFNγ production and a chronic myelogenous leukemia-like syndrome that is apparent even in the heterozygotes, suggesting haploin sufficiency (319). Because this phenotype is not shared with the IRF1 or PKR knockouts, ICSBP has a unique role in regulating hematopoiesis.

> **Visit the *Annual Reviews* home page at**
> **http://www.AnnualReviews.org.**

Literature Cited

1. Darnell JE Jr, Kerr IM, Stark GR. 1994. *Science* 264:1415–21
2. Schindler C, Darnell JE Jr. 1995. *Annu. Rev. Biochem.* 64:621–51
3. Levy DE. 1995. *Virology* 6:181–89
4. Leaman DW, Leung S, Li XX, Stark GR. 1996. *FASEB J.* 10:1578–88
5. Leaman DW. 1998. *Prog. Mol. Subcell. Biol.* 22: In press
6. Ihle JN, Kerr IM. 1995. *Trends Genet.* 11:69–74
7. Ihle JN. 1996. *Cell* 84:331–34
8. Darnell JE Jr. 1996. *Proc. Natl. Acad. Sci. USA* 93:6221–24
9. Darnell JE Jr. 1998. *Science* 277:1630–35
10. Sen GC, Ransohoff RM. 1997. *Transcriptional Regulation in the Interferon System.* Georgetown, TX: Landes Bio-Sci.
11. Bach EA, Aguet M, Schreiber RD. 1997. *Annu. Rev. Immunol.* 15:563–91
12. Farrar MA, Schreiber RD. 1993. *Annu. Rev. Immunol.* 11:571–611
13. Soh J, Donnelly RJ, Kotenko S, Mariano TM, Cook JR, et al. 1994. *Cell* 76:793–802
14. Hemmi S, Bohni R, Stark G, DiMarco F, Aguet M. 1994. *Cell* 76:803–10
15. Bach EA, Tanner JW, Marsters SA, Ashkenazi A, Aguet M, et al. 1996. *Mol. Cell. Biol.* 16:3214–21
16. Kotenko SV, Izotova LS, Pollack BP, Mariano TM, Donnelly RJ, et al. 1995. *J. Biol. Chem.* 270:20915–21
17. Sakatsume M, Igarashi K, Winestock

KD, Garotta G, Larner AC, Finbloom DS. 1995. *J. Biol. Chem.* 270:17528–34

18. Kaplan DH, Greenlund AC, Tanner JW, Shaw AS, Schreiber RD. 1996. *J. Biol. Chem.* 271:9–12

19. Fountoulakis M, Zulauf M, Lustig A, Garotta G. 1992. *Eur. J. Biochem.* 208: 781–87

20. Greenlund AC, Schreiber RD, Goeddel DV, Pennica D. 1993. *J. Biol. Chem.* 268:18103–10

21. Walter MR, Windsor WT, Nagabhushan TL, Lundell DJ, Lunn CA, et al. 1995. *Nature* 376:230–35

22. Marsters S, Pennica D, Bach E, Schreiber RD, Ashkenazi A. 1995. *Proc. Natl. Acad. Sci. USA* 92:5401–5

23. Briscoe J, Rogers NC, Witthuhn BA, Watling D, Harpur AG, et al. 1996. *EMBO J.* 15:799–809

24. Kotenko SV, Izotova LS, Pollack BP, Muthukumaran G, Paukku K, et al. 1996. *J. Biol. Chem.* 271:17174–82

25. Kohlhuber F, Rogers NC, Watling D, Feng J, Guschin D, et al. 1997. *Mol. Cell. Biol.* 17:695–706

26. Greenlund AC, Farrar MA, Viviano BL, Schreiber RD. 1994. *EMBO J.* 13:1591–600

27. Igarashi K, Garotta G, Ozmen L, Ziemieckl A, Wilks AF, et al. 1994. *J. Biol. Chem.* 269:14333–36

28. Greenlund AC, Morales MO, Viviano BL, Yan H, Krolewski J, Schreiber RD. 1995. *Immunity* 2:677–87

29. Heim MH, Kerr IM, Stark GR, Darnell JE Jr. 1995. *Science* 267:1347–49

30. Schindler C, Shuai K, Prezioso VR, Darnell JE Jr. 1992. *Science* 257:809–13

31. Shuai K, Schindler C, Prezioso VR, Darnell JE Jr. 1992. *Science* 258:1808–12

32. Shuai K, Stark GR, Kerr IM, Darnell JE Jr. 1993. *Science* 261:1744–46

33. Sekimoto T, Nakajima K, Tachibana T, Hirano T, Yoneda Y. 1997. *J. Biol. Chem.* 271:31017–20

34. Wen ZL, Zhong Z, Darnell JE Jr. 1995. *Cell* 82:241–50

35. David M, Petricoin E III, Benjamin C, Pine R, Weber MJ, Larner AC. 1995. *Science* 269:1721–23

36. Pernis A, Gupta S, Gollob KJ, Garfein E, Coffman RL, et al. 1995. *Science* 269:245–47

37. Bach EA, Szabo SJ, Dighe AS, Ashkenazi A, Aguet M, et al. 1995. *Science* 270:1215–18

38. Starr R, Willson TA, Viney EM, Murray LJL, Rayner JR, et al. 1997. *Nature* 387:917–21

39. Endo TA, Masuhara M, Yokouchi M, Suzuki R, Sakamoto H, et al. 1997. *Nature* 387:921–24

40. Naka T, Narazaki M, Hirata M, Matsumoto T, Minamoto S, et al. 1997. *Nature* 387:924–29

41. Meraz MA, White JM, Sheehan KCF, Bach EA, Rodig SJ, et al. 1996. *Cell* 84:431–42

42. Durbin JE, Hackenmiller R, Simon MC, Levy DE. 1996. *Cell* 84:443–50

43. Reid LE, Brasnett AH, Gilbert CS, Porter ACG, Gewert DR, et al. 1997. *Proc. Natl. Acad. Sci. USA* 86:840–44

44. Bluyssen HAR, Muzaffar R, Vlieststra RJ, van der Made ACJ, Leung S, et al. 1995. *Proc. Natl. Acad. Sci. USA* 92:5645–49

45. Look DC, Pelletier MR, Tidwell RM, Roswit WT, Holtzman MJ. 1997. *J. Biol. Chem.* 270:30264–67

46. Perez C, Wietzerbin J, Benech PD. 1997. *Mol. Cell. Biol.* 13:2182–92

47. Perez C, Coeffier E, Moreau-Gachelin F, Wietzerbin J, Benech PD. 1997. *Mol. Cell. Biol.* 14:5023–31

48. Novick D, Cohen B, Rubinstein M. 1994. *Cell* 77:391–400

49. Pfeffer LM, Basu L, Pfeffer SR, Yang CH, Murti A, et al. 1997. *J. Biol. Chem.* 272:11002–5

50. Lutfalla G, Holland SJ, Cinato E, Monneron D, Reboul J, et al. 1995. *EMBO J.* 14:5100–8

51. Russell-Harde D, Pu HF, Betts M, Harkins RN, Perez HD, Croze E. 1995. *J. Biol. Chem.* 270:26033–36

52. Cohen B, Novick D, Barak S, Rubinstein M. 1995. *Mol. Cell. Biol.* 15:4208–14

53. Colamonici O, Yan H, Domanski P, Handa R, Smalley D, et al. 1994. *Mol. Cell. Biol.* 14:8133–42

54. Colamonici OR, Uyttendaele H, Domanski P, Yan H, Krolewski JJ. 1994. *J. Biol. Chem.* 269:3518–22

55. Gauzzi MC, Velazquez L, McKendry R, Mogensen KE, Fellous M, Pellegrini S. 1996. *J. Biol. Chem.* 271:20494–500

56. Pellegrini S, John J, Shearer M, Kerr IM, Stark GR. 1989. *Mol. Cell. Biol.* 9:4605–12

57. Velazquez L, Mogensen KE, Barbieri G, Fellous M, Uzé G, Pellegrini S. 1995. *J. Biol. Chem.* 270:3327–34

58. Krishnan K, Yan H, Lim JTE, Krolewski JJ. 1996. *Oncogene* 13:125–33

59. Li XX, Leung S, Kerr IM, Stark GR. 1997. *Mol. Cell. Biol.* 17:2048–56

60. Stancato LF, David M, Carter-Su C, Larner AC, Pratt WB. 1996. *J. Biol. Chem.* 271:4134–37

61. Yan H, Krishnan K, Greenlund AC, Gupta S, Lim JTE, et al. 1996. *EMBO J.* 15:1064–74
62. David M, Zhou GC, Pine R, Dixon JE, Larner AC. 1996. *J. Biol. Chem.* 271: 15862–65
63. John J, McKendry R, Pellegrini S, Flavell D, Kerr IM, Stark GR. 1991. *Mol. Cell. Biol.* 11:4189–95
64. Bluyssen HAR, Muzaffar R, Vlieststra RJ, van der Made ACJ, Leung S, et al. 1995. *Proc. Natl. Acad. Sci. USA* 92: 5645–49
65. Harada H, Matsumoto M, Sato M, Kashiwazaki Y, Kimura T, et al. 1996. *Genes Cells* 1:995–1005
66. Haque SJ, Williams BRG. 1994. *J. Biol. Chem.* 269:19523–29
67. Li XX, Leung S, Qureshi S, Darnell JE Jr, Stark GR. 1996. *J. Biol. Chem.* 271:5790–94
68. Dale TC, Rosen JM, Guille MJ, Lewin AR, Porter AGC, et al. 1989. *EMBO J.* 8:831–39
69. Imam AMA, Ackrill AM, Dale TC, Kerr IM, Stark GR. 1990. *Nucleic Acids Res.* 18:6573–80
70. Qureshi SA, Salditt-Georgieff M, Darnell JE Jr. 1995. *Proc. Natl. Acad. Sci. USA* 92:3829–33
71. Bluyssen HAR, Levy DE. 1997. *J. Biol. Chem.* 272:4600–5
72. Horvath CM, Wen ZL, Darnell JE Jr. 1995. *Genes Dev.* 9:984–94
73. Horvath CM, Stark GR, Kerr IM, Darnell JE Jr. 1996. *Mol. Cell. Biol.* 16: 6957–64
74. Decker T, Kovarik P, Meinke A. 1997. *J. Interferon Cytokine Res.* 17:121–34
75. Xu XA, Sun Y-L, Hoey T. 1996. *Science* 273:794–97
76. Vinkemeier U, Cohen SL, Moarefi I, Chait BT, Kuriyan J, Darnell JE Jr. 1996. *EMBO J.* 15:5616–26
77. Guyer NB, Severns CW, Wong P, Feghali CA, Wright TM. 1995. *J. Immunol.* 155:3472–80
78. Porter ACG, Chernajovsky Y, Dale TC, Gilbert CS, Stark GR, Kerr IM. 1988. *EMBO J.* 7:85–92
79. Zhang JJ, Vinkemeier U, Gu W, Chakravarti D, Horvath CM, Darnell JE Jr. 1996. *Proc. Natl. Acad. Sci. USA* 93: 15092–96
80. Horvai AE, Xu L, Korzus E, Brard G, Kalafus D, et al. 1997. *Proc. Natl. Acad. Sci. USA* 94:1074–79
81. Rutherford MN, Kumar A, Haque SJ, Ghysdael J, Williams BRG. 1997. *J. Interferon Cytokine Res.* 17:1–10
82. Bhattacharya S, Eckner R, Grossman S, Oldread E, Arany Z, et al. 1996. *Nature* 383:344–47
83. Müller M, Laxton C, Briscoe J, Schindler C, Improta T, et al. 1993. *EMBO J.* 12:4221–28
84. Stancato LF, Sakatsume M, David M, Dent P, Dong F, et al. 1997. *Mol. Cell. Biol.* 17:3833–40
85. Flati V, Haque SJ, Williams BRG. 1996. *EMBO J.* 15:1566–71
86. Uddin S, Yenush L, Sun X-J, Sweet ME, White MF, Platanias LC. 1995. *J. Biol. Chem.* 270:15938–41
87. Burfoot MS, Rogers NC, Watling D, Smith JM, Pons S, et al. 1997. *J. Biol. Chem.* 272:24183–90
88. Diaz M, Bohlander S, Allen G. 1993. *J. Interferon Res.* 13:243–44
89. Weck P, Apperson S, May L, Stebbing N. 1981. *J. Gen. Virol.* 57:233–37
90. Evinger M, Rubinstein M, Pestka S. 1981. *Arch. Biochim. Biophys.* 210:319–29
91. Foster GR, Rodrigues O, Ghouze F, Schulte-Frohlinde E, Testa D, et al. 1996. *J. Interferon Cytokine Res.* 16: 1027–33
92. Pestka S, Langer JA, Zoon KC, Samuel CE. 1987. *Annu. Rev. Biochem.* 56:727–77
93. Platanias LC, Uddin S, Colamonici OR. 1994. *J. Biol. Chem.* 269:17761–64
94. Abramovich C, Shulman LM, Ratovitski E, Harroch S, Tovey M, et al. 1994. *EMBO J.* 13:5871–77
95. Platanias LC, Uddin S, Domanski P, Colamonici OR. 1996. *J. Biol. Chem.* 271:23630–33
96. Croze E, Russell-Harde D, Wagner TC, Pu HF, Pfeffer LM, Perez HD. 1996. *J. Biol. Chem.* 271:33165–68
97. Rani MRS, Foster GR, Leung S, Leaman D, Stark GR, Ransohoff RM. 1996. *J. Biol. Chem.* 271:22878–84
98. Harada H, Fujita T, Miyamoto M, Kimurak Y, Murayama M, et al. 1989. *Cell* 58:729–39
99. Nelson N, Marks MS, Driggers PH, Ozato K. 1993. *Mol. Cell. Biol.* 13:588–99
100. Yamagata T, Nishida J, Tanaka S, Sakai R, Mitani K, et al. 1996. *Mol. Cell. Biol.* 16:1283–94
101. Friedman RL, Manly SP, McMahon M, Kerr IM, Stark GR. 1984. *Cell* 38:745–55
102. Larner AC, Chaudhuri A, Darnell JE Jr. 1986. *J. Biol. Chem.* 261:453–59
103. David M, Grimley PM, Finbloom DS, Larner AC. 1993. *Mol. Cell. Biol.* 13: 7515–21

104. Haspel RL, Salditt-Georgieff M, Darnell JE Jr. 1996. *EMBO J.* 15:6262–68
105. Kim TK, Maniatis T. 1996. *Science* 273:1717–19
106. Igarashi K-I, David M, Larner AC, Finbloom DS. 1993. *Mol. Cell. Biol.* 13:3984–89
107. Haque SJ, Flati V, Deb A, Williams BRG. 1995. *J. Biol. Chem.* 270:25709–14
108. David M, Chen HE, Goelz S, Larner AC, Neel BG. 1995. *Mol. Cell. Biol.* 15:7050–58
109. Jiao HY, Berrada K, Yang WT, Tabrizi M, Platanias LC, Yi TL. 1996. *Mol. Cell. Biol.* 16:6985–92
110. Petricoin EF III, David M, Igarashi K, Benjamin C, Ling L, et al. 1996. *Mol. Cell. Biol.* 16:1419–24
111. Petricoin EF III, Hackett RH, Akai H, Igarashi K, Finbloom DS, Larner AC. 1992. *Mol. Cell. Biol.* 12:2486–95
112. Kimura T, Kadokawa Y, Harada H, Matsumoto M, Sato M, et al. 1996. *Genes Cells* 1:115–24
113. Isaacs A, Lindenmann J. 1957. *Proc. R. Soc. London Ser. B* 147:258–67
114. van den Broek MF, Muller U, Huang S, Aguet M, Zinkernagel RM. 1995. *J. Virol.* 69:4792–96
115. Vilcek J, Sen GC. 1996. In *Fields Virology*, ed. BN Fields, DM Knipe, PM Howley, pp. 375–400. Philadelphia: Lippincott-Raven. 3rd ed.
116. Meurs E, Chong K, Galabru J, Thomas NS, Kerr IM, et al. 1990. *Cell* 62:379–90
117. McMillan NAJ, Williams BRG. 1996. In *Protein Phosphorylation in Cell Growth Regulation*, ed. MJ Clemens, pp. 225–54. London: Harwood Acad.
118. Carpick BW, Graziano V, Schneider D, Maitra RK, Lee X, Williams BRG. 1997. *J. Biol. Chem.* 272:9510–16
119. Meurs EF, Watanabe Y, Kadereit S, Barber GN, Katze MG, et al. 1992. *J. Virol.* 66:5805–14
120. Der SD, Lau AS. 1995. *Proc. Natl. Acad. Sci. USA* 92:8841–45
121. Yang YL, Reis LFL, Pavlovic J, Aguzzi A, Schafer R, et al. 1995. *EMBO J.* 14:6095–106
122. Kerr IM, Brown RE. 1978. *Proc. Natl. Acad. Sci. USA* 75:256–60
123. Silverman RH, Cirino NM. 1997. In *mRNA Metabolism and Post-Transcriptional Gene Regulation*, ed. DR Morris, JB Harford, pp. 295–309. New York: Wiley & Sons
124. Wreschner DH, McCauley JW, Skehel JJ, Kerr IM. 1981. *Nature* 289:414–17
125. Floyd-Smith G, Slattery E, Lengyel P. 1981. *Science* 212:1030–32
126. Carroll SS, Chen E, Viscount T, Geib J, Sardana MK, et al. 1996. *J. Biol. Chem.* 271:4988–92
127. Chebath J, Benech P, Hovanessian AG, Galabru J, Revel M. 1987. *J. Biol. Chem.* 262:3852–57
128. Ghosh SK, Kusari J, Bandyopadhyay SK, Samanta H, Kumar R, Sen GC. 1991. *J. Biol. Chem.* 266:15293–99
129. Rutherford MN, Kumar A, Nissim A, Chebath J, Williams BRG. 1991. *Nucleic Acids Res.* 19:1917–24
130. Marie I, Hovanessian AG. 1992. *J. Biol. Chem.* 267:9933–39
131. Dong BH, Silverman RH. 1995. *J. Biol. Chem.* 270:4133–37
132. Cole JL, Carroll SS, Kuo LC. 1996. *J. Biol. Chem.* 271:3979–81
133. Cole JL, Carroll SS, Blue ES, Viscount T, Kuo LC. 1997. *J. Biol. Chem.* 272:19187–92
134. Carroll SS, Cole JL, Viscount T, Geib J, Gehman J, Kuo L. 1997. *J. Biol. Chem.* 272:19193–98
135. Zhou AM, Hassel BA, Silverman RH. 1993. *Cell* 72:753–65
136. Hassel BA, Zhou AM, Sotomayor C, Maran A, Silverman RH. 1993. *EMBO J.* 12:3297–304
137. Dong B, Silverman RH. 1997. *J. Biol. Chem.* 272:22236–42
138. Sidrauski C, Walter P. 1997. *Cell* 90:1031–39
139. Diaz-Guerra M, Rivas C, Esteban M. 1997. *Virology* 227:220–28
140. Zhou A, Paranjape J, Brown TL, Nie H, Naik S, et al. 1997. *EMBO J.* 16:6355–63
141. Torrence PF, Maitra RK, Lesiak K, Khamnei S, Zhou A, Silverman RH. 1993. *Proc. Natl. Acad. Sci. USA* 90:1300–4
142. Cirino NM, Li GY, Xiao W, Torrence PF, Silverman RH. 1997. *Proc. Natl. Acad. Sci. USA* 94:1937–42
143. Mitra A, Higgins DW, Langenberg WG, Nie HQ, Sengupta DN, Silverman RH. 1996. *Proc. Natl. Acad. Sci. USA* 93:6780–85
144. Ogawa T, Hori T, Ishida I. 1996. *Nat. Biotechnol.* 14:1566–69
145. Horisberger MA. 1995. *Am. J. Respir. Crit. Care Med.* 152:S67–S71
146. Arnheiter H, Frese M, Kambadur R, Meier E, Haller O. 1996. *Curr. Top. Microbiol. Immunol.* 206:119–47
147. Takei K, McPherson PS, Schmid SL, De Camilli P. 1995. *Nature* 374:186–90

148. Hinshaw JE, Schmid SL. 1995. *Nature* 374:190–92
149. Nakayama M, Yazaki K, Kusano A, Nagata K, Hanai N, Ishihama A. 1993. *J. Biol. Chem.* 268:15033–38
150. Richter MF, Schwemmle M, Herrmann C, Wittinghofer A, Staeheli P. 1995. *J. Biol. Chem.* 270:13512–17
151. Ponten A, Sick C, Weeber M, Haller O, Kochs G. 1997. *J. Virol.* 71:2591–99
152. Haller O, Frese M, Rost D, Nuttal PA, Kochs G. 1995. *Virology* 4:2596–601
153. Schnorr J-J, Schneider-Schaulies S, Simon-Jodicke A, Pavlovic J, Horisberger MA, ter Meulen V. 1993. *J. Virol.* 67:4760–68
154. Thimme R, Frese M, Kochs G, Haller O. 1995. *Virology* 211:296–301
155. Frese M, Kochs G, Feldmann H, Hertkorn C, Haller O. 1996. *J. Virol.* 70: 915–23
156. Zhao H, De BP, Das T, Banerjee AK. 1996. *Virology* 220:330–38
157. Frese M, Kochs G, Meier-Dieter U, Siebler J, Haller O. 1995. *J. Virol.* 69:3904–9
158. Pitossi F, Blank A, Schroder A, Schwarz A, Hussi P, Schwemmle M. 1993. *J. Virol.* 67:6726–32
159. Schwemmle M, Weining KC, Richter MF, Schumacher B, Staeheli P. 1995. *Virology* 206:545–54
160. Stranden AM, Staeheli P, Pavlovic J. 1993. *Virology* 197:642–51
161. Pavlovic J, Haller O, Staeheli P. 1992. *J. Virol.* 66:2564–69
162. Floyd-Smith GJ. 1988. *Cell. Biochem.* 38:13–21
163. Staeheli P, Horisberger MA, Haller O. 1984. *Virology* 132:456–61
164. Karupiah G, Xie QW, Buller RM, Nathan C, Duarte C, MacMicking JD. 1993. *Science* 261:1445–48
165. Alber D, Staeheli P. 1996. *J. Interferon Cytokine Res.* 16:375–80
166. Basinga M. 1992. *Science* 258:1730–31
167. Gooding LR. 1992. *Cell* 71:5–7
168. McFadden G, Graham K, Ellison K, Barry M, Macen J, et al. 1995. *J. Leukocyte Biol.* 57:731–38
169. Moore PS, Boshoff C, Weiss RA, Chang Y. 1996. *Science* 274:1739–44
170. Smith GL. 1996. *Curr. Opin. Immunol.* 8:467–71
171. Cayley PJ, Knight M, Kerr IM. 1982. *Biochem. Biophys. Res. Commun.* 104: 376–82
172. Bisbal C, Martinand C, Silhol M, Lebleu B, Salehzada T. 1995. *J. Biol. Chem.* 270:13308–17
173. Kitajewski J, Schneider RJ, Safer B, Munemitsu SM, Samuel CE, et al. 1986. *Cell* 45:195–200
174. Anderson KP, Fennie EH. 1987. *J. Virol.* 61:787–95
175. Clarke PA, Schwemmle M, Schikinger J, Hilse K, Clemens M. 1991. *Nucleic Acids Res.* 19:243–48
176. Swaminathan S, Honeycut BS, Reiss CS, Kieff E. 1992. *J. Virol.* 66:5133–36
177. Imani F, Jacobs BL. 1988. *Proc. Natl. Acad. Sci. USA* 85:7887–91
178. Beattie E, Denzler KL, Tartaglia J, Perkus ME, Paoletti E, Jacobs BL. 1995. *J. Virol.* 69:499–505
179. Roy S, Katze MG, Parkin NT, Edery I, Hovanessian AG, Sonenberg N. 1990. *Science* 247:1216–20
180. McMillan NA, Chun RF, Siderovski DP, Galabru J, Toone WM, et al. 1995. *Virology* 213:413–24
181. Gale MJ Jr, Korth MJ, Tang NM, Tan SL, Hopkins DA, et al. 1997. *Virology* 230:217–27
182. Black TL, Safer B, Hovanessian AG, Katze MG. 1989. *J. Virol.* 63:2244–51
183. Dubois MF, Hovanessian AG. 1990. *Virology* 179:591–98
184. Melville MW, Hansen WJ, Freeman BC, Welch WJ, Katze MG. 1997. *Proc. Natl. Acad. Sci. USA* 94:97–102
185. Swaminathan S, Rajan P, Savinova O, Jagus R, Thimmapaya B. 1996. *Virology* 219:321–23
186. Shen Y, Shenk TE. 1995. *Curr. Biol.* 5:105–11
187. Der SD, Yang Y-L, Weissmann C, Williams BRG. 1997. *Proc. Natl. Acad. Sci. USA* 94:3279–83
188. Ray CA, Black RA, Kronheim SR, Greenstreet TA, Sleath PR, et al. 1992. *Cell* 69:597–604
189. Thome M, Schneider P, Hofmann K, Fickenscher H, Meinl E, et al. 1997. *Nature* 386:517–21
190. Henderson S, Huen D, Rowe M, Dawson C, Johnson G, Rickinson A. 1993. *Proc. Natl. Acad. Sci. USA* 90:8479–83
191. Reich N, Pine R, Levy D, Darnell JE. 1988. *J. Virol.* 62:114–19
192. Ackrill AM, Foster GR, Laxton CD, Flavell DM, Stark GR, Kerr IM. 1991. *Nucleic Acids Res.* 19:4387–93
193. Leonard GT, Sen GC. 1997. *J. Virol.* 71:5095–101
194. Perea SE, Lopez-Ocejo O, Von Gabain A, Arana MDJ. 1997. *Int. J. Oncol.* 11: 169–73
195. Symons JA, Alcami A, Smith GL. 1995. *Cell* 81:551–60

196. Stuart AD, Stewart AP, Arrand JR, Mac-Kett M. 1995. *Oncogene* 11:1711–19
197. Deiss LP, Feinstein E, Berissi H, Cohen O, Kimchi A. 1995. *Genes Dev.* 9:15–30
198. Higuchi T, Hannigan GE, Malkin D, Yeger H, Williams BRG. 1991. *Cancer Res.* 51:3958–64
199. Nason-Burchenal K, Gandini D, Bott M, Allpenna J, Seale JR, Cross NC, et al. 1996. *Blood* 88:3926–36
200. Kumar R, Atlas I. 1992. *Proc. Natl. Acad. Sci. USA* 89:6599–603
201. Resnitzky D, Tiefenbrun N, Berissi H, Kimchi A. 1992. *Proc. Natl. Acad. Sci. USA* 89:402–6
202. Melamed D, Tiefenbrun N, Yarden A, Kimchi A. 1993. *Mol. Cell. Biol.* 12:5255–65
203. Tiefenbrun N, Melamed D, Levy N, Resnitzky D, Hoffmann I, et al. 1996. *Mol. Cell. Biol.* 16:3934–44
204. Raveh T, Hovanessian AG, Meurs EF, Sonenberg N, Kimchi A. 1996. *J. Biol. Chem.* 271:25479–84
205. Deng C, Zhang P, Harper JW, Elledge SJ, Leder P. 1995. *Cell* 82:675–84
206. Hobeika AC, Subramaniam P, Johnson HM. 1997. *Oncogene* 14:1165–70
207. Grawunder U, Melchers F, Rolink A. 1993. *Eur. J. Immunol.* 23:544–51
208. Rojas R, Roman J, Torres A, Ramirez R, Carracedo J. 1996. *Leukemia* 10:1782–88
209. Buschle M, Campana D, Carding SR, Richard C, Hoffbrand AV, Breener MK. 1993. *J. Exp. Med.* 177:213–18
210. Novelli F, Di Pierro F, di Celle PF, Bertini S, Affaticati P, et al. 1994. *J. Immunol.* 152:496–504
211. Ghayur T, Banerjee S, Hugunin M, Butler D, Herzog L, et al. 1997. *Nature* 386:619–23
212. Gu Y, Kuida K, Tsutsui H, Ku G, Hsiao K, et al. 1997. *Science* 275:206–9
213. Deiss LP, Kimchi A. 1991. *Science* 252:117–20
214. Kissil JL, Deiss LP, Bayewitch M, Raveh T, Khaspekov G, Kimchi A. 1995. *J. Biol. Chem.* 270:27932–36
215. Feinstein E, Kimchi A, Wallach D, Boldin M, Varfolomeev E. 1995. *Trends Biochem. Sci.* 20:342–44
216. Deiss LP, Galinka H, Berissi H, Cohen O, Kimchi A. 1996. *EMBO J.* 15:3861–70
217. Levy-Strumpf N, Deiss LP, Berissi H, Kimchi A. 1997. *Mol. Cell. Biol.* 17:1615–25
218. Cohen O, Feinstein E, Kimchi A. 1997. *EMBO J.* 16:998–1008
219. Kissil JL, Feinstein E, Cohen O, Jones PA, Tsai YC, et al. 1997. *Oncogene* 15:403–7
219a. Inbal B, Cohen O, Polak-Charcon S, Kopolovic J, Vadai E, et al. 1997. *Nature* 390:180–84
220. Uzé G, Lutfalla G, Mogensen KE. 1995. *J. Interferon Res.* 15:3–26
221. Boehm U, Klamp T, Groot M, Howard JC. 1997. *Annu. Rev. Immunol.* 15:749–95
222. Reis LFL, Harada H, Wolchok JD, Taniguchi T, Vilcek J. 1992. *EMBO J.* 11:185–93
223. Chang C-H, Hammer J, Loh JE, Fodor WL, Flavell RA. 1992. *Immunogenetics* 35:378–84
224. Mach B, Steimle V, Martinez-Soria E, Reith W. 1996. *Annu. Rev. Immunol.* 14:301–31
225. Mond JJ, Carman J, Sarma C, Ohara J, Finkelman FD. 1986. *J. Immunol.* 137:3534–37
226. York IA, Rock KL. 1996. *Annu. Rev. Immunol.* 14:369–96
227. Boes B, Hengel H, Ruppert T, Multhaup G, Koszinowski UH, Kloetzel PM. 1994. *J. Exp. Med.* 179:901–9
228. Groettrup M, Soza A, Eggers M, Keuhn L, Dick TP, et al. 1996. *Nature* 381:166–68
229. Trowsdale J, Hanson I, Mockridge I, Beck S, Townsend A, Kelly A. 1990. *Nature* 348:741–44
230. Epperson DE, Arnold E, Spies T, Cresswell P, Pober JS, Johnson DR. 1992. *J. Immunol.* 149:3297–301
231. Abbas AK, Murphy KM, Sher A. 1997. *Nature* 383:787–93
232. Hsieh C-S, Macatonia S, Tripp CS, Wolf SF, O'Garra A, Murphy KM. 1993. *Science* 260:547–49
233. Murphy TL, Cleveland MG, Kulesza P, Magram J, Murphy KM. 1995. *Mol. Cell. Biol.* 15:5258–67
234. Dighe AS, Campbell D, Hsieh C-S, Clarke S, Greaves DR, et al. 1995. *Immunity* 3:657–66
235. Flesch IEA, Hess JH, Huang S, Aguet M, Rothe J, et al. 1995. *J. Exp. Med.* 181:1615–21
236. Trinchieri G. 1995. *Annu. Rev. Immunol.* 13:251–76
237. Szabo SJ, Dighe AS, Gubler U, Murphy KM. 1997. *J. Exp. Med.* 185:817–24
238. Szabo SJ, Jacobson NG, Dighe AS, Gubler U, Murphy KM. 1995. *Immunity* 2:665–75
239. Gajewski TF, Fitch FW. 1988. *J. Immunol.* 140:4245–52
240. Adams DO, Hamilton TA. 1984. *Annu. Rev. Immunol.* 2:283–318

241. Buchmeier NA, Schreiber RD. 1985. *Proc. Natl. Acad. Sci. USA* 82:7404–8
242. Dalton DK, Pitts-Meek S, Keshav S, Figari IS, Bradley A, Stewart TA. 1993. *Science* 259:1739–42
243. Huang S, Hendriks W, Althage A, Hemmi S, Bluethmann H, et al. 1993. *Science* 259:1742–45
244. Newport MJ, Huxley CM, Huston S, Hawrylowicz CM, Oostra BA, et al. 1996. *N. Engl. J. Med.* 335:1941–49
245. Jouanguy E, Altare F, Lamhamedi S, Revy P, Newport M, et al. 1996. *N. Engl. J. Med.* 335:1956–61
246. Klebanoff SJ. 1992. In *Inflammation: Basic Principles and Clinical Correlates*, ed. JI Gallin, pp. 541–89. New York: Raven. 2nd ed.
247. MacMicking J, Xie Q-W, Nathan C. 1997. *Annu. Rev. Immunol.* 15:323–50
248. Snapper CM, Peschel C, Paul WE. 1988. *J. Immunol.* 140:2121–27
249. Snapper CM, McIntyre TM, Mandler R, Pecanha LMT, Finkelman FD, et al. 1992. *J. Exp. Med.* 175:1367–71
250. Snapper CM, Paul WE. 1987. *Science* 236:944–47
251. Finkelman FD, Svetic A, Gresser I, Snapper C, Holmes J, et al. 1991. *J. Exp. Med.* 174:1179–88
252. Wang Y-D, Wong K, Wood WI. 1995. *J. Biol. Chem.* 270:7021–24
253. Fujitani Y, Hibi M, Fukada T, Takahashi-Tezuka M, Yoshida H, et al. 1997. *Oncogene* 14:751–61
254. Lobie PE, Ronsin B, Silvennoinen O, Haldosen LA, Norstedt G, Morel G. 1996. *Endocrinology* 137:4037–45
255. Conway G, Margoliath A, Wongmadden S, Roberts RJ, Gilbert W. 1997. *Proc. Natl. Acad. Sci. USA* 94:3082–87
256. Russell SM, Tayebi N, Nakajima H, Riedy MC, Roberts JL, et al. 1995. *Science* 270:797–800
257. Nosaka T, van Deursen JM, Tripp RA, Thierfelder WE, Witthuhn BA, et al. 1995. *Science* 270:800–2
258. Smit LS, Meyer DJ, Billestrup N, Norstedt G, Schwartz J, Carter-Su C. 1996. *Mol. Endocrinol.* 10:519–33
259. Wang XY, Fuhrer DK, Feng GS, Marshall MS, Yang Y-C. 1995. *Blood* 86:608
260. Chauhan D, Kharbanda SM, Ogata A, Urashima M, Frank D, et al. 1995. *J. Exp. Med.* 182:1801–6
261. Han YL, Leaman DW, Watling D, Rogers NC, Groner B, et al. 1996. *J. Biol. Chem.* 271:5947–52
262. Winston LA, Hunter T. 1996. *Curr. Biol.* 6:668–71
263. Yin TG, Shen R, Feng GS, Yang YC. 1997. *J. Biol. Chem.* 272:1032–37
264. Uddin S, Sweet M, Colamonici OR, Krolewski JJ, Platanias LC. 1997. *FEBS Lett.* 403:31–34
265. Uddin S, Sher DA, Alsayed Y, Pons S, Colamonici OR, et al. 1997. *Biochem. Biophys. Res. Commun.* 235:83–88
266. Takahashitezuka M. 1997. *Oncogene* 14:2273–82
267. Danial NN, Pernis A, Rothman PB. 1995. *Science* 269:1875–77
268. Uddin S, Gardziola C, Dangat A, Yi T, Platanias LC. 1996. *Biochem. Biophys. Res. Commun.* 225:833–38
269. Leaman DW, Pisharody S, Flickinger TW, Commane MA, Schlessinger J, et al. 1996. *Mol. Cell. Biol.* 16:369–75
270. Hirano T. 1998. *Int. Rev. Immunol.* 16:249–84
271. Kishimoto T, Akira S, Narazaki M, Taga T. 1995. *Blood* 86:1243–54
272. Argetsinger LS, Hsu GW, Myers MG Jr, Billestrup N, White MF, Carter-Su C. 1995. *J. Biol. Chem.* 270:14685–92
273. Wang HY, Zamorano J, Yoerkie JL, Paul WE, Keegan AD. 1997. *J. Immunol.* 158:1037–40
274. Novick D, Cohen B, Kim HS, Levy Y, Rubinstein M. 1996. *Eur. Cytokine Netw.* 7:523
275. Takeshita T, Arita T, Higuchi M, Asao H, Endo K, et al. 1997. *Immunity* 6:449–57
275a. Kumar A, Commane M, Flickinger TW, Horvath CM, Stark GR. 1997. *Science* 278:1630–32
276. Williams BRG. 1995. *Semin. Virol.* 6:191–202
277. Kumar A, Haque J, Lacoste J, Hiscott J, Williams BR. 1994. *Proc. Natl. Acad. Sci. USA* 91:6288–92
278. Maran A, Maitra RK, Kumar A, Dong BH, Xiao W, et al. 1994. *Science* 265:789–92
279. Kumar A, Yang Y-L, Flati V, Der S, Kadereit S, et al. 1997. *EMBO J.* 16:406–16
280. DiDonato JA, Hayakawa M, Rothwarf DM, Zandi E, Karin M. 1997. *Nature* 388:548–53
281. Regnier CH, Song HY, Gao X, Goeddel DV, Cao Z, Rothe M. 1997. *Cell* 90:373–83
282. Offerman MK, Simring J, Mellits KH, Hagan MK, Shaw R, et al. 1995. *Eur. J. Biochem.* 232:28–36
283. Koromilas AE, Cantin C, Craig AWB, Jagus R, Hiscott J, Sonenberg N. 1995. *J. Biol. Chem.* 270:25426–34

284. Wong AH-T, Tam NWN, Yang Y-L, Cuddihy AR, Li S, et al. 1997. *EMBO J.* 16:1291–1304

285. Mundschau LJ, Faller DV. 1995. *J. Biol. Chem.* 270:3100–6

286. Chong KL, Feng L, Schappert K, Meurs E, Donahue TF, et al. 1992. *EMBO J.* 11:1553–62

287. Dever TE, Chen J-J, Barber GN, Cigan AM, Feng L, et al. 1993. *Proc. Natl. Acad. Sci. USA* 90:4616–20

288. Koromilas AE, Roy S, Barber GN, Katze MG, Sonenberg N. 1992. *Science* 257:1685–89

289. Meurs EF, Galabru J, Barber GN, Katze MG, Hovanessian AG. 1993. *Proc. Natl. Acad. Sci. USA* 90:232–36

290. Barber GN, Wambach M, Thompson S, Jagus R, Katze MG. 1995. *Mol. Cell. Biol.* 15:3138–46

291. Lee SB, Esteban M. 1994. *Virology* 199: 491–96

292. Lee SB, Rodriguez D, Rodriguez JR, Esteban M. 1997. *Virology* 231:81–88

293. Yeung MC, Liu J, Lau AS. 1996. *Proc. Natl. Acad. Sci. USA* 93:12451–55

294. Wada N, Matsumura M, Oba Y, Kobayashi N, Takizawa T, Nakanishi Y. 1995. *J. Biol. Chem.* 270:18007–12

295. Takizawa T, Ohashi K, Nakanishi Y. 1996. *J. Virol.* 70:8128–32

296. Stark GR, Dower WJ, Schimke RT, Brown RE, Kerr IM. 1979. *Nature* 278:471–73

297. Cohrs RJ, Goswami BB, Sharma OK. 1988. *Biochemistry* 27:3246–52

298. Reid TR, Hersh CL, Kerr IM, Stark GR. 1984. *Anal. Biochem.* 136:136–41

299. Ilson DH, Torrence PF, Vilcek J. 1986. *J. Interferon Res.* 6:5–12

300. Hovanessian AG, Laurent AG, Chebath J, Galabru J, Robert N, Svab J. 1987. *EMBO J.* 6:1273–80

301. Hovanessian AG, Svab J, Marie I, Robert N, Chamaret S, Laurent AG. 1988. *J. Biol. Chem.* 263:4945–49

302. Marie I, Blanco J, Rebouillat D, Hovanessian AG. 1997. *Eur. J. Biochem.* 248:558–66

303. Ball LA, White CN. 1979. *Regulation of Macromolecular Synthesis by Low Molecular Weight Mediators,* ed. H Koch, D Richter, pp. 303–17. New York: Academic

304. Justesen J, Ferbus D, Thang MN. 1980. *Proc. Natl. Acad. Sci. USA* 77:4618–22

305. Cayley PJ, Kerr IM. 1982. *Eur. J. Biochem.* 122:601–8

306. Justesen J, Worm-Leonhard H, Ferbus D, Petersen HU. 1985. *Biochimie* 67: 651–55

307. Turpaev K, Hartmann R, Kisselev L, Justesen J. 1997. *FEBS Lett.* 408:177–81

308. Cayley PJ, Davies JA, McCullagh KG, Kerr IM. 1984. *Eur. J. Biochem.* 143: 165–74

309. Matsuyama T, Grossman A, Mittrucker HW, Siderovski DP, Kiefer F, et al. 1995. *Nucleic Acids Res.* 23:2127–36

310. Eisenbeis CF, Singh H, Storb U. 1995. *Genes Dev.* 9:1377–87

311. Briken V, Ruffner H, Schultz U, Schwarz A, Reis LFL, et al. 1995. *Mol. Cell. Biol.* 15:975–82

312. Matsuyama T, Kimura T, Kitagawa M, Pfeffer K, Kawakami T, et al. 1993. *Cell* 75:83–97

313. Reis LF, Ruffner H, Stark G, Aguet M, Weissmann C. 1994. *EMBO J.* 13:4798–806

314. White LC, Wright KL, Felix NJ, Ruffner H, Reis LF, et al. 1996. *Immunity* 4:365–76

315. Lohoff M, Ferrick D, Mittrucker HW, Duncan GS, Bischof S, et al. 1997. *Immunity* 6:681–89

316. Taki S, Sato T, Ogasawara K, Fukuda T, Sato M, et al. 1997. *Immunity* 6:673–79

317. Kamijo R, Harada H, Matsuyama T, Bosland M, Gerecitano J, et al. 1994. *Science* 263:1612–15

318. Tamura T, Ishihara M, Lamphier MS, Tanaka N, Oishi I, et al. 1995. *Nature* 376:596–99

319. Holtschke T, Lohler J, Kanno Y, Fehr T, Giese N, et al. 1996. *Cell* 87:307–17

Annu. Rev. Biochem. 1998. 67:265–306

NUCLEOCYTOPLASMIC TRANSPORT: The Soluble Phase

Iain W. Mattaj and Ludwig Englmeier

European Molecular Biology Laboratory, Meyerhofstrasse 1, D-69117 Heidelberg, Germany; e-mail: Mattaj@embl-heidelberg.de

KEY WORDS: nuclear import, nuclear export, Ran GTPase, nuclear pore complex

ABSTRACT

Active transport between the nucleus and cytoplasm involves primarily three classes of macromolecules: substrates, adaptors, and receptors. Some transport substrates bind directly to an import or an export receptor while others require one or more adaptors to mediate formation of a receptor-substrate complex. Once assembled, these transport complexes are transferred in one direction across the nuclear envelope through aqueous channels that are part of the nuclear pore complexes (NPCs). Dissociation of the transport complex must then take place, and both adaptors and receptors must be recycled through the NPC to allow another round of transport to occur. Directionality of either import or export therefore depends on association between a substrate and its receptor on one side of the nuclear envelope and dissociation on the other. The Ran GTPase is critical in generating this asymmetry. Regulation of nucleocytoplasmic transport generally involves specific inhibition of the formation of a transport complex; however, more global forms of regulation also occur.

CONTENTS

265

0066-4154/98/0701-0265$08.00

INTRODUCTION: WHAT'S THE PROBLEM?

The compartmentation of eukaryotic cells gives rise to a need for intercompartmental transport of macromolecules. Specialized systems have evolved that allow proteins to be imported into membrane-bound organelles such as mitochondria, chloroplasts, lysosomes, the endoplasmic reticulum, and nuclei. Nuclear transport is unusual in that both import into and export out of the organelle are major processes, whereas in other organelles transport is largely unidirectional. All nuclear proteins are made in the cytoplasm and must be imported to the nucleus. In cells with an open mitosis these proteins must be reimported after each nuclear division. RNAs transcribed in the nucleus are almost all exported to the cytoplasm, in the form of ribonucleoproteins (RNPs). Many proteins shuttle continuously between the nucleus and cytoplasm. In total, this gives rise to an enormous level of nucleocytoplasmic traffic. Even a conservative estimate suggests that more than 1 million macromolecules are transferred between the two compartments each minute in a growing mammalian cell (1).

The nuclear envelope (NE) consists of a double lipid bilayer with an intervening lumen. The lumen and the outer bilayer are continuous with the endoplasmic reticulum and, thus, with the cellular secretory system (2). The NE is penetrated by nuclear pore complexes (NPCs). These huge structures, 125 million Daltons in vertebrates (3), form aqueous channels through which all nucleocytoplasmic transport is thought to occur. The NPC is composed of between 50 and 100 distinct polypeptides (4) that are often called nucleoporins. Molecules of up to approximately 9 nm in diameter, corresponding to a globular protein of approximately 60 kDa, can in principle enter or leave the nucleus by diffusion through the NPC, although in practice very few proteins and no known RNAs do so. Rather, nucleocytoplasmic transport is an active, signal-mediated process.

Very large complexes, like ribosomal subunits or even larger RNPs (5), are actively transported through the NPC. The functional pore size for active transport is somewhat greater than 25 nm (6). The difference between the size of the diffusion and active transport channels means that active transport must be accompanied by large conformational changes in the NPC. How this happens and, more generally, the mechanism of active translocation through the NPC are fascinating topics. Many recent reviews have covered NPC composition, assembly, structure, and function (7–13). The size of the diffusion pores means that ions and small metabolites must, in the free state, be able to cross the NPC unhindered. This does not mean that they all equilibrate between the nucleus and cytoplasm, since their distribution will be determined by the location of the macromolecules to which they bind on either side of the nuclear envelope.

Until recently, it was not clear whether nucleocytoplasmic transport represented active movement against a chemical concentration gradient or facilitated movement through the NPC followed by binding of the transported substrate and consequent retention in the target compartment. This was obviously a critical point in interpreting mechanistic studies of transport (see 14). Experiments in both mammalian and yeast cells have recently confirmed that at least one form of nucleocytoplasmic transport conforms to the active transport paradigm. This involved showing that molecules that were free to diffuse, and thus not retained in the nucleus by binding interactions, could be pumped into the nucleus against a concentration gradient when energy was provided. Substrates that were small enough to diffuse between the nucleus and cytoplasm and that carried a signal for nuclear import were constructed. When cells were depleted of energy, either by using a combination of energy poisons and low temperature or by low-temperature treatment alone, these substrates diffused throughout the nuclear and cytoplasmic compartments. Restoring ATP production or increasing the temperature caused the proteins to reaccumulate in the nucleus (15, 16). These studies demonstrate that the class of nuclear import signal tested does not bind tightly to an immobile nuclear phase, but rather is recognized by an active transport system capable of carrying the substrate against a concentration gradient. A further logical consequence is that the signal acts in one direction only, from the cytoplasm to the nucleus.

Given the similarities among the transport systems described in this review, it is likely that this conclusion will hold for most transport substrates. This review examines our current understanding of nuclear import and export signals, the transport mediators that recognize them, and the regulation of signal-mediator interactions. During import or export, transport mediators are found at the NPC, but they also spend some time free in the cytoplasm or nucleoplasm, and that soluble phase is the focus of this review.

NUCLEAR IMPORT

Import Signals and Receptors: How Much Diversity?

Signal sequences involved in targeting proteins into either the endoplasmic reticulum or mitochondria are generally removed during transit. As noted above, in many cell types nuclear proteins have to reaccumulate in the nucleus after each mitotic division. This means that nuclear targeting signals must be part of the mature nuclear protein rather than being removed on use (17, 18). The definition of what came to be called nuclear localization signals (NLSs) began with the study of proteolytic fragments of nucleoplasmin (19), but DNA-based technology soon took over.

The two best defined NLSs are those of SV40 large T antigen (SV40 TAg) and nucleoplasmin (20, 21; see Table 1). In this review we use the term NLS to refer only to the class of nuclear import signal represented by these two examples. They are both short and contain several critical basic amino acids. The discovery that a sequence as short as seven amino acids could direct nuclear import (20) allowed the synthesis of artificial import substrates by chemical cross-linking of the SV40 TAg NLS to human serum albumin (22). These conjugates, at high concentration, competitively inhibited nuclear import of NLS-bearing proteins, demonstrating that NLS-protein import is saturable. Together with earlier data on the saturability of tRNA export (23) and on the energy dependence of both NLS-protein import and tRNA export (23, 24) these results confirmed that nucleocytoplasmic transport processes are not only active but require saturable mediators.

NLS conjugates have also been used to determine whether all nuclear import substrates require the same saturable mediators. Initially, it seemed that most nuclear proteins might do so since the import of many members of a radio-labeled mixture of nuclear proteins was affected by saturating levels of SV40 TAg NLS (25). In contrast, the NLS conjugate had no effect on import of U snRNPs (26) that assemble in the cytoplasm and are imported to the nucleus (27). Various techniques were used to demonstrate that saturation of either NLS-protein import or U snRNP import did not affect the transport of the other karyophile and thus to suggest that these two substrates did not require the same saturable import mediator (26, 28).

More recent work has revealed the existence of a variety of protein import signals whose activity is not affected by saturation of the NLS import mediator (Table 1). The stage to which these signals have been characterized varies and is discussed later. In general, Table 1 shows that earlier ideas on the limited diversity of import pathways for proteins may have to be revised. A significant and growing number of well-defined import signals interact with import mediators that do not recognize the classical type of NLS. In addition, the signals listed at the foot of Table 1 do not appear to correspond to any of the better-characterized examples described in the body of the table. This should alert us to the possibility that the diversity of signals and transport mediators remaining to be discovered may still be extensive. Limitations to these arguments are discussed in a later section. There is at present a single example of nuclear uptake

Protein/signal	Nature of signal	Import receptor (plus adaptor)
SV40 large T antigen (simple basic NLS)	PKKKRKV Short sequences containing a single cluster of basic amino acids, often preceded by an acidic amino acid or a proline residue	Importin β together with members of the importin α family
Nucleoplasmin (bipartite basic NLS)	KRPAATKKAGQAKKKK Two interdependent clusters of basic amino acids separated by a flexible spacer (21); neutral and acidic residues flanking the motif contribute (282)	Importin β together with members of the importin α family
hnRNP A1 (M9-domain)	Amino acids 265–303 of human hnRNP A1, a region rich in glycines and aromatic residues. Also an export signal in mammalian cells	Transportin
STAT 1	Not known; formed after tyrosine phosphorylation and dimerization (263, 283)	Importin β together with one of the importin α family, NPI-1
hnRNP K (KNS)	Amino acids 323–361 of human hnRNP K. HnRNP K also carries a bipartite NLS. The KNS is not conserved in evolution. Also an export signal in mammalian and avian cells	Unknown, but experimental evidence excluding importin α and β and transportin
U snRNPs	Complex: RNA-bound Sm core proteins and the trimethyl cap structure of the RNA	Importin β is necessary; Snurportin binds to the m₃GpppN cap
Some ribosomal proteins	Accumulation of basic amino acids. Currently presumed to be distinct from the simple basic or bipartite NLS	In yeast Kap123p/Yrb4 or Pse1p
U1A	Amino acids 94–204 of human U1A protein	Importin α involvement unlikely
TFIIIA	Not yet well-defined; located within the zinc fingers (284)	Importin α involvement unlikely
Lamin B receptor	Within the N-terminal 204 amino acids. This signal targets membrane proteins to the inner nuclear membrane (285)	Importin α involvement unlikely (286)
Suppressor of white apricot	SR domain, i.e. a domain containing serine-arginine repeats (143, 287)	Not characterized
Glycoconjugates	Glucose, fucose, or mannose residues (288)	Importin α involvement unlikely; physiological significance unclear

[a]Examples of NLSs that do not conform to the classes above have been found in the HTLV-1 Tax transactivator (289), the yeast Mat α2 protein (290, 291), the yeast Gal 4 transactivator (292), Adenovirus E1A protein (293), and the c-myc protein (282). It is unclear whether these signals use as yet unknown import pathways, are divergent members of one of the classes above, or lead to nuclear accumulation via a piggyback mechanism, as has been shown for E2F/D1 transcription factors (294), the CBP20 protein (248), yeast Cdc2 protein kinase (295), and others. References not present in the table are given in the text.

of a protein that requires a mediator but does not seem to be energy dependent. This involves calmodulin, whose nuclear accumulation is saturable but appears to require no energy in the form of nucleotide triphosphate hydrolysis (29) and can thus be considered facilitated, rather than active, transport.

Recognition of the NLS by Importin

The NLS-conjugate saturation experiments provided early evidence for the existence of saturable mediators of import, the import receptors. Several such receptors have been characterized molecularly (Table 1). The initial experimental approach used to identify them was to search for proteins that would bind to the well-characterized SV40 TAg NLS (reviewed in 30). Concurrently, in vitro assays for protein import were established (24, 31, 32). These assays were based either on nuclei added to or assembled in *Xenopus* egg extracts or on mammalian cultured cells that had been gently extracted with detergent to permeabilize the plasma membrane and remove soluble cytoplasmic components but leave the nucleus intact. Import in the latter system was dependent on the readdition of cytosol to the permeabilized cells (32). Fractionation of the added cytosol was the major technique used in the identification of the soluble factors required for NLS-protein import.

Researchers identified a protein from reticulocyte lysate (the NLS receptor) that could be cross-linked to the SV40 TAg NLS and functioned in NLS import together with a second protein (p97) (33–35). The NLS receptor from *Xenopus* was cloned and called importin α (36). Importin α was related in sequence to the previously identified yeast Srp1p protein (37), and Srp1p was subsequently shown to be the functional homologue of importin α (38). Although yeast has a single importin α protein, cloning of homologues from other species (e.g. 39–43) demonstrated that multicellular eukaryotes encode a family of closely related importin α–like proteins (reviewed in 1, 44). In contrast, the second subunit of the heterodimer that functions in NLS-protein import, importin β, is unique (38, 45–48). There are several alternative nomenclatures for the importin subunits, such as nuclear pore targeting complex (PTAC)58 and PTAC97, Karyopherin α and β, and NLS receptor and p97. In addition, some individual members of the importin α family have been given several names (1, 44). Where possible, we use the importin α and β nomenclature to avoid confusion.

NLS-protein import has been grossly divided into two stages: energy-independent docking at the cytoplasmic face of the NPC and energy-dependent NPC translocation (49, 50). In the permeabilized cell assay, the importin α/β heterodimer is both required and sufficient for the docking step (45, 51–53). The two importin subunits have specialized functions in docking. Importin α binds the substrate protein through recognition of the NLS, whereas importin β interacts with the NPC (33–35, 41, 51–53). Although this has been shown formally only for the importin-NLS protein complex, all import receptors are likely capable of docking at the NPC with their substrate.

The primary docking sites for importin were identified by electron microscopic examination of the docking of NLS substrates that had been conjugated to colloidal gold (6). The docking sites were found on fibers that extend from the NPC into the cytoplasm. Although there may be only one high-affinity docking site per fiber (89), considerable arrays of colloidal gold particles, apparently attached to the NPC fibers, were seen when the substrate was present at high concentrations (6, 49). This observation suggests that the importin complex, via importin β, can bind to multiple sites on the fibers. The existence of multiple docking sites could concentrate the import complex close to the site of NPC translocation.

NPC translocation of the importin-NLS protein complex requires two additional soluble proteins, the Ran GTPase and p10/NTF2 (see below). Dissection of importin α led to the important conclusion that the only functions of α in NLS-protein import are its binding to the NLS and importin β (54, 55). The N-terminal basic region of importin α that is responsible for binding to β (the importin β–binding or IBB domain) includes sequences that look remarkably like a bipartite basic NLS (Table 1). In spite of this, the IBB domain binds to importin β, not to α, and when fused to a reporter protein, is sufficient to target that protein to the nucleus in a β-dependent and α-independent manner (54, 55). Thus it is possible to consider importin β as the genuine import mediator or receptor, and α as an adaptor that joins β and the NLS substrate.

Other Import Signals and Receptors

An interesting mechanistic variation concerns the STAT1 protein, one member of the signal transducer and activator of transcription protein family. STATs are transcription regulators that are cytoplasmic in the resting state but are activated by various extracellular stimuli to move to the nucleus (56, 57). STAT1 activation involves tyrosine phosphorylation–dependent dimerization, which is required to generate a nuclear targeting signal. Unlike proteins that contain an NLS, which seem capable of utilizing any of the importin α family members for import (1, 58), STAT1 import is more selective. STAT1 can be imported by one human importin α, namely hSRP1/NPI-1, but not by another, hSRP1α/Rch1 (59). NPI-1 can also import NLS proteins, but the region of NPI-1 required for NLS binding is distinct from that needed for STAT1 binding (59).

This observation has two major implications. The first implication is practical. We should not expect to be able to recognize all substrates of a particular import receptor by looking at the sequence of their nuclear import signals since STAT1 has no obvious NLS-like sequence. Thus examples of divergent import signals do not necessarily imply the existence of novel import receptors.

Indeed, it is unclear if saturation by NLS-conjugates blocks STAT1 import, and it may be that this test does not unambiguously assign a substrate to a specific transport receptor or adaptor.

The second implication is evolutionary. Given the existence of multiple import (and export) receptors that are capable of shuttling between the nucleus and cytoplasm (see below), it might be relatively easy to generate a domain on a potential substrate protein that can bind somewhere on the surface of one of the receptors, and thus to target a new protein for import or export. Import or export requires both association with a receptor on one side of the NPC and dissociation from the receptor on the other side, and evolution of new substrates therefore is made more difficult.

U snRNP import was mentioned previously. The nuclear import signal of these RNPs is bipartite (Figure 1; Table 1). The first feature consists of an ill-defined component that entails binding of the so-called Sm core proteins, a collection of eight proteins, to an RNA element found only in these imported U snRNAs (27, 60). Assembly of the Sm core structure in vivo may not occur spontaneously, but rather appears to require the mediation of at least two assembly factors, one of which is encoded by the gene whose mutation causes the common inherited genetic disorder spinal muscular atrophy (61, 62). In addition to the Sm core structure, the trimethylguanosine cap that is formed when the Sm core proteins bind the snRNA is also part of the U snRNP import signal (63–67).

As mentioned above, U snRNP uptake was the first example of import proposed to require a different saturable mediator than did NLS proteins. The saturable adaptor that binds to the NLS is importin α, and direct study of in vitro import of U snRNPs has confirmed that importin α is not required (68). However, a variety of approaches led to the conclusion that importin β is needed for U snRNP import. Most directly, depletion of importin β blocked U snRNP import, and readdition of recombinant β reversed this defect (68). The protein that recognizes the trimethyl cap structure during import, Snurportin 1, is a distant relative of importin α and, like α, has an N-terminal IBB domain (J Huber, C Marshallsay, U Crontzhagen, M Sekine, R Lührmann, manuscript submitted). The role of importin β in U snRNP import may therefore involve only its interaction with Snurportin 1. However, Snurportin 1 does not appear to interact with the Sm core structure (J Huber, C Marshallsay, U Crontzhagen, M Sekine, R Lührmann, manuscript submitted). This means that the U snRNP–containing import complex contains minimally one additional essential component, and this may also interact with importin β (68). The advantage of having two or more separate adaptors that interact with the same import receptor, as exemplified by the mutually exclusive interactions of importin α and Snurportin 1

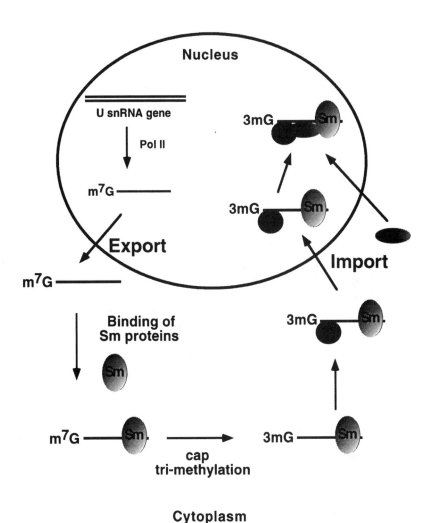

Figure 1   A simplified scheme of U snRNP assembly. This scheme applies to RNA polymerase II–transcribed spliceosomal U snRNAs. They are transcribed in the nucleus and obtain a mono-methylguanosine cap cotranscriptionally. After nuclear export, they bind the Sm core proteins. Maturation steps occur, including formation of the trimethylguanosine cap. The U snRNP can then be imported to the nucleus. Other U snRNP–specific proteins can join the RNP either in the cytoplasm or, more likely (79, 300–302), in the nucleus.

Table 2 Importin β-like proteins[a]

Mammalian transport factor	Yeast homologue	Function
Importin β	Importin β/Kap95p	Import of NLS-proteins; import of U snRNPs in vertebrates
Transportin	Transportin/Kap104p	Import of mRNA-binding proteins
Ran BP5 (296)/ Karyopherin β3 (297)	Probably Pse1p	Pse1p thought to mediate import of ribosomal proteins in yeast
?	Yrb4p/Kap123p	Import of ribosomal proteins in yeast
RanBP7 (71)	Possibly Sxm1p or Nmd5p	Unknown
CRM1	Crm1p/Xpo1p	Export of NES-containing proteins
CAS	Cse1p	Export of importin α

[a]Sxm1p, Mtr10p, and Los1p are additional importin β–like proteins for which experimental evidence of an involvement in nucleocytoplasmic transport exists (see text). Other members of the importin β family (70, 71) have not been further characterized. References not present in the table are given in the text.

with importin β, is presumably to allow separate regulation of import of the two distinct classes of import substrates.

Either by coincidence or for unknown functional reasons, the remaining well-defined examples of import signal-receptor pairs in Table 1 all involve RNA-binding proteins. The receptors involved in the transport of hnRNP and ribosomal proteins characterized to date are distantly related to importin β (Table 2; 13, 70, 71) but do not follow the importin paradigm of division of function between two subunits. Rather, they are either known or thought to bind their substrates directly rather than via an α-like adaptor.

Transportin is the import receptor for hnRNP A1 (72, 73). The hnRNP A1 import signal, the M9 domain, is much longer than the SV40 TAg NLS at 38 amino acids (74, 75) and is rich in glycine and aromatic amino acids rather than being basic in character. Transportin binds directly to the M9 domain (72, 73). Mutation of yeast transportin, Kap104p (76, 77), leads to defective import of one hnRNP protein, Nab2p, and yeast transportin has also been implicated in the import of a second one, Hrp1p/Nab4p (76). The import signals of the yeast transportin substrate proteins have not been delineated, but they are not closely related in sequence to the M9 domain. Similarly, human transportin interacts with a variety of hnRNP proteins, not all of which contain regions that resemble M9 (77a).

The human hnRNP K protein has two import signals. One is a bipartite basic NLS and the second an unrelated element called the KNS (78). KNS-mediated in vitro import is not inhibited by saturating amounts of either the IBB domain (which blocks both NLS-protein and U snRNP import) or the

M9 domain and thus is not likely to be dependent on either importin β or transportin (78). Similarly, the U1A protein of U1 snRNP has a complex nuclear import signal of roughly 100 amino acids in length (79). Import driven by the U1A import signal is not affected by saturation with the IBB, M9, or KNS domains (M Hetzer, unpublished data). These two examples, plus others in Table 1, offer strong evidence that not all the import mediators have been defined.

Almost all ribosomal proteins have to be imported into the nucleoplasm, and then to the nucleolus, for ribosomal subunit assembly. It is unclear whether all ribosomal proteins are imported individually or if some may enter the nucleus as preassembled complexes. Studies of a few yeast and human ribosomal proteins have revealed that some proteins carry more than one functional nuclear import signal (e.g. 80–85). This is probably easier to rationalize by proposing that complexes containing more than one ribosomal protein assemble prior to import. If this were the case, some of these signals would in fact represent interaction domains with other ribosomal proteins required to form a complex. Each complex of ribosomal proteins would require only one "real" import signal, that is, one sequence capable of interaction with an import receptor. All ribosomal protein import signals will presumably have to be covered up on assembly of a ribosomal subunit to prevent their recognition by the import machinery in the cytoplasm.

The best-characterized ribosomal protein import signals either include, or consist of, short, basic peptides (80–85) and therefore were initially presumed to be imported via the NLS-importin pathway. The first indication that this might not be the case was the observation that import directed by the signal derived from the yeast L25 protein (81) was not affected in cells carrying a mutant of the yeast NPC protein, or nucleoporin, Nsp1p (86). In the same cells, import mediated by signals from other yeast nuclear proteins was inhibited (86).

More recently, proteins carrying the L25 import signal have been shown to associate with two members of the importin β family in yeast cells, namely Pse1p and Yrb4p/Kap123p (Tables 1 and 2; 87, 88). Genetic evidence for roles of these proteins in nuclear uptake mediated by the L25 import signal has been presented (87, 88). Pse1p is essential in yeast, while Yrb4p/Kap123p is not. These two import receptors have therefore been proposed to act in a partially redundant way to bring about ribosomal protein import. Many interesting questions remain regarding ribosomal protein import. Apart from the uncertainty about the mono- or oligomeric state of the proteins during import, it also remains to be seen whether all ribosomal proteins will use the same import receptors. In vitro studies with pure proteins and receptors would help to resolve this issue.

Ran

The energy-dependent step of NPC translocation occurs through the center of the membrane-embedded part of the NPC, the so-called transporter or central plug (6, 89, 90). Translocation of substrate-receptor complexes requires the Ran GTPase. Ran is a critical component of almost all known nucleocytoplasmic transport pathways (91–97) and is discussed at several points in this review. But first, we provide some essential background information (see also 98, 99).

Ran is extremely abundant and at steady state is mainly nuclear, although it is believed to move between the nucleus and cytoplasm. Like many other regulatory GTPases, Ran has low intrinsic activity. It has to interact with both a GTPase-activating protein (RanGAP1 in humans, Rna1p in yeast) and a small Ran-binding protein (RanBP1/Yrb1) to achieve maximal GTPase activity. Once hydrolysis has occurred, Ran also needs a cofactor to dissociate from the GDP formed, the guanine nucleotide exchange factor, or GEF, RCC1/Prp20p. Although other Ran-binding proteins exist, these are the four major components involved in the parts of Ran function that are elucidated thus far. A critical aspect of Ran's function relies on the fact that RCC1 is nuclear and is stably bound to chromatin in the nucleus (100), while both RanBP1 and RanGAP1 are found either on the cytoplasmic side of the NPC or in the cytosol (98, 99). This distribution predicts that RanGTP concentration will be high in the nucleus and low in the cytoplasm. Treatments that collapse this RanGTP concentration inequality, such as increasing the cytoplasmic RanGTP:RanGDP ratio or decreasing the nuclear RanGTP concentration, block nucleocytoplasmic transport (91–97, 101, 102).

The mechanism of the block of import when the cytoplasmic RanGTP: RanGDP ratio is increased seems straightforward. RanGTP binds directly to import receptors like importin β or transportin. In these cases, and apparently also for the Yrb4p-L25 import complex, the result of RanGTP binding is to cause substrate-receptor complex disassembly (77a, 88, 97, 102–105). This interaction may cause dissociation of import complexes in the nucleus after NPC translocation (1, 99, 102, 107), although other possibilities have also been raised (103). It therefore seems logical that cytoplasmic RanGTP would inhibit nuclear import, as it will cause import complexes to disassemble before they reach the NPC. In regulating the formation of import complexes by allowing assembly to occur in the cytoplasm and causing disassembly in the nucleus, Ran provides asymmetry, a property that is needed for transport across the NPC against a concentration gradient. Ran thus helps impart directionality to the process of import. This is the first critical function that Ran is proposed to play in nuclear import.

Crucial though this is, it does not seem to be the only function of Ran in import. Energy-dependent import is inhibited by nonhydrolyzable GTP analogues, but not by nonhydrolyzable ATP analogues, and is also inhibited in a dominant way by Ran mutants that can bind GTP but are unable to perform hydrolysis (91–93, 95, 108, 109). Although all these treatments will generate cytoplasmic RanGTP, and thus cause inhibition via disruption of import receptor-substrate complexes, the results have often been interpreted as suggesting that GTP hydrolysis by Ran would play a direct role in nuclear import.

The best evidence for this came from the use of a mutant form of Ran that binds to and hydrolyzes xanthine triphosphate (XTP) rather than GTP. This Ran XTPase supported importin-mediated nuclear transport when tested in permeabilized cells. Instead of cytosol, the cells were supplemented only with the soluble factors required for NLS-protein import in recombinant form (101). Import in this case required both XDP and small amounts of XTP and was not inhibited by nonhydrolyzable GTP (or ATP) analogues in quantities that inhibited cytosol-mediated import (101). The inescapable conclusion of this experiment seems to be that if nucleotide triphosphate hydrolysis is at all required for the nuclear import observed, then XTP (or GTP) hydrolysis by Ran can fulfill the energy requirement.

It should be borne in mind that other studies with the XTPase Ran mutant, also carried out with permeabilized cells but using Ran-depleted cytosol as a source of import factors, led to the proposal that at least one additional GTPase plays a role in NLS-protein import (110). Resolution of this issue will require either functional characterization of the second putative GTPase activity or the development of a simpler in vitro system for NLS-protein import. A candidate for the putative second GTPase is the yeast GTP-binding protein, and possible GTPase, Gtr1p. A mutant form of Gtr1p suppressed mutations in both yeast RanGAP1 (Rna1p) and RanGEF (Prp20p) suggesting that this protein may be involved in nuclear transport (111).

Several other factors have been proposed to play a role in nuclear import. We next discuss additional proteins whose function in relation to import is either poorly understood or has not been firmly established.

p10/NTF2

This small protein is required for efficient NLS-protein nuclear import in permeabilized cells (112, 113). The yeast homologue Ntf2p is essential (114–116), and genetic evidence supports its role in nuclear import (115–117). p10/NTF2 interacts with Ran in the GDP-bound state (114, 118, 119), and it also binds both to a number of nucleoporins and to the Ran-binding protein Yrb2p/RanBP3, which is located in the nucleus (see below) (113, 114). Overexpression of Ntf2p suppresses certain gsp1p (Ran) mutants in yeast (117). The interesting result of the converse experiment, namely that the lethality caused by deletion of the

NTF2 gene can be suppressed by Gsp1p (Ran) overexpression (119) further suggests that the critical role of the protein is to increase the efficiency of one or more of the functions of Ran. For example, this role could involve localizing Ran to a particular place on the NPC, although more information is required before drawing firm conclusions about p10/NTF2 function.

RanBP1/Yrb1p

RanBP1 binds to RanGTP and increases the rate of RanGAP1-induced GTP hydrolysis (120). Both RanBP1 and RanGAP1 are cytoplasmic and therefore maintain cytoplasmic RanGTP at low levels, allowing interaction between import receptors and their adaptors and substrates (101, 102, 104, 107, 121, 122).

Some properties of RanBP1 suggest an additional function, although this function is not yet well characterized. RanBP1 is capable of forming a stable complex with RanGDP and importin β. RanGDP sits between the two larger proteins and is needed to mediate complex formation since either RanBP1 or importin β alone binds well to RanGTP but very poorly to RanGDP (104, 105). Unlike the RanGTP–importin β complex, the RanBP1–RanGDP-β heterotrimer is capable of binding importin α and α-associated NLS proteins (104, 105, 122). Indirect evidence suggests a role for the ternary complex in NLS-protein import (105).

There are some problems with this theory (see also 99). First, RanBP1 contains a signal that retains it in the cytoplasm, presumably through binding to a nontransportable partner. Only when this retention signal is mutated can RanBP1 be detected in the nucleus (123). Further, RanBP1 also contains a signal for active nuclear export (123, 124). If one RanBP1 molecule were carried into the nucleus (as the ternary complex) with each importin β, then an equal and opposite amount of export, and export receptor, would be necessary for each round of import. This is possible, but it is not economical. Third, nuclear RanBP1 is toxic, at least in part because it inhibits many types of nuclear export (97, 124). Given the nuclear toxicity of RanBP1, and the dominance of the cytoplasmic retention signal over RanBP1 nuclear import (123), the nuclear export signal in RanBP1 may be required to prevent trapping of even small amounts of RanBP1 in the reforming nucleus after mitosis.

Yrb2p/RanBP3

There is no direct evidence that this protein is involved in nucleocytoplasmic transport. On the basis of its sequence it was originally designated Nup36p, for nucleoporin of 36 kDa, and shown to interact with p10/Ntf2p (114). However, both the yeast (Yrb2p) and human (RanBP3) homologues are located in the nucleoplasm rather than at the NPC (125; L Mueller, VC Cordes,

FR Bischoff, H Ponstingl, manuscript submitted). Yrb2p makes a Ran-dependent interaction with Prp20p, and their human counterparts, the RanBP3 and RCC1 proteins, behave in a similar way (125; L Mueller, VC Cordes, FR Bischoff, H Ponstingl, manuscript submitted). Yrb2p/RanBP3 also binds to RanGTP on its own, although this interaction is weak (125; L Mueller, VC Cordes, FR Bischoff, H Ponstingl, manuscript submitted). Analysis of Yrb2p mutants has provided no evidence to date of a role in transport, although yeast cells lacking Yrb2p are cold sensitive for growth (125).

RanBP2/Nup358

RanBP2/Nup358 is an extremely large and interesting nucleoporin. It is located on the cytoplasmic fibers of the NPC (126, 127). It includes four RanBP1-like Ran-binding domains that function in a similar way to RanBP1 in in vitro assays that measure interaction with and modification of Ran activity (121, 126). A fraction of RanGAP1 is found at the NPC rather than in the cytoplasm because it binds to RanBP2 (128, 129). This fraction is modified by the addition of a small, ubiquitin-like peptide, and this modification is required for tight RanBP2-RanGAP1 interaction (128–130). RanBP2 therefore potentially provides a site on the NPC to which both RanGTP and RanGAP1 are bound. Furthermore, the RanGTP, through binding to the RanBP1-like domains, would be in its most sensitive configuration for GTPase activation.

Two possibilities exist for the function of RanBP2 and the bound proteins. First, RanBP2 might bind RanGTP as it leaves the nucleus (perhaps together with export receptors; see below) and hydrolyze it, thus increasing the efficiency by which the cytoplasmic RanGTP concentration is maintained at a low level (99, 107). Alternatively, RanBP2 could function in coupling RanGTP hydrolysis to NPC translocation (128, 129). Since the latter function should be an essential one, the fact that the soluble, cytoplasmic Yrb1p is the only essential RanBP1-like RanGTP-binding protein in yeast speaks against this idea. The only other two RanBP1-like proteins in yeast are Nup2p, a nucleoporin (131, 132), and Yrb2p. They both bind RanGTP poorly (99, 125; L Mueller, VC Cordes, FR Bischoff, H Ponstingl, manuscript submitted), and they are not essential for yeast growth. Even a strain lacking both proteins is viable (133). Unfortunately, it seems as if the function of RanBP2 will have to be elucidated without the benefit of comparative yeast genetics.

Hsp70/hsc70

A final factor with a rather ill-defined role in nuclear import is the heat shock protein, hsp70, and its constitutively expressed cognate, hsc70. Inhibition experiments involving either microinjection of anti-hsc70 antibodies into cells or their addition to in vitro assays provided first evidence for a role of hsc70 in

NLS-protein import (134, 135). More persuasively, depletion of the cytosolic extract used in an import assay with ATP agarose, which would remove hsc70 among other things, inactivated the extract, and this effect could be reversed by addition of recombinant hsp70 or hsc70 (135). More recent experiments in yeast showed that overexpression of hsp70 increased import rate (16).

Although it is hard to rule out nonspecific effects of the hsps in all of these experiments—because they are able to help keep all the proteins required for import in an active conformation—these results may indicate a direct role for hsps in import. This would not have been an a priori expectation since, given both the size of the active NPC transport channel and the fact that RNP substrates such as ribosomal subunits cross the NPC intact, proteins clearly do not need to unfold during NPC translocation. One additional finding that might help to illuminate the role of hsps in import is the observation that not all NLS substrates are affected by hsc70 depletion (136), leading to the suggestion that hsc70 might be required in some cases to allow presentation of the NLS to importin α. A final problem in proposing a role for hsp/hsc70 in nuclear import is that although the known functions of these proteins require ATP hydrolysis, import, at least in vitro, does not (101, 108).

NUCLEAR EXPORT

Nuclear Export and Nuclear Retention

Many substrates for protein import are individual proteins that carry a nuclear import signal and, thus, are relatively small and simple. This is probably not the case for nuclear export substrates, many of which are ribonucleoprotein particles (RNPs) that can be both large and complex in composition. Ribosomal subunits are one obvious example and mRNPs another (137). A favorite experimental system for microscopic study illustrates this point. The Balbiani Ring (BR) mRNAs of *Chironomus tentans* encode secreted proteins of roughly 10^6 Daltons that are made in larval salivary gland cells. The BR transcripts are 35–40 kb in length, and each is estimated to associate with roughly 500 protein molecules (reviewed by 5, 138). The BR RNPs are large enough to allow easy identification in electron microscope sections, and their passage through the NPC has been studied in detail. The particles, which form during transcription and fold into regularly shaped structures of 50-nm diameter, have to partially unfold at the NPC to permit translocation (139, 140). The unfolded RNP that is translocated has a diameter of 25 nm, roughly equivalent to the maximal size of substrates for nuclear import.

Although different in size from other mRNPs, studies of the protein composition of the BR RNPs have revealed that in this respect they seem to be representative of mRNPs as a class. For example, BR mRNAs are associated with hnRNP proteins, SR proteins, and the nuclear cap-binding complex, CBC (5). The SR proteins are a family of splicing factors containing repeated serine-arginine dipeptides. They share many properties with hnRNP proteins, being highly abundant, binding to RNA with relatively weak sequence discrimination, and including both members that are constitutively nuclear as well as ones that shuttle between nucleus and cytoplasm (141–143). Further, the BR mRNAs contain introns that must be removed before they are exported from the nucleus; thus, the mRNAs must also associate with the pre-mRNA splicing machinery. Conclusions from the study of BR RNPs are likely to be generally applicable to mRNPs.

One of the most significant observations made thus far is that individual proteins leave the BR RNP at different stages during its formation and passage through the nucleoplasm and NPC into the cytoplasm. Some splicing factors, such as the U snRNPs, are associated with only the nascent or newly mature RNPs (144) and thus probably dissociate soon after splicing has occurred. Other proteins whose mammalian homologues have roles in splicing behave differently. The homologue of hnRNP A1 remains with the RNA at all steps in its maturation and transport and is still detected on the mRNA even when translation is occurring (145). The nuclear CBC stays on the RNA through NPC translocation, but is then removed (146). This is particularly interesting since BR RNPs always exit the nucleus in a 5′ to 3′ orientation, that is, with the capped end, to which CBC is bound, in the lead (139, 140). Finally some proteins, such as the homologue of the SR splicing factor ASF/SF2, dissociate prior to NPC translocation and thus remain in the nucleoplasm (147).

Removal of some RNP proteins is likely to be a general prerequisite for RNP export, due to the phenomenon of nuclear retention. Certain RNAs do not leave the nucleus (except during mitosis), and these RNAs are actively retained in the nucleus by binding to specific, saturable factors (148–151). Similarly, splicing factors that bind to the splice sites of introns in pre-mRNAs prevent these RNAs from leaving the nucleus until the introns have been removed and the mature mRNA released (152–154). How spliced mRNAs are released from the retaining splicing complexes is not understood, although Prp22, a putative RNA helicase of the DEAD/H box class, is required for this process in mammalian cells (155).

With one exception, protein sequences directly involved in nuclear retention are poorly characterized. This exception involves the hnRNP C protein, which is constitutively nuclear. Since hnRNP C associates with many polyA-containing mRNAs and pre-mRNAs in the nucleus (156, 157), it must be removed from them before mRNA export. The nuclear retention sequence (NRS; 158) of the protein plays a role in this process. The NRS has been defined as a 78-amino-acid segment of hnRNP C (158). The NRS is dominant when fused to a protein carrying a nuclear export signal (see below), but it is unclear how it functions. Two classes of explanation for NRS activity seem possible. An NRS might interact with some untransportable structure in the nucleus, such as the nuclear lamina, or an NRS might block the activity of, or physically mask, export signals and thus need to be removed to allow transport to occur. It is possible that NRSs may be used for the regulation of export of specific mRNAs since the adenovirus E4 34K protein has an hnRNP C-like NRS that is somehow involved in the selective export of viral RNA, as opposed to cellular mRNAs, from the nuclei of infected cells (159).

It will be interesting to learn how hnRNP C, the SF2/ASF homologue on the BR RNP, and other nuclearly retained mRNA-binding proteins are removed from mRNPs before export, and if there is indeed a single mechanism that achieves this for the different retained proteins. Retention and export factors can be easily distinguished experimentally since saturation of a factor required for retention allows export to happen while saturation of a factor required for export competitively inhibits export (see e.g. 151). However, an important corollary of the seeming generality of nuclear retention is that factors that can be shown experimentally to be required for active export may be needed to release retention rather than in a more direct positive sense.

Export Signals on RNA Are Generally Adaptor-Binding Sites

As noted previously, the export of tRNA was the first nucleocytoplasmic transport process shown to be saturable and thus receptor mediated (23). Later work revealed that other classes of RNA that are exported from the nucleus, like ribosomal RNA (in the form of ribosomal subunits), U snRNA, 5S rRNA, and mRNAs, also require saturable mediators (160–162). These mediators appear to be class specific. For example, saturable factor(s) required for tRNA export are different from those needed for mRNAs, U snRNAs, 5S rRNA, or ribosomal subunits (161, 162). It has been shown (see below), or is believed, that the class-specific saturable proteins detected in these experiments are not export receptors per se, but rather are RNA-binding proteins that recognize the specific RNAs and mediate their interaction with the actual export receptors. In other words, they are adaptors that are analogous to importin α, which recognizes NLS proteins and mediates their interaction with the actual import receptor, importin β.

Before discussing these adaptors we describe a few features of RNAs that are recognized during export (Table 3). Some of these features are simple: For example, the export of U snRNAs depends upon recognition of their monomethyl-guanosine cap structures (154, 161, 163, 164). Others are complex and redundant: For example, nonoverlapping regions of yeast heat-shock mRNAs are

Table 3 Export signals[a]

Export signals on proteins protein/signal	Nature of signal	Export pathway
HIV-1 Rev, PKI (leucine-rich NES)	$L-X_{2-3}-(F,I,L,V,M)-X_{2-3}-L-X-(L,I)$; consensus is not exclusive[b]	Receptor: CRM1
Importin α	Unknown	Receptor: CAS
HnRNP A1 (M9-domain)	See Table 1	Receptor unknown; tested in mammalian cells
HnRNP K (KNS)	See Table 1	Receptor unknown; tested in mammalian and avian cells
CBC	Unknown, formed when CBC binds capped RNA	Receptor: CRM1; CBC conserved in evolution, but evidence for transport function so far in vertebrates and insects
Export signals on RNA		
Rev responsive element of HIV-1 (RRE)	A 234 nt region within the HIV *env* gene with a complicated secondary structure	RRE is bound by Rev, which carries the NES
Constitutive transport element (CTE) of simple retroviruses	Mapped in the Mason-Pfizer monkey virus (MPMV); a 154 nt sequence that forms a long, imperfectly paired stem	Interaction with a component of the cellular mRNA export pathway
U snRNAs	Monomethyl guanosine cap structure	m_7G cap is bound by CBC
5SRNA	Binding sites for TFIIIA or ribosomal protein L5	Ability to bind to either TFIIIA or L5 is necessary for 5SRNA export
Histone mRNA	Acquired during 3' end formation; a second signal in the mature histone mRNA	

[a]An increasing number of proteins are known to shuttle between the nucleus and cytoplasm, including transcription factors (258, 259), nucleolar proteins (299), other RNA-binding proteins (see text), and protein phosphatases and kinases (185, 258). Whether these proteins all carry signals for active export and what these signal are remains to be elucidated.

[b]Like bipartite basic NLSs, leucine-rich NESs can also be divergent from the consensus, e.g. the equine infectious anemia virus (EIAV) Rev protein NES, GPLESGQWCRVLRQSLPE (298). References to other data in the table are in the text.

recognized during their export (165), and multiple features of polyadenylated mRNAs including the cap structure, the polyA tail, and the "body" of the mRNA (161) seem to contribute to their efficient export and to combine to produce maximal export rates. Similar conclusions with respect to recognition of several RNA features by export factors were obtained in studies of non-polyadenylated histone mRNA export (166; Table 3).

As mentioned above, intron-containing cellular mRNA precursors (pre-mRNAs) are generally retained in the nucleus. For nuclear viruses, for example, the retroviruses, this presents a problem since some viral mRNAs, as well as genomic RNAs, contain sequences that either are or are not spliced out of different individual transcripts to produce mRNAs with different coding capacities (167, 168). Thus these viruses need to export unspliced as well as spliced RNAs. Simple retroviruses like Mason-Pfizer monkey virus (MPMV) or Rous sarcoma virus (RSV) have evolved RNA structures called constitutive transport elements (CTEs; Table 3). These elements allow export of the unspliced mRNAs to which they are attached (169–173). Experiments carried out in *Xenopus* oocytes have shown that saturating quantities of CTE-containing RNAs prevent the export of mRNAs but do not inhibit export of the other classes of RNA tested (U snRNAs, tRNA), suggesting that CTEs are bound by a cellular factor required for mRNA export (172, 173). The identity of this factor is of considerable interest, not least because the CTE acts in a dominant manner to direct the export of an excised intron that would normally be retained in the nucleus (172) whereas mRNA sequences do not have this property. The CTE binds to a number of cellular proteins in vitro (173, 174), but there is no direct evidence that these proteins are transport factors.

Complex retroviruses have developed a more complete solution to the problem of exporting nonspliced RNAs. The best studied example is HIV-1. The RNA element in this case (see Table 3) is predicted to be a complicated collection of hairpin loops called the Rev Response Element, or RRE (175). The RRE, as its name suggests, binds directly to the Rev protein (176).

Export Adaptors

Rev is the best-understood example of an RNA export adaptor. A direct demonstration that Rev promoted RNA export from the nucleus came from experiments in *Xenopus* oocytes (177). As previously discussed, introns are retained in the nucleus, even after being spliced out of pre-mRNAs. When the RRE was inserted into an intron, Rev-dependent export of the intron took place, that is, like the CTE, the RRE-Rev complex is dominant over nuclear retention mechanisms. Rev binds to the RRE through an arginine-rich motif (178) and requires an additional peptide region located C-terminal to this RNA-binding domain for function (179). Initially this region was called the activation domain, and it was defined by extensive mutagenic analyses as an eight-amino-acid leucine-rich peptide (179, 180). Later work showed that peptides corresponding to the activation domain of the Rev protein (Table 3), when cross-linked to BSA, directed nuclear export of the conjugate, identifying the leucine-rich peptide as an NES, or nuclear export signal (181). Simultaneously, a cellular protein called PKI (protein kinase A inhibitor) was shown to contain a similar leucine-rich NES (182).

PKI is an example of the use of protein nuclear export for regulation. The cyclic AMP–dependent protein kinase (PKA) exists in unstimulated cells as an inactive heterotetramer of four subunits, two catalytic (C) and two regulatory (R). On stimulation, the complex dissociates and the active C subunits enter the nucleus by diffusion and can phosphorylate substrate proteins there (183, 184). PKI is a small protein that binds to the C subunit, and its NES directs the C subunit–PKI complex back to the cytoplasm, thereby limiting the time in which the C subunit can remain active in the nucleus (182). Mutations in several components of the nucleocytoplasmic transport machinery give rise to complex phenotypes, including the common observation of cell-cycle-related defects. We therefore suspect that transport-determined activation or inactivation of regulatory molecules like kinases or protein-degrading enzymes, including those involved in regulating cell cycle progression (see e.g. 185), may be rather common.

Leucine-rich NESs have been implicated in transport of RNA substrates other than those that carry an RRE. The first evidence for this involvement was that saturation of NES-dependent export with BSA-NES peptide conjugates blocked the export of both 5S rRNA and U snRNAs in *Xenopus* oocytes (181); however, the conjugates had no effect on mRNA export.

To be transported out of the nucleus of *Xenopus* oocytes, 5S has to be able to bind to either TFIIIA or ribosomal protein L5 (186). TFIIIA contains a leucine-rich sequence that, as an isolated peptide, is a functional NES (181, 187). The block to 5S rRNA export caused by saturating levels of BSA-NES conjugates, and the presence of a NES in TFIIIA, further suggest that TFIIIA might mediate 5S rRNA export. However, since 5S mutants that do not bind TFIIIA but do bind to the L5 protein are also exported (186), this cannot be a complete explanation for the export of this RNA.

Although the BSA-NES conjugate experiments indicate that leucine-rich NESs do not play an essential role in mRNA export from oocyte nuclei, some data suggest that the situation may be different in yeast cells. Two *S. cerevisiae* proteins, Gle1p (or Rss1p) and Mex67p, were identified by their genetic interactions with the nucleoporins Nup100p and Nup85p, respectively, and are located at the NPC (188–190). Mutation of either of these proteins can cause the accumulation of polyA in yeast nuclei, an assay that was developed specifically to allow identification of yeast genes involved in some aspect of mRNA export (e.g. see 191–193). This assay is extremely useful, although its major weakness is that polyA does not necessarily equate with mRNA. Conditional (temperature-sensitive) mutants of both Gle1p and Mex67p were used to show that polyA accumulation, that is, the presumed mRNA export defect, occurred rapidly after inactivation of the proteins (188–190).

Further evidence that Mex67p could be involved directly in RNA export came from the observation that the protein can be cross-linked to RNA in vivo and that it interacts with RNA-binding proteins (190). Both Mex67p and Gle1p contain sequences that resemble a leucine-rich NES (188, 190), although these putative NESs are not conserved in their vertebrate homologues (190; L Englmeier, unpublished data). Nevertheless, tested as short peptides, the NES-like sequences from both Gle1p and Mex67p do signal export on microinjection into vertebrate cell nuclei (188, 190). Moreover, deletion of these peptides or point mutations in the NES-like sequences gives rise to either nonfunctional or only conditionally functional proteins, respectively, with consequent accumulation of polyA in the nucleus. This observation suggests that NESs may be involved in yeast mRNA export. To confirm this conclusion, it must be shown that the sequences function as NESs in the context of either Gle1p or Mex67p in yeast cells, and that this function is required for mRNA export.

The hnRNP family of proteins is implicated in mRNA export (157, 194). HnRNP proteins associate with polyA-containing RNA in the nucleus. There are roughly 20 proteins in the human hnRNP family, although neither the size of the family, nor the sequences of individual members, are highly conserved in evolution. Since hnRNP proteins are located in mammalian cell nuclei at steady state, they were not initially considered as potential export mediators. The demonstration that some hnRNP proteins shuttle continuously between the nucleus and cytoplasm (195, 196) changed this perception. The best-studied member of this protein family to date is hnRNP A1. Both the number of hnRNP A1 molecules per HeLa cell nucleus (10^8) and the amount of A1 shuttling (10^5 molecules per min) is enormous (197). As noted above, the *C. tentans* hnRNP A1 homologue leaves the nucleus bound to the BR mRNA (145). These results suggested a role for hnRNP A1 in mRNA export.

This proposal was further strengthened when a region of hnRNP A1, the M9 domain, was found not only to direct nuclear import of the protein (Tables 1 and 3) but also to be sufficient for nuclear export (15). Note that a different shuttling signal has since been identified in hnRNP K (Tables 1 and 3), which must also be considered a candidate mRNA export mediator (78). Microinjection of saturating amounts of hnRNP A1 into *Xenopus* oocyte nuclei blocked export of some mRNAs (198; see also 97, 172). The export block presumably reflects titration of an export mediator by hnRNP A1, and the variation between RNAs may be explained by the fact that different mRNAs preferentially bind different hnRNP proteins (199, 200). Some mRNAs might, for example, have sufficient hnRNP K bound not to need the factor that recognizes hnRNP A1 for their export.

Evidence suggests that the export mediator titrated by hnRNP A1 might be distinct from transportin, the M9 import receptor (Table 1). A mutation of the M9 domain that prevents its recognition by transportin and blocks hnRNP A1 nuclear import (15, 72, 198) does not abrogate the effect of saturating levels of hnRNP A1 on mRNA export, although deletion of the whole M9 domain does so (198). Further, transportin is dissociated from M9-containing proteins by RanGTP (77a, 97). RanGTP is found at high concentration in the nucleus; therefore, it might prevent transportin-M9 interaction from occurring there. Definitive proof that transportin is not involved in mRNA export will require identification of the export receptor that recognizes hnRNP A1.

Yeast also contains hnRNP-like proteins: Npl3p/Nop3p/Nab1p (201–205), Nab2p (205), Nab3p (201), and Hrp1p (206). Npl3p is the most extensively studied. It bears a distant resemblance to hnRNP A1, as does Hrp1p, and its mutation results in polyA accumulation in the nucleus (202, 204). Like hnRNP A1, Npl3p shuttles between the nucleus and the cytoplasm (203, 207). These characteristics indicate a level of functional homology between Npl3p and hnRNP A1; however, critical differences exist. First, although hnRNP A1 export is transcription independent (195), movement of Npl3p to the cytoplasm requires ongoing transcription and is blocked by mutation in an RNA-binding domain of Npl3p (207). This observation might suggest that Npl3p is carried to the cytoplasm by RNA rather than the reverse, although Npl3p might carry an export signal that is exposed only on RNA binding. Further, mutants of Npl3p prevent NLS-mediated protein import (208), whereas hnRNP A1 has no obvious connection to this import pathway. The implication of this finding for Npl3p function is not obvious. Again, as for hnRNP A1, identification of the hnRNP export receptors in yeast will help to resolve Npl3p's functional role and further clarify whether the apparent differences with hnRNP A1 are significant.

Although the earliest RNA export studies were with tRNA, the export of this RNA is still not well-understood. Evidence suggests that tRNA export in vertebrates exhibits differences from that of other RNA species, not only in terms of the saturable export mediators with which it interacts but also in relation to NPC proteins involved in its export (161, 209, 210; but see below). However, no proteins specifically involved in tRNA export have been identified. Yeast genetic studies have demonstrated interaction between the nucleoporin Nsp1p and two proteins that are involved in tRNA production (211). The proteins are a pseudouridine synthase (Pus1p) that modifies certain uridine nucleotides in tRNAs to pseudouridine and Los1p. Los1p localizes to the NPC (211) and had previously been implicated in pre-tRNA splicing (212). Los1p is distantly related to importin β (Table 2; 70, 71). These facts suggest that Los1p could be involved in tRNA export and, more generally, that tRNA maturation and export could be coupled events.

The last part of this section is devoted to CBC, the nuclear cap-binding complex that functions in U snRNA export. This heterodimeric complex, composed of the CBP80 and CBP20 proteins (163, 213–217), binds to capped nuclear RNAs. CBC translocates through the NPC attached to the 5' end of BR mRNPs (146). Since these mRNPs always traverse the pore 5' end first, CBC might have a role in orienting the RNPs for efficient translocation. However, it does not appear that the role of CBC in mRNA export is more than a minor one (161, 215).

In contrast, preventing the interaction of CBC with newly synthesized U snRNAs disrupts their nuclear export (154, 161, 163, 215) and thus prevents their assembly with U snRNP proteins in the cytoplasm (Figure 1). A likely explanation for the difference in the relative importance of the role of CBC for U snRNA and mRNA export is that additional proteins that bind to the latter, such as the hnRNP proteins, are sufficient to allow their export even in the absence of CBC. CBC is multifunctional, with roles in pre-mRNA processing as well as in RNA export (214, 215, 218, 219). Although CBC is also found in yeast (220–222) and functions in pre-mRNA processing there, no evidence exists to indicate that yeast CBC plays a direct role in U snRNA or mRNA export.

Export Receptors, the Exportins

The previous section discussed adaptors that we proposed would mediate interaction between RNAs and "real" export receptors. Two such receptors have just been identified, although only one, CRM1, has thus far been shown to mediate RNP export. The second, CAS1, was initially identified as the human homologue of an *S. cerevisiae* gene, *CSE1*, whose mutation caused abnormal chromosome segregation in mitosis (223, 224). Mutants of *S. pombe* CRM1, in contrast, caused abnormal chromosome morphology (225). These phenotypes underline the previously discussed problem of identifying transport factors on the basis of the visible defects they cause when mutated. Clearly, blocking nucleocytoplasmic transport of a class of macromolecules can affect virtually any aspect of cell growth and morphology.

Other information on the CRM1 and CAS proteins provided more direct evidence of their function. Both are distantly related to importin β, particularly in its N-terminal Ran-binding region (70, 71). Mammalian CRM1 had furthermore been identified as an NPC-associated protein that binds directly to at least one nucleoporin, CAN/Nup214 (70). Therefore, there were good reasons to expect a functional relationship between the proteins and importin β.

This turned out to be the case. CAS binds importin α in a RanGTP-dependent way (226), that is, under conditions thought to exist in the nucleus. An in vitro assay showed that importin α re-export from the nucleus depends on the presence of CAS (226). Although the region of importin α that interacts with CAS is not defined, previous work had shown that the N-terminal importin β–binding (IBB) domain was sufficient to direct nuclear import but not export (54, 55), indicating that CAS binding will require other regions of importin α.

CRM1 also binds to the substrates it transports cooperatively with RanGTP (227). These substrates are proteins like Rev and PKI that carry a leucine-rich NES (Table 3) (227–229). Leptomycin B was identified as a potential fungicide and later shown to be an inhibitor of CRM1 function (227, 230, 231). As expected from the role of CRM1 as an export receptor for NES proteins, leptomycin B inhibits Rev export from the nucleus (227, 232). Consistent with the fact that saturation of the NES pathway also blocks U snRNA export, leptomycin β also inhibits export of these RNAs in *Xenopus* oocytes (227), although it is unclear whether CRM1 interacts directly with CBC, the export adaptor of these RNAs.

One mutant form of *S. cerevisiae* Crm1p blocks mRNA export (229), which recalls our prior discussion of the role of proteins containing NES-like sequences in mRNA export in yeast and somewhat strengthens the argument that leucine-rich NESs may be involved in mRNA transport in this organism. Some caution is required, however. Mutation of importin β, an import receptor, can give rise to dominant negative effects that block transport of many substrates whose import or export does not require importin β (233). At least one mutation in human CRM1 (one that changes only two amino acids of the protein) blocks mRNA export when expressed in *Xenopus* oocytes even though CRM1 is not required for this form of export (M Ohno, unpublished data). Furthermore, several additional yeast Crm1p mutants that block Rev NES-mediated transport in *S. cerevisiae* do not affect mRNA export (234).

Proteins other than CRM1 had previously been found to interact with the Rev NES. Might they also have a role in export? One of these proteins was eukaryotic initiation factor (eIF)5A (235, 236). This protein was thought to be involved in translation initiation, but its role was poorly defined (see 235). We must admit to skepticism about the involvement of eIF5A in Rev function. First, it was identified by a five-step procedure that would require co-migration on anion exchange chromatography, (isoelectric) chromatofocusing, and two-dimensional gel electrophoresis of the native 16.7-kDa eIF5A protein and of the protein after its cross-linking to a Rev NES peptide of 2 kDa in mass and net charge of -3. Furthermore, the most direct experimental evidence for a role of eIF5A in Rev function was based on the authors' contention that overexpression of eIF5A was required for Rev to show export activity in *Xenopus* oocytes (235). This was later shown not to be the case (172, 177).

The second pair of candidate Rev NES-binding proteins were yeast Rip1p and the human Rip/Rab protein, which are distantly related in sequence (237–239). The proposed interaction between these proteins and the Rev NES was based almost entirely on the results of two-hybrid screens. As previously discussed (227), the Rip proteins are related to nucleoporins and in particular bear a striking resemblance to CAN/Nup214, the protein with which mammalian CRM1 interacts in vivo (70). The resemblance with Rip1p/Rip/Rab is in the domain of CAN/Nup214 required for CRM1 binding (240). It was therefore proposed that the two-hybrid data might be explicable if endogenous yeast Crm1p mediated the interaction between the Rev NES and Rip proteins. Indeed, no two-hybrid interaction was seen between Rip1p and the Rev NES in yeast strains carrying mutant CRM1 genes (234). This observation strongly supports the view that Crm1p mediates NES-Rip1p interaction. Thus, as suggested by the original studies (237–239), Rip1p may well have a function in Rev NES export, but this function will involve interaction with Crm1p.

We end this section with a short description of three additional importin β–related proteins for which there is genetic evidence of a role in nuclear transport (Table 2). Los1p (211) was already discussed, and the indirect evidence for its possible role in tRNA export described. Los1p is unlikely to be the only import or export receptor for specific substrates, because it is not an essential gene (211, 212).

The proposal (87, 88) that Pse1p and Yrb4p/Kap123p may be redundant import receptors for ribosomal proteins (Table 2) is also a topic treated previously. Surprisingly, combination of a YRB4/KAP123 gene deletion that is viable with a conditional allele of PSE1 leads to polyA accumulation in the nucleus in nonpermissive conditions, that is, presumably to a defect in mRNA export (241). This result may indicate additional functions for either of these importin β relatives. The alternative explanation raised earlier in relation to CRM1, that is, the possibility of dominant effects of the mutant Pse1p protein on transport events not directly mediated by Pse1p, also applies here. Nevertheless, further study of the basis of these effects is clearly of interest. There was no defect in NLS protein or hnRNP protein import in these strains (241), consistent with the conclusion that the yeast importin and transportin homologues are their required import receptors (Table 1).

Possession of the doubly mutant yeast strain allowed the identification of Sxm1p as a high-copy suppressor of the phenotype caused by combination of the conditional Pse1p mutant and the deletion of YRB4/KAP123 (241). Sxm1p could also rescue the lethality caused by PSE1 deletion (241). This is evidence for a functional relationship between Sxm1p and Pse1p. Like *YRB4/KAP123*, *SXM1* is not an essential gene (241).

Finally, Mtr10p was identified by a mutation that caused polyA accumulation in the nucleus (193). Further study revealed abnormal 18S rRNA production in the mutant strain, indicative of pleiotropic defects (193). Nevertheless, the correlation between relatedness to importin β and function in nucleocytoplasmic transport (Table 2) means that Mtr10p and the other importin β family members (70, 71) are valuable objects of further study.

ASYMMETRY AND RECYCLING

Ran and the Real Difference Between the Nucleus and the Cytoplasm

We now discuss the interaction between Ran and exportins, emphasizing the likely importance of this interaction for our understanding of nucleocytoplasmic transport. Two previously discussed points bear repeating here: First, RanGTP interacts with at least three import receptor-substrate complexes and causes their dissociation (88, 97, 102, 103, 105, 107; Figure 2); and second, high nuclear and low cytoplasmic RanGTP concentrations are predicted (98, 99). These points make obvious the significance of the observation that the two identified export receptors, CAS and CRM1, bind tightly to their substrates only through cooperative binding involving both the substrates and RanGTP (Figure 2). Export complexes form in the nucleus in the presence of RanGTP. Their dissociation in the cytoplasm is likely to involve the hydrolysis of GTP by the receptor-bound Ran under the influence of the RanBP1, RanBP2, and RanGAP1 proteins that are either cytosolic or associated with the cytoplasmic face of the NPC (99, 107, 226, 242). In contrast, import complexes are stable only in the absence of receptor-bound RanGTP, that is, in the cytoplasm, and they dissociate in its presence (Figure 2). Therefore, the differential RanGTP concentrations in the nucleus and cytoplasm are likely to be a fundamental cause of directionality in receptor-mediated nucleocytoplasmic transport events and, thus, of the difference between the macromolecular composition of the nucleus and the cytoplasm. Ran is unlikely to be the only determinant of transport asymmetry; other contributions to directionality are likely to include those of the structurally asymmetric NPC.

Although the nuclear and cytoplasmic concentrations of RanGTP are unknown, strong indirect evidence indicates the importance of the RanGTP concentration gradient across the NPC. The studies reviewed here focus on vertebrate systems, but studies in yeast (see 98, 243) have also contributed greatly to this conclusion.

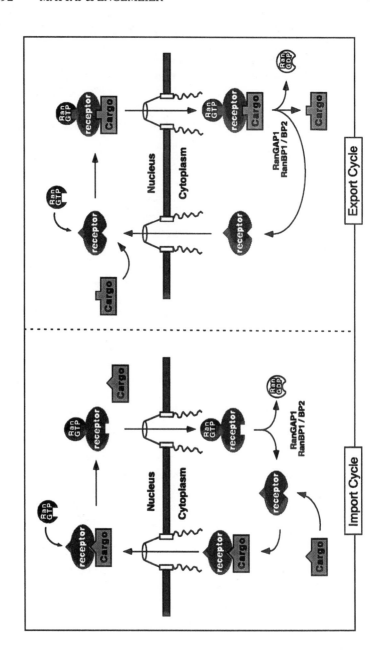

A first line of evidence has come from the use of the tsBN2 Chinese hamster ovary cell line. In these cells RCC1, the nuclear Ran guanine nucleotide exchange factor, is temperature sensitive and unstable at the nonpermissive temperature (244). The loss of RCC1 protein at high temperature will result in reduced regeneration of RanGTP from RanGDP in the nucleus and also results in relocation of Ran to the cytoplasm. As a consequence, tsBN2 cells at the nonpermissive temperature are defective in NLS-protein import (245), in the export of polyA-containing (m)RNA and U snRNAs (210, 246), and in ribosomal RNA production, perhaps as a result of defective ribosomal protein import (210). Some forms of transport, probably those that are either most rapid or have the lowest requirement for RanGTP, and are thus most difficult to inhibit by RanGTP depletion, still go on in these cells. Examples are tRNA export and Rev export (96, 210). However, Rev export is inhibited by further reduction of nuclear RanGTP in the nuclei of these cells induced by direct injection of RanGAP1 (96).

Similarly, a series of experiments designed to decrease nuclear RanGTP concentration in the nuclei of *Xenopus* oocytes caused a direct block of all tested forms of nuclear export of proteins and RNPs, including those containing tRNA (97). That the effect was direct, and not an indirect consequence of a block of import of adaptor proteins, was shown by the fact that export could be inhibited in conditions that did not block protein import (97). The conclusion of these studies is that high nuclear RanGTP concentration is critical for all forms of nucleocytoplasmic transport examined. The one apparent exception, involving heat shock, is discussed later.

The vesicular stomatitis virus M protein, when injected into or expressed in vertebrate cells (247), results in a complex phenotype similar to that seen in tsBN2 cells at the nonpermissive temperature or in *Xenopus* oocytes into whose nuclei high concentrations of RanGTP in a nonhydrolyzable form have been injected (97, 247). If the hypothesis (247) can be confirmed that M protein directly affects some aspect of the RanGTP cycle, it will be an extremely useful tool in the study of Ran function.

←——

Figure 2   Simplified scheme of the interactions between import and export receptors, their cargo, and RanGTP. RanGTP causes assembly of export complexes and disassembly of import complexes in the nucleus. The cooperative action of either cytosolic RanBP1 and RanGAP1 or NPC-associated RanBP2 and RanGAP1 combine to cause hydrolysis of Ran-bound GTP and thus release export cargo or regenerate import receptor in the active conformation. In many cases, adaptors (such as importin α) are required to mediate receptor-substrate interaction. For simplicity's sake we have considered these adaptors as part of the cargo. Note, however, that like the receptors these adaptors must also be recycled to their original starting compartment. This involves receptor-mediated transport. See text for references.

If the RanGTP distribution across the NPC is so important, how is it set up and maintained through mitotic or meiotic division? Here, little concrete evidence has been presented, but it seems probable that the tight association of RCC1 with chromosomes (100, 244) may be critical. Since chromosomes are first separately surrounded by nuclear envelopes at telophase to form individual mini-nuclei before fusing to form the nucleus (2), chromosome-associated RCC1 might be incorporated directly into nuclei at the first assembly stage, leaving RanBP1 and RanGAP1 in the cytoplasm and thus automatically leading to the re-establishment of the RanGTP gradient.

An interesting speculation arises from the differential effects of RanGTP on the binding between import and export receptors and their substrates (Figure 2). In theory, a single receptor could bind to an import substrate in the absence of RanGTP, that is, in the cytoplasm, and release it into the nucleoplasm on binding RanGTP. At the same time, RanGTP could induce a conformation in the receptor that would allow association with an export substrate, which would be carried to the cytoplasm and released on hydrolysis of the Ran-bound GTP. Thus a single receptor could function both as an import and an export receptor. There seems no obvious logical reason why this should not occur, even if there are currently no examples of such behavior.

Recycling of Adaptors and Receptors

We have discussed both import and export as unidirectional events. In order to function, however, both adaptors and receptors have to be recycled in the empty state to the compartment (nucleus or cytoplasm) where they will pick up a new cargo molecule. Current evidence suggests that adaptors function as adaptors in one direction and as substrates in the other. For example, importin α mediates interaction between NLS proteins and importin β during import. For export, importin α binds to CAS in a RanGTP-dependent way (226). CAS prefers to bind importin α in the absence of a bound NLS (226). Whether this, and the lower affinity of importin α for the NLS after separation from β (102–105), is enough to explain α-NLS dissociation in the nucleus remains to be seen. An active dissociation mechanism might be required. In any event, CAS is the receptor that recycles the "empty" importin α adaptor back to the cytoplasm.

HnRNP A1, proposed to mediate mRNA export, can be imagined in a similar way. It leaves the nucleus bound to mRNA (145, 195, 196) and after dissociation is recycled to the nucleus via transportin-mediated import (72, 73). The CBC cycle is both more complex than this and economical in a rather elegant way. CBC leaves the nucleus bound to capped RNA (146; M Ohno, unpublished data). This export is mediated by CRM1 (227), and CBC is presumably bound either directly or indirectly to CRM1. The association between CBC and CRM1 is thought to be broken in the cytoplasm via hydrolysis of the GTP on the CRM1-bound Ran (226, 227). However, CBC also makes an unusual interaction with importin. CBC, through its NLS-containing CBP80 component, remains stably bound to importin α in the nucleus rather than dissociating like other NLS proteins (220). Thus the CBC-U snRNA-CRM1 export complex also contains importin α. When this complex leaves the nucleus and, thus, the region of high RanGTP concentration, importin β can interact with the CBC-bound α molecule. This interaction simultaneously causes CBC to release the bound RNA (220) and targets CBC for import back into the nucleus by the normal importin α/β-mediated mechanism (248). In this way, two receptors, CRM1 and importin β, combine to ensure the directionality of CBC-mediated U snRNA export.

On purely logical grounds, the adaptor-recycling mechanism seems extremely unlikely to apply to the receptors, the importin β family, themselves. The reason for this is simple: If each molecule of importin β that moves into the nucleus needed a receptor, say CRM1, to carry it back out, then CRM1 would need an import receptor (X) to carry it back in, and so on in a never-ending cycle of receptors in one direction turning into cargo in the other. This scenario is clearly absurd, as an infinite number of transport events and an infinite amount of energy would be needed to import a single NLS protein. Although an NES-like sequence in yeast importin β is required for its function (249), we consider it very unlikely that this sequence will direct CRM1-dependent export. Rather, the export of importin β and the import of CRM1, as examples, should follow a different principle. It seems most likely that the receptors, devoid of adaptors and transport substrates, will be recycled directly via interactions with the NPC.

In a formal sense, receptor recycling is the reverse of the functional import or export step but in the absence of bound cargo (adaptors plus substrate). Thus it represents a form of the next critical problem that needs to be solved in the field of nucleocytoplasmic transport, the mechanism by which NPC translocation occurs. Although many interactions between importin β family members and individual nucleoporins have been reported (47, 70, 76, 87, 103, 250, 251), little evidence exists for the functional significance of most of these (discussed in 13, 251).

Ran is also involved in the control of receptor-nucleoporin interactions. RanGTP binding to importin β dissociates nucleoporin–importin β complexes in vitro (103). In addition, a mutant form of importin β from which the RanGTP-binding site is deleted is capable of NPC translocation but appears to bind very tightly to specific sites on the nuclear side of the NPC (102). This mutant is a dominant inhibitor of all tested forms of active nucleocytoplasmic transport (233). This shows that importin β can be translocated through the NPC without bound Ran and suggests that RanGTP is required for clearance of importin β from a critical site or sites that it occupies once nucleus-directed translocation is complete. The dominant effect on other transport events could indicate that these sites are commonly used by different transport receptors, but other explanations are also possible (discussed in 233). Perhaps the most interesting possibility is the idea that the site to which the mutant form of β binds is some kind of "checkpoint," whose occupancy would in some way close down the pore for the transport of other receptor complexes. If true, this might reflect a mechanism to prevent traffic jams in the pore (233).

It is very likely that RanGTP binding to importin β at the sites occupied by the mutant is not only the last (or termination) step of NLS-protein import but also the first step in importin β re-export (102). Thus importin β would be expected to exit the nucleus bound to RanGTP. Conversely, export receptors would be expected to exit the nucleus bound to both cargo and RanGTP, but would return to the nucleus without either. Further details on the mechanism of these recycling processes should be forthcoming.

It is possible that the NPC clearance or termination step of importin-mediated transport is the only step of the process that requires RanGTP, in other words, that RanGTP is needed to dissociate importin β from the inner side of the NPC and from importin α. Similarly, there is no definitive argument against the possibility that GTP hydrolysis by Ran is required only to prevent the build-up of cytoplasmic RanGTP, which would prevent interaction between importin β and the α-NLS protein complex. Thus more data are required to prove that NPC translocation per se is an energy-requiring process.

A mutant form of Ran that cannot hydrolyze GTP is able to support several forms of nuclear export (97). This observation again raises the possibility that NPC translocation is not obligatorily coupled to hydrolysis of GTP by Ran. This does not mean that we hold the view that translocation will be found to be uncoupled from the provision of energy. It seems unlikely that the aqueous NPC channel could be opened from 9 to 25 nm without energy input. Similarly, transport of large substrates such as the BR RNPs seems on the one hand inconceivable without energy and on the other not fundamentally different from transport of smaller mRNP substrates. Where and how energy is needed for translocation is a critical area for further study. It may be that receptor recycling, since it involves simpler transport substrates, is the experimental system of choice to answer these questions.

REGULATION OF NUCLEOCYTOPLASMIC TRANSPORT

This section describes examples of regulated nuclear transport, involving both protein and RNA substrates. The examples, while not comprehensive, illustrate themes in transport regulation, such as changes in transport that are responses to growth state, extracellular signals, developmental stage, or stress. Furthermore, they illustrate the theme that the many possible targets of regulation in the transport machinery are all likely to be used in one case or the other. Feldherr and colleagues (reviewed in 252) have observed that cells have different nuclear "import capacities" when they are in different states. This observation reveals itself in two ways: First, rates of import change; and second, when import substrates are coupled to colloidal gold particles of different sizes, the ability of cells to import large- versus small-sized conjugates varies. For example, proliferating cells import faster and have a larger apparent active NPC diameter than do quiescent cells (253). Similar differences are seen between transformed and nontransformed cells or between cells immediately after mitosis and at later times in the cell cycle (252). Although the molecular basis of these effects is still unknown, it is hard to see how they could be explained without proposing changes in the NPC itself. Further progress in understanding this global regulation is eagerly awaited.

At a more specific level, the nuclear accumulation of many proteins, including many transcription regulators, is controlled in response to signals that result in a change in their phosphorylation. One of the first examples discovered was the yeast Swi5p protein, which helps control expression of the endonuclease responsible for mating-type switching, a prerequisite for sexual reproduction in *S. cerevisiae*. Swi5p is nuclear in G1 but cytoplasmic at other times in the cell cycle. A 50-amino-acid fragment from Swi5p confers this regulated import behavior on a reporter protein (254). The 50 amino acids contain a bipartite basic NLS and two serine residues whose phosphorylation regulates the activity of the NLS. Their mutation to alanine results in a constitutively nuclear protein (254). Thus the activity of the Swi5p NLS is negatively regulated by the phosphorylation of critical serine residues.

A similar vertebrate example is the NF-AT family of proteins, involved in regulating genes involved early in mounting an immune response. These proteins are competent for both DNA binding and transcriptional activation when isolated from cytosol and, thus, appear to depend on control of nuclear import to regulate their activity (255). NF-AT activation occurs as a result of signal-mediated changes in calcium concentration. Elevated intracellular calcium ion concentration activates calcineurin, a phosphatase that dephosphorylates critical serine residues in the NF-AT proteins. Dephosphorylation exposes two basic NLSs, allowing nuclear accumulation and target gene activation (256). An interesting facet of this regulation is that calcineurin enters the nucleus with NF-ATs and must continuously dephosphorylate the proteins, antagonizing a nuclear kinase, to maintain their nuclear localization (257, 258). Thus, phosphorylated NF-ATs must be substrates for (phosphate-regulated) nuclear export.

The best-characterized example of regulated protein export concerns yeast Pho4p, a transcription activator involved in turning on genes involved in phosphate metabolism. These genes are switched off rapidly when yeast cells are exposed to extracellular phosphate. The nuclear localization of Pho4p is regulated by the cyclin/cyclin-dependent kinase pair Pho80p-Pho85p (259) as part of this phosphate response. Phosphorylated Pho4p is cytoplasmic and therefore inactive as a transcription activator, whereas the dephosphorylated, functional form of the protein is nuclear. It was recently shown that the Pho80p-Pho85p kinase is constitutively nuclear, and that Pho4p is nuclear until Pho80p-Pho85p is activated in response to extracellular phosphate, at which time Pho4p is phosphorylated and rapidly exported (E O'Shea, personal communication).

Another common strategy for controlling the nuclear import of transcription factors is to regulate interactions with inhibitory factors that cause their cytoplasmic retention. Examples are the *Drosophila* heat shock transcription factor (260), the mammalian glucocorticoid receptor (261) and STAT factors (262, 263). The best-characterized case is NFκB, which is ubiquitously expressed but seems to have a particularly significant role in the immune system. In unstimulated cells NFκB is present in trimeric cytoplasmic complexes like the one that consists of the p50 and p65 subunits of NFκB bound to IκBα. I stands for inhibitor, and IκB prevents nuclear entry of the active heterodimeric NFκB by occluding its NLS (264). One pathway of NFκB activation involves tyrosine phosphorylation of IκB. This is sufficient to cause IκB to dissociate from NFκB and to allow NFκB nuclear translocation (265). The more commonly used pathway is more complex. It involves phosphorylation of two serine residues on IκB (266, 267). This targets IκB degradation by proteasome-mediated hydrolysis (268–271), releasing NFκB for nuclear entry.

Transport regulation is not confined to proteins, and examples of RNA export regulation have been described. Several viruses have evolved mechanisms that allow selective export of viral mRNAs (reviewed in 272), but for a long time the only example of regulated cellular mRNA export was that of maternal histone mRNAs in certain sea urchin species (273, 274). In this case, maternal histone messages are stored in the nuclei of the eggs and released shortly before maturation to allow the bulk histone protein synthesis required during the first, very short, cleavage-stage cell cycles of embryo development. The basis of this retention and release mechanism has not been studied.

Recently, a second example of regulated RNA export has been discovered that points to the existence of a transport mechanism quite different from any of those discussed here so far. The regulated mRNAs are those encoding the *S. cerevisiae* heat shock proteins. Despite their name, these mRNAs are not only synthesized in response to excessive heat but also in response to signals induced by multiple forms of cellular stress. The discussion that follows is relevant to the stress response in general, not just to heat shock. We focus on the effect of stringent heat shock, for *S. cerevisiae* at 42°C, as the situation appears different after a milder heat shock at 37°C (165, 275).

The basic observation on which this work was based was the finding that heat shock (hs) prevents bulk polyA (i.e. mRNA) export from the nucleus (276). Cole and colleagues realized, however, that since induction of some hs genes involved turning on their transcription, the newly made hs mRNAs must be exported to allow hs protein synthesis (165). By comparing the export of bulk polyA and specific hs mRNAs they proved that hs mRNA was indeed exported in stress conditions (165). The specific mRNA they examined, encoded by the *SSA4* gene, carries at least two different nonoverlapping *cis*-acting signals for export under stress conditions. The signals were defined by fusing them to reporter mRNAs and demonstrating export of the fusion transcripts after heat shock (165). The conclusion of this experiment was that hs mRNAs might bind different proteins than bulk polyA-containing RNA.

Further evidence for this conclusion came from studying cellular factors required for hs mRNA export. As discussed above, mutants of the hnRNP-like Npl3p protein are dominant inhibitors of polyA RNA export in yeast (reviewed in 277); however, hs mRNA export was normal in *npl3* mutant strains (275). This observation might indicate that Npl3p does not interact with hs mRNAs. An even greater surprise came from analysis of hs mRNA export in strains mutant in the Gsp1p (Ran), Prp20p (RanGEF), or Rna1p (RanGAP) components of the yeast Ran system. As described above, these three components of the Ran system are essential for all other studied forms of nuclear import and export. In contrast, hs mRNA export was not affected in these mutant strains (165, 275), suggesting a fundamental difference between hs mRNA export and the better understood nucleocytoplasmic transport pathways described in this review.

A first glimpse of what is required for hs mRNA export came from the observation that Rip1p, the nonessential nucleoporin previously implicated in Rev-NES-mediated nuclear export (237), is required for hs mRNA export (275, 278). The relationship between hs mRNA export and NES-dependent export is still unclear (275, 278), but further studies of this fascinating form of transport are to be expected. In our opinion, the major question is, What substitutes for Ran in this export pathway?

PERSPECTIVES

The field of nucleocytoplasmic transport is booming. We expect further rapid progress in identification of the functions of the new import and export receptors that are members of the importin β family. A further expansion in the collection of well-defined import and export signals is also expected. Further examination of energy consumption during different import and export processes, using in vitro transport systems and yeast genetics, is another area with open questions and experimental possibilities. Comparison of conventional transport pathways with unusual examples such as the heat shock mRNAs should be productive. The specific roles of the different Ran-binding proteins that are not related to importin β also need to be studied further. We expect such studies to lead to the discovery of Ran functions that are not related directly to nucleocytoplasmic transport. Some transport substrates, specifically mRNPs, may not leave the nucleus isometrically, but rather might be exported out of the nucleus in a directional way (reviewed in 279). If this is true, transport could be studied in terms of nuclear architecture. The atomic structures of both p10/NTF2 (280) and Ran in its GDP-bound state (281) have been elucidated, but several fascinating problems of protein structure that relate to nucleocytoplasmic transport remain unsolved. These problems include how proteins such as importin α, transportin, or CRM1 can accurately bind to such diverse, but seemingly simple, transport signals, and the nature of the conformational changes induced by RanGTP binding to transport receptors.

The critical current question in the field, in our opinion, concerns the NPC translocation step of import and export. We have not discussed this topic in depth in this review because less is known about translocation than about other aspects of nucleocytoplasmic transport. Nevertheless, we are confident that the translocation mechanism will be elucidated through the enormous effort of the various scientists investigating this area.

ACKNOWLEDGMENTS

We wish to thank the many colleagues who helped in the preparation of this review by providing us with unpublished information. We also thank members of our laboratory, specifically Isabel Palacios, Puri Fortes, Martin Hetzer, Maarten Fornerod, and Mutsuhito Ohno for critical comments on the manuscript and Bertrand Séraphin and Maarten Fornerod for the figures.

> Visit the *Annual Reviews home page* at
> http://www.AnnualReviews.org.

Literature Cited

1. Görlich D, Mattaj IW. 1996. *Science* 271: 1513–18
2. Wilson KL, Wiese C. 1996. *Semin. Cell Dev. Biol.* 7:487–96
3. Reichelt R, Holzenburg A, Buhle EL Jr, Jarnik M, Engel A, Aebi U. 1990. *J. Cell Biol.* 110:883–94
4. Rout MP, Blobel G. 1993. *J. Cell Biol.* 123:771–83
5. Daneholt B. 1997. *Cell* 88:585–88
6. Feldherr CM, Kallenbach E, Schultz N. 1984. *J. Cell Biol.* 99:2216–22
7. Forbes DJ. 1992. *Annu. Rev. Cell Biol.* 8:495–527

8. Davis LI. 1995. *Annu. Rev. Biochem.* 64: 865–96
9. Panté N, Aebi U. 1993. *J. Cell Biol.* 122: 977–84
10. Doye V, Hurt E. 1997. *Curr. Opin. Cell Biol.* 9:401–11
11. Rout MP, Wente SR. 1994. *Trends Cell Biol.* 4:357–65
12. Macaulay C, Forbes DJ. 1996. *Semin. Cell Dev. Biol.* 7:475–86
13. Ohno M, Fornerod M, Mattaj IW. 1998. *Cell.* 92:327–36
14. Dingwall C, Laskey RA. 1986. *Annu. Rev. Cell Biol.* 2:367–90
15. Michael WM, Choi M, Dreyfuss G. 1995. *Cell* 83:415–22
16. Shulga N, Roberts P, Gu ZY, Spitz L, Tabb MM, et al. 1996. *J. Cell Biol.* 135:329–39
17. Bonner WM. 1975. *J. Cell Biol.* 64:421–30
18. De Robertis EM, Longthorne RF, Gurdon JB. 1978. *Nature* 272:254–56
19. Dingwall C, Sharnick SV, Laskey RA. 1982. *Cell* 30:449–58
20. Kalderon D, Roberts BL, Richardson WD, Smith AE. 1984. *Cell* 39:499–509
21. Robbins J, Dilworth SM, Laskey RA, Dingwall C. 1991. *Cell* 64:615–23
22. Goldfarb DS, Gariepy J, Schoolnik G, Kornberg RD. 1986. *Nature* 322:641–44
23. Zasloff M. 1983. *Proc. Natl. Acad. Sci. USA* 80:6436–40
24. Newmeyer DD, Lucocq JM, Burglin TR, De Robertis EM. 1986. *EMBO J.* 5:501–10
25. Michaud N, Goldfarb DS. 1993. *Exp. Cell Res.* 208:128–36
26. Michaud N, Goldfarb DS. 1991. *J. Cell Biol.* 112:215–23
27. Mattaj IW. 1988. In *Structure and Function of Major and Minor Small Nuclear Ribonucleoprotein Particles*, ed. ML Birnstiel, pp. 100–14. Berlin: Springer-Verlag
28. Fischer U, Darzynkiewicz E, Tahara SM, Dathan NA, Lührmann R, Mattaj IW. 1991. *J. Cell Biol.* 113:705–14
29. Pruschy M, Ju Y, Spitz L, Carafoli E, Goldfarb DS. 1994. *J. Cell Biol.* 127: 1527–36
30. Powers MA, Forbes DJ. 1994. *Cell* 79: 931–34
31. Newmeyer DD, Finlay DR, Forbes DJ. 1986. *J. Cell Biol.* 103:2091–102
32. Adam SA, Marr RS, Gerace L. 1990. *J. Cell Biol.* 111:807–16
33. Adam SA, Lobl TJ, Mitchell MA, Gerace L. 1989. *Nature* 337:276–79
34. Adam SA, Gerace L. 1991. *Cell* 66:837–47

35. Adam EJH, Adam SA. 1994. *J. Cell Biol.* 125:547–55
36. Görlich D, Prehn S, Laskey RA, Hartmann E. 1994. *Cell* 79:767–78
37. Yano R, Oakes M, Yamaghishi M, Dodd JA, Nomura M. 1992. *Mol. Cell Biol.* 12:5640–51
38. Enenkel C, Blobel G, Rexach M. 1995. *J. Biol. Chem.* 270:16499–502
39. Imamoto N, Shimamoto T, Takao T, Tachibana T, Kose S, et al. 1995. *EMBO J.* 14:3617–26
40. Moroianu J, Blobel G, Radu A. 1995. *Proc. Natl. Acad. Sci. USA* 92:2008–11
41. Weis K, Mattaj IW, Lamond AI. 1995. *Science* 268:1049–53
42. Török I, Strand D, Schmitt R, Tick G, Török T, et al. 1995. *J. Cell Biol.* 129: 1473–89
43. Küssel P, Frasch M. 1995. *J. Cell Biol.* 129:1491–1507
44. Nigg EA. 1997. *Nature* 386:779–87
45. Görlich D, Kostka S, Kraft R, Dingwall C, Laskey RA, et al. 1995. *Curr. Biol.* 5:383–92
46. Imamoto N, Shimamoto T, Kose S, Takao T, Tachibana T, et al. 1995. *FEBS Lett.* 368:415–19
47. Radu A, Blobel G, Moore MS. 1995. *Proc. Natl. Acad. Sci. USA* 92:1769–73
48. Chi NC, Adam EJ, Adam SA. 1995. *J. Cell Biol.* 130:265–74
49. Richardson WD, Mills AD, Dilworth SM, Laskey RA, Dingwall C. 1988. *Cell* 52:655–64
50. Newmeyer DD, Forbes DJ. 1988. *Cell* 52:641–53
51. Moore MS, Blobel G. 1992. *Cell* 69:939–50
52. Görlich D, Vogel F, Mills AD, Hartmann E, Laskey RA. 1995. *Nature* 377:246–48
53. Moroianu J, Hijikata M, Blobel G, Radu A. 1995. *Proc. Natl. Acad. Sci. USA* 92: 6532–36
54. Görlich D, Henklein P, Laskey RA, Hartmann E. 1996. *EMBO J.* 15:1810–17
55. Weis K, Ryder U, Lamond A. 1996. *EMBO J.* 15:1818–25
56. Darnell JE Jr. 1996. *Recent Prog. Horm. Res.* 51:391–403
57. Ihle JN. 1996. *Cell* 34:331–43
58. Adam SA, Adam EJH, Chi NC, Visser GD. 1995. *Cold Spring Harbor Symp. Quant. Biol.* 60:687–94
59. Sekimoto T, Imamoto N, Nakajima K, Hirano T, Yoneda Y. 1997. *EMBO J.* 16: 7067–77
60. Lührmann R, Kastner B, Bach M. 1990. *Biochim. Biophys. Acta* 1087:265–92
61. Liu Q, Fischer U, Wang F, Dreyfuss G. 1997. *Cell* 90:1013–22

62. Fischer U, Liu Q, Dreyfuss G. 1997. *Cell* 90:1023–30
63. Mattaj IW, De Robertis EM. 1985. *Cell* 40:111–18
64. Mattaj IW. 1986. *Cell* 46:905–11
65. Hamm J, Darzynkiewicz E, Tahara SM, Mattaj IW. 1990. *Cell* 62:569–77
66. Fischer U, Lührmann R. 1990. *Science* 249:786–90
67. Fischer U, Sumpter V, Sekine M, Satoh T, Lührmann R. 1993. *EMBO J.* 12:573–83
68. Palacios I, Hetzer M, Adam SA, Mattaj IW. 1997. *EMBO J.* 16:6783–92
69. Deleted in proof
70. Fornerod M, van Deursen J, van Baal S, Reynolds A, Davis D, et al. 1997. *EMBO J.* 16:807–16
71. Görlich D, Dabrowski M, Bischoff FR, Kutay U, Bork P, et al. 1997. *J. Cell Biol.* 138:65–80
72. Pollard VW, Michael WM, Nakielny S, Siomi MC, Wang F, Dreyfuss G. 1996. *Cell* 86:985–94
73. Fridell RA, Truant R, Thorne L, Benson RE, Cullen BR. 1997. *J. Cell Sci.* 110:1325–31
74. Siomi H, Dreyfuss G. 1995. *J. Cell Biol.* 129:551–60
75. Weighardt F, Biamonti G, Riva S. 1995. *J. Cell Sci.* 108:545–55
76. Aitchison JD, Blobel G, Rout MP. 1996. *Science* 274:624–27
77. Nakielny S, Siomi MC, Siomi H, Michael WM, Pollard V, Dreyfuss G. 1996. *Exp. Cell Res.* 229:261–66
77a. Siomi MC, Eder PS, Kataoka N, Wan L, Liu Q, Dreyfuss G. 1997. *J. Cell Biol.* 138:1–12
78. Michael WM, Eder PS, Dreyfuss G. 1997. *EMBO J.* 16:3587–98
79. Kambach C, Mattaj IW. 1992. *J. Cell Biol.* 118:11–21
80. Moreland RB, Nam HG, Hereford LM, Fried HM. 1985. *Proc. Natl. Acad. Sci. USA* 82:6561–65
81. Schaap PJ, van't Riet J, Woldringh CL, Raue HA. 1991. *J. Mol. Biol.* 221:225–37
82. Schmidt C, Lipsius E, Kruppa J. 1995. *Mol. Biol. Cell* 6:1875–85
83. Quaye IK, Toku S, Tanaka T. 1996. *Eur. J. Cell Biol.* 69:151–55
84. Russo G, Ricciardelli G, Pietropaolo C. 1997. *J. Biol. Chem.* 272:5229–35
85. Underwood MR, Fried HM. 1990. *EMBO J.* 9:91–99
86. Nehrbass U, Fabre E, Dihlmann S, Herth W, Hurt EC. 1993. *Eur. J. Cell Biol.* 62:1–12
87. Rout MP, Blobel G, Aitchison JD. 1997. *Cell* 89:715–25

88. Schlenstedt G, Smirnova E, Deane R, Solsbacher J, Kutay U, et al. 1997. *EMBO J.* 16:6237–49
89. Panté N, Aebi U. 1996. *Science* 273: 1729–32
90. Akey CW, Goldfarb DS. 1989. *J. Cell Biol.* 109:971–82
91. Moore MS, Blobel G. 1993. *Nature* 365:661–63
92. Melchior F, Paschal B, Evans E, Gerace L. 1993. *J. Cell Biol.* 123:1649–59
93. Corbett AH, Koepp DM, Schlenstedt G, Lee MS, Hopper AK, Silver PA. 1995. *J. Cell Biol.* 130:1017–26
94. Schlenstedt G, Saavedra C, Loeb JDJ, Cole CN, Silver PA. 1995. *Proc. Natl. Acad. Sci. USA* 92:225–29
95. Palacios I, Weis K, Klebe C, Mattaj IW, Dingwall C. 1996. *J. Cell Biol.* 133:485–94
96. Richards SA, Carey KL, Macara IG. 1997. *Science* 276:1842–44
97. Izaurralde E, Kutay U, von Kobbe C, Mattaj IW, Görlich D. 1997. *EMBO J.* 16:6535–47
98. Koepp DM, Silver PA. 1996. *Cell* 87:1–4
99. Görlich D. 1997. *Curr. Opin. Cell Biol.* 9:412–19
100. Ohtsubo M, Okazaki H, Nishimoto T. 1989. *J. Cell Biol.* 109:1389–97
101. Weis K, Dingwall C, Lamond AI. 1996. *EMBO J.* 15:7120–28
102. Görlich D, Panté N, Kutay U, Aebi U, Bischoff FR. 1996. *EMBO J.* 15:5584–94
103. Rexach M, Blobel G. 1995. *Cell* 83:683–92
104. Chi NC, Adam EJH, Visser GD, Adam SA. 1996. *J. Cell Biol.* 135:559–69
105. Chi NC, Adam EJH, Adam SA. 1997. *J. Biol. Chem.* 272:6818–22
106. Deleted in proof
107. Lounsbury KM, Macara IG. 1997. *J. Biol. Chem.* 272:551–55
108. Moore MS, Blobel G. 1995. *Cold Spring Harbor Symp. Quant. Biol.* 60:701–5
109. Carey KL, Richards SA, Lounsbury KM, Macara IG. 1996. *J. Cell Biol.* 133:985–96
110. Sweet DJ, Gerace L. 1996. *J. Cell Biol.* 133:971–83
111. Nakashima N, Hayashi N, Noguchi E, Nishimoto T. 1996. *J. Cell Sci.* 109:2311–18
112. Moore MS, Blobel G. 1994. *Proc. Natl. Acad. Sci. USA* 91:10212–16
113. Paschal BM, Gerace L. 1995. *J. Cell Biol.* 129:925–37
114. Nehrbass U, Blobel G. 1996. *Science* 272:120–22
115. Corbett AH, Silver PA. 1996. *J. Biol. Chem.* 271:18477–84

116. Paschal BM, Fritze C, Guan T, Gerace L. 1997. *J. Biol. Chem.* 272:21534–39
117. Wong DH, Corbett AH, Kent HM, Stewart M, Silver PA. 1997. *Mol. Cell Biol.* 17:3755–67
118. Clarkson WD, Kent HM, Stewart M. 1996. *J. Mol. Biol.* 263:517–24
119. Paschal BM, Delphin C, Gerace L. 1996. *Proc. Natl. Acad. Sci. USA* 93:7679–83
120. Bischoff FR, Krebber H, Smirnova E, Dong WH, Ponstingl H. 1995. *EMBO J.* 14:705–15
121. Lounsbury KM, Beddow AL, Macara IG. 1994. *J. Biol. Chem.* 269:11285–90
122. Chi NC, Adam SA. 1997. *Mol. Biol. Cell* 8:945–56
123. Richards SA, Lounsbury KM, Carey KL, Macara IG. 1996. *J. Cell Biol.* 134:1157–68
124. Zolotukhin AS, Felber BK. 1997. *J. Biol. Chem.* 272:11356–60
125. Taura T, Schlenstedt G, Silver PA. 1997. *J. Biol. Chem.* 272:31877–84
126. Yokoyama N, Hayashi N, Seki T, Pante N, Ohba T, et al. 1995. *Nature* 376:184–88
127. Wu J, Matunis MJ, Kraemer D, Blobel G, Coutavas E. 1995. *J. Biol. Chem.* 270:14209–13
128. Matunis MJ, Coutavas E, Blobel G. 1996. *J. Cell Biol.* 135:1457–70
129. Mahajan R, Delphin C, Guan T, Gerace L, Melchior F. 1997. *Cell* 88:97–107
130. Saitoh H, Pu R, Cavenagh M, Dasso M. 1997. *Proc. Natl. Acad. Sci. USA* 94:3736–41
131. Loeb JD, Davis LI, Fink GR. 1993. *Mol. Biol. Cell* 4:209–22
132. Dingwall C, Kandels-Lewis S, Sérpahin B. 1995. *Proc. Natl. Acad. Sci. USA* 92:7525–29
133. Noguchi E, Hayashi N, Nakashima N, Nishimoto T. 1997. *Mol. Cell Biol.* 2235–46
134. Imamoto N, Matsuoka Y, Kurihara T, Kohno K, Miyagi M, et al. 1992. *J. Cell Biol.* 119:1047–61
135. Shi Y, Thomas JO. 1992. *Mol. Cell Biol.* 12:2186–92
136. Yang J, DeFranco DB. 1994. *Mol. Cell Biol.* 14:5088–98
137. Franke WW, Scheer U. 1974. *Symp. Soc. Exp. Biol.* 28:249–82
138. Mehlin H, Daneholt B. 1993. *Trends Cell Biol.* 3:443–47
139. Mehlin H, Daneholt B, Skoglund U. 1992. *Cell* 69:605–13
140. Mehlin H, Daneholt B, Skoglund U. 1995. *J. Cell Biol.* 129:1205–16
141. Manley JL, Tacke R. 1996. *Genes Dev.* 10:1569–79

142. Zahler AM, Lane WS, Stolk JA, Roth MB. 1992. *Genes Dev.* 6:837–47
143. Cáceres JF, Misteli T, Screaton GR, Spector DL, Krainer AR. 1997. *J. Cell Biol.* 138:225–38
144. Kiseleva E, Wurtz T, Visa N, Daneholt B. 1994. *EMBO J.* 13:6052–61
145. Visa N, Alzhanova-Ericsson AT, Sun X, Kiseleva E, Björkroth B, et al. 1996. *Cell* 84:253–64
146. Visa N, Izaurralde E, Ferreira J, Daneholt B, Mattaj IW. 1996. *J. Cell Biol.* 133:5–14
147. Alzhanova-Ericsson AT, Sun X, Visa N, Kiseleva E, Wurtz T, Daneholt B. 1996. *Genes Dev.* 10:2881–93
148. Vankan P, McGuigan C, Mattaj IW. 1990. *EMBO J.* 9:3397–404
149. Terns MP, Dahlberg JE. 1994. *Science* 264:959–61
150. Terns MP, Grimm C, Lund E, Dahlberg JE. 1995. *EMBO J.* 14:4860–71
151. Boelens WC, Palacios I, Mattaj IW. 1995. *RNA* 1:273–83
152. Legrain P, Rosbash M. 1989. *Cell* 57:573–83
153. Chang DD, Sharp PA. 1989. *Cell* 59:789–95
154. Hamm J, Mattaj IW. 1990. *Cell* 63:109–18
155. Ohno M, Shimura Y. 1996. *Genes Dev.* 10:997–1007
156. Dreyfuss G, Matunis MJ, Piñol-Roma S, Burd CG. 1993. *Annu. Rev. Biochem.* 62:289–321
157. Nakagawa TY, Swanson MS, Wold BJ, Dreyfuss G. 1986. *Proc. Natl. Acad. Sci. USA* 83:2007–11
158. Nakielny S, Dreyfuss G. 1996. *J. Cell Biol.* 134:1365–73
159. Dobbelstein M, Roth J, Kimberly WT, Levine AJ, Shenk T. 1997. *EMBO J.* 16:4276–84
160. Bataillé N, Helser T, Fried HM. 1990. *J. Cell Biol.* 111:1571–82
161. Jarmolowski A, Boelens WC, Izaurralde E, Mattaj IW. 1994. *J. Cell Biol.* 124:627–35
162. Pokrywka NJ, Goldfarb DS. 1995. *J. Biol. Chem.* 270:3619–24
163. Izaurralde E, Stepinski J, Darzynkiewicz E, Mattaj IW. 1992. *J. Cell Biol.* 118:1287–95
164. Terns MP, Dahlberg JE, Lund E. 1993. *Genes Dev.* 7:1898–908
165. Saavedra C, Tung KS, Amberg DC, Hopper AK, Cole CN. 1996. *Genes Dev.* 10:1608–20
166. Eckner R, Ellmeier W, Birnstiel ML. 1991. *EMBO J.* 10:3513–22
167. Cullen BR, Malim MH. 1991. *Trends Biochem. Sci.* 16:346–50

168. Hammarskjöld ML. 1997. *Semin. Cell Dev. Biol.* 8:83–90
169. Bray M, Prasad S, Dubay JW, Hunter E, Jeang KT, et al. 1994. *Proc. Natl. Acad. Sci. USA* 91:1256–60
170. Ogert RA, Lee LH, Beemon KL. 1996. *J. Virol.* 70:3834–43
171. Tabernero C, Zolotukhin AS, Valentin A, Pavlakis GN, Felber BK. 1996. *J. Virol.* 70:5998–6011
172. Saavedra C, Felber B, Izaurralde E. 1997. *Curr. Biol.* 7:619–28
173. Pasquinelli AE, Ernst RK, Lund E, Grimm C, Zapp ML, et al. 1997. *EMBO J.* 16:7500–10
174. Tang H, Gaietta GM, Fischer WH, Ellisman MH, Wong-Staal F. 1997. *Science* 276:1412–15
175. Malim MH, Hauber J, Le SY, Maizel JV, Cullen BR. 1989. *Nature* 338:254–57
176. Zapp ML, Green MR. 1989. *Nature* 342:714–16
177. Fischer U, Meyer S, Teufel M, Heckel C, Lührmann R, Rautmann G. 1994. *EMBO J.* 13:4105–12
178. Tan RY, Chen L, Buettner JA, Hudson D, Frankel AD. 1993. *Cell* 73:1031–40
179. Malim MH, Bohnlein S, Hauber J, Cullen BR. 1989. *Cell* 58:205–14
180. Malim MH, McCarn DF, Tiley LS, Cullen BR. 1991. *J. Virol.* 65:4248–54
181. Fischer U, Huber J, Boelens WC, Mattaj IW, Lührmann R. 1995. *Cell* 82:475–83
182. Wen W, Meinkoth JL, Tsien RY, Taylor SS. 1995. *Cell* 82:463–73
183. Nigg EA, Hilz H, Eppenberger HM, Dutly F. 1985. *EMBO J.* 4:2801–6
184. Harootunian AT, Adams SR, Wen W, Meinkoth JL, Taylor SS, Tsien RY. 1993. *Mol. Biol. Cell* 4:993–1002
185. Pines J, Hunter T. 1991. *J. Cell Biol.* 115:1–17
186. Guddat U, Bakken AH, Pieler T. 1990. *Cell* 60:619–28
187. Fridell RA, Fischer U, Lührmann R, Meyer BE, Meinkoth JL, et al. 1996. *Proc. Natl. Acad. Sci. USA* 93:2936–40
188. Murphy R, Wente SR. 1996. *Nature* 383:357–60
189. Del Priore V, Snay CA, Bahr A, Cole CN. 1996. *Mol. Biol. Cell* 7:1601–21
190. Segref A, Sharma K, Doye V, Hellwig A, Huber J, et al. 1997. *EMBO J.* 16:3256–71
191. Amberg DC, Goldstein AL, Cole CN. 1992. *Genes Dev.* 6:1173–89
192. Kadowaki T, Zhao YM, Tartakoff AM. 1992. *Proc. Natl. Acad. Sci. USA* 89:2312–16
193. Kadowaki T, Chen S, Hitomi M, Jacobs E, Kumagai C, et al. 1994. *J. Cell Biol.* 126:649–59

194. Nakielny S, Dreyfuss G. 1997. *Curr. Opin. Cell Biol.* 9:420–29
195. Piñol-Roma S, Dreyfuss G. 1991. *Science* 253:312–14
196. Piñol-Roma S, Dreyfuss G. 1992. *Nature* 355:730–32
197. Michael WM, Siomi H, Choi M, Piñol-Roma S, Nakielny S, et al. 1995. *Cold Spring Harbor Symp. Quant. Biol.* 60:663–68
198. Izaurralde E, Jarmolowski A, Beisel C, Mattaj IW, Dreyfuss G. 1997. *J. Cell Biol.* 137:27–35
199. Matunis EL, Matunis MJ, Dreyfuss G. 1993. *J. Cell Biol.* 121:219–28
200. Bennett M, Piñol-Roma S, Staknis D, Dreyfuss G, Reed R. 1992. *Mol. Cell Biol.* 12:3165–75
201. Wilson SM, Datar KV, Paddy MR, Swedlow JR, Swanson MS. 1994. *J. Cell Biol.* 127:1173–84
202. Russell I, Tollervey D. 1995. *Eur. J. Cell Biol.* 66:293–301
203. Flach J, Bossie M, Vogel J, Corbett A, Jinks T, et al. 1994. *Mol. Cell Biol.* 14:8399–407
204. Singleton DR, Chen SP, Hitomi M, Kumagai C, Tartakoff AM. 1995. *J. Cell Sci.* 108:265–72
205. Anderson JT, Wilson SM, Datar KV, Swanson MS. 1993. *Mol. Cell Biol.* 13:2730–41
206. Henry M, Borland CZ, Bossie M, Silver PA. 1996. *Genetics* 142:103–15
207. Lee MS, Henry M, Silver PA. 1996. *Genes Dev.* 10:1233–46
208. Bossie MA, DeHoratius C, Barcelo G, Silver P. 1992. *Mol. Biol. Cell* 3:875–93
209. Powers MA, Douglass JF, Dahlberg JE, Lund E. 1997. *J. Cell Biol.* 136:241–50
210. Cheng Y, Dahlberg JE, Lund E. 1995. *Science* 267:1807–10
211. Simos G, Tekotte H, Grosjean H, Segref A, Sharma K, et al. 1996. *EMBO J.* 15:2270–84
212. Hopper AK, Schultz LD. 1980. *Cell* 19:741–51
213. Ohno M, Kataoka N, Shimura Y. 1990. *Nucleic Acids Res.* 18:6989–95
214. Izaurralde E, Lewis J, McGuigan C, Jankowska M, Darzynkiewicz E, Mattaj IW. 1994. *Cell* 78:657–68
215. Izaurralde E, Lewis J, Gamberi C, Jarmolowski A, McGuigan C, Mattaj IW. 1995. *Nature* 376:709–12
216. Kataoka N, Ohno M, Kangawa K, Tokoro Y, Shimura Y. 1994. *Nucleic Acids Res.* 22:3861–65
217. Kataoka N, Ohno M, Moda I, Shimura Y. 1995. *Nucleic Acids Res.* 23:3638–41

218. Lewis JD, Izaurralde E, Jarmolowski A, McGuigan C, Mattaj IW. 1996. *Genes Dev.* 10:1683–98
219. Flaherty SM, Fortes P, Izaurralde E, Mattaj IW, Gilmartin GM. 1997. *Proc. Natl. Acad. Sci. USA* 94:11893–98
220. Görlich D, Kraft R, Kostka S, Vogel F, Hartmann E, et al. 1996. *Cell* 87:21–32
221. Colot HV, Stutz F, Rosbash M. 1996. *Genes Dev.* 10:1699–708
222. Lewis JD, Görlich D, Mattaj IW. 1996. *Nucleic Acids Res.* 24:3332–36
223. Xiao ZX, McGrew JT, Schroeder AJ, Fitzgerald-Hayes M. 1993. *Mol. Cell Biol.* 13:4691–702
224. Brinkmann U, Brinkmann E, Gallo M, Pastan I. 1995. *Proc. Natl. Acad. Sci. USA* 92:10427–31
225. Adachi Y, Yanagida M. 1989. *J. Cell Biol.* 108:1195–207
226. Kutay U, Bischoff FR, Kostka S, Kraft R, Görlich D. 1997. *Cell* 90:1061–71
227. Fornerod M, Ohno M, Yoshida M, Mattaj IW. 1997. *Cell* 90:1051–60
228. Fukuda M, Asano S, Nakamura T, Adachi M, Yoshida M, et al. 1997. *Nature* 390:308–11
229. Stade K, Ford CS, Guthrie C, Weis K. 1997. *Cell* 90:1041–50
230. Hamamoto T, Gunji S, Tsuji H, Beppu T. 1983. *J. Antibiot.* 36:639–45
231. Nishi K, Yoshida M, Fujiwara D, Nishikawa M, Horinouchi S, Beppu T. 1994. *J. Biol. Chem.* 269:6320–24
232. Wolff B, Sanglier JJ, Wang Y. 1997. *Chem. Biol.* 4:139–47
233. Kutay U, Izaurralde E, Bischoff FR, Mattaj IW, Görlich D. 1997. *EMBO J.* 16:1153–63
234. Neville M, Lee L, Stutz F, Davis LI, Rosbash M. 1997. *Curr. Biol.* 7:767–75
235. Ruhl M, Himmelspach M, Bahr GM, Hammerschmid F, Jaksche H, et al 1993. *J. Cell Biol.* 123:1309–20
236. Bevec D, Jaksche H, Oft M, Wöhl T, Himmelspach M, et al. 1996. *Science* 271:1858–60
237. Stutz F, Neville M, Rosbash M. 1995. *Cell* 82:495–506
238. Bogerd HP, Fridell RA, Madore S, Cullen BR. 1995. *Cell* 82:485–94
239. Fritz CC, Zapp ML, Green MR. 1995. *Nature* 376:530–33
240. Fornerod M, Boer J, van Baal S, Morreau H, Grosveld G. 1996. *Oncogene* 13:1801–8
241. Seedorf M, Silver PA. 1997. *Proc. Natl. Acad. Sci. USA* 94:8590–95
242. Floer M, Blobel G, Rexach M. 1997. *J. Biol. Chem.* 272:19538–46

243. Izaurralde E, Mattaj IW. 1995. *Cell* 81:153–59
244. Nishitani H, Ohtsubo M, Yamashita K, Iida H, Pines J, et al. 1991. *EMBO J.* 10:1555–64
245. Tachibana T, Imamoto N, Seino H, Nishimoto T, Yoneda Y. 1994. *J. Biol. Chem.* 269:24542–45
246. Kadowaki T, Goldfarb D, Spitz LM, Tartakoff AM, Ohno M. 1993. *EMBO J.* 12:1929–37
247. Her LS, Lund E, Dahlberg JE. 1997. *Science* 276:1845–48
248. Izaurralde E, McGuigan C, Mattaj IW. 1995. *Cold Spring Harbor Symp. Quant. Biol.* 60:669–75
249. Iovine MK, Wente SR. 1997. *J. Cell Biol.* 137:797–811
250. Radu A, Moore MS, Blobel G. 1995. *Cell* 81:215–22
251. Iovine MK, Watkins JL, Wente SR. 1995. *J. Cell Biol.* 131:1699–713
252. Feldherr CM, Akin D. 1994. *Int. Rev. Cytol.* 151:183–228
253. Feldherr CM, Akin D. 1993. *Exp. Cell Res.* 205:179–86
254. Moll T, Tebb G, Surana U, Robitsch H, Nasmyth K. 1991. *Cell* 66:743–58
255. Flanagan WM, Corthésy B, Bram RJ, Crabtree GR. 1991. *Nature* 352:803–7
256. Beals CR, Clipstone NA, Ho SN, Crabtree GR. 1997. *Genes Dev.* 11:824–34
257. Timmerman LA, Clipstone NA, Ho SN, Northrop JP, Crabtree GR. 1996. *Nature* 383:837–40
258. Shibasaki F, Price ER, Milan D, McKeon F. 1996. *Nature* 382:370–73
259. O'Neill EM, Kaffman A, Jolly ER, O'Shea EK. 1996. *Science* 271:209–12
260. Zandi E, Tran TNT, Chamberlain W, Parker CS. 1997. *Genes Dev.* 11:1299–314
261. Yamamoto KR. 1995. *Harvey Lect.* 91:1–19
262. Shuai K, Stark GR, Kerr IM, Darnell JE Jr. 1993. *Science* 261:1744–46
263. Shuai K, Horvath CM, Huang LHT, Qureshi SA, Cowburn D, Darnell JE Jr. 1994. *Cell* 76:821–28
264. Henkel T, Zabel U, van Zee K, Muller JM, Fanning E, Baeuerle PA. 1992. *Cell* 68:1121–33
265. Imbert V, Rupec RA, Livolsi A, Pahl HL, Traenckner EBM, et al. 1996. *Cell* 86:787–98
266. Traenckner EBM, Pahl HL, Henkel T, Schmidt KN, Wilk S, Baeuerle PA. 1995. *EMBO J.* 14:2876–83
267. Brown K, Gerstberger S, Carlson L, Franzoso G, Siebenlist U. 1995. *Science* 267:1485–88

268. Traenckner EBM, Wilk S, Baeuerle PA. 1994. *EMBO J.* 13:5433–41
269. Alkalay I, Yaron A, Hatzubai A, Orian A, Ciechanover A, Ben-Neriah Y. 1995. *Proc. Natl. Acad. Sci. USA* 92:10599–603
270. Palombella VJ, Rando OJ, Goldberg AL, Maniatis T. 1994. *Cell* 78:773–85
271. DiDonato JA, Hayakawa M, Rothwarf D, Zandi E, Karin M. 1997. *Nature* 388:548–54
272. Krug RM. 1993. *Curr. Opin. Cell Biol.* 5:944–49
273. Showman RM, Wells DE, Anstrom J, Hursh DA, Raff RA. 1982. *Proc. Natl. Acad. Sci. USA* 79:5944–47
274. DeLeon DV, Cox KH, Angerer LM, Angerer RC. 1983. *Dev. Biol.* 100:197–206
275. Saavedra CA, Hammell CM, Heath CV, Cole CN. 1997. *Genes Dev.* In press
276. Tani T, Derby RJ, Hiraoka Y, Spector DL. 1995. *Mol. Biol. Cell* 6:1515–34
277. Corbett AH, Silver PA. 1997. *Microbiol. Mol. Biol. Rev.* 61:193–211
278. Stutz F, Kantor J, Zhang D, McCarthy T, Neville M, Rosbash M. 1997. *Genes Dev.* 11:2857–68
279. Davis I. 1997. *Semin. Cell Dev. Biol.* 8:99–97
280. Bullock TL, Clarkson WD, Kent HM, Stewart M. 1996. *J. Mol. Biol.* 260:422–31
281. Scheffzek K, Klebe C, Fritz-Wolf K, Kabsch W, Wittinghofer A. 1995. *Nature* 374:378–81
282. Makkerh JPS, Dingwall C, Laskey RA. 1996. *Curr. Biol.* 6:1025–27
283. Schindler C, Shuai K, Prezioso VR, Darnell JE Jr. 1992. *Science* 257:809–13
284. Rudt F, Pieler T. 1996. *EMBO J.* 15:1383–91
285. Soullam B, Worman HJ. 1993. *J. Cell Biol.* 120:1093–100
286. Soullam B, Worman HJ. 1995. *J. Cell Biol.* 130:15–27
287. Li H, Bingham PM. 1991. *Cell* 67:335–42
288. Duverger E, Pellerin-Mendes C, Mayer R, Roche AC, Monsigny M. 1995. *Cell Sci.* 108:1325–32
289. Smith MR, Greene WC. 1992. *Virology* 187:316–20
290. Hall MN, Hereford L, Herskowitz I. 1984. *Cell* 36:1057–65
291. Hall MN, Craik C, Hiraoka Y. 1990. *Proc. Natl. Acad. Sci. USA* 87:6954–68
292. Silver PA, Keegan LP, Ptashne M. 1984. *Proc. Natl. Acad. Sci. USA* 81:5951–55
293. Standiford DM, Richter JD. 1992. *J. Cell Biol.* 118:991–1002
294. De la Luna S, Burden MJ, Lee CW, La Thangue NB. 1996. *J. Cell Sci.* 109:2443–52
295. Booher RN, Alfa CE, Hyams JS, Beach DH. 1989. *Cell* 58:485–97
296. Deane R, Schäfer W, Zimmermann HP, Mueller L, Görlich D, et al. 1997. *Mol. Cell Biol.* 17:5087–96
297. Yaseen NR, Blobel G. 1997. *Proc. Natl. Acad. Sci. USA* 94:4451–56
298. Meyer BE, Meinkoth JL, Malim MH. 1996. *J. Virol.* 70:2350–59
299. Meier UT, Blobel G. 1992. *Cell* 70:127–38
300. Kambach C, Mattaj IW. 1994. *J. Cell Sci.* 107:1807–16
301. Romac JM, Graff DH, Keene JD. 1994. *Mol. Cell Biol.* 14:462–70
302. Klein Gunnewiek JMT, van Aarssen Y, van der Kemp A, Nelissen R, Pruijn GJM, van Venrooij WJ. 1997. *Exp. Cell Res.* 235:265–73

Annu. Rev. Biochem. 1998. 67:307–33

ROLE OF SMALL G PROTEINS IN YEAST CELL POLARIZATION AND WALL BIOSYNTHESIS[1]

Enrico Cabib, Jana Drgonová, and Tomás Drgon
National Institute of Diabetes and Digestive and Kidney Diseases, National Institutes
of Health, Bethesda, Maryland 20892; e-mail: enricoc@bdg10.niddk.nih.gov,
janad@bdg10.niddk.nih.gov, tomasd@bdg10.niddk.nih.gov

KEY WORDS: budding, Bud1/Rsr1, Cdc42, Rho1, $\beta(1 \rightarrow 3)$glucan synthase

ABSTRACT

In the vegetative (mitotic) cycle and during sexual conjugation, yeast cells display polarized growth, giving rise to a bud or to a mating projection, respectively. In both cases one can distinguish three steps in these processes: choice of a growth site, organization of the growth site, and actual growth and morphogenesis. In all three steps, small GTP-binding proteins (G proteins) and their regulators play essential signaling functions. For the choice of a bud site, Bud1, a small G protein, Bud2, a negative regulator of Bud1, and Bud5, an activator, are all required. If any of them is defective, the cell loses its ability to select a proper bud position and buds randomly. In the organization of the bud site or of the site in which a mating projection appears, Cdc42, its activator Cdc24, and its negative regulators play a fundamental role. In the absence of Cdc42 or Cdc24, the actin cytoskeleton does not become organized and budding does not take place. Finally, another small G protein, Rho1, is required for activity of $\beta(1 \rightarrow 3)$glucan synthase, the enzyme that catalyzes the synthesis of the major structural component of the yeast cell wall. In all of the above processes, G proteins can work as molecular switches because of their ability to shift between an active GTP-bound state and an inactive GDP-bound state.

CONTENTS

INTRODUCTION

During both vegetative proliferation and sexual conjugation, the yeast cell undergoes localized morphogenetic changes that are preceded by cell polarization. Polarization and morphogenesis occur at specific stages of these processes and thus require temporal and spatial regulation. How does the cell dictate the time and place for new morphological changes to occur? Evidence accumulated in the past decade indicates that the timing and localization involve the concerted interactions of a large number of molecules. Most prominent among the regulatory factors that control these processes are small GTP-binding proteins of the Ras superfamily. Small G proteins are especially suited to function as molecular switches because of their ability to shift between a GTP-bound active form and a GDP-bound inactive form. These changes are regulated by other proteins: GTPase-activating proteins (GAPs), which enhance the intrinsic GTPase activity of the G protein, thus stimulating the transition from GTP- to GDP-bound state; GTP-GDP exchange factors (GEFs), which at high GTP-GDP ratios found in the cell lead to an increase in the GTP-bound form of the protein; and GDP dissociation inhibitors (GDIs). In the small G protein systems, GAPs function as negative regulators and GEFs as activators, whereas GDIs tend to keep the G protein in the cytoplasm in an inactive state (1). The participation of small G proteins in cell polarization and in subsequent morphogenesis is the subject of this review.

In both the formation of a new bud and the construction of the pointed projection that precedes sexual conjugation, three consecutive steps can be recognized: (*a*) choice of a site at the cell cortex where the new growth will occur;

(*b*) organization of the site, including assembly of the machinery required for subsequent growth; (*c*) actual building of the new structure, the morphogenetic step. These three stages are considered below. Lack of space precludes discussion of pseudohyphal growth, although small G proteins have recently been implicated in regulation of this morphogenetic process (2).

WHERE TO START

Yeast cells, which are usually ovoid, develop polarity and display polarized growth in two different modes of their life cycle: budding and mating. In the vegetative cycle, yeast cells divide by budding, and the position of bud emergence is predetermined. *Saccharomyces cerevisiae* cells have long been known to exhibit two different budding patterns (3–6), depending on their ploidy. Haploid yeast cells exist in two mating types: **a** and *α*. Both **a** and *α* cells exhibit so-called axial budding, in which a new bud always emerges at the cell pole where budding occurred previously. Diploid cells, on the other hand, display so-called polar budding. In this case, a new bud emerges at either of the cell poles (Figure 1).

Cells of each mating type secrete a distinct mating pheromone, and at the same time they sense the pheromone of their mating partner. When cells of the two mating types come together, they stop dividing (arrest) and develop a polarized growth projection toward their mating partner. Eventually, cells fuse at the tips of these projections, giving rise to a diploid zygote.

In recent years, much has been learned through genetic studies about the mechanisms underlying the budding pattern (7–10a, 11). Genes isolated with a variety of screens were systematized, and four gene classes were identified with respect to bud site selection:

1. Genes that are responsible for establishment of nonrandom (either axial or polar) budding patterns [*BUD1* (12), *BUD2* (13–15), and *BUD5* (16, 17); for a list of most of the genes mentioned in this review, see Table 1]. Mutants in these genes exhibit a random budding pattern but do not show any growth impairment. Genes in this group are required for proper function of axial and polar genes in classes 2 and 3 (12, 13, 17–19).

2. Genes that are responsible for development of the axial budding pattern displayed by **a** or *α* haploid cells; i.e. *BUD3*, *BUD4* (12, 20, 21), *AXL1* (22), *AXL2 /BUD10* (23, 24). Mutants in these genes, despite being haploid, exhibit a polar budding pattern without any significant growth defect.

3. Genes required for the polar budding pattern in diploids [*ACT1* (25, 28), *SPA2* (26), *RVS161*, *RVS167* (27, 29), *BNI1*, *BUD6*, *BUD7*, *BUD8*, *BUD9*

haploid cells
axial budding

diploid cells
bipolar budding

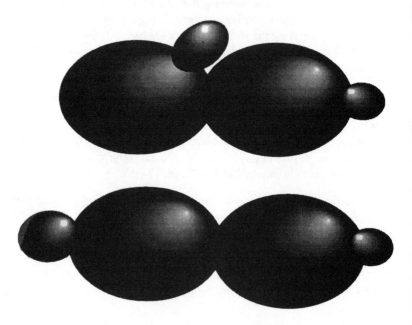

(30, 31)]. Diploids mutated in these genes bud randomly or manifest a bias for one of the poles in bud site selection. The budding pattern of haploid cells is not affected by mutations in these genes.

4. Genes required for organization of the bud site. Mutants in these genes are unable to bud; they are discussed in a later section.

The molecular principles underlying development of cell polarity are conserved among many eukaryotic organisms, including mammals and plants. Yeast proteins involved in these processes have close homologues in other eukaryotic cells (32–35, 35a, and references therein).

Bud Site Selection

Chronologically, the first step in polarity development in yeast is to mark the site where growth will occur (Figure 2). That a physical landmark exists at this site is supported by the finding that in haploid cells each succeeding bud site is immediately adjacent to the preceding one, as determined from the distribution of bud scars remaining after cell division (36). In addition to septin proteins (discussed later in this section), the landmark may consist of Bud3 (20), Bud4 (21), and Axl2/Bud10 (23, 24). These proteins form a ring around the mother-bud neck that splits during cytokinesis and is retained on both the mother and daughter cells (8, 20, 21, 23, 24). In diploid cells, preferred polar bud sites exist, but their nature is not well understood (12). Candidate genes that may participate in polar bud site marking in diploid cells (*ACT1*, *SPA2*, *BNI1*, *BUD6*, *BUD7*, *BUD8*, *BUD9*) are discussed in reference 19.

The next step, recruiting the budding machinery to the site delimited by the landmark, is common to both axial and polar budding. Execution of this step is dependent on a small G protein, Bud1, and its regulators, Bud2 and Bud5 (12, 13, 17, 18). If any of these proteins is defective or absent, cells establish a bud site at random (Figure 2; 13, 16–18). Nevertheless, even if the bud site is chosen randomly, it is functional and indistinguishable from a nonrandomly selected one.

Site selection is followed by accumulation of various structures at the selected site, one of which is a microfilament ring composed of at least four proteins now called septins (37): Cdc3 (38), Cdc10 (39), Cdc11 (40), Cdc12 (39). Septins are often referred to as GTP-binding proteins on the basis of protein sequence analysis. However, no data showing actual GTPase activity or

←

Figure 1 Schematic representation of different budding patterns in yeast. Haploid cells bud axially, the new bud emerging always adjacent to the site of previous budding. Diploid cells bud polarly, the new bud emerging at either of the cell poles.

Table 1   Genes involved in bud site selection and organization and in cell wall synthesis

Gene	Proposed function of encoded protein	References
ACT1	Actin; cytoskeleton, bipolar bud site selection	25, 28
AXL1	Similar to insulin-degrading enzymes; axial bud site selection	22
AXL2/BUD10	Axial bud site selection	23, 24
BEE1/LAS17	Homologous to Wiskott-Aldrich syndrome protein; nucleation of actin	66, 67
BEM1	SH3 domains; cell polarization	54
BEM2	GTPase-activating protein (GAP) for Rho1	69, 135
BEM3	GAP for Cdc42	71
BEM4	Chaperone for small G proteins	100
BNI1, BNR1	Formins; organization of actin cytoskeleton	73–75, 102
BOI1, BOI2	Interact with Bem1	101
BUD1/RSR1	Small Ras-like G protein; general bud site selection	12, 18
BUD2	GAP for Bud1	13–15
BUD3, BUD4	Axial landmark proteins; axial budding	12, 20, 21
BUD5	GTP-GDP exchange factor (GEF) for Bud1	16, 17
BUD6, BUD7, BUD8, BUD9	Bipolar budding	30, 31
CDC3, CDC10, CDC11, CDC12	Septins; components of the neck filament ring	37–40
CDC24	GEF for Cdc42	71
CDC42	Rho-like small G protein; bud site organization, pheromone signaling	49, 65, 88, 89
CDC43	Geranylgeranyl transferase subunit; prenylation of Cdc42	92, 93
CLA4	Protein kinase; homologous to Ste20, cell morphogenesis	84
FKS1, FKS2	$\beta(1 \rightarrow 3)$glucan synthase subunits	124–126
LRG1	GAP for Rho proteins	137
PCA1	Nucleation of actin	66
PKC1	Protein kinase C; triggers MAP kinase cascade involved in maintenance of cell wall integrity	139–144
RGA1/DBM1	GAP for Cdc42	95, 136
RHO1	Small G protein; activator of $\beta(1 \rightarrow 3)$glucan synthase and protein kinase C; actin organization	74, 122, 123, 139, 140
RHO2	RHO1 homolog of uncertain function	120
RHO3, RHO4	Small G proteins; actin organization, polarity maintenance?	121, 150–152
ROM1, ROM2	GEF for Rho1	134
RVS161, RVS167	Bipolar budding	27, 29
SAC7	GAP for Rho1	98
SPA2	Bipolar budding	26
STE20	Protein kinase; pheromone signaling	83, 86, 90
TOR2	Phosphatidylinositol kinase; actin organization	97

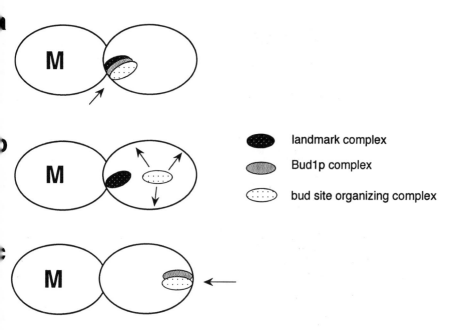

Figure 2 Bud site selection and construction. (*a*) The participation of different protein complexes in bud site selection may be rationalized by assuming that Bud1, together with Bud2 and Bud5, acts as a mediator between the proteins involved in establishing a landmark (Bud3, Bud4, Axl1, and Axl2) and those that organize the bud site. (*b*) In the absence of Bud1, the bud site organizers are not able to recognize any landmark, and budding occurs randomly. (*c*) If the landmark is defective, Bud1 and associated proteins are still able to guide the organizers to a default site and polar budding occurs. *Arrows* point to positions where a new bud will emerge. *M*, mother cell.

GTP binding are available. Analysis of mutants showed that the microfilament ring is necessary for normal growth of the bud; however, septins may also be involved in establishment of the landmark that delineates the site for budding in the next division. Thus, some temperature-sensitive mutants in septin genes, when grown as haploids at a restrictive temperature, do not form buds at their normal axial location (20, 25, 41). Moreover, in a *cdc12^{ts}* mutant, the typical localization of Bud3 (20, 38), Bud4 (21), and Bud10/Axl2 (23, 24) in a double ring at the bud neck is lost or disturbed. Bud10/Axl2, however, still localizes to nascent and small buds in this mutant. Therefore, it is possible that in axially budding cells, septins serve as an anchoring site for other proteins or structures, such as Bud3, Bud4, and Bud10/Axl2.

In this review, we focus on those components of the bud site selection machinery that have been shown to operate on the basis of GTP-GDP cycling.

Components of the Bud Site Selection Apparatus

RSR1/BUD1 was isolated as a multicopy suppressor of a *cdc24* mutation that causes inability to establish cell polarity and to bud (18). Sequence analysis revealed that *BUD1* codes for a Ras-related small GTPase protein. Because *BUD1* deletion itself does not have any effect on growth rate, but only causes the cells to bud randomly, it is suggested that Bud1 acts only at the level of nonrandom bud site selection and not on further polarity development (17, 18). Interestingly, dominant positive (GTPase deficient) as well as dominant negative (GTP-binding deficient) mutants in *BUD1* also randomize their budding pattern with a strong tendency to form the first bud at the distal pole of the cell (42). In addition, a dominant positive mutation in *BUD1* suppresses the ability of nitrogen-starved diploid cells to undergo pseudohyphal growth (43).

BUD2 was isolated by screening a yeast genomic library for complementation of a *bud2* mutation that causes random budding (13). Chromosomal deletion or overexpression of *BUD2* again caused a random budding pattern without affecting growth rates of the strains. Bud2 can function as a GTPase-activating protein for Bud1 in vitro (13) and in vivo (44). *BUD2* was also isolated as *CLA2* (14) and *ERC25* (15) in screens for genes required for budding in the absence of G1 cyclins Cln1 and Cln2, proteins that regulate the transition from G1 to S phase in the cell cycle. Cln1 and Cln2 might play a role in bud site selection (14, 15), but this requires further experimentation.

BUD5, like *BUD1* and *BUD2*, is required for nonrandom budding (16, 17). Bud5 shares significant homology with Cdc25, a GTP-GDP exchange factor for Ras2 (16), and is required for suppression of a *cdc24* mutation by wild-type *BUD1* but not by a GTPase-deficient *bud1* mutant (18). This and other genetic evidence (16) suggests that Bud5 is a GDP-GTP exchange factor (activator) of Bud1 (44).

Bud1 Cycling and the Assembly of the Budding Complex

As mentioned above, the phenotype of mutants in the *BUD* genes suggests that Bud1 is critical for bringing together the proteins necessary for bud formation and the proteins that mark the incipient bud site (Figure 2). In fact, it has been shown recently (45) that Bud1 interacts with Cdc24 and Bem1, two components of the protein complex required for bud development (see below). Cycling of Bud1 between the GTP- and the GDP-bound forms affects these interactions: In the GTP-bound state, Bud1 binds preferentially to Cdc24, whereas in the GDP-bound state, binding to Bem1 is favored (45). We return to this point when discussing the protein complex that organizes bud formation.

PREPARING TO GROW

Once the site for bud emergence is defined, polarity establishment proteins are recruited to the chosen site. These proteins form a structure that organizes actin filaments into mobile cortical patches at the presumptive bud site and at the tip of the growing bud; this structure orients actin cables toward the growing bud (46–48). Orientation of the actin cytoskeleton is necessary for polarized delivery of building materials and for restriction of cell surface growth to the bud (11, 35a, 49, 50). Known members of the polarity establishment complex are the products of the *CDC42* (49), *CDC24* (52, 53), and *BEM1* genes (54).

CDC42

Cdc42 is a member of the *rho* (*ras* homologous) family of small GTP-binding proteins. *CDC42* is an essential gene originally isolated in a screen for mutants unable to bud at high temperature (49). The arrested cells grow in volume and continue to carry out DNA replication and nuclear division, hence becoming multinucleate (49). Such cells also display delocalized chitin deposition in the cell wall, a phenotype associated with loss of actin polarization (49). Immunofluorescence and immunoelectron microscopy demonstrated that Cdc42 localizes to the plasma membrane near secretory vesicles that accumulate at the site of bud emergence and at the tips and sides of enlarging buds. The Cdc42 staining was most pronounced near plasma membrane invaginations where cortical actin also was found (55); however, its overall staining pattern was different from that of actin (56). Although at permissive temperature $cdc42\text{-}1^{ts}$ cells bud axially, overexpression of *CDC42*, as well as expression of another ts allele, $cdc42^{W97R}$ in single copy, randomizes the budding pattern (57, 58), suggesting that Cdc42 may participate in recognition of the landmark that defines the incipient bud site.

Close homologues of Cdc42 are found in cells of other eukaryotes, such as *Homo sapiens*, *Schizosaccharomyces pombe*, and *Caenorhabditis elegans* (59–63). Some of these homologues are able to complement the *S. cerevisiae cdc42* mutation, which suggests that they may participate in a similar process.

Organization of Actin

A large amount of data has accumulated on proteins involved in polarity establishment and on actin and actin-binding proteins, but a huge gap remains in our knowledge of the regulation of actin polarization. In a newborn daughter cell that must grow isotropically to reach its mature size, actin is randomly distributed around the cell cortex as patches (46). Shortly before emergence of a bud, these patches congregate at a specifically selected site where the bud will emerge. During bud development, the actin patches are localized almost exclusively within the bud itself, and actin cables that form in the mother cell orient along the mother-bud axis (46). As mentioned earlier, these cables are believed to serve as "highways" for delivery of new material carried by secretory vesicles to the growing site (64). When the bud is mature, the patches become randomized again. At cytokinesis, the patches reassemble at the mother-bud neck, where the division septum is constructed (46).

An elegant experiment showed direct involvement of Cdc42 in regulating actin assembly (65). Rhodamine-labeled actin monomers added to permeabilized yeast cells accumulated in buds to form cortical patches similar to those observed in vivo. Actin incorporation into the bud was stimulated by GTP-γS and was reduced by a mutation in *CDC42*. The impaired actin nucleation activity in the *cdc42* mutant was restored when a constitutively active (GTPase-deficient) Cdc42 protein was added to the assay (65). Lechler & Li (66) recently modified the assay to identify two sequentially acting protein factors required for actin nucleation: Bee1 (also called Las17), a protein homologous to mammalian Wiskott-Aldrich syndrome protein (67), and a new protein, Pca1. These results also show that the permeabilized cell assay is a fruitful approach to obtain more information about the function of other genes implicated in actin regulation.

One such gene with an elusive function is *BEM1* (bud emergence), which encodes a protein without any obvious enzymatic activity but containing two SH3 (*src* homology) domains found in proteins interacting with actin (54, 68). Bem1 localizes to the bud site at an early stage of bud assembly (unpublished data of Corrado & Pringle cited in reference 19) and binds to GDP-Bud1/Rsr1 (45). It also binds, in a Ca^{2+}-sensitive manner, to Cdc24 (69, 70), which is a GEF for Cdc42 (see below; 71). Bem1 also interacts with the protein kinase Ste20 (see below; 72) and, most interestingly, with actin (72). The physiological importance of the interaction between Bem1 and actin is not clear, but it seems that Bem1 does not directly regulate formation of actin structures. Rather, its function may be to localize the components of the polarity establishment complex represented by Cdc24 and Cdc42 to the preselected bud site (see above) and to stabilize their interaction with actin.

Mutants in the *BNI1* gene show defects in cytokinesis during vegetative growth (73, 74) and in polarized morphogenesis during mating (75). Bni1 is the yeast homologue of mammalian proteins called formins, which participate in morphogenesis (75). The NH_2-terminal portion of Bni1 binds to the activated form of Cdc42 and other rho-type G proteins, including Rho1, Rho3, and Rho4 (74, 75), all implicated in organization of the actin cytoskeleton. The *RHO1* product is a yeast homologue of human RhoA. The latter regulates actin-dependent cell functions such as cell motility, cytokinesis, cell adhesion, and smooth muscle contraction (76); at least the latter two functions are performed through regulation of myosin phosphorylation (77). Mammalian RhoA can complement some functions of yeast *RHO1* (78). The COOH-terminal portion of Bni1 interacts with profilin (75), a highly conserved protein that stimulates actin polymerization (79). Thus Bni1 may be the common mediator through which rho-type G proteins induce actin polymerization, as predicted by Narumiya (76).

Reviews are available concerning actin structure and function (80), its regulation during the yeast cell cycle (64), and the role of rho-type G proteins in actin organization in higher eukaryotes (76, 81).

STE20

To date, the best-characterized target of Cdc42 is the protein kinase Ste20. Its homologue from mammalian cells, PAK (p21-activated protein kinase), is stimulated by binding of the mammalian homologue of Cdc42 (82). Loss of *STE20* function does not affect budding or the establishment of cell polarity in yeast (83); however, activation of Ste20 or its close homologue, Cla4, by GTP-Cdc42 is essential for localization of cell growth with respect to the septin ring and for cytokinesis (84). A *cla4 ste20* double mutant cannot undergo cytokinesis and is inviable (84, 85). In addition, Ste20 functions during mating in both the pheromone signaling pathway (83, 86) and the cell wall integrity signaling pathway (87) as well as in the transition to pseudohyphal and invasive growth (2). Ste20 mutants that can no longer bind Cdc42 are defective in pseudohyphal growth and are unable to restore growth of *ste20 cla4* mutant cells (85). Paradoxically, although the necessity for both Cdc42 (88, 89) and Ste20 (83, 86, 90) in the pheromone signaling pathway has been well documented, mutations in either of their mutual binding domains did not abolish transduction of the mating signal (85). Perhaps, in the mating response, the contact between Ste20 and its partners is facilitated by the scaffolding protein, Ste5 (72, 90). Alternatively, during pheromone signaling, Ste20 may be activated by a factor different from Cdc42. The latter is consistent with the observation that overexpression of dominant negative or constitutively active alleles of *CDC42* has no effect on the basal activity of the pheromone signaling pathway (89, 90), although overexpression of the constitutively active allele potentiates the response to the mating pheromone (89, 90).

Regulators

CDC24, CDC43, BEM2, and *BEM3* genes were also found to be necessary for the assembly of the bud (49, 71, 91). Cdc24 is a GEF for Cdc42, as mentioned above (71). It also binds GTP-bound Bud1 (45, 70), thus perhaps linking bud site selection and bud site formation.

CDC43 encodes a β-subunit of the geranylgeranyl transferase I that prenylates Cdc42, thus facilitating its attachment to the membrane (92, 93).

Mutations in *CDC42* that eliminate the isoprenylation site result in a nonfunctional product, which suggests that proper membrane localization is necessary for Cdc42 function (56, 94).

BEM2 and *BEM3* contain a rho-GTPase-activating protein (rho-GAP) homology domain, but only Bem3 is able to stimulate hydrolysis of GTP by Cdc42 in vitro (71). Another protein with rho-GAP activity for Cdc42 is Rga1 (Rho GTPase activating protein) (95). Loss of Rga1 activity causes cells to bud polarly and slightly increases activity of the pheromone signaling pathway, whereas its overexpression dampens it (95). Although genetic interactions between *RGA1* and *BEM3* with $cdc24^{ts}$ support the possibility that they both function as negative regulators of Cdc42 in vivo (95), mutation of *BEM3* has not been reported to activate the pheromone pathway. At least one more protein with GAP activity on Cdc42 is expected to function in yeast because the *rga1 bem3* double mutation has a less severe phenotype than the constitutively active $CDC42^{G12V}$ mutation (95). Several other genes encoding putative Rho-GAPs were identified in the yeast genome, but their function is unknown.

Phosphatidylinositol 4,5-bisphosphate (PIP$_2$) can regulate mammalian Cdc42 and Rho by functioning as their GEF in vitro (96). In *S. cerevisiae*, the putative phosphatidylinositol kinase Tor2, involved in organization of actin (97), appears to act upstream of Rho1 and Rho2 (98). However, in mammalian cells, phosphatidylinositol 4-phosphate-5-kinase was found to act downstream of Rho (99).

Other Players

BEM4 encodes a protein required for bud emergence at higher temperature (100). Because it interacts with constitutively active and dominant negative forms of Cdc42 as well as with Rho1, Rho2, and Rho4, its role in cell morphogenesis may be more general: e.g. it may play the role of a chaperone for small GTPases (100).

Boi1 and Boi2 (Bem one interacting) is another pair of proteins with unknown function that interact with both Bem1 and the GTP-bound form of Cdc42 (101). The stoichiometry of the components in this complex is crucial because overexpression of *BOI1* inhibits bud emergence, but this inhibition is counteracted by co-overexpression of *CDC42* (101). This complex may link the functions of Cdc42 and Rho3, because *RHO3* is an efficient dosage suppressor of *boi1 boi2* mutant, and, at the same time, overexpression of *RHO3* exacerbates the effect of overexpressed *BOI1* (101).

How the Machine Works

After seeing only a few parts of a complex apparatus, it is difficult to determine how it functions and how each component contributes. But a model, even if tentative, may help to orient us through the mass of data (Figure 3).

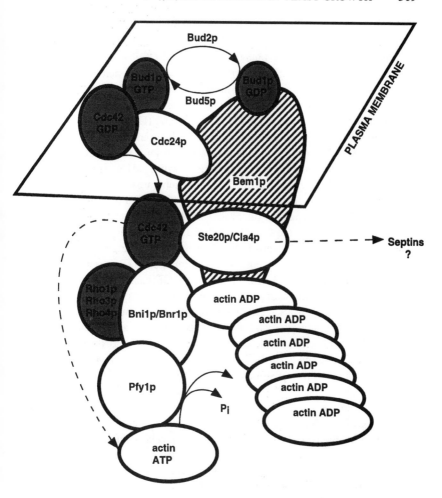

Figure 3 Schematic model for the organization of a bud site. *Solid arrows* indicate known or suspected conformational changes, and *dashed arrows* indicate hypothetical interactions. Small G proteins participating in the complex are shaded. Bee1/Las17, and Pca1, which probably function downstream of Cdc42 (66), are not depicted. For simplicity, proteins of the landmark complex have been omitted.

In the model depicted in Figure 3, the polarity establishment complex is targeted to the landmark defining the bud site by the activity of the small GTPase Bud1 (12). Because Bud1 itself is distributed rather uniformly over the cell surface, the landmark may bring about localized activation of Bud1 by affecting the positions and activities of its GEF Bud5 (16, 18) or its GAP Bud2 (13, 44). A loss of function of either Bud2 or Bud5, as well as any mutation of Bud1 that locks it in one state (either GTP- or GDP-bound) results in random budding.

Thus, for proper assembly of the polarity establishment complex, Bud1 must cycle between its GTP- and GDP-bound forms. Each of these forms has a different target: GTP-Bud1 binds to Cdc24, and GDP-Bud1 binds to Bem1 (Figure 3; 45, 70). In addition, Bem1 and Cdc24 bind to each other (Figure 3; 69, 70). What is the physiological consequence of all these interactions? Park et al (45) proposed the following model: In the first step, GTP-Bud1 would bind Cdc24, which in turn binds GDP-Cdc42 and Bem1 (Figure 3), thus recruiting the whole complex to the future bud site. In the second step, Bud2-induced hydrolysis of GTP-Bud1 to GDP-Bud1 would reorient Bud1 to Bem1. Consequently, Cdc24 released from the interactions with Bud1 and Bem1 would be able to activate Cdc42. Finally, Bud5 would regenerate GTP-Bud1 from GDP-Bud1, thus making it ready to start another cycle with a new load of Cdc24, Cdc42, and Bem1. Repeating the cycle would lead to an increased concentration of Bem1 and activated Cdc42 at the bud site. This model is at odds with results of Zheng et al (70), who observed that in vitro binding of Cdc24 to GTP-Bud1 inhibits both its intrinsic and Bud2p-stimulated GTPase activity and that binding of Bem1 or of Bud1 does not affect the GEF activity of Cdc24 toward Cdc42. If these behaviors are manifest in vivo, the cycle proposed by Park et al (45) would be blocked after the first step. Moreover, Bud1 seems not to be required for activation of Cdc42, only for its proper localization (see above). More work is necessary to understand the mechanism of bud site organization.

Activated Cdc42 binds Bni1 (Figure 3; 75), which through interactions with both actin and profilin can stimulate actin polymerization (Figure 3; 75, 79). Because it also binds actin, Bem1 may be positively involved in this process (Figure 3; 71). However, Bni1 is essential for actin polarization only during mating (75). During vegetative growth, its role may be at least partially supplanted by Bnr1 (102) or other proteins with related function. Bni1 interacts with the G protein Rho1 in vivo and in vitro and is a potential target of Rho1 (Figure 3; 74). At this point, it is not clear whether Rho1, similarly to mammalian Rho, functions downstream of Cdc42 (103, 104) or whether it acts independently.

Other targets of GTP-Cdc42 are the protein kinases, Ste20 and Cla4, which are required in a later stage of bud development for proper localization of cell growth with respect to the septin ring (Figure 3; 84).

Mating

As already mentioned, when a yeast cell is near another yeast cell of opposite mating type, it will grow a polarized projection along the pheromone gradient produced by the mating partner (105, 106). As in bud formation, the actin cytoskeleton, secretion, and new cell wall construction are polarized toward the tip of the projection (107) by the same polarity establishment molecules, which include *CDC42, CDC24,* and *BEM1* (10, 54, 88, 108, 109). In this case, genetically determined instructions for axis formation are usually ignored, and the spatial cues are probably provided by the highest concentration of the pheromone-bound receptor, by Far1, (110, 111), and undoubtedly by other molecules. If these cues are missing because of the *far1-s* mutation, or because of the absence of the pheromone gradient when a cell is incubated in an isotropic solution of pheromone, the cell orients actin toward a regular (axial) bud site but still produces a mating projection and not a bud (112). It seems that the components of the polarity establishment complex that are common for both budding and mating can always recognize the bud site, but the presence of pheromone signal changes the situation inside the cell in such a way that only a projection can be formed.

BUILDING NEW CELL WALL

Once the budding machinery has been organized, actual growth of the daughter cell can begin, including addition of new cell surface (membrane and cell wall) and of intracellular material, both in soluble form and as organelles. We concentrate here on growth of the cell wall because it is directly involved in morphogenesis. The wall determines cell shape: Enzymatic digestion of the cell wall leads to the formation of spherical protoplasts; conversely, cell walls isolated after mechanical breakage of cells maintain the shape of the intact cell. Cell wall synthesis must be regulated in synchrony with the cell cycle. Before bud emergence, the mother cell wall is in a quiescent state. As the bud starts growing, wall synthesis is switched on; at daughter cell maturation, it is switched off. The composition of the cell wall is relatively simple, consisting of a few polysaccharides and of mannoproteins (113). The synthesis of each of these components must be under the control of the cell wall growth switches. Thus, control of wall growth can be studied at the molecular level by following biosynthesis of one of the main constituents. This approach has led to the finding that Rho1 is an essential regulator of the synthesis of the main structural component of the *S. cerevisiae* cell wall, $\beta(1 \rightarrow 3)$glucan.

A GTP Requirement for β(1 → 3)Glucan Synthase

Although in vitro biosynthesis of $\beta(1 \rightarrow 3)$glucan in *S. cerevisiae* was reported as early as 1975 (114), conditions for an efficient transfer of glucose from UDP-glucose to an accepting glucan chain in the presence of membrane preparations were not determined until 1980 (115, 116). In these studies, GTP or some of its analogs were found to be potent stimulators of the enzymatic activity. The nucleotides were active at concentrations in the micromolar or submicromolar range, a level that may easily be present in vivo, suggesting the possibility that they were physiological regulators of the synthase. Further progress was slow because localization of the synthase in membranes hindered its dissection and characterization. However, a later study of the effect of GTP on a number of fungal $\beta(1 \to 3)$glucan synthases (117) culminated in the finding that the enzyme could be dissociated by extraction with a mixture of salt and detergent into two components, one soluble (fraction A) and the other still membrane bound (118). Reconstruction of enzymatic activity required both fractions plus GTP. Thermal stability experiments in the absence or presence of the nucleotide suggested that GTP was interacting with the solubilized fraction. Similar results were later obtained with membrane preparations from *S. cerevisiae*. In this case, further extraction of the insoluble fraction with other detergents resulted in the solubilization of another fraction (fraction B), which, when added to fraction A, supported polysaccharide synthesis in the presence of GTP (119).

A GTP-Binding Protein Is a Component of $\beta(1 \to 3)$Glucan Synthase

Purification of fraction A from yeast by ion exchange and gel filtration chromatography led to cofractionation of GTP-binding activity, as measured by adsorption on nitrocellulose membranes, and ability to complement fraction B in a glucan synthase assay (119). The component of fraction A required for glucan synthesis was a GTP-binding protein (119). The elution profile of a sizing column suggested a molecular weight around 25,000, whereas a band at 20 kDa was photolabeled in the presence of $[\gamma\text{-}^{32}P]GTP$ (119). However, microsequence analysis of the latter protein showed that this band was unrelated to the glucan synthase system (T Drgon and E Cabib, unpublished results). The labeling appears to have been spurious.

These results suggested that a small GTP-binding protein was involved in the activity of $\beta(1 \to 3)$glucan synthase and provided a rationale for the switching on and off of the enzyme during the cell cycle, presumably by alternating between the active GTP-bound and the inactive GDP-bound states (119). A genetic approach was needed to determine whether this GTP-binding protein was a new small G protein or one of those already detected in yeast.

Rho1 Is the Regulatory Subunit of $\beta(1 \to 3)$Glucan Synthase

In an attempt to identify the small G protein involved in glucan synthase activity, the reported phenotypes of mutants in known yeast G proteins were scrutinized. Only mutants in two pairs of closely related genes, those in the *RHO1-RHO2* pair (120) and in the *RHO3-RHO4* pair (121), showed a phenotype compatible with the expected defect. In both cases, conditional mutants lysed when brought to a nonpermissive condition, and the majority of cells susceptible to the lysis bore a small bud. It was reasoned that a cell incapable of switching glucan synthase on at bud emergence would give rise to a bud with a defect in the wall, leading to cell lysis. Measurements of glucan synthase activity showed a normal enzyme in $rho4\Delta rho3^{ts}$ mutants. However, membrane preparations from $rho1^{ts}$ mutants were clearly defective even at permissive temperatures and showed almost no stimulation by GTP (122). Addition of purified fraction A or recombinant Rho1 restored both activity and GTP stimulation (122, 123). Glucan synthase activity was also reconstituted by addition of recombinant Rho1 to fraction B, showing that Rho1 is the only active component of purified fraction A (122). Furthermore, when purified preparations of fraction A from wild-type and $rho1^{ts}$ cells were subjected to SDS-polyacrylamide gel electrophoresis, a band at 24 kDa was found to be absent in the mutant. The same band was labeled by ADP-ribosylation with *Clostridium botulinum* C-3 exoenzyme, as was recombinant Rho1 (122). Rho1 is the only Rho protein from *S. cerevisiae* that is ADP-ribosylated by the C-3 exoenzyme (78, 124). From all this evidence it was concluded that Rho1 is the G protein present in fraction A that is necessary for glucan synthase activity.

On the Nature of the Direct Target of Rho1 in the Glucan Synthase System

To act as an activator of glucan synthase, Rho1 presumably must interact directly with the catalytic subunit of glucan synthase or with a protein associated with it. Fraction B is a crude preparation that contains many proteins. However, glucan synthase has been purified extensively from solubilized membranes by the product entrapment procedure (125). These preparations were enriched in the product of the *FKS1* gene, which, together with its homologue, *FKS2* (126), has been implicated in the activity of glucan synthase (127, 128). Mutants in *FKS1* were first isolated in a screen for hypersensitivity to immunosuppressants (129), and again, under the name *etg1*, in a screen for strains resistant to glucan synthase inhibitors (127). Null (*fks1*Δ) mutants show a decrease in glucan synthase activity (128) and in incorporation of glucose into β $(1 \rightarrow 3)$glucan in vivo (130); double null (*fks1*Δ*fks2*Δ) mutants are inviable (126). Furthermore, immunoprecipitation of Fks1 or Fks2 from purified preparations of the synthase resulted in coprecipitation of enzymatic activity (126, 131). These results led to the conclusion that Fks1 and Fks2, large hydrophobic proteins with 16 putative transmembrane domains (126, 128), are essential components of the glucan synthase system. It could not be established conclusively, however, whether they represent the catalytic subunit of the synthase, because even the most purified preparations contained other proteins. Fks1 and Fks2 seem to have different, if overlapping, functions because they are regulated differently. During vegetative growth, Fks1 is preferentially expressed (126), whereas Fks2 is required for sporulation (126). Also, Fks2 is induced by Ca^{2+} as well as by pheromones, in both cases in a calcineurin-dependent fashion (126).

Although it has not been shown that Fks1 or Fks2 interacts directly with Rho1, they appear to be part of the same complex. Rho1 copurifies with Fks1 upon product entrapment of glucan synthase (123, 131), and immunoprecipitation of Fks1 results in coprecipitation of Rho1 (123, 131). Consequently, although it has not been proven that Fks1 and Fks2 are direct targets of Rho1, they are strong candidates.

Localization of the Glucan Synthase Components and a Scheme for Their Interaction

It has long been known that $\beta(1 \to 3)$glucan synthase activity is localized to the plasma membrane (115). Because Rho1 is essential for activity, both the G protein and the catalytic components must be at that location. The presence of a prenyl group on Rho1 (132) favors its binding to the membrane. At bud emergence, the glucan synthase complex would be expected to be at the bud organization site, to start cell wall formation in the new bud. In fact, Yamochi & coworkers (132) detected Rho1 at the bud site, in the same general area where actin is localized, by immunofluorescence, whereas Qadota et al (123) used a similar methodology to show Fks1 at the same site. Both proteins appear to relocalize to the neck between mother and daughter cell at septum formation (123, 132). These findings suggest a simple scheme for the interaction of the glucan synthase components (Figure 4). The catalytic portion may be present in the plasma membrane in an inactive form (Figure 4, where the closed "gate" symbolizes the inactive state). When recruited to the bud site (Figure 4), it finds Rho1, which by exchanging GDP for GTP has undergone a conformational change that enables it to bind to the catalytic complex. This binding, in turn, changes the conformation of the catalytic subunit (gate open), uncovering the active site, to which UDP-glucose can now attach. Glucan synthesis ensues, with simultaneous extrusion of the polysaccharide into the extracellular space, as was formerly demonstrated for chitin (133). The possibility of an indirect interaction with actin, as discussed in a previous section, is also indicated (Figure 4).

The Regulation of Rho1

Candidate regulators of Rho1 have been found in yeast, although there is evidence for only some of these proteins that they have the expected physiological function (Figure 5). Two GEFs with apparently overlapping functions were detected: Rom1 and Rom2 (134). A double null (rom1Δ rom2Δ) mutant showed a phenotype similar to that of rho1Δ strains, i.e. the mother cell arrests with a small bud (134). The question of which of several candidates may be physiological GAPs for Rho1 is not yet resolved. The GAP domain of Bem2 is active on Rho1 in vitro (69), but the phenotype of cells containing a bem2Δ mutation

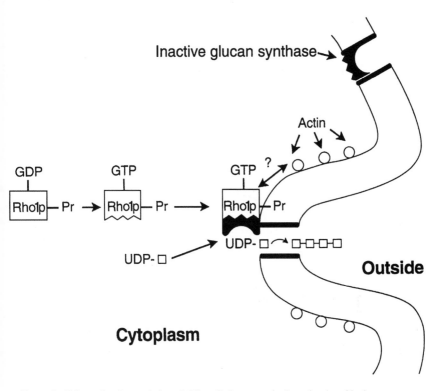

Figure 4 Scheme for the regulation of $\beta(1 \to 3)$glucan synthesis at the site of bud emergence. Actin filaments are shown in section around an invagination of the plasma membrane encompassing a small portion of the bud site (55). The GTP-GDP exchange in Rho1 is shown as taking place in the cytoplasm, but it could occur on the membrane. The small squares attached to UDP are glucosyl units. *Pr*, prenyl group. Adapted from reference 122.

(large and multinucleate) (135) is difficult to interpret as the consequence of an inability to inactivate Rho1. It is also unclear why overexpression of *RHO1* and *RHO2* would suppress rather than exacerbate a *bem2* defect (135). Another candidate GAP is Sac7, although in this case the demonstration of in vitro activity is not convincing (98). In the course of purification of Rho1 (fraction A), a fraction with apparent GAP activity on Rho1 was isolated (119). The activity is still present in extracts from strains with mutations in putative GAPs (J Drgonová, CSM Chan, E Cabib, unpublished data), such as Bem2 (135), Bem3 (71), Dbm1/Rga1 (136), and Lrg1 (137). Finally, a gene for a GDI was cloned and the corresponding protein purified and shown to act on Rho1 (138). There is, however, no phenotype associated with disruption of this gene (138).

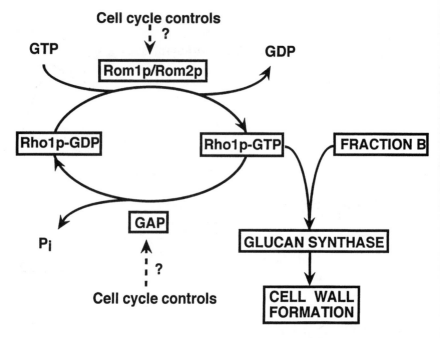

Figure 5   Scheme for the regulation of Rho1 activity and $\beta(1 \to 3)$glucan synthesis. For explanations see text.

The effects of the different regulators are summarized in Figure 5. Clearly, the opposing actions of a GEF and a GAP can switch Rho1 activity and therefore glucan synthase and cell wall biosynthesis, on and off. That in the glucan synthase system only GTP-Rho1 is active was demonstrated with the help of the above-mentioned purified GAP activity (119). Furthermore, in cells containing a dominant active allele of *RHO1*, glucan synthase was active in the absence of added GTP (123). However, GAPs and GEFs represent only one aspect of regulation. Glucan synthase activity must be strictly regulated in time, in synchrony with the cell cycle. Therefore, the cell cycle machinery may modulate the GAPs or GEFs or both (Figure 5).

Other Functions of Rho1

In addition to its direct effect on $\beta(1 \to 3)$glucan synthase, Rho1 has other functions. Both genetic (139) and biochemical (140) evidence demonstrates that Rho1 can interact with and activate the protein kinase Pkc1. This enzyme in turn activates a MAP kinase cascade, which functions to maintain integrity of the cell wall (141–144). The nature of the ultimate target(s) of the cascade is not clear; however, one of its effects is to increase transcription of *FKS2* at high temperatures (140). A *pkc1* mutant shows a partial defect in components of the cell wall and exhibits synthetic lethality with some members of the *KRE* gene family (145). The latter are involved in the synthesis of $\beta(1 \rightarrow 6)$glucan, which plays a central role in the cross-linking of the different components of the yeast cell wall (146). A recent report (147) identified SBF, a transcription factor that regulates the transition between G1 and S in the cell cycle, as another target of the Pkc1-regulated MAP kinase cascade. In this signaling pathway, the Cdc28 kinase seems to work as an upstream regulator of Pkc1 (148, 148a). So far, however, it is not known whether Rho1 mediates the effect of the Cdc28 kinase on Pkc1.

A *pkc1Δ* null mutant lyses with a small bud, a phenotype similar to that of *rho1^ts* mutants (132, 144, 148b). However, the *pkc1* defect is suppressed by osmotic protectors (141), whereas that of *rho1* mutants is not (132; E Cabib, J Drgonová, unpublished data). If the only other target of Rho1, in addition to Pkc1, were glucan synthesis, osmotic protectors should prevent lysis, just as they do for protoplasts that completely lack a cell wall. A possible function of Rho1 in actin organization (see above) might explain these results, because actin mutants show some lysis that is not suppressed by sorbitol (149). Another issue with the phenotype of *pkc1* mutants arises in the above-mentioned case of SBF. Because the latter is required for the G1 to S transition, and if the function of SBF depended solely on the activity of Pkc1, *pkc1* mutants should be blocked in G1. However, they proceed to bud before they lyse. Together with the finding that overexpression of Pkc1 suppresses a null mutation of *SWI4*, which encodes a component of SBF (148a), these results suggest the existence of two pathways for budding, one that requires SBF and the other Pkc1 activity. Despite these complications, it seems safe to conclude that Rho1 regulates cell wall synthesis both directly through glucan synthase and indirectly through the Pkc1-dependent MAP kinase cascade.

Rho3 and Rho4, Two Proteins Implicated in Bud Growth

RHO3 and *RHO4* encode small G proteins that appear to have overlapping functions (121). Disruption of *RHO3* results in very slow growth, whereas null mutants of *RHO4* have no discernible phenotype. However, *RHO4* is a dosage suppressor of the *rho3* null defect, and a double deletion of both genes is lethal. The phenotypes of mutants are somewhat variable. When a strain harboring deletions of both *RHO3* and *RHO4* as well as a plasmid carrying *RHO4* under control of a *GAL7* promoter was shifted from galactose to glucose medium, after many hours most of the cells were found to be lysed with a small bud (150). However, when cells with a *rho4* deletion and a temperature-sensitive *rho3* mutation were shifted to the nonpermissive temperature, they ended up as large cells in which actin was delocalized (151). Matsui and his coworkers concluded that *RHO3* and *RHO4* are involved in maintaining cell polarity during bud growth (150, 151), but it is not clear how they accomplish this function. There are tantalizing interactions with other genes. Thus, overexpression of *BEM1* or *CDC42*, both participants in bud site organization (see above), suppresses the *rho3/rho4* defect (150). As mentioned above, overexpression of *RHO3* or *RHO4* suppresses the *boi1* or *boi2* mutation, which has a similar phenotype to *rho3* (101, 152). Finally, *rho3* is synthetically lethal with mutants of *SEC4*, whose product is a Rab-type G protein involved in the fusion of vesicles with the plasma membranes (151). To understand these interactions, it is necessary to determine the targets of Rho3 and Rho4.

SMALL G PROTEINS MUSINGS

As has been described here, both the selection of a bud site and its organization are intricate processes that require the formation of complexes that include multiple proteins. Most of these proteins have been identified by the use of screens for interacting genes, such as suppression of a mutation by high-copy expression of another gene, two-hybrid interaction, and synthetic lethality. Whereas these techniques have been invaluable for the identification of proteins potentially participating in a physiological function, they are not without flaws. For instance, it may happen that the two-hybrid procedure recognizes two proteins that are part of the same complex but do not interact functionally or physically or that the simultaneous presence of two unrelated mutations makes the cell so sick that "synthetic" lethality ensues. Furthermore, genetic methods do not indicate the nature of the interaction among different proteins. Their use has often led to very complex schemes in which the putative participants in a cellular process are connected by arrows whose meaning is difficult to assess. One striking example of this situation is found in the recent cloning of two genes found in so many screens that they were named *ZDS* for zillion different screens (153). Some of these difficulties are caused at least in part by the fact that such complex processes as cell polarization and morphogenesis are not strictly sequential or only so to a certain extent (and in those cases genetic determination of epistasis is of great help). In contrast, metabolic pathways are usually linear processes in which each product is the precursor of the next one, which explains the spectacular success of genetics in the elucidation of metabolic routes. Therefore it is very important to complement genetic data with results obtained by other methodologies. These strategies include localization of gene products by immunofluorescence; in vitro binding of heterologously expressed proteins; and, in the case of G proteins, biochemical measurement of the activity of a putative GAP, GEF, or GDI. A more ambitious methodology is needed, however—specifically, in vitro reconstruction of the system under study. This approach has been very successful in the analysis of vesicular transport and secretion (154), which involve events of a complexity comparable to those discussed in this review. A beginning has been made by Lee and coworkers (65) with the use of permeabilized cells to study actin organization (see above). Further development of such systems, though difficult, could be very productive.

Despite the problems and uncertainties, great progress has been made in understanding the mechanism of bud site selection and organization, especially in the identification of major players in these complex events. Two of them are small G proteins, Bud1 for bud site selection and Cdc42 for bud site assembly. In their absence, the corresponding process does not take place. What determines their localization and how they interact with their presumptive targets is not clear. More work is also needed to establish how their cycling between active (GTP-bound) and inactive (GDP-bound) forms is integrated in the processes they control.

The function of Rho1 in cell wall morphogenesis as represented by the synthesis of $\beta(1 \rightarrow 3)$glucan is easier to understand because here the immediate target, the enzyme $\beta(1 \rightarrow 3)$glucan synthase, is known. In fact, so far this case is unique because a small G protein directly signals the system that gives rise to the ultimate product of the regulatory cascade, here the cell wall. This mechanism for regulation of $\beta(1 \rightarrow 3)$glucan synthase seems to be quite general in fungi, because fractions with similar activity to fraction A from *S. cerevisiae* were isolated from several organisms (117). In *Schizosaccharomyces pombe*, a homologue of *S. cerevisiae* Rho1 is the synthase regulator (155). On the other hand, we know very little about the upstream regulation of Rho1 that ultimately must synchronize cell wall synthesis with other events of the cell cycle.

Homologues of the three G proteins discussed here are known to exist in mammalian cells, where they participate in many cellular processes, such as formation of stress fibers (156), lamellipodia (157), and filopodia (158, 159). The mechanisms of action of these proteins may be expected to share similarities in both mammalian organisms and yeast. The latter is more amenable to genetic approaches, whereas mammalian cells are accessible to certain techniques, such as microinjection, that cannot be used with yeast. The results obtained with one type of cell may therefore complement those found with the other and help in the elucidation of mechanisms for polarization and morphogenesis, which are processes common to all organisms.

ACKNOWLEDGMENTS

We thank JG Gray, I Herskowitz, D Johnson, Y Matsui, M Snyder, and Y Takai for communicating results before publication.

Visit the *Annual Reviews home page* at
http://www.AnnualReviews.org.

Literature Cited

1. Boguski MS, McCormick F. 1993. *Nature* 366:643–54
2. Mösch H-U, Roberts RL, Fink GR. 1996. *Proc. Natl. Acad. Sci. USA* 93: 5352–56
3. Winge O. 1935. *C. R. Lab. Carlsberg. Ser. Physiol.* 21:77–111
4. Freifelder D. 1960. *J. Bacteriol.* 124: 511–23
5. Streiblova E. 1970. *Can. J. Microbiol.* 16:827–31
6. Hicks J, Strathern JN. 1977. *Brookhaven Symp. Biol.* 29:233–42
7. Herskowitz I, Park HO, Sanders S, Valtz N, Peter M. 1995. *Cold Spring Harbor Symp. Quant. Biol.* 60:717–27
8. Sanders SL, Field CM. 1995. *Curr. Biol.* 5:1213–15
9. Chant J, Pringle JR. 1991. *Curr. Opin. Genet. Dev.* 1:342–50
10. Roemer T, Vallier LG, Snyder M. 1996. *Trends Cell Biol.* 6:434–41
10a. Cid VJ, Duran A, del Rey F, Snyder MP, Nombela C, Sanchez M. 1995. *Microbiol. Rev.* 59:345–86
11. Chant J. 1996. *Curr. Opin. Cell Biol.* 8: 557–65
12. Chant J, Herskowitz I. 1991. *Cell* 65:1203–12
13. Park H-O, Chant J, Herskowitz I. 1993. *Nature* 365:269–74
14. Cvrckova F, Nasmyth K. 1993. *EMBO J.* 12:5277–86
15. Benton BK, Tinklenberg AH, Jean D, Plump SD, Cross FR. 1993. *EMBO J.* 12:5267–75
16. Powers S, Gonzales E, Christensen T, Cubert J, Broek D. 1991. *Cell* 65:1225–31
17. Chant J, Corrado K, Pringle JR, Herskowitz I. 1991. *Cell* 65:1213–24
18. Bender A, Pringle JR. 1989. *Proc. Natl. Acad. Sci. USA* 86:9976–80
19. Pringle JR, Bi E, Harkins HA, Zahner JE, De Virgilio C, et al. 1995. *Cold Spring Harbor Symp. Quant. Biol.* 60:729–44
20. Chant J, Mischke M, Mitchell E, Herskowitz I, Pringle JR. 1995. *J. Cell Biol.* 129:767–78
21. Sanders SL, Herskowitz I. 1996. *J. Cell Biol.* 134:413–27
22. Fujita A, Oka C, Arikawa Y, Katagai T, Tonouchi A, et al. 1994. *Nature* 372:567–70
23. Roemer T, Madden K, Chang JT, Snyder M. 1996. *Genes Dev.* 10:777–93
24. Halme A, Michelitch M, Mitchell EL, Chant J. 1996. *Curr. Biol.* 6:570–79
25. Yang S, Ayscough KR, Drubin DG. 1997. *J. Cell Biol.* 136:111–23
26. Snyder M. 1989. *J. Cell Biol.* 108:1419–29
27. Bauer F, Urdaci M, Aigle M, Crouzet M. 1993. *Mol. Cell. Biol.* 13:5070–84
28. Drubin DG, Jones HD, Wertman KF. 1993. *Mol. Biol. Cell* 4:1277–94
29. Sivadon P, Bauer F, Aigle M, Crouzet M. 1995. *Mol. Gen. Genet.* 246:485–95
30. Durrens P, Revardel E, Bonneu M, Aigle M. 1995. *Curr. Genet.* 27:213–16
31. Zahner JE, Harkins HA, Pringle JR. 1996. *Mol. Cell. Biol.* 16:1857–70
32. White GC, Crawford N, Fischer TH. 1993. *Adv. Exp. Med. Biol.* 344:187–94
33. Ruggieri R, Bender A, Matsui Y, Powers S, Takai Y, et al. 1992. *Mol. Cell. Biol.* 12:758–66
34. Neufeld TP, Rubin GM. 1994. *Cell* 77:371–79
35. McCabe PC, Haubruck H, Polakis P, McCormick F, Innis MA. 1992. *Mol. Cell. Biol.* 12:4084–92
35a. Drubin DG, Nelson JW. 1996. *Cell* 84:335–44
36. Chant J, Pringle JR. 1995. *J. Cell Biol.* 129:751–65
37. Chant J. 1996. *Cell* 84:187–90
38. Kim HB, Haarer BK, Pringle JR. 1991. *J. Cell Biol.* 112:535–44
39. Haarer BK, Pringle JR. 1987. *Mol. Cell. Biol.* 7:3678–87
40. Ford SK, Pringle JR. 1991. *Dev. Genet.* 12:281–92
41. Flescher EG, Madden K, Snyder M. 1993. *J. Cell Biol.* 122:373–86
42. Michelitch M, Chant J. 1996. *Curr. Biol.* 6:446–54
43. Gimeno CJ, Ljungdahl PO, Styles CA, Fink GR. 1992. *Cell* 68:1077–90
44. Bender A. 1993. *Proc. Natl. Acad. Sci. USA* 90:9926–29
45. Park H-O, Bi EF, Pringle JR, Herskowitz I. 1997. *Proc. Natl. Acad. Sci. USA* 94: 4463–68

46. Kilmartin JV, Adams AEM. 1984. *J. Cell Biol.* 98:922–33
47. Waddle JA, Karpova TS, Waterston RH, Cooper JA. 1996. *J. Cell Biol.* 132:861–70
48. Doyle T, Botstein D. 1996. *Proc. Natl. Acad. Sci. USA* 93:3886–91
49. Adams AEM, Johnson DI, Longnecker RM, Sloat BF, Pringle J. 1990. *J. Cell Biol.* 111:131–42
50. Preuss D, Mulholland J, Franzusoff A, Segev N, Botstein D. 1992. *Mol. Biol. Cell* 3:789–803
51. Deleted in proof
52. Sloat BF, Pringle JR. 1978. *Science* 200:1171–73
53. Sloat BF, Adams A, Pringle JR. 1981. *J. Cell Biol.* 89:395–405
54. Chenevert J, Corrado K, Bender A, Pringle J, Herskowitz I. 1992. *Nature* 356:77–79
55. Mulholland J, Preuss D, Moon A, Wong A, Drubin D, Botstein D. 1994. *J. Cell Biol.* 125:381–91
56. Ziman M, Preuss D, Mulholland J, O'Brien JM, Botstein D, Johnson DI. 1993. *Mol. Biol. Cell.* 4:1307–16
57. Johnson DI, Pringle JR. 1990. *J. Cell Biol.* 111:143–52
58. Miller PJ, Johnson DI. 1997. *Yeast* 13:561–72
59. Shinjo K, Koland JG, Hart MJ, Narasimhan V, Johnson DI, et al. 1990. *Proc. Natl. Acad. Sci. USA* 87:9853–57
60. Munemitsu S, Innis MA, Clark R, McCormick F, Ullrich A, Polakis P. 1990. *Mol. Cell. Biol.* 10:5977–82
61. Fawell E, Bowden S, Armstrong J. 1992. *Gene* 114:153–54
62. Miller PJ, Johnson DI. 1994. *Mol. Cell. Biol.* 14:1075–83
63. Chen WN, Lim HH, Lim L. 1993. *J. Biol. Chem.* 268:13280–85
64. Lew DJ, Reed SI. 1995. *Curr. Opin. Genet. Dev.* 5:17–23
65. Li R, Zheng Y, Drubin DG. 1995. *J. Cell Biol.* 128:599–615
66. Lechler T, Li R. 1997. *J. Cell Biol.* 138:95–103
67. Li R. 1997. *J. Cell Biol.* 136:649–58
68. Drubin DG, Mulholland J, Zhu ZM, Botstein D. 1990. *Nature* 343:288–90
69. Peterson J, Zheng Y, Bender L, Myers A, Cerione R, Bender A. 1994. *J. Cell Biol.* 127:1395–406
70. Zheng Y, Bender A, Cerione RA. 1995. *J. Biol. Chem.* 270:626–30
71. Zheng Y, Cerione R, Bender A. 1994. *J. Biol. Chem.* 269:2369–72
72. Leeuw T, Fourest-Lieuvin A, Wu CL,
73. Chenevert J, Clark K, et al. 1995. *Science* 270:1210–13
74. Jansen R-P, Dowzer C, Michaelis C, Galova M, Nasmyth K. 1996. *Cell* 84:687–97
75. Kohno H, Tanaka K, Mino A, Umikawa M, Imamura H, et al. 1996. *EMBO J.* 15:6060–68
76. Evangelista M, Blundell K, Longtine MS, Chow CJ, Adames N, et al. 1997. *Science* 276:118–22
77. Narumiya S. 1996. *J. Biochem.* 120:215–28
78. Kimura K, Ito M, Amano M, Chihara K, Fukata Y, et al. 1996. *Science* 273:245–48
79. Qadota H, Anraku Y, Botstein D, Ohya Y. 1994. *Proc. Natl. Acad. Sci. USA* 91:9317–21
80. Pantaloni D, Carlier MF. 1993. *Cell* 75:1007–14
81. Ayscough KR, Drubin DG. 1996. *Annu. Rev. Cell Dev. Biol.* 12:129–60
82. Tapon N, Hall A. 1997. *Curr. Opin. Cell Biol.* 9:86–92
83. Manser E, Leung T, Salihuddin H, Zhao ZS, Lim L. 1994. *Nature* 367:40–46
84. Leberer E, Dignard D, Harcus D, Thomas DY, Whiteway M. 1992. *EMBO J.* 11:4815–24
85. Cvrckova F, De Virgilio C, Manser E, Pringle JR, Nasmyth K. 1995. *Genes Dev.* 9:1817–30
86. Peter M, Neiman AM, Park H-O, van Lohuizen M, Herskowitz I. 1996. *EMBO J.* 15:7046–59
87. Ramer SW, Davis RW. 1993. *Proc. Natl. Acad. Sci. USA* 90:452–56
88. Zarzov P, Mazzoni C, Mann C. 1996. *EMBO J.* 15:83–91
89. Zhao Z-S, Leung T, Manser E, Lim L. 1995. *Mol. Cell. Biol.* 15:5246–57
90. Simon M-N, De Virgilio C, Souza B, Pringle JR, Abo A, Reed SI. 1995. *Nature* 376:702–5
91. Akada R, Kallal L, Johnson DI, Kurjan J. 1996. *Genetics* 143:103–17
92. Bender A, Pringle JR. 1991. *Mol. Cell. Biol.* 11:1295–305
93. Finegold AA, Johnson DI, Farnsworth CC, Gelb MH, Judd SR, et al. 1991. *Proc. Natl. Acad. Sci. USA* 88:4448–52
94. Ohya Y, Qadota H, Anraku Y, Pringle JR, Botstein D. 1993. *Mol. Biol. Cell* 4:1017–25
95. Ziman M, O'Brien JM, Ouellette LA, Church WR, Johnson DI. 1991. *Mol. Cell. Biol.* 11:3537–44
96. Stevenson BJ, Ferguson B, De Virgilio C, Bi E, Pringle JR, et al. 1995. *Genes Dev.* 9:2949–63

96. Zheng Y, Glaven JA, Wu WJ, Cerione RA. 1996. *J. Biol. Chem.* 271:23815–19
97. Schmidt A, Kunz J, Hall MN. 1996. *Proc. Natl. Acad. Sci. USA* 93:13780–85
98. Schmidt A, Bickle M, Beck T, Hall MN. 1997. *Cell* 88:531–42
99. Chong LD, Traynor-Kaplan A, Bokoch GM, Schwartz MA. 1994. *Cell* 79:507–13
100. Mack D, Nishimura K, Dennehey BK, Arbogast T, Parkinson J, et al. 1996. *Mol. Cell. Biol.* 16:4387–95
101. Bender L, Lo HS, Lee H, Kokojan V, Peterson J, Bender A. 1996. *J. Cell Biol.* 133:879–94
102. Imamura H, Tanaka K, Hirata T, Umikawa M, Kamei T, et al. 1997. *EMBO J.* 16:2745–55
103. Chant J, Stowers L. 1995. *Cell* 81:1–4
104. Symons M. 1996. *Trends Biochem. Sci.* 21:178–81
105. Tkacz JS, MacKay VL. 1979. *J. Cell Biol.* 80:326–33
106. Segall JE. 1993. *Proc. Natl. Acad. Sci. USA* 90:8332–36
107. Hasek J, Rupes I, Svobodova J, Streiblova E. 1987. *J. Gen. Microbiol.* 133:3355–63
108. Chenevert J. 1994. *Mol. Biol. Cell* 5:1169–75
109. Chenevert J, Valtz N, Herskowitz I. 1994. *Genetics* 136:1287–96
110. Jackson CL, Konopka JB, Hartwell LH. 1991. *Cell* 67:389–402
111. Valtz N, Peter M, Herskowitz I. 1995. *J. Cell Biol.* 131:863–73
112. Dorer R, Pryciak PM, Hartwell LH. 1995. *J. Cell Biol.* 131:845–61
113. Klis FM. 1994. *Yeast* 10:851–69
114. Sentandreu R, Elorza MV, Villanueva JR. 1975. *J. Gen. Microbiol.* 90:13–20
115. Shematek EM, Braatz JA, Cabib E. 1980. *J. Biol. Chem.* 255:888–94
116. Shematek EM, Cabib E. 1980. *J. Biol. Chem.* 255:895–902
117. Szaniszlo PJ, Kang MS, Cabib E. 1985. *J. Bacteriol.* 161:1188–94
118. Kang MS, Cabib E. 1986. *Proc. Natl. Acad. Sci. USA* 83:5808–12
119. Mol PC, Park HM, Mullins JT, Cabib E. 1994. *J. Biol. Chem.* 269:31267–74
120. Madaule P, Axel R, Myers AM. 1987. *Proc. Natl. Acad. Sci. USA* 84:779–83
121. Matsui Y, Toh-e A. 1992. *Gene* 114:43–49
122. Drgonová J, Drgon T, Tanaka K, Kollár R, Chen GC, et al. 1996. *Science* 272:277–79
123. Qadota H, Python CP, Inoue SB, Arisawa M, Anraku Y, et al. 1996. *Science* 272:279–81
124. McCaffrey M, Johnson JS, Goud B, Myers AM, Rossier J, et al. 1991. *J. Cell Biol.* 115:309–19
125. Inoue SB, Takewaki N, Takasuka T, Mio T, Adachi M, et al. 1995. *Eur. J. Biochem.* 231:845–54
126. Mazur P, Morin N, Baginsky W, El-Sherbeini M, Clemas JA, et al. 1995. *Mol. Cell. Biol.* 15:5671–81
127. Douglas CM, Marrinan JA, Li W, Kurtz MB. 1994. *J. Bacteriol.* 176:5686–96
128. Douglas CM, Foor F, Marrinan JA, Morin N, Nielsen JB, et al. 1994. *Proc. Natl. Acad. Sci. USA* 91:12907–11
129. Parent SA, Nielsen JB, Morin N, Chrebet G, Ramadan N, et al. 1993. *J. Gen. Microbiol.* 139:2973–84
130. Castro C, Ribas JC, Valdivieso MH, Varona R, del Rey F, Durán A. 1995. *J. Bacteriol.* 177:5732–39
131. Mazur P, Baginsky W. 1996. *J. Biol. Chem.* 271:14604–9
132. Yamochi W, Tanaka K, Nonaka H, Maeda A, Musha T, Takai Y. 1994. *J. Cell Biol.* 125:1077–93
133. Cabib E, Bowers B, Roberts RL. 1983. *Proc. Natl. Acad. Sci. USA* 80:3318–21
134. Ozaki K, Tanaka K, Imamura H, Hihara T, Kameyama T, et al. 1996. *EMBO J.* 15:2196–207
135. Kim YJ, Francisco L, Chen GC, Marcotte E, Chan CSM. 1994. *J. Cell Biol.* 127:1381–94
136. Chen GC, Zheng L, Chan CSM. 1996. *Mol. Cell. Biol.* 16:1376–90
137. Muller L, Xu G, Wells R, Hollenberg CP, Piepersberg W. 1994. *Nucleic Acids Res.* 22:3151–54
138. Masuda T, Tanaka K, Nonaka H, Yamochi W, Maeda A, Takai Y. 1994. *J. Biol. Chem.* 269:19713–18
139. Nonaka H, Tanaka K, Hirano H, Fujiwara T, Kohno H, et al. 1995. *EMBO J.* 14:5931–38
140. Kamada Y, Qadota H, Python CP, Anraku Y, Ohya Y, Levin DE. 1996. *J. Biol. Chem.* 271:9193–96
141. Levin DE, Bartlett-Heubusch E. 1992. *J. Cell Biol.* 116:1221–29
142. Lee KS, Irie K, Gotoh Y, Watanabe Y, Araki H, et al. 1993. *Mol. Cell. Biol.* 13:3067–75
143. Errede B, Levin DE. 1993. *Curr. Opin. Cell Biol.* 5:254–60
144. Levin DE, Bowers B, Chen CY, Kamada Y, Watanabe M. 1994. *Cell. Mol. Biol. Res.* 40:229–39

145. Roemer T, Paravicini G, Payton MA, Bussey H. 1994. *J. Cell Biol.* 127:567–79
146. Kollár R, Reinhold BB, Petráková E, Yeh HJC, Ashwell G, et al. 1997. *J. Biol. Chem.* 272:17762–75
147. Madden K, Sheu YJ, Baetz K, Andrews B, Snyder M. 1997. *Science* 275:1781–84
148. Marini NJ, Meldrum E, Bueher B, Hubberstey AV, Stone DE, et al. 1996. *EMBO J.* 15:3040–52
148a. Gray JV, Ogas JP, Kamada Y, Stone M, Levin DE, Herskowitz I. 1997. *EMBO J.* 16:4924–37
148b. Levin DE, Fields FU, Kurusawa R, Bishop JM, Thorner J. 1990. *Cell* 62:213–24
149. Novick P, Botstein D. 1985. *Cell* 40:405–16
150. Matsui Y, Toh-e A. 1992. *Mol. Cell. Biol.* 12:5690–99
151. Imai J, Toh-e A, Matsui Y. 1996. *Genetics* 142:359–69
152. Matsui Y, Matsui R, Akada R, Toh-e A. 1996. *J. Cell Biol.* 133:865–78
153. Yu YX, Jiang YW, Wellinger RJ, Carlson K, Roberts JM, Stillman DJ. 1996. *Mol. Cell. Biol.* 16:5254–63
154. Rothman JE, Wieland FT. 1996. *Science* 272:227–34
155. Arellano M, Durán A, Perez P. 1996. *EMBO J.* 15:4584–91
156. Ridley AJ, Hall A. 1992. *Cell* 70:380–99
157. Ridley AJ, Paterson HF, Johnston CL, Diekmann D, Hall A. 1992. *Cell* 70:401–10
158. Kozma R, Ahmed S, Best A, Lim L. 1995. *Mol. Cell. Biol.* 81:1159–70
159. Nobes CD, Hall A. 1995. *Cell* 81:1–20

Annu. Rev. Biochem. 1998. 67:335–94

RNA LOCALIZATION IN DEVELOPMENT

Arash Bashirullah,[1,3] *Ramona L. Cooperstock,*[1,2] *and Howard D. Lipshitz*[1,2]

[1]Program in Developmental Biology, Research Institute, The Hospital for Sick Children, Toronto, Canada M5G 1X8; [2]Department of Molecular & Medical Genetics, University of Toronto, Toronto, Canada; [3]Division of Biology, California Institute of Technology, Pasadena, California; e-mail: lipshitz@sickkids.on.ca

KEY WORDS: *Drosophila, Xenopus, Saccharomyces,* oocyte, embryo, RNA-binding protein, cytoskeleton, microtubules, microfilaments, 3′-untranslated region (3′-UTR)

ABSTRACT

Cytoplasmic RNA localization is an evolutionarily ancient mechanism for producing cellular asymmetries. This review considers RNA localization in the context of animal development. Both mRNAs and non-protein-coding RNAs are localized in *Drosophila, Xenopus,* ascidian, zebrafish, and echinoderm oocytes and embryos, as well as in a variety of developing and differentiated polarized cells from yeast to mammals. Mechanisms used to transport and anchor RNAs in the cytoplasm include vectorial transport out of the nucleus, directed cytoplasmic transport in association with the cytoskeleton, and local entrapment at particular cytoplasmic sites. The majority of localized RNAs are targeted to particular cytoplasmic regions by *cis*-acting RNA elements; in mRNAs these are almost always in the 3′-untranslated region (UTR). A variety of *trans*-acting factors— many of them RNA-binding proteins—function in localization. Developmental functions of RNA localization have been defined in *Xenopus, Drosophila,* and *Saccharomyces cerevisiae.* In *Drosophila,* localized RNAs program the anteroposterior and dorso-ventral axes of the oocyte and embryo. In *Xenopus,* localized RNAs may function in mesoderm induction as well as in dorso-ventral axis specification. Localized RNAs also program asymmetric cell fates during *Drosophila* neurogenesis and yeast budding.

CONTENTS

INTRODUCTION

Since the early days of experimental embryology it has been suggested that the asymmetric distribution of substances in the egg cytoplasm might confer particular fates to cells that receive that cytoplasm (reviewed in 1). However, it is only in the past 13 years that specific maternally synthesized, asymmetrically distributed RNA and protein molecules have been identified in oocytes and early embryos of *Xenopus, Drosophila*, ascidians, zebrafish, and echinoderms. This review focuses largely on RNAs that are localized to specific cytoplasmic regions in eggs and early embryos. It addresses both the mechanisms of cytoplasmic RNA localization and the developmental functions of this localization. Some consideration is also given to RNA localization later in development, in differentiating or differentiated cells. However, since both the mechanisms and the functions of this later localization are not well understood, the emphasis here is on RNA localization in oocytes.

This review considers only RNAs that are asymmetrically distributed in the cytoplasm. Examples of RNAs that are localized to and within the nucleus—even to specific chromosomes or regions of chromosomes (2–4)—are covered elsewhere. The first maternally synthesized cytoplasmically localized RNAs were identified in *Xenopus* in a molecular screen for RNAs enriched in either the vegetal (*Vg* RNAs) or animal (*An* RNAs) hemisphere of the *Xenopus* oocyte (5). Shortly thereafter, an RNA was discovered that is localized to the anterior pole of the *Drosophila* oocyte and early embryo (6). This RNA is encoded by the *bicoid* maternal effect locus (7), which plays a crucial role in specifying cell fates in the anterior half of the early *Drosophila* embryo (8). The facile combination of genetics and molecular biology in *Drosophila* led to *bicoid* becoming the first case in which it was demonstrated that RNA localization per se was important for normal development. Delocalization of *bicoid* RNA led to defects in anterior cell fate specification (7). Over 75 cytoplasmically localized RNAs have now been identified, and many of these are localized in eggs, early embryos, or differentiating cells (Table 1).

To date, it has been possible to address both the mechanisms and the developmental functions of RNA localization almost exclusively in *Drosophila* and *Xenopus*. The large size of the *Xenopus* oocyte has allowed mapping of sequences necessary and sufficient for RNA localization through injection of in vitro synthesized transcripts engineered to contain an exogenous reporter sequence and part or all of the localized RNA. Further, in some cases inactivation, delocalization, or degradation of specific RNAs has been induced through microinjection of antisense RNA or DNA. The ability to manipulate *Xenopus* oocytes and to apply various cytoskeleton-destabilizing drugs or other inhibitors has demonstrated the importance of the cytoskeleton in RNA localization.

Drosophila oocytes and early embryos are also large and have also been used for drug and inhibitor studies. In contrast to *Xenopus*, however, the ability to generate transgenic lines that express reporter-tagged transcripts during oogenesis has obviated the need for micronjection studies, although some of these have been conducted. Finally, the ability to obtain mutations in the endogenous gene that encodes the localized RNA or in factors that function in trans in its localization or in its translational regulation, has facilitated analyses of the

Table 1   Localized RNAs

Species	Transcript name	Protein product	Localization pattern	Cell	Reference
Ascidians					
	Actin	Cytoskeletal component	Myoplasm and ectoplasm	Oocyte	249
	PCNA	Auxilary protein of DNA polymerase	Ectoplasm	Oocyte	100
	Ribosomal protein L5	Ribosomal component	Myoplasm	Oocyte	102
	YC RNA	Noncoding RNA	Myoplasm	Oocyte	101
Drosophila					
	Add-hts	Cytoskeletal component	Anterior	Oocyte and embryo	16, 17
	Bicaudal-C	Signal transduction/ RNA-binding protein	Anterior	Oocyte	18
	Bicaudal-D	Cytoskeleton interacting protein (?)	Anterior	Oocyte	19
	bicoid	Transcription factor	Anterior	Oocyte and embryo	7, 32
	crumbs	Transmembrane protein	Apical	Cellular blastoderm	67
	Cyclin B	Cell cycle regulator	Posterior and perinuclear	Oocyte and embryo	59, 62
	egalitarian	Novel	Anterior	Oocyte	20
	even-skipped	Transcription factor	Apical	Cellular blastoderm	69
	fushi tarazu	Transcription factor	Apical	Cellular blastoderm	64
	germ cell-less	Nuclear pore associated protein	Posterior	Oocyte and embryo	57, 239
	gurken	Secreted growth factor	Anterior-dorsal	Oocyte	21
	hairy	Transcription factor	Apical	Cellular blastoderm	65
	Hsp83	Molecular chaperone	Posterior	Embryo	63
	inscuteable	Novel	Apical	Neuroblast	127
	K10	Novel	Anterior	Oocyte	22
	mtlrRNA	Noncoding RNA	Posterior	Oocyte and embryo	61, 134

(Continued)

Table 1   (*Continued*)

Species	Transcript name	Protein product	Localization pattern	Cell	Reference
	nanos	RNA binding protein	Posterior	Oocyte and embryo	28, 43
	orb	RNA binding protein	Posterior	Oocyte and embryo	23
	oskar	Novel	Posterior	Oocyte and embryo	24, 25
	Pgc	Noncoding RNA	Posterior	Oocyte and embryo	26
	prospero	Transcription factor	Apical/basal	Neuroblast	127
	pumilio	RNA binding protein	Posterior	Embryo	250
	runt	Transcription factor	Apical	Cellular blastoderm	66
	sevenless	Transmembrane receptor	Apical	Eye imaginal Epithelial cells	126
	tudor	Novel	Posterior	Oocyte	27
	wingless	Secreted ligand	Apical	Cellular blastoderm	68
	yemanuclein-α	Transcription factor	Anterior	Oocyte	251
Echinoderms					
	SpCOUP-TF	Hormone receptor	Lateral to animal-vegetal axis	Oocyte	103
Mammals					
	β-actin	Cytoskeletal component	Specialized periphery	Fibroblasts, myoblasts, and epithelial cells	123–125
	Arc	Cytosketal component	Somatodendritic	Neurons	104
	BC-1	Noncoding RNA	Somatodentritic and axonal	Neurons	107, 111
	BC-200	Noncoding RNA	Somatodendritic	Neurons	108
	CaMKIIα	Signalling component	Somatodendritic	Neurons	109
	F1/GAP43	PKC substrate	Somatodendritic	Neurons	104
	InsP3 receptor	Integral membrane receptor	Somatodendritic	Neurons	110

(*Continued*)

Table 1   (*Continued*)

Species	Transcript name	Protein product	Localization pattern	Cell	Reference
	MAP2	Cytosketal component	Somatodendritic	Neurons	106
	MBP	Membrane protein	Myelinating membrane	Oligodendrocyte and Schwann cells	120
	Myosin heavy chain	Cytoskeletal component	Peripheral	Muscle	252
	OMP/odorant receptors	Integral membrane receptor	Axonal	Neurons	118, 119
	Oxytocin	Neuropeptide	Axonal	Neurons	115
	Prodynorphin	Neuropeptide	Axonal	Neurons	116, 117
	RC3	PKC substrate	Somatodendritic	Neurons	104
	tau	Cytoskeletal component	Axon hillock	Neurons	112
	Tropomyosin-5	Cytoskeletal component	Pre-axonal pole	Neurons	113
	V-ATPase subunits	Membrane protein	Specialized membrane	Osteoclasts	122
	Vassopressin	Neuropeptide	Axonal	Neurons	253
Xenopus					
	Actin	Cytoskeletal component	Periplasmic	Oocyte	254
	Anl (a and b)	Cyoplasmic protein (ubiquitin-like)	Animal	Oocyte	5, 81
	An2	mt ATPase subunit	Animal	Oocyte	5, 82
	An3	RNA binding protein	Animal	Oocyte	5
	An4 (a and b)	Novel	Animal	Oocyte	83
	βTrCP	Signaling molecule	Animal	Oocyte	83
	βTrCP-2	Signaling molecule	Vegetal	Oocyte	83
	βTrCP-3	Signaling molecule	Vegetal	Oocyte	83
	B6	NR[a]	Vegetal	Oocyte	70
	B7	NR	Vegetal	Oocyte	70
	B9	NR	Vegetal	Oocyte	70
	B12	NR	Vegetal	Oocyte	70
	C10	NR	Vegetal	Oocyte	70

(*Continued*)

Table 1   (*Continued*)

Species	Transcript name	Protein product	Localization pattern	Cell	Reference
	G-proteins	Signaling molecule	Animal	Oocyte	84
	Oct60	Transcription factor	Animal	Oocyte	85
	PKCα	Signaling molecule	Animal	Oocyte	86
	α-tubulin	Cytoskeletal component	Periplasmic	Oocyte	254
	VegT (*Antipodean*)	Transcription factor	Vegetal	Oocyte	91, 92
	Vgl	Signaling molecule	Vegetal	Oocyte	5
	Xcat-2	RNA-binding protein	Vegetal	Oocyte	78, 79
	Xcat-3	RNA-binding protein	Vegetal	Oocyte	95
	Xcat-4	NR	Vegetal	Oocyte	70
	xl-21	Transcription factor (?)	Animal	Oocyte	87
	Xlan4	P-rich and PEST sequences	Animal	Oocyte	88
	Xlcaax-1	Membrane protein	Animal	Oocyte	89
	Xlsirt	Noncoding RNA	Vegetal	Oocyte	97
	Xwnt-11	Secreted ligand	Vegetal	Oocyte	98
Yeast					
	ASH1	Transcription factor	Budding site	Mother cell	128, 129
Zebrafish					
	Vasa	RNA-binding protein	Cleavage plane	Early embryo	209

[a]NR, Not reported.

mechanisms of RNA localization, the developmental functions of these RNAs, and of their localization per se.

This review begins with a description of patterns of cytoplasmic RNA localization with an emphasis on *Xenopus* and *Drosophila*. To help explain the patterns and their significance, brief descriptions of the structure and development of *Xenopus* and *Drosophila* oocytes and/or early embryos are included. After considering the patterns of RNA localization, the focus switches to mechanisms. First, the dynamics of RNA localization are considered, including the role of the cytoskeleton in RNA transport and anchoring. Then specific components of the localization mechanism are dissected; these include *cis*-acting sequences and *trans*-acting factors that function either in localization per se or in control of RNA stability or translation during and after localization. Finally, developmental functions of RNA localization are discussed.

PATTERNS OF RNA LOCALIZATION

All cells are nonhomogeneous since they are compartmentalized into organelles with distinct functions and locations. These inhomogeneities can result in several forms of cellular symmetry and asymmetry. For example, positioning of the nucleus in the center of an otherwise quite homogeneous spherical cell produces spherical symmetry. In such a cell (and there are few if any examples, with the possible exception of some oocytes), certain RNAs might be localized close to the nucleus (perinuclear) while others might be positioned more peripherally. More complex cellular asymmetries result from variations in cell shape and the position of the nucleus and other subcellular organelles. Cells can be radially symmetric or even further polarized in two or three axes. In these cases, RNA localization can occur relative to one, two, or three axes (e.g. to the dorsal anterior pole). Regardless of cell shape or size, RNA distribution patterns are usually based on preexisting asymmetries and can, in turn, lead to the establishment of further asymmetries.

This section describes the dynamics and patterns of subcellular distribution of cytoplasmically localized RNAs. It provides a cellular and developmental context for consideration of the mechanisms and functions of RNA localization in subsequent sections. The emphasis here is on the best understood of the examples listed in Table 1.

Drosophila Oocytes and Early Embryos

STRUCTURE AND DEVELOPMENT OF THE NURSE CELL–OOCYTE COMPLEX The two bilaterally symmetric *Drosophila* ovaries each consist of about 16 ovarioles. At the anterior tip of the ovariole is the germarium. Here the oogonial stem cells divide asymmetrically producing a stem cell and a commited cell, which is called a cystoblast. Each cystoblast divides four times with incomplete cytokinesis to form 16 cystocyte cells interconnected by cytoplasmic bridges that run through specialized membrane cytoskeletal structures called ring canals. Only 1 of the 16 cystocytes becomes the oocyte, and the remaining 15 become nurse cells. Each 16-cell germarial cyst becomes surrounded by somatically derived follicle cells to form a stage 1 egg chamber. The more posterior part of the ovariole comprises a connected series of progressively older egg chambers ordered such that the youngest is most anterior and the oldest (stage 14) most posterior relative to the body axis of the female. It takes three days for an egg chamber to produce a mature egg. Except during the final six hours, the nurse cells synthesize large amounts of RNA and protein that are transported into the developing oocyte. Many of these molecules are required during the first two hours of embryonic development prior to the onset of zygotic transcription.

Selection of the oocyte from among the 16 cystocytes is not random. Of the 16 cells, 2 are connected to 4 others and 1 of these always becomes the oocyte (9). A large cytoplasmic structure—called the fusome—containing several cytoskeletal proteins, runs through the ring canals that connect cystocytes and has been implicated in oocyte determination (10–13). The only microtubule organizing center (MTOC) in the 16-cell complex is localized to the pro-oocyte, and microtubule arrays connect all 16 cells through the ring canals (reviewed in 14, 15). Because the MTOC nucleates the minus ends of the microtubules, the microtubule-based cytoskeleton that connects the 16 cells is polarized. This has important consequences for RNA transport from the nurse cells into the oocyte as well as for RNA localization within the oocyte itself.

RNA LOCALIZATION IN STAGE 1–6 EGG CHAMBERS   Many RNAs that are later localized within the growing oocyte are first transcribed in all 16 germline cells but accumulate specifically in the pro-oocyte. These include the *Adducin like-huli tai shao* (*Add-hts*) (16, 17), *Bicaudal-C* (18), *Bicaudal-D* (19), *egalitarian* (20), *gurken* (21), *K10* (22), *orb* (23), *oskar* (24, 25), *Polar granule component* (*Pgc*) (26), and *tudor* (27) transcripts. Other RNAs that will later be localized within the oocyte are transcribed at very low levels in the nurse cells at this early stage and so cannot be visualized easily. For example, *nanos* is transcribed at low levels, and oocyte accumulation can be seen only later (28). Additional transcripts, such as *ovarian tumor* (29) and *cytoplasmic tropomyosin II* (*cTmII*) (30), also accumulate in the oocyte at this and later stages but are not localized. Therefore oocyte-specific accumulation is not unique to RNAs that will be localized within the oocyte during later stages of oogenesis but is a property of many RNAs synthesized in the germline of early egg chambers. The fact that many RNAs that accumulate in the early oocyte appear to do so with higher concentrations at the posterior cortex—the site of the only MTOC in the egg chamber (31)—is an early indication of the role of the polarized microtubule network in RNA transport and localization (see below).

The exact stage at which transcription of different localized RNAs commences (or at least the stage at which the transcripts can first be detected) varies. For example, *bicoid* is first transcribed in the nurse cells of stage 5 egg chambers and then accumulates in the oocyte (32), whereas *oskar* (24, 25) and *K10* (22) RNAs already accumulate in the oocyte in the germarium over a day earlier. Interestingly, the dynamics of accumulation of these RNAs is identical if they are intentionally transcribed at the same time (33). Thus temporal differences in patterns of oocyte accumulation of different RNAs are a consequence of variation in time of transcription and are not indicative of a difference in underlying transport mechanism, which in fact is similar for different RNAs transcribed at distinct stages.

REORGANIZATION OF THE CYTOSKELETON AND RNA LOCALIZATION DURING STAGE 7   The process of nurse cell transcription and oocyte accumulation of RNAs continues through stage 6. During this time oocyte-follicle cell interactions establish anterior-posterior polarity within the oocyte and in the surrounding follicle cells (34, 35). Reciprocal signaling between the follicle cells and the oocyte results in a reorganization of the cytoskeleton such that, by the end of stage 7, the MTOC disappears from the posterior of the oocyte and microtubules become concentrated at the anterior oocyte margin (31, 34, 35). Concomitant with this change in cytoskeletal organization, RNAs that previously accumulated at the posterior of the oocyte localize in a ring-like pattern at the anterior oocyte margin (Figure 1) (e.g. *Bicaudal-C*, *Bicaudal-D*, *bicoid*, *egalitarian*, *gurken*, *K10*, *nanos*, *orb*, *oskar*, and *Pgc*) (18–26). At this stage several proteins are also seen in an anterior ring-like pattern (e.g. Egalitarian and Bicaudal-D) (20). This redistribution of RNA and protein is likely a consequence of the reorganization of the microtubule network such that the minus ends of microtubules move from the posterior pole to the anterior. Consistent with this, a β-galactosidase fusion to the minus-end-directed microtubule motor, Nod, relocalizes from the posterior pole of the oocyte at stage 6 to a ring around the anterior margin by the end of stage 7 (36). Thus RNAs that are transported into and within the oocyte by minus-end-directed microtubule motors would be expected to accumulate at the anterior rather than the posterior pole. As expected, the transient anterior localization of transcripts (e.g. *Bicaudal-D*, *bicoid*, *K10*, *orb*) is colchicine sensitive (37), and microtubules are required for anterior Egalitarian protein localization (20).

Mutations in genes involved in oocyte-follicle cell signaling during stages 6 and 7 cause defects in oocyte polarity (see below) and in the microtubule-based cytoskeleton (e.g. *Delta*, *gurken*, *Notch*, *PKA*) (34, 35, 38, 39). For example, double-anterior oocytes can form in which microtubules have their minus ends at both oocyte poles and their plus ends at its center. Mutations in *homeless* cause a similar disorganization of microtubules (40). Such disorganization results in *bicoid* RNA localization at both poles of the oocyte while *oskar* RNA and plus-end-directed kinesin-β-galactosidase fusion protein localize in the middle (38, 41, 42). These data emphasize that microtubule polarity directs intraoocyte transcript localization.

bicoid   gurken   nanos   oskar

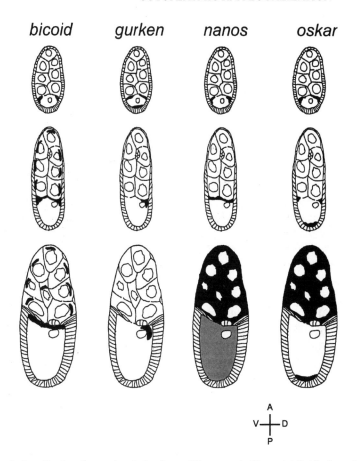

A
V——D
P

Figure 1 Localization of transcripts during *Drosophila* oogenesis. Transcript distribution patterns
are shown in black or gray shading for *bicoid, gurken, nanos,* and *oskar* RNAs in stage 8 (*top row*),
9 (*middle row*), and 10B (*bottom row*) egg chambers. A, anterior; P, posterior; D, dorsal; V, ventral.
Drawings of egg chambers are after King (9).

RNA LOCALIZATION AT STAGES 8–9   Further changes in RNA localization patterns occur at stages 8 and 9. Certain RNAs localize in a general anterior
cortical pattern. These include *Bicaudal-C* (18), *Bicaudal-D* (19), *bicoid* (7),
K10 (22), *nanos* (28, 43), *orb* (23), and Pgc (26). In contrast, *Add-hts* RNA
shifts from a general cortical pattern to an anterior cortical pattern (16, 44, 45;
KL Whittaker, D Ding, WW Fisher, HD Lipshitz, manuscript in preparation)
whereas *gurken* RNA localizes in a dorso-anterior pattern around the nucleus
(21). Others such as *egalitarian* RNA delocalize from the anterior and become

uniformly distributed (20). Still others begin to relocalize to the posterior: *Pgc* RNA spreads posteriorly along the oocyte cortex (26) while *oskar* RNA gradually moves away from the anterior and begins to accumulate at the posterior pole of the oocyte (24, 25).

During this same period a kinesin-β-galactosidase fusion protein localizes to the posterior (41). This observation suggests that, even though it has not been possible to visualize microtubules that traverse antero-posterior axis of the oocyte at this stage (31), they must be present. RNAs that enter the oocyte at this stage or that were previously localized to the anterior of the oocyte in association with minus-end-directed microtubule motors, must dissociate from these motors and associate with plus-end-directed motors in order to be translocated to the posterior pole. One such RNA is *oskar*, which by the end of stage 9, is present only at the posterior pole of the oocyte (24, 25). *oskar* RNA plays a key role in nucleating formation of the polar granules at the posterior pole of the oocyte. The polar granules, which are involved in germ-cell specification (see below), gradually assemble during stages 9–14 of oogenesis (reviewed in 47).

NURSE CELL "DUMPING" AND RNA LOCALIZATION FROM STAGES 10–14   During stages 9 and 10 nurse cells increase synthesis of RNA and protein, dump their contents into the oocyte starting at stage 10B, and degenerate by the end of stage 12. This massive transfer of material is aided by contraction of the actin cortex of nurse cells. "Dumpless" mutants affect this process as well as ring canal structure. These genes, *chickadee* (encoding a profilin homolog), *singed* (encoding a fascin homolog), and *quail* (encoding a vilin homolog), are involved in F-actin crosslinking, indicating a major role for the actin-based cytoskeleton (48–50). Moreover, these studies demonstrate that the actin-based cytoskeleton is involved in anchoring the nurse cell nuclei so that they do not plug the ring canals during the dumping process. Interestingly, *bicoid* is localized apically in nurse cells during this phase (32). This apical distribution of *bicoid* RNA indicates a preexisting asymmetry within the nurse cells, but whether this transient *bicoid* localization in nurse cells serves any function is unclear. A similar apical nurse cell RNA localization pattern is observed for ectopically expressed *oskar* and *K10* transcripts at this stage (33).

During stage 10B, microtubules rearrange into subcortical parallel arrays in the oocyte, and a microtubule-based process called ooplasmic streaming begins. Capuccino and Spire proteins are required for control of ooplasmic streaming (51, 52). During stages 10B–12 the dumping of large amounts of RNA into the oocyte along with ooplasmic streaming make it difficult to distinguish delocalization of previously localized RNA on the one hand, from a transient stage during which the dumped RNA is becoming localized (i.e. is joining the previously localized RNA at its intracellular target site) on the other. This is further complicated by the release of many transiently anteriorly localized RNAs, followed by their gradual translocation toward the posterior in some cases, or their complete delocalization in others. As a result, many RNAs appear to be generally distributed in the stage 11 oocyte. These include *Bicaudal-C* (18), *Bicaudal-D* (19), *egalitarian* (20), *nanos* (28), and *orb* (23). Other previously localized transcripts such as *K10* and *gurken* disappear by stage 11 (21, 22).

In contrast, the anterior localization pattern of *bicoid* and *Add-hts* transcripts is maintained throughout these stages (7, 16, 32). Maintenance of this pattern is likely the result of two factors: (*a*) anteriorly localized RNAs are trapped as they enter the oocyte from the nurse cells (WE Theurkauf, TI Hazelrigg, personal communication) and (*b*) previously anchored *bicoid* and *Add-hts* RNA is not released from the anterior pole during dumping and so does not have the opportunity to become generally distributed throughout the oocyte (16, 32).

After ooplasmic streaming is completed (stage 12), subcortical microtubules are replaced by randomly oriented short cytoplasmic filaments, and F-actin reorganizes from a dense cortical filament network to an extensive deep cytoplasmic network (31, 54, 55). At this stage, several newly localized RNAs can be seen at the posterior of the oocyte. These include *nanos* (28, 56), *germ cell-less* (57), and probably *orb* (23). In addition, *oskar* and *Pgc* transcripts exhibit a posteriorly enriched pattern (24–26). *bicoid* and *Add-hts* transcripts remain localized at the anterior pole of late oocytes (7, 16, 32).

RNA LOCALIZATION IN EARLY EMBRYOS   After egg activation, the cytoskeleton reorganizes once again with actin and tubulin concentrated in the cortex and deeper filamentous networks of microtubules (31, 54, 55). Some longitudinal actin fibers may also be present in the early embryo. The *Drosophila* zygote undergoes 13 synchronous nuclear divisions without cytokinesis, forming a syncytial embryo containing several thousand nuclei that share the same cytoplasm (58). This syncytial state persists until the end of the 14th cell cycle when approximately 6000 nuclei reside at the cortex. Subsequently, invagination of membranes forms individual cells to give the cellular blastoderm. At this point the antero-posterior and dorso-ventral positional fates of the cells are specified.

The two anteriorly localized maternal RNAs—*bicoid* (Figures 1 and 2) and *Add-hts*—persist in early cleavage embryos. *Add-hts* is released and diffuses posteriorly (16), while *bicoid* appears to remain anchored at the anterior cortex (7, 32). By the cellular blastoderm stage both RNAs are gone.

Three RNAs that are posteriorly localized in the oocyte—*oskar*, *nanos* (Figures 1 and 2), and *Pgc*—retain posterior localization in early cleavage stage embryos. By nuclear cycle 6/7 *oskar* RNA is gone (24), whereas *Pgc* and

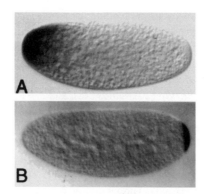

Figure 2   Localized maternal transcripts in early *Drosophila* embryos. A: *bicoid* RNA; B: *nanos* RNA. Images are of whole mount RNA tissue in situ hybridizations to stage 2 embryos using digoxigenin-labeled probes. Anterior is to the left and dorsal toward the top of the page.

nanos RNAs are associated with polar granules and are taken up into pole cells together with these granules (26, 56). Some maternal RNAs do not become posteriorly localized until late in oogenesis or early embryogenesis. Examples are *Cyclin B* (59–62) and *Hsp83* (60, 61, 63) transcripts. By the cellular blastoderm stage, maternal transcripts of *Cyclin B, germ cell-less, Hsp83, nanos, orb,* and *Pgc* can be detected only in pole cells.

During the syncytial and cellular blastoderm stages, zygotic synthesis of RNA commences. Several of these zygotically synthesized transcripts, including *crumbs, even-skipped, fushi tarazu, hairy, runt,* and *wingless* are apically localized in the blastoderm (64–69), an epithelium surrounding the syncytial yolk mass of the zygote.

Xenopus Oocytes

As in *Drosophila*, maternally synthesized gene products play a key role in the development of the *Xenopus* embryo (reviewed in 70). Zygotic transcription initiates at the 4000-cell mid blastula stage. Unlike in *Drosophila*, however, synthesis of maternal molecules occurs in the oocyte itself. Thus the issue of transport into the oocyte from interconnected nurse cells does not arise.

DEVELOPMENT OF THE OOCYTE   The *Xenopus* oocyte is initially a small spherical cell of 30 μm diameter when it is produced by mitosis of a stem cell, the oogonium (71). However, even at this stage, its nucleus and organelles are asymmetrically distributed (72, 73). Unlike *Drosophila* oogenesis, which lasts just over a week, *Xenopus* oogenesis lasts three years, although most of the synthesis of oocyte contents occurs in the third year (71). The oocyte reaches a final diameter of approximately 1.5 mm (71).

An early indicator of asymmetry is the mitochondrial cloud or Balbiani body (71). It is composed of clumps of mitochondria, rough endoplasmic reticulum, and dense granules that initially are evenly distributed around the periphery of the germinal vesicle in early stage I oocytes ($\sim$80 μm diameter). By the end of stage I these components condense on one side of the germinal vesicle as a cap-like structure that grows and assumes a spherical shape. Beginning in stage II, the mitochondrial cloud moves toward the future vegetal pole initially changing shape to become disk-like and then reorganizing into a wedge-like shape (late stage II/early stage III). Subsequently its components become localized to the vegetal cortex of the oocyte during stage III/IV.

The mitochondrial cloud probably functions in the accumulation and localization of material needed for the formation of germ plasm at the vegetal pole of the early embryo (reviewed in 70, 74). In structure and function the *Xenopus* germ plasm is comparable to the *Drosophila* posterior polar plasm, and it contains germinal granules that function in germ-cell determination. However, while *Drosophila* polar granules are sufficient for the induction of germ cells (75, 76), *Xenopus* primordial germ cells are not irreversibly determined (77). Preliminary evidence indicates that certain RNA and protein components of the *Drosophila* and *Xenopus* germinal granules are evolutionarily conserved (70, 78, 79) and that in both cases RNA localization is an important mechanism used to locally assemble these structures.

The oocyte cytoskeleton is symmetric early in oogenesis. Until stage II the germinal vesicle appears to serve as the only MTOC with microtubules emanating radially toward the plasma membrane (80). This array loses its symmetry as the germinal vesicle moves toward the future animal pole and the mitochondrial cloud starts condensing at the opposite (vegetal) side. At this time microtubules start to concentrate at the vegetal side of the germinal vesicle, colocalizing with the wedge-shaped mitochondrial cloud by late stage II (77). The germinal vesicle completes movement to the animal hemisphere by stage V/VI and at this time the microtubule array disappears.

RNA LOCALIZATION PATTERNS   The first collection of localized RNAs was reported for *Xenopus* oocytes. Three animal hemisphere–enriched RNAs (*An1, An2, An3*) and one vegetal hemisphere–localized RNA (*Vg1*; Figures 3 and 4) were identified in a molecular screen for cDNAs representing mRNAs differentially distributed along the animal-vegetal axis (5).

The animal hemisphere (*An*)–enriched RNAs are not tightly localized within the animal hemisphere but are at least fourfold enriched in this hemisphere

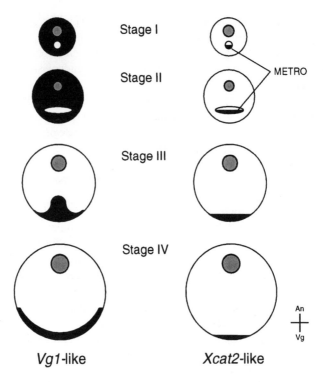

Figure 3 Localization of transcripts during *Xenopus* oogenesis. Transcript distribution patterns are shown in black, on the left for transcripts that show a *Vg1*-like pattern and on the right for transcripts that show an *Xcat-2*-like distribution pattern. Oocytes from stages I–IV are schematized. An, animal pole; Vg, vegetal pole. The germinal vesicle (oocyte nucleus) is shown in gray and the METRO as a white circle (stage I) or ellipsoid (stage II). Drawings of oocytes are after Kloc & Etkin (90).

relative to their vegetal concentration. Generally distributed maternal RNAs are twice as abundant in the animal than in the vegetal hemisphere (5). Thus the *An* RNAs are enriched in the animal hemisphere at least twofold versus other maternal RNAs. There are over a dozen known examples of *An* RNAs, all with similar distribution patterns: *An1a* (5, 81), *An1b* (5, 81), *An2* (5, 82), *An3* (5), *An4a* (83), *An4b* (83), *β-TrCP* (83), *G protein* (84), *oct60* (85), *PKC-α* (86), *xl-21* (87), *Xlan-4* (88), and *Xlcaax-1* (89). Unfortunately, reported in situ RNA hybridization data are lacking, making detailed comparisons of distribution patterns impossible.

In contrast to the *An* RNAs, the vegetally localized RNAs exhibit highly restricted distribution patterns. Careful observations of the patterns of localization

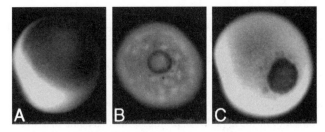

Figure 4   Localized transcripts in *Xenopus* oocytes. A: *Vg1* RNA; B: *Xcat-2* RNA; C: *Xlsirt* RNA. All images are of whole mount RNA tissue in situ hybridizations to either late stage III (A) or early stage II (B, C) oocytes. In A, the vegetal pole points toward the top right; in B and C the vegetal tip of the oocyte is at the center of the transcript distribution (i.e. the oocytes are viewed from the vegetal side). The small patches of *Xcat-2* RNA that are radially symmetrically distributed around the vegetal ring of transcripts are in the islands of germ plasm.

in situ in whole mounts have suggested two or three distinct patterns (70, 90–92) exemplified by *Vg1*, *Xcat-2*, and, possibly, *VegT/Antipodean*.

Four RNAs show the *Vg1*-like pattern of localization: *Vg1* (Figures 3 and 4); and *Xcat-4*, *B12*, and *B9* (70). *Vg1* RNA is initially distributed uniformly in the oocyte cytoplasm and is excluded from the mitochondrial cloud during stages I and II (93). During late stage II/early stage III, *Vg1* transcripts accumulate in a wedge-like pattern toward the vegetal pole before becoming restricted to the vegetal cortex by late stage III (93–95). Finally, at stage IV, *Vg1* RNA is tightly localized to the cortex of the entire vegetal hemisphere (94, 96). This localization pattern correlates with the dispersion of the mitochondrial cloud (see above).

The second pattern of RNA localization is exhibited by seven RNAs: *Xcat-2*, *Xwnt-11*, *Xcat-3*, *B6*, *B7*, *C10*, and the *Xlsirts* (Figures 3 and 4) (70, 78, 79, 95, 97, 98). This localization process includes passage through what has been called a message transport organizer (METRO) within the mitochondrial cloud (Figure 3) and occurs in three distinct steps (90): (*a*) movement from the germinal vesicle to the mitochondrial cloud, (*b*) sorting within the METRO, and (*c*) translocation and anchoring to vegetal cortex. Localization begins during early oocyte stage I when transcripts appear to be distributed throughout the cytoplasm at low levels with slightly higher concentrations around the germinal vesicle. As the mitochondrial cloud condenses into a sphere during mid stage I, these transcripts are transported to the METRO. For *Xcat-2* this change in pattern is not the result of a reduction of RNA elsewhere, as the total amount of RNA in the oocyte remains constant or even increases (79). Within the METRO the transcripts are sorted such that, for example, *Xcat-2* is localized first, followed by *Xlsirts*, and then *Xwnt-11* (70, 79, 90, 99). By late stage I, when the METRO is disc-like, *Xcat-2* RNA is in the periphery of the disc, *Xwnt-11* RNA is in the center, and *Xlsirts* reside between these two (Figure 3). The METRO moves vegetally such that by stage IV, all three RNAs are localized at the tip of the vegetal pole (70, 79, 90, 99). At this stage, the *Vg1* RNA pattern is quite distinct as it is distributed throughout the vegetal cortex (70, 79, 90, 99).

The third type of localization, typified by *VegT* (*Antipodean*), appears to deviate from both of these patterns, although there is some discrepancy in the distributions described in two reports (91, 92). *VegT/Apod* maternal transcripts are initially uniform in the cytoplasm in stage I. Later they appear to move to the vegetal pole with a timing and pattern similar to *Xlsirts*, *Xcat-2*, and *Xwnt-11* transcripts. One report describes a distribution at late stage III/early stage IV similar to that of *Vg1* (92), while another describes a novel final pattern predominantly in the vegetal yolk (91). The latter distribution would imply a target of localization that is distinct from all other vegetally localized transcripts.

After fertilization, the *B7* transcripts disappear, the *Xlsirt*, *Xcat-2*, and *Xcat-3* transcripts are associated with the germ plasm in the primordial germ cells, and the other transcripts (*B9, B12, C10, Vg1, Xcat-4,* and *Xwnt-11*) are all located in the vegetal blastomeres (70, 79, 90, 93, 98, 99; L Etkin, personal communication).

Ascidian Oocytes

Several RNAs are localized in ascidian (*Styela*) oocytes (Table 1). These include *YC* RNA, *PCNA* mRNA, and *ribosomal protein L5* mRNA. *PCNA* mRNA is initially uniformly distributed throughout the previtellogenic oocyte (100). During maturation, *PCNA* mRNA becomes concentrated in the central ectoplasm and cortical regions surrounding the myoplasm but is absent from the myoplasm per se (100). While *PCNA* mRNA can be observed in both the ectoplasm and the myoplasm after the first phase of ooplasmic segregation, which restricts the ectoplasm and myoplasm to the vegetal hemisphere, *PCNA* mRNA is absent from the myoplasm in the two-cell embryo (100). *YC* RNA is distributed throughout the cytoplasm of previtellogenic oocytes (101). However, during early vitellogenesis, *YC* transcripts are localized around the nucleus. They gradually move away from the nucleus as the oocyte increases in size, until they become restricted to the cortex of postvitellogenic oocytes (101). After the first phase of ooplasmic segregation shortly after fertilization, *YC* transcripts are localized in the vegetal cap of myoplasm (101). A third pattern of localization is exhibited by *L5* transcripts, which are concentrated in the cortical myoplasm (except at the animal pole) during oogenesis (102).

After fertilization, *L5* transcripts become restricted to the vegetal myoplasm (102).

Echinoderm Oocytes

There is one example of a maternally synthesized mRNA that is localized in the echinoderm (*Strongylocentrotus purpuratus*) oocyte and early embryo (103). This *SpCOUP-TF* mRNA, which encodes an "orphan" nuclear steroid hormone receptor, is localized subcortically in one hemisphere of the sea urchin oocyte and mature egg. Since there are no markers of the animal-vegetal axis of the egg, the location of *SpCOUP-TF* transcripts in the egg was inferred from their distribution in cleavage stage embryos where the animal-vegetal and oral-aboral axes are evident morphologically. From this inference it was concluded that *SpCOUP-TF* transcripts are localized such that they are restricted to one of the two cells produced by the first cleavage (i.e. lateral to the animal-vegetal axis) and, thus, are fixed at a 45° angle relative to the future oral-aboral axis (103).

Zebrafish Embryos

There is one example of a maternally synthesized mRNA that is localized within the cells of the zebrafish (*Brachydanio rerio*) embryo (209). This mRNA encodes a fish homolog of *Drosophila* Vasa, a DEAD-box RNA helicase that is known to function in germ plasm assembly (see below). Maternally synthesized zebrafish *vasa* transcripts localize to the inner (yolk-most) edges of the cleavage furrows at the first embryonic cell division (209). This localization pattern is maintained through the four-cell stage. From the 8- to the 1000-cell stage, the *vasa* transcripts remain in only four cells (the presumptive primordial germ cells) and are found in intracellular clumps that likely represent the assembling germ plasm. Subsequently, *vasa* transcripts are found in all primordial germ cells and germ cells.

Polarized Somatic Cells

Many cells in addition to oocytes are polarized. Epithelial cells have an apical-basal polarity. Differentiated neurons have dendritic arbors and an axon. Fibroblasts have specialized moving membranes (lamellopodia) at defined surfaces. These classes of cells also show asymmetric distributions of RNAs. However, in most cases, the developmental significance of RNA localization is unknown or, alternatively, transcript localization serves a function in the fully differentiated cell rather than during its development or differentiation. Several RNA localization patterns in these polarized cells are described here.

Although neurons can exhibit very complex and quite varied cytoarchitectures, they are classic examples of polarized cells and generally have an axon on one side of the cell body (soma) and dendrites on the other. Most neuronal RNAs are present only in the soma and are excluded from dendrites and axons. At least 12 localized neuronal RNAs have been reported (1, 104, 105). These RNAs can be classified into two different patterns. The first is somatodendritic: *MAP2* (106); *BC-1* (107); *BC-200* (108); *CaMKII-α* (109); *IP3 receptor* (110); and *Arc, F1/GAP43*, and *RC3* (104) RNAs. The second is axonal: *BC-1* (111), *tau* (112), *tropomyosin-5* (113), *vasopressin/oxytocin* (114, 115), *prodynorphin* (116, 117), and *odorant receptor* (118, 119) RNAs. In almost all cases these RNAs are localized in differentiated neurons. The exception, *tropomyosin-5* (*Tm5*) RNA, is localized prior to any structural polarities at the future axonal pole of differentiating neurons (113). In mature neurons *Tm5* RNA is present only in the soma (113).

Several types of cells have defined areas of membrane devoted to a particular function. Myelinating membranes of oligodendrocytes and Schwann cells have associated myelin basic protein (MBP) mRNA (120, 121). mRNA for vacuolar H^+-ATPase subunits is localized to the bone resorption membranes of osteoclasts (122). Lamellopodia of fibroblasts (123) and myoblasts (124) contain localized cytoplasmic *β-actin* mRNA. Apical ends of villar epithelial cells also have high concentrations of *actin* mRNA (125). In these cases, the distribution of mRNA likely follows the differentiation of these cell types rather than playing a role during their differentiation.

In *Drosophila* the location of plus-end (kinesin)- and minus-end (Nod)-directed microtubule motors provides a readout of the polarities of various cell types. Localization of these motors within oocytes was mentioned earlier. In addition, these motors localize to opposite ends of polarized cells (36): epithelia (Nod is apical and kinesin is basal), mitotic spindles (Nod is at the poles), neurons (Nod is dendritic and kinesin is axonal), and muscle (Nod is at the center and kinesin at attachment sites). As mentioned previously, several mRNAs (e.g. *crumbs, even-skipped, fushi tarazu, hairy, runt,* and *wingless*) are apically localized within undifferentiated epithelia such as the embryonic blastoderm (64–69). In addition, mRNAs (e.g. *sevenless*) are localized within the developing epithelium of imaginal discs such as the eye disc (126).

It has been reported that *prospero* and *inscuteable* mRNAs are localized within embryonic *Drosophila* neuroblasts (127). The *inscuteable* transcripts are apically localized during interphase of the neuroblast cell divisions, while *prospero* transcripts are apically localized at interphase but are basal from prophase to telophase. Basal *prospero* RNA is segregated into one daughter cell (the ganglion mother cell).

The *S. cerevisiae ASH1* mRNA is localized within yeast cells at the site of the future bud and is segregated into the daughter cell (128, 129).

MECHANISMS OF RNA LOCALIZATION

This section focuses on general classes of RNA localization mechanisms. Specific details of *cis*-acting sequences and *trans*-acting factors that function in RNA localization are reviewed in the following section.

Nucleo-Cytoplasmic Transport

An obvious way to achieve cytoplasmic RNA localization is to export transcripts vectorially from only one side of the nucleus and then to transport or anchor them in the cytoplasm on that side of the nucleus. Substantial progress has been made recently in understanding the mechanisms of nucleo-cytoplasmic transport (130); however, studies of vectorial aspects of transport from the nucleus are in their infancy.

In general, it has been difficult to establish vectorial nucleo-cytoplasmic transport for particular transcripts due to experimental limitations. An exception is the case of pair-rule gene transcripts (*hairy* and *fushi tarazu*) in the cellularizing blastoderm of *Drosophila* (65, 131). Here it was possible to use mutations to produce two layers of nuclei (or displaced nuclei) in the cortex of the syncytial blastoderm and, thus, to show—for the inner nuclei—that transcripts are vectorially exported even in the absence of normal apical cytoskeletal structures. The fact that this is possible suggests that the nuclei themselves have a polarity independent of the cytoplasmic cytoskeleton. Moreover, this directed export depends on the 3′-UTR (65), suggesting that it is specific to these transcripts (see below for a discussion of the role of 3′-UTRs).

A second example of vectorial nucleo-cytoplasmic export may be the *Drosophila gurken* mRNA that is localized dorso-anterior to the nucleus in stage 8 oocytes. The *gurken* transcripts are synthesized in the oocyte nucleus itself (R Cohen, personal communication), and the K10 and Squid proteins may function in vectorial transport of the *gurken* mRNA (see below).

Transport from One Cell Type into Another

A second class of localization mechanism applies during *Drosophila* oogenesis. As outlined previously, RNAs are transported from the nurse cells into the oocyte through intercellular bridges known as ring canals. Nurse cells connect only to the presumptive anterior pole of the oocyte, so that the imported RNAs first arrive at the oocyte's anterior pole. It is likely that anteriorly localized RNAs (e.g. *bicoid*) are trapped at the anterior pole when they enter the oocyte (WE Theurkauf, TI Hazelrigg, personal communication). The fact that mutants in which nurse cells connect to the oocyte at both poles result in bipolar transport into the oocyte and trapping of *bicoid* RNA at both poles (42) supports this hypothesis. Recent experiments in which so-called localization particles were followed by time-lapse confocal microscopy supports this hypothesis further

(WE Theurkauf, TI Hazelrigg, personal communication). In contrast to the entrapment seen for anterior-localized RNAs, those that are localized to the posterior pole are actively transported there in association with the cytoskeleton or are localized there by other mechanisms such as degradation-protection (see below).

Transport Out of Mitochondria

The posterior polar plasm of *Drosophila* oocytes and early embryos contains large, non-membrane-bound organelles known as polar granules, which are involved in germ cell formation and specification (see below). Mitochondria are found in close association with the polar granules. One of the more remarkable examples of a localized RNA is the *Drosophila 16S mitochondrial large ribosomal RNA* (*mtlrRNA*), which is encoded by the mitochondrial genome (132). This RNA appears to be exported from the mitochondria into the cytoplasm within the posterior polar plasm and to be associated with polar granules (61, 133–137). Indeed, given the apposition of polar granules and mitochondria, the *mtlrRNA* may in fact be exported vectorially out of mitochondria directly into or onto the polar granules. The function of the *mtlrRNA* in the polar granules is unclear, although it has been implicated in pole cell formation (133). There is some disagreement, however, about whether a high local concentration of *mtlrRNA* is indeed necessary for pole cell formation (135–137).

Generalized Degradation with Localized Protection

It was postulated several years ago that one mechanism by which a generalized RNA distribution could be converted to a restricted pattern was through degradation of the RNA throughout the cell except at the site of localization (138). Several *Drosophila* transcripts represent variants of this type of process. For example, while the bulk of maternally synthesized *nanos* and *cyclin B* transcripts are concentrated in the posterior polar plasm of the early embryo, a subset of these transcripts remains unlocalized (62, 139–141). The posteriorly localized transcripts are taken up into the pole cells when they bud, while the unlocalized transcripts are degraded (62, 139–141). Similarly, maternally synthesized *Hsp83* transcripts are generally distributed in the early embryo (61, 63, 142; SR Halsell, A Bashirullah, RL Cooperstock, WW Fisher, A Karaiskakis, HD Lipshitz, manuscript in preparation). *Hsp83* transcripts in the posterior polar plasm also are taken up into the pole cells when they bud, while the remaining transcripts are degraded (61, 63, 142; SR Halsell, A Bashirullah, RL Cooperstock, WW Fisher, A Karaiskakis, HD Lipshitz, manuscript in preparation). There is a close correlation between translational repression of unlocalized *nanos* transcripts and their degradation (reviewed in 144). Under normal conditions, the polar granules are necessary and sufficient for protection of *nanos*, *cyclin B*, and *Hsp83* transcripts from degradation at the posterior (56, 60, 63).

Another example of transcript destabilization and localization may be maternally synthesized *PCNA* mRNA in the ascidian oocyte. There is sequence complementarity between a non–protein-coding RNA, *YC*, whose 3' end is complementary to the 3'-UTR of the *PCNA* RNA (100) as well as to the 5'-UTR of *ribosomal protein L5* RNA (Sc*YC26a*) (102). During oogenesis the *YC* RNA is perinuclear, gradually moving to the cortex, and after fertilization the RNA segregates to the myoplasm and associates with the cytoskeleton (101). Uniformly expressed maternal *PCNA* RNA initially overlaps with *YC* RNA but later becomes depleted in the myoplasm (100). Investigators have suggested that the double-stranded *YC-PCNA* RNA hybrid in the myoplasm might somehow destabilize *PCNA* RNA, thus representing an example of RNA localization by degradation-protection. The hypothesis that such double-stranded RNA hybrids destabilize specific RNAs in the myoplasm is, however, confounded by the data for the *L5 ribosomal protein* maternal RNA. Although there is substantial sequence complementarity between *YC* and *L5* RNAs, *L5* RNA is concentrated in myoplasm along with *YC* RNA, rather than destabilized there (102). In this case, the *YC* RNA has been postulated to aid the anchoring of *L5* RNA. Whether *PCNA* RNA localizes by a mechanism that involves hybrid-induced degradation remains an open question.

Directed Cytoplasmic Transport of RNA

Asymmetries in cytoskeletal organization have been described earlier for both *Xenopus* and *Drosophila* oocytes. Further, there is colocalization of specific RNAs with either a minus-end-directed microtubule motor (Nod) or a plus-end-directed motor (kinesin), in particular regions of the *Drosophila* oocyte's cytoplasm (see above). There is now substantial evidence that cytoplasmic RNA transport to specific intracellular destinations is accomplished by both the microtubule- and the microfilament-based cytoskeleton. The following section reviews evidence for a role of the cytoskeleton in anchoring localized RNAs at their intracellular destinations. Here the role of the cytoskeleton in directed cytoplasmic transport is reviewed.

Analysis of intracellular transport mechanisms requires the ability to systematically perturb normal cytoskeletal function. These studies have been aided in *Xenopus* and *Drosophila* by drugs that specifically perturb either the microtubule-based (colchicine, nocodazole, or taxol) or the microfilament-based (cytochalasins) cytoskeleton. In addition, mutations that affect components of the cytoskeleton have led to informative results in *Drosophila* and *Saccharomyces*.

THE ROLE OF THE CYTOSKELETON IN RNA TRANSPORT DURING DROSOPHILA OOGENESIS   Localized RNAs have several characteristic and sequential patterns of expression during *Drosophila* oogenesis that correlate with particular aspects of the cytoskeleton, particularly the microtubules (see above).

Over a dozen transcripts are synthesized in the nurse cells and specifically accumulate in the oocyte within early egg chambers prior to their localization [*bicoid* (7), *nanos* (28), *orb* (23), *oskar* (24, 25), *Add-hts* (16), *Bicaudal-C* (18), *Bicaudal-D* (19), *gurken* (21), *Pgc* (26), *K10* (22), *egalitarian* (20), and *tudor* (27)]. Transport of these RNAs into the oocyte is likely to be carried out by minus-end-directed microtubule motors since the MTOC is located in the oocyte during these stages (see above). Although no specific motors have been demonstrated to be involved in this process, the kinesin-like minus-end-directed motor—Nod—localizes first to the oocyte and then to its posterior at the same stages as many of these RNAs are transported into the oocyte and then accumulate at its posterior pole (36). Dynein (a minus-end-directed motor) is also localized to the oocyte at these stages (145) but does not appear to be involved in RNA transport and localization (146). During these stages, several RNAs are present in detergent insoluble fractions (e.g. *bicoid*, *oskar*, *Bicaudal-D*, *K10*, *orb*) indicating association with the cytoskeleton (147). Moreover, the association of *bicoid*, *oskar*, and *Bicaudal-D* RNAs with the cytoskeleton is sensitive to colchicine and not to cytochalasins, indicating that microtubules but not microfilaments are involved in their transport and localization (37, 147).

The phenotypes of *orb*, *egalitarian*, and *Bicaudal-D* mutants suggest a role in oocyte specific RNA accumulation. Egalitarian and Bicaudal-D proteins are made in nurse cells and are transported to the posterior of the oocyte presumably along minus-directed microtubules (20). The distribution of Egalitarian and Bicaudal-D proteins parallels that of the RNAs that are transported into the oocyte at these stages. Microtubule inhibitors result in delocalization of Egalitarian protein (20). Moreover, in *egalitarian* mutants *oskar* and *orb* RNA are no longer associated with the cytoskeleton (147). This indicates that Egalitarian and Bicaudal-D proteins may be involved—directly or indirectly—in transporting localized RNAs along microtubule networks into and to the posterior of the oocyte.

Mutations in genes required for early localization also perturb oocyte polarity; *egalitarian* and *Bicaudal-D* mutations cause all 16 cells of the cyst to become polyploid nurse cells; thus oocyte-specific accumulation of transcripts cannot occur because there is no oocyte (20, 148). Orb is required for oocyte polarity, and in *orb* mutants, oocytes are located at ectopic positions within the egg chamber (149). In *orb* mutant oocytes certain RNAs (*orb*, *oskar*) are still localized—albeit at abnormal positions—whereas others are not localized at all (*Add-hts*, *Bicaudal-D*, *K10*) (149). It is possible that Orb protein is required to establish microtubule polarity, whereas Egalitarian and Bicaudal-D are necessary for its maintenance.

As described above, the majority of RNAs transported into and localized in the oocyte have in common early transport from the nurse cells (stage 1–5), transient localization to the posterior (stage 6), and subsequent localization to the anterior (stages 7–8). In addition, *oskar* and *Pgc* transcripts move back to the posterior pole of the oocyte at stage 9, whereas the *bicoid* and *Add-hts* RNAs never show the early posterior localization but are either always anteriorly localized (*bicoid*) or are initially localized throughout the cortex and subsequently localize to the anterior (*Add-hts*). These data suggest that a default transport and localization mechanism is carried out by minus-end-directed microtubule motors, and that certain RNAs (*bicoid, Add-hts, oskar, Pgc*) initially use this mechanism to enter the oocyte but then engage a different localization machinery. Since *bicoid* and *oskar* RNA transport and localization are best understood and exemplify distinct localization mechanisms, they are discussed below.

bicoid RNA is localized at the anterior pole of the oocyte by stage 8 of oogenesis and remains anterior until the late cleavage stage of embryogenesis when it is degraded. During its translocation from the nurse cells into the oocyte, it is apically localized within the nurse cells (32). Both apical nurse cell localization and anterior oocyte localization are sensitive to microtubule depolymerizing drugs (colchicine, nocodazole, tubulozole C) but not to inhibitors of F-actin polymerization (cytochalasin D and B) (150). Recent evidence suggests that transport of *bicoid* RNA into the oocyte actually involves several distinct steps that might be mediated by distinct localization mechanisms; for example, two distinct microtubule-dependent steps drive *bicoid* RNA localization particles within the nurse cell cytoplasm (WE Theurkauf, TI Hazelrigg, personal communication), but transport through the ring canals into the oocyte is resistant to both microtubule and actin filament inhibitors (WE Theurkauf, TI Hazelrigg, personal communication). The Exuperantia protein may mediate this microtubule-independent transport (151). The cytoskeletal association of *bicoid* transcripts is stage specific (147). During early oocyte accumulation, *bicoid* transcripts are associated with the cytoskeleton (i.e. the detergent-insoluble fraction). However, during stages 8–11, when *bicoid* transcripts are at the anterior margin of the oocyte, they are not cytoskeleton associated. This observation supports the idea that anchoring at the anterior pole at these stages is accomplished by some other structures. Later, during stage 14, *bicoid* RNA is localized in a tight cap at the anterior, is again associated with the microtubule-based cytoskeleton, and its localization is again sensitive to colchicine. In early embryos, *bicoid* RNA is no longer restricted to the cortex and is not associated with the cytoskeleton (147).

A kinesin-β-galactosidase fusion protein localizes to the posterior (41) at stages 8–9, indicating that microtubules traverse the antero-posterior axis of the oocyte at these stages and are oriented with their plus ends at the posterior (41). This localization coincides with the initiation of *oskar* and *Pgc* translocation from the anterior pole of the oocyte to its posterior. Thus it is likely that *oskar* and *Pgc* switch from the use of minus-end-directed microtubule motors to the use of plus-end-directed ones in order to achieve transport to the posterior. Posterior localization of the kinesin-β-galactosidase fusion protein is lost once ooplasmic streaming begins (stage 10) and is absent in *capuccino* and *spire* mutant oocytes that undergo premature streaming (41, 52). In *capuccino* and *spire* mutants *oskar* RNA is not localized to the posterior during stages 8 and 9, but instead *oskar* RNA is uniformly distributed throughout the oocyte (24, 25). These mutants cause an early cytoplasmic streaming during stage 7 and 8 instead of 10B (52), suggesting that premature assembly of microtubules into the parallel arrays in the subcortex drives cytoplasmic streaming (52). In other words, *oskar* RNA does not localize to the posterior in *capuccino* and *spire* mutants because these mutants omit the stage during which antero-posterior axial organization of microtubules is used for directed transport of *oskar* RNA to the posterior.

Evidence suggests a role for the actin-based cytoskeleton at the anterior of the oocyte in transfer of RNAs to the microtubules that run from the anterior to the posterior pole. In oocytes that are mutant for a component of the actin-based cytoskeleton—cytoplasmic (nonmuscle) tropomyosin II (cTmII)—*oskar* RNA remains anteriorly localized at stage 9 and never localizes posteriorly (152, 153). This observation suggests a role for cTmII—and possibly the actin-based cytoskeleton—in transfer of *oskar* RNA to the axial microtubules. Staufen protein similarly fails to translocate posteriorly in *cTmII* mutants (152), suggesting that the entire transport particle containing Staufen protein and *oskar* RNA fails to be transferred to the posterior translocation apparatus.

THE ROLE OF THE CYTOSKELETON IN RNA TRANSPORT DURING XENOPUS OOGENESIS   As discussed above, the dynamics of transcript localization to the vegetal pole of *Xenopus* oocytes can largely be classified into two different patterns exemplified by *Vg1* and *Xcat-2* RNAs.

Xcat-2 transcripts are distributed uniformly in early *Xenopus* oocytes (79) and are then sequestered into the METRO region of the mitochondrial cloud along with *Xwnt-11*, *Xlsirt*, and other RNAs (70, 90). This step appears to be mediated by selective entrapment of these RNAs, possibly similar to posterior polar-granule-localized RNAs in *Drosophila* (see below). Once localized to the METRO, *Xcat-2* accompanies the mitochondrial cloud to the vegetal pole (90). *Xcat-2* RNA then relocates to form a disc-like pattern at the tip of the vegetal pole (90).

Vg1 RNA is initially generally distributed in the oocyte and later localizes in the wedge-shaped pattern that overlaps but differs from that of *Xcat-2* RNA at the vegetal pole (90). Accumulation of *Vg1* to the vegetal pole requires functional microtubules but not actin microfilaments (94). Later *Vg1* RNA is found throughout the cortex of the vegetal hemisphere, unlike *Xcat-2*, which is localized to a more restricted area at the vegetal pole (90). During this late stage, *Vg1* RNA is enriched 30- to 50-fold in the detergent-insoluble fraction (96). Moreover, this association and cortical *Vg1* RNA localization are not sensitive to microtubule-depolymerizing drugs (nocodazole and colchicine) but rather to microfilament-disrupting agents (cytochalasin B) (94). This is an indication of a two-step localization mechanism for *Vg1* RNA where microtubules are required for translocation and actin filaments for anchoring (94). Recent evidence indicates that *Vg1* RNA is associated with the endoplasmic reticulum (ER) and that *Vg1* RNA-ER complexes move to the vegetal pole along with the mitochondrial cloud in a microtubule-dependent fashion (154).

Xcat-2 RNA injected into later oocytes is able to localize cortically without prior association with the METRO (99). This localization is dependent on microtubules and cannot occur in late oocytes (stage VI) when microtubules are no longer present (99). Moreover, the *cis*-acting elements within the *Xcat-2* 3'-UTR that are required for METRO localization and cortical localization are different but overlapping (79, 99) (see below). In addition, injected *Xcat-2* transcripts that localize to the vegetal cortex without METRO do so in a pattern similar to *Vg1* (throughout the vegetal hemisphere) but different from that of endogenous *Xcat-2* transcripts (99). Thus the differences in *Vg1* and *Xcat-2* localization patterns are a consequence of the fact that *Xcat-2* is normally associated with the METRO, rather than in some inherent difference in their ability to associate with the microtubule-based cytoskeleton.

ROLE OF THE CYTOSKELETON IN RNA TRANSPORT IN OTHER CELL TYPES   Observations of localized RNAs in living neurons in culture have suggested that they are present in particles composed of several RNAs and proteins including polyribosomes (155). These particles translocate inside the cell in a microtubule- but not microfilament-dependent manner (155). Similar studies with *MBP* RNA injected into oligodendrocytes in culture indicate that the initially homogeneous *MBP* RNA becomes organized into granules that align on microtubule tracks in the peripheral processes (156). Endogenous *MBP* RNA is seen in granules and fractionates with the insoluble fraction in cell extracts, consistent with an association of "transport" granules with the cytoskeleton (156).

Mammalian *tau* mRNA is localized to the proximal hillock of axons (112). This localization is mediated by the *tau* 3'-UTR (see below) and depends on microtubules (157). Interestingly, in vitro synthesized *tau* 3'-UTR injected into *Xenopus* oocytes localizes to the vegetal pole in a pattern identical to *Vg1* RNA (158). Moreover, this localization is dependent first on microtubules and then on actin microfilaments, as for *Vg1* RNA (158). This observation demonstrates that once an RNA associates with the cytoskeletal transport apparatus, it localizes according to the type of cell in which it is. This occurs even if that RNA normally would not be present in this cell type. Thus the role of the cytoskeleton in RNA localization is highly conserved across evolution (see below).

RNA localization also occurs in the yeast *S. cerevisiae*. In this budding yeast, *ASH1* mRNA is localized first to the future bud site and then to the daughter cell by a mechanism involving actin microfilaments (128, 129). The role of microfilaments was demonstrated genetically using mutants in actin, myosin, profilin, and tropomyosin (which form part of the microfilament network). In contrast, disruption of microtubules by tubulin mutants, or disruption of the process of budding with *MYO2* mutants, has no effect on *ASH1* transcript localization.

In various somatic cells (e.g. fibroblasts and myoblasts) *β-actin* mRNA is localized to moving membranes (123, 124). This localization is not dependent on microtubules or intermediate filaments but on microfilaments (123). Both RNA transport and anchoring are dependent on the actin cytoskeleton (123).

The data described above indicate a key role for microtubules in directed mRNA transport, especially in *Drosophila* and *Xenopus* oocytes but also in polarized cells such as neurons and glia. In several cases, microfilaments also play a crucial role in RNA localization. However, it is unclear whether the instances of microfilament-based RNA localization indicate a role in directed RNA transport versus in entrapment/anchoring of the RNA at the site of localization (see below).

Entrapment/Anchoring of RNA at the Site of Localization

At stage 10B of *Drosophila* oogenesis, the microtubules in the oocyte assemble into subcortical arrays that direct circumferential ooplasmic streaming (see above). Thus at these stages, RNAs cannot be localized to the posterior by directed transport on axial microtubules. Localized RNAs most likely are trapped at the posterior as they circulate through the oocyte cortex along with unlocalized components of the cytoplasm. In support of this hypothesis, it has been shown that, while long-range transport of injected *oskar* RNA to the posterior requires microtubules (159), local injection of *oskar* RNA near the posterior pole of large (stage 10–11) oocytes or anywhere in smaller (stage 9) oocytes results in posterior localization in a microtubule-independent process. This observation suggests that short-range RNA transport and/or local entrapment of RNA is not dependent on microtubules. Trapping of injected *oskar* transcripts at the posterior pole fails in *cTmII* mutant oocytes (159), consistent with a role for the actin-based cytoskeleton in this process.

Egalitarian and Bicaudal-D may also be required for posterior anchoring or trapping of RNAs such as *oskar*. In a situation where Bicaudal-D is eliminated during late oogenesis (160), *oskar* RNA is initially localized normally but then is lost from the posterior pole. In these mutant oocytes kinesin-β-galactosidase protein localizes normally to the posterior; therefore, the microtubule-based cytoskeleton involved in posterior transport is not disrupted. Posterior RNA localization occurs independent of Bicaudal-D, but maintenance of localization is dependent on Bicaudal-D function. Bicaudal-D may be involved in anchoring of RNA at the posterior pole of the oocyte, where it may function in a complex with Egalitarian (20).

Anchoring of RNAs at the posterior pole of the *Drosophila* oocyte also requires the integrity of the polar granules. These are large non-membrane-bound organelles composed of RNA and protein. Many posteriorly localized RNAs are either components of the polar granules or are associated with the granules (reviewed in 142). Mutations that disrupt the posterior polar granules (e.g. *capuccino*, *oskar*, *spire*, *staufen*, *tudor*, *valois*, and *vasa*) cause delocalization of posteriorly localized RNAs while ectopic assembly of polar granules at the anterior pole of the oocyte results in anterior localization of RNAs normally localized to the posterior (reviewed in 142), with the possible exception of the *mtlrRNA* (137). Developmental functions of the posterior polar granules and their associated localized RNAs are considered below.

In *Xenopus* oocytes, anchoring of *Vg1* RNA to the vegetal cortex of stage IV oocytes requires microfilaments (94), which as for *Drosophila*, implicates the actin-based cytoskeleton in anchoring of RNA. Anchoring of *Vg1* RNA also involves other—non-protein-coding—localized RNAs, *Xlsirts* (161). Injection of oligonucleotides complementary to *Xlsirt* transcripts causes delocalization of anchored *Vg1* RNA in a manner similar to delocalization by F-actin-disrupting drugs (161). This observation indicates a connection between *Xlsirt* RNA, *Vg1* RNA, and F-actin; however, no direct interactions have been demonstrated. A role for the noncoding *YC* RNA in anchoring *L5* RNA in the myoplasm of the ascidian oocyte and early embryo has also been postulated (102). However, no experimental evidence supports this hypothesis.

Finally, for *ASH1* RNA in yeast (128, 129) as well as for *β-actin* mRNA in mammalian somatic cells (123), the actin-based cytoskeleton is necessary for localization (see above). However, the mechanism of localization of these RNAs is not clear; it has not yet been determined whether localization is by selective entrapment or by directed transport.

RNA Transport/Anchoring Particles

Specific *trans*-acting factors that interact with localized RNAs are discussed later in a separate section. Large ribonucleoprotein RNA transport particles have been visualized at the light microscope level in several systems. For example, *bicoid* RNA is transported from the nurse cells into the oocyte in particles that contain Exuperantia protein (162; WE Theurkauf, TI Hazelrigg, personal communication). Maintenance of *bicoid* transcripts at the anterior pole, as well as transport of *oskar* RNA to the posterior pole, likely involves particles that contain Staufen protein (163, 164). In *Xenopus*, RNA-containing transport particles have been visualized (90, 99). In neurons and glia, large transport particles containing *tau* RNA (157) and *MBP* RNA (156), respectively, have been identified. Particles containing *ASH1* mRNA have been detected in *S. cerevisiae* (129). The polar granules of *Drosophila* and the germinal granules of *Xenopus* are very large (organelle-sized) ribonucleoprotein (RNP) particles that are involved in anchoring localized RNAs at the posterior and vegetal poles of *Drosophila* and *Xenopus* oocytes, respectively. Several of the particles mentioned here contain not just proteins and mRNAs, but also non-protein-coding RNAs (26, 90, 97, 134, 155).

CIS-ACTING ELEMENTS THAT TARGET RNAs FOR LOCALIZATION

In principle, there are two mechanisms by which an RNA molecule could be targeted for cytoplasmic localization. An RNA element or elements could be recognized by the localization machinery. Alternatively (only in the case of mRNAs), the polypeptide product of the RNA might be recognized and the RNA translocated intracellularly along with the polypeptide. The latter mechanism—through recognition of the signal peptide by the SRP apparatus—may be used to bring mRNAs that encode secreted or transmembrane proteins into association with the endoplasmic reticulum (reviewed in 165). All other defined cytoplasmic RNA localization mechanisms involve recognition of RNA elements, particularly in the 3'-UTR of mRNAs. This section summarizes the methods used to map *cis*-acting sequence elements within localized RNAs. Generalities about the location of such elements within the RNAs are then drawn, and specific features of these elements are discussed. A subsequent section focuses on *trans*-acting factors that interact with these elements.

Mapping of Cis-Acting Elements in Localized RNAs

The general method used to map *cis*-acting sequences that function in RNA localization is to produce hybrid RNAs that include an exogenous reporter sequence (e.g. part of the *E. coli* β-*galactosidase* RNA) and part or all of the RNA under study. These hybrid transcripts are introduced into the cell type of interest, and the reporter sequence is used as a tag to assay localization of the hybrid transcript. Initially the location of the transcripts was assayed by isolating parts of cells and carrying out RNase protection assays for the presence of the transcripts (166). However, this method was replaced by the use of in situ hybridization with antisense reporter RNA probes (e.g. antisense *β-galactosidase* RNA probes) either in tissue sections or whole mounts. More recently—in the case of injected in vitro transcribed RNAs that have been fluorescently tagged—it has been possible to visualize the injected RNAs directly by fluorescence microscopy (159). Elements involved in various aspects of localization are then mapped further by testing a series of RNAs that carry deletions or mutations in the region that confers localization. Using these methods, *cis*-acting sequences sufficient (i.e. that are capable of conferring localization) or necessary (i.e. that when deleted result in failure of localization) can be mapped. Additional information may be gained by performing sequence alignments of the region of interest among different species. Conserved domains may represent important *cis*-acting elements required for localization (167).

Methods available for introducing the hybrid RNA into the cell-type of interest vary depending on the system. In *Drosophila* transgenic flies can be obtained that synthesize the hybrid transcripts in the correct cell type and often at the correct developmental stage. This can be accomplished using an inducible promoter, a cell-type-specific promoter, or the promoter of the relevant endogenous gene. Thus there is seldom any question as to the in vivo significance of the results obtained. In many cases mutations exist that eliminate or reduce the function of the endogenous gene. This then enables one to test whether a transgene carrying all of the endogenous regulatory and protein coding sequences rescues the mutant phenotype. If so, then transgenes in which specific RNA elements are deleted can be assayed for phenotypic rescue or lack thereof; thus both the effects on RNA localization, and the phenotypic consequences of disrupting that localization, can be tested simultaneously.

A second common method for introducing the hybrid transcripts into the host cytoplasm is by injection of in vitro transcribed RNA. This has been done occasionally in *Drosophila* (60, 159) but is most common in *Xenopus* oocytes where transgenic technology is still rather primitive. In both *Drosophila* and *Xenopus*, the large size of the oocyte enables the injected RNA to be introduced at a specific location within the cell.

A third method, used mostly for analyses of somatic cells in culture, is to transfect expression plasmids carrying the hybrid transcription unit into the cells, to wait for transcription and possible localization to occur, and then to assay the distribution of the hybrid RNAs.

Cis-Acting Sequences for Localization Map
to the 3'-Untranslated Region of mRNAs

Table 2 lists localized RNAs in which *cis*-acting localization elements have been mapped. For all mRNAs studied to date, sequences that are necessary for localization are found in the 3'-untranslated region (3'-UTR). In many cases,

Table 2   *Cis*-acting localization elements

Species	Transcript name	Localization signal(s)	Sufficient subregions (subelements)	Reference
Drosophila				
	Add-hts	3′-UTR (345 nt)	345 nt sufficient (ALE1 = 150 nt)	—[a]
	bicoid	3′-UTR (817 nt)	625 nt sufficient (BLE1 = 53 nt)	166, 176
	Cyclin B	3′-UTR (776 nt)	94 nt + 97 nt (TCE = 39 nt)	182
	even-skipped	3′-UTR (190 nt)	163 nt (124 nt in UTR)	65
	fushi tarazu	3′-UTR (455 nt)	Not defined	65
	hairy	3′-UTR (816 nt)	Not defined	65
	Hsp83	3′-UTR (407 nt)	107 nt sufficient (protection)	—[b]
	K10	3′-UTR (1400 nt)	44 nt (TLS) sufficient	181
	nanos	3′-UTR (849 nt)	543 nt sufficient (localization) (TCE = 90 nt translational control) (SRE = 60 nt translational control)	43, 141, 185
	orb	3′-UTR (1200 nt)	280 nt sufficient	169
	oskar	3′-UTR (1043 nt)	924 nt sufficient (localization) (BRE = 71 nt translational control)	168, 184
	wingless	3′-UTR (1083 nt)	363 nt sufficient	—[c]
Mammals				
	β-actin	3′-UTR (591 nt)	54 nt or 43 nt sufficient (43 nt less active)	171
	BC1	5′ region (152 nt)	62 nt sufficient	173
	CaMKIIα	3′-UTR (3200 nt)	Not defined	109
	tau	3′-UTR (3847 nt)	1395 nt sufficient (VgRBP-binding region = 624 nt)	157, 158
Xenopus				
	Vg1	3′-UTR (1300 nt)	340 nt (VgLE) sufficient (85 nt repeat)	170, 183
	TGF β-5	3′-UTR (1102 nt)	Not defined	198
	Xcat-2	3′-UTR (410 nt)	150 nt (mitochondrial cloud) 120 nt (vegetal cortex)	99
	Xlsirt	3–12 repeat sequences (79–81 nt)	Two copies of repeat sufficient	97
Yeast				
	ASH1	3′-UTR	250 nt sufficient	128

[a] KL Whittaker, D Ding, WW Fisher, HD Lipshitz, manuscript in preparation.

[b] SR Halsell, A Bashirullah, RL Cooperstock, WW Fisher, A Karaiskakis, HD Lipshitz, manuscript in preparation.

[c] H Krause, personal communication.

these 3'-UTR elements are also sufficient for localization. Examples of mRNAs that contain such localization elements in *Drosophila* include anteriorly localized transcripts such as *bicoid*, *Add-hts*, and *K10* (22, 166; KL Whittaker, D Ding, WW Fisher, HD Lipshitz, manuscript in preparation), posteriorly localized transcripts such as *oskar*, *nanos*, *orb*, *Cyclin B*, and *Hsp83* (43, 60, 142, 168, 169; SR Halsell, A Bashirullah, RL Cooperstock, WW Fisher, A Karaiskakis, HD Lipshitz, manuscript in preparation), and apically localized pair-rule gene transcripts such as *even-skipped*, *fushi-tarazu*, *hairy*, and *wingless* (65). *Xenopus* also provides examples such as *Vg1*, *TGFβ-5*, and *Xcat-2* (99, 170). Interestingly, the rat *tau* RNA's 3'-UTR, which confers localization to the proximal hillock of rat axons, also mediates vegetal localization in *Xenopus* oocytes (158). The chicken *β-actin* 3'-UTR contains a *cis*-acting element sufficient for peripheral localization in chicken embryonic fibroblasts and myoblasts (171). The presence of localization tags in the 3'-UTR adds to the list of 3'-UTR *cis*-acting elements involved in posttranscriptional mRNA regulation via control of stability, cytoplasmic polyadenylation, and translation (172).

Cis-Acting Sequences for Localization Map Within Non-Protein-Coding RNAs

Non-protein-coding RNAs also contain discrete elements that target the RNAs for localization (Table 2). Such elements have been mapped in *Xenopus Xlsirt* (97) and in neuronal *BC-1* RNAs (173).

Alternative Splicing Can Generate Localized vs Unlocalized RNA Isoforms

Drosophila Cyclin B transcripts exemplify the importance of alternative splicing of sequences that target localization (60). Alternative splicing within the 3'-UTR generates two *Cyclin B* mRNA isoforms that differ by 393 nucleotides (nt). The shorter splice variant is synthesized preferentially during early oogenesis and is present throughout the pro-oocyte until stages 7–8. The longer splice variant is synthesized in the nurse cells later in oogenesis, during stages 9–11. It is then transported into the oocyte, with an initially uniform distribution and is later concentrated at the posterior pole (60, 61). The transcript also exhibits perinuclear localization in the synctial embryo. Posterior localization of the long *Cyclin B* mRNA isoform is directed by the additional sequences spliced into its 3'-UTR relative to the short mRNA isoform (which is unable to localize) (60).

Alternative splicing also plays a role in the localization of *Add-hts* transcripts. Three classes of *Add-hts* contain unique 3'-UTRs introduced by alternative splicing (KL Whittaker, D Ding, WW Fisher, HD Lipshitz, manuscript in preparation). This alternative splicing also introduces variability in the carboxy-terminal regions of the encoded Adducin-like protein isoforms (174; KL Whittaker, D Ding, WW Fisher, HD Lipshitz, manuscript in preparation). Only one of the mRNA variants, N4, exhibits transport into and localization within the oocyte (16; KL Whittaker, D Ding, WW Fisher, HD Lipshitz, manuscript in preparation). The N4 3′-UTR is necessary and sufficient for this transport and localization, suggesting that use of alternative 3′-UTRs is one mechanism by which different *Add-hts* protein isoforms are restricted to different subsets of the nurse cell–oocyte complex (KL Wittaker, D Ding, WW Fisher, HD Lipshitz, manuscript in preparation).

Although not a consequence of alternative slicing, two *actin* RNA isoforms, encoding α- and β-actin, possess isoform-specific 3′-UTRs that can confer differential intracellular targeting (175). *β-actin* transcripts are localized to the leading lamellae in both differentiating myoblasts and small myotubes, while *α-actin* transcripts associate with a perinuclear compartment.

Discrete Localization Elements

In the case of *bicoid* RNA, discrete elements within the 3′-UTR have been defined that confer distinct aspects of the RNA localization pattern. A decade ago a 625-nt subset of the *bicoid* 3′-UTR was found to be sufficient for anterior localization (166). At that time it was suggested that the secondary structure of the 3′-UTR, which can be folded into several long stem-loops, might be recognized by the localization machinery. Subsequent evolutionary sequence comparisons supported this hypothesis since the secondary structure appears to be conserved in distant *Drosophila* species (*melanogaster, teissieri,* and *virilis*) despite the fact that the primary sequence of these 3′-UTRs has diverged by up to 50% (167). Further, the 3′-UTR from one species can direct anterior localization in a distant species (167). There is complementarity between two single-stranded regions predicted in the secondary structure, implying that tertiary base-pairing interactions might also be important (167).

Subsequent analyses followed two different routes. In one set of experiments, deletions were used to define an approximately 50-nt region, called *bicoid* localization element 1 (BLE1), which is necessary and sufficient (when present in two copies) to direct nurse cell–oocyte transport and anterior transcript localization during mid-oogenesis (176). However, anterior localization is lost later. BLE1 interacts with Exl protein, which might function in localization to the anterior of the oocyte (177) (see below). In addition, linker scanning and point mutational analyses were used to define regions of the 3′-UTR that are important for anterior localization late in oogenesis and in the early embryo (163, 164). These regions interact with the double-stranded RNA-binding protein Staufen (178, 179), which functions to anchor *bicoid* RNA at the anterior of the late-stage oocyte and early embryo (see below). This is accomplished in part by promoting quaternary (inter-3'-UTR) interactions via the complementary single-stranded regions mentioned earlier (164).

Although not yet as well characterized as *bicoid*, it has similarly been possible to map discrete elements in other 3'-UTRs that direct subsets of the localization pattern. For example, the N4 isoform of *Add-hts* mRNA is transported from the nurse cells into the oocyte starting in the germarium, then localized cortically in the oocyte (stages 7–8) and to the anterior pole (stage 9) (16, 44, 45, 180). In this case, a central (100–150 nt) element of the 3'-UTR, *Add-hts* localization element 1 (ALE1), is necessary and sufficient for nurse cell–oocyte transport as well as for cortical localization within the oocyte (KL Whittaker, D Ding, WW Fisher, HD Lipshitz, manuscript in preparation). The region that includes ALE1 comprises the most extensive predicted secondary structure within the *Add-hts* N4 3'-UTR (KL Whittaker, D Ding, WW Fisher, HD Lipshitz, manuscript in preparation). When additional, adjacent parts of the N4-3'-UTR are added to ALE1, anterior localization is conferred starting at stage 9 (KL Whittaker, D Ding, WW Fisher, HD Lipshitz, manuscript in preparation). The *K10* 3'-UTR reveals several long inverted repeats, suggesting that it forms extensive secondary structure (22). A short region within the *K10* 3'-UTR, TLS (transport/localization sequence, 44 nt in length) is predicted to form a stem-loop structure and is necessary and sufficient for nurse cell–oocyte transport and anterior localization (181). Mutations that disrupt this structure block transport and localization, while compensatory mutations that preserve the structure restore these processes.

With respect to posteriorly localized RNAs, it has also been possible in some cases to map discrete localization elements. For example, as mentioned above, a 181-nt element in the long isoform of *Cyclin B* mRNA is necessary for posterior localization (182). Similarly, a 107-nt element in the *Hsp83* 3'-UTR is necessary and sufficient for association with the posterior polar plasm (SR Halsell, A Bashirullah, RL Cooperstock, WW Fisher, A Karaiskakis, HD Lipshitz, manuscript in preparation).

Repeated/Redundant Localization Elements

In the examples described above it was possible to define discrete, relatively small (<150 nt) localization elements that confer specific aspects of localization. In other instances, while discrete elements have been identified that direct localization, there is some redundancy in the system such that more than one localization element capable of conferring a particular aspect of the localization pattern, is present in the RNA.

An example comes from the *Xenopus Vg1* 3'-UTR. A 340-nt region within the 3'-UTR is necessary and sufficient for vegetal localization of *Vg1* RNA (170). Deletion analysis indicates that there is considerable redundancy within this region but that critical elements can be defined that lie at each end (183). An 85-nt subelement from the 5' end of the region, when duplicated, is sufficient to direct vegetal localization (183).

A second example is in the *orb* 3'-UTR (169). In early oogenesis, *orb* transcripts accumulate preferentially in the pro-oocyte (stage 1). They localize transiently to the oocyte posterior (stages 2–7) and then to the oocyte anterior (stages 8–10). A 280-nt element is sufficient to confer oocyte accumulation, posterior localization, and then anterior localization. Further analysis has shown that when the element is split in two, each half on its own confers oocyte accumulation, although the level of accumulation is reduced relative to the intact element (169). Several possibilities could account for this result. Each element may constitute an independent binding site for the localization machinery, and the presence of both elements might recruit more localization factors. Alternatively, the two elements may interact with each other to present a better binding site and recruit a single binding factor.

A third example comes from analysis of the chicken β-*actin* 3'-UTR (171). β-*actin* mRNA is localized to the leading edge of lamellae in chicken embryonic fibroblasts and myoblasts. A so-called peripheral zipcode element consisting of the first 54 nt of the 3'-UTR is sufficient to direct localization of a heterologous transcript. When this element is deleted from the full-length 3'-UTR, the transcript is still able to localize, suggesting the presence of a redundant element. An inspection of the remainder of the 3'-UTR revealed a region of homology to the 54-nt zipcode within a more 3'-located 43-nt sequence. When this 43-nt region is present on its own, it is able to direct localization, albeit less effectively than the 54-nt element. These data suggest functional redundancy, but a functional analysis gave complicated results. Transfection of oligonucleotides complementary to each element individually significantly reduced localization (171). This observation may suggest that both elements are required in their natural context to mediate localization and, therefore, that they are not fully redundant. However, each element can mediate localization in isolation. In addition, antisense oligonucleotides used may have recognized both elements (since they share sequence homology) and thus simultaneously inactivated both localization elements.

A final case of repetition comes from the noncoding *Xlsirts*. These RNAs include repeated sequence elements flanked by unique sequences. The repeated element is 79–81 nt long and is tandemly repeated 3–13 times (97). Vegetal localization can be conferred by as few as two of the 79-nt sequence elements (97).

Dispersed/Nonredundant Localization Elements

In contrast to the aforementioned examples, in some cases it has been difficult to map discrete localization elements. For example, the *nanos* 3'-UTR contains a 547-nt region that is necessary and sufficient to confer localization (43). Two overlapping subregions map within this larger region, either of which are capable of conferring localization. However, these subregions are 400 and 470 nt in length, respectively, and cannot be subdivided without disrupting localization (43). Another possible example is the 1043-nt *oskar* 3'-UTR (168). In this case deletions have been used to define elements necessary for distinct aspects of *oskar* RNA localization. However, it has been difficult to demonstrate sufficiency of individual elements for any specific aspect of localization (168).

In principle, several scenarios might prevent definition of discrete and small (<150 nt) elements sufficient for localization. For example, there might be several dispersed elements, each necessary for localization, distributed over a large region. Deletion of any one of these elements would disrupt localization. Alternatively, the arrangement of specific localization elements within the larger region or the secondary structure of the RNA might preclude the use of gross deletional studies to define subregions sufficient for localization.

Additive Function of Localization Elements

Implicit in much of the preceding discussion is the fact that different subsets of the RNA confer different aspects of the localization pattern. In other words, localization elements act additively. Examples already mentioned are the *bicoid* 3'-UTR, which has distinct elements for early (BLE1) vs later (Staufen-mediated) localization to the anterior pole, and the *Add-hts* N4 3'-UTR, which has an early transport and cortical localization element (ALE1) and distinct elements that function in anterior localization. In each case, the combination of the defined elements (with possible contributions from other undefined elements) directs localization with the correct spatial and temporal dynamics.

Elements That Function in Translational Control During or After Localization

Several localized RNAs in the *Drosophila* oocyte are not translated until they are localized. For example, *oskar* RNA is not translated until it is localized to the posterior pole of the stage 9 oocyte. Translation of unlocalized *oskar* RNA leads to major developmental defects. Thus there is an intimate and important link between localization and translational control of *oskar* RNA. Sequence elements for translational control are separable from those that function in localization per se. Elements known as Bruno response elements (BREs) have been mapped within the *oskar* 3'-UTR and are necessary and sufficient for preventing translation of unlocalized RNA. Three discrete segments (A, B, and C) within the 3'-UTR bind an 80-kDa protein (Bruno) that mediates translational repression (discussed in the next section) (184). These segments share a conserved 7- to 9-nt sequence [U(G/A)U(A/G)U(G/A)U] that is present as a single copy in elements A and B, and in two copies in element C. Mutation of this sequence abolishes binding of Bruno and thus translational repression of unlocalized *oskar* RNA (184).

A second example comes from the *nanos* RNA. Unlike *oskar* RNA, which becomes tightly localized to the posterior pole of the oocyte, some unlocalized *nanos* RNA always exists in the embryo even after most of it is localized to the posterior. This unlocalized *nanos* RNA must be translationally repressed in order to prevent pattern defects in the early embryo (43, 139–141, 185). This translational repression is mediated by a 184-nt translational control element (TCE) in the *nanos* 3'-UTR (140) that contains two separable Smaug recognition elements (SREs), which bind a translational repressor to be described in the next section (141). These elements map within evolutionarily conserved regions of the *nanos* 3'-UTR (185). SRE1 lies between nucleotides (nt) 25 and 40 (141) within a 90-nt region (nt 1–90 of the 3'-UTR), which was shown independently to confer translational repression (185). SRE2 maps to nt 130–144, downstream in the 3'-UTR (141), within an adjacent 88-nt region (nt 91–178), which independent analyses showed has limited ability to repress translation (185). The SREs can form stem-loops (14–23 nt in length) with a highly conserved loop sequence (CUGGC) while the stem sequence is not conserved. Point mutations in the loops abolish binding of Smaug protein and eliminate translational repression of *nanos* RNA.

The BREs in the *oskar* 3'-UTR and the SREs in the *nanos* 3'-UTR are clear examples of discrete elements that are repeated within the 3'-UTRs and are largely functionally redundant.

A translational control element has also been mapped within the *Cyclin B* 3'-UTR, to a 39-nt region distinct from the posterior localization element (see above) (60, 182). In this case, the translational control element represses translation of localized maternal *Cyclin B* mRNA until late stage 14, about 11 h after fertilization. Deletion of the element results in premature translation, starting an hour after fertilization (182). The functional significance of this translational control has not been determined.

Primary, Secondary, Tertiary, and Quaternary Structures

A final question to be addressed here is the nature of the localization and translational control elements themselves. That is, is it the primary, secondary, tertiary, or quaternary structure that is recognized by the localization and translational control machinery? Instances of each of these possibilities have been mentioned in the preceding discussion. Several conserved primary sequence elements have been defined. For example, the SREs in the *nanos* 3'-UTR include a highly conserved loop sequence (CUGGC) recognized by Smaug (141), and the BREs in the *oskar* 3'-UTR contain a highly conserved sequence

[U(G/A)U(A/G)U(G/A)U] important for Bruno binding (184, 186). Similarly, a conserved sequence in *Vg1* and several other 3'-UTRs (*TGFβ-5*, *Xwnt-11*, *Xlsirt*, *tau*, *oskar*, *nanos*, and *gurken*) has been implicated in localization (187). As mentioned above, the chicken *β-actin* 3'-UTR contains two regions that confer peripheral localization (171); these regions each contain two conserved motifs (GGACT and AATGC).

The importance of RNA secondary structure in localization has been clear for some time (e.g. *bicoid* 3'-UTR) (166, 167). Conserved primary sequence elements are often parts of a stem-loop (e.g. SREs) and are likely to be bound by factors that interact with single-stranded RNA. Other examples of important stem-loop structures are the *K10* TLE (181) and, possibly, the *Add-hts* ALE1 (KL Whittaker, D Ding, WW Fisher, HD Lipshitz, manuscript in preparation).

Initially it was proposed that tertiary structure is also important for localization based on the discovery of complementarity between two loops within stem-loop III of the *bicoid* 3'-UTR (166, 167). These complementary regions actually undergo quaternary interactions (i.e. between different 3'-UTR molecules) mediated by the double-stranded RNA-binding Staufen protein (see below) (163, 164).

TRANS-ACTING FACTORS INVOLVED IN RNA LOCALIZATION AND TRANSLATIONAL CONTROL OF LOCALIZED RNAs

The previous section outlined the *cis*-acting sequences that function in RNA localization. It also discussed the sequences involved in localization as well as the elements involved in translational control of mRNAs during or after localization. Here, the focus is on *trans*-acting factors that function during RNA localization. These factors can function in RNA localization per se or in translational control during or after localization. The latter class of factors is included only if translational control is related directly either to the localization process or to the functional significance of RNA localization.

Identification of Trans-Acting Factors

Three strategies have led to the identification of *trans*-acting factors that function in RNA localization and/or translational control of localized RNAs. The first strategy, genetic definition of genes involved in RNA localization followed by molecular cloning of the genes and molecular biological and biochemical analyses of their encoded products, is restricted to *D. melanogaster* and *S. cerevisiae*. Examples are *Drosophila* Bicaudal-C, Bicaudal-D, Exuperantia, Homeless, K10, Squid, Staufen, Swallow, and Vasa. Although this strategy ensures that the gene product is involved in RNA localization, it cannot be determined at the outset whether the effects on RNA localization are direct or indirect.

A second strategy proceeds in the opposite direction by starting with biochemical searches for factors that bind defined RNA elements involved in localization and/or translational control. Ultimately, the gene encoding the identified factor is cloned and, in *Drosophila*, also mutated in order to assay function of the endogenous protein. With this method, it is known from the outset that the factor interacts directly with the target RNA; however, one has no assurance that the identified protein will indeed function in RNA localization/translational control rather than in some other aspect of RNA metabolism. Examples of *trans*-acting factors identified in this way are Bruno, Smaug, and Exl in *Drosophila*, and Vg1 RBP and Vera in *Xenopus*.

A final approach (not so much a strategy) that has led to the identification of factors involved in RNA localization has derived from molecular screens for gene products (RNA or protein) with interesting intracellular distributions (e.g. localization) or with interesting molecular homologies (e.g. RNA-binding motifs). In this case, as for the second strategy described above, one has no prior indication that the gene product functions in localization or translational control. However, if it is present in the right place at the right time, and possesses the appropriate molecular properties, the gene product may have a function in the process of interest. Examples of gene products identified in this way are Oo18 RNA-binding protein (Orb) and *Pgc* RNA in *Drosophila*, the *Xlsirt* RNAs in *Xenopus*, and the *YC* RNA in ascidians.

Factors That Interact Directly with Defined RNA Elements

STAUFEN PROTEIN *(DROSOPHILA)*   Alleles of the *staufen* gene were first recovered as maternal effect mutations with defects in anterior and posterior (abdominal) pattern in the *Drosophila* embryo (188). It was implicated in *bicoid* RNA localization to the anterior since *bicoid* RNA is partially delocalized in early embryos produced by mutant mothers (7, 32, 189). In *staufen* mutants *oskar* RNA is maintained at the anterior of the oocyte until stage 10 when it delocalizes (25) (in wild-type oocytes *oskar* RNA is transported from the anterior pole to the posterior by stage 9). Therefore, *staufen* is essential for initiation of posterior transport of *oskar* RNA. Moreover, weak *staufen* alleles show normal posterior localization of *oskar* RNA at stage 9, but the posterior localization is lost in later oocytes (stage 11) indicating that Staufen is also required to maintain posteriorly localized *oskar* RNA (190). During oogenesis Staufen protein first appears uniformly in stage 3–4 egg chambers, and by stage 8 it is present in a ring at the oocyte anterior as well as at the posterior of the oocyte (179). By stage 10B Staufen protein is at the posterior pole of oocyte; therefore, *oskar*

RNA and Staufen protein colocalize. In the early embryo, Staufen protein colocalizes with *oskar* RNA at the posterior pole and with *bicoid* RNA at the anterior pole (179).

Staufen is a double-stranded RNA-binding protein (178). Varying the amount of the Staufen target RNAs (*bicoid* or *oskar*), or of Staufen protein, indicates that the amount of Staufen protein recruited to the anterior or posterior pole depends on the amount of *bicoid* or *oskar* RNA present; thus the RNA targets rather than Staufen protein are limiting. Injected in vitro transcribed *bicoid* 3'-UTR RNA recruits Staufen protein into particles that colocalize with microtubules near the site of injection in the early embryo (163). The formation of these particles requires specific 3'-UTR elements previously defined as important for *bicoid* RNA localization (see above). Specifically, evidence suggests that two single-stranded regions of stem-loop III within the *bicoid* 3'-UTR form intermolecular double-stranded RNA hybrids (i.e. via quaternary interactions) that are bound by Staufen protein (164). Staufen protein also interacts later in embryogenesis with the *prospero* 3'-UTR in neuroblasts and is necessary for basal localization of *prospero* transcripts (127).

Given that Staufen interacts with the best-characterized 3'-UTRs of localized RNAs (those in *bicoid* and *oskar* RNAs), that it is an RNA-binding protein, and that mutations exist both in the *staufen* gene and in the genes that encode its target RNAs, Staufen is by far the best-understood *trans*-acting factor involved in RNA localization.

EXL PROTEIN (*DROSOPHILA*)   BLE1 is a 53-nt element in the *bicoid* 3'-UTR that is sufficient (when present in two copies but not one) to direct early nurse cell–oocyte transport and anterior localization of RNA (see above) (176). 2xBLE1 was used in UV-crosslinking assay to search for directly interacting proteins (177). A single protein of 115 kDa (Exl) was found to bind 2xBLE1 but not 1xBLE1, consistent with a role in localization. Definition of Exl-binding sites within BLE1 and mutation of these sites gave results consistent with a role for Exl in BLE1-mediated anterior localization. Exl might inter-act directly with BLE1 in the *bicoid* 3'-UTR during localization, and it might mediate Exuperantia protein interaction with *bicoid* RNA in the localization particles (177) (see below for discussion of Exuperantia). To date, the gene encoding Exl has not been cloned, and mutations in the gene have not been identified.

BRUNO PROTEIN (*DROSOPHILA*)   Bruno, an 80-kDa protein, was identified in a UV-crosslinking screen for *trans*-acting factors that bind the *oskar* 3'-UTR (184). There are three binding sites (BREs) in the *oskar* 3'-UTR that share a 7- to 9-nt motif. Mutations in these elements abolish Bruno binding in vitro

(184). Endogenous *oskar* mRNA is translated only after the transcript is localized to the posterior of the oocyte in stages 8–9 (184). An *oskar* [BRE⁻] transgene results in premature translation of *oskar* RNA during stages 7–8, prior to *oskar* RNA localization, producing gain-of-function phenotypes (double-posterior or posteriorized embryos) (184), consistent with spatially inappropriate Oskar protein expression (75, 76). BRE elements can confer translational repression on a heterologous transcript (184). Bruno-mediated translational control of other transcripts is suggested by the ability of Bruno protein to bind *gurken* RNA (184).

The gene encoding Bruno was recently cloned (186). The sequence reveals RNP/RRM-type ribonucleoprotein RNA-binding domains consistent with direct interaction with RNA. The RNP/RRM-domain RNA-binding motif was first defined in yeast mRNA poly(A)-binding protein and mammalian hnRNP protein A1 (reviewed in 191). The gene encoding Bruno is a previously identified genetic locus, *arrest*, which is necessary for female as well as male fertility (192, 193). This observation, together with the fact that *arrest* mutants display defects in oogenesis prior to the time that Bruno binds *oskar* RNA, suggests that Bruno regulates other transcripts in addition to *oskar* RNA.

Bruno also interacts with Vasa protein as assayed by far-Western analysis (186). Vasa is itself an RNA helicase related to eIF-4A, which functions in translation initiation (194–196). This suggests a possible mechanism for Bruno; Bruno may protect *oskar* RNA from premature translational activation by the Vasa protein (186).

Identification and cloning of Bruno, and its correlation with a previously identified genetic locus, is the first instance in which a *trans*-acting factor involved in translational control of a localized RNA has been identified biochemically and then studied both molecularly and genetically.

SMAUG PROTEIN (*DROSOPHILA*)   As described above, translational repression of unlocalized *nanos* mRNA (43, 140, 141) is accomplished through two elements that bind in vitro to Smaug, a 135-kDa protein (141). Mutation of these SREs abolishes Smaug binding in vitro. Embryos from *nanos* [SRE⁻] transgenic mothers lose head structures, which is consistent with translation of *nanos* RNA throughout the embryo rather than solely at the posterior (141). Smaug binding alone is sufficient to mediate translational repression since a transgene containing three SREs but no other part of the *nanos* 3'-UTR is translationally repressed (141). The gene encoding Smaug has not been cloned.

VG1 RNA-BINDING PROTEIN (RBP) (*XENOPUS*)   UV crosslinking has also been used to identify *trans*-acting factors that bind to the *Vg1* 3'-UTR, which functions in vegetal localization of *Vg1* RNA in *Xenopus* oocytes (197). The crosslinking experiments led to the identification of a 69-kDa protein called Vg1

RNA-binding protein (RBP). Vg1 RBP may also bind in vitro to the 3'-UTR of a second vegetal pole–localized RNA, *TGFβ-5* (187, 197), but not to *An2* RNA, which is localized to the animal hemisphere. The precise role of Vg1 RBP in vegetal RNA localization is not clear. Evidence indicates that it mediates the association of *Vg1* RNA with microtubules (199), which are necessary for vegetal RNA localization (see above). Based on the results of UV-crosslinking experiments Vg1 RBP is likely to be a member of an RNP complex containing up to six proteins plus *Vg1* mRNA (200). Six protein bands (one of which is likely to be Vg1 RBP) were identified from stage II–III oocyte extracts, corresponding to the period during which *Vg1* RNA is localized (200). Fewer proteins were labeled in earlier and later stage oocyte extracts, at times during which *Vg1* RNA is not being localized.

VERA PROTEIN (*XENOPUS*)   In a search for *trans*-acting factors that bind the *Vg1* localization element a new 75-kDa protein, Vera, was purified (154). A mutant form of the *Vg1* localization element (deleted for three out of four repeated sequence motifs) that does not bind Vera in vitro exhibits impaired localization in vivo (154). Vera protein co-sediments with Trap-α, an integral membrane protein associated with the protein translocation machinery of the endoplasmic reticulum (ER) (154). Vera may link *Vg1* mRNA to the vegetal ER subcompartment while the ER (along with *Vg1* RNA) is transported via microtubules to the vegetal pole.

Other Factors That Function in RNA Localization

EXUPERANTIA PROTEIN (*DROSOPHILA*)   The genetic analyses that led to the identification of *staufen* as functioning in *bicoid* RNA localization also led to the identification of *exuperantia* (188, 189). The first observable *bicoid* RNA localization defect in *exuperantia* mutants is loss of apical transcript localization in nurse cells (32) at a stage at which Exuperantia protein is localized around the nurse cell nuclei (201, 202). Exuperantia protein is highly concentrated in the anterior cortex of the oocyte between stages 8 and 10 (162, 201). In stage 10 egg chambers mutant for *exuperantia*, *bicoid* RNA delocalizes from the anterior of the oocyte (7, 150). In late oocytes (stage 14) mutant for *exuperantia*, *bicoid* RNA is released from the microtubule-based cytoskeleton (147). However, Exuperantia protein is not present in late oocytes or in embryos, indicating that Exuperantia is involved in establishing but not in maintaining anterior *bicoid* RNA localization (201, 202).

Visualization of an Exuperantia-GFP fusion protein in live oocytes (162) demonstrates that Exuperantia is present in large particles that are transported from the nurse cells into the oocyte through the ring canals (see above). Transport of these Exuperantia-containing particles appears to be a multistep process: Colchicine-sensitive steps are transport within the nurse cells and anchoring at the anterior of the oocyte; whereas transport through the ring canals into the oocyte is insensitive to both microtubule and microfilament inhibitors (162; WE Theurkauf, TI Hazelrigg, personal communication). Exuperantia may function in this microtubule-independent transport of *bicoid* transcripts through the ring canals into the oocyte (151).

Taxol stabilizes microtubules and causes aberrant microtubule bundles in the oocyte that contain ectopically localized *bicoid* transcripts (150). Taxol-treated *exuperantia* mutant oocytes do not exhibit ectopic *bicoid* transcript localization (150). The microtubule networks in *exuperantia* oocytes are normal. These results are consistent with *bicoid* RNA association with ectopic microtubules requiring Exuperantia protein.

Deletion of the *cis*-acting element BLE1 from the *bicoid* 3′-UTR mimics the *bicoid* transcript delocalization defects caused by *exuperantia* mutants (176). However, BLE1 specifically binds a protein called Exl (see above) and Exuperantia alone can bind RNA only nonspecifically (177). Exuperantia protein may interact with Exl protein in *bicoid* RNA localization particles. While Exuperantia does not interact specifically with *bicoid* RNA, it does function specifically in *bicoid* RNA localization; other anteriorly localized RNAs such as *Add-hts*, *Bicaudal-D*, *K10*, and *orb* are not delocalized in *exuperantia* mutant oocytes (16, 37).

SWALLOW PROTEIN (*DROSOPHILA*)   The third genetic locus initially shown to be necessary for *bicoid* transcript localization during oogenesis is *swallow* (7, 189). Subsequent analyses implicated it in the cortical and anterior localization of a second RNA, *Add-hts* (16, 174; KL Whittaker, D Ding, WW Fisher, HD Lipshitz, manuscript in preparation). The *swallow* locus is not necessary for the transient anterior positioning of other RNAs such as *Bicaudal-D*, *K10*, and *orb* (37). Swallow protein maintains cortical localization of transcripts such as *bicoid* and *Add-hts* as well as of RNPs such as the polar granules (44, 45). With regard to *bicoid*, Swallow appears to be involved in maintaining rather than establishing *bicoid* transcript association with microtubules since taxol-treated *swallow* oocytes exhibit ectopic *bicoid* transcript localization (150). The Swallow protein may possess a highly divergent RNP/RRM motif suggestive of a direct interaction with RNA (203). Swallow protein is distributed throughout the oocyte at stages 5–7 and at its anterior cortex at stages 8–10, consistent with localization of *Add-hts* and *bicoid* RNAs to these regions (44, 45, 204).

OSKAR PROTEIN (*DROSOPHILA*)   Nonsense mutants such as *oskar*[54] show abnormal *oskar* RNA localization (24, 25). In these mutants *oskar* RNA localizes to the posterior in stage 8 but subsequently delocalizes, indicating a role for Oskar protein in anchoring *oskar* RNA at the posterior. Directly or indirectly,

Oskar protein's role in *oskar* transcript localization is mediated by the *oskar* 3'-UTR since a chimeric *oskar* 3'-UTR construct localizes to the posterior and is maintained there in wild type but fails to be maintained there in an *oskar*[54] mutant (190). Further support for this scenario is the observation that in *D. melanogaster*, a transgenic *D. virilis oskar* RNA is able to localize to the posterior but fails to maintain its position at the oocyte posterior in the absence of *D. melanogaster* Oskar protein (205). *D. virilis* Oskar can direct posterior abdominal patterning but not pole-cell formation. This observation indicates that pole-cell formation may require a high concentration of Oskar protein and that this is provided for by anchoring high concentrations of *oskar* RNA at the posterior. Oskar protein nucleates the formation of the posterior polar granules, organelles that function in germ-cell formation and specification (see below). Oskar protein may interact directly with the *oskar* RNA's 3'-UTR to maintain transcript localization. Alternatively, Oskar function may be indirect in nucleating formation of polar granules that in turn are needed for *oskar* RNA localization.

VASA PROTEIN (*DROSOPHILA*) Vasa protein is a component of the perinuclear "nuage" material in the nurse cells and also of the polar granules at the posterior pole of the *Drosophila* oocyte (194, 196, 206–208). Loss of *vasa* function results in destabilization of the polar granules and delocalization of posteriorly localized RNAs (188, 194, 196, 206–208). Vasa protein is a founding member of the DEAD-box family of ATP-dependent RNA helicases (194, 207), and it binds duplex RNA (196). There is no evidence for specific binding of Vasa to any localized RNA; however, far-Western analysis has demonstrated that Vasa protein interacts directly with Bruno protein, which in turn binds directly to the BRE translational control elements in the *oskar* RNA's 3'-UTR (see above) (186). Transcripts encoding the zebrafish homolog of Vasa are localized subcellularly beginning at the two-cell stage and segregate into the primordial germ cells (209). This observation suggests likely evolutionary conservation of Vasa function in metazoan germ plasm.

HOMELESS PROTEIN (*DROSOPHILA*) Homeless is another member of the DEAD/DE-H family of RNA-binding proteins (40). Its amino terminal portion contains a region that bears homology to yeast splicing factors PRP2 and PRP16 and to the *Drosophila* Maleless protein (40). Transport and localization of *gurken*, *oskar*, and *bicoid* transcripts are severely disrupted in *homeless* mutant ovaries, which also contain reduced amounts of *K10* and *orb* transcripts (40). In contrast, *Bicaudal-D* and *Add-hts* transcript localization is unaffected (40). It is unknown whether Homeless protein interacts directly with any of the affected transcripts.

ORB PROTEIN (*DROSOPHILA*)   Orb protein contains two RNP/RRM-type RNA-binding domains and functions in antero-posterior and dorso-ventral patterning during *Drosophila* oogenesis (23, 149, 210). *orb* mutants affect the transport and localization of several RNAs including *Add-hts*, *Bicaudal-D*, *K10*, *gurken* and *oskar* (149, 210). For example, in wild-type oocytes during stages 8–10, *gurken* mRNA is usually restricted to the dorsal side of the nucleus at the antero-dorsal pole. In *orb* mutants, however, *gurken* transcripts are present throughout the entire anterior of the oocyte. In the wild type, *oskar* mRNA is localized to the posterior pole of the oocyte. In *orb* mutants *oskar* transcripts are distributed throughout the oocyte. Direct binding of Orb protein to either RNA has not been demonstrated.

SQUID PROTEIN (*DROSOPHILA*)   Squid is a member of the hnRNP family of RNA-binding proteins and was identified as functioning in dorso-ventral axis formation (211). There are three protein isoforms that share a common amino terminus containing two RNA-binding motifs. Two of these isoforms, Squid-A and Squid-B, are present in the oocyte. Squid is required for the correct dorso-anterior localization of *gurken* mRNA in the oocyte. In *squid* mutant ovaries, *gurken* transcripts are localized throughout the anterior of the oocyte rather than just antero-dorsally (211). Since *gurken* is unusual in that it is transcribed in the oocyte nucleus (R Cohen, personal communication), Squid's function in *gurken* transcript localization may initiate in the oocyte nucleus. It is unknown whether Squid interacts directly with the *gurken* mRNA or with the *gurken* pre-mRNA. As for K10 (below), Squid may function in vectorial transport of *gurken* transcripts out of the oocyte nucleus.

K10 PROTEIN (*DROSOPHILA*)   *K10* gene function is required during stage 8 of oogenesis for localization of *gurken* mRNA adjacent to the oocyte nucleus at the antero-dorsal tip of the oocyte (21, 212). K10 protein is restricted to the oocyte nucleus (213) but does not regulate *gurken* transcript production or stability (212). Since *gurken* is transcribed in the oocyte nucleus (R Cohen, personal communication) and becomes restricted to the oocyte cytoplasm dorso-anteriorly to the oocyte nucleus, K10 protein might function specifically in vectorial nucleo-cytoplasmic transport of *gurken* transcripts (R Cohen, personal communication). Whether this function is through direct interaction with the *gurken* mRNA is not known at present.

BICAUDAL-C PROTEIN (*DROSOPHILA*)   Mutations that reduce *Bicaudal-C* gene dosage result in defects in RNA localization in the oocyte: Most *oskar* RNA remains at the anterior pole of the oocyte and early embryo instead of being transported to the posterior by stage 9 (18). Possibly as a consequence, in these *Bicaudal-C* mutants, *nanos* RNA is localized ectopically near the anterior pole in patches on the dorsal and ventral sides rather than at the posterior pole as in wild type (18). These embryos develop a bicaudal (double-abdomen) phenotype. *Bicaudal-C* mutations have no effect on *gurken* or *orb* transcript localization to the anterior pole. Females homozygous for strong *Bicaudal-C* alleles produce oocytes that do not form anterior chorion as a consequence of defects in follicle cell migration over the oocyte anterior. Bicaudal-C is a transmembrane protein that has two conserved cytoplasmic domains (18): an Eph domain that is present in transmembrane receptor tyrosine kinases and is involved in signal transduction, and a KH domain that has been implicated in binding of single-stranded DNA or RNA. Since *Bicaudal-C* RNA is localized to the anterior of the oocyte, it is reasonable to assume that the Eph domain functions in the intercellular signaling from oocyte to follicle cells that programs their migration (18). The KH domain might bind to and interact with the *oskar* transcripts during their localization, possibly functioning in their transfer to the machinery that transports RNA to the posterior (see above) (18).

BICAUDAL-D PROTEIN (*DROSOPHILA*)   Bicaudal-D function has been covered with respect to RNA transport and the oocyte cytoskeleton above. The Bicaudal-D protein includes a region with homology to the coiled-coil domains of several cytoskeletal proteins and is required for maintenance of *oskar* RNA localization at the posterior pole of the oocyte (160).

EGALITARIAN PROTEIN (*DROSOPHILA*)   In *egalitarian* mutants the *Bicaudal-D*, *orb*, and *K10* RNAs do not accumulate in the oocyte (25). In addition, these RNAs no longer coprecipiate with the cytoskeletal fraction of oocytes as in the wild type (147). Egalitarian protein contains regions homologous to c10G6.1 from *Caenorhabditis elegans*, an EST from *Arabidopsis thaliana*, and ribonuclease D from *Haemophilus influenzae* (20). It is also predicted to include a coiled-coil region. Egalitarian protein localization to and within the oocyte is dependent on microtubules (20). Further, the Egalitarian and Bicaudal-D proteins copurify (20). Egalitarian and Bicaudal-D may be components of the cytoskeletal apparatus involved in RNA localization.

BULLWINKLE (*DROSOPHILA*)   Mutations in the *bullwinkle* gene have several defects in posterior body patterning (214, 215). The *bullwinkle* gene is required to localize *oskar* transcripts to the posterior of the oocyte, to maintain *oskar* RNA at posterior, and to regulate the level of *oskar* protein (214, 215). Cloning of the *bullwinkle* gene has not been reported.

AUBERGINE (*DROSOPHILA*)   While previously *aubergine* had been implicated in dorso-ventral body patterning (192), two new alleles were identified in a recent genetic screen for genes involved in posterior body patterning (214, 216).

Aubergine functions to enhance the translation of *oskar* mRNA in the ovary, mediated through the *oskar* 3'-UTR (216). *Aubergine*'s enhancement of *oskar* translation is independent of Bruno-mediated repression of translation since an *oskar* [BRE⁻] transgenic RNA still requires *aubergine* function for its translation (216). Cloning and molecular analysis of *aubergine* have not been reported.

XLSIRT RNA *(XENOPUS)*   *Xlsirt*s are a family of nontranslatable, interspersed repeat transcripts that localize to the vegetal cortex of *Xenopus* oocytes (see above) (97, 161). These RNAs are involved in the localization of *Vg1* but not *Xcat-2* transcripts. It is unknown whether *Xlsirt* and *Vg1* RNAs interact directly or whether *Xlsirt* RNA function in *Vg1* transcript localization is indirect via the cytoskeletal network or as part of a localization particle/organelle.

PGC RNA *(DROSOPHILA)*   In *Drosophila* a non-protein-coding RNA, *Pgc*, is a component of the posterior polar granules and is required for pole cells to migrate normally and to populate the gonad (26). Reduction of *Pgc* RNA at the posterior pole results in a reduction in the amount of posteriorly localized *nanos* and *germ cell-less* RNA and Vasa protein (26). It is unknown whether *Pgc* RNA interacts with other posteriorly localized RNAs and/or with protein components of the polar granules.

YC RNA *(STYELA)*   As described above, the noncoding *YC* RNA is localized in the yellow crescent of Ascidian eggs and embryos (101). The 3'-UTR of *PCNA* RNA contains a 521-nt region of complementarity to *YC* RNA (100), while the 5'-UTR of *ribosomal protein L5* mRNA exhibits 789 nt of complementarity to *YC* RNA (102). *YC* RNA may interact directly with these two RNAs in vivo (100–102); however, no such interaction has been demonstrated, and its function remains unclear (see above).

Summary

In summary, genetic and molecular strategies have identified a host of *trans*-acting factors that function in transcript localization. To date, although many of these are homologous to known RNA-binding proteins, only a handful have been shown to interact directly with specific localized RNAs. Many localized RNAs have collections of discrete *cis*-acting localization elements that mediate distinct aspects of their localization. It might therefore be predicted that many different *trans*-acting factors will function during localization of any one RNA, each binding to a different type of element and each possibly functioning at a different time and intracellular location. RNA localization particles are likely to consist of these directly interacting factors in addition to numerous others that are involved in linking the RNA to the cytoplasmic translocation machinery as well as anchoring it at its intracellular target site. It would be surprising if evolution has not also utilized the base-pairing capability of RNA in the translocation and anchoring process.

DEVELOPMENTAL FUNCTIONS OF RNA LOCALIZATION

The cellular functions of RNA localization have been reviewed extensively (see e.g. 1, 217) and so are considered here only briefly. First, for mRNAs, localization directs high-level synthesis of the encoded protein at the site of localization. Thus if protein function requires a high concentration (e.g. *Drosophila* Oskar in the polar granules), this requirement can be met through mRNA localization. Second, localizing an mRNA also excludes protein synthesis from other parts of the cell, thus reducing the amount of protein present in those regions. Often, not only is local protein synthesis directed by mRNA localization, but translational control mechanisms actually prevent translation of unlocalized RNA either during RNA transit to the target site (e.g. *Drosophila oskar* RNA) or of unlocalized RNA that remains after transcript localization is complete (e.g. *Drosophila nanos* RNA).

A third postulated function of mRNA localization is to direct specific protein isoforms to particular regions of the cell. Often this is accomplished by alternative pre-mRNA splicing such that different 3'-UTRs direct different isoforms to different cytoplasmic domains (e.g. *Drosophila Add-hts* RNA). Fourth, intracellular localization is used as a mechanism to segregate RNAs unequally between the products of cell division, particularly when these divisions are asymmetric (e.g. *ASH1* transcripts during yeast budding, *prospero* transcripts during *Drosophila* neuroblast division). Fifth, certain non-protein-coding RNAs are localized (e.g. *Xenopus Xlsirts* or *Drosophila Pgc* RNAs). The detailed role of these RNAs and of their localization is currently under intensive study. These RNAs may serve as structural components of localization particles or organelles such as the germinal granules. Alternatively, they may function in the RNA localization or anchoring process, possibly through sequence complementarity to mRNAs that are being localized.

Specification of the Anterior-Posterior and Dorsal-Ventral Axes of the Drosophila Oocyte

As mentioned above, the *gurken* mRNA is unusual in that it is synthesized in the oocyte nucleus (R Cohen, personal communication). It is localized to the posterior pole of the oocyte at stage 7, then to both the anterior and posterior poles at stage 8, and finally to the antero-dorsal pole from late stage 8 through stage 10 (21, 218). Gurken protein is a TGFα-like secreted growth factor (21). Establishment of the antero-posterior and dorso-ventral axes of the oocyte is accomplished between stages 7 and 9 of oogenesis through signaling between the oocyte and the surrounding follicle cells (34, 35). The Gurken protein functions as a key signal from the oocyte to the follicle cells in both of these processes. First, due to posterior-localized *gurken* RNA, local production of Gurken protein at the posterior of the oocyte signals to the posterior follicle cells. This signaling is essential for establishing the antero-posterior oocyte axis and the polarization of the oocyte microtubule-based cytoskeleton that plays a crucial role in RNA localization. Subsequently, antero-dorsal localized *gurken* RNA directs local Gurken protein synthesis, enabling oocyte–nurse cell signaling that establishes the dorso-ventral axis of the egg chamber. The dorso-anterior localization of *gurken* mRNA depends on anterior migration of the oocyte nucleus on the polarized microtubule cytoskeleton. Thus the antero-posterior axis is primary and the establishment of the dorso-ventral axis secondary (35). Both axes depend on localization of *gurken* mRNA for localized signaling. If *gurken* mRNA is mislocalized or delocalized, for example, by mutating the *K10*, *squid*, or *orb* genes, severe defects in the formation of both axes result.

Specification of Anterior Cell Fates in the Drosophila Embryo

Shortly after fertilization, *bicoid* mRNA is translated (219, 220). Since the embryo is syncytial, Bicoid protein diffuses away from its site of translation at the anterior pole, forming an antero-posterior protein gradient with its peak at the anterior tip (219, 220). Since Bicoid is a homeodomain-containing transcription factor, its function is to activate zygotic transcription of pattern-specifying genes in the syncytial nuclei in the anterior half of the embryo. It does this in a concentration-dependent fashion (221–223). For example, the *hunchback* gene contains high-affinity Bicoid-binding sites in its transcriptional control region, so its transcription is activated by low as well as high Bicoid concentrations throughout the anterior half of the embryo. In contrast, genes such as *orthodenticle* and *empty spiracles* have lower affinity Bicoid-binding sites and so are activated only by higher Bicoid protein concentrations in the more anterior part of the embryo. In this way, different combinations of zygotic pattern genes are activated in different subsets of the anterior part of the embryo leading to different cell fates within this region (e.g. head more anteriorly, thorax more posteriorly). *Bicoid* mRNA localization controls the amount of Bicoid transcription factor in different regions and thus specifies distinct cell fates. Delocalization of *bicoid* mRNA can be accomplished by mutating genes that encode *trans*-acting factors that function in *bicoid* transcript localization (7, 32, 189) (e.g. *exuperantia*, *swallow*, *staufen*). Delocalization results in lower levels of Bicoid protein at the anterior pole than in the wild type. As a consequence, acronal and head structures cannot be specified.

Localization of *bicoid* mRNA serves two additional functions. First, mislocalization of *bicoid* RNA to the posterior pole can result in developmental defects in the posterior part of the embryo through cells mistakenly adopting anterior fates (224). Thus a corollary function of anterior *bicoid* transcript localization is to prevent Bicoid protein synthesis in the posterior of the embryo. Translational control mechanisms also prevent Bicoid protein synthesis at the posterior (141, 205, 225). Second, the Bicoid homeodomain protein can function not only as a transcription factor but also can directly bind RNA and translationally repress mRNAs such as *caudal* through interaction with the *caudal* 3'-UTR (226, 227). Thus anterior localization of *bicoid* mRNA and the resultant Bicoid protein gradient creates a reverse gradient of Caudal protein with its peak at the posterior pole. Caudal protein is involved in specifying pattern in the posterior of the early embryo (228, 229).

Specification of Abdominal Cell Fates in the Drosophila Embryo

A key player in abdominal cell fate specification is Nanos, a Zinc-finger-containing protein (56, 230). The *nanos* mRNA is localized at the posterior pole of the late oocyte and early embryo, although some unlocalized RNA is present throughout the embryo (56, 231). After fertilization, the posteriorly localized *nanos* RNA in the syncytial embryo is translated (there is repression of unlocalized *nanos* RNA translation by Smaug protein). This translation leads to a gradient of Nanos protein with a peak at the posterior pole (139–141, 185, 231). Unlike *bicoid*, which controls anterior cell fates by a combination of transcriptional control of target genes and direct translational repression of *caudal* mRNA in the anterior, all of the Nanos protein's effects in abdominal patterning derive from its translational repression of target RNAs. One target, *hunchback* maternal RNA, is distributed uniformly in the early embryo (232, 233) and encodes a Zinc-finger transcription factor that specifies anterior cell fates (234–236). Thus, if Hunchback protein were synthesized in the posterior of the embryo, posterior cells would mistakenly adopt anterior fates. Nanos protein in the posterior of the embryo prevents this by translationally repressing *hunchback* RNA. The Pumilio protein, previously shown to be important for abdominal patterning (237), specifically binds to Nanos response elements (NREs) in the *hunchback* 3'-UTR, recruiting Nanos through protein-protein interactions (140). If *nanos* RNA is misexpressed throughout the embryo by mutation of its SREs, head defects result, probably through repression of *bicoid* translation by Nanos protein since the *bicoid* 3'-UTR also contains NREs (141). Misexpression of high levels of Nanos protein in the anterior results in bicaudal embryos (231). Thus the combination of posterior localization of *nanos* mRNA and the translational repression of unlocalized *nanos* transcripts plays a crucial role in patterning the abdomen of *Drosophila*.

Assembly of Polar Granules and Specification of Germ Cells in the Drosophila Embryo

oskar mRNA is localized to the posterior pole of the stage 9 oocyte. Translation of *oskar* RNA at this site nucleates the formation of the posterior polar granules and polar plasm (25, 75, 76). The polar granules and posterior polar plasm serve two functions. The first is the anchoring of *nanos* transcripts at the posterior (75, 76). Disruption of the polar granules results in delocalization of *nanos* RNA, translational repression by Smaug, and ultimately the production of embryos without abdomens. The second role of polar granules is to specify the formation of germ (pole) cells and to restrict their formation to the posterior tip of the embryo (25, 75, 76). Disruption of *oskar* function results in an inability to nucleate polar granules and, consequently, absence of pole cells (238). Alternatively, misexpression of Oskar protein throughout the oocyte during *oskar* RNA localization (184, 186), overexpression of *oskar* RNA throughout the oocyte (75), or mislocalization of *oskar* RNA and protein to the anterior pole of the oocyte (76) all result in severe pattern defects. In the latter two situations ectopic pole cells form at or near the anterior of the embryo. The mechanisms by which the polar granules specify the formation and function of the pole cells in the early embryo are not yet fully understood but appear to require the function of several other posteriorly localized RNAs such as *Pgc*, *mtlrRNA*, *nanos*, and *germ cell-less* (26, 57, 133, 134, 136, 137, 239, 240). Thus, both the establishment of the polar granules, and their function in pole cell formation, require RNA localization.

Signaling of Dorso-Ventral Axis and Mesoderm Induction in the Xenopus Embryo

The animal-vegetal axis of the *Xenopus* embryo is established during oogenesis. The three germ layers of the early embryo (ectoderm, mesoderm, and endoderm) are established along the animal-vegetal axis (72, 73). The darkly pigmented animal hemisphere of the oocyte gives rise to ectoderm while the vegetal hemisphere cells become endoderm. The mesoderm is derived from animal hemisphere cells that lie adjacent to the vegetally derived mesoderm. Mesodermal development is not autonomous but is a result of inductive interactions from the endoderm (72, 73).

Asymmetrically distributed RNAs localized to the vegetal hemisphere that encode secreted growth factors such as Vg1 and TGFβ-5 have been implicated in mesoderm induction (5, 241–243). The secreted growth factor TGF-β1 can act synergistically with another growth factor, bFGF, to induce mesoderm, whereas antibodies against TGF-β2 can reduce mesoderm induction when injected into *Xenopus* embryos. Since *Vg1* and *TGF-β5* RNAs are localized to the vegetal hemisphere from which the inducing signal derives, and since Vg1 and TGF-β5 proteins are related, respectively, to TGF-β1 and 2, these proteins are strong candidates for mesoderm-inducing signals (5, 241–243). Indeed, engineered processed Vg1 protein (in the form of BMP2/4-Vg1 fusion protein) can function as a mesoderm inducer when ectopically expressed in embryos (241, 242).

The orientation of the dorso-ventral axis of the early embryo is not established prior to fertilization. Rather, the sperm entry point in the animal hemisphere establishes this axis in part by causing an oriented cytoplasmic rearrangement (72, 73). This rearrangement relocates interior cytoplasm (endoplasm) relative to the stationary cortical cytoplasm. Treatments, such as UV-irradiation, prevent cytoplasmic rearrangement and ventralize the embryo (i.e. prevent formation of the dorsal-most tissues). Maternal *Xwnt-11* mRNA is localized vegetally in oocytes and early embryos (98). Injection of *Xwnt-11* RNA into embryos that have been ventralized by UV-irradiation substantially rescues the UV-induced defect by inducing formation of dorsal tissues such as somitic muscle and neural tube (98). This observation suggests that Xwnt-11 protein functions during normal embryogenesis in dorso-ventral axis formation and that localization of *Xwnt-11* mRNA and protein may play a role in induction of this axis.

The inductive events discussed previously are complex both at the level of inducing signals and at the level of mesodermal cell fate outcomes. The inability to genetically inactivate genes in *Xenopus* has been a major drawback in defining endogenous factors necessary (rather than sufficient) for induction. Thus the functions during normal development of *Xwnt-11*, *Vg1*, and *TGFβ-5* RNA localization and of their encoded proteins remain to be determined.

Specification of Cell Fates During Asymmetric Cell Divisions

During *Drosophila* neurogenesis, a stem cell called a neuroblast divides asymmetrically to form a ganglion mother cell (GMC) and another neuroblast. The GMC then divides to form neurons. The Prospero nuclear protein is required for neuronal differentiation (244) and axonal pathfinding (245). The *prospero* mRNA and the Prospero protein are intially apically localized in the neuroblast at interphase but relocalize basally from prophase through telophase, thus segregating into the GMC (127, 246, 247). Basal localization of *prospero* RNA requires Inscuteable and Staufen proteins (127), and Staufen binds directly to the *prospero* 3'-UTR (127). The Miranda protein functions as an adapter that links Prospero protein to the basal cell membrane during the asymmetric neuroblast division (248). Thus RNA and protein localization are used to segregate the Prospero protein into one product of an asymmetric cell division, conferring appropriate neuronal fates upon that cell and its progeny. Asymmetric segregation of cell fate determinants through mRNA localization has also been described in the budding yeast *S. cerevisiae*. In this case, during cell division the *ASH1* mRNA is localized to the site of the bud and then into the daughter cell that forms there (128, 129). The *ASH1* protein acts as a repressor of the HO endonuclease, which is responsible for mating-type switching (128, 129). Thus localization of the *ASH1* mRNA and its asymmetric segregation into the daughter cell ensures that the daughter cell cannot switch mating type while the mother cell (which does not inherit *ASH1* mRNA) can switch.

EVOLUTIONARY CONSIDERATIONS

Many features of mRNA localization appear to have been conserved during evolution, suggesting that RNA localization is an ancient mechanism for producing cytoplasmic asymmetry. For example, large stereotypic secondary structures in 3'-UTRs that function in *bicoid* transcript localization are evolutionarily conserved and functionally interchangeable between *Drosophila* species separated by over 60 million years (167). Further, the *bicoid* 3'-UTR, which directs anterior RNA localization in the oocyte, can also direct apical transcript localization in epithelia such as the blastoderm (65). Consistent with this observation, *bicoid* RNA is localized apically in the nurse cells prior to its transport into the oocyte during normal development. Thus, at least within the same species, different polarized cell types appear to share localization signals and factors.

More remarkable is the fact that the mammalian *tau* 3'-UTR, which directs *tau* transcript localization to the axons of neurons, can also direct vegetal transcript localization in *Xenopus* oocytes with a pattern and dynamics indistinguishable from *Vg1* transcripts (158). This result suggests that RNA targeting elements and localization machinery are conserved from *Xenopus* to mammals and from oocytes to neurons. Whether this functional conservation extends to the primary RNA sequence level remains to be seen; however, a small sequence element has been reported to be conserved in the 3'-UTR of *tau*, *Vg1*, and several other localized RNAs in mammals, *Xenopus*, and even *Drosophila* (187).

Recently, RNA localization has been reported in the budding yeast, *S. cerevisiae* (128, 129), and functions during budding to confer asymmetric fates on the mother and daughter cells. mRNA localization (e.g. of *prospero* transcripts) can serve a similar function in higher eukaryotes (127). This suggests that the process of RNA localization dates at least to the invention of single-celled organisms with specialized cytoplasmic domains and/or that undergo asymmetric cell divisions. The demonstration that the yeast *ASH1* mRNA's 3′-UTR carries information for intracellular targeting implies that the position of *cis*-acting localization elements may be conserved in mRNAs from yeast to mammals. Future studies that focus on the identification and analysis of *trans*-acting factors that target RNAs for localization are likely to uncover additional conserved components of the cytoplasmic RNA localization mechanism.

ACKNOWLEDGMENTS

The following kindly shared unpublished or in press results with us: P Macdonald, R Cohen, P Lasko, W Theurkauf, T Hazelrigg, H Krause, R Long, R Singer, P Takizawa, H Tiedge, L Etkin, and K Mowry. Special thanks to L Etkin, University of Texas, for providing the images of *Xenopus* localized RNAs used in Figure 4. We thank S Lewis, L Etkin, and M Kloc for comments on the manuscript. Our research on RNA localization has been funded over the past decade by research grants to HDL from the National Institutes of Health, the National Science Foundation, and currently the Medical Research Council of Canada.

Visit the *Annual Reviews home page* at
http://www.AnnualReviews.org.

Literature Cited

1. Lipshitz HD, ed. 1995. *Localized RNAs.* Austin, TX: Landes/Springer-Verlag
2. Meller VH, Wu K-H, Roman G, Kuroda MI, Davis RL. 1997. *Cell* 88:445–57
3. Herzing LB, Romer JT, Horn JM, Ashworth A. 1997. *Nature* 386:272–75
4. Penny GD, Kay GF, Sheardown SA, Rastan S, Brockdorff N. 1996. *Nature* 379:131–37
5. Rebagliati MR, Weeks DL, Harvey RP, Melton DA. 1985. *Cell* 42:769–77
6. Frigerio G, Burri M, Bopp D, Baumgartner S, Noll M. 1986. *Cell* 47:735–46
7. Berleth T, Burri M, Thoma G, Bopp D, Richstein S, et al. 1988. *EMBO J.* 7:1749–56
8. Frohnhöfer HG, Nüsslein-Volhard C. 1986. *Nature* 324:120–25
9. King RC. 1970. *Ovarian Development in Drosophila melanogaster.* New York: Academic
10. McKearin D, Ohlstein B. 1995. *Development* 121:2937–47
11. Lin HF, Yue L, Spradling AC. 1994. *Development* 120:947–56
12. Lin HF, Spradling AC. 1995. *Dev. Genet.* 16:6–12
13. de Cuevas M, Lee JK, Spradling AC. 1996. *Development* 122:3959–68
14. Stephenson EC. 1995. See Ref. 1, pp. 63–76
15. Cooley L, Theurkauf WE. 1994. *Science* 266:590–96
16. Ding D, Parkhurst SM, Lipshitz HD. 1993. *Proc. Natl. Acad. Sci. USA* 90:2512–16
17. Yue L, Spradling AC. 1992. *Genes Dev.* 6:2443–54
18. Mahone M, Saffman EE, Lasko PF. 1995. *EMBO J.* 14:2043–55
19. Suter B, Romberg LM, Steward R. 1989. *Genes Dev.* 3:1957–68
20. Mach JM, Lehmann R. 1997. *Genes Dev.* 11:423–35
21. Neuman-Silberberg FS, Schüpbach T. 1993. *Cell* 75:165–74
22. Cheung H-K, Serano TL, Cohen RS. 1992. *Development* 114:653–61
23. Lantz V, Ambrosio L, Schedl P. 1992. *Development* 115:75–88
24. Kim-Ha J, Smith JL, Macdonald PM. 1991. *Cell* 66:23–35
25. Ephrussi A, Dickinson LK, Lehmann R. 1991. *Cell* 66:37–50

26. Nakamura A, Amikura R, Mukai M, Kobayashi S, Lasko PF. 1996. *Science* 274:2075–79
27. Golumbeski GS, Bardsley A, Tax F, Boswell RE. 1991. *Genes Dev.* 5:2060–70
28. Wang C, Dickinson LK, Lehmann R. 1994. *Dev. Dyn.* 199:103–15
29. Tirronen M, Lahti V-P, Heino TI, Roos C. 1995. *Mech. Dev.* 52:65–75
30. Hales KH, Meredith JE, Storti RV. 1994. *Dev. Biol.* 165:639–53
31. Theurkauf WE. 1994. *Dev. Biol.* 165:352–60
32. St. Johnston D, Driever W, Berleth T, Richstein S, Nüsslein-Volhard C. 1989. *Development* 107(Suppl.):13–19
33. Karlin-McGinnes M, Serano TL, Cohen RS. 1996. *Dev. Genet.* 19:238–48
34. Roth S, Neuman-Silberberg FS, Barcelo G, Schüpbach T. 1995. *Cell* 81:967–78
35. González-Reyes A, Elliot H, Johnston DS. 1995. *Nature* 375:654–58
36. Clark IE, Jan LY, Jan YN. 1997. *Development* 124:461–70
37. Pokrywka N-J, Stephenson EC. 1995. *Dev. Biol.* 167:363–70
38. Lane ME, Kalderon D. 1994. *Genes Dev.* 8:2986–95
39. Ruohola H, Bremer KA, Baker D, Swedlow JR, Jan LY, Jan YN. 1991. *Cell* 66:433–49
40. Gillespie DE, Berg CA. 1995. *Genes Dev.* 9:2495–508
41. Clark I, Giniger E, Ruohola-Baker H, Jan LY, Jan YN. 1994. *Curr. Biol.* 4:289–300
42. González-Reyes A, Johnston DS. 1994. *Science* 266:639–41
43. Gavis ER, Curtis D, Lehmann R. 1996. *Dev. Biol.* 176:36–50
44. Zaccai M, Lipshitz HD. 1996. *Dev. Genet.* 19:249–57
45. Zaccai M, Lipshitz HD. 1996. *Zygote* 4:159–66
46. Deleted in proof
47. Lasko PF. 1994. *Molecular Genetics of Drosophila oogenesis.* Austin, TX: Landes
48. Mahajan-Miklos S, Cooley L. 1994. *Cell* 78:291–301
49. Cooley L, Verheyen E, Ayers K. 1992. *Cell* 69:173–84
50. Cant K, Knowles BA, Mooseker MS, Cooley L. 1994. *J. Cell Biol.* 125:369–80
51. Theurkauf WE, Smiley S, Wong ML, Alberts BM. 1992. *Development* 115:923–36
52. Theurkauf WE. 1994. *Science* 265:2093–96
53. Deleted in proof
54. Karr TL, Alberts BM. 1986. *J. Cell Biol.* 102:1494–509
55. Sullivan W, Fogarty P, Theurkauf W. 1993. *Development* 118:1245–54
56. Wang C, Lehmann R. 1991. *Cell* 66:637–47
57. Jongens TA, Hay B, Jan LY, Jan YN. 1992. *Cell* 70:569–84
58. Foe VA, Alberts BM. 1983. *J. Cell Sci.* 61:31–70
59. Raff JW, Whitfield WGF, Glover DM. 1990. *Development* 110:1249–61
60. Dalby B, Glover DM. 1992. *Development* 115:989–97
61. Ding D, Lipshitz HD. 1993. *Zygote* 1:257–71
62. Whitfield WGF, Gonzalez C, Sanchez-Herrero E, Glover DM. 1989. *Nature* 338:337–40
63. Ding D, Parkhurst SM, Halsell SR, Lipshitz HD. 1993. *Mol. Cell. Biol.* 13:3773–81
64. Edgar BA, O'Dell GM, Schubiger G. 1987. *Genes Dev.* 1:1226–37
65. Davis I, Ish-Horowicz D. 1991. *Cell* 67:927–40
66. Gergen JP, Butler BA. 1988. *Genes Dev.* 2:1179–93
67. Tepass U, Speicher SA, Knust E. 1990. *Cell* 61:787–99
68. Baker NE. 1988. *Dev. Biol.* 125:96–108
69. Macdonald PM, Ingham PW, Struhl G. 1986. *Cell* 47:721–34
70. King ML. 1995. See Ref. 1, pp. 137–48
71. Dumont JN. 1972. *J. Morphol.* 136:153–80
72. Gerhart J, Keller R. 1986. *Annu. Rev. Cell Biol.* 2:201–29
73. Gerhart J, Danilchik M, Doniach T, Roberts S, Rowning B, Stewart R. 1989. *Development* 107(Suppl.):37–51
74. Kloc M, Etkin LD. 1995. See Ref. 1, pp. 149–56
75. Smith JL, Wilson JE, Macdonald PM. 1992. *Cell* 70:849–59
76. Ephrussi A, Lehmann R. 1992. *Nature* 358:387–92
77. Wylie CC, Heasman J, Snape A, O'Driscoll M, Holwill S. 1985. *Dev. Biol.* 112:66–72
78. Mosquera L, Forristall C, Zhou Y, King ML. 1993. *Development* 117:377–86
79. Zhou Y, King M-L. 1996. *Development* 122:2947–53
80. Wylie CC, Brown D, Godsave SF, Quarmby J, Heasman J. 1985. *J. Embryol. Exp. Morphol.* 89(Suppl.):1–15
81. Linnen JM, Bailey CP, Weeks DL. 1993. *Gene* 128:181–88

82. Weeks DL, Melton DA. 1987. *Proc. Natl. Acad. Sci. USA* 84:2798–802
83. Hudson JW, Alarcón VB, Elinson RP. 1996. *Dev. Genet.* 19:190–98
84. Otte AP, McGrew LL, Olate J, Nathanson NM, Moon RT. 1992. *Development* 116:141–46
85. Hinkley CS, Martin JF, Leibham D, Perry M. 1992. *Mol. Cell. Biol.* 12:638–49
86. Otte AP, Moon RT. 1992. *Cell* 68:1021–29
87. Kloc M, Reddy BA, Miller M, Eastman E, Etkin L. 1991. *Mol. Reprod. Dev.* 28:341–45
88. Reddy BA, Kloc M, Etkin LD. 1992. *Mech. Dev.* 39:1–8
89. Kloc M, Miller M, Carrasco A, Eastman E, Etkin L. 1989. *Development* 107:899–907
90. Kloc M, Etkin LD. 1995. *Development* 121:287–97
91. Stennard F, Carnac G, Gurdon JB. 1996. *Development* 122:4179–88
92. Zhang J, King M-L. 1996. *Development* 122:4119–29
93. Melton DA. 1987. *Nature* 328:80–82
94. Yisraeli JK, Sokol S, Melton DA. 1990. *Development* 108:289–98
95. Elinson RP, King M-L, Forristall C. 1993. *Dev. Biol.* 160:554–62
96. Pondel MD, King ML. 1988. *Proc. Natl. Acad. Sci. USA* 85:7612–16
97. Kloc M, Spohr G, Etkin LD. 1993. *Science* 262:1712–14
98. Ku M, Melton DA. 1993. *Development* 119:1161–73
99. Zhou Y, King ML. 1997. *Dev. Biol.* 179:173–83
100. Swalla BJ, Jeffery WR. 1996. *Dev. Biol.* 178:23–34
101. Swalla BJ, Jeffery WR. 1995. *Dev. Biol.* 170:353–64
102. Swalla BJ, Jeffrey WR. 1996. *Dev. Genet.* 19:258–67
103. Vlahou A, Gonzalez-Rimbau M, Flytzanis CN. 1996. *Development* 122:521–26
104. Steward O, Kleiman R, Banker G. 1995. See Ref. 1, pp. 235–55
105. Van Minnen J. 1994. *Histochem. J.* 26:377–91
106. Garner CC, Tucker RP, Matus A. 1988. *Nature* 336:674–76
107. Tiedge H, Fremeau RT, Weinstock PH, Arancio O, Brosius J. 1991. *Proc. Natl. Acad. Sci. USA* 88:2093–97
108. Tiedge H, Chen W, Brosius J. 1993. *J. Neurosci.* 13:2382–90
109. Mayford M, Baranes D, Podsypanina K, Kandel ER. 1996. *Proc. Natl. Acad. Sci. USA* 93:13250–55
110. Furuichi T, Simon-Chazottes D, Fujino I, Yamada N, et al. 1993. *Recept. Channels* 1:11–24
111. Tiedge H, Zhou A, Thorn NA, Brosius J. 1993. *J. Neurosci.* 13:4214–19
112. Litman P, Barg J, Rindzoonski L, Ginzburg I. 1993. *Neuron* 10:627–38
113. Hannan AJ, Schevzov G, Gunning P, Jeffrey PL, Weinberger RP. 1995. *Mol. Cell. Neurosci.* 6:397–411
114. Murphy D, Levy A, Lightman S, Carter D. 1989. *Proc. Natl. Acad. Sci. USA* 86:9002–5
115. Jirikowski GF, Sanna PP, Bloom FE. 1990. *Proc. Natl. Acad. Sci. USA* 87:7400–4
116. Mohr E, Fehr S, Richter D. 1991. *EMBO J.* 10:2419–24
117. Mohr E, Richter D. 1992. *Eur. J. Neurosci.* 9:870–76
118. Ressler KJ, Sullivan SL, Buck LB. 1994. *Cell* 79:1245–55
119. Vassar R, Chao SK, Sitcheran R, Nuñez JM, Vosshall LB, Axel R. 1994. *Cell* 79:981–91
120. Trapp BD, Moench T, Pully M, Barbosa E, Tennekoon G, Griffin J. 1987. *Proc. Natl. Acad. Sci. USA* 84:7773–77
121. Colman DR, Kriebich G, Frey AB, Sabatini DD. 1982. *J. Cell Biol.* 95:598–608
122. Laitala-Leinonen T, Howell ML, Dean GE, Väänänen HK. 1996. *Mol. Biol. Cell* 7:129–42
123. Sundell CL, Singer RH. 1991. *Science* 253:1275–77
124. Lawrence JB, Singer RH. 1986. *Cell* 45:407–15
125. Cheng H, Bjerknes M. 1989. *J. Mol. Biol.* 210:541–49
126. Banerjee U, Renfranz PJ, Pollock JA, Benzer S. 1987. *Cell* 49:281–91
127. Li P, Yang XH, Wasser M, Cai Y, Chia W. 1997. *Cell* 90:437–47
128. Long RM, Singer RH, Meng XH, Gonzalez I, Nasmyth K, Jansen R-P. 1997. *Science* 277:383–87
129. Takizawa P, Sil A, Swedlow JR, Herskowitz I, Vale RD. 1997. *Nature* 389:90–93
130. Daneholt B. 1997. *Cell* 88:585–88
131. Francis-Lang H, Davis I, Ish-Horowicz D. 1996. *EMBO J.* 15:640–49
132. Kobayashi S, Okada M. 1990. *Nucleic Acids Res.* 18:4592
133. Kobayashi S, Okada M. 1989. *Development* 107:733–42
134. Kobayashi S, Amikura R, Okada M. 1993. *Science* 260:1521–24
135. Kobayashi S, Amikura R, Okada M. 1994. *Int. J. Dev. Biol.* 38:193–99

136. Kobayashi S, Amikura R, Nakamura A, Saito H, Okada M. 1995. *Dev. Biol.* 169: 384–86
137. Ding D, Whittaker KL, Lipshitz HD. 1994. *Dev. Biol.* 163:503–15
138. Gottlieb E. 1990. *Curr. Opin. Cell Biol.* 2:1080–86
139. Gavis ER, Lehmann R. 1994. *Nature* 369:315–18.
140. Dahanukar A, Wharton RP. 1996. *Genes Dev.* 10:2610–20
141. Smibert CA, Wilson JE, Kerr K, Macdonald PM. 1996. *Genes Dev.* 10:2600–9
142. Halsell SR, Lipshitz HD. 1995. See Ref. 1, pp. 9–39
143. Deleted in proof
144. Cooperstock RL, Lipshitz HD. 1998. *Semin. Cell Dev. Biol.* 8(6):541–49
145. Li MG, McGrail M, Serr M, Hays TS. 1994. *J. Cell Sci.* 126:1475–94
146. McGrail M, Hays TS. 1997. *Development* 124:2409–19
147. Pokrywka N-J, Stephenson EC. 1994. *Dev. Biol.* 166:210–19
148. Suter B, Steward R. 1991. *Cell* 67:917–26
149. Lantz V, Chang JS, Horabin JI, Bopp D, Schedl P. 1994. *Genes Dev.* 8:598–613
150. Pokrywka NJ, Stephenson EC. 1991. *Development* 113:55–66
151. Macdonald PM, Kerr K. 1997. *RNA* 3: 1413–1420
152. Erdélyi M, Michon A-M, Guichet A, Glozter JB, Ephrussi A. 1995. *Nature* 377:524–27
153. Tetzlaff MT, Jäckle H, Pankratz MJ. 1996. *EMBO J.* 15:1247–54
154. Deshler JO, Highett MI, Schnapp BJ. 1997. *Science* 276:1128–31
155. Knowles RB, Sabry JH, Martone ME, Deerinck TJ, Ellisman MH, et al. 1996. *J. Neurosci.* 16:7812–20
156. Ainger K, Avossa D, Morgan F, Hill SJ, Barry C, et al. 1993. *J. Cell Biol.* 123:431–41
157. Behar L, Marx R, Sadot E, Barg J, Ginzburg I. 1995. *Int. J. Dev. Neurosci.* 13:113–27
158. Litman P, Behar L, Elisha Z, Yisraeli JK, Ginzburg I. 1996. *Dev. Biol.* 176:80–94
159. Glotzer JB, Saffrich R, Glotzer M, Ephrussi A. 1997. *Curr. Biol.* 7:326–37
160. Swan A, Suter B. 1996. *Development* 122:3577–86
161. Kloc M, Etkin LD. 1994. *Science* 265: 1101–3
162. Wang SX, Hazelrigg T. 1994. *Nature* 369:400–3
163. Ferrandon D, Elphick L, Nüsslein-Volhard C, St. Johnston D. 1994. *Cell* 79:1221–32
164. Ferrandon D, Koch I, Westhof E, Nüsslein-Volhard C. 1997. *EMBO J.* 16: 1751–58
165. Walter P, Johnson AE. 1994. *Annu. Rev. Cell Biol.* 10:87–119
166. Macdonald PM, Struhl G. 1988. *Nature* 336:595–98
167. Macdonald PM. 1990. *Development* 110:161–72
168. Kim-Ha J, Webster PJ, Smith JL, Macdonald PM. 1993. *Development* 119: 169–78
169. Lantz V, Schedl P. 1994. *Mol. Cell Biol.* 14:2235–42
170. Mowry KL, Melton DA. 1992. *Science* 255:991–94
171. Kislauskis EH, Zhu X-C, Singer RH. 1994. *J. Cell Biol.* 127:441–51
172. Jackson RJ. 1993. *Cell* 74:9–14
173. Muslimov IA, Santi E, Homel P, Perini S, Higgins D, Tiedge H. 1997. *J. Neurosci.* 17:4722–33
174. Whittaker KL, Lipshitz HD. 1995. See Ref. 1, pp. 41–61
175. Kislauskis EH, Li ZF, Singer RH, Taneja KL. 1993. *J. Cell Biol.* 123:165–72
176. Macdonald PM, Kerr K, Smith JL, Leask A. 1993. *Development* 118:1233–43
177. Macdonald PM, Leask A, Kerr K. 1995. *Proc. Natl. Acad. Sci. USA* 92:10787–91
178. St. Johnston D, Brown NH, Gall JG, Jantsch M. 1992. *Proc. Natl. Acad. Sci. USA* 89:10979–83
179. St. Johnston D, Beuchle D, Nüsslein-Volhard C. 1991. *Cell* 66:51–63
180. Ding D, Lipshitz HD. 1993. *BioEssays* 10:651–58
181. Serano T, Cohen R. 1995. *Development* 121:3809–18
182. Dalby B, Glover DM. 1993. *EMBO J.* 12:1219–27
183. Gautreau D, Cote CA, Mowry KL. 1997. *Development* 124:5013–20
184. Kim-Ha J, Kerr K, Macdonald PM. 1995. *Cell* 81:403–12
185. Gavis ER, Lunsford L, Bergsten SE, Lehmann R. 1996. *Development* 122: 2791–800
186. Webster PJ, Liang L, Berg CA, Lasko P, Macdonald PM. 1997. *Genes Dev.* 11:2510–21
187. Yisraeli JK, Oberman F, Pressman Schwartz S, Havin L, Elisha Z. 1995. See Ref. 1, pp. 157–71
188. Schüpbach T, Wieschaus E. 1986. *Roux's Arch. Dev. Biol.* 195:302–17
189. Frohnhöfer HG, Nüsslein-Volhard C. 1987. *Genes Dev.* 1:880–90

190. Rongo C, Gavis E, Lehmann R. 1995. *Development* 121:2737–46
191. Bandziulis R, Swanson MS, Dreyfuss G. 1989. *Genes Dev.* 3:431–37
192. Schüpbach T, Wieschaus E. 1991. *Genetics* 129:1119–36
193. Castrillon DH, Gönczy P, Alexander S, Rawson R, Eberhart CG, et al. 1993. *Genetics* 135:489–505
194. Lasko PF, Ashburner M. 1988. *Nature* 335:611–16
195. Hay B, Jan LY, Jan YN. 1988. *Cell* 55:577–87
196. Liang L, Diehl-Jones W, Lasko P. 1994. *Development* 120:1201–11
197. Schwartz SP, Aisenthal L, Elisha Z, Oberman F, Yisraeli JK. 1992. *Proc. Natl. Acad. Sci. USA* 89:11895–99
198. Perry-O'Keefe H, Kintner C, Yisraeli J, Melton DA. 1990. In *In situ Hybridization and the Study of Development and Differentiation*, ed. N Harris, D Wilkenson, pp. 115–30. Cambridge: Cambridge Univ. Press
199. Elisha Z, Havin L, Ringel I, Yisraeli JK. 1995. *EMBO J.* 14:5109–14
200. Mowry KL. 1996. *Proc. Natl. Acad. Sci. USA* 93:14608–13
201. Marcey D, Watkins WS, Hazelrigg T. 1991. *EMBO J.* 10:4259–66
202. Macdonald PM, Luk SKS, Kilpatrick M. 1991. *Genes Dev.* 5:2455–66
203. Chao Y-C, Donahue KM, Pokrywka NJ, Stephenson EC. 1991. *Dev. Genet.* 12:333–41
204. Hegdé J, Stephenson EC. 1993. *Development* 119:457–70
205. Webster P, Suen J, Macdonald P. 1994. *Development* 120:2027–37
206. Lasko PF, Ashburner M. 1990. *Genes Dev.* 4:905–21
207. Hay B, Ackerman L, Barbel S, Jan LY, Jan YN. 1988. *Development* 103:625–40
208. Hay B, Jan LY, Jan YN. 1990. *Development* 109:425–33
209. Yoon C, Kawakami K, Hopkins N. 1997. *Development* 124:3157–66
210. Christerson LB, McKearin DM. 1994. *Genes Dev.* 8:614–28
211. Kelley RL. 1993. *Genes Dev.* 7:948–60
212. Serano TL, Karlin-McGinnes M, Cohen RS. 1995. *Mech. Dev.* 51:183–92
213. Prost E, Derychere F, Roos C, Haenlin M, Pantesco V, Mohier E. 1988. *Genes Dev.* 2:891–900
214. Wilson JE, Connell JE, Schlenker JD, Macdonald PM. 1996. *Dev. Genet.* 19:199–209
215. Rittenhouse KR, Berg CA. 1995. *Development* 121:3032–33

216. Wilson JE, Connell JE, Macdonald PM. 1996. *Development* 122:1631–39
217. St. Johnston D. 1995. *Cell* 81:161–70
218. Neuman-Silberberg FS, Schupbach T. 1994. *Development* 120:2457–63
219. Driever W, Nüsslein-Volhard C. 1988. *Cell* 54:83–93
220. Driever W, Nüsslein-Volhard C. 1988. *Cell* 54:95–104
221. Struhl G, Struhl K, Macdonald PM. 1989. *Cell* 57:1259–73
222. Driever W, Thoma G, Nüsslein-Volhard C. 1989. *Nature* 340:363–67
223. Driever W, Nüsslein-Volhard C. 1989. *Nature* 337:138–43
224. Driever W, Siegel V, Nüsslein-Volhard C. 1990. *Development* 109:811–20
225. Wharton RP, Struhl G. 1991. *Cell* 67:955–67
226. Rivera-Pomar R, Niessing D, Schmidt-Ott U, Gehring WJ, Jäckle H. 1996. *Nature* 379:746–49
227. Dubnau J, Struhl G. 1996. *Nature* 379:694–99
228. Mlodzik M, Fjose A, Gehring WJ. 1985. *EMBO J.* 4:2961–69
229. Macdonald PM, Struhl G. 1986. *Nature* 324:537–45
230. Lehmann R, Nüsslein-Volhard C. 1991. *Developmen* 112:679–93
231. Gavis ER, Lehmann R. 1992. *Cell* 71:301–13
232. Hülskamp M, Schröder C, Pfeifle C, Jäckle H, Tautz D. 1989. *Nature* 338:629–32
233. Irish V, Lehmann R, Akam M. 1989. *Nature* 338:646–48
234. Lehmann R, Nüsslein-Volhard C. 1987. *Dev. Biol.* 119:402–17
235. Tautz D, Lehmann R, Schnürch H, Schuh R, Seifert E, et al. 1987. *Nature* 327:383–89
236. Stanojevic D, Hoey T, Levine M. 1989. *Nature* 341:331–35
237. Lehmann R, Nüsslein-Volhard C. 1987. *Nature* 329:167–70
238. Lehmann R, Nüsslein-Volhard C. 1986. *Cell* 47:141–52
239. Jongens TA, Ackerman LD, Swedlow JR, Jan LY, Jan YN. 1994. *Genes Dev.* 8:2123–36
240. Kobayashi S, Yamada M, Asaoka M, Kitamura T. 1996. *Nature* 380:708–11
241. Thomsen GH, Melton DA. 1993. *Cell* 74:433–41
242. Dale L, Matthews G, Colman A. 1993. *EMBO J.* 12:4471–80
243. Kondaiah P, Sands M, Smith JM, Fields A, Roberts AB, et al. 1990. *J. Biol. Chem.* 265:1089–93

244. Doe CQ, Chu-LaGraff Q, Wright DM, Scott MP. 1991. *Cell* 65:451–64
245. Vaessin H, Grell E, Wolff E, Bier E, Jan LY, Jan YN. 1991. *Cell* 67:941–53
246. Hirata J, Nakagoshi H, Nabeshima Y-I, Matsuzuka F. 1995. *Nature* 377:627–30
247. Knoblich JA, Jan LY, Jan YN. 1995. *Nature* 377:624–27
248. Shen C-P, Jan LY, Jan YN. 1997. *Cell* 90:449–58
249. Jeffery WR, Tomlinson CR, Brodeur RD. 1983. *Dev. Biol.* 99:408–20
250. Macdonald PM. 1992. *Development* 114:221–32
251. Aït-Ahmed O, Thomas-Cavallin M, Rosset R. 1987. *Dev. Biol.* 122:153–62
252. Pomeroy ME, Lawrence JB, Singer RH, Billings-Gagliardi S. 1991. *Dev. Biol.* 143:58–67
253. Trembleau A, Calas A, Fèvre-Montange M. 1990. *Mol. Brain Res.* 8:37–45
254. Perry BA. 1988. *Cell Differ. Dev.* 25:99–108

Annu. Rev. Biochem. 1998. 67:395–424

BIOCHEMISTRY AND GENETICS OF VON WILLEBRAND FACTOR

J. Evan Sadler

Howard Hughes Medical Institute, Department of Medicine and Department of
Biochemistry and Molecular Biophysics, Washington University School of Medicine,
Saint Louis, Missouri 63110; e-mail: esadler@im.wustl.edu

KEY WORDS: von Willebrand disease, hemostasis, platelet adhesion, factor VIII,
             Weibel-Palade body

ABSTRACT

Von Willebrand factor (VWF) is a blood glycoprotein that is required for normal
hemostasis, and deficiency of VWF, or von Willebrand disease (VWD), is the most
common inherited bleeding disorder. VWF mediates the adhesion of platelets to
sites of vascular damage by binding to specific platelet membrane glycoproteins
and to constituents of exposed connective tissue. These activities appear to be
regulated by allosteric mechanisms and possibly by hydrodynamic shear forces.
VWF also is a carrier protein for blood clotting factor VIII, and this interaction
is required for normal factor VIII survival in the circulation. VWF is assembled
from identical ≈250 kDa subunits into disulfide-linked multimers that may be
>20,000 kDa. Mutations in VWD can disrupt this complex biosynthetic process
at several steps to impair the assembly, intracellular targeting, or secretion of
VWF multimers. Other VWD mutations impair the survival of VWF in plasma
or the function of specific ligand binding sites. This growing body of information
about VWF synthesis, structure, and function has allowed the reclassification of
VWD based upon distinct pathophysiologic mechanisms that appear to correlate
with clincial symptoms and the response to therapy.

CONTENTS

INTRODUCTION

Von Willebrand factor (VWF) is a large, multimeric glycoprotein that is found in blood plasma, platelet α-granules, and subendothelial connective tissue. VWF performs two essential functions in hemostasis: it mediates the adhesion of platelets to subendothelial connective tissue, and it binds blood clotting factor VIII (the protein missing in hemophilia A). In the absence of VWF, factor VIII is rapidly removed from the circulation. Consequently, patients who lack VWF have a severe bleeding disorder because they have profound defects both in blood clotting and in the formation of platelet plugs at sites of vascular injury. As is the case for many hemostatic proteins, the first clue to the existence of VWF came from the discovery of bleeding patients who lacked it.

HISTORICAL ASPECTS

In 1924, Erik von Willebrand, a Finnish physician, began to study an inherited bleeding disorder affecting consanguineous families from the Åland Islands, which lie in the narrow Gulf of Bothnia that separates Sweden and Finland (1). The proband was a five-year-old girl with severe bleeding since birth. Three of her sisters had died before the age of four with bleeding from various sites; one living sister, aged three, also was severely affected. Two sisters and three brothers were free of symptoms. Both parents, two brothers, and many other relatives of both sexes had milder bleeding problems, suggesting autosomal dominant transmission of a mild disorder that became severe in the homozygous or compound heterozygous proband and four of her siblings. The bleeding usually was from skin or mucous membranes, and hemorrhage into joints or deep tissues was rare. Von Willebrand distinguished the condition from thrombocytopenic purpura and from other known congenital bleeding diseases, including hemophilia A (factor VIII deficiency) and Glanzmann thrombasthenia (deficiency of platelet integrin $\alpha IIb\beta 3$). By the time his first account was published in 1926 (1), the affected sister also had died of massive gastrointestinal bleeding. The proband herself died with her fourth menstrual period at the age of 13 (2). Von Willebrand was not able to determine whether this serious bleeding tendency was caused by a defect in platelets, blood plasma, or the vasculature.

The pathophysiology of this disorder, now called von Willebrand disease (VWD), has been clarified during the past forty years. In the 1950s, patients with VWD were shown to have decreased plasma levels of blood coagulation factor VIII (3–5). Both the skin bleeding tendency and factor VIII deficiency were corrected by transfusion of a partially purified factor VIII preparation, indicating that VWD is caused by deficiency of a blood protein. Most remarkably, these abnormalities also were corrected by transfusion of a similar concentrate prepared from the plasma of patients with severe hemophilia (6, 7). This provided the first evidence for a hemostatic "von Willebrand factor" distinct from factor VIII and other clotting factors in the blood. VWF was first purified in the early 1970s and its complete sequence was reported in 1986 [reviewed in (8)]. The last decade has seen considerable progress in understanding VWF assembly, the structural requirements for its function, and the molecular basis of VWD.

BIOSYNTHESIS AND SECRETION OF VWF

The assembly of this complex multimeric protein requires many separate steps, each of which may be disrupted by mutations. In addition, the largest VWF multimers are hemostatically active, whereas small multimers are not. The structure and biosynthesis of VWF therefore provides an essential framework for understanding the molecular pathology of VWD.

Subunit Structure and Domain Organization

VWF is made by endothelial cells throughout the body (9) and by megakaryocytes (10). The primary translation product consists of 2813 amino acids and includes a signal peptide of 22 residues, a large propeptide of 741 residues, and the mature subunit of 2050 residues (Figure 1). At least 95% of the protein sequence consists of repeated domains or motifs that are shared with other proteins. These are arranged in the sequence: D1-D2-D'-D3-A1-A2-A3-D4-B1-B2-B3-C1-C2-CK. The protein is remarkably rich in cysteine, which comprises 234 of the 2813 residues in preproVWF, or 8.3 percent. Cysteine is abundant in all of the domains except the triplicated A domains, which together contain only six cysteines. In the secreted protein, all cysteine residues appear to be paired in disulfide bonds (11). The mature subunit is extensively glycosylated

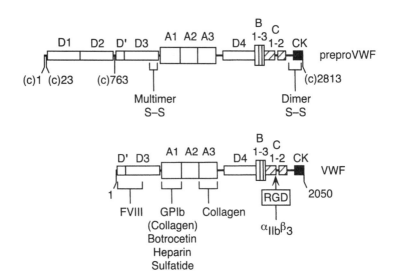

Figure 1 Structure of von Willebrand factor. The structure of preproVWF is shown above the structure of the mature VWF subunit. The locations of five kinds of conserved structural domains (A, B, C, D, CK) are indicated. *preproVWF*: Amino acid residues of preproVWF are numbered consecutively by codon number (c)1-(c)2813, beginning with the initiation codon. *Amino acid residues (c)1-(c)22*, the cleaved signal peptide; *residues (c)23-(c)763*, the propeptide; *residues (c)764-(c)2813*, the mature subunit polypeptide. Intersubunit disulfide bonds are formed near the carboxyl-termini of proVWF dimers in the ER. Additional intersubunit disulfide bonds are formed near the amino-terminus of the mature subunits to assemble multimers in the Golgi. *VWF*: Amino acid residues of the mature VWF subunit are numbered 1–2050. Binding sites within the subunit have been localized for several macromolecules as indicated. Activated platelet integrin $\alpha IIb\beta 3$ binds VWF through a segment that includes the tripeptide sequence Arg-Gly-Asp (RGD) in VWF domain C1.

with 12 N-linked and 10 O-linked oligosaccharides, and the propeptide has three more potential N-glycosylation sites. One or both of the oligosaccharides at Asn384 and Asn468 of the mature subunit are sulfated (12). The N-linked oligosaccharides of VWF are unusual compared to those of other plasma glycoproteins because they contain ABO blood group oligosaccharides (13).

Common practice is to number the amino acid residues of the mature VWF subunit from 1 to 2050, beginning at the amino-terminal serine, and this convention will be followed herein. This may be confusing when it becomes necessary to discuss specific residues of both the mature subunit and the propeptide. In this article, amino acid residues of the signal peptide or propeptide are designated by codon number and are distinguished by the prefix (c) (for "codon"). For example, the propeptide (Figure 1) extends from (c)23 to (c)763.

Dimerization in the Endoplasmic Reticulum

After translocation into the endoplasmic reticulum (ER), proVWF subunits dimerize through disulfide bonds near their carboxyl termini. This "tail-to-tail" dimerization function requires only sequences within the last 150 residues (11, 14). The carboxyl-terminal 90 residues comprise the "CK" domain that is homologous to the "cystine knot" superfamily of proteins. This family includes nerve growth factor (NGF), transforming growth factor β (TGF-β), platelet-derived growth factor (PDGF), chorionic gonatotrophin, Norrie disease protein (NDP), and several epithelial mucins (15–17). These family members share a tendency to dimerize, often through disulfide bonds. Alignment of the VWF cystine knot with its nearest homologues shows the cysteine that forms an interchain disulfide in TGF-β is conserved in VWF, and suggests that Cys2010 of the mature VWF subunit may contribute to dimerization of proVWF in the ER. This conclusion is supported by the phenotype of patients with heterozygous Cys2010Arg mutations, who cannot form large VWF multimers. In addition, recombinant carboxyl-terminal fragments of VWF that bear this mutation do not dimerize covalently (18).

Exit from the ER appears to be regulated by both glycosylation and dimerization. Inhibition of N-linked glycosylation with tunicamycin causes proVWF monomers to accumulate in the ER, suggesting glycosylation must occur before dimerization (19). Normal proVWF monomers with fully processed N-linked oligosaccharides are not found within the cell and therefore do not appear to leave the ER, but deletion of the carboxyl-terminal 20 kDa of the subunit permits the transport of monomers to the Golgi (20). Thus, sequences near the carboxyl-terminus may be responsible for the retention of monomers within the ER until they are dimerized or degraded.

Multimer Assembly in the Golgi

"Tail-to-tail" proVWF dimers are transported to the Golgi and there form additional "head-to-head" disulfide bonds near the amino-termini of the subunits, yielding multimers that may exceed 20 million Da in size. Additional modifications in the Golgi include the proteolytic removal of the large VWF propeptide, possibly by the propeptide processing protease furin (21); the completion of N-linked and O-linked glycosylation; and sulfation of certain N-linked oligosaccharides. Pulse-chase studies indicate that propeptide cleavage and multimer formation follow sulfation. Since sulfation appears to occur in the trans-Golgi network, multimerization and propeptide cleavage also occur in the trans-Golgi network or a later compartment (22). Deletion of the carboxyl-terminal dimerization region is compatible with transport to the Golgi, the formation of "head-to-head" disulfide bonds characteristic of multimers, and secretion of the resultant unnatural dimers (20, 23). Thus, "tail-to-tail" dimerization and "head-to-head" multimerization reactions are independent.

The VWF propeptide plays an essential role in the assembly of multimers. Deletion of the propeptide does not prevent transport to the Golgi, but abolishes multimerization (24). The isolated propeptide also can promote the intracellular multimerization of VWF subunits when expressed in trans rather than in cis (25). Propeptide cleavage is not necessary for multimerization, because the mutation Arg763Gly prevents propeptide cleavage and results in the assembly of multimers composed of proVWF subunits (26).

The oxidation of thiols is facilitated at alkaline pH, and the acidic Golgi apparatus is an unusual site for disulfide bond formation. However, the VWF propeptide appears to promote multimer formation under these hostile conditions. In fact, an acidic Golgi environment may be necessary, since treatment of cells with a weak base, ammonium chloride or chloroquine, prevents multimerization without inhibiting propeptide cleavage (19). Isolated proVWF dimers form multimers optimally in vitro at pH <6.0 in the presence of calcium ions, conditions like those felt to prevail in the trans-Golgi (27).

Examination of the propeptide structure has suggested a possible mechanism for this function (28). The D1 and D2 domains of the VWF propeptide contain CXXC sequences, Cys(c)159-Gly-Leu-Cys(c)162 and Cys(c)521-Gly-Leu-Cys(c)524, that resemble the functional sites of thiol:disulfide oxidoreductases. Insertion of an extra glycine into either of these sequences was compatible with dimer formation in the ER, but the dimers were transported to the Golgi and secreted without forming multimers. Based on these results, the VWF propeptide was proposed to facilitate interchain disulfide bond formation by catalyzing protein disulfide interchange, although disulfide isomerase activity has not been demonstrated directly for the VWF propeptide (28).

Some "head-to-head" intersubunit disulfide bonds have been localized. Digestion of VWF with V8 protease and trypsin yields a dimeric fragment containing residues Lys273 and Lys728. Both flanking fragments are monomeric, so all "head-to-head" disulfide bonds must lie within this segment (11). Characterization of a smaller dimeric trypsin fragment indicates that interchain disulfides involve at least one of the three cysteine residues at positions 459, 462, and 464 (29). In addition, a Cys379-Cys379 disulfide bond was demonstrated between subunits (30). There are 23 cysteine residues between Lys273 and Lys448, among which only six have been shown to form intrasubunit disulfide bonds (11); therefore, 16 other cysteines in this interval are candidates to be involved in VWF multimer assembly.

Intracellular Storage

VWF is partitioned between two pathways in the Golgi of endothelial cells. The majority, perhaps up to 95%, is secreted constitutively, whereas the remainder is stored in cytoplasmic granules called Weibel-Palade bodies that are specific for endothelium (31, 32). These are rod-shaped membrane enclosed organelles 0.1–0.2 μm wide and up to 4 μm long (33). They have longitudinal striations and in cross-section consist of closely packed 150 Å to 200 Å tubules (33) that appear to be composed of VWF multimers (31) (Figure 2). Small clusters of similar tubules that contain VWF are found at the periphery of platelet α-granules (34). Weibel-Palade bodies have been found in endothelium of all vertebrate classes examined, including mammals, amphibians (33), birds, reptiles, and fish (35).

The signals that direct VWF to the storage compartment are partly elucidated. Wild-type VWF is stored in a variety of cells that have granules, including AtT-20 mouse pituituary cells, RIN5F rat insulinoma cells (20), CV-1 monkey kidney cells (36), and MDCK cells (37), indicating that the signal for VWF storage is recognized by cells other then endothelial cells and megakaryocytes. Cells without a regulated secretory pathway such as COS-1, 3T3, and CHO cells, do not form granules when induced to express VWF (20). Targeting appears to require components of both the propeptide and the mature subunit, since neither is directed to granules when expressed individually (20, 36, 37). However, targeting does not require the propeptide to be colinear with the mature subunit, because coexpression in trans from separate plasmids restores both multimer formation (25) and targeting to granules (37).

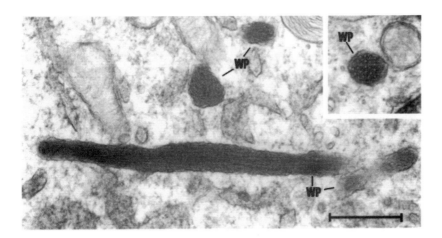

Figure 2 Weibel-Palade bodies in endothelial cells visualized by electron microscopy. Longitudinal striations are clearly seen in several membrane-delimited Weibel-Palade bodies (WP). A cross-section (*inset*) shows that the striations represent closely packed tubules. These are composed principally of VWF. The scale bar indicates 0.5 μm. Figure courtesy of Elisabeth Cramer, Hôpital Henri Mondor, Creteil, France.

The relationship of multimer assembly to intracellular storage remains poorly understood. In endothelial cells, the VWF within granules contains very large multimers made almost exclusively of mature VWF subunits, whereas the VWF that is secreted constitutively contains mostly small multimers and a high proportion of unprocessed proVWF subunits (32). Since pulse-chase analysis indicates that propeptide cleavage and multimer formation precede the formation of Weibel-Palade bodies (22), fully processed and highly multimerized VWF may be targeted preferentially to these granules. However, deletion of the carboxyl-terminal 20 kDa or 80 kDa does not impair targeting even though it allows the formation of only "head-to-head" dimers in the Golgi (20). Similarly, single amino acid insertions in the propeptide that allow the formation of only "tail-to-tail" dimers in the ER are compatible with targeting to granules (28). The effect of mutating the propeptide cleavage site has varied, suggesting that targeting machinery may differ among cell lines. The mutation Arg763Gly appeared to prevent storage in RIN 5F and AtT-20 cells (38), but not in CV-1 or MDCK cells (36, 37). Thus, targeting to the regulated secretory compartment appears to require components of both the propeptide and the mature subunit, but under some circumstances targeting can be dissociated from propeptide cleavage and the formation of covalent intersubunit bonds.

After cleavage, the propeptide remains noncovalently associated with VWF multimers. This association is promoted by the high calcium concentration and acidic pH of the trans-Golgi network, and may explain the observed 1:1 stoichiometry of mature VWF subunit and its propeptide in Weibel Palade bodies (22). Besides VWF and its propeptide, only two other proteins are known to reside in Weibel-Palade bodies, and both are found in its membrane rather than its matrix. These are the cell adhesion molecule P-selectin (39, 40), and the lysosomal membrane sialoglycoprotein CD63 (41). P-selectin also is a membrane constituent of the platelet α-granule, the other major storage site for VWF. P-selectin is targeted to Weibel-Palade bodies at least partly by its cytoplasmic tail, and heterologous proteins can be redirected to the storage compartment when fused to this segment of P-selectin (42). The targeting of CD63 to Weibel-Palade bodies is felt to be specific rather than due to missorting of a lysosomal signal, since other lysosomal membrane proteins are not found there (41). No obvious structural motif is shared by VWF, P-selectin, and CD63 that might be a candidate for a common targeting signal.

VWF Secretion and Catabolism

In cultured endothelial cells, VWF secretion is stimulated by a variety of agents including histamine, thrombin, fibrin, β-adrenergic agonists, calcium ionophore A23187, and phorbol myristate acetate [reviewed in (43)]. This induced secretion of VWF is associated with the loss of Weibel-Palade bodies.

The magnitude of the acute secretory response to epinephrine correlates with intracellular cyclic AMP level (44).

In vivo, the plasma level of VWF is increased acutely by adrenergic stress, thrombin generation, or treatment with the vasopressin analog 1-desamino-8-D-arginine vasopressin (DDAVP). These responses are presumed to reflect secretion from Weibel-Palade bodies, although this has been demonstrated directly only for thrombin generation (45). The ability of DDAVP to raise plasma VWF levels has made it a major drug for the treatment of VWD and hemophilia (46). The mechanism of DDAVP action in vivo seems to be indirect since endothelial cells in culture do not respond to it (47, 48).

The VWF multimers secreted into the blood consist of flexible ≈2 nm thick strands with regularly spaced nodules (Figure 3). The molecules usually appear as tangled, condensed coils, but occasionally are extended so that the structure is seen to have ≈120 nm periodicity. One impressive multimer observed by electron microscopy was >1,800 nm long and contained 17 repeats (49). The repeating units have two-fold symmetry, with a small ≈5 nm central nodule connected by 30–34 nm flexible rods to ≈35 × 6.5 nm globular ends. The large globular regions are dimeric and contain domains D'-D3-A1-A2-A3; the rod and central nodule contain domains D4-B1-B2-B3-C1-C2-CK (50). The small central nodules may represent dimeric CK domains.

VWF multimers and VWF propeptide are secreted together in 1:1 stoichiometric amounts, but subsequently have different fates. After secretion, the propeptide dissociates from VWF multimers (51) and circulates independently as a noncovalent homodimer with a very short halflife of ≈2 h (52). The plasma level of VWF propeptide is ≈1 μg/ml. In contrast, VWF multimers are cleared

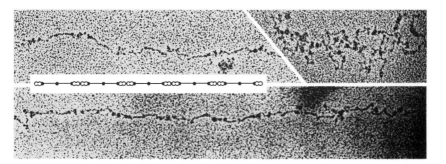

Figure 3 Electron micrographs of VWF molecules rotary shadowed with platinum. The long, straight molecules are shown approximately on register with a ≈120 nm repeat, as illustrated in the schematic representation of VWF multimer structure (*inset*). Tangled molecules (*upper right*) were observed most frequently and probably reflect the conformation in solution. This figure is adapted from Fowler et al (49), courtesy of Harold Erickson, Duke University.

more slowly with a halflife of $\approx$12 h, and the plasma concentration averages $\approx$10 μg/ml (52). The propeptide is not known to have an independent physiological function after it is secreted, although it can bind collagen (53) and also can support integrin $\alpha4\beta1$-mediated adhesion of certain cell lines (54).

BIOLOGICAL ACTIVITIES AND STRUCTURE-FUNCTION RELATIONSHIPS

VWF is not an enzyme, but performs its hemostatic functions through binding to factor VIII, to platelet surface glycoproteins, and to constituents of connective tissue. Binding sites for several of these physiologically important ligands have been localized in the VWF subunit sequence. The binding of VWF to platelets appears to be regulated by its initial interaction with connective tissue, and also by shear stress in flowing blood. Two major goals of current VWF research are to explain these regulatory mechanisms in molecular detail.

Platelet Adhesion to Connective Tissue

In the absence of injury, VWF does not appear to interact with circulating platelets. However, damage to the endothelium allows VWF to bind constituents of subendothelial connective tissue, and this enables VWF to bind platelets with affinity sufficient to snare them from the rapidly flowing blood and retain them at the site of injury. Thus, the binding of VWF to platelets behaves as though it is regulated, perhaps allosterically.

VWF BINDING TO COLLAGENS   VWF binds specifically to several collagens in vitro, including types I, II, III, IV, V, and VI, but it does not bind denatured gelatin (55–57). Immobilized fibrillar collagens (e.g. types I or III) also support platelet adhesion under high shear conditions (58). However, the significance of fibrillar collagen interactions is challenged by reports that VWF binds normally to extracellular matrix that is depleted of collagen either by collagenase digestion or by growth of cells in the presence of α,α'-dipyridyl, a collagen synthesis inhibitor (59). Also, monoclonal antibodies to VWF have been characterized that prevent VWF binding to either extracellular matrix or fibrillar collagens, but not both (60). These findings have focused attention on other candidates. For example, the nonfibrillar collagen type VI binds VWF (57) and is resistant to collagenase (61) or α,α'-dipyridyl (62). VWF also appears to colocalize with collagen type VI but not fibrillar collagens in subendothelial matrix (63). However, surfaces coated with collagen type VI support VWF-dependent platelet adhesion at low shear rates ($100\ s^{-1}$) rather than the high shear rates ($\geq1000\ s^{-1}$) that are optimal for platelet adhesion to vessel walls or fibrillar collagens (64).

These various results suggest one or more subendothelial collagens may be physiological ligands for VWF, but other connective tissue components also may be important.

Binding sites for fibrillar collagens have been identified within VWF domains A1 and A3 (65–67), although mutagenesis studies suggest the major collagen binding site of multimeric VWF is within domain A3. Recombinant VWF lacking domain A1 (ΔA1-VWF) bound normally to collagen III, whereas the binding of VWF lacking domain A3 (ΔA3-VWF) was reduced $\approx$40-fold. Competition studies indicated the residual collagen binding of ΔA3-VWF was mediated by domain A1 because it was inhibited by excess VWF but not by excess ΔA1-VWF. These data also suggest that the major site in VWF domain A3 and the minor site in domain A1 interact with different targets on collagen (68).

The structure of domain A3 determined by X-ray crystallography was reported recently (69, 70). As expected based upon homology to A domains found in the α-chains of integrins $\alpha M\beta 2$ (Mac-1) (71) and $\alpha L\beta 2$ (LFA1) (72), the VWF A3 domain consists of a central hydrophobic parallel β-sheet surrounded by seven amphipathic α-helices in an open α/β sheet or dinucleotide-binding fold (73). The single disulfide bond between Cys923 and Cys1109 is at the base of the domain. A divalent metal ion binding site is present in this region of the A domains of αM and αL. In the VWF A3 domain structure this site is not conserved and is filled instead by water molecules, and the binding of VWF to collagen does not depend on metal ions. The collagen binding surface has not yet been localized.

BINDING TO PLATELET GLYCOPROTEIN IB   Once VWF is immobilized in subendothelial connective tissue, platelets recognize and adhere to it. This interaction requires the glycoprotein (GP) Ib-IX-V complex, the only receptor on a nonactivated platelet with significant affinity for VWF. GPIb-IX-V consists of at least four polypeptide chains encoded by separate genes: GPIbα, GPIbβ, GPIX, and GPV. GPIbα and GPIbβ are disulfide-linked and associate noncovalently with GPIX and GPV (74, 75). The binding site for VWF is in the amino-terminal 293 residues of the GPIbα chain, and optimal binding requires sulfation of tyrosine residues at positions 276, 278, and 279 (76).

VWF-dependent platelet adhesion occurs optimally under conditions of high fluid shear stress, but such conditions are cumbersome to duplicate in vitro and several more convenient surrogate assays are commonly used. The small bacterial glycopeptide antibiotic ristocetin dimerizes and binds to both VWF and platelets (77), and it causes VWF-dependent platelet aggregation by inducing VWF to bind GPIbα (78). Consequently, ristocetin-induced platelet aggregation has been used to assay VWF and to investigate the structural requirements for VWF binding to GPIb. Botrocetin, a ≈25 kDa venom protein from the viper *Bothrops jararaca*, also binds VWF and causes VWF-dependent platelet aggregation by a mechanism slightly different from that of ristocetin (79). Botrocetin is thought to activate the endogenous GPIb binding site of VWF, although this has not been demonstrated directly.

The GPIb binding site on VWF has been localized to domain A1 (29), and the requirements for binding have been studied by scanning mutagenesis (80–82). Interpretation of the results is facilitated by reference to a molecular model of part of the VWF A1 domain (83) that is based on the homologous A domains of integrins αM (71) and αL (72) (Figure 4). The model is limited to amino acid residues Cys509-Cys695 because the flanking portions of the A1 domain are not represented in either reference structure. The model consists of a central 5-stranded parallel β-sheet with a short sixth antiparallel strand on one edge. This hydrophobic β-sheet is enclosed by seven amphipathic α-helices. Unlike the integrin A domains, VWF domain A1 does not appear to have a

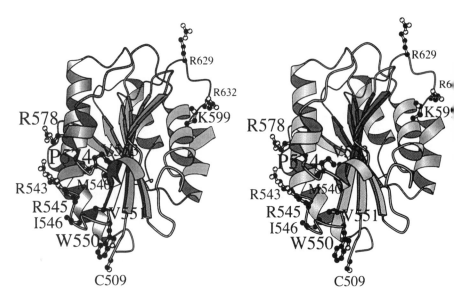

Figure 4 Stereo view of a molecular model of the VWF A1 domain based upon the structures of homologous A domains in integrins αM and αL (83). Selected amino acids are shown by their side chains (ball-and-stick) and residue numbers. Mutagenesis studies suggest that Lys599, Arg629 and Arg632 at the upper right may interact with a binding site on platelet GPIbα. *Lower left*, clustered residues mark the location of a regulatory site that inhibits binding to GPIbα until VWF first interacts with connective tissue or certain soluble modulators.

metal binding site at its apex because the residues that coordinate metal ions in αM and αL are not conserved; this is consistent with the lack of metal ion dependence for VWF binding to platelet GPIbα.

The segment between His463 and Gly716 was targeted for mutagenesis because it is similar to a proteolytic fragment of VWF that binds GPIbα in a ristocetin-dependent or botrocetin-dependent manner (84, 85). The 68 charged amino acids in this segment were changed to alanine in clusters of up to five residues. If the resultant protein had decreased affinity for platelets, the corresponding single alanine substutitions were made. All mutant proteins were secreted, but mutations at seven positions appeared to cause global misfolding as judged by failure to react with any of nine conformation-dependent monoclonal antibodies. Forty-six mutant proteins were characterized further for binding to GPIbα and botrocetin.

Both ristocetin-induced and botrocetin-induced binding to GPIbα were decreased by mutations at Lys599, Arg629, and Arg632. The Lys599Ala mutant was unique among this group because it bound normally to radiolabeled botrocetin, suggesting that Lys599 may bind directly to GPIbα. The effects of ristocetin and botrocetin were dissociated in other mutant proteins. Alanine substitutions at eight positions caused loss of only ristocetin-induced binding to GPIbα, and mutations at Arg636 or Lys667 selectively decreased botrocetin-induced binding to GPIbα (as well as direct binding of botrocetin to VWF) (80–82). The results suggest botrocetin binds to the VWF A1 domain in part through residues R636 and K667, thereby activating the adjacent GPIb binding site that contains residue K599 (Figure 4).

One of the most interesting results of these studies was the identification of mutations that appear to localize a regulatory function within domain A1. Several clustered alanine substitutions in the segments Glu497-Arg511, and Arg687-Glu689, dramatically increased binding to platelets induced by ristocetin (80). The substitution Arg545Ala exhibited increased sensitivity to both ristocetin and botrocetin, and also bound spontaneously in the absence of either modulator. This gain-of-function phenotype is similar to that caused by mutations in VWD type 2B, as discussed in a later section. Both scanning and natural mutations that cause this phenotype cluster in a surface patch of $\approx$30 Å $\times$ 20 Å near the base of the A1 domain (Figure 4) that may correspond to a negative regulatory site.

These results suggest an "allosteric" model in which VWF has at least two conformations; one interacts weakly or not at all with platelet GPIbα, and the second binds with much higher affinity. The low affinity state prevails in the blood because the regulatory site inhibits the GPIb binding site on the opposite side of the A1 domain. This inhibition is relieved and a high affinity state is induced by binding to connective tissue, by interaction with certain soluble modulators (e.g. ristocetin, botrocetin), or by constitutive gain-of-function mutations in the regulatory site.

Such an allosteric model is consistent with the behavior of the A domain of integrin $\alpha M\beta 2$. Wild-type $\alpha M\beta 2$ supports cell adhesion to fibrinogen only after the integrin has been activated. However, when mutations were made in the segments of αM that are homologous to the proposed regulatory domain of VWF domain A1, the mutant integrin supported spontaneous cell adhesion to fibrinogen (86).

In one alternative "entropic" model, VWF in solution does not bind tightly to platelets because the intrinsic affinity of each binding site is low. Multivalent binding is difficult in solution because of the conformational flexibility of VWF. Immobilization of VWF in connective tissue could restrict the motion of multiple binding sites, distributing them favorably to bind several GPIb receptors and retain platelets at the vessel wall. This mechanism would avoid the requirement for local allosteric activation of each binding site. However, "allosteric" and "entropic" models need not be mutually exclusive.

The crystal structure of the recombinant VWF A1 domain was presented recently (87, 88). The structure predicted by modeling appears to be similar to that determined by X-ray crystallography, although there certainly will be significant differences in some features. The A1 domain structure will provide a valuable framework for understanding how VWF binds to its ligands and for interpreting the effects of mutations on these interactions.

BINDING OF VWF TO PLATELET INTEGRIN $\alpha_{IIb}\beta_3$ The platelet GPIIb-IIIa complex, or $\alpha IIb\beta 3$, is a member of the integrin family of cell-surface receptors. On resting platelets, $\alpha IIb\beta 3$ does not bind to VWF. However, after platelets are activated by thrombin or other agonists, $\alpha IIb\beta 3$ becomes able to bind fibrinogen (89), fibronectin (90), or VWF with high affinity (91). The binding site on VWF includes the tetrapeptide motif Arg-Gly-Asp-Ser near the carboxyl-terminal end of domain C1. This interaction alone is not sufficient for platelet adhesion, but specific inhibition of GPIIb-IIIa function impairs the adhesion of platelets to surfaces coated with either VWF (92) or collagen (93, 94). Therefore, GPIIb-IIIa may contribute to platelet adhesion that is initiated through GPIb binding to VWF.

Fluid Shear Stress and VWF Function

A remarkable feature of VWF-mediated platelet adhesion is its dependence on fluid shear stress. As blood flows through a vessel, velocity is maximal in the center and falls to zero at the wall. Conversely, the gradient of velocity, or shear rate (expressed in units of cm/s per cm, or inverse seconds, s^{-1}), is maximal at the vessel wall and zero at the center [reviewed in (95)]. To form a thrombus, platelets must bind to the vessel wall despite these opposing shear forces. Part of the effect of shear is to increase the transport of platelets to the vessel wall, but a more important effect is to increase the reactivity of VWF. At the low shear rates of veins and normal arteries ($\approx$100 s^{-1}), platelet adhesion is not stimulated by VWF. However, at shear rates >1,000 s^{-1}, platelet adhesion is strongly dependent on VWF. In vivo, shear rates in this range occur in small arterioles of 10–50 μm diameter, where the shear rates are estimated to vary between $\approx$470 and $\approx$4,700 s^{-1}, with a median of 1,700 s^{-1} (96). Above atherosclerotic plaques in partially occluded arteries, higher shear rates of up to 11,000 s^{-1} have been calculated (96).

Under such conditions of hydrodynamic stress, almost all platelets that contact surface-immobilized VWF become attached (92). At shear rates of 1,500 s^{-1} to 6,000 s^{-1}, the adherent platelets translate along the surface slowly (<10 μm/s) in the direction of flow. The exceptional efficiency of platelet capture, and the halting "stop-start" motion, probably depend on rapid association and dissociation rates for the interaction of the VWF A1 domain with GPIbα (92). With increasing time at high shear, the translocating platelets are activated, spread on the surface, and become immobile. These later reactions depend on high-affinity interactions between αIIbβ3 and the Arg-Gly-Asp-Ser site in VWF domain C1. The kinetics of αIIbβ3 binding are too slow to mediate platelet adhesion directly at high shear rates (92). Therefore, the initial reversible tethering mediated by GPIb-VWF binding may provide sufficient time to make the additional tight integrin-dependent bonds that are necessary for irreversible platelet binding.

These observations are consistent with a two-step model of platelet thrombus formation at high shear rates. First, the kinetically rapid interaction between GPIbα and VWF promotes the reversible tethering of platelets that approach the vessel wall. Second, the slowly moving tethered platelets adhere irreversibly through the kinetically slow binding of cell surface αIIbβ3 (or other platelet integrins) to ligands in connective tissue. The transition from GPIb-dependent to integrin-dependent binding is accelerated by platelet activation, which can be mediated by constituents of connective tissue such as collagen or by soluble platelet agonists such as ADP or epinephrine (92, 95). Leucocytes use an analogous mechanism to overcome the shear stresses in flowing blood and bind to venular endothelial cells. As a leukocyte speeds past, cell surface "selectins" interact with their cognate membrane bound ligands, enabling the leukocyte to roll slowly across the endothelium. Subsequent integrin-dependent interactions then allow the leukocyte to stop completely, flatten, and exit the vasculature through an endothelial cell junction (97).

The puzzle remains that VWF does not appear to support platelet adhesion at low shear rates, yet is extremely effective at high shear rates. No satisfactory explanation has been proposed for this phenomenon. However, the hydrodymamic properties of VWF multimers may be an important part of the mechanism. In the absence of shear stress, VWF deposited on a hydrophobic surface had a globular, condensed conformation by atomic force microscopy (98). Above a critical shear stress value, VWF adopted an extended chain conformation with the exposure of intra-molecular globular domains (Figure 5). The shear stress required to induce this conformational transition appears comparable to the threshold of shear stress for VWF-dependent platelet adhesion to vessel segments. Thus, shear-induced conformational changes in VWF may contribute to the regulation of VWF binding to platelet GPIbα (98).

Stabilization of Blood Coagulation Factor VIII

Patients with hemophilia have low levels of factor VIII and normal levels of VWF, whereas patients with severe VWD have undetectable levels of VWF and factor VIII levels <10% of normal. This behavior reflects the dependence of normal factor VIII survival on the formation of noncovalent factor VIII-VWF complexes, and it illustrates the biological importance of the factor VIII-VWF interaction [reviewed in (99)].

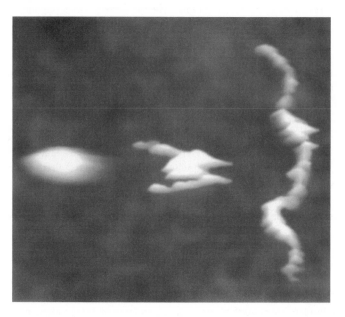

Figure 5   Effect of shear stress on VWF conformation. Atomic force microscopy (AFM) in tapping mode exerts negligible shear force on the sample and images VWF as a globular protein (*left*). AFM in contact mode applies shear forces of a few nanoNewtons and can slightly extend the VWF molecule (*center*). Shear force of 35 dyn/cm^2 applied by a rotating disk before imaging extends the VWF molecule in the direction of the shear field, with molecular lengths ranging from 146 to 774 nm. This figure is from Siedlecki et al (98).

Factor VIII binds to a single class of sites on VWF binds with a Kd of 200 to 400 pM (100, 101), and all sizes of VWF multimers bind factor VIII with similar affinity (101). In solution, almost all VWF subunits can bind factor VIII, but adsorption of VWF to surfaces markedly decreases the stoichiometry of binding (100). The rate of factor VIII-VWF association is relatively rapid (5.9×10^6 $M^{-1} s^{-1}$), so that exogenously administered factor VIII may bind endogenous VWF in vivo with a half-life for complex formation of ≈ 2 seconds (101).

The factor VIII binding site on VWF is within the amino-terminal 272 amino acid residues of the mature subunit (102; Figure 1). The corresponding VWF binding site on factor VIII is near the amino-terminus of the 80 kDa light chain. The factor VIII light chain consists of residues 1649-2332 of the factor VIII precursor. The VWF binding site involves amino acids in the segment between residues 1669–1689 (103, 104), and optimal binding requires sulfation of Tyr1680 (104). When factor VIII is activated during blood coagulation, thrombin cleaves it after Arg1689. This cleavage destroys the VWF binding site and releases factor VIIIa. Thus, VWF adsorbed to connective tissue at a site of platelet thrombus formation may deliver factor VIII for participation in additional blood coagulation reactions.

Other Binding Activities

Several other compounds have been shown to bind VWF, although the physiological significance of these additional interactions is not known. For example, VWF binds heparin through domain A1 (105). Heparin competitively inhibits the binding of VWF to platelet GPIb, and this could affect platelet adhesion during heparin therapy (106). VWF binds to some sulfated glycolipids through domain A1, and the sulfatide binding site appears to be distinct from the heparin binding site (107). A proteolytic fragment of VWF corresponding to amino acid residues 1–272 also binds heparin (50), but this second heparin site appears to be unavailable in native VWF (108).

GENETICS OF VWF AND VWD

The VWF Gene and Pseudogene

The VWF gene is located near the tip of the short arm of human chromosome 12 at 12p13.2 (109, 110). It contains 52 exons and spans approximately 180 kb (111). Exon 28 is exceptionally large, comprising 1379 base pairs that encode all of domains A1 and A2. This has facilitated the cloning of $\approx 16\%$ of the VWF coding sequence by PCR of exon 28 for more than 30 species of placental mammals representing 15 orders (112, 113).

In addition to the VWF gene, a partial unprocessed VWF pseudogene is located on human chromosome 22q11.2 (114). The pseudogene is 21 to 29 kb in length and corresponds to exons 22–34 of the VWF gene, which encode domains A1-A2-A3 (115). The pseudogene has accumulated several missense, nonsense, and splice-site mutations that would prevent the generation of a functional transcript. The VWF pseudogene and gene have diverged only ≈3.1% in nucleotide sequence, suggesting the pseudogene arose relatively recently by partial gene duplication (115). This is consistent with the presence of the VWF pseudogene in great apes (bonabo, chimpanzee, gorilla, orangutan) but not in more distantly related primates such as rhesus or spider monkey (LA Westfield, JE Sadler, unpublished information).

Recombination as a Possible Cause of VWD

Some of the sequences characteristic of the VWF pseudogene have been found in the VWF gene of a few patients with VWD, and this has been proposed to result from localized gene conversion (116). For example, a silent mutation in the codon for Ser500 and a Pro503Leu mutation were found in two patients with VWD type 2B; these two pseudogene-like mutations also were found with a third, Val516Ile, in a patient with VWD type 1. Similarly, the mutations Val566Gly and Asn468Thr are present in the VWF pseudogene and were found also in the VWF gene of another patient with VWD type 2B. The gene segments proposed to be affected by gene conversion are relatively short: depending on the patient, the lower limit is 7 to 47 nt, and the upper limit is 95 to 195 nt. Gene conversion is consistent with these findings, but additional evidence is needed to support or refute this mechanism [reviewed in (116)].

Evolution of Repeated Motifs in VWF

Homologues of VWF repeated domains occur in many otherwise unrelated proteins, indicating that they have been dispersed throughout the genome by duplication and exon shuffling. At least 46 VWF A domains are known to be distributed among 22 distinct human genes (Figure 6) [reviewed in (117, 118)]. VWF A domains are found inserted into the α-subunits of certain integrins including five leukocyte adhesion receptors ($\alpha M\beta 2$, $\alpha L\beta 2$, $\alpha x\beta 2$, $\alpha d\beta 2$, and $\alpha E\beta 7$) (119, 120) and two collagen receptors ($\alpha 1\beta 1$ and $\alpha 2\beta 1$). VWF A domains also occur in several nonfibrillar collagens (types VI, VII, XII, and XIV); cartilage matrix protein and the closely related matrilin-2 (121) and matrilin-3 (122); the noninhibitory α2 and α3 subunits of an inter-α-trypsin Kunitz-type protease inhibitor; the α2 subunit of a dihydropyridine-sensitive calcium channel; and the serine proteases complement factor B and complement component C2. Many of these proteins contain several A domains. For example, the collagen α1(VI) and α2(VI) chains each have three A domains, and the α3(VI) chain contains up to 12 A domains.

The A domains of VWF, complement factor B and component C2, and several integrin α-chains have been shown to bind macromolecular ligands, and their

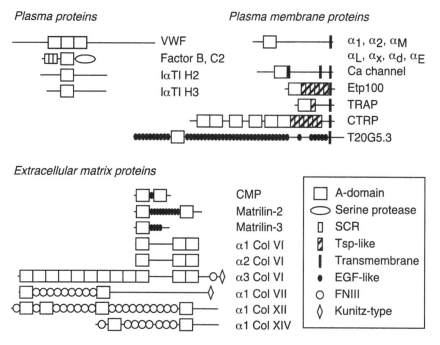

Figure 6 Schematic structure of proteins with conserved VWF A domains. Plasma proteins include VWF, complement factor B, complement component C2, and inter-α-trypsin inhibitor (IαTI) noninhibitory subunits H2 and H3. Plasma membrane proteins include the α-subunits of integrins $\alpha 1 \beta 1$, $\alpha 2 \beta 1$, $\alpha M \beta 2$, $\alpha L \beta 2$, $\alpha x \beta 2$, $\alpha d \beta 2$, and $\alpha E \beta 7$; the $\alpha 2$ subunit of a dihydropyridine-sensitive calcium channel (Ca channel); *E. tenella* microneme protein Etp100; *P. falciparum* thrombospondin-related anonymous protein (TRAP) and circumsporozoite TRAP-related protein (CTRP); and *C. elegans* hypothetical protein T20G5.3. Extracellular matrix proteins include cartilage matrix protein (CMP) and the related matrilin-2 and matrilin-3; and collagens (Col) VI, VII, XII, and XIV. The symbols for structural motifs are shown and include VWF A domains, serine protease domains, complement consensus repeats (SCR), thrombospondin-like domains (Tsp-like), transmembrane domains, epidermal growth factor-like domains (EGF-like), fibronectin type III repeats (FNIII), and Kunitz-type protease inhibitor domains. See text for references.

binding functions appear to be regulated. The A domains of factor B and several integrins also have divalent metal ion binding sites that must be occupied to bind ligands. By analogy to these examples, some VWF A domains in other proteins, for which functions have not been demonstrated, are likely to mediate binding interactions that are regulated allosterically and that may be metal ion-dependent [reviewed in (71)].

The VWF B, C, D, and CK domains near the carboxyl-terminal end of the VWF subunit are shared with certain epithelial mucins, often in association with VWF D domains and, sometimes, B domains [reviewed in (17, 123)]

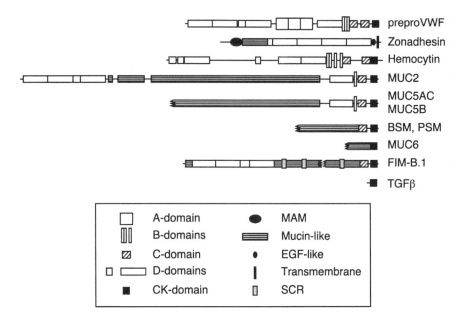

Figure 7 Schematic structure of proteins with conserved VWF B, C, D, and CK domains. In addition to preproVWF, these include zonadhesin; *B. mori* hemocytin; the mammalian epithelial mucins MUC2, MUC5AC, MUC5B, and MUC6; bovine and porcine submaxillary mucins (BSM, PSM); *X. laevis* integumentary mucin FIM-B.1; and the large family of neurotrophins represented here by transforming growth factor-β (TGFβ). Portions of certain mucins have not yet been cloned and these are indicated by jagged interruptions in mucin-like domains. Symbols for structural motifs include VWF A, B, C, and D domains; CK domains; MAM domains named for a common motif found in meprin, *X. laevis* neuronal protein A5, and protein tyrosine phosphatase-μ (153); mucin-like domains, EGF-like domains, transmembrane domains, and complement consensus repeats (SCR). See text for other references.

(Figure 7). These proteins form homodimers or oligomers. Dimerization of porcine submaxillary mucin is mediated by disulfide bonds between carboxyl-terminal CK domains (124), and this mechanism may apply to other structurally similar mucins. The VWF CK domains also are homologous to the neurotrophin family of growth factors and are particularly similar to transforming growth factor-β (16). Like mucins that contain CK domains, growth factors of the neurotrophin family usually are dimeric (15), suggesting that CK domains can function as dimerization motifs in several contexts.

The detritus of VWF evolution is scattered even more broadly in the animal kingdom. Five VWF D domains are found in zonadhesin, a mammalian sperm membrane glycoprotein that appears to function in the species-specific recognition of egg extracellular matrix (125). VWF A domains occur in protozoan parasite and nematode proteins, including *Plasmodium falciparum* malarial thrombospondin-related anonymous protein (TRAP), and circumsporozoate TRAP-related protein (CTRP); *Eimeria tenella* microneme protein Etp100; and *Caenorhabditis elegans* 338 kDa (partial) hypothetical protein T20G5.3 (126– 128). An especially remarkable homologue of VWF is found in the silkworm moth, *Bombyx mori*. The inducible defense protein, hemocytin, is very similar in structure to VWF except that it lacks type A domains (Figure 7; 129). Thus, VWF is a mosaic protein with a complex ancestry. Representatives of all VWF domains are found throughout the vertebrates and in several invertebrate orders, indicating VWF has deep evolutionary roots that predate the development of vertebrate blood circulation and its role in hemostasis.

Prevalence and Medical Significance of VWD

VWD appears to be the most common inherited human bleeding disorder. In Scandinavia, where VWD was discovered, its prevalence is estimated to be approximately 100 per million population (130). Clinically significant VWD appears to be less common elsewhere, although the method of patient identification strongly affects the apparent prevalence. For example, screening of asymptomatic schoolchildren in Italy suggests that approximately 8000 per million population have inherited defects in VWF function, but most persons identified in this way are asymptomatic (131).

The symptoms of VWD usually are those of platelet dysfunction and include nose bleeding, skin bruises and hematomas, prolonged bleeding from trivial wounds, oral cavity bleeding, and excessive menstrual bleeding [reviewed in (8)]. Gastrointestinal bleeding appears to be relatively rare, but may be very serious when it occurs. Severe deficiency of VWF, or a specific defect in the interaction of VWF with factor VIII, causes a secondary moderate deficiency of factor VIII. These patients may have symptoms that are more characteristic of hemophilia, such as bleeding into joints or soft tissues including muscle and brain. Bleeding in VWD may be exacerbated by the use of aspirin or corticosteroids, and decreased by the use of oral contraceptives.

Classification and Molecular Defects in VWD

The wealth of information about VWF biosynthesis and function has led to a simplified classification for VWD (132). The three major categories are partial quantitative deficiency (type 1), qualitative deficiency (type 2), and total deficiency (type 3). Qualititative type 2 VWD is divided further into four variants (2A, 2B, 2M, and 2N) based upon details of the phenotype. These six categories correspond to distinct pathophysiologic mechanisms and they are intended to correlate with distinct clinical features and therapeutic requirements.

VWD TYPE 1   The distribution of VWF multimers is normal in VWD type 1, and all molecular forms are decreased proportionately (Figure 8). About 70% of patients with VWD appear to have VWD type 1, but many factors conspire to make this common variant a diagnostic problem. The bleeding symptoms of VWD are not specific and may occur with any cause of platelet dysfunction. Some symptoms, such as bruising and menorrhagia, occur frequently in apparently normal persons. Low VWF levels in the blood can be hard to evaluate because the normal range of VWF concentration is very broad. One determinant of VWF level is ABO blood type, and the mean VWF:Ag level is ≈25% lower for persons of blood type O compared to other blood types (133). Additional variation in VWF level is caused by unknown factors, so that for normal type O blood donors, the mean level of VWF:Ag is reported to be 0.75 U/ml with a range (±2 SD) of 0.36−1.57 U/ml, where the mean level in the population is 1.0 U/ml (133). Therefore, bleeding symptoms and low VWF levels are individually common and may occur together by chance. The case for VWD is strengthened by evidence that bleeding symptoms and VWF deficiency are coinherited, but in practice the diagnosis often is uncertain.

VWD type 1 frequently is caused by frameshifts, nonsense mutations, or deletions that overlap with those identified in VWD type 3, discussed below.

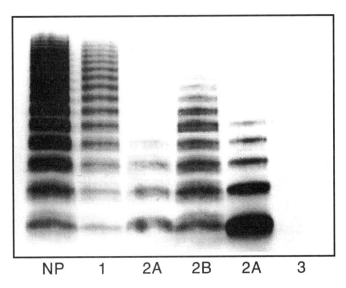

Figure 8   Multimer patterns in VWD. Samples of plasma from a normal individual (N) and from patients with VWD types 1, 2A, 2B, and 3 were electrophoresed through a 1.3% agarose/SDS gel and VWF was detected by Western blotting. [This figure is from (154)].

Possession of a single such mutant allele does not consistently cause symptoms. Patients with VWD type 1 often have mild bleeding, and in their families the disease may seem to exhibit low penetrance. In some cases, extreme variation in symptoms appears to be explained by the chance coinheritance of a mutation for a distinct VWD subtype. For example, compound heterozygosity for VWD type 1 and VWD type 2N was found in patients with significant bleeding, whereas relatives with either single mutant allele were asymptomatic (134).

Occasionally, VWD type 1 appears to be inherited as a dominant trait with very high penetrance and exceptionally low VWF levels. In one family this phenotype was associated with a Cys386Arg mutation in the VWF subunit that inhibited the secretion of wild-type VWF subunits, providing a plausible biochemical explanation for the strongly dominant transmission of symptoms (135). The mutant proVWF subunits were retained intracellularly in the ER. If "tail-to-tail" pairing of proVWF subunits were random, retention of mutant homodimers and heterodimers would reduce the amount of normal proVWF dimers arriving in the Golgi to 25% of the total. Any tendency to favor homodimerization or heterodimerization in the ER would, respectively, ameliorate or exacerbate the phenotype. This dominant-negative mechanism appears to provide a reasonable explanation for at least some unexpectedly severe VWD type 1 phenotypes.

VWD TYPE 2A   Only large VWF multimers have significant hemostatic activity, and the selective absence of these large multimers (Figure 8) predictably causes bleeding. This class of qualitative defects is referred to as VWD type 2A and is the most common subtype of VWD type 2.

VWD type 2A usually is transmitted as a dominant disorder caused by single amino acid substitutions within repeated domain A2 (residues 717–909) of the mature VWF subunit (Figure 9; 136). These mutations usually act by either of two mechanisms. Some impair multimer assembly within the Golgi so that only small, nonfunctional multimers are secreted. Others are compatible with normal multimer assembly and secretion but lead to rapid clearance from the circulation, probably by increased proteolytic degradation (137). A distinct set of mutations within the VWF propeptide causes a recessive form of VWD type 2A, also by impairing multimer assembly in the Golgi (138). The basis for the recessive rather than dominant clinical phenotype is not understood. Finally, the mutation Cys2011Arg has been described in a patient with a recessive form of VWD type 2A (originally designated type IID). Cys2011 is proposed to form an intersubunit disulfide bond between proVWF subunits in the ER, and substitution by Arg at this site leads to the formation of only nonfunctional "head-to-head" dimers in the Golgi (18). For all of these VWD type 2A mutations, the common pathophysiologic mechanism involves the absence of

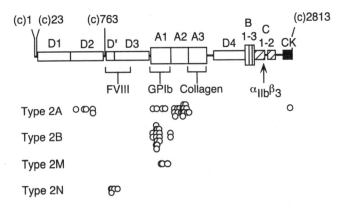

Figure 9 Mutations in VWD type 2. References to the individual mutations can be found in the database of VWD mutations (136) maintained by David Ginsburg (University of Michigan) and accessible at http://mmg2.im.med.umich.edu/vWF.

large VWF multimers and the consequent failure of VWF-dependent platelet adhesion in vivo.

VWD TYPE 2B   VWD type 2B is characterized by increased affinity of the mutant VWF for platelets (139). This apparently leads to the spontaneous binding of large VWF multimers to platelets in vivo, followed by clearance of both the VWF and the platelets. Paradoxically, this exaggerated function does not cause thrombosis. Instead, the remaining small VWF multimers (Figure 8) are not hemostatically effective, and patients have a bleeding diathesis that may seem severe for the degree of VWF deficiency shown by laboratory testing. Thrombocytopenia in VWD type 2B may be intermittent and often is worsened by stress such as infection or pregnancy.

At least 14 distinct mutations have been characterized in patients with VWD type 2B (136, 140, 141); one is a single amino acid insertion and the remainder are single amino acid substitutions (136). These cluster in a discrete region on the A1 domain (Figure 9), and appear to mark the location of a regulatory site that normally inhibits the binding of domain A1 to platelet GPIb (Figure 4). Mutations in the site relieve this inhibition and cause constitutive binding, thereby causing the observed dominant, gain-of-function phenotype.

VWD TYPE 2M   VWD type 2M ("M" for "multimer") includes variants in which platelet adhesion is impaired but the VWF multimer distribution is normal. In principle, this loss-of-function phenotype might be produced by mutations that inactivate specific binding sites for ligands on platelets or in connective tissue. So far, the few known VWD type 2M mutations are located in domain

A1 and disrupt binding to platelet GPIb (Figure 9). Screening laboratory results were similar to those in VWD type 2A, but multimer gel analysis showed that large multimers were present. The patients have had significant bleeding symptoms despite factor VIII levels in the normal range, thereby confirming that VWF-mediated platelet adhesion is biologically important, independent of the role of VWF in factor VIII stabilization.

VWD TYPE 2N   Several missense mutations in VWF have been reported that selectively inactivate the factor VIII binding site (Figure 9). These mutations produce an autosomal recessive VWD phenotype in which the platelet-dependent functions of VWF are preserved and the multimer pattern appears normal, but factor VIII levels are low, often <10% (142, 143). This fascinating autosomal mimic of hemophilia A is named VWD "type 2N" after "Normandy," the birth province of one index case [reviewed in (144)]. The correct diagnosis is made by demonstrating that the patient's VWF has markedly decreased affinity for factor VIII.

The prevalence of VWD type 2N is not known, but it probably is much higher than currently appreciated. For example, 31 affected families were identified between 1990 and 1995 in France, where only two centers perform VWF-factor VIII binding assays (145). Some of these cases had been misdiagnosed as hemophilia A patients or carriers. Screening of patients with hemophilia A has shown that a few percent of those with apparent mild hemophilia A actually have VWD type 2N (146).

Although VWD type 2N superficially resembles mild hemophilia A, or a hemophilia carrier state with low factor VIII levels resulting from extreme lyonization, the therapeutic requirements in VWD type 2N are quite distinct. In particular, highly purified factor VIII preparations often give poor factor VIII recovery and survival in vivo, because the patient's endogenous VWF cannot stabilize it normally. Furthermore, genetic counseling for an autosomal recessive disease (VWD type 2N) is very different from that for an X chromosome-linked disease (hemophilia A). Because of these important points, the differential diagnosis of VWD type 2N versus hemophilia A should be considered in any patient with low factor VIII levels in whom the evidence for X-linked inheritance is incomplete, or in whom initial therapy with pure factor VIII gives unexpectedly poor results.

VWD TYPE 3   VWD type 3 is a recessive disorder in which there is virtually no detectable VWF protein (Figure 8). The pathophysiology of VWD type 3 is not fundamentally different from that of VWD type 1, but complete VWD deficiency receives a distinct label because the symptoms are so severe and the therapeutic requirements are distinct. The absence of VWF causes a moderate secondary deficiency of factor VIII, and such patients have a combined defect in platelet adhesion and blood clotting that leads to severe bleeding symptoms from birth. The plasma factor VIII is reduced to levels similar to those in VWD type 2N, and patients with VWD type 3 may have hemarthroses and soft tissue bleeding that are more typical of hemophilia A. VWD type 3 is the diagnosis of the proband in the first family characterized by von Willebrand (1). Almost 70 years later, the mutation was identified in surviving members of this family and shown to be a frameshift in exon 18 that leads to a premature termination within the propeptide (147).

The parents of patients with VWD type 3 may be phenotypically normal, or may have bleeding symptoms and meet criteria for VWD type 1. Such variability was evident in the first report by von Willebrand (1). The proband was one of five siblings with VWD type 3 and both parents had mild symptoms, but several relatives who could transmit the disease appeared to be unaffected. Similar variability in symptoms is evident in many other families for which mutations have been characterized. VWD type 3 and VWD type 1 coexist in the same pedigrees, and both appear to be caused mainly by deletion, frameshift, and nonsense mutations (148–150).

Questions regarding the "expressivity" or "penetrance" of VWD in the relatives of type 3 patients are linked inextricably to questions regarding the pathophysiology of VWD type 1. Only a small fraction of what might be termed "carriers" of type 3 mutations are symptomatic. For example, the prevalence of VWD type 3 is reportedly about 2.5 to 3 per million population in Scandinavian countries, and 0.1 to 1.6 per million in many counties of Europe and the Middle East (151). An intermediate prevalence of $\approx$1.5 per million was reported for the United States and Canada (152). These figures represent approximate values for q^2, where q is the gene frequency for type 3 alleles; therefore, the prevalence of heterozygous persons (2pq) is in the range of 660 to 3400 per million. The prevalence of heterozygotes may be even higher because of the shortened life expectancy of VWD type 3. However, the prevalence of symptomatic VWD type 1 appears to be only $\approx$10% of these values. Thus, a single nonfunctional VWF allele may sometimes cause VWD type 1, but is not always sufficient.

CONCLUSIONS

The biosynthesis of VWF is unusual because it involves the intracellular assembly and storage of enormous multimers, and it requires the formation of disulfide bonds in the acidic environment of the Golgi apparatus. Some of the structural requirements for targeting and disulfide assembly have been characterized. Depending on which step in this complex process is affected, VWF mutations can have a dominant effect and produce several distinct VWD phenotypes.

Retention of mutant subunits in the ER appears able to cause relatively severe quantitative deficiency (VWD type 1), whereas mutant subunits that make it as far as the Golgi can cause qualitative defects. Mutations that interfere with multimer assembly or stability cause VWD type 2A. Mutations that inactivate specific ligand binding sites, or disrupt the regulation of binding to platelets, cause distinct VWD phenotypes that clearly illustrate the biological importance of VWF-dependent platelet adhesion and factor VIII stabilization. This recent progress in understanding the structure-function relationships of VWF has led to a clinically useful classification of VWD based on pathophysiologic mechanism.

> **Visit the *Annual Reviews* home page at
> http://www.AnnualReviews.org.**

Literature Cited

1. von Willebrand EA. 1926. *Fin. Laekare-saellsk. Hand.* 68:87–112
2. von Willebrand EA, Jürgens R. 1933. *Dtsch. Arch. Klin. Med.* 175:453–83
3. Alexander B, Goldstein B. 1953. *J. Clin. Invest.* 32:551
4. Larrieu MJ, Soulier JP. 1953. *Rev. Hematol.* 8:361–70
5. Quick AJ, Hussey CV. 1953. *J. Lab. Clin. Med.* 42:929–30
6. Nilsson IM, Blombäck M, von Francken I. 1957. *Acta Med. Scand.* 159:35–37
7. Nilsson IM, Blombäck M, Jorpes E, Blombäck B, Johansson S-A. 1957. *Acta Med. Scand.* 159:179–88
8. Sadler JE. 1995. In *The Metabolic and Molecular Basis of Inherited Disease*, ed. CR Scriver, AL Beaudet, WS Sly, D Valle, pp. 3269–87. New York: McGraw-Hill
9. Jaffe EA, Hoyer LW, Nachman RL. 1974. *Proc. Natl. Acad. Sci. USA* 71:1906–9
10. Nachman R, Levine R, Jaffe EA. 1977. *J. Clin. Invest.* 60:914–21
11. Marti T, Rösselet SJ, Titani K, Walsh KA. 1987. *Biochemistry* 26:8099–109
12. Carew JA, Browning PJ, Lynch DC. 1990. *Blood* 76:2530–39
13. Matsui T, Titani K, Mizuochi T. 1992. *J. Biol. Chem.* 267:8723–31
14. Voorberg J, Fontijn R, Calafat J, Janssen H, van Mourik JA, Pannekoek H. 1991. *J. Cell Biol.* 113:195–205
15. McDonald NQ, Hendrickson WA. 1993. *Cell* 73:421–24
16. Meitinger T, Meindl A, Bork P, Rost B, Sander C, et al. 1993. *Nat. Genet.* 5:376–80
17. Desseyn J-L, Aubert JP, Van Seuningen I, Porchet N, Laine A. 1997. *J. Biol. Chem.* 272:16873–83
18. Schneppenheim R, Brassard J, Krey S, Budde U, Kunicki TJ, et al. 1996. *Proc. Natl. Acad. Sci. USA* 93:3581–86
19. Wagner DD, Mayadas T, Marder VJ. 1986. *J. Cell Biol.* 102:1320–24
20. Wagner DD, Saffaripour S, Bonfanti R, Sadler JE, Cramer EM, et al. 1991. *Cell* 64:403–13
21. Rehemtulla A, Kaufman RJ. 1992. *Blood* 79:2349–55
22. Vischer UM, Wagner DD. 1994. *Blood* 83:3536–44
23. Voorberg J, Fontijn R, van Mourik JA, Pannekoek H. 1990. *EMBO J.* 9:797–803
24. Verweij CL, Hart M, Pannekoek H. 1987. *EMBO J.* 6:2885–90
25. Wise RJ, Pittman DD, Handin RI, Kaufman RJ, Orkin SH. 1988. *Cell* 52:229–36
26. Verweij CL, Hart M, Pannekoek H. 1988. *J. Biol. Chem.* 263:7921–24
27. Mayadas TN, Wagner DD. 1989. *J. Biol. Chem.* 264:13497–503
28. Mayadas TN, Wagner DD. 1992. *Proc. Natl. Acad. Sci. USA* 89:3531–35
29. Fujimura Y, Titani K, Holland LZ, Russell SR, Roberts JR, et al. 1986. *J. Biol. Chem.* 261:381–85
30. Dong Z, Thoma RS, Crimmins DL, McCourt DW, Tuley EA, Sadler JE. 1994. *J. Biol. Chem.* 269:6753–58
31. Wagner DD, Olmsted JB, Marder VJ. 1982. *J. Cell Biol.* 95:355–60
32. Sporn LA, Marder VJ, Wagner DD. 1986. *Cell* 46:185–90

33. Weibel ER, Palade GE. 1964. *J. Cell Biol.* 23:101–12
34. Cramer EM, Meyer D, le Menn R, Breton-Gorius J. 1985. *Blood* 66:710–13
35. Santolaya RC, Bertini F. 1970. *Z. Anat. Entwickl.-Gesch.* 131:148–55
36. Voorberg J, Fontijn R, Calafat J, Janssen H, van Mourik JA, Pannekoek H. 1993. *EMBO J.* 12:749–58
37. Hop C, Fontijn R, van Mourik JA, Pannekoek H. 1997. *Exp. Cell Res.* 230:352–61
38. Journet AM, Saffaripour S, Cramer EM, Tenza D, Wagner DD. 1993. *Eur. J. Cell Biol.* 60:31–41
39. McEver RP, Beckstead JH, Moore KL, Marshall-Carlson L, Bainton DF. 1989. *J. Clin. Invest.* 84:92–99
40. Bonfanti R, Furie BC, Furie B, Wagner DD. 1989. *Blood* 73:1109–12
41. Vischer UM, Wagner DD. 1993. *Blood* 82:1184–91
42. Disdier M, Morrissey JH, Fugate RD, Bainton DF, McEver RP. 1992. *Mol. Biol. Cell* 3:309–21
43. Wagner DD. 1990. *Annu. Rev. Cell Biol.* 6:217–46
44. Vischer UM, Wollheim CB. 1997. *Thromb. Haemost.* 77:1182–88
45. Richardson M, Tinlin S, De Reske M, Webster S, Senis Y, Giles AR. 1994. *Arterioscler. Thromb.* 14:990–99
46. Mannucci PM, Ruggeri ZM, Pareti FI, Capitanio A. 1977. *Lancet* 1:869–72
47. Moffat EH, Giddings JC, Bloom AL. 1984. *Br. J. Haematol.* 57:651–62
48. Booth F, Allington MJ, Cederholm-Williams SA. 1987. *Br. J. Haematol.* 67:71–78
49. Fowler WE, Fretto LJ, Hamilton KK, Erickson HP, McKee PA. 1985. *J. Clin. Invest.* 76:1491–500
50. Fretto LJ, Fowler WE, McCaslin DR, Erickson HP, McKee PA. 1986. *J. Biol. Chem.* 261:15679–89
51. Wagner DD, Fay PJ, Sporn LA, Sinha S, Lawrence SO, Marder VJ. 1987. *Proc. Natl. Acad. Sci. USA* 84:1955–59
52. Borchiellini A, Fijnvandraat K, ten Cate JW, Pajkrt D, van Deventer SJ, et al. 1996. *Blood* 88:2951–58
53. Takagi J, Kasahara K, Sekiya F, Inada Y, Saito Y. 1989. *J. Biol. Chem.* 264:10425–30
54. Isobe T, Hisaoka T, Shimizu A, Okuno M, Aimoto S, et al. 1997. *J. Biol. Chem.* 272:8447–53
55. Santoro SA. 1981. *Thromb. Res.* 21:689–91
56. Morton LF, Griffin B, Pepper DS, Barnes MJ. 1983. *Thromb. Res.* 32:545–56
57. Rand JH, Patel ND, Schwartz E, Zhou SL, Potter BJ. 1991. *J. Clin. Invest.* 88:253–59
58. Houdijk WP, Sakariassen KS, Nievelstein PF, Sixma JJ. 1985. *J. Clin. Invest.* 75:531–40
59. Wagner DD, Urban-Pickering M, Marder VJ. 1984. *Proc. Natl. Acad. Sci. USA* 81:471–75
60. de Groot PG, Ottenhof-Rovers M, van Mourik JA, Sixma JJ. 1988. *J. Clin. Invest.* 82:65–73
61. von der Mark H, Aumailley M, Wick G, Fleischmajer R, Timpl R. 1984. *Eur. J. Biochem.* 142:493–502
62. Colombatti A, Bonaldo P. 1987. *J. Biol. Chem.* 262:14461–66
63. Wu XX, Gordon RE, Glanville RW, Kuo HJ, Uson RR, Rand JH. 1996. *Am. J. Pathol.* 149:283–91
64. Ross JM, McIntire LV, Moake JL, Rand JH. 1995. *Blood* 85:1826–35
65. Roth GJ, Titani K, Hoyer LW, Hickey MJ. 1986. *Biochemistry* 25:8357–61
66. Kalafatis M, Takahashi Y, Girma JP, Meyer D. 1987. *Blood* 70:1577–83
67. Pareti FI, Niiya K, McPherson JM, Ruggeri ZM. 1987. *J. Biol. Chem.* 262:13835–41
68. Lankhof H, van Hoeij M, Schiphorst ME, Bracke M, Wu YP, et al. 1996. *Thromb. Haemost.* 75:950–58
69. Bienkowska J, Cruz M, Atiemo A, Handin R, Liddington R. 1997. *J. Biol. Chem.* 272:25162–67
70. Huizinga EG, van der Plas RM, Sixma JJ, Kroon J, Gros P. 1997. *Structure* 5:1147–56
71. Lee JO, Rieu P, Arnaout MA, Liddington R. 1995. *Cell* 80:631–38
72. Qu AD, Leahy DJ. 1995. *Proc. Natl. Acad. Sci. USA* 92:10277–81
73. Branden CI. 1980. *Q. Rev. Biophys.* 13:317–38
74. Du X, Beutler L, Ruan C, Castaldi PA, Berndt MC. 1987. *Blood* 69:1524–27
75. Modderman PW, Admiraal LG, Sonnenberg A, Borne AEGKV. 1992. *J. Biol. Chem.* 267:364–69
76. Marchese P, Murata M, Mazzucato M, Pradella P, De Marco L, et al. 1995. *J. Biol. Chem.* 270:9571–78
77. Scott JP, Montgomery RR, Retzinger GS. 1991. *J. Biol. Chem.* 266:8149–55
78. Berndt MC, Du X, Booth WJ. 1988. *Biochemistry* 27:633–40
79. Andrews RK, Booth WJ, Gorman JJ, Castaldi PA, Berndt MC. 1989. *Biochemistry* 28:8317–26
80. Matsushita T, Sadler JE. 1995. *J. Biol. Chem.* 270:13406–14

81. Kroner PA, Frey AB. 1996. *Biochemistry* 35:13460–68
82. Matsushita T, Sadler JE. 1997. *Thromb. Haemost.* Suppl.:364
83. Hillery CA, Mancuso DJ, Sadler JE, Ponder JW, Jozwiak ML, et al. 1998. *Blood* 91:1572–81
84. Andrews RK, Gorman JJ, Booth WJ, Corino GL, Castaldi PA, Berndt MC. 1989. *Biochemistry* 28:8326–36
85. Miura S, Fujimura Y, Sugimoto M, Kawasaki T, Ikeda Y, et al. 1994. *Blood* 84:1553–58
86. Zhang L, Plow EF. 1996. *J. Biol. Chem.* 271:29953–57
87. Celikel R, Varughese KI, Madhusudan, Yoshioka A, Ware J, Ruggeri ZM. 1997. *Thromb. Haemost.* Suppl.:769
88. Cruz MA, Emsley J, Diacovo T, Liddington R, Handin RI. 1997. *Blood* 90(Suppl. 1):430a
89. Marguerie GA, Plow EF, Edgington TS. 1979. *J. Biol. Chem.* 254:5357–63
90. Plow EF, Ginsberg MH. 1981. *J. Biol. Chem.* 256:9477–82
91. Fujimoto T, Ohara S, Hawiger J. 1982. *J. Clin. Invest.* 69:1212–22
92. Savage B, Saldivar E, Ruggeri ZM. 1996. *Cell* 84:289–97
93. Fressinaud E, Baruch D, Girma JP, Sakariassen KS, Baumgartner HR, Meyer D. 1988. *J. Lab. Clin. Med.* 112:58–67
94. Fressinaud E, Girma J-P, Sadler JE, Baumgartner HR, Meyer D. 1990. *Thromb. Haemost.* 64:589–93
95. Ruggeri ZM. 1997. *J. Clin. Invest.* 99:559–64
96. Back LD, Radbill JR, Crawford DW. 1977. *J. Biomech.* 10:339–53
97. Konstantopoulos K, McIntire LV. 1996. *J. Clin. Invest.* 98:2661–65
98. Siedlecki CA, Lestini BJ, Kottke-Marchant KK, Eppell SJ, Wilson DL, Marchant RE. 1996. *Blood* 88:2939–50
99. Sadler JE, Davie EW. 1994. In *The Molecular Basis of Blood Diseases*, ed. G Stamatoyannopoulos, AW Nienhuis, PW Majerus, H Varmus, pp. 657–700. Philadelphia: WB Saunders
100. Vlot AJ, Koppelman SJ, van den Berg MH, Bouma BN, Sixma JJ. 1995. *Blood* 85:3150–57
101. Vlot AJ, Koppelman SJ, Meijers JCM, Damas C, van den Berg HM, et al. 1996. *Blood* 87:1809–16
102. Foster PA, Fulcher CA, Marti T, Titani K, Zimmerman TS. 1987. *J. Biol. Chem.* 262:8443–46
103. Lollar P, Hill-Eubanks DC, Parker CG. 1988. *J. Biol. Chem.* 263:10451–55
104. Leyte A, van Schijndel HB, Niehrs C, Huttner WB, Verbeet MP, et al. 1991. *J. Biol. Chem.* 266:740–46
105. Fujimura Y, Titani K, Holland LZ, Roberts JR, Kostel P, et al. 1987. *J. Biol. Chem.* 262:1734–39
106. Sobel M, McNeill PM, Carlson PL, Kermode JC, Adelman B, et al. 1991. *J. Clin. Invest.* 87:1787–93
107. Christophe O, Obert B, Meyer D, Girma JP. 1991. *Blood* 78:2310–17
108. Sixma JJ, Schiphorst ME, Verweij CL, Pannekoek H. 1991. *Eur. J. Biochem.* 196:369–75
109. Ginsburg D, Handin RI, Bonthron DT, Donlon TA, Bruns GAP, et al. 1985. *Science* 228:1401–6
110. Kuwano A, Morimoto Y, Nagai T, Fukushima Y, Ohashi H, et al. 1996. *Hum. Genet.* 97:95–98
111. Mancuso DJ, Tuley EA, Westfield LA, Worrall NK, Shelton-Inloes BB, et al. 1989. *J. Biol. Chem.* 264:19514–27
112. Porter CA, Goodman M, Stanhope MJ. 1996. *Mol. Phylogenet. Evol.* 5:89–101
113. Springer MS, Cleven GC, Madsen O, de Jong WW, Waddell VG, et al. 1997. *Nature* 388:61–64
114. Patracchini P, Calzolari E, Aiello V, Palazzi P, Banin P, et al. 1989. *Hum. Genet.* 83:264–66
115. Mancuso DJ, Tuley EA, Westfield LA, Lester-Mancuso TL, Le Beau MM, et al. 1991. *Biochemistry* 30:253–69
116. Eikenboom JCJ, Vink T, Briët E, Sixma JJ, Reitsma PH. 1994. *Proc. Natl. Acad. Sci. USA* 91:2221–24
117. Colombatti A, Bonaldo P. 1991. *Blood* 77:2305–15
118. Perkins SJ, Smith KF, Williams SC, Haris PI, Chapman D, Sim RB. 1994. *J. Mol. Biol.* 238:104–19
119. Shaw SK, Cepek KL, Murphy EA, Russell GJ, Brenner MB, Parker CM. 1994. *J. Biol. Chem.* 269:6016–25
120. Van der Vieren M, Le Trong H, Wood CL, Moore PF, St. John T, et al. 1995. *Immunity* 3:683–90
121. Deák F, Piecha D, Bachrati C, Paulsson M, Kiss I. 1997. *J. Biol. Chem.* 272:9268–74
122. Wagener R, Kobbe B, Paulsson M. 1997. *FEBS Lett.* 413:129–34
123. Toribara NW, Ho SB, Gum E, Gum JR Jr, Lau P, Kim YS. 1997. *J. Biol. Chem.* 272:16398–403
124. Perez-Vilar J, Eckhardt AE, Hill RL. 1996. *J. Biol. Chem.* 271:9845–50
125. Hardy DM, Garbers DL. 1995. *J. Biol. Chem.* 270:26025–28
126. Tomley FM, Clarke LE, Kawazoe U,

Dijkema R, Kok JJ. 1991. *Mol. Biochem. Parasitol.* 49:277–88

127. Wilson R, Ainscough R, Anderson K, Baynes C, Berks M, et al. 1994. *Nature* 368:32–38

128. Trottein F, Triglia T, Cowman AF. 1995. *Mol. Biochem. Parasitol.* 74:129–41

129. Kotani E, Yamakawa M, Iwamoto S, Tashiro M, Mori H, et al. 1995. *Biochim. Biophys. Acta* 1260:245–58

130. Holmberg L, Nilsson IM. 1992. *Eur. J. Haematol.* 48:127–41

131. Rodeghiero F, Castaman G, Dini E. 1987. *Blood* 69:454–59

132. Sadler JE. 1994. *Thromb. Haemost.* 71:520–25

133. Gill JC, Endres-Brooks J, Bauer PJ, Marks WJ, Montgomery RR. 1987. *Blood* 69:1691–95

134. Eikenboom JC, Reitsma PH, Peerlinck KMJ, Briët E. 1993. *Lancet* 341:982–86

135. Eikenboom JCJ, Reitsma PH, Matsushita T, Tuley EA, Castaman G, et al. 1996. *Blood* 88:2433–41

136. Ginsburg D, Sadler JE. 1993. *Thromb. Haemost.* 69:177–84

137. Lyons SE, Bruck ME, Bowie EJW, Ginsburg D. 1992. *J. Biol. Chem.* 267:4424–30

138. Gaucher C, Diéval J, Mazurier C. 1994. *Blood* 84:1024–30

139. Ruggeri ZM, Pareti FI, Mannucci PM, Ciavarella N, Zimmerman TS. 1980. *N. Engl. J. Med.* 302:1047–51

140. Holmberg L, Dent JA, Schneppenheim R, Budde U, Ware J, Ruggeri ZM. 1993. *J. Clin. Invest.* 91:2169–77

141. Ribba AS, Christophe O, Derlon A, Cherel G, Siguret V, et al. 1994. *Blood* 83:833–41

142. Nishino M, Girma J-P, Rothschild C, Fressinaud E, Meyer D. 1989. *Blood* 74:1591–99

143. Mazurier C, Dieval J, Jorieux S, Delobel J, Goudemand M. 1990. *Blood* 75:20–26

144. Mazurier C. 1992. *Thromb. Haemost.* 67:391–96

145. Mazurier C, Meyer D. 1996. *Thromb. Haemost.* 76:270–74

146. Schneppenheim R, Budde U, Krey S, Drewke E, Bergmann F, et al. 1996. *Thromb. Haemost.* 76:598–602

147. Zhang ZP, Blombäck M, Nyman D, Anvret M. 1993. *Proc. Natl. Acad. Sci. USA* 90:7937–40

148. Zhang ZP, Lindstedt M, Falk G, Blombäck M, Egberg N, Anvret M. 1992. *Am. J. Hum. Genet.* 51:850–58

149. Zhang ZP, Blombäck M, Egberg N, Falk G, Anvret M. 1994. *Genomics* 21:188–93

150. Schneppenheim R, Krey S, Bergmann F, Bock D, Budde U, et al. 1994. *Hum. Genet.* 94:640–52

151. Mannucci PM, Bloom AL, Larrieu MJ, Nilsson IM, West RR. 1984. *Br. J. Haematol.* 57:163–69

152. Weiss HJ, Ball AP, Mannucci PM. 1982. *N. Engl. J. Med.* 307:127

153. Beckmann G, Bork P. 1993. *Trends Biochem. Sci.* 18:40–41

154. Sadler JE. 1994. In *Haemostasis and Thrombosis*, ed. AL Bloom, CD Forbes, DP Thomas, EGD Tuddenham, pp. 843–57. Edinburgh: Churchill-Livingstone

Annu. Rev. Biochem. 1998. 67:425–79

THE UBIQUITIN SYSTEM

Avram Hershko and Aaron Ciechanover

Unit of Biochemistry, Faculty of Medicine and the Rappaport Institute for Research in the Medical Sciences, Technion-Israel Institute of Technology, Haifa 31096, Israel

KEY WORDS: ubiquitin, protein degradation, proteasome, proteolysis

ABSTRACT

The selective degradation of many short-lived proteins in eukaryotic cells is carried out by the ubiquitin system. In this pathway, proteins are targeted for degradation by covalent ligation to ubiquitin, a highly conserved small protein. Ubiquitin-mediated degradation of regulatory proteins plays important roles in the control of numerous processes, including cell-cycle progression, signal transduction, transcriptional regulation, receptor down-regulation, and endocytosis. The ubiquitin system has been implicated in the immune response, development, and programmed cell death. Abnormalities in ubiquitin-mediated processes have been shown to cause pathological conditions, including malignant transformation. In this review we discuss recent information on functions and mechanisms of the ubiquitin system. Since the selectivity of protein degradation is determined mainly at the stage of ligation to ubiquitin, special attention is focused on what we know, and would like to know, about the mode of action of ubiquitin-protein ligation systems and about signals in proteins recognized by these systems.

CONTENTS

425

INTRODUCTION

The past few years have witnessed a dramatic increase in our knowledge of the important functions of ubiquitin-mediated protein degradation in basic biological processes. The selective and programmed degradation of cell-cycle regulatory proteins, such as cyclins, inhibitors of cyclin-dependent kinases, and anaphase inhibitors are essential events in cell-cycle progression. Cell growth and proliferation are further controlled by ubiquitin-mediated degradation of tumor suppressors, protooncogenes, and components of signal transduction systems. The rapid degradation of numerous transcriptional regulators is involved in a variety of signal transduction processes and responses to environmental cues. The ubiquitin system is clearly involved in endocytosis and down-regulation of receptors and transporters, as well as in the degradation of resident or abnormal proteins in the endoplasmic reticulum. There are strong indications for roles of the ubiquitin system in development and apoptosis, although the target proteins involved in these cases have not been identified. Dysfunction in several ubiquitin-mediated processes causes pathological conditions, including malignant transformation.

The role of ubiquitin in protein degradation was discovered and the main enzymatic reactions of this system elucidated in biochemical studies in a cell-free system from reticulocytes (reviewed in 1). In this system, proteins are targeted for degradation by covalent ligation to ubiquitin, a 76-amino-acid-residue protein. The biochemical steps in the ubiquitin pathway have been reviewed previously (2, 3) and are illustrated in Figure 1A. Briefly, ubiquitin-protein ligation requires the sequential action of three enzymes. The C-terminal Gly residue of ubiquitin is activated in an ATP-requiring step by a specific activating enzyme, E1 (Step 1). This step consists of an intermediate formation of ubiquitin adenylate, with the release of PP_i, followed by the binding of ubiquitin to a Cys residue of E1 in a thiolester linkage, with the release of AMP. Activated ubiquitin is next transferred to an active site Cys residue of a ubiquitin-carrier protein, E2 (Step 2). In the third step catalyzed by a ubiquitin-protein ligase or E3 enzyme, ubiquitin is linked by its C-terminus in an amide isopeptide linkage to an ε-amino group of the substrate protein's Lys residues (Figure 1A, Step 3).

Usually there is a single E1, but there are many species of E2s and multiple families of E3s or E3 multiprotein complexes (see below). Specific E3s appear to be responsible mainly for the selectivity of ubiquitin-protein ligation (and, thus, of protein degradation). They do so by binding specific protein substrates that contain specific recognition signals. In some cases, binding of the

A.

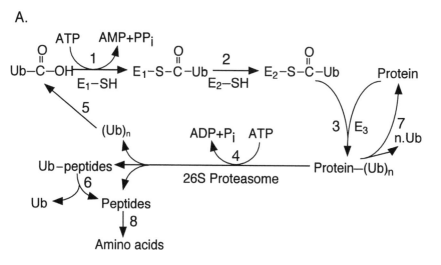

B.

1.

$$E_3-S-\overset{\overset{O}{\|}}{C}-Ub \longrightarrow E_3-SH + Protein-\overset{\overset{H}{|}}{N}-\overset{\overset{O}{\|}}{C}-Ub$$

Protein–NH$_2$

2.

$$E_3 \cdot E_2-S-\overset{\overset{O}{\|}}{C}-Ub \longrightarrow E_3+E_2-SH + Protein-\overset{\overset{H}{|}}{N}-\overset{\overset{O}{\|}}{C}-Ub$$

Protein–NH$_2$

Figure 1 Enzymatic reactions of the ubiquitin system. *A.* Sequence of reactions in the proteolytic pathway. *B.* Possible mechanisms of ubiquitin transfer by different types of E3 enzymes. See the text. *Ub*, ubiquitin.

substrate protein to an E3 is indirect, via an adaptor protein. Different types of E3s may carry out the transfer of ubiquitin to the substrate protein by two different mechanisms. In some cases, such as with the *Hect*-domain family of E3 enzymes (see below), ubiquitin is first transferred from an appropriate E2 to an active site Cys residue of the E3 enzyme. This E3-ubiquitin thiolester is the donor for amide bond formation with the protein substrate (Figure 1*B*.1). In other families of E3 enzymes, E3-ubiquitin thiolester formation cannot be demonstrated. Since E3 enzymes bind cognate E2s tightly (see below) and they also bind their appropriate protein substrate, ubiquitin can be transferred directly from E2 to the protein substrate (Figure 1B.2). After the linkage of ubiquitin to the substrate protein, a polyubiquitin chain is usually formed, in which the C-terminus of each ubiquitin unit is linked to a specific Lys residue (most commonly Lys[48]) of the previous ubiquitin.

Proteins ligated to polyubiquitin chains are usually degraded by the 26S proteasome complex (reviewed in 4) that requires ATP hydrolysis for its action. The 26S proteasome is formed by an ATP-dependent assembly of a 20S proteasome, a complex that contains the protease catalytic sites, with 19S "cap" or regulatory complexes (5). The 19S complexes contain several ATPase subunits and other subunits that are presumably involved in the specific action of the 26S proteasome on ubiquitinylated proteins. The roles of ATP in the assembly of the 26S proteasome complex and in its proteolytic action are not understood. The action of the 26S proteasome presumably generates several types of products: free peptides, short peptides still linked to ubiquitin via their Lys residues, and polyubiquitin chains (Figure 1A, Step 4). The latter two products are converted to free and reusable ubiquitin by the action of ubiquitin-C-terminal hydrolases or isopeptidases (Steps 5 and 6). Some isopeptidases may also disassemble certain ubiquitin-protein conjugates (Step 7) and thus prevent their proteolysis by the 26S proteasome (see below). The latter type of isopeptidase action may have a correction function to salvage incorrectly ubiquitinylated proteins or may have a regulatory role. Short peptides formed by the above processes can be further degraded to free amino acids by cytosolic peptidases (Figure 1A, Step 8).

In the five years since our last review on the ubiquitin system in this series (2), there has been an exponential increase of information on the subject. The reader is referred to reviews on the 20S and 26S proteasomes, including their subunit composition and crystal structure of the 20S proteasome from the *Thermoplasma* archaebacterium (4, 5). Ubiquitin-C-terminal hydrolases and isopeptidases are described elsewhere (6, 7). Hochstrasser's review (7) also provides a catalog of the known components of the ubiquitin system in the yeast *Saccharomyces cerevisiae*. This review discusses these subjects only briefly, focusing instead on selected examples that illustrate the mode of action and basic functions of the ubiquitin system.

ENZYMES OF UBIQUITIN-PROTEIN LIGATION

Ubiquitin Carrier Proteins (E2s)

A large number of E2s (also called Ubiquitin-conjugating enzymes or Ubcs) have been identified. In the relatively small genome of *S. cerevisiae* 13 genes encode E2-like proteins (7), so more are likely to be found in higher eukaryotes. Some E2s have overlapping functions, whereas others have more specific roles. For example, in *S. cerevisiae*, Ubc2/Rad6 is required for DNA repair and proteolysis of so-called N-end rule substrates, Ubc3/Cdc34 is required for the G1 to S-phase transition in the cell cycle, and Ubc4 and Ubc5 are needed for the degradation of many abnormal and short-lived normal proteins (reviewed in 7, 8). Specific functions of some E2s in higher organisms have been reported. For example, *Drosophila* UbcD1[1] is needed for proper detachment of telomeres in mitosis and meiosis (9). Some mutant alleles of UbcD1 cause abnormal attachment between telomeres of sister chromatids or fusion of chromosomes through their telomere ends. UbcD1-dependent degradation of some telomere-associated proteins may be required for telomere detachment (9). Another interesting example of a specific lesion caused by a mutation in an E2 enzyme is that of the *Drosophila bendless* gene, which is required for the establishment of synaptic connectivity in development (10; see below). The inactivation of HRB6B, one of the two mouse homologs of the yeast Ubc2/Rad6 E2 enzyme, causes male sterility due to decreased spermatogenesis (11; see below). Disruption of the gene of UbcM4, a mouse E2 homologous to yeast Ubc4/Ubc5, causes embryonic lethality possibly owing to impairment of the placenta's development (12).

Because of the specific effects of mutations in some E2 genes, it was proposed that E2s may participate in the recognition of the protein substrate, either directly or in combination with an E3 enzyme (7, 8). However, not much experimental evidence exists for the direct binding of E2s to protein substrates, with the notable exception of the interactions of E2-like Ubc9 with many proteins (see below) and that of E2-25 kDa with Huntingtin, the product of the gene affected in Huntington's disease (13). Specific functions of some E2s may be the result of their association with specific E3s, which in turn bind their specific protein substrates. For example, E2-14 kDa and its yeast homolog Ubc2/Rad6 specifically bind to E3α (14) or to its yeast counterpart Ubr1p (15). Ubc2/Rad6 also binds strongly to Rad18, a yeast DNA-binding protein involved in DNA repair (16). The biochemical function of Rad18 is not known, but it may be part of an E3 complex that directs it to the site of DNA repair. The Ubc3/Cdc34 E2 protein in the budding yeast specifically associates with Cdc53p and Cdc4p, which are involved in the degradation of cell-cycle regulators necessary for the G1 to-S-phase transition (17, 18; see below).

Another specific E2 involved in cell-cycle regulation is E2-C, which was first observed as a novel E2 required for the ubiquitinylation of cyclin B in a

---

[1] According to the currently used nomenclature, the different E2s/Ubcs are numbered according to the chronological order of their discovery in each organism.

reconstituted system from clam oocytes (19). E2-C acts in concert with the cyclosome/APC, a large complex that has cell-cycle–regulated ubiquitin ligase activity specific for mitotic cyclins and some other cell-cycle regulators that contain the so-called destruction box degradation signal (20; see next section). E2-C from clam has a 30-amino-acid N-terminal extension and several unique internal sequences (21). Homologs of E2-C were found in *Xenopus* (22), human (23), and fission yeast (24). Expression of a dominant-negative derivative of human E2-C arrests cells in mitosis (23), as is the case with a temperature-sensitive mutant of the fission yeast homolog (24), suggesting the conservation of its cell-cycle function in evolution. However, there is no homolog of E2-C in the budding yeast, even though the subunits of the cyclosome/APC are strongly conserved in this organism (see below). The budding yeast cyclosome may act with a nonspecific E2. In a cell-free system from *Xenopus* eggs (but not in that from clam oocytes), E2-C can be replaced by the nonspecific E2 Ubc4 (25). Though the interaction of the cyclosome with E2-C has not been defined, this interaction may be less stringent in some species than in others.

Stringency of E2-E3 interactions depends not only on species but also, or mainly, on the identity of the E2 and E3 enzymes. Some E2s (for example, Ubc4) can act with more than one E3 enzyme, and some E3s can act with several E2s. For example, the ubiquitinylation of proteins by the E6-AP E3 enzyme (see next section) can be supported by UbcH5, a human homolog of yeast Ubc4 (26, 27), as well as by the closely related UbcH5B and UbcH5C (28) and the less related UbcH7 (29) (previously described as E2-F1; see 30) or UbcH8 (31). By using a yeast two-hybrid assay, researchers were able to detect interaction of E6-AP with UbcH7 and UbcH8 but not with UbcH5 (31). In contrast, UbcH5B interacts with E6-AP in an in vitro binding assay (32). These E2s may bind to E6-AP with different affinities, in which case, the strength of the binding would determine whether the association could be detected by a certain assay.

A mysterious case of an E2-like protein, Ubc9, has been solved recently. Ubc9 was originally described as an essential yeast protein required for cell-cycle progression at the G2- or early M-phase and for the degradation of B-type cyclins (33). It was proposed that the proteolytic pathway that degrades B-type cyclins involves Ubc9 (33); however, subsequent work showed that the conjugation of cyclin B to ubiquitin in a cell-free system from *Xenopus* eggs could not be supported by a *Xenopus* homolog of Ubc9 (25). Furthermore, no formation of thiolester of ubiquitin with Ubc9 could be observed following incubation with E1 and ATP (T Hadari & A Hershko, unpublished results), and the crystal structure of mammalian Ubc9 showed significant differences in the region of the active site as compared to other E2s (34).

Still, Ubc9 has important functions, as indicated by its strong conservation in many eukaryotes (see 34 and references therein). Ubc9 was identified as an interacting protein in yeast two-hybrid searches with a surprisingly large number of proteins, including Rad51 (35) and Rad52 (36) human recombination proteins, a negative regulatory domain of the Wilms' tumor suppressor gene product (37), subunits of the CBF-3 DNA-binding complex of the yeast centromere (38), papillomavirus E1 replication protein (39), adenovirus-transforming E1A protein (40), poly (ADP-ribose) polymerase (41), transcription regulatory E2A proteins (42), the Fas (CD95) receptor of the tumor necrosis family (43, 44), and the RanBP2/RanGAP1 complex of proteins required for the action of Ran GTPase in nuclear transport (45). This last observation provided a clue to the function of Ubc9, owing to the recent discovery of the covalent modification of RanGAP1 with a small ubiquitin-like protein (46, 47). This protein has been termed UBL1 (36), sentrin (48), and SUMO-1 (47). We use the term UBL1.

The covalent ligation of UBL1 to RanGAP1 is required for its association with RanBP2, which appears to be important for the localization of the GTPase activator at the nuclear pore complex (46, 47). It was observed that a thiolester is formed between UBL1 and Ubc9, following incubation with a crude extract and ATP (M Dasso, personal communication). The reaction is analogous to the charging of E2s with activated ubiquitin and presumably involves an E1-like UBL1-activating enzyme provided by the extract. It thus appears that Ubc9 is an E2-like enzyme specific for the ligation of UBL1 to proteins. Since nuclear transport is essential for cell-cycle progression and for the degradation of mitotic cyclins (49), it was suggested that Ubc9 affects cyclin degradation indirectly, by modifying the function of RanGAP1 by ligation to UBL1 (45).

The discovery of the function of Ubc9 illustrates the importance of combining biochemical work with molecular genetic studies. It remains to be seen which other proteins are modified by ligation to UBL1 and whether at least some of the many proteins that interact with Ubc9 are also substrates for ligation to UBL1. In this system, Ubc9 may bind directly to the proteins ligated to UBL1; however, since most of the interactions of Ubc9 with various proteins were not studied with purified preparations, some of these interactions may be mediated by other proteins such as E3-like enzymes. Nonspecific interactions of some proteins with a positively charged surface of Ubc9 (34) are also possible. In vitro studies on the ligation of UBL1 to specific proteins, using purified proteins and enzymes, should resolve these questions.

Ubiquitin-Protein Ligases (E3s)

Though ubiquitin-protein ligases have centrally important roles in determining the selectivity of ubiquitin-mediated protein degradation, our knowledge of these enzymes remains limited. The difficulty in identifying new E3 enzymes is due, in part, to the lack of sequence homologies between different types of E3s, except for sequence similarities between members of the same E3 family.

In addition, some E3s are associated with large multisubunit complexes, and it is unclear which subunits of these complexes are responsible for their ubiquitin-protein ligase activities. There is even some confusion in the literature about the properties that define an E3 enzyme. This confusion has resulted from the variety of mechanisms by which different types of E3s promote ubiquitin-protein ligation. In some cases, the protein substrate is bound directly to an E3, while in others the substrate is bound to the ligase via an adaptor molecule (see below). The mechanisms of the transfer of activated ubiquitin from a thiolester intermediate to the amino group of a protein appear to differ in various types of E3s. In some cases, E3 accepts the activated ubiquitin from an E2 and binds it as a thiolester intermediate prior to transfer to protein, while in others a ligase may help to transfer ubiquitin directly from E2 to a protein, by tight binding of E2 and the protein substrate (Figure 1*B*). The first E3 discovered, E3α, was originally defined operationally as a third enzyme component required, in addition to E1 and E2, for the ligation of ubiquitin to some specific proteins (50). We can now replace this operational definition by a more mechanistic but broad definition. We define E3 as an enzyme that binds, directly or indirectly, specific protein substrates and promotes the transfer of ubiquitin, directly or indirectly, from a thiolester intermediate to amide linkages with proteins or polyubiquitin chains.

According to this definition, four types of ubiquitin-protein ligases are known (Figure 2). The main N-end rule E3, E3α (and its yeast counterpart, Ubr1p), is still among the best-characterized ubiquitin ligases (reviewed in 2). It is an approximately 200-kDa protein that binds N-end rule protein substrates that have basic (Type I) or bulky-hydrophobic (Type II) N-terminal amino acid residues to separate binding sites specific for such residues (Figure 2*A*). Some protein substrates that do not have N-end rule N-terminal amino acid residues, such as unfolded proteins and some N-α-acetylated proteins (51), bind to this enzyme at a putative "body" site that has not been well characterized. E3α also binds a specific E2 [E2-14 kDa (14) or its yeast homolog, Ubc2p/Rad6 (15)], thus facilitating the transfer of activated ubiquitin from E2 to the substrate protein. Thus E3α is responsible for the recognition of some N-end rule protein substrates for ubiquitin ligation and degradation. A related enzyme appears to be E3β, which has been only partially purified and characterized, and which may be specific for proteins with small and uncharged N-terminal amino acid residues (52). Though the N-end rule recognition mechanism is strongly conserved in eukaryotic evolution, its main physiological functions and substrates are still not known (see below).

A second major family of E3 enzymes is the *hect* (homologous to E6-AP C-terminus) domain family. The first member of this family, E6-AP (E6-associated protein), was discovered as a 100-kDa cellular protein that was

A. N-end rule E3 (E3α)

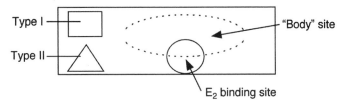

Type I — "Body" site

Type II —

E$_2$ binding site

B. Hect-domain E3 (Rsp5p)

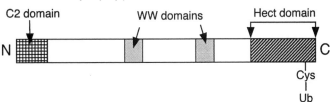

C2 domain   WW domains   Hect domain

N   C

Cys
|
Ub

C. Cyclosome/APC

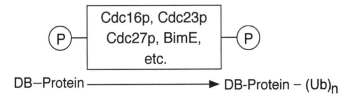

(P) Cdc16p, Cdc23p
Cdc27p, BimE,
etc. (P)

DB–Protein ⟶ DB-Protein – (Ub)$_n$

D. Phosphoprotein-ubiquitin ligase complexes

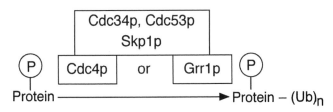

(P) Cdc34p, Cdc53p
Skp1p

Cdc4p   or   Grr1p (P)

Protein ⟶ Protein – (Ub)$_n$

Figure 2   Different types of E3 enzymes or E3 enzyme complexes. See the text. *DB*, destruction box.

required, together with papillomavirus E6 oncoprotein, for the ubiquitinylation and degradation of p53 in reticulocyte lysates (53). In contrast to E3α, E6-AP does not bind directly to p53 but rather binds indirectly via E6, which binds to both p53 and E6-AP. In other cases, however, E6-AP can promote the transfer of ubiquitin to some cellular proteins in the absence of E6. The action of E6-AP involves an intermediary ubiquitin transfer reaction, in which activated ubiquitin is transferred from an appropriate E2 to form a thiolester with a specific Cys residue near its N-terminus (54). This thiolester is apparently the donor of ubiquitin for amide linkage with the protein substrate, because mutation of this Cys residue of E6-AP abolishes its activity in protein ubiquitinylation.

A large family of proteins that contain an approximately 350-amino-acid C-terminal region homologous to that of E6-AP, the *hect*-domain family, has been identified in many eukaryotic organisms (55; see Figure 2*B*). All *hect* proteins contain a conserved active site Cys residue near the C-terminus. In contrast to the conservation of the C-terminal domain, the N-terminal regions of the different *hect* proteins are highly variable. The N-terminal domains may be involved in the recognition of specific protein substrates (55); this has been proven in some cases (see below). Most *hect*-domain proteins are likely E3 enzymes or parts of multiprotein complexes that contain E3-like activities. At present, only fragmentary information exists about possible functions of some *hect* proteins. Some cases of Angelman syndrome, a human hereditary disease characterized by mental retardation and seizures, are due to mutations in the E6-AP gene (56, 57). This observation suggests that E6-AP-mediated protein ubiquitinylation is required for brain development (see below). More specific functions were identified for Rsp5p, one of the five *hect* proteins of the yeast *S. cerevisiae*. Rsp5p specifically binds and ubiquitinylates in vitro several yeast cellular proteins, including the large subunit of RNA polymerase II (58). The N-terminal domain of Rsp5p binds the polymerase subunit while the C-terminal (*hect*) domain does not bind, suggesting the role of the N-terminal domain in substrate binding. The relevance of these in vitro findings to similar processes occurring in vivo was suggested by the finding that inhibition of the expression of Rsp5p caused a fivefold increase in the steady-state levels of the RNA polymerase subunit (58). This subunit is usually a long-lived protein, so it is possible that it is degraded rapidly only under special conditions. In mammalian cells, the large subunit of RNA polymerase is ubiquitinylated following DNA damage induced by UV irradiation or cisplatin treatment (59) and is degraded by a proteasome-mediated process (DB Bregman, personal communication).

Two interesting problems are (*a*) How does DNA damage expose the RNA polymerase subunit to the action of the ubiquitin ligase? and (*b*) Does this process plays a role in DNA repair? Pub1, a close homolog of Rsp5p found in fission yeast, is involved in the degradation of a different protein, the Cdc25 phosphatase (60). This phosphatase activates protein kinase Cdk1 by the removal of an inhibitory phosphate group from a tyrosine residue and thus plays an essential role in the entry of cells into mitosis. The levels of Cdc25 oscillate in the cell cycle (20). The degradation of Cdc25 is apparently mediated by the Pub1 ubiquitin ligase, as indicated by observations that disruption of *pub1* markedly increases Cdc25 levels, and *pub1* interacts with genes that control Cdc25 function. In addition, by using a mutant defective in a subunit of the 26S proteasome, researchers showed that ubiquitin conjugates of Cdc25 accumulate in pub+ but not in *pub1*-deleted cells (60).

Although in the above cases ubiquitin ligation by Rsp5p/Pub1 is apparently followed by proteasome-mediated degradation, in other instances ubiquitin ligation by the same E3 protein is involved in endocytosis (see below). Thus the general amino acid permease of the budding yeast, GAP1p, is rapidly inactivated and degraded by the addition of NH_4^+ ions. The *NPI1* gene, which is required for this process, is similar to that of the Rsp5 ubiquitin ligase (61). Rsp5p is also required for the degradation (61) and ubiquitinylation (62) of the Fur4p uracil permease. The degradation of these permeases is the result of endocytosis into the vacuole, as indicated by the finding that their degradation was inhibited in mutants of vacuolar proteases but not in mutants of proteasome subunits (see below). The action of Rsp5p on these membrane proteins may be mediated by a calcium-lipid-binding domain (CaLB/C2), which is located near the N-terminus of Rsp5p and of its homologs from other organisms (see 61 and references therein).

Another motif found in the N-terminal region of Rsp5p and of its homologs is the WW domain, an approximately 30-amino-acid region thought to be involved in interactions with proline-rich sequences containing an XPPXY (or PY) motif (see 63 and references therein). Several WW domains exist in yeast Rsp5p (see Figure 2B) and in Nedd4, its mammalian homolog. The rat Nedd4 was isolated as a protein that interacts with subunits of an epithelial sodium channel (63). The C-terminal tails of these channel subunits contain PY motifs. Deletion of these C-terminal tails in a human hereditary disease called Liddle's syndrome causes hypertension owing to hyperactivation of the sodium channel. Using yeast two-hybrid and in vitro binding assays, researchers showed that Nedd4 binds through its WW domains to the PY motifs of the sodium channel's subunits. It was suggested (63) and subsequently demonstrated (64) that Nedd4 suppresses the epithelial sodium channel by its ubiquitin-mediated degradation. Though Rsp5p and its homologs can directly bind at least some protein substrates, it is unknown whether they act in a monomeric form or in multiprotein complexes. In yeast, Rsp5p is associated with a protein designated Bul1, which is not a substrate for degradation (65). A part of Rsp5p molecules is associated with Bul1 in a high-molecular-weight complex, and Bul1 may be a modulator of the Rsp5p ubiquitin ligase (65).

A high-molecular-weight complex, called the cyclosome (66) or anaphase promoting complex (APC) (25), has a ubiquitin ligase activity specific for cell-cycle regulatory proteins that contain a nine-amino-acid degenerate motif called the destruction box (Figure 2C; see also below). Its substrates are mitotic cyclins, some anaphase inhibitors, and spindle-associated proteins, all of which are degraded at the end of mitosis (see below). The cyclosome/APC was discovered by biochemical studies in early embryonic cell-free systems that reproduce cell-cycle-related processes. Fractionation of extracts of clam oocytes first showed that the system that ligates cyclin B to ubiquitin contained a particle-associated E3-like activity that was cell-cycle regulated. This complex was inactive in the interphase but became active at the end of mitosis, when cyclin B was degraded (19). It was dissociated from particles by extraction with high salt and was found to be an approximately 1,500-kDa complex containing destruction box–specific cyclin-ubiquitin ligase activity. The complex was named the cyclosome, to denote its large size and important roles in cell-cycle regulation (66). In the early embryonic cell cycles, the cyclosome is converted to the active form by phosphorylation (66, 67; see also below). A similar complex, called the APC, was purified from *Xenopus* eggs by immunoprecipitation (25, 68). The *Xenopus* complex has eight subunits, three of which are homologous to *S. cerevisiae* Cdc16, Cdc23, and Cdc27 proteins, which are required for exit from mitosis and for the degradation of B-type cyclins in yeasts (69). These three cyclosome subunits contain tetratrico-peptide motifs, proposed to be involved in protein-protein interactions (70). A fourth subunit is homologous to *Aspergillus nidulans* BimE protein, essential for the completion of mitosis (68, 71). These four cyclosome subunits are strongly conserved in evolution, from yeast to humans (reviewed in 72). Partial sequences obtained from four other subunits of the *Xenopus* cyclosome/APC are not homologous to proteins with known functions (68). The subunits of the cyclosome involved in its ubiquitin ligase functions, such as those responsible for specific binding to destruction box–containing substrates, and of its E2 partner, E2-C (21), have yet to be identified. Other aspects of cyclosome/APC involvement in the degradation of different cell-cycle regulators, and of the control of its activity in the cell cycle, are described in a subsequent section.

A different type of multisubunit ubiquitin ligase is involved in the degradation of some other cell-cycle regulators, such as the Sic1p Cdk inhibitor or the G1 cyclin Cln2p. In these cases, phosphorylation of the substrate converts it to a form susceptible to the action of the ubiquitin ligase complex. We designate these complexes phosphoprotein-ubiquitin ligase complexes (PULCs) (Figure 2D).

It appears that different PULCs exist, although present information is incomplete. These PULCs share some common components but may also have other components specific for certain protein substrates. Thus the degradation of the Cdk inhibitor Sic1p, a process essential for the G → S transition in the budding yeast, requires its phosphorylation by a G1 cyclin–activated protein kinase (73) as well as the products of *CDC34, CDC53, CDC4* (74), and *SKP1* (75). Cdc34p is an E2 protein (8), but the other gene products do not resemble proteins with known functions. Some of these components are required for the ligation of ubiquitin to Sic1p in vitro (76). Cdc34p, Cdc53p, and Cdc4p are physically associated, as indicated by their co-purification from yeast lysates (17, 18). It thus appears that a complex containing the above-mentioned components may be responsible for the ubiquitinylation of phosphorylated Sic1.

The ubiquitinylation and degradation of the yeast G1 cyclin Cln2 also requires its phosphorylation and the actions of Cdc34p, Cdc53p (17), and Skp1p (75). However, Cdc4p is not required for the degradation of G1 cyclins (S Sadis & D Finley, personal communication). Instead, the product of the *GRR1* gene is required for the degradation of the yeast G1 cyclins Cln1p and Cln2p (77). Both Cdc4p and Grr1p contain a motif called the F-box, which is present in a variety of proteins that bind to Skp1p (75). It was proposed that Skp1p is a component of ubiquitin-protein ligase complexes that connect them to specific "adaptor" proteins, such as Cdc4p and Grr1p, which would in turn bind their specific protein substrates, such as phosphorylated Sic1p and Cln2p, respectively (75). Figure 2D shows this model for the mode of action of different PULCs, which still has to be examined. It also remains to be seen what other specific features in protein substrates (in addition to the phosphorylated residues) are recognized by the different PULC complexes.

While the above-described information on phosphoprotein-ubiquitin ligase complexes is based on studies in yeast, it seems that at least some components of these machineries are conserved in evolution. Numerous homologs of yeast Cdc53, called cullins, were found in many eukaryotes. One of these, Cul-1, is a negative regulator of cell proliferation in *Caenorhabditis elegans* (78). A human Cdc53 homolog, Cul-2, binds to the von Hippel-Lindau tumor suppressor (79). Skp1 is also strongly conserved, and close homologs were found in many eukaryotic organisms (75). These findings suggest that similar ubiquitin ligase complexes may be involved in the degradation of a variety of regulators in higher organisms.

In addition to the four types of ubiquitin-protein ligases described above, several other E3s have been partially characterized. An approximately 550-kDa E3, designated E3L, was partially purified from rabbit reticulocytes (80). It acts on some non-N-end rule substrates, such as actin, troponin T, and MyoD. The physiological substrates of this enzyme and the signals it recognizes are unknown. A 280-kDa E3, which ligates ubiquitin to c-fos, was purified approximately 350-fold from Fraction 2 of reticulocytes (81). An approximately 140-kDa protein was tentatively identified as a subunit of this enzyme, but the preparation was not homogenous. The formation of a thiolester between ubiquitin and the putative E3 subunit was demonstrated (81). The cloning of this E3 is necessary to examine whether it is a novel member of the *hect* family of E3 enzymes. An approximately 320-kDa E3 from reticulocytes promotes the ligation of ubiquitin to the p105 precursor of NF-κB (82).

Much remains to be learned about the identity, specificity, and regulation of E3 enzymes or E3 complexes. The lack of sequence similarity between the different types of E3 enzymes necessitates the identification of new types of E3s by biochemical methods. Because of the variety of mechanisms by which E3 enzymes carry out their two basic functions of protein substrate recognition and ubiquitin transfer, these mechanisms have to be characterized for each type of E3 enzyme. Because of these variable mechanisms, different families of E3s, specific for the recognition of different classes of protein substrates, may have evolved that do not have many features in common. The only similarity between various E3s may be the binding of E2s, but since different E3s bind different E2s, it may not be easy to recognize similarities in the various E2-binding sites. One of the major challenges in the ubiquitin field is the identification and elucidation of the mode of action of different E3s that recognize specific signals in cellular proteins.

SIGNALS IN PROTEINS FOR UBIQUITINYLATION AND DEGRADATION

Our knowledge of different signals in proteins that mark them for ubiquitinylation is also limited. Recent results indicate that many proteins are targeted for degradation by phosphorylation. It was observed previously that many rapidly degraded proteins contain PEST elements, regions enriched in Pro, Glu, Ser, and Thr residues (83, 84). More recently, it was pointed out that PEST elements are rich in S/TP sequences, which are minimum consensus phosphorylation sites for Cdks and some other protein kinases (85). Indeed, it now appears that in several (though certainly not all) instances, PEST elements contain phosphorylation sites necessary for degradation. Thus multiple phosphorylations within PEST elements are required for the ubiquitinylation and degradation of the yeast G1 cyclins Cln3 (85) and Cln2 (86), as well as the Gcn4 transcriptional activator (87). Other proteins, such as the mammalian G1 regulators cyclin E (88) and cyclin D1 (89), are targeted for ubiquitinylation by phosphorylation at specific, single sites. In the case of the IκBα inhibitor of the NF-κB transcriptional regulator, phosphorylation at two specific sites, Ser32 and Ser36, is required for ubiquitin ligation (see below). β-Catenin, which is targeted for ubiquitin-mediated degradation by phosphorylation (see below), has a sequence motif similar to that of IκBα around these phosphorylation sites (90). However, the homology in phosphorylation patterns of these two proteins is not complete, because phosphorylation of other sites of β-catenin is also required for its degradation (90; see below).

Other proteins targeted for degradation by phosphorylation include the Cdk inhibitor Sic1p (73) and the STAT1 transcription factor (91). Though different patterns of phosphorylation target different proteins for degradation, a common feature appears to be that the initial regulatory event is carried out by a protein kinase, while the role of a ubiquitin ligase would be to recognize the phosphorylated form of the protein substrate. It further appears that different ubiquitin ligases recognize different phosphorylation patterns as well as additional motifs in the various protein substrates. However, the identity of such E3s is unknown, except for some PULC-type ubiquitin ligases that act on some phosphorylated cell-cycle regulators in the budding yeast (see previous section). The multiplicity of signals that target proteins for ubiquitin-mediated degradation (and of ligases that have to recognize such signals) is underscored by observations that the phosphorylation of some proteins actually prevents their degradation. Thus the phosphorylation of the c-Mos protooncogene on Ser3 (92) and the multiple phosphorylations of c-Fos (93) and c-Jun (94) protooncogenes at multiple sites by MAP kinases suppress their ubiquitinylation and degradation (see also below).

Among degradation signals inherent in primary protein structure, the best characterized is still the N-end rule system, in which the ubiquitinylation and degradation of a protein is determined by the nature of its N-terminal amino acid residue (reviewed in 95). However, there are few known physiological protein substrates of this system, presumably because of the specificity of methionine aminopeptidases that do not remove initiating Met residues from nascent proteins when the second amino acid residue is an N-end rule destabilizing residue (96). An important function of this pathway may be to remove from the cytosol erroneously transported or compartmentalized proteins, in which a destabilizing N-terminal residue is produced in cleavage by a signal peptidase. Because the few known physiological substrates of the N-end rule system, such as the Gα subunit of G-protein (97) or the CUP9 transcriptional repressor of peptide import in yeast (95), do not have destabilizing N-terminal residues, they are presumably recognized by some other internal signal.

A signal important for the degradation of mitotic cyclins and certain other cell-cycle regulators is the destruction box. It was first discovered as a partially conserved, 9-amino-acid sequence motif usually located approximately 40–50 amino acid residues from the N-terminus of mitotic cyclins and is necessary for their ubiquitinylation and degradation in extracts of *Xenopus* eggs (98). Compilation of destruction box sequences from nearly 40 B-type and A-type cyclins from various organisms (99) showed that they have the following general structure:

R	(A/T)	(A)	L	(G)	x	(I/V)	(G/T)	(N)
1	2	3	4	5	6	7	8	9

Amino acid residues, or combinations of two residues, that appear in parentheses in the above structure occur in more than 50% of known destruction sequences. Thus the only invariable residues are R and L in positions 1 and 4, respectively; the rest of the destruction box sequence is quite degenerate.

Still, the destruction box signal is absolutely necessary for the ubiquitinylation and degradation of mitotic cyclins in vitro (66, 100) and in vivo (see 99 and references therein), as shown by the prevention of these processes by deletion of the destruction box region or by point mutations in its conserved residues. Moreover, the destruction box–containing N-terminal fragments of mitotic cyclins act as transferable signals in vitro and in vivo, as indicated by the cell-cycle-stage-specific degradation of reporter proteins fused to such fragments (98, 100, 101). Similar destruction box motifs are required for the degradation of certain non-cyclin cell-cycle regulators that are degraded at late mitosis, such as anaphase inhibitors and the spindle-associated protein Ase1p (see below).

All presently known destruction box–containing cell-cycle regulators are ligated to ubiquitin by the cyclosome/APC and are degraded after the conversion of the cyclosome to the active form at late mitosis. However, some destruction box–containing proteins are degraded at slightly different times during the cell cycle, indicating additional levels of regulation (see below). Thus the cyclosome-mediated ubiquitinylation of destruction box–containing proteins may be an example (thus far, unique) of a strategy by which a limited set of proteins that perform related functions and share a common degradation signal are substrates for a common ubiquitin ligase. Some proteins that are not related to cell-cycle regulation, such as the budding yeast Ras exchange factor Cdc25p (102) and uracil permease (61), have been reported to be degraded in a destruction box–related manner, but it is unknown whether cyclosome action is involved in these cases.

Much less is known about signals or domains recognized for degradation in other cellular proteins. Truncations or deletions of several rapidly degraded proteins cause their stabilization (eg, see 103, 104), but since it has not been shown that these regions contain transferable degradation signals, stabilization may be the result of secondary effects on protein structure. An exception is the case of c-Jun, in which the δ-domain of a sequence of 27 amino acid residues near the N-terminus is a transferable ubiquitinylation signal (105; see also below).

DEGRADATION OF UBIQUITIN-PROTEIN CONJUGATES

The 20S and 26S Proteasome Complexes

The structure and function of the 20S and 26S proteasome complexes have been reviewed elsewhere (see 4, 6, 7, 106–109). In this section, we update the reader on some important recent developments.

Important progress has been made in the resolution at 2.4 Å of the crystal structure of the eukaryotic (yeast) 20S proteasome (110). This study corroborated previous observations on the structure of the complex from the archeabacterium *Thermoplasma acidophilum* but also revealed some unexpected features. Like the *T. acidophilum* proteasome, the yeast complex is also arranged as a stack of four rings, each containing seven subunits, $\alpha_7\beta_7\beta_7\alpha_7$. The catalytic sites reside in some of the β rings. However, the composition of the eukaryotic 20S proteasome is more complicated than that of the archaeal complex. While each of the *T. acidophilum* proteasome rings is composed of identical subunits, seven identical α subunits for each of the two α rings and seven identical β subunits for each of the two β rings, the rings of the yeast enzyme are composed of seven distinct subunits. Thus, the 20S proteasome of yeast is composed of 14 pairs of protein subunits, 7 different α and 7 different β subunits organized as $\alpha_{1-7}\beta_{1-7}\beta_{1-7}\alpha_{1-7}$.

Resolution of the crystal structure enabled better understanding of the biogenesis of the different chains. Five β-type subunits are synthesized as proproteins with N-terminal extensions of up to 75 residues and are cleaved during proteasome maturation (reviewed in 4, 110). Three of the subunits (β1/PRE3, β2/PUP1, and β5/PRE2) undergo cleavage between the last Gly residue of the pro-peptide and Thr^1 of the mature subunit that also constitutes the catalytic site. The enzymes use the side chain of the Thr residue as a nucleophile in a catalytic attack at the carbonyl carbon. Activation of the side chain occurs by transfer of its proton to the free N-terminus. The Thr residue occupies an unusual fold (common also to other aminohydrolases such as glutamine PRPP amidotransferase, the penicillin acylase, and the aspartylglucosaminidase), which provides the capacity for both the nucleophilic attack and autocatalytic processing.

Several other adjacent preserved residues in β-type subunits (Gly^{-1}, Asp^{17}, Lys^{33}, Ser^{129}, Asp^{166}, and Ser^{169}) are also important for the structural integrity of the catalytic site (111–113). Topological analysis of the location of the different subunits has revealed that for the three distinct proteolytic activities—the trypsin-like, the chymotrypsin-like, and the post-glutamyl peptidyl hydrolytic (PGPH) activities—the active sites are generated by adjacent pairs of identical β-type subunits residing in different β rings. These findings have been corroborated independently by genetic analysis (114), as well as by immunoelectronmicroscopy and chemical cross-linking of neighboring sub-units (115). The crystal structure has also shown that the α chains, although catalytically inactive, play an essential role in stabilizing the two-ring structure of the β chains. They must also play a role in the binding of the 19S cap or regulatory complexes, but the structure of the contacts and mechanisms of bind-ing will be elucidated only when the structure of the 26S complex is resolved. The crystal structure has revealed a distance of 28 Å between the Thr[1] active sites of adjacent active β subunits. This distance may determine the length of the peptides generated during the proteolytic process ($\sim$8 amino acid residues) and may explain the role of the proteasome in generation of antigenic peptides presented on class I MHC molecules (109, 116, 117; see also below).

An unresolved problem involves the entry of protein substrates into, and exit of proteolysis products from, the proteasome. In the *T. acidophilum* pro-teasome, there are two putative entry pores of approximately 13 Å at the two ends of the cylinder surrounded by defined segments of the seven α subunits (118). In striking and rather surprising contrast, these pores do not exist in the yeast 20S proteasome, and entry to the inter-β rings catalytic chamber is not possible from the ends of the complex. The N-terminal domains of α1/C7, α2/Y7, α3/Y13, α6/PRE5, and α7/C1 protrude toward each other and fill the space in several layers of tightly interacting side chains (110). Thus, entry from the ends may be possible only after substantial rearrangement that can occur after association with the 19S regulatory complex. Such a rearrange-ment may also require energy that can be provided by the ATPase activity of the 19S regulatory complex. Also, unlike the *T. acidophilum* proteasome, the yeast complex displays some narrow side orifices, particularly at the interface between the α and β rings. These openings lead directly to the Thr[1] active sites. They are coated with polar residues that can potentially rearrange to gen-erate $\sim$10-Å apertures through which unfolded and extended protein substrates may enter.

Substrate recognition by the 26S proteasome is probably mediated by the interaction of specific subunits of the 19S regulatory complex with polyubiq-uitin chains. Indeed, such subunits have been described both in humans (S5a; see 119) and in plants (MBP1; see 120). These subunits bind at high-affinity polyubiquitin chains, in particular those that contain more than four moieties, but they also bind ubiquitin markers. The association of these subunits with the 19S complex and their preference for polyubiquitinylated tagged substrates suggests a crucial role for these subunits in ubiquitin-mediated protein degra-dation. *Mcb1*, the yeast gene encoding the homologous subunit was cloned recently. Surprisingly, $\Delta mcb1$ deletion mutants do not display any growth de-fect and degrade normally ubiquitinylated proteins, except for the fusion model protein ubiquitin-Pro-β-Gal. These mutants do display a slight sensitivity to stress, such as exposure to amino acid analogs (121). A possible explanation for these results is that ubiquitinylated proteins are recognized by additional, as-yet-undefined proteasomal subunits.

Specific inhibitors of the proteasome have proved to be important research tools, probing the structure and function of the proteasome and establishing the involvement of the ubiquitin-proteasome pathway in the degradation of specific proteins. The initial inhibitors were derivatives of the calpain inhibitors I [N-acetyl-Leu-Leu-norleucinal (ALLN)] and II [N-acetyl-Leu-Leu-methioninal (ALLM)]. These inhibitors block degradation of most cellular proteins, both short and long lived (122). They modify covalently and irreversibly the Thr[1] in the catalytically active β subunits. While they are quite specific toward the proteasome, at higher concentrations, they also inhibit calpains. By contrast, the *Streptomyces* metabolite lactacystin appears to be a specific inhibitor of the proteasome (123). It modifies covalently the active site Thr[1] residues and strongly inhibits the trypsin- and chymotrypsin-like activities of the complex and, less efficiently, the PGPH activity. A recently developed derivative of the calpain inhibitors, carboxybenzyl-Leu-Leu-Leu-vinyl sulfone (Z-L$_3$VS) inhibits efficiently and specifically all three activities of the proteasome (124). It is cell permeable and inhibits the activity of the complex in vivo as well. Although vinyl sulfone derivatives were described originally as cysteine protease inhibitors, like all other known inhibitors of the proteasome, these derivatives covalently modify the Thr[1] residues in active β subunits.

Ubiquitin-C-Terminal Hydrolases and Isopeptidases

The subject of ubiquitin-C-terminal hydrolases (UCHs) and ubiquitin-specific proteases (UBPs) (also called isopeptidases and de-ubiquitinating enzymes) is reviewed elsewhere (6, 7, 125), and we discuss here only some recent developments. Genes of 16 different UBPs are found in the yeast genome (7). The large number of hydrolases suggests that some of them may have specific functions, such as the recognition of different types of ubiquitin conjugates. Thus a family of low-molecular-mass (25- to 28-kDa) UCHs specifically act on adducts of ubiquitin with small molecules or peptides (126). The crystal structure of one of these, UCH-L3, has been solved at 1.8 Å resolution (127). The enzyme comprises a central antiparallel β-sheet flanked on both sides by α helices. The β-sheet and one of the helices are similar to those observed in the thiol protease cathepsin B. The similarity includes the three amino acid residues that comprise the active site, Cys[95], His[169], and Asp[184]. The active site appears to fit the binding of ubiquitin that may anchor also at an additional site. The catalytic site in the free enzyme is masked by two different segments of the molecule that limit nonspecific hydrolysis and must undergo conformational rearrangement after substrate binding.

Another hydrolase, isopeptidase T (IsoT), acts preferentially on free, unanchored polyubiquitin chains and stimulates protein breakdown by the disassembly of such chains that inhibit the action of the 26S proteasome (128). IsoT acts by a sequential *exo* mechanism, starting from the end of the polyubiquitin chain that contains a free C-terminus of ubiquitin (125). This free C-terminus can be exposed following the action of the 26S proteasome on the protein moiety of polyubiquitin-protein conjugates. A recent report (129) describes the characterization of Ubp14, the yeast homolog of IsoT. Like IsoT, Ubp14 is involved in disassembly of free, unanchored polyubiquitin chains. A Δ*Ubp14* mutant, as well as a yeast expressing a dominant-negative mutant form of the enzyme, display a lowered rate of general protein degradation accompanied by accumulation of free ubiquitin chains, probably bound to the proteasome. Unexpectedly, overexpression of the wild-type protein also results in inhibition of proteolysis of certain proteins. It is possible that certain substrates are tagged by direct transfer of polyubiquitin chains, and the low level of such chains resulting from overexpression of the wild-type enzyme leads to inhibition of their degradation. Complementation experiments have revealed that Ubp14 and IsoT are functional homologs, confirming in vivo the initial characterization of the enzyme carried out in a cell-free system using a model substrate.

The action of the UCHs and IsoT stimulates protein breakdown by the removal of inhibitory polyubiquitin chains and by the regeneration of free and reusable ubiquitin. In other cases, the action of an isopeptidase may inhibit protein breakdown. For example, a mutation in the *Drosophila faf facets* (*faf*) gene, which encodes an isopeptidase affecting eye development (see below) is suppressed by another mutation in a proteasome subunit (130). These results indicate that the *faf* isopeptidase stabilizes some unidentified proteins, which are also stabilized by the proteasome mutation. It is possible that certain isopeptidases can stabilize particular proteins by the removal of ubiquitin from conjugates that would be otherwise targeted for degradation by the 26S proteasome. An editing function for some isopeptidases was proposed a long time ago (131). Recently, Lam et al (132) reported that the 19S regulatory complex of the 26S proteasome contains a 37-kDa ubiquitin-aldehyde-sensitive but ATP-independent isopeptidase that removes single ubiquitin moieties from the distal end of short polyubiquitin chains. The authors proposed that this isopeptidase is involved in editing and in rescue of poorly ubiquitinated or slowly degraded proteins from degradation, which differs from the function of isopeptidase Doa4 (135) and the ATP-dependent but Ubal-insensitive isopeptidase (136) involved mostly in recycling of ubiquitin and maintenance of free ubiquitin levels in the cell.

Low concentrations of ubiquitin aldehyde, an inhibitor of some isopeptidases (133) stimulates the degradation of excess globin α-chains in reticulocytes of thalassemic patients (134). This observation suggests that ubiquitin conjugates of α-globin are disassembled by an isopeptidase that prevents its degradation by the 26S proteasome.

Different ubiquitin-C-terminal isopeptidases affect a variety of other basic processes, including development (see below), gene silencing (137), and long-term memory (138). In none of these cases were the target proteins identified, nor was the mode of action of the isopeptidase characterized in the degradation or stabilization of target proteins. Another interesting function of specific isopeptidases is the regulation of cell proliferation. It was observed that cytokines induced in T-cells specific de-ubiquitinating enzymes (DUBs), termed DUB-1 (139) and DUB-2 (140). DUB-1 is induced by stimulation of the cytokine receptors for IL-3, IL-5, and GM-CSF, suggesting a role in its induction for the β-common (betac) subunit of the interleukin receptors. Overexpression of a dominant negative mutant of JAK2 inhibits cytokine induction of DUB-1 (141), suggesting that the regulation of the enzyme is part of the cell response to the JAK/STAT signal transduction pathway. Continued expression of DUB-1 arrests cells at G_1; therefore, the enzyme appears to regulate cellular growth via control of the G_0-G_1 transition. The catalytic conserved Cys residue of the enzyme is required for its activity. DUB-2 is induced by IL-2 as an immediate early (IE) gene that is down-regulated shortly after the initiation of stimulation. The function of this enzyme is also obscure. It may stimulate or inhibit the degradation of a critical cell-cycle regulator (see below).

CELLULAR PROTEINS DEGRADED BY THE UBIQUITIN SYSTEM

Cell-Cycle Regulators

Progress in the eukaryotic cell-cycle is driven by oscillations in the activities of cyclin-dependent kinases (Cdks). Cdk activity is controlled by periodic synthesis and degradation of positive regulatory subunits, cyclins, as well as by fluctuations in levels of negative regulators, by Cdk inhibitors (Ckis), and by reversible phosphorylation (reviewed in 142). The different cyclins, specific for the G1, S-, or M-phases of the cell cycle, accumulate and activate Cdks at the appropriate times during the cell cycle and then are degraded, causing kinase inactivation. Levels of some Ckis, which specifically inhibit certain cyclin/Cdk complexes, also rise and fall at specific times during the cell cycle. Selective, ubiquitin-mediated degradation of cyclins, Ckis, and other cell-cycle regulators appear to play centrally important roles in cell-cycle control, as described below.

MITOTIC CYCLINS   Though all cyclins are degraded by ubiquitin-mediated processes, the systems that carry out their ligation to ubiquitin, and the mode by which these systems are connected to the cell-cycle regulatory phosphorylation network, are different for mitotic and G1 cyclins. Mitotic B-type cyclins and some S-phase cyclins such as cyclin A (66) are ligated to ubiquitin by the cyclosome, while G1 cyclins are ubiquitinylated by PULC-type E3 enzymes (see previous section). In the former case, the activity of the ligase is regulated in the cell cycle, whereas the latter process is triggered by the phosphorylation of the G1 cyclin substrate. The mitotic cyclin B was the first cyclin discovered, by its striking degradation at the end of each mitosis in early sea urchin embryos (143). Cyclin B combines with Cdk1 (also called Cdc2 or Cdc28 in the fission or budding yeasts, respectively) to form the major mitotic kinase MPF (M-phase promoting factor). MPF causes entry of cells into mitosis and, after a lag, activates the system that degrades its cyclin subunit (reviewed in 20). MPF inactivation, caused by the degradation of cyclin B, is required for exit from mitosis, as shown by observations that cells expressing nondegradable forms of cyclin B arrest in late anaphase (144).

Initial evidence indicating that the degradation of cyclin B is carried out by the ubiquitin system was based on correlations between the degradation of cyclin B and its ubiquitinylation in extracts of *Xenopus* eggs (98), and on the inhibition of cyclin degradation by methylated ubiquitin (an inhibitor of polyubiquitin chain formation; see 145) in extracts of clam oocytes (146). Fractionation of these extracts (19) led to the identification of the specific components of this system, the novel ubiquitin-carrier protein E2-C (21), and the cyclosome complex that has cyclin-ubiquitin ligase activity (66; see also previous section). E2-C is constitutively active, but the activity of the cyclosome is cell-cycle regulated: It is inactive in the interphase of the embryonic cell cycle and is converted to the active form at the end of mitosis by phosphorylation (19, 25, 66; see also below). Thus the regulation of the degradation of mitotic cyclins in the cell cycle is carried out mainly by the modulation of the cyclin-ubiquitin ligase (E3) activity of the cyclosome (67).

Molecular genetic studies in intact cells, mainly in yeasts, corroborated the results of biochemical studies on the mode of ubiquitin-mediated degradation of mitotic cyclins. The degradation of B-type cyclins in yeasts requires functional subunits of the 26S proteasome (147, 148). Most significantly, the discovery that products of the budding yeast genes CDC16 and CDC23 are required for cyclin B proteolysis in vivo (69) led to the identification of their homologs as subunits of the cyclosome/APC (68). Other cyclosome subunits, such as homologs of BimE of *A. nidulans*, are also required for the degradation of B-type cyclins in both the budding (71) and fission (149) yeasts.

The molecular mechanisms regulating the machinery that degrades cyclin B are not well understood. In the relatively simple early embryonic cell cycles, it seems that the activity of the cyclosome is mainly regulated by its reversible phosphorylation, as indicated by the observations that the inactive, interphase form of the cyclosome can be converted in vitro to the active form by incubation with MPF (19, 66) and that the active, mitotic form of the cyclosome can be converted to the inactive form by treatment with an okadaic acid–sensitive phosphatase (67). Conversion of the cyclosome to the active form by MPF, previously observed with partially purified preparations, has been confirmed using highly purified preparations of cyclosome from clam oocytes (M Shteinberg & A Hershko, unpublished results), indicating that activation is due to direct phosphorylation of the cyclosome by MPF.

Cyclosome activation by MPF-dependent phosphorylation may involve the action of the suc1/cks family of proteins. These proteins were discovered in yeasts as gene products that interact with Cdk1 and were subsequently found in higher organisms (reviewed in 150). In yeasts, suc1/cks proteins are required at several stages of the cell cycle, including entry into mitosis, exit from mitosis, and the degradation of B-type cyclins. Immunodepletion experiments in extracts of *Xenopus* eggs also indicated that suc1/cks has multiple roles in the cell cycle, including the degradation of cyclin B (151). The requirement of cyclin degradation for suc1/cks may be explained by the recent finding that the active, phosphorylated form of the cyclosome binds to p13[suc1] beads (152). Several lines of evidence indicated that the cyclosome does not bind to the Cdk-binding site but rather to a phosphate-binding site of suc1. Thus the cyclosome could be eluted from suc1-Sepharose beads by phosphate-containing compounds, an observation used to develop a procedure for the affinity purification of the cyclosome (152).

A conserved phosphate-binding site was found by x-ray crystallography (in addition to the Cdk-binding site) in all suc1/cks proteins, and researchers suggested that this site directs Cdks to some phosphorylated proteins (150). If multiple phosphorylations are required for cyclosome activation, initial slow phosphorylations may cause tighter binding of MPF to the cyclosome via suc1/cks, thus accelerating additional phosphorylations. Such a model may explain, at least in part, the lag kinetics of interphase cyclosome activation by MPF. This lag, which can be reproduced in vitro, presumably plays an important role in preventing premature self-inactivation of MPF prior to the end of mitosis. This model of the possible role of suc1/cks proteins in the kinetics of cyclosome activation remains to be investigated.

Information on the regulation of cyclin B degradation is based on studies in relatively simple early embryonic cell-cycle systems, which consist of rapidly alternating S- and M-phases, without any intervening G1 and G2 phases. The regulation of mitotic cyclin degradation is more complicated in the more complex cell cycles of somatic cells and unicellular eukaryotes, which contain many additional events in the G1 and G2 phases and have to respond to a variety of extracellular stimuli. Thus the activation of the B-type cyclin proteolysis machinery in yeasts also occurs at the end of mitosis but requires several gene products (reviewed in 153), including a protein phosphatase (154). This process is very different from the regulation of the early embryonic cyclosome, which is inactivated by phosphatase action (67). It is unknown whether the phosphatase acts on the cyclosome directly, removing an inhibitory phosphate group, or whether it is a part of a signal transduction system that affects the cyclosome indirectly.

The inactivation of the cyclin-degrading machinery in yeast or somatic cells is also very different from that of early embryos. In early embryos the cyclin-degrading system is active for only a few minutes at the end of mitosis (155), whereas in yeast it remains active until the end of the G1 phase of the next cell cycle, when it is turned off by the action of G1 cyclins (156). It is unknown how the B-type cyclin proteolytic machinery is turned off by G1 cyclins, but some phosphorylation event by G1 cyclin/Cdk complexes likely inhibits cyclosome activity directly or indirectly. Similarly, in cultured mammalian cells the proteolysis of mitotic cyclins is activated shortly before anaphase and is turned off only at the end of the G1 phase of the next cell cycle (101). The continued activity of this degradation machinery during the G1 phase of the cell cycle may be needed to prevent the premature accumulation of S-phase cyclin substrates of the cyclosome. More research is needed on the mechanisms by which cyclosome activity is regulated in the somatic type of cell cycles.

G1 CYCLINS   Cyclins specific for the G1 phase of the cell cycle are also highly unstable proteins. G1 cyclins do not contain a destruction box motif but are targeted for degradation by phosphorylation. The phosphorylation of yeast G1 cyclins Cln2p (86) and Cln3p (85) is required for their degradation. Available information indicates that phosphorylated G1 cyclins may be ligated to ubiquitin by PULC-type complexes. Thus the degradation of both Cln2p (17) and Cln3p (85) requires the Cdc34p E2 enzyme. Ligation of Cln2 to ubiquitin and its degradation both require Cdc53p, and the phosphorylated form of Cln2p co-purifies with Cdc53p (17). The degradation of Cln2p requires the action of *SKP1* (75), and the degradation of Cln1p and Cln2p require *GRR1* (77). Although evidence for the involvement of PULCs in the degradation of yeast G1 cyclins is much less complete than that available for the Sic1p Cdk inhibitor (see below), similar complexes likely carry out the ubiquitinylation of yeast G1 cyclins.

Available information on the mode of degradation of mammalian G1 cyclins is much more limited, but it appears that in this case, too, phosphorylation of the cyclin substrate is required for its ubiquitin-mediated degradation. However, in contrast to the multiple phosphorylations required for the degradation of yeast Cln2p (86), phosphorylation of specific, single sites has a strong influence on the degradation of mammalian G1 cyclins. Thus the rapid degradation of human cyclin E is markedly slowed by mutation of a specific Cdk phosphorylation site at T380 (88, 157). Residue T380 of cyclin E is phosphorylated in vivo and autophosphorylated in vitro (88). When cells were treated with proteasome inhibitors, the accumulation of ubiquitinylated derivatives of wild-type cyclin E (88, 157), but not of the T480A mutant of cyclin E, was observed. It was suggested that autophosphorylation of cyclin E initiates its ubiquitinylation and degradation (88). The degradation of another mammalian G1 cyclin, cyclin D1, also requires its specific phosphorylation at T286 (89). In this case, too, proteasome inhibitors caused the accumulation of ubiquitinylated derivatives of wild-type cyclin D1, but not of the T286A mutant of cyclin D1, indicating that phosphorylation is required for ubiquitinylation (89). The ubiquitinylation systems that act on phosphorylated mammalian G1 cyclins remain to be identified.

CDK INHIBITORS   The activities of some Cdks are controlled tightly by fluctuations in the levels of their negative regulatory proteins, Ckis. Thus a cyclin/Cdk complex cannot act until the inhibitor is removed by selective proteolysis. A well-studied case is that of the Sic1p Cdk inhibitor of the budding yeast, *S. cerevisiae*. Sic1p inhibits the activity of complexes of Cdk1 with B-type cyclins but does not inhibit the activity of G1 cyclin/Cdk1 complexes (74). Levels of Sic1p are high in the G1 phase of the cell cycle, but the inhibitor is degraded rapidly at the G1 to S-phase transition. The degradation of Sic1p is a centrally important event in the transition from G1 to the S-phase, because it permits the action of S-phase-promoting B-type cyclins to initiate DNA replication (74). The degradation of Sic1p requires its prior phosphorylation by G1 cyclin/Cdk1 complexes, as indicated by the findings that Sic1p is phosphorylated at multiple sites at the end of G1, and this phosphorylation depends on the activities of G1 cyclins (73).

Sic1p degradation also requires the action of the Cdc34p E2 enzyme, as well as that of the products of the *CDC4*, *CDC53* (74), and *SKP1* (75) genes. Biochemical experiments in which the ligation of Sic1p to ubiquitin was reconstituted in vitro indicated that the products of most of these genes are directly required for this process (76). Most of the components required for Sic1p ubiquitinylation have been shown to be assembled in a multiprotein complex (18). Thus phosphorylated Sic1p is likely targeted for degradation by a PULC-type complex (Figure 2D); however, the action of PULCs may not be specific for cell-cycle regulators. For example, the degradation of the Gcn4 transcriptional activator of amino acid biosynthesis in yeast, which takes place throughout the cell cycle, requires its phosphorylation and the action of *CDC34* (87) as well as *CDC53*, *CDC4*, and *SKP1* genes (D Kornitzer, personal communication). In contrast to the regulation of the cyclosome, the activity of which is turned on and off at specific points of the cell cycle, PULCs may be constitutively active. The precise timing of the degradation of Sic1p is apparently determined by the accumulation of G1 cyclin/Cdk1 complexes in late G1, which phosphorylate Sic1p and thus trigger its degradation.

In the fission yeast *Schizosaccharomyces pombe*, a similar function is carried out by the rum1 protein, a specific inhibitor of cyclin B/Cdk1 complexes (158). In this case, too, levels of rum1 are high in G1, but the inhibitor is degraded upon transition to the S-phase. The mode of the degradation of rum1 is not known, except that it requires pop1, the fission yeast homolog of Cdc4p (159). A different function is carried out by another budding yeast Cdk inhibitor, Far1p: It accumulates in response to mating pheromone and arrests cells in early G1 owing to the inhibition of G1 cyclin/Cdk1 complexes (160). Thus Far1p action mediates response to extracellular signals, as is the case with some vertebrate Cdk inhibitors (see below). Upon release of yeasts from arrest in early G1, Far1p is degraded rapidly (161). The degradation of Far1p is preceded by its phosphorylation (161) and requires the function of Cdk1 (162). As a result, the phosphorylation of Far1p is likely required for its degradation. The machinery that carries out the degradation of Far1p is not known, except that Far1p accumulates in *cdc34* or *cdc4* mutants (161), and it is markedly stabilized by the deletion of its 50 N-terminal amino acids (162).

Many Ckis have been identified in mammalian cells, and these can be divided into two families based on sequence similarities: The KIP/CIP family contains p21, p27, and p57, and the INK family includes p15, p16, p18, and p19 (163). All mammalian Ckis inhibit G1 cyclin/Cdk complexes (though with different specificities) and thus mediate cell-cycle arrest in response to a variety of growth-inhibiting conditions. For example, p21 is induced by DNA damage in a process mediated by p53 (163), p27 levels are increased greatly in cells arrested by deprivation of growth factors or contact inhibition (164), and p18 levels are elevated in terminal differentiation associated with permanent cell-cycle arrest (165).

Several mammalian Ckis are unstable proteins, the levels of which may be modulated by alterations in the rates of their degradation. Thus the high levels of p27 in quiescent cells result, at least in part, from decreased degradation (166, 167). After growth stimulation, p27 levels decrease rapidly owing to degradation by the ubiquitin system, as indicated by the accumulation of its ubiquitinylated derivatives following treatment of cells with proteasome inhibitor (166). Furthermore, rates of ubiquitinylation of p27 in vitro are higher in extracts of proliferating cells than in those of quiescent cells (166). Co-overexpression of different derivatives of p27 with Cdk2/cyclin E led to the suggestion that phosphorylation of p27 by cdk2/cyclin E on T187 causes its degradation (168). Since p27 inhibits the action of cdk2/cyclin E, it was further proposed that the phosphorylation of p27 by the kinase is faster than the formation of the inhibited complex (168). It is unclear whether the ratio of p27 to Cdk2/cyclin E in these overexpression experiments is comparable to the physiological situation. Thus, the identity of the system that ligates p27 to ubiquitin and the mode of the regulation of this process remain to be identified.

The p21 Cki is also a rapidly degraded protein that is ubiquitinylated in vivo (169). Another interesting case is that of a p15 Cki, the degradation of which is inhibited by transforming factor β, thus causing its accumulation (170). Based on these findings, the regulation of Cdk inhibitor–degradation in animal cells may be involved in connecting to the basic cell-cycle machinery a variety of signals that affect cell proliferation.

ANAPHASE INHIBITORS AND OTHER CELL-CYCLE REGULATORS   In addition to cyclins and Ckis, the levels of numerous other cell-cycle regulators oscillate during the cell cycle. Some of these regulators are targeted for degradation by cyclosome-mediated ubiquitin ligation, some by PULC-type mechanisms, and for others, the mechanisms are unknown. In extracts of *Xenopus* eggs, the addition of an N-terminally truncated, nondegradable derivative of cyclin B caused arrest at late anaphase, while the addition of the N-terminal fragment of cyclin B caused an earlier arrest at the metaphase (171). This observation indicates that the machinery that degrades mitotic cyclins is also involved in the degradation of other cell-cycle regulators. Since the N-terminal fragment of cyclin B contains the destruction box region, it was suggested that the degradation of some other destruction box–containing protein (which is competed by the N-terminal fragment of cyclin B) is required for the separation of sister chromatids that takes place in the metaphase-anaphase transition (171).

These in vitro observations were confirmed in vivo (99); however, the identity of the putative anaphase inhibitors remained unknown until recently. One of these inhibitors is the Cut2 protein of *S. pombe*, which is localized on the mitotic spindle and is degraded rapidly at the end of the anaphase (172). The degradation of Cut2 requires the presence of two destruction box regions at its N-terminal domain (173) and the activity of the cyclosome (172). Most significantly, the expression of nondegradable derivatives of Cut2 prevented the separation of sister chromatids. As a result, destruction box–dependent, cyclosome-mediated degradation of Cut2 may be required for the onset of the anaphase (172). Essentially similar results were observed on Pdsp1p, another anaphase inhibitor in *S. cerevisiae* (174).

The mode of action of anaphase inhibitors is unknown. They may act as "molecular glues" that hold sister chromatids together until they are degraded, or they may act by much more indirect mechanisms. Another destruction box–containing protein in the budding yeast, Ase1p, is bound to the midzone of the mitotic spindle and is degraded at the end of mitosis by a mechanism that requires the activity of the cyclosome (175). This process may be involved in the disassembly of the mitotic spindle, since the expression of a nondegradable derivative of Ase1 caused a delay in spindle disassembly (175). It seems that the cyclosome is responsible for the degradation of various cell-cycle regulators at the end of mitosis, all of which share the destruction box–recognition determinant.

Another important unstable cell-cycle regulator is the Cdc6 protein of the budding yeast, which is required for the initiation of DNA replication (reviewed in 153, 176). Cdc6p is synthesized in G1, associates with the origin replication complex, and promotes its conversion to a form competent for replication. After the initiation of DNA replication, Cdc6p is degraded rapidly (177). A similar function is carried out in the fission yeast by a homologous protein, Cdc18 (178). In the fission yeast, the elimination of Cdc18 after the initiation of DNA replication may prevent replication of DNA more than once in a cell cycle, since massive overexpression of Cdc18 caused repeated rounds of DNA synthesis, without mitosis (178). In contrast, ectopic expression of Cdc6p in the budding yeast after G1 does not cause re-replication (177). In the latter case, the prevention of re-replication appears to be tightly controlled by other mechanisms; thus the degradation of Cdc6p may be an additional safeguard to ensure that DNA is replicated only once in a cell cycle. The mechanisms of Cdc6p degradation are unknown, although the action of *CDC4* is required (177), suggesting the possible involvement of a PULC-type complex. Likewise, the ubiquitinylation and degradation of Cdc18 in the fission yeast requires the action of *pop1+*, a homolog of *CDC4* (159).

Transcription Factors, Tumor Suppressors, and Oncoproteins

The activity of many short-lived regulatory proteins of the transcription factor, tumor suppressor, and oncoprotein classes is controlled by proteolysis via the ubiquitin pathway. This section discusses some selected cases.

NF-κB AND IκBα NF-κB (nuclear factor κB) is a ubiquitous inducible transcription factor involved in central immune, inflammatory, stress, and developmental processes (reviewed in 179, 180). The best-studied transcription factor is a heterodimer, NF-κB1, that is composed of p50 and p65 (RelA). Similar complexes containing other subunits such as p52 (p52-p65 complex is NF-κB2) and RelB also exist. p50 and p52 are the processing products of larger precursor molecules, p105 and p100, respectively, and are derived from the N-terminal domain of the precursor molecules: The C-terminal domain is degraded. The processed transcription factor is retained in a latent form in the cytoplasm of nonstimulated cells via association with inhibitory molecules collectively termed IκBs (inhibitors of κB). Following exposure of the cell to a variety of extracellular stimuli such as cytokines, viral and bacterial products, and stress, IκBs are degraded rapidly and the active heterodimer is translocated into the nucleus where it exerts its transcriptional activity. The inhibitors fulfill their task via a dual action: They sterically hinder the nuclear localization signal of the NF-κB proteins, thus retaining them in the cytosol, and they also inhibit their DNA binding and transactivation capacity. Recent evidence implicates the ubiquitin system both in the processing of p105 (and probably p100) and in the signal-induced degradation of IκBα (and probably IκBβ).

Fan & Maniatis (181) showed that p50 is generated in vivo by processing of the precursor protein p105. Reconstitution of a cell-free system using a truncated form of the precursor protein, p60, has revealed that the process requires ATP and is sensitive to N-ethylmaleimide, which inactivates E1, E2, and certain E3 enzymes. These findings suggest that the process may be mediated by the ubiquitin system. Palombella & colleagues (182) showed that COS cells transfected with p105 process the precursor protein to p50 and that processing can be inhibited by proteasome inhibitors. Experiments with yeast mutant cells defective in Pre1, one of the proteasomal subunits, provided direct evidence that processing is mediated by the proteasome. Reconstitution of a cell-free system has demonstrated that the processing requires ubiquitin (182), E2-F1 (or UbcH5 or UbcH7), and a novel species of E3 (82; see also above).

These findings establish a role for the ubiquitin system in the processing of p105. Since p105 is the only molecule known to be processed by the ubiquitin system rather than being completely destroyed, an obvious problem concerns the underlying mechanism(s) involved. Lin & Ghosh (183) have reported that a Gly-rich region (GRR) that spans amino acid residues 376–404, and that contains 19 (out of 29) Gly residues, constitutes an independent stop-transfer signal that prevents processing of p105. Removal of GRR inhibits processing. A chimera protein constituted of p50, GRR, and IκBα was processed to yield p50 and IκBα, suggesting that generation of p50 from p105 proceeds in a two-step mechanism, a single endoproteolytic cleavage of p105 to generate p50 and the C-terminal domain, followed by degradation of the C-terminal part by an as-yet-unknown mechanism. Insertion of the motif between two unrelated proteins, gp10 and GST, yielded two authentic processing products, suggesting that the GRR can serve as a transferable stop processing signal.

Other experiments (A Orian & A Ciechanover, unpublished results), however, could not substantiate all of these conclusions. For example, insertion of GRR into Dorsal, the noncleavable *Drosophila melanogaster* p105 homolog, did not lead to processing of the protein. Insertion of GRR between p50 and ornithine decarboxylase (ODC) results in processing, but the ODC moiety is completely degraded even in the absence of antizyme. Thus it appears that certain steric constraints determine recognition for processing and its mechanism, and that GRR clearly is not a universal transferable processing element. As for the mechanism of action of GRR, it may generate a structure that cannot enter the 26S proteasome. For example, insertion of a similar repeat at the C-terminal domain of Epstein-Barr nuclear antigen 4 (EBNA4) prevented its degradation but not its conjugation (184; see also below).

The IκB family of proteins contains, among other proteins, four inhibitors: IκBα, IκBβ, IκBε, and Cactus (179, 185). Most stimuli that induce NF-κB activation target the inhibitors for degradation. Signal-induced phosphorylation at specific sites directs IκB proteins for degradation via the ubiquitin system. When either Ser^{32} or Ser^{36} of IκBα were mutated, the inhibitor did not undergo stimulation-induced phosphorylation, was not degraded, and NF-κB could not be activated in response to a broad array of stimuli (186, 187). The S32E/S36E double mutant was constitutively unstable and NF-κB was constitutively hyperactive, suggesting a role for the negatively charged phosphate groups in the recognition process.

Additional studies have further indicated that phosphorylation targets the protein for degradation by the ubiquitin system. Incubation of TNF-α or phorbol ester/ionomycin-stimulated Jurkat T-cells in the presence of proteasome inhibitors results in stabilization of the otherwise rapidly degraded phosphorylated IκBα and in accumulation of high-molecular-mass ubiquitin conjugates of the protein (188–190). Reconstitution of a cell-free system revealed an excellent correlation with the in vivo findings. Only the doubly phosphorylated form, but not the S32A/S36A double mutant or singly mutated S32A or S36A species, could be ubiquitinylated in vitro. Similarly, the S32E/S36E double mutant could be ubiquitinylated in vitro (188). Reconstitution of degradation revealed that the process requires ATP, ubiquitin, and the E2 enzymes E2-F1, which is the rabbit homolog of human UbcH7 (see above) or Ubc5.

In ts20 cells that harbor a thermolabile E1, phosphorylated IκBα was stable at the nonpermissive temperature (189). Addition of purified 26S proteasome to immunopurified IκBα ubiquitin conjugates led to their degradation (188). The ubiquitinylation sites have been localized to Lys residues 21 and 22 (191). These findings strongly link stimulation-induced phosphorylation at Ser residues 32 and 36 to recognition by the ubiquitin-conjugating machinery and to subsequent degradation by the proteasome. It is important to note that phosphorylation does not release IκBα from the NF-κB complex: it is degraded while still bound to the complex (188, 192).

An interesting problem involves the mechanism of phosphorylation-dependent targeting of IκBα. Is the phosphorylated domain [32-S(P)GLDS(P)-36] recognized directly by the IκBα-ubiquitin ligase, or does phosphorylation affect the 3-D structure of the inhibitor indirectly in a manner that exposes a remote E3-binding site? Yaron et al (193) have shown that phosphorylated peptides that span the phosphorylation domain, but not their unmodified or mutated counterparts, specifically inhibit conjugation and degradation of $I\kappa B\alpha$ in a cell-free reconstituted system. A seven-residue peptide [31-DS(P)GLDS(P)M-37] is sufficient to exert the inhibitory effect. Notably, Lys residues 21 and 22 are dispensable and do not constitute a part of the recognition signal. Incubation of a crude extract with immobilized peptide leads to specific binding of a conjugating activity that can be complemented only by the addition of an E3-rich fraction, but not by E1 and E2. Microinjection of these phosphopeptides into cells leads to inhibition of translocation of NF-κB to the nucleus and, consequently, to inhibition of expression of the NF-κB-dependent gene, *E-selectin*. These results suggest that the phosphorylated domain serves as recognition motif for an E3 that acts on $I\kappa B\alpha$ and that inhibition of this E3 by the mimetic peptides can inhibit the biological functions of NF-κB.

Two kinases that phosphorylate $I\kappa B\alpha$ have been cloned recently and were identified as the previously known Ser-Thr kinase of unknown function, CHUK, and a close homolog of this protein (194, 195, 195a). CHUK associates directly with $I\kappa B\alpha$ and phosphorylates it on Ser residues 32 and 36. The direct or indirect role(s) of other kinases that have been reported to be involved in NF-κB activation via phosphorylation of $I\kappa B\alpha$, such as the mitogen-activated protein kinase/ERK kinase kinase 1 (MEKK1; 196), the protein kinase A catalytic (PKAc) subunit (197), and an unidentified novel kinase that appears to be activated by ubiquitinylation (198) are not clear. At least for CHUK, there is no evidence that it is activated by ubiquitinylation.

P53   The tumor suppressor protein p53 is degraded by the ubiquitin system both in vitro (199) and in vivo (200). Its degradation in vitro requires E1 and is blocked in intact cells by inhibitors of the proteasome with the concomitant accumulation of high-molecular-mass p53-ubiquitin adducts. Degradation is accelerated dramatically by the high-risk human papilloma virus (HPV) onco-protein E6 (53). The E6-dependent degradation is mediated by the E3 enzyme E6-AP and by one of several E2 enzymes (see previous section).

Since wild-type p53 is short lived in all cells examined, and since it is targeted for degradation by the ubiquitin system, an important problem involves the identity of the ubiquitin pathway enzymes (E2 and E3), which conjugate p53 in cells that do not express E6. Antisense targeting of E6-AP results in elevated p53 levels in HPV-infected cells but not in normal cells (201), suggesting that E6-AP plays a role in targeting of p53 only in the presence of E6. In most cells that are not transformed by HPV, ubiquitinylation of p53 is mediated by an unidentified species of E3.

Haupt et al (202) and Kubbutat et al (203) have reported that Mdm2 promotes rapid, ubiquitin-dependent degradation of p53. The oncoprotein Mdm2 is a potent inhibitor of p53. It binds to the transcriptional activation domain of p53 and inhibits its ability to activate target genes, to exert its antiproliferative effects, to mediate cell-cycle arrest following exposure to DNA damage, and to fulfill its apoptotic functions. p53 regulates the transcription of Mdm2 in an autoregulatory feedback loop (202 and references therein). The transient stabilization of p53 that occurs following UV irradiation and DNA damage, for example, leads to an increase in its level, which enables the tumor suppressor protein to curb the damage. Concomitantly, increased, p53-dependent transcription of Mdm2 ensures, by targeting p53 for degradation, release from the cell-cycle arrest and other untoward effects caused by elevated p53 level. Thus the targeting of p53 for degradation by Mdm2 provides another way to remove p53 after its repair functions have been fulfilled.

The interval between p53 stabilization and activation and its inactivation and targeting by induced Mdm2 defines a time window for its activities. Induction of degradation is accompanied by accumulation of high-molecular-mass p53-ubiquitin adducts and can be inhibited by lactacystin. The destabilizing effect of Mdm2 requires physical interaction between the two proteins. This interaction is mediated by a small region in the N-terminal domain of p53. Fusion of amino acid residues 1–42 of p53 to Gal4, an otherwise stable protein, renders the chimeric protein susceptible to Mdm2-mediated degradation (202). Thus the N-terminal domain of p53 involved in Mdm2 binding is necessary and sufficient to confer upon the tumor suppressor Mdm2-dependent proteolytic sensitivity. It remains to be seen whether, like E6, Mdm2 is part of an E3 complex that targets p53 for degradation.

JUN   Treier et al (105) reported that c-Jun, but not its transforming counterpart v-Jun, is multiply ubiquitinylated and rapidly degraded in cells. Detailed analysis of the differential sensitivity to the ubiquitin system revealed that the δ domain of c-Jun, an amino acid sequence that spans residues 31–57 and that is missing in the retrovirus-derived molecule, confers instability on the normal, cellular protein. Deletion of the domain stabilizes c-Jun. This requirement for the δ domain raised the question of whether it contains either the ubiquitinylation site or a recognition signal for the conjugation machinery. Alteration of all the Lys residues in the domain did not alter the protein's sensitivity to degradation, and no single Lys residue in the remaining parts of the molecule was essential for ubiquitinylation. Thus, the δ domain is a *cis*-acting signal required for ubiquitinylation and subsequent degradation of c-Jun. Transfer of this element to β-Gal, an otherwise stable protein, rendered the protein susceptible to multiple ubiquitinylation and degradation.

The lack of the δ domain from v-Jun, a protein that is otherwise highly homologous to c-Jun, provides a mechanistic explanation for the stability and the possible resulting transforming activity of v-Jun. The loss of the δ domain during the retroviral transduction is another example of the sophisticated diverse mechanisms developed by viruses to ensure replication and continuity of infection (see above for the effect of E6 on targeting p53 and below for the effect of the cytomegalovirus proteins US2 and US11 on targeting MHC class I heavy chains). Musti et al (94) showed that phosphorylation by mitogen-activated protein kinases (MAPKs), such a Jun kinase 1 (JNK1), reduces c-Jun ubiquitinylation and increases its stability. c-Jun^{Asp}, in which one of the Ser targets of JNK1 was substituted by the phosphate-mimetic residue, Asp, acts as a gain of function form of c-Jun. It is not ubiquitinylated and, consequently, remains stable. In contrast, c-Jun^{Ala} is as efficiently ubiquitinylated as the wild-type protein. Thus, it appears that phosphorylation-dependent stabilization contributes to the efficient activation of target genes following exposure to growth factors, stress, or other stimulators of c-Jun activity. In another study, Fuchs et al (204) showed that JNK2 can, in addition, stimulate ubiquitinylation of c-Jun, underscoring the complexity of the signals and extracellular events that govern c-Jun stability.

β-CATENIN β-Catenin and its *D. melanogaster* homolog, Armadillo, are multifunctional proteins involved in cell-cell adhesion complexes, signal transduction along the wingless (*D. Melanogaster*)/Wnt-1 (mammals) pathway, and regulation of transcription (reviewed in 205). Accordingly, the protein is found in the plasma membrane, cytoplasm, and nucleus. The wingless/Wnt-1 pathway is involved in several key developmental processes, such as in determining anterior-posterior patterning in *Drosophila* segments and axis formation in *Xenopus*. Its specific expression in early mouse embryos implicates the protein in the development of mammals as well. Genetic and biochemical analyses showed that β-catenin plays a major role in signal transduction and differentiation in the mammalian colorectal epithelium, and that aberrations in the process are important in the multistep development of colorectal tumors.

In the absence of signaling by wingless/Wnt-1 ligands, the downstream *Drosophila* Ser/Thr kinase zeste white 3 (zw3), or its mammalian homolog glycogen synthase kinase-3 (GSK-3), is active and promotes degradation of β-catenin, most probably by ubiquitinylation (90). The cadherin-associated β-catenin in the cell-cell junctions remains stable. Stimulation by wingless/Wnt-1 promotes activation of the cytosolic protein disheveled (dsh), which antagonizes the function of zw3/GSK-3, leading to stabilization and accumulation of β-catenin. Free β-catenin binds and activates transcription factors Tcf (T-cell transcription factor) and Lef (lymphoid enhancer factor). The translocation of the active complexes to the nucleus initiates transcription of specific, yet-to-be-identified target genes.

β-Catenin interacts with both GSK-3 and an additional protein, the ~300-kDa tumor suppressor APC (adenomatous polyposis coli), which appears to regulate β-catenin intracellular level (206). GSK-3 phosphorylates β-catenin in three conserved Ser residues and one Thr residue in the N-terminal domain (33-SGIHSGATTTAPS-45). Mutations of these residues (S33Y, S37F, S45Y), or deletions of the N-terminal domain containing these residues in different malignant melanoma cell lines, result in increased stability and consequent increased activity of the protein (207, 208). The constitutive high levels of β-catenin-Tcf and β-catenin-Lef complexes in these conditions may result in persistent *trans*-activation of the target gene(s) and may play a crucial role in malignant transformation of these cells.

Phosphorylation-dependent degradation of β-catenin is another example in which this posttranslational modification destabilizes a protein (see above for other examples). It is unknown how APC regulates the stability of β-catenin; however, APC interacts with β-catenin via two distinct sets of binding sites. β-catenin accumulates in colon cancer cells that do not express the protein (APC$^{-/-}$) or that harbor APC proteins that lack one of the binding clusters. The accumulated β-catenin associates with Tcf or Lef and leads, most probably, to overexpression of their dependent genes (209). Expression of full-length APC in these cells leads to degradation of excess β-catenin and to abrogation of the *trans*-activation effect. Although the mechanism(s) that underlie the function of APC in regulating β-catenin stability have not been elucidated, it may serve as a ubiquitin ligase or as part of a larger complex that serves as a ligase. Indeed, wild-type but not ΔN1-89 β-catenin is part of a large complex that contains, among other proteins, APC (210).

E2F-1   The E2F family of transcription factors plays an important role in regulating cell-cycle progression at the G_1-S transition. One of the best studied members of this family is E2F-1, a 437-amino-acid residue protein that has a half life of approximately 2–3 h. E2F-1 can act under different circumstances as an oncogene or as an inducer of apoptosis and is controlled by multiple mechanisms, among which are binding to the retinoblastoma tumor suppressor (Rb) protein, activation by Cdk3, and S-phase-dependent down-regulation of its DNA-binding capacity by cyclin A–dependent kinase. E2F-1 is degraded in a regulated manner by the ubiquitin system (211, 212). In the presence of proteasome inhibitors, the protein was stabilized with the concomitant accumulation of high-molecular-mass E2F-1-ubiquitin adducts. Co-expression of Rb with E2F-1 results in a marked stabilization of E2F-1. The stabilization is the result of direct binding of Rb to the transcription factor, as deletion of either the E2F-1-Rb- or the Rb-E2F-1-binding sites abrogated the stabilizing effect of Rb.

Mechanistically, binding of Rb prevents ubiquitinylation of E2F-1. The dephosphorylated form of Rb binds E2F-1, as a mutant form of Rb that could not undergo phosphorylation was much more efficient in stabilizing E2F-1 than its wild-type counterpart. An interesting problem involves the cell-cycle regulation of E2F-1 and its linkage to Rb binding. Hofmann et al (211) noted a cell-cycle oscillation in the level of E2F-1. In particular, there appeared to be a sharp decrease during S-phase. Whether the changes reflect cell-cycle-dependent dissociation from Rb and degradation of free E2F-1 remains to be examined.

Membrane Proteins

It has been generally accepted that the ubiquitin system is involved in selective degradation of cytosolic and nuclear proteins. However, it is clear now that the system is also involved in two distinct pathways of degradation of membrane proteins. A new and rather unexpected function of ubiquitin tagging was found in targeting some membrane receptors and transporters for endocytosis and degradation in the lysosome (or the vacuole in yeast). In a distinct pathway, native or misfolded proteins in the endoplasmic reticulum (ER) are targeted to the proteasome and degraded in the cytosol (or on the cytosolic surface of the membrane), either with or without prior ubiquitinylation. These novel findings raise important, yet unresolved, mechanistic issues. For membrane proteins an obvious problem relates to the role of ubiquitin modification: Is it required for endocytosis of the tagged protein or in a later stage, for its specific targeting and uptake by the lysosome? For membrane ER proteins degraded by the cytosolic proteasome, important questions involve the mechanisms that underlie retrieval of these proteins across the membrane back into the cytosol.

Analysis of the fate of the *S. cerevisiae* Ste2p, the G-protein-coupled plasma membrane receptor of the α factor involved in the mating response pathway, has shown that binding of the ligand leads to ubiquitinylation of the receptor that is essential for endocytosis of the receptor-ligand complex (213, 214). The complex is targeted for degradation in the vacuole: In mutant yeast species lacking vacuolar proteolytic enzymes, the receptor is stable. These results suggest that both the ectoplasmic and cytoplasmic portions of the molecule are degraded within the vacuole. Binding of the α factor leads to phosphorylation of the receptor on Ser residues that reside on a well-defined internalization signal, SINNDAKSS, that is both necessary and sufficient for receptor endocytosis (215, 216). The Lys residue in this sequence is ubiquitinylated, though it is dispensable: Other adjacent Lys residues can also bind ubiquitin. Detailed structural analysis of the SINNDAKSS sequence showed that receptor ubiquitinylation is necessary for internalization of the ligand-receptor complex: Inhibition of ubiquitinylation results in stabilization of the receptor on the cell surface as shown by radiolabeled ligand binding assays. Interestingly, endocytosis of the Ste6 ABC transporter also requires ubiquitinylation and a DAKTI motif (217) that is similar to the essential SINNDAKSS sequence in the Ste2p receptor. In addition, Pdr5, the yeast multidrug transporter, is also targeted to the vacuole by ubiquitinylation (218).

Using a Chinese hamster cell-cycle mutant cell that harbors a thermolabile E1, Strous et al (219) showed that following binding of growth hormone (GH) at the permissive temperature, the receptor of GH undergoes rapid ubiquitinylation and subsequent degradation in the lysosome: Inhibitors of endosomal/lysosomal function, such as NH_4Cl and bafilomycin A1, significantly inhibited ligand-induced degradation of the receptor. In contrast, at the non-permissive temperature, the receptor was not ubiquitinylated and remained stable. Similar to the case of the Ste2p receptor, these findings establish a direct linkage between ubiquitinylation and early events in endocytosis of ligand-receptor complexes in mammalian cells. The exact role of ubiquitinylation in the endocytic process is unclear. It may serve, for example, as an anchoring site for adaptor molecules or cytoskeletal elements involved in vesicle budding and movement. It can also signal limited proteolysis of the cytosolic tail by the proteosome that is necessary for initiation of the endocytic process to occur.

The yeast Gap1p amino acid permease and the Fur4 uracil permease are also targeted for degradation in the vacuole following ubiquitinylation that is mediated by the E3 enzyme Npi1p/Rsp5 (61,62; see also previous section). Ubiquitinylation of the two proteins is triggered by NH_4^+ or stress conditions (such as approach to the stationary growth phase or inhibition of protein synthesis), respectively. Recent findings demonstrate that ubiquitinylation of the Fur4 protein, and probably of other membrane proteins, is mediated by a mechanism that is distinct from that of tagging soluble proteins. Polyubiquitinylation and endocytosis of the protein were inhibited in a yeast mutant lacking the deubiquitinylating enzyme Doa4 that prevents the regeneration of free ubiquitin. Interestingly, both processes could be rescued by overexpression of ubiquitin mutants carrying Lys → Arg mutations at Lys^{29} and Lys^{48}. By contrast, a ubiquitin mutated at Lys^{63} did not restore Fur4 polyubiquitinylation but interfered only slightly with endocytosis (220).

Similar findings were reported for the endocytosis of the α factor receptor (220a). It appears, therefore, that ubiquitin-Fur4 and ubiquitin-Ste2p conjugates are extended via Lys^{63}, but this process is not essential for endocytosis: Monoubiquitinylation is sufficient to drive the receptors into the vacuolar system of the cell. Another yeast membrane protein, the galactose transporter

Gal2p, is ubiquitinylated, endocytosed, and targeted to the vacuole following transfer of cells from a galactose- to a glucose-containing medium (221). Down-regulation of the Kit receptor by its ligand, the soluble steel factor (SSF), is also mediated by ubiquitinylation that targets the endocytosed receptor to the lysosome (222). Down-regulation of this receptor, involved in hematopoiesis, melanogenesis, and gametogenesis, requires the kinase activity of the molecule, although surprisingly, substitution of Tyr^{821}, which undergoes autophosphorylation, does not affect the process (223). Thus the role of the receptor kinase in the endocytic process is still obscure. It was reported that angiotensin II down-regulates the inositol triphosphate receptors via the proteasome, but it is unclear whether ubiquitinylation is required in this process (224).

The involvement of the ubiquitin-proteasome pathway in different aspects of membrane protein degradation has highlighted the role the system plays in a variety of pathophysiological processes in the immune system. The ζ subunit of the T-cell receptor (TCR), a disulfide-linked homodimer that plays an important role in TCR-mediated signal transduction, also undergoes polyubiquitinylation in response to receptor engagement (225). This modification can occur on multiple Lys residues (226) and requires phosphorylation of the cytosolic tail of the chain by the tyrosine kinase $p56^{lck}$ (226). A tyrosine phosphatase that regulates the activity of the kinase is also required in the process (227). The TCR is a long-lived protein, and it is unclear whether ubiquitinylation plays a role in modulating TCR activity via degradation of a single subunit, or whether it is required for the function of the receptor in signal transduction.

While most cell surface receptors appear to be targeted to the lysosome/vacuole, several exceptional cases were reported in which ubiquitinylation targets membrane proteins to the proteasome. For example, the platelet-derived growth factor receptor is ubiquitinylated following ligand binding and is degraded by a process partially inhibited by proteasome inhibitors (228). In addition, the Met tyrosine kinase receptor involved in pleiotropic cellular response following activation by its ligand, the hepatocyte growth factor/scatter factor (HGF/SF; see 229), and the gap junction protein connexin43 (230) are ubiquitinylated and degraded in a similar fashion. Ubiquitinylation of other cell surface receptors such as the high-affinity IgE receptor (231), the prolactin receptor (232), and the EGF receptor (233), has been reported; however, the function of the modification in regulating the level and activity of these molecules and the underlying mechanisms involved are still obscure.

Thus, an emerging view is that rapid ligand-induced endocytosis of certain cell surface receptors requires ubiquitinylation of their cytosolic tails. The role of ubiquitinylation in the endocytic process is not known and may involve proteolysis of the tail: Formation of multivesicular, multilayered bodies as transport intermediates en route to the lysosome appears to be essential for degradation of both the ectoplasmic and the cytosolic domains of these proteins in the vacuole/lysosome system. Unlike the degradation of lumenal ER proteins, proteolysis of membrane-anchored proteins in the lysosome does not require their transport across the membrane into the cytosol.

The ER is the port of entry of most compartmentalized, membrane-anchored, and secretory proteins. It is also the site of folding and modification of nascent chains and of assembly of multisubunit complexes. Therefore, it must be endowed with a "quality control," editing machinery to remove proteins that fail to fold properly or to oligomerize. Such proteins can include aberrant, mutated proteins or excess subunits of large complexes. Some resident ER proteins are under tight physiological control, and their degradation must also be regulated by a highly specific machinery, distinct from the lysosome. Until recently, degradation of ER proteins was thought to involve an unidentified ER-localized protease(s)/proteolytic system (234). It has become clear, however, that this is not the case, and that ER degradation represents a novel type of ubiquitin-mediated degradation of membrane proteins, distinct from that of mature cell surface proteins. Degradation of these proteins occurs in the cytosol and is mediated by the 26S proteasome, with or without prior ubiquitinylation (for recent reviews, see 235, 236). Unlike ubiquitin-induced targeting of cell surface proteins to lysosomes, degradation of ER proteins requires retrograde transport of these proteins back into the cytosol. Here the function of ubiquitinylation, when modification occurs, is known, and it most likely serves in targeting to the proteasome.

An important ER membrane protein is 3-hydroxy-3-methylglutaryl-coA reductase (HMG-R), a key regulatory enzyme in sterol synthesis. The mammalian enzyme has several transmembrane domains and is tightly regulated by the final product of the biosynthetic pathway, cholesterol. Regulation is mediated by both feedback inhibition and rapid targeting for degradation. Yeast has two isozymes, Hmg1p and Hmg2p. The regulation of the first enzyme occurs mostly at the translational level, whereas that of Hmg2p is mediated in an manner identical to the regulation of the mammalian enzyme, via end-product inhibition and induced degradation. Recent evidence implicates the ubiquitin-proteasome pathway in the degradation of Hmg2p in yeast. Degradation proceeds independently of vacuolar proteases (237). Genetic analysis has revealed that the proteolytic process requires the product of *HRD2* (HMG-R degradation), which is a subunit of the 19S regulatory complex of the proteasome and is homologous to the p97/TRAP2 19S subunit in the mammalian proteasome (238).

A recent study showed that the enzyme undergoes Ubc7-mediated polyubiquitinylation prior to its degradation (239). In contrast, while the mammalian enzyme is also degraded by the proteasome as shown by inhibition of sterol-induced degradation by lactacystin, it does not appear to require prior ubiquitinylation (240). However, genetic analysis in mammalian cells is more difficult than in yeast, and the conclusion that ubiquitinylation is not involved in the degradation of a certain protein is based solely on experiments with a temperature-sensitive mutant of E1, which is difficult to inactivate completely (241).

The cystic fibrosis transmembrane conductance regulator (CFTR) is synthesized as an approximately 140-kDa core glycosylated protein that is associated initially with the cytosolic chaperone Hsc70 and then with the ER chaperone calnexin. Only 25–50% of the wild-type protein matures to the cell surface, whereas most of the protein molecules do not fold properly and are degraded in the ER (242, 243). A single mutation in the protein, deletion of Phe[508], is the underlying cause of most cases of cystic fibrosis. The mutated protein is never released from its complex with the chaperone, fails to acquire carbohydrate moieties (a hallmark of transit through the Golgi apparatus), and is degraded. Recent evidence suggests that the process is mediated by the ubiquitin-proteasome pathway. Lactacystin inhibits degradation of both the wild-type and the mutant forms of CFTR. Degradation is also inhibited in a mutant cell line harboring a thermolabile mutation in E1 and in cells that express a dominant negative form of ubiquitin (244, 245). These data suggest that ubiquitinylation precedes recognition and degradation of the wild-type and mutant CFTR proteins by the proteasome.

Degradation of TCR subunits in the ER constitutes an important mechanism that ensures that only correctly assembled receptor complexes will reach the cell surface. The CD3-δ subunit of the TCR is degraded in the ER following ubiquitinylation (246). Inhibition of the proteasome by lactacystin leads to accumulation of an insoluble, membrane-bound form of CD3-δ, suggesting that the ubiquitin system selectively degrades a misfolded denatured form of the protein. The misfolded form is generated, most probably, because of the inability of the excess unassembled chains to incorporate into the mature TCR complex. A more detailed study of the mechanism revealed that the process initiates with the trimming of mannose residues from an N-linked oligosaccharide, generation of membrane-bound ubiquitin conjugates, and removal of these conjugates by the proteasome. The α chain of the TCR is degraded via a similar mechanism, except that ubiquitinylation does not appear to mediate recognition by the proteasome (247, 248). Again, similar to the case of mammalian HMG-R, the lack of requirement for ubiquitinylation is not firmly substantiated.

Antigenic peptides that are processed in the cytosol are presented to cytotoxic T-cells on MHC class I complexes. These complexes are heterodimers constituted of transmembrane heavy chains (HCs) associated noncovalently with β_2-microglobulin (β_2-m). The proper folding of the complex is aided by the ER chaperones calnexin and calreticulin. If β_2-m is absent, the complex will not be assembled and antigenic peptides will not be transported and presented on the cell surface. Inhibition of the proteasome in β_2-m-deficient cells leads to accumulation of de-glycosylated HCs in the cytosol (249). These molecules that were transported into the ER membrane and glycosylated cannot fold properly and are transported back to the cytosol, de-glycosylated, and degraded by the proteasome.

An interesting observation involving the stability of MHC class I complexes relates to the multiple mechanisms by which viruses evade the immune surveillance machinery. One such mechanism involves a viral matrix protein–directed phosphorylation of the immediate early (IE) protein of the cytomegalovirus (CMV) that prevents its processing to peptide antigens and, consequently, their presentation to cytotoxic T-cells (CTLs; see 250). Since the proteasome is involved in processing of antigens presented via class I MHC molecules (see below), it is assumed that inhibition of processing involves interference with recognition by the proteasome.

In a different mechanism, researchers noted that the human CMV encodes two ER resident proteins, US2 and US11, that down-regulate the expression of MHC class I HC molecules. The MHC molecules are synthesized on membrane-bound ribosomes, transported to the ER where they are glycosylated, but shortly thereafter, in cells expressing US2 or US11, are transported back to the cytosol, de-glycosylated by N-glycanase, and degraded by the proteasome (251, 252). It appears that the viral products bind to the MHC molecules and escort/dislocate them to the translocation machinery where they are translocated back into the cytosol in an ATP-dependent process. The mechanism of action of the viral proteins is not known. They may diffuse laterally in the membrane, interact with the emerging nascent MHC chain, and not allow insertion of the stop-transfer signal and proper anchoring of the molecule in the membrane. Alternatively, they may compete with the binding of the ER chaperone Bip/kar2, which may be necessary for proper folding of the HC molecule. In a different case, ICP467, an IE cytosolic protein encoded by the herpes simplex virus, prevents transport of cytosolic peptides into the ER lumen. Expression of this viral protein also leads to rapid degradation of MHC HCs (249). The lack of peptides can lead to changes in the conformation of otherwise intact and properly folded MHC complexes. Consequently, they are retrieved from the ER and degraded in the cytosol by the proteasome.

Carboxypeptidase Y (CPY) is a yeast vacuolar enzyme that, like all other vacuolar enzymes, traverses the ER lumen en route to the vacuole. Mutated CPY is degraded rapidly and never matures to the vacuole (253). Recent evidence suggests that it is ubiquitinylated and degraded by the proteasome in the cytosol (254). In clear distinction from the MHC molecules that are membrane-anchored proteins, CPY is a soluble lumenal ER protein. Thus, following synthesis, glycosylation, and *complete* transport to the ER lumen, it has to bind to the lumenal face of the ER membrane and be translocated into the cytosol for ubiquitinylation and degradation.

The notion that ER proteins are translocated in a retrograde manner raises several important mechanistic problems. One problem involves the specificity of recognition and the ability of the dislocation machinery to distinguish between properly folded and misfolded substrates. It is assumed that the proper folding of a protein requires a chaperone, and if the chaperone "fails" to fold a protein in repeated ATP-dependent association-dissociation cycles, it presents it to the ubiquitin proteolytic machinery (the "refold or degrade" function of chaperones; see 255). Such a mechanism has been described for soluble proteins (256, 257). The process in the ER probably involves chaperones as well. Indeed, the dislocation of CPY requires the Bip/Kar2 chaperone (258). The chaperones involved in folding of normal proteins must distinguish between normal proteins in the process of folding and misfolded proteins destined for degradation. The underlying mechanisms involved in signaling one population from the other are not known. The viral proteins US2 and US11 probably act as chaperones as well; however, they are different from the cellular native chaperones in the sense that they target a normal, properly folded (or in the process of folding) protein for degradation.

Another problem involves the machinery through which transport occurs. It appears that the retrograde transport is mediated by the Sec61 complex, which is also involved in anterograde insertion and transport of proteins into the ER membrane and lumen. In US2-expressing cells, de-glycosylated MHC molecules were immunoprecipitated along with components of Sec61 (252). Similarly, retrograde transport of mutated CPY and other misfolded secretory soluble proteins also require the yeast Sec61 translocation complex (258, 259). Interestingly, degradation of misfolded mutant components of the Sec61 complex requires the E2 enzymes Ubc6 and Ubc7 and is mediated by the proteasome (260). Since Ubc7 is a soluble cytosolic protein, it has to be recruited to the membrane, a process mediated by a novel protein, Cue1p (260a). Finally, dislocation requires energy. Indeed, US2-mediated targeting of the HCs of the MHC complex is ATP dependent. The cytosolic face of the ER membrane contains several candidate ATPases that could provide the energy necessary for dislocation. These include the cytosolic chaperone Hsc70 and the 19S regulatory complex of the 26S proteasome.

DIVERSE FUNCTIONS OF THE UBIQUITIN SYSTEM

Recent evidence implicates the involvement of the ubiquitin pathway in a variety of basic cellular processes. In many cases evidence is still circumstantial, and neither the cellular targeted substrates nor the underlying mechanisms involved have been elucidated. Yet, the processes appear to be important and therefore deserve special attention.

DEVELOPMENT   The involvement of the ubiquitin system in human brain development is indicated by a defect in a gene coding for the E3 enzyme E6-AP, which has been implicated as the cause of Angelman syndrome, a disorder characterized by mental retardation, seizures, and abnormal gait (56, 57; see also above). Other evidence links the ubiquitin system to developmental processes in the central nervous system (CNS). The *Drosophila bendless* (*ben*) gene encodes an E2 that appears to be restricted to the CNS during development. Mutations in this gene lead to morphological abnormalities within the visual system that involve, for example, impairment of synaptogenesis between photoreceptor cells and other elements of the system (10, 261). The mutation was described initially as a behavioral defect affecting the escape jump response and was ascribed to a lesion affecting the connectivity of the giant fiber to the motor neuron innervating the tergotrochanter muscle. It appears that the function of the gene product is broader and is involved in other developmental processes of the neuromuscular system as well.

Another gene that encodes a developmentally regulated de-ubiquitinating enzyme from the UBP (see above) family is the *Drosophila faf facets* (*faf*; see 130, 262). The *faf* gene is specifically required for eye development, and mutant *faf* flies have more than eight photoreceptors in each of the compound eye units, the facets. The *faf facets* protein is probably involved in generating the inhibitory signal sent by the photoreceptor cells to undifferentiated surrounding cells to stop differentiation and migration to the facet unit. The only other defect found in *faf* mutants affects the eggs that do not reach cellularization during early embryogenesis. The target protein(s) of *faf facets* have not been identified so far (however, see below).

Another recently described link between the ubiquitin system and *Drosophila* eye development is related to the function of *Tramtrak* (TTK88), *Phyllopod* (PHYL), and *Seven in Absentia* (SINA). TTK88 is a zinc-finger embryonic transcription factor whose expression represses neuronal cell fate determination in the developing eye. PHYL, in a mechanism that also requires SINA, antagonizes the activity of TTK88. Activation of the Sevenless Ras/MAPK signaling cascade by the Sevenless receptor tyrosine kinase, known to be involved in cell differentiation, leads to transcriptional induction of PHYL. The induced PHYL, TTK88, and SINA generate a heterotrimeric complex that targets TTK88 for degradation (263). SINA was found to interact with UbcD1, a *D. melanogaster* E2 enzyme, and a mutation in *UbcD1* serves as a dominant suppressor of the effects caused by overexpression of the SINA protein. A mutation in *faf facets* serves as a dominant enhancer for reduced *sina* activity (263). The role of SINA may be as an adaptor protein that provides the link between the TTK88/PHYL/SINA heterotrimer and the ubiquitin system (see above), whereas TTK88 may well be one substrate that is targeted by *faf facets*: *faf* may be an editing isopeptidase, protecting TTK88 from untimed degradation (see above). In another study, Li et al (264) showed that SINA and PHYL promote ubiquitinylation of TTK88, which is then degraded rapidly by the proteasome.

Roest et al (11) demonstrated that inactivation of HR6B, a human homolog of the yeast E2 RAD6/Ubc2 involved in DNA repair and targeting of N-end rule substrates (see above), leads to a single defect, male sterility: Knockout females are completely normal. The defect is specific to the development of sperm and does not involve the general process of meiosis. The lack of more severe effects may be due to the presence of a highly homologous enzyme, HR6A. The protein target of HR6B in the spermatids has not been identified. The authors hypothesized that HR6B is involved in polyubiquitinylation and degradation of histones. This process is critical for chromatin remodeling, which involves replacement of histones with transition protamines and, subsequently, with protamines. Histones are N-α-acetylated proteins, and HR6B may act on these proteins either via the non-N-end rule recognition site of E3 (see above) or via a novel, yet-to-be-identified species of E3. Another ubiquitin system–related gene involved in sex differentiation, oogenesis, or spermatogenesis is the *hyperplastic disc* (*hyd*) gene, which appears to play a major role in *D. melanogaster* development. The null phenotype appears to be lethality at an early embryonic stage; however, adults obtained by crosses of temperature-sensitive alleles and maintained at the permissive temperature are sterile and display, in addition to imaginal disc overgrowth, morphological abnormalities in the germ tissue (265). The gene encodes an approximately 280-kDa protein that belongs to the HECT family of E3 enzymes (see above). The target proteins of this E3 enzyme have not been identified.

APOPTOSIS   During development, a large number of cells die in a predicted spatial and temporal pattern known as programmed cell death (PCD), or apoptosis. This process is crucial for differentiation and involves programmed regulation of gene expression. One of the first genes shown to be involved in PCD is the polyubiquitin gene that is up-regulated during the metamorphosis of the hawkmoth *Manduca sexta* (266). During the 30-h period that precedes the adult moth's emergence from the pupal cuticle, there is a rapid degradation of the large mass of intersegmental muscles that served the pupa. The resulting amino acids are probably used in building the wing muscles and also as a source of energy for the butterfly's short life span. The strong induction of transcription of the polyubiquitin gene during the metamorphotic process is accompanied by a parallel increase in the level of ubiquitin. Qualitative and quantitative changes occur concurrently in the 26S proteasome (267, 268). Pools of ubiquitin conjugates increase 10-fold during this period, accompanied by an increase in the activity of E1, several E2 enzymes, and E3 enzyme(s) of unknown specificity (269). This coordinated induction of ubiquitin conjugation and degradation pathways is stimulated by the programmed decline in the steroid-like molting hormone, 20-hydroxyecdisone. All the changes can be blocked by administration of the hormone.

Several other studies have linked the ubiquitin system to apoptosis in different systems; however, because of the complexity and variety of the apoptotic pathways, it is unclear whether the system plays a primary causative or only a secondary role in the process. Also, the system's protein targets in these processes remain unidentified. The complexity of the processes is also reflected in the apparently conflicting reports as to the system's role in apoptosis in different experimental systems. In some model systems, PCD requires the activity of the ubiquitin pathway, whereas in others PCD is invoked following inhibition of the ubiquitin system.

γ-irradiation-induced apoptosis in human lymphocytes is accompanied by increased ubiquitin mRNA and ubiquitinylated nuclear proteins. Expression of ubiquitin sequence-specific antisense oligonucleotides leads to a significant decrease in the proportion of cells displaying the apoptotic phenotype (270). Similarly, lactacystin prevents ionizing irradiation-induced cell death of thymocytes (271). Inhibition of the ubiquitin system leads to prevention of apoptosis induced by NGF deprivation in sympathetic neurons (272). These findings suggest an active role for the ubiquitin system in PCD. However, in leukemic cells (273), activated T-cells (274), and some neuronal cells (275), inhibition of the ubiquitin system stimulates apoptosis. Thus, the involvement of the system appears to be cell- and environment-specific. In some cases inhibition of the system may lead to the accumulation of abnormal proteins with the possible consequent induction of apoptosis, whereas in others the system may play a direct role in the destructive process and its inhibition leads to an inhibition or a delay in the onset of the apoptotic chain of events.

ANTIGEN PROCESSING   Peptide epitopes presented to cytotoxic lymphocytes (CTLs) on class I MHC molecules are generated in the cytosol by limited processing of antigenic proteins. Although it was reported that the process can be mediated by ubiquitinylation- (276, 277) or proteasome-independent (278) mechanisms, it appears that many MHC class I antigens are processed by the ubiquitin-proteasome pathway (for review, see 4, 279).

Rock et al showed that both the less specific peptide aldehyde inhibitors (122) and the more specific proteasome inhibitor, lactacystin (280), inhibit processing and presentation of class I MHC antigens. The cytokine γ-interferon (γ-IFN) that stimulates antigen presentation leads also to induction and exchange of three proteasomal subunits in human cells: LMP2 for X (MB1, ε), LMP7 for Y (δ), and MECL1 for Z (281, 282). These exchanges lead to alteration in the cleavage site preferences of the proteasome: The tryptic- and chymotryptic-like activities are stimulated, whereas the PGPH activity is decreased. The changes in activities, which are mostly the result of the newly incorporated subunits LMP2 and LMP7, result in peptides that terminate mostly with basic and hydrophobic residues, similar to the vast majority of known peptides presented on MHC class I molecules. These C-termini may be required for selective uptake by the ER TAP transporter (283) and for better binding to the MHC molecule (284).

Initial reports showed that LMP2 and LMP7 are not essential for antigen presentation (285, 286). However, more detailed quantitative analyses showed that cells lacking these subunits have a decreased efficiency rather than an absolute inability to present antigens (287, 288). Macrophages and spleen cells derived from knockout mice lacking LMP2 exhibit a reduced capacity to stimulate a T-cell specific for a nucleoprotein epitope of the *Haemophilus influenza* A virus (288). The mutant mice themselves have significantly reduced levels of CD8+ T lymphocytes and also generate five- to six-fold fewer CTLs to the viral antigen. Similarly, mice lacking LMP7 demonstrate reduced cell surface levels of MHC class molecules (empty MHC molecules that do not carry antigenic peptides are unstable) and also demonstrate reduced T-cell response to the viral antigen HY (287). Thus the γ-IFN-induced alterations in the composition of the proteasomal subunits may increase the efficiency of antigen presentation and, consequently, of the immune surveillance.

Although the proteasome clearly is involved in cleavage at the C-terminal residue of antigenic peptides, the mechanism involved in specific cleavage at the N-terminal residue has yet to be identified. Recent evidence implicates the PA28 regulator of the proteasome in coordinated dual cleavages that lead to generation of the final antigenic peptides (289). The PA28 (designated also as the 11S regulator) strongly activates hydrolysis of peptides by the 20S proteasome in an ATP-independent manner, but it does not appear to be involved in degradation of intact proteins (290, 291). It is composed of two γ-IFN-inducible, approximately 27.5-kDa subunits, α and β. The chains form hexameric rings of approximately 200 kDa, composed of alternating subunits that associate loosely with the α endplates of the 20S proteasome and dissociate from it in low salt.

Overexpression of the PA28α (which is sufficient to stimulate the peptide-hydrolyzing activity) in mouse or human fibroblasts, along with the murine CMV pp89 protein, results in enhanced presentation to the appropriate CTL of an antigen derived from the viral protein. Enhanced presentation was observed also when an influenza nucleoprotein was expressed along with the α subunit. When purified PA28α was incubated along with purified 20S proteasome and a long peptide that contains the sequence of two antigens, one derived from the JAK1 protein of a mastocytoma cell line and the other derived from pp89, the protease trimmed the peptides on both sides to generate the authentic antigenic epitopes (289). However, because of the rarity of free peptides in the cytosol, the physiological role of PA28 is not clear.

Interestingly, it was reported that vaccination against Hsp70 derived from a sarcoma cell line (292) or from several autologous carcinomas, such as lung carcinoma, melanoma, and colon carcinoma, (292a) can render mice immune against the tumors. The Hsp associates with the antigenic peptide, and the function of the complex may be to carry the peptide from the site of its generation at the proteasome to the ER transport machinery. An Hsp-chaperoned peptide was shown to be channeled from the exogenous, MHC class II pathway to the endogenous, MHC class I pathway (293). A fraction of the proteasomes in the cell may exist as PA28/20S/19S heterotrimers. In this case, the proteasome will degrade intact proteins into large peptides, and the PA28 complex will trim the peptides to generate the antigenic epitopes.

Michalek et al (294) were the first to report that ubiquitinylation must precede processing of ovalbumin for presentation of the antigenic peptide SIINFEKL (amino acid residues 257–264) to the appropriate CTL. Using a mutant cell that harbors a thermolabile E1, the researchers showed that at the nonpermissive temperature, the peptide is not presented. Incubation at the high temperature, however, did not affect presentation of a peptide expressed in these cells from a minigene. This control experiment ruled out any defect in the transport or presentation machineries of the cell. Manipulation of genes that encode for antigenic proteins in a manner that renders the proteins more susceptible to ubiquitin-mediated degradation (such as conversion of the N-terminal amino acid residue into a "destabilizing" moiety) demonstrated that ubiquitin conjugation is a rate-limiting step in antigen presentation (295). Ben-Shahar et al (296) showed that production of SIINFEKL in a cell-free system from lymphocytes is mediated by the ubiquitin system: The process required ATP and ubiquitin, and could be reversibly inhibited by methylated ubiquitin. The in vitro experimental system may allow analysis of other components that may be involved in the process, such as molecular chaperones, which may be required to escort the generated peptides to the ER transport machinery, thus protecting them from cellular peptidases.

An interesting question involves recognition of protein antigens by the ubiquitin system. The Epstein-Barr virus (EBV) nuclear antigen 1 (EBNA1) persists in healthy virus carriers for life and is the only viral protein regularly detected in all EBV-associated malignancies. Unlike EBNAs 2–4, which are strong immunogens, EBNA1 is not processed and cannot elicit a CTL response. The persistence of EBNA1 contributes, most probably, to some of the pathologies caused by the virus. An interesting structural feature common to all EBNA1 proteins derived from different EBV strains is a relatively long Gly-Ala repeat of a variable length at the C-terminus domain of the molecule. Transfer of a strong antigenic epitope from EBNA4 (residues 416–424) to EBNA1 prevented its presentation, while its insertion in an EBNA1 deletion mutant that lacks the Gly-Ala repeat resulted in its presentation to the appropriate CTL. Similarly, insertion of the Gly-Ala repeat downstream to the 416–424 epitope in EBNA4 inhibited CTL recognition of the EBNA4 chimeric protein (297). Thus, the Gly-Ala repeat constitutes a *cis*-acting element that inhibits antigen processing and subsequent presentation of the resulting antigenic epitopes.

A more recent study carried out in a cell-free system showed that EBNA4 is degraded in an ATP-, ubiquitin-, and proteasome-dependent manner, whereas EBNA1 is resistant to proteolysis. However, EBNA1 is degraded by the ubiquitin cell-free system following deletion of the Gly-Ala repeat. Transfer of the signal to EBNA4 prevented, as expected, its degradation by the ubiquitin system (184). A short Gly-Ala repeat (38 amino acid residues) as well as a short Pro-Ala repeat have similar effects. Interestingly, the presence of the Gly-Ala repeat does not prevent ubiquitin conjugation to Gly-Ala-containing EBNA4. This finding suggests that the Gly-Ala repeat, like the Gly-rich region in p105 (see above), interferes with processing of the protein by the 26S proteasome. Since a Gly-rich region as well as Gly-Ala or Gly-Pro repeats of various lengths have similar inhibitory effects on processing, it appears that a formation of domains composed of small uncharged/hydrophobic residues prevent entry into the proteasome, thus inhibiting degradation of an already ubiquitin-tagged protein.

CONCLUDING REMARKS

Spectacular progress has been achieved in the past few years in our understanding of the important functions of selective, ubiquitin-mediated protein degradation in a variety of cellular processes. Ubiquitin-mediated protein degradation may be comparable to protein phosphorylation in the variety of its regulatory functions. This seemingly wasteful mechanism of using disposable protein regulators may be essential to ensure irreversibility of temporally controlled processes, such as cell cycle or development. The rapid degradation of protein regulators is especially important when the regulator should act for a short period of time (e.g., cyclins or some transcriptional regulators) or when a process is initiated by the degradation of an inhibitor (e.g. degradation of Ckis or IκBs). Further examples of ubiquitin-mediated regulatory processes will likely be discovered in the near future. For instance, it remains to be seen whether the rapid degradation of components of the circadian clock (298) is carried out by the ubiquitin system, and what controls the precise timing of its degradation.

We still know very little about the ubiquitin ligases (E3 enzymes), which are responsible mainly for the selectivity and regulation of ubiquitin-mediated protein degradation. As discussed above, this is partly because of the divergence of different families of E3 enzymes and should be tackled by a combination of biochemical and molecular genetic approaches. For example, work with cell-free systems is needed to determine whether Mdm2, which is required for the degradation of p53 (202, 203), or the APC protein required for the degradation of β-catenin (206, 208) are parts of E3 enzyme complexes that act in the ubiquitinylation of the respective proteins. Of equal importance is the identification of signals in proteins that are recognized by the different E3 enzymes. Significant progress has been made in this respect, such as the identification of the role of phosphorylation in the degradation of many proteins, but this is just the tip of the iceberg. Different phosphorylated degradation signals are likely recognized by different E3 enzymes, and many other degradation signals probably exist. The elucidation of other types of interplay between protein phosphorylation and protein degradation, as illustrated by the regulation of cyclosome activity in the cell cycle, are major challenges for the future.

ACKNOWLEDGMENTS

We thank many colleagues for communicating results prior to publication. The skillful and devoted secretarial assistance of Mary Williamson is gratefully acknowledged. Research in the laboratories of AH and AC is supported by grants from the Israel Science Foundation founded by the Israeli Academy of Sciences and Humanities, US-Israel Binational Science Foundation, and the Israeli Ministry of Sciences and the Arts. AC is also supported by the Council for Tobacco Research, Inc.; the German-Israeli Foundation for Scientific Research and Development; the Israel Cancer Research Fund; a TMR grant from the European Community and the UK-Israel Foundation for Scientific Research and Biotechnology. This article was written during the stay of AH (on sabbatical leave) at the Institute for Cancer Research, Fox Chase Cancer Center, Philadelphia, PA.

Literature Cited

1. Hershko A. 1996. *Trends Biochem. Sci.* 21:445–49
2. Hershko A, Ciechanover A. 1992. *Annu. Rev. Biochem.* 61:761–807
3. Ciechanover A. 1994. *Cell* 79:13–21
4. Coux O, Tanaka K, Goldberg AL. 1996. *Annu. Rev. Biochem.* 65:801–47
5. Rechsteiner M. 1997. In *Ubiquitin and the Biology of the Cell*, ed. D Finley, J-M Peters. New York/London: Plenum. In press
6. Wilkinson KD. 1995. *Annu. Rev. Nutr.* 15:161–89
7. Hochstrasser M. 1996. *Annu. Rev. Genet.* 30:405–39
8. Jentsch S. 1992. *Annu. Rev. Genet.* 26:179–207
9. Cenci G, Rawson RB, Belloni G, Castrillon DH, Tudor M, et al. 1997. *Genes Dev.* 11:863–75
10. Muralidhar MG, Thomas JB. 1993. *Neuron* 11:253–66
11. Roest HP, van Klaveren J, de Wit J, van Gurp CG, Koken MHM, et al. 1996. *Cell* 86:799–810
12. Harbers K, Muller U, Grams A, Li E, Jaenisch R, Franz T. 1996. *Proc. Natl. Acad. Sci. USA* 93:12412–17
13. Kalchman MA, Graham RK, Xia G, Koide HB, Hodgson JG, et al. 1996. *J. Biol. Chem.* 271:19385–94
14. Reiss Y, Heller H, Hershko A. 1989. *J. Biol. Chem.* 264:10378–83
15. Dohmen RJ, Madura K, Bartel B, Varshavsky A. 1991. *Proc. Natl. Acad. Sci. USA* 88:7351–55
16. Bailly V, Prakash S, Prakash L. 1997. *Mol. Cell. Biol.* 17:4536–43
17. Willems AR, Lanker S, Patton EE, Craig KL, Nason TF, et al. 1996. *Cell* 86:453–63
18. Mathias N, Johnson SL, Winey M, Adams AEM, Goetsch L, et al. 1996. *Mol. Cell. Biol.* 16:6634–43
19. Hershko A, Ganoth D, Sudakin V, Dahan A, Cohen LH, et al. 1994. *J. Biol. Chem.* 269:4940–46
20. Hershko A. 1997. *Curr. Opin. Cell Biol.* 9:788–99
21. Aristarkhov A, Eytan E, Moghe A, Admon A, Hershko A, Ruderman JV. 1996. *Proc. Natl. Acad. Sci. USA* 93:4294–99
22. Yu H, King RW, Peters J-M, Kirschner MW. 1996. *Curr. Biol.* 6:455–66
23. Townsley FM, Aristarkhov A, Beck S, Hershko A, Ruderman JV. 1997. *Proc. Natl. Acad. Sci. USA* 94:2362–67
24. Osaka F, Seino H, Seno T, Yamao F. 1997. *Mol. Cell. Biol.* 17:3388–97
25. King RW, Peters J-M, Tugendreich S, Rolfe M, Hieter P, et al. 1995. *Cell* 81:279–88
26. Scheffner M, Huibregtse JM, Howley PM. 1994. *Proc. Natl. Acad. Sci. USA* 91:8797–801
27. Rolfe M, Beer-Romero P, Glass S, Eckstein J, Berdo I, et al. 1995. *Proc. Natl. Acad. Sci. USA* 92:3264–68
28. Jensen JP, Bates PW, Yang M, Vierstra RD, Weissman AM. 1995. *J. Biol. Chem.* 270:30408–14
29. Nuber U, Schwarz S, Kaiser P, Schneider R, Scheffner M. 1996. *J. Biol. Chem.* 271:2795–800
30. Blumenfeld N, Gonen H, Mayer A, Smith CE, Siegel NR, et al. 1994. *J. Biol. Chem.* 269:9574–81
31. Kumar S, Kao WH, Howley PM. 1997. *J. Biol. Chem.* 272:13548–54
32. Hatakeyama S, Jensen JP, Weissman AM. 1997. *J. Biol. Chem.* 272:15085–92
33. Seufert W, Futcher B, Jentsch S. 1995. *Nature* 373:78–81
34. Tong H, Hateboer G, Perrakis A, Bernards R, Sixma TK. 1997. *J. Biol. Chem.* 272:21381–87
35. Kovalenko OV, Plug AW, Haaf T, Gonda DK, Ashley T, et al. 1996. *Proc. Natl. Acad. Sci. USA* 93:2958–63
36. Shen ZY, Pardington-Purtymun PE, Comeaux JC, Moyzis RK, Chen DJ. 1996. *Genomics* 37:183–86
37. Wang Z-Y, Qiu Q-Q, Seufert W, Taguchi T, Testa JR, et al. 1996. *J. Biol. Chem.* 271:24811–16
38. Jiang W, Koltin Y. 1996. *Mol. Gen. Genet.* 251:153–60
39. Yasugi T, Howley PM. 1996. *Nucleic Acids Res.* 24:2005–10
40. Hateboer G, Hijmans EM, Nooij JBD, Schlenker S, Jentsch S, et al. 1996. *J. Biol. Chem.* 271:25906–11
41. Masson M, deMurcia JM, Mattei M-G, deMurcia G, Niedergang CP. 1997. *Gene* 190:287–96
42. Kho C-J, Huggins GS, Endege WO, Hsieh C-M, Lee M-E, et al. 1997. *J. Biol. Chem.* 272:3845–51
43. Wright DA, Futcher B, Ghosh P, Geha RS. 1996. *J. Biol. Chem.* 271:31037–43
44. Becker K, Schneider P, Hofmann K, Mattman C, Tschopp J. 1997. *FEBS Lett.* 412:102–6
45. Saito H, Pu R, Cavenagh M, Dasso M.

1997. *Proc. Natl. Acad. Sci. USA* 94: 3736–41
46. Matunis MJ, Coutavas E, Blobel G. 1996. *J. Cell Biol.* 135:1457–70
47. Mahajan R, Delphin C, Guan T, Gerace L, Melchior F. 1997. *Cell* 88:97–107
48. Okura T, Gong L, Kamitani T, Wada T, Okura I, Wei CF, et al. 1996. *J. Immunol.* 157:4277–81
49. Loeb JDJ, Schlenstedt G, Pellman D, Kornitzer D, Silver PA, Fink GR. 1995. *Proc. Natl. Acad. Sci. USA* 92:7647–51
50. Hershko A, Heller H, Elias S, Ciechanover A. 1983. *J. Biol. Chem.* 258: 8206–14
51. Gonen H, Schwartz AL, Ciechanover A. 1991. *J. Biol. Chem.* 266:19221–31
52. Heller H, Hershko A. 1990. *J. Biol. Chem.* 265:6532–35
53. Scheffner M, Huibregtse JM, Vierstra RD, Howley PM. 1993. *Cell* 75:495–505
54. Scheffner M, Nuber U, Huibregtse JM. 1995. *Nature* 373:81–83
55. Huibregtse JM, Scheffner M, Beaudenon S, Howley PM. 1995. *Proc. Natl. Acad. Sci. USA* 92:2563–67
56. Kishino T, Lalande M, Wagstaff J. 1997. *Nat. Genet.* 15:70–73
57. Matsuura T, Sutcliffe JS, Fang P, Galjaard R-J, Jiang Y-H, et al. 1997. *Nat. Genet.* 15:74–77
58. Huibregtse JM, Yang JC, Beaudenon SL. 1997. *Proc. Natl. Acad. Sci. USA* 94:3656–61
59. Bregman DB, Halaban R, van Gool AJ, Henning KA, Friedberg EC, et al. 1996. *Proc. Natl. Acad. Sci. USA* 93:11586–90
60. Nefsky B, Beach D. 1996. *EMBO J.* 15:1301–12
61. Hein C, Springael J-Y, Volland C, Haguenauer-Tsapis R, Andre B. 1995. *Mol. Microbiol.* 18:77–87
62. Galan JM, Moreau V, Andre B, Volland C, Haguenauer-Tsapis R. 1996. *J. Biol. Chem.* 271:10946–52
63. Staub O, Dho S, Henry PC, Correa J, Ishikawa T, et al. 1996. *EMBO J.* 15: 2371–80
64. Staub O, Gautschi I, Ishikawa T, Breitschopf K, Ciechanover A, et al. 1977. *EMBO J.* 16:6325–36
65. Yashiroda H, Oguchi T, Yasuda Y, Toh-e A, Kikuchi Y. 1996. *Mol. Cell. Biol.* 16: 3255–63
66. Sudakin V, Ganoth D, Dahan A, Heller H, Hershko J, et al. 1995. *Mol. Biol. Cell* 6:185–98
67. Lahav-Baratz S, Sudakin V, Ruderman JV, Hershko A. 1995. *Proc. Natl. Acad. Sci. USA* 92:9303–7
68. Peters J-M, King RW, Höög C,

Kirschner MW. 1996. *Science* 274: 1199–201
69. Irniger S, Piatti S, Michaelis C, Nasmyth K. 1995. *Cell* 81:269–77
70. Lamb JR, Michaud WA, Sikorski RS, Hieter PA. 1994. *EMBO J.* 13:4321–28
71. Zachariae W, Shin TH, Galova M, Obermeier B, Nasmyth K. 1996. *Science* 274:1201–4
72. King RW, Deshaies RJ, Peters J-M, Kirschner MW. 1996. *Science* 274: 1652–59
73. Schneider BL, Yang Q-H, Futcher AB. 1996. *Science* 272:560–62
74. Schwob E, Böhm T, Mendenhall MD, Nasmyth K. 1994. *Cell* 79:233–44
75. Bai C, Sen P, Hofmann K, Ma L, Goebl M, et al. 1996. *Cell* 86:263–74
76. Verma R, Feldman RMR, Deshaies RJ. 1997. *Mol. Biol. Cell.* 8:1427–37
77. Barral Y, Jentsch S, Mann C. 1995. *Genes Dev.* 9:399–409
78. Kipreos ET, Lander LE, Wing JP, He WW, Hedgecock EM. 1996. *Cell* 85:829–39
79. Pause A, Lee S, Worrell RA, Chen DYT, Burghess WH, et al. 1997. *Proc. Natl. Acad. Sci. USA* 94:2156–61
80. Gonen H, Stancovski I, Shkedy D, Hadari T, Bercovich B, et al. 1996. *J. Biol. Chem.* 271:302–10
81. Stankovski I, Gonen H, Orian A, Schwartz AL, Ciechanover A. 1995. *Mol. Cell. Biol.* 15:7106–16
82. Orian A, Whiteside S, Israël A, Stankovski I, Schwartz AL, et al. 1995. *J. Biol. Chem.* 270:21707–14
83. Rogers S, Wells R, Rechsteiner M. 1986. *Science* 234:364–68
84. Rechsteiner M, Rogers SW. 1996. *Trends Biochem. Sci.* 21:267–71
85. Yaglom J, Linskens MHK, Sadis S, Rubin DM, Futcher B, et al. 1995. *Mol. Cell. Biol.* 15:731–41
86. Lanker S, Valdivieso MH, Wittenberg C. 1996. *Science* 271:1597–600
87. Kornitzer D, Raboy B, Kulka RG, Fink GR. 1994. *EMBO J.* 13:6021–30
88. Won K-A, Reed SI. 1996. *EMBO J.* 15:4182–93
89. Diehl JA, Zindy F, Sherr CJ. 1997. *Genes Dev.* 11:957–72
90. Aberle H, Bauer A, Stappert J, Kispert A, Kemler R. 1997. *EMBO J.* 16:3797–804
91. Kim TK, Maniatis T. 1996. *Science* 273: 1717–19
92. Nishizawa M, Okazaki K, Furuno N, Watanabe N, Sagata N. 1992. *EMBO J.* 11:2433–46

93. Okazaki K, Sagata N. 1995. *EMBO J.* 14:5048–59
94. Musti AM, Treier M, Bohmann D. 1997. *Science* 275:400–2
95. Varshavsky A. 1996. *Proc. Natl. Acad. Sci. USA* 93:12142–49
96. Sherman F, Stewart JW, Tsunasawa S. 1985. *BioEssays* 3:27–31
97. Madura K, Varshavsky A. 1994. *Science* 265:1454–58
98. Glotzer M, Murray AW, Kirschner MW. 1991. *Nature* 349:132–38
99. Yamano H, Gannon J, Hunt T. 1996. *EMBO J.* 15:5268–79
100. King RW, Glotzer M, Kirschner MW. 1996. *Mol. Biol. Cell* 7:1343–57
101. Brandeis M, Hunt T. 1996. *EMBO J.* 15:5280–89
102. Kaplon T, Jacquet M. 1995. *J. Biol. Chem.* 270:20742–47
103. Hochstrasser M, Varshavsky A. 1990. *Cell* 61:697–708
104. Bies J, Wolff L. 1997. *Oncogene* 14:203–12
105. Treier M, Staszewski LM, Bohmann D. 1994. *Cell* 78:787–98
106. Baumeister W, Lupas A. 1997. *Curr. Opin. Struct. Biol.* 7:273–78
107. Hilt W, Wolf DH. 1996. *Trends Biochem. Sci.* 21:96–102
108. Rubin DM, Finley D. 1995. *Curr. Biol.* 5:854–58
109. Stock D, Nederlof PM, Seemüller E, Baumeister W, Huber R, Löwe J. 1996. *Curr. Opin. Biotechnol.* 7:376–85
110. Groll M, Ditzel L, Löwe J, Stock D, Bochtler M, Bartunik HD, Huber R. 1997. *Nature* 386:463–71
111. Brannigan JA, Dodson G, Duggleby HJ, Moody PCE, Smith JL, et al. 1996. *Nature* 378:416–19
112. Schmidtke G, Kraft R, Kostka S, Henklein P, Frömmel K, et al. 1996. *EMBO J.* 15:6887–98
113. Chen P, Hochstrasser M. 1996. *Cell* 86:961–72
114. Arendt CS, Hochstrasser M. 1997. *Proc. Natl. Acad. Sci. USA* 94:7156–61
115. Kopp F, Hendil KB, Dahlmann B, Kristensen P, Sobek A, Uerkvitz W. 1997. *Proc. Natl. Acad. Sci. USA* 94:2939–44
116. Cerundolo V, Kelly A, Elliott T, Trowsdale J, Townsend A. 1995. *Eur. J. Immunol.* 25:554–62
117. Dick LR, Moomaw CR, DeMartino GN, Slaughter CA. 1991. *Biochemistry* 30:2725–34
118. Löwe J, Stock D, Jap B, Zwickl P, Baumeister W, Huber R. 1995. *Science* 268:533–39
119. Deveraux Q, Ustrell V, Pickart C, Rechsteiner M. 1994. *J. Biol. Chem.* 269:7059–61
120. van Nocker S, Deveraux Q, Rechsteiner M, Vierstra RD. 1996. *Proc. Natl. Acad. Sci. USA* 93:856–60
121. van Nocker S, Sadis S, Rubin DM, Glickman M, Fu H, et al. 1996. *Mol. Cell. Biol.* 16:6020–28
122. Rock KL, Gramm C, Rothstein L, Clark K, Stein R, et al. 1994. *Cell* 78:761–71
123. Fenteany G, Standaert RF, Lane WS, Choi S, Corey EJ, Schreiber SL. 1995. *Science* 268:726–31
124. Bogyo M, McMaster JS, Gaczynska M, Tortorella D, Goldberg AL, Ploegh H. 1997. *Proc. Natl. Acad. Sci. USA* 94:6629–34
125. Wilkinson KD, Tashayev VL, O'Connor LB, Larsen CN, Kasperek E, et al. 1995. *Biochemistry* 34:14535–46
126. Pickart CM, Rose IA. 1986. *J. Biol. Chem.* 261:10210–17
127. Johnston SC, Larsen CN, Cook WJ, Wilkinson KD, Hill CP. 1997. *EMBO J.* 16:3787–96
128. Hadari T, Warms JVB, Rose IA, Hershko A. 1992. *J. Biol. Chem.* 267:719–27
129. Amerik AYu, Swaminathan S, Krantz BA, Wilkinson KD, Hochstrasser M. 1997. *EMBO J.* 16:4826–38
130. Huang Y, Baker RT, Fischer-Vize JA. 1995. *Science* 270:1828–31
131. Hershko A, Ciechanover A, Heller H, Haas AL, Rose IA. 1980. *Proc. Natl. Acad. Sci. USA* 77:1783–86
132. Lam YA, Xu W, DeMartino GN, Cohen RE. 1997. *Nature* 385:737–40
133. Hershko A, Rose IA. 1987. *Proc. Natl. Acad. Sci. USA* 84:1829–33
134. Shaeffer JR, Cohen RE. 1997. *Blood* 90:1300–8
135. Papa FR, Hochstrasser M. 1993. *Nature* 366:313–19
136. Eytan E, Armon T, Heller H, Beck S, Hershko A. 1993. *J. Biol. Chem.* 268:4668–74
137. Moazed D, Johnson AD. 1996. *Cell* 86:667–77
138. Hegde AN, Inokuchi K, Pei W, Casadio A, Ghirardi M, et al. 1997. *Cell* 89:115–26
139. Zhu Y, Carroll M, Papa FR, Hochstrasser M, D'Andrea AD. 1996. *Proc. Natl. Acad. Sci. USA* 93:3255–79
140. Zhu Y, Lambert K, Corless C, Copeland NG, Gilbert DJ, et al. 1997. *J. Biol. Chem.* 272:51–57
141. Jaster R, Zhu Y, Pless M, Bhattacharya S, Mathey-Prevot B, D'Andrea AD. 1997. *Mol. Cell. Biol.* 17:3364–72

142. Nigg EA. 1995. *BioEssays* 17:471–80
143. Evans T, Rosenthal ET, Youngblom J, Distel D, Hunt T. 1983. *Cell* 33:389–96
144. Surana U, Amon A, Dowzer C, McGrew J, Byers B, et al. 1993. *EMBO J.* 12:1969–78
145. Hershko A, Heller H. 1985. *Biochem. Biophys. Res. Commun.* 128:1079–86
146. Hershko A, Ganoth D, Pehrson J, Palazzo RE, Cohen LH. 1991. *J. Biol. Chem.* 266:16376–79
147. Gordon C, McGurk G, Dillon P, Rosen C, Hastle ND. 1993. *Nature* 366:355–57
148. Ghislain M, Udvardy A, Mann C. 1993. *Nature* 366:358–62
149. Yamashita YM, Nakaseko Y, Samejima I, Kumada K, Yamada H, et al. 1996. *Nature* 384:276–79
150. Pines J. 1996. *Curr. Biol.* 6:1399–402
151. Patra D, Dunphy WG. 1996. *Genes Dev.* 10:1503–15
152. Sudakin V, Shteinberg M, Ganoth D, Hershko J, Hershko A. 1997. *J. Biol. Chem.* 272:18051–59
153. Nasmyth K. 1996. *Trends Genet.* 12:405–12
154. Ishii K, Kumada K, Toda T, Yanagida M. 1996. *EMBO J.* 15:6629–40
155. Hunt T, Luca FC, Ruderman JV. 1992. *J. Cell Biol.* 116:707–24
156. Amon A, Irniger S, Nasmyth K. 1994. *Cell* 77:1037–50
157. Clurman BE, Sheaff RJ, Thress K, Groudine M, Roberts JM. 1996. *Genes Dev.* 10:1979–90
158. Correa-Bordes J, Nurse P. 1995. *Cell* 83:1001–9
159. Kominami K-I, Toda T. 1997. *Genes Dev.* 11:1548–60
160. Peter M, Herskowitz I. 1994. *Science* 265:1228–31
161. McKinney JD, Chang F, Heintz N, Cross FR. 1993. *Genes Dev.* 7:833–43
162. McKinney JD, Cross FR. 1995. *Mol. Cell. Biol.* 15:2509–16
163. Elledge SJ, Winston J, Harper JW. 1996. *Trends Cell Biol.* 6:388–92
164. Sherr CJ, Roberts JM. 1995. *Genes Dev.* 9:1149–63
165. Franklin DS, Xiong Y. 1996. *Mol. Biol. Cell* 7:1587–99
166. Pagano M, Tam SW, Theodoras AM, Beer-Romeno P, Del Sal G, et al. 1995. *Science* 269:682–85
167. Hengst L, Reed SI. 1996. *Science* 271:1861–64
168. Sheaff RJ, Groudine M, Gordon M, Roberts JM, Clurman BE. 1997. *Genes Dev.* 11:1464–78
169. Maki CG, Howley PM. 1997. *Mol. Cell. Biol.* 17:355–63
170. Sandhu C, Garbe J, Bhattacharya N, Daksis J, Pan C-H, et al. 1997. *Mol. Cell. Biol.* 17:2458–67
171. Holloway SL, Glotzer M, King RW, Murray AW. 1993. *Cell* 73:1393–402
172. Funabiki H, Yamano H, Kumada K, Nagao K, Hunt T, et al. 1996. *Nature* 381:438–41
173. Funabiki H, Yamano H, Nagao K, Tanaka H, Yasuda H, et al. 1997. *EMBO J.* 16:5977–87
174. Cohen-Fix O, Peters J-M, Kirschner MW, Koshland D. 1996. *Genes Dev.* 10:3081–93
175. Juang Y-L, Huang J, Peters J-M, McLaughlin ME, Tai C-Y, et al. 1997. *Science* 275:1311–14
176. Stillman B. 1996. *Science* 274:1659–64
177. Piatti S, Böhm T, Cocker JH, Diffley JFX, Nasmyth K. 1996. *Genes Dev.* 10:1516–31
178. Nishitani H, Nurse P. 1995. *Cell* 83:397–405
179. Baeuerle PA, Baltimore D. 1996. *Cell* 87:13–20
180. Baldwin AS. 1996. *Annu. Rev. Immunol.* 14:649–83
181. Fan C-M, Maniatis T. 1991. *Nature* 354:395–98
182. Palombella VJ, Rando OJ, Goldberg AL, Maniatis T. 1994. *Cell* 78:773–85
183. Lin L, Ghosh S. 1996. *Mol. Cell. Biol.* 16:2248–54
184. Levitskaya J, Shapiro A, Leonchiks A, Ciechanover A, Masucci MG. 1997. *Proc. Natl. Acad. Sci. USA* 94:12616–21
185. Whiteside ST, Epinat J-C, Rice NR, Israël A. 1997. *EMBO J.* 16:1413–26
186. Brown K, Gerstberger S, Carlson L, Franzoso G, Siebenlist U. 1995. *Science* 267:1485–88
187. Brockman JA, Scherer DC, McKinsey TA, Hall SM, Qi X, et al. 1995. *Mol. Cell. Biol.* 15:2809–18
188. Chen ZJ, Hagler J, Palombella VJ, Melandri F, Scherer D, et al. 1995. *Genes Dev.* 9:1586–97
189. Alkalay I, Yaron A, Hatzubai A, Orian A, Ciechanover A, Ben Neriah Y. 1995. *Proc. Natl. Acad. Sci. USA* 92:10599–603
190. Traenckner E B-M, Pahl HL, Henkel T, Schmidt KN, Wilk S, Baeuerle PA. 1995. *EMBO J.* 14:2876–83
191. Scherer DC, Brockman JA, Chen Z, Maniatis T, Ballard DW. 1995. *Proc. Natl. Acad. Sci. USA* 92:11259–63
192. Alkalay I, Yaron A, Hatzubai A, Jung S, Avraham A, et al. 1995. *Mol. Cell. Biol.* 15:1294–301

193. Yaron A, Gonen H, Alkalay I, Hatzubai A, Jung S, et al. 1997. *EMBO J.* 16: 6486–94

194. Régnier CH, Song HY, Gao X, Goeddel DV, Cao Z, Rothe M. 1997. *Cell* 90:373–83

195. DiDonato JA, Hayakawa M, Rothwarf DM, Zandi E, Karin M. 1997. *Nature* 388:548–54

195a. Mercurio F, Zhu HY, Murray BW, Shevchenko A, Bennett BL, et al. 1997. *Science* 278:860–66

196. Lee FS, Hagler J, Chen ZJ, Maniatis T. 1997. *Cell* 88:213–22

197. Zhong HH, Su Yang H, Erdjument-Bromage H, Tempst P, Ghosh S. 1997. *Cell* 89:413–24

198. Chen ZJ, Parent L, Maniatis T. 1996. *Cell* 84:853–62

199. Ciechanover A, DiGiuseppe JA, Bercovich B, Orian A, Richter JD, et al. 1991. *Proc. Natl. Acad. Sci. USA* 88: 139–43

200. Maki CG, Huibregtse JM, Howley PM. 1996. *Cancer Res.* 56:2649–54

201. Beer-Romero P, Glass S, Rolfe M. 1997. *Oncogene* 14:595–602

202. Haupt Y, Maya R, Kazaz A, Oren M. 1997. *Nature* 387:296–99

203. Kubbutat MHG, Jones SN, Vousden KH. 1997. *Nature* 387:299–303

204. Fuchs SY, Dolan L, Davis RJ, Ronai Z. 1996. *Oncogene* 13:1531–35

205. Miller JR, Moon RT. 1996. *Genes Dev.* 10:2527–39

206. Yost C, Torres M, Miller JR, Huang E, Kimelman D, Moon RT. 1996. *Genes Dev.* 10:1443–54

207. Morin PJ, Sparks AB, Korinek V, Barker N, Clevers H, et al. 1997. *Science* 275: 1787–90

208. Rubinfeld B, Robbins P, El-Gamil M, Albert I, Porfiri E, Polakis P. 1997. *Science* 275:1790–92

209. Korinek V, Barker N, Morin PJ, van Wichen D, de Weger R, et al. 1997. *Science* 275:1784–87

210. Munemitsu S, Albert I, Souza B, Rubinfeld B, Polakis P. 1995. *Proc. Natl. Acad. Sci. USA* 92:3046–50

211. Hofmann F, Martelli F, Livingston DM, Wang ZY. 1996. *Genes Dev.* 10:2949–59

212. Campanero MR, Flemington EK. 1997. *Proc. Natl. Acad. Sci. USA* 94:2221–26

213. Hicke L, Riezman H. 1996. *Cell* 84:277–87

214. Roth AF, Davis NG. 1996. *J. Cell Biol.* 134:661–74

215. Rohrer J, Bénédetti H, Zanolari B, Riezman H. 1993. *Mol. Biol. Cell* 4:511–21

216. Hicke L, Zanolari B, Riezman H. 1998. *J. Cell Biol.* In press

217. Kölling R, Losko S. 1997. *EMBO J.* 16:2251–61

218. Egner R, Kuchler K. 1996. *FEBS Lett.* 378:177–81

219. Strous GJ, van Kerkhof P, Govers R, Ciechanover A, Schwartz AL. 1996. *EMBO J.* 15:3806–12

220. Galan J-M, Haguenauer-Tsapis R. 1997. *EMBO J.* 16:5847–54

220a. Terrell J, Shih S, Dunn R, Hicke L. 1998. *Mol. Cell* 1:193–202

221. Horak J, Wolf DH. 1997. *J. Bacteriol.* 179:1541–49

222. Miyazawa K, Toyoma K, Gotoh A, Hendrie PC, Mantel C, Broxmeyer HE. 1994. *Blood* 83:137–45

223. Yee NS, Hsiau CWM, Serve H, Vosseller K, Besmer P. 1994. *J. Biol. Chem.* 269:31991–98

224. Bokkala S, Joseph SK. 1997. *J. Biol. Chem.* 272:12454–61

225. Cenciarelli C, Hou D, Hsu K-C, Rellahan BL, Wiest DL, et al. 1992. *Science* 257:795–97

226. Hou D, Cenciarelli C, Jensen JP, Nguyen HB, Weissman AM. 1994. *J. Biol. Chem.* 269:14244–47

227. Cenciarelli C, Wilhelm KG Jr, Guo A, Weissman AM. 1996. *J. Biol. Chem.* 271:8709–13

228. Mori S, Tanaka K, Omura S, Saito Y. 1995. *J. Biol. Chem.* 270:29447–52

229. Jeffers M, Taylor GA, Weidner KM, Omura S, Vande Woude GF. 1997. *Mol. Cell. Biol.* 17:799–808

230. Laing JG, Beyer EC. 1995. *J. Biol. Chem.* 270:26399–403

231. Paolini R, Kinet JP. 1993. *EMBO J.* 12:779–86

232. Cahoreau C, Garnier L, Djiane J, Devauchelle G, Cerutti M. 1994. *FEBS Lett.* 350:230–34

233. Galcheva-Gargova Z, Theroux SJ, Davis RJ. 1995. *Oncogene* 11:2649–55

234. Bonifacino JS, Lippincott-Schwartz J. 1991. *Curr. Opin. Cell Biol.* 3:592–600

235. Kopito RR. 1997. *Cell* 88:427–30

236. Brodsky JL, McCracken AA. 1997. *Trends Cell Biol.* 7:151–56

237. Hampton RY, Rine J. 1994. *J. Cell Biol.* 125:299–312

238. Hampton RY, Gardner RG, Rine J. 1996. *Mol. Biol. Cell* 7:2029–44

239. Hampton RY, Bhakta H. 1997. *Proc. Natl. Acad. Sci. USA* 94:12944–48

240. McGee TP, Cheng HH, Kumagai H, Omura S, Simoni RD. 1996. *J. Biol. Chem.* 271:25630–38

241. Mayer A, Gropper R, Schwartz AL, Ciechanover A. 1989. *J. Biol. Chem.* 264:2060–68
242. Lukacs GL, Mohamed A, Kertner N, Chang X-B, Riordan JR, Grinstein S. 1994. *EMBO J.* 13:6076–86
243. Ward CL, Kopito RR. 1994. *J. Biol. Chem.* 269:25710–18
244. Ward CL, Omura S, Kopito RR. 1995. *Cell* 83:121–27
245. Jensen TJ, Loo MA, Pind S, Williams DB, Goldberg AL, Riordan JR. 1995. *Cell* 83:129–35
246. Yang M, Omura S, Bonifacino JS, Weissman AM. 1998. *J. Exp. Med.* 187:835–46
247. Huppa JB, Ploegh HL. 1997. *Immunity* 7:113–22
248. Yu H, Kaung G, Kobayashi S, Kopito RR. 1997. *J. Biol. Chem.* 272:20800–4
249. Hughes EA, Hammond C, Cresswell P. 1997. *Proc. Natl. Acad. Sci. USA* 94:1896–901
250. Gilbert MJ, Riddell SR, Plachter B, Greenberg PD. 1996. *Nature* 383:720–22
251. Wiertz EJHJ, Jones TR, Sun L, Bogyo M, Geuze HJ, Ploegh H. 1996. *Cell* 84:769–79
252. Wiertz EJHJ, Tortorella D, Bogyo M, Yu J, Mothes W, et al. 1996. *Nature* 384:432–38
253. Knop M, Finger A, Braun T, Hellmuth K, Wolf DH. 1996. *EMBO J.* 15:753–63
254. Hiller MM, Finger A, Schweiger M, Wolf DH. 1996. *Science* 273:1725–28
255. Craig EA, Baxter BK, Becker J, Halladay J, Ziegelhoffer D. 1994. In *The Biology of Heat Shock Proteins and Molecular Chaperones*, ed. RI Morimoto, A Tissières, C Georgopoulos, pp. 31–52. New York: Cold Spring Harbor Lab. Press
256. Lee DH, Sherman MY, Goldberg AL. 1996. *Mol. Cell. Biol.* 16:4773–81
257. Bercovich B, Stancovski I, Mayer A, Blumenfeld N, Laszlo A, et al. 1997. *J. Biol. Chem.* 272:9002–10
258. Plemper RK, Böhmler S, Bordallo J, Sommer T, Wolf DH. 1997. *Nature* 388:891–95
259. Pilon M, Schekman R, Romisch K. 1997. *EMBO J.* 16:4540–48
260. Biederer T, Volkwein C, Sommer T. 1996. *EMBO J.* 15:2069–76
260a. Biederer T, Volkwein C, Sommer T. 1997. *Science* 278:1806–9
261. Oh CE, McMahon R, Benzer S, Tanouye MA. 1994. *J. Neurosci.* 14:3166–79
262. Huang Y, Fischer-Vize JA. 1996. *Development* 122:3207–16

263. Tang AH, Neufeld TP, Kwan E, Rubin GM. 1997. *Cell* 90:459–67
264. Li S, Li Y, Carthew RW, Lai Z-C. 1997. *Cell* 90:469–78
265. Mansfield E, Hersperger E, Biggs J, Shearn A. 1994. *Dev. Biol.* 165:507–26
266. Schwartz LM, Myer A, Kosz L, Engelstein M, Maier C. 1990. *Neuron* 5:411–19
267. Jones ME, Haire MF, Kloetzel PM, Mykles DL, Schwartz LM. 1995. *Dev. Biol.* 169:436–47
268. Dawson SP, Arnold JE, Mayer NJ, Reynolds SE, Billett MA, et al. 1995. *J. Biol. Chem.* 270:1850–58
269. Haas AL, Baboshina O, Williams B, Schwartz LM. 1995. *J. Biol. Chem.* 270:9407–12
270. Delic J, Morange M, Magdelenat H. 1993. *Mol. Cell. Biol.* 13:4875–83
271. Grimm LM, Goldberg AL, Poirier GG, Schwartz LM, Osborne BA. 1996. *EMBO J.* 15:3835–44
272. Sadoul R, Fernandez PA, Quiquerez AL, Martinou I, Maki M, et al. 1996. *EMBO J.* 15:3845–52
273. Drexler HCA. 1997. *Proc. Natl. Acad. Sci. USA* 94:855–60
274. Cui H, Matsui K, Omura S, Schauer SL, Matulka RA, et al. 1997. *Proc. Natl. Acad. Sci. USA* 94:7515–20
275. Lopes UG, Erhardt P, Yao RJ, Cooper GM. 1997. *J. Biol. Chem.* 272:12893–96
276. Cox JH, Galardy P, Bennink JR, Yewdell JW. 1995. *J. Immunol.* 154:511–19
277. Dick LR, Aldrich C, Jameson SC, Moomaw CR, Pramanik BC, et al. 1994. *J. Immunol.* 152:3884–94
278. Vinitsky A, Antón LC, Snyder HL, Orlowski M, Bennink JR, Yewdell JW. 1997. *J. Immunol.* 159:554–64
279. Groettrup M, Soza A, Kuckelkorn U, Kloetzel PM. 1996. *Immunol. Today* 17:429–35
280. Craiu A, Gaczynska M, Akopian T, Gramm CF, Fenteany G, et al. 1997. *J. Biol. Chem.* 272:13437–45
281. Gaczynska M, Rock KL, Goldberg AL. 1993. *Nature* 365:264–67
282. Driscoll J, Brown MG, Finley D, Monaco JJ. 1993. *Nature* 365:262–64
283. Heemels MT, Schumacher TNM, Wonigeit K, Ploegh HL. 1993. *Science* 262:2059–63
284. Rammensee HG, Falk K, Rotzschke O. 1993. *Annu. Rev. Immunol.* 11:213–44
285. Arnold D, Driscoll J, Androlewicz M, Hughes E, Cresswell P, Spies T. 1992. *Nature* 360:171–74

286. Yewdell J, Lapham C, Bacik I, Spies T, Bennink J. 1994. *J. Immunol.* 152:1163–70

287. Fehling HJ, Swat W, Laplace C, Kuhn R, Rajewsky K, et al. 1994. *Science* 265:1234–37

288. van Kaer L, Ashton-Rickardt PG, Eichelberger M, Gaczynska M, Nagashima K, et al. 1994. *Immunity* 1:533–41

289. Dick TP, Ruppert T, Groettrup M, Kloetzel PM, Kuehn L, et al. 1996. *Cell* 86:253–62

290. Ma CP, Slaughter CA, DeMartino GN. 1992. *J. Biol. Chem.* 267:10515–23

291. Dubiel W, Pratt G, Ferell K, Rechsteiner M. 1992. *J. Biol. Chem.* 267:22369–77

292. Udono H, Srivastava PK. 1993. *J. Exp. Med.* 178:1391–96

292a. Tamura Y, Peng P, Liu K, Daou M, Srivastava PK. 1997. *Science* 278:117–20

293. Suto R, Srivastava PK. 1995. *Science* 269:1585–88

294. Michalek MT, Grant EP, Gramm C, Goldberg AL, Rock KL. 1993. *Nature* 363:552–54

295. Grant EP, Michalek MT, Goldberg AL, Rock KL. 1995. *J. Immunol.* 155:3750–58

296. Ben-Shahar S, Cassuto B, Novak L, Porgador A, Reiss Y. 1997. *J. Biol. Chem.* 272:21060–66

297. Levitskaya J, Coram M, Levitsky V, Imreh S, Steigerwald-Mullen PM, et al. 1995. *Nature* 375:685–888

298. Zeng H, Qian Z, Myers MP, Rosbash M. 1996. *Nature* 380:129–35

Annu. Rev. Biochem. 1998. 67:481–507

PHOSPHOINOSITIDE KINASES

David A. Fruman, Rachel E. Meyers, and Lewis C. Cantley
Division of Signal Transduction, Department of Medicine, Beth Israel Deaconess
Medical Center, Boston, Massachusetts 02215, and Department of Cell Biology,
Harvard Medical School, Boston, Massachusetts 02115;
e-mail: dfruman@bidmc.harvard.edu, rmeyers@bidmc.harvard.edu,
cantley@helix.mgh.harvard.edu

KEY WORDS: lipid kinase, phosphatidylinositol, signal transduction, second messengers,
              phospholipids

ABSTRACT

Phosphatidylinositol, a component of eukaryotic cell membranes, is unique among
phospholipids in that its head group can be phosphorylated at multiple free hydroxyls. Several phosphorylated derivatives of phosphatidylinositol, collectively
termed phosphoinositides, have been identified in eukaryotic cells from yeast
to mammals. Phosphoinositides are involved in the regulation of diverse cellular processes, including proliferation, survival, cytoskeletal organization, vesicle
trafficking, glucose transport, and platelet function. The enzymes that phosphorylate phosphatidylinositol and its derivatives are termed phosphoinositide kinases.
Recent advances have challenged previous hypotheses about the substrate selectivity of different phosphoinositide kinase families. Here we re-examine the
pathways of phosphoinositide synthesis and the enzymes involved.

CONTENTS

481

INTRODUCTION

Although phosphatidylinositol (PtdIns) represents only a small percentage of total cellular phospholipids, it plays a crucial role in signal transduction as the precursor of several second-messenger molecules. The inositol head group contains five free hydroxyls with the potential to become phosphorylated (Figure 1a). Thus, numerous derivatives of PtdIns[1] could exist in cells, each with a unique function. To date, the following phosphoinositides have been identified in cells: PtdIns-3-phosphate (hereafter termed PtdIns-3-P), PtdIns-4-phosphate (PtdIns-4-P), PtdIns-5-phosphate (PtdIns-5-P), PtdIns-3,4-*bis*phosphate (PtdIns-3,4-P_2), PtdIns-3,5-*bis*phosphate (PtdIns-3,5-P_2), PtdIns-4,5-*bis*phosphate (PtdIns-4,5-P_2), and PtdIns-3,4,5-*tris*phosphate (PtdIns-3,4,5-P_3) (Figure 1b). Three general functions of these lipids could be imagined: (*a*) to serve as phospholipase substrates for the generation of soluble inositol phosphate (and membrane-associated diacylglycerol) second messengers; (*b*) to interact directly with intracellular proteins, affecting their location and/or activity; (*c*) to alter local membrane topology by electrostatic interactions. The first two functions are well established for certain phosphoinositides, and some data provide evidence for the third role. The specific cellular functions of the individual phosphoinositides have recently been reviewed in detail (1–4). Our intention here is to present our current understanding of the pathways of synthesis of each phosphorylated lipid and how different steps are regulated. Although some phosphoinositides can be generated by lipid-specific phosphatases acting upon more highly phosphorylated forms, the lipid kinases are the focus of this review.

Many phosphoinositide kinases were described initially as enzymatic activities capable of transferring a phosphate to a particular position on the inositol ring of PtdIns or one or more of its phosphorylated derivatives. Studies of these purified enzymes led to the categorization of PtdIns or phosphoinositide kinases into three general families: phosphoinositide 3-kinases (PI3Ks), PtdIns 4-kinases (PtdIns4Ks), and PtdIns-P (PIP) kinases (PIP5Ks). Research in this area has been accelerated by the cloning of genes encoding several of these enzyme classes. Members of each phosphoinositide kinase family have been identified in yeast and other lower eukaryotes; each shares substantial protein-sequence homology with its mammalian counterparts. This evolutionary

---

[1] This review uses nomenclature that distinguishes phosphatidylinositol from phosphoinositides, the phosphorylated derivatives of PtdIns.

Phosphatidylinositol (PtdIns)

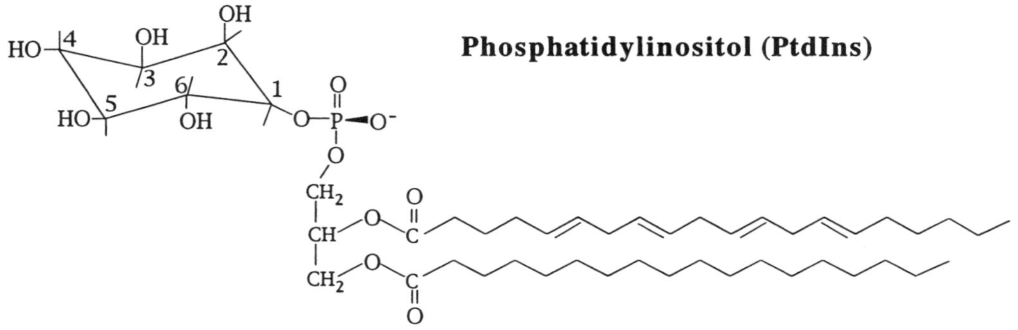

b.

Monophosphoinositides: PtdIns-3-P, PtdIns-4-P, PtdIns-5-P

Bisphosphoinositides: PtdIns-3,4-P$_2$, PtdIns-3,5-P$_2$, PtdIns-4,5-P$_2$

Trisphosphoinositides: PtdIns-3,4,5-P$_3$

Figure 1 (*a*) Chemical structure of phosphatidylinositol (PtdIns). Note the free hydroxyls at positions 2–6 of the inositol head group. The *myo*-D-enantiomer of inositol is shown, in which the 2'-hydroxyl is axial and the other hydroxyls are equatorial. (*b*) List of phosphoinositides known to exist in mammalian cells. PtdIns-4-P and PtdIns-4,5-P$_2$ represent approximately 60% and 30%, respectively, of the total phosphoinositides in cells. It is important to note that PtdIns itself is not considered a phosphoinositide. Some of the enzymes discussed in this review are not strictly phosphoinositide kinases because they phosphorylate only PtdIns (see text). However, they belong to the phosphoinositide kinase superfamily based on sequence homology.

conservation underscores the importance of these enzymes in the physiology of all eukaryotic cells. Sequence homology generally supports the separate classification of PI3Ks, PtdIns4Ks, and PIP kinases (for reviews of phosphoinositide kinase phylogeny, see References 5–7). However, studies of the recombinant enzymes have revealed that certain phosphoinositide kinases have different or broader activities than realized previously.

Each section that follows begins with a review of the molecular biology of a particular phosphoinositide kinase family and concludes with a discussion of the enzymology and regulation of the family members. For some enzymes we suggest new nomenclature that reflects our current understanding of substrate selectivity. We conclude with an overview of phosphoinositide synthesis pathways, using the proposed nomenclature where appropriate. The reader is referred to several recent reviews that detail the cellular functions of the enzymes and their lipid products (1–4, 8, 9).

PI 3-KINASES

PI3Ks have been studied intensively since the discovery of a PI3K activity associated with two viral oncoproteins: polyoma middle T (mT) antigen and pp60^{v-src} (10). Subsequent work has confirmed a role for PI3Ks and their products not only in growth regulation but also in various other cellular responses. Interest in these enzymes has further increased following recent findings that PI3K activation prevents cell death (9), that PI3K is a retrovirus-encoded oncogene (11), and that PI3K mutations increase lifespan in *Caenorhabditis elegans* (12). We refer to the different PI3K classes using nomenclature proposed recently by Domin & Waterfield (7).

Class I PI3Ks

MOLECULAR BIOLOGY   Class I PI3Ks are heterodimeric proteins, each of which consists of a catalytic subunit of 110–120 kDa and an associated regulatory subunit. Three mammalian PI3Ks sharing 42–58% amino acid sequence identity have been cloned and designated p110α, p110β, and p110δ (13–15). Each of these proteins contains an N-terminal region that interacts with regulatory subunits, a domain that binds to the small G protein ras, a "PIK domain" homologous to a region found in other phosphoinositide kinases, and a C-terminal catalytic domain (Figure 2). p110-related genes have been cloned from a range of eukaryotes including *C. elegans*, *Drosophila melanogaster*, and *Dictyostelium discoideum* (12, 16, 17). Together these gene products are termed class I$_A$ PI3Ks.

The regulatory subunits that associate with class I$_A$ PI3Ks are often called p85 proteins, based on the molecular weight of the first two isoforms to be purified (18) and cloned (19–21): p85α and p85β. p85 proteins do not possess any

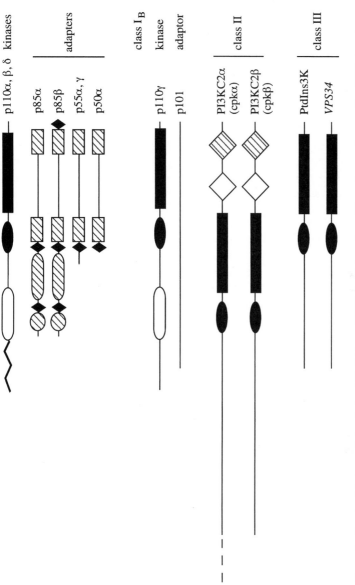

Figure 2 Structural features of phosphoinositide 3-kinase (PI3K) family members. Three classes of mammalian PI3Ks have been defined; class I is further subdivided based on the associated adapter (regulatory) subunit. The only yeast PI3K, Vps34, is homologous to the human phosphatidylinositol (PtdIns)-specific class III enzyme (denoted PtdIns 3-kinase). The protein domains are as follows: catalytic domain (*solid rectangles*), phosphoinositide kinase (PIK) domain (*open ovals*), ras-binding domain (*open ovals*), rhoGAP-homology domain (*hatched oval*), SH2 domain (*hatched rectangles*), SH3 domain (*hatched circle*), proline-rich motif (*solid diamonds*), C2 domain (*hatched diamonds*), PX domain (*open diamonds*), p85-interaction region in class IA enzymes (*sawtoothed line*).

known enzymatic activity but are composed of several domains with homology to those found in other signaling proteins. These domains are termed modular because they can be separated functionally and spatially from the rest of the protein in which they reside. p85α and p85β contain an N-terminal src-homology 3 (SH3) domain, two or three proline-rich segments, a region of homology to GTPase-activating proteins for the rho family of small G proteins (rho-GAPs), and two src-homology 2 (SH2) domains (Figure 2). The function of each of these modules is discussed below. Between the two SH2 domains is a region that is necessary and sufficient for interaction with the N terminus of p110 catalytic subunits. Termed the inter-SH2 domain, this region contains sequences that are predicted to form α-helices that fold into a coiled-coil structure.

The p85α gene has several splice variants, two of which encode the smaller proteins p50α and p55α (22–25). These proteins have unique N termini of 6 and 35 amino acids, respectively, and share the C terminus of p85α, including the second proline-rich motif, the SH2 domains, and the inter-SH2 domain (Figure 2). A third gene, p55γ, encodes a protein with similar overall structure to p55α (26). There is no evidence to date that different regulatory subunit isoforms pair preferentially with different p110 isoforms (15). However, some data suggest that different p85 subunits may associate with different subsets of intracellular proteins (27–29). *Drosophila* contain a p85-related protein of 60 kDa with conserved SH2 domains and an inter-SH2 domain with probable coiled-coil structure (30).

A protein with 36% identity to p110α was cloned and designated p110γ (31–33). p110γ contains a PIK domain, a kinase domain, and a ras-binding domain but diverges from class I$_A$ PI3Ks at its N terminus (Figure 2) and does not interact with p85 proteins. It has therefore been designated a class I$_B$ PI3K. p101, a putative regulatory subunit for p110γ, possesses no recognizable homology to other proteins (33). The regions of interaction between p101 and p110γ have not been mapped.

ACTIVITY AND REGULATION   In vitro, all the class I PI3Ks are able to phosphorylate PtdIns, PtdIns-4-P, or PtdIns-4,5-P$_2$ on the free 3-position (Table 1). Class I$_A$ PI3Ks also phosphorylate PtdIns-5-P in vitro (34). However, agonists that stimulate these enzymes in vivo cause increases mainly in the cellular levels of PtdIns-3,4-P$_2$ and PtdIns-3,4,5-P$_3$. In addition, kinetic studies in ^{32}P-labeled cells suggest that PtdIns-3,4-P$_2$ may be produced in part by the action of a 5-phosphatase on PtdIns-3,4,5-P$_3$ (35). Thus, the class I PI3Ks may be selective for PtdIns-4,5-P$_2$ in vivo. Of note, the inter-SH2 domain of p85α binds to PtdIns-4-P and PtdIns-4,5-P$_2$ in vitro (36); this property may provide a mechanism for presenting these substrates to the catalytic subunit or to concentrate the enzyme at membranes rich in these lipids.

Table 1 Properties of mammalian phosphoinositide kinases

Enzyme	Catalytic subunits		Regulatory subunits	In vitro substrates	Inhibitors[b]
	Isoforms	MW (kDa)[a]			
Class I$_A$ PI3K	p110α	123 (110)	p85α, p55α,	PtdIns-4,5-P$_2$,	Detergent,
	p110β	123 (119)	p50α, p85β,	PtdIns-4-P,	wortmannin (1–10),
	p110δ	119 (115)	p55γ[c]	PtdIns,[d]	Ly294002 (1000)
				PtdIns-5-P	
Class I$_B$ PI3K	p110γ	120 (110)	p101	PtdIns-4,5-P$_2$,	Detergent,
				PtdIns-4-P,	wortmannin (1–10),
				PtdIns[d]	Ly294002 (1000)
Class II PI3K	P13KC2α	170, 210		PtdIns,	Wortmannin
	P13KC2β	180		PtdIns-4-P,	(50–450),
				(PtdIns-4,5-P$_2$)[e]	Ly294002 (19000)
Class III PI3K (PtdIns3K)		101	p150	PtdIns	Wortmannin (2–10)
PtdIns4Kα		100, 230[f]		PtdIns	Adenosine 4C5G antibody Wortmannin?[f]
PtdIns4Kβ		90 (110)		PtdIns	Wortmannin (150)
PIP4K	α and β	47 (53)		PtdIns-5-P, PtdIns-3-P	PtdIns-4,5-P$_2$ Heparin
PIP5K	α and β	61 (68)		PtdIns-4-P, PtdIns-3,4-P$_2$, PtdIns-3-P[g]	PtdIns-4,5-P$_2$

[a]The deduced molecular weight (MW) of each phosphoinositide kinase is shown. Numbers in parentheses represent cases where the apparent molecular weight determined by sodium dodecyl sulfate-polyacrylamide gel electrophoresis (SDS-PAGE) is significantly different.
[b]Values in parentheses indicate the 50% inhibitory concentration (IC$_{50}$) or range of IC$_{50}$ values (in nanomolar concentration) of the inhibitor.
[c]Most of the possible combinations of class I$_A$ catalytic and regulatory subunits have been demonstrated to occur in vivo (15).
[d]Each class I enzyme can utilize all three lipid substrates in vitro, although phosphatidylinositol-4,5-bisphosphate (PtdIns-4,5-P$_2$) appears to be the preferred substrate in vivo.
[e]PtdIns is the preferred substrate of class II enzymes in vitro. PtdIns-4,5-P$_2$ is phosphorylated only when phosphatidylserine is used as a carrier.
[f]The two PtdIns4Kα proteins appear to be products of alternative splicing. The higher-molecular-weight form has been reported to be inhibited by ∼200 nM wortmannin (116).
[g]PtdIns-3-P can be converted to PtdIns-3,4,5-P$_3$ in an apparent concerted reaction (133).

Class I PI3Ks possess intrinsic protein kinase activity that is inseparable from their lipid kinase activity (15, 37–39). In fact, all phosphoinositide kinases contain several key residues conserved in the catalytic domains of classical protein kinases. The major substrates of the protein kinase activity of a class I PI3K are serine residues within the catalytic subunit itself and/or its associated regulatory subunit. p110α phosphorylates p85α at serine 608; interestingly, this phosphorylation results in down-regulation of the lipid kinase activity of p110α (37, 38). p110δ prefers autophosphorylation to intersubunit phosphorylation, but this autophosphorylation similarly down-regulates enzyme activity (15). p110γ also autophosphorylates, but without demonstrably affecting enzyme activity (39).

p85/p110 complexes can phosphorylate insulin receptor substrate-1 (IRS-1) in vitro and possibly in vivo (40, 41); however, other protein substrates for class I PI3Ks in vivo have not been established.

The fungal metabolite wortmannin is a potent inhibitor of the lipid (and protein) kinase activities of class I PI3Ks. The 50% inhibitory concentration (IC_{50}) values for inhibition of the isolated enzymes are all in the range of 1–10 nM (Table 1). Similar concentrations are required to inhibit p85/p110 PI3K in vivo, as judged by effects on the activity of the enzyme immunoprecipitated from cells treated with the drug. Wortmannin irreversibly inhibits p110α by reacting covalently with lysine-802 (42), a residue required for catalytic activity that is conserved in all phosphoinositide kinases (and in protein kinases). A second pharmacological PI3K inhibitor, Ly294002, is a reversible inhibitor of class I enzymes with IC_{50} values of approximately 1 μM (43). Wortmannin and Ly294002 have been used extensively to study the physiological role of class I PI3Ks in various cellular responses. However, some of these studies should be interpreted with caution owing to the emerging evidence that at somewhat higher concentrations these compounds inhibit other signaling enzymes, including PtdIns4Kβ (44) and the related protein kinases TOR (45) and DNA-PK (46).

Class I_A PI3Ks are regulated by interaction with the small G protein ras. Many extracellular stimuli activate ras by increasing the ratio of bound GTP to GDP. In turn, ras-GTP interacts with a number of downstream "effectors." p110α has been established as a ras effector: Its activity is increased in vitro and in vivo by ras-GTP, dominant negative forms of ras can interfere with 3'-phosphorylated phosphoinositide production, and ras effector domain mutants that fail to interact with PI3K are defective in certain ras-dependent cellular responses (47–50).

The activities and subcellular locations of Class I_A PI3K catalytic subunits are also regulated by p85 proteins. p85 and its relatives are sometimes referred to as adapter subunits because they possess several modular domains with the capacity to interact with other signaling proteins. The SH2 domains of p85 have been studied in detail. Like SH2 domains in other signaling proteins, they bind selectively to phosphotyrosyl (pTyr) residues within specific sequence contexts. In all known p85 proteins, both the N-terminal SH2 (N-SH2) and C-terminal SH2 (C-SH2) domains bind preferentially to polypeptides containing a p-Tyr-X1-X2-Met motif (51). A second methionine or valine at the X1 position increases binding affinity, particularly for the N-SH2 domain (51). A crystal structure of the N-SH2 domain bound to a phosphopeptide explains the binding selectivity (52). A solution structure of the C-SH2 domain has also been determined by nuclear magnetic resonance (53). Synthetic peptides containing tandem pTyr-Met-X-Met (pYMXM) motifs separated by a spacer region bind with high affinity to p85 proteins and, importantly, increase the catalytic activity of the associated p110 subunits two- to threefold in vitro (54, 55).

Many stimuli trigger phosphorylation of YMXM motifs, which recruits p85-p110 complexes and thereby enhances PI3K activity (discussed in detail in 8). Most of the agonists that activate p110 via p85/pYMXM interactions also activate ras, itself a p110 activator, as discussed above. For example, polyoma mT has a YMXM sequence to interact with PI3K adapter proteins and other tyrosine residues that mediate interactions leading to activation of ras. Variant polyoma mT proteins that fail to activate ras are unable to increase production of cellular 3'-phosphorylated phosphoinositides, even if the YMXM motif is intact (56, 57). Many YMXM-containing proteins are membrane associated, as is ras. Therefore, recruitment of p85/p110 complexes not only increases catalytic activity but also brings the PI3K from the cytoplasm to the membrane, where its substrates and a potential activator (ras) reside.

Interestingly, the SH2 domains of p85 proteins also bind to phosphoinositides in vitro and exhibit marked selectivity for PtdIns-3,4,5-P_3 (58). Lipid binding to the SH2 domains competes with the binding of pTyr-containing peptide, which suggests that the production of PtdIns-3,4,5-P_3 by activated PI3K causes dissociation of the p85/phosphopeptide complex from its pYMXM docking sites. This model is supported by the finding that wortmannin treatment stabilizes p85/pTyr interactions in cells (58). It is also possible that lipid binding regulates enzyme activity allosterically or by influencing its membrane attachment.

SH3 domains are known to interact with proline-rich sequences with the consensus motif φ–P-p-φ-P, where P is an invariant proline, p is a weakly conserved proline, and φ is an aliphatic amino acid (59). This motif forms a left-handed type II polyproline helix that fits into a hydrophobic platform in the SH3 domain (60). Solution structures have been determined for the p85α SH3 domain alone or bound to a high-affinity peptide ligand (61, 62). A GST fusion protein of p85α SH3 selected several candidate partner proteins from bovine brain extracts, including the GTPase dynamin (61). In addition, the p85α-SH3 domain may interact with proline-rich motifs within the p85 protein itself (63). Intramolecular association of these two modules may prevent either of them from finding other, higher-affinity partners in unstimulated cells.

The proline-rich motifs of p85α mediate binding to SH3 domains of src family kinases including src itself, lck, lyn, and fyn (63–69). The SH3 domain of the cytoplasmic tyrosine kinase abl also associates with p85α (63). For binding to lyn and fyn, the N-terminal proline-rich motif of p85α is more effective than the C-terminal motif (69). The proline-rich regions of p85β diverge considerably in sequence from the analogous regions of p85α. In addition, p85β possesses a third PPXP motif between the C-SH2 domain and the C terminus. Thus, p85β may select a different set of SH3-containing proteins in vivo. The p50α, p55α, and p55γ regulatory subunits contain only a single polyproline motif and lack the N-terminal motif that is selective for binding src.

The binding of protein kinases of the src family via their SH3 domains to the proline-rich motifs of p85 does not depend on any posttranslational modification. This fact suggests that the p85-kinase interaction is regulated by a different mechanism. The crystal structures of src and hck in their inactive states shows that the endogenous SH3 domain makes intramolecular contact with a cryptic polyproline-like helix formed by the spacer sequence between the SH2 and kinase domains (70, 71). Thus, intermolecular association of src family kinases and p85 proteins may first require release of intramolecular contacts within each protein. Association of p85α with the phosphoprotein cbl is also consistent with this type of model. Binding of pTyr residues on cbl to the SH2 domains of p85α appears to expose the SH3 domain to allow high-affinity interactions with proline-rich regions of cbl (72).

The rho-GAP homology region—originally termed the breakpoint cluster region (bcr)-homology domain—of p85α lacks GTPase-promoting activity in vitro toward the small G proteins rho, rac, and cdc42. The crystal structure of the p85α rho-GAP domain is similar to bona fide rho-GAPs but lacks five conserved residues likely to be important for catalysis (73–75). Nevertheless, the p85α rho-GAP domain binds to rac and cdc42 in vitro in a GTP-dependent manner (76, 77). It is possible that rac and/or cdc42 regulates p85/p110 complexes or helps locate these complexes at specific regions of the cell (or vice versa). Because the rho-GAP homology region of p85α lies between its SH3 domain and its second proline-rich motif in the primary structure, binding of a small G protein may either disrupt or enhance the intramolecular interactions discussed above. The rho-GAP region of p85β is only 42% identical to the corresponding domain of p85α, and it similarly lacks residues thought to be important for GTPase-promoting activity (20, 73).

p110γ differs from other class I enzymes because of its ability to be directly activated by $\beta\gamma$ subunits of heterotrimeric G proteins (31, 32, 78, 79). Many G protein–coupled serpentine receptors increase the levels of various 3′-phosphorylated PtdIns species upon ligand binding. Recent evidence has shown that these increases are mediated by p110γ (78). One group found that G$\beta\gamma$ subunits could activate p110γ directly (32), whereas another reported that the associated p101 subunit binds G$\beta\gamma$ and is required for significant activation (33). Interestingly, a p85/p110-type PI3K activity was also shown to be activated by G$\beta\gamma$ subunits, but only in the presence of pTyr peptides that bind to the SH2 domains of the adapter subunit (79, 80). p110γ/p101 activity is not affected by phosphotyrosyl peptides. p110γ also contains a ras-binding domain and associates with ras in vitro (80a). However, a role for ras in p110γ activation has not been demonstrated.

Class II PI3Ks

MOLECULAR BIOLOGY   Class II PI3Ks are large (170–210 kDa) proteins that contain a PIK domain and a catalytic domain 45–50% similar to class I PI3Ks (Figure 2). Class II PI3K genes have been cloned from humans and mice as well as from *D. melanogaster*, *D. discoideum*, and *C. elegans* (17, 81–84). Each of these proteins contains a C-terminal region with homology to C2 domains; indeed, the class II PI3Ks have been termed PI3KC2 (used herein) or cpk (for C2-containing phosphoinositide kinase). Other proteins with C2 domains include certain protein kinase C (PKC) isoforms and the synaptic vesicle membrane protein synaptotagmin. C2 domains have been implicated in the Ca^{2+}-dependent binding of proteins to lipid vesicles. However, certain Asp residues important for Ca^{2+} binding are absent in the C2 domains of class II PI3Ks. The *Drosophila* PI3KC2 was found to bind acidic phospholipids in a Ca^{2+}-independent manner (81). Mammalian PI3KC2α and the novel human PI3KC2β have an additional sequence motif termed the PX domain (see Figure 2) (S Volinia, personal communication). Homologous domains have been noted in several signaling proteins, including the NADPH oxidase-associated proteins phox-40 and phox-47. The function of PX domains is unknown.

ACTIVITY AND REGULATION   In vitro, class II PI3Ks preferentially phosphorylate PtdIns and PtdIns-4-P (Table 1), although human PI3KC2α phosphorylates PtdIns-4,5-P_2 in the presence of phosphatidylserine (84). The enzymes from *Drosophila*, mouse, and human differ significantly in sensitivity to inhibition by wortmannin (IC_{50} values of 5, 50, and 450 nM, respectively) (81, 83, 84) (Table 1). It has not yet been determined which lipids are produced by class II PI3Ks in vivo and how their activities are regulated. If these enzymes make PtdIns-3,4-P_2 in vivo, they are not likely to be active in resting cells where this lipid is undetectable. It is possible that class II PI3Ks contribute to the buildup of PtdIns-3,4-P_2 observed in stimulated cells. There is evidence for rapid and transient tyrosine phosphorylation of PI3KC2α and -β following mitogen stimulation (82; S Volinia, personal communication), but the effect on enzyme activity is unknown.

Class III PI3Ks

MOLECULAR BIOLOGY   The prototype class III PI3K was first identified in yeast in a screen for mutants conditionally defective in vacuolar protein sorting (85). The corresponding gene, *VPS34*, was cloned and found to be essential for accurate transport of newly synthesized proteins from the Golgi to the vacuole, an organelle similar to the lysosome of higher eukaryotes. The lipid kinase activity of this yeast protein was appreciated only after cloning of the p110 subunit of mammalian class I PI3Ks revealed that the proteins shared extensive sequence homology (86) (Figure 2). Bona fide *VPS34* homologues have now been cloned from humans, *Dictyostelium*, and *Drosophila* (17, 87, 88). Each of the proteins also possesses a PIK domain (Figure 2).

ACTIVITY AND REGULATION   Class III PI3Ks phosphorylate only PtdIns (Table 1); therefore they should be called PtdIns 3-kinases to differentiate them from the phosphoinositide 3-kinases with broader substrate specificity. Although the yeast Vps34 is relatively insensitive to wortmannin (IC_{50} ~2.5 μM), human and *Drosophila* Vps34 homologues are inhibited with IC_{50} values of 2–10 nM (Table 1). In addition, wortmannin treatment lowers the level of PtdIns-3-P in platelets (89, 90).

The yeast protein Vps34 associates with another protein, Vps15, that is also required for vesicle sorting (91). Vps15 is a serine/threonine kinase that recruits Vps34 to membranes and enhances its lipid kinase activity. A human DNA encoding of protein with 30% sequence identity to the *VPS15* gene product was cloned. This protein was found to act as an adapter for the human class III PI3K (92). The human heterodimeric class III PI3K was found to associate with phosphatidylinositol transfer protein, which stimulated lipid kinase activity (92). That PtdIns-3-P levels are comparable in both resting and stimulated cells (90) suggests that class III PI3Ks induce local increases in PtdIns-3-P that may be required for agonist-independent membrane trafficking processes.

PtdIns 4-KINASES

The existence of PtdIns 4-kinases (PtdIns4Ks) has been appreciated for some 30 years, yet our understanding of these enzymes has been limited. The genes encoding the enzymes have only recently been cloned. PtdIns4Ks convert PtdIns to PtdIns-4-P by phosphorylating the inositol ring at the D4 position. The PtdIns-4-P thus generated can be further phosphorylated by both PI3Ks and PtdIns-4-P 5-kinases (PIP5Ks) to yield PtdIns-3,4-P_2 or PtdIns-4,5-P_2, respectively. In this way, the PtdIns4Ks play a central role in signaling by feeding into multiple pathways. These enzymes are ubiquitously expressed and are most abundant in cellular membranes including the Golgi, lysosomes, endoplasmic reticulum (ER), plasma membrane, and a variety of vesicles that include Glut 4–containing vesicles, secretory vesicles, and coated pits (93). Unlike the functionally related PI3Ks and the PIP5Ks, these enzymes appear to use only PtdIns as a substrate and cannot phosphorylate either the singly or doubly phosphorylated lipids generated by the other enzymes.

The PtdIns4Ks were classically subdivided into two types (II and III) based on biochemical differences of the partially purified enzymes (94, 95). Recently, several PtdIns4Ks have been cloned from yeast, mammals, and other species.

Characterization of the enzymes that have been cloned from both yeast and mammalian cells has revealed additional subtypes of PtdIns 4-kinases. In the next section we describe the properties of the purified enzymes and then describe the structure and enzymology of the cloned PtdIns 4-kinases.

Purification of the 4-Kinases

Type II PtdIns4Ks were defined biochemically as single subunit, membrane-bound enzymes whose lipid kinase activity is stimulated by nonionic detergent and inhibited by adenosine (95). They were purified from a variety of mammalian sources, including fibroblasts, bovine brain, A431 cells, and erythrocytes (94–97). The type II PtdIns4K was characterized as a 55-kDa protein that could be renatured from a sodium dodecyl sulfate (SDS) gel (96, 98) and could be inhibited by the monoclonal antibody 4C5G (99). Although this activity is the most potent PtdIns4K in most cell types, the gene has not yet been cloned.

The p55 enzyme has been implicated in many signaling pathways through its association with a variety of molecules and subcellular compartments, including epidermal growth factor (EGF) receptor (100), chromaffin granules (101), CD4-p56^{lck} (102), integrins (103), and PKCs. Unfortunately, a more detailed analysis of the role of this enzyme in these pathways awaits the cloning of the gene encoding the p55 type II enzyme.

Type III PtdIns4Ks were defined as membrane-bound enzymes whose lipid kinase activity is unaffected by high concentrations of adenosine and is maximally active in nonionic detergent (95). The enzyme, purified from bovine and rat brain, has an apparent molecular weight of 220 kDa by sucrose gradients and gel filtration (95, 104), and it is resistant to the monoclonal antibody 4C5G (99).

The prototype PtdIns4Ks were first cloned from yeast and designated *PIK1* (114) and *STT4* (106). Subsequently, cDNAs for two mammalian PtdIns4Ks were cloned; their properties are described below. The encoded enzymes do not strictly follow the typing described above and are therefore called PtdIns4Kα and PtdIns4Kβ, based on the order in which they were isolated from mammalian sources.

PtdIns4Kα/STT4

MOLECULAR BIOLOGY   PtdIns4Kα is homologous to a yeast PtdIns4K, the *STT4* gene product that was identified as a staurosporine- and temperature-sensitive mutant in yeast (105). Stt4 is a ~210-kDa protein with a conserved C-terminal catalytic domain that is ~35% identical to PtdIns4Kβ and ~27% identical to the PI3Ks (106). Null mutants (*stt4Δ*) are lethal unless supplied with an exogenous osmotic support (1M sorbitol). Under these conditions, the cells display decreased doubling times and a fivefold decrease in the level of PtdIns-4-P (106). Additionally, overexpression of a protein kinase C gene

(*PKC1*) can complement the staurosporine sensitivity of an *stt4-1* mutant, which suggests that *PKC1* is downstream of *STT4* in yeast (107). However, PKC enzymes bind staurosporine, so this rescue could be indirect, because *PKC1* overexpression does not rescue any of the other phenotypes of *sst4Δ* mutants.

PtdIns4Kα was cloned from a human placenta cDNA library. It encodes an 854-amino-acid protein (p97) that is 50% identical to Stt4 in the catalytic domain and is also more similar to Stt4 than to Pik1 in the noncatalytic region (108). Homologues have now been cloned from rat and bovine brain (109, 110). Interestingly, the rat brain protein shares 98% identity with human p97 PtdIns4Kα over the region of overlap but encodes a protein of 2041 amino acids (p220) with a distinct N-terminal half. We refer to these genes as p97 PtdIns4Kα and p220 PtdIns4Kα. In addition to the C-terminal catalytic domain, p97 and p220 PtdIns4Kα contain an amino-proximal PIK domain (Figure 3) as well as regions with possible homology to ankyrin repeats. The unique N-terminal portion of p220 PtdIns4Kα contains a proline-rich segment and an SH3 domain (111) (Figure 3). The p97 PtdIns4Kα enzyme is likely to be a spliced variant of the larger protein because antibodies against p97 cross-react with a larger protein in human cells (112). By RNA hybridization and immunoblotting, most cells appear to express predominantly p220 PtdIns4Kα with the noted exception of platelets, which express significant amounts of p97 PtdIns4Kα (K Wong, personal communication).

ACTIVITY AND REGULATION   In vitro, the PtdIns4Kα/Stt4 isoforms phosphorylate only PtdIns. Although PtdIns4Kα is relatively resistant to wortmannin (K Wong, personal communication), recent data suggest that the p220 PtdIns4Kα protein can be completely inhibited at a high concentration of

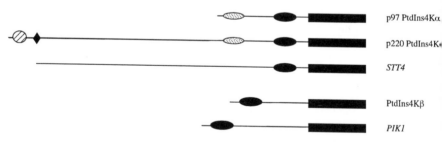

Figure 3   Structural features of phosphatidylinositol 4-kinases (PtdIns4K) family members. Two types of PtdIns4K genes have been cloned from yeast and mammals. Mammalian PtdIns4Kα is most closely related to yeast Stt4, whereas mammalian PtdIns4Kβ is more homologous to yeast Pik1. The protein domains are as follows: catalytic domain (*solid rectangles*), phosphoinositide kinase (PIK) domain (*solid ovals*), SH3 domain (*hatched circle*), proline-rich motif (*solid diamond*), ankyrin-like repeats (*stippled ovals*).

wortmannin (10 μM) (109). By contrast, the yeast Stt4 is >90% inhibited by 10 nM wortmannin in vitro (113). In most mammalian cells, the levels of PtdIns-4-P do not appreciably change in response to growth factors or osmotic stress, which suggests that if PtdIns4Ks are regulated by extracellular signals, the regulation is likely to be at specific regions of the cell. Indeed, recent studies have localized the mammalian PtdIns4Ks—PtdIns4Kα and PtdIns4Kβ—to distinct intracellular membranes (ER and Golgi, respectively), where they may regulate the local levels of PtdIns-4-P (111, 112).

PtdIns4Kβ/PIK1

MOLECULAR BIOLOGY   Protein purification (117) and peptide sequencing led to the cloning of the first PtdIns4K, *PIK1*, from yeast (114). Pik1 is a 125-kDa enzyme that is essential for *Saccharomyces cerevisiae* survival. It has a catalytic domain that is 30% identical to the catalytic domains of the PI3K family members, and it has a PIK domain (Figure 3). Null mutants (*pik1Δ*) are not viable (114), and temperature-sensitive (ts) alleles show defects in cytokinesis, with most cells appearing as pairs with fully separated nuclei (115). Additionally, it has been reported that *pik1* ts mutants show an approximately threefold decrease in PtdIns4K activity in the nuclear fraction (115).

Recently, cDNAs encoding homologues of *PIK1* have been cloned from rat (109), human (44), and bovine (116) tissues. These enzymes, variously named 92kPtdIns4K, PtdIns4Kβ (used herein), and PtdIns4KIIIβ, are highly related and almost certainly are the same gene product. A *D. discoideum* gene highly related to this family has also been cloned (17). The PtdIns4Kβ/*PIK1* family of enzymes is distinguished from the PtdIns4Kα/*STT4* family and the PI3K family in that the PIK domain is at the N terminus rather than adjacent to the catalytic domain in sequence (Figure 3).

Both Pik1 and PtdIns4Kβ are mostly soluble, cytosolic enzymes (112, 117), although some reports suggest that Pik1 is nuclear (115), and a significant fraction of mammalian PtdIns4Kβ is associated with the Golgi (112).

ACTIVITY AND REGULATION   PtdIns4Kβ/Pik1 enzymes phosphorylate only PtdIns to generate PtdIns-4-P (Table 1). The mammalian enzyme is inhibited by wortmannin with an IC_{50} of $\sim$50–100 nM (44, 118, 119), whereas the yeast enzyme is resistant to as much as 5 μM wortmannin (R Meyers, unpublished data; J Thorner, personal communication). Interestingly, wortmannin treatment of adrenal glomerulosa cells leads to a decrease in the hormone-stimulated production of PtdIns-4-P and PtdIns-4,5-P_2 (118). Furthermore, 10 μM but not 300 nM wortmannin inhibits the sustained increase in both IP_3 production and cytoplasmic Ca^{2+} in agonist-stimulated adrenal glomerulosa and Jurkat cells. These results suggest a role for a wortmannin-sensitive PtdIns4K (but not PI3K) in regulation of agonist-sensitive pools of PtdIns-4,5-P_2 in these cells. Because the p55 type II PtdIns4K is not affected by high concentrations of wortmannin, these results suggest that PtdIns4Kβ (or PtdIns4Kα—see above) may account for a significant fraction of the lipid pool utilized in agonist-stimulated phosphoinositide turnover in these cells (118, 119). In contrast, B cells treated with 25 nM wortmannin (which inhibits PI3K but not PtdIns4K) showed inhibition of the initial as well as the sustained increases in both IP_3 production and cytoplasmic Ca^{2+}, thus implicating PI3Ks in regulation of IP_3 production in this cell type (120).

OTHER PROTEINS WITH HOMOLOGY TO PtdIns KINASES AND PHOSPHOINOSITIDE KINASES

The conserved catalytic domain present in PI3Ks and PtdIns4Ks is shared by a more distantly related family of proteins. These molecules include a group of gene products that control cell cycle progression in response to DNA damage (*RAD3, MEC1, TEL 1*, ATM, and ATR) as well as the DNA-activated protein kinase (DNA-PK) and the rapamycin/FKBP-binding proteins *TOR*/FRAP/ RAFT1 (121–123). These enzymes are more closely related to each other than to the PI3K and PtdIns4K genes and lack a definitive PIK element. Several of these proteins, including DNA-PK, FRAP, and ATM, have demonstrable protein kinase activity (46, 124, 125). There are reports that the yeast protein Tor2 and its mammalian homologue RAFT1 display PtdIns4K activity on purification from cells (126, 127). The nature of this PtdIns4K is unclear because it shares properties with both type II and type III PtdIns4K. Although immunoprecipitates of a ts mutant of yeast Tor2 showed greatly reduced levels of PtdIns4K activity (127), PtdIns4K activity in Tor2/RAFT1/FRAP is more likely to be the result of the presence of an associated lipid kinase. Consistent with this idea, rapamycin does not affect the lipid kinase activity of Tor2/RAFT1/FRAP or alter the levels of lipids in [^{3}H]-inositol-labeled cells (126–128). Additionally, equal amounts of PtdIns4K activity are detected from immunoprecipitates of wild-type and kinase-dead recombinant FRAP/RAFT1 isolated from SF9 insect cells or mammalian cells (124). More work is needed to address the catalytic activities of these important enzymes.

PtdIns-P KINASES

Two distinct families of enzymes have been purified based on their ability to phosphorylate PIP to produce PtdIns-4,5-P_2. Genes encoding these two families of PIP kinases have now been cloned and found to contain significant sequence

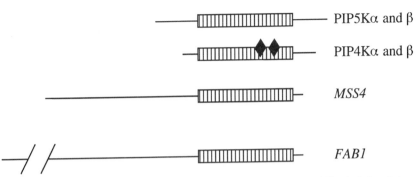

Figure 4   Structural features of PtdIns-4-P 5-kinases (PIP5Ks) and phosphatidylinositol-phosphate 4-kinases (PIP4Ks). Two isoforms of both PIP5K (formerly termed type I PIP5K) and PIP4K (formerly type II PIP5K) have been cloned. Despite their different enzymatic activities, they share significant homology in their putative catalytic domains (*striped rectangles*). The PIP4Ks contain polyproline motifs (*solid diamonds*). Two large proteins in yeast, Mss4 and Fab1, possess domains related to the mammalian enzymes, but their enzymic activities have not been determined. Both PIP4Ks and PIP5Ks are more closely related to Mss4 than to Fab1 (131).

similarity (129–132). Although these enzymes were previously termed type I and type II PtdIns-4-P 5-kinases (PIP5Ks), recent studies have revealed that the two families of enzymes selectively phosphorylate different positions on the inositol ring (34, 133). To reflect the distinct sites of phosphorylation, the two families have been designated PIP5Ks and PIP4Ks, as discussed below.

PIP5Ks

MOLECULAR BIOLOGY   Two human PIP5Ks (PIP5Kα and -β) have been cloned. Two groups cloned these genes but used the α and β suffix in a reciprocal manner (130, 131) (Figure 4). We use the designation of Ishihara et al (130). PIP5Kα and -β are 61-kDa proteins migrating at 68 kDa upon SDS-polyacrylamide gel electrophoresis (PAGE) and sharing 64% identity overall and 83% identity in their catalytic domains. Interestingly, type I PIP5Ks exhibit little homology to PI3Ks and PtdIns4Ks. Splice variants of the PIP5Kβ gene have been identified by cDNA cloning and RNA hybridization (131); some of these variants may represent 90-kDa and 110-kDa proteins detected by an anti-PIP5K antiserum (6).

The kinase domains of PIP5Ks are related to two proteins in *S. cerevisiae*: Mss4 and Fab1 (Figure 4). The enzymatic activities of these yeast proteins have not been determined. *MSS4* is an essential gene in yeast, whereas mutations in *FAB1* cause defects in morphology and vacuolar function (107, 134). Genetic evidence places *MSS4* downstream of the yeast PtdIns 4-kinase *STT4*, consistent with Mss4 acting as a PtdIns-4-P kinase (107). Genetic relationships between the *FAB1* gene and either *STT4* or *PIK1*, the other yeast PtdIns4K, have not been established. The human PIP5Ks are more closely related to Mss4 (40–44% identity) than to Fab1 (27–29% identity) (131). A *C. elegans* gene homologous to *FAB1* (40% identity) has also been identified (131), which suggests that a more highly related mammalian homologue may also exist.

ACTIVITY AND REGULATION   The best-studied reaction catalyzed by PIP5Ks is the conversion of PtdIns-4-P to PtdIns-4,5-P$_2$. Using enzyme purified from erythrocytes, the K_m for PtdIns-4-P in micelles, liposomes, or native membranes ranges from 1 to 10 μM (6). This reaction is stimulated by heparin and spermine and inhibited by the product, PtdIns-4,5-P$_2$. Importantly, PIP5Ks are stimulated as much as 50-fold by phosphatidic acid (135, 136). This lipid can be generated from phosphatidylcholine by phospholipase D and from diacylglycerol (DAG) by DAG-kinases.

PIP5Ks have been shown to phosphorylate other phosphoinositides in vitro. The α and β isoforms convert PtdIns-3,4-P$_2$ to PtdIns-3,4,5-P$_3$ with K_{cat}/K_m ratios that are 3-fold and 100-fold lower, respectively, than those observed using PtdIns-4-P as the substrate (133). Type I PIP5Ks also utilize PtdIns-3-P as a substrate in vitro (133; K Tolias, personal communication). Zhang et al (133) identified the PtdIns-*bis*phosphate product as PtdIns-3,4-P$_2$; in contrast, work in our laboratory has shown that the *bis*phosphorylated product includes PtdIns-3,5-P$_2$ (K Tolias, personal communication), a lipid that was recently demonstrated to exist in vivo (137, 137a). Pulse-labeling experiments indicated that most of this lipid is generated by phosphorylation of PtdIns-3-P at the 5′ position (137), a reaction that could be catalyzed by a PIP5K.

Remarkably, PtdIns-3,4,5-P$_3$ is also produced when recombinant PIP5Ks are mixed with PtdIns-3-P (133; K Tolias, personal communication). The amount of PtdIns-3,4,5-P$_3$ produced by PIP5Kα is significant, approaching 50% as much as the PtdIns *bis*phosphate produced in a 10-min reaction (133). The mechanism of this reaction and its relevance for lipid production in vivo are unclear. It is not known whether the enzyme generates PtdIns-3,4,5-P$_3$ by a true concerted reaction in which the substrate remains bound to the enzyme during both phosphorylation steps or whether a PtdIns-*bis*phosphate is released and bound again. One possibility is that the enzyme makes both PtdIns-3,5-P$_2$ and PtdIns-3,4-P$_2$ from PtdIns-3-P; these initial products could then be phosphorylated again to produce PtdIns-3,4,5-P$_3$, although PtdIns-3,5-P$_2$ may be a very poor substrate for further phosphorylation. In any case, it now appears that the enzymes known as type I PIP5Ks do more than just phosphorylate PtdIns-4-P.

Type I PIP5K activity in vivo may be influenced by small G proteins. The nonhydrolyzable GTP analog GTPγS increases PIP5K activity in several cell types (6). PIP5K activity associates constitutively with rac1 (77) and rhoA (138). The interaction with rac1 requires a carboxyterminal basic region of this small G protein (139). Addition of rac-GTP to permeabilized platelets or to neutrophil extracts causes increased PtdIns-4,5-P_2 synthesis (140, 141). RhoA was also reported to increase PtdIns-4,5-P_2 production (138). Like PI3Ks, PIP5Ks may be regulated by both small G proteins and heterotrimeric G proteins. For example, a cholera toxin-sensitive G protein was found to regulate a 5-kinase activity in rat liver membranes (142). The availability of cDNA clones for the different PIP5Ks should facilitate the mapping of regulatory interaction sites and the mutational analysis of these regions and their physiological relevance.

PIP4Ks

MOLECULAR BIOLOGY   Based on the ability to phosphorylate commercial PtdIns-4-P to produce PtdIns-4,5-P_2, a second form of PIP-kinase was purified (98, 143) and cloned (129, 144). This enzyme was thought to be a PtdIns-4-P 5-kinase and was named type IIα PIP5K. However, as discussed below, this PIP-kinase is actually a PtdIns-5-P 4-kinase (PIP4Kα). A second highly related gene (78% identical) has also been cloned (132). Previously termed type IIβ PIP5K, this enzyme is designated PIP4Kβ herein. Both cDNAs encode 47-kDa proteins whose kinase domains contain proline-rich sequences in contexts that match the consensus for binding to SH3 domains (Figure 4). The PIP4Ks are about 35% identical to the PIP5Ks in their kinase domains but are dissimilar in their amino- and carboxyterminal extensions (Figure 4). PIP4Ks also share homology with Mss4 and Fab1, and, like PIP5Ks, are more closely related to Mss4 (43%) than to Fab1 (28%) (131).

ACTIVITY AND REGULATION   Recent work from our laboratory has established that in the presence of commercial PtdIns-4-P and [γ-^{32}P]-labeled ATP, the PIP4Kα produces PtdIns-4,5-P_2 radiolabeled at the 4' position rather than at the 5' position (34). This paradoxical observation is readily explained by the presence of contaminating PtdIns-5-P in the PtdIns-4-P used as substrate. Many commercial preparations of PtdIns-4-P are isolated from brain lipid extracts, so they also contain significant amounts of PtdIns-5-P, which was only recently found to exist in vivo (34) and is difficult to separate from PtdIns-4-P. The failure to phosphorylate the 5' position of PtdIns-4-P may explain why red blood cell membranes could not be significantly phosphorylated by this enzyme (143); presumably, red cell membranes contain little PtdIns-5-P. Through the use of pure preparations of PtdIns-4-P and PtdIns-5-P, it has been found that PIP4Kα and β can phosphorylate only the latter (34; L Rameh, personal communication). Thus, these enzymes are PtdIns-5-P 4-kinases.

PIP4Kα also phosphorylates the 4'-OH of PtdIns-3-P in vitro (34, 133), although PtdIns-5-P is a better substrate (34). These enzymes probably account for the PtdIns-3-P 4-kinase activity that was purified from platelets (145) and erythrocytes (97) using PtdIns-3-P as a substrate. PIP4Ks do not synthesize PtdIns-3,4,5-P_3 when PtdIns-3,4-P_2 is presented as a substrate (133), which strengthens the argument that these enzymes are not 5-kinases. In addition, they do not significantly phosphorylate PtdIns (L Rameh, personal communication). In view of their ability to phosphorylate the 4′ position of both PtdIns-3-P and PtdIns-5-P, we refer to these enzymes not as PtdIns-5-P 4-kinases but by the more general term PIP4K (Table 1) (34).

Little is known about the regulation of PIP4K activity. In vitro activity is inhibited by PtdIns-4,5-P_2 and by heparin. PIP4Kβ (previously termed type IIβ PIP5K) associates directly with the p55 subunit of the tumor necrosis factor (TNF) receptor, and indirect evidence suggests that its activity increases following TNF treatment of cells (132). An activity that converts PtdIns-3-P to PtdIns-3,4-P_2 is stimulated by an integrin-mediated pathway in platelets (145a). It is possible that the proline-rich motifs within the kinase domains of PIP4Ks mediate binding to proteins containing SH3 domains. Because PtdIns-3,4-P_2 is an important second messenger that participates in the activation of Akt protein kinase and certain PKCs (3), the PIP4Ks may play a role upstream of these kinases (145a).

PHOSPHOINOSITIDE SYNTHESIS: A CURRENT MODEL

Based on data discussed in this review, a model can be proposed for the synthesis of each of the known phosphoinositides. Although PtdIns has free hydroxyls at positions 2–6, only positions 3–5 are known to be phosphorylated in vivo (Figure 1). Figure 5 shows the phosphorylation steps reported to be catalyzed by purified, heterologously expressed PtdIns kinases and phosphoinositide kinases in vitro. Some of these pathways have been shown to occur in vivo, based on pulse-labeling studies and steady-state analysis of lipid levels in cells overexpressing constitutively active or dominant negative forms of the enzymes. Confirmation that other pathways occur in vivo awaits similar investigations.

Three different monophosphorylated forms of PtdIns have been shown to exist in vivo (Figure 1*b*). PtdIns is converted to PtdIns-3-P by PI3Ks (Figure 5). In vitro, all known types of PI3K catalyze this reaction. However, it is thought that most of the PtdIns-3-P in cells is synthesized by class III PI3Ks. Synthesis of PtdIns-4-P is catalyzed by PtdIns4Ks (Figure 5). PtdIns-5-P was recently identified in cells, when chromatographic separation from PtdIns-4-P was achieved (34). It is not yet known whether a phosphoinositide kinase generates PtdIns-5-P in vivo or whether it is produced only by the action of lipid

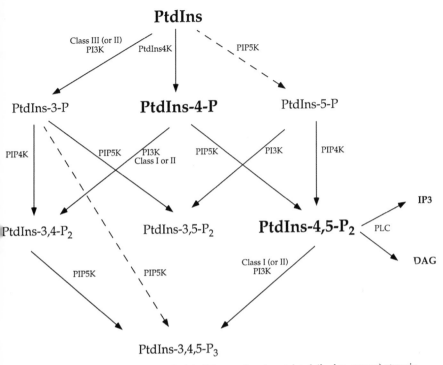

Figure 5 Flow chart of the established (*solid arrows*) and postulated (*broken arrows*) steps in phosphoinositide synthesis in mammalian cells. *Vertical arrows* indicate reactions catalyzed by phosphatidylinositol 4-kinases (PtdIns4Ks) or phosphatidylinositol-phosphate 4-kinases (PIP4Ks). *Diagonal-left arrows* are mediated by phosphoinositide 3-kinases (PI3Ks), *diagonal-right arrows* by phosphatidylinositol-phosphate 5-kinases (PIP5Ks). As stated in Figure 1, PtdIns-4-phosphate (PtdIns-4-P) and PtdIns-4,5-*bis*phosphate (PtdIns-4,5-P_2) (in larger type to emphasize abundance) together represent approximately 90% of the total phosphoinositides. PtdIns-3-phosphate (PtdIns-3-P) and PtdIns-5-phosphate (PtdIns-5-P) each represent approximately 2–5% of the total. PtdIns-3,4-P_2 and PtdIns-3,4,5-P_3 are barely detectable in quiescent cells and rise to about 1–3% of the total in stimulated cells. PtdIns-3,5-P_2 represents about 2% of phosphoinositides in fibroblasts. The levels of PtdIns, PtdIns-3-P, and PtdIns-4-P remain fairly constant in quiescent and activated cells, whereas relative amounts of PtdIns-3,4-P_2, PtdIns-4,5-P_2, and PtdIns-3,4,5-P_3 can change substantially under different conditions.

phosphatases. However, PIP5Ks phosphorylate the 5′-OH of PtdIns in vitro (Figure 5) (K Tolias, personal communication).

Three different *bis*phosphorylated forms of PtdIns have also been shown to exist in vivo (Figure 1*b*). PtdIns-4,5-P_2 is the predominant PtdIns *bis*phosphate in cells; its synthesis and hydrolysis have been studied in great detail. It is now clear that there are two pathways for synthesis of PtdIns-4,5-P_2: phosphorylation of PtdIns-4-P at the 5 position by PIP5Ks and of PtdIns-5-P at the 4 position by PIP4Ks (Figure 5). When cells are pulse labeled with ^{32}P, most of the radioactivity in PtdIns-4,5-P$_2$ is found at the 5 position, which suggests that the 5′ phosphate is usually added after the 4′-phosphate (146). This result is consistent with the greater abundance of PtdIns-4-P relative to PtdIns-5-P. It will be interesting to determine whether the ratio of labeled phosphates can be altered by overexpression of PIP5Ks versus PIP4Ks or in response to specific stimuli.

PtdIns-3,4-P$_2$ can be generated from PtdIns-4-P by class I and class II PI3Ks (Figure 5). In addition, it can be synthesized by phosphorylation of PtdIns-3-P by PIP4Kα (34, 133) (Figure 5). Although the PIP5Ks were reported to generate PtdIns-3,4-P$_2$ from PtdIns-3-P (133), a more recent study concluded that these enzymes preferentially phosphorylate the 5 position of PtdIns-3-P (K Tolias, personal communication). PtdIns-3,5-P$_2$ has been found in cells (137, 137a), and pulse-labeling studies show that the 5-phosphate is added last. This finding suggests that most of this lipid is produced by phosphorylation of the 5 position of PtdIns-3-P (137). However, this lipid can also be generated from PtdIns-5-P by PI3Ks in vitro (34).

It is well established that PtdIns-3,4,5-P$_3$ is synthesized by class I PI3Ks using PtdIns-4,5-P$_2$ as substrate (147) (Figure 5). In addition, the PIP5Ks can phosphorylate PtdIns-3,4-P$_2$ at the 5 position in vitro (133) (Figure 5). Interestingly, the PIP5Ks were also found to generate PtdIns-3,4,5-P$_3$ from the monophosphorylated PtdIns-3-P in a concerted reaction (133). There has been some controversy about the order of phosphate addition in the production of PtdIns-3,4,5-P$_3$ in vivo. Two early studies indicated that, in thrombin-stimulated platelets and platelet-derived growth factor (PDGF)-stimulated fibroblasts, the 5′ phosphate is added last, consistent with a role for a PtdIns-3,4-*bis*phosphate 5-kinase (148, 149). However, in later studies of formyl peptide-activated neutrophils (35), PDGF-stimulated fibroblasts (150), and thrombin-stimulated platelets (151), evidence supported addition of a 3′ phosphate to PtdIns-4,5-P$_2$. In view of the recent in vitro results, it would not be surprising if multiple pathways exist in vivo for the synthesis of PtdIns-3,4,5-P$_3$.

It is important to note that overall levels and local concentrations of the different phosphoinositides are also regulated by lipid phosphatases and by phospholipases. PtdIns-4,5-P$_2$ is the preferred substrate of phospholipase C (PLC) enzymes (Figure 5). Lipolysis of PtdIns-4,5-P$_2$ generates two second messengers with well-characterized actions: the lipid DAG, which activates some protein kinase C (PKC) isozymes, and the water-soluble inositol-1,4,5-*tris*phosphate, which promotes a rise in intracellular Ca^{2+} concentration (152). This process is sometimes referred to as the classical phosphoinositide-turnover pathway. Unlike PtdIns-4,5-P$_2$, the 3′-OH phosphoinositides, namely PtdIns-3-P,

PtdIns-3,4-P_2, and PtdIns-3,4,5-P_3, are poor substrates for PLCs (153). It is possible that these lipids are hydrolyzed by phospholipases that have yet to be discovered. Nevertheless, it is clear that 3'-OH phosphoinositides act directly as second messengers to regulate proteins at the membrane (3).

CONCLUDING REMARKS

The study of phosphoinositide kinases has progressed rapidly since PI3Ks were first cloned from yeast (85) and mammals (13). Molecular tools are now available to dissect the complex regulation of phosphoinositide synthesis. Several challenges remain. It will be important to determine the three-dimensional structures of all these enzymes. The protein kinase activity of PI3Ks and their detectable sequence homology with classical protein kinases suggest that their catalytic domains might fold similarly. It will be especially interesting to compare the PI3Ks and PtdIns4Ks with apparently unrelated families of lipid kinases, including the PIP4Ks and PIP5Ks and the TOR/ATM/DNA-PK group. These structural studies will be crucial for designing compounds to differentially modulate the activities of these enzymes for research and clinical applications. Another important task will be to compare expression of the different isoforms in different tissues as well as within subcellular compartments. Most studies of phosphoinositide levels have measured lipids in total cell extracts; a more precise understanding of lipid metabolism will require analysis of intracellular pools of the products of the different enzymes we have discussed here. Furthermore, investigation of phosphoinositide biology must be complemented by the study of lipid phosphatases. These enzymes may be just as important as lipid kinases in determining the balance of different phosphoinositides and lipid second messengers in the cell. These avenues of study are certain to increase our knowledge of how the cell generates and breaks down phosphoinositides to regulate downstream signaling.

ACKNOWLEDGMENTS

We thank Lucia Rameh, Kimberly Tolias, Stefano Volinia, and Karen Wong for sharing unpublished results, and Lucia Rameh, Kimberly Tolias, Stefano Volinia, and Kay Lee-Fruman for critical reading of the manuscript. DAF is a Fellow of the Cancer Research Fund of the Damon Runyon–Walter Winchell Foundation. REM and LCC are supported by National Institutes of Health Grants CA68719 and GM36624.

Visit the *Annual Reviews home page* at
http://www.AnnualReviews.org.

Literature Cited

1. Divecha N, Irvine RF. 1995. *Cell* 80: 269–78
2. Rittenhouse SE. 1996. *Blood* 88:4401–14
3. Toker A, Cantley LC. 1997. *Nature* 387: 673–76
4. Vanhaesebroeck B, Leevers SJ, Panayotou G, Waterfield MD. 1997. *Trends Biochem. Sci.* 22:267–72
5. Zvelebil MJ, MacDougall L, Leevers S, Volinia S, Vanhaesebroeck B, et al. 1996. *Philos. Trans. R. Soc. London Ser. B* 351:217–23
6. Loijens JC, Boronenkov IV, Parker GJ, Anderson RA. 1996. *Adv. Enzyme Regul.* 36:115–40
7. Domin J, Waterfield MD. 1997. *FEBS Lett.* 410:91–95
8. Duckworth BC, Cantley LC. 1996. In *Handbook of Lipid Research: Lipid Second Messengers*, ed. RM Bell, JH Exton, 8:125–75. New York: Plenum
9. Franke TF, Kaplan DR, Cantley LC. 1997. *Cell* 88:435–37
10. Cantley LC, Auger KR, Carpenter C, Duckworth B, Graziani A, et al. 1991. *Cell* 64:281–302
11. Chang HW, Aoki M, Fruman D, Auger KR, Bellacosa A, et al. 1997. *Science* 276:1848–50
12. Morris JZ, Tissenbaum HA, Ruvkun G. 1996. *Nature* 382:536–39
13. Hiles ID, Otsu M, Volinia S, Fry MJ, Gout I, et al. 1992. *Cell* 70:419–29
14. Hu P, Mondino A, Skolnik EY, Schlessinger J. 1993. *Mol. Cell. Biol.* 13: 7677–88
15. Vanhaesebroeck B, Welham MJ, Kotani K, Stein R, Warne PH, et al. 1997. *Proc. Natl. Acad. Sci. USA* 94:4330–35
16. Leevers SJ, Weinkove D, MacDougall LK, Hafen E, Waterfield MD. 1996. *EMBO J.* 15:6584–94
17. Zhou K, Takegawa K, Emr SD, Firtel RA. 1995. *Mol. Cell. Biol.* 15:5645–56
18. Carpenter CL, Duckworth BC, Auger KR, Cohen B, Schaffhausen BS, et al. 1990. *J. Biol. Chem.* 265:19704–11
19. Escobedo JA, Navankasattusas S, Kavanaugh WM, Milfay D, Fried VA, et al. 1991. *Cell* 65:75–82
20. Otsu M, Hiles I, Gout I, Fry MJ, Ruiz-Larrea F, et al. 1991. *Cell* 65:91–104
21. Skolnik EY, Margolis B, Mohammadi M, Lowenstein E, Fischer R, et al. 1991. *Cell* 65:83–90
22. Fruman DA, Cantley LC, Carpenter CL. 1996. *Genomics* 37:113–21

23. Antonetti DA, Algenstaedt P, Kahn CR. 1996. *Mol. Cell. Biol.* 16:2195–203
24. Inukai K, Anai M, Van Breda E, Hosaka T, Katagiri H, et al. 1996. *J. Biol. Chem.* 271:5317–20
25. Inukai K, Funaki M, Ogihara T, Katagiri H, Kanda A, et al. 1997. *J. Biol. Chem.* 272:7873–82
26. Pons S, Asano T, Glasheen E, Miralpeix M, Zhang YT, et al. 1995. *Mol. Cell. Biol.* 15:4453–65
27. Reif K, Gout I, Waterfield MD, Cantrell DA. 1993. *J. Biol. Chem.* 268:10780–88
28. Baltensperger K, Kozma LM, Jaspers SR, Czech MP. 1994. *J. Biol. Chem.* 269:28937–46
29. Shepherd PR, Nave BT, Rincon J, Nolte LA, Bevan AP, et al. 1997. *J. Biol. Chem.* 272:19000–7
30. Weinkove D, Leevers SJ, MacDougall LK, Waterfield MD. 1997. *J. Biol. Chem.* 272:14606–10
31. Stephens L, Smrcka A, Cooke FT, Jackson TR, Sternweis PC, et al. 1994. *Cell* 77:83–93
32. Stoyanov B, Volinia S, Hanck T, Rubio I, Loubtchenkov M, et al. 1995. *Science* 269:690–93
33. Stephens LR, Eguinoa A, Erdjument-Bromage H, Lui M, Cooke F, et al. 1997. *Cell* 89:105–14
34. Rameh LE, Tolias KF, Duckworth BC, Cantley LC. 1997. *Nature* 390:192–96
35. Stephens LR, Hughes KT, Irvine RF. 1991. *Nature* 351:33–39
36. End P, Gout I, Fry MJ, Panayotou G, Dhand R, et al. 1993. *J. Biol. Chem.* 268:10066–75
37. Carpenter CL, Auger KR, Duckworth BC, Hou WM, Schaffhausen B, et al. 1993. *Mol. Cell. Biol.* 13:1657–65
38. Dhand R, Hiles I, Panayotou G, Roche S, Fry MJ, et al. 1994. *EMBO J.* 13:522–33
39. Stoyanova S, Bulgarelli-Leva G, Kirsch C, Hanck T, Klinger R, et al. 1997. *Biochem. J.* 324:489–95
40. Lam K, Carpenter CL, Ruderman NB, Friel JC, Kelly KL. 1994. *J. Biol. Chem.* 269:20648–52
41. Freund GG, Wittig JG, Mooney RA. 1995. *Biochem. Biophys. Res. Commun.* 206:272–78
42. Wymann MP, Bulgarelli-Leva G, Zvelebil MJ, Pirola L, Vanhaesebroeck B, et al. 1996. *Mol. Cell. Biol.* 16:1722–33
43. Vlahos CJ, Matter WF, Hui KY, Brown RF. 1994. *J. Biol. Chem.* 269:5241–48

44. Meyers R, Cantley LC. 1997. *J. Biol. Chem.* 272:4384–90
45. Brunn GJ, Williams J, Sabers C, Wiederrecht G, Lawrence JC Jr, Abraham RT. 1996. *EMBO J.* 15:5256–67
46. Hartley KO, Gell D, Smith GCM, Zhang H, Divecha N, et al. 1995. *Cell* 82:849–56
47. Rodriguez-Viciana P, Warne PH, Dhand R, Vanhaesebroeck B, Gout I, et al. 1994. *Nature* 370:527–32
48. Rodriguez-Viciana P, Warne PH, Vanhaesebroeck B, Waterfield MD, Downward J. 1996. *EMBO J.* 15:2442–51
49. Rodriguez-Viciana P, Warne PH, Khwaja A, Marte BM, Pappin D, et al. 1997. *Cell* 89:457–67
50. Khwaja A, Rodriguez-Viciana P, Wennstrom S, Warne PH, Downward J. 1997. *EMBO J.* 16:2783–93
51. Zhou SY, Shoelson SE, Chaudhuri M, Gish G, Pawson T, et al. 1993. *Cell* 72:767–78
52. Nolte RT, Eck MJ, Schlessinger J, Shoelson SE, Harrison SC. 1996. *Nat. Struct. Biol.* 3:364–74
53. Booker GW, Breeze AL, Downing AK, Panayotou G, Gout I, et al. 1992. *Nature* 358:684–87
54. Backer JM, Myers MJ, Shoelson SE, Chin DJ, Sun XJ, et al. 1992. *EMBO J.* 11:3469–79
55. Carpenter CL, Auger KR, Chanudhuri M, Yoakim M, Schaffhausen B, et al. 1993. *J. Biol. Chem.* 268:9478–83
56. Druker BJ, Ling LE, Cohen B, Roberts TM, Schaffhausen BS. 1990. *J. Virol.* 64:4454–61
57. Ling LE, Druker BJ, Cantley LC, Roberts TM. 1992. *J. Virol.* 66:1702–8
58. Rameh LE, Chen C-S, Cantley LC. 1995. *Cell* 83:1–20
59. Sparks AB, Rider JE, Hoffman NG, Fowlkes DM, Quilliam LA, et al. 1996. *Proc. Natl. Acad. Sci. USA* 93:1540–44
60. Feng SB, Chen JK, Yu HT, Simon JA, Schreiber SL. 1994. *Science* 266:1241–47
61. Booker GW, Gout I, Downing AK, Driscoll PC, Boyd J, et al. 1993. *Cell* 73:813–22
62. Yu HT, Chen JK, Feng SB, Dalgarno DC, Brauer AW, et al. 1994. *Cell* 76:933–45
63. Kapeller R, Prasad KVS, Janssen O, Hou W, Schaffhausen BS, et al. 1994. *J. Biol. Chem.* 269:1927–33
64. Liu XQ, Marengere LE, Koch CA, Pawson T. 1993. *Mol. Cell. Biol.* 13:5225–32
65. Vogel LB, Fujita DJ. 1993. *Mol. Cell. Biol.* 13:7408–17
66. Prasad KVS, Janssen O, Kapeller R, Raab M, Cantley LC, et al. 1993. *Proc. Natl. Acad. Sci. USA* 90:7366–70
67. Prasad KVS, Kapeller R, Janssen O, Duke-Cohan JS, Repke H, et al. 1993. *Philos. Trans. R. Soc. London Ser.* 342:35–42
68. Karnitz LM, Sutor SL, Abraham RT. 1994. *J. Exp. Med.* 179:1799–808
69. Pleiman CM, Hertz WM, Cambier JC. 1994. *Science* 263:1609–12
70. Xu WQ, Harrison SC, Eck MJ. 1997. *Nature* 385:595–602
71. Sicheri F, Moarefi I, Kuriyan J. 1997. *Nature* 385:602–9
72. Soltoff SP, Cantley LC. 1996. *J. Biol. Chem.* 271:563–67
73. Musacchio A, Cantley LC, Harrison SC. 1996. *Proc. Natl. Acad. Sci. USA* 93:14373–78
74. Barrett T, Xiao B, Dodson EJ, Dodson G, Ludbrook SB, et al. 1997. *Nature* 385:458–61
75. Rittinger K, Walker PA, Eccleston JF, Nurmahomed K, Owen D, et al. 1997. *Nature* 388:693–97
76. Zheng Y, Bagrodia S, Cerione RA. 1994. *J. Biol. Chem.* 269:18727–30
77. Tolias KF, Cantley LC, Carpenter CL. 1995. *J. Biol. Chem.* 270:17656–59
78. Lopez-Ilasaca M, Crespo P, Pellici PG, Gutkind JS, Wetzker R. 1997. *Science* 275:394–97
79. Tang XW, Downes CP. 1997. *J. Biol. Chem.* 272:14193–99
80. Okada T, Hazeki O, Ui M, Katada T. 1996. *Biochem. J.* 317:475–80
80a. Rubio I, Rodriguez-Viciana P, Downward J, Wetzker R. 1997. *Biochem. J.* 326:891–5
81. MacDougall LK, Domin J, Waterfield MD. 1995. *Curr. Biol.* 5:1404–15
82. Molz L, Chen Y-W, Hirano M, Williams LT. 1996. *J. Biol. Chem.* 271:13892–99
83. Virbasius JV, Guilherme A, Czech MP. 1996. *J. Biol. Chem.* 271:13304–7
84. Domin J, Pages F, Volinia S, Rittenhouse SE, Zvelebil MJ, et al. 1997. *Biochem. J.* 326:139–47
85. Herman PK, Emr SD. 1990. *Mol. Cell. Biol.* 10:6742–54
86. Schu PV, Takegawa K, Fry MJ, Stack JH, Waterfield MD, et al. 1993. *Science* 260:88–91
87. Volinia S, Dhand R, Vanhaesebroeck B, MacDougall LK, Stein R, et al. 1995. *EMBO J.* 14:3339–48
88. Linassier C, MacDougall LK, Domin J, Waterfield MD. 1997. *Biochem. J.* 321:849–56

89. Kovacsovics TJ, Bachelot C, Toker A, Vlahos CJ, Duckworth B, et al. 1995. *J. Biol. Chem.* 270:11358–66
90. Toker A, Bachelot C, Chen C-S, Falck JR, Hartwig JH, et al. 1995. *J. Biol. Chem.* 270:29525–31
91. Stack JH, Herman PK, Schu PV, Emr SD. 1993. *EMBO J.* 12:2195–204
92. Panaretou C, Domin J, Cockcroft S, Waterfield MD. 1997. *J. Biol. Chem.* 272:2477–85
93. Pike LJ. 1992. *Endocr. Rev.* 13:692–706
94. Whitman M, Kaplan D, Roberts T, Cantley L. 1987. *Biochem. J.* 247:165–74
95. Endemann G, Dunn SN, Cantley LC. 1987. *Biochemistry* 26:6845–52
96. Walker DH, Dougherty N, Pike LJ. 1988. *Biochemistry* 27:6504–11
97. Graziani A, Ling LE, Endemann G, Carpenter CL, Cantley LC. 1992. *Biochem. J.* 284:39–45
98. Ling LE, Schulz JT, Cantley LC. 1989. *J. Biol. Chem.* 264:5080–88
99. Endemann GC, Graziani A, Cantley LC. 1991. *Biochem. J.* 273:63–66
100. Thompson DM, Cochet C, Chambaz EM, Gill GN. 1985. *J. Biol. Chem.* 260: 8824–30
101. Wiedemann C, Schafer T, Burger MM. 1996. *EMBO J.* 15:2094–101
102. Prasad KVS, Kapeller R, Janssen O, Repke H, Duke-Cohan JS, et al. 1993. *Mol. Cell. Biol.* 13:7708–17
103. Berditchevski F, Tolias KF, Wong K, Carpenter CL, Hemler ME. 1997. *J. Biol. Chem.* 272:2595–98
104. Li YS, Porter FD, Hoffman RM, Deuel TF. 1989. *Biochem. Biophys. Res. Commun.* 160:202–9
105. Yoshida S, Ikeda E, Uno I, Mitsuzawa H. 1992. *Mol. Gen. Genet.* 231:337–44
106. Yoshida S, Ohya Y, Goebl M, Nakano A, Anraku Y. 1994. *J. Biol. Chem.* 269: 1166–72
107. Yoshida S, Ohya Y, Nakano A, Anraku Y. 1994. *Mol. Gen. Genet.* 242:631–40
108. Wong K, Cantley LC. 1994. *J. Biol. Chem.* 269:28878–84
109. Nakagawa T, Goto K, Kondo H. 1996. *Biochem. J.* 320:643–49
110. Gehrmann T, Vereb G, Schmidt M, Klix D, Meyer HE, et al. 1996. *Biochim. Biophys. Acta* 1311:53–63
111. Nakagawa T, Goto K, Kondo H. 1996. *J. Biol. Chem.* 271:12088–94
112. Wong K, Meyers R, Cantley LC. 1997. *J. Biol. Chem.* 272:13236–41
113. Cutler NS, Heitman J, Cardenas ME. 1997. *J. Biol. Chem.* 272:27671–77
114. Flanagan CA, Schnieders EA, Emerick AW, Kunisawa R, Admon A, et al. 1993. *Science* 262:1444–48
115. Garcia BJ, Marini F, Stevenson I, Frei C, Hall MN. 1994. *EMBO J.* 13:2352–61
116. Balla T, Downing GJ, Jaffe H, Kim S, Zolyomi A, et al. 1997. *J. Biol. Chem.* 272:18358–66
117. Flanagan CA, Thorner J. 1992. *J. Biol. Chem.* 267:24117–25
118. Nakanishi S, Catt KJ, Balla T. 1995. *Proc. Natl. Acad. Sci. USA* 92:5317–21
119. Downing GJ, Kim S, Nakanishi S, Catt KJ, Balla T. 1996. *Biochemistry* 35:3587–94
120. Hippen KL, Buhl AM, D'Ambrosio D, Nakamura K, Persin C, et al. 1997. *Immunity* 7:49–58
121. Keith CT, Schreiber SL. 1995. *Science* 270:50–51
122. Carr AM. 1996. *Science* 271:314–15
123. Hoekstra MF. 1997. *Curr. Opin. Genet. Dev.* 7:170–75
124. Brown EJ, Beal PA, Keith CT, Chen J, Shin TB, et al. 1995. *Nature* 377:441–46
125. Jung M, Kondratyev A, Lee SA, Dimtchev A, Dritschilo A. 1997. *Cancer Res.* 57:24–27
126. Sabatini DM, Pierchala BA, Barrow RK, Schell MJ, Snyder SH. 1995. *J. Biol. Chem.* 270:20875–78
127. Cardenas ME, Heitman J. 1995. *EMBO J.* 14:5892–907
128. Zheng XF, Fiorentino D, Chen J, Crabtree GR, Schreiber SL. 1995. *Cell* 82: 121–30
129. Boronenkov IV, Anderson RA. 1995. *J. Biol. Chem.* 270:2881–84
130. Ishihara H, Shibasaki Y, Kizuki N, Katagiri H, Yazaki Y, et al. 1996. *J. Biol. Chem.* 271:23611–14
131. Loijens JC, Anderson RA. 1996. *J. Biol. Chem.* 271:32937–43
132. Castellino AM, Parker GJ, Boronenkov IV, Anderson RA, Chao MV. 1997. *J. Biol. Chem.* 272:5861–70
133. Zhang X, Loijens JC, Boronenkov IV, Parker GJ, Norris FA, et al. 1997. *J. Biol. Chem.* 272:17756–61
134. Yamamoto A, DeWald DB, Boronenkov IV, Anderson RA, Emr SD, et al. 1995. *Mol. Biol. Cell* 6:525–39
135. Moritz A, De Graan PNE, Gispen WH, Wirtz KWA. 1992. *J. Biol. Chem.* 267: 7207–10
136. Jenkins GH, Fisette PL, Anderson RA. 1994. *J. Biol. Chem.* 269:11547–54
137. Whiteford CC, Brearley CA, Ulug ET. 1997. *Biochem. J.* 323:597–601

137a. Dove SK, Cooke FT, Douglas MR, Sayers LG, Parker PJ, Michell RH. 1997. *Nature* 390:187–92
138. Chong LD, Traynor-Kaplan A, Bokoch GM, Schwartz MA. 1994. *Cell* 79:507–13
139. Tolias KF, Couvillon AD, Cantley LC, Carpenter CL. 1998. *Mol. Cell. Biol.* 18:762–70
140. Hartwig JH, Bokoch GM, Carpenter CL, Janmey PA, Taylor LA, et al. 1995. *Cell* 82:1–20
141. Zigmond SH, Joyce M, Borleis J, Bokoch GM, Devreotes PN. 1997. *J. Cell Biol.* 138:363–74
142. Urumow T, Wieland OH. 1988. *Biochim. Biophys. Acta* 972:232–38
143. Bazenet CE, Ruano AR, Brockman JL, Anderson RA. 1990. *J. Biol. Chem.* 265:18012–22
144. Divecha N, Truong O, Hsuan JJ, Hinchliffe KA, Irvine RF. 1995. *Biochem. J.* 309:715–19
145. Yamamoto K, Graziani A, Carpenter C, Cantley LC, Lapetina EG. 1990. *J. Biol. Chem.* 265:22086–89
145a. Banfic H, Tang X, Batty IH, Downes CP, Chen C, Rittenhouse SE. 1998. *J. Biol. Chem.* 273:13–16
146. Carpenter CL, Cantley LC. 1990. *Biochemistry* 29:11147–56
147. Auger KR, Serunian LA, Soltoff SP, Libby P, Cantley LC. 1989. *Cell* 57:167–75
148. Cunningham TW, Lips DL, Bansal VS, Caldwell KK, Mitchell CA, et al. 1990. *J. Biol. Chem.* 265:21676–83
149. Cunningham TW, Majerus PW. 1991. *Biochem. Biophys. Res. Commun.* 175:568–76
150. Hawkins PT, Jackson TR, Stephens LR. 1992. *Nature* 358:157–59
151. Carter AN, Huang RS, Sorisky A, Downes CP, Rittenhouse SE. 1994. *Biochem. J.* 301:415–20
152. Berridge MJ. 1987. *Annu. Rev. Biochem.* 56:159–93
153. Serunian LA, Haber MT, Fukui T, Kim JW, Rhee SG, et al. 1989. *J. Biol. Chem.* 264:17809–15

Annu. Rev. Biochem. 1998. 67:509–44

THE GREEN FLUORESCENT PROTEIN

Roger Y. Tsien
Howard Hughes Medical Institute; University of California, San Diego; La Jolla,
CA 92093-0647

KEY WORDS: *Aequorea*, mutants, chromophore, bioluminescence, GFP

ABSTRACT

In just three years, the green fluorescent protein (GFP) from the jellyfish *Aequorea victoria* has vaulted from obscurity to become one of the most widely studied and exploited proteins in biochemistry and cell biology. Its amazing ability to generate a highly visible, efficiently emitting internal fluorophore is both intrinsically fascinating and tremendously valuable. High-resolution crystal structures of GFP offer unprecedented opportunities to understand and manipulate the relation between protein structure and spectroscopic function. GFP has become well established as a marker of gene expression and protein targeting in intact cells and organisms. Mutagenesis and engineering of GFP into chimeric proteins are opening new vistas in physiological indicators, biosensors, and photochemical memories.

CONTENTS

509

0066-4154/98/0701-0509$08.00

NATURAL AND SCIENTIFIC HISTORY OF GFP

Discovery and Major Milestones

Green Fluorescent Protein was discovered by Shimomura et al (1) as a companion protein to aequorin, the famous chemiluminescent protein from *Aequorea* jellyfish. In a footnote to their account of aequorin purification, they noted that "a protein giving solutions that look slightly greenish in sunlight through only yellowish under tungsten lights, and exhibiting a very bright, greenish fluorescence in the ultraviolet of a Mineralite, has also been isolated from squeezates." This description of the appearance of GFP solutions is still accurate. The same group (2) soon published the emission spectrum of GFP, which peaked at 508 nm. They noted that the green bioluminescence of living *Aequorea* tissue also peaked near this wavelength, whereas the chemiluminescence of pure aequorin was blue and peaked near 470 nm, which was close to one of the excitation peaks of GFP. Therefore the GFP converted the blue emission of aequorin to the green glow of the intact cells and animals. Morin & Hastings (3) found the same color shift in the related coelenterates *Obelia* (a hydroid) and *Renilla* (a sea pansy) and were the first to suggest radiationless energy transfer as the mechanism for exciting coelenterate GFPs in vivo. Morise et al (4) purified and crystallized GFP, measured its absorbance spectrum and fluorescence quantum yield, and showed that aequorin could efficiently transfer its luminescence energy to GFP when the two were coadsorbed onto a cationic support. Prendergast & Mann (5) obtained the first clear estimate for the monomer molecular weight. Shimomura (6) proteolyzed denatured GFP, analyzed the peptide that retained visible absorbance, and correctly proposed that the chromophore is a 4-(*p*-hydroxybenzylidene)imidazolidin-5-one attached to the peptide backbone through the 1- and 2-positions of the ring.

Aequorea and *Renilla* GFPs were later shown to have the same chromophore (7); and the pH sensitivity, aggregation tendency (8), and renaturation (9) of *Aequorea* GFP were characterized. But the crucial breakthroughs came with the cloning of the gene by Prasher et al (10) and the demonstrations by Chalfie et al (11) and Inouye & Tsuji (12) that expression of the gene in other organisms creates fluorescence. Therefore the gene contains all the information necessary for the posttranslational synthesis of the chromophore, and no jellyfish-specific enzymes are needed.

Occurrence, Relation to Bioluminescence, and Comparison with Other Fluorescent Proteins

Green fluorescent proteins exist in a variety of coelenterates, both hydrozoa such as *Aequorea*, *Obelia*, and *Phialidium*, and anthozoa such as *Renilla* (3, 13). In this review, GFP refers to the *Aequorea* protein except where another genus name is specifically indicated. These GFPs seem to be partners with chemiluminescent proteins and to control the color of the emission in vivo. Despite interesting speculations, it remains unclear why these coelenterates glow, why green emission should be ecologically so superior to the blue of the primary emitters, and why the animals synthesize a separate GFP rather than mutate the chemiluminescent protein to shift its wavelengths. Other than *Aequorea* GFP, only *Renilla* GFP has been biochemically well characterized (14). Despite the apparent identity of the core chromophore in *Renilla* and *Aequorea* GFP, *Renilla* GFP has a much higher extinction coefficient, resistance to pH-induced conformational changes and denaturation, and tendency to dimerize (7).

Unfortunately, *Aequorea* GFP genes are the only GFP genes that have been cloned. Several other bioluminescent species also have emission-shifting accessory proteins, but so far the chromophores all seem to be external cofactors such as lumazines (15) or flavins (16), which diminish their attractiveness as biotechnological tags and probes. Likewise phycobiliproteins (17) and peridinin-chlorophyll-a protein (18), which are highly fluorescent and attractively long-wavelength accessory pigments in photosynthesis, use tetrapyrrole cofactors as their pigments. Correct insertion of the cofactors into the apoproteins has not been demonstrated in foreign organisms, so these proteins are not ready to compete with *Aequorea* GFP. A variety of marine organisms fluoresce, but the biochemistry of the fluorophores is almost completely unknown. Painstaking research like that undertaken by the pioneers of *Aequorea* and *Renilla* GFP would be needed before cloning efforts could begin. It is unclear whether any investigators or granting agencies are still patient enough to undertake and fund such long-term groundwork.

PRIMARY, SECONDARY, TERTIARY, AND QUATERNARY STRUCTURE

Primary Sequence from Cloning

The sequence of wild-type *Aequorea* GFP (10) is given in Figure 1. Sequences of at least four other isoforms are known (19), though none of the mutations seem to be in positions known to influence protein behavior. Most cDNA constructs derived from the original sequence contain the innocuous mutation Q80R, probably resulting from a PCR error (11). Also, the gene has been resynthesized with altered codons and improved translational initiation sequences (see section on "Promoters, Codon Usage, and Splicing").

The chromophore is a p-hydroxybenzylideneimidazolinone (10, 20) formed from residues 65–67, which are Ser-Tyr-Gly in the native protein. Figure 2 shows the currently accepted mechanism (21–23) for chromophore formation. First, GFP folds into a nearly native conformation, then the imidazolinone is formed by nucleophilic attack of the amide of Gly67 on the carbonyl of residue 65, followed by dehydration. Finally, molecular oxygen dehydrogenates the α-β bond of residue 66 to put its aromatic group into conjugation with the imidazolinone. Only at this stage does the chromophore acquire visible absorbance and fluorescence. This mechanism is based on the following arguments: (*a*) Atmospheric oxygen is required for fluorescence to develop (21, 24). (*b*) Fluorescence of anaerobically preformed GFP develops with a simple exponential time course after air is readmitted (21, 25), which is essentially unaffected by the concentration of the GFP itself or of cellular cofactors. (*c*) Analogous imidazolinones autoxidize spontaneously (26). (*d*) The proposed

-->

Figure 1 GFP sequences. (*Line 1*) The wild-type (WT) *gfp10* gene as originally cloned and sequenced by Prasher et al (10). (*Line 2*) A popular humanized version (EGFP, Clontech Laboratories, Palo Alto, CA) (64) incorporating (*a*) an optimal sequence for translational initiation (66), including insertion of a new codon GTG; (*b*) mutation of Phe64 to Leu to improve folding at 37°C; (*c*) mutation of Ser65 to Thr to promote chromophore ionization; and (*d*) mutation of His231 to Leu, which was probably inadvertent and neutral. (*Line 3*) WT amino acid sequence. (*Lines 4* and *5*) Numbering of amino acids and differences between EGFP and WT. The inserted Val is numbered 1a to maintain correspondence with the WT numbering. Note that constructs derived from the natural *gfp10* cDNA contain an apparently neutral mutation Gln80 → Arg (Q80R) caused by a PCR error that changed the CAG codon to CGG (11). In some genes artificially resynthesized with different codons, this error was corrected (e.g. 63 and Clontech's EGFP) but was left as an arginine codon in other instances (40, 62).

```
   AGT AAA GGA GAA GAA CTT TTC ACT GGA GTT GTC CCA ATT CTT GTT GAA TTA GAT GGT  60
   GTG AGC AAG GGC GAG GAG CTG TTC ACC GGG GTG GTG CCC ATC CTG GTC GAG CTG GAC GGC
   Val Ser Lys Gly Glu Glu Leu Phe Thr Gly Val Val Pro Ile Leu Val Glu Leu Asp Gly
       Val          5               10               15               20
(1a)
   GAT GTT AAT GGG CAC AAA TTT TCT GTC AGT GGA GAG GGT GAA GGT GAT GCA ACA TAC GGA 120
   GAC GTA AAC GGC CAC AAG TTC AGC GTG TCC GGC GAG GGC GAG GGC GAT GCC ACC TAC GGC
   Asp Val Asn Gly His Lys Phe Ser Val Ser Gly Glu Gly Glu Gly Asp Ala Thr Tyr Gly
                    25               30               35               40

   AAA CTT ACC CTT AAA TTT ATT TGC ACT ACT GGA AAA CTA CCT GTT CCA TGG CCA ACA CTT 180
   AAG CTG ACC CTG AAG TTC ATC TGC ACC ACC GGC AAG CTG CCC GTG CCC TGG CCC ACC CTC
   Lys Leu Thr Leu Lys Phe Ile Cys Thr Thr Gly Lys Leu Pro Val Pro Trp Pro Thr Leu
                    45               50               55               60

   GTC ACT ACT TTC TCT TAT GGT GTT CAA TGC TTT TCA AGA TAC CCA GAT CAT ATG AAA CAG 240
   GTG ACC ACC CTG ACC TAC GGC GTG CAG TGC TTC AGC CGC TAC CCC GAC CAC ATG AAG CAG
   Val Thr Thr Phe Ser Tyr Gly Val Gln Cys Phe Ser Arg Tyr Pro Asp His Met Lys Gln
                Leu Thr          70               75               80
                    65

   CAT GAC TTT TTC AAG AGT GCC ATG CCC GAA GGT TAT GTA CAG GAA AGA ACT ATA TTT TTC 300
   CAC GAC TTC TTC AAG TCC GCC ATG CCC GAA GGC TAC GTC CAG GAG CGC ACC ATC TTC TTC
   His Asp Phe Phe Lys Ser Ala Met Pro Glu Gly Tyr Val Gln Glu Arg Thr Ile Phe Phe
                    85               90               95               100

   AAA GAT GAC GGG AAC TAC AAG ACA CGT GCT GAA GTC AAG TTT GAA GGT GAT ACC CTT GTT 360
   AAG GAC GAC GGC AAC TAC AAG ACC CGC GCC GAG GTG AAG TTC GAG GGC GAC ACC CTG GTG
   Lys Asp Asp Gly Asn Tyr Lys Thr Arg Ala Glu Val Lys Phe Glu Gly Asp Thr Leu Val
                    105              110              115              120

   AAT AGA ATC GAG TTA AAA GGT ATT GAT TTT AAA GAA GAT GGA AAC ATT CTT GGA CAC AAA 420
   AAC CGC ATC GAG CTG AAG GGC ATC GAC TTC AAG GAG GAC GGC AAC ATC CTG GGG CAC AAG
   Asn Arg Ile Glu Leu Lys Gly Ile Asp Phe Lys Glu Asp Gly Asn Ile Leu Gly His Lys
                    125              130              135              140

   TTG GAA TAC AAC TAT AAC TCA CAC AAT GTA TAC ATC ATG GCA GAC AAA CAA AAG AAT GGA 480
   CTG GAG TAC AAC TAC AAC AGC CAC AAC GTC TAT ATC ATG GCC GAC AAG CAG AAG AAC GGC
   Leu Glu Tyr Asn Tyr Asn Ser His Asn Val Tyr Ile Met Ala Asp Lys Gln Lys Asn Gly
                    145              150              155              160

   ATC AAA GTT AAC TTC AAA ATT AGA CAC AAC ATT GAA GAT GGA AGC GTT CAA CTA GCA GAC 540
   ATC AAG GTG AAC TTC AAG ATC CGC CAC AAC ATC GAG GAC GGC AGC GTG CAG CTC GCC GAC
   Ile Lys Val Asn Phe Lys Ile Arg His Asn Ile Glu Asp Gly Ser Val Gln Leu Ala Asp
                    165              170              175              180

   CAT TAT CAA CAA AAT ACT CCA ATT GGC GAT GGC CCT GTC CTT TTA CCA GAC AAC CAT TAC 600
   CAC TAC CAG CAG AAC ACC CCC ATC GGC GAC GGC CCC GTG CTG CTG CCC GAC AAC CAC TAC
   His Tyr Gln Gln Asn Thr Pro Ile Gly Asp Gly Pro Val Leu Leu Pro Asp Asn His Tyr
                    185              190              195              200

   CTG TCC ACA CAA TCT GCC CTT TCG AAA GAT CCC AAC GAA AAG AGA GAC CAC ATG GTC CTT 660
   CTG AGC ACC CAG TCC GCC CTG AGC AAA GAC CCC AAC GAG AAG CGC GAT CAC ATG GTC CTG
   Leu Ser Thr Gln Ser Ala Leu Ser Lys Asp Pro Asn Glu Lys Arg Asp His Met Val Leu
                    205              210              215              220

   CTT GAG TTT GTA ACA GCT GCT GGG ATT ACA CAT GGC ATG GAT GAA CTA TAC AAA TAA TAA 720
   CTG GAG TTC GTG ACC GCC GCC GGG ATC ACT CTC GGC ATG GAC GAG CTG TAC AAG TAA
   Leu Glu Phe Val Thr Ala Ala Gly Ile Thr His Gly Met Asp Glu Leu Tyr Lys Stop
                    225              230 Leu            235
```

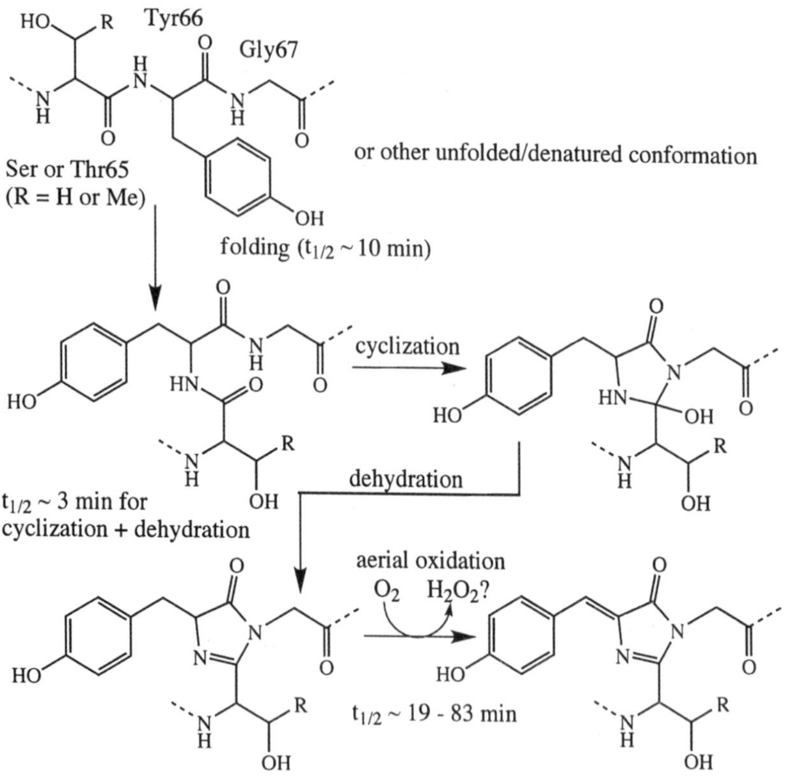

Figure 2 Mechanism proposed by Cubitt et al (22) for the intramolecular biosynthesis of the GFP chromophore, with rate constants estimated for the Ser65 → Thr mutant by Reid & Flynn (23) and Heim et al (25).

cyclization is isosteric with the known tendency for Asn-Gly sequences to cyclize to imides (27). Glycine is by far the best nucleophile in such cyclizations because of its minimal steric hindrance, and Gly67 is conserved in all known mutants of GFP that retain fluorescence. (*e*) Electrospray mass spectra indicate that anaerobically preformed GFP loses only 1 ± 4 Da upon exposure to air, consistent with the predicted loss of two hydrogens (22). This implies that the dehydration (-18 Da) must already have occurred anaerobically and must precede oxidation. (*f*) Reid & Flynn (23) have extensively characterized the kinetics of in vitro refolding of GFP from bacterial inclusion bodies with no chromophore, urea-denatured protein with a mature chromophore, and denatured protein with a chromophore reduced by dithionite. Renaturation was measured by development of fluorescence and resistance to trypsin attack. Their results support the sequential mechanism and provide the rate constants shown in Figure 2. However, many other aspects of the maturation mechanism remain obscure, such as the steric and catalytic roles of neighboring residues, the means by which mutations can improve folding efficiency, and the dependence of oxidation rate on the oxygen concentration and the protein sequence.

One predicted consequence of oxidation by O_2 is that hydrogen peroxide, H_2O_2, is presumably released in 1:1 stoichiometry with mature GFP. This byproduct might explain occasions when high-level expression of GFP can be deleterious. Perhaps catalase could be useful in such cases. Some difficult GFPs seem to express most readily when targeted to peroxisomes and mitochondria (28; R Rizzuto, T Pozzan, personal communication). Is it a coincidence that these organelles are the best at coping with reactive oxygen species?

Crystal Structures; Tolerance of Truncations

Although GFP was first crystallized in 1974 (4) and diffraction patterns reported in 1988 (29), the structure was first solved in 1996 independently by Ormö et al (30), Protein Data Bank accession number 1EMA, and by Yang et al (31), accession number 1GFL. Both groups relied primarily on multiple anomalous dispersion of selenomethionine groups to obtain phasing information from recombinant protein. Subsequent structures of other crystal forms and mutants (32–34a) have been solved by molecular replacement from the 1EMA coordinates. GFP is an 11-stranded β-barrel threaded by an α-helix running up the axis of the cylinder (Figure 3). The chromophore is attached to the α-helix and is buried almost perfectly in the center of the cylinder, which has been called a β-can (31, 34a). Almost all the primary sequence is used to build the β-barrel and axial helix, so that there are no obvious places where one could design large deletions and reduce the size of the protein by a significant fraction. Residues 1 and 230–238 were too disordered to be resolved; these regions correspond closely to the maximal known amino- and carboxyl-terminal deletions that still permit fluorescence to develop (35). A surprising number of polar groups and structured water molecules are buried adjacent to the chromophore (Figure 4). Particularly important are Gln69, Arg96, His148, Thr203, Ser205, and Glu222.

Dimerization

The excitation spectrum of wild-type GFP changes its shape as a function of protein concentration, implying some form of aggregation (8). The spectroscopic effects of such aggregation are discussed in the section on "Absorbance and Fluorescence Properties." In the Yang et al structure for wild-type GFP (31), the GFP is dimeric. The dimer interface includes hydrophobic residues Ala206, Leu221, and Phe 223 as well as hydrophilic contacts involving Tyr39, Glu142, Asn 144, Ser147, Asn149, Tyr151, Arg168, Asn170, Glu172, Tyr200,

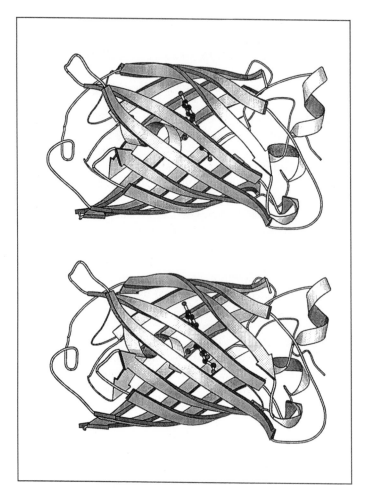

Figure 3 Stereoview of the three-dimensional structure of GFP (30), showing 11 β-strands forming a hollow cylinder through which is threaded a helix bearing the chromophore, shown in ball-and-stick representation. The drawing was prepared by the program MOLSCRIPT and is intended for viewing with uncrossed eyes. Figure courtesy of SJ Remington, University of Oregon.

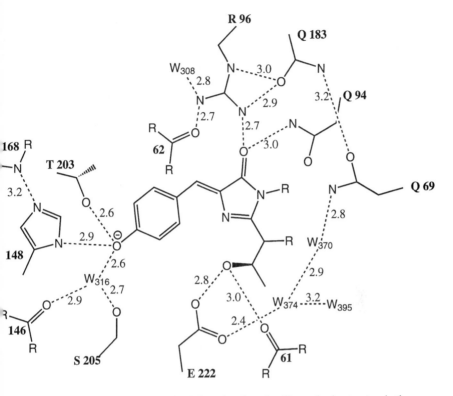

Figure 4 Amino acid side chains, main chain carbonyls and amides, and solvent waters in the immediate vicinity of the chromophore of S65T GFP (30). Side chains are labeled with the one-letter code for the amino acid and the residue number. Main chain groups are labeled with the residue number. Water oxygens are denoted by W and the corresponding serial number within Protein Data Bank structure 1EMA. Probable hydrogen bonds are shown as dotted lines labeled with the distance between the heteroatoms in angstroms. Obviously the true three-dimensional relationships cannot be depicted accurately in this two-dimensional schematic. Figure courtesy of SJ Remington, University of Oregon.

Ser202, Gln204, and Ser208. However, the same wild-type GFP could also crystallize as a monomer (32), isomorphous to the monomeric crystals formed by the S65T mutant (30). Even though GFP can hardly be more concentrated than in a crystal, the formation of dimers seems to be highly dependent on crystal growth conditions rather than an obligatory feature of GFP (33). The dissociation constant for the homodimer has been estimated as 100 μM (34a). By contrast, *Renilla* GFP is an obligate dimer, which is dissociated only under denaturing conditions (14).

ABSORBANCE AND FLUORESCENCE PROPERTIES

Classification of GFPs

The currently known GFP variants may be divided into seven classes based on the distinctive component of their chromophores: class 1, wild-type mixture of neutral phenol and anionic phenolate; class 2, phenolate anion; class 3, neutral phenol; class 4, phenolate anion with stacked π-electron system; class 5, indole; class 6, imidazole; and class 7, phenyl. Each class has a distinct set of excitation and emission wavelengths (Table 1). Classes 1–4 are derived from polypeptides with Tyr at position 66, whereas classes 5–7 result from Trp, His, and Phe at that position. Structures of the resulting chromophores are shown in Figure 5, together with typical fluorescence spectra.

WILD-TYPE MIXTURE OF NEUTRAL PHENOL AND ANIONIC PHENOLATE (CLASS 1) The wild-type *Aequorea* protein has the most complex spectra of all the GFPs. It has a major excitation peak at 395 nm that is about three times higher in amplitude than a minor peak at 475 nm. In normal solution, excitation at 395 nm gives emission peaking at 508 nm, whereas excitation at 475 nm gives a maximum at 503 nm (21). The fact that the emission maximum depends on the excitation wavelength indicates that the population includes at least two chemically distinct species, which do not fully equilibrate within the lifetime of the excited state. At pH 10–11, when the protein is on the verge of unfolding, increasing pH increases the amplitude of the 475-nm absorbance or excitation peak at the expense of the 395-nm peak (8). The simplest interpretation is that the 475-nm peak arises from GFP molecules containining deprotonated or anionic chromophores, whereas the 395-nm peak represents GFPs containing protonated or neutral chromophores (21, 22). The latter would be expected to deprotonate in the excited state, because phenols almost always become much more acidic in their excited states. Light-induced ionization to the anion would explain why excitation of the neutral chromophores gives emission at greater than 500 nm, similar to but not quite identical to the direct excitation of anionic chromophores. Picosecond spectroscopy gives direct evidence for such excited-state proton transfer (36). After a flash at 395 nm, the emission shifts from a 460- to a 508-nm peak over about 10 ps. These kinetics can be slowed greatly by cooling to 77°K and increasing viscosity, or by deuterium substitution, which argue strongly for excited-state proton transfer.

During most light absorption/emission cycles, the proton transfer eventually reverses. However, occasionally the proton does not return to the chromophore, so the neutral chromophore is photoisomerized to the anionic form. Thus on

Table 1 Spectral characteristics of the major classes of green fluorescent proteins (GFPs)

Mutation[a]	Common name	λ_{exc} (ϵ)[b]	λ_{em} (QY)[c]	Rel. fl.[d] @ 37°C	References[e]
Class 1, wild-type					
None or Q80R	Wild type	395–397 (25–30) 470–475 (9.5–14)	504 (0.79)	6	43, 45
F99S, M153T, V163A	Cycle 3	397 (30) 475 (6.5–8.5)	506 (0.79)	100	43, 45
Class 2, phenolate anion					
S65T		489 (52–58)	509–511 (0.64)	12	43–45
F64L, S65T	EGFP	488 (55–57)	507–509 (0.60)	20	43–45
F64L, S65T, V163A		488 (42)	511 (0.58)	54	44
S65T, S72A, N149K, M153T, I167T	Emerald	487 (57.5)	509 (0.68)	100	44
Class 3, neutral phenol					
S202F, T203I	H9	399 (20)	511 (0.60)	13	44
T203I, S72A, Y145F	H9–40	399 (29)	511 (0.64)	100	44
Class 4, phenolate anion with stacked π-electron system (yellow fluorescent proteins)					
S65G, S72A, T203F		512 (65.5)	522 (0.70)	6	44
S65G, S72A, T203H		508 (48.5)	518 (0.78)	12	44
S65G, V68L, Q69K S72A, T203Y	10C Q69K	516 (62)	529 (0.71)	50	44
S65G, V68L, S72A, T203Y	10C	514 (83.4)	527 (0.61)	58	44
S65G, S72A, K79R, T203Y	Topaz	514 (94.5)	527 (0.60)	100	44
Class 5, indole in chromophore (cyan fluorescent proteins)					
Y66W		436	485	—	21
Y66W, N146I, M153T, V163A	W7	434 (23.9) 452	476 (0.42) 505	61	44
F64L, S65T, Y66W, N146I, M153T, V163A	W1B or ECFP	434 (32.5) 452	476 (0.4) 505	80	44
S65A, Y66W, S72A, N146I, M153T, V163A	W1C	435 (21.2)	495 (0.39)	100	44
Class 6, imidazole in chromophore (blue fluorescent proteins)					
Y66H	BFP	384 (21)	448 (0.24)	18	44
Y66H, Y145F	P4–3	382 (22.3)	446 (0.3)	52	44
F64L, Y66H, Y145F	EBFP	380–383 (26.3–31)	440–447 (0.17–0.26)	100	43, 44
Class 7, phenyl in chromophore					
Y66F		360	442	—	22

[a]Substitutions from the primary sequence of GFP (see Figure 1) are given as the single-letter code for the amino acid being replaced, its numerical position in the sequence, and the single-letter code for the replacement. Note that many valuable mutants have been left out of this table for reasons of brevity and because quantitative spectral and brightness data were not available; therefore omission does not imply denigration. Phenotypically neutral substitutions such as Q80R, H231L, and insertion of residue 1a (see Figure 1) have also been omitted.

[b]λ_{exc} is the peak of the excitation spectrum in units of nanometers. ϵ in parentheses is the absorbance extinction coefficient in units of $10^3\,M^{-1}\,cm^{-1}$. Estimates of extinction coefficients have tended to increase as expression and purification are optimized; obsolete older values have been omitted. Two numbers separated by a dash indicate a range of estimates from different authors working under slightly different conditions. Two numbers on separate lines indicate two distinct peaks in the excitation spectrum.

[c]λ_{em} is the peak of the emission spectrum in units of nanometers. QY in parentheses is the fluorescence quantum yield, which is dimensionless. The best figure of merit for the overall brightness of properly matured GFPs is the product of ϵ and QY. See footnote b for explanation of pairs of values.

[d]Relative fluorescence intensities for proteins expressed in *Escherichia coli* at 37°C from the same vector background under similar conditions. These numbers include not only the intrinsic brightnesses measured by $\epsilon \cdot$ QY but also the folding efficiencies at 37°C. They are only rough estimates, which will change under different expression conditions. They have been arbitrarily normalized to 100 for the brightest member of each class and cannot be used to compare different classes.

[e]References only for the quantitative spectral and brightness data. References to the origin and use of the mutants have been omitted for lack of space.

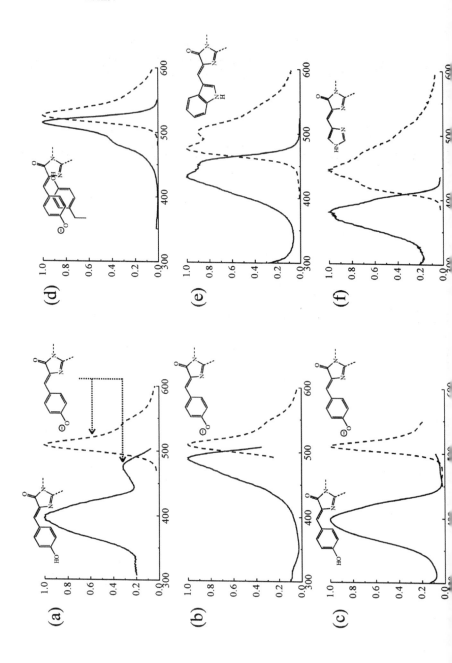

intense UV illumination, the 395-nm absorbance and excitation peak of the neutral form gradually declines and the 470-nm peak of the chromophore anion increases (22, 36, 37). Before illumination, wild-type GFP contains about a 6:1 ratio of neutral-to-anionic forms, but with enough UV the percentage of anionic form can increase several-fold. The probable mechanism (32, 33) is that proton transfer occurs via the hydrogen bonds of a buried water and Ser205 to Glu222. Meanwhile the side chain of Thr203 rotates to solvate and stabilize the phenolate oxyanion. In the crystal structure of monomeric wild-type GFP, Thr203 exists in two conformations: approximately 85% with the OH facing away from the phenol oxygen, and 15% with the OH rotated toward it (32). This proportion agrees well with the spectroscopic estimate for the ratio of neutral to anionic chromophores at equilibrium (36).

Wild-type GFP folds fairly efficiently when expressed at or below room temperature, but its folding efficiency declines steeply at higher temperatures. Presumably this natural temperature sensitivity is of no consequence to the jellyfish, which would never encounter warm water in the Pacific Northwest. Temperature sensitivity is restricted to the folding process. GFP that has matured properly at low temperature is stable and fluorescent at temperatures up to at least 65°C. The poor ability of GFP to mature in warm temperatures has been used in pulse-chase experiments in which the fate of fluorescent protein made at low temperatures is followed after restoration of normal warmth and simultaneous suppression of new fluorescence (38, 39). However, for other applications it would be desirable to have a GFP that works well at 37°C.

The most extensive attempt to develop such a mutant while preserving the complex wild-type spectrum utilized DNA shuffling (40), a technique for recombining various mutations while creating new ones. This approach produced a triple mutant, F99S, M153T, V163A, which improved 37°C-folding, reduced aggregation at high concentrations, and increased the diffusibility of the protein inside cells (37). The latter two mutations had already been found by more conventional mutagenesis procedures (41, 42). Such folding mutations do not

←――

Figure 5 Fluorescence excitation and emission spectra (solid and dashed lines, respectively) for typical members of the six major classes of GFP mutants, together with the chromophore structures believed to be responsible for the spectra. Spectra have been normalized to a maximum amplitude of 1. For comparison of absolute brightnesses, see the extinction coefficients and quantum yields in Table 1. When only one structure is drawn, both excitation and emission spectra arise from the same state of chromophore protonation. The actual GFPs depicted are (*a*) wild-type, (*b*) Emerald, (*c*) H9-40, (*d*) Topaz, (*e*) W1B, and (*f*) P4-3. The detailed substitutions within each of these variants are listed in Table 1.

increase the intrinsic brightness of properly matured GFP molecules. Such brightness is measured by the product of extinction coefficient and fluorescence quantum yield (Table 1). The folding mutations merely increase the percentage of molecules that mature properly under adverse conditions, such as 37°C and high GFP concentrations that promote aggregation (33, 43–45). Although the folding mutations are highly valuable and should be incorporated routinely into new constructs, they produce less dramatic or no improvements at lower temperatures and levels of expression. Also the increments in brightness achieved by compounding such mutations will be limited by the obvious fact that folding cannot exceed 100% efficiency.

The coexistence of neutral and anion chromophores giving two excitation peaks in the wild-type spectrum has a few advantages and many disadvantages for cell biological applications. If the GFP fluorescence is to be detected by the naked eye, UV excitation is convenient (40) because UV is inherently invisible. However, because intense UV can damage the eye, an external excitation-blocking filter would be advisable even for visual inspection. Also, scattering, autofluorescence, and the possibility of tissue damage are more severe with UV excitation. Excitation at the 470-nm peak would reduce these problems but is inefficient because only 15% of the protein has the anionic chromophore that absorbs there. The photoisomerization is a major hindrance to quantitation of images, but it also permits the diffusion or trafficking of GFP-labeled proteins to be monitored by locally irradiating a cell with a point or stripe of intense UV and then imaging the subsequent fate of the photoisomerized protein with 470-nm excitation (37).

PHENOLATE ANION IN CHROMOPHORE (CLASS 2)   GFPs with phenolate anions in the chromophore have become the most widely used class for routine cell biological use because they were the first group to combine high brightness with simple excitation and emission spectra peaking at wavelengths very similar to fluorescein, the most popular small-molecule fluorophore. The most commonly used mutation to cause ionization of the phenol of the chromophore is a replacement of Ser65 by Thr, or S65T (25), though several other aliphatic residues such as Gly, Ala, Cys, and Leu have roughly similar effects (25, 46, 47). The triple mutation F64M, S65G, Q69L, found by random mutagenesis around the chromophore, has achieved considerable popularity under the name RSGFP4 (46). In both S65T and RSGFP4, the wild-type 395-nm excitation peak due to the neutral phenol is suppressed, and the 470- to 475-nm peak due to the anion is enhanced five- to sixfold in amplitude and shifted to 489–490 nm (25, 46, 48). The oxidation to the mature fluorophore was about fourfold faster in S65T than in the wild type (25). Like wild-type GFP, S65T folds fairly efficiently when expressed at room temperature or below but tends to misfold and produce mostly nonfluorescent aggregates at higher temperatures. Because of the obvious interest in expression at 37°C, much effort has been devoted to finding additional mutations that give greater brightness at warmer temperatures. The most often used of these have been F64L (47) and V163A (42), though other mutations such as S72A, N149K, M153T, and I167T (33) (Table 1) can also be helpful alone or in combination. As with GFPs of wild-type spectra, such mutations improve only folding efficiency, not the brightness of properly folded molecules.

The probable mechanism by which replacement of Ser65 promotes chromophore ionization (30, 32) is that only Ser65 can donate a hydrogen bond to the buried side chain of Glu222 to allow ionization of that carboxylate, which is within 3.7 Å of the chromophore. Gly, Ala, and Leu cannot donate hydrogen bonds, and Thr and Cys are too large to adopt the correct conformation in the crowded interior of the protein. Such residues at position 65 force the carboxyl of Glu222 to remain neutral. The other polar groups solvating the chromophore are then sufficient to promote its ionization to an anion, whereas if Glu222 is an anion, electrostatic repulsion forbids the chromophore from becoming an anion as well. This hypothesis explains why mutation of Glu222 to Gly gives the same spectral shape and wavelengths (49) as Ser65 mutations. However, practical applications of E222G have not been reported.

NEUTRAL PHENOL IN CHROMOPHORE (CLASS 3)   Ionization of the chromophore cannot only be favored but be repressed. Mutation of Thr203 to Ile (21, 49) largely suppresses the 475-nm excitation peak, leaving only the shorter wavelength peak at 399 nm. Presumably a chromophore anion cannot be adequately solvated once the OH of Thr203 is gone, so the chromophore is neutral in almost all the ground-state molecules. However, the emission is still at 511 nm because the excited state remains acidic enough to eject a proton. This mutant and its folding-optimized descendants (Table 1) could be valuable alternatives for UV-excited green fluorescence without the complicated photochemistry of wild-type (class 1) GFPs. Because they lack an excitation maximum near 479–490 nm, they could be used in conjunction with the phenolate anion (class 2) GFPs for double-labeling. Images taken with the two different excitation bands near 400 and 480 nm but the same greater-than-500-nm emission would be compared. Even though the spectral contrast between the two GFPs would not be as great as when both excitation and emission wavelengths are varied, the use of two excitation wavelengths fits the many imaging systems designed for excitation-ratioing indicators and avoids any image registration problems created by alternating emission filters. The neutral phenol GFPs also have the largest gap in wavelengths between excitation and emission peaks of any of the GFPs. This large Stokes' shift could be advantageous in supporting laser action, where it is important that the dye should have as little absorbance as possible at the wavelengths of fluorescence and lasing.

PHENOLATE ANION WITH STACKED π-ELECTRON SYSTEM (CLASS 4)   The longest wavelengths currently available by mutation result from stacking an aromatic ring next to the phenolate anion of the chromophore. So far the aromatic ring has always come from the side chain of residue 203, and residue 65 is Gly or Thr instead of Ser, to promote ionization of the chromophore. All four aromatic residues at that position 203 (His, Trp, Phe, and Tyr) increase the excitation and emission wavelengths by up to 20 nm, with the shifts increasing in the stated order (30). These mutants were rationally designed from the crystal structure of S65T GFP in the expectation that the additional polarizability around the chromophore and π-π interaction would reduce the excited state energy, that is, increase both the excitation and emission wavelengths. The mutants would have been nearly impossible to find by random mutagenesis, because all three bases in the original codon (ACA) encoding Thr203 would have to be replaced to encode an aromatic amino acid, and any random mutation rate high enough to give a significant probability of changing three bases in one codon would mutate so many other residues as to kill the protein.

The actual crystal structure of a mutant containing Tyr203 has verified that its aromatic ring stacks next to the chromophore (RM Wachter, GT Hanson, AB Cubitt, K Kallio, RY Tsien, SJ Remington, manuscript in preparation). Mutation of Gln69 to Lys (Q69K) gives an additional shift of about 1–2 nm, resulting in an emission peak around 529 nm, the longest now known (Table 1). Although 529 nm itself is rather greenish, the tail at longer wavelengths is sufficient to give the fluorescence an overall yellowish appearance, which is clearly distinguishable by eye from the more greenish emission of GFP classes 1–3. Therefore class 4 GFPs have been called YFPs for yellowish fluorescent proteins (50), though this name has also been used for a fluorescent protein from *Vibrio fischeri* (16). The so-called BioYellow variety marketed by Pharmingen (51) is identical to RSGFP4 (46), a class 2 GFP with emission maximum at 505 nm, the same as the wild type.

INDOLE IN CHROMOPHORE DERIVED FROM Y66W (CLASS 5)   Substitution of Trp for Tyr66 produces a new chromophore with an indole instead of a phenol or phenolate (21). Excitation and emission wavelengths are 436 and 476 nm, intermediate between neutral phenol and anionic phenolate chromophores. The increased bulk of the indole requires many additional mutations to restore reasonable brightness (41), but when such mutations are provided, the over-all performance is fairly good (Table 1). These proteins are called cyan fluo-rescent proteins, or CFPs, because of their blue-green or cyan emission. One curious and so far unexplained feature (Figure 5) is that most have double-humped rather than conventional single excitation and emission peaks. The origin of the doubled emission peaks must be vibrational levels or other quan-tum states that equilibrate within the lifetime of the excited state, because their shapes and relative amplitudes are the same regardless of the excitation wavelength.

IMIDAZOLE IN CHROMOPHORE DERIVED FROM Y66H (CLASS 6)   Substitution of His for Tyr 66 puts an imidazole in the chromophore (21) and shifts the wave-lengths yet shorter than Trp66. The excitation and emission peaks are around 383 and 447 nm (Table 1), so the emission is blue. A convenient abbreviation is therefore BFP, although other blue fluorescent proteins [e.g. spent aequorin and lumazine-containing proteins from *Photobacterium phosphoreum* (15)] have previously shared the same acronym. Crystal structures for several BFPs have been solved (33, 34). As usual, these proteins benefit considerably from fold-ing mutations (33, 41). BFP and a UV-excitable GFP permit double-labeling of cellular structures with two emission colors arising from a common excitation wavelength near 390 nm (28). However, even with folding improvements BFP still suffers from a relatively low fluorescence quantum yield and relatively easy bleaching (28). A functional dye laser has been constructed with purified BFP as the gain medium, but the duration of lasing at 450 nm is limited by protein bleaching (SJ Remington, D Alavi, M Raymer, RY Tsien, manuscript in preparation).

PHENYL IN CHROMOPHORE DERIVED FROM Y66F (CLASS 7)   The very shortest wavelengths are obtained with Phe at 66 (22). This mutant has been little in-vestigated because no obvious practical use for proteins requiring such short wavelength excitation has been proposed. Nevertheless it proves that any aro-matic residue at position 66 can form a chromophore.

General Relation of Structure to Spectra

The denatured wild-type protein absorbs maximally at 384 nm at neutral or acidic pH and at 448 nm at alkaline pH, with a pK_a of 8.1 (7). This rough similarity to the absorbance and excitation maxima of the intact protein was a primary motivation for assigning the 395- and 470-nm excitation peaks of the latter to the neutral and anionic chromophores. Denatured GFPs or small proteolytic fragments carrying the chromophore are essentially totally nonfluo-rescent, presumably because the chromophore is unprotected from quench-ing by jostling water dipoles, paramagnetic oxygen molecules, or *cis-trans* isomerization (7, 52). The slight difference in absorbance wavelengths between denatured and intact proteins is not unreasonable for the structured environment of the latter. In particular, Arg96 puts a positively charged guanidinium quite close to the carbonyl group of the imidazolinone. This cation would electrostatically stabilize increased electron-density on the carbonyl oxygen in the chromophore's excited state. This electrostatic attraction would explain much of the red shift of intact protein relative to denatured protein. Indeed, mutation of Arg96 to Cys in S65T blue-shifts the excitation maximum from 489 to 472 nm and the emission maximum from 511 to 503 nm (R Ranganathan, personal communication), supporting a major role for Arg96 in lowering the energy of the excited state. Theoretical calculations of the energy levels of the chromophore in vacuo have led to the proposal that the imidazolinone-ring nitrogen adjacent to the hydroxybenzylidene must be protonated (53). However, the large effects on the chromophore of buried water molecules and the microenvironment supplied by the protein (52) would seem to provide a chemically more plausible explanation.

Two-Photon Excitation

One of the most promising new techniques in high-resolution fluorescence microscopy is two-photon excitation (54), in which two infrared photons hit a fluorophore within a few femtoseconds of each other and sum their energy to simulate a single photon of half the wavelength, that is, ultraviolet to blue. Such coincidence of infrared photons requires extremely high fluxes and therefore occurs to a significant extent only at the focus of a microscope objective of high numerical aperture, illuminated by a pulsed laser. Because other regions of the specimen are effectively not excited, they neither emit fluorescence nor are subject to photobleaching or photodynamic damage. As in confocal microscopy, the image is built up by scanning the focus point in a raster, but unlike confocal microscopy, out-of-focus planes are protected from bleaching, which is a tremendous advantage for two-photon excitation.

GFPs are quite good fluorophores for two-photon excitation. Wild-type GFP is readily excited with 780- to 800-nm pulses (43, 54–56), which are in the optimal output range for commercial mode-locked titanium-sapphire lasers. However, the photoisomerization proceeds just as with 390- to 400-nm single-photon excitation (55). Class 3 (neutral phenol) GFP mutants have not been tried but should be better because they disfavor photoisomerization. S65T, the prototypic class 2 GFP mutant, is optimally excited near 910 nm and has a slightly higher two-photon cross-section than the wild type (54). Two-photon excitation is also effective on class 6 (imidazole) blue mutants (43) as well as class 5 (indole) cyan mutants (H Fujisaki, G Fan, A Miyawaki, RY Tsien, unpublished observations).

Effects of pH

As noted above, wild-type GFP at high pH (11–12) loses absorbance and ex-citation amplitude at 395 nm and gains amplitude at 470 nm (8, 57). Such pH values, though mechanistically revealing, are almost never encountered in bi-ology. Wild-type GFP is also quenched by acidic pH values with an apparent pK_a near 4.5. Several of the mutants with enhanced spectral properties at pH 7 are actually more acid sensitive than is the wild type; thus EGFP is 50% quenched at pH 5.5 (43). pK_as as high as 6.8 are found in some of the class 4 mutants with Thr203 replaced by an aromatic residue (J Llopis, RY Tsien, manuscript in preparation; R Wachter, SJ Remington, personal communica-tion). The mechanistic explanation for these relatively high pK_as is not entirely clear. Loss of the Thr203 hydroxyl would indeed be expected to destabilize the phenolate form of the chromophore. However, the effect of acid is to quench the fluorescence altogether rather than simply shift it toward the short wavelengths expected of a protonated chromophore. The sensitivity of some GFPs to mildly acidic pH values carries both advantages and disadvantages. Such GFPs could be quenched to a major extent in acidic organelles such as lysosomes, endo-somes, and Golgi compartments. The pH sensitivity of some GFPs can also be put to good use to measure organellar pH (J Llopis, RY Tsien, manuscript in preparation; R Wachter, SJ Remington, personal communication) by targeting appropriate GFPs to those locations.

Effects of Temperature and Protein Concentrations

Higher GFP concentrations amplify the main excitation peak at 395 nm at the expense of the subsidiary peak at 470 nm (8). Because the 395- and 470-nm peaks are believed to result from neutral and anionic fluorophores, respectively, aggregation probably inhibits ionization of the fluorophores. Increasing temper-ature from 15 to 65°C modestly decreases the 395-nm and increases the 470-nm excitation peak of mature wild-type GFP. Yet higher temperatures cause denat-uration, with 50% of fluorescence lost at 78°C (8). As already mentioned, much more modest temperature increases from 20 to 37°C can profoundly decrease maturation efficiency of GFPs lacking mutations to improve folding.

Effects of Prior Illumination

GFPs have a variety of remarkable abilities to undergo photochemical trans-formations, which enables visualization of the diffusion or trafficking of GFP-tagged proteins. A defined zone within a cell or tissue is momentarily exposed to very bright illumination, which initiates the photochemistry. The subsequent fate of the photoconverted protein is imaged over time. At least four distinct types of semipermanent photochemical transformation have been reported from one or more GFPs: (a) simple irreversible photobleaching, (b) conversion from a 395- to 475-nm excitation maximum, (c) loss of 488-nm-excited fluorescence, reversible by illumination at 406 nm, and (d) generation of rhodamine-like orange or red fluorescence upon illumination at 488 nm under strictly anaerobic conditions (58).

IRREVERSIBLE PHOTOBLEACHING   Photobleaching is the simplest and most universal behavior of fluorophores. Most GFPs are relatively resistant to photobleaching (22, 59), perhaps because the fluorophore is well shielded from chemical reactants such as O_2. The bleach rate of the prototypic class 2 GFP, the S65T mutant, was reported to be relatively indifferent to equilibration with 0–100% oxygen or addition of quenchers of triplet states, singlet oxygen, and radicals (59). Nevertheless, with sufficient laser power, photobleaching is easily observed and exploited for measurements of fluorescence recovery (37, 59, 60). The class 6 mutants (BFPs) are generally more photosensitive than classes 1–5 (28). Cell-permeant antioxidants may be helpful in protecting such GFPs from bleaching. An example is Trolox, 6-hydroxy-2,5,7,8-tetramethylchroman-2-carboxylic acid, a water-soluble vitamin E analog commercially available from Aldrich Chemical Co., Milwaukee, WI.

SHIFTING TO A LONGER-WAVELENGTH EXCITATION PEAK   This behavior is characteristic of wild-type and other class 1 GFPs (11, 22, 36). As discussed previously, the mechanism is probably a light-driven proton transfer from the neutral chromophore to the carboxylate of Glu222, yielding an anionic chromophore and a protonated Glu222 (32). This UV-induced enhancement of the blue excitation peak has been exploited to measure lateral diffusion of GFP-tagged proteins (37). Because the proton transfer is mediated by Thr203 and Ser205, mutation of those residues might be a promising way to enhance this photochromic effect. Indeed, UV irradiation of the double mutant T203S, S205T increases the amplitude of its long-wavelength excitation peak 11.8-fold, whereas wild-type GFP under the same conditions increases by at most 3.6-fold (R Heim, RY Tsien, unpublished information).

REVERSIBLE LOSS OF LONGER-WAVELENGTH EXCITATION PEAK; SINGLE MOLECULE DETECTION   The opposite behavior, a shift from a longer to a shorter excitation wavelength, seems to occur in class 4 mutants. Upon intense laser illumination and observation of the fluorescence from single molecules immobilized in a polyacrylamide gel, such mutants both blink reversibly on a time scale of seconds and switch the fluorescence off over tens of seconds (61). However, the apparent bleaching can be reversed by illumination at short wavelengths such as 406 nm. Probably the chromophore, which is normally mostly anionic, can eventually be driven into a protonated state with an excitation maximum near 405 nm, whereupon it appears nonfluorescent and bleached to the probe laser at 488 nm. However, excitation of the protonated state then restores the normal anionic state. Such cycling can be repeated many times with apparently no fatigue, so that it potentially represents a basis for an optical memory at the single molecule level. It might also be particularly advantageous for multiple determinations of diffusion or trafficking on the same region of interest.

ANAEROBIC PHOTOCONVERSION TO A RED FLUORESCENT SPECIES   A variety of GFPs, including wild-type, S65T, and EGFP, undergo a remarkable photo-conversion to a red fluorescent species under rigorously anaerobic conditions, for example, in microorganisms that have exhausted the oxygen in the medium, or in the presence of oxygen scavengers such as glucose plus glucose oxidase and catalase (58). The nature of this species emitting at 600 nm remains to be clarified. This effect has been used to measure the diffusibility of GFP in live bacteria. One complication is that the red emission develops with an exponential time constant of about 0.7 s after the illuminating flash (58).

EXPRESSION, FORMATION, MATURATION, RENATURATION, AND OBSERVATION

Promoters, Codon Usage, and Splicing

The expression level and detectability of GFP depend on many factors, the most important of which are summarized in Table 2. Obviously the more copies of the gene and the stronger the promoters/enhancers driving its transcription, the more protein that will be made per cell. In plants, it has been important to alter the original codon usage to eliminate a cryptic splice site (62). Codons have also been altered to conform to those preferred in mammalian systems (63, 64) and in the pathogenic yeast *Candida albicans* (65). Some authors have found such codon alterations to improve expression levels in mammalian systems (44, 63, 64), whereas others have found little improvement (43). Because the mammalianized genes are now widely available and may well be beneficial, they might as well be incorporated into all new GFP constructs for use in vertebrates. Our impression is that mammalian codons do not hurt expression levels in bacteria. Yet another improvement for mammalian systems is the inclusion of an optimal ribosome-binding site, also known as a Kozak sequence for translational initiation (66). Such a sequence requires insertion of an additional codon immediately after the starting methionine, as shown in Figure 1. The additional valine or alanine (40) does not seem to interfere with protein function; we prefer to number it 1a so that the numbering of the subsequent amino acids continues to correspond with the wild-type numbering. GFP can be expressed with reasonable efficiency in a cell-free in vitro translation system (67), a finding that confirms the protein can fold autonomously.

Table 2 Factors affecting the detectability of green fluorescent protein (GFP)

Total amount of GFP (picked out by antibodies, or by position on gel if GFP is abundant enough)
  Number of copies of gene, duration of expression
  Strength of transcriptional promotors and enhancers
  Efficiency of translation including Kozak sequence and codon usage
  Absence of mRNA splicing, protein degradation and export
Efficiency of posttranslational fluorophore formation
  Solubility vs. formation of inclusion bodies
  Availability of chaperones
  Hindrance to folding because of unfortunate fusions to host proteins
  Time, temperature, availability of O_2, and intrinsic rate of cyclization/oxidation
Molecular properties of mature GFP
  Wavelengths of excitation and emission
  Extinction coefficient and fluorescence quantum yield
  Susceptibility to photoisomerization/bleaching
  Dimerization
Competition with noise and background signals
  Autofluorescence of cells or culture media at preferred wavelengths
  Location of GFP, diffuse vs. confined to small subregions of cells or tissues
  Quality of excitation and emission filters and dichroic mirrors
  Sensitivity, noise, and dark current of photodetector

Folding Mutations and Thermotolerance

As mentioned previously, several mutations improve the ability of GFP to fold at temperatures above those to which the jellyfish would have been exposed. Most of them replace bulky residues with smaller ones. Their scattered locations throughout the three-dimensional structure of the mature protein (Figure 6) give little hint as to why they should help folding and maturation. Of course, the X-ray structures are all determined on well-folded mature proteins. At the stage when the mutations are needed, the protein is presumably less well ordered. As an alternative to mutating the GFP, the presence of chaperones can also help GFP fold (68). Aside from the potential technical importance of providing chaperones, this finding makes GFP a useful substrate for testing chaperone function (69), since GFP provides a continuous nondestructive assay for successful folding.

Requirement for O_2

The requirement for O_2 to dehydrogenate the α,β bond of residue 66 (21, 23, 24) means that GFP probably cannot become fluorescent in obligate anaerobes. So far this is the only fundamental limitation on the range of systems in which GFP can be expressed. Once GFP is matured, O_2 has no further effect (59). The oxidation seems to be the slowest step (Figure 2) in the maturation of GFP (23),

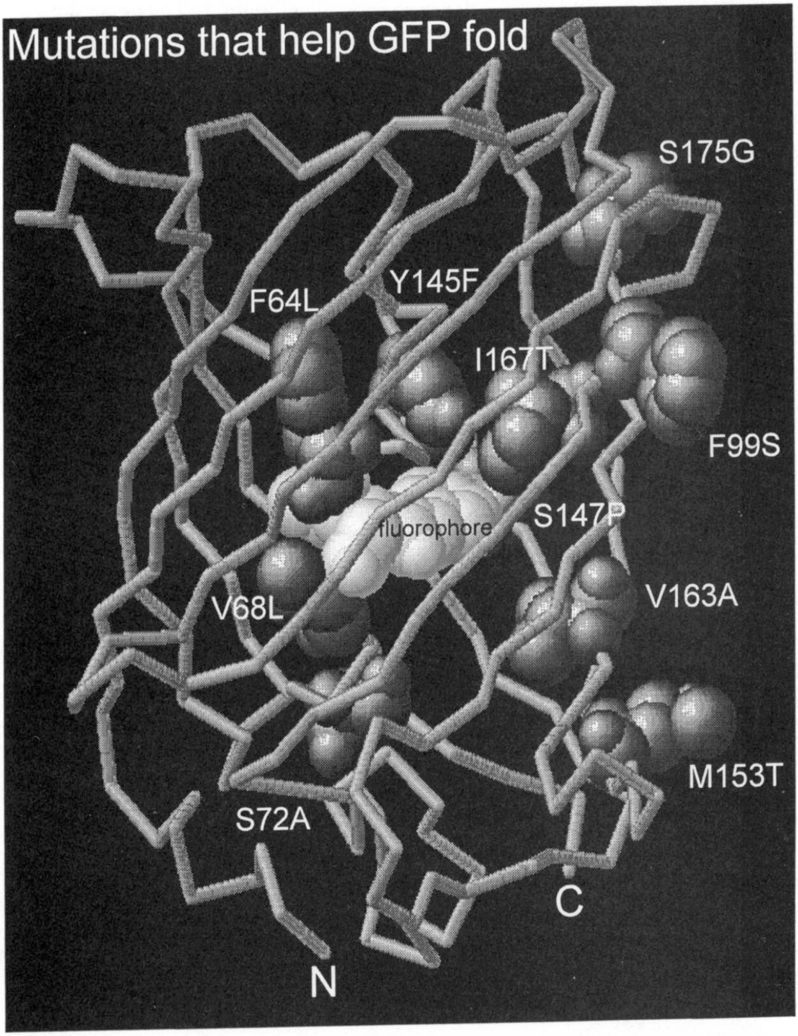

Figure 6 Location on the GFP crystal structure (30) of the most important sites that improve folding at 37°C. The amino acids shown in space-filling representation are the wild-type residues that are replaced by the mutations listed.

so it imposes the ultimate limit on the ability of GFP fluorescence to monitor rapid changes in gene expression. Considering its importance, surprisingly little work has been done on how to accelerate this oxidation. The dependence of rate on oxygen pressure has not been characterized, so it is unknown whether pO_2 higher than ordinary atmospheric would speed up the reaction. The mutant S65T has been reported to oxidize with an exponential time constant of 0.5 h, 4 times faster than the 2 h for wild-type protein under parallel conditions (25). Strong reductants such as dithionite can decolorize mature GFP (24, 52), probably by rehydrogenating the chromophore. Such reduction may also require that the GFP become denatured to allow access to the buried chromophore (23).

Histology in Fixed Tissues

Although the prime advantage of GFP is its ability to generate fluorescence in live tissue, its fluorescence does survive glutaraldehyde and formaldehyde fixatives (11). Occasional problems in maintaining fluorescence during fixation may result from uncontrolled acidity of the fixative solution or the use of excessive organic solvents, which denature the protein and destroy the fluorescence (7).

PASSIVE APPLICATIONS OF GFP

Cell biological applications of GFP may be divided into uses as a tag or as an indicator. In tagging applications, the great majority to date, GFP fluorescence merely reflects levels of gene expression or subcellular localizations caused by targeting domains or host proteins to which GFP is fused. As an indicator, GFP fluorescence can also be modulated posttranslationally by its chemical environment and protein-protein interactions.

Reporter Gene, Cell Marker

The first proposed application of GFP was to detect gene expression in vivo (11), especially in the nematode *Caenorhabditis elegans*, whose cuticle hinders access of the substrates required for detecting other reporter genes. GFP was particularly successful at confirming the pattern of expression of the *mec-7* promoter, which drives the formation of β-tubulin in a limited number of mechanosensory neurons. GFP's independence from enzymatic substrates is likewise particularly promising in intact transgenic embryos and animals (70–76) and for monitoring the effectiveness of gene transfer (48, 77, 78). However, GFP seems to need rather strong promoters to drive sufficient expression for detection, especially in mammalian cells. Most published examples, even those using brightened GFPs with mutations to promote folding at 37°C, have used constitutive promoters from viruses such as cytomegalovirus (CMV), SV40, or HIV long terminal repeat (79), or strong exogenous regulators such as the tetracycline transactivator system (80, 81), rather than native genetic response elements modulated by endogenous signals.

The somewhat disappointing sensitivity of GFP as a so-called gene-tag is probably an inherent result of its lack of amplification. GFP is not an enzyme that catalytically processes an indefinite number of substrate molecules. Instead, each GFP molecule produces at most one fluorophore. It has been estimated that 1 μM well-folded wild-type GFP molecules are required to equal the endogenous autofluorescence of a typical mammalian cell (55), that is, to double the fluorescence over background. Mutant GFPs with improved extinction coefficients might improve this detection limit six- to tenfold (43) (see also Table 1), but 0.1 μM GFP is still approximately 10^5 copies per typical cell of 1–2 pL volume. This estimate already assumes perfect GFP maturation; imperfect or incomplete maturation would raise the threshold copy number even further. The ultimate sensitivity limit is set not by instrumentation but by cellular autofluorescence.

If cytosolic GFP is inadequately sensitive as a reporter gene, two alternatives should be considered. If the gene product can be detected by microscopic imaging with subcellular resolution, then targeting the GFP to a defined subcompartment of the cell can greatly reduce the number of molecules required. The GFP becomes highly concentrated, and the surrounding unlabeled region of the cell provides an internal reference for the autofluorescence background, which is usually diffuse in each cell. It is far easier to see local contrast within a cell than to quantitate a cell's average fluorescence relative to unlabeled standards. Thus as few as 300–3000 GFPs packed into a centrosome are readily visible as a green dot inside a cell (82). However, compartmentation of the GFP does not help with nonimaging detection methods such as fluorometry in cuvets or microtiter plates or fluorescence-activated cell sorting (FACS). A different solution is to use reporter gene products that can enzymatically catalyze a large change in the fluorescence of substrates that can be loaded into intact, fully viable cells. For example, the bacterial enzyme β-lactamase can be detected at levels as low as 60 pM in single mammalian cells (50 molecules per cell) with substrates loaded as membrane-permeant esters (83).

Fusion Tag

The most successful and numerous class of GFP applications has been as a genetic fusion partner to host proteins to monitor their localization and fate. The gene encoding a GFP is fused in frame with the gene encoding the endogenous protein and the resulting chimera expressed in the cell or organism of interest. The ideal result is a fusion protein that maintains the normal functions and localizations of the host protein but is now fluorescent. The range of successful fusions is now much greater than previously tabulated (22). Not all fusions are successful, but the failures are almost never published, so it is difficult to assess the overall success rate. GFP has been targeted successfully to practically every major organelle of the cell, including plasma membrane (37, 84–87), nucleus (28, 38, 87–92), endoplasmic reticulum (50, 60, 93), Golgi apparatus (93), secretory vesicles (39, 94), mitochondria (28, 89, 95, 96), peroxisomes (97), vacuoles (98), and phagosomes (99). Thus the size and shape of GFP and the differing pHs and redox potentials of such organelles do not seem to impose any serious barrier. Even specific chromosomal loci can be tagged indirectly by inserting multiple copies of Lac operator sites and decorating them with a fusion of GFP with the Lac repressor protein (100). In general, fusions can be attempted at either the amino or carboxyl terminus of the host protein, sometimes with intervening spacer peptides. However, the crystal structures of GFP (30, 31) show that the N- and C-terminii of its core domain are not far apart, so it might be possible to splice GFP into a noncritical exterior loop or domain boundary of the host protein. For example, residues 2–233 of GFP have been inserted between the last transmembrane segment and the long cytoplasmic tail of a Shaker potassium channel (100a).

GFP AS AN ACTIVE INDICATOR

The rigid shell in GFP surrounding the chromophore enables it to be fluorescent and protects it from photobleaching but also hinders environmental sensitivity. Nevertheless, GFPs that act as indicators of their environment have been created by combinations of random and directed mutagenesis. The pH sensitivity of certain mutants and their potential application to measure organellar pH have already been mentioned. It is possible to engineer phosphorylation sites into GFP such that phosphorylation produces major changes in fluorescence under defined conditions (AB Cubitt, personal communication). The engineered fusion of GFP within the Shaker potassium channel is the first genetically encoded optical sensor of membrane potential (100a). Depolarization causes at most a 5% decrease in fluorescence with a time constant of approximately 85 ms, but both the amplitude and speed may well improve in future versions. But the most general way to make biochemically sensitive GFPs is to exploit fluorescence resonance energy transfer (FRET) between GFPs of different color. FRET is a quantum-mechanical phenomenon that occurs when two fluorophores are in molecular proximity (<100 Å apart) and the emission spectrum of one fluorophore, the donor, overlaps the excitation spectrum of the second fluorophore, the acceptor. Under these conditions, excitation of the donor can produce emission from the acceptor at the expense of the emission from the donor that would normally occur in the absence of the acceptor. Any biochemical signal that changes the distance between the fluorophores or relative orientation of their transition dipoles will modulate the efficiency of FRET (101–103). Because FRET is a through-space effect, it is not necessary to perturb either GFP alone but rather only the linkage or spatial relationship between them. The potential utility of FRET between GFPs was the main motivation for the development of most of the mutations in Table 1. The change in ratio of acceptor to donor emissions is nearly ideal for cellular imaging and flow cytometry because the two emissions can be obtained simultaneously and their ratio cancels out variations in the absolute concentration of the GFPs, the thickness of the cell, the brightness of the excitation source, and the absolute efficiency of detection. Because the sample need be excited at only one wavelength, which should preferentially excite the donor, FRET is ideal for laser-scanning confocal microscopy and FACS (103). FRET also causes changes in donor fluorescence lifetime and bleaching rate (104), but detection of those signals either requires much more sophisticated instrumentation or is destructive.

Protease Action

The simplest and first-demonstrated way to achieve and modulate FRET between GFPs was to fuse a blue-emitting (class 6) GFP mutant (i.e. a BFP) to a phenolate-containing (class 2) green GFP via an intervening protease-sensitive spacer (41, 105). The broad emission spectrum of the donor BFP, peaking at 447 nm, overlaps fairly well with the excitation spectrum of the class 2 GFP, peaking at 489 nm. Wild-type GFP would not be satisfactory, because the 383 nm used to excite the BFP would directly and efficiently excite the 395-nm excitation peak of wild-type GFP even in the absence of FRET. The tandem fusion exhibits FRET, which is then disrupted when a protease is added to cleave the spacer and let the GFPs diffuse apart. Heim & Tsien (41) used P4-3 and S65C or S65T and a trypsin- or enterokinase-sensitive 25-residue linker, and achieved a 4.6-fold increase in the ratio of blue to green emissions resulting from protease action. Control experiments verified that the two GFPs were unaffected by the proteases at the concentrations used, so the spectral change reflected cleavage of the linker. Mitra et al (105) used BFP5 (F64M, Y66H) and RSGFP4 with a Factor X_a-sensitive linker and obtained a 1.9-fold increase in the analogous ratio. Although FRET-based assays for proteases are well known (106, 107), synthetic peptide substrates are limited in length and useful only in vitro. The special advantages of GFP-based constructs are that they could incorporate full-length protein substrates and could be expressed and assayed inside live cells or organisms.

Transcription Factor Dimerization

A static homodimerization of the transcription factor Pit-1 has been detected by coexpression of BFP-Pit-1 and GFP-Pit-1 fusions in HeLa cells. Homodimerization is inherently more difficult than heterodimerization to demonstrate by FRET, because at most 50% of the complexes will combine BFP- and GFP-labeled proteins, while nonproductive BFP-BFP and GFP-GFP complexes will each account for 25% of the homodimers. Nevertheless, careful spectral analysis indicated that homodimerization was detectable by FRET (108). Unfortunately, no modulation of the Pit-1 interaction or new biological conclusions were reported.

Ca^{2+} Sensitivity

The first dynamically responsive biochemical indicators based on GFP are Ca^{2+} sensors, independently developed almost simultaneously by Romoser et al (109) and by Miyawaki et al (50). Romoser et al linked commercially available class 6 BFP and class 2 GFP mutants with a 26-residue spacer containing the calmodulin (CaM)-binding domain from avian smooth muscle myosin light chain kinase. This spacer allowed FRET to occur from the BFP to the GFP, perhaps because it was long and flexible enough for the two GFPs to dimerize. Addition of Ca^{2+}-CaM disrupted FRET, presumably by binding to and straightening the linker so that the two GFPs were unable to dimerize. Such binding of Ca^{2+}-CaM decreased the 505 nm emission by 65% and the ratio of 505- to 440-nm emissions by sixfold in vitro, an impressive spectral change for a reversible conformational change less drastic than proteolytic cleavage. The bacterially expressed recombinant protein was then microinjected into individual HEK-293 cells. In such intact cells, elevations of cytosolic free Ca^{2+} produced much more modest decreases (5–10%) in 510-nm emission, which could be amplified to about 30% decreases if exogenous calmodulin was co-injected. Thus the response of the indicator in cells was limited by CaM availability, implying that the indicator is responsive to cellular Ca^{2+}-CaM rather than Ca^{2+} per se. Because the heterologously expressed protein had to be microinjected, the unique ability of GFP to be continuously synthesized by the target cell was not exploited, and only cytosolic signals could be monitored.

Miyawaki et al (50) fused BFP or class 5 cyan-fluorescent protein (CFP) to the N-terminus of CaM, and class 2 GFP or class 4 yellow-fluorescent protein (YFP) to the C-terminus of M13, the CaM-binding peptide from skeletal muscle myosin light chain kinase. The CFP-CaM and the M13-YFP could either be fused via two glycines (in which case all four protein domains were joined into a 76-kDa tandem chimera) or left separate. In either case, binding of Ca^{2+} to the CaM caused it to grab the M13, thus increasing FRET, the opposite spectral effect from that of Romoser et al (109). By using GFPs with mutations to optimize mammalian expression, the indicators were bright enough to be introduced into cells by DNA transfection rather than protein microinjection. Because the four-domain chimeras were expressed in situ, they could readily be targeted to organelles such as the nucleus or endoplasmic reticulum by addition

Table 3   Advantages and disadvantages of GFP-based Ca^{2+} indicators[a]

Advantages
  Applicable to nearly all organisms; no need for ester permeation and hydrolysis
  Can be targeted to specific tissues, cells, organelles, or proteins
  Unlikely to diffuse well enough to blur spatial gradients
  Modular construction is readily modified/improved by mutagenesis
  Good optical properties: visible excitation, emission ratioing, high photostability
  cDNAs or improved sequences are cheap to replicate and distribute
  Should be generalizable to measure many bioactive species other than Ca^{2+},
    as long as a conformationally sensitive receptor is available

Disadvantages
  Gene transfection required
  The maximum change in emission ratio is currently less than for small-molecule dyes
  The binding kinetics are somewhat slower
  The CaM or M13 might have some additional biological activity

[a]GFP, green fluorescent proteins.

of appropriate targeting sequences. Ca^{2+} affinities were readily adjustable by mutation of the CaM. Thus free Ca^{2+} concentrations in the endoplasmic reticulum were measured to be 60–400 μM in unstimulated cells, decreasing to 1–50 μM in cells treated with Ca^{2+}-mobilizing agonists. Advantages and disadvantages of the GFP-based Ca^{2+} indicators (called cameleons) compared to conventional Ca^{2+} indicators are summarized in Table 3.

The constructs with separate CFP-CaM and M13-YFP proved that FRET between GFP mutants can dynamically monitor protein-protein interaction in single living cells. Binding of the two host proteins to each other brings their fused GFPs into proximity and enhances FRET. By comparison the mechanism of Romoser et al (109) requires that mutual binding must substantially change the distance between N- and C-terminii of at least one of the partners.

In principle the use of FRET offers some major advantages and disadvantages over other current methods for detecting protein-protein interaction (Table 4). The most unique advantages are the spatial and temporal resolution and the ability to observe the proteins in any compartment of the cell. The biggest disadvantage is that even in the absence of any protein interaction, a substantial background signal is present when illuminating at the donor's excitation maximum and observing at the acceptor's emission maximum. This background signal arises because the donor emission has a tail that extends into the acceptor's emission band, and the acceptor excitation has a tail that extends into the donor's excitation band. For these reasons, FRET is probably not a suitable method for detecting trace interactions or fishing for unknown partners but

Table 4   Advantages and disadvantages of FRET between (GFPs) to monitor protein interactions.

Advantages
  Works in vitro and in living mammalian cells, not just yeast
  Can respond dynamically to posttranslational modifications
  Has high temporal (milliseconds) and spatial (submicron) resolution
  Interacting proteins can be anywhere in the cell and do not need to be sent to nucleus
  Degree of association can be quantified, if 0% and 100% binding can be established in situ
  Efficiency of FRET at 100% complexation gives some structural information

Disadvantages of intermolecular FRET
  Must express fusion proteins, in which host protein and GFPs must both remain functional
  If GFPs are too far from each other ($\gg$80 Å) or unluckily oriented, FRET will fail
  Even with no association, spectral overlap contributes some signal at the FRET wavelengths
  Trace or rare interactions will be hard to detect
  Need negative and positive controls, i.e. reference conditions of 0% and 100% association
  Homodimerization is more difficult to monitor than heterodimerization

is best used at a later stage when the two host proteins are molecularly well characterized and the detailed spatiotemporal dynamics of their interaction is to be determined in live cells.

What Are the Best FRET Partners?

The efficiency of FRET is given by the expression $R_0^6/(R_0^6 + r^6)$, where r is the actual distance between the centers of the chromophores and R_0 is the distance at which FRET is 50% efficient. R_0 depends on the quantum yield of the donor, the extinction coefficient of the acceptor, the overlap of the donor emission and acceptor excitation spectra, and the mutual orientation of the chromophores (102). The early attempts to obtain FRET between GFPs all used BFPs (class 6 mutants) as donors and class 2 (phenolate anion) GFPs as acceptors because these were the first available pairs with sufficiently distinct wavelengths. For such pairs, R_0 is calculated to range from 40 to 43 Å. However, the poor extinction coefficients, quantum yields, and photostabilities of the BFPs have convinced us (50) that cyan mutants (class 5) are much better donors. The acceptor correspondingly must become a class 4 yellow mutant so that its excitation spectrum overlaps the donor emission as much as possible, whereas the two emissions remain as distinct as possible. Combinations of cyan donors with yellow acceptors have R_0 values of 49–52 Å and are our currently preferred donor-acceptor pairs. So far, the highest value of R_0 between two GFPs is 60 Å for class 3 mutant H9-40 as donor to Topaz, but unfortunately these mutants' emission spectra are too close to each other for good discrimination.

The above calculations have assumed that the GFPs are randomly oriented or tumbling with respect to one another, which is the conventional assumption made in calculating R_0 (102). If instead the mutual orientation of the two chromophores were the same as in the crystal structure for the wild-type dimer (31), the R_0s would be about 72% of the previously calculated values, for example, 35–37 Å for cyan-to-yellow FRET. For comparison, the actual distance r between the centers of the chromophores in the dimer is about 25 Å, which would predict that FRET would occur with about 90% efficiency in such a direct heterodimer between cyan and yellow mutants.

OUTLOOK FOR FUTURE RESEARCH

Despite all that has been learned about how GFP works and how it can be exploited as a research tool, enormous challenges and opportunities remain. Listed below are some unanswered general questions about GFP that are among the most intriguing, excluding problems related to narrow applications in cell biology:

Cloning of Related GFPs

What are the genetic sequences and structures of GFP homologs from bioluminescent organisms other than *Aequorea*? This information would illuminate the evolution of fluorescent proteins, reveal the essential conserved elements of the structure, and provide the genetic raw material for combinatorial mixing and matching to produce hybrid proteins with new phenotypes. *Renilla* GFP is the most obvious next cloning target, but even more bioluminescent organisms should be investigated.

Protein Folding and Chromophore Folding

We need to know much more about how GFP folds into its β-barrel conformation and synthesizes its internal chromophore. Now that many of the steps have been kinetically resolved (23), the effect of mutations on each of the steps needs to be determined. The most informative mutants will not be the majority that completely prevent the formation of fluorescence, because those could act anywhere in the entire cascade including disruption of the final state. Instead, mutants or chaperonins that affect the rates but not the final extent of fluorescence development are likely to be most valuable. The molecular mechanism, kinetics, and byproducts of chromophore formation by O_2 are particularly critical questions.

Altered Wavelengths of Fluorescence

Yet longer wavelengths of excitation and emission than are currently available from the class 4 (π-stacked phenolate) mutants (Table 1) would be useful for multiple labels and reporters and to serve as resonance energy transfer acceptors.

For example, an increase in excitation maximum to 540–550 nm would permit efficient energy transfer from terbium chelates, whose millisecond excited-state lifetimes make them useful as energy transfer donors (110). The 560–570 nm emission from such longer-wavelength GFP mutants would also be distinct enough from the standard class 2 (phenolate) mutants to make such green-orange pairs useful as FRET partners. Another approach to improving FRET would be to reduce the emission bandwidth of the class 5 (tryptophan-based) cyan mutants and thereby improve the quantitative separation of cyan and yellow emissions. Emission spectral alterations should be most easily screened by fluorescence-activated cell sorting (FACS). The chemical structure of the red fluorescent species formed by intense illumination in anaerobic conditions (58) must be determined as the first step in making this extraordinary photochemical reaction more general and useful.

Altered Chemical and Photochemical Sensitivities

The sensitivity of GFP spectra to environmental factors such as pH and past illumination is valuable if pH indication or photochemical tagging is desired but is a nuisance for most other applications. Therefore we need to understand the molecular mechanisms of such environmental modulations and to find mutations that enhance or eliminate those mechanisms. In many cases, it will be important to use screening methods that, unlike FACS, permit longitudinal comparison of individual cells or colonies before and after a chemical or actinic challenge. Digital imaging of colonies on plates (e.g. 111) is likely to be advantageous.

Fusions Other Than at N- or C-Terminus

Almost all fusions of host proteins with GFP have been simple tandem fusions in which the C-terminus of one protein is genetically concatenated to the N-terminus of the other. Because not all such fusions work, general rules are needed in order to predict when fluorescence will be intact and when host protein function will be preserved. Sometimes neither order of simple concatenation produces functional chimeras. Could splicing of GFP into the middle of the host protein be made easier and more general? It might be helpful to engineer GFP to move its N- and C-terminii as close to each other as possible, perhaps by addition of spacers or by circular permutation.

Alternatives to Fluorescence

Chromophores can be harnessed to perform many functions other than fluorescence, such as phosphorescence (emission from the triplet state), generation of reactive oxygen species such as singlet oxygen or hydroxyl radical, and photochemical cleavage. Can GFP be engineered to do such tricks? Phosphorescence typically gives lifetimes in microseconds rather than nanoseconds and therefore permits exploration of protein dynamics on longer time scales. Controlled generation of singlet oxygen can be useful to polymerize diaminobenzidine locally into a polymer visible by electron microscopy, so that the location of fluorophores can be verified at ultrastructural resolution (112). Laser pulses can be used to kill proteins within a few nanometers of a suitable chromophore that generates hydroxyl radicals or other reactive species (113). Photochemical cleavage is the basis of important methods to produce sudden changes in the concentration of signaling molecules (114). These techniques would be revolutionized if their crucial molecules could be synthesized or at least localized in situ under molecular biological control, in the same way as GFP. Of course, the rigid shell protecting the chromophore of GFP from the environment may intrinsically prevent even mutagenized GFPs from fulfilling such alternative functions, so that completely different proteins may need to be found or devised. Perhaps GFP will become just one prototype of a collection of genetically encoded, light-driven macromolecular reagents.

ACKNOWLEDGMENTS

I thank the many collaborators and colleagues who allowed me to cite their unpublished results or manuscripts in preparation. The Howard Hughes Medical Institute and the National Institutes of Health (NS27177) provided essential financial support.

> Visit the *Annual Reviews home page* at
> http://www.AnnualReviews.org.

Literature Cited

1. Shimomura O, Johnson FH, Saiga Y. 1962. *J. Cell. Comp. Physiol.* 59:223–39
2. Johnson FH, Shimomura O, Saiga Y, Gershman LC, Reynolds GT, Waters JR. 1962. *J. Cell. Comp. Physiol.* 60:85–103
3. Morin JG, Hastings JW. 1971. *J. Cell. Physiol.* 77:313–18
4. Morise H, Shimomura O, Johnson FH, Winant J. 1974. *Biochemistry* 13:2656–62
5. Prendergast FG, Mann KG. 1978. *Biochemistry* 17:3448–53
6. Shimomura O. 1979. *FEBS Lett.* 104:220–22
7. Ward WW, Cody CW, Hart RC, Cormier MJ. 1980. *Photochem. Photobiol.* 31:611–15
8. Ward WW, Prentice HJ, Roth AF, Cody CW, Reeves SC. 1982. *Photochem. Photobiol.* 35:803–8
9. Ward WW, Bokman SH. 1982. *Biochemistry* 21:4535–40
10. Prasher DC, Eckenrode VK, Ward WW, Prendergast FG, Cormier MJ. 1992. *Gene* 111:229–33
11. Chalfie M, Tu Y, Euskirchen G, Ward WW, Prasher DC. 1994. *Science* 263:802–5
12. Inouye S, Tsuji FI. 1994. *FEBS Lett.* 341:277–80
13. Ward WW. 1979. In *Photochemical and Photobiological Reviews*, ed. KC Smith, 4:1–57. New York: Plenum
14. Ward WW, Cormier MJ. 1979. *J. Biol. Chem.* 254:781–88
15. Lee J, Gibson BG, O'Kane DJ, Kohnle A, Bacher A. 1992. *Eur. J. Biochem.*

210:711–19

16. Macheroux P, Schmidt KU, Steinerstauch P, Ghisla S. 1987. *Biochem. Biophys. Res. Commun.* 146:101–6
17. Glazer AN. 1989. *J. Biol. Chem.* 264:1–4
18. Song P-S, Koka P, Prézelin BB, Haxo FT. 1976. *Biochemistry* 15:4422–27
19. Tsien RY, Prasher DC. 1997. In *GFP: Green Fluorescent Protein Strategies and Applications*, ed. M Chalfie, S Kain. New York: Wiley & Sons
20. Cody CW, Prasher DC, Westler WM, Prendergast FG, Ward WW. 1993. *Biochemistry* 32:1212–18
21. Heim R, Prasher DC, Tsien RY. 1994. *Proc. Natl. Acad. Sci. USA* 91:12501–4
22. Cubitt AB, Heim R, Adams SR, Boyd AE, Gross LA, Tsien RY. 1995. *Trends Biochem. Sci.* 20:448–55
23. Reid BG, Flynn GC. 1997. *Biochemistry* 36:6786–91
24. Inouye S, Tsuji FI. 1994. *FEBS Lett.* 351:211–14
25. Heim R, Cubitt AB, Tsien RY. 1995. *Nature* 373:663–64
26. Kojima S, Hirano T, Niwa H, Ohashi M, Inouye S, Tsuji FL. 1997. *Tetrahedron Lett.* 38:2875–78
27. Wright HT. 1991. *Crit. Rev. Biochem. Mol. Biol.* 26:1–52
28. Rizzuto R, Brini M, DeGiorgi F, Rossi R, Heim R, et al. 1996. *Curr. Biol.* 6:183–88
29. Perozzo MA, Ward KB, Thompson RB, Ward WW. 1988. *J. Biol. Chem.* 263:7713–16
30. Ormö M, Cubitt AB, Kallio K, Gross LA, Tsien RY, Remington SJ. 1996. *Science* 273:1392–95
31. Yang F, Moss LG, Phillips GN Jr. 1996. *Nat. Biotechnol.* 14:1246–51
32. Brejc K, Sixma TK, Kitts PA, Kain SR, Tsien RY, et al. 1997. *Proc. Natl. Acad. Sci. USA* 94:2306–11
33. Palm GJ, Zdanov A, Gaitanaris GA, Stauber R, Pavlakis GN, Wlodawer A. 1997. *Nat. Struct. Biol.* 4:361–65
34. Wachter RM, King BA, Heim R, Kallio K, Tsien RY, et al. 1997. *Biochemistry* 36:9759–65
34a. Phillips GN Jr. 1997. *Curr. Opin. Struct. Biol.* 7:821–27
35. Dopf J, Horiagon TM. 1996. *Gene* 173:39–44
36. Chattoraj M, King BA, Bublitz GU, Boxer SG. 1996. *Proc. Natl. Acad. Sci. USA* 93:8362–67
37. Yokoe H, Meyer T. 1996. *Nat. Biotechnol.* 14:1252–56
38. Lim CR, Kimata Y, Oka M, Nomaguchi K, Kohno K. 1995. *J. Biochem.* 118:13–17

39. Kaether C, Gerdes HH. 1995. *FEBS Lett.* 369:267–71
40. Crameri A, Whitehorn EA, Tate E, Stemmer WPC. 1996. *Nat. Biotechnol.* 14:315–19
41. Heim R, Tsien RY. 1996. *Curr. Biol.* 6:178–82
42. Kahana J, Silver P. 1996. In *Current Protocols in Molecular Biology*, ed. FM Ausabel, R Brent, RE Kingston, DD Moore, JG Seidman, et al, 9(7):22–28 New York: Wiley & Sons
43. Patterson GH, Knobel SM, Sharif WD, Kain SR, Piston DW. 1997. *Biophys. J.* 73:2782–90
44. Cubitt AB, Heim R, Woollenweber LA. 1997. *Methods Cell Biol.* In press
45. Ward WW. 1997. In *Green Fluorescent Protein: Properties, Applications, and Protocols*, ed. M Chalfie, S Kain. New York: John Wiley & Sons
46. Delagrave S, Hawtin RE, Silva CM, Yang MM, Youvan DC. 1995. *Bio-Technology* 13:151–14
47. Cormack BP, Valdivia RH, Falkow S. 1996. *Gene* 173:33–38
48. Cheng LZ, Fu J, Tsukamoto A, Hawley RG. 1996. *Nat. Biotechnol.* 14:606–9
49. Ehrig T, O'Kane DJ, Prendergast FG. 1995. *FEBS Lett.* 367:163–66
50. Miyawaki A, Llopis J, Heim R, McCaffery JM, Adams JA, et al. 1997. *Nature* 388:882–87
51. Wu C, Liu HZ, Crossen R, Gruenwald S, Singh S. 1997. *Gene* 190:157–62
52. Niwa H, Inouye S, Hirano T, Matsuno T, Kojima S, et al. 1996. *Proc. Natl. Acad. Sci. USA* 93:13617–22
53. Voityuk AA, Michel-Beyerle ME, Rösch N. 1997. *Chem. Phys. Lett.* 272:162–67
54. Xu C, Zipfel W, Shear JB, Williams RM, Webb WW. 1996. *Proc. Natl. Acad. Sci. USA* 93:10763–68
55. Niswender KD, Blackman SM, Rohde L, Magnuson MA, Piston DW. 1995. *J. Microsc.* 180:109–16
56. Potter SM, Wang C-M, Garrity PA, Fraser SE. 1996. *Gene* 173:25–31
57. Bokman SH, Ward WW. 1981. *Biochem. Biophys. Res. Commun.* 101:1372–80
58. Elowitz MB, Surette MG, Wolf E, Stock J, Leibler S. 1997. *Curr. Biol.* 7:809–12
59. Swaminathan KS, Hoang CP, Verkman AS. 1997. *Biophys. J.* 72:1900–7
60. Subramanian K, Meyer T. 1997. *Cell* 89:963–71

61. Dickson RM, Cubitt AB, Tsien RY, Moerner WE. 1997. *Nature* 388:355–58
62. Haseloff J, Siemering KR, Prasher DC, Hodge S. 1997. *Proc. Natl. Acad. Sci. USA* 94:2122–27
63. Zolotukhin S, Potter M, Hauswirth WW, Guy J, Muzyczka N. 1996. *J. Virol.* 70:4646–54
64. Yang T-T, Cheng L, Kain SR. 1996. *Nucleic Acids Res.* 24:4592–93
65. Cormack BP, Bertram G, Egerton M, Gow NAR, Falkow S, Brown AJP. 1997. *Microbiology* 143:303–11
66. Kozak M. 1989. *J. Cell Biol.* 108:229–41
67. Kahn TW, Beachy RN, Falk MM. 1997. *Curr. Biol.* 7:R207–8
68. Weissman JS, Rye HS, Fenton WA, Beechem JM, Horwich AL. 1996. *Cell* 84:481–90
69. Makino Y, Amada K, Taguchi H, Yoshida M. 1997. *J. Biol. Chem.* 272:12468–74
70. Ikawa M, Kominami K, Yoshimura Y, Tanaka K, Nishimune Y, Okabe M. 1995. *Dev. Growth Differ.* 37:455–59
71. Zernicka-Goetz M, Pines J, Hunter SM, Dixon JPC, Siemering KR, et al. 1997. *Development* 124:1133–37
72. Chiocchetti A, Tolosano E, Hirsch E, Silengo L, Altruda F. 1997. *Biochim. Biophys. Acta* 1352:193–202
73. Gagneten S, Le Y, Miller J, Sauer B. 1997. *Nucleic Acids Res.* 25:3326–31
74. Takada T, Iida K, Awaji T, Itoh K, Takahashi R, et al. 1997. *Nat. Biotechnol.* 15:458–61
75. Ikawa M, Kominami K, Yoshimura Y, Tanaka K, Nishimune Y, Okabe M. 1995. *FEBS Lett.* 375:125–28
76. Amsterdam A, Lin S, Moss LG, Hopkins N. 1996. *Gene* 173:99–103
77. Muldoon RR, Levy JP, Kain SR, Kitts PA, Link CJJ. 1997. *BioTechniques* 22:162–67
78. Zhang GH, Gurtu V, Kain SR. 1996. *Biochem. Biophys. Res. Commun.* 227:707–11
79. Gervaix A, West D, Leoni LM, Richman DD, Wong-Staal F, Corbeil J. 1997. *Proc. Natl. Acad. Sci. USA* 94:4653–58
80. Anderson MT, Tjioe IM, Lorincz MC, Parks DR, Herzenberg LA, Nolan GP. 1996. *Proc. Natl. Acad. Sci. USA* 93:8508–11
81. Mosser DD, Caron AW, Bourget L, Jolicoeur P, Massie B. 1997. *BioTechniques* 22:150–61

82. Shelby RD, Hahn KM, Sullivan KF. 1996. *J. Cell Biol.* 135:545–57
83. Zlokarnik G, Negulescu PA, Knapp TE, Santiso-Mere D, Burres N, et al. 1998 *Science* 279:84–88
84. Marshall J, Molloy R, Moss GWJ, Howe JR, Hughes TE. 1995. *Neuron* 14:211–15
85. Barak LS, Ferguson SSG, Zhang J, Martenson C, Meyer T, Caron MG. 1997. *Mol. Pharmacol.* 51:177–84
86. Moriyoshi K, Richards LJ, Akazawa C, O'Leary DDM, Nakanishi S. 1996. *Neuron* 16:255–60
87. Hanakam F, Albrecht R, Eckerskorn C, Matzner M, Gerisch G. 1996. *EMBO J.* 15:2935–43
88. Grebenok RJ, Pierson E, Lambert GM, Gong F-C, Afonso CL, et al. 1997. *Plant J.* 11:573–86
89. DeGiorgi F, Brini M, Bastianutto C, Marsault R, Montero M, et al. 1996. *Gene* 173:113–17
90. Corbett AH, Koepp DM, Schlenstedt G, Lee MS, Hopper AK, Silver P. 1995. *J. Cell Biol.* 130:1017–26
91. Carey KL, Richards SA, Lounsbury KM, Macara IG. 1996. *J. Cell Biol.* 133:985–96
92. Chatterjee S, Stochaj U. 1996. *BioTechniques* 21:62–63
93. Presley JF, Cole NB, Schroer TA, Hirschberg K, Zaal KJM, Lippincott-Schwartz J. 1997. *Nature* 389:81–85
94. Lang T, Wacker I, Steyer J, Kaether C, Wunderlich I, et al. 1997. *Neuron* 18:857–63
95. Murray AW, Kirschner MW. 1989. *Science* 246:614–21
96. Yano M, Kanazawa M, Terada K, Namchai C, Yamaizumi M, et al. 1997. *J. Biol. Chem.* 272:8459–65
97. Wiemer EAC, Wenzel T, Deerinck TJ, Ellisman MH, Subramani S. 1997. *J. Cell Biol.* 136:71–80
98. Cowles CR, Odorizzi G, Payne GS, Emr SD. 1997. *Cell* 91:109–18
99. Maniak M, Rauchenberger R, Albrecht R, Murphy J, Gerisch G. 1995. *Cell* 83:915–24
100. Straight AF, Marshall WF, Sedat JW, Murray AW. 1997. *Science* 277:574–78
100a. Siegel MS, Isacoff EY. 1997. *Neuron* 19:735–41
101. Stryer L. 1978. *Annu. Rev. Biochem.* 47:819–46
102. Lakowicz JR. 1983. *Principles of Fluorescence Spectroscopy.* New York: Plenum

103. Tsien RY, Bacskai BJ, Adams SR. 1993. *Trends Cell Biol.* 3:242–45
104. Jovin TM, Arndt-Jovin DJ. 1989. *Annu. Rev. Biophys. Biophys. Chem.* 18:271–308
105. Mitra RD, Silva CM, Youvan DC. 1996. *Gene* 173:13–17
106. Krafft GA, Wang GT. 1994. *Methods Enzymol.* 241:70–86
107. Knight CG. 1995. *Methods Enzymol.* 248:18–34
108. Periasamy A, Kay SA, Day RN. 1997. In *Functional Imaging and Optical Manipulation of Living Cells, Proc. SPIE 2983,* ed. DL Farkas, BJ Tromberg. Bellingham, WA: SPIE
109. Romoser VA, Hinkle PM, Persechini A. 1997. *J. Biol. Chem.* 272:13270–74
110. Selvin PR. 1996. *IEEE J. Sel. Top. Quant. Electron.* 2:1077–87
111. Youvan DC, Goldman ER, Delgrave S, Yang MM. 1995. *Methods Enzymol.* 246:732–48
112. Deerinck TJ, Martone ME, Lev-Ram V, Green DPL, Tsien RY, et al. 1994. *J. Cell Biol.* 126:901–10
113. Linden KG, Liao JC, Jay DG. 1992. *Biophys. J.* 61:956–62
114. Adams SR, Tsien RY. 1993. *Annu. Rev. Physiol.* 55:755–84

Annu. Rev. Biochem. 1998. 67:545–79

ALTERATION OF NUCLEOSOME STRUCTURE AS A MECHANISM OF TRANSCRIPTIONAL REGULATION

J. L. Workman

Howard Hughes Medical Institute and Department of Biochemistry and Molecular Biology, The Pennsylvania State University, University Park, Pennsylvania 16802; e-mail: jlw10@psu.edu

R. E. Kingston

Department of Molecular Biology, Massachusetts General Hospital, Boston, Massachusetts 02114; e-mail: kingston@frodo.mgh.harvard.edu

KEY WORDS: acetylation, chromatin, DNA-protein interactions

ABSTRACT

The nucleosome, which is the primary building block of chromatin, is not a static structure: It can adopt alternative conformations. Changes in solution conditions or changes in histone acetylation state cause nucleosomes and nucleosomal arrays to behave with altered biophysical properties. Distinct subpopulations of nucleosomes isolated from cells have chromatographic properties and nuclease sensitivity different from those of bulk nucleosomes. Recently, proteins that were initially identified as necessary for transcriptional regulation have been shown to alter nucleosomal structure. These proteins are found in three types of multi-protein complexes that can acetylate nucleosomes, deacetylate nucleosomes, or alter nucleosome structure in an ATP-dependent manner. The direct modification of nucleosome structure by these complexes is likely to play a central role in appropriate regulation of eukaryotic genes.

CONTENTS

545

INTRODUCTION

It has become increasingly apparent that modulation of chromatin structure plays an important role in the regulation of transcription in eukaryotes. Chromatin structures can inhibit the binding and function of the numerous proteins that collaborate to produce appropriate levels of transcription. Recent studies have identified a large group of proteins whose primary function is to help activate transcription by altering chromatin so that its DNA sequences become more transparent to the transcriptional apparatus. Conversely, different proteins help repress transcription by making chromatin structure less transparent.

The nucleosome core is the target of many of these activities. Once thought of as a static building block of chromatin structure, the nucleosome core is now clearly understood as a dynamic structure whose stability can be regulated by posttranslational modification and by enzymatic function. In addition, regulatory proteins can bind directly to nucleosomal histones to create altered structures. Thus, the state of the nucleosome core plays a central role in determining the transcriptional competence of any region of chromatin.

The purpose of this review is to examine what is known about nucleosome core structure and how that structure might be altered during gene regulation. The nucleosome core contains about 146 bp of DNA and two copies each of the histones H2A, H2B, H3, and H4. We discuss what is known about histone-histone interactions and histone-DNA interactions within this particle, with an emphasis on how these interactions might contribute to regulation. Of particular importance are the N-terminal tails of the core histones, which are less structured and extend out of the central structured portions of the nucleosome core in a manner that frees them for interaction with DNA or with other proteins. We discuss the enzymatic acetylation and deacetylation activities that modify these tails, and we relate these modifications to changes in the structural properties of nucleosomes and to changes in gene regulation. We conclude with a discussion of ATP-dependent protein complexes that alter nucleosome structure to increase the ability of transcription factors to interact with their DNA sequences.

STEPS IN TRANSCRIPTIONAL REGULATION

The nucleosome can inhibit several processes that must occur for a eukaryotic gene to be appropriately regulated (Figure 1). The initial event in activation is thought to be binding of sequence-specific activators to both enhancer and promoter regions. To measure the effects of nucleosomes on activator binding, investigators have assembled binding sites into mononucleosomes by salt dialysis of purified histones, or they have used either *Xenopus* or *Drosophila* chromatin assembly extracts to assemble sites into arrays of nucleosomes. These experiments demonstrate that nucleosomes inhibit the binding of most transcriptional activators; the glucocorticoid receptor is the only example of an activator that binds with similar affinity to naked and nucleosomal DNA (1). Estimation of the dissociation constants for binding indicates that the degree of inhibition for other activators varies between one and greater than four orders of magnitude and that one activator can facilitate binding of an adjacent activator (see below). Therefore the characteristics of nucleosomal structure that cause inhibition of activator binding, and the mechanisms that enhance activator binding, are critical to understanding transcriptional regulation.

Nucleosome structures also inhibit transcriptional initiation. This inhibition most likely occurs via inhibition of the binding of general transcription factors, which must associate with the template to enable specific initiation by RNA polymerase. Formation of the preinitiation complex, which includes general transcription factors and RNA polymerase, is inhibited by nucleosome formation. Prebinding the template with fractions that include TFIID, or prebinding with purified TATA-binding protein, partially relieves this inhibition (2). TBP binding is inhibited by nucleosomal formation (3), so it is possible that precisely the same mechanisms that affect access of activators to nucleosomal DNA also affect access of the preinitiation complex.

The final steps in transcription are elongation and termination. Nucleosomes create a significant topological problem for transcriptional elongation, as a megadalton RNA polymerase must proceed around 1.65 turns of DNA wrapped around the histone octamer. In vitro studies have demonstrated that purified phage RNA polymerase can elongate through a nucleosome and that eukaryotic polymerases are either significantly slowed or even stopped by a nucleosome, depending on the template and reaction conditions (4–9). Given the varied nature of the reactions that must occur, the ability of RNA polymerase to elongate through a nucleosome is likely to be affected by aspects of nucleosome structure that differ from those that affect sequence-specific DNA binding. In

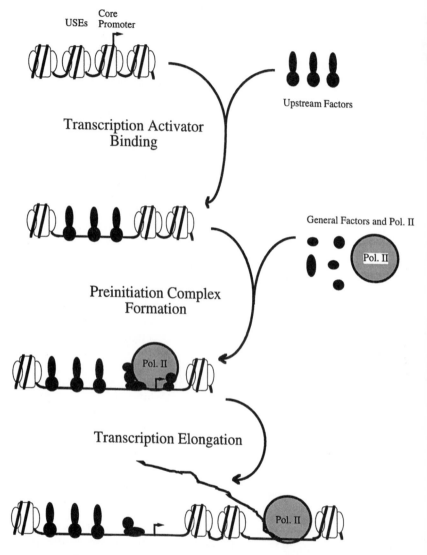

Figure 1 Nucleosomes can inhibit multiple steps required for gene transcription. The binding of upstream regulatory factors requires accessing nucleosomal DNA and may result in displacement or rearrangement of the histone octamers. Similarly the formation of preinitiation complexes at the TATA-box and transcription start site is also suppressed by the presence of nucleosomes. Finally, the elongation of RNA polymerase II is inhibited by nucleosome arrays, resulting in increasing pausing of engaged polymerases.

reviewing nucleosome structure and the activities that are believed to modify that structure, we focus on the characteristics that are likely to be important to each of the above steps in transcriptional regulation.

THE NUCLEOSOME

The discovery of the nucleosome as the repeating subunit of chromatin was a seminal event in chromatin research. The nucleosome was discovered through (a) its appearance in the electron microsope as the beads (nu-bodies) in the extended beads-on-a-string structure of the decondensed chromatin fiber (10–13), (b) by the generation of approximately 200-bp DNA ladders upon digestion of chromatin with endogenous (14) or exogenous (15) nucleases, and (c) by the isolation of 11.5S nucleoprotein complexes (16). These observations, histone/histone cross-linking patterns (17), and X-ray diffraction data led Kornberg to propose that the nucleosome comprised approximately 200 bp of DNA and two copies of each of the four core histones (18). The nucleosome proper is composed of (a) the nucleosome core (146 bp of DNA wrapped 1.65 turns around the histone octamer), (b) the linker histone H1 (or an H1 variant), and (c) the linker DNA between nucleosome cores (19). A subnucleosome particle containing the nucleosome core, histone H1, and 168 bp of DNA (i.e. the nucleosome core and 20 base pairs interacting with histone H1) has been termed a chromatosome (20). Numerous details of nucleosome structure have been revealed over the past two and a half decades. These discoveries culminated with the recent solution of a high-resolution crystal structure of the nucleosome core (21).

Nucleosome Structure

The histone octamer is a cylindrical wedge with an outer diameter of 6.5 nm (22, 23) that is composed of a central tetramer of two copies each of histone H3 and H4 that is flanked by two H2A/H2B heterodimers (17, 24–26). The H3/H4 tetramer is assembled by the association of two heterologous dimers of H3 and H4. Each of the core histones shares a common motif consisting of two short alpha helixes and a long central helix seperated by beta bridges. These structures constitute most of the ordered portion of the histones and have been termed the histone fold (27). The histone fold provides sites of histone DNA interactions as well as provides for the dimerization of histones via a handshake motif, in which each monomer clasps its partner in a head-to-tail arrangement (21, 27).

Homology searches have revealed that several proteins with functions related to DNA also contain putative histone fold motifs including TAFs (TBP-associated factors), which are subunits of the general transcription initiation factor TFIID (28). The histone fold motifs in TAFs also appear to participate in dimerization of subunits (29–31). Indeed, heterodimers between TAFs and core histones can be formed in vitro (30). These observations raise the possibility that there is a histone octamer–like core structure within TFIID that might similarly wrap DNA. However, these TAFs lack arginine side chains, which are found in core histones; these side chains insert into the minor groove of the DNA helix. This difference suggests that any DNA binding by these TAFs must be by different protein-DNA interactions (21).

In addition to assisting dimerization, histone folds contribute to further oligomerization of histone proteins within the histone octamer. The central and third alpha helices of the histone folds of H3, H4, and H2A are the primary, but not only, determinants for assembly of the H3/H4 tetramer and the octamer through the formation of four helix bundles (21). An H3-H3 interaction between H3/H4 dimers drives the formation of the tetramer, and an interaction between each H4 of the tetramer with an H2B of an H2A/H2B dimer drives formation of the octamer. It is important that the H4-H2B interface is considerably more hydrophobic than the H3-H3 interfaces and thus expected to be less stable at low ionic strength (21). This characteristic may account for the early observation that whereas the H3/H4 tetramer is stable at physiological salt concentrations, the complete octamer (i.e. including binding of the H2A/H2B dimers) is stable only at high salt concentrations or when wrapped with DNA in the form of a nucleosome core (26, 32, 33). As discussed below, the lability of the H2A/H2B dimer–H3/H4 tetramer interactions may have significance for conformational changes and/or disassembly of the nucleosome core.

Within the nucleosome core, 146 bp of DNA are wrapped 1.65 times around the histone octamer in a left-handed superhelix with about 7.6 turns of the DNA helix/superhelical turn (22). The path of the central 12 turns of DNA follows a path of repeating histone-positive charges on the surface of the histone octamer (34). Each histone heterodimer within the octamer forms a cresent shape that arcs 27–28 bp of DNA along their axis in a 140° bend, leaving 4 bp linkers between them (21). The interactions of the histone folds organize the central 121 bp of DNA with additional H3 interactions, extending the DNA superhelix. Further DNA interactions are provided by the N-terminal tails of H3 and H2B, which pass through minor groove channels between the DNA helices (one tail each 20 bp) and by the binding of the H2A tail to the minor groove outside the superhelix (21). The path of the DNA helix around the histone octamer is not uniform: It is distorted by the local structure of the histone DNA–binding surface. Most notably are major bends at 10–15 and 40 bp from the dyad axis (center of the DNA superhelix), where histone interactions bulge or buckle the DNA outward (21, 22).

NUCLEOSOME DYNAMICS

Unfolding or Disassembling Nucleosome Cores

A number of physical and biochemical studies have investigated potential structural transitions in the nucleosome core when challenged with low or high ionic strengths, changes in pH, binding of intercalating agents, or denaturation caused by urea. These studies have revealed that although the nucleosome core is stable under physiological conditions, it is susceptible to unfolding and/or unwrapping of the DNA under a variety of conditions (35). The altered nucleosomal structures that form under these nonphysiological conditions are candidates for structures that might be induced in vivo by ATP-dependent remodeling complexes or by the action of acetylation or deacetylation activities. Histone-binding proteins might also be involved in stabilizing structures such as these in vivo. Only limited data address the relationship of these altered structures to structures that might be stabilized in vivo by regulatory proteins; however, it is important to consider the types of structural changes that nucleosomes can undergo when formulating models for action of regulatory proteins (Table 1).

Table 1 Proposed nucleosome conformation changes

Transition (references)	Disruption of histone-histone contacts	Disruption of histone-DNA contacts	Features
Low salt transition (36, 39–41)	+	−	Partial unfolding of the histone octamer
Loss of H2A/H2B dimer(s) (43, 44, 49, 51)	+	+	Increased factor access Decreased fiber folding
Unfolded H3-SH accessible (54, 56, 58, 60)	+	−	Enriched in transcribed sequences and acetylated histones
Transient unwrapping of nucleosome DNA (65, 67, 69, 70)	−	+	Increased factor access Facilitates cooperative factor binding
Short-range octamer movement in *cis* (74–77)	−	+	Temperature-enhanced redistribution
Long-range octamer movement in *cis* (71, 72, 80, 81)	−	+	Enhanced by high salt or ATP-dependent remodeling complexes
Histone octamer transfer (4, 7, 8, 50)	−	+	Transfers intact octamers to other DNA sites in *cis* or *trans*

The nucleosome core undergoes a structural transition when salt concentrations are dropped below 1 mM. Modeling of the low-salt nucleosome conformation has suggested an elongated partially unfolded structure (36), whereas other studies disagree with this view (37). It is interesting to note that the extent of the low-salt nucleosome transition as well as a reversible thermal transition (see below) is affected by the histone subtypes contained in the nucleosome core particle (38). The low-salt transition is thought to involve unfolding of the histone octamer in some manner because it appears to break H4-H2A contacts (suggesting weakening of the H3/H4 tetramer–H2A/H2B dimer interface) and is inhibited by histone-histone cross-linking (36, 39–42).

The interaction of the H2A/H2B dimers with the H3/H4 tetramer has been suggested to modulate many processes. For example, RNA polymerase II preferentially associates with nucleosomes deficient in one H2A/H2B dimer (43). The in vitro binding of TFIIIA adjacent to an H3/H4 tetramer reconstituted on a 5S RNA gene occurs readily but is inhibited by the association of H2A/H2B dimers that occupy its binding site (44). Acidic nucleosome assembly factors—nucleoplasmin and NAP-1, which preferentially interact with H2A/H2B(45–48)—stimulate the binding of transcription factors to nucleosome cores and can lead to the depletion of H2A/H2B dimers from factor-bound nucleosomes (49, 50). As with the low-salt nucleosome transition, stimulation of factor binding and histone removal by nucleoplasmin and NAP-1 is inhibited by cross-linking the histone octamer (50). It has also been suggested that depletion of H2A/H2B dimers can disrupt the ability of nucleosome arrays to fold into higher-order chromatin structures (51).

Genetic evidence supports an important role for association of H2A/H2B dimers within the nucleosome core. The SWI/SNF complex can remodel nucleosome core structure (see below), and depletion of one of the H2A/H2B gene pairs in yeast alters the transcription of particular genes in vivo, in a manner that supresses defects in the SWI/SNF complex (52). Moreover, single amino acid substitutions for H4 tyrosines that interact with H2A/H2B dimers cause defects in the transcription of several genes, including crucial regulators of cell cycle progression (53).

Extended unfolded nucleosomes have been described by electron spectroscopic imaging of a nucleosome subfraction isolated by Hg-affinity chromatography (54). These U-shaped particles are retained on Hg columns, in part because of accessibility of the single H3 thiol, and are enriched in acetylated histones and transcribed DNA sequences (55–58). This H3 thiol (H3-C110) is in the center of the H3-H3 four-helix bundle that binds the two H3/H4 dimers into a tetramer (21). Presumably, reactivity of this thiol to Hg would require a substantial alteration in the conformation of the tetramer, consistent with the unfolded structures described above. This conformation appears to be different from that of the low-salt nucleosome transition. Early studies that measured energy transfer between fluorescent dyes attached to the H3-cysteins did not detect a substantial increase in the distance between the cysteins at low salt (59), suggesting that a major conformation change in the tetramer did not take place at this location. Studies in yeast have described a transcription-induced split nucleosome structure based on nuclease sensitivities that may be related to the U-shaped thiol-reactive nucleosome conformation (60). The appearance of these altered nucleosome structures is induced by transcription in vivo. These structures accumulate at the 3' end of the gene and may result from RNA polymerase–induced positive supercoiling (60, 61). Template supercoiling may induce these structures, but in vivo nuclease digestion experiments indicate that template supercoiling does not appear to be necessary for efficient transcription in yeast (62).

Regardless of the precise relationship between split nucleosomes and nucleosomes that can be bound on Hg columns, the association of these apparently altered nucleosomal structures with transcribed chromatin implies that some process connected to gene activation causes enrichment of these structures. They might be a by-product of transcriptional activation; for example, they might be induced by RNA polymerase movement through chromatin. Alternatively, they might be induced by remodeling activities prior to transcription and might therefore represent an altered nucleosomal state that is formed as a prerequisite for appropriate activation of genes. Thus, while altered structures of the nucleosome can be observed in vivo and in vitro, neither the genesis of these structures nor the role that they play in regulatory processes is clear.

Unwrapping of Nucleosomal DNA

Another reversible conformational change in the nucleosome core particle is the unwrapping of DNA from the histone octamer as revealed by thermal denaturation studies. Given that most DNA-binding proteins do not easily form a ternary complex on a nucleosome, unwrapping of nucleosomal DNA may be an important mechanism for increasing access of regulatory proteins to their target DNA sequence. With increasing temperature, denaturation of nucleosomal DNA occurs in two phases at temperatures that are higher than required for melting of naked DNA. The first phase is reversible and represents the melting of approximately 40 bp of DNA, while the second phase is irreversible and represents melting of the remaining base pairs (63). Further experiments illustrated that the first phase of nucleosomal DNA denaturation represented the melting of approximately 20 bp at each end of the nucleosome core following their release from the octamer. Moreover, this melting was not inhibited by histone-histone cross-linking, indicating that it did not require dissociation of histone protein (64). Thus, these studies demonstrated that the first 20 bp into the nucleosome core are particularly able to reversibly dissociate from the histone octamer.

Further support for the notion that the DNA at the edge of the nucleosome core is less tightly constrained comes from analysis of transcription factor–nucleosome interactions. Numerous transcription factors interact with nucleosomal DNA albeit usually with affinities that are reduced relative to naked DNA (1). In most instances the affinity is greatest when the recognition site for the factor is located near the edge of the nucleosome core (65–67). When a single nucleosome has multiple recognition sites for the same factor, those near the edge are bound first (65). However, the binding of factors to the more accessible sites at the edge of the nucleosome core enhances the affinity of other sites deeper into the core particle (65, 68). This leads to a cooperative effect on the binding of multiple factors to a single nucleosome core that does not require direct interaction between the factors but instead appears to result from the cumulative disruption of histone-DNA interactions (67).

These data have been used to develop a model for cooperative binding of factors to nucleosomal target sites (69). The model posits that factors access their binding sites during transient unpeeling of DNA from the histone octamer (which initiates at the edge of the nucleosome core) (70). The binding of two factors may then be linked in a thermodynamic cycle governed by the equilibrium constants for exposure of each recognition site. Cooperativity could be attributable to the binding of each individual factor stabilizing the otherwise transient release of DNA sequences from the surface of the histone octamer and thereby increasing the probability of exposure of sites for additional factors.

Histone Octamer Movement

The above studies demonstrate possible mechanistic pathways for altering the nucleosome to potentially increase the ability of regulatory proteins to interact with their recognition sequences. A second type of mechanism to increase factor interaction is for the core histones to slide or jump along the DNA sequence to expose important sequences. Indeed, histone octamers are able to relocate to other positions on a DNA strand in *cis* (71, 72). This long-range movement of octamers along DNA generally requires elevated salt concentrations (~0.5 M). By contrast, short-range movement of histone octamers (i.e. tens of base pairs) can occur at physiological salt concentrations especially at 37°C. Several studies have used tandem repeats of sea urchin 5S RNA gene sequences (73) to measure nucleosome mobility at physiological ionic strength. These studies revealed that on these sequences, nucleosomes adopt a primary translational frame surrounded by others located at intervals of 10 bp (74, 75). These data illustrate a cluster of octamer positions on the 5S sequences that are in equilibrium at low ionic strength but maintain the rotational phasing of the DNA helix (i.e. the direction of DNA bending around the octamer). Similar localized mobility of histone octamers was found to occur on other DNA sequences, suggesting that it is not solely a property of 5S RNA gene sequences (76). This localized histone octamer mobility is reduced by the binding of histone H1 (77, 78). Importantly, histone octamer sliding has been suggested to be enhanced by ATP-dependent nucleosome remodeling activities present in *Drosophila* embryo extracts (79–81).

Histone octamer transfer is another form of histone movement that appears to be mediated by the direct transfer of histones from one region or strand of DNA to another. This pathway has been illustrated by studies of phage RNA polymerase elongation through nucleosome cores. Studies by Felsenfeld and colleagues have suggested that a histone octamer can step around an elongating polymerase (4, 7, 8). The model suggests that as the elongating polymerase peels DNA off a histone octamer in front of it, the same octamer begins to reassociate with DNA behind the polymerase. That histone-histone cross-linking within the histone octamer does not interfere with elongation (82) suggested that the octamer most likely remains intact during transfer. Thus, the histone octamer appears to be directly transferred from DNA in front of the polymerase to DNA behind it. A similar pathway of histone octamer transfer might occur upon destabilization of nucleosomes by the binding of multiple factors. In vitro studies have shown that the nucleosome binding by five dimers of GAL4 derivatives can induce transhistone displacement from GAL4 sites onto competitor DNA (49, 83, 84). This transfer does not require dissociation of the histone octamer (50) and thus may also occur by direct contact between the donor and recipient DNA strands. Importantly, histone octamer transfer differs from disassembly of the histone octamer mediated by histone-binding proteins, nucleoplasmin, or NAP-1, which requires initial removal of H2A/H2B dimers (50; see above).

HISTONE ACETYLATION

The Histone Amino Terminal Tails

The N termini of core histones are central to the processes that modulate nucleosome structure. The N termini of histones H3 and H4 are the most conserved portions of these highly evolutionarily conserved proteins, and genetic studies in *Saccharomyces cerevisiae* have shown that small deletions or point mutations in the N termini lead to a remarkable breadth and severity of phenotypes (85, 86). N termini can be modified posttranslationally by acetylation and by phosphorylation. Both modifications alter the charge distribution on the N termini: Acetylation neutralizes the positive charge of N-terminal lysine residues, while phosphorylation introduces a negative charge at a conserved serine (position 10) of histone H3. Each histone can be acetylated at multiple conserved positions; for example, histone H4 can be acetylated at four positions, and the ratio of acetylation of each position varies depending on whether H4 is newly deposited on chromatin (87, 88). This has important implications for potential regulatory mechanisms, as acetylation of multiple sites on each of the eight histone monomers in a nucleosome would dramatically alter the overall charge of the nucleosome, and hundreds of different nucleosomal charge distribution states might exist in vivo as a result of different combinations of acetylation patterns on the eight tails in each nucleosome.

Histone N termini have a disordered structure in solution and in crystals, as evidenced by nuclear magnetic resonance (NMR) structural studies (89) and by the lack of order in each increasingly resolved crystal structure (however one portion of the H4 tail is structured in the most recent crystal structure) (21; see below). These tails therefore do not appear to have a static interaction within the particle in which they are found; instead they may interact with other DNA sequences or with other proteins in a manner that contributes to changed functional properties of the nucleosome. The role of the N termini has been explored by characterizing nucleosomes with increased levels of acetylation or nucleosomes that have been treated with trypsin to remove the tails.

Treatment of nucleosome cores with limited amounts of trypsin removes the histone N termini while leaving the nucleosome core structure intact (90, 91). Initial studies revealed surprisingly few changes in the properties of mononucleosomes following trypsinization (92, 93). The extended structure that mononucleosomes form as salt increases was not significantly altered by removal of the tails; nor were there significant changes in circular dichroism (CD) spectra. Trypsinized mononucleosomes appear to be more accessible to proteins, as DNAse cleavage and binding of sequence-specific DNA-binding proteins are substantially enhanced by trypsinization (65, 93–95). Thus, the tails on purified mononucleosomes can impede access of proteins to nucleosomal DNA even though these domains are not required for the formation of nucleosomes. These studies with single nucleosome cores implicate a function of the histone tails in modulating access of factors to DNA sequences within the primary level of chromatin structure. One possibility is that removal of the histone tails enhances the transient unpeeling of DNA from the histone octamer (see above), thus facilitating factor binding. Indeed, by stimulating the binding of individual factors, removal of the core histone tails reduces the apparent cooperativity of multiple factors binding to nucleosomes (65).

In addition to the effects of the histone tails observed with mononucleosomes, studies with arrays of nucleosomes highlight additional functions of the core histone tails in chromatin structure. Arrays of nucleosomes form a compacted structure, as measured by sedimentation rate, as the NaCl concentration increases from less than 10 mM to 150 mM or as Mg concentration increases to

2 mM (96–98). Increased compaction under these conditions can also be seen by electron microscopy and by nondenaturing gel electrophoresis. Trypsinized cores do not form this compacted structure (99, 100). Each nucleosome in a trypsinized array is more senstive than normal nucleosomes to micrococcal nuclease cleavage at low salt, creating a 106-bp protected fragment instead of a 146-bp fragment. This sensitivity decreases as salt is increased, but the fully compacted structure is not formed. This sensitivity to micrococcal nuclease is similar to the increased sensitivity of acetylated mononucleosomes to DNase (101) and might reflect the same alteration. Although the effect of the histone tails on the condensation of nucleosome arrays may be indirect (e.g. contributing to necessary charge shielding of the DNA backbone), these experiments raise the intriguing possibility of important interactions between the tails and linker DNA and/or adjacent core particles that participate in formation of a higher-order structure. Thus, interactions involving the histone tails may not be constrained to their own nucleosome, and histone tails may be available for interaction with other proteins (see below).

Similar conclusions can be drawn from comparisons of nucleosomes with high and low levels of acetylation. There are no dramatic changes in biophysical properties of mononucleosomes upon acetylation. However, there are changes in access of DNA-binding proteins to the mononucleosome: DNase cleavage is enhanced at a position 60 bp from one end (73, 102, 103), and transcription factor access can be increased by acetylation (94, 104). Acetylation does have a pronounced effect on supercoiling of arrays of nucleosomes (105, 106). In a normal nucleosomal array, each nucleosome introduces approximately one negative supercoil into a closed circular plasmid. In arrays formed with hyperacetylated nucleosomes, each nucleosome introduces approximately 0.8 negative supercoils, and this 20% decrease appears to be due largely to acetylation on histones H3 and H4 (105). Similar effects can be seen on a subpopulation of plasmids in intact cells following butyrate treatment to increase intracellular acetylation levels (107); however, no changes in linking number were seen in simian virus 40 (SV40) minichromosomes as a result of increased acetylation (108). Thus, in a purified in vitro system, or in vivo at very high template concentration, acetylation induces a change in topology. This change is not seen in vivo at lower template concentration. One possible interpretation is that tails can alter nucleosome structure to affect long-range interactions (as above), and acetylation can affect that process. However, in vivo other proteins might interact with tails to alter these changes.

Histone tails might interact with DNA, with other proteins, or with both. Nonacetylated tails have a strong positive charge, suggesting an ability to bind tightly to DNA. The ability of tails to interact with DNA has been examined by chemical strategies designed to map regions of histones that contact DNA and by thermal denaturation experiments. A peptide representing the N-terminal tail of histone H4 is able to interact with DNA, as shown by an increase in the denaturation temperature of DNA when the peptide is present (109). However, this effect is seen only at low salt concentrations and is not seen at physiological levels of salt. Acetylation of tails reduces this interaction, a result that is consistent with the decreased positive charge of acetylated tails.

An analysis of the interaction of H2A N terminus with nucleosomal DNA was obtained by placing a cross-linking agent at specific residues on the H2A tail and assembling the modified histone into mononucleosomes (110). This site-specific cross-linking agent interacts with DNA at two specific sites on the nucleosomal DNA that are symmetrically disposed about 40 nucleotides from the dyad. This is consistent with a specific interaction between each of the two H2A tails in a nucleosome and two specific, symmetric sites. The specificity of the observed interaction is most simply interpreted as showing a specific localization of the H2A tail in these mononucleosomes. The histone H4 N terminus can be cross-linked to regions near the dyad (111). Chemical modification studies have been used to examine how the sites where H2B and H3 contacts DNA change in sea urchin sperm chromatin when linker DNA is present. The modified H2B that is found in this organism was shown to contact linker DNA (112). These data show that histone tails can interact with DNA. However, all of these studies have been performed using either purified mononucleosomes or purified arrays of nucleosomes. It is unclear how many of these contacts would occur in a nuclear environment where many other proteins might interact with the tails.

Whereas histone tails are not ordered, the high-resolution crystal structure of the nucleosome shows that the tails of histones H3 and H2B emerge from the core by tracking through a tunnel made by the minor grooves of adjacent DNA helices; thus they have significant DNA contact (21). These contact points are not near the points of acetylation, however, and there is no observed contact between DNA and the acetylated regions of the tails that are not visible in the crystals. The nucleosome preparation used in the crystal structure study has no linker DNA, so it is possible that an ordered interaction between tails and DNA would be seen in the presence of linker DNA. It seems likely, however, from the biochemical data discussed above and from genetic studies, that the tails are not bound tightly to DNA but rather might have transient interactions with DNA that make them available for interactions with nuclear proteins.

One such interaction is seen in the crystal structure (21). Amino acids 16–24 of histone H4 are bound to a negatively charged face of histones H2A and H2B on an adjacent nucleosome. There might be an artifactual contribution of crystal packing to this interaction. The interaction is particularly interesting because this region of histone H4 has been shown by genetic and biochemical studies to interact with the gene-silencing protein SIR3 of *S. cerevisiae* (113–117). The SIR proteins are believed to form a stable structure that coats nucleosomal arrays, with SIR3 functioning via a direct contact with the histone H4 tail. SIR3, therefore, might alter long-range nucleosomal stability by removing one interaction between nucleosomes and replacing it with a presumably more stable structure. This interaction site is directly adjacent to where acetylation occurs, and it indeed contains lysine 16, one of the sites of acetylation. Acetylation might be expected to weaken the interaction between the H4 N-terminal tail and this negatively charged region of H2A and H2B by decreasing the positive charge on H4. This structural information provides a possible explanation for a paradoxical observation: Mutation of a yeast deacetylase complex increases silencing at telomeres, the opposite of the expected effect (deacetylation most normally being associated with repression, thus mutating a deacetylase complex should decrease silencing) (115). Silencing at telomeres requires SIR3 interaction, so it is possible that the increased acetylation of H4 in mutant deacetylase strains increases the access of SIR3 to the H4 tail by decreasing H4 interactions with the adjacent nucleosome or with DNA.

The basic hypothesis that underlies the above example can be stated as follows: Histone tails are able to form transient interactions with various elements of nucleosome structure in a manner that leaves them sufficiently weakly bound (with a sufficiently high off-rate) that they are available for interactions with other proteins, including repressing proteins such as the SIR complex, or enzymatic activities such as the acetylase and deacetylase activities discussed in the following section. This hypothesis is consistent with the biochemical and genetic data and implies that the tails form critical handles that can be used to alter nucleosome structure in a manner that facilitates appropriate regulation. A central role for tails, particularly H3 and H4 tails, in modulating nucleosome structure might be the underlying reason that the sequence of the tails has been so strictly conserved in evolution.

Histone Acetyltransferases

Since the initial observation that core histones are posttranslationally modified by acetylation (118), strong correlations between histone acetylation and gene transcription (119–121) and histone deposition during chromatin assembly (122–126) have been developed. This generated a long-standing interest in the enzymes that catalyzed this modification. For example, numerous studies have used fractionated cell extracts to identify and characterize histone acetyltransferases (HATS) from organisms as diverse as rat (127–130), calf (131, 132), *Drosophila* (133), yeast (134–137), *Tetrahymena* (121), corn (138, 139), and *Physarum* (140). These studies made the important observation that there are multiple forms of cytoplasmic and nuclear histone acetyltransferases and that they often differed with regard to the histones they acetylated (127, 129, 135, 138, 140). Thus, HATs are grouped into two broad categories. These are the nuclear type A HATs, which acetylate chromosomal histones (131, 141–144), and cytoplasmic type B HATs, which acetylate free cytoplasmic histones prior to chromatin assembly (133, 141, 144). The recent purification and cloning of type A and type B HATs have revealed remarkable insights regarding their functions.

CYTOPLASMIC, TYPE B, HATS   Newly synthesized histones H3 and H4 are modified by acetylation (123) and are subsequently deacetylated after their assembly into chromatin (124). The newly sythesized and acetylated H3 and H4 are assembled into a chromatin assembly complex (CAC), which also contains the chromatin assembly factor CAF-1 (145). CAF-1 comprises three subunits (p160, p60, and p48) and promotes nucleosome assembly during in vitro DNA replication and DNA repair (146–149). Analysis of CAC with H4 acetyl-lysine–specific antibodies suggests that it contains H4 proteins acetylated at lysines 5, 8, and 12 (145). H4, which is acetylated at lysines 5 and 12, is also found in newly assembled chromatin (150), partially consistent with the premise that it is deposited by CAC. The small subunit of CAF-1, p48, is identical to RbAp48 (retinoblastoma-associated protein 48) (145), previously identified through interactions with retinoblastoma protein in vitro (151, 152). Drosophila CAF-1 also contains an RbAp48 homologue (153). p48 appears to play multiple roles in histone metabolism. It is also tightly associated with HD1, a human histone deacetylase (154, also see below). Its S. cerevisiae homologue, Hat2p, is a subunit of a yeast type B HAT (155). Moreover, that p48 and Hat2p bind directly to histone H4 (145, 155) suggests that p48 may funtionally link histone H4 acetylation with assembly into chromatin and maturation/deacetylation (156).

Whereas little is known about cytoplasmic HATs that modify H3, type B HATs that acetylate H4 have been purified from yeast and corn (155, 157). Both enzymes are specific to free histones (i.e. not associated with DNA) and contain two subunits of approximately 45 and 50 kDa. Sequencing of the large subunit of the yeast enzyme, Hat1p, led to its identification as the catalytic subunit and product of the HAT1 gene (155). The HAT1 gene was also identified by enzymatic screens of mutants defective in H4 acetylation (158). Yeast Hat1p is the predominant cytoplasmic HAT and acetylates H4 primarily on lysine 12 (155, 158). Moreover, the activity of the corn enzymes generates diacetylated H4, suggesting that it also acetylates a second lysine, presumably lysine 5 (157). Thus, the higher eukaryotic enzymes may have expanded specificity, generating H4 products more closely resembling those found in human CAC (145) or newly assembled chromatin (150). As mentioned above, the small subunit of the yeast

HATB, Hat2p, was identified as the histone-binding subunit and the homologue of RbAp48, leading to the proposal that the Hat2p/RbAp48 family of proteins serve as escorts of histone metabolism enzymes (155).

NUCLEAR, TYPE A, HATS   A major breakthrough in the study of type A HATs came with the initial purification and cloning of a catalytic subunit. Brownell & Allis used an in gel activity assay to identify the polypeptide, p55, corresponding to the catalytic subunit of a type A HAT from *Tetrahymena* (159). Purification and cloning of p55 led to its identification as a homologue of the yeast transcriptional adaptor protein Gcn5 (160). Gcn5 was originally identified through genetic screens in yeast as a protein that functionally interacts with the transcriptional activator Gcn4 (161) and with the activation domain of the herpes simplex virus protein VP16 (162). Thus, the discovery of Brownell and colleagues that Gcn5 is a type A HAT provided a direct link between pathways of transcription activation and histone acetylation (160). Indeed, recombinant yeast Gcn5 was shown to acetylate lysines in histones H3 (lysine 14) and H4 (lysines 8 and 16) that are thought to be sites of transcription-linked acetylation in vivo (163). Moreover, the function of Gcn5 in vivo requires its identified HAT domain indicating that this catalytic activity is crucial to the function of the Gcn5 transcriptional adaptor (164, 165).

Additional genetic screens in yeast for mutants that relieved GAL4-VP16–mediated toxicity in vivo revealed additional genes whose products were thought to be functionally related to that of Gcn5. These include ADA2 (166), ADA3 (167), ADA5/SPT20 (168, 169), and ADA1 (170). Ada2, Ada3, and Gcn5 were shown to physically interact with each other (162, 171, 172). High molecular weight complexes of 170–200 kDa, 800–900 kDa, and 1.8–2.0 MDa containing Ada and Gcn5 proteins have been identified in yeast (170, 173–175). Moreover, analysis of mutant strains has shown that Gcn5 is the primary if not only catalytic HAT subunit of each of these HATA complexes (174, 175).

Mutants in ADA1 and ADA5 show more severe phenotypic defects than mutants in the other ADA genes and also display *spt*-phenotypes (168–170). In fact, ADA5 was simultaneously discovered in a screen for SPT mutants and named SPT20 (169). SPT genes have been isolated in genetic screens for suppressors of Ty and delta promoter insertion mutations (176). Genetic and biochemical studies illustrated interactions between Spt3, Spt7, Spt8, Spt20, and TBP (177–179), suggesting that these proteins might constitute a multiprotein complex. These observations prompted testing of the 0.8-MDa and 1.8-MDa ADA/GCN5 HAT complexes for the presence of Spt proteins. Both complexes contain Ada2, Ada3, and Gcn5, and the larger complex was found to also contain Spt3, Spt7, Spt8, and Spt20/Ada5 (174). Although both complexes contain several yet unidentified proteins, neither complex appears to contain the TATA-binding protein, TBP, which is encoded by SPT15. Thus, the 0.8-MDa complex has been termed the ADA complex while the 1.8-MDa complex has been termed the SAGA complex (Spt-Ada-Gcn5-Acetyltransferase) (174).

The SAGA complex provides a link between histone acetylation by Gcn5p and the function of the Spt and Ada proteins. This complex contains proteins that interact with acidic activation domains and TBP (179–182) and appears to be a primary transcriptional regulator in yeast. Further genetic analysis indicates that SAGA carries out multiple functions, a subset of which are dependent on Gcn5 and thus presumably histone acetylation (179). Moreover, mutations in components of the SWI/SNF chromatin remodeling complex (see below) and SRB/Mediator coactivator complexes cause lethality in spt20/Ada5 deletion mutants. Thus, the SAGA, SWI/SNF, and SRB/Mediator complexes appear to represent a trio of partially redundant activities that provide a mutual backup system for appropriate regulation of RNA polymerase II transcription (179).

In addition to the Gcn5-containing ADA and SAGA adaptor HAT complexes, there are other HAT complexes in yeast that neither contain the Spt or Ada proteins described above nor use Gcn5 as a catalytic subunit (174). However, GCN5 belongs to a diverse superfamily of genes (GNATs) that share four regions of sequence homology, including domains implicated in enyzmatic activity (183). In addition, screens for genes enhancing mating defects in Saccharomyces have revealed a separate family of tentative nuclear histone acetyltransferases (the MYST family). Of these, the SAS2 and SAS3 (SAS, something about silencing) genes contain regions of homology to GCN5, HAT1, and other acetyltransferases as well as extensive homology to each other throughout the rest of their sequence (184). SAS2 and SAS3 are also closely homologous to the HIV Tat-interacting protein, Tip60 (185), and the MOZ gene product, which is a translocation partner in the MOZ-CBP chimaeric oncogene (186). Thus, the GNAT and MYST gene families represent potential catalytic subunits of yet uncharacterized nuclear HAT complexes.

The discovery of nuclear proteins that function as type A HATs from mammals has strengthened the link between histone acetylation and transcription activation and provided a possible link to cellular transformation. Both p300 and p/CAF can function as HATs. p300 and its closely related functional homologue, CBP, are transcriptional adaptors that interact with numerous transcription factors including CREB, Jun, Myb, Fos, MyoD, and nuclear hormone receptors (187–196). Moreover, p300/CBP interacts with the adenovirus E1A transforming protein (197, 198). P/CAF binds to p300 and CBP (199) and is also closely homologous to yeast GCN5 and the human homologue hGCN5 (200). Recombinant p/CAF acetylates primarily H3 while recombinant p300/CBP acetylates all four core histones, suggesting that the two catalytic subunits might function within a single adaptor complex to efficiently acetylate nucleosomal histones.

In addition, recombinant forms of p/CAF and p300/CBP effectively acetylate histones within nucleosomes (199, 201). By contrast, recombinant yeast Gcn5 does not modify nucleosomal histones (199), but Gcn5 does so within native HAT complexes (174).

While these data suggest that histone acetylation might be one mechanism through which these proteins regulate gene expression, other nuclear proteins may be crucial acetylation targets. For example, in addition to histones, p300 will acetylate the tumor suppressor protein, p53, enhancing its in vitro DNA-binding activity (202). In this instance it is not clear whether histones, p53, and/or other nuclear proteins are the critical targets for p300 action in vivo, an issue that also pertains to other proteins that have been identified as histone acetyltransferases.

Acetylation might be involved in other regulatory events in mammals. p300/CBP and pCAF are recruited to ligand-bound nuclear hormone receptors by an additional coactivator, ACTR, which also has HAT activity on both free histones and nucleosomes (203). Thus, during transcriptional activation by ligand-bound nuclear hormone receptors it appears that a multiprotein complex (204) containing three distinct HAT activities (ACTR, p300/CBP, and p/CAF) is assembled by the receptor to modify nucleosomal histones. Importantly, the recruitment of these HATs by the receptors, and presumably the acetylation of nearby nucleosomes, is ligand dependent (203). The importance of histone acetylation during activation by the receptors is further illustrated by the fact that the unliganded receptors instead recruit corepressor complexes that carry histone deacetylase activity (see below). Steroid receptor–responsive promoters are sufficiently defined in terms of nucleosomal structure that it should be possible to examine in detail how the acetylation state of these promoters is changed as various receptor agonists and antagonists are introduced.

In addition to the yeast ADA and SAGA complexes and mammalian p/CAF-p300/CBP, further coactivator complexes might carry out histone acetylation as part of their functions. For example, human $TAF_{II}250$ and its yeast homologue also contain HAT activity (205). $TAF_{II}250$ is part of the TFIID complex that also contains TBP, the TATA-binding protein. The presence of a HAT activity in TFIID suggests that histone acetylation may be particularly important at the core promoter (i.e. TATA-box) to mediate transcription factor interactions with nucleosomal DNA. In addition to HAT activity, $TAF_{II}250$ preparations are able to phosphorylate the basal transcription factor RAP74 (206). $TAF_{II}250$ is therefore a multifunctional enzyme that catalyzes at least two distinct reactions as well as serving as a core subunit of TFIID. Thus, $TAF_{II}250$ provides an example of a protein that can function as a HAT but that has other activities relevant to gene regulation. This is likely to be true of other HATs and HAT complexes as well (e.g. see 179).

Histone Deacetylases

For histone acetylation to play an important role in signaling pathways controlling transcription, efficient reversal of the modification (i.e. histone deacetylation) is crucial (Figure 2). The importance of histone deacetylase enzymes (HDs) in the turnover of the acetate modification became clear from early studies that illustrated that inhibition of HD activity led to an accumulation of acetylated histones in vivo (207). Multiple forms of HDs have been characterized in several organisms (139, 140, 208–210), and like HATs, they appear to be found in high molecular weight complexes (211–214). As discussed below, some histone deacetylases have been linked to RNA polymerase II transcription; however, there are likely to be additional HDs that perform other functions in chromatin. For example, purification of a Maize histone deacetylase, HD2, has led to its identification as an acidic nucleolar phosphoprotein that may regulate ribosomal gene chromatin structure and function (215).

A direct link between histone deacetylases and transcription resulted from the purification and cloning of a mammalian HD, called HDAC1, which was found to be homologous to the yeast Rpd3 protein (154). A related protein, HDAC2, was found in screens for corepressors that interact with the YY1 transcription repressor/activator (216). Rpd3 was found to be the catalytic subunit of a 600-kDa yeast HD complex, HDB (213, 217). A second yeast HD complex, HDA, is distinct from HDB (350 kDa) and contains the Hda1 protein, which is similar to Rpd3 (213, 217). Moreover, the presence of three additional ORFs in yeast (HOS1, HOS2, HOS3) that are similar to RPD3 and HDA1 suggest that there may be additional HD complexes (217).

Rpd3 and Hda1 may act as both negative and positive regulators of transcription. Deletions of Rpd3 or Hda1 increase repression at telomeric loci in yeast, and mutations in a *Drosophila* RPD3 homologue increase position effect variegation (115, 217). These observations suggest that in these contexts, Rpd3 and Hda1 counteract repression mechanisms. However, it is not yet known in this instance whether Rpd3 and Hda1 act directly or indirectly in relief of repression.

Figure 2 Tentative model of the action of HD and HAT complexes. Sequence-specific DNA-binding repressor proteins recruit HD complexes to the promoter region of a gene, maintaining nucleosomal histones in the deacetylated state (black nucleosomes), which is refractory to transcription. The binding of sequence-specific activator proteins recruits HAT complexes, which leads to acetylation of the nucleosomal histones (white nucleosomes). This acetylation in turn leads to conformational changes within the nucleosomal array. The acetylated nucleosomes permit the recruitment of general transcription factors and RNA polymerase II, which bind the promoter and initiate transcription. The rebinding of a sequence-specific repressor recruits HD complexes, which deacetylates the nucleosomal histones and thus leads to the cessation of transcription.

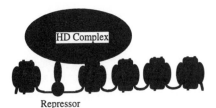

**Repressor and HD bound
Histones Deacetylated**

Repressor

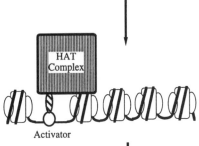

**Activator binding
Recruitment of HAT
Histones Acetylated**

Activator

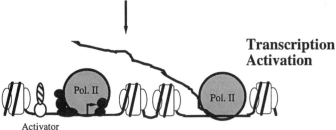

**Transcription
Activation**

Activator

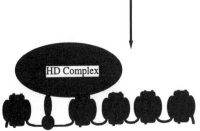

**Repressor binding
Recruitment of HD
Histones Deacetylated**

Repressor

For example, Rpd3 might counteract silencing by acting as a repressor, reducing expression of genes encoding other repressor proteins.

The possibility that Rpd3 functions directly as a repressor protein is supported by numerous biochemical and genetic studies. Rpd3 acts as a negative regulator of genes that are also negatively regulated by the Sin3 protein (218–222). Sin3 is physically associated with Rpd3; both are contained in a very large, 2-MDa, complex (214, 223). The relationship between this 2-MDa complex and the 600-kDa HDB complex described above is not clear (other than that they both contain Rpd3). One possibility is that the 600-kDa complex is a subcomplex that associates with additional proteins to form the larger complex. Alternatively, Rpd3 may function in two distinct complexes. Neither Sin3 nor Rpd3 binds DNA directly; however, LexA fusion proteins containing either Sin3 or Rpd3 repress transcription from reporter genes bearing lexA sites, suggesting that they act as corepressors (223, 224).

The Sin3/Rpd3 corepressor complex interacts with sequence-specific DNA-binding proteins, which suggests that this interaction is the mechanism of its targeting to specific promoters. Sin3 binds directly to Ume6, a sequence-specific DNA-binding protein that recognizes URS1, an upstream repression sequence found in several yeast promoters (223). Thus, the picture emerging from the studies in yeast is that a specific DNA-binding protein that binds repressor elements recruits the Sin3/Rpd3 corepressor complex, which mediates transcriptional repression and histone deacetylation. It will be interesting to see whether HDA is similarly targeted to yeast promoters. It is not yet clear whether histone deacetylation is the primary mechanism of transcriptional repression mediated by the Sin3/Rpd3 complex. It is likely that the complex performs multiple functions, some of which may play a more prominent role in transcription repression. However, the catalytic deacetylase subunit, Rpd3, is clearly required for repression by Sin3 (223). Generation of mutants in Rpd3 that reduce its deacetylase activity but leave the Sin3/Rpd3 corepressor complex intact will help address the importance of histone deacetylation in its repression function.

Simultaneous studies of corepressor complexes in mammals have strengthened the connection between transcriptional repression and histone deacetylases. Mad/Max and Mxi/Max heterodimers and unliganded retinoid and thyroid hormone receptors function as sequence-specific transcriptional repressors (225–228). Mad represses transcription via interactions with mSin3A or mSin3B, mammalian homologues of yeast Sin3 protein (229, 230). Repression of the unliganded thyroid and retinoid receptors is mediated via SMRT and N-CoR corepressors (227, 228). N-CoR and SMRT interact directly with mSin3A/B (231–233). mSin3A/B interacts with the Rpd3 homologues HDAC1 and/or HDAC2 (231–234) to form high molecular weight corepressor complexes that also contain RbAP 46/48 and several novel proteins (235, 236). Transcriptional repression by the mSin3/HDAC complexes is blocked by histone deacetylase inhibitors (Tricostatin A, trapoxin), implicating deacetylase activity in the transcriptional repression mechanism. As mentioned above, RbAp48 is also the histone-binding subunit (HAT2) of the yeast cytoplasmic type B HAT (155), suggesting that it might similarly connect the corepressor/deacetylase complex to histone substrates.

As noted above in discussing acetylase activities, a caveat in interpreting these data is that there may be further protein substrates for these deacetylase complexes. The finding that tumor suppressor protein, p53, is acetylated by p300 makes this possibility more intriguing (202). However, the model of recruitment of corepressor/deacetylase complexes by sequence-specific DNA-binding repressor proteins fits nicely with the recruitment of coactivator/HAT complexes by sequence-specific DNA-binding activators. This is illustrated most clearly by the switch for nuclear hormone receptors, which in their unliganded form recruit mSin3-HDAC repressor complexes (231–233) and when bound by hormone recruit the p300/pCAF coactivator/HAT complex (195, 196). Moreover, the recruitment of histone-modifying enzymes (HATs, HDs) for both activation and active repression strongly implicate nucleosome functions in the control of transcription.

ATP-DEPENDENT REMODELING COMPLEXES

Genetic and biochemical studies have identified complexes that are able to alter nucleosome structure in an ATP-dependent manner. These studies imply that at least some, if not all, of these complexes play important roles in regulating the activation of transcription by increasing access of the transcription machinery (237). Five related complexes of this type have been identified (the SWI/SNF, NURF, RSC, CHRAC, and ACF complexes), and each of these complexes can increase either transcription factor binding or restriction enzyme access to nucleosomal DNA (238–246). Although the mechanism(s) by which these complexes function has not yet been elucidated, it appears that they increase accessibility to the DNA of the nucleosome without removing histones. This suggests that they must alter histone-DNA contacts either by remaining in contact with the nucleosome to actively disrupt these contacts or by catalyzing the formation of a remodeled configuration of histones and DNA that has weakened histone-DNA contacts. Four important questions regarding function of these complexes are as follows: (a) Does each of the ATP-dependent remodeling complexes function by the same mechanism? (b) What is the configuration of histones and DNA in a remodeled nucleosome? (c) Will a remodeled nucleosome remain remodeled following the removal of the remodeling complex

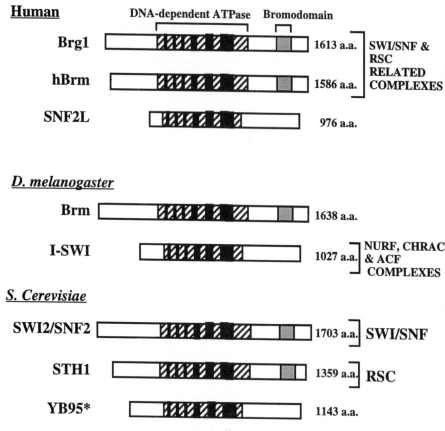

Figure 3 The ATPase subunits of known ATP-dependent chromatin remodeling complexes. SWI/SNF2 and STH1 are homologous to BRG1 and hBRM in humans and Brm in *Drosophila*. ISWI is homologous to YB95 (and several other potential homologues) in yeast and hSNF2L in humans. Complexes are indicated where they have been identified and shown to alter some aspect of chromatin structure in an ATP-dependent manner.

(i.e. how stable is the remodeled state?) (*d*) Are these complexes targeted to specific regions of chromatin, and if so, what does the targeting?

There are likely to be significant differences in the mechanisms used by these complexes to remodel nucleosomes. The five complexes can be divided into two families based on protein composition (Figure 3). The SWI/SNF and RSC complexes each contain numerous subunits: SWI/SNF has 11 subunits in yeast (240, 245, 247), and the putative homologue has 8 subunits in humans (238, 248, 249), while the RSC complex has approximately 15 subunits (246).

The DNA-dependent ATPase subunits of SWI/SNF and RSC are similar to each other (see Figure 3), as are at least three other subunits (246). This similarity suggests that SWI/SNF and RSC might alter nucleosome structure using similar mechanisms, and it further suggests that these complexes are evolutionarily related. In humans, the hBRM and BRG1 genes (250–252) appear to be homologues of the SWI2/SNF2 and STH1 genes of yeast, so it is possible that the hBRM and BRG1 proteins organize two separate complexes that correspond to SWI/SNF and RSC. Both BRG1 and hBRM proteins have been associated with complexes that can alter nucleosome structure in an ATP-dependent manner, and immunoprecipitation experiments imply that complexes exist that contain only hBRM or only BRG1 (248, 249, 253).

Genetic studies suggest that there are differences between the roles that SWI/SNF and RSC play in vivo. Deletion of genes that encode components of the SWI/SNF complex is not lethal to growth, whereas deletion of RSC components is lethal (239, 254, 255). Temperature-sensitive mutations in subunits of RSC affect cell cycle progression, whereas there are no known effects of SWI/SNF mutations on cell cycle (256). In human cells, the BRG1 gene cannot be deleted unless BRG1 is expressed from an ectopic source, suggesting that the BRG1-based complex might be essential to cell viability, similar to RSC (257). Deletion of the hBrm gene is not lethal. These considerations, in addition to the fact that BRG1 is slightly more similar to STH1 than to BRG1, raise the possibility that the BRG1-based complex and RSC have similar functions in yeast and humans. These differences between SWI/SNF and RSC function in vivo might reflect differences in their mechanism of action. Alternatively, these phenotypic differences might reflect differential requirements for SWI/SNF and RSC in cellular processes and different abilities of these complexes to receive signals and be targeted to specific regions of chromatin.

The NURF (241, 242, 258), CHRAC (243), and ACF (244) remodeling complexes have been isolated from *Drosophila* extracts. Each contains the ISWI protein, a DNA-dependent ATPase that has homology to SWI2/SNF2 only in the ATPase domain. In addition to ISWI, these complexes contain either three or four additional subunits, so these complexes are smaller than the SWI/SNF family of complexes. Another difference is that the SWI/SNF ATPase is stimulated by bare DNA or by nucleosomal DNA whereas the NURF ATPase is stimulated only by nucleosomal DNA (242). These data suggest that there might be differences in the function between the ISWI-based complexes and the SWI/SNF family of complexes.

The potential for differences in function between ATP-dependent remodeling complexes is further substantiated by the finding that the different ISWI family complexes have different characteristics in vitro. CHRAC will remodel nucleosome arrays to increase restriction enzyme access in a manner that is not seen with NURF when the two activities are compared side by side (243). In addition, CHRAC contains Topoisomerase II as a subunit, whereas NURF does not. Although it is not clear what activity is contributed to CHRAC by Topoisomerase II, because inhibitors of Topoisomerase II function do not alter the basic remodeling activity of CHRAC, it seems likely that Topoisomerase II plays an important role in the mechanism of action of CHRAC. Whereas ACF and CHRAC share similar abilities in promoting appropriate spacing of nucleosomes on plasmid templates, ACF does not contain Topoisomerase II and thus is likely to have some mechanistic differences when compared to CHRAC (244).

Two types of models explain how any of these remodeling complexes might use the energy of ATP hydrolysis. This energy might be used in a power-stroke mechanism that catalyzes an isomerization of the nucleosome structure into a different structure with the same components, but with altered histone-DNA contacts (Figure 4b,c), or ATP hydrolysis might be used to promote continual movement of the complex around the nucleosomal DNA (259), with the complex physically prying the DNA away from the histone core as it moves (Figure 4a). These two models currently cannot be rigorously evaluated for either the SWI/SNF or ISWI families of complexes. The latter model is argued against in the case of SWI/SNF function by the observation that ATP is not required to maintain a disrupted state of a nucleosome (260), although it is conceivable that removal of ATP freezes the nucleosome in a disrupted state with SWI/SNF still present. The central distinguishing feature of these models is that the first model posits the existence of an altered nucleosome structure that might be stable in the absence of the remodeling complex.

Functional analysis of the NURF complex implies that NURF contacts the histone N termini to alter nucleosomal structure. Truncation of the N termini from nucleosomes blocks remodeling by NURF and also inhibits the ability of these nucleosomes to stimulate ATPase activity (261). This inhibition appears to be caused by a direct interaction between the N termini and NURF, because isolated tails of histones can act as inhibitors of NURF ATPase activity. These data are most consistent with a model in which NURF alters structure of the nucleosomes and uses the tails as part of a handle to accomplish this alteration. These data are less consistent with a model in which NURF continually translocates around the nucleosome, because it is not clear why histone tail contact would be required for this translocation.

The ability of chromatin remodeling complexes to catalyze formation of an altered nucleosomal state might be the first step of a mechanistic pathway that leads to a stable alteration in chromatin structure. For example, the combined action of SWI/SNF and the binding of multiple GAL4 derivatives can mediate displacement of histone octamers from nucleosome cores and generate

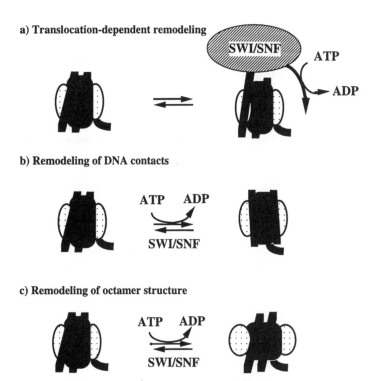

Figure 4   Three different mechanisms that might be used to alter nucleosomal structure by ATP-dependent remodeling complexes (using SWI/SNF as an example): (*a*) Continual ATP hydrolysis might be required to separate DNA from histones (259); (*b*) histone-DNA contacts might be altered by a mechanism that alters the path of DNA around an unaltered histone octamer; and (*c*) histone-DNA contacts might be altered by a mechanism that changes the conformation of the histone octamer in a manner that weakens bonds to DNA. An ATP-dependent movement such as that shown in (*a*) could lead to more stable structural changes such as those shown in (*b*) or (*c*). It is also possible that both the path of the DNA and the histone octamer might be altered—that is, a combination of (*b*) and (*c*).

DNAse 1 hypersensitive sites that persist after removal of the SWI/SNF complex (262). It is not clear whether this persistent chromatin remodeling requires the SWI/SNF complex and the binding of the transcription factors to happen in concert or whether SWI/SNF first catalyzes a remodeled state, and GAL4 binds only after the large SWI/SNF complex has dissociated from the remodeled nucleosome. The conformation of nucleosomes that have been remodeled by these ATP-dependent remodeling activities may resemble one of the altered nucleosome states that has been identified by varying solution conditions (discussed above). These biophysical data emphasize that nucleosome structure can be dynamic in solution, providing precedence for the possibility that these complexes use the energy of ATP hydrolysis to drive the nucleosome into an altered structure.

The SWI/SNF complex was initially identified via genetic studies in *Saccharomyces cerevisiae*, and mutations in histones have been isolated that suppress mutations in the SWI/SNF genes (263, 264). The location of these mutations on the crystal structure of the nucleosome informs possible models for SWI/SNF action (21). Two of the six isolated point mutations of histone H4 mutate an Arginine residue (R45) that inserts into the minor groove of DNA and is one of the key contact points between histones and DNA. Two other point mutations affect amino acids that contribute to the shape of the loop that contains R45, or directly contact R45 and DNA, indicating that weakening a normal histone-DNA contact confers a phenotype that alleviates the lack of SWI/SNF function. This further validates the belief that the primary role of this remodeling complex is to weaken histone-DNA contacts. Perhaps more intriguing are the locations of the other two point mutations. One (H3 R116) is buried in the structure and thus might contribute to altering the overall structure to favor that of a remodeled nucleosome. Another mutant (H3 E105) mutates a surface negative charge to a positive charge in a region that does not contact DNA in the solved structure. It is intriguing to speculate that this region of the surface might somehow be involved in redirecting histone-DNA contacts in a remodeled nucleosome structure (the increased positive charge stabilizing the remodeled interactions). Another possibility, however, is that altered contacts between this region and adjacent nucleosomes cause the phenotype.

Regardless of how these complexes are able to remodel nucleosomes, a critical question concerning function of the complexes in vivo concerns targeting of these activities. Several different remodeling complexes are found in cells; thus each is anticipated to play a different role in regulating nuclear processes. It is therefore important to understand the mechanism(s) that target each complex to a specific region of chromatin and those that regulate activity at specific stages of cellular development. Very little concrete information is available on this issue.

The clearest example of targeting comes from immunoprecipitation studies that show that the glucocorticoid receptor can interact with the SWI/SNF complex (265). An analysis of the possible function of this interaction demonstrated that binding the glucocorticoid receptor to a nucleosome increased the ability of SWI/SNF to disrupt the nucleosome, a result that is most simply interpreted as arguing that SWI/SNF was targeted to the nucleosome by the receptor (266). Although it is not clear from these experiments whether SWI/SNF directly contacts the glucocorticoid receptor, or whether intermediary proteins are involved, these experiments suggest that targeting of a remodeling complex might be an important aspect of activation by the glucocorticoid receptor. Taken together with the observations that many steroid receptors bind nucleosomal DNA (267–269) and that many have also been linked to acetylation and deacetylation activities (see above), these results all point toward the possibility that regulation of chromatin structure is a general mechanism used by receptors to affect gene expression. A different mechanism of targeting is suggested by the association of SWI/SNF with yeast RNA polymerase holoenzyme (270). This might allow targeting of SWI/SNF via the targeting of RNA polymerase, although the potency of this interaction has been questioned (246). It will be informative to determine whether associations exist between SWI/SNF and RNA polymerase in mammalian systems and whether this association results in the targeting of remodeling activities when polymerase is bound to a template.

SUMMARY

Descriptive studies of chromatin structure in vitro and in vivo have led to the unequivocal conclusion that altered states of the nucleosome exist. More recent functional analysis of the role that chromatin structure plays in gene regulation has led to the possibility that these altered nucleosomal states play an important role in establishing the proper chromatin structure on a gene. Generation of a chromatin structure that is more accessible to regulatory proteins is believed to be an important mechanism in gene activation. Conversely, a more tightly ordered chromatin structure is believed to be important in repression. Stable alterations in nucleosome structure might be an important means of creating an altered chromatin state, or alterations in nucleosome structure might be transient but be an essential step in a pathway that creates a stably altered chromatin state.

The discovery of complexes of proteins that acetylate nucleosomes, that deacetylate nucleosomes, and that remodel nucleosome structure in an ATP-dependent manner provides an essential advance in understanding the role that chromatin structure plays in regulating nuclear events. The purification and characterization of these complexes will allow several questions to be addressed: How do these complexes alter nucleosome structure? What are the functional consequences of these alterations for regulation of nuclear processes? How are these activities spatially and temporally targeted? The answers to these questions are vital to an understanding of regulatory mechanisms in the eukaryotic nucleus.

Visit the *Annual Reviews home page* at
http://www.AnnualReviews.org.

Literature Cited

1. Owen-Hughes T, Workman JL. 1994. *Crit. Rev. Eukaryot. Gene Exp.* 4:403–41
2. Felsenfeld G. 1992. *Nature* 355:219–24
3. Imbalzano AN, Kwon H, Green MR, Kingston RE. 1994. *Nature* 370:481–85
4. Clark DJ, Felsenfeld G. 1992. *Cell* 71:11–22
5. Izban MG, Luse DS. 1992. *J. Biol. Chem.* 267:13647–55
6. Chang CH, Luse DS. 1997. *J. Biol. Chem.* 272:23427–34
7. Studitsky VM, Clark DJ, Felsenfeld G. 1995. *Cell* 83:19–27
8. Studitsky VM, Clark DJ, Felsenfeld G. 1994. *Cell* 76:371–82
9. Lorch Y, LaPointe JW, Kornberg RD. 1987. *Cell* 49:203–10
10. Olins AL, Olins DE. 1974. *Science* 183:330–32
11. Olins AL, Carlson D, Olins DE. 1975. *J. Cell Biol.* 64:528–37
12. Oudet P, Gross-Bellard M, Chambon P. 1975. *Cell* 4:281–300
13. Woodcock CLF. 1973. *J. Cell Biol.* 59:A368
14. Hewish DR, Burgoyne LA. 1973. *Biochem. Biophys. Res. Commun.* 52:504–10
15. Noll M. 1974. *Nature* 251:249–51
16. Sahasrabuddhe CG, Van Holde KE. 1974. *J. Biol. Chem.* 249:152–56
17. Kornberg RD, Thomas JO. 1974. *Science* 184:865–68
18. Kornberg RD. 1974. *Science* 184:868–71
19. Van Holde K, Zlatanova J, Arents G, Moudrianakis E. 1995. In *Chromatin Structure and Gene Expression*, ed. SCR Elgin, pp. 1–21. Cambridge, UK: Oxford Univ. Press
20. Simpson RT. 1978. *Biochemistry* 17:5524–31
21. Luger K, Mader AW, Richmond RK, Sargent DF, Richmond TJ. 1997. *Nature* 389:251–60
22. Richmond TJ, Finch JT, Rhodes D, Klug A. 1984. *Nature* 311:532–37
23. Arents G, Burlingame RW, Wang BC, Love WE, Moudrianakis EN. 1991. *Proc. Natl. Acad. Sci. USA* 88:10148–52
24. Camerini-Otero RD, Sollner-Webb B, Felsenfeld G. 1976. *Cell* 8:333–47
25. Sollner-Webb B, Camerini-Otero RD, Felsenfeld G. 1976. *Cell* 9:179–93
26. Eickbush TH, Moudrianakis EN. 1978. *Biochemistry* 17:4955–64
27. Arents G, Moudrianakis EN. 1995. *Proc. Natl. Acad. Sci. USA* 92:11170–74
28. Baxevanis AD, Arents G, Moudrianakis EN, Landsman D. 1995. *Nucleic Acids Res.* 23:2685–91
29. Xie X, Kokubo T, Cohen SL, Mirza UA, Hoffmann A, et al. 1996. *Nature* 380:316–22
30. Hoffmann A, Chiang C-M, Oelgeschlager T, Xie X, Burley SK, et al. 1996. *Nature* 380:356–59
31. Nakatani Y, Bagby S, Ikura M. 1996. *J. Biol. Chem.* 271:6575–78
32. Thomas JO, Butler PJ. 1977. *J. Mol. Biol.* 116:769–81
33. Ruiz-Carrillo A, Jorcano JL. 1979. *Biochemistry* 18:760–68
34. Arents G, Moudrianakis EN. 1993. *Proc. Natl. Acad. Sci. USA* 90:10489–93
35. McGhee JD, Felsenfeld G. 1980. *Annu. Rev. Biochem.* 49:1115–56
36. Wu HM, Dattagupta N, Hogan M, Crothers DM. 1979. *Biochemistry* 18:3960–65
37. Uberbacher EC, Remakrishnan V, Olins DE, Bunick GJ. 1983. *Biochemistry* 22:4916–23
38. Simpson RT. 1981. *Proc. Natl. Acad. Sci. USA* 78:6803–7
39. Gordon VC, Knobler CM, Olins DE, Schumaker VN. 1978. *Proc. Natl. Acad. Sci. USA* 75:660–63
40. Gordon VC, Schumaker VN, Olins DE, Knobler CM, Horwitz J. 1979. *Nucleic Acids Res.* 6:3845–58
41. Martinson HG, True RJ, Burch JBE. 1979. *Biochemistry* 18:1082–89
42. Burch JBE, Martinson HG. 1980. *Nucleic Acids Res.* 80:4969–87
43. Baer BW, Rhodes D. 1983. *Nature* 301:482–88
44. Hayes JJ, Wolffe AP. 1992. *Proc. Natl. Acad. Sci. USA* 89:1229–33
45. Dilworth SM, Black SJ, Laskey RA. 1987. *Cell* 51:1009–18
46. Ishimi Y, Kojima M, Yamada M, Hanaoka F. 1987. *Eur. J. Biochem.* 162:19–24
47. Kleinschmidt JA, Fortkamp E, Krohne G, Zentgraf H, Franke WW. 1985. *J. Biol. Chem.* 260:1166–76
48. Kleinschmidt JA, Seiter A, Zentgraf H. 1990. *EMBO J.* 9:1309–18
49. Chen H, Li BY, Workman JL. 1994. *EMBO J.* 13:380–90
50. Walter PP, Owen-Hughes TA, Côté J, Workman JL. 1995. *Mol. Cell. Biol.* 15:6178–87
51. Hansen JC, Wolffe AP. 1994. *Proc. Natl. Acad. Sci. USA* 91:2339–43
52. Hirschhorn JN, Brown SA, Clark CD, Winston F. 1992. *Genes Dev.* 6:2288–98

53. Santisteban MS, Arents G, Moudrianakis EN, Smith MM. 1997. *EMBO J.* 16:2493–506
54. Bazett-Jones DP, Mendez E, Czarnota GJ, Ottensmeyer FP, Allfrey VG. 1996. *Nucleic Acids Res.* 24:321–29
55. Sterner R, Boffa LC, Chen TA, Allfrey VG. 1987. *Nucleic Acids Res.* 15:4375–91
56. Allegra P, Sterner R, Clayton DF, Allfrey VG. 1987. *J. Mol. Biol.* 196:379–88
57. Walker J, Chen TA, Sterner R, Berger M, Winston F, Allfrey VG. 1990. *J. Biol. Chem.* 265:5736–46
58. Chen TA, Sterner R, Cozzolino A, Allfrey VG. 1990. *J. Mol. Biol.* 212:481–93
59. Eshaghpour H, Dieterich AE, Cantor CR, Crothers DM. 1980. *Biochemistry* 19:1797–805
60. Lee M-S, Garrard WT. 1991. *EMBO J.* 10:607–15
61. Lee M-S, Garrard WT. 1991. *Proc. Natl. Acad. Sci. USA* 88:9675–79
62. Liang CP, Garrard WT. 1997. *Mol. Cell. Biol.* 17:2825–34
63. Weischet WO, Tatchell K, Van Holde KE, Klump H. 1978. *Nucleic Acids Res.* 5:139–60
64. Simpson RT. 1979. *J. Biol. Chem.* 254:10123–27
65. Vettese-Dadey M, Walter P, Chen H, Juan L-J, Workman JL. 1994. *Mol. Cell. Biol.* 14:970–81
66. Li Q, Wrange Ö. 1993. *Genes Dev.* 7:2471–82
67. Adams CC, Workman JL. 1995. *Mol. Cell. Biol.* 15:1405–21
68. Taylor ICA, Workman JL, Schuetz TJ, Kingston RE. 1991. *Genes Dev.* 5:1285–98
69. Polach KJ, Widom J. 1996. *J. Mol. Biol.* 258:800–12
70. Polach KJ, Widom J. 1995. *J. Mol. Biol.* 254:130–49
71. Beard P. 1978. *Cell* 15:955–67
72. Weischet WO. 1979. *Nucleic Acids Res.* 7:291–304
73. Simpson RT, Thoma F, Brubaker JM. 1985. *Cell* 42:799–808
74. Pennings S, Meersseman G, Bradbury EM. 1991. *J. Mol. Biol.* 220:101–10
75. Meersseman G, Pennings S, Bradbury EM. 1991. *J. Mol. Biol.* 220:89–100
76. Meersseman G, Pennings S, Bradbury EM. 1992. *EMBO J.* 11:2951–59
77. Pennings S, Meersseman G, Bradbury EM. 1994. *Proc. Natl. Acad. Sci. USA* 91:10275–79
78. Ura K, Hayes JJ, Wolffe AP. 1995. *EMBO J.* 14:3752–65
79. Wall G, Varga-Weisz PD, Sandaltzopou-

los R, Becker PB. 1995. *EMBO J.* 14:1727–36
80. Varga-Weisz PD, Blank TA, Becker PB. 1995. *EMBO J.* 14:2209–16
81. Pazin MJ, Bhargava P, Geiduschek EP, Kadonaga JT. 1997. *Science* 276:809–12
82. O'Neill TE, Smith JG, Bradbury EM. 1993. *Proc. Natl. Acad. Sci. USA* 90:6203–7
83. Workman JL, Kingston RE. 1992. *Science* 258:1780–84
84. Owen-Hughes T, Workman JL. 1996. *EMBO J.* 15:4702–12
85. Csordas A. 1990. *Biochem. J.* 265:23–38
86. Grunstein M. 1997. *Nature* 389:349–52
87. Brownell JE, Allis CD. 1996. *Curr. Opin. Genet. Dev.* 6:176–84
88. Turner BM, O'Neill LP. 1995. *Semin. Cell Biol.* 6:229–36
89. Smith RM, Rill RL. 1989. *J. Biol. Chem.* 264:10574–81
90. Cary PD, Moss T, Bradbury EM. 1978. *Eur. J. Biochem.* 89:475–82
91. Bohm L, Sautiere P, Cary PD, Crane-Robinson C. 1982. *Biochem. J.* 203:577–82
92. Dong F, Nelson C, Ausio J. 1990. *Biochemistry* 29:10710–16
93. Ausio J, Dong F, van Holde K. 1989. *J. Mol. Biol.* 206:451–63
94. Lee DY, Hayes JJ, Pruss D, Wolffe AP. 1993. *Cell* 72:73–84
95. Juan L-J, Utley RT, Adams CC, Vettese-Dadey M, Workman JL. 1994. *EMBO J.* 13:6031–40
96. Hansen JC, Ausio J, Stanik VH, van Holde K. 1989. *Biochemistry* 28:9129–36
97. Fletcher TM, Krishnan U, Serwer P, Hansen JC. 1994. *Biochemistry* 33:2226–33
98. Hansen JC, Lohr D. 1993. *J. Biol. Chem.* 268:5840–48
99. Fletcher TM, Hansen JC. 1995. *J. Biol. Chem.* 270:25359–62
100. Garcia RM, Dong F, Ausio J. 1992. *J. Biol. Chem.* 267:19587–95
101. Simpson RT. 1978. *Cell* 13:691–99
102. Garcia RM, Rocchini C, Ausio J. 1995. *J. Biol. Chem.* 270:17923–28
103. Ausio J, van Holde K. 1986. *Biochemistry* 25:1421–28
104. Vettese DM, Grant PA, Hebbes TR, Crane RC, Allis CD, Workman JL. 1996. *EMBO J.* 15:2508–18
105. Norton VG, Marvin KW, Yau P, Bradbury EM. 1990. *J. Biol. Chem.* 265:19848–52
106. Norton VG, Imai BS, Yau P, Bradbury EM. 1989. *Cell* 57:449–57
107. Thomsen B, Bendixen C, Westergaard O. 1991. *Eur. J. Biochem.* 201:107–11

108. Lutter LC, Judis L, Paretti RF. 1992. *Mol. Cell. Biol.* 12:5004–14
109. Hong L, Schroth GP, Matthews HR, Yau P, Bradbury EM. 1993. *J. Biol. Chem.* 268:305–14
110. Lee K-M, Hayes JJ. 1997. *Proc. Natl. Acad. Sci. USA* 94:8959–64
111. Ebralidse KK, Grachev SA, Mirzabekov AD. 1988. *Nature* 331:365–67
112. Hill CS, Thomas JO. 1990. *Eur. J. Biochem.* 187:145–53
113. Hecht A, Laroche T, Strahl-Bolsinger S, Gasser SM, Grunstein M. 1995. *Cell* 80:583–92
114. Johnson LM, Kayne PS, Kahn ES, Grunstein M. 1990. *Proc. Natl. Acad. Sci. USA* 87:6286–90
115. De Rubertis F, Kadosh D, Henchoz S, Pauli D, Reuter G, et al. 1996. *Nature* 384:589–91
116. Moretti P, Freeman K, Coodly L, Shore D. 1994. *Genes Dev.* 8:2257–69
117. Renauld H, Aparicio OM, Zierath PD, Billington BL, Chhablani SK, Gottschling DE. 1993. *Genes Dev.* 7:1133–45
118. Allfrey VG, Faulkner R, Mirsky AE. 1964. *Proc. Natl. Acad. Sci. USA* 51:786–94
119. Mathis D, Oudet P, Chambon P. 1980. *Proc. Natl. Acad. Sci. USA* 24:1–55
120. Vavra KJ, Allis CD, Gorovsky MA. 1982. *J. Biol. Chem.* 257:2591–98
121. Chicoine LG, Richman R, Cook RG, Gorovsky MA, Allis CD. 1987. *J. Cell Biol.* 105:127–35
122. Louie AJ, Dixon GH. 1972. *Proc. Natl. Acad. Sci. USA* 1972:1975–79
123. Ruiz-Carrillo A, Wangh LJ, Allfrey VG. 1975. *Science* 190:117–28
124. Jackson V, Shires A, Tanphaichitr N, Chalkley R. 1976. *J. Mol. Biol.* 104:471–83
125. Allis CD, Chicoine LG, Richman R, Schulman IG. 1985. *Proc. Natl. Acad. Sci. USA* 82:8048–52
126. Annunziato AT, Seale RL. 1983. *J. Biol. Chem.* 258:12675–84
127. Yukioka M, Sasaki S, Qi S-L, Inoue A. 1984. *J. Biol. Chem.* 259:8372–77
128. Fukushima M, Ota K, Fujimoto D, Horiuchi K. 1980. *Biochem. Biophys. Res. Commun.* 92:1409–14
129. Gallwitz D, Sures I. 1972. *Biochim. Biophys. Acta* 263:315–28
130. Libby PR. 1968. *Biochem. Biophys. Res. Commun.* 31:59–65
131. Belikoff E, Wong L-J, Alberts BM. 1980. *J. Biol. Chem.* 255:11448–53
132. Bohm J, Schlaeger EJ, Knippers R. 1980. *Eur. J. Biochem.* 112:353–62

133. Wiegand RC, Brutlag DL. 1981. *J. Biol. Chem.* 256:4578–83
134. Travis GH, Colavito-Shepanski M, Grunstein M. 1984. *J. Biol. Chem.* 259:14406–12
135. Lopez-Rodas G, Tordera V, Sanchez del Pino MM, Franco L. 1989. *J. Biol. Chem.* 264:9028–33
136. Lopez-Rodas G, Perez-Ortin JE, Tordera V, Salvador ML, Franco L. 1985. *Arch. Biochem. Biophys.* 239:184–90
137. Lopez-Rodas G, Tordera V, Sanchez del Pino MM, Franco L. 1991. *Biochemistry* 30:3728–32
138. Lopez-Rodas G, Georgieva EI, Sendra R, Loidl P. 1991. *J. Biol. Chem.* 266:18745–50
139. Grabher A, Brosch G, Sendra R, Lechner T, Eberharter A, et al. 1994. *Biochemistry* 33:14887–95
140. Lopez-Rodas G, Brosch G, Golderer G, Lindner H, Grobner P, Loidl P. 1992. *FEBS Lett.* 296:82–86
141. Garcea RL, Alberts BM. 1980. *J. Biol. Chem.* 255:11454–63
142. Sures I, Gallwitz D. 1980. *Biochemistry* 19:943–51
143. Wiktorowicz JE, Bonner J. 1982. *J. Biol. Chem.* 257:12893–900
144. Kelner DN, McCarty KS Sr. 1984. *J. Biol. Chem.* 259:3413–19
145. Verreault A, Kaufman PD, Kobayashi R, Stillman B. 1996. *Cell* 87:95–104
146. Stillman B. 1986. *Cell* 45:555–65
147. Smith S, Stillman B. 1989. *Cell* 58:15–25
148. Kaufman PD, Kobayashi R, Kessler N, Stillman B. 1995. *Cell* 81:1105–14
149. Gaillard P-H, Martini EM-D, Kaufman PD, Stillman B, Moustacchi E, Almouzni G. 1996. *Cell* 86:887–96
150. Sobel RE, Cook RG, Perry CA, Annunziato AT, Allis CD. 1995. *Proc. Natl. Acad. Sci. USA* 92:1237–41
151. Qian YW, Wang YCJ, Hollingsworth REJ, Jones D, Ling N, Lee EYHP. 1993. *Nature* 364:648–52
152. Qian YW, Lee EYHP. 1995. *J. Biol. Chem.* 270:25507–13
153. Tyler JK, Bulger M, Kamakaka RT, Kobayashi R, Kadonaga JT. 1996. *Mol. Cell. Biol.* 16:6149–59
154. Taunton J, Hassig CA, Schreiber SL. 1996. *Science* 272:408–11
155. Parthun MR, Widom J, Gottschling DE. 1996. *Cell* 87:85–94
156. Roth SY, Allis CD. 1996. *Cell* 87:5–8
157. Eberharter A, Lechner T, Goralik-Schramel M, Loidl P. 1996. *FEBS Lett.* 386:75–81
158. Kleff S, Andrulis ED, Anderson CW,

Sternglanz R. 1995. *J. Biol. Chem.* 270: 24674–77

159. Brownell JE, Allis CD. 1995. *Proc. Natl. Acad. Sci. USA* 92:6364–68

160. Brownell JE, Zhou J, Ranalli T, Kobayashi R, Edmondson DG, et al. 1996. *Cell* 84:843–51

161. Georgakopoulos T, Thireos G. 1992. *EMBO J.* 11:4145–52

162. Marcus GA, Silverman N, Berger SL, Horiuchi J, Guarente L. 1994. *EMBO J.* 13:4807–15

163. Kuo M-H, Brownell JE, Sobel RE, Ranalli TA, Cook RG, et al. 1996. *Nature* 383: 269–72

164. Wang LA, Mizzen C, Ying C, Candau R, Barlev N, et al. 1997. *Mol. Cell. Biol.* 17: 519–27

165. Candau R, Zhou JX, Allis CD, Berger SL. 1997. *EMBO J.* 16:555–65

166. Berger SL, Pina B, Silverman N, Marcus GA, Agapite J, et al. 1992. *Cell* 70:251–65

167. Pina B, Berger S, Marcus GA, Silverman N, Agapite J, Guarente L. 1993. *Mol. Cell. Biol.* 13:5981–89

168. Marcus FA, Horiuchi J, Silverman N, Guarente L. 1996. *Mol. Cell. Biol.* 16: 3197–205

169. Roberts SM, Winston F. 1996. *Mol. Cell. Biol.* 16:3206–13

170. Horiuchi J, Silverman N, Pina B, Marcus GA, Guarente L. 1997. *Mol. Cell. Biol.* 17:3220–28

171. Horiuchi J, Silverman N, Marcus GA, Guarente L. 1995. *Mol. Cell. Biol.* 15: 1203–9

172. Candau R, Berger SL. 1996. *J. Biol. Chem.* 271: 5237–45

173. Saleh A, Lang V, Cook R, Brandl CJ. 1997. *J. Biol. Chem.* 272:5571–78

174. Grant PA, Duggan L, Cote J, Roberts SM, Brownell JE, et al. 1997. *Genes Dev.* 11:1640–50

175. Ruiz-Garcia AB, Sendra R, Pamblanco M, Tordera V. 1997. *FEBS Lett.* 403:186–90

176. Winston F, Carlson M. 1992. *Trends Genet.* 8:387–91

177. Eisenmann DM, Arndt KM, Ribupero SL, Rooney JW, Winston F. 1992. *Genes Dev.* 6:1319–31

178. Eisenmann DM, Chapon C, Roberts SM, Dollard C, Winston F. 1994. *Genetics* 137:647–57

179. Roberts SM, Winston F. 1997. *Genetics* 147:451–65

180. Silverman N, Agapite J, Guarente L. 1994. *Proc. Natl. Acad. Sci. USA* 91: 11665–68

181. Barlev NA, Candau R, Wang L, Darpino

P, Silverman N, Berger SL. 1995. *J. Biol. Chem.* 270:9337–44

182. Madison JM, Winston F. 1997. *Mol. Cell. Biol.* 17:287–95

183. Neuwald AF, Landsman D. 1997. *Trends Biochem. Sci.* 22:154–55

184. Reifsnyder C, Lowell J, Clarke A, Pillus L. 1996. *Nat. Genet.* 14:42–49

185. Kamine J, Elangovan B, Subramanian T, Coleman D, Chinnadurai G. 1996. *Virology* 216:357–66

186. Borrow J, Stanton VPJ, Andresen JM, Becher R, Behm FG, et al. 1996. *Nat. Genet.* 14:33–41

187. Chrivia JC, Kwok RP, Lamb N, Hagiwara M, Montiminy MR, Goodman RH. 1993. *Nature* 365:855–59

188. Kwok RP, Lundblad JR, Chrivia JC, Richards JP, Bachinger HP, Brennan RG, et al. 1994. *Nature* 370:223–26

189. Arany Z, Sellers WR, Livingston DM, Eckner R. 1994. *Cell* 77:799–800

190. Arias J, Alberts AS, Brindle P, Claret FX, Smeal T, et al. 1994. *Nature* 370:226–29

191. Bannister AJ, Oehler T, Wilhelm D, Angel P, Kouzarides T. 1995. *Oncogene* 11: 2509–14

192. Dai P, Akimaru H, Tanaka Y, Hou DX, Yasukawa T, et al. 1996. *Genes Dev.* 10:528–40

193. Oelgeschlager M, Janknecht R, Krieg J, Schreek S, Luscher B. 1996. *EMBO J.* 15:2771–80

194. Yuan WC, Condorelli G, Caruso M, Felsani A, Giordano A. 1996. *J. Biol. Chem.* 271:9009–13

195. Chakravarti D, La Morte VJ, Nelson MC, Nakajima T, Schulman IG, et al. 1996. *Nature* 383:99–103

196. Kamei Y, Xu L, Heinzel T, Torchia J, Kurokawa R, et al. 1996. *Cell* 85:403–14

197. Eckner R, Ewen ME, Newsome D, Gerdes M, DeCaprio JA, et al. 1994. *Genes Dev.* 8:869–84

198. Arany Z, Newsome D, Oldread E, Livingston DM, Eckner R. 1995. *Nature* 374:81–84

199. Yang X-J, Ogryzko VV, Nishikawa J-I, Howard BH, Nakatani Y. 1996. *Nature* 382:319–24

200. Candau R, Moore PA, Wang L, Barlev N, Ying CY, et al. 1996. *Mol. Cell. Biol.* 16:593–602

201. Ogryzko VV, Schiltz RL, Russanova V, Howard BH, Nakatani Y. 1996. *Cell* 87: 953–59

202. Gu W, Roeder RG. 1997. *Cell* 90:595–606

203. Chen HW, Lin RJ, Schiltz RL, Chakravarti D, Nash A, et al. 1997. *Cell* 90:569–80

204. Fondell JD, Ge H, Roeder RG. 1996. *Proc. Natl. Acad. Sci. USA* 93:8329–33
205. Mizzen CA, Yang X-J, Kokubo T, Brownell JE, Bannister AJ, et al. 1996. *Cell* 87:1261–70
206. Dikstein R, Ruppert S, Tjian R. 1996. *Cell* 84:781–90
207. Boffa L, Vidali G, Mann R, Allfrey V. 1978. *J. Biol. Chem.* 253:3364–66
208. Georgieva EI, Lopez-Rodas G, Sendra R, Grobner P, Loidl P. 1991. *J. Biol. Chem.* 266:18751–60
209. Brosch G, Georgieva EI, Lopez-Rodas G, Lindner H, Loidl P. 1992. *J. Biol. Chem.* 267:20561–64
210. Lechner T, Lusser A, Brosch G, Eberharter A, Goralik-Schramel M, Loidl P. 1996. *Biochim. Biophys. Acta* 1296:181–88
211. Hay CW, Candido EP. 1983. *Biochemistry* 22:6175–80
212. Sanchez del Pino MM, Lopez-Rodas G, Sendra R, Tordera V. 1994. *Biochem. J.* 303:723–29
213. Carmen AA, Rundlett SE, Grunstein M. 1996. *J. Biol. Chem.* 271:15837–44
214. Kasten MM, Dorland S, Stillman DJ. 1997. *Mol. Cell. Biol.* 17:4852–58
215. Lusser A, Brosch G, Loidl A, Haas H, Loidl P. 1997. *Science* 277:88–91
216. Yang WM, Inouye C, Zeng YY, Bearss D, Seto E. 1996. *Proc. Natl. Acad. Sci. USA* 93:12845–50
217. Rundlett SE, Carmen AA, Kobayashi R, Bavykin S, Turner BM, Grunstein M. 1996. *Proc. Natl. Acad. Sci. USA* 93:14503–8
218. Bowdish KS, Mitchell AP. 1993. *Mol. Cell. Biol.* 13:2172–81
219. McKenzie EA, Kent NA, Dowell SJ, Moreno F, Bird LE, Mellor J. 1993. *Mol. Gen. Genet.* 240:374–86
220. Stillman DJ, Dorland S, Yu YX. 1994. *Genetics* 136:781–88
221. Vidal M, Strich R, Esposito RE, Gaber RF. 1991. *Mol. Cell. Biol.* 11:6306–16
222. Vidal M, Gaber RF. 1991. *Mol. Cell. Biol.* 11:6317–27
223. Kadosh D, Struhl K. 1997. *Cell* 89:365–71
224. Wang HM, Stillman DJ. 1993. *Mol. Cell. Biol.* 13:1805–14
225. Amati B, Dalton S, Brooks MW, Littlewood TD, Evan GI, Land H. 1992. *Nature* 359:423–26
226. Kretzner L, Blackwood EM, Eisenman RN. 1992. *Nature* 359:426–29
227. Chen JD, Evans RM. 1995. *Nature* 377:454–57
228. Horlein AJ, Naar AM, Heinzel T, Torchia J, Gloss B, et al. 1995. *Nature* 377:397–404
229. Ayer DE, Lawrence QA, Eisenman RN. 1995. *Cell* 80:767–76
230. Schreiber-Agus N, Chin L, Chen K, Torres R, Rao G, et al. 1995. *Cell* 80:777–86
231. Nagy L, Kao H-Y, Chakravarti D, Lin RJ, Hassig CA, et al. 1997. *Cell* 89:373–80
232. Alland L, Muhle R, Hou HJ, Potes J, Chin L, et al. 1997. *Nature* 387:49–55
233. Heinzel T, Lavinsky RM, Mullen T-M, Soderstrom M, Laherty CD, et al. 1997. *Nature* 387:43–48
234. Laherty CD, Yang W-M, Sun J-M, Davie JR, Seto E, Eisenman RN. 1997. *Cell* 89:349–56
235. Hassig CA, Fleischer TC, Billin AN, Schreiber SL, Ayer DE. 1997. *Cell* 89:341–47
236. Zhang Y, Iratni R, Erdjument-Bromage H, Tempst P, Reinberg D. 1997. *Cell* 89:357–64
237. Kingston RE, Bunker CA, Imbalzano AN. 1996. *Genes Dev.* 10:905–20
238. Kwon H, Imbalzano AN, Khavari PA, Kingston RE, Green MR. 1994. *Nature* 370:477–81
239. Peterson CL, Herskowitz I. 1992. *Cell* 68:573–83
240. Côté J, Quinn J, Workman JL, Peterson CL. 1994. *Science* 265:53–60
241. Tsukiyama T, Becker PB, Wu C. 1994. *Nature* 367:525–31
242. Tsukiyama T, Wu C. 1995. *Cell* 83:1011–20
243. Varga WP, Wilm M, Bonte E, Dumas K, Mann M, Becker PB. 1997. *Nature* 388:598–602
244. Ito T, Bulger M, Pazin MJ, Kobayashi R, Kadonaga JT. 1997. *Cell* 90:145–55
245. Cairns BR, Kim Y-J, Sayre MH, Laurent BC, Kornberg RD. 1994. *Proc. Natl. Acad. Sci. USA* 91:1950–54
246. Cairns BR, Lorch Y, Li Y, Zhang M, Lacomis L, et al. 1996. *Cell* 87:1249–60
247. Peterson CL, Dingwall A, Scott MP. 1994. *Proc. Natl. Acad. Sci. USA* 91:2905–8
248. Wang WD, Côté J, Xue Y, Zhou S, Khavari PA, et al. 1996. *EMBO J.* 15:5370–82
249. Wang WD, Xue YT, Zhou S, Kuo A, Cairns BR, Crabtree GR. 1996. *Genes Dev.* 10:2117–30
250. Muchardt C, Yaniv M. 1993. *EMBO J.* 12:4279–90
251. Khavari PA, Peterson CL, Tamkun JW, Crabtree GR. 1993. *Nature* 366:170–74
252. Chiba H, Muramatsu M, Nomoto A, Kato H. 1994. *Nucleic Acids Res.* 22:1815–20
253. Muchardt C, Reyes JC, Bourachot B, Leguoy E, Yaniv M. 1996. *EMBO J.* 15:3394–402
254. Laurent BC, Yang XL, Carlson M. 1992.

Mol. Cell. Biol. 12:1893–902

255. Neigeborn L, Carlson M. 1984. *Genetics* 108:845–58

256. Cao Y, Cairns BR, Kornberg RD, Laurent BC. 1997. *Mol. Cell. Biol.* 17:3323–34

257. Sumi IC, Ichinose H, Metzger D, Chambon P. 1997. *Mol. Cell. Biol.* 17:5976–86

258. Tsukiyama T, Daniel C, Tamkun J, Wu C. 1995. *Cell* 83:1021–26

259. Pazin MJ, Kadonaga JT. 1997. *Cell* 88:737–40

260. Imbalzano AN, Schnitzler GR, Kingston RE. 1996. *J. Biol. Chem.* 271:20726–33

261. Georgel PT, Tsukiyama T, Wu C. 1997. *EMBO J.* 16:4717–26

262. Owen-Hughes T, Utley RT, Cote J, Peterson CL, Workman JL. 1996. *Science* 273:513–16

263. Kruger W, Peterson CL, Sil A, Coburn C, Arents G, et al. 1995. *Genes Dev.* 9:2770–79

264. Wechser MA, Kladde MP, Alfieri JA, Peterson CL. 1997. *EMBO J.* 16:2086–95

265. Yoshinaga SK, Peterson CL, Herskowitz I, Yamamoto KR. 1992. *Science* 258:1598–604

266. Ostlund FA, Blomquist P, Kwon H, Wrange Ö. 1997. *Mol. Cell. Biol.* 17:895–905

267. Pina B, Bruggemeier U, Beato M. 1990. *Cell* 60:719–31

268. Perlmann T, Wrange Ö. 1991. *Mol. Cell. Biol.* 11:5259–65

269. Wong JM, Shi Y-B, Wolffe AP. 1995. *Genes Dev.* 9:2696–711

270. Wilson CJ, Chao DM, Imbalzano AN, Schnitzler GR, Kingston RE, Young RA. 1996. *Cell* 84:235–44

Annu. Rev. Biochem. 1998. 67:581–608

STRUCTURE AND FUNCTION IN GroEL-MEDIATED PROTEIN FOLDING

Paul B. Sigler,[1,2] Zhaohui Xu,[1,2] Hays S. Rye,[2,3] Steven G. Burston,[3] Wayne A. Fenton,[3] and Arthur L. Horwich[2,3]

[1]Department of Molecular Biophysics and Biochemistry, [2]Howard Hughes Medical Institute, and [3]Department of Genetics, School of Medicine, Yale University, New Haven, Connecticut 06510; e-mail: sigler@csb.yale.edu

KEY WORDS: ATP, chaperonin, GroES, Hsp60

ABSTRACT

Recent structural and biochemical investigations have come together to allow a better understanding of the mechanism of chaperonin (GroEL, Hsp60)–mediated protein folding, the final step in the accurate expression of genetic information. Major, asymmetric conformational changes in the GroEL double toroid accompany binding of ATP and the cochaperonin GroES. When a nonnative polypeptide, bound to one of the GroEL rings, is encapsulated by GroES to form a *cis* ternary complex, these changes drive the polypeptide into the sequestered cavity and initiate its folding. ATP hydrolysis in the *cis* ring primes release of the products, and ATP binding in the *trans* ring then disrupts the *cis* complex. This process allows the polypeptide to achieve its final native state, if folding was completed, or to recycle to another chaperonin molecule, if the folding process did not result in a form committed to the native state.

CONTENTS

PURPOSE

The main purpose of this review is to put forth an integrated structure/function analysis of chaperonin-assisted protein folding that coordinates recent high-resolution crystallographic results with functional studies in solution. To provide an appropriate context, however, we first review where molecular chaperones fit in the scheme of biological catalysis and the special problems presented by the cell's need for quick and accurate expression of genetic information, ultimately in the form of properly folded functional proteins and their assemblies.

THE ROLE OF MOLECULAR CHAPERONES IN THE CELL

Biological Catalysts

Biological systems are necessarily metastable. They are created, modulated, and destroyed according to a temporal plan that meets the survival needs of the cell, organism, and species. This metastability applies over a wide time and spatial frame, ranging from the femtosecond and Ångstrom scale of molecular dynamics to the millennia lifetimes and 200-foot cellulose scaffold of the giant sequoia. Clearly, no biological system is close to true equilibrium or it would be dead—thus systems survive by consuming free energy and regulating rates. For example, a human turns over 40 kg of ATP daily, and 90% of the information in a genome encodes biological catalysts. To properly coordinate life processes, a system must have the capability to control rates of reactions independently; that is, the catalysts must be specific and carry out their tasks precisely. For interesting philosophical reasons beyond the scope of this review, one can justify the fact that biological catalysts almost always speed up reactions. Herein lies an important trade-off: The more quickly something is done, the more difficult it is to do it accurately. Accuracy and specificity often are sacrificed in the name of speed and vice versa. This balance is especially important in optimizing the rate of information transfer from the gene to the tertiary and quaternary structures of functional nucleic acids and proteins.

Biological catalysts can be categorized by the nature of the reactions they catalyze. *Enzymes* make and break covalent bonds. They address rates that alter the primary structure, where the components of the free energy of the activation barriers are both enthalpic (disruption of the electronic structure) and entropic (restricting the orientation and trajectory of the reacting compounds). *Channels and small-molecule transporters* facilitate the passage of ions and polar molecules through restrictive membrane barriers, and like enzymes that catalyze thermodynamically unfavorable reactions, they often couple free energy input to the movement of a substrate against a gradient.

Molecular chaperones address a problem on a different scale, that of achieving the correct tertiary structure (and in an indirect way quaternary structure) of proteins. Here the notion of an activation barrier cannot be viewed simply as a high free energy point along a linear reaction coordinate. Rather, a free energy hypersurface must be envisioned, defined by the polypeptide itself and its ultimate cellular destination, such as the bacterial cytosol or periplasm, a eukaryotic secretory vesicle, or an organelle. This folding hypersurface is replete with false minima, some of which are deep enough to entrap the wayward nonnative polypeptide for a period that can be considered irreversible on the cell's time scale (see Reference 1, for example). Just as enzymes guide chemical reactions along reaction paths with minimal free energy barriers, molecular chaperones catalyze productive folding by escorting nonnative polypeptides across the folding free energy surface, avoiding and, if necessary, reversing states that lead to the truly dead-end pitfalls of aggregation and/or proteolysis. Finally there are *scaffolding or assembly proteins*, which facilitate the efficient formation of quaternary structures. These are encountered, for example, in the assembly of a virion, the DNA replication machinery, or a preinitiation transcription complex.

The Entropic Barrier to Expressing Genes Accurately

Consider the likelihood that a gene will be replicated, transcribed, and translated nearly error free. If the cell completely lost its mechanisms for ensuring fidelity in replicating or transcribing the 600 nucleotides that encode the expressed component of a 200–amino acid protein, then the polymerases, in principle, might synthesize any one of $4^{600}-1$ incorrect sequences. Moreover, translation of the polypeptide has a potential for 20^{200} errors. It would take about 10^{80} universes to accommodate such a collection of faulty proteins, assuming they would be completely folded and packed effectively. Even if the cell could tolerate polypeptide products with an average of one mistake per chain, the error rate in translation alone would have to be less than 10^{-4} at each step. If the process were slow and deliberate enough, such a low error rate might be achievable, but aminoacylation and encoded ribosomal synthesis of polypeptides proceeds at very high speed; therefore, the likelihood of "noise-free" information transfer is negligible. The same considerations apply to replicating and transcribing the gene. How does the cell accomplish this seemingly impossible task? The answer is analogous to the process by which the writers of this review and the editors reduce errors in this chapter to a tolerable limit: They proofread and edit. Thus, the entropic activation barrier to prompt and accurate genetic expression at the level of enzymatic polymerization is overcome by the lavish use of nucleoside triphosphate to proofread and edit the polymer sequences. The same is true in the final steps of protein folding and assembly: Molecular chaperones, especially chaperonins, consume ATP, sometimes lavishly, to correct the inevitable and potentially irreversible mistakes in folding.

Categories of Molecular Chaperones

Studies of the past decade have focused intensively on a number of different types of conformational "editors"—families of molecular chaperones. Many of these were first identified by their increased expression during heat shock, where the workload of editing goes far beyond the "noise" of normal growth conditions. Each chaperone family appears to recognize specific nonnative protein conformations and acts on them in a characteristic way. However, all the families appear to share the ability to recognize hydrophobic surfaces exposed in nonnative proteins, surfaces that ultimately become buried in the interior of native proteins. The Hsp70 class appears to prefer hydrophobic regions in extended polypeptide chains (2–4), for example, those transiting cellular membranes (5). Through the action of ATP and cochaperones, polypeptides are released from the Hsp70 proteins, retaining their extended state in preparation for further steps of biogenesis. The Hsp20 class chaperones appear to act as small globular collectives like sponges, binding a multitude of nonnative species at their outer surface during heat shock (6, 7). On return to normal temperature, the substrate proteins are transferred to other chaperones and return to native form. The Hsp90 class chaperones appear to generally function in large multimeric complexes, recognizing a host of important signal transduction proteins in forms that may be near native and that in many cases await the appearance/binding of a ligand, such as a steroid hormone, for final conversion to an active conformation (8, 9).

But perhaps the most interesting of these is the chaperonin class (comprising the Hsp60/GroEL and TF55/CCT families) of ring-shaped complexes that recognize exposed hydrophobic surfaces of a wide range of globular nonnative conformations and bind them in the central cavity. Here, binding confers stability upon exposed, aggregation-prone, hydrophobic surfaces, preventing irreversible aggregation. Multivalent binding by the high density of neighboring hydrophobic sites lining this machine's channel may also serve to untangle misfolded structures. Most dramatic, however, is that when a chaperonin binds ATP and, in some cases, a cochaperonin, it is converted to an active device that may further unfold a polypeptide and then releases it into a new, shielded environment—encapsulated, expanded in volume, and now hydrophilic—favorable to folding to the native state. This type of assistance probably represents an ultimate level of general conformational editing to be found in the cell. We describe below the structural and functional workings of this machine, focusing largely on the *Escherichia coli* chaperonin, GroEL (10–12).

CHAPERONINS

Chaperonin Architecture

From the outset, all chaperonins (cpn60), irrespective of their cellular or subcellular origin, were seen to be double toroids in the electron microscope (13, 14). The need for this architecture is now better understood and is discussed more fully below. The bacterial and organellar chaperonins are assisted by single-ring cochaperonins (cpn10) (15, 16). The *E. coli* cochaperonin, GroES, is bound as a cap at one end of GroEL to form a bulletlike structure (17–19) or, sometimes, at both ends to form a football-like structure (20, 21). As shown in Figure 1 (opposite p. 586), the asymmetric bullets represent well-defined, biochemically accessible states in the folding cycle. The football structures probably are transient states whose presence can be enriched by nonnatural magnesium ion analogs such as Mn^{2+} or other conditions (22, 23) but which are not obligatory participants in productive folding reactions (22). Whether they constitute an intermediate form that occurs in vivo remains unclear.

Early electron microscopy (EM) studies of both unlabeled (18) and gold-tagged (24) polypeptides showed nonnative polypeptide bound in the central channel of the double toroid. This finding has been confirmed by single-particle correlation reconstructions of cryoelectron micrographs (25) and low-angle neutron-scattering curves (26). The latter suggest that the bound peptide distributes itself in the shape of a champagne cork adhering to and protruding from the end of the central cavity.

The symmetry of the chaperonin assembly depends on the number and uniqueness of the subunits comprising the rings. Bacterial (GroEL) and mitochondrial (Hsp60) chaperonins are composed of only one type of subunit. Hence, the two rings can be considered identical and, extrapolating from the crystallographic studies of GroEL (and in agreement with earlier EM studies), they are all stacked back to back with twofold rotational symmetry. The number of subunits in each ring varies. Those from eubacteria, mitochondria, and chloroplasts contain seven. Archaeal members of the TF55/CCT family have eight (27) or nine (28); the eight-membered archaeal thermosome contains two types of subunits that alternate within a ring to give fourfold rotational symmetry (27). Finally, the eukaryotic cytosolic chaperonin rings contain eight different gene products, which, cross-linking studies have indicated, are arranged in an explicit permutation (29). Unlike the bacterial chaperonins and their counterparts in endosymbiotic organelles, archaeal and eukaryal cytosolic chaperonins appear to function without cochaperonins (30–32). A possible structural basis for this is described below.

GroEL and GroES Structures

In 1994, the first high-resolution structure of GroEL was determined to 2.8 Å (33). The structure is a porous, thick-walled cylinder that is slightly taller than it is wide and contains a substantial central cavity or channel (Figure 1). As indicated by the electron microscope, it is composed of two rings of seven subunits arranged with nearly exact sevenfold rotational symmetry. The rings are arranged back to back, contacting one another through an extensive, yet remarkably flat, equatorial interface that contains seven molecular dyads (so exact that one of them coincides with a crystallographic dyad).

The GroEL subunit (547 amino acids) folds into three distinctive domains (Figure 2):

1. A well-ordered, highly α-helical, "equatorial" domain that forms a solid foundation around the waist of the assembly. In addition to providing most of the intersubunit contacts as described above, the equatorial domain contributes most of the residues that constitute the ATPase site.

3. An "apical" domain that surrounds the opening at the ends of the central channel. The apical domain is considerably less well ordered and shows local flexibility within the domain as well as en bloc movements around a hinge that connects it to the intermediate domain.

3. A small, slender "intermediate" domain at the periphery of the cylinder that links the equatorial domain to the apical domain.

The crystal structures of isolated GroES (a heptamer of 10 kDa subunits) at 2.8 Å (34) and cpn10 from *Mycobacterium leprae* at 3.5 Å (35) show a sevenfold rotationally symmetric, dome-shaped architecture, about 75 Å in diameter and 30 Å high. Each of the seven subunits has a core β-barrel structure (Figure 2C) with two β-hairpin loops. One arches upward and inward from the top aspect, collectively forming the top of the dome. The other, at the lower lateral aspect, contains a structure that is disordered in all but one subunit, where

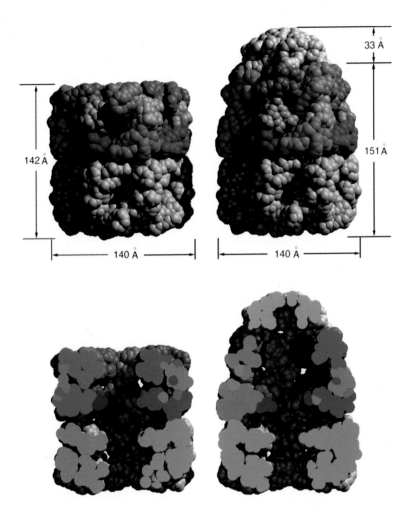

Figure 1   Overall architecture and dimensions of GroEL and GroEL-GroES-(ADP)$_7$. van der Waals space-filling models (6Å spheres around Cα) of GroEL (*left*) and GroEL-GroES-(ADP)$_7$ (*right*). *Upper panels* are outside views, showing outer dimensions; *lower panels* show the insides of the assemblies and were generated by slicing off the front half with a vertical plane that contains the cylindrical axis. Various colors are used to distinguish the subunits of GroEL in the upper ring. The domains are indicated by shading: equatorial, *dark hue*; apical, *medium hue*; intermediate, *light hue*. The lower GroEL ring is *uniformly yellow*. GroES is *uniformly gray*.

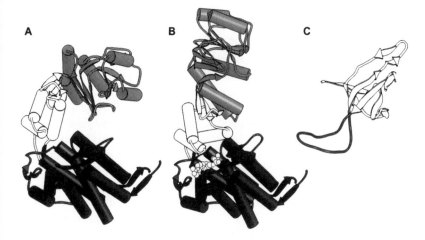

Figure 2 The subunit structures of GroEL and GroES. The subunit structure of GroEL as seen from within (*A*) an unliganded GroEL ring, as well as the *trans* ring of a GroEL-GroES-(ADP)$_7$ complex, and (*B*) the *cis* ring of a GroEL-GroES-(ADP)$_7$ complex. The *shading* corresponds to the three GroEL domains: equatorial, *dark gray*; intermediate, *white*; and apical, *medium gray*. Note the nucleotide (*white*) in the equatorial binding site in (*B*). (*C*) The subunit structure of GroES as seen in the GroEL-GroES-(ADP)$_7$ complex in the orientation required for interaction with the GroEL subunit in (*A*). The mobile loop is *dark gray*. The en bloc conformational changes relating the structures in (*A*) and (*B*) are shown diagrammatically in Figure 6.

it is stabilized by crystal contacts. Earlier proton nuclear magnetic resonance (NMR) studies indicated that this loop (Glu16 to Ala32) was mobile (hence the term mobile loop) but became better ordered on interaction with GroEL (36). These results have been confirmed in the GroEL-GroES-(ADP)$_7$ complex (see below).

The Chaperonin Reaction Cycle

As noted above, the function of the GroEL-GroES chaperonin system as a biological catalyst is to facilitate the folding of proteins within the cellular environment—an enviroment that a variety of critical proteins, or at least some of their folding intermediates, find hostile during their journey to the native state. At its simplest, the chaperonin reaction consists of the cyclic binding and release of target polypeptides. When one considers the potentially tangled and aggregated confusion in which folding polypeptides might find themselves, it seems remarkable that such a simple mechanism can be so effective. Yet GroEL-GroES can routinely rescue greater than 80% of a denatured protein population that would generate no more than a few percent of native molecules without chaperonin (37). As the devil is always in the details, the secrets behind the success of this deceptively simple mechanism lie in the remarkable structure of the GroEL-GroES protein complex, how it changes during the ATP-consuming reaction cycle, and exactly what effects the conformational states of the chaperonin have on the energetics of target polypeptides. Essentially, GroEL modulates its affinity for folding intermediates through the binding and hydrolysis of ATP. The highly coordinated act of binding and releasing substrate proteins then becomes sufficient to drive an otherwise dead-end folding reaction all the way to the native state. This reaction cycle can be operationally divided into four phases (Figure 3): (*a*) binding of polypeptide by GroEL (not shown); (*b*) release of polypeptide into the central chamber and initiation of folding, accomplished by binding GroES and ATP to form the high-energy *cis*-active state (Figure 3*A*); (*c*) decay of the high-energy *cis* state by hydrolysis of ATP, priming the *cis* assembly for the release of bound peptide, GroES, and ADP (Figure 3*B*); and (*d*) binding of ATP to the *trans* ring, providing the trigger to discharge GroES and entrapped polypeptide from the opposite side (Figure 3*C*).

Phase I: Polypeptide Binding

The central channel of GroEL functions as two separate cavities, one in each ring, that are separated from each other by the confluence of the crystallographically disordered 24–amino acid C-terminal segments of the seven subunits. In electron micrographs and by small-angle neutron scattering, these segments appear to coalesce and block the central channel at the level of the equatorial domain (26, 38). Because polypeptide substrates cannot escape through the equatorial segments of single-ring GroEL mutants that are otherwise competent in refolding reactions (39, 40), the pair of cavities in double-ring GroEL do not appear able to exchange substrate across the equatorial plane. The total volume of each cavity is measured in the crystal structure to be ~85,000 Å^3, just large enough for a native protein with a molecular size of 70 kDa, assuming a perfect fit. The size of a more loosely packed, nonnative polypeptide that could fit completely inside the cavity would be much smaller. Nevertheless, because the channel is open at this stage of binding, polypeptides can protrude slightly

---→

Figure 3 Model of the pathway of chaperonin-mediated protein folding. Phase I is not shown. *Panel A* represents phase II; *panel B*, phase III; *panel C*, phase IV; the *final panel* reflects current uncertainty about the details of the pathway by which the complex relaxes and recycles into a new peptide-binding state. The orientation of the intermediate and apical domains of the bottom ATP-bound ring simply reflects an ATP-driven change; neither the character nor the extent of this change has been established. See the text for details. GroES is *shaded gray*; folding polypeptide is the *cross-hatched circle*; misfolded or folded (or committed) polypeptide is shown as the *open circles*. The domains of GroEL are indicated by *rectangles* (equatorial), *small ovals* (intermediate), and *large ovals* (apical).

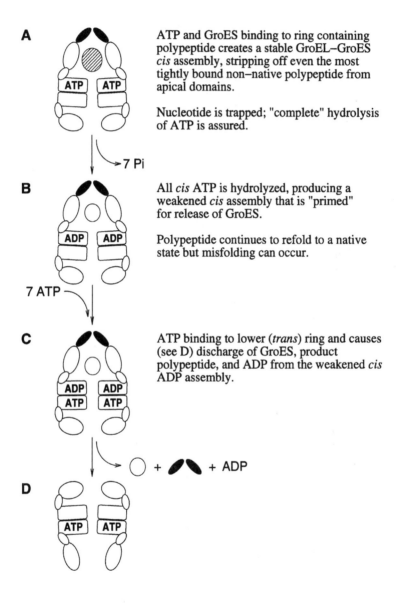

A — ATP and GroES binding to ring containing polypeptide creates a stable GroEL–GroES *cis* assembly, stripping off even the most tightly bound non–native polypeptide from apical domains.

Nucleotide is trapped; "complete" hydrolysis of ATP is assured.

→ 7 Pi

B — All *cis* ATP is hydrolyzed, producing a weakened *cis* assembly that is "primed" for release of GroES.

Polypeptide continues to refold to a native state but misfolding can occur.

7 ATP

C — ATP binding to lower (*trans*) ring and causes (see D) discharge of GroES, product polypeptide, and ADP from the weakened *cis* ADP assembly.

+ ADP

D

from the cavity, as cryo-EM and small-angle neutron-scattering studies have shown.

Nine residues (located on helices H and I and a loop between strands 6 and 7) of the apical domains have been identified by mutagenesis as important for polypeptide binding (41). These residues, eight of which have nonpolar side chains (the ninth is serine), face the central cavity in the unliganded GroEL structure (33, 41), closely matching the position of bound polypeptides seen in EM images of GroEL-substrate binary complexes. Thus, in the polypeptide acceptor state, GroEL presents a ring of hydrophobic binding surface along the inner edge of its apical cavities, poised to interact with the excess hydrophobic surface typically presented by folding intermediates. This strong hydrophobic contribution to the binding of substrate polypeptides by GroEL has been examined and confirmed with a number of substrates (42, 43), although polar and ionic contributions may also play a role (44, 45).

Despite these general characteristics of substrate interaction with GroEL, the exact stereochemistry of substrate binding has not been well characterized. Because GroEL evolved to interact with a wide variety of folding intermediates encompassing a large diversity of sequence information, its promiscuity would seem to preclude a specific and universal binding interaction. For example, both α-helical and extended secondary structures have been observed associated with chaperonin. In NMR studies, a rhodanese peptide formed an α-helical structure when bound to GroEL (46). On the other hand, the recent crystal structure of an isolated GroEL apical domain containing a 17-residue N-terminal tag showed a well-resolved interaction between this N-terminal segment, in an extended conformation, and the apical binding surface of a neighbor in the crystal lattice (47). Seven of the nine mutagenically implicated GroEL residues were found in contact with the peptide segment, as were several other hydrophobic residues, most of which are located on helices H and I. The binding mode of small isolated peptides, especially those presented by a neighboring molecule of a crystal lattice, may be different from those of nonnative substrate proteins because the binding of the latter to a ring of seven apical domains is subject to restraints that do not apply to small peptides. Thus, it is difficult to extrapolate from current structural information to a general model of GroEL-substrate interaction, if indeed one exists.

As pointed out above, the vast conformational space available to a folding polypeptide can lead to significant local energetic minima, outside the global (or near-global) minimum for which the protein is searching. These kinetic traps can represent off-pathway misarrangements of the individual chain (misfolding) or potentially irreversible nonproductive interactions between chains (aggregation). Because GroEL must interact with individual folding intermediates and productively promote movement along a folding trajectory toward the native state, one or more aspects of the interaction must be capable of driving the substrate polypeptide out of these kinetic traps. One way for GroEL to accomplish this goal is through unfolding, either globally or locally, the polypeptide that it binds (48). The potentially cooperative nature of binding a folding intermediate to the multiple, neighboring apical sites of the GroEL toroid make this an especially attractive mechanism. Once bound, the polypeptide may be further unfolded by the subsequent elevating and twisting movements of the apical domains that occur on binding of nucleotide and GroES and ultimately lead to its release into the central cavity (49). Thus, the energy required for unfolding could come from either or both of two sources, namely the energy produced by the interaction between the polypeptide and the hydrophobic channel face and the energy generated by binding ATP and GroES to form the *cis* folding-active complex.

The coupling of substrate binding to substrate unfolding on GroEL has been examined in a number of studies, either by characterizing the point at which GroEL interacts with a polypeptide along its folding pathway or by examining rates of hydrogen-deuterium exchange of bound substrates. In general, it appears that in the absence of GroES and ATP, GroEL interacts rapidly (10^6–10^7 M^{-1} s^{-1}) with relatively early, collapsed folding intermediates (48; MS Goldberg, unpublished results). Moreover, GroEL is capable of shifting the natural equilibrium between the native and unfolded states of several proteins toward the unfolded state by stably binding only the unfolded protein (50–52). These results do not in themselves constitute a demonstration that binding causes unfolding, however. Evidence for direct participation of GroEL in polypeptide unfolding has been more mixed. Studies with a small protein, barnase, seem to indicate that global unfolding of this polypeptide occurs concomitant with binding (48). Similar, but less certain, conclusions were reached by an examination of cyclophilin bound to GroEL (53). On the other hand, deuterium exchange experiments with three-disulfide α-lactalbumin intermediates bound to GroEL showed a low degree of overall protection, consistent with the retention of some secondary structure (54). Moreover, a pair of recent studies with dihydrofolate reductase (DHFR), using NMR analysis of hydrogen-deuterium exchange (55) and mass spectrometry (56), could find no evidence for large-scale global unfolding of the stably bound DHFR intermediates. Indeed, the protection afforded against hydrogen-deuterium exchange was consistent with a highly structured and native-like, albeit unstable, form of DHFR bound to GroEL.

Active unfolding of a substrate protein by GroEL needs to be invoked only if the kinetic traps are operating at the level of misfolding within individual polypeptide chains. The alternative off-pathway reaction that can prevent the acquisition of the native state is aggregation, which may be dominant in some instances. Strong binding by GroEL of aggregation-prone folding intermediates would be sufficient to block this nonproductive reaction. Furthermore, provided there was timely and efficient release of the bound polypeptide, the overall result of this mechanism would be to catalyze the production of the native state without directly affecting the unfolding of the polypeptide itself. Ranson and colleagues have shown that the folding of mitochondrial malate dehydrogenase by GroEL involves such a mechanism (57). Here, even at stoichiometries as low as 1:20 relative to substrate, GroEL can catalyze the acquisition of the native state by increasing the proportion of monomeric folding intermediates available to the normal folding pathway. It is important to point out, however, that it is not known to what extent aggregation or intramolecular misfolding may contribute to the off-pathway fate of any substrate of GroEL, particularly in the context of the intact cell. Indeed, it seems likely that both mechanisms may be operative at different points on the folding pathway, and it may be a critical feature of chaperonin-mediated folding that both can be dealt with by the same machinery.

Phase II: Nucleotide and GroES Binding

Since the first in vitro GroEL-assisted refolding experiments were performed, it has been noted that whereas GroEL alone inhibits refolding, GroEL in the presence of K^+ ions, Mg-ATP, and GroES promotes efficient folding to the native state (37, 58–61). This led investigators to propose the existence of at least two distinct conformations of the complex: one that binds unfolded polypeptides tightly and ATP weakly and another in which the binding properties are reversed (60, 62), suggesting that nucleotide-modulated conformational changes of the chaperonin are inherent to the protein folding cycle. Initial equilibrium measurements of ATP binding and hydrolysis by GroEL revealed a high degree of cooperativity, which was enhanced further upon the addition of GroES (63–66). Moreover, GroES binding depends on the binding of nucleotide. The order of these events and their rates were studied by following changes in the fluorescence of pyrene-labeled GroEL. Weak binding of ATP to GroEL triggers a rapid conformational change (half-time $\sim$40 ms) (65), which precedes the fast association of GroES ($> 4 \times 10^7 \, M^{-1} \, s^{-1}$) (67). Conformational changes have been observed directly upon the addition of ATP, using cryo-EM and three-dimensional single-particle reconstruction. They have been interpreted, assuming the en bloc domain movements demonstrated by the crystal structure, to show that the apical domains of one ring open out by about 5–10° relative to the equatorial axis and also twist, causing a slight elongation of the GroEL cylinder (25, 68). The relative effect of the various nucleotides upon the apical domain movements followed the order ATP > AMP-PNP > ADP, which is consistent with the molecular mechanism outlined below.

The allosteric behavior of GroEL with respect to its ATPase function does not fit well into a simple model of cooperativity. Rather, it has been quantitatively described in a nested model that combines positive allostery within each ring with a negative effect between the rings (69). This model provides a context within which to appreciate the molecular details of the ligand-induced conformational changes that underlie the allosteric effects, as revealed by the crystal structures (see below), and to rationalize the effects of various mutational changes that independently disrupt one or the other allosteric interaction. However, the remarkable structure/function relationships of the folding cycle are difficult to completely capture with the compact formalisms originally developed to describe the behavior of hemoglobin and the regulation of oligomeric enzymes. For example, GroEL-bound ATP is hydrolyzed on only one ring at a time, even in the presence of excess ATP (67), whereas the binding of seven ADP molecules to one ring of GroEL is virtually noncooperative and only marginally asymmetric between rings. In the presence of GroES, however, ADP binding is complete and essentially irreversible within one ring and nonexistent in the opposite one. It would appear that there is inherent positive and negative cooperativity within the GroEL tetradecamer that is driven by GroES into the discrete intermediate states that characterize the stages of the folding cycle.

The structure of nucleotide in its binding site at the top of the equatorial domain, facing the central cavity, was first observed in the crystal structure of a variant GroEL (R13G/A126V) fully complexed with 14 ATPγS molecules, solved to a resolution of 2.4 Å (70) (Figure 4A). The site contains residues 87–91 (DGTTT), which interact with the β- and γ-phosphates of ATP and lie in a loop region (between helices C and D) that is highly conserved among chaperonins (71). Two metal ions are present in the nucleotide binding site. One is a magnesium ion, which chelates an oxygen from each of the three nucleotide phosphates, the carboxylate of Asp87, and two water molecules. Surprisingly, the overall architecture of the ATPγS complex was largely superimposable over the X-ray structure of unliganded GroEL (33, 72), in contrast to the significant domain movements seen in the cryo-EM image reconstructions. It was suggested that the discrepancy between the cryo-EM and crystal structures might be due in part to a decreased negative cooperativity observed for the variant protein (73), allowing the binding of ATPγS to all 14 subunits and creating a highly symmetrical molecule that is able to form a well-defined crystal lattice. This appears not to be the case, however, because further crystallographic studies with wild-type GroEL and ATPγS also reveal an isomorphous structure (D Boisvert, unpublished observations), leaving open the possibility that lattice forces have prevailed over movements of a flexible apical domain. In contrast to the crystalline specimens, which demand a uniform state, the noncrystalline complexes in the cryo-EM studies are not subject to such constraints on

(A)

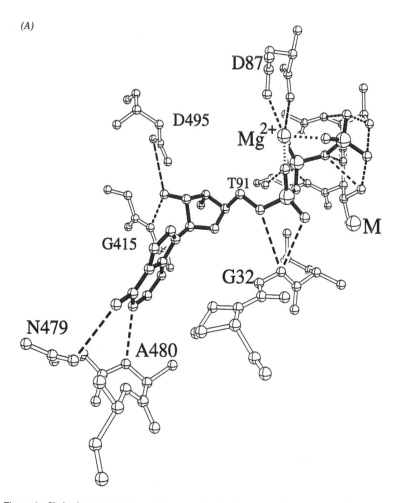

Figure 4 Skeletal representations of the nucleotide binding sites. The view is similar to that in Figure 2. The protein has *unshaded* bonds, and the nucleotide has *gray* bonds. The metal ions are *large spheres*. (A) ATPγS bound to GroEL in the GroEL-(ATPγS)$_{14}$ structure. (B) ADP bound to GroEL in the GroEL-GroES-(ADP)$_7$ structure. Two residues from the M helix of the intermediate domain have *black* bonds.

symmetry. Nonetheless, two significant local substructure shifts were observed in the ATPγS crystal structure: a large axial translation of helix C and a movement of a stem loop (Lys34 to Asp52), whose antiparallel stem forms an essential parallel β-contact with the neighboring subunit within the ring through a β-strand near the C terminus (strand 19) (Figure 5A). The importance of these two substructure movements is now evident in light of the recent GroEL-GroES-(ADP)$_7$ complex structure (74).

(B)

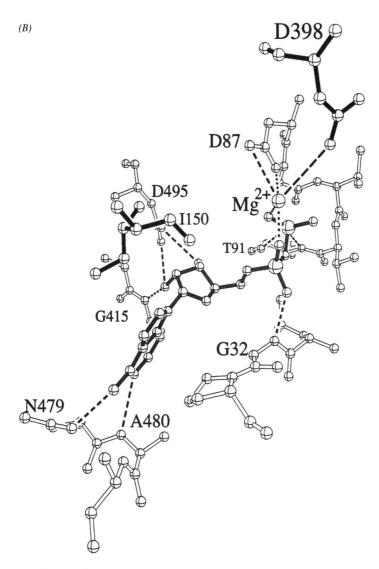

Figure 4   (Continued)

(A)

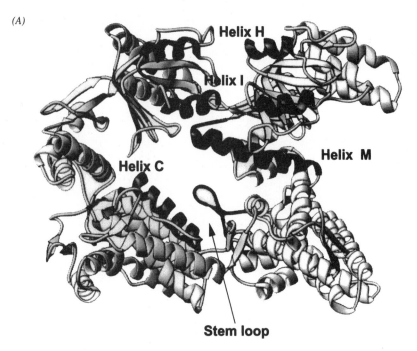

Figure 5   Substructures involved in stabilizing the *cis* assembly. Ribbon drawings for two adjacent subunits in *trans* (A) and *cis* (B) GroEL rings of the GroEL-GroES-(ADP)₇ complex, viewed from the inside of the ring. Both are oriented similarly with respect to the equatorial interface (at the bottom). The substructures that form the new interface between equatorial and intermediate domains (helix C, helix M, and the stem loop), as well as those that form the GroEL-GroES interface (helix H and helix I), are *dark gray*. Helix C and the stem loop are shifted by nucleotide to provide new stabilizing contacts for the reoriented helix M, as shown in *panel B*.

The Asymmetrical GroEL-GroES-ADP Complex Structure

OVERALL STRUCTURE AND DOMAIN MOVEMENTS   The overall structure of the GroEL-GroES-(ADP)₇ asymmetrical complex is the expected bullet-shaped image (Figure 1, opposite p. 586), resulting from the smoothness of the union between GroEL and GroES (Figure 5B). GroES caps one end of GroEL, which is elongated and tapered toward the GroEL-GroES interface. The change in the shape of GroEL is due mostly to the change in the *cis* GroEL ring. Large en bloc movements of the apical and intermediate domains in the *cis* ring widen and elongate the *cis* cavity. The GroES ring, assembled as in its stand-alone structure, caps the apical surface of the *cis* ring, anchoring the elevated orientation of the apical domains and closing off the end of the cavity. The net result is a dome-shaped chamber that has the elevated apical domains as its

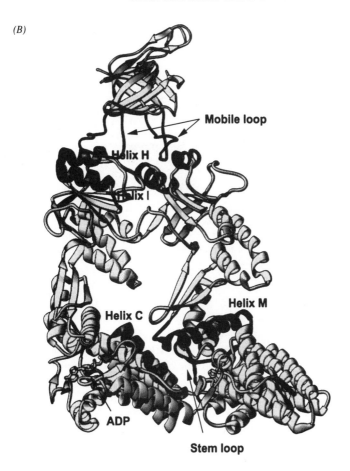

(B)

Figure 5   (Continued)

walls and the GroES cap as its roof. The GroEL and GroES rings share one nearly exact sevenfold rotational axis (Figure 1, opposite p. 586). In contrast to the dramatic changes in the *cis* ring, the *trans* ring (the empty ring) closely resembles that of unliganded GroEL.

The dramatic reshaping of the *cis* ring is due to rearrangements involving both intermediate and apical domains (Figures 4, 5, and 6). First, the intermediate domain swings down toward the equatorial domain and the central channel, pivoting approximately 25° around Pro137 and Gly410. The movement closes the occupied nucleotide binding site, located on the top inner surface of the equatorial domain, and generates numerous new interactions with the bound

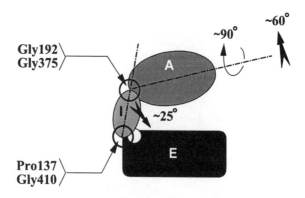

Figure 6 A schematic drawing showing direction and magnitude of the domain movement within the *cis* ring. The *shading* corresponds to the three GroEL domains: equatorial, *dark gray*; intermediate, *white*; and apical, *medium gray*. The *small disc* on the top of the equatorial domain represents the nucleotide binding site.

nucleotide and the equatorial domain. Second, the apical domain swings up 60° relative to the equator and twists around the long axis of the domain by about 90°, forming new interfaces with neighboring apical domains and leading to an interaction with the mobile loop of GroES. Both intermediate and apical domain movements are largely en bloc. The *cis* equatorial domains also show an en bloc movement that is small in magnitude compared to those of the intermediate and apical domain but every bit as important. The equatorial domains of the *cis* assembly tilt inward toward the cylindrical axis by 4° (Figure 9). Because the strong interface between the rings is conserved, there is a complementary outward tilt of the equatorial domains in the *trans* ring. This imposes a further asymmetry between the two rings and underlies the negative allostery that relates them (see below).

NUCLEOTIDE SITE   The asymmetry of the GroEL-GroES-(ADP)$_7$ complex is most obvious in its ligand binding: GroES binds to one end of GroEL, and the seven *cis* nucleotide binding sites are fully occupied, while the seven *trans* sites are completely empty. Except for the absence of a second metal ion–mediated interaction with the α-phosphate of ATPγS, the specific interactions of ADP with the equatorial domain largely mirror those in the GroEL-ATPγS structure (70) (Figure 4*B*). In the GroEL-ATPγS structure (as well as in unliganded GroEL), the nucleotide binding site is largely open (compare panels *A* and *B* in both Figure 2 and 5), so the nucleotide can enter and exit without much steric hindrance. Upon GroES binding, however, residues of the helices F and M of the reoriented intermediate domain clamp onto the nucleotide, the Mg^{2+} cofactor, and residues of the equatorial domain, thereby closing the nucleotide binding site. Thus, nucleotide is trapped in the *cis* ring and will remain there until the *cis* complex is disassembled, even at nucleotide concentrations low enough to completely empty the *trans* ring binding sites.

Some of the interactions clarify previously unexplained observations. For example, Ile150 of helix F forms a van der Waals interaction with the sugar moiety of the ADP, which is probably why mutation of Ile150 is lethal (41). Helix M contributes the carboxylate oxygen of Asp398 to the Mg^{2+} ion coordination cage, explaining why the D398A mutant GroEL retains only 2% of the wild-type ATPase activity, even though its affinity for ATP is unaffected (40). Furthermore, if Asp398 is prevented from assuming this new active-site position through restriction of domain movements by covalent cross-linking, GroEL is unable to hydrolyze bound ATP (75). These observations reaffirm the conclusion drawn from the fully liganded ATPγS-bound structure; namely, that binding of ATP to GroEL does not require shifts of the intermediate or apical domains but that subsequent hydrolysis of ATP does.

GroES BINDING   The binding of ATP or ADP supports the binding of GroES (16), albeit with different rates of GroES association to the different GroEL-nucleotide binary complexes. For example, GroES binds rapidly ($> 4 \times 10^7$ $M^{-1} s^{-1)}$) to GroEL-(ATP)$_7$ after the ATP-induced conformational change (67), whereas GroES associates more slowly ($1 \times 10^5 M^{-1} s^{-1)}$) with the GroEL-(ADP)$_7$ state (65). The stereochemical bases for the interdependence of GroES and nucleotide binding are threefold. First, the equatorial domains of the *cis* ring of the GroEL-GroES-(ADP)$_7$ complex show the same nucleotide-induced substructural shifts seen in the fully saturated ATPγS structure: an axial shift of helix C and displacement of the Lys34-Asp52 stem loop. These shifts provide stabilizing contacts for the radically reoriented intermediate domain (Figure 5). Second, the Mg^{2+} complex of the bound nucleotide provides additional direct stabilizing contacts for the reoriented intermediate domain. These interactions are the same as those that trap nucleotide in the *cis* assembly. Third, the nucleotide-shifted intermediate domain has now repositioned the hinge connecting it to the apical domain, so that intermediate domain/apical domain contacts between subunits of the unliganded structure are disrupted and the ensuing stabilizing interactions with GroES are sterically feasible. Thus, GroES binding is enabled by the structural transition initiated by nucleotide binding.

The extent of the ligand-induced conformational changes follows the order ATP > AMPPNP > ADP, implying highly stereo-explicit interactions with the β-γ phosphoanhydride of the triphosphate moiety. The same order of functional effectiveness is also observed in the apparent rates and affinities of GroES binding to the different GroEL-nucleotide complexes and in the increased stability of the GroEL-GroES-(ATP)$_7$ complex over its ADP counterpart (40), suggesting a similar dependence on the terminal diphosphate. However, the specific interactions that account for this ranking cannot be derived with confidence from the current structures.

The movement of Ile150 and Asp398 into the ATPase active site is locked into place by the binding of GroES, thereby sequestering the nucleotide and precluding the loss of ATP (or the hydrolysis product ADP) or its exchange with free ligand. Because of its matching sevenfold rotational symmetry, GroES imposes the same restriction on all subunits of the *cis* ring simultaneously. Therefore, only in the presence of GroES are the seven bound ATP nucleotides of the *cis* ring committed to hydrolysis as a unit. Thus, GroES markedly increases the positive cooperativity of ATP hydrolysis within one ring, producing the quantized behavior of the ATPase activity of GroEL, as described by Viitanen and coworkers (76) and quantitated by Burston and colleagues (67).

In contrast to the major structural changes occurring in GroEL, the GroES heptamer ring in the GroEL-GroES complex is similar to that in the stand-alone structure (34) except for the mobile-loop residues, which now become structured. As expected, the mobile loop forms the interface with the elevated and twisted H and I helices of the GroEL apical domain through small aliphatic side chains, including Ile25, Val26, and Leu27 of GroES.

Several observations indicate that certain aspects of the role of the cochaperonin in the folding reaction remain incompletely defined. First, archaeal and eukaryal cytosolic chaperonins (TF55 and CCT, respectively) do not have a cochaperonin. Instead the thermosome appears to have an extended loop in its apical domain that serves to cap the *cis* folding chamber (76a). Perhaps the need to assist the folding of multidomain substrate proteins requires a less restricted apical hole. Second, mutations in the hinge regions of GroEL (V174F, V190I, and G375S) suppress the disruption of GroES binding caused by certain mutations in the GroES mobile loop (77). Current crystal structures do not offer a straightforward explanation for this observation. Third, GroES binding appears to limit the size of polypeptides whose folding can be assisted by the GroEL-GroES complex. For example, although T4 bacteriophage utilizes the host chaperonin, it encodes its own version of cochaperonin, known as gp31. The recent crystal structure of gp31 (78) suggests that it can form a larger folding cavity in the *cis* complex (Anfinsen cage) to accommodate the >50 kDa phage head protein (gp23), which may not fit comfortably under GroES.

Phase III: Polypeptide Release and Folding

The crystal structure of the GroEL-GroES complex suggests how the substrate polypeptide can be stripped from its binding sites on the channel walls and released into the cavity of the *cis* assembly (Figure 7, opposite pp. 600 and 601). Helices H and I of the apical domain, bearing peptide-binding residues Leu234, Leu237, Val259, Leu263, and Val264, move to the very top of the GroEL

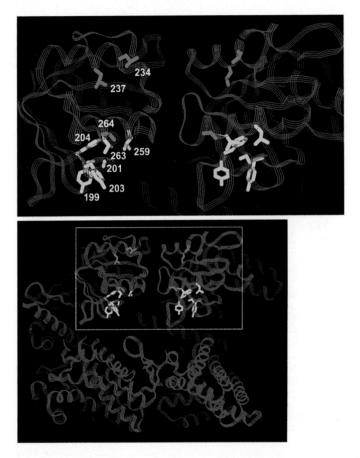

Figure 7A The change in the central cavity. Coiled-line ribbon drawing of two neighboring subunits of the *trans* ring viewed from the central cavity, oriented with the equatorial plane at the bottom; a magnified view of the polypeptide-interacting region (*rectangular area*) is shown at the top of the panel. Skeletal side chains denote residues involved in polypeptide binding, derived from mutagenesis studies; these residues, with the exception of S201, have hydrophobic side chains.

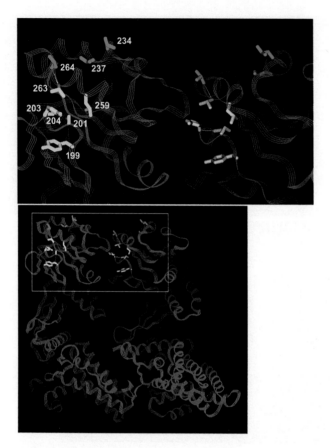

Figure 7B Two neighboring subunits of the *cis* ring, viewed in the same orientation and highlighted as in Figure 7A. The polypeptide-binding residues have moved away from the cavity surface, either to form the GroES interface [L234, L237, V264 (*green*)] or to contribute to the new interfaces between the apical domains of the *cis* GroEL ring [Y199, S201, Y203, F204, L259, V263 (*yellow*)].

cavity to form part of the GroEL-GroES interface. The loop between strands 6 and 7, bearing peptide-binding residues Tyr199, Ser201, Phe203, and Tyr204, is elevated and rotated into the *cis* ring's newly formed interface between the reoriented apical domains. Thus, the binding of GroES and nucleotide deprives the substrate polypeptide of its binding elements, which are now involved either in binding GroES directly (helices H and I) or in supporting GroES binding indirectly by stabilizing the interface between elevated and rotated apical domains. That the nine polypeptide-binding residues support the folding-active *cis* assembly explains the observation that a drastic mutagenic change of just one residue can prevent GroES binding and, indeed, be lethal to the mutant strain (41). As the apical domains move into their new positions, the corresponding peptide binding elements on the apical domains separate, possibly putting the substrate "on the rack" and helping to unfold it, either locally or globally (49). This might be sufficient to pull a bound polypeptide out of a conformation that is in a kinetic folding trap and place it back on the folding landscape just prior to or concomitant with its release into the cavity. Hydrophobic residues, which originally bound the nonnative polypeptide (presumably through hydrophobic interaction) in the cavity of the otherwise unliganded ring, are now buried in the *cis* ring assembly and have been replaced on the cavity walls with mostly polar residues [Figure 8 (opposite p. 602), compare *cis* and *trans* rings]. The released polypeptide is now free to reinitiate folding as an isolated molecule in a much-enlarged cavity whose hydrophilic lining is conducive to burial of hydrophobic residues and folding into a native structure.

Phase IV: Protein Folding and Release of Ligands

COMMUNICATION TO THE *TRANS* RING AND NEGATIVE COOPERATIVITY   Superimposition of the equatorial domains of the *cis* GroEL ring on those of unliganded GroEL shows that the plane of the *cis* ring is slightly deformed (Figure 9). In the *cis* ring, each subunit tilts about 4° toward the cylinder axis, so that the inside of the ring is 3 Å lower than the original plane and the outside is 5 Å higher. Some of the largest shifts are observed for residues that are involved in cross-ring interactions: For example, the Cα of Glu434 moves 4.9 Å and the Cα of Ala109 moves 3.8 Å away from equatorial plane. Despite these shifts, the chemical details of the interface are maintained. To preserve the inter-ring interface, the *trans* ring must shift in a complementary direction. This causes each *trans* subunit to tilt in the opposite direction, that is, away from the central axis, by about 2°. Thus, the formation of the *cis* GroEL-GroES assembly favors a structural change in the opposite ring that is opposed to the formation of a second *cis* assembly (a symmetrical football). Binding events in the *cis* ring compete against similar events in the *trans* ring, explaining the transmission of negative allosteric effects across the equatorial plane. Like the positive effects

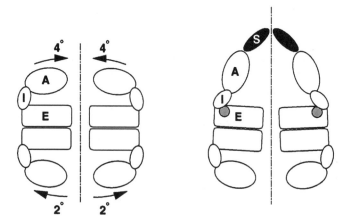

Figure 9 Negative cooperativity between rings. The presence of nucleotide (ATP or ADP) (*shaded circle*) and GroES causes the top of the *cis* equatorial domains to tilt inward by about 4°. Preservation of the equatorial interface forces the top of the *trans* equatorial domains to splay outward 2°, in a direction opposing the formation of a second GroES/nucleotide assembly on this ring. The *shading* corresponds to the three GroEL domains: equatorial, *dark gray*; intermediate, *white*; and apical, *medium gray*. S indicates GroES.

within a ring, this transmission is primarily through en bloc movements rather than through conformational shifts within the domains. Unlike the mechanisms in most other allosteric systems, however, this expression of negative allostery depends on the preservation, rather than alteration, of the quaternary interfacial contacts across the equatorial plane.

The winner in the competition between the two rings is decided by the type of adenine nucleotide it binds. Functional experiments show that ATP is dominant over ADP because it provides a stronger stabilizing force for the complex. First, binding of GroES with ATP, but not ADP, will release the most tightly bound nonnative polypeptides into the domed folding cavity of the *cis* ring, permitting them to initiate and, if sufficient time elapses, complete folding to a native protein (40). Second, bound ATP, but not ADP, causes an ATPase-deficient assembly (containing the D398A mutation) to maintain its structural integrity when challenged by low temperature and 0.5 M guanidinium HCl (40). Finally, ATPase-deficient rings of D398A maintain their domed *cis* assembly and will not release nucleotide, GroES, and folded polypeptide upon exposure of the *trans* ring to ATP, unless the bound ATP in the *cis* ring is permitted to hydrolyze to ADP. As noted earlier, it appears that the β-γ diphosphate/Mg^{2+} portion of the nucleoside triphosphate contributes additional strong contacts beyond those available from ADP that ultimately stabilize the changes that underlie the formation of a domed *cis* assembly. Thus, the *cis* ring association

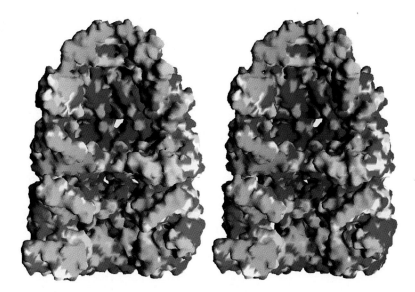

Figure 8 Stereo view of the solvent-accessible surface of the central cavity of the GroEL-GroES-(ADP)$_7$ complex. The three subunits from each of the rings nearest the reader were removed to show the inside of the assembly. Colors represent the type of surface: all backbone atoms, *white*; all hydrophobic side-chain atoms (A, V, L, I, M, F, P, Y), *yellow*; all polar and charged side-chain atoms (S, T, H, C, N, Q, K, R, D, E), *blue*; all solvent-excluded surface at subunit interfaces, *gray*. Most of the *yellow* hydrophobic patches on the surface of the *trans* GroEL cavity are replaced by *blue* polar patches on the surface of *cis* GroEL cavity.

of GroES is weakened when the nucleotide loses its γ-phosphate by hydrolysis, and the complex becomes susceptible to disassembly (and release of GroES, polypeptide, and ADP) when subjected to the outward tilting stress imposed by the binding of ATP on the opposite ring (Figure 3). This model is fully consistent with the role of nucleotides in driving other molecular mechanical systems such as G-proteins and contractile systems. Moreover, it explains why the GroEL-assisted dynamic protein folding cycle requires a double-toroid structure.

RECYCLING CHAPERONIN AND NONNATIVE POLYPEPTIDE   The timing of the formation and dissociation of the folding-active GroEL-GroES intermediates can be gauged from several experiments (at 23°C). Following binding of GroES and ATP to a GroEL-polypeptide binary complex to form a *cis* ATP complex, polypeptide is released into the central channel within a second, as revealed by fluorescence anisotropy measurements (39, 40). Interestingly, during this initial second, there is a sharp drop of anisotropy, reflecting increased mobility of the reporting groups and possibly the process of precipitous unfolding mentioned above, as the polypeptide is initially stretched upon the mobilizing apical binding sites. After polypeptide release into the channel, there is a period of ∼6–8 s before ATP hydrolysis in the *cis* complex converts it to the primed ADP state (67). Productive folding appears to continue during this time and after formation of the *cis* ADP state, as evidenced both by the continuous anisotropy changes and by ongoing accretion of biological activity of the folding substrate (39, 40). After ∼5–10 s in the *cis* ADP state, discharge of GroES and polypeptide is prompted by ATP binding in the *trans* ring (40). The dissolution of the complex appears likely to be a fast step, as suggested by recent dynamic fluorescence experiments (HS Rye, unpublished results), consistent with the quantized cooperativeness within a ring being coupled to negative cooperativeness between the rings, as described above. What likely follows for the former *cis* ring is that its apical domains relax and move back down to the conformation with high affinity for polypeptide, as seen in the unliganded GroEL structures. What happens to the former *trans* ring remains poorly understood. For example, it has not been clearly established whether nonnative polypeptide or GroES becomes bound on the *trans* ring during folding and/or ATP-triggered discharge of the *cis* complex. In a study with substrate bound in a *trans* ternary ADP complex, the addition of excess GroES and ATP was associated with nonproductive release instead of productive folding, suggesting that binding a second GroES is disfavored (WA Fenton, unpublished data). Thus, although it is attractive to consider that the chaperonin machine could simultaneously operate two different productive reactions within opposite rings, shifted 180° in the cycle from each other, this may not be its working mode. It remains possible that polypeptide and GroES must both be completely cleared from GroEL before a next round of productive binding (to rings in a fully relaxed state) can ensue. In this regard, a study with a mixed-ring GroEL assembly, able to bind polypeptide and GroES on only one of its rings, exhibited kinetics of refolding of rhodanese that were identical to those of refolding by wild-type GroEL (79). This supports the idea that, at least for rhodanese, GroEL progresses through one entire cycle before commencing the next; otherwise, one would have expected the kinetics of the wild-type chaperonin, with two competent cavities, to be faster. Further studies may resolve how the machine cycles between active states.

Given the schedule of events on GroEL dictated by ATP binding and hydrolysis, it becomes clear that nonnative proteins encapsulated in the *cis* ternary complex have a set amount of time, ~15 s, in which to reach the native state prior to disruption of the folding-active chaperonin complex. It appears that only a certain fraction of bound, nonnative polypeptide molecules can reach the native state or a conformation that is committed to the native state (that is, no longer recognizable by GroEL) during this time. But the remaining fraction fails to reach native form and requires either another round of interaction with GroEL or, in the cell, interaction with other chaperones or with protease. The latter possibility seems particularly important for removal of damaged proteins to prevent the clogging of the chaperone machinery. This nonnative fraction appears in large part to be discharged from GroEL, as revealed by a number of experiments, in particular those using a "trap" molecule—a mutant or modified form of GroEL that can bind but not release nonnative proteins. Addition of such molecules to a folding reaction has reproducibly revealed a rate of release of rhodanese, for example, that is approximately 10 times its rate of refolding. Thus, a substantial portion of the nonnative forms, if not all, are discharged from chaperonin during a round of dissociation of the *cis* complex (76, 79–83). The question has been raised as to whether rebinding of such forms occurs in a conformation that has progressed toward the native state or whether this is an all-or-none process, with each round of folding starting from the same or similar nonnative conformation. The evidence at hand would argue for the latter, as the conformation of nonnative protein bound after transfer to other GroEL molecules appears to correspond to that initially bound to chaperonin (56, 80).

CONCLUSIONS AND PROSPECTS

Certainty Versus Uncertainty in the Stereochemistry of the Folding Cycle

We now understand in molecular detail how the binding of nucleotides potentiates the large conformational changes in the *cis* ring that lead to the release of bound nonnative polypeptide into the shielded folding chamber of the *cis* assembly. We can also speculate with some confidence on how these large domain shifts might further unfold the polypeptide prior to or just at the time of its release. We also know that the mechanism of this release is coupled to the stabilization of the *cis* complex, as residues of the apical domain implicated in peptide binding form interfaces that support both the walls and the roof of the *cis* assembly. Bound GroES, acting as a multivalent keystone of the dome, stabilizes these enormous shifts. In doing so, it not only holds all the nucleotides of the *cis* complex in place but also maintains the catalytic machinery necessary to hydrolyze ATP effectively. This latter property may exemplify the general theme of extreme induced fit, wherein large domain shifts, triggered by nucleoside triphosphate binding, are stabilized by specific interactions with partner proteins that promote hydrolysis and productive return of the nucleotide-driven cycle to its ground state. Here, the partner protein is GroES; in the myosin headpiece, it is actin; and in G proteins, they are GTPase activating proteins (GAPs) or regulators of G-protein signaling (RGS) proteins. We also appreciate that for the cycle to proceed with productive release of polypeptide, along with GroES and ADP, ATP hydrolysis in *cis* (priming disruption) must be followed by ATP binding in *trans* (triggering disassembly). This mechanism underlies the need for a double toroid for efficient folding.

Although we understand the cooperative, quantized action of the *cis* system in convincing stereochemical detail, the mechanism of negative cooperativity between rings is less well understood. The inward tilting of the *cis* equatorial domains, coupled to the preservation of the interface between rings, causes outward tilting in the *trans* ring. This presents an attractive mechanism for negative cooperativity, in which formation of a *cis* complex on one ring opposes a similar event on the opposite ring. Thus, ATP binding and generation of a *cis* assembly on what was originally the *trans* ring could lead to the disruption of the original, but now weakened, ADP-primed *cis* complex, with productive release of GroES, polypeptide, and ADP. However, this opposing-tilt mechanism is derived from a structure in which bound GroES stabilizes a fully developed *cis* assembly, and there is no evidence that GroES is required, in addition to bound ATP, to discharge the products sequestered in the opposing ring. Indeed, a variety of experiments indicate that ATP binding alone will suffice. It will be necessary to visualize—in the absence of GroES—the mechanism by which ATP binding to one ring disrupts a *cis* assembly on the other before these questions can be answered.

The puzzle of the negative allosteric interplay between rings extends to the binding of nonnative polypeptides to otherwise unliganded GroEL. When presented with an excess of nonnative polypeptide, the chaperonin rarely binds polypeptide to both rings. Indeed, it is still not clear whether a nonnative polypeptide substrate can bind to the empty *trans* ring during the interval when protected folding and ATP hydrolysis is taking place in the *cis* complex. Prudence would dictate that we retreat—at least until there are supporting data—from the tempting two-cylinder metaphor, in which the expulsion of products from one ring is coupled to the intake of reactants in the other. Rather, at this stage, we must restrict ourselves to the more conservative, albeit potentially less efficient, model shown in Figure 3.

While sorting out the exact sequence of events in the folding cycle, it will also be necessary to establish the molecular basis of the unique capacity of ATP to drive the cycle by binding to the catalytic center of the equatorial domain. This has been a central focus of research on the other nucleotide-driven molecular machines mentioned above. We have only inferential hints derived from the structures of the symmetrical GroEL(ATPγS)$_{14}$ complex and the asymmetrical GroEL-GroES-(ADP)$_7$ one. Direct visualization of an ATP complex may be necessary to address this question.

A special feature of chaperonins is the wide range of nonnative folding intermediates that can be captured in the central cavity, even though specificity is apparent, at least in vivo. This raises several questions that require further study. To what extent are polypeptides bound by a unique surface of the apical domains, as exemplified in the binding of peptides to Hsp90 and the antigen presentation system? Are multiple alternative binding modes used to grasp nonnative polypeptides? And if so, to what extent and by what mechanism do they share common residues on the apical surface? What are the principal determinants of affinity in the polypeptide-GroEL interface? What fraction of subunits in a ring are needed to capture and stably bind a full-length nonnative polypeptide? Despite the analysis of folding dynamics from deuterium exchange and fluorescence studies, we still have only glimpses of the structural basis for these events.

Finally, can we extrapolate from the extensive studies of the GroEL-GroES folding cycle to the mechanisms of other chaperonin-assisted folding pathways? There are clear architectural differences between GroEL-GroES and the chaperonins of chloroplasts and, in particular, those of archaea and the eukaryotic cytosol that may require different mechanisms. This is especially true for events involving GroES, because there is no cochaperonin in archaea or the eukaryotic cytosol. On the other hand, sequence conservation in the nucleotide-binding segments of the equatorial domain suggests that the initial nucleotide-driven allosteric events in all chaperonins may be well represented by those seen in GroEL. The symmetry and subunit uniformity of GroEL and GroES, coupled to the amenity of *E. coli* to mutagenesis and recombinant technologies, have made the GroE system the ideal starting point for these inquiries into structure/function relationships in chaperonins. Even if we were to understand the

GroE system in detail, we would still be only near the end of the introductory chapter of the long and essential story of the final steps in the accurate expression of the genetic code.

ACKNOWLEDGMENTS

We thank Dr. Patrick Fleming for invaluable assistance with the illustrations.

Visit the *Annual Reviews* home page at
http://www.AnnualReviews.org.

Literature Cited

1. Dill KA, Chan HS. 1997. *Nat. Struct. Biol.* 4:10–19
2. Flynn GC, Chappell TG, Rothman JE. 1989. *Science* 245:385–90
3. Zhu XT, Zhao X, Burkholder WF, Gragerov A, Ogata CM, et al. 1996. *Science* 272:1606–14
4. Rudiger S, Germeroth L, Schneider-Mergener J, Bukau B. 1997. *EMBO J.* 16:1501–7
5. Rudiger S, Buchberger A, Bukau B. 1997. *Nat. Struct. Biol.* 4:342–49
6. Ehrnsperger M, Gräber S, Gaestel M, Buchner J. 1997. *EMBO J.* 16:221–29
7. Lee GJ, Roseman AM, Saibil HR, Vierling E. 1997. *EMBO J.* 16:659–71
8. Sullivan W, Stensgard B, Caucutt G, Bartha B, McMahon N, et al. 1997. *J. Biol. Chem.* 272:8007–12
9. Bohen SP, Kralli A, Yamamoto KR. 1995. *Science* 268:1303–4
10. Fenton WA, Horwich AL. 1997. *Protein Sci.* 6:743–60
11. Ellis RJ, ed. 1996. *The Chaperonins.* San Diego, CA: Academic
12. Hartl F-U. 1996. *Nature* 381:571–80
13. Hendrix RW. 1979. *J. Mol. Biol.* 129:375–92
14. Hohn T, Hohn B, Engel A, Wurtz M, Smith PR. 1979. *J. Mol. Biol.* 129:359–73
15. Tilly K, Murialdo H, Georgopoulos C. 1981. *Proc. Natl. Acad. Sci. USA* 78:1629–33
16. Chandrasekhar GN, Tilly K, Woolford C, Hendrix R, Georgopoulos C. 1986. *J. Biol. Chem.* 261:12414–19
17. Saibil H, Dong Z, Wood S, auf der Mauer A. 1991. *Nature* 353:25–26
18. Langer T, Pfeifer G, Martin J, Baumeister W, Hartl F-U. 1992. *EMBO J.* 11:4757–65
19. Ishii N, Taguchi H, Sumi M, Yoshida M. 1992. *FEBS Lett.* 299:169–74
20. Azem A, Kessel M, Goloubinoff P. 1994. *Science* 265:653–56
21. Schmidt M, Rutkat K, Rachel R, Pfeifer G, Jaenicke R, et al. 1994. *Science* 265:656–59
22. Engel A, Hayer-Hartl MK, Goldie KN, Pfeifer G, Hegerl R, et al. 1995. *Science* 269:832–36
23. Azem A, Diamant S, Kessel M, Weiss C, Goloubinoff P. 1995. *Proc. Natl. Acad. Sci. USA* 92:12021–25
24. Braig K, Simon M, Furaya F, Hainfeld JF, Horwich AL. 1993. *Proc. Natl. Acad. Sci. USA* 90:3978–82
25. Chen S, Roseman AM, Hunter AS, Wood SP, Burston SG, et al. 1994. *Nature* 371:261–64
26. Thiyagarajan P, Henderson SJ, Joachimiak A. 1996. *Structure* 4:79–88
27. Phipps BM, Hoffmann A, Stetter KO, Baumeister W. 1991. *EMBO J.* 10:1711–22
28. Trent JD, Nimmesgern E, Wall JS, Hartl F-U, Horwich AL. 1991. *Nature* 354:490–93
29. Liou AKF, Willison KR. 1997. *EMBO J.* 16:4311–16
30. Horwich AL, Willison KR. 1993. *Philos. Trans. R. Soc. London Ser. B* 339:313–26
31. Kubota H, Hynes G, Carne A, Ashworth A, Willison K. 1994. *Curr. Biol.* 4:89–99
32. Lewis SA, Tian G, Vainberg IE, Cowan NJ. 1996. *J. Cell Biol.* 132:1–4
33. Braig K, Otwinowski Z, Hegde R, Boisvert DC, Joachimiak A, et al. 1994. *Nature* 371:578–86
34. Hunt JF, Weaver AJ, Landry SJ, Gierasch L, Deisenhofer J. 1996. *Nature* 379:37–45
35. Mande SC, Mehra V, Bloom BR, Hol WGJ. 1996. *Science* 271:203–7
36. Landry SJ, Zeilstra-Ryalls J, Fayet O, Georgopoulos C, Gierasch LM. 1993. *Nature* 364:255–58

37. Goloubinoff P, Christeller JT, Gatenby AA, Lorimer GH. 1989. *Nature* 342:884–89
38. Saibil HR, Zheng D, Roseman AM, Hunter AS, Watson GMF, et al. 1993. *Curr. Biol.* 3:265–73
39. Weissman JS, Rye HS, Fenton WA, Beechem JM, Horwich AL. 1996. *Cell* 84:481–90
40. Rye HS, Burston SG, Fenton WA, Beechem JM, Xu Z, et al. 1997. *Nature* 388:792–98
41. Fenton WA, Kashi Y, Furtak K, Horwich AL. 1994. *Nature* 371:614–19
42. Lin ZL, Schwarz FP, Eisenstein E. 1995. *J. Biol. Chem.* 270:1011–14
43. Itzhaki LS, Otzen DE, Fersht AR. 1995. *Biochemistry* 34:14581–87
44. Katsumata K, Okazaki A, Tsurupa GP, Kuwajima K. 1996. *J. Mol. Biol.* 264:643–49
45. Perrett S, Zahn R, Sternberg G, Fersht AR. 1997. *J. Mol. Biol.* 269:892–901
46. Landry SJ, Gierasch LM. 1991. *Biochemistry* 30:7359–62
47. Buckle AM, Zahn R, Fersht AR. 1997. *Proc. Natl. Acad. Sci. USA* 94:3571–75
48. Zahn R, Perrett S, Stenberg G, Fersht AR. 1996. *Science* 271:642–45
49. Lorimer G. 1997. *Nature* 388:720–23
50. Zahn R, Pluckthun A. 1994. *J. Mol. Biol.* 242:165–74
51. Walter S, Lorimer GH, Schmid FX. 1996. *Proc. Natl. Acad. Sci. USA* 93:9425–30
52. Viitanen PV, Donaldson GK, Lorimer GH, Lubben TH, Gatenby AA. 1991. *Biochemistry* 30:9716–23
53. Zahn R, Spitzfaden C, Ottiger M, Wuthrich K, Pluckthun A. 1994. *Nature* 368:261–65
54. Robinson CV, Groß M, Eyles SJ, Ewbank JJ, Mayhew M, et al. 1994. *Nature* 372:646–51
55. Goldberg MS, Zhang J, Sondek S, Matthews CR, Fox RO, Horwich AL. 1997. *Proc. Natl. Acad. Sci. USA* 94:1080–85
56. Groß M, Robinson CV, Mayhew M, Hartl F-U, Radford SE. 1996. *Protein Sci.* 5:2506–13
57. Ranson NA, Dunster NJ, Burston SG, Clarke AR. 1995. *J. Mol. Biol.* 250:581–86
58. Laminet AA, Ziegelhoffer T, Georgopoulos C, Pluckthun A. 1990. *EMBO J.* 9:2315–19
59. Viitanen PV, Lubben TH, Reed J, Goloubinoff P, O'Keefe DP, Lorimer GH. 1990. *Biochemistry* 29:5665–71
60. Martin J, Langer T, Boteva R, Schramel A, Horwich AL, Hartl F-U. 1991. *Nature* 352:36–42
61. Buchner J, Schmidt M, Fuchs M, Jaenicke R, Rudolph R, et al. 1991. *Biochemistry* 30:1586–91
62. Badcoe IG, Smith CJ, Wood S, Halsall DJ, Holbrook JJ, et al. 1991. *Biochemistry* 30:9195–9200
63. Gray TE, Fersht AR. 1991. *FEBS Lett.* 292:254–58
64. Bochkareva ES, Lissin NM, Flynn GC, Rothman JE, Girshovich AS. 1992. *J. Biol. Chem.* 267:6796–800
65. Jackson GS, Staniforth RA, Halsall DJ, Atkinson T, Holbrook JJ, et al. 1993. *Biochemistry* 32:2554–63
66. Todd MJ, Viitanen PV, Lorimer GH. 1993. *Biochemistry* 32:8560–67
67. Burston SG, Ranson NA, Clarke AR. 1995. *J. Mol. Biol.* 249:138–52
68. Roseman AM, Chen SX, White H, Braig K, Saibil HR. 1996. *Cell* 87:241–51
69. Yifrach O, Horovitz A. 1995. *Biochemistry* 34:9716–23
70. Boisvert DC, Wang JM, Otwinowski Z, Horwich AL, Sigler PB. 1996. *Nat. Struct. Biol.* 3:170–77
71. Kim S, Willison KR, Horwich AL. 1994. *Trends Biochem. Sci.* 19:543–48
72. Braig K, Adams PD, Brunger AT. 1995. *Nat. Struct. Biol.* 2:1083–94
73. Aharoni A, Horovitz A. 1996. *J. Mol. Biol.* 258:732–35
74. Xu Z, Horwich AL, Sigler PB. 1997. *Nature* 388:741–50
75. Murai N, Makino Y, Yoshida M. 1996. *J. Biol. Chem.* 271:28229–34
76. Todd MJ, Viitanen PV, Lorimer GH. 1994. *Science* 265:659–66
76a. Klumpp M, Baumeister W, Essen L-O. 1977. *Cell* 91:263–70
77. Zeilstra-Ryalls J, Fayet O, Georgopoulos C. 1996. *FASEB J.* 10:148–52
78. Hunt JF, van der Vies SM, Henry L, Deisenhofer J. 1997. *Cell* 90:361–71
79. Burston SG, Weissman JS, Farr GW, Fenton WA, Horwich AL. 1996. *Nature* 383:96–99
80. Weissman JS, Kashi Y, Fenton WA, Horwich AL. 1994. *Cell* 78:693–702
81. Smith KE, Fisher MT. 1995. *J. Biol. Chem.* 270:21517–23
82. Taguchi H, Yoshida M. 1995. *FEBS Lett.* 359:195–98
83. Mayhew M, da Silva ACR, Martin J, Erdjument-Bromage H, Tempst P, Hartl F-U. 1996. *Nature* 379:420–26

Annu. Rev. Biochem. 1998. 67:609–52

MATRIX PROTEOGLYCANS: From Molecular Design to Cellular Function

Renato V. Iozzo

Department of Pathology, Anatomy and Cell Biology, and the Kimmel Cancer Center, Jefferson Medical College, Thomas Jefferson University, Philadelphia, Pennsylvania 19107-6799; e-mail: iozzo@lac.jci.tju.edu

KEY WORDS: glycosaminoglycan, heparan sulfate, dermatan sulfate, chondroitin sulfate, leucine-rich repeat

---

ABSTRACT

The proteoglycan superfamily now contains more than 30 full-time molecules that fulfill a variety of biological functions. Proteoglycans act as tissue organizers, influence cell growth and the maturation of specialized tissues, play a role as biological filters and modulate growth-factor activities, regulate collagen fibrillogenesis and skin tensile strength, affect tumor cell growth and invasion, and influence corneal transparency and neurite outgrowth. Additional roles, derived from studies of mutant animals, indicate that certain proteoglycans are essential to life whereas others might be redundant.

The review focuses on the most recent genetic and molecular biological studies of the matrix proteoglycans, broadly defined as proteoglycans secreted into the pericellular matrix. Special emphasis is placed on the molecular organization of the protein core, the utilization of protein modules, the gene structure and transcriptional control, and the functional roles of the various proteoglycans. When possible, proteoglycans have been grouped into distinct gene families and subfamilies offering a simplified nomenclature based on their protein core design. The structure-function relationship of some paradigmatic proteoglycans is discussed in depth and novel aspects of their biology are examined.

CONTENTS

609

0066-4154/98/0701-0609$08.00

INTRODUCTION

This review focuses on areas of proteoglycan biology that are coming under the scrutiny of molecular biology, genetics, and mutant animal studies. Progress has been made in understanding the major biosynthetic pathways by which cells produce proteoglycans, some of the largest and most complex molecular structures in mammalian cells. However, the function of many of these compounds is not understood. Because of space limitations, only the matrix proteoglycans are discussed here. These can be separated into three groups: the basement membrane proteoglycans, the hyalectans—proteoglycans interacting with hyaluronan and lectins, and the small leucine-rich proteoglycans. The structure and function of these and other proteoglycan gene families have been covered in recent reviews (1–11).

The references cited here are not intended to be exhaustive but rather to direct the reader to pertinent primary literature where additional details can be found. In some instances, I have speculated rather freely, while in others I tried to confine such speculations to areas that can be tested experimentally. When possible, I have attempted to connect work on the biology of a specific proteoglycan, or family of proteoglycans, to fundamental cellular processes and pathology. To what extent are common molecular mechanisms involved? What are the underlying mechanisms that regulate the intrinsic function of a given proteoglycan? Is there any common theme in transcriptional regulation of proteoglycan gene expression? Why are so many genes expressed for seemingly identical functions? What is the level of redundancy? What kinds of mechanisms dictate the developmental expression of a specific proteoglycan? These and other important questions are answered in the context of our current knowledge of developmental processes and the extracellular matrix at large.

BASEMENT MEMBRANE PROTEOGLYCANS

Basement membranes are biochemically complex and heterogeneous structures containing laminin, collagen type IV, nidogen, and at least one type of

Table 1   General properties of basement membrane proteoglycans

Proteoglycan	Gene	Chromosomal mapping		Protein core ($\sim$kDa)[a]	Glycosaminoglycan type (number)
		Human	Mouse		
Perlecan	HSPG2	1p36	4, distal	400–467	Heparan/chondroitin sulfate (3)
Agrin	AGRN	1p32-pter	4, distal	250	Heparan sulfate (3)
Bamacan[b]				138	Chondroitin sulfate (3)

[a]The size of individual protein core does not include any posttranslational modification.
[b]We have no information regarding chromosomal assignment and designation of the bamacan gene. Southern analysis indicates the presence of a single-copy gene in eukaryotic cells (14).

proteoglycan. Three proteoglycans are characteristically present in vascular and epithelial basement membranes of mammalian organisms: perlecan (12), agrin (13), and bamacan (14) (Table 1). The first two carry primarily heparan sulfate side chains, whereas the latter carries primarily chondroitin sulfate. The chimeric structural design of these proteoglycans suggests that they may be involved in numerous biological processes. It is unclear why only these three seemingly diverse molecules are associated with the basement membranes. The following sections discuss the structural and functional properties of these gene products and propose additional functional roles predicted from structural affinities.

General Structural Features

PERLECAN   The name derives from its rotary shadowing appearance suggesting a string of pearls. The three glycosaminoglycan side chains are located at one end of the molecule, which also contains numerous globular regions interlinked by rod-like segments (15). The general structural features are shown in Figure 1 and summarized in Table 2. This multidomain proteoglycan is one of the most complex gene products because of its enormous dimensions and number of posttranslational modifications (9, 16–19). It comprises five domains that harbor protein modules used by disparate proteins involved in lipid uptake and metabolism, cell adhesion, and cellular growth. The N-terminal domain I contains three SGD tripeptides, the attachment sites for heparan sulfate chains. This region, which has no internal repeats and is devoid of cysteine residues, is enriched in acidic amino acid residues that facilitate heparan sulfate polymerization. Recombinant domain I can accept either heparan or chondroitin sulfate chains, and this selection of glyconation appears to be cell specific (20–23).

The distal portion of domain I encompasses a SEA module, named after the three proteins—sperm protein, enterokinase, and agrin—in which it was first identified. This module has been proposed to regulate binding to neighboring

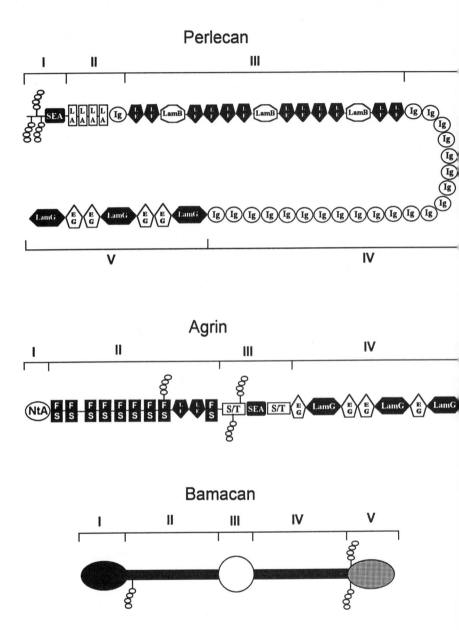

Table 2   Structural and functional motifs in the human perlecan proteoglycan

Domain	Coding exons	Domain features	Homology	Proposed function
Signal peptide	1	Hydrophobic	Common type	Signal peptide
I	5	SEA	Sperm protein, enterokinase, agrin	Regulate binding to neighboring carbohydrate moieties
II	3	4 cysteine-rich repeats	LDL receptor	Binding to lipids (?)
IIa	2	1 Ig-like repeat	N-CAM	Spacer
III	27	3 globular and 4 cysteine-rich repeats	Short arm of laminin-1	Binding to fibronectin, interaction with cell surface integrin[a]
IV	40	21 Ig-like repeats[b]	N-CAM	Homophilic interaction (?)
V	16	3 globular and 4 EGF-like repeats[c]	G-domain of laminin-1	Oligomerization, cell binding, neurite outgrowth

[a]The murine species has an Arg-Gly-Asp (RGD) motif that could mediate the binding of perlecan protein core to integrins. However, the human perlecan lacks such a sequence.
[b]Apparently, the murine species has only 14 Ig-like repeats with the possibility of alternatively splice variants in this region. However, no formal evidence for splice variants has been presented so far.
[c]EGF, epidermal growth factor.

carbohydrate moieties (24) and can enhance heparan sulfate synthesis in mouse perlecan (22). Notably, all SEA-containing proteins are glycoproteins or proteoglycans with glycosaminoglycans attached proximally to SEA (24). Secondary structure prediction suggests a conformation of alternating β sheets and α helices (24). Circular dichroism studies of recombinant domain I demonstrate a distinct α helix/β sheet structure of approximately 20 and 60%, respectively

←————————————————————————————————————

Figure 1   Schematic representation of three major proteoglycans found in basement membranes. The Roman numerals indicate the proposed domains. The symbols and designations of individual protein modules for perlecan and agrin are according to Bork & Patthy (24) with minor modifications: SEA, a module first identified in sperm protein, enterokinase, and agrin; LA, LDL receptor class A module; Ig, immunoglobulin-like repeat typically found in N-CAM; LE, laminin-1 EGF-like; LamB, a globular module similar to that found in the short arm of the α-1 chain of laminin-1; LamG, first identified as the G domain in long arm of the α-1 chain of laminin-1; EG, EGF-like; NtA, N-terminal domain that binds laminin; FS, follistatin-like; ST, serine/threonine rich. The glycosaminoglycan side chains are designated by strings of small circles.

(23). The N-terminal region of domain I contains six Ser/Thr residues that are substituted with galactosamine-containing oligosaccharides and one N-substituted Asn residue (23).

Domain II comprises four repeats homologous to the ligand-binding portion of the low-density lipoprotein (LDL) receptor (LA module) with six perfectly conserved cysteine residues and a pentapeptide, DGSDE, that mediates ligand binding by the LDL receptor (19). Rotary shadowing electron microscopy of recombinant domain II demonstrates a globular domain connected to a short segment, suggesting that the tandem arrays of LDL receptor class A modules form rod-like elements (25). Further analysis of recombinant domain II shows that it represents an autonomously folding unit within the perlecan structure (25). The function of this domain remains conjectural, and whether this molecule can indeed bind LDL is unknown. However, its location immediately following the heparan sulfate-attachment domain may favor the interaction of perlecan with either lipid bilayers or soluble lipids, or perhaps this region may direct the proteoglycan to the basolateral surfaces of epithelial cells for vectorial deposition into basement membranes (5, 15). Nidogen, a known ligand for perlecan (26, 27), binds to recombinant domain II, though less avidly than to the whole perlecan molecule (25).

Domain III, most homologous to the subdomain IVa and IVb of the short arm of the α-1 chain of laminin-1 (28), harbors three distinct globular domains (LamB modules) connected by short rods, the cysteine-rich epidermal growth factor (EGF)-like regions (LE modules) (29, 30). This domain contains an RGD tripeptide that can promote integrin-mediated cell attachment (31). However, because this sequence is not conserved in the human (32, 33), it is not clear whether additional regions in the protein core may mediate cell attachment.

Domain IV is the largest and the most repetitive since it contains 14 and 21 Ig-like repeats in the murine and human species, respectively. These repeats are similar to those found in members of the Ig gene superfamily such as those described in the neural cell adhesion molecule N-CAM. Thus this domain may be implicated in dimerization or intermolecular self-association.

Domain V has homology to the so-called G domain of the long arm of laminin-1 and has three distinct globular regions (LamG modules) connected by EGF-like repeats (EG modules). Domain V is responsible for self-assembly and may be important for basement membrane formation in vivo (34). This domain contains an LRE tripeptide that is a cell adhesion site for laminin-3 (S-laminin) (35).

The complexity of the perlecan protein core is mirrored by a series of posttranslational modifications and additional not-well-understood events that lead to multiple isoforms of the parent molecule. Such biosynthetic events include fatty acylation of the protein core (36), attachment of both chondroitin and heparan sulfate side chains (37–39), substitution with undersulfated or totally unsulfated heparan sulfate side chains (40), and secretion of the protein core without addition of any glycosaminoglycan side chains (41). These forms of perlecan may be tissue or cell specific; however, their function remains to be fully elucidated.

AGRIN   A constituent of the basement membrane that causes aggregation of acetylcholine receptors, agrin is a major heparan sulfate proteoglycan of neuro-muscular junctions (13, 42) and renal tubular basement membranes (43). Agrin is a multidomain protein that shares similarities with perlecan and laminin and can be divided into four distinct domains (Figure 1). Domain I harbors the first 130 amino acid residues following the signal peptide. This region has been identified in the chick as a novel module that binds laminin-1 and is thus called NtA, for N-terminal in agrin (44). This domain has been also identified in mouse and human expressed sequence tags and shows a very high conservation among species, up to 90% (44). Thus such a domain is likely to be operational also in mammalian agrin. Alternatively spliced variants may exist that harbor an insertion of approximately 21 amino acid residues (42, 45).

Domain II is characterized by nine follistatin-like repeats, the last of which is interrupted by the insertion of two cysteine-rich, EGF-like modules typically found in laminin β- and γ-chains. This domain may function as a protease in-hibitor or may mediate growth-factor binding in vivo (46). If the two EGF-like repeats behave as in laminin, they might also mediate binding of agrin to nido-gen. Domain III is highly glycosylated and is characterized by a central SEA module, as in perlecan, flanked by two Ser/Thr-rich regions. In this domain, there are two conserved glycosaminoglycan attachment sites and potential for several O-linked oligosaccharide attachments. Domain IV is the most similar to perlecan and comprises three G modules found in laminin α-chains interrupted by EGF-like repeats. By analogy to laminin and perlecan, this domain may self-assemble. Alternatively spliced variants of domain IV with insertion of small peptide sequences, 4–19 amino acid residues long, have been described (47). These short peptides can significantly influence the binding of agrin to heparin, α-dystroglycan, and the cell surface (46). Moreover, this proteogly-can harbors at least six potential sites for glycosaminoglycan attachment and five N-glycosylation sites in addition to the two Ser/Thr-rich regions described above. Thus agrin may be heterogeneous in tissues.

BAMACAN   The presence of chondroitin sulfate–containing proteoglycans in the basement membranes of various tissues is well documented (5). Bamacan was cloned and sequenced recently and is the product of a single gene different from any other basement membrane proteoglycan (14). It is probably the same proteoglycan that was synthesized in organ cultures of rat parietal yolk sac, the so-called Reichert's membrane (48), which also contained chondroitinase ABC-sensitive glycosaminoglycans (49).

Structurally, bamacan can be subdivided into five domains (Figure 1). Domain I is devoid of cysteine residues, is largely hydrophilic, and has several β turns. Domain II likely assumes a coiled-coil configuration owing to the absence of proline residues. A possible glycosaminoglycan attachment site is also present in this domain. Domain III is a rod-like region with four cysteine residues and a VTxG sequence, which may mediate cell attachment as in thrombospondin-1. Domain IV comprises the second coiled-coil domain very similar in size and overall structure to domain II. As in perlecan, it contains an LRE tripeptide that is a cell adhesion site for laminin-3 (S-laminin) (35). Domain V is also hydrophilic and contains four SG sequences that are potential binding sites for glycosaminoglycan side chains. This region, in addition to potentially interacting with domain I, may undergo heterotypic interactions to incorporate bamacan into the basement membrane (14).

Genomic Organization and Transcriptional Control

Among the basement membrane proteoglycans, only the genomic organization of human perlecan (50) and mouse agrin (51) have been described to date. By comparing these two gene products with other modular proteins, some interesting observations can be made regarding their evolutionary development. The human perlecan (HSPG2) is a single-copy gene located on the short arm of human chromosome 1 at 1p36 (52) and on a syntenic region of mouse chromosome 4 (53). It comprises 94 exons and spans at least 120 kb of continuous DNA (9). The gene duplication theory of molecular evolution is supported strongly by the remarkable conservation of the intron-exon junctions in the various modules of perlecan. For instance, the exon sizes of the LDL receptor–like repeats in domain II are identical to those encoding the ligand-binding region of the LDL receptor with only a few base-pair differences. In other parts of the perlecan gene, some of the repeats of domain IV are almost identical, differing by only two or three nucleotides. Moreover, there is a striking conservation of intron phases among the units encoding the Ig folds. Domain III shows no correlation between exon arrangement and either nominal domain or repeat boundaries. Notably, a similar comparison between the laminin-$\beta 1$ and -$\beta 2$ shows considerable divergence between these molecules with no conservation of exon structure or domain location. We can thus assume that the laminin-like region of the perlecan gene might have evolved from an ancestral gene that has undergone extensive rearrangement. As in the case of the other laminin-like region, the genomic organization of perlecan domain V lacks correlation between domain boundaries and exon structure.

The human agrin gene has been mapped to the distal region of the short arm of human chromosome 1, at 1p32-pter, relatively close to the perlecan gene and to the same syntenic region of mouse chromosome 4 (51). The intron-exon pattern of the agrin gene displays a remarkable correspondence to the proposed domain structure of the protein. Once again, the common theme of exon shuffling and duplication seems to have prevailed in the evolution of the agrin molecule. The follistatin repeats 1–7 in the N terminus are all encoded by single exons flanked by phase I introns. These follistatin repeats could have evolved from unequal crossover events after the deletion of the internal intron found between repeats 8 and 9. As in the case of laminin and perlecan, the C-terminal part of agrin shows no significant correlation between exon-intron organization of the gene and the modular organization of the protein. In summary, whereas at the amino acid level these two proteoglycans are related, genomic analysis shows a remarkable divergence.

To date, only the promoter of human perlecan has been sequenced (50) and tested for functional activity (54). Structurally, the human perlecan promoter is enriched in $G + C$ nucleotides, lacks canonical TATA or CAAT boxes, and contains several cognate *cis*-acting elements and palindromic inverted repeats. These features are observed typically in housekeeping and growth factor–encoding genes that are generally devoid of TATA or CAAT boxes and contain multiple transcription initiation sites. The proximal promoter region contains four GC boxes and 15 consensus hexanucleotide-binding sites for the zinc-finger transcription factor Sp1, five of which are located in the first exon.

Another striking feature of the perlecan promoter is the presence of numerous AP2 motifs, eight residing in the first 1.5 kb and two in the most distal areas (54). Notably, the AP2 transcription factors can be suppressed by SV40 T antigen and can confer phorbol ester and cAMP induction (55). These cognate *cis*-acting elements are very likely to be operational in vivo since SV40 T antigen inhibits transcription of perlecan in renal tubular epithelial cells (56), whereas perlecan expression is markedly upregulated by phorbol ester in colon cancer (57) and in erythroleukemia K562 (58) cells, respectively. The perlecan promoter also contains several motifs that bind transcription factors involved in hematopoiesis, including two PEA3 motifs and nine GATA-1 motifs that are involved in erythrocyte differentiation (59). Because perlecan is expressed abundantly in the hematopoietic system (60), these transcription factors may also play a role in regulating perlecan gene expression during bone marrow development and lymphoid organ formation (61).

In the distal promoter region, perlecan contains a binding site for NF-κB, a factor that has been involved in interleukin-induced transcription of several genes (55). The full-length promoter is quite active in a variety of human cells with various histogenetic backgrounds, as well as in mouse fibroblastic and melanoma cells (54). Collectively, these data are consistent with the fact that perlecan is ubiquitously expressed (60, 62). The perlecan promoter contains a transforming growth factor (TGF)-β-responsive element with a $5'$-TGGCC.N$_{3-5}$.GCC-$3'$ consensus sequence (54) resembling that described in rat and mouse α2(I) collagen (63, 64), elastin (65), type I plasminogen activator inhibitor (66), and growth hormone (67) genes. This sequence binds to NF-1-like members of transcription factors and is transcriptionally activated by TGF-β (54), thereby validating previous results obtained in human colon carcinoma (68) and murine uterine epithelial (69) cells. In both cases, TGF-β induces the mRNA and protein levels of perlecan. Thus this TGF-β-responsive element may regulate the expression of this proteoglycan at sites of tissue remodeling and tumor stroma formation.

Other factors that regulate perlecan gene transcription are beginning to be elucidated. In F-9 embryonal carcinoma cells, induction of differentiation into the parietal endoderm phenotype occurs with a combination of cAMP and retinoic acid, which results in an induction of perlecan transcription (70). Notably, cAMP alone downregulates perlecan expression in glomerular epithelial cells (71). Another potential regulator of perlecan gene expression is glucose (72). Increased concentrations of glucose correlate with an inhibition of de novo proteoglycan synthesis by glomerular (73) and mesangial cells (72), and these effects may be posttranscriptionally controlled (74). In contrast, long-term exposure of human mesangial cells to elevated concentrations of D-glucose induces perlecan gene expression (75). Moreover, D-glucose can induce dysmorphogenesis of embryonic kidneys at least in part by reducing the expression of perlecan, which would thus act as an essential morphogenetic regulator of extracellular matrix (76, 77).

Expression and Functional Properties

PERLECAN   A vast body of evidence that links perlecan to cell differentiation and tissue morphogenesis indicates that this gene product may play an important role in embryogenesis. Earlier studies identified perlecan in preimplantation embryos prior to the formation of a basement membrane (78). In addition to being deposited in the blastocyst interior, perlecan epitopes can be detected on the outer surface of the trophectoderm cells at the time of attachment competence (79), suggesting that perlecan may play a role in attachment of the embryo to the uterine linings.

Increased perlecan expression during blastocyst attachment competence is regulated at multiple levels. It can derive from increased transcription as in normal implantation or from increased translation of preexisting mRNA as in delayed implantation (80). A systematic study of murine embryogenesis has shown that perlecan expression appears early in tissues of vasculogenesis such as the heart primordium and major blood vessels (81). Subsequently it accumulates in a number of mesenchymal tissues, especially in cartilage undergoing endochondral ossification, where it persists throughout the developmental stages and into adulthood (81). Perlecan expression correlates universally with tissue maturation and is always prominent in the endothelial cell basement membrane of all vascularized organs, particularly the liver, lung, spleen, pancreas, and kidney. Thus perlecan may play important roles not only in the early steps of blood vessel development but also in the maturation and maintenance of a variety of differentiated epithelial and mesenchymal tissues, among which cartilage is prominent.

The distribution of FGF-2 in various basement membranes (82) parallels that observed for perlecan in the mouse embryo (81). This suggests that perlecan plays a key role as a regulator of FGF-2 signaling (see also below) and as a gatekeeper to limit access of growth factors to subjacent target cells (82). Indeed, vascular heparan sulfate proteoglycan can activate or block FGF-2 activity (83) and can control the access of FGF-2 to the underlying vascular smooth muscle cells (84). Moreover, various isoforms of TGF-β (85) as well as platelet factor 4 (86) bind to specific heparan sulfate sequences.

The developmental timing of perlecan expression further suggests that this proteoglycan may play a role in controlling smooth muscle cell replication in vasculogenesis. Heparin-like molecules have long been implicated in vascular smooth muscle cell proliferation (87), and perlecan is a potent growth inhibitor of such proliferation (88). The pattern of perlecan mRNA and protein expression correlates inversely with the degree of smooth muscle cell replication during rat aortic development (89). Therefore, perlecan could modify the behavior of replicative cells by controlling the amount of growth factors involved in vascular morphogenesis (81). The report of a perlecan splice variant that activates FGFs during early neuronal development (90) supports this concept.

Mouse perlecan is capable of downregulating Oct-1, a transcription factor involved in vascular smooth muscle cell growth control (91). Because the addition of soluble heparin does not elicit the same response, it is plausible that the ability of perlecan to alter smooth muscle cell function resides in the coordinated binding of the protein core and heparan sulfate chains (91). The human system is more complex since a variety of proteoglycans, including perlecan synthesized by arterial endothelial cells, do not directly suppress the growth of vascular smooth muscle cells (92). Additional studies are needed to identify the signals involved in perlecan-mediated growth inhibition and changes in gene expression.

The strategic location of perlecan immediately suggests that this gene product may be involved directly in the modulation of cell surface events known to be altered in the multistep process of invasion and metastasis. A cardinal role for cell-associated heparan sulfate proteoglycans was revealed by the finding that high-affinity receptor binding of FGF-2 is abrogated in mutant cell lines defective in heparan sulfate (93) and in myoblasts depleted of sulfated glycosaminoglycans (94). Bone marrow heparan sulfate proteoglycans, including perlecan, bind growth factors and present them to hematopoietic progenitor cells (95), whereas highly O-sulfated oligosaccharide sequences are required for FGF-2 binding and receptor activation (96, 97). Perlecan is involved directly in this coupling by acting as a low-affinity receptor and as an angiogenic modulator (98).

An emerging body of evidence further supports the notion that perlecan is involved directly in promoting the growth and invasion of tumor cells through its ability to capture and store growth factors (99) by entrapping them within both the basement membrane (100) and the tumor stroma (101). In melanomas a predominance of heparan sulfate proteoglycans at the cell surface is a marker of a more aggressive phenotype (102), and perlecan mRNA and protein levels are notably increased in the metastatic neoplasms (103). Purified perlecan enhances invasiveness of human melanoma cells (104), whereas contact with basement membrane perlecan augments the growth of transformed endothelial cells but suppresses that of their normal counterparts (105). Stable overexpression of an antisense perlecan cDNA in NIH-3T3 cells as well as in human metastatic melanomas leads to reduced levels of perlecan and concurrent suppression of cellular responses to FGF-2 (106).

In contrast, in fibrosarcoma cells, antisense expression of perlecan cDNA causes enhanced tumorigenesis characterized by heightened growth in vitro and in soft agar, increased cellular invasion into a collagenous matrix, and faster appearance of tumor xenografts in nude mice (107). Thus the cellular context is important in mediating perlecan's functions. Indeed, perlecan can behave as either an adhesive or an antiadhesive protein for endothelial and bone marrow cells, respectively (108). Perlecan also inhibits mesangial cell adhesion to fibronectin (109) and is antiadhesive for polymorphonuclear granulocytes (110). Though antiadhesive for hematopoietic and fibrosarcoma cells, perlecan still binds granulocyte/macrophage colony stimulating factor and presents it to the hematopoietic progenitor cells (108).

Large deposits of immunoreactive perlecan are present in the newly vascularized stroma of colon, breast, and prostate carcinomas (19). In tumor xenografts induced by subcutaneous injection of human prostate carcinoma PC3 cells into nude mice, though perlecan was actively synthesized by the human tumor cells, it was clearly deposited along the newly formed blood vessels of murine origin (19). Hence, perlecan deposited by growing tumor cells may act as a scaffold upon which proliferating capillaries grow and eventually form functional blood vessels. The observation that FGF-2 binds to heparan sulfate chains located in the N-terminal domain of perlecan synthesized by human endothelial cells (111) reinforces the hypothesis that perlecan represents a major storage site for FGF-2 in the blood vessel wall. The release of growth factor/heparan sulfate complexes via controlled proteolytic processing (111, 112) is a physiologic mechanism that disengages biologically active molecules at the site of remodeling and tumor invasion (18).

Another function of perlecan is in regulating the permeability of the glomerular basement membrane (16). It has long been known that removal of heparan sulfate chains increases glomerular permeability to proteins and leads to proteinuria (113). Injection of monoclonal antibodies against heparan sulfate chains derived from glomerular proteoglycans is nephritogenic and can induce a selective proteinuria, likely due to the neutralization of anionic sites on the heparan sulfate (114). Alterations in expression patterns are also noted in both protein core and side chains in various glomerulopathies (115), indicating that perlecan may be involved in several renal pathologies. The urine of patients who have end-stage renal failure and are undergoing hemodialysis contains a fragment of human perlecan derived from the carboxyl end of the protein core (116), further suggesting a role in glomerular filtration.

AGRIN   Agrin is among the best-characterized molecules of the synaptic basement membrane. First isolated as a glycoprotein from the basement membrane of the *Torpedo californica* electric organ on the basis of its ability to aggregate acetylcholine receptors, it was subsequently shown to be a key organizer of the postsynaptic apparatus at the neuromuscular junction (117). When added to cultured muscle cells, agrin causes aggregation of acetylcholine receptors and other proteins that are enriched at the neuromuscular junctions (46). It is now recognized that agrin is a heparan sulfate proteoglycan (13) and that the "functionally active" agrin investigated in the past is a proteolytically processed form of the parent molecule, essentially a large C-terminal fragment.

The N-terminal extension of agrin is required for the proper secretion of agrin, and this region is substituted with heparan sulfate side chains (42). Moreover, this N-terminal domain, the NtA domain (Figure 1), mediates binding to laminin-1 (44) as well as to laminin-2 and -4, the predominant laminin isoforms of muscle basement membranes; however, it has no affinity for either perlecan or collagen type IV (44). Thus the specific binding of agrin to laminin may provide the basis for its localization within basement membranes at the neuromuscular junctions (117) and in the renal tubules (43).

Agrin exists as a heparan sulfate proteoglycan in a variety of species including mouse, rat, cow, and human (118). Because nonglycosylated splice variants were not detected in any of the tissues investigated (118), it is highly likely that the agrin molecule occurs primarily as a proteoglycan. Moreover, agrin epitopes have been found in basement membranes of skin, gastrointestinal tract, and heart (119), with a distribution that has no association with synapses. Thus agrin has molecular functions beyond its role in synaptogenesis.

A novel function for agrin is a proposed role in maintaining cerebral microvascular impermeability insofar as agrin accumulates in the brain microvascular basement membranes during development of the blood-brain barrier (120). In both avian and murine brain, agrin accumulates on vessels around the time the vasculature becomes impermeable. Moreover, the agrin isoform that accumulates at such loci lacks the 8- and 11-amino-acid sequences known to confer on agrin high potency in acetylcholine receptor clustering (120).

Of special interest are sites in agrin that undergo alternate mRNA splicing. One site, recently described in the avian form of agrin, is located in the N-terminal end of the molecule (45). In the developing chicken, this variant is expressed primarily by nonneuronal cells such as astrocytes, smooth muscle, and cardiac muscle cells. Its upregulation is consistent with the proportional increase in glial cells during brain development (45). Whereas motor neurons of chicken spinal cord express primarily this splice variant of agrin, muscle cells synthesize primarily agrin lacking the extra sequence (42).

Two additional sites of alternate splicing called a and b in avian species (y and z in rat) are positioned within domain IV of agrin in the second LamG domain and distal to the terminal EGF-like module (Figure 1). When small peptides are inserted in these regions, the overall properties of agrin are modified significantly. Insertion of a four-amino-acid sequence at the y site confers heparin-binding properties (47). When the insertion is present, the binding of agrin to α-dystroglycan is inhibited by heparin.

Considerable excitement followed the finding that agrin is bound by α-dystroglycan, a molecule belonging to the complex that links dystrophin to the cell surface (117). However, it has since been shown that the region of the agrin molecule by which it binds to α-dystroglycan is not essential for acetylcholine aggregating activity. The fact that agrin activity needs protein phosphorylation suggests that agrin may activate a protein kinase.

A significant advance in our understanding of how agrin functions has come from genetic analysis of mutant mice. Researchers have examined mice in which the differentially spliced exons required for the acetylcholine receptor clustering of agrin are deleted. These animals die perinatally and possess no acetylcholine receptor clusters associated with the motor nerves (121). In addition, axons of the motor neurons do not stop growing, and their growth cones fail to differentiate into presynaptic nerve terminals. These genetic experiments establish agrin as a key regulator of synaptogenesis at the neuromuscular junction.

A similar phenotype was obtained in mice harboring a targeted disruption of a gene encoding a muscle-specific receptor tyrosine kinase, called MuSK

(122). The phenotypic similarities between the two conditions indicate strongly that agrin acts via the MuSK receptor. Indeed, agrin induces autophosphorylation of MuSK within minutes, and this phosphorylation is observed only with spliced variants that are also active in acetylcholine receptor aggregation. However, agrin does not appear to bind directly to the MuSK receptor (123), requiring instead an accessory component that has not been identified. Collectively, these results indicate that signal-activated protein phosphorylation, a widespread component of growth-controlled systems, also plays a key part in synapse formation (124).

BAMACAN   Among the basement membrane proteoglycans, bamacan's functions are the least well-known. As mentioned previously, bamacan likely represents the proteoglycan that was originally isolated and partially characterized from organ cultures of rat Reichert's membrane (48), the embryonic basement membrane sandwiched between the parietal endoderm and the trophoblast. Biosynthetic experiments revealed the presence of a large, high-density chondroitin sulfate proteoglycan with a core protein of about 135 kDa (48). Subsequent studies using immunohistochemical and histochemical techniques coupled with chondroitinase ABC digestion revealed the presence of two major proteoglycans (49): one sensitive and one resistant to chondroitinase ABC digestion. The latter is probably perlecan.

The concept that basement membranes can contain two or more proteoglycans is supported by the finding that in the EHS tumor matrix, both perlecan and bamacan exist as intrinsic constituents (39). Antibodies against the chondroitin sulfate proteoglycan of Reichert's membrane recognize bamacan in a variety of tissues (5, 125). The deduced protein core structure of bamacan (14) reveals a unique molecule with coiled-coil regions (Figure 1). Because of these unique structural features, bamacan is likely to be an important functional molecule within specialized basement membranes. For example, tissues undergoing morphogenesis appear to lack or express lower amounts of bamacan (5).

Bamacan is immunologically unrelated to perlecan, is apparently regulated during embryonic development, and has been implicated in the pathogenesis of several disease processes in which basement membranes are affected (5). Bamacan is expressed in nearly all of the basement membranes investigated so far (125, 126); however, there are some important exceptions. It is localized in the basement membrane of Bowman's capsule and within the mesangial matrix but is not found in the capillary glomerular basement membrane (125). This observation agrees with the fact that mesangial cells synthesize a spectrum of proteoglycans species that include bamacan (127).

In contrast, perlecan immunostaining is associated with all of the major basement membranes of the kidney. Immunoelectron microscopic studies have further localized bamacan epitopes to subregions of the mesangium, consistently being found directly subjacent to the lamina of the perimesangial portion of the glomerular basement membrane and the juxtamesangial portion of the capillary endothelial cells. From developmental studies of skin, hair follicle, and kidney, one can infer that this proteoglycan may play a role in basement membrane stability (5). During these developmental processes, the presence of bamacan may inhibit, while the presence of perlecan may enhance, branching morphogenesis. How bamacan interacts specifically with other basement membrane constituents and how it imparts stability is not known.

HYALECTANS: PROTEOGLYCANS INTERACTING WITH HYALURONAN AND LECTINS

Molecular cloning has enabled the identification of a family of proteoglycans that share structural and functional similarities at both the genomic and protein levels. This family currently contains four distinct genes, namely versican, aggrecan, neurocan, and brevican (Figure 2 and Table 3). A common feature of these proteoglycans is their tridomain structure: an N-terminal domain that binds hyaluronan, a central domain that carries the glycosaminoglycan side chains, and a C-terminal region that binds lectins. On this basis, the term hyalectans, an acronym for hyaluronan- and lectin-binding proteoglycans, has been proposed (9). Alternate exon usage occurs extensively, and various degrees of glycanation and glycosylation make these proteoglycans appropriate molecular bridges between cell surfaces and extracellular matrices (128, 129).

General Structural Features

VERSICAN   The largest member of the hyalectan gene family, versican (130), is the mammalian counterpart of the so-called PG-M isolated from avian tissue (131). Domain I (Figure 2) contains one Ig repeat followed by two consecutive modules, the link protein modules, which are involved in mediating the binding of proteins to hyaluronan (132). The entire link module is approximately 100 amino acids in length and has a characteristic consensus sequence with

Figure 2   Schematic representation of four members of the hyalectans, the hyaluronan and lectin-binding proteoglycans. The Roman numerals indicate the proposed domains. The symbols and designations are as follows: Ig, immunoglobulin-type repeat; LP, link-protein type module; GAG, glycosaminoglycan-binding domain, where α and β refer to the two alternatively spliced variants encoded by individual exons; EG, EGF-like module; Lectin, C-type lectin-like module; CR, complement regulatory protein; KS, keratan sulfate-attachment module. The glycosaminoglycan side chains are indicated by strings of small circles. The GPI-anchor in brevican is denoted by two filled circles and a curved tail.

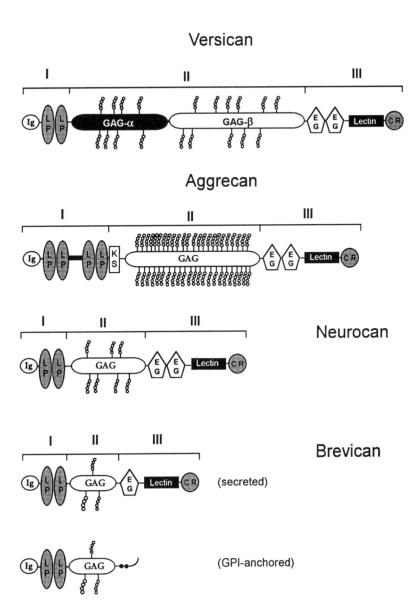

Table 3 General properties of the hyalectans, proteoglycans interacting with hyaluronan and lectins

Proteoglycan	Gene	Chromosomal mapping		Protein core ($\sim$kDa)[a]	Glycosaminoglycan type (number)
		Human	Mouse		
Versican	CSPG2	5q13.2	13	265–370	Chondroitin/dermatan sulfate (10–30)
Aggrecan	AGC1	15q26	7	$\sim$220	Chondroitin sulfate ($\sim$100)
Neurocan	NCAN		8	$\sim$136	Chondroitin sulfate (3–7)
Brevican	BCAN	1q25–q31[b]	3	$\sim$100	Chondroitin sulfate (1–3)

[a]The size of individual protein core does not include any posttranslational modification.
[b]The mouse brevican gene is closely linked to the osteocalcin gene on chromosome 3 and is flanked by markers that are all located in a syntenic region of the long arm of human chromosome 1, at 1q25–31 (150). Thus, it is likely that the human BCAN homologue also maps to this region of human chromosome 1.

four disulfide-bonded cysteine residues. The solution structure of a homologous link module was elucidated recently and consists of two α helices and two antiparallel β sheets arranged around a large hydrophobic core (133). A hydrophobic/hydrophilic region is proposed to mediate hyaluronan binding in the various members of the link module superfamily since there is conservation of the hydrophobic and some of the charged residues. Notably, there is a striking similarity between the link module and the C-type lectin domain (133). These domains have identical topologies and a similar organization of their secondary structure. The major difference is in the length of the loop. The long loop that contains the Ca^{2+}-binding residues in the lectin domain is missing in the link module.

Because there is no evidence that the binding of hyaluronan to the link protein module requires Ca^{2+}, it is quite likely that this structural difference is involved directly in dictating the differential affinity of the two domains for carbohydrates. The structural similarities between the link module and the C-type lectin domain suggest that they have a common evolutionary origin, even though the sequence similarity between the two is not as evident (132). Both recombinant versican and a truncated form of versican containing domain I bind to hyaluronan with a K_d of approximately 4 nM, in the same range as aggrecan (134). This observation suggests that versican may form a molecular link between lectin-containing glycoproteins at the cell surface and extracellular hyaluronan. Because hyaluronan is bound to the cell surface via its CD44 receptor, versican may also stabilize a large supramolecular complex at the plasma membrane zone.

Domain II consists of two large subdomains, designated GAG-α and GAG-β (135), that are encoded by two alternatively spliced exons (136). These regions lack cysteine residues and contain as many as 30 potential consensus sequences for glycosaminoglycan attachment as well as several binding sites for N- and O-linked oligosaccharides.

There are at least four possible splice variants of mammalian versican. The largest one, designated V0, contains both GAG-α and GAG-β and has an estimated M_r of approximately 370 excluding the signal peptide and the possible posttranslational modifications. The other three variants contain only GAG-β (V1 variant), GAG-α (V2 variant), or neither region (V3 variant). In birds, there is an additional exon, designated PLUS, in the N-terminal region corresponding to the mammalian GAG-α (137). This exon can be alternatively spliced giving rise to two additional isoforms. No corresponding region was found in the mammalian genome, suggesting a divergence of this domain during evolution. Sequence homology, however, indicates that the PLUS domain of avian versican may correspond to the keratan sulfate attachment region of aggrecan (137).

Domain III contains a series of structural motifs including two EGF-like repeats, a C-type lectin domain, and a complement regulatory protein-like module. These motifs are characteristically observed in the selectins, adhesion receptors regulating leukocyte homing and extravasation at inflammatory sites. The C-type animal lectins harbor a Ca^{2+}-dependent carbohydrate-recognition motif that is highly conserved among species (9, 138). The recombinant C-terminal portion of avian versican can bind heparin, heparan sulfate, and simple carbohydrates (139), whereas the recombinant human lectin domain can bind fucose and GlcNAc as well as tenascin-R (140). The binding is Ca^{2+} dependent and is abrogated by deglycosylation of tenascin-R. Other C-type lectin domains may have different saccharide-binding specificity, a mechanism that may provide additional specialized and refined functions for the hyalectans (138).

AGGRECAN   The overall organization of aggrecan, the main proteoglycan of cartilaginous tissues, is similar to that of versican with a few exceptions. Domain I contains four link protein-like modules in addition to the Ig-like repeat (Figure 2). These modules form two globular domains also known as G1 and G2 (141). An interglobular region has a rod-like structure and contains cleavage sites for proteases involved in the degradation of aggrecan (8). The function of the G2 domain is poorly understood insofar as this region does not mediate the binding to hyaluronan. Immediately following the G2 domain is a relatively small region that contains numerous keratan sulfate consensus sequences. This domain has few similarities among species, and its size also varies in different species (142).

Domain II is the largest domain of aggrecan and contains the glycosamino-glycan-binding region. This domain is similar in size to the GAG-β of versican but harbors many more consensus sequences for glyconation. Thus a fully glycosylated aggrecan may contain up to 100 chondroitin sulfate chains. The human glycosaminoglycan-binding domain harbors a polymorphism that depends on a high degree of sequence conservation (143). A variable number of tandem repeats generates at least 13 different alleles in the general population with repeats numbering between 13 and 33. This could lead to a great variation in the degree of glycosylation and ultimately charge (sulfation) of the parent molecule within cartilage.

The structural motifs in domain III are very similar to those of versican; however, there is evidence that in both the human and murine species, the EGF repeats can be alternatively spliced (144). As in the case of versican, the lectin-like domain of aggrecan can bind simple sugars in a Ca^{2+}-dependent manner (145). Thus aggrecan may also be bridging or interconnecting various constituents of the cell surface and extracellular matrix via its terminal domains.

NEUROCAN   The third member of the hyalectan gene family is neurocan, a chondroitin sulfate proteoglycan originally cloned from rat brain (146). Rotary shadowing electron microscopy of tissue-derived neurocan reveals two globular domains connected by a central rod of 60–90 nm (147), in agreement with the organization derived from protein sequencing and cDNA cloning (Figure 2). As with other members of the hyalectan gene family, neurocan has an N-terminal region with all the typical arrangements found in link protein. Recombinant neurocan domain I interacts with hyaluronan in gel permeation assays, and the isolated, retarded preparations contain supramolecular complexes of hyaluronan and globular profiles of domain I (147). Thus all domain Is likely mediate hyaluronan binding in vivo. Domain II contains at least seven potential binding sites for glycosaminoglycans and has no significant homology to any other protein.

Recombinant studies have demonstrated that glycanation is restricted solely to this central domain even though other portions of the molecule contain SG repeats (147). The C-terminal domain is again very similar to that of the other members of the hyalectan gene family with approximately 60% identity between the rat neurocan and human versican and aggrecan. Although not formally demonstrated, this domain may mediate the binding of neurocan to a variety of brain glycoproteins including Ng-CAM, N-CAM, and tenascin (discussed below).

BREVICAN   Brevican is the most recently discovered member of this class of proteoglycans and takes its eponym from the Latin word *brevis* (short) because it has the shortest glycosaminoglycan-binding region (Figure 2) (128, 148). Within conserved domains, sequence homology with the other members is relatively uniform (55–60%). However, domain II shows little homology to the corresponding regions of the other hyalectans. This domain is also characterized by a relatively high content of glutamic acid, including a sequence of eight consecutive residues. Such a cluster of acidic residues, which is also present in the link protein-like module of versican, may mediate binding to cationic proteins and, perhaps, minerals.

Similarly to neurocan, brevican exists in vivo either as a full-length proteoglycan or as a proteolytically processed form lacking the GAG-binding region and the N-terminal domain I. Proteolytic cleavage of rat brevican occurs at the same site as bovine brevican within an amino acid sequence that is similar to the aggrecanase cleavage site in aggrecan (149); this site is also conserved in mouse brevican (150). Domain III is also organized similarly to the other hyalectans. However, unlike the other members, brevican contains only one EGF-like repeat, which shows high sequence similarity to the second EGF repeat of versican and neurocan (128).

In addition to the secreted species of brevican, an isoform of rat brevican encoded by a shorter 3.3 kb mRNA is bound to the plasma membrane via a GPI anchor (151). Immunochemical and biochemical data demonstrate that both soluble and GPI-anchored forms of rat brevican exist and that the latter derives from alternatively processed transcripts. The glypiation signal is encoded by a DNA segment that is removed as an intron from the 3.6 kb-transcript encoding the secreted (larger) form of brevican (151). The GPI-anchored form contains no EGF, lectin, or CRP motifs but contains a stretch of hydrophobic amino acids resembling the GPI-anchor. When transfected into human epithelial cells, this form of brevican localizes to the cell surface and is cleavable by the PI-specific phospholipase C (151). Interestingly, the nucleotide sequence of the rat GPI-anchored species is nearly identical to a sequence immediately distal to exon 8 of the mouse brevican gene (150), which is followed by a potential polyadenylation signal. Collectively, these data indicate that the GPI-anchored brevican variant derives from lack of splicing at the 3' end of exon 8.

Genomic Organization and Transcriptional Control

Analysis of the genomic organization of the hyalectan genes indicates that they are modular and have utilized exon shuffling and duplication during evolution to permit progressive refinement of protein function. The versican (CSPG2) gene (136) maps to human chromosome 5q13.2 (152) and to a syntenic region of mouse chromosome 13 (153). The rat (154), mouse (142, 155), and human (156) aggrecan (AGC1) genes have been fully sequenced and assigned to human chromosome 15q26 (157) and mouse chromosome 7 (158), whereas neurocan (NCAN) and brevican (BCAN) have been mapped to mouse chromosomes 8 (159) and 3 (150), respectively. This diverse chromosomal location not only suggests an early divergence of the hyalectan genes during evolution but also indicates a significant evolutionary pressure to maintain the overall structure of these modular proteoglycans.

A salient feature of the hyalectans is a remarkable conservation of the exon/intron junctions. For example, the hyaluronan-binding region in these genes is encoded by four exons with identical conservation of the exon size and exon/intron phases (9). All of the introns flanking the individual modules are phase I introns, thereby allowing alternative splicing to occur without altering the reading frame. A typical example is the splicing of the GAG-α and GAG-β modules of versican. Alternate exon usage is indeed a widely used mechanism for increasing coding diversity within genes coding for extracellular matrix proteins (160, 161).

Another intriguing property of versican and aggrecan is the large size of the spliced exons encoding the central domain. They vary from 3 to 5.3 kb. This observation is in contrast to the average size of exons ($\sim$150 bp) for a variety of genes and suggests that these genes have evolved a way to bypass the rules of exon definition. In neurocan, the central nonhomologous domain is encoded by two distinct exons of relatively smaller dimension ($\sim$0.6 and 1.2 kb) (159). An established notion is that intervening sequences function by limiting amplification of genomic DNA that needs to maintain a constant size. Thus genes that code for proteins in which the strict dimension and copies of the repeats are not demanded are likely to harbor large exons. A potential advantage for large exons is an improved RNA processing (161).

Another striking conservation of the exon/intron organization occurs in the 3' end of the hyalectan genes. The selectin region is encoded by six exons with identical size and phasing in versican, aggrecan, and neurocan. This arrangement should allow alternative splicing of the EGF and CRP repeats. However, splicing events of this region have been observed only in aggrecan (162). Whether this specialized splicing within the selectin domain has any biological significance awaits further experimentation. Notably, murine brevican is the first hyalectan where an intron bordering a domain module is not a phase I intron, thereby preventing the alternative splicing of the CRP domain (150).

The promoter sequences of the human versican (136), rat and mouse aggrecan (142, 154), and mouse neurocan (159) and brevican (150) genes have been sequenced. However, functional studies are available for only the human versican and rat aggrecan promoters. The human versican promoter is active in both squamous carcinoma cells and in embryonic lung fibroblasts (136). It contains a TATA box and numerous cognate cis-acting elements that could drive the tissue-specific expression of versican. For example, a cluster of AP2-binding sites acts as an enhancer element since the presence of AP2 markedly increases the level of versican promoter activity (61). AP2 is expressed in the neural crest and its derivatives, and versican, and perhaps other members of the hyalectan gene family, may be regulated by AP2 in the central nervous system. Also present in the human versican promoter are binding sites for CCAAT-binding transcription factor, Sp1, CEBP, and several cAMP-responsive elements. Of note, a CCAAT-binding transcription factor has been previously involved in TGF-β-induced upregulation of type I plasminogen activator (66), and it may mediate the established upregulation of versican by TGF-β (163).

In contrast to versican, both the rat (154) and mouse (142) aggrecan promoters lack a TATA box and display multiple transcription start sites. They are relatively enriched in G + C, similar to many housekeeping and growth factor–encoding genes. Accordingly, several Sp1-binding sites are scattered throughout the promoter. There are, however, some similarities with versican insofar as the proximal promoter region of rat aggrecan also harbors several AP2-binding sites. There are conserved sequences common to link protein, aggrecan, and type II collagen (142, 154). These sequences include a potential binding site for NF-κB, a factor known to interact with various cytokines that may also play a role in cartilage differentiation. Since these genes are expressed primarily in cartilaginous tissues, these regions may universally mediate chondrocyte gene expression.

The neurocan putative promoter region contains a TATA box and several cognate *cis*-acting elements commonly found in the other hyalectan genes (159) including several Sp1, AP2, AP1, and glucocorticoid-responsive elements. Partial characterization of the 5' flanking region of mouse brevican shows no TATA or CAAT boxes but reveals a relatively high G + C content, numerous transcription start sites, and several *cis*-acting elements potentially involved in the regulation of neural gene expression (150), consistent with its restricted tissue distribution.

Expression and Functional Roles

Based on the combination of the structural domains summarized above, the hyalectans' major functional roles would be to bind complex carbohydrates such as hyaluronan at their N termini and less complex sugars at their C termini. Less clear is the role of the central nonhomologous regions insofar as they can be substituted with as few as 2 or as many as 100 glycosaminoglyan chains. Via multiple isoforms generated by alternative splicing, these central domains would provide a means to introduce glycosaminoglycans into various extracellular matrices. The different spliced variants of hyalectans are often expressed in distinct spatial and temporal patterns. For example, the V0 and V1 variants of versican are present in fibroblasts, chondrocytes, hepatocytes, and smooth muscle cells of the aorta and myometrium (135). In contrast, keratinocytes express only versican V1.

Chondroitin sulfate proteoglycans have been implicated in the regulation of cell migration and pattern formation in the developing peripheral nervous system. Versican is selectively expressed in embryonic tissues that act as barriers to neural crest cell migration and axonal outgrowth (164). Versican interferes with the attachment of embryonic fibroblasts to various substrata, including fibronectin, laminin, and collagen. Therefore, the expression of versican within barrier tissues may be linked to guidance of migratory neural crest cells and outgrowing axons (164). In contrast, a report suggests that aggrecan, but not versican, inhibits avian neural crest cell migration (165). In muscular arteries versican-specific epitopes are restricted to the tunica adventitia (166), but in the media and the split elastic interna of atherosclerotic lesions, versican is prominent. The latter observation supports a role for versican in the development of atherosclerotic lesions (167). Of relevance, matrilysin, a matrix metalloproteinase, is expressed by macrophages at sites of potential rupture in atherosclerotic lesions and specifically degrades versican (168). Versican has also been implicated in retaining hyaluronan in mouse cumulus cell-oocyte complexes (169) and could play a role in hair follicle development and cycling (170).

Versican is upregulated in smooth muscle cells treated with platelet-derived growth factor (PDGF) and TGF-β (163, 171, 172). The PDGF-mediated induction of versican gene expression is abrogated by genistein, a broad inhibitor of tyrosine kinase activity (173). Versican can also be induced by a cocktail of growth factors including EGF, TGF-β, and PDGF (174), suggesting that there are independent and synergistic signal-transducing pathways regulating its expression. Abnormal versican expression has been demonstrated in human colon cancer where it occurs as a consequence of hypomethylation of its control genomic regions (175, 176), and enhanced deposits of versican have been observed in the stroma of various tumors, particularly in hyaluronan-rich regions (177, 178). These findings support the observation that abrogation of versican expression by stable antisense transfection can revert the malignant phenotype (179). A link between versican expression and cellular growth is provided by the observation that ectopic expression of the retinoblastoma (RB) protein, a potent tumor suppressor, in RB-negative mammary carcinoma cells induces a relatively small number of genes including the two proteoglycans versican and decorin (180). The RB-induced versican transcript was approximately 5 kb, a size compatible with a selective upregulation of the V2 splice variant containing the GAG-α module (136, 181).

The functional properties of aggrecan reside in the two structural features summarized above: the high concentration of chondroitin sulfate side chains and the formation of large supramolecular aggregates with hyaluronan (8). In cartilage, each aggrecan monomer occupies a large hydrodynamic volume, and when subjected to compressive forces, water is displaced from individual monomers. This swelling of the tissue is dissipated readily when the compressive forces are removed and the water molecules are siphoned back into the tissue. The function of aggrecan in cartilage development and homeostasis is demonstrated by the phenotype of two mutant animals—the nanomelic chicken (182, 183) and the cartilage matrix-deficient (*cmd*) mouse (158). Nanomelia, a recessively inherited connective tissue disorder of the chicken in which very low levels of aggrecan mRNA are found, affects cartilage formation. A single base mutation leads to a premature truncation of the protein core, which lacks the C-terminal globular (G3) domain, and the truncated precursor accumulates in the endoplasmic reticulum (182). In agreement with these findings, aggrecan mRNA is detectable in the nuclei of nanomelic chondrocytes but is greatly reduced in the cytoplasm (183).

In the *cmd* mice, a 7-bp deletion in exon 5 of the aggrecan gene has been discovered (158). The mutation occurs in the G1 domain and leads to a termination codon within exon 6, resulting in a truncated polypeptide of approximately 36 kDa. Although heterozygous *cmd* mice appear normal, the homozygous mice die soon after birth due to respiratory failure. Both the nanomelic chicken and the *cmd* mouse are characterized by shortened limbs attesting to the importance of aggrecan in the space-filling role in cartilage.

The functional roles of the two brain proteoglycans, neurocan and brevican, are less well understood. Neurocan binds with a relatively high affinity ($K_d \sim 6$ nM) to the neural cell adhesion molecules Ng-CAM and N-CAM, inhibits their homophilic interactions, and blocks neurite outgrowth (184, 185). Moreover, neurocan interacts with tenascin (186) and axonin-1 (187). Some of these interactions may be confined to restricted areas or to a relatively brief developmental stage, and the multiplicity of potential ligands may provide a mechanism for fine tuning of various regulatory processes of neurocan (187).

Neurocan is developmentally regulated (188, 189), and astrocytes, in addition to neurons, may be a cellular source of neurocan in the brain (190). In situ hybridization and immunohistochemical studies have demonstrated that neurocan transcripts are distributed widely in pre- and postnatal brain but not in other organs (191). The adult form of neurocan is generated by a developmentally regulated proteolytic processing of the larger species predominant in the early postnatal brain (146). The adult protein core, mostly devoid of glycosaminoglycan side chains, is formed by proteolytic cleavage of the C terminus.

Brevican is probably the most abundant hyalectan in the adult brain (128). Two species, a 145-kDa and an 80-kDa band representing the full-length and the N-terminally truncated isoforms of brevican, respectively, represent the largest fraction of whole brain extracts that binds to DEAE-cellulose columns (128). The content of the 80-kDa species increases in the later stages of development, suggesting that the proteolytic processing of brevican may also be developmentally regulated.

Finally, what is the function of a GPI-anchored form of brevican? The obvious answer is that this isoform may link hyaluronan to the cell surface of a specific subset of neurons, or it may induce vectorial insertion of brevican into finite subdomains of the plasma membrane, since the axonal membranes of neurons are equivalent to the apical membranes of polarized epithelia (128).

SMALL LEUCINE-RICH PROTEOGLYCANS

The family of small leucine-rich proteoglycans (SLRPs) contains at least nine distinct products encoded by separate genes (Table 4). They were previously termed nonaggregating or small dermatan-sulfate proteoglycans because of their inability to interact with hyaluronan or because of their type of glycosaminoglycans, respectively. Based on their protein and genomic organization, three classes of SLRPs can be easily identified (Table 4 and Figure 3). Essentially, they are all characterized by a central domain containing leucine-rich

Table 4   General properties of small leucine-rich proteoglycans

Proteoglycan	Gene	Chromosomal mapping Human	Chromosomal mapping Mouse	Protein core (~kDa)[a]	Glycosaminoglycan type (number)
Class I					
Decorin	DCN	12q23	10	40	Dermatan/chondroitin sulfate (1)[b]
Biglycan	BGN	Xq28	X	40	Dermatan/chondroitin sulfate (2)
Class II					
Fibromodulin	FMOD	1q32		42	Keratan sulfate (2–3)
Lumican	LUM	12q21.3–22	10	38	Keratan sulfate (3–4)
Keratocan				38	Keratan sulfate (3–5)
PRELP[c]	PRELP	1q32		44	Keratan sulfate (2–3)
Osteodherin				42	Keratan sulfate (2–3)
Class III					
Epiphycan	DSPG3	12q21		35	Dermatan/chondroitin sulfate (2–3)
Osteoglycin	OG			35	Keratan sulfate (2–3)

[a]The size of individual protein core does not include any posttranslational modification.

[b]Adult chicken cornea contains decorin with keratan sulfate side chains. Also in avian decorin there is the possibility of two glycosaminoglycan side chains.

[c]The inclusion of PRELP in this category is still preliminary because in most cases PRELP appears as a glycoprotein rather than a proteoglycan.

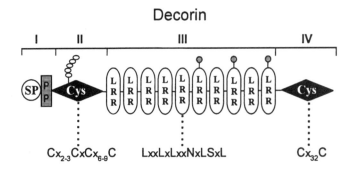

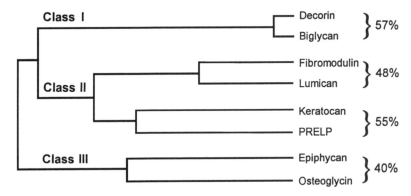

Figure 3   Schematic representation of the structural features of decorin, the prototype member of the small leucine-rich proteoglycans (*top*) and sequence-based evolutionary tree (*bottom*) of the various members of this gene family. The Roman numerals at the top indicate the proposed domains of decorin. The symbols and designations are as follows: SP, signal peptide; PP, propeptide; Cys, cysteine-rich region; LRR, leucine-rich repeat. The consensus sequences for the N-terminal and C-terminal cysteine-rich regions, as well for the leucine-rich repeats are also shown. The glycosaminoglycan side chain and potential N-linked oligosaccharides are indicated by strings of small circles or by a single circle, respectively. In the dendrogram obtained with the program CLUSTAL, branch lengths (*horizontal lines*) are proportional to evolutionary distances. The percent identity in amino acid sequence for each pair of related proteoglycans is indicated in the right margins.

repeats (LRR) flanked at either side by small cysteine-clusters. The prototype member, decorin, is shown in Figure 3 (*top*).

General Structural Features: Three Distinct Classes

Analysis by multiple sequence alignments identifies three classes of SLRPs and two subfamilies. Class I comprises decorin and biglycan, which show the highest homology (57% identity). Class II includes fibromodulin (192), lumican (193), keratocan (194), and PRELP (195). Two subclasses can also be detected (Figure 3): Fibromodulin and lumican comprise the first subclass (48% identity), and keratocan and PRELP comprise the second subfamily (55% identity). The inclusion of PRELP as a Class II proteoglycan is still preliminary because there is only weak evidence that PRELP is indeed a proteoglycan. In most tissues, PRELP is a glycoprotein. Recently, a novel member of the Class II SLRP gene family has been cloned and named osteoadherin (RD Heinegård, personal communication). As in the case of the other members of Class II, osteoadherin is substituted with keratan sulfate and contains tyrosine sulfate in its N-terminal end. However, osteoadherin has a larger C-terminal extension and is primarily expressed in bone.

Class III comprises epiphycan and osteoglycin. Epiphycan (196, 197), also known as PG-Lb, derives its name from its highly selected expression in epiphyseal cartilage. Based on the dendrogram shown in Figure 3, Class III proteoglycans diverged from the common precursors of Class I and II before the presumed gene duplication and independent evolution of the latter two subfamilies.

The overall structural characteristics of the SLRPs are similar although there are some interesting variations. In the prototype SLRP decorin, four domains can be identified: domain I, which contains the signal peptide and a propeptide; domain II, which contains four evenly spaced cysteine residues and the glycosaminoglycan attachment site; domain III, which contains the LRRs; and domain IV, which contains a relatively large loop with two cysteine residues. Interestingly, only decorin and biglycan contain a propeptide (198, 199). While the signal peptide targets the nascent core protein to the rough endoplasmic reticulum, less clear is the function of the propeptide. This sequence is highly conserved across species (200), and the propeptide may function as a recognition signal for the first enzyme (xylosyltransferase) involved in the biosynthesis of glycosaminoglycans (201).

When constructs containing deletions of the decorin propeptide are transfected into mammalian cells, the secreted proteoglycans are substituted with shorter glycosaminoglycan chains (202). Thus deletion of the propeptide may lower the affinity for xylosyltransferase or may induce a more rapid transition through the Golgi compartment. This situation may also occur in biglycan because a recombinant molecule that lacks the propeptide results in a protein core devoid of glycosaminoglycans (203). The presence of the pro-biglycan species has been documented in keratinocytes (204) and endothelial cells (205). In contrast to Class I, Class II SLRPs appear to be proteolytically processed following removal of their signal peptide and thus are devoid of the propeptide (206, 207).

Domain II is the negatively charged region carrying sulfated glycosaminoglycans in decorin, biglycan, and epiphycan and carrying sulfotyrosine in all of the rest, with the exception of PRELP (10). Decorin and biglycan both have a region that contains a series of acidic amino acids followed by a consensus GAG attachment domain with potential attachment sites at Ser[4], or Ser[5] and Ser[11], respectively (208). Site-directed mutagenesis of the serine residue has demonstrated the requirement of a specific amino acid sequence to direct proper O-glycosylation of decorin (209). Biglycan and decorin can occur as either a monoglycanated (208) or biglycanated (210) species. All Class II and III SLRPs contain consensus sequences for tyrosine sulfation, with the exception of PRELP. This sequence is characterized by at least one tyrosine residue followed by an acidic amino acid (Asp or Glu).

Domain II also contains a cluster of highly conserved cysteine residues with a general consensus $Cx_{2-3}CxCx_{6-9}C$, where x is any amino acid and the subscripts denote the number of intervening residues (Figure 3). Interestingly, within each subclass there is identical spacing of the intervening amino acid residues. For example, in Class I the four cysteine residues are spaced by 3, 1, and 6 residues; in Class II, the spacing is a 3, 1, 9 pattern; whereas Class III follows a 2, 1, 6 pattern (10). Not only is the spacing of the cysteine residues conserved within each class of SLRPs, but also the nature of the intervening amino acid is maintained.

Domain III comprises 10 tandem LRRs, with the exception of epiphycan and osteoglycin, which contain only six repeats. In the LRR consensus sequence, that is, LxxLxLxxNxLSxL, L is leucine, isoleucine, or valine, and S is serine or threonine (10). If the consensus for LRRs is interpreted with less stringency, then there could be two additional LRRs, one in the N-terminal and one in the C-terminal end of the molecule. As discussed below, the major function of these LRRs is to bind and interact with other proteins.

The level of complexity increases if one considers the substitution of the protein core with several N-linked oligosaccharides. For example, human decorin has three potential sites and biglycan has two. The variability in the number and complexity of oligosaccharide chains, which are of both the high-mannose and the complex type, could modulate some of the functions of these gene products. A proposed role of the N-linked oligosaccharides in decorin and biglycan is to retard self-aggregation, thereby favoring interactions with cell surface proteins and/or other extracellular matrix constituents.

In Class II proteoglycans, one to several asparagine residues can be substituted with N-linked keratan sulfate. At least in the case of fibromodulin, all four Asn residues in domain III can be acceptors (211). Moreover, polylactosamine, essentially an unsulfated keratan sulfate, can be found in both fibromodulin (212) and keratocan (194). Lumican and keratocan have a characteristic unlike other SLRP members: In the cornea, they are both keratan sulfate–carrying proteoglycans; in other tissues, these molecules occur as poorly sulfated or unsulfated glycoproteins (194).

The function of the C-terminal domain of the SLRPs is the least characterized so far. It comprises about 50 amino acid residues with considerable similarity among the various members. However, there are several stretches that show a profound divergence. Domain IV contains two cysteines spaced by 32 intervening amino acids, except in keratocan and PRELP, where an insertion of 7 and 8 amino acids, respectively, has occurred. For bovine biglycan, a disulfide bond has been demonstrated in this region (208). Thus a large loop of approximately 34 amino acid residues would be formed at the C end of the SLRPs. Notably, reduction and alkylation of decorin and lumican (213, 214) abolishes their ability to interact with collagen. Thus the disulfide bonding at the C-terminal end may also be crucial in maintaining the fibrillogenesis-controlling activity of SLRPs.

Genomic Organization and Transcriptional Control

The various members of the SLRP gene family map to relatively few chromosomes (Table 4). Decorin, lumican, and epiphycan map to the long arm of chromosome 12 between 12q21 and 12q23 (215–217) and in the corresponding syntenic regions of mouse chromosome 10 (200). In contrast, biglycan maps to Xq28 (218), and fibromodulin and PRELP map to the long arm of chromosome 1 at 1q32 (219). The pairwise similarity at the protein level is also reflected at the genomic level. For example, both members of Class I, decorin and biglycan, are encoded by eight distinct exons with very similar intron/exon boundaries (216, 220), and these features are very well conserved in the mouse (200, 221). A unique feature of the human decorin gene is the presence of alternatively spliced leader exons, Ia and Ib, encoding 5'-untranslated sequences (216). The region corresponding to exon Ib is not found in the mouse (200), suggesting that the decorin gene has undergone significant recombination during evolution (10).

In contrast to Class I, members of Class II SLRPs are encoded by three exons, with all ten LRRs encoded by a single large exon (222, 223). In these genes, the introns are positioned at identical sites, just proximal to the translation initiation and termination codons, respectively (223). There is no published information regarding the genomic organization of class III SLRPs; however, a preliminary study suggests that their intron/exon structure follows a distinct pattern (224), further stressing their belonging to a distinct subfamily.

So far, only the promoter regions of decorin and biglycan have been sequenced and shown to be functionally active (61). Decorin has a complex promoter in the region flanking exon Ib (225). The discovery of two leader exons in the 5'-untranslated region has suggested that a two-promoter system and alternative splicing could be responsible for the presence of heterogeneous transcripts from a single gene (216). However, no functional activity for the region flanking exon Ia was found (225). In contrast, strong basal and inducible promoter activity was detected using the approximately 1-kb region 5' to exon Ib (225). This promoter can be divided arbitrarily into two main regions: a proximal promoter of approximately 188 bp and a distal promoter of approximately 800 bp. The proximal promoter region contains two functional TATA boxes and a CAAT box (225). Moreover, it contains two tumor necrosis factor-α (TNF-α)-responsive elements that mediate the binding of TNF-α-induced nuclear proteins and a consequent transcriptional repression of the decorin gene (226). The proximal promoter region contains also a canonical and a functional AP1–binding site, a bimodal regulator of decorin gene expression, which allows both repression by TNF-α and induction by interleukin-1 (IL-1) (227).

These transcriptional data thus provide a molecular mechanism for the previous observations that decorin expression can be induced by either IL-1 (228) or IL-4 (229). The distal promoter of decorin harbors a number of cognate cis-acting factors including AP1, AP5, and NF-κB; several direct repeats; and a TGF-β-negative element. The latter element has been found in a variety of proteinases that are involved in the degradation of collagens and proteoglycans during remodeling and suggests a link between suppression of matrix-degrading enzymes and downregulation of decorin gene expression (230). The distal promoter also contains a long stretch of homopurine/homopyrimidine residues. When contained in a supercoiled plasmid, this sequence is sensitive to endonuclease S1, an enzyme that digests preferentially single-stranded DNA. Moreover, this sequence can upregulate a minimal heterologous promoter (225). Thus this region may adopt an intramolecular hairpin triplex and may regulate in vivo the transcription of decorin.

In contrast to decorin, the promoter of the human biglycan gene contains neither a TATA nor a CAAT box but is G + C rich (220) with a 66% overall G + C content and two clusters of G + C that reach 73 and 87%, respectively. These features have been conserved in the mouse promoter, which contains an overall G + C content of approximately 60% (221). The human biglycan promoter contains numerous Sp1-binding sites and several transcription initiation sites. In contrast to decorin, the biglycan promoter is highly conserved in the mouse, particularly in the proximal region where two AP2 and two Sp1-binding sites are perfectly maintained (221). In both the human and mouse biglycan promoters there are numerous motifs that could potentially bind members of the Ets family of oncogenes (PU-boxes and PEA3 motifs). These factors can transcriptionally activate B cells and macrophages (231).

The human promoter contains at least five IL-6-responsive elements, TNF-α-responsive elements similar to those found in the decorin gene promoter, and a binding site for the liver-specific transcription factor C/EBP (232). The latter is interesting because it may explain why biglycan expression is induced in the liver during the transformation of the fat-storing cells into myofibroblast-like cells (233). Biglycan can be transcriptionally induced in MG-63 osteosarcoma cells by forskolin and 8-Bromo-cAMP (234), and the biglycan promoter is upregulated by exogenous IL-6 and downregulated by TNF-α (232). That the putative binding sites for the corresponding cytokines are functionally active is demonstrated in human endothelial cells whose biglycan expression is down-regulated by TNF-α (235).

Decorin is generally downregulated by TGF-β, whereas biglycan is upregulated in various cells and organisms (61, 163, 236–239). In contrast, dexamethasone increases decorin production and also prevents the TGF-β-elicited downregulation of decorin gene expression (240). In contrast to decorin where TNF-α and TGF-β are additive in their inhibitory action, TNF-α counteracts the effects of TGF-β and IL-6 in controlling biglycan gene expression. Thus these two important members of the SLRP gene family can be diversely regulated at the site of injury by a finely balanced release of active cytokines. A unique feature of the biglycan gene, being located on the X chromosome, is that it should follow the rules of X chromosomal inactivation. However, biglycan does not follow the expected conventional correlation between gene dosage and expression rate (241). That is, biglycan behaves as a pseudoautosomal gene even though in somatic cell hybrid experiments, biglycan undergoes X chromosomal inactivation. These data suggest that there may be an additional regulatory gene(s) that controls the transcriptional activity of biglycan (241).

Control of Collagen Fibrillogenesis

Collagen fibril formation is a self-assembly process that has been investigated in vitro for over four decades. Although it is clear from fibril reconstitution experiments that information to build periodic fibrils resides in the amino acid sequence of the fibril-forming collagens, other molecules have been identified that modulate the assembly process (242). Both the kinetics of assembly and the ultimate fibril diameters are modulated by these factors, and both acceleration and inhibition of fibril formation have been reported.

After the discovery that the interaction of dermatan sulfate proteoglycans with collagen causes increased stability of collagen fibrils and a change in their solubility (243, 244), it was demonstrated that various members of the SLRP gene family interact directly with fibrillar collagens (245). The orthogonal position of the proteoglycans would facilitate proper spacing of the collagen fibrils during axial growth or perhaps during lateral fusion (see below). At least three members of the SLRP gene family, namely decorin (246), fibromodulin (247, 248), and lumican (214), can delay fibril formation in a dynamic fibrillogenesis assay. Neither removal of the glycosaminoglycan chain nor the 17-amino acid N-terminal portion of the decorin protein core significantly alters collagen fibrillogenesis (249), indicating that the decorin/collagen interactions require the remaining protein core (250). Corneal and scleral SLRPs also retard fibrillogenesis in vitro and reduce the size of the formed fibrils (251), in agreement with data generated from analysis of developing avian tendon where a decrease in fibril-associated decorin is necessary for fibril growth during tissue maturation (252).

The three-dimensional (3D) model of human decorin (253) and the ultra-structural observation that several SLRPs are horseshoe shaped (254) predict a close interaction between decorin and fibrillar collagen. This close interaction would help stabilize fibrils and orient fibrillogenesis. The 3D model of decorin, based on the crystal structure of the porcine ribonuclease inhibitor (255), predicts an arch-shaped structure with the inner concave surface lined by β-strands and the outer convex surface formed by α-helices. This model would allow easy access of interactive proteins to the inner surface. Another interesting feature of the decorin model is that the three N-linked oligosaccharides and the single glycosaminoglycan chain at Ser^7 are all positioned on one side of the arch-shaped structure, and the glycosaminoglycan chain is relatively free to project away from the protein core. This arrangement would also account for the proposed function of the glycosaminoglycan chain to maintain inter-fibrillary space (256). The inner surface of the arch-shaped decorin molecule contains a series of charged residues that would complement the charges in a corresponding amino acid stretch within the collagen type I triple helix, the proposed binding site for decorin (257).

Of course, the data generated with the decorin model should be interpreted cautiously because the homology between ribonuclease inhibitor and decorin is relatively low, particularly at the N- and C-terminal ends. However, even if the protein is folding quite differently in these regions, it has been shown that the central domain, which should fold as the ribonuclease inhibitor does, is the essential part for modulating collagen fibrillogenesis (258–260).

Genetic evidence for a role of decorin in maintaining collagen fibrillogenesis has been provided by the phenotype of decorin null animals. The mice carrying a homozygous disruption of the decorin gene grow normally to adulthood; however, they manifest a phenotype characterized by increased skin fragility (261). When samples of skin from the wild-type and decorin null animals were subjected to biomechanical testing, the latter samples exhibited a markedly reduced tensile strength that could be associated with an abnormal collagen fiber formation. Ultrastructural analysis of skin revealed bizarre and irregular collagen morphology with coarser and irregular fiber outlines in the decorin null specimens (Figure 4). Although the mean cross-sectional diameter of the fibers did not vary significantly between the wild-type and decorin null animals, the latter showed a markedly increased range with profiles varying between 40 and 260 nm. Scanning-transmission electron microscopy of isolated collagen fibers revealed that the wide variation in range was not the result of multiple

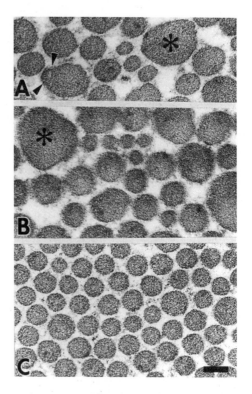

Figure 4   Ultrastructural appearance of dermal collagen from the skin of decorin null (*A* and *B*) and wild-type (*C*) mice. Notice the larger and irregular cross-sectional profiles in the decorin null collagen fibers (*asterisks*) with evidence of lateral fusion (*A, arrowheads*). Bar: 90 nm.

populations of fibrils of different diameter but rather was due to a single population of fibrils with markedly irregular profiles along their axes (261). These studies establish a functional role for decorin in maintaining the structural integrity of the cutis. While decorin may affect other functions (see below), this genetic evidence unequivocally demonstrates that the tensile strength of skin is under the control of so-called modifiers of collagen fibrillogenesis and that decorin is an important member of these modifier proteins.

Interactions with TGF-β and the Control of Cell Proliferation and Corneal Transparency

Increased TGF-β production is a hallmark of a variety of fibrotic states, including cirrhosis, pulmonary fibrosis, and glomerular sclerosis. One of the most important characteristics of decorin is a high affinity for TGF-β (262). TGF-β1, -β2, and -β3 isoforms all bind to the decorin core protein with similar efficiency

(262), allowing decorin to function as a reservoir for these growth factors in the extracellular milieu. This line of research derives from the observation that Chinese hamster ovary (CHO) cells, which do not constitutively synthesize decorin, are growth suppressed when decorin is ectopically expressed (263). Investigators later showed that this growth inhibition was caused by a blocking of TGF-β-activity by decorin (264). Because CHO cells require TGF-β for their growth, it was concluded that decorin-induced growth inhibition is secondary to the blockage of this growth factor. Administration of recombinant decorin or decorin purified from bovine tissues prevents glomerular sclerosis in a rat model of glomerulonephritis (265). The lesions caused by this kind of glomerular injury can be ameliorated by gene therapy with a decorin-expressing vector directly transfected into skeletal muscle (266).

Although this evidence implicates decorin in blocking the action of TGF-β, controversy remains as to whether decorin is a universal TGF-β inhibitor. For example, in quiescent fibroblasts that produce up to 40 times more decorin than cells in the logarithmic phase of growth, nanomolar amounts of TGF-β are still fully active, that is, they can induce the endogenous expression of biglycan (226). Intuitively, the amount of decorin in the medium would have been sufficient to block the exogenous TGF-β. One possibility is that decorin binds avidly to fibrillar collagen, thereby preventing further interaction with TGF-β. Moreover, in the presence of 10,000-fold molar excess of exogenous decorin, TGF-β effects could still be detected in human monocytic cells (267). The addition of decorin to osteoblastic cells enhances the binding of radiolabeled TGF-β to its receptors (268).

Therefore, it appears that in certain cellular systems, decorin blocks the activity of TGF-β, whereas in others its binding augments the bioactivity of the cytokine. How can we reconcile these seemingly conflicting observations? One possibility is that the decorin/TGF-β complexes may still be capable of interacting with at least one signal-transducing pathway of TGF-β signaling. That is, under certain conditions, these complexes may activate rather than repress the cytokine activity, a phenomenon that would be cell specific.

An emerging function of decorin is its ability to inhibit cellular proliferation. When a full-length cDNA driven by the potent cytomegalovirus promoter is introduced into colon carcinoma cells (which do not synthesize decorin), the cells become quiescent, form small colonies in soft agar, and do not generate tumors in immunocompromised hosts (269). All of these effects are independent of TGF-β. Interestingly, a number of clones are arrested in the G_1 phase of the cell cycle, and their growth suppression can be restored by treatment with decorin antisense oligodeoxynucleotides. The decorin-induced growth arrest is associated with an induction of p21, a potent inhibitor of cyclin-dependent kinase activity (270). Ectopic expression of decorin proteoglycan or protein core, a mutated form lacking any glycosaminoglycan side chains, induces growth suppression in neoplastic cells of various histogenetic origins (271). Even when recombinant decorin is added to tumor cells, all react by slowing their proliferative status and by inducing p21 (271). Thus it appears that decorin is an important inhibitor of growth that can act directly on a signal-transduction pathway that leads to activation of cyclin-dependent kinase inhibitors and ultimately to arrest in G_1. Decorin causes rapid phosphorylation of the EGF receptor and a concurrent activation of the mitogen-activated protein (MAP) kinase signal pathway (271a). This leads to a protracted induction of endogenous p21 and ultimate cell cycle arrest. Moreover, recombinant decorin causes a rapid increase in intracellular Ca^{2+} levels in A431 cells (271b). The effects of decorin persist in the absence of extracellular Ca^{2+} but are blocked by AG1478, an EGF-specific tyrosine kinase inhibitor and by downregulation of the EGF receptor. These results indicate a novel action of decorin on the EGF receptor, which results in mobilization of intracellular Ca^{2+} and activation of a signal-transducing pathway that culminates in growth suppression by blocking the cell cycle machinery. The discovery that two tyrosine kinase orphan receptors, the discoidin domain receptors DDR1 and DDR2, are the receptors for fibrillar collagen (Types I, III, and V) opens a novel perspective in extracellular matrix research (271c, 271d). Both fibrillar collagen and decorin converge on similar tyrosine kinase receptors, and their activation may allow a specific cross talk betwen cells and the extracellular matrix.

The ability of secreted decorin to induce growth suppression gives further support to the concept that abnormal production of this proteoglycan around invading carcinomas represents a specialized biological response of the host designed to counterbalance the invading tumor cells (176, 272, 273).

Decorin, as well as other SLRP members, may also modulate the remodeling of the extracellular matrix since decorin can induce collagenase (274). When vascular endothelial cells, which usually synthesize neither type I collagen nor decorin, initiate the formation of tubes or cords, they begin to synthesize both molecules (275). In contrast, when endothelial cells are wounded in vitro, biglycan is induced at the edge of the migrating endothelial cells, a process that is apparently mediated by release of endogenous FGF-2 (276). Thus the same theme emerges, with decorin and biglycan, though structurally related, serving diverse functions when cells are migrating to form new blood vessels. The high-affinity interaction of decorin and biglycan with important mediators of the inflammatory process, such as the C1q component of the C1 complex (277, 278), and the affinity of the other SLRP members for a variety of extracellular matrix constituents suggest that these molecules play a primary role in repair processes.

Biglycan has also been directly implicated in regulating hemopoiesis. For example, biglycan was identified as a potent factor stimulating monocytic activity from thymic myoid cells (279) and as one of several products that have affinity for precursors of B lymphocytes (280). Notably, while biglycan is incapable of stimulating the growth of myeloid cells, it markedly enhances the cloning of IL-7-responsive precursors (280).

An important function of members of the SLRP gene family is related to the maintenance of corneal transparency. In cornea, the diameter of the collagen fibrils and the interfibrillary gaps must be kept constant to prevent corneal opacity (256). Lateral growth of collagen segments isolated from bovine (214) or avian (281) corneas can be retarded by addition of decorin or lumican extracted from their respective corneal stromas. Moreover, treatment of the developing avian cornea with β-D-xyloside, an inhibitor of glycosaminoglycan attachment (which presumably affects only dermatan/chondroitin sulfate glycanation but not the addition of keratan sulfate) does not alter collagen fibril diameter (282).

The developmental changes of lumican, primarily those occurring in the first two weeks of corneal development, are fundamental to the maintenance of corneal transparency (283). In fact, the fully sulfated species of lumican is not abundant before day 15 of avian corneal development. Before that time, there is a progressive accumulation of lumican substituted with polylactosamine, suggesting that its subsequent sulfation could play a role during the acquisition of corneal transparency (283). This notion is also supported by the finding that in macular corneal dystrophy, the polylactosamine form of lumican is the primary species (284). A recent study using synchrotron X-ray diffraction has demonstrated a unique 4.6-C periodicity in macular corneal dystrophy (285). After observing digestion of proteoglycans with various polysaccharide lyases, researchers concluded that dermatan sulfate or keratan sulfate proteoglycan hybrids would be also present in this disease.

PERSPECTIVES

Our appreciation of the molecular design of the matrix proteoglycans is leading to a better understanding of their cellular functions. The application of modern gene technology in this field has been slower than in others, but initial successes are now being reported. Current work aims at understanding the mechanisms that govern proteoglycan diversity, especially in regard to tissue-specific constraints that favor the expression of specialized variants. Contemporary areas of active research also focus on clarifying the mechanisms that regulate the generation of proteoglycan isoforms—containing a specific set of glycosaminoglycans or a proteolytically processed species—and of biologically active growth factor/proteoglycan complexes.

Some of the major tasks that lie ahead relate to the delineation of the primary mechanisms that trigger a specific cellular response to various proteoglycans, and the determinants of cellular responsiveness to them. Ultimately, proteogly-can-dependent induction of transcription factors and their modulation will be an area of vibrant research. How do proteoglyans induce so many pleiotropic responses? How do diverse proteoglycans activate specific signal-transducing pathways? What is the molecular cross-talk between a given proteoglycan family and the transcriptional machinery? How do cells regulate their complex biosynthetic pathways?

In part, some of these questions will be answered by the analysis of mutant animals, whereas others will be answered by resolving the various biochemical steps of proteoglycan assembly and metabolism. The correlation of structure and function may soon be possible for several actively investigated members of the matrix proteoglycans by using modern genetic approaches that involve site-directed mutagenesis, transfection into appropriate recipients, and analysis of their functional consequences. One can anticipate continued rapid progress and await with anticipation the further advances in this exciting area of research.

ACKNOWLEDGMENTS

I am grateful to many colleagues for helpful discussion and apologize to those whose work was not referenced because of space limitations. I thank J Esko for his friendship and generosity; W Halfter, G Cole, D Heinegård, J Gallagher, and U Rauch for sharing data before publication; CC Clark for critical reading of this review; and I Eichstetter for help with the illustrations. The original work was supported by National Institutes of Health grants 5RO1 CA39481-14 and 5RO1 CA47282-08.

> Visit the *Annual Reviews home page* at
> http://www.AnnualReviews.org.

Literature Cited

1. Gallagher JT. 1989. *Curr. Opin. Cell Biol.* 1201–18
2. Kjellén L, Lindahl U. 1991. *Annu. Rev. Biochem.* 60:443–75
3. Esko JD. 1991. *Curr. Opin. Cell Biol.* 3: 805–16
4. Bernfield M, Kokenyesi R, Kato M, Hinkes MT, Spring J, et al. 1992. *Annu. Rev. Cell Biol.* 8:365–93
5. Couchman JR, Woods A. 1993. In *Cell Surface and Extracellular Glycoconjugates*, ed. DD Roberts, RP Mecham, pp. 33–81. San Diego: Academic
6. Humphries DE, Stevens RL. 1992. In *Heparin and Related Polysaccharides*, ed. DA Lane, I Bjork, U Lindahl, pp. 59–67. New York: Plenum
7. Elenius K, Jalkanen M. 1994. *J. Cell Sci.* 107:2975–82
8. Roughley PJ, Lee ER. 1994. *Micro. Res. Tech.* 28:385–97
9. Iozzo RV, Murdoch AD. 1996. *FASEB J.* 10:598–614
10. Iozzo RV. 1997. *Crit. Rev. Biochem. Mol. Biol.* 32:141–74
11. Rosenberg RD, Shworak NW, Liu J, Schwartz JJ, Zhang L. 1997. *J. Clin. Invest.* 99:2062–70

12. Noonan DM, Fulle A, Valente P, Cai S, Horigan E, et al. 1991. *J. Biol. Chem.* 266:22939–47
13. Tsen G, Halfter W, Kröger S, Cole GJ. 1995. *J. Biol. Chem.* 270:3392–99
14. Wu R-R, Couchman JR. 1997. *J. Cell Biol.* 136:433–44
15. Noonan DM, Hassell JR. 1993. *Kidney Int.* 43:53–60
16. Murdoch AD, Iozzo RV. 1993. *Virchows Arch. A* 423:237–42
17. Timpl R. 1993. *Experientia* 49:417–28
18. Iozzo RV. 1994. *Matrix Biol.* 14:203–8
19. Iozzo RV, Cohen IR, Grässel S, Murdoch AD. 1994. *Biochem. J.* 302:625–39
20. Kokenyesi R, Silbert JE. 1995. *Biochem. Biophys. Res. Commun.* 211:262–67
21. Groffen AJA, Buskens CAF, Tryggvason K, Veerkamp JH, Monnens LAH, et al. 1996. *Eur. J. Biochem.* 241:827–34
22. Dolan M, Horchar T, Rigatti B, Hassell JR. 1997. *J. Biol. Chem.* 272:4316–22
23. Costell M, Mann K, Yamada Y, Timpl R. 1997. *Eur. J. Biochem.* 243:115–21
24. Bork P, Patthy L. 1995. *Protein Sci.* 4: 1421–25
25. Costell M, Sasaki T, Mann K, Yamada Y, Timpl R. 1996. *FEBS Lett.* 396:127–31
26. Battaglia C, Mayer U, Aumailley M, Timpl R. 1992. *Eur. J. Biochem.* 208: 359–66
27. Reinhardt D, Mann K, Nischt R, Fox JW, Chu M-L, et al. 1993. *J. Biol. Chem.* 268:10881–87
28. Timpl R, Brown JC. 1994. *Matrix Biol.* 14:275–81
29. Schulze B, Mann K, Battistutta R, Wiedemann H, Timpl R. 1995. *Eur. J. Biochem.* 231:551–56
30. Schulze B, Sasaki T, Costell M, Mann K, Timpl R. 1996. *Matrix Biol.* 15:349–57
31. Chakravarti S, Horchar T, Jefferson B, Laurie GW, Hassell JR. 1995. *J. Biol. Chem.* 270:404–9
32. Kallunki P, Tryggvason K. 1992. *J. Cell Biol.* 116:559–71
33. Murdoch AD, Dodge GR, Cohen I, Tuan RS, Iozzo RV. 1992. *J. Biol. Chem.* 267: 8544–57
34. Yurchenco PD, Cheng Y-S, Ruben GC. 1987. *J. Biol. Chem.* 262:17668–76
35. Hunter DD, Porter BE, Bulock JW, Adams SP, Merlie JP, et al. 1989. *Cell* 59:905–13
36. Iozzo RV, Kovalszky I, Hacobian N, Schick PK, Ellingson JS, et al. 1990. *J. Biol. Chem.* 265:19980–89
37. Danielson KG, Martinez-Hernandez A, Hassell JR, Iozzo RV. 1992. *Matrix* 11: 22–35
38. SundarRaj N, Fite D, Ledbetter S, Chakravarti L, Hassell JR. 1995. *J. Cell Sci.* 108:2601–4
39. Couchman JR, Kapoor R, Sthanam M, Wu R-R. 1996. *J. Biol. Chem.* 271:9595–602
40. Iozzo RV. 1989. *J. Biol. Chem.* 264: 2690–99
41. Iozzo RV, Hassell JR. 1989. *Arch. Biochem. Biophys.* 269:239–49
42. Denzer AJ, Gesemann M, Schumacher B, Ruegg MA. 1995. *J. Cell Biol.* 131: 1547–60
43. Hagen SG, Michael AF, Butkowski RJ. 1993. *J. Biol. Chem.* 268:7261–69
44. Denzer AJ, Brandenberger R, Gesemann M, Chiquet M, Ruegg MA. 1997. *J. Cell Biol.* 137:671–83
45. Tsen G, Napier A, Halfter W, Cole GJ. 1995. *J. Biol. Chem.* 270:15934–37
46. Denzer AJ, Gesemann M, Ruegg MA. 1996. *Semin. Neurosci.* 8:357–66
47. O'Toole JJ, Deyst KA, Bowe MA, Nastuk MA, McKechnie BA, et al. 1996. *Proc. Natl. Acad. Sci. USA* 93:7369–74
48. Iozzo RV, Clark CC. 1986. *J. Biol. Chem.* 261:6658–69
49. Iozzo RV, Clark CC. 1987. *Histochemistry* 88:23–29
50. Cohen IR, Grässel S, Murdoch AD, Iozzo RV. 1993. *Proc. Natl. Acad. Sci. USA* 90:10404–8
51. Rupp F, Özçelik T, Linial M, Peterson K, Francke U, et al. 1992. *J. Neurosci.* 12:3535–44
52. Dodge GR, Kovalszky I, Chu M-L, Hassell JR, McBride OW, et al. 1991. *Genomics* 10:673–80
53. Wintle RF, Kisilevsky R, Noonan D, Duncan AMV. 1990. *Cytogenet. Cell Genet.* 54:60–61
54. Iozzo RV, Pillarisetti J, Sharma B, Murdoch AD, Danielson KG, et al. 1997. *J. Biol. Chem.* 272:5219–28
55. Mitchell PJ, Tjian R. 1989. *Science* 245: 371–78
56. Piédagnel R, Prié D, Cassingéna R, Ronco PM, Lelong B. 1994. *J. Biol. Chem.* 269:17469–76
57. McBain JA, Pettit GR, Mueller GC. 1990. *Cell Growth Differ.* 1:281–91
58. Grässel S, Cohen IR, Murdoch AD, Eichstetter I, Iozzo RV. 1995. *Mol. Cell. Biochem.* 145:61–68
59. Orkin SH. 1995. *J. Biol. Chem.* 270: 4955–58
60. Murdoch AD, Liu B, Schwarting R, Tuan RS, Iozzo RV. 1994. *J. Histochem. Cytochem.* 42:239–49

61. Iozzo RV, Danielson KG. 1998. *Prog. Nucleic Acids Res. Mol. Biol.* In press
62. Couchman JR, Ljubimov AV, Sthanam M, Horchar T, Hassell JR. 1995. *J. Histochem. Cytochem.* 43:955–63
63. Ritzenthaler JD, Goldstein RH, Fine A, Lichtler A, Rowe DW, et al. 1991. *Biochem. J.* 280:157–62
64. Rossi P, Karsenty G, Roberts AB, Roche NS, Sporn MB, et al. 1988. *Cell* 52:405–14
65. Marigo V, Volpin D, Vitale G, Bressan GM. 1994. *Biochem. Biophys. Res. Commun.* 199:1049–56
66. Riccio A, Pedone PV, Lund LR, Olesen T, Olsen HS, et al. 1992. *Mol. Cell. Biol.* 12:1846–55
67. Courtois SJ, Lafontaine DA, Lemaigre FP, Durviaux SM, Rousseau GG. 1990. *Nucleic Acids Res.* 18:57–64
68. Dodge GR, Kovalszky I, Hassell JR, Iozzo RV. 1990. *J. Biol. Chem.* 265:18023–29
69. Morris JE, Gaza G, Potter SW. 1994. *Cell. Dev. Biol.* 30A:120–28
70. Chakravarti S, Hassell JR, Phillips SL. 1993. *Dev. Dyn.* 197:107–14
71. Ko CW, Bhandari B, Yee J, Terhune WC, Maldonado R, et al. 1996. *Mol. Cell. Biochem.* 162:65–73
72. Kasinath BS, Grellier P, Ghosh-Choudhury G, Abboud SL. 1996. *J. Cell. Physiol.* 167:131–36
73. Kanwar YS, Rosenzweig LJ, Linker A, Jakubowski ML. 1983. *Proc. Natl. Acad. Sci. USA* 80:2272–75
74. Templeton DM, Fan M-Y. 1996. *Metabolism* 45:1136–46
75. Wahab NA, Harper K, Mason RM. 1996. *Biochem. J.* 316:985–92
76. Kanwar YS, Liu ZZ, Kumar A, Usman MI, Wada J, et al. 1996. *J. Clin. Invest.* 98:2478–88
77. Kanwar YS, Liu ZZ, Wallner EI. 1997. *Lab. Invest.* 76:671–81
78. Dziadek M, Fujiwara S, Paulsson M, Timpl R. 1985. *EMBO J.* 4:905–12
79. Carson DD, Tang J-P, Julian J. 1993. *Dev. Biol.* 155:97–106
80. Smith SE, French MM, Julian J, Paria BC, Key SK, et al. 1997. *Dev. Biol.* 184:38–47
81. Handler M, Yurchenco PD, Iozzo RV. 1997. *Dev. Dyn.* 210:130–45
82. Friedl A, Chang Z, Tierney A, Rapraeger AC. 1997. *Am. J. Pathol.* 150:1443–55
83. Nugent MA, Karnovsky MJ, Edelman ER. 1993. *Circ. Res.* 73:1051–60
84. Forsten KE, Courant NA, Nugent MA. 1997. *J. Cell. Physiol.* 172:209–20
85. Lyon M, Rushton G, Gallagher JT. 1997. *J. Biol. Chem.* 272:18000–6
86. Stringer SW, Gallagher JT. 1997. *J. Biol. Chem.* 272:20508–14
87. Clowes AW, Karnovsky MJ. 1977. *Nature* 265:625–26
88. Benitz WE, Kelley RT, Anderson CM, Lorant DE, Bernfield M. 1990. *Am. J. Respir. Cell Mol. Biol.* 2:13–24
89. Weiser MCM, Belknap JK, Grieshaber SS, Kinsella MG, Majack RA. 1996. *Matrix Biol.* 15:331–40
90. Joseph SJ, Ford MD, Barth C, Portbury S, Bartlett PF, et al. 1996. *Development* 122:3443–52
91. Weiser MCM, Grieshaber NA, Schwartz PE, Majack RA. 1997. *Mol. Biol. Cell* 8:999–1011
92. Whitelock J, Mitchell S, Underwood PA. 1997. *Cell Biol. Int.* 21:181–89
93. Yayon A, Klagsbrun M, Esko JD, Leder P, Ornitz DM. 1991. *Cell* 64:841–48
94. Rapraeger AC, Krufka A, Olwin BB. 1991. *Science* 22:1705–8
95. Roberts R, Gallagher J, Spooncer E, Allen TD, Bloomfield F, et al. 1988. *Nature* 332:376–78
96. Turnbull JE, Fernig DG, Ke Y, Wilkinson MC, Gallagher JT. 1992. *J. Biol. Chem.* 267:10337–41
97. Ornitz DM, Herr AB, Nilsson M, Westman J, Svahn C, et al. 1995. *Science* 268:432–36
98. Aviezer D, Hecht D, Safran M, Eisinger M, David G, et al. 1994. *Cell* 79:1005–13
99. Folkman J, Klagsbrun M, Sasse J, Wadzinski M, Ingber D, et al. 1988. *Am. J. Pathol.* 130:393–400
100. Vigny M, Ollier-Hartmann MP, Lavigne M, Fayein N, Jeanny JC, et al. 1988. *J. Cell. Physiol.* 137:321–28
101. Ohtani H, Nakamura S, Watanabe Y, Mizoi T, Saku T, et al. 1993. *Lab. Invest.* 68:520–27
102. Timar J, Ladanyi A, Lapis K, Moczar M. 1992. *Am. J. Pathol.* 141:467–74
103. Cohen IR, Murdoch AD, Naso MF, Marchetti D, Berd D, et al. 1994. *Cancer Res.* 54:5771–74
104. Marchetti D, Menter D, Jin L, Nakajima M, Nicolson GL. 1993. *Int. J. Cancer* 55:692–99
105. Imamura T, Tokita Y, Mitsui Y. 1991. *Cell Struct. Funct.* 16:225–30
106. Aviezer D, Iozzo RV, Noonan DM, Yayon A. 1997. *Mol. Cell. Biol.* 17:1938–46
107. Mathiak M, Yenisey C, Grant DS, Sharma B, Iozzo RV. 1997. *Cancer Res.* 57:2130–36

108. Klein G, Conzelmann S, Beck S, Timpl R, Müller CA. 1995. *Matrix Biol.* 14: 457–65
109. Gauer S, Schulzelohoff E, Schleicher E, Sterzel RB. 1996. *Eur. J. Cell Biol.* 70:233–42
110. Frieser M, Hallmann R, Johansson S, Vestweber D, Goodman SL, et al. 1996. *Eur. J. Immunol.* 26:3127–36
111. Whitelock JM, Murdoch AD, Iozzo RV, Underwood PA. 1996. *J. Biol. Chem.* 271:10079–86
112. Saksela O, Rifkin DB. 1990. *J. Cell Biol.* 110:767–75
113. Kanwar YS, Linker A, Farquhar MG. 1980. *J. Cell Biol.* 86:688–93
114. van den Born J, van den Heuvel LPWJ, Bakker MAH, Veerkamp JH, Assmann KJM, et al. 1992. *Kidney Int.* 41:115–23
115. van den Born J, van den Heuvel LPWJ, Bakker MAH, Veerkamp JH, Assmann KJM, et al. 1993. *Kidney Int.* 43:454–63
116. Oda O, Shinzato T, Ohbayashi K, Takai I, Kunimatsu M, et al. 1996. *Clin. Chim. Acta* 255:119–32
117. Ruegg MA. 1996. *Curr. Opin. Neurobiol.* 6:97–103
118. Halfter W, Schurer B, Yip J, Yip L, Tsen G, et al. 1997. *J. Comp. Neurol.* 381:1–17
119. Godfrey EW, Dietz ME, Morstad AL, Wallskog PA, Yorde DE. 1988. *J. Cell Biol.* 106:1263–72
120. Barber AJ, Lieth E. 1997. *Dev. Dyn.* 208:62–74
121. Gautam M, Noakes PG, Moscoso L, Rupp F, Scheller RH, et al. 1996. *Cell* 85:525–35
122. DeChiara TM, Bowen DC, Valenzuela DM, Simmons MV, Poueymirou WT, et al. 1996. *Cell* 85:501–12
123. Glass DJ, Bowen DC, Stitt TN, Radziejewski C, Bruno J, et al. 1996. *Cell* 85:513–23
124. Slater CR. 1996. *Nature* 381:478–79
125. McCarthy KJ, Abrahamson DR, Bynum KR, St. John PL, Couchman JR. 1994. *J. Histochem. Cytochem.* 42:473–84
126. McCarthy KJ, Couchman JR. 1990. *J. Histochem. Cytochem.* 38:1479–86
127. Thomas GJ, Shewring L, McCarthy KJ, Couchman JR, Mason RM, et al. 1995. *Kidney Int.* 48:1278–89
128. Yamaguchi Y. 1996. *Perspect. Dev. Neurobiol.* 3:307–17
129. LeBaron RG. 1996. *Perspect. Dev. Neurobiol.* 3:261–71
130. Zimmermann DR, Ruoslahti E. 1989. *EMBO J.* 8:2975–81
131. Shinomura T, Nishida Y, Ito K, Kimata K. 1993. *J. Biol. Chem.* 268:14461–69
132. Neame PJ, Barry FP. 1993. *Experientia* 49:393–402
133. Kohda D, Morton CJ, Parkar AA, Hatanaka H, Inagaki FM, et al. 1996. *Cell* 86:767–75
134. LeBaron RG, Zimmermann DR, Ruoslahti E. 1992. *J. Biol. Chem.* 267:10003–10
135. Dours-Zimmermann MT, Zimmermann DR. 1994. *J. Biol. Chem.* 269:32992–98
136. Naso MF, Zimmermann DR, Iozzo RV. 1994. *J. Biol. Chem.* 269:32999–3008
137. Zako M, Shinomura T, Kimata K. 1997. *J. Biol. Chem.* 272:9325–31
138. Kishore U, Eggleton P, Reid KBM. 1997. *Matrix Biol.* 15:583–92
139. Ujita M, Shinomura T, Ito K, Kitagawa Y, Kimata K. 1994. *J. Biol. Chem.* 269:27603–9
140. Aspberg A, Binkert C, Ruoslahti E. 1995. *Proc. Natl. Acad. Sci. USA* 92: 10590–94
141. Doege KJ, Sasaki M, Kimura T, Yamada Y. 1991. *J. Biol. Chem.* 266:894–902
142. Watanabe H, Gao L, Sugiyama S, Doege K, Kimata K, et al. 1995. *Biochem. J.* 308:433–40
143. Doege KJ, Coulter SN, Meek LM, Maslen K, Wood JG. 1997. *J. Biol. Chem.* 272:13974–79
144. Fülöp C, Walcz E, Valyon M, Glant TT. 1993. *J. Biol. Chem.* 268:17377–83
145. Drickamer K. 1993. *Curr. Opin. Struct. Biol.* 3:393–400
146. Rauch U, Karthikeyan L, Maurel P, Margolis RU, Margolis RK. 1992. *J. Biol. Chem.* 267:19536–47
147. Retzler C, Wiedemann H, Kulbe G, Rauch U. 1996. *J. Biol. Chem.* 271: 17107–13
148. Yamada H, Watanabe K, Shimonaka M, Yamaguchi Y. 1994. *J. Biol. Chem.* 269:10119–26
149. Yamada H, Watanabe K, Shimonaka M, Yamasaki M, Yamaguchi Y. 1995. *Biochem. Biophys. Res. Commun.* 216: 957–63
150. Rauch U, Meyer H, Brakebusch C, Seidenbecher C, Gundelfinger ED, et al. 1997. *Genomics* 44:15–21
151. Seidenbecher CI, Richter K, Rauch U, Fässler R, Garner CC, et al. 1995. *J. Biol. Chem.* 270:27206–12
152. Iozzo RV, Naso MF, Cannizzaro LA, Wasmuth JJ, McPherson JD. 1992. *Genomics* 14:845–51
153. Naso MF, Morgan JL, Burchberg AM, Siracusa LD, Iozzo RV. 1995. *Genomics* 29:297–300

154. Doege KJ, Garrison K, Coulter SN, Yamada Y. 1994. *J. Biol. Chem.* 269: 29232–40
155. Walcz E, Deák F, Erhardt P, Coulter SN, Fülöp C, et al. 1994. *Genomics* 22:364–71
156. Valhmu WB, Palmer GD, Rivers PA, Ebara S, Cheng J-F, et al. 1995. *Biochem. J.* 309:535–42
157. Korenberg JR, Chen XN, Doege K, Grover J, Roughley PJ. 1993. *Genomics* 16:546–48
158. Watanabe H, Kimata K, Line S, Strong D, Gao L-y, et al. 1994. *Nat. Genet.* 7:154–57
159. Rauch U, Grimpe B, Kulbe G, Arnold-Ammer I, Beier DR, et al. 1995. *Genomics* 28:405–10
160. Ushkaryov YA, Südhof TC. 1993. *Proc. Natl. Acad. Sci. USA* 90:6410–14
161. Boyd CD, Pierce RA, Schwarzbauer JE, Doege K, Sandell LJ. 1993. *Matrix* 13:457–69
162. Grover J, Roughley PJ. 1993. *Biochem. J.* 291:361–67
163. Kähäri V-M, Larjava H, Uitto J. 1991. *J. Biol. Chem.* 266:10608–15
164. Landolt RM, Vaughan L, Winterhalter KH, Zimmermann DR. 1995. *Development* 121:2303–12
165. Perris R, Perissinotto D, Pettway Z, Bronner-Fraser M, Mörgelin M, et al. 1996. *FASEB J.* 10:293–301
166. Bode-Lesniewska B, Dours-Zimmermann MT, Odermatt BF, Briner J, Heitz PU, et al. 1996. *J. Histochem. Cytochem.* 44:303–12
167. Wight TN, Kinsella MG, Qwarnström EA. 1992. *Curr. Opin. Cell Biol.* 4:793–801
168. Halpert I, Sires UI, Roby JD, Potter-Perigo S, Wight TN, et al. 1996. *Proc. Natl. Acad. Sci. USA* 93:9748–53
169. Camaioni A, Salustri A, Yanagishita M, Hascall VC. 1996. *Arch. Biochem. Biophys.* 325:190–98
170. du Cros DL, LeBaron RG, Couchman JR. 1995. *J. Invest. Dermatol.* 105:426–31
171. Schönherr E, Järveläinen HT, Sandell LJ, Wight TN. 1991. *J. Biol. Chem.* 266: 17640–47
172. Häkkinen L, Westermarck J, Kähäri V-M, Larjava H. 1996. *J. Dent. Res.* 75:1767–78
173. Schönherr E, Kinsella MG, Wight TN. 1997. *Arch. Biochem. Biophys.* 339: 353–61
174. Tiedemann K, Malmström A, Westergren-Thorsson G. 1996. *Matrix Biol.* 15:469–78

175. Adany R, Iozzo RV. 1990. *Biochem. Biophys. Res. Commun.* 171:1402–13
176. Iozzo RV. 1995. *Lab. Invest.* 73:157–60
177. Isogai Z, Shinomura T, Yamakawa N, Takeuchi J, Tsuji T, et al. 1996. *Cancer Res.* 56:3902–8
178. Nara Y, Kato Y, Torii Y, Tsuji Y, Nakagaki S, et al. 1997. *Histochem. J.* 29:21–30
179. Yamagata M, Kimata K. 1994. *J. Cell Sci.* 107:2581–90
180. Rohde M, Warthoe P, Gjetting T, Lukas J, Bartek J, et al. 1996. *Oncogene* 12: 2393–401
181. Zimmermann DR, Dours-Zimmermann MT, Schubert M, Bruckner-Tuderman L. 1994. *J. Cell Biol.* 124:817–25
182. Li H, Schwartz NB, Vertel BM. 1993. *J. Biol. Chem.* 268:23504–11
183. Primorac D, Stover ML, Clark SH, Rowe DW. 1994. *Matrix Biol.* 14:297–305
184. Grumet M, Flaccus A, Margolis RU. 1993. *J. Cell Biol.* 120:815–24
185. Friedlander DR, Milev P, Karthikeyan L, Margolis RK, Margolis RU, et al. 1994. *J. Cell Biol.* 125:669–80
186. Grumet M, Milev P, Sakurai T, Karthikeyan L, Bourdon M, et al. 1994. *J. Biol. Chem.* 269:12142–46
187. Milev P, Maurel P, Häring M, Margolis RK, Margolis RU. 1996. *J. Biol. Chem.* 271:15716–23
188. Engel M, Maurel P, Margolis RU, Margolis RK. 1996. *J. Comp. Neurology* 366:34–43
189. Meyer-Puttlitz B, Milev P, Junker E, Zimmer I, Margolis RU, et al. 1995. *J. Neurochem.* 65:2327–37
190. Watanabe E, Aono S, Matsui F, Yamada Y, Naruse I, et al. 1995. *Eur. J. Neurosci.* 7:547–54
191. Margolis RU, Margolis RK. 1994. *Methods Enzymol.* 245:105–26
192. Oldberg C, Antonsson P, Lindblom K, Heinegård D. 1989. *EMBO J.* 8:2601–4
193. Blochberger TC, Vergnes J-P, Hempel J, Hassell JR. 1992. *J. Biol. Chem.* 267:347–52
194. Corpuz LM, Funderburgh JL, Funderburgh ML, Bottomley GS, Prakash S, et al. 1996. *J. Biol. Chem.* 271:9759–63
195. Bengtsson E, Neame PJ, Heinegård D, Sommarin Y. 1995. *J. Biol. Chem.* 270:25639–44
196. Deere M, Johnson J, Garza S, Harrison WR, Yoon S-J, et al. 1996. *Genomics* 38:399–404
197. Johnson J, Rosenberg L, Choi HU, Garza S, Höök M, et al. 1997. *J. Biol. Chem.* 272:18709–17

198. Krusius T, Ruoslahti E. 1986. *Proc. Natl. Acad. Sci. USA* 83:7683–87
199. Fisher LW, Termine JD, Young MF. 1989. *J. Biol. Chem.* 264:4571–76
200. Scholzen T, Solursh M, Suzuki S, Reiter R, Morgan JL, et al. 1994. *J. Biol. Chem.* 269:28270–81
201. Sawhney RS, Hering TM, Sandell LJ. 1991. *J. Biol. Chem.* 266:9231–40
202. Oldberg C, Antonsson P, Moses J, Fransson L-C. 1996. *FEBS Lett.* 386:29–32
203. Hocking AM, McQuillan DJ. 1996. *Glycobiology* 6:717
204. Bianco P, Riminucci M, Fisher LW. 1993. In *Dermatan Sulphate Proteoglycans*, ed. JE Scott, pp. 193–205. London: Portland
205. Yeo TK, Torok MA, Kraus HL, Evans SA, Zhou Y, et al. 1995. *J. Vasc. Res.* 32:175–82
206. Plaas AHK. 1992. *Trends Glycosci. Glycotech.* 4:445–55
207. Funderburgh JL, Funderburgh ML, Brown SJ, Vergnes J-P, Hassell JR, et al. 1993. *J. Biol. Chem.* 268:11874–80
208. Neame PJ, Choi HU, Rosenberg LC. 1989. *J. Biol. Chem.* 264:8653–61
209. Mann DM, Yamaguchi Y, Bourdon MA, Ruoslahti E. 1990. *J. Biol. Chem.* 265:5317–23
210. Blaschke UK, Hedbom E, Bruckner P. 1996. *J. Biol. Chem.* 271:30347–53
211. Plaas AHK, Neame PJ, Nivens CM, Reiss L. 1990. *J. Biol. Chem.* 265:20634–40
212. Plaas AHK, Wong-Palms S. 1993. *J. Biol. Chem.* 268:26634–44
213. Scott PG, Winterbottom N, Dodd CM, Edwards E, Pearson CH. 1986. *Biochem. Biophys. Res. Commun.* 138:1348–54
214. Rada JA, Cornuet PK, Hassell JR. 1993. *Exp. Eye Res.* 56:635–48
215. McBride OW, Fisher LW, Young MF. 1990. *Genomics* 6:219–25
216. Danielson KG, Fazzio A, Cohen I, Cannizzaro LA, Eichstetter I, et al. 1993. *Genomics* 15:146–60
217. Chakravarti S, Stallings RL, SundarRaj N, Cornuet PK, Hassell JR. 1995. *Genomics* 27:481–88
218. Traupe H, van den Ouweland AMW, van Oost BA, Vogel W, Vetter U, et al. 1992. *Genomics* 13:481–83
219. Grover J, Chen X-N, Korenberg JR, Recklies AD, Roughley PJ. 1996. *Genomics* 38:109–17
220. Fisher LW, Heegaard A-M, Vetter U, Vogel W, Just W, et al. 1991. *J. Biol. Chem.* 266:14371–77
221. Wegrowski Y, Pillarisetti J, Danielson KG, Suzuki S, Iozzo RV. 1995. *Genomics* 30:8–17
222. Antonsson P, Heinegård D, Oldberg C. 1993. *Biochim. Biophys. Acta* 1174:204–6
223. Grover J, Chen X-N, Korenberg JR, Roughley PJ. 1995. *J. Biol. Chem.* 270:21942–49
224. Ujita M, Shinomura T, Kimata K. 1995. *Gene* 158:237–40
225. Santra M, Danielson KG, Iozzo RV. 1994. *J. Biol. Chem.* 269:579–87
226. Mauviel A, Santra M, Chen YQ, Uitto J, Iozzo RV. 1995. *J. Biol. Chem.* 270:11692–700
227. Mauviel A, Korang K, Santra M, Tewari D, Uitto J, et al. 1996. *J. Biol. Chem.* 271:24824–29
228. Heino J, Kähäri V, Mauviel A, Krusius T. 1988. *Biochem. J.* 252:309–12
229. Wegrowski Y, Paltot V, Gillery P, Kalis B, Randoux A, et al. 1995. *Biochem. J.* 307:673–78
230. Iozzo RV, Cohen I. 1993. *Experientia* 49:447–55
231. Klemsz MJ, McKercher SR, Celada A, Van Beveren C, Maki RA. 1990. *Cell* 61:113–24
232. Ungefroren H, Krull NB. 1996. *J. Biol. Chem.* 271:15787–95
233. Meyer DH, Krull N, Dreher KL, Gressner AM. 1992. *Hepatology* 16:204–16
234. Ungefroren H, Cikós T, Krull NB, Kalthoff H. 1997. *Biochem. Biophys. Res. Commun.* 235:413–17
235. Nelimarkka L, Kainulainen V, Schönherr E, Moisander S, Jortikka M, et al. 1997. *J. Biol. Chem.* 272:12730–37
236. Westergren-Thorsson G, Antonsson P, Malmström A, Heinegård D, Oldberg C. 1991. *Matrix* 11:177–83
237. Romaris M, Heredia A, Molist A, Bassols A. 1991. *Biochim. Biophys. Acta* 1093:229–33
238. Vogel KG, Hernandez DJ. 1992. *Eur. J. Cell Biol.* 59:304–13
239. Roughley PJ, Melching LI, Recklies AD. 1994. *Matrix Biol.* 14:51–59
240. Kähäri VM, Hakkinen L, Westermarck J, Larjava H. 1995. *J. Invest. Dermatol.* 104:503–8
241. Geerkens C, Vetter U, Just W, Fedarko NS, Fisher LW, et al. 1995. *Hum. Genet.* 96:44–52
242. Kadler KE, Holmes DF, Trotter JA, Chapman JA. 1996. *Biochem. J.* 316:1–11
243. Toole BP, Lowther DA. 1968. *Biochem. J.* 109:857–66
244. Toole BP. 1969. *Nature* 222:872–73

245. Scott JE, Orford CR. 1981. *Biochem. J.* 197:213–16
246. Vogel KG, Paulsson M, Heinegård D. 1984. *Biochem. J.* 223:587–97
247. Hedbom E, Heinegård D. 1993. *J. Biol. Chem.* 268:27307–12
248. Nurminskaya MV, Birk DE. 1996. *Biochem. J.* 317:785–89
249. Vogel KG, Koob TJ, Fisher LW. 1987. *Biochem. Biophys. Res. Commun.* 148: 658–63
250. Vogel KG, Trotter JA. 1987. *Collagen Rel. Res.* 7:105–14
251. Chandrasekhar S, Kleinman HK, Hassell JR, Martin GR, Termine JD, et al. 1984. *Collagen Rel. Res.* 4:323–38
252. Birk DE, Nurminskaya MV, Zycband EI. 1995. *Dev. Dyn.* 202:229–43
253. Weber IT, Harrison RW, Iozzo RV. 1996. *J. Biol. Chem.* 271:31767–70
254. Scott JE. 1996. *Biochemistry* 35:8795–99
255. Kobe B, Deisenhofer J. 1993. *Nature* 366:751–56
256. Scott JE. 1995. *J. Anat.* 187:259–69
257. Yu L, Cummings C, Sheehan JK, Kadler KE, Holmes DF, et al. 1993. In *Dermatan Sulphate Proteoglycans*, ed. JE Scott, pp. 183–92. London: Portland
258. Spiro RC, Countaway JL, Gaarde WA, Garcia JA, Leisten J, et al. 1994. *Mol. Biol. Cell* 55:A303
259. Svensson L, Heinegård D, Oldberg C. 1995. *J. Biol. Chem.* 270:20712–16
260. Schönherr E, Hausser H, Beavan L, Kresse H. 1995. *J. Biol. Chem.* 270: 8877–83
261. Danielson KG, Baribault H, Holmes DF, Graham H, Kadler KE, et al. 1997. *J. Cell Biol.* 136:729–43
262. Hildebrand A, Romaris M, Rasmussen LM, Heinegård D, Twardzik DR, et al. 1994. *Biochem. J.* 302:527–34
263. Yamaguchi Y, Ruoslahti E. 1988. *Nature* 336:244–46
264. Yamaguchi Y, Mann DM, Ruoslahti E. 1990. *Nature* 346:281–84
265. Border WA, Noble NA, Yamamoto T, Harper JR, Yamaguchi Y, et al. 1992. *Nature* 360:361–64
266. Isaka Y, Brees DK, Ikegaya K, Kaneda Y, Imai E, et al. 1996. *Nat. Med.* 2:418–23
267. Kresse H, Hausser H, Schönherr E, Bittner K. 1994. *Eur. J. Clin. Chem. Clin. Biochem.* 32:259–64
268. Takeuchi Y, Kodama Y, Matsumoto T. 1994. *J. Biol. Chem.* 269:32634–38
269. Santra M, Skorski T, Calabretta B, Lattime EC, Iozzo RV. 1995. *Proc. Natl. Acad. Sci. USA* 92:7016–20
270. De Luca A, Santra M, Baldi A, Giordano A, Iozzo RV. 1996. *J. Biol. Chem.* 271:18961–65
271. Santra M, Mann DM, Mercer EW, Skorski T, Calabretta B, et al. 1997. *J. Clin. Invest.* 100:149–57
271a. Moscatello DK, Santra M, Mann DM, McQuillan DJ, Wong AJ, et al. 1998. *J. Clin. Invest.* 101:406–12
271b. Patel S, Santra M, McQuillan DJ, Iozzo RV, Thomas AP. 1998. *J. Biol. Chem.* 273:3121–24
271c. Vogel W, Gish GD, Alves F, Pawson T. 1997. *Mol. Cell.* 1:13–23
271d. Shrivastava A, Radziejewski C, Campbell E, Kovac L, McGlynn M, et al. 1997. *Mol. Cell.* 1:25–34
272. Adany R, Heimer R, Caterson B, Sorrell JM, Iozzo RV. 1990. *J. Biol. Chem.* 265:11389–96
273. Adany R, Iozzo RV. 1991. *Biochem. J.* 276:301–6
274. Huttenlocher A, Werb Z, Tremble P, Huhtala P, Rosenberg L, et al. 1996. *Matrix Biol.* 15:239–50
275. Järveläinen HT, Iruela-Arispe ML, Kinsella MG, Sandell LJ, Sage EH, et al. 1992. *Exp. Cell Res.* 203:395–401
276. Kinsella MG, Tsoi CK, Järveläinen HT, Wight TN. 1997. *J. Biol. Chem.* 272: 318–25
277. Krumdieck R, Höök M, Rosenberg LC, Volanakis JE. 1992. *J. Immunol.* 149:3695–701
278. Hocking AM, Strugnell RA, Ramamurthy P, McQuillan DJ. 1996. *J. Biol. Chem.* 271:19571–77
279. Kamo I, Kikuchi A, Nonaka I, Yamada E, Kondo J. 1993. *Biochem. Biophys. Res. Commun.* 195:1119–26
280. Oritani K, Kincade PW. 1996. *J. Cell Biol.* 134:771–81
281. Birk DE, Hahn RA, Linsenmayer C, Zycband EI. 1996. *Matrix Biol.* 15:111–18
282. Hahn RA, Birk DE. 1992. *Development* 115:383–93
283. Cornuet PK, Blochberger TC, Hassell JR. 1994. *Invest. Ophthalmol. Vis. Sci.* 35:870–77
284. Nakazawa K, Hassell JR, Hascall VC, Lohmander LS, Newsome DA, et al. 1984. *J. Biol. Chem.* 259:13751–57
285. Quantock AJ, Klintworth GK, Schanzlin DJ, Capel MS, Lenz ME, et al. 1996. *Biophys. J.* 70:1966–72

Annu. Rev. Biochem. 1998. 67:653–92

G PROTEIN–COUPLED RECEPTOR KINASES

Julie A. Pitcher,[1] *Neil J. Freedman,*[1] *and Robert J. Lefkowitz*

Howard Hughes Medical Institute, Departments of Medicine (Cardiology) and
Biochemistry, Duke University Medical Center, Durham, North Carolina 27710;
e-mail: pitch001@mc.duke.edu, freed002@mc.duke.edu, lefko001@mc.duke.edu

KEY WORDS: signaling, regulation, phosphorylation, desensitization, β-arrestin

ABSTRACT

G protein–coupled receptor kinases (GRKs) constitute a family of six mammalian
serine/threonine protein kinases that phosphorylate agonist-bound, or activated, G
protein–coupled receptors (GPCRs) as their primary substrates. GRK-mediated
receptor phosphorylation rapidly initiates profound impairment of receptor signaling, or desensitization. This review focuses on the regulation of GRK activity by a variety of allosteric and other factors: agonist-stimulated GPCRs,
βγ subunits of heterotrimeric GTP-binding proteins, phospholipid cofactors, the
calcium-binding proteins calmodulin and recoverin, posttranslational isoprenylation and palmitoylation, autophosphorylation, and protein kinase C–mediated
GRK phosphorylation. Studies employing recombinant, purified proteins, cell
culture, and transgenic animal models attest to the general importance of GRKs
in regulating a vast array of GPCRs both in vitro and in vivo.

CONTENTS

[1] These authors contributed equally to this work and may be cited in interchangeable order.

0066-4154/98/0701-0653$08.00

THE BIOLOGICAL CONTEXT OF DESENSITIZATION AND RESENSITIZATION

As if to respond selectively to new or crescendo stimuli, biological systems consistently diminish their responses to persistent or stable stimuli in a process termed desensitization, or adaptation. Desensitization manifests itself at the cellular level in biological processes as diverse as bacterial chemotaxis (1), mating responses in yeast (2), light perception in *Drosophila* (3), and neurotransmission in mammals (4). This review focuses on G protein–coupled receptor kinases (GRKs), a eukaryotic class of enzymes important for desensitization of signaling mediated by G protein–coupled receptors, the largest known family of signal-transducing proteins (5).

The classical G protein–coupled receptors comprise an extracellular N-terminal domain; seven membrane-spanning domains connected by three extracellular and three intracellular polypeptide loops; and an intracellular, C-terminal cytoplasmic tail (5). Upon binding its agonist, the receptor is stabilized in its active conformation (6) and, with its intracellular domains, stimulates heterotrimeric guanine nucleotide-binding regulatory (G) proteins. These G proteins dissociate into α and $\beta\gamma$ subunits, which modulate the activity of one or more effectors: adenylyl cyclases, phospholipases, phosphodiesterases, and ion channels for calcium or potassium. The activity of these effector enzymes and ion channels regulates the intracellular concentration of "second-messenger" molecules or ions, which elicit cellular responses to the agonist initially bound by the receptor. Although desensitization of a G protein–coupled signaling system can involve the receptor, the G protein, and/or the effector, impairment of the receptor's ability to activate its G protein appears to account for most desensitization, especially within minutes of agonist stimulation (7, 8).

Characterizing desensitization by time course and stimulus specificity provides perspective on the role of GRKs in this phenomenon. Within milliseconds to minutes of agonist challenge, cells can diminish or virtually eliminate (7–10) their agonist-evoked responses, in a process that involves phosphorylation of the receptors on one or more intracellular domains. After several hours of agonist exposure, the short-term desensitization just described is augmented by receptor down-regulation, a process in which the cellular complement of stimulated receptors is decreased by a combination of protein degradation, transcriptional, and posttranscriptional mechanisms (11). The cellular response to a given agonist may be desensitized by cellular exposure to that agonist itself, in a process descriptively dubbed *homologous* desensitization. Alternatively, desensitization of the response to a given agonist may be engendered by cellular exposure to agonists for distinct receptor signaling systems, in a process termed *heterologous* desensitization. Potentially affecting multiple receptor systems, heterologous desensitization involves phosphorylation of GPCRs by second-messenger–dependent kinases, such as cyclic AMP (cAMP)–dependent protein kinase and protein kinase C (PKC). Receptor phosphorylation by these kinases, as an isolated event, substantially impairs the ability of purified receptors to stimulate their G proteins (12, 13).

The discovery of GRKs emerged from investigations of mechanisms responsible for short-term, homologous desensitization of the β_2-adrenergic receptor (β_2AR) and rhodopsin, the prototypic "light receptor." With the β_2AR, agonist-induced receptor phosphorylation associated with homologous desensitization (14) was found to occur even in cells genetically lacking cAMP-dependent protein kinase (15). The enzyme responsible for this activity was purified from these cells (16) as well as from bovine brain (17) and was named β-adrenergic receptor kinase (later, GRK2). Rhodopsin kinase (GRK1) was identified earlier (18) as the enzyme responsible for phosphorylating light-bleached (agonist-activated) rhodopsin in rod outer segments. Subsequently, GRK1-mediated phosphorylation of rhodopsin was associated with desensitization of the rhodopsin/G_T/cGMP phosphodiesterase system (19). Identification of the GRK family followed the molecular cloning of cDNAs for GRK2 (20) and then GRK1 (21). Homology (22–25) or positional (26) cloning strategies subsequently identified the other four known mammalian GRKs (see below). The GRK family of serine/threonine kinases shares the unusual feature of phosphorylating specifically the agonist-occupied, or activated, conformation of G protein–coupled receptors.

The current model of GRK action (Figure 1) proposes that a receptor phosphorylated by a GRK can subsequently bind stoichiometrically to one of a family of cytoplasmic inhibitory proteins known as arrestin isoforms (27, 28) in the retina or β-arrestin-1 (29–31) and β-arrestin-2 (8, 31) isoforms in extraretinal tissues. As a result of arrestin or β-arrestin binding, the receptor is prevented from activating its G protein and, therefore, its effector(s). This two-step process of GRK-initiated desensitization can reduce by as much as 70–80% the ability of fully activated β_2ARs or rhodopsin to activate their respective G proteins (7, 8). Furthermore, the binding of nonretinal arrestins to GRK-phosphorylated receptors is believed to initiate G protein–coupled receptor endocytosis, or sequestration (32), into recycling endosomes (33). In this subcellular location, G protein–coupled receptors appear to be dephosphorylated by a membrane-associated phosphatase (34), in the process of resensitization, before returning to the cell surface in signaling-competent form.

The importance of GRKs in homologous desensitization of G protein–coupled receptors has been highlighted over the past 12 years by the convergence of several lines of evidence. First, at a time when cAMP-dependent kinase was believed to be important in β_2AR desensitization, homologous desensitization and

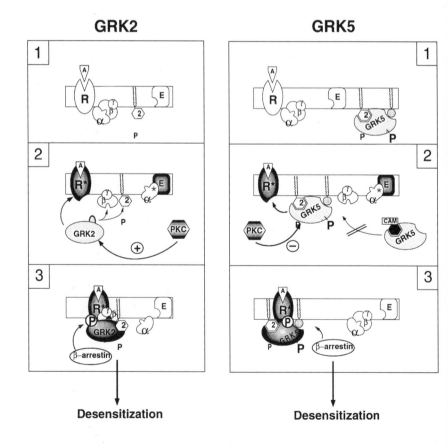

agonist-induced phosphorylation of the β_2AR was shown to transpire—via a GRK2-initiated process (35)—in mutant cells that lacked the cAMP-dependent protein kinase (15). Second, diminished agonist-induced receptor phosphorylation and/or desensitization was observed when cloned receptors were mutated (10, 36, 37) at serine and threonine residues thought—or subsequently demonstrated (38, 39)—to be sites for GRK phosphorylation. Third, chemical or genetic inhibition of GRK action in cells or transgenic animals has either blunted receptor desensitization or potentiated receptor function. Lastly, augmentation of GRK activity by enzyme overexpression has amplified receptor desensitization or attenuated receptor function in transfected cells and transgenic animals.

THE FAMILY OF GRKs

Structure, Evolution, and Enzymology

Six mammalian cDNAs encoding members of the GRK subfamily of serine/threonine kinases (EC.2.7.1-) have been identified to date: GRK1 (rhodopsin kinase) (21); GRK2 (β-adrenergic receptor kinase-1) (20); GRK3 (β-adrenergic receptor kinase-2) (22); GRK4 (IT-11) (26); GRK5 (23, 24); and GRK6 (25). (The original names of these enzymes are given in parentheses.) With the exception of GRK1 [which is found almost exclusively in retina (21)] and GRK4 [which is expressed at significant levels only in testes (26, 40)], GRKs

←————————————————————————

Figure 1 GRK-mediated G protein–coupled receptor phosphorylation, schematically depicted in three arbitrary stages. Separate schemata are presented for GRK2 (*left*) and GRK5 (*right*). Stage 1: All signaling proteins are quiescent: receptor (R), which has not yet bound agonist (A); heterotrimeric G protein ($\alpha\beta\gamma$); and effector (E). GRK5 adheres to the cytosolic surface of the plasma membrane (*stippled rectangle*) via binding to phosphatidylinositol-bisphosphate (P_2) with its N-terminal domain and any phospholipid (*stippled circle*) with its C-terminal domain. Phospholipid-bound GRK5 undergoes autophosphorylation (P), which is required for receptor-phosphorylating activity. Stage 2 (activation): Activated by agonist, the receptor (R^*) in turn activates the G protein, which dissociates into $\beta\gamma$ and active, GTP-bound α (α^*) subunits. The GTP-bound α subunit stimulates the effector(s) (E) to produce second messenger(s), which in turn can activate protein kinase C (PKC) isoforms, among other proteins. $G_{\beta\gamma}$ subunits, along with PIP_2 (P_2), facilitate the translocation of GRK2 to the plasma membrane, where GRK2 binds to the activated receptor. Phospholipid-bound GRK5, by contrast, binds to the activated receptor without the assistance of $G_{\beta\gamma}$ subunits. Phospholipid binding by GRK5 is inhibited by calcium/calmodulin (CAM). Whereas the receptor-phosphorylating activity of GRK2 is enhanced if the GRK2 is phosphorylated by PKC, the receptor-phosphorylating activity of GRK5 is diminished if the GRK5 is phosphorylated by PKC. Stage 3 (signal termination): The receptor-bound GRKs phosphorylate (P) the activated receptors, which can then bind stoichiometrically to a β-arrestin molecule. Bound to a β-arrestin (or arrestin) molecule, the phosphorylated, activated receptor can no longer activate G proteins. Second-messenger signaling terminates as the G_α subunit hydrolyzes GTP to GDP, thereby inactivating itself, and again binds to the $G_{\beta\gamma}$ subunits. Throughout the schematic, *dark shading* is used to denote activated proteins.

are ubiquitously expressed (41). GRK4 is the only member of this family shown to undergo alternative splicing (40, 42). Four splice variants exist, with the alternatively spliced exons occurring in the amino and carboxyl termini (40).

Evolutionary conservation among GRKs isolated from nematodes, insects, and mammals emphasizes their biological importance. GPRK1 and wj283.2 encode, respectively, *Drosophila* (43) and *Caenorhabditis elegans* (44) proteins with 62% and 51% homology to bovine GRK2. Similarly, GPRK2 and f19c6.1 encode *Drosophila* and *C. elegans* proteins with 50–53% homology to bovine GRK5 (43, 44). Among the six mammalian GRKs, amino acid sequence similarity is 53–93%, with GRK1 and GRK2 being the most divergent.

GRKs are most closely related to PKC and cAMP-dependent protein kinase families (PKAs), although sequences within some GRK catalytic subdomains (I, II, VI, VII, and VIII) (45) are strikingly different from these enzymes. Most notable among these differences is that found in the ATP-binding region (46) of subdomain VII, where the DFG (Asp Phe Gly) sequence of most serine/threonine kinases (45) is replaced by DLG (Asp Leu Gly) in GRKs. Based on sequence and functional similarities, the GRK family has been divided into three subfamilies: (*a*) GRK1, (*b*) GRK2 or β-adrenergic receptor kinase (βARK) (GRK2 and GRK3), and (*c*) GRK4 (GRK4, GRK5, and GRK6) (47). Members of the GRK2 subfamily share approximately 84% sequence similarity, and members of the GRK4 subfamily share approximately 70% sequence similarity.

Structurally, GRKs contain a centrally located 263–266 amino acid catalytic domain flanked by large amino- and carboxyl-terminal regulatory domains. The amino-terminal domains of GRKs share a common size (∼185 amino acids) and demonstrate a fair degree of structural homology. These observations have prompted the speculation that amino-terminal domains may perform a common function in all GRKs, potentially that of receptor recognition (48). In contrast, the carboxyl-terminal domain of GRKs is highly variable in both length and structure. GRK1 contains a carboxyl terminus of approximately 100 amino acid residues, the GRK2 subfamily approximately 230 amino acid residues, and the GRK4 subfamily approximately 130 amino acid residues. This region of GRKs represents the site of several posttranslational modifications and regulatory protein/protein interactions. It appears to be involved in the membrane and receptor targeting of these enzymes.

Functionally, GRKs exhibit several hallmark characteristics:

1. They preferentially phosphorylate activated (agonist-occupied) rather than inactive or antagonist-occupied G protein–coupled receptor (GPCR) substrates.

2. Interaction of GRKs with their activated receptor substrates potently activates these enzymes.

3. GRK-mediated GPCR phosphorylation requires the participation of regulatory mechanisms responsible for the membrane localization and receptor targeting of these enzymes.

Catalytically, GRKs have demonstrated no clear consensus sequence in their receptor substrates, even though GRK-mediated GPCR phosphorylation has been demonstrated to occur on specific residues in the cytoplasmic carboxyl tail [rhodopsin (39), β_2AR (38)] or the third intracellular loop [m2 muscarinic acetylcholine receptor (49), α_{2A}-adrenergic receptor (50)] of several receptors. However, in some cases, the putative or sequenced sites of GRK phosphorylation have pairs of acidic residues located at the amino-terminal side of the most amino-terminal phosphorylated residue (38, 50–52). These findings contrast somewhat with peptide phosphorylation studies, which have revealed apparent GRK substrate specificity. While GRK1 and GRK2 most actively phosphorylate peptides containing acidic residues flanking serines or threonines on the carboxyl- or amino-terminal sides, respectively (53), GRK5 and GRK6 most actively phosphorylate peptides containing basic residues amino terminal to the serine target residues (54, 55). GRK-mediated phosphorylation of peptides may proceed in a sequential manner (51, 52, 56)—that is, phosphorylation of a residue adjacent to acidic amino acids appears to be required for the GRK-mediated phosphorylation of more distal serine/threonine residues, in a manner analogous to the behavior of casein kinase II and glycogen synthase kinase 3 (57, 58). Since the stoichiometry of GRK-mediated phosphorylation of most GPCRs in intact cells approximates 1 mole phosphate/mole receptor (59–62), however, the relevance to receptors of sequential substrate phosphorylation remains uncertain.

REGULATION OF GRKS

GRK-Mediated Phosphorylation of GPCR Substrates: The Agonist-Occupied Receptor/GRK Ternary Complex

A defining feature of the GRK family is its substrate specificity. GRKs phosphorylate GPCRs, but only when the receptors are in their activated (agonist-occupied) state. What mechanisms underlie the specific recognition and phosphorylation of activated receptors by GRKs? Agonist occupancy of GPCRs is accompanied by a change in receptor conformation (6). Thus, GRKs may simply interact with and phosphorylate regions of the receptor exposed following receptor activation. This simple model of GRK/GPCR complex formation has been abandoned in the face of a considerable body of evidence suggesting a highly complex and specific kinase/substrate interaction that has functional consequences not only for the receptor but also for the kinase itself.

Several lines of evidence indicate that GRKs interact with receptor substrates at sites distinct from their sites of phosphorylation:

1. Compared with agonist-occupied receptors, synthetic peptides serve as extremely poor substrates for these enzymes. $V_{max}:K_m$ ratio values for synthetic peptide phosphorylation are 100–100,000-fold lower than values for agonist-occupied GPCRs (53–55, 63).

2. GRK-mediated receptor phosphorylation can be inhibited by peptides derived from rhodopsin or β_2AR intracellular domains that are remote from the actual sites of receptor phosphorylation (63–65). Moreover, receptor peptides specifically inhibit phosphorylation of receptor, as opposed to peptide, substrates (63). These results indicate that receptor peptides specifically block the GRK/GPCR interaction rather than directly inhibit the GRKs.

3. GRK-mediated peptide phosphorylation is markedly enhanced in the presence of activated receptors (66–68).

The three lines of evidence outlined above support a model in which receptor domains other than the site of GRK phosphorylation are primarily responsible for mediating the interaction between the GRK and its GPCR substrate.

What regions of a GPCR substrate interact with the GRK? And what functional consequences does this multisite interaction have for GRK function? Insight into the functional consequences that such multisite interactions may have for GRK activity first came from studies with rhodopsin, when it was demonstrated that GRK1-mediated phosphorylation of a peptide substrate is significantly enhanced (>100-fold) specifically in the presence of activated rhodopsin (66, 67). This effect was achieved with an enzymatically digested form of rhodopsin, which lacked the carboxyl-terminal domain residues phosphorylated by GRK1 (66). These results confirm that GRK1 interacts with receptor domains distinct from those that serve as phosphate acceptors. Furthermore, these results demonstrate that an activated GPCR serves not only as a GRK substrate but also as a GRK activator. The interaction of GRK2 with either activated rhodopsin or β_2AR (68) has also been demonstrated to lead to allosteric activation of this enzyme. Notably, however, PKA is not allosterically activated by occupied GPCR substrates (68). Enhancement of GRK activity by activated GPCRs is thus proposed to be a feature common among, and specific to, members of the GRK family of enzymes. The ability of activated GPCRs to enhance GRK activity toward peptide substrates raises the possibility that in a cellular setting, GRKs may phosphorylate nonreceptor substrates following receptor activation.

The third intracellular loop of GPCRs has been implicated in GRK activation, because proteolysis of this region in rhodopsin prevents allosteric activation of GRK1 (66). Regions of the third intracellular loop in close proximity to the transmembrane domains of GPCRs are also proposed to represent sites participating in the interaction with, and activation of, heterotrimeric G proteins (69). In this respect, it is of particular interest that mastoparan, a peptide activator of G proteins, has been shown to stimulate GRK1-mediated phosphorylation of peptide substrates: The $V_{max}:K_m$ ratio increases 10-fold (66). Similarly, GRK2-mediated phosphorylation of a fusion protein encompassing the third loop of the m2 muscarinic acetylcholine receptor is dramatically enhanced in the presence of mastoparan: The $V_{max}:K_m$ ratio increases 1000-fold (70). These results are consistent with the hypothesis that domains within the receptor responsible for G-protein activation following agonist occupancy may also bind to and activate GRKs. Thus the same regions of the GPCR responsible for initiating signal transduction may also play a role in facilitating signal termination mediated by GRKs.

The Pleckstrin Homology Domain of GRK2 and GRK3

GRK2 and GRK3 contain within their carboxyl termini an approximately 100–amino acid protein module termed a pleckstrin homology (PH) domain. This recently described region of sequence homology has been identified in at least 70 different proteins, many of which can be clustered into groups of functionally related molecules. PH domains are found in serine/threonine-specific kinases (e.g. GRK2, GRK3, RAC, and Nrk); tyrosine-specific kinases (such as Btk, Tec-A, and tlk); all of the known mammalian phospholipase Cs; regulators of small GTP-binding proteins [i.e. GTPase-activating proteins (GAPs) and guanine nucleotide–releasing factors (GRFs)]; and a number of cytoskeletal proteins (including β-spectrin) (71, 72). Sequence identity between PH domains is limited. Only a single amino acid, a tryptophan residue, is invariant in all sequences. However, PH domain alignment shows six blocks of homology containing conserved hydrophobic residues; the blocks are separated by variable-length inserts (71).

The classification of PH domains as discrete protein modules is supported by analysis of their molecular structures. With nuclear magnetic resonance (NMR) and X-ray crystal analysis, the structures for several PH domains from different proteins (73–80) have been solved. The conservation of structural features between PH domains is remarkable (72), suggesting that this domain may subserve a common function in different proteins. Although the function or functions of PH domains remain to be determined, PH domains have been proposed to act as mediators of protein/protein and protein/membrane inter-actions. Several different PH domain ligands have been identified, including protein PKC, phosphatidylinositol-4,5-bisphosphate (PIP_2), the $\beta\gamma$ subunits of heterotrimeric G proteins ($G_{\beta\gamma}$), and inositol 1,4,5-trisphosphate (IP_3) (71, 72). The PH domain of GRK2 represents one of the most extensively character-ized members of the PH domain family. It remains to be determined, however, whether the ligands and ligand-binding characteristics of the GRK2 PH domain are shared by domains in other molecular settings.

THE ROLE OF $\beta\gamma$ SUBUNITS OF HETEROTRIMERIC G PROTEINS AND PHOS-PHATIDYLINOSITOL-4,5-BISPHOSPHATE IN REGULATING GRK ACTIVITY   In cells, GRK2 and GRK3 associate with plasma membranes in an agonist-depen-dent fashion (81, 82). A potential molecular basis for this agonist-stimulated redistribution of GRK2 emerged from in vitro studies demonstrating that the $G_{\beta\gamma}$ binds the carboxyl terminus of this enzyme (83, 84). G-protein γ subunits are modified by isoprenylation, in which a 15-carbon farnesyl [in the case of $\gamma 1$ (retinal G_γ)], or more commonly a 20-carbon geranylgeranyl isoprenoid moi-ety, is covalently attached to the cysteine residue of a CAAX motif. Isoprenyl-ation of γ subunits is responsible for the membrane targeting of $G_{\beta\gamma}$ (85). Association of $G_{\beta\gamma}$ with the carboxyl terminus of GRK2 promotes in vitro the association of GRK2 with lipid vesicles and rod outer-segment membranes (83, 86, 87). Membrane association dramatically enhances GRK2-mediated phosphorylation of activated receptors reconstituted in phospholipid vesicles, such as the β_2AR and the m2 muscarinic acetylcholine receptor, as well as rhodopsin in rod outer-segment membranes (83, 86–89). On the basis of these observations, a model has been proposed in which agonist occupancy of a GPCR leads to the activation of heterotrimeric G proteins and the release of the free $G_{\beta\gamma}$ dimer. The membrane-localized $G_{\beta\gamma}$ subsequently interacts with GRK2 and serves to target this enzyme to its membrane-incorporated receptor substrate. Evidence for the operation of such a mechanism in intact cells has recently been obtained with the demonstration by coimmunoprecipitation of an agonist-specific association between GRK2 or GRK3 and $G_{\beta\gamma}$ (82).

Compared with other GRKs, members of the GRK2 subfamily have extended carboxyl termini. The presence of the $G_{\beta\gamma}$-binding site within this region is consistent with the identification of these enzymes as the only $G_{\beta\gamma}$-regulated GRKs to be identified to date. $G_{\beta\gamma}$ binds to an approximately 125–amino acid region (residues 546–670), the distal end of which is located 19 residues from the carboxyl terminus of GRK2 (84). This region encompasses the PH do-main of GRK2 (residues 553–656). Specifically it is the carboxyl-terminal half of the PH domain, and residues extending beyond the end of the recognized PH domain, that are involved in $G_{\beta\gamma}$ binding (90, 91). Consistent with this observation, $G_{\beta\gamma}$ will bind to a 28–amino acid GRK2 peptide that comprises 9 amino acids from the carboxyl terminus of the PH domain and 19 amino acids from the remaining carboxyl-terminal domain (84). Mapping the $G_{\beta\gamma}$-binding domain to regions including and extending beyond the carboxyl-terminal region of the PH domain suggests that $G_{\beta\gamma}$ binding may be a property common to a subset of PH domain–containing proteins—those containing the appropriate sequence determinants adjacent to the PH domain (90). The amino terminus of PH domains forms a hydrophobic β-barrel pocket (72). This structure is similar to that of the retinoid-binding protein family, all of which bind lipophilic molecules (92, 93). Indeed, in vitro, isolated PH domains display specific binding to the charged lipid phosphatidylinositol-4,5-bisphosphate (PIP2) (94). PIP_2 binds to the amino terminus of a GRK2 PH domain lacking carboxyl-terminal sequences required for $G_{\beta\gamma}$ binding (91, 94). This observation immediately suggests that the PH domain of GRK2 is the site of multiple ligand interactions.

What role does PIP_2 play in regulating GRK2 activity? An initial survey of published literature reveals a confusing picture, with reports of both enhancement and inhibition of GRK2 activity by PIP_2 (87, 95–97). Differences in experimental procedures, principally the concentration of PIP_2 used, probably account for these disparate findings. A clearer understanding of the regulatory role of PIP_2 is obtained when experiments performed at physiological concentrations of PIP_2 (98) are examined.

GRK2 activity has traditionally been assessed in vitro with purified GPCRs reconstituted in vesicles composed of multiple lipids, or alternatively with rhodopsin in rod outer-segment membranes—again, a heterogeneous lipid environment. Under these conditions, GRK2 catalyzes agonist-dependent GPCR phosphorylation, which is enhanced on addition of $G_{\beta\gamma}$ subunits (83, 84, 86, 88, 89). To assess the role of PIP_2 in the regulation of GRK2, this traditional approach required modification such that receptor substrates are reconstituted in lipid vesicles or detergent/lipid micelles of defined lipid composition. This change in experimental approach revealed a previously unappreciated aspect of GRK regulation: that GRKs are lipid-dependent enzymes. Indeed, GRK2 fails to phosphorylate activated GPCRs when these receptor substrates are reconstituted in vesicles composed of pure phosphatidylcholine (PC) (87, 95, 97). This null background facilitated the demonstration that both PIP_2 and $G_{\beta\gamma}$ are required for effective GRK2-mediated βAR (87) and m2 muscarinic acetylcholine receptor phosphorylation (97). The synergistic activation of GRK2 observed in the presence of both PH domain ligands suggests their cooperative binding to the PH domain of GRK2. It remains to be determined whether the coordinate binding of multiple ligands is a feature shared with other PH domains.

The synergistic enhancement of GRK2-mediated GPCR phosphorylation observed in the presence of $G_{\beta\gamma}$ and PIP_2 cannot be accounted for by $G_{\beta\gamma}$-

and/or PIP_2-mediated direct activation of the enzyme, because GRK2-mediated phosphorylation of soluble substrates is at best only modestly (approximately twofold) affected by these ligands (70, 86, 87, 95). How then does PH domain ligand binding facilitate GRK2-mediated receptor phosphorylation? The coordinated presence of $G_{\beta\gamma}$ and PIP_2 promotes membrane association of GRK2 (87). By restricting diffusion of GRK2 from three dimensions in solution to two dimensions at the surface of the membrane, PH domain ligands are proposed to facilitate the interaction of GRK2 with its receptor substrates. $G_{\beta\gamma}$- and PIP2-mediated membrane localization may thus play an important role in promoting GRK2-mediated GPCR phosphorylation.

$G_{\beta\gamma}$ appears to play an additional role above and beyond the simple targeting of GRK2 to the membrane. $G_{\beta\gamma}$ binding specifically targets GRK2 to its activated receptor substrate and promotes the GPCR-mediated allosteric activation of this enzyme (70, 86, 99). Thus, although $G_{\beta\gamma}$ does not significantly affect the rate of GRK2-mediated peptide phosphorylation in the absence of additional activators, $G_{\beta\gamma}$ addition dramatically promotes peptide phosphorylation observed in the presence of activated GPCRs or mastoparan (a peptide mimic of activated receptors) (70, 86, 99). In other words, GPCR-mediated allosteric activation of GRK2 occurs more effectively in the presence of $G_{\beta\gamma}$.

THE $\beta\gamma$ SUBUNIT SPECIFICITY OF THE PH DOMAINS OF GRK2 AND GRK3
Although GRK2 and GRK3 are highly homologous proteins, they are only 39% identical across a span of 18 amino acids (residues 648–665) within the $G_{\beta\gamma}$-binding domain (100). Mammalian cDNAs encoding 6 G_β subunits (101) and 12 G_γ subunits (102) have been identified, combining to form a multitude of potential $G_{\beta\gamma}$ combinations. The presence of a variable domain within the $G_{\beta\gamma}$-binding site of GRK2 and GRK3 suggests the potential for specificity of $G_{\beta\gamma}$ isoform binding. This hypothesis has been confirmed in solution phase and in cellular studies (82). Incubation of fusion proteins encompassing the carboxyl termini of either GRK2 or GRK3 with a diverse mixture of $G_{\beta\gamma}$'s purified from bovine brain reveals that $G_{\beta 3}$ binds to the carboxyl terminus of GRK3 but not of GRK2. In contrast, $G_{\beta 1}$ and $G_{\beta 2}$ bound with apparently equal affinities to both GRK2 subfamily members (82). Notably, in a separate study, peptides derived from GRK2 or GRK3 displayed no selectivity of G_β binding (100), suggesting that determinants outside of the 18–amino acid region of these enzymes are responsible for conferring specificity.

The specificity of the carboxyl terminus of GRK3 for $G_{\beta 3}$ was confirmed through the use of purified recombinant $G_{\beta 1 \gamma 5}$ and $G_{\beta 3 \gamma 5}$ (82). Although $G_{\beta 1 \gamma 5}$ bound to both GRK2 and GRK3 fusion proteins, $G_{\beta 3 \gamma 5}$ bound exclusively to the GRK3 (82). The nature of the G_β subunit appears to be the primary determinant regulating the specificity of the GRK2 and GRK3 $G_{\beta\gamma}$ interactions.

The G_γ subunits of $G_{\beta\gamma}$ have been proposed to play a role in determining the affinity of the GRK/$G_{\beta\gamma}$ interaction. Thus purified recombinant $G_{\beta1}$ or $G_{\beta2}$ complexed with $\gamma2$, $\gamma5$, or $\gamma7$ is more efficacious at promoting GRK2- and GRK3-mediated phosphorylation of activated receptor substrates than when $G_{\beta1}$ or $G_{\beta2}$ is complexed to $G_{\gamma3}$ (103, 104). Although these data may indicate a lower affinity of $G_{\beta1\gamma3}$ and $G_{\beta2\gamma3}$ for GRKs, they may also reflect an impaired ability of these $G_{\beta\gamma}$ combinations to interact with activated receptor substrates.

That GRK2 and GRK3 exhibit specificity for distinct $G_{\beta\gamma}$ combinations in intact cells has been shown in studies that monitored GRK/$G_{\beta\gamma}$ complex formation following agonist occupancy of GPCRs. It has also been shown by the specific blockade of GPCR-mediated desensitization by peptides derived from the $G_{\beta\gamma}$-binding domain of GRK2 and GRK3 (105). The formation of a GRK/$G_{\beta\gamma}$ complex in cells overexpressing either GRK2 or GRK3 following activation of GPCRs was monitored by probing GRK immunoprecipitates for the presence of G_β. The complexing of $G_{\beta\gamma}$ with both GRK2 and GRK3 was observed following agonist activation of either the lysophosphatidic acid receptor or the β_2AR. In marked contrast, complexing of $G_{\beta\gamma}$ with only GRK3 was observed following activation of thrombin receptors (82). This observation suggests that in the cell line tested, GRK3 preferentially interacts with the specific $G_{\beta\gamma}$ combination prevalent in the G protein that couples to thrombin receptors. The coupling of specific receptors to distinct $G_{\beta\gamma}$ combinations has been suggested from experiments utilizing antisense mRNAs directed against G_β and G_γ subunits (106). Because $G_{\beta3}$ binds to GRK3, but not to GRK2, it is tempting to speculate whether this G_β subunit plays a role in thrombin-mediated signal transduction. The inhibition of neuronal voltage-dependent Ca^{2+} channels mediated by α_2-adrenergic receptors desensitizes, via a GRK3-specific mechanism, following prolonged exposure to neurotransmitter (105). A peptide derived from the $G_{\beta\gamma}$-binding domain of GRK3, but not of GRK2, blocks this desensitization (105). These results suggest that GRK3 specifically interacts with the $G_{\beta\gamma}$ combination released following α_2-adrenergic receptor activation. The differential binding of specific $G_{\beta\gamma}$ isoforms to GRK2 and GRK3 would be predicted to be of primary importance in determining the substrate specificity of these enzymes.

The Lipid Dependence of GRKs

As discussed above, in the context of PIP$_2$, specific experimental procedures have revealed the lipid dependence of GRKs. Lipids other than PIP$_2$ have also been shown to bind to and regulate GRK2. Negatively charged phospholipids such as phosphatidylserine (PS) promote GRK2-mediated phosphorylation of the β_2AR and the m2 muscarinic acetylcholine receptor (95, 97, 107). Several characteristics distinguish the enhanced GRK2 activity observed in the presence of PS from that observed in the presence of PIP_2. These differences, summarized below, suggest that PS and PIP_2 represent members of two distinct classes of GRK2 lipid ligands and that their effects may be mediated by binding to distinct sites on GRK2.

PS and PIP_2 bind to GRK2 with very different affinities. At saturating $G_{\beta\gamma}$ (~120 nM), concentrations of PIP_2 ranging from 1 to 20 mM dramatically stimulate GRK2-mediated GPCR phosphorylation (87, 97). In marked contrast, the same concentrations of other negatively charged phospholipids have no effect on GRK2 activity (87, 97). Effective stimulation of GRK2 activity by PS requires the presence of 10- to 20-fold higher concentrations of lipid. The difference in affinity of these two ligands for GRK2 is congruent with the difference in their estimated plasma membrane concentrations. Whereas the mole fraction percent for PS is estimated to be approximately 10% (108), that for PIP_2 is estimated to be 1–3% (98). Both lipids may thus represent physiologically relevant regulators of GRK2 activity.

PS, but not PIP_2, directly activates GRK2. In contrast to PIP_2, PS and other negatively charged phospholipids enhance GRK2-mediated phosphorylation of soluble peptide substrates (95, 97). Consistent with this observation, PS, but not PIP_2, appears to promote a comformational change in GRK2, as assessed by the lipid-promoted incorporation of a cross-linking agent into the enzyme (95).

The PS-binding site of GRK2 has been mapped to the carboxyl-terminal domain (97). In contrast to PIP_2, however, the GRK2 PH domain has not been directly implicated as the site of PS interaction. Resolution of the crystal structures of several PH domains complexed to IP_3 indicates that the 4' and 5' phosphates of the inositol ring are critical determinants for binding (74, 76, 109). PS lacks these determinants. This observation, coupled with the distinct GRK2-binding properties of PS and PIP_2, suggests that PS may potentially bind to a carboxyl-terminal GRK2 site distinct from the PH domain.

Like GRK2, GRK5 is regulated by multiple lipid ligands. Two lipid-binding sites have been identified on GRK5: one in the carboxyl terminus and one in the amino terminus of this enzyme (110, 111). The carboxyl-terminal lipid-binding site resides in the last 100 amino acids of GRK5 and displays no lipid specificity. Occupancy of this site stimulates GRK5 autophosphorylation and has been shown to promote GRK5-mediated phosphorylation of receptor substrates (110). Although lipid binding stimulates autophosphorylation, the binding of lipid to GRK5 is apparently independent of the autophosphorylation status of GRK5 (110). In contrast to the carboxyl-terminal site, the amino-terminal lipid-binding site exhibits a high degree of specificity for PIP_2 (111). However, PIP_2 binding to this site does not affect GRK5 autophosphorylation (110). When incorporated in PC vesicles, physiological concentrations of PIP_2

promote the vesicle association of GRK5 and restore the ability of this ki-nase to phosphorylate activated β_2AR (111). As is the case for GRK2, PIP_2 has no direct effect on the catalytic activity of GRK5: GRK5-mediated phos-phorylation of soluble peptide substrates is unaffected by the presence of this lipid. Autophosphorylation-deficient mutants of GRK5 reveal that the interac-tion with PIP_2 occurs independently of the autophosphorylation status of this enzyme. Mutation to alanine of six basic amino acid residues (Lys-22, 24, 26, 28, and Arg-23) in the amino terminus of GRK5 ablates PIP_2-dependent GRK5 activity (111).

The PIP_2-binding site on GRK5 coincides with the GRK5-binding site for calcium/calmodulin (111, 112), which inhibits the enzyme. The binding of PIP_2 and calcium/calmodulin to GRK5 would thus be predicted to be mutu-ally exclusive. The sequence encompassing the PIP_2/calmodulin-binding site is highly conserved among GRKs 4–6 and is divergent in GRK1, GRK2, and GRK3. Consistent with this observation, GRK4α, GRK5, and GRK6 all display PIP_2-dependent β_2AR phosphorylation (111). Both GRK5 and GRK6 bind cal-cium/calmodulin (112). Although it remains to be determined experimentally, it seems likely that this regulatory feature is also shared by GRK4.

Posttranslational Lipid Modifications of GRKs

Phosphorylation of activated GPCRs by GRKs necessitates the membrane lo-calization of these enzymes. For three members of the GRK family (GRK1, GRK4, and GRK6), membrane localization is accomplished, at least in part, by the covalent attachment of lipids to their carboxyl-terminal domains. The amino acid sequence of GRK1 terminates in a CAAX motif (where C is cysteine, A is a small aliphatic residue, and X is an uncharged amino acid). For GRK1 this motif is CVLS, which directs the farnesylation (C15 isoprenylation) and carboxylmethylation of this protein (21). The presence of the farnesyl group is essential for light-dependent membrane association of GRK1 (113). A mutant unfarnesylated form of the kinase remains in the soluble fraction following light exposure and displays a reduced ability to phosphorylate rhodopsin. In contrast, a mutant kinase bearing a more hydrophobic geranylgeranyl (C20) isoprenoid moiety is constitutively associated with the membrane but phos-phorylates rhodopsin at a rate comparable to wild-type (farnesylated) GRK1 (113). The specific modification found in vivo, farnesylation, ensures that membrane association of GRK1 occurs only in the presence of its activated receptor substrate.

GRK4 and GRK6 are palmitoylated (40, 114). Palmitoylation is the acyla-tion of a protein with a 16-carbon saturated fatty acid (palmitic acid) through a thioester or an oxyester bond. In the most prevalent form of palmitoylation, cysteine residues are modified through a thioester bond. One or more of the cysteines located 12–17 amino acids from the carboxyl terminus represent the sites of palmitolylation in GRK6 (114). By analogy with this enzyme, the probable site or sites of GRK4 palmitoylation are the carboxyl-terminal cysteine of GRK4 and a cysteine located 17 amino acids away from the carboxyl terminus of this enzyme (40). Palmitoylated GRK4 and GRK6 are found exclusively associated with membranes (40, 114). Palmitoylation is a reversible posttranslational modification. For several proteins, GPCR activation has been shown to be accompanied by a change in palmitoylation status. These proteins include certain GPCRs (115, 116) as well as nitric oxide synthetase (117) and the α subunits of heterotrimeric G proteins (118, 119). For $G\alpha_s$ the loss of palmitic acid following GPCR activation is accompanied by the release of this protein from the membrane (120). It remains to be determined whether the palmitoylation status, and thus potentially the subcellular localization, of GRK4 and GRK6 is regulated by agonist occupancy of GPCRs.

Regulation of GRKs by Protein Kinase C

GRK2 (a member of the GRK2 subfamily) and GRK5 (a member of the GRK4 subfamily) are phosphorylated in vitro and in intact cells by protein kinase C (PKC) (121–123). The functional consequences of this phosphorylation event, however, are very different for these two enzymes. In vitro, PKC-mediated phosphorylation of GRK2 leads, albeit somewhat slowly, to the incorporation of between 0.5 and 0.9 moles P_i/mole GRK2 and is accompanied by an approximately twofold activation of the enzyme toward activated rhodopsin (121, 122). Notably, the PKC-dependent activation of GRK2 is specific to receptor substrates (as opposed to peptides), suggesting that PKC phosphorylation promotes membrane and/or receptor association of GRK2 rather than directly enhancing the catalytic activity of this enzyme (122). In support of this hypothesis, direct activation of PKC by phorbol ester treatment of cells overexpressing GRK2 leads to a decrease in cytosolic GRK2 immunoreactivity, a result suggesting a PKC-dependent association of GRK2 with the plasma membrane (122). In mononuclear leukocytes, GRK-mediated β_2AR desensitization is potentiated in phorbol ester–treated cells, suggesting that PKC may play a physiologically relevant role in the regulation of GRK2 (121). Although the PKC phosphorylation site or sites on GRK2 are unknown, a fusion protein encompassing the last 137 amino acids of GRK2 serves as a PKC substrate (122). This fusion protein encompasses the PH domain of GRK2, and in this respect it is interesting to note that the PH domains of two other protein kinases [Bruton tyrosine kinase and RAC (related to A- and C-kinase)] bind PKC (124, 125).

GRK5 serves as an excellent substrate for PKC in vitro; it is phosphorylated rapidly to a stoichiometry of approximately 2 moles P_i/mole GRK5 (123). In contrast to GRK2, GRK5 dramatically loses enzymatic activity–on both receptor and soluble substrates–consequent to phosphorylation by PKC (123). The V_{max}:K_m ratio for GRK5-mediated phosphorylation of rhodopsin is decreased approximately 13-fold following PKC phosphorylation. Although the direct inhibition of catalytic activity appears to be the principal effect of PKC phosphorylation, phosphorylated GRK5 may also exhibit a reduced affinity for rhodopsin, as evidenced by a decreased ability to associate with rod outer-segment membranes but not lipid vesicles (123). Mapping the two PKC phosphorylation sites on GRK5 reveals that both reside within the carboxyl-terminal 26 amino acids of GRK5 (123) at sites distinct from the sites of GRK5 autophosphorylation (110).

Thus, members of two GRK subfamilies are regulated by PKC-mediated phosphorylation. The functional consequences of this phosphorylation event appear subfamily specific: GRK2 is activated, while GRK5 is inhibited. Alterations of PKC activity in a cellular setting would thus be predicted to regulate the cellular complement of active GRKs. Agonist occupancy of receptors coupled to $G_{q/11}$ and phospholipase C leads to PKC activation. Desensitization of such receptors would thus be predicted to be mediated by GRK2 (a PKC-activated enzyme) rather than by GRK5 (an enzyme potently inhibited by PKC).

Olfactory Receptor Desensitization: A Novel Paradigm of GPCR Desensitization

Activation of odorant receptors in the cilia of olfactory neurons can lead to the generation of the second messengers cAMP and inositol-1,4,5-trisphosphate (IP$_3$) (126). A characteristic feature of the signal transduction pathways in this tissue is the extent and rapidity of the desensitization process (126–128). In a permeabilized rat olfactory cilia preparation, odorants induce an approximately fourfold increase in second-messenger (cAMP and IP$_3$) concentrations, which peak 25–50 ms after agonist exposure (9, 129, 130). Olfactory receptor desensitization occurs extremely rapidly such that 100 ms after agonist exposure, second-messenger levels have returned to near basal values (9, 129, 130).

Subtype-specific antibodies to GRK2 and GRK3 reveal that in marked contrast to most other tissues, olfactory epithelium expresses exclusively GRK3 (9, 129). That GRK3 participates in olfactory receptor desensitization was demonstrated through the use of inhibitors of this enzyme: heparin (9); anti-GRK3 antibodies (9, 129); and a fusion protein encompassing the carboxyl terminus of the GRK3, a $G_{\beta\gamma}$ sequestrant (130). Inhibition of GRK3 activity led to a complete loss of desensitization for both the cAMP and IP$_3$ signals. Interestingly, a similar pattern was observed when inhibitors of the second messenger–dependent protein kinases were used (9). Thus, inhibition of either GRK3 or PKA/PKC completely blocked olfactory receptor desensitization. This pattern of desensitization is in marked contrast to that observed with the β_2AR, where

GRK and PKA inhibitors partially and additively inhibit receptor desensitization (131). The complete inhibition of olfactory receptor desensitization observed in the presence of either GRK3 inhibitors or the second messenger–dependent kinase inhibitors suggests that in the olfactory epithelium, these enzymes act in a coordinated rather than an independent fashion. Notably, inhibition of PKA completely blocks both desensitization and receptor phosphorylation (9). In contrast, inhibition of GRK3 completely blocks olfactory receptor desensitization but only partially blocks receptor phosphorylation. Because second-messenger phosphorylation events are operative in the presence of GRK inhibitors, these results suggest that PKA-mediated phosphorylation of olfactory receptors does not induce receptor desensitization. Several alternative regulatory mechanisms could explain the pattern of olfactory receptor desensitization. Second messenger–dependent kinases may directly phosphorylate and activate GRKs. Alternatively, PKA and PKC may phosphorylate and inactivate a GRK3 inhibitor.

In olfactory cilia, as described above, the PKA inhibitor peptide inhibits olfactory receptor desensitization. The PKA inhibitor peptide inhibits not only receptor desensitization but also translocation of GRK3 to the membrane (132). Consistent with this observation, cAMP addition promotes the membrane localization of GRK3 (132). These results suggest that an important effect of PKA activity may be to facilitate membrane, and thus receptor, targeting of GRK3. The ability of PKA to modulate GRK3 membrane targeting could be mediated by phosducin, a $G_{\beta\gamma}$-binding PKA substrate found in olfactory cilia. Once phosphorylated by PKA, phosducin loses its affinity for $G_{\beta\gamma}$, which could then recruit GRK3 to the odorant-activated receptors (133–135). Whether termination of the odorant-induced generation of IP_3 is controlled by a similar cascade is unclear, because PKC-mediated phosphorylation of phosducin has not been demonstrated. Notably, however, PKC has been shown to phosphorylate and activate GRK2 in vitro (121, 122). If PKC similarly regulates GRK3, then for receptors coupled to IP_3 generation, the specific desensitization pattern observed in olfactory cilia may be explained by the PKC-mediated direct activation of this enzyme.

Regulation of GRKs by Ca^{2+}-Binding Proteins

Many Ca^{2+}-sensing transducers are members of the EF hand superfamily of Ca^{2+}-binding proteins (135a). Two members of this superfamily (recoverin and calmodulin) bind to and inhibit GRKs (112, 136–138). Recoverin specifically inhibits GRK1 activity ($IC_{50} \sim 2~\mu M$) (136). In contrast, calmodulin inhibits GRK2, GRK5, and GRK6 (112, 137, 138). Although not directly assessed, GRK3 and GRK4 are also presumed to represent calmodulin targets by virtue of their similarities to GRKs 2 and 5, respectively. Calmodulin inhibition of GRKs exhibits a selectivity for GRK5 ($IC_{50} \sim 50$ nM) > GRK6 $\gg$ GRK2 ($IC_{50} \sim 2$ μM), with inhibition of GRK1 activity observed only at supraphysiological concentrations of calmodulin (112, 137, 138). Both recoverin- and calmodulin-mediated inhibition of GRKs are Ca^{2+} dependent.

RECOVERIN-MEDIATED INHIBITION OF GRK1   Recoverin is present in vertebrate photoreceptors, certain retinal cone bipolar cells, and pineal glands (139, 140), a distribution that closely mimics that of GRK1, its presumed physiological target. Recoverin inhibits GRK1 by binding directly to the enzyme in a Ca^{2+}-dependent fashion, and it does so more potently in its myristoylated form (136). Ca^{2+}-complexed, myristoylated recoverin is membrane associated (141–143). Similarly, the recoverin/GRK1 complex would be predicted to be membrane localized but inactive against receptor substrates. Recoverin would thus appear to inhibit GRK1 activity by preventing the interaction between this enzyme and activated photolysed rhodopsin.

In addition to potentiating the inhibitory effect of recoverin on GRK1, the N-myristoyl moiety of recoverin also confers cooperativity to the Ca^{2+}-dependent inhibition of rhodopsin phosphorylation (136, 144). Recoverin-mediated inhibition of GRK1 is thus extremely sensitive to Ca^{2+} across a narrow range of concentrations (136, 144). In the vertebrate photoreceptor system, Ca^{2+} levels are high in the dark and drop following illumination (145). Thus when rhodopsin is in its inactive state, GRK1 would be predicted to be complexed to recoverin, membrane associated, and inactive. Following rhodopsin activation, a decrease in intracellular Ca^{2+} would be accompanied by the release of GRK1 from recoverin. The release of uninhibited GRK1 in close proximity to its activated receptor substrate may facilitate rapid rhodopsin phosphorylation and desensitization. Although this model is attractive, the inhibitory effects of recoverin in vitro require Ca^{2+} concentrations significantly higher than the bulk intracellular free Ca^{2+} levels in vertebrate photoreceptors (200–600 nM in the dark) (145). What accounts for this apparent discrepancy? One explanation is that recoverin may function in a local environment; that is, close to the membrane, where Ca^{2+} concentrations may be significantly higher than the concentration of the bulk free Ca^{2+}. Alternatively, additional proteins absent from the reconstituted systems used in vitro may be required for recoverin to operate in the physiological range of free Ca^{2+} concentrations.

CALMODULIN-MEDIATED INHIBITION OF GRK2 AND GRK5   As with recoverin-mediated inhibition of GRK1, calmodulin binds directly, though with different affinities, to both GRK2 ($IC_{50} \sim 2$ μM) and GRK5 ($IC_{50} \sim 50$ nM) to inhibit the interaction of these enzymes with their agonist-occupied receptor substrates (112, 137, 138). Inhibiting the interaction of the GRKs with activated receptor substrates would be predicted to inhibit GPCR-mediated allosteric activation of these enzymes. Indeed, for GRK2 or GRK2/$G_{\beta\gamma}$, calmodulin in the presence of Ca^{2+} inhibits the agonist-dependent stimulation of peptide phosphorylation observed in the presence of a nonphosphorylatable mutant of the m2 muscarinic acetylcholine receptor (138). GRK5's calmodulin binding, in addition to inhibiting the interaction of GRK5 with GPCRs, inhibits the association of this kinase with lipid vesicles (112). Calmodulin is not modified by fatty acylation. The GRK5/calmodulin complex would thus be predicted to differ from the GRK1/recoverin complex in that it would not be membrane associated.

The binding site for calmodulin has been mapped on GRK5 and shown to reside within the amino terminus of the enzyme (residues 20–39) (112). This region also constitutes the PIP_2-binding site of GRK5 (111), an observation that may explain the inhibition of GRK5 binding to lipid vesicles observed in the presence of calmodulin (112). The calmodulin-binding site of GRK5 displays properties characteristic of an amphiphilic helix and is thus similar to calmodulin-binding sites identified in other proteins (146). The peptide sequence surrounding the PIP_2/calmodulin-binding site is highly conserved between members of the GRK4 subfamily, but is divergent in GRK1 and GRK2. This observation suggests (a) that calmodulin binding, like PIP_2 binding, is likely to be a property shared by all the members of the GRK4 subfamily and (b) that GRK2 contains a calmodulin-binding site structurally distinct from that mapped in GRK5. Structurally distinct calmodulin-binding domains would explain the very different affinities of GRK2 and GRK5 for calmodulin. GRK2 and GRK5 bind PIP_2 via structurally distinct domains, and an analogous situation may thus also exist for calmodulin.

Calmodulin binding to GRK5 has an additional effect that is likely to be specific to this enzyme: It promotes GRK5 autophosphorylation at sites distinct from those phosphorylated in the absence of this protein (112). Although not primarily responsible for mediating calmodulin-dependent inhibition, Ca^{2+}-stimulated autophosphorylation of GRK5 prevents the interaction between GRK5 and rhodopsin in the absence of calmodulin (112). The GRK5 inhibitory effects of intracellular calcium transients might thus be prolonged by calmodulin-stimulated GRK5 autophosphorylation (112). Interestingly, an inverse regulatory mechanism has been proposed for CaM-kinase II, an enzyme for which calmodulin binding and calmodulin-stimulated autophosphorylation activate rather than inhibit activity (147).

GRK2- and GRK5-mediated desensitization of GPCRs would be predicted to be inhibited in the presence of Ca^{2+}/calmodulin. The relatively low affinity of GRK2 for calmodulin suggests that inhibition of GRK2 would occur only at locations specifically enriched in calmodulin. One such locale is the brain, where concentrations of calmodulin are reported to range between 1 and 10 μM (147).

GRK2 is particularly abundant at the synapse (4), and a calmodulin-dependent inhibition of GRK2 activity may be of physiological relevance under these conditions. Alternatively, an EF hand–binding protein distinct from calmodulin may represent the true physiological ligand for GRK2. This hypothesis is particularly attractive because a number of neuronal proteins highly homologous to recoverin, and capable of inhibiting GRK1 activity in vitro, have recently been identified (148). Interestingly, one such protein, neuronal calcium sensor (NCS-1), can bind to and regulate a number of calmodulin targets (149). It will be of interest to determine whether NCS-1 is also capable of regulating GRKs 2–6. The high affinity of GRK5 for calmodulin suggests that this regulatory mechanism may be operative in multiple cell types. Coupled with the inhibitory effect of PKC on GRK5 activity, the marked inhibition of GRK5 by calmodulin strongly suggests that this kinase is probably not involved in the phosphorylation and desensitization of $G_{q/11}$- and phospholipase-C-coupled receptors, such as the angiotensin II receptor, in vivo.

Regulatory Role of GRK Autophosphorylation

Of the six GRKs, only GRK1 and GRK5 undergo significant autophosphorylation (23, 24, 65, 110, 150). GRK1 autophosphorylation occurs rapidly to a stoichiometry of approximately 3 moles P_i/mole GRK1 (65, 150). The autophosphorylation sites have been mapped to Ser-488 and Thr-489, with Ser-21 representing a minor autophosphorylation site (151). Autophosphorylation of GRK1 does not affect the ability of this enzyme to bind to or phosphorylate rhodopsin (65, 150). However, as compared to an autophosphorylation-deficient mutant of the kinase, the autophosphorylated form of GRK1 displays impaired binding to phosphorylated, light-activated rhodopsin (150). GRK1 autophosphorylation has thus been proposed to play a role in facilitating dissociation of GRK1 from its receptor substrate following phosphorylation.

The effect of autophosphorylation on GRK5 function appears distinct from that on GRK1. Autophosphorylation of GRK5 is stimulated nonspecifically by lipids to a stoichiometry of approximately 2 moles P_i/mole GRK5 (110). Comparison of the GRK1 and GRK5 sequences reveals that the two major autophosphorylation sites in GRK1 are conserved in GRK5 (residues Ser-484 and Thr-485). Mutation of these sites to alanine produces a form of GRK5 that does not autophosphorylate (110). Compared with wild-type GRK5, this autophosphorylation mutant displays a dramatically impaired ability to phosphorylate membrane-incorporated receptor substrates (110). However, the catalytic activity of GRK5 seems minimally affected by autophosphorylation, because GRK5-mediated phosphorylation of soluble peptide substrates is only modestly stimulated (approximately twofold) by autophosphorylation (110). Furthermore, autophosphorylation does not affect the lipid binding of GRK5, because an autophosphorylation-deficient mutant of GRK5 can inhibit phospholipid-stimulated autophosphorylation of the native enzyme (110). Together, these observations suggest that autophosphorylation specifically promotes the interaction of GRK5 with its activated receptor substrates, thereby facilitating GPCR-mediated allosteric activation of this enzyme.

The GRK5 autophosphorylation stimulated by calcium/calmodulin occurs at sites distinct from Ser-484 and Thr-485 (112) and impairs rather that facilitates GRK5/receptor binding. Thus, the nature of the ligand binding to GRK5—phospholipid or calcium/calmodulin—determines which sites become autophosphorylated. This autophosphorylation in turn provides another level of regulation for the interaction between GRK5 and its receptor substrates.

Regulation of GRK Activity by Other Targeting Proteins

Significant amounts of GRK2 have been shown to be specifically associated with microsomal membranes (152, 153). This GRK2/microsomal membrane interaction is reversible, of high affinity (nanomolar), and, as judged by protease sensitivity, mediated via a protein/protein interaction (152). GRK2 binding to microsomal membranes is blocked upon addition of a GST-fusion protein containing residues from the amino (residues 50–145) but not the carboxyl terminus of the kinase (153). Although membranes containing the GRK2-binding protein inhibit GRK2-mediated rhodopsin phosphorylation, this inhibition is relieved in the presence of activators of the heterotrimeric G proteins (153). The binding of GRK2 to the inhibitory GRK2-binding protein of microsomal membranes provides a mechanism for regulating, in a GTP-dependent fashion, the activity of membrane-associated GRK2. These findings raise the possibility that additional, unidentified proteins direct GRK2, as well as other members of the GRK family, to distinct subcellular compartments and thus potentially to different receptor substrates.

CELLULAR STUDIES

From the time they led to the discovery of GRK2 (15), cultured cells have served as fundamental tools for investigations of GRK function. These investigations can be categorized broadly as studies that (a) evaluate the contribution of GRK action, among other mechanisms, to the desensitization of a particular receptor; (b) investigate the kinetics of GRK-initiated desensitization; (c) define GRK subcellular localization; (d) assay factors affecting GRK expression or activity; and (e) evaluate candidate receptor substrates. Although the preponderance of these studies employ immortalized cell lines, several of the conclusions derived from these cell lines have been corroborated by studies performed in primary cultures of cells (105, 154–158), the constituents of which more closely model in vivo conditions. In immortalized cell lines (159–162) as well as in primary-culture models (9, 105, 129, 163, 164), there may be considerable cell type–specific variation in relative and even absolute GRK expression. For certain receptors, this cell type–specific GRK expression may affect susceptibility to GRK-mediated phosphorylation and desensitization (165).

The Contribution of GRKs to Homologous Desensitization

To evaluate the importance of GRK action in cellular systems, various methods of inhibiting GRK action in cells have been devised. In permeabilized human epidermoid carcinoma A431 cells, the GRK-selective inhibitor heparin (24, 54, 55, 166) has been used to estimate that GRK activity contributes ~60% of the agonist-induced β_2AR phosphorylation and desensitization observed (the remainder was attributed to cAMP-dependent protein kinase) (167). In the same permeabilized cell system, inhibiting GRK2 with a peptide corresponding to the first intracellular loop of the β_2AR suggested that 52% of agonist-induced β_2AR desensitization derived from GRK action (64). In consonance with these results, when A431 cells were treated with antisense oligodeoxynucleotides to inhibit GRK2 synthesis, the amount of β_2AR desensitization attributed to GRK activity was also ~50% (160). For β_2AR (160) and H_2-histamine receptor (168) desensitization in other cell lines, the antisense approach has shown the relative contribution of GRK2 to range from 0% to 100%, an observation that attests to cell type–specific variability in desensitization mechanisms. In permeabilized rabbit cardiac myocytes, GRK-specific antibodies were used to demonstrate that agonist-induced β_2AR phosphorylation is totally GRK2-dependent (158). In transfected cells, the dominant negative (K220R or equivalent) mutant of GRK2 (169) has been used to inhibit specifically GRK-mediated agonist-induced phosphorylation and/or desensitization of several receptors (60–62, 170–173), with results varying from 40% to 60% inhibition of phosphorylation and 50% to 100% inhibition of desensitization (61, 62, 171).

The relative importance of GRK-initiated receptor desensitization has also been addressed with the use of mutant receptors in transfected cells. After the β_2AR Ser and Thr residues phosphorylated by purified GRK2 were localized to the cytoplasmic tail of the receptor (174), mutations of 11 Ser and Thr residues in this region to Ala or Gly yielded a receptor that sustained only 50% as much agonist-induced phosphorylation and desensitization as the wild-type β_2AR (37). The observation of a reduction in agonist-induced phosphorylation and/or homologous desensitization consequent to mutating presumptive GRK phosphorylation sites has subsequently been used to implicate GRKs in the regulation of many receptors (50, 175–178). This strategy should be used cautiously, however, because results obtained with site-directed mutagenesis can sometimes prove misleading, as the prototypical β_2AR has illustrated.

A mutant β_2AR resistant to homologous desensitization and agonist-induced phosphorylation was created by mutating to Ala or Gly 4 of the 11 cytoplasmic tail Ser/Thr residues described above: Ser-355, Ser-356, Thr-360, and Ser-364 (179). However, subsequent amino acid sequencing of GRK2-phosphorylated β_2AR peptides (38) demonstrated that the only β_2AR Ser and Thr residues phosphorylated by GRK2 or GRK5 lie in the cytoplasmic tail C-terminal to the Ser-364 mutated in the study cited above.

The Kinetics of GRK-Mediated Events

The rapid intracellular kinetics of GRK-initiated desensitization have been studied best in two permeabilized cell systems: (a) A431 cells (180) and (b) dendrites of rat olfactory neurons (9, 129). Representative of somatic cells, A431 cells demonstrated GRK-mediated β_2AR phosphorylation and desensitization to transpire with a $t_{1/2}$ of ~15 sec. Representative of more specialized neuronal cells, olfactory dendrites demonstrated GRK3-mediated odorant receptor desensitization to transpire considerably more rapidly—within 200 ms. Because GRKs 2 and 3 have been shown to localize to receptor-rich pre- and postsynaptic densities in the rat central nervous system (4), it is not surprising that GRK-mediated receptor regulation proceeds most rapidly in neuronal cells (105).

Subcellular Localization

The subcellular localization of GRKs is of interest primarily because the receptor substrates for these enzymes are integral membrane proteins. From studies of unstimulated, disrupted cells, it seems that GRKs 4–6 are tightly adherent to cell membranes (24, 40, 114), whereas GRKs 1–3 are primarily cytosolic and translocate to the membrane fraction of the disrupted cells when the cells are stimulated with an appropriate receptor agonist (81, 130, 159, 181, 182). The agonist-dependent GRK2 translocation from cytosol to plasma membrane can be seen as well in whole cells, with confocal microscopy (182). It is also possible, however, to demonstrate GRK2 in the plasma membrane, cytosolic, and microsomal membrane compartments of whole cells with immunofluorescence or density-gradient centrifugation (152).

Factors Affecting GRK Expression or Activity

The factors affecting cellular expression and activity of GRKs are only beginning to manifest themselves in experimental systems. Perhaps surprisingly, elevating cellular cAMP levels appears to have no effect on GRK2 expression (30) and appears to decrease GRK5 expression in rat thyroid cells (183). Despite these results in cultured cells, β-adrenergic antagonist treatment of swine has been shown to decrease assayable GRK activity in ventricular myocardium (184). Chronic activation of protein kinase C (PKC) with phorbol ester in lymphocytes (157) appears to up-regulate GRK2 expression two- to three-fold. Interestingly, short-term PKC activation appears to increase the activity of GRK2 in these same cells, as judged by β_2AR desensitization (121). Under somewhat different conditions, phorbol ester can also apparently up-regulate GRK6 and down-regulate GRK2 in lymphocytes (161). Polyclonal activation of human lymphocytes with phytohemagglutinin (157, 161) or interleukin-2 (161) has also been shown to up-regulate GRK2 and GRK6 expression. Whereas a twofold up-regulation of GRK6 accompanies myeloid differentiation of HL-60 leukemia cells, a twofold up-regulation of GRK2 accompanies monocytic differentiation of these same cells (161). Undoubtedly, many of the factors that modulate the cellular expression of GRKs will, like the cellular expression of GRKs, prove cell type–specific.

RECEPTOR SUBSTRATES OF GRKS

Approaches Employed to Identify GRK Substrates

To portray the continually enlarging cadre of receptors that appear to be regulated by a GRK mechanism, we have created a data summary in Table 1. Attempts to determine whether a particular G protein–coupled receptor is or can be phosphorylated by a particular GRK have evolved in several directions, and Table 1 seeks to accommodate considerable diversity among the experimental data. Below, we discuss the types and quality of evidence used to implicate GRKs in homologous receptor desensitization.

Because agonist-induced receptor phosphorylation constitutes the hallmark of GRK activity, direct studies of receptor phosphorylation have provided the most straightforward and convincing evidence for identifying GRK substrates. Studies of homologous receptor desensitization, though more difficult to interpret than phosphorylation studies, have also provided strong evidence for identifying GRK substrates. The combination of phosphorylation and desensitization data militates most cogently for the identity of a particular GRK substrate.

To discern that a particular GRK can phosphorylate a particular receptor, the earliest phosphorylation assays involved purified kinases and purified receptors reconstituted in phospholipid vesicles, or rod outer segments. The demonstration of desensitization in these assays has utilized heterotrimeric G proteins, in agonist-dependent GTPase assays with (7, 8, 29, 185, 186) or without (99, 186–189) the addition of arrestin or β-arrestin proteins. Because many G protein–coupled receptors lose ligand-binding capacity in detergent solution (190), however, this approach has been limited to a small number of model receptors. Moreover, because phospholipids (95, 97, 110, 111) and G proteins (83, 88) are important for GRK-mediated receptor phosphorylation, other experimental systems seemed desirable. Accordingly, GRK-mediated receptor

Table 1   G protein–coupled receptor substrates of GRKs

Receptor	Kinase(s)	Experimental Evidence		References
		Purified proteins	Cellular assays	
Adenosine A$_1$	GRK2	P, S		188
Adenosine A$_{2a}$	GRK unknown		D$_{DN}$	191
Adenosine A$_{2b}$	GRK unknown		D$_{DN}$	191
Adenosine A$_3$	GRK2	P	IP	192
α_{1B}-adrenergic	GRKs 2, 3, 6		P, IP$_{DN}$, S (GRK2, 3)	173
α_{2A}-adrenergic	GRKs 2, 3	P	P, IP, D$_{hep}$	193–196
α_{2B}-adrenergic	GRKs 2, 3	P	P	194, 197
β_1-adrenergic	GRKs 1, 2, 3, 5	P (GRK2, 5)	P, IP, IP$_{DN}$, S	60
β_2-adrenergic	GRKs 1, 2, 3, 4, 5, 6	P, S (GRK2)	IP$_{hep}$, IP$_{DN}$, IP$_{Ab}$, P, D$_{hep}$, S, T1	13, 16, 17, 22, 24 29, 40, 81, 158, 160, 167, 172, 198, 199
Angiotensin II type 1A	GRKs 2, 3, 5		P, IP$_{DN}$, D$_{DN}$, S	61, 62
B$_2$ Bradykinin	GRK2	P	IP	200
C5a	GRK unknown		IP	201, 202
CCR-5 (chemokine)	GRKs 2, 3, 5, 6		P, S	165
CCR-2B (chemokine)	GRK3		S	203
Cholecystokinin	GRK unknown		IP$_{hep}$	156
CXCR1 (chemokine)	GRK unknown		IP	178
CXCR2 (chemokine)	GRK unknown		IP	206
Dopamine D$_{1A}$	GRKs 2, 3, 5		P, S	204
Endothelin A	GRKs 2, 3, 5, 6		P, IP$_{DN}$, D$_{DN}$, S	62
Endothelin B	GRKs 2, 3, 5, 6		P, IP, S	62
N-Formyl peptide	GRK unknown		IP	201
H$_2$ Histamine	GRK2 (not GRK6)		D$_{AS}$	168
5-Hydroxytryptamine$_4$	GRK unknown		D$_{hep}$	205
Lutropin/choriogona- dotropin	GRKs 2, 4		S	40
m1 acetylcholine	GRK2	P		207
m2 acetylcholine	GRKs 2, 3, 5, 6	P, S (GRK2, 3)	IP$_{DN}$	54, 55, 170, 187
m3 acetylcholine	GRKs 2, 3	P		96
Odorant	GRK3		P$_{hep}$, P$_{Ab}$, P$_{pep}$, D$_{hep}$, D$_{Ab}$, D$_{pep}$	9, 129, 130
δ-opioid	GRKs 2, 5		P, IP$_{DN}$, S	171
κ-opioid	GRK unknown		D$_{DN}$	208
Platelet-activating factor	GRK unknown		IP, T1, D$_{hep}$	159, 176, 209
Prostaglandin E1	GRK unknown		T1	210
Parathyroid hormone	GRK unknown		IP	211
Rhodopsin	GRKs 1, 2, 3, 4, 5, 6	P		18, 22, 23, 25 164, 198
Somatostatin	GRK unknown		T1	212
Substance P	GRKs 2, 3	P		213
Thrombin (PAR-1)	GRK3 > GRK2		P, IP, S	62, 177
Thyrotropin	GRKs 2, 5		D$_{AS}$, S	183, 214

Notes: GRK, G protein–coupled receptor kinase; P, receptor *phosphorylation* mediated by a specific GRK; S, desensitization or inhibition of receptor *signaling* mediated by a specific GRK, in a time course consistent with GRK action; IP, *immunoprecipitation* of receptors phosphorylated by a cellular kinase(s) resistant to inhibitors of second messenger–dependent kinases; IP$_{DN}$, like IP, but showing GRK activity by virtue of inhibiting agonist-induced receptor phosphorylation with a *dominant negative* GRK2 construct; D$_{DN}$, receptor *desensitization* inhibited by cellular expression of a *dominant negative* GRK2; D$_{hep}$, receptor *desensitization* inhibited by treating per-meabilized cells with *heparin*; D$_{AS}$, receptor *desensitization* inhibited by GRK-specific *antisense* oligodeoxynucleotides or cDNAs, with demonstrated decrease in GRK expression; T1, agonist-induced *translocation* of GRK activity from the cytosolic to membrane fraction of disrupted cells.

phosphorylation was demonstrated with membranes from cells expressing high receptor numbers, either with highly purified plasma membrane preparations (96, 197) or with crude cell membranes subjected to receptor immunoprecipitation subsequent to phosphorylation (192). To evaluate GRK-mediated phosphorylation in a cellular milieu, several laboratories (60–62, 170–173) have used cells transfected with receptors and a dominant negative (K220R or equivalent) mutant of GRK2, which specifically (60), but incompletely (169), inhibits GRK activity. Inhibition of agonist-induced receptor phosphorylation by this dominant negative GRK2, assessed by receptor immunoprecipitation, has been interpreted to imply that a cellular GRK not only can, but also does, phosphorylate the receptor. To attribute activity on certain receptors to a particular GRK, permeabilized cells have been treated with either GRK-specific antibodies (9, 158) or GRK2- or GRK3-specific antagonist peptides (105, 130) to inhibit agonist-induced receptor phosphorylation (9, 130, 158) and desensitization (9, 105, 130, 158). Yet another method employed to attribute intracellular agonist-induced receptor phosphorylation to a particular GRK has involved cells transfected with both receptor and individual GRK cDNAs. In these assays, transfected cells express levels of GRK protein manyfold higher than endogenous cellular levels. Augmentation of agonist-induced receptor phosphorylation in cells overexpressing a GRK has therefore been interpreted as evidence for that GRK's activity, an inference confirmed with several receptors in assays using purified proteins (17, 60, 172, 193, 194). Negative results from these assays may be difficult to interpret, however, because the receptor phosphorylation observed derives from the activity of both transfected and endogenous cellular GRKs (usually GRK2).

Agonist-induced receptor phosphorylation studies in intact cells have also generated data that could be classified as evidence suggestive of GRK activity. These receptor immunoprecipitation studies have demonstrated agonist-induced receptor phosphorylation transpiring within the rapid time course characteristic of GRK-mediated processes and furthermore demonstrated the receptor phosphorylation to be resistant to inhibitors of second messenger–dependent protein kinases (60–62, 156, 167, 171, 177, 178, 192, 200, 201, 206, 209, 211). A last category of suggestive phosphorylation evidence comprises early experiments in which cells were stimulated with specific agonists and then disrupted and separated into membrane and cytosolic fractions, which were assayed for GRK activity. If GRK activity translocated from the cytosolic to the membrane fraction of the cell on agonist stimulation, the receptor for that agonist was inferred to be a GRK substrate (81, 159, 212). We have omitted data based on only GRK-mediated phosphorylation of receptor-derived peptide fragments (52, 176), because peptide phosphorylation studies involve no allosteric activation of the GRK by agonist-occupied receptors.

Cellular signaling studies roughly parallel the receptor phosphorylation assays. The inhibition of homologous receptor desensitization in permeabilized or intact cells has provided evidence for GRK substrates. This evidence ranges in quality from suggestive, with heparin (9, 167, 176, 196, 205), to convincing, with specific antibodies (9, 158), peptides (105, 130), or antisense oligonucleotides or mRNAs (160, 168, 215). The ability of dominant negative GRK2 to attenuate desensitization assessed in membrane assays has accorded well with phosphorylation studies employing this construct (61, 62). Suppression of agonist-evoked second-messenger responses in intact cells transfected with GRKs, on the other hand, has provided data discordant with phosphorylation studies (61, 62, 173). The suppression of receptor signaling by a transfected GRK in these overexpression systems appears to correlate not with GRK-mediated receptor phosphorylation but with agonist-induced binding of the overexpressed GRK to the activated receptor, as assessed by coimmunoprecipitation (62).

Substrate Specificity

Increasingly, it seems that the vast majority of G protein–coupled receptors will prove to be substrates for GRKs. Because the preponderance of mammalian cells and tissues tested thus far express more than one GRK, divining which GRKs regulate which receptors has proved daunting. Only with GRK1 and GRK4, with severely restricted cell-type expression, can we confidently infer—or reasonably speculate—about in vivo receptor substrates. GRK1, restricted primarily to retinal photoreceptor cells (21), certainly phosphorylates rhodopsin in vivo, and GRK4, restricted to the spermatogonia cell lineage (42), may prove to phosphorylate the olfactory-like sperm receptors (216) in vivo. With the surfeit of receptors and relative scarcity of GRKs, it seems overwhelmingly likely that the other GRKs each regulate innumerable receptors, probably in a cell type–specific manner. That is, regulation of a given receptor in a particular cell will be determined not only by which GRK or GRKs are expressed in that cell, but also by the relative and absolute expression levels of each GRK in that cell (165). The inhibition of specific GRKs in cellular systems, described above (9, 129, 130, 158, 160, 168, 215), has offered valuable insights into cell type–specific GRK substrate specificity.

Intimations of true enzymatic GRK substrate specificity have emanated from phosphorylation assays with purified proteins (17, 54, 96, 197, 198) and peptides (53–55). By examining the kinetics and stoichiometry of receptor phosphorylation, these studies suggest that certain receptors are not equally well phosphorylated by each of the purified GRKs (54, 96, 197, 198). In interpreting these studies, however, we must consider the possibilities that the absence of GRK isoform activators or the presence of inadvertently copurified GRK isoform inhibitors (207) have affected the results.

Finally, let us consider GRK substrate specificity more globally, by asking what G protein–coupled receptors, subjected to the assays described above, have proved refractory to GRK-mediated phosphorylation. To our knowledge, there are only two: the β_3-adrenergic (217) and the α_{2c}-adrenergic (194, 195) receptors. Neither of these receptors appears to undergo agonist-induced desensitization, and neither appears susceptible to GRK2-mediated phosphorylation in the sort of intact cell systems described above. These receptors therefore provide valuable negative controls supporting the validity of the approaches taken to implicate GRKs in the homologous desensitization of other GPCRs.

STUDIES WITH GENETICALLY ALTERED MICE

Most information about GRKs has been obtained from in vitro studies, many of which utilized purified and recombinant molecules. Recently, however, a number of papers have been published that investigate the roles of kinases in living animals. These studies, which have utilized both transgenic and knockout mice, have both confirmed predictions from earlier in vitro work and provided a number of interesting surprises. Most of the data pertain to the roles of GRKs in cardiac function and development (218, 219).

Transgenic Animals

Transgenic mice were created with cardiac-specific overexpression of either GRK2 or a GRK2 antagonist, comprising the carboxy terminal 195–amino acid segment of GRK2 (residues 495–689) (220) previously shown to function as a $G_{\beta\gamma}$-sequestering protein in vitro (84). In each case the transgene was driven by the murine alpha myosin heavy-chain promoter (221). Animals overexpressing GRK2 showed attenuation of isoproterenol-stimulated left ventricular contractility in vivo, reduced sarcolemmal adenylyl cyclase activity, and diminished functional coupling of β-adrenergic receptors (220). Conversely, mice expressing the GRK2 inhibitor displayed enhanced cardiac contractility in vivo with or without isoproterenol (220). These studies were the first to demonstrate the ability of a GRK to desensitize G protein–coupled receptors in vivo. Moreover, they demonstrated the important role of GRK2 in modulating in vivo myocardial function, even in the absence of exogenous catecholamines. Consonant with these results, isolated cardiac myocytes from these transgenic mice demonstrate perturbations in contractility similar to those observed in vivo (222).

GRK5 is also abundantly expressed in the mammalian heart (23, 24). A transgenic mouse overexpressing GRK5 in the heart also showed marked β-adrenergic receptor desensitization as judged by in vivo isoproterenol-stimulated contractility (223). The dampening of catecholamine-stimulated contractility was even more striking than that observed in the GRK2-overexpressing animals; however, the relative overexpression of GRK5 was also much greater than that of GRK2. Interestingly, whereas angiotensin II–stimulated contractility was unaffected in these GRK5-overexpressing mice, it was blunted in the GRK2-overexpressing mice (223). Because activation of angiotensin II type 1 receptors increases intracellular calcium, the consequent activation of calmodulin may inhibit myocardial GRK5 action more than it does GRK2 action (112, 137) and explain the apparent GRK specificity. These data suggest that myocardial overexpression of GRK5 results in selective uncoupling of G protein–coupled receptors and demonstrate that receptor specificity of GRKs may be important in determining physiological phenotype.

Knockout Animals

When the mouse gene for GRK2 was disrupted by homologous recombination, no homozygous embryos survived beyond gestational day 15.5 (224). Prior to this time, −/− embryos displayed pronounced hypoplasia of the ventricular myocardium essentially identical to the thin myocardium syndrome observed on gene inactivation of several transcription factors (RXRα, N-myc, TEF-1, WT-1) (225). Embryonic death of the −/− embryos appears to be due to heart failure, as cardiac ejection fraction is markedly reduced (224). These results, along with the virtual absence of endogenous GRK activity in GRK2 −/− whole embryos, demonstrate that GRK2 is the predominant GRK in early embryogenesis and that it plays a fundamental role in cardiac development. The nature of this role remains obscure. The data, however, hint at an as yet unappreciated interaction between GRK2 and transcription factor pathways.

Whereas GRK2−/− embryos all die in utero, +/− embryos not only survive but also appear normal with regard to gross morphology, growth, and development. As adults they have a ∼50% reduction in tissue (cardiac) GRK2 levels (224). Interestingly, this 50% reduction in cardiac GRK2 levels is associated with significant increases in in vivo contractile responses to isoproterenol (H Rockman, W Koch, S Akhter, RJ Lefkowitz, MG Caron, submitted manuscript). When GRK2 +/− animals are bred with animals overexpressing the GRK2 inhibitor peptide, the doubly transgenic animals show even greater contractility enhancement and a further lowering of myocardial $G_{\beta\gamma}$-stimulated GRK activity to ∼25% of wild-type animals (H Rockman, W Koch, S Akhter, RJ Lefkowitz, MG Caron, submitted manuscript). Similar alterations in contractility could be observed in ventricular myocytes isolated from these various animal lines. These data further strengthen the notion that cardiac function in vivo is strongly modulated by the levels of GRK2 activity.

In contrast to the embryonic lethality of the GRK2 gene knockout, deletion of the GRK3 gene in mice allows for normal embryonic and postnatal development (227). GRK3 is known to be highly expressed in olfactory epithelium and to function in desensitization of olfactory second-messenger signaling

(9, 129, 130). In accordance with these observations, olfactory cilia preparations derived from GRK3-deficient mice lack the fast agonist-induced desensitization normally seen after odorant stimulation (227). In addition, olfactory cilia (but not other tissues) of these mice demonstrate a dampening of the G protein–adenylyl cyclase system that may compensate for the desensitization defect observed. The findings confirm the requirement of GRK3 for odorant-induced desensitization of cAMP responses.

POTENTIAL INVOLVEMENT OF GRKS IN HUMAN DISEASE

Because the array of receptors regulated by GRKs (Table 1) affect so many vital functions, it seems likely that disorders of GRK-mediated regulation would contribute to, if not engender, disease. Alternatively, alterations in GRK expression and activity might help to compensate for excessive stimulation of certain receptor signaling systems. These possibilities have manifested themselves thus far in five disease areas.

1. *Opiate addiction*:   In rats treated chronically with morphine, GRK2 levels increased in the locus coeruleus. This increased GRK2 activity (perhaps involved in phosphorylating the μ-opioid receptor) may both compensate for hyperstimulation of central nervous system opioid receptors and contribute to the problem of opiate tolerance (228).

2. *Retinitis pigmentosa*:   Failure to desensitize rhodopsin signaling leads to photoreceptor cell death (3, 10), as is observed in retinitis pigmentosa. This disease has been associated with many mutations in rhodopsin, including several in the cytoplasmic tail domain (229, 230), where GRK1-mediated phosphorylation leads to termination of receptor signaling (47).

3. *Hypertension*:   Elevated GRK2 levels have recently been demonstrated in peripheral blood lymphocytes of a subgroup of hypertensive patients with impaired β_2AR-mediated vasodilation (231). Because β_2AR regulation in lymphocytes parallels that observed in vascular smooth muscle cells in hypertensive subjects, the GRK2 up-regulation seen in these lymphocytes may underlie the attenuation of β_2AR-mediated vasodilation in the hypertensive subjects studied (231).

4. *Myocardial ischemia*:   Up-regulation of GRK2 manifests itself in rat heart muscle deprived of oxygen for prolonged periods. This GRK2 up-regulation correlates temporally with diminishing responsiveness of β-adrenergic receptor-stimulated cyclase activity (232).

5. *Chronic heart failure*:   A reproducible constellation of abnormalities in myocardial β-adrenergic receptor signaling has been observed both in material from human hearts (233) and in various animal models of heart failure (234). These include decreased contractile and adenylyl cyclase responses to β-agonists, decreased numbers and G_s coupling of myocardial β_1-adrenergic receptors, and elevated levels of myocardial GRK activity (presumably GRK2 but possibly including GRK5) (233, 235–237). The extent of GRK2 up-regulation observed in human chronic heart failure approximates that achieved in the GRK2-overexpressing transgenic mice described above (220, 236).   As we have seen, a growing body of data appears to indicate that myocardial GRK2 levels correlate directly with and may significantly control myocardial contractility. These findings have suggested a testable hypothesis: that lowering myocardial GRK2 levels might offer a therapeutic strategy for improving myocardial contractile performance in the setting of the failing heart (238).

This hypothesis has been tested only in isolated cardiac myocytes from rabbits with pacing-induced congestive heart failure, characterized by a diminished myocardial maximum dP/dT (239).   The pacing-induced heart failure was associated with reduction of the number of β-adrenergic receptors and increases in membrane GRK2 activity in the isolated ventricular myocytes (239). Adenoviral-mediated gene transfer of the GRK2 carboxy terminus peptide inhibitor into these failing myocytes led to the restoration of βAR signaling (239), much as this GRK2 inhibitor enhanced isoproterenol-stimulated cAMP in ventricular myocytes from normal rabbits (238). These in vitro studies raise the possibility that successful gene transfer or chemical inhibitors of GRK2 might have therapeutic utility in the setting of human heart failure.

FUTURE DIRECTIONS AND PERSPECTIVES

The initial observations that phosphorylation of activated rhodopsin or β_2AR initiates receptor desensitization led to the identification of the GRK family of serine/threonine kinases, currently comprising six members. In the years since their discovery, the GRKs have been shown to phosphorylate and initiate desensitization of an impressive array of GPCR substrates (see Table 1). The critical importance of GRKs in regulating GPCR function has been revealed by a combination of studies in vitro using purified proteins, investigations of GPCR phosphorylation and desensitization in cellular systems, and, more recently, examination of the effects of deletion or overexpression of GRK genes in intact animals. Furthermore, burgeoning investigations portray a family of enzymes exquisitely and differentially regulated by a variety of different signaling molecules. Thus, for GRKs, the interaction with GPCR substrates; posttranslational modifications; and the binding of distinct structural domains to $G_{\beta\gamma}$, calmodulin, and a variety of different lipids serve to link GPCR activation inextricably to GRK-initiated desensitization.

Recent data hint at further, hitherto unsuspected roles for these enzymes and for GPCR phosphorylation in general. In addition to receptor desensitization and attenuation of cellular signaling, GRKs and cAMP-dependent protein kinase (PKA) appear to play an additional role: that of signal switching.

The β_2AR, like many GPCRs, couples to several different classes of G proteins (240–243). While many β_2AR processes occur consequent to β_2AR/G_s coupling and activation of adenylyl cyclase (37, 244), other β_2AR-stimulated processes occur consequent to G_i activation (240, 242). β_2AR-mediated G_i activation has recently been shown to depend on the phosphorylation status of the receptor (245). PKA-mediated phosphorylation of the β_2AR appears to switch the coupling of this receptor from G_s to G_i (245). Thus, PKA-mediated β_2AR phosphorylation attenuates β_2AR-stimulated cAMP production both by inhibiting the β_2AR's coupling to G_s and by promoting its coupling to G_i.

In many systems, the $G_{\beta\gamma}$ subunits of G_i can initiate mitogen-activated protein (MAP) kinase cascades (246). The β_2AR has also been shown to initiate this effector pathway (245), but in a manner that is dependent on phosphorylation of the β_2AR not only by PKA but also by a GRK (247). Whereas PKA-mediated receptor phosphorylation facilitates coupling to G_i, GRK-mediated receptor phosphorylation facilitates the receptor's binding to a β-arrestin isoform. This binding event both uncouples the receptor from G proteins and targets the receptor to clathrin-coated pits, where receptor endocytosis begins. β_2AR endocytosis appears essential for the activation of MAP kinase by this receptor (247). β_2AR-mediated MAP kinase activation is thus intimately dependent on the phosphorylation status of the receptor. These observations suggest a new paradigm for GRK- and PKA-mediated phosphorylation of GPCRs in which these enzymes, rather than acting merely as signal attenuators, act as signal switches determining the nature of the signal transduction cascade initiated by a single receptor.

In addition to terminating and switching signal transduction pathways via receptor phosphorylation, GRKs may participate directly in GPCR-mediated signal transduction. The interaction of GRKs with activated receptor substrates leads to the allosteric activation of these enzymes, dramatically enhancing peptide substrate phosphorylation. GRKs thus represent enzymes that are specifically activated following GPCR occupancy and may participate in signal transduction cascades by phosphorylating non-GPCR substrates. This model of GRK action is particularly attractive in light of the recent observation that tubulin, the building block of microtubules, serves as an excellent substrate for GRK2 (248). The kinetic parameters of tubulin phosphorylation are similar to those of agonist-occupied β_2AR, and GRK2 associates with microtubules in intact cells. Furthermore, in cells, tubulin phosphorylation is potentiated following GPCR activation (248). It is tempting to speculate whether GRK2-mediated tubulin phosphorylation mediates, at least in part, the reported effects of GPCRs on the cytoskeleton (249, 250). The search for additional nonreceptor GRK substrates is in progress.

The binding of kinases to specific proteins with discrete subcellular distributions provides an attractive mechanism for regulating their activity and specificity. Such targeting proteins have been identified for PKA and PKC: A-kinase anchoring proteins (AKAPs) (251) and receptors for activated PKC (RACKs) (252), respectively. That similar proteins exist for GRKs is supported by the observation that GRK2 binds with high affinity to a microsomal protein, dubbed protein X (152, 153). When bound to protein X, GRK2 is inhibited, but activators of heterotrimeric G proteins (GTPγS or A1F$_4^-$) can relieve this inhibition, even though protein X does not appear to be a heterotrimeric G protein (153). In this respect it is interesting to note that the GRKs contain within their amino termini a region that is homologous to a family of proteins termed regulators of G protein signaling (RGS) (253). RGS proteins bind to the α subunits of activated G proteins to facilitate GTP hydrolysis (254, 255). Could protein X be homologous to G protein α subunits? The characterization of protein X and the potential identification of other GRK-targeting proteins are likely to provide considerable insight into the mechanisms of regulation as well as the substrate specificity of GRKs.

Most of the studies utilizing GRKs have centered on their role as mediators of GPCR desensitization. Recent evidence, however, points to a role for these enzymes as propagators or initiators of signal transduction. The second decade of GRK investigations will likely witness studies designed to address the generality of the signal switch model of GPCR phosphorylation and the existence of novel GRK substrates and binding proteins. Such studies have the potential to greatly expand our understanding of the cellular functions of GRKs.

ACKNOWLEDGMENTS

This work was supported, in part, by National Institutes of Health Grants HL03008 (to NJF) and HL16037 (to RJL).

We wish to thank Dr. Richard T Premont for valuable discussions, and M Holben, D Addison, and D Sawyer for expert assistance in manuscript preparation.

Literature Cited

1. Springer MS, Goy MF, Adler J. 1979. *Nature* 280:279–84
2. Reneke JE, Blumer KJ, Courschesne W, Thorner J. 1988. *Cell* 55:221–34
3. Dolph PJ, Ranganathan R, Colley NJ, Hardy RW, Socolich M, Zuker CS. 1993. *Science* 260:1910–16
4. Arriza JL, Dawson TM, Simerly RB, Martin LJ, Caron MG, et al. 1992. *J. Neurosci.* 12:4045–55
5. Watson S, Arkinstall A. 1994. *The G-Protein Linked Receptor Facts Book.* San Diego: Academic Press
6. Samama P, Cotecchia S, Costa T, Lefkowitz RJ. 1993. *J. Biol. Chem.* 268:4625–36
7. Lohse MJ, Andexinger S, Pitcher J, Trukawinski S, Codina J, et al. 1992. *J. Biol. Chem.* 267:8558–64
8. Attramadal H, Arriza JL, Aoki C, Dawson TM, Codina J, et al. 1992. *J. Biol. Chem.* 267:17882–90
9. Schleicher S, Boekhoff I, Arriza J, Lefkowitz RJ, Breer H. 1993. *Proc. Natl. Acad. Sci. USA* 90:1420–24
10. Chen J, Makino CL, Peachey NS, Baylor DA, Simon MI. 1995. *Science* 267:374–77
11. Hausdorff WP, Caron MG, Lefkowitz RJ. 1990. *FASEB J.* 4:2881–89
12. Benovic JL, Pike LJ, Cerione RA, Staniszewski C, Yoshimasa T, et al. 1985. *J. Biol. Chem.* 260:7094–101
13. Pitcher J, Lohse MJ, Codina J, Caron MG, Lefkowitz RJ. 1992. *Biochemistry* 31:3193–97
14. Stadel J, Nambi P, Shorr R, Sawyer D, Caron M, Lefkowitz R. 1983. *Proc. Natl. Acad. Sci. USA* 80:3173–77
15. Strasser R, Sibley D, Lefkowitz R. 1986. *Biochemistry* 25:1371–77
16. Benovic JL, Strasser RH, Caron MG, Lefkowitz RJ. 1986. *Proc. Natl. Acad. Sci. USA* 83:2797–801
17. Benovic JL, Mayor F Jr, Staniszewski C, Lefkowitz RJ, Caron MG. 1987. *J. Biol. Chem.* 262:9026–32
18. Shichi H, Somers RL. 1978. *J. Biol. Chem.* 253:7040–46
19. Sitaramayya A, Liebman P. 1983. *J. Biol. Chem.* 258:12106–9
20. Benovic JL, DeBlasi A, Stone WC, Caron MG, Lefkowitz RJ. 1989. *Science* 246:235–40
21. Lorenz W, Inglese J, Palczewski K, Onorato JJ, Caron MG, Lefkowitz RJ. 1991. *Proc. Natl. Acad. Sci. USA* 88:8715–19
22. Benovic JL, Onorato JJ, Arriza JL, Stone WC, Lohse M, et al. 1991. *J. Biol. Chem.* 266:14939–46
23. Kunapuli P, Benovic JL. 1993. *Proc. Natl. Acad. Sci. USA* 90:5588–92
24. Premont RT, Koch WJ, Inglese J, Lefkowitz RJ. 1994. *J. Biol. Chem.* 269:6832–41
25. Benovic JL, Gomez J. 1993. *J. Biol. Chem.* 268:19521–27
26. Ambrose C, James M, Barnes G, Lin C, Bates G, et al. 1992. *Hum. Mol. Genet.* 1:697–703
27. Smith WC, Milam AH, Dugger D, Arendt A, Hargrave PA, Palczewski K. 1994. *J. Biol. Chem.* 269:15407–10
28. Craft CM, Whitmore DH. 1995. *FEBS Lett.* 362:247–55
29. Lohse MJ, Benovic JL, Codina J, Caron MG, Lefkowitz RJ. 1990. *Science* 248:1547–50
30. Parruti G, Peracchia F, Sallese M, Ambrosini G, Masini M, et al. 1993. *J. Biol. Chem.* 268:9753–61
31. Sterne-Marr R, Gurevich VV, Goldsmith P, Bodine RC, Sanders C, et al. 1993. *J. Biol. Chem.* 268:15640–48
32. Ferguson SS, Downey WE Jr, Colapietro AM, Barak LS, Menard L, Caron MG. 1996. *Science* 271:363–66
33. von Zastrow M, Kobilka BK. 1992. *J. Biol. Chem.* 267:3530–38
34. Pitcher JA, Payne ES, Csortos C, DePaoli-Roach AA, Lefkowitz RJ. 1995. *Proc. Natl. Acad. Sci. USA* 92:8343–47
35. Hughes RJ, Anderson KL, Kiel D, Insel PA. 1996. *Am. J. Physiol.* 270:C885–91
36. Bouvier M, Hausdorff WP, De Blasi A, O'Dowd BF, Kobilka BK, et al. 1988. *Nature* 333:370–73
37. Hausdorff WP, Bouvier M, O'Dowd BF, Irons GP, Caron MG, Lefkowitz RJ. 1989. *J. Biol. Chem.* 264:12657–65
38. Fredericks ZL, Pitcher JA, Lefkowitz RJ. 1996. *J. Biol. Chem.* 271:13796–803
39. McDowell JH, Nawrocki JP, Hargrave PA. 1993. *Biochemistry* 32:4968–74

40. Premont RT, Macrae AD, Stoffel RH, Chung N, Pitcher JA, et al. 1996. *J. Biol. Chem.* 271:6403–10
41. Freedman NJ, Lefkowitz RJ. 1996. *Recent Prog. Horm. Res.* 51:319–51
42. Sallese M, Lombardi MS, De Blasi A. 1994. *Biochem. Biophys. Res. Commun.* 199:848–54
43. Cassill JA, Whitney M, Joazeiro CA, Becker A, Zuker CS. 1991. *Proc. Natl. Acad. Sci. USA* 88:11067–70
44. Wilson R, Ainscough R, Anderson K, Baynes C, Berks M, et al. 1994. *Nature* 368:32–38
45. Hanks SK, Quinn AM, Hunter T. 1988. *Science* 241:42–52
46. Brenner S. 1987. *Nature* 329:21
47. Premont RT, Inglese J, Lefkowitz RJ. 1995. *FASEB J.* 9:175–82
48. Inglese J, Freedman NJ, Koch WJ, Lefkowitz RJ. 1993. *J. Biol. Chem.* 268:23735–38
49. Nakata H, Kameyama K, Haga K, Haga T. 1994. *Eur. J. Biochem.* 220:29–36
50. Eason MG, Moreira SP, Liggett SB. 1995. *J. Biol. Chem.* 270:4681–88
51. Ohguro H, Palczewski K, Ericsson LH, Walsh KA, Johnson RS. 1993. *Biochemistry* 32:5718–24
52. Prossnitz ER, Kim CM, Benovic JL, Ye RD. 1995. *J. Biol. Chem.* 270:1130–37
53. Onorato JJ, Palczewski K, Regan JW, Caron MG, Lefkowitz RJ, Benovic JL. 1991. *Biochemistry* 30:5118–25
54. Kunapuli P, Onorato JJ, Hosey MM, Benovic JL. 1994. *J. Biol. Chem.* 269:1099–105
55. Loudon RP, Benovic JL. 1994. *J. Biol. Chem.* 269:22691–97
56. Giannini E, Brouchon L, Boulay F. 1995. *J. Biol. Chem.* 270:19166–72
57. Litchfield DW, Arendt A, Lozeman FJ, Krebs EG, Hargrave PA, Palczewski K. 1990. *FEBS Lett.* 261:117–20
58. Fiol CJ, Haseman JH, Wang YH, Roach PJ, Roeske RW, et al. 1988. *Arch. Biochem. Biophys.* 267:797–802
59. Sibley D, Strasser R, Caron M, Lefkowitz R. 1985. *J. Biol. Chem.* 260:3883–86
60. Freedman NJ, Liggett SB, Drachman DE, Pei G, Caron MG, Lefkowitz RJ. 1995. *J. Biol. Chem.* 270:17953–61
61. Oppermann M, Freedman NJ, Alexander RW, Lefkowitz RJ. 1996. *J. Biol. Chem.* 271:13266–72
62. Freedman NJ, Ament AS, Oppermann M, Stoffel RH, Exum ST, Lefkowitz RJ. 1997. *J. Biol. Chem.* 272:17734–43
63. Benovic JL, Onorato J, Lohse MJ, Dohlman HG, Staniszewski C, et al.
1990. *Br. J. Clin. Pharmacol.* 30:S3–12
64. Lohse MJ, Lefkowitz RJ, Caron MG, Benovic JL. 1989. *Proc. Natl. Acad. Sci. USA* 86:3011–15
65. Kelleher DJ, Johnson GL. 1990. *J. Biol. Chem.* 265:2632–39
66. Palczewski K, Buczylko J, Kaplan MW, Polans AS, Crabb JW. 1991. *J. Biol. Chem.* 266:12949–55
67. Brown NG, Fowles C, Sharma R, Akhtar M. 1993. *Eur. J. Biochem.* 212:840
68. Chen CY, Dion SB, Kim CM, Benovic JL. 1993. *J. Biol. Chem.* 268:7825–31
69. Dohlman HG, Thorner J, Caron MG, Lefkowitz RJ. 1991. *Annu. Rev. Biochem.* 60:653–88
70. Haga K, Kameyama K, Haga T. 1994. *J. Biol. Chem.* 269:12594–99
71. Shaw G. 1996. *BioEssays* 18:35–46
72. Lemmon MA, Ferguson KM, Schlessinger J. 1996. *Cell* 85:621–24
73. Ferguson KM, Lemmon MA, Schlessinger J, Sigler PB. 1994. *Cell* 79:199–209
74. Ferguson KM, Lemmon MA, Schlessinger J, Sigler PB. 1995. *Cell* 83:1037–46
75. Downing AK, Driscoll PC, Gout I, Salim K, Zvelebil MJ, Waterfield MD. 1994. *Curr. Biol.* 4:884–91
76. Hyvonen M, Macias MJ, Nilges M, Oschkinat H, Saraste M, Wilmanns M. 1995. *EMBO J.* 14:4676–85
77. Timm D, Salim K, Gout I, Guruprasad L, Waterfield M, Blundell T. 1994. *Nat. Struct. Biol.* 1:782–88
78. Fushman D, Cahill S, Lemmon MA, Schlessinger J, Cowburn D. 1995. *Proc. Natl. Acad. Sci. USA* 92:816–20
79. Zhang P, Talluri S, Deng H, Branton D, Wagner G. 1995. *Structure* 3:1185–95
80. Macias MJ, Musacchio A, Ponstingl H, Nilges M, Saraste M, Oschkinat H. 1994. *Nature* 369:675–77
81. Strasser RH, Benovic JL, Caron MG, Lefkowitz RJ. 1986. *Proc. Natl. Acad. Sci. USA* 83:6362–66
82. Daaka Y, Pitcher JA, Richardson M, Stoffel RH, Robishaw JD, Lefkowitz RJ. 1997. *Proc. Natl. Acad. Sci. USA* 94:2180–85
83. Pitcher JA, Inglese J, Higgins JB, Arriza JL, Casey PJ, et al. 1992. *Science* 257:1264–67
84. Koch WJ, Inglese J, Stone WC, Lefkowitz RJ. 1993. *J. Biol. Chem.* 268:8256–60
85. Simonds WF, Butrynski JE, Gautam N, Unson CG, Spiegel AM. 1991. *J. Biol. Chem.* 266:5363–66

86. Kim CM, Dion SB, Benovic JL. 1993. *J. Biol. Chem.* 268:15412–18
87. Pitcher JA, Touhara K, Payne ES, Lefkowitz RJ. 1995. *J. Biol. Chem.* 270:11707–10
88. Haga K, Haga T. 1990. *FEBS Lett.* 268:43–47
89. Haga K, Haga T. 1992. *J. Biol. Chem.* 267:2222–27
90. Touhara K, Inglese J, Pitcher JA, Shaw G, Lefkowitz RJ. 1994. *J. Biol. Chem.* 269:10217–20
91. Touhara K, Koch WJ, Hawes BE, Lefkowitz RJ. 1995. *J. Biol. Chem.* 270:17000–5
92. Cowan SW, Newcomer ME, Jones TA. 1990. *Proteins* 8:44–61
93. Yoon HS, Hajduk PJ, Petros AM, Olejniczak ET, Meadows RP, Fesik SW. 1994. *Nature* 369:672–75
94. Harlan JE, Hajduk PJ, Yoon HS, Fesik SW. 1994. *Nature* 371:168–70
95. Onorato JJ, Gillis ME, Liu Y, Benovic JL, Ruoho AE. 1995. *J. Biol. Chem.* 270:21346–53
96. DebBurman SK, Kunapuli P, Benovic JL, Hosey MM. 1995. *Mol. Pharmacol.* 47:224–33
97. DebBurman SK, Ptasienski J, Benovic JL, Hosey MM. 1996. *J. Biol. Chem.* 271:22552–62
98. Brockman JL, Anderson RA. 1991. *J. Biol. Chem.* 266:2508–12
99. Kameyama K, Haga K, Haga T, Moro O, Sadee W. 1994. *Eur. J. Biochem.* 226:267–76
100. Chuang TT, Pompili E, Paolucci L, Sallese M, Degioia L, et al. 1997. *Eur. J. Biochem.* 245:533–40
101. Watson AJ, Aragay AM, Slepak VZ, Simon MI. 1996. *J. Biol. Chem.* 271:28154–60
102. Ray K, Kunsch C, Bonner LM, Robishaw JD. 1995. *J. Biol. Chem.* 270:21765–71
103. Muller S, Hekman M, Lohse MJ. 1993. *Proc. Natl. Acad. Sci. USA* 90:10439–43
104. Muller S, Straub A, Lohse MJ. 1997. *FEBS Lett.* 401:25–29
105. Diverse-Pierluissi M, Inglese J, Stoffel RH, Lefkowitz RJ, Dunlap K. 1996. *Neuron* 16:579–85
106. Gudermann T, Schoneberg T, Schultz G. 1997. *Annu. Rev. Neurosci.* 20:399–427
107. DebBurman SK, Ptasienski J, Boetticher E, Lomasney JW, Benovic JL, Hosey MM. 1995. *J. Biol. Chem.* 270:5742–47
108. Quinn PJ. 1976. The *Molecular Biology of Cell Membranes.* Baltimore: Univ. Park Press
109. Lemmon MA, Ferguson KM, O'Brien R, Sigler PB, Schlessinger J. 1995. *Proc. Natl. Acad. Sci. USA* 92:10472–76
110. Kunapuli P, Gurevich VV, Benovic JL. 1994. *J. Biol. Chem.* 269:10209–12
111. Pitcher JA, Fredericks ZL, Stone WC, Premont RT, Stoffel RH, et al. 1996. *J. Biol. Chem.* 271:24907–13
112. Pronin AN, Satpaev DK, Slepak VZ, Benovic JL. 1997. *J. Biol. Chem.* 272:18273–80
113. Inglese J, Koch WJ, Caron MG, Lefkowitz RJ. 1992. *Nature* 359:147–50
114. Stoffel RH, Randall RR, Premont RT, Lefkowitz RJ, Inglese J. 1994. *J. Biol. Chem.* 269:27791–94
115. Mouillac B, Caron M, Bonin H, Dennis M, Bouvier M. 1992. *J. Biol. Chem.* 267:21733–37
116. Kennedy ME, Limbird LE. 1994. *J. Biol. Chem.* 269:31915–22
117. Robinson LJ, Busconi L, Michel T. 1995. *J. Biol. Chem.* 270:995–98
118. Wedegaertner PB, Bourne HR. 1994. *Cell* 77:1063–70
119. Degtyarev MY, Spiegel AM, Jones TL. 1994. *J. Biol. Chem.* 269:30898–903
120. Levis MJ, Bourne HR. 1992. *J. Cell Biol.* 119:1297–307
121. Chuang TT, LeVine H III, De Blasi A. 1995. *J. Biol. Chem.* 270:18660–65
122. Winstel R, Freund S, Krasel C, Hoppe E, Lohse MJ. 1996. *Proc. Natl. Acad. Sci. USA* 93:2105–9
123. Pronin AN, Benovic JL. 1997. *J. Biol. Chem.* 272:3806–12
124. Yao L, Kawakami Y, Kawakami T. 1994. *Proc. Natl. Acad. Sci. USA* 91:9175–79
125. Konishi H, Kuroda S, Kikkawa U. 1994. *Biochem. Biophys. Res. Commun.* 205:1770–75
126. Schleicher S, Boekhoff I, Arriza J, Lefkowitz RJ, Breer H. 1993. *Proc. Natl. Acad. Sci. USA* 90:1420–24
127. Boekhoff I, Inglese J, Schleicher S, Koch WJ, Lefkowitz RJ, Breer H. 1994. *J. Biol. Chem.* 269:37–40
128. Dawson TM, Arriza JL, Jaworsky DE, Borisy FF, Attramadal H, et al. 1993. *Science* 259:825–29
129. Dawson TM, Arriza JL, Jaworsky DE, Borisy FF, Attramadal H, et al. 1993. *Science* 259:825–29
130. Boekhoff I, Inglese J, Schleicher S, Koch WJ, Lefkowitz RJ, Breer H. 1994. *J. Biol. Chem.* 269:37–40
131. Lohse MJ, Benovic JL, Caron MG, Lefkowitz RJ. 1990. *J. Biol. Chem.* 265:3202–11
132. Boekhoff I, Touhara K, Danner S, Inglese J, Lohse MJ, et al. 1997. *J. Biol. Chem.* 272:4606–12

133. Kuo CH, Akiyama M, Miki N. 1989. *Brain Res. Mol. Brain Res.* 6:1–10
134. Lee RH, Brown BM, Lolley RN. 1990. *J. Biol. Chem.* 265:15860–66
135. Hawes BE, Touhara K, Kurose H, Lefkowitz RJ, Inglese J. 1994. *J. Biol. Chem.* 269:29825–30
135a. Moncrief ND, Kretsinger RH, Goodman M. 1990. *J. Mol. Evol.* 30:522–62
136. Chen CK, Inglese J, Lefkowitz RJ, Hurley JB. 1995. *J. Biol. Chem.* 270:18060–66
137. Chuang TT, Paolucci L, De Blasi A. 1996. *J. Biol. Chem.* 271:28691–96
138. Haga K, Tsuga H, Haga T. 1997. *Biochemistry* 36:1315–21
139. Korf HW, White BH, Schaad NC, Klein DC. 1992. *Brain Res.* 595:57–66
140. Milam AH, Dacey DM, Dizhoor AM. 1993. *Vis. Neurosci.* 10:1–12
141. Zozulya S, Stryer L. 1992. *Proc. Natl. Acad. Sci. USA* 89:11569–73
142. Dizhoor AM, Chen CK, Olshevskaya E, Sinelnikova VV, Phillipov P, Hurley JB. 1993. *Science* 259:829–32
143. Sanada K, Shimizu F, Kameyama K, Haga K, Haga T, Fukada Y. 1996. *FEBS Lett.* 384:227–30
144. Calvert PD, Klenchin VA, Bownds MD. 1995. *J. Biol. Chem.* 270:24127–29
145. Gray-Keller MP, Polans AS, Palczewski K, Detwiler PB. 1993. *Neuron* 10:523–31
146. O'Neil KT, DeGrado WF. 1990. *Science* 250:646–51
147. Gnegy ME. 1993. *Annu. Rev. Pharmacol. Toxicol.* 33:45–70
148. De Castro E, Nef S, Fiumelli H, Lenz SE, Kawamura S, Nef P. 1995. *Biochem. Biophys. Res. Commun.* 216:133–40
149. Schaad NC, De Castro E, Nef S, Hegi S, Hinrichsen R, et al. 1996. *Proc. Natl. Acad. Sci. USA* 93:9253–58
150. Buczylko J, Gutmann C, Palczewski K. 1991. *Proc. Natl. Acad. Sci. USA* 88:2568–72
151. Palczewski K, Buczylko J, Van Hooser P, Carr SA, Huddleston MJ, Crabb JW. 1992. *J. Biol. Chem.* 267:18991–98
152. Garcia-Higuera I, Penela P, Murga C, Egea G, Bonay P, et al. 1994. *J. Biol. Chem.* 269:1348–55
153. Murga C, Ruiz-Gomez A, Garcia-Higuera I, Kim CM, Benovic JL, Mayor F Jr. 1996. *J. Biol. Chem.* 271:985–94
154. Garcia-Higuera I, Mayor F Jr. 1994. *J. Clin. Invest.* 93:937–43
155. Shui Z, Boyett MR, Zang WJ, Haga T, Kameyama K. 1995. *J. Physiol.* 487:359–66
156. Ozcelebi F, Miller LJ. 1995. *J. Biol. Chem.* 270:3435–41
157. De Blasi A, Parruti G, Sallese M. 1995. *J. Clin. Invest.* 95:203–10
158. Oppermann M, Diverse-Pierluissi M, Drazner MH, Dyer SL, Freedman NJ, et al. 1996. *Proc. Natl. Acad. Sci. USA* 93:7649–54
159. Chuang TT, Sallese M, Ambrosini G, Parruti G, De Blasi A. 1992. *J. Biol. Chem.* 267:6886–92
160. Shih M, Malbon CC. 1994. *Proc. Natl. Acad. Sci. USA* 91:12193–97
161. Loudon RP, Perussia B, Benovic JL. 1996. *Blood* 88:4547–57
162. Menard L, Ferguson SSG, Zhang J, Lin FT, Lefkowitz RJ, et al. 1997. *Mol. Pharmacol.* 51:800–8
163. McGraw DW, Liggett SB. 1997. *J. Biol. Chem.* 272:7338–44
164. Sallese M, Mariggio S, Collodel G, Moretti E, Piomboni P, et al. 1997. *J. Biol. Chem.* 272:10188–95
165. Aramori I, Zhang J, Ferguson S, Bieniasz P, Cullen B, Caron M. 1997. *EMBO J.* 16:4606–16
166. Benovic JL, Stone WC, Caron MG, Lefkowitz RJ. 1989. *J. Biol. Chem.* 264:6707–10
167. Lohse MJ, Benovic JL, Caron MG, Lefkowitz RJ. 1990. *J. Biol. Chem.* 265:3202–11
168. Nakata H, Kinoshita Y, Kishi K, Fukuda H, Kawanami C, et al. 1996. *Digestion* 57:406–10
169. Kong G, Penn R, Benovic JL. 1994. *J. Biol. Chem.* 269:13084–87
170. Tsuga H, Kameyama K, Haga T, Kurose H, Nagao T. 1994. *J. Biol. Chem.* 269:32522–27
171. Pei G, Kieffer BL, Lefkowitz RJ, Freedman NJ. 1995. *Mol. Pharmacol.* 48:173–77
172. Ferguson SS, Menard L, Barak LS, Koch WJ, Colapietro AM, Caron MG. 1995. *J. Biol. Chem.* 270:24782–89
173. Diviani D, Lattion AL, Larbi N, Kunapuli P, Pronin A, et al. 1996. *J. Biol. Chem.* 271:5049–58
174. Dohlman HG, Bouvier M, Benovic JL, Caron MG, Lefkowitz RJ. 1987. *J. Biol. Chem.* 262:14282–88
175. Lattion A-L, Diviani D, Cotecchia S. 1994. *J. Biol. Chem.* 269:22887–93
176. Takano T, Honda Z, Sakanaka C, Izumi T, Kameyama K, et al. 1994. *J. Biol. Chem.* 269:22453–58
177. Ishii K, Chen J, Ishii M, Koch WJ, Freedman NJ, et al. 1994. *J. Biol. Chem.* 269:1125–30
178. Richardson RM, DuBose RA, Ali H,

Tomhave ED, Haribabu B, Snyderman R. 1995. *Biochemistry* 34:14193–201

179. Hausdorff WP, Campbell PT, Ostrowski J, Yu SS, Caron MG, Lefkowitz RJ. 1991. *Proc. Natl. Acad. Sci. USA* 88: 2979–83

180. Roth NS, Campbell PT, Caron MG, Lefkowitz RJ, Lohse MJ. 1991. *Proc. Natl. Acad. Sci. USA* 88:6201–4

181. Kuhn H. 1978. *Biochemistry* 17:4389–95

182. Ruiz-Gomez A, Mayor F Jr. 1997. *J. Biol. Chem.* 272:9601–4

183. Nagayama Y, Tanaka K, Namba H, Yamashita S, Niwa M. 1996. *Thyroid* 6:627–31

184. Ping P, Gelzer-Bell R, Roth DA, Kiel D, Insel PA, Hammond HK. 1995. *J. Clin. Invest.* 95:1271–80

185. Wilden U, Hall S, Kuhn H. 1986. *Proc. Natl. Acad. Sci. USA* 83:1174–78

186. Benovic JL, Kuhn H, Weyand I, Codina J, Caron MG, Lefkowitz RJ. 1987. *Proc. Natl. Acad. Sci. USA* 84:8879–82

187. Richardson RM, Kim C, Benovic JL, Hosey MM. 1993. *J. Biol. Chem.* 268: 13650–56

188. Ramkumar V, Kwatra M, Benovic JL, Stiles GL. 1993. *Biochim. Biophys. Acta* 1179:89–97

189. Wilden U. 1995. *Biochemistry* 34:1446–54

190. Murphy TJ, Alexander RW, Griendling KK, Runge MS, Bernstein KE. 1991. *Nature* 351:233–36

191. Mundell SJ, Benovic JL, Kelly E. 1997. *Mol. Pharmacol.* 51:991–98

192. Palmer TM, Benovic JL, Stiles GL. 1995. *J. Biol. Chem.* 270:29607–13

193. Benovic JL, Regan JW, Matsui H, Mayor F Jr, Cotecchia S, et al. 1987. *J. Biol. Chem.* 262:17251–53

194. Kurose H, Lefkowitz RJ. 1994. *J. Biol. Chem.* 269:10093–99

195. Jewell-Motz EA, Liggett SB. 1996. *J. Biol. Chem.* 271:18082–87

196. Liggett SB, Ostrowski J, Chesnut LC, Kurose H, Raymond JR, et al. 1992. *J. Biol. Chem.* 267:4740–46

197. Pei G, Tiberi M, Caron MG, Lefkowitz RJ. 1994. *Proc. Natl. Acad. Sci. USA* 91: 3633–36

198. Benovic JL, Mayor F Jr, Somers RL, Caron MG, Lefkowitz RJ. 1986. *Nature* 321:869–72

199. Pippig S, Andexinger S, Daniel K, Puzicha M, Caron MG, et al. 1993. *J. Biol. Chem.* 268:3201–8

200. Blaukat A, Alla SA, Lohse MJ, Muller-Esterl W. 1996. *J. Biol. Chem.* 271: 32366–74

201. Ali H, Richardson RM, Tomhave ED, Didsbury JR, Snyderman R. 1993. *J. Biol. Chem.* 268:24247–54

202. Giannini E, Boulay F. 1995. *J. Immunol.* 154:4055–64

203. Franci C, Gosling J, Tsou CL, Coughlin SR, Charo IF. 1996. *J. Immunol.* 157:5606–12

204. Tiberi M, Nash SR, Bertrand L, Lefkowitz RJ, Caron MG. 1996. *J. Biol. Chem.* 271:3771–78

205. Ansanay H, Sebben M, Bockaert J, Dumuis A. 1992. *Mol. Pharmacol.* 42:808–16

206. Mueller SG, Schraw WP, Richmond A. 1995. *J. Biol. Chem.* 270:10439–48

207. Haga K, Kameyama K, Haga T, Kikkawa U, Shiozaki K, Uchiyama H. 1996. *J. Biol. Chem.* 271:2776–82

208. Raynor K, Kong H, Hines J, Kong G, Benovic J, et al. 1994. *Mol. Pharmacol.* 270:1381–86

209. Ali H, Richardson RM, Tomhave ED, DuBose RA, Haribabu B, Snyderman R. 1994. *J. Biol. Chem.* 269:24557–63

210. Strasser RH, Benovic JL, Lefkowitz RJ, Caron MG. 1988. *Adv. Exp. Med. Biol.* 231:503–17

211. Blind E, Bambino T, Nissenson RA. 1995. *Endocrinology* 136:4271–77

212. Mayor F Jr, Benovic JL, Caron MG, Lefkowitz RJ. 1987. *J. Biol. Chem.* 262:6468–71

213. Kwatra MM, Schwinn DA, Schreurs J, Blank JL, Kim CM, et al. 1993. *J. Biol. Chem.* 268:9161–64

214. Iacovelli L, Franchetti R, Masini M, De Blasi A. 1996. *Mol. Endocrinol.* 10: 1138–46

215. Nagayama Y, Tanaka K, Hara T, Namba H, Yamashita S, et al. 1996. *J. Biol. Chem.* 271:10143–48

216. Parmentier M, Libert F, Schurmans S, Schiffmann S, Lefort A, et al. 1992. *Nature* 355:453–55

217. Liggett SB, Freedman NJ, Schwinn DA, Lefkowitz RJ. 1993. *Proc. Natl. Acad. Sci. USA* 90:3665–69

218. Rockman H, Koch W, Lefkowitz R. 1997. *Am. J. Physiol.* 272:H1553–59

219. Koch WJ, Milano CA, Lefkowitz RJ. 1996. *Circ. Res.* 78:511–16

220. Koch WJ, Rockman HA, Samama P, Hamilton RA, Bond RA, et al. 1995. *Science* 268:1350–53

221. Subramaniam A, Jones WK, Gulick J, Wert S, Neumann J, Robbins J. 1991. *J. Biol. Chem.* 266:24613–20

222. Korzick DH, Xiao RP, Ziman BD, Koch WJ, Lefkowitz RJ, Lakatta EG. 1997. *Am. J. Physiol.* 41:H590–96

223. Rockman HA, Choi DJ, Rahman NU, Akhter SA, Lefkowitz RJ, Koch WJ. 1996. *Proc. Natl. Acad. Sci. USA* 93: 9954–59
224. Jaber M, Koch WJ, Rockman H, Smith B, Bond RA, et al. 1996. *Proc. Natl. Acad. Sci. USA* 93:12974–79
225. Rossant J. 1996. *Circ. Res.* 78:349–53
226. Deleted in proof.
227. Peppel KC, Boekhoff I, McDonald PH, Breer H, Caron MG, Lefkowitz RJ. 1997. *J. Biol. Chem.* 272:25425–28
228. Terwilliger RZ, Ortiz J, Guitart X, Nestler EJ. 1994. *J. Neurochem.* 63: 1983–86
229. McInnes R, Bascom R. 1992. *Nat. Genet.* 1:155–57
230. Restagno G, Maghtheh M, Bhattacharya S, Ferrone M, Garnerone S, et al. 1993. *Hum. Mol. Genet.* 2:207–8
231. Gros R, Benovic JL, Tan CM, Feldman RD. 1997. *J. Clin. Invest.* 99:2087–93
232. Ungerer M, Kessebohm K, Kronsbein K, Lohse MJ, Richardt G. 1996. *Circ. Res.* 79:455–60
233. Bristow M, Herschberger R, Port J, Gilbert E, Sandoval A, et al. 1990. *Circulation* 82:I12–25
234. Hongo M, Ryoke T, Ross J Jr. 1997. *Trends Cardiovasc. Med.* 7:161–67
235. Bristow M, Kantrowitz N, Ginsburg R, Fowler M. 1985. *J. Mol. Cell. Cardiol.* 17:41–52
236. Ungerer M, Bohm M, Elce JS, Erdmann E, Lohse MJ. 1993. *Circulation* 87:454–63
237. Ungerer M, Parruti G, Bohm M, Puzicha M, DeBlasi A, et al. 1994. *Circ. Res.* 74:206–13
238. Drazner MH, Peppel KC, Dyer S, Grant AO, Koch WJ, Lefkowitz RJ. 1997. *J. Clin. Invest.* 99:288–96
239. Akhter SA, Skaer CA, Kypson AP,

McDonald PH, Peppel KC, et al. 1997. *Proc. Natl. Acad. Sci. USA* 94:12100–5
240. Abramson SN, Martin MW, Hughes AR, Harden TK, Neve KA, et al. 1988. *Biochem. Pharmacol.* 37:4289–97
241. Herrlich A, Kuhn B, Grosse R, Schmid A, Schultz G, Gudermann T. 1996. *J. Biol. Chem.* 271:16764–72
242. Xiao RP, Ji XW, Lakatta EG. 1995. *Mol. Pharmacol.* 47:322–29
243. Laugwitz KL, Allgeier A, Offermanns S, Spicher K, Van Sande J, et al. 1996. *Proc. Natl. Acad. Sci. USA* 93:116–20
244. Strulovici B, Cerione RA, Kilpatrick BF, Caron MG, Lefkowitz RJ. 1984. *Science* 225:837–40
245. Daaka Y, Luttrell LM, Lefkowitz RJ. 1997. *Nature* 390:88–91
246. van Biesen T, Luttrell LM, Hawes BE, Lefkowitz RJ. 1996. *Endocr. Rev.* 17:698–714
247. Daaka Y, Luttrell LM, Ahn S, Della Rocca GJ, Ferguson SSG, et al. 1998. *J. Biol. Chem.* 273:685–88
248. Pitcher JA, Hall RA, Daaka Y, Zhang J, Ferguson SSG, et al. 1998. *J. Biol. Chem.* In press
249. Keller H, Niggli V, Zimmermann A. 1991. *Adv. Exp. Med. Biol.* 297:23–37
250. Jalink K, Eichholtz T, Postma FR, van Corven EJ, Moolenaar WH. 1993. *Cell Growth Differ.* 4:247–55
251. Dell'Acqua ML, Scott JD. 1997. *J. Biol. Chem.* 272:12881–84
252. Mochly-Rosen D, Smith BL, Chen CH, Disatnik MH, Ron D. 1995. *Biochem. Soc. Trans.* 23:596–600
253. Siderovski DP, Hessel A, Chung S, Mak TW, Tyers M. 1996. *Curr. Biol.* 6:211–12
254. Koelle MR. 1997. *Curr. Opin. Cell Biol.* 9:143–47
255. Dohlman HG, Thorner J. 1997. *J. Biol. Chem.* 272:3871–74

Annu. Rev. Biochem. 1998. 67:693–720

ENZYMATIC TRANSITION STATES AND TRANSITION STATE ANALOG DESIGN

Vern L. Schramm

Department of Biochemistry, Albert Einstein College of Medicine of Yeshiva
University, Bronx, New York 10461; email: vern@aecom.yu.edu

KEY WORDS: catalysis, isotope effects, inhibitor, enzymes, inhibitor design

ABSTRACT

All chemical transformations pass through an unstable structure called the transition state, which is poised between the chemical structures of the substrates and products. The transition states for chemical reactions are proposed to have lifetimes near 10^{-13} sec, the time for a single bond vibration. No physical or spectroscopic method is available to directly observe the structure of the transition state for enzymatic reactions. Yet transition state structure is central to understanding catalysis, because enzymes function by lowering activation energy. An accepted view of enzymatic catalysis is tight binding to the unstable transition state structure. Transition state mimics bind tightly to enzymes by capturing a fraction of the binding energy for the transition state species. The identification of numerous transition state inhibitors supports the transition state stabilization hypothesis for enzymatic catalysis. Advances in methods for measuring and interpreting kinetic isotope effects and advances in computational chemistry have provided an experimental route to understand transition state structure. Systematic analysis of intrinsic kinetic isotope effects provides geometric and electronic structure for enzyme-bound transition states. This information has been used to compare transition states for chemical and enzymatic reactions; determine whether enzymatic activators alter transition state structure; design transition state inhibitors; and provide the basis for predicting the affinity of enzymatic inhibitors. Enzymatic transition states provide an understanding of catalysis and permit the design of transition state inhibitors. This article reviews transition state theory for enzymatic reactions. Selected examples of enzymatic transition states are compared to the respective transition state inhibitors.

693

CONTENTS

INTRODUCTION

Enzymes catalyze chemical reactions at rates that are astounding relative to uncatalyzed chemistry at the same conditions. Typical enzymatic rate enhancements are 10^{10} to 10^{15}, accomplishing in 1 sec that which would require 300 to 30,000,000 years in the absence of enzymes (1). Each catalytic event requires a minimum of three or often more steps, all of which occur within the few milliseconds that characterize typical enzymatic reactions. According to transition state theory, the smallest fraction of the catalytic cycle is spent in the most important step, that of the transition state. However, this step cannot occur without participation of the enzymatic forces that occur as the Michaelis complex is transformed into the transition state by precise alignment of catalytic groups by enzymatic and substrate conformational changes (Figure 1).

The original proposals of absolute reaction rate theory for chemical reactions defined the transition state as a distinct species in the reaction coordinate that determined the absolute reaction rate (2). Soon thereafter, Linus Pauling proposed that the powerful catalytic action of enzymes could be explained by specific tight binding to the transition state species (3). Because reaction rate is proportional to the fraction of the reactant in the transition state complex, the enzyme was proposed to increase the concentration of the reactive species. This proposal was formalized by Wolfenden and coworkers, who hypothesized that the rate increase imposed by enzymes is proportional to the affinity of

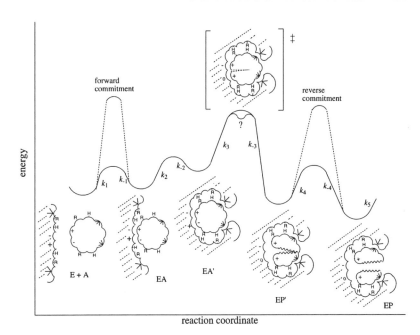

reaction coordinate

Figure 1   Reaction coordinate diagram for conversion of substrate (A) to enzyme-bound products (EP). Symbols are R = H-bond acceptor, H = H-bond donor, + and − are ionic charges, and > represents hydrophobic sites. The *solid line* is an example of fully rate-limiting transition state formation, providing intrinsic kinetic isotope effects. The energetic barriers (*dashed lines*) labeled "forward commitment" and "reverse commitment" make substrate binding and/or product release rate limiting and suppress kinetic isotope effects. The transition state has unique properties of charge and optimal H-bond alignment not found in any of the reactant species. In this example, charge repulsion serves to clear the catalytic site and restore the enzyme to the open form (E) after k_5.

the enzyme for the transition state structure relative to the Michaelis complex (4; Figure 2). Because enzymes typically increase the noncatalyzed reaction rate by factors of 10^{10}–10^{15}, and Michaelis complexes often have dissociation constants in the range of 10^{-3}–10^{-6} M, it is proposed that transition state complexes are bound with dissociation constants in the range of 10^{-14}–10^{-23} M. Analogs that resemble the transition state structures should therefore provide the most powerful noncovalent inhibitors known, even if only a small fraction of the transition state energy is captured.

Reviews from 1976, 1988, and 1995 listed enzymes known to interact with transition state inhibitors, defined as tight binding inhibitors that resemble the hypothetical transition states or intermediates for various enzymes (5–7). Between 1988 and 1995, the list grew from 33 to 132 enzymes. In 1976 and 1988, most of the inhibitors were natural products. The 1995 list of enzymes and transition state inhibitors is dominated by intentionally synthesized inhibitors

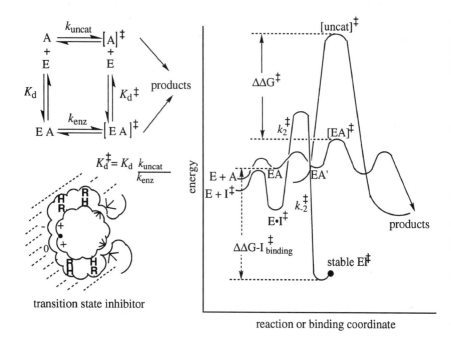

Figure 2   The thermodynamic box that predicts transition state binding affinity (*upper left*) compares the rates of uncatalyzed (k_{uncat}) and enzymatic (k_{enz}) reactions and assumes that the transmission coefficients from $[A]^{\ddagger}$ and $[EA]^{\ddagger}$ are equal. The dissociation constants for EA and $[EA]^{\ddagger}$ are given by K_d and $K_d^{\ddagger}$. Transition state inhibitors (*lower left*) invoke the unique ionization and conformational structure found exclusively in the transition state (see Figure 1). Catalysis is prevented by a stable bond at the reaction center (represented by the *solid dot*). The energetics of catalysis and binding an energetically perfect transition state analog are compared in the reaction or binding coordinate diagram. Energy of enzymatic transition state stabilization ($\Delta\Delta G^{\ddagger}$) is converted into binding energy for the transition state inhibitor ($\Delta\Delta G\text{-}I^{\ddagger}$) to form the stable $EI^{\ddagger}$ complex, which cannot escape by the product release pathway. The slow-onset inhibition common with transition state inhibitors occurs after formation of a readily reversible $E{\cdot}I^{\ddagger}$ complex. The rate of onset for tight-binding inhibitors is $k_2^{\ddagger}$ and the rate of escape is $k_{-2}^{\ddagger}$. Note the unfavorable energetic barrier for escape from the stable $EI^{\ddagger}$ complex.

in response to inhibitor development for the AIDS protease, β-lactamases, metalloproteinases, cyclooxygenases, and a growing list of enzymes that are targets for pharmaceutical intervention. Because many inhibitor developments are proprietary, the list for 1995 is likely to be an incomplete representation of the known list of transition state inhibitors.

In 1947, an experimental approach to chemical transition states was discovered by Bigeleisen and coworkers, who, along with others, established the relationships among isotopic substitution, altered reaction rates, and altered bonding between reactants and transition states (8–10). A few applications of individual kinetic isotope effects were made to enzymatic reaction mechanisms before 1970. Qualitative results from these studies indicated whether the bond changes to the isotopically labeled atom occurred in the rate-limiting step (e.g. Figure 1). The Steenbock Symposium of 1976, "Isotope Effects on Enzyme-Catalyzed Reactions," provided the impetus for additional studies on enzymes (11). The 1978 book *Transition States of Biochemical Processes* included the provocative and fundamentalist position of RL Schowen that "the entire and sole source of catalytic power is the stabilization of the transition state; that reactant-state interactions are by nature inhibitory and only waste catalytic power" (12, p. 78). These works bridged the gap between chemical and biological transition states, recognizing that even complex biological transformations involve formation of one or more defined transition states and are therefore susceptible to transition state analysis based on the measurement of kinetic isotope effects. Development of steady state and pre–steady state kinetic methods to reveal intrinsic isotope effects permitted interpretation of results in terms of transition state structure (13–17).

Although the focus of this review is enzymatic transition states and related inhibitors, transition state similarity is not necessary for tight-binding inhibition of enzymes. Any combination of multiple favorable hydrogen, ionic, or hydrophobic bonds between enzyme and substrate can provide the summation of binding interactions leading to tight-binding inhibition and is the basis for inhibitory screening from chemical and combinatorial libraries. Examples of such inhibitors are found in nature as antibiotics. Streptomycin, erythromycin, and rifampicin are examples of complex natural products that have no known similarity to the transition states involved in protein and RNA synthesis. Inhibitors designed to match contacts in the catalytic site or those that are stable analogs of the substrate can bind tightly but do not qualify as transition state inhibitors. One example is the tight binding of methotrexate to dihydrofolate reductase, in which the analog is bound upside down with respect to substrate in the catalytic site (18). Another example is 9-deaza-9-phenyl-guanine, a powerful inhibitor of purine nucleoside phosphorylase. It was designed from the X-ray crystal structure of substrate and product complexes and does not resemble the transition state (19, 20) (Figure 3).

TRANSITION STATE THEORY FOR ENZYME-CATALYZED REACTIONS

Nature of the Transition State

As substrate progresses from the Michaelis complex to product, chemistry occurs by enzyme-induced changes in electron distribution in the substrate. Enzymes alter the electronic structure by protonation, proton abstraction, electron

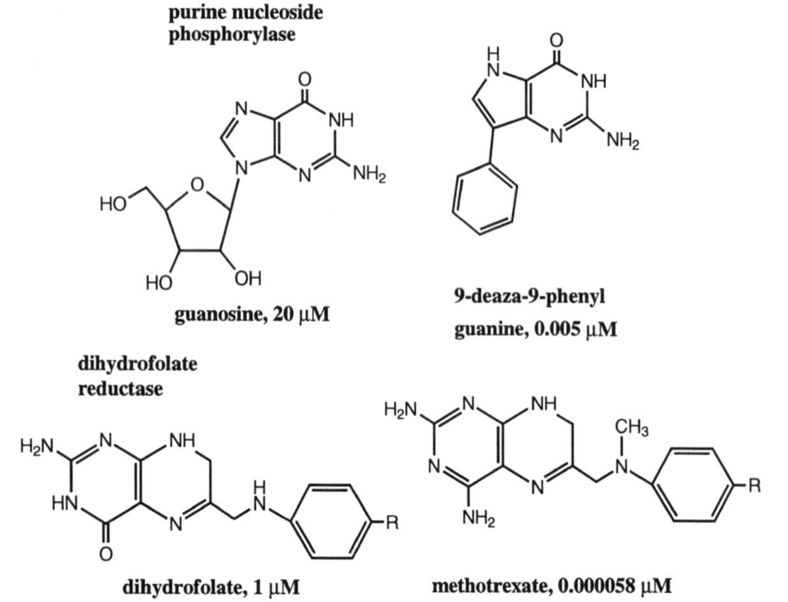

Figure 3 Tight-binding inhibitors of purine nucleoside phosphorylase and dihydrofolate reductase that are not transition state analogs despite high-affinity binding. The K_m values for substrates and equilibrium K_i values for the inhibitors are shown.

transfer, geometric distortion, hydrophobic partitioning, and interaction with Lewis acids and bases. These are accomplished by sequential protein and substrate conformational changes (Figure 1). When a constellation of individually weak forces are brought to bear on the substrate, the summation of the individual energies results in large forces capable of relocating bonding electrons to cause bond-breaking and bond-making. Substrates of modest molecular size (for example, glucose), bound in the active sites of proteins, typically interact with two H-bonds at each donor/acceptor site. These interactions alone, in typical H-bond energies of 3 kcal/mol/H-bond, can provide over 30 kcal/mol of energy toward redistribution of electrons and therefore toward ionizations and covalent bond changes. The restricted geometry and hydrophobic environment allow short H-bonds to form that would be unfavorable in solution (21, 22).

The lifetime for chemical transition states is short, approximately 10^{-13} sec, the time for conversion of a bond vibrational mode to a translational mode (23, 24). This theory requires reexamination for enzymes, because the protein domain motion resulting in transition state formation may stabilize the altered bond lengths of the bound transition state for a lifetime sufficient for 10^1–10^6

vibrations. This hypothesis is indicated by the dimpled transition state feature labeled with a question mark in Figure 1; it remains unexplored except by computational theory. Enzymes that form transient covalent intermediates require two distinct transition states. These transition states are surrounded by lower energy complexes and should be distinguished from the altered vibrational state implied by the question mark in Figure 1. The classic interpretation of the transition state supposes a short lifetime during which an infinitesimal force toward product or substrate leads to EP' and EA' respectively.

Induced Protein Conformational Change

The energetic problem that must be solved by enzymes is to bind tightly only to the unstable transition state structure while avoiding tight binding to the substrate and products. Enzymes often bind to the substrate at diffusion-controlled rates, and subsequent conformational or electronic changes are mandatory for catalysis. Placing the enzyme-bound substrate in the solvent-restricted environment of the closed catalytic site permits the subsequent events to occur in the altered solvent of the catalytic site.

Presenting the enzyme with a transition state mimic results in a mismatch of the substrate-recognizing features. Many transition state inhibitors are slow-onset, typified by a rapid weak binding followed by a slow tight-binding interaction. The energetics of this interaction (Figure 2) show rapid formation of the encounter complex $E \cdot I^{\ddagger}$ (similar to the EA complex of Figure 1) followed by a difficult (high-energy, slow) entry to the stable $EI^{\ddagger}$ complex. The bound inhibitor does not induce the conformational change with the efficiency of substrate and requires time to permit the transition state conformational change to occur on the protein. This time corresponds to the slow onset of tight-binding inhibition commonly observed with transition state mimics (6). These data argue that the substrate actively induces the catalytic conformational change, because the rate of catalysis is substantially greater than the rate at which most tight-binding inhibitors induce enzymes into the transition state configuration. This view of transition state inhibitor interaction predicts that near-perfect inhibitors would still exhibit slow-onset inhibition because the enzyme is designed to recognize the ground state of the substrate. However, some tightly bound inhibitors are reported to achieve inhibition on the time scale of catalysis. For example, the inhibition of adenosine deaminases by purine riboside and of cytidine deaminase by pyrimidin-2-one riboside is fast (25–26). These inhibitors are estimated to bind with dissociation constants of 10^{-13} and 10^{-12} M respectively, approaching the hypothetical 10^{-16} M dissociation constant for the actual transition states. Rapid onset occurs because these are half-reaction substrates for the enzymes. The deaminases contain a tightly bound zinc that acts as the catalytic site base to ionize a water molecule to the hydroxide and position it near

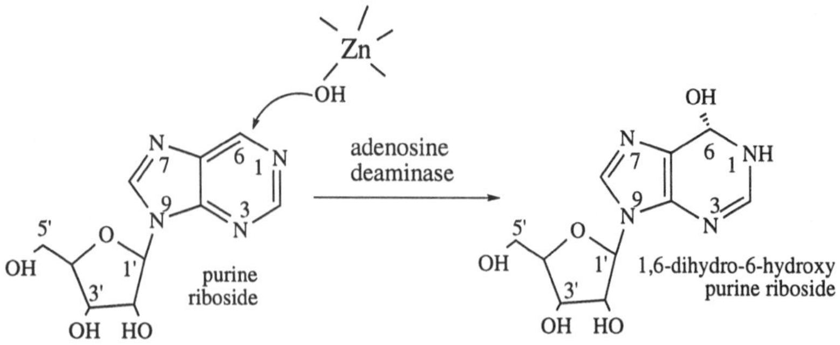

Figure 4 Hydration of purine riboside by a catalytic site hydroxyl generated at the tightly bound Zn^{2+}. The hydrated purine is bound tightly. An analogous reaction occurs with cytidine deaminase.

the reactive carbon of the aromatic rings (27; Figure 4). Enzymatic protonation of the adjacent aromatic ring nitrogen assists in the required rehybridization of the rings. These inhibitors take advantage of the substrate-induced transition state configuration to cause rapid hydration and tight binding of the hydrated species. The sp^3 hybridization at the reactive carbon is a transition state feature, and without the amino leaving group, a stable complex is formed.

Tight Binding of the Transition State Structure

Binding energies of enzymatic transition states are generated by the realignment of substrate (Michaelis) contacts as the enzyme and substrate mutually change their structures toward the transition state (Figure 1). The strong dependence of hydrogen and ionic bond energy on bond distance, angle, solvent environment, and relative pK_a values can be invoked to explain the increases in binding forces of the transition state complex relative to the Michaelis complex (21, 28). Structural rearrangements tighten the protein around the catalytic site to exclude solvent and to make stronger electrostatic contacts. These are shown as well-aligned H-bonds at the transition state and as ionic attraction and repulsion as catalytic forces (Figure 1). Enzymatic reactions usually demonstrate distinct pK_a values for substrate binding and k_{cat}, testifying to the ionic changes between the enzyme substrate and enzyme transition state complexes (29, 30). A common mechanism for enzymatic catalysis is to generate differential charge between substrate and transition state, permitting electrostatic interactions to specifically stabilize the transition state (28). The imperfect match between the enzyme and the transition state inhibitor is inevitable because it is impossible to recreate perfectly the nonequilibrium bond lengths of the transition state with stable compounds. The $\Delta\Delta G\text{-}I^{\ddagger}_{binding}$ energy of Figure 2 is shown for a perfect transition state inhibitor, with full $K^{\ddagger}_d$ of the transition state. For the necessarily imperfect transition state inhibitors, the $k_2^{\ddagger}$ barrier is the time to fit the imperfect inhibitor into the lowest-energy structure of the analog complex.

Relaxation of the Transition State Complex

Catalysis requires rapid relaxation of the transition state energy to allow the products to dissociate. The relative enzymatic affinity for the transition state and products decreases by about 12 orders of magnitude within milliseconds. For catalytically efficient enzymes, the Michaelis complex is a diffusional event with a rate constant near 10^9 $M^{-1}sec^{-1}$. Transition state formation, product formation, and release are as rapid as this, so the diffusional event defines the maximal catalytic rate (31). The change in electron distribution as bonds are broken creates a repulsive interaction that opens the catalytic site and expels products (see Figure 1). This step is the opposite of what occurs when the Michaelis complex is converted into the transition state complex.

Computer modeling of well-understood enzymatic reactions has allowed examination of the factors important in catalysis. The resulting conclusion is that electrostatic stabilization of the transition state is the most significant factor (28). Understanding the electrostatic nature of the transition state therefore provides a template for the synthesis of transition state inhibitors. The experimental approach that provides the most direct information for enzymatic transition state structure is analysis of kinetic isotope effects (32–34).

EXPERIMENTAL APPROACHES TO ENZYMATIC TRANSITION STATE STRUCTURE

Chemical Precedent

Chemical reactivity series with leaving groups of different pK_a values or electron withdrawing capability can indicate the position of the transition state in the reaction coordinate. Brønstead and Hammett plots have had wide application in chemical mechanisms, but less in enzymology because the specificity of enzymes often does not permit the use of a variety of substrates (35, 36). Behavior of model chemistry in solution for the reaction of interest provides a transition state benchmark for comparison with the transition state structure imposed by the enzyme.

Transition State Inhibitors

Inhibitors that bind tightly, are slow-onset, and resemble the expected transition states have been used to predict state features (4–6). For example, the inhibition of deaminases and proteases by analogs with sp^3 reaction centers was useful in establishing the nature of these transition states and intermediates (37, 38). Inhibitor specificity is subject to the vagaries of enzymatic binding

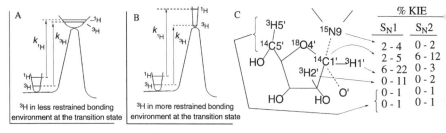

Figure 5 Simplified bond vibrational basis for the relationship between a kinetic isotope effect and transition state structure. The parabolae represent the bonding environment (restoring force) for the vibrational mode of a bonded atom (^{1}H) and its isotope (^{3}H) in the reactant and transition states. The *arrows* are the activation energies for ^{1}H- and ^{3}H-labeled reactants. In *panel A*, ^{1}H-substrate reacts more rapidly (has lower activation barrier). In *panel B*, ^{3}H-substrate reacts more rapidly (has lower activation barrier). In practice, all significant modes are considered. In *panel C* are the anticipated values of kinetic isotope effects for C1'-N9 N-ribosidic bond scission with S_N1 or S_N2 character. The ranges arise from variability in ring geometry and the degree of the associated O' nucleophile at transition states with predominant S_N1 or S_N2 character. The percent kinetic isotope effect (%KIE) = [(unlabeled rate/labeled rate) − 1.00] × 100%.

sites. Transition state properties cannot always be predicted, as demonstrated by methotrexate in the catalytic site of dihydrofolate reductase (18; Figure 3). Direct information on transition state structure is available from kinetic isotope effect studies.

Kinetic Isotope Effects

The only method available for the direct determination of enzymatic transition state structure is the measurement of kinetic isotope effects. Excellent accounts of the theory and its development can be found in the literature (39–43) and are beyond the scope of this review. Kinetic isotope effects compare the enzymatic reaction rates of isotopically labeled and unlabeled substrates (Figure 5). Isotopically labeled molecules have molecular energy different from that of unlabeled molecules and thus require a different amount of energy to reach the transition state, where the molecular energies may also be perturbed by the isotope. If the bonding environment for the labeled atom is less restricted in the transition state than in the reactant, the isotope effect will be normal (the heavy isotope substrate reacting more slowly than the unlabeled substrate). Likewise, if the bonding environment at the transition state is more restricted than for the reactant, the substrate with the higher mass isotope will react more rapidly (an inverse kinetic isotope effect). Isotope effects also provide quantitative information because the magnitude of the isotope effect indicates the extent of bond change. For some isotope effects, bond geometry can be determined because of the dihedral angular dependence for conjugative isotope effects β to the bond being broken (44, 45). By measuring kinetic isotope effects at every position in a substrate molecule that might be expected to be perturbed at the transition state, a unique description of the transition state can be deduced if intrinsic isotope effects are being measured (13, 46). The steps involved in transition state analysis by kinetic isotope effects are to

1. synthesize substrates with appropriate isotopic labels

2. measure kinetic isotope effects to an accuracy of better than 0.5%

3. compute a truncated transition state with bond lengths and angles matching the isotope effects (limited to 25 atoms), using normal-mode bond-energy bond-order vibrational analysis (47, 48)

4. restore the complete molecular structure for the transition state by fixing the bonds established from kinetic isotope effects and optimizing the remaining structure using semiempirical methods (49)

5. determine the electron wave function for the molecule to determine electron distribution at the van der Waals surface (50).

EXAMPLES OF ENZYMATIC TRANSITION STATES

AMP Nucleosidase

AMP nucleosidase catalyzes the hydrolysis of AMP to adenine and ribose 5-phosphate under the allosteric control of MgATP and inorganic phosphate, which serve as allosteric activator and inhibitor, respectively (51, 52):

AMP + H_2O → adenine + ribose 5-phosphate.

The enzyme is found only in prokaryotes and is thought to play a role in the regulation of energy metabolism. However, genetic ablation of the enzyme from *Escherichia coli* had only minor effects on growth rates (53). The enzyme from *Azotobacter vinelandii* requires MgATP as a k_{cat} activator. In the absence of activator, the enzyme binds substrate, but k_{cat} is 10^{-3} of that with the allosteric activator. This provides the opportunity to determine whether allosteric activation is capable of changing the chemical nature of the transition state or only serves to change energies of activation (altered rate constants) for steps through the reaction cycle. Systematic determination of kinetic isotope effects is made possible by the combined chemical and enzymatic synthesis of AMP (Figure 6) with the isotopic labels indicated in Figure 7 (54–56).

Kinetic isotope effects for the enzyme were intrinsic based on the catalytic mechanism and by comparison with the isotope effects for acid-catalyzed

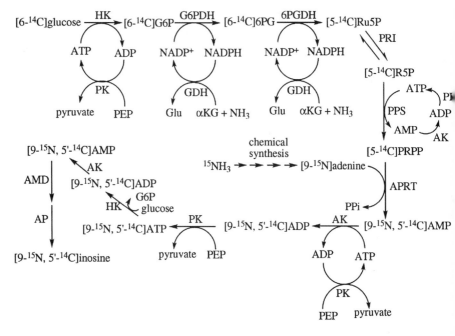

Figure 6 Combined enzymatic and chemical synthesis of specifically labeled nucleosides and nucleotides (synthesis of [9-^{15}N, 5′-^{14}C]ATP, AMP, and inosine). All steps from glucose to ATP are coupled in a single reaction mixture. Separate reaction mixtures are used to convert ATP to AMP and to inosine. The enzymes are HK = hexokinase, PK = pyruvate kinase, G6PDH = glucose 6-phosphate dehydrogenase, 6PGDH = 6-phosphogluconate dehydrogenase, GDH = glutamate dehydrogenase, PRI = phosphoriboisomerase, PPS = 5-phosphoribosyl-1-pyrophosphate synthetase, AK = adenylate kinase, APRT = adenine phosphoribosyltransferase, AMD = AMP deaminase, AP = alkaline phosphatase. Using specific labels in glucose, ribose and adenine can yield the desired labels at any position of the nucleosides and nucleotides.

solvolysis of AMP (56). The transition state for the acid-catalyzed reaction was compared to those determined for the enzyme in the presence and absence of allosteric activator (Figure 7). Because kinetic isotope effects report directly on the bond environment at the transition state, comparison of the experimental kinetic isotope effects established that the transition states differ for the enzyme-catalyzed and acid-catalyzed hydrolysis. The allosteric activator, which causes a 10^3-fold increase in catalytic rate, also changes the nature of the transition state. Quantitiation of the isotope effects in terms of bond orders for reactants and transition states indicated that the enzymatic catalyst permitted the reaction to reach the transition state earlier, when there was more bond order remaining

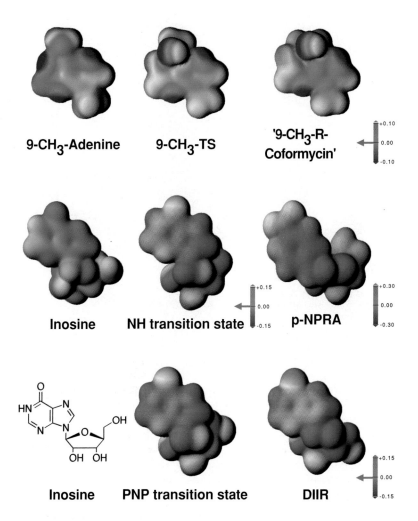

Figure 10 Molecular electrostatic potential at the van der Waals surfaces for substrates, transition state, and transition state inhibitors for AMP deaminase (*top*), nucleoside hydrolase (NH) (*center*), and purine nucleoside phosphorylase (PNP) (*bottom*). The chemical structure of inosine is shown. Other structures are shown in Figures 9, 11, and 12. Details of the calculations are in References 82, 89, and 92. In every case, the electrostatic potential similarity between the transition state and the transition state inhibitor is greater than between the substrate and the transition state. The attacking water and phosphate nucleophiles are omitted from the transition state structures for NH and PNP. The color codes for the molecular electrostatic potentials are shown to the *right*.

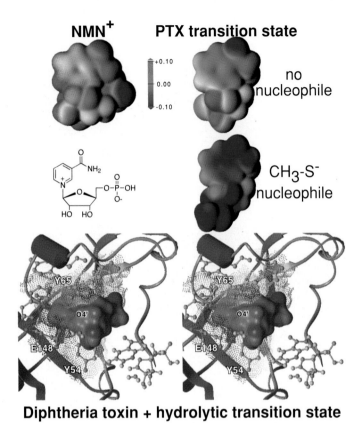

Diphtheria toxin + hydrolytic transition state

Figure 17 Electrostatic potential surfaces for NMN^+, the transition state for NAD^+ hydrolysis by pertussis toxin (PTX) showing only the NMN^+ portion, and the transition state for ADP-ribosylation of the C-terminal 20–amino acid peptide from $G_{i\alpha 3}$ by pertussis toxin showing only the atoms of NMN^+ and a CH_3S^- to replace the peptide. A docked model of the transition state for diphtheria toxin in the crystal structure of diphtheria toxin is shown in stereo. The chemical and electrostatic potential surface for NMN^+ is shown in the *upper left*. At the transition state, the nicotinamide positive charge migrates into the ribosyl, forming the ribooxocarbenium. In the no-nucleophile structure, the charge (*red*) is clearly visible. Participation of the thiolate anion at the transition state causes the charge gradient to be lost across the susceptible bond, so that both the nicotinamide and ribosyl are partial negative (*blue*), causing C1-N1' bond scission with lagging attack of the thiolate anion (CH_3S^-). Docking the no-nucleophile transition state into the crystal structure for diphtheria toxin shows electrostatic complementarity, especially at E148, the essential catalytic site carboxylate that interacts with the positive charge (*red*) of the oxocarbenium ion. The stick model to the *lower right* of the stereo figure shows NMN^+ and the attacking water nucleophile in the same relative orientation of the transition state embedded into the protein.

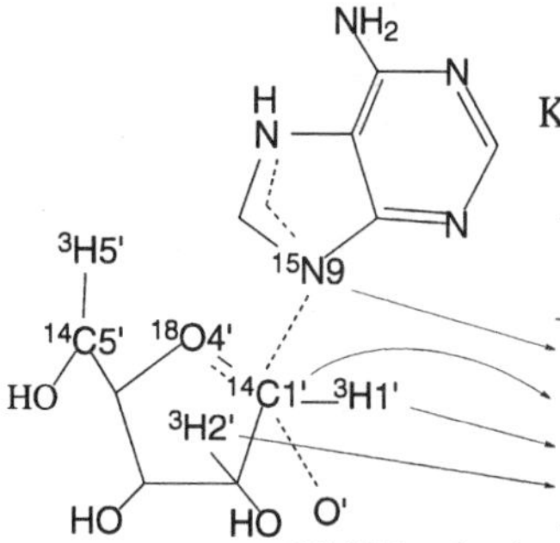

Figure 7 shows a chemical structure diagram on the left and a data table. The structure is AMP with labeled atoms: NH_2, $^{15}N9$, $^{3}H5'$, $^{14}C5'$, $^{18}O4'$, $^{14}C1'$, $^{3}H1'$, $^{3}H2'$, HO, HO, HO, O'.

Kinetic Isotope Effects for AMP Nucleosidases

		% KIE			
		native		mutant	
	acid	-MgATP	+MgATP	-MgATP	+MgATP
	3.0	3.0	2.5	3.4	2.1
	4.4	3.5	3.2	4.1	4.1
	21.6	6.9	4.7	8.6	9.4
	7.7	6.1	4.3	8.9	8.6
C1'-N9 bond order	0.02	0.16	0.21	0.09	0.18
C1'-O' bond order	0.02	0.03	0.03	0.03	0.01

Figure 7 Kinetic isotope effects (KIEs) and bond orders at the transition states for the C1'-N9 bond hydrolysis of AMP by acid and AMP nucleosidases. All isotope effects were measured with a mixture of two AMP molecules, one labeled near and one distant from the C'1-N9 bond. Both native and mutant enzymes were activated by a factor of $\sim 10^3$ by the addition of MgATP. Kinetic isotope effects were measured with average standard deviations of 0.4%. Bond orders for the C1'-N9 and C1'-O' bonds were determined from bond vibrational analysis (48, 56). When bonds surrounding C1' are weakest at the transition state, the $^3H1'$ kinetic isotope effect is greatest. Bond order (n) is based on the Pauling rule: $r_n = r_1 - 0.3 \ln n$, where r_1 is the single bond length (58).

to the leaving group (Figure 7). The protein structural change induced by the allosteric activator caused the transition state to occur even earlier in the reaction coordinate. Changes in the transition state structure also resulted from the forced evolution of a β-galactosidase in *E. coli*, causing the enzyme to become catalytically more efficient (57).

Substrate and inhibitor specificity studies revealed that formycin A 5'-phosphate binds $>10^3$ times tighter to AMP nucleosidases than does substrate (59, 60). The X-ray crystal structure of this inhibitor helped to establish the properties of the transition state (61). The inhibitor has a *syn*-ribosyl torsion angle, preferred in all nucleotide inhibitors of the enzyme (59). The molecular electrostatic potential at the van der Waals surface revealed similar electronic structures of inhibitor and transition state because of the common protonation at N7, which is not present in the substrate (62).

Random mutagenesis of the organism expressing AMP nucleosidase provided an enzyme with a k_{cat} 2% that of the native enzyme (Figure 7) (63). The purpose of the mutagenesis was to determine if catalytic site mutation changes transition state structure. The results established substantially different kinetic isotope effects for the mutant enzyme, demonstrating differences in bond structure at the transition state. The mutant enzyme was less effective in stabilizing

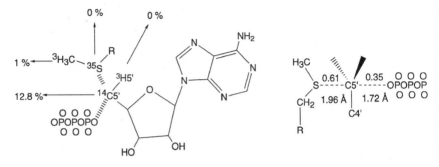

Figure 8   Kinetic isotope effects for the SAM-synthetase reaction (*left*), where the ^{35}S-molecule represents the attacking methionine sulfur nucleophile. The isotope effects were measured in individually labeled substrates but are illustrated here as a single molecule. Only symmetrical S_N2 displacements give primary ^{14}C isotope effects of 12.8% (see Figure 5). The transition state structure consistent with all isotope effects is shown on the *right*. The bond orders were 0.61 and 0.35, which correspond to bond lengths of 1.96 and 1.72 Å for S-C5' and C5'-O bonds, respectively. The mass difference between sulfur and oxygen accounts for the differences, which are equivalent in bond vibrational energy. C5' is the center of reaction coordinate motion, giving the large ^{14}C isotope effect.

the ribooxocarbenium ion of the transition state and in protonating the adenine leaving group (64).

S-Adenosylmethionine Synthetase

The chemical reaction of S-adenosylmethionine (SAM) synthetase involves displacement of the tripolyphosphate of ATP by reaction of the sulfur of methionine with C5' of ATP (Figure 8). The reaction results in the further hydrolysis of the phosphate chain to phosphate and pyrophosphate, to release the products SAM, phosphate, and pyrophosphate from the catalytic site (65–66). This is one of the few reactions of ATP in which all phosphates are displaced in a single reaction. The reaction is irreversible under physiological conditions as a consequence of the tripolyphosphorolysis. Potassium ion is required for efficient catalysis, activating the enzyme approximately 100-fold. Kinetic isotope effect studies of the enzyme were designed to establish the nature of the transition state under optimal catalytic conditions and to establish the effect of K$^+$ ion activation. Monovalent cations are common in kinase activation and are thought to provide a positive charge to neutralize the transition state complex at highly charged reaction centers typified by phosphoryl transfers (67). Activation can be achieved either by changing a rate constant or by changing the transition state structure.

Measurement of kinetic isotope effects with limiting K$^+$ gave intrinsic isotope effects (68). The ^{14}C5' kinetic isotope effect was 12.8%, near the theoretical limit for a primary ^{14}C isotope effect (33). This value is realized only when carbon dominates reaction coordinate motion in a symmetric nucleophilic displacement reaction. The result provided a ready solution for the reaction mechanism. In nucleophilic displacements, neighboring α- and β-secondary ^{3}H isotope effects are expected to be insignificant, and this was confirmed for both [5'-^{3}H]ATP and methyl-[^{3}H$_3$]methionine. These kinetic isotope effects provide proof that the chemical mechanism of SAM synthetase at limiting K$^+$ is a symmetric S$_N$2 reaction. Addition of K$^+$ to activate the rate by 100-fold decreased the [5'-^{14}C]ATP isotope effect, consistent with either a change in transition state structure or increased forward commitment (Figure 1).

Substrate trapping experiments, pioneered by I Rose (69), permitted the direct measurement of forward commitment for SAM synthetase (the probability that a bound substrate molecule will be transformed to product relative to release from the catalytic site). Only substrate molecules in equilibrium with free substrate contribute to the observed isotope effect (13). The results established that K$^+$ improved catalytic efficiency by increasing the forward commitment. The intrinsic isotope effect (corrected for commitment) was unchanged, thus the structure of the transition state is unchanged by K$^+$. This example of enzymatic activation demonstrates lowering the transition state energetic barrier without a measurable change in the bond environment of the transition state (Figure 1).

AMP Deaminase and Adenosine Deaminase

The deaminations of AMP and adenosine are aromatic nucleophilic substitutions at carbon, a reaction type well known in chemistry and characterized by rate-limiting formation of an unstable tetrahedral intermediate that decomposes rapidly to yield products (70). AMP deaminase is found only in eukaryotes, and it is proposed to be involved in the regulation of the adenine nucleotide pool through the purine nucleotide cycle (71). Activation of AMP deaminase in times of energy deficiency leads to IMP formation and a decrease in the total adenylate pool size (72). Humans deficient in specific muscle isozymes of AMP deaminase suffer moderate difficulty in muscular work, but deficiencies in the erythrocyte/heart isoform have no known phenotype (73). It has been proposed that inhibition of the heart isozyme during recovery from heart attacks might speed recovery by preserving the adenylates and preventing oxygen radical damage associated with the conversion of hypoxanthine to uric acid (74). Adenosine deaminase deficiency leads to B- and T-cell immunodeficiency, presumably as a result of the accumulation of dATP in the progenitor cells preventing clonal expansion in response to an immune challenge (75).

The natural products coformycin, deoxycoformycin, and their 5'-phosphates are tight-binding transition state inhibitors for adenosine and AMP deaminases, binding approximately 10^7-fold more tightly than substrates (76–78). The

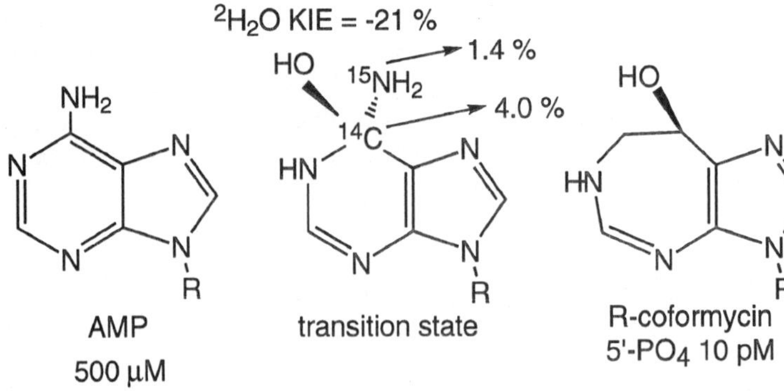

Figure 9 Substrate, transition state, and transition state inhibitor for yeast AMP deaminase. The kinetic isotope effects (KIEs) are superimposed on a diagram of the transition state. The R-substituent is ribose 5'-phosphate. The isotope effects were measured using the natural abundance of ^{15}N, specifically labeled ^{14}C-AMP, and with solvent deuterium for the ^2H_2O KIE.

transition state structures of these enzymes are of interest for comparing transition state structure deduced from kinetic isotope effects to the structures of transition state inhibitors. Specific questions about the nature of the transition states include the degree of hydroxyl attack (sp^3 hybridization) at C6 in the transition states; whether protonation of the leaving amino group has occurred; whether amino group departure has begun; and whether the enzymatic transition state structures can explain the tight binding of the coformycins.

Kinetic isotope effects from yeast AMP deaminase were similar to those measured for adenosine deaminase, indicating that both have similar transition states (Figure 9; 79, 80). The conserved protein motif for the catalytic site zinc also supports common catalytic site chemistry (81). The enzyme-stabilized transition state is reached prior to formation of the tetrahedral intermediate, at a bond order of 0.8 to the incoming water nucleophile, while retaining full bond order to the leaving amino group. The increased bond order to the attacking water oxygen requires that the purine ring lose its conjugation and that N1 become protonated to reach the transition state. Before the N6 amino group is competent for departure, it must be protonated, a process that occurs in a rapid step after transition state formation. The transition state is early, preceding formation of the intermediate. Formation of the intermediate, protonation, and loss of the exocyclic amino group are all fast relative to the initial addition of water to form the transition state.

Molecular electrostatic potential surface analysis of the transition state and substrate reveals substantial changes and compelling similarity between the transition state and coformycin (Figure 10, color plate) (82). The striking correspondence indicates that the molecular electrostatic potential surface of this transition state is closely related to that for the transition state inhibitor.

Nucleoside Hydrolase

Protozoan parasites are purine auxotrophs, using salvage enzymes to recover purines from their environment (83). The nucleoside hydrolases cleave the N-ribosidic bonds of nucleosides to liberate the purine and pyrimidine bases and to generate ribose (84–88):

nucleosides + $H_2O \rightarrow$ base + ribose.

Liberated bases are salvaged by phosphoribosyltransferases. Ribosyl nucleoside substrates and substrate analogs bind poorly to the nonspecific nucleoside hydrolase from *Crithidia fasciculata* with dissociation constants from 0.2 to 10 mM (84). The nucleoside hydrolases are not found in mammalian genomes; thus the enzymes have been targeted for antiprotozoan agents (89, 90). A goal of this study was to characterize the transition state for the nonspecific nucleoside hydrolase from *C. fasciculata* and use the information to design transition state inhibitors. At the transition state, the N9-C1' ribosidic bond cleavage is well advanced, with 0.22 bond order remaining (Figure 11). The attacking water

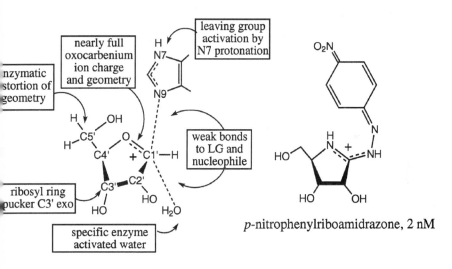

p-nitrophenylriboamidrazone, 2 nM

Figure 11   Features of the transition state for the nonspecific nucleoside hydrolase from *Crithidia fasciculata*. The substrates were labeled inosines (Figure 12) synthesized as in Figure 6, and the isotope effects were interpreted according to normal-mode and ab initio methods (89, 90). A tight-binding transition state inhibitor is shown on the *right*. The resonance structure of the inhibitor bound to the enzyme was determined by resonance Raman spectroscopy.

nucleophile lags well behind N-ribosidic bond breaking with approximately 0.03 bond order from the attacking oxygen to C1'. The loss of most of the N-ribosidic bond before water attacks leaves the ribose as an oxocarbenium cation at the transition state with C1' rehybridized nearly completely to sp^2. This flattens the ribose ring around C1', causing the atoms of the ribosyl ring to be coplanar with the exception of C3'. The kinetic isotope effect from ^{3}H2' is dependent on the dihedral angle to the C1'-N9 bond. Its value required that C3' lies below the plane of the ring, in the C3'-*exo* ribosyl configuration, to provide near eclipse in the H2'-C2'-C1'-N9 dihedral angle. Solvent D$_2$O and ^{15}N9 isotope effects established that the hypoxanthine leaving group is protonated, most likely at N7, prior to reaching the transition state. This creates a neutral, planar, and hydrophobic leaving group. A suprising remote 5'-^{3}H isotope effect of 5.1% was observed in the enzyme but not in acid-catalyzed hydrolysis of inosine, establishing enzyme-induced distortion (89, 90). The interpretation that the 5'-oxygen is rotated above the oxocarbenium ion is shown in Figure 11 and was recently confirmed by X-ray crystallogaphic studies with a bound transition state inhibitor (90a).

Based on the transition state structure, iminoribitol and riboamidrazone inhibitors with K_m/K_i to 200,000 have been synthesized and characterized (90–94). The best inhibitor (*p*-nitrophenylriboamidrazone; Figure 11) incorporates most of the electrostatic traits of the transition state but is structurally distinct from both the substrate and transition states. The similarity to the transition state is apparent in the comparison of the molecular electrostatic potentials of substrate, transition state, and transition state inhibitor in Figure 10 (color plate) (95). Thus, the molecular electrostatic potential surface matching of inhibitor and transition states leads to powerful binding energy.

Purine Nucleoside Phosphorylase

Scission of the N-ribosidic bonds of the purine nucleosides and deoxynucleosides in mammals is accomplished almost exclusively by the phosphorolysis reaction of purine nucleoside phosphorylase (PNP). Inosine, guanosine, and 2'-deoxyguanosine are the major substrates. The genetic deficiency of human PNP causes a T-cell deficiency as the major physiological defect (96). Apoptosis in T cells that are not immunologically stimulated to divide results in recycling of DNA in these thymic cells. When PNP is absent, 2'-deoxyguanosine is converted to dGTP instead of being degraded. Imbalance of deoxynucleotides prevents clonal expansion of normal T-cell populations. Several human disorders may involve T-cell action, including T-cell lymphoma, lupus, psoriasis, rheumatoid arthritis, and transplant tissue rejection. Specific inhibitors for PNP may provide useful agents for these disorders. The transition state structure of

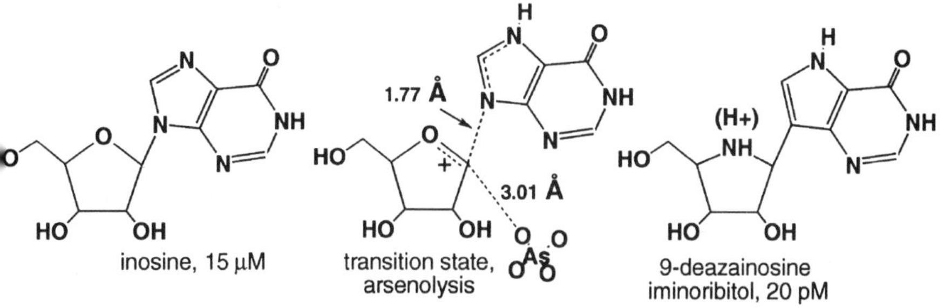

Figure 12   Substrate, transition state, and transition state inhibitor for purine nucleoside phosphorylase. At the transition state, N7 is protonated and the ribosyl is positively charged. These features are incorporated into the transition state inhibitor.

PNP has been determined to permit the development of inhibitors related to the transition state.

Transition state analysis was accomplished using arsenate (AsO_4) because in the presence of phosphate, commitment factors prevented measurement of intrinsic isotope effects (20). In the absence of phosphate or arsenate, the enzyme catalyzes the slow hydrolysis of inosine at one of the trimeric catalytic sites but retains tightly bound hypoxanthine (97). Under these conditions, pre–steady state isotope effects were intrinsic and permitted the hydrolytic and arsenolysis transition states to be compared (98). A fundamental question for the transition state of PNP is the degree to which the transition state is nucleophilic (extent of arsenate or phosphate bonding to C1′ at the transition state), because the reaction occurs with inversion of configuration, as expected for nucleophilic displacements. The results shown in Figure 12 establish that the transition state for arsenolysis is oxocarbenium ion in character, with bond order 0.4 to the leaving group but bond order <0.04 to the incoming water or arsenate oxygens. The enzyme forms an enzyme-bound ribooxocarbenium ion without bound phosphate or arsenate, and participation of the attacking nucleophile occurs late in the reaction coordinate. Formation of the ribooxocarbenium ion requires activation of the purine leaving group, and isotope effects indicate that N7 is protonated at the transition state. The ribose ring accumulates positive charge at the transition state. A transition state inhibitor was designed with a H-bond acceptor at N7 of the purine ring analog, and a positive charge to mimic the ribooxocarbenium ion in the ribosyl analog. 9-Deazainosine iminoribitol is a tight-binding inhibitor binding 10^6-fold tighter than substrate (R Miles, PC Tyler, RH Furneaux, VL Schramm, unpublished data; Figure 12). The molecular electrostatic potential surfaces are compared in Figure 10 (color plate).

Orotate Phosphoribosyl Transferase

Phosphoribosyl transferases were investigated in the pioneering kinetic isotope effects measured by Goiten et al in 1978 (99). The measurement of kinetic isotope effects from both [1-^{3}H]- and [1-^{14}C]5-phosphoribosyl-1-pyrophosphate suggested a pattern consistent with a dissociative mechanism, with pyrophosphate departure being well developed prior to the attack of the nitrogen from the purine. However, many of the phosphoribosyl transferases demonstrated substantial commitment factors, causing the kinetic isotope effects to be obscured.

Orotate phosphoribosyl transferase (OPRT), similar to other enzymes in this group, does not give useful kinetic isotope effects under normal assay conditions. However, the use of phosphonoacetic acid as a pyrophosphate analog made it possible to study the conversion of orotidine 5'-phosphate to orotate and 5-phosphoribosyl-1-phosphonoacetic acid under conditions that gave intrinsic kinetic isotope effects (100). In reactions with a concerted chemical step, the transition state structure can be determined from kinetic isotope effect measurements in either the forward or reverse directions because the same transition state defines both reactions. Differing chemical reactivity with slow substrates may result in a transition state that differs from the normal reaction. However, these differences are expected to be small because the template of the catalytic site is expected to favor a specific transition state geometry for the reaction.

At the transition state for OPRT, the N1-C1' bond has a residual bond order of 0.28 and the attacking oxygen from the pyrophosphate analog is just beginning to participate with a bond order of <0.02 (Figure 13). The kinetic isotope effect from [2'-^{3}H]orotidine 5'-phosphate is relatively large at 14%, establishing that the dihedral angle to the leaving group is nearly eclipsed. The remote 5'-^{3}H isotope effect is significant at 2.8%, confirming that ribosyl-transferases commonly use geometric distortion from the ribosyl 5'-position to form the transition state geometry. Features of this transition state require a ribooxocarbenium ion structure. The X-ray crystal structure of the enzyme has been solved with orotidine 5'-monophosphate or 5-phosphoribosyl-1-pyrophosphate at the catalytic site (101, 102). Although the structure is not in a transition state conformation, Arg156 is shown to interact with the 4-carbonyl group of orotic acid and can be proposed to activate the orotic acid leaving group.

NAD$^+$ and the ADP-Ribosylating Toxins: Cholera, Diphtheria, and Pertussis

Early investigations of the hydrolysis of nicotinamide from NAD$^+$ and NMN$^+$ in solution and by NAD$^+$ hydrolases used [1'-^{2}H]- in NAD$^+$ or NMN$^+$ as the isotopic probes (103–104). Based on the cases where isotope effects were observed, the transition states were characterized as dissociative and electropositive.

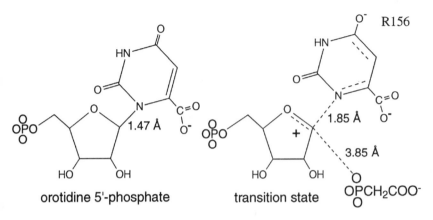

orotidine 5'-phosphate   transition state

Figure 13 Substrate and transition state for orotidine phosphoribosyl transferase (OPRT) based on kinetic isotope effects. Arg156 appears to be the primary group for stabilizing the leaving group based on the crystal structure. The major enzymatic contacts to the ribosyl are to the 5'-phosphate. Leaving group activation is therefore implicated in forming the transition state from bound orotidine 5'-phosphate.

Recent studies have fully characterized the transition states for NAD^+ hydrolysis in solution and by cholera, diphtheria, and pertussis toxins. These studies have been made possible only by the development of enzymatic synthetic methods to provide specifically labeled NAD^+ (105; Figure 14).

The physiological reaction catalyzed by these bacterial toxins is covalent ADP-ribosylation of regulatory GTP-binding proteins. The covalent modification disrupts the normal functions (106; Figure 15). Because the toxins are a primary cause of tissue damage from these bacterial infections, understanding the transition state structures is intended to guide the design of transition state inhibitors to intervene at the site of tissue damage. An advantage of toxin intervention is that it would not be expected to lead to genetic selection for resistant strains and could be used to limit tissue damage in populations at risk in these infections.

A question addressed by these studies is the nature of the transition state as substrates of differing nucleophilicity are involved in displacing nicotinamide. The results (Figure 16) indicate that the transition states for hydrolysis are similar for pH-independent hydrolysis (pH values from 3 to 7) and for the slow NAD^+ hydrolysis reactions catalyzed by all three enzymes when acceptor proteins are not present (107–110). In every case the ribosyl group has well-developed oxocarbenium ion character, with nearly full positive charge at the transition state. The attacking water nucleophile in all cases has a low bond order, more characteristic of a "spectator" nucleophile than one actively

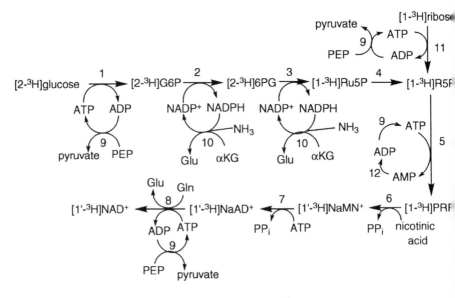

Figure 14 Enzymatic synthesis of specifically labeled NAD⁺. Enzymes not identified in Figure 6 include 6 = nicotinate phosphoribosyl transferase, 7 = NAD⁺ pyrophosphorylase, 8 = NAD⁺ synthetase, and 11 = ribokinase.

participating in bond formation. However, the magnitude of the kinetic isotope effects requires a small degree of nucleophilic participation, less than 0.01 bond order at the transition state. The results establish that the water nucleophile does not play a role in forming the transition state, as occurs in nucleophilic displacement reactions. Rather, the enzymes catalyze formation of the ribooxocarbenium ion, and the neighboring water molecule then reacts with the electron-deficient center.

Transition state structures have been solved for pertussis toxin catalyzing the ADP-ribosylation of the C-terminal 20–amino acid peptide from $G_{i\alpha 3}$ and the G-protein $G_{i\alpha 1}$ (111, 112; Figure 16). In both cases, the ADP-ribosyl acceptor is a specific Cys sulfhydryl. The reaction for peptide ADP-ribosylation gave intrinsic kinetic isotope effects, and the transition state structure shows increased nucleophilic participation by the thiolate anion (Figure 16). Electrostatic potential surfaces of the transition state for ADP-ribosylation of a thiolate anion are shown in Figure 17 (color plate). In NAD⁺ the positive charge is distributed nearly equally in the ribosyl and nicotinamide, giving rise to an elongated, weak N-ribosidic bond. At the transition state the leaving group is neutral and the full positive charge has migrated into the ribose. Attack by the thiolate anion neutralizes the oxocarbenium ion charge to form the ion-paired transition state (111).

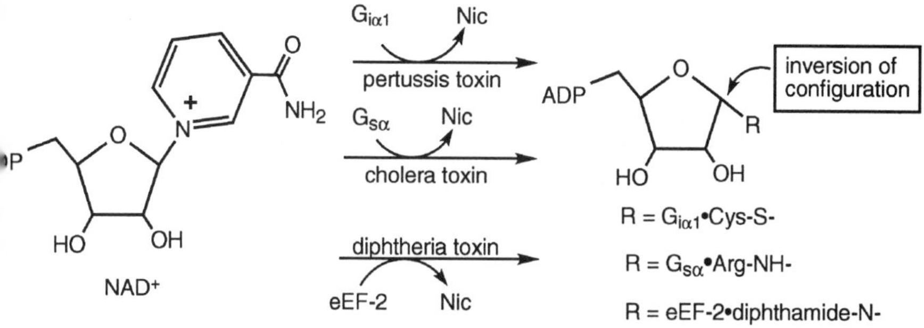

Figure 15 Reactions catalyzed by the ADP-ribosylating agents: pertussis toxin, cholera toxin, and diphtheria toxin. In the absence of the G-protein acceptors, all toxins catalyze the slow hydrolysis of NAD^+ to form nicotinamide (Nic) and ADP-ribose (where R = OH). All of the known ADP-ribosylation reactions involve inversion of configuration at C1' of the ribosyl.

The ADP-ribosylation reaction for the protein $G_{i\alpha 1}$ involves the concerted interactions of the G protein, NAD^+, and pertussis toxin A-chain enzyme. To express intrinsic kinetic isotope effects, the chemical bond change must be the slowest step in the reaction. It was previously unknown whether protein covalent modifications could be investigated by kinetic isotope effect methods. Of concern were the slow conformational changes that dominate most protein-protein interactions and the potentially rate-limiting release of a protein product

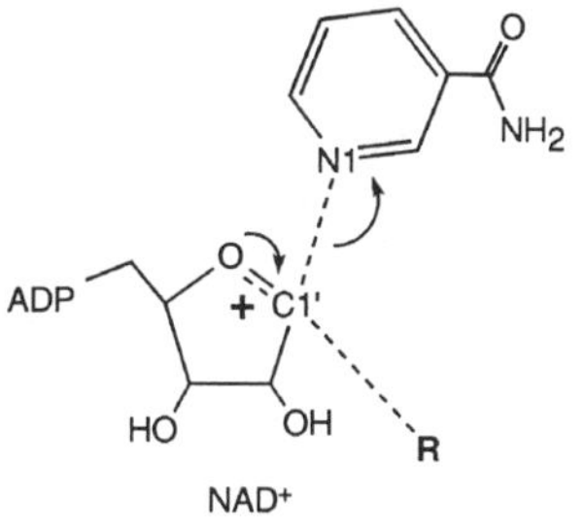

catalyst - reaction	bond order at transition state	
	C1'-N1	C1'-**R**
solvolysis	0.02	0.005
cholera-hydrolysis	0.11	0.002
pertussis-hydrolysis	0.11	0.001
diphtheria-hydrolysis	0.02	0.03
pertussis-peptide-ADPR	0.14	0.11
pertussis-$G_{i\alpha 1}$-ADPR	0.11	0.09

Figure 16 Summary of transition state structures for NAD^+ hydrolysis or ADP-ribosyl transfer. Solvolysis was at pH 4 in buffered H_2O. Hydrolysis by the three toxins is in the absence of ADP-ribose acceptors. For the pertussis-peptide-ADP-ribose (ADPR) reaction, the C-terminal 20–amino acid peptide from $G_{i\alpha 3}$ was used as the acceptor. Cys16 of the peptide is ADP-ribosylated. For the pertussis-$G_{i\alpha 1}$-ADPR reaction, purified recombinant $G_{i\alpha 1}$ was the acceptor and ADP-ribosylation occurs at a Cys located 4 amino acids from the C terminal of the protein.

from the catalytic site. Pertussis toxin binds tightly to the trimeric $\alpha\beta\gamma$-G-protein complex (0.8 μM K_m), but when $G_{i\alpha 1}$ is used alone, the K_m increases 1000-fold to 800 μM K_m with no change in k_{cat}, and intrinsic kinetic isotope effects are expressed (112). Similar methods may be applicable to other co-valent modifications because the catalytic rates for covalent modifications are usually slow, a condition that favors expression of intrinsic kinetic isotope effects.

The site of ADP-ribosylation for pertussis toxin is a specific Cys residue lo-cated four amino acids from the C terminus of several G_i proteins. The attacking Cys thiol has been ionized to form the thiolate anion (RS$^-$) prior to reach-ing the transition state. Thiolate anions are efficient nucleophiles. As the en-zyme converts NAD$^+$ to the enzyme-stabilized ribooxocarbenium ion, the jux-taposition of the thiolate results in a transition state with an increase in the bond order from the attacking nucleophile (Figure 16). The transition states increase in nucleophilic character when the physiological peptide substrates are present. This change is likely to reflect an ability of the enzyme to position the attacking thiolate nucleophile for optimum interaction as the ribooxocarbenium ion is formed.

Docking the NAD$^+$ hydrolytic transition state formed by diphtheria toxin into the X-ray crystal structure determined with NAD$^+$ in the catalytic site provides revealing information for the formation of the ribooxocarbenium transition state (Figure 17, color plate). The crystal structure of the Michaelis complex of diph-theria toxin-NAD$^+$ reveals the nicotinamide ring in a hydrophobic pocket (113). The ring is not inserted full depth into the pocket, and the essential catalytic site carboxylate, Glu148, is poorly positioned with respect to the ribooxocar-benium ion of the transition state. Removing NAD$^+$ from the crystal structure and replacing it with the transition state also provides poor enzymatic contacts. The ribooxocarbenium ion does not contact the protein until the nicotinamide group is inserted more deeply, about an additional 1 Å, into the hydrophobic pocket. Following this translation, the catalytic site Glu148 is well positioned to stabilize the positive charge that develops at C1$'$ in the transition state. The hydrophobic pocket is energetically unfavorable for the positive charge on the reactant nicotinamide and forces it toward the ribose, where it is accommo-dated in the oxocarbenium ion and stabilized by Glu148. Water is an inefficient nucleophile to attack the ribooxocarbenium ion, and hydrolysis occurs slowly. When the acceptor G protein is present, the reaction occurs more rapidly be-cause the toxin ionizes the Cys ADP-ribosylation site to form the nucleophilic thiolate anion. These features of the transition state are adequate to predict the structure of stable compounds that mimic the transition state. It is hoped that this information will lead to new agents to intervene in these infections.

INHIBITOR PREDICTION FROM TRANSITION STATE INHIBITORS

Similarity Measures

A goal of transition state investigation has been to use the knowledge to design transition state inhibitors and to predict inhibitory strength. Transition states stabilized by enzymes bind strongly by virtue of their geometric and electrostatic fit to the enzyme conformation that stabilizes the unstable electronic structure. Geometric similarity of a transition state inhibitor is necessary to provide volumetric access to the catalytic site and to establish the correct distance to the enzymatic contacts. Electronic similarity is essential to make the correct H-bond, ionic, and hydrophobic contacts found in the transition state interaction. If every detail of the transition state structure could be reproduced in a stably bonded transition state analog, it would approach the transition state binding energy of 10^{10}–10^{15} times tighter than the substrate. Despite the problem of exact matching, striking similarity is apparent in molecular electrostatic potential surfaces of transition state structures and those of transition state inhibitors (Figure 10, color plate). It is possible to establish a predictive relationship between the binding affinity of an inhibitor and its similarity to the enzymatic transition state.

Prediction of Inhibitory Strength

Bagdassarian et al (114, 115) examined this relationship using three enzymes for which the transition state structures had been characterized by kinetic isotope effects. AMP deaminase, adenosine deaminase, and AMP nucleosidase were considered by comparing the geometric and molecular electrostatic potential surfaces of the experimentally determined transition states with the substrates and a series of inhibitors exhibiting classic competitive inhibition and slow-onset, tight-binding inhibition (e.g. 78). An algorithm was used that considers the electrostatic potential and its spatial distribution at the van der Waals surface of test inhibitors. These parameters were compared to those for the transition state structure, and an electronic similarity index (S_e) was assigned, with a value of 1.0 indicating a perfect match for molecular electrostatic potential (a molecule compared to itself gives $S_e = 1.0$). Comparison of the transition state S_e with those from substrate and inhibitors that bind better than substrate gave a strong correlation in electrostatic similarity and binding energy ($\Delta G/RT$) over a large range of binding energies encompassing the substrate and the transition state. Approaches that match the features of experimentally determined transition states to those of proposed inhibitors have substantial potential for the prediction of new inhibitors prior to synthetic efforts.

CONCLUSIONS

Experimental access to enzymatic transition state structures has provided novel information about the nature of catalysis. It is reasonable to anticipate that the continued application of this information will permit the design and synthesis of powerful transition state inhibitors that incorporate desired pharmacologic properties. The ability to solve transition state structures has been made possible only by the development of kinetic methods to establish the rate-limiting steps in reactions, synthetic methods for labeled compounds, and computational chemistry and theory to quantitate kinetic isotope effects and to predict electron distribution in partially bonded molecules. The next advance in this area will see these technologies applied to the development of new inhibitors.

ACKNOWLEDGMENTS

Preparation of this work and research in this laboratory have been made possible by the support of research and training grants from the National Institutes of Health, the United States Army, The G. Harold and Leila Y. Mathers Charitable Foundation, the American Cancer Society, and the National Sciences and Engineering Research Council (Canada). I thank Dr. Paul Berti for assistance in figure preparation.

**Visit the *Annual Reviews* home page at
http://www.AnnualReviews.org.**

Literature Cited

1. Radzicka A, Wolfenden R. 1995. *Science* 267:90–93
2. Glasstone S, Laidler KJ, Eyring HK. 1941. *The Theory of Rate Processes.* New York: McGraw-Hill
3. Pauling L. 1948. *Am. Sci.* 36:50–58
4. Wolfenden R. 1972. *Acc. Chem. Res.* 5:10–18
5. Wolfenden R. 1976. *Annu. Rev. Biophys. Bioeng.* 5:271–306
6. Morrison JF, Walsh CT. 1988. *Adv. Enzymol. Relat. Areas Mol. Biol.* 61:201–301
7. Radzicka A, Wolfenden R. 1995. *Methods Enzymol.* 249:284–312
8. Bigeleisen J, Mayer MG. 1947. *J. Chem. Phys.* 15:261–67
9. Bigeleisen J, Wolfsberg M. 1958. *Adv. Chem. Phys.* 1:15–76
10. Streitwiser A Jr, Jagow RH, Fahey RC, Suzuki S. 1958. *J. Am. Chem. Soc.* 80: 2326–32
11. Cleland WW, O'Leary MH, Northrop DB, eds. 1977. *Isotope Effects on Enzyme-Catalyzed Reactions.* Baltimore, MD: Univ. Park Press
12. Gandour RD, Schowen RL, eds. 1978. *Transition States of Biochemical Processes.* New York: Plenum
13. Northrop DB. 1981. *Annu. Rev. Biochem.* 50:103–31
14. Cook PF, Cleland WW. 1981. *Biochemistry* 20:1790–96
15. Cook PF, Oppenheimer NJ, Cleland WW. 1981. *Biochemistry* 20:1817–25
16. Cleland WW. 1982. *Methods Enzymol.* 87:625–41
17. Scharschmidt M, Fisher MA, Cleland WW. 1984. *Biochemistry* 23:5471–78
18. Bolin JT, Filman DJ, Matthews DA, Hamlin RC, Kraut J. 1982. *J. Biol. Chem.* 257:13650–62
19. Kimble E, Hadala J, Ludewig R, Peters P, Greenberg G, et al. 1995. *Inflamm. Res.* 44:S181–82
20. Kline PC, Schramm VL. 1993. *Biochemistry* 32:13212–19

21. Shan S-O, Herschlag D. 1996. *Proc. Natl. Acad. Sci. USA* 93:14474–79
22. Cleland WW, Kreevoy MM. 1994. *Science* 264:1887–90
23. Truhlar DG, Hase WL, Hynes JT. 1983. *J. Phys. Chem.* 87:2664–82
24. Albery WJ. 1993. *Adv. Phys. Org. Chem.* 28:139–70
25. Kurz LC, Weitkamp E, Frieden C. 1987. *Biochemistry* 26:3027–32
26. Frick L, Yang C, Marquez VE, Wolfenden R. 1989. *Biochemistry* 28:9423–30
27. Wilson DK, Rudolph FB, Quiocho FA. 1991. *Science* 252:1278–84
28. Warshel A. 1991. *Computer Modeling of Chemical Reactions in Enzymes and Solutions.* New York: Wiley & Sons
29. Cleland WW. 1977. *Adv. Enzymol. Relat. Areas Mol. Biol.* 45:273–387
30. Parkin DW, Schramm VL. 1995. *Biochemistry* 34:13961–66
31. Albery WJ, Knowles JR. 1977. *Angew. Chem.* 16:285–93
32. Schowen RL. 1978. In *Transition States of Biochemical Processes,* ed. RD Gandour, RL Schowen, pp. 77–114. New York: Plenum
33. Melander L, Saunders WJ Jr. 1980. *Reaction Rates of Isotopic Molecules.* New York: Wiley & Sons
34. Cleland WW. 1995. *Methods Enzymol.* 249:341–73
35. Jencks WP. 1987. In *Catalysis in Chemistry and Enzymology,* pp. 170–82. New York: Dover
36. Hammond GS. 1955. *J. Am. Chem. Soc.* 77:334–38
37. Agarwal RP, Spector T, Parks RE Jr. 1977. *Biochem. Pharmacol.* 26:359–67
38. Bachovin WW, Wong WYL, Farr-Jones S, Shenvi AB, Kettner CA. 1988. *Biochemistry* 27:12839–46
39. Rodgers J, Femec DA, Schowen RL. 1982. *J. Am. Chem. Soc.* 104:3263–68
40. Schramm VL, Horenstein BA, Kline PC. 1994. *J. Biol. Chem.* 269:18259–62
41. Huskey WP. 1991. See Ref. 116, pp. 37–72
42. Suhnel J, Schowen RL. 1991. See Ref. 116, pp. 3–35
43. Cleland WW. 1987. *Bioorg. Chem.* 15:282–302
44. Sunko DE, Szele I, Hehre WJ. 1977. *J. Am. Chem. Soc.* 99:5000–4
45. Bennet AJ, Sinnott ML. 1986. *J. Am. Chem. Soc.* 108:7287–94
46. Northrop DB. 1975. *Biochemistry* 14:2644–51
47. Sims LB, Burton GW, Lewis DE. 1977. *BEBOVIB-IV, QCPE No. 337.* Blooming-

ton, IN. Quantum Chem. Program Exch., Dep. Chem., Univ. Indiana
48. Sims LB, Lewis DE. 1984. In *Isotopes in Organic Chemistry,* ed. E Buncel, CC Lee, 6:161–259. New York: Elsevier
49. Stewart JJP. 1989. *Comput. Chem.* 10:209–20
50. Frisch MJ, Trucks GW, Schlegel HB, Gill PMW, Johnson BG, et al. 1995. *Gaussian 94, Rev. C. 2, 1995.* Pittsburgh, PA: Gaussian Inc.
51. Schramm VL. 1976. *J. Biol. Chem.* 251:3417–24
52. Schramm VL. 1974. *J. Biol. Chem.* 249:1729–36
53. Leung HB, Schramm VL. 1984. *J. Biol. Chem.* 259:6972–78
54. Parkin DW, Leung HB, Schramm VL. 1984. *J. Biol. Chem.* 259:9411–17
55. Parkin DW, Schramm VL. 1987. *Biochemistry* 26:913–20
56. Mentch F, Parkin DW, Schramm VL. 1987. *Biochemistry* 26:921–30
57. Srinivasan K, Konstantinidis A, Sinnott ML, Hall BG. 1993. *Biochem. J.* 291:15–17
58. Pauling L. 1960. *The Nature of the Chemical Bond.* Ithaca, NY: Cornell Univ. Press. 3rd ed.
59. DeWolf WE Jr, Fullin FA, Schramm VL. 1979. *J. Biol. Chem.* 254:10868–75
60. Leung HB, Schramm VL. 1980. *J. Biol. Chem.* 255:10867–74
61. Giranda VL, Berman HM, Schramm VL. 1988. *Biochemistry* 27:5813–18
62. Ehrlich JI, Schramm VL. 1994. *Biochemistry* 33:8890–96
63. Leung HB, Schramm VL. 1981. *J. Biol. Chem.* 256:12823–29
64. Parkin DW, Mentch F, Banks GA, Horenstein BA, Schramm VL. 1991. *Biochemistry* 30:921–30
65. Parry RJ, Minta A. 1982. *J. Am. Chem. Soc.* 104:871–72
66. Markham GD, Hafner EW, Tabor CW, Tabor H. 1980. *J. Biol Chem.* 255:9082–92
67. Larsen TM, Laughlin LT, Holden HM, Rayment I, Reed GH. 1994. *Biochemistry* 33:6301–9
68. Markham GD, Parkin DW, Mentch F, Schramm VL. 1987. *J. Biol. Chem.* 262:5609–15
69. Rose IW. 1980. *Methods Enzymol.* 64:47–59
70. Carey FA, Sundberg RF. 1990. *Advanced Organic Chemistry. Part A: Structure and Mechanism,* pp. 579–83. New York: Plenum. 3rd ed.
71. Merkler DJ, Wali AS, Taylor J, Schramm

VL. 1989. *J. Biol. Chem.* 264:21422–30

72. Lowenstein JM. 1972. *Physiol. Rev.* 52:382–414
73. Sabina RL, Holmes EW. 1995. See Ref. 117, pp. 1769–80
74. Xia Y, Khatchikian G, Zweier JL. 1996. *J. Biol. Chem.* 271:10096–102
75. Hershfield MS, Mitchell BS. 1995. See Ref. 117, pp. 1725–68
76. Frieden C, Kurz LC, Gilbert HR. 1980. *Biochemistry* 19:5303–9
77. Kati WM, Wolfenden R. 1989. *Science* 243:1591–93
78. Merkler DJ, Brenowitz M, Schramm VL. 1990. *Biochemistry* 29:8358–64
79. Merkler DJ, Kline PC, Weiss P, Schramm VL. 1993. *Biochemistry* 32:12993–3001
80. Weiss PM, Cook PF, Hermes JD, Cleland WW. 1987. *Biochemistry* 26:7378–84
81. Merkler DJ, Schramm VL. 1993. *Biochemistry* 32:5792–99
82. Kline PC, Schramm VL. 1994. *Biochemistry* 34:1153–62
83. Hammond DJ, Gutteridge WE. 1984. *Mol. Biochem. Parasitol.* 13:243–61
84. Parkin DW, Horenstein BA, Abdulah DR, Estupiñán B, Schramm VL. 1991. *J. Biol. Chem.* 266:20658–65
85. Estupiñán B, Schramm VL. 1994. *J. Biol. Chem.* 269:23068–73
86. Parkin DW. 1996. *J. Biol. Chem.* 271:21713–19
87. Horenstein BA, Parkin DW, Estupiñán B, Schramm VL. 1991. *Biochemistry* 30:10788–95
88. Pelle R, Schramm VL, Parkin DW. 1998. *J. Biol. Chem.* 273:2118–26
89. Horenstein BA, Schramm VL. 1993. *Biochemistry* 32:7089–97
90. Horenstein BA, Schramm VL. 1993. *Biochemistry* 32:9917–25
90a. Degano M, Almo SC, Sacchettini JC, Schramm VL. 1998. *Biochemistry.* In press
91. Horenstein BA, Zabinski RF, Schramm VL. 1993. *Tetrahedron Lett.* 34:7213–16
92. Boutellier M, Horenstein BA, Semenyaka A, Schramm VL, Ganem B. 1994. *Biochemistry* 33:3994–4000
93. Parkin DW, Schramm VL. 1995. *Biochemistry* 34:13961–66
94. Furneaux RH, Limberg G, Tyler PC, Schramm VL. 1997. *Tetrahedron* 53:2915–30

95. Deng H, Chan AWY, Bagdassarian CK, Estupiñán B, Ganem B, et al. 1996. *Biochemistry* 35:6037–47
96. Markert ML, Finkel BD, McLaughlin TM, Watson TJ, Collard HR, et al. 1997. *Hum. Mutat.* 9:118–21
97. Kline PC, Schramm VL. 1992. *Biochemistry* 31:5964–73
98. Kline PC, Schramm VL. 1995. *Biochemistry* 34:1153–62
99. Goiten RK, Chelsky D, Parsons SM. 1978. *J. Biol. Chem.* 253:2963–71
100. Tao W, Grubmeyer C, Blanchard JS. 1996. *Biochemistry* 35:14–21
101. Scapin G, Grubmeyer C, Sacchettini JC. 1994. *Biochemistry* 33:1287–94
102. Scapin G, Ozturk DH, Grubmeyer C, Sacchettini JC. 1995. *Biochemistry* 34:10744–54
103. Bull HG, Ferraz JP, Cordes EH, Ribbi A, Apitz-Castro R. 1978. *J. Biol. Chem.* 253:5186–92
104. Ferraz JP, Bull HG, Cordes EH. 1978. *Arch. Biochem. Biophys.* 191:431–36
105. Rising KA, Schramm VL. 1994. *J. Am. Chem. Soc.* 116:6531–36
106. Moss J, Vaughan M, eds. 1990. *ADP-Ribosylating Toxins and G-Proteins. Insights into Signal Transduction.* Washington, DC: Am. Soc. Microbiol.
107. Rising KA, Schramm VL. 1997. *J. Am. Chem. Soc.* 119:27–37
108. Berti PJ, Schramm VL. 1997. *J. Am. Chem. Soc.* 119:12069–78
109. Scheuring J, Schramm VL. 1997. *Biochemistry* 36:4526–34
110. Berti PJ, Blanke SR, Schramm VL. 1997. *J. Am. Chem. Soc.* 119:12079–88
111. Scheuring J, Schramm VL. 1997. *Biochemistry* 36:8215–23
112. Scheuring J, Berti PJ, Schramm VL. 1998. *Biochemistry.* 37:2748–58
113. Bell CE, Eisenberg D. 1996. *Biochemistry* 35:1137–49
114. Bagdassarian CK, Braunheim BB, Schramm VL, Schwartz DD. 1996. *Int. J. Quantum Chem.* 60:73–80
115. Bagdassarian CK, Schramm VL, Schwartz SD. 1996. *J. Am. Chem. Soc.* 118:8825–36
116. Cook PF, ed. 1991. *Enzyme Mechanism from Isotope Effects.* Boca Raton, FL: CRC
117. Scriver CR, Beaudet AL, Sly WS, Valle D, eds. 1995. *The Metabolic and Molecular Basis of Inherited Disease.* New York: McGraw-Hill. 7th ed.

Annu. Rev. Biochem. 1998. 67:721–51

THE DNA REPLICATION FORK IN EUKARYOTIC CELLS

Shou Waga and Bruce Stillman
Cold Spring Harbor Laboratory, P.O. Box 100, Cold Spring Harbor, New York 11724

KEY WORDS: polymerase switching, Simian virus 40, cell cycle, Okazaki fragments, replisome proteins

ABSTRACT

Replication of the two template strands at eukaryotic cell DNA replication forks is a highly coordinated process that ensures accurate and efficient genome duplication. Biochemical studies, principally of plasmid DNAs containing the Simian Virus 40 origin of DNA replication, and yeast genetic studies have uncovered the fundamental mechanisms of replication fork progression. At least two different DNA polymerases, a single-stranded DNA-binding protein, a clamp-loading complex, and a polymerase clamp combine to replicate DNA. Okazaki fragment synthesis involves a DNA polymerase-switching mechanism, and maturation occurs by the recruitment of specific nucleases, a helicase, and a ligase. The process of DNA replication is also coupled to cell-cycle progression and to DNA repair to maintain genome integrity.

CONTENTS

INTRODUCTION

In a proliferating eukaryotic cell, the duplication of the genetic complement occurs every S phase of the cell cycle and must occur with high accuracy only once per cell cycle. Furthermore, DNA replication in a single cell must be coordinated with other cell-cycle processes such as mitosis and cytokinesis and with DNA replication in the surrounding cells in tissue. Much of this regulation occurs at the level of initiation of DNA replication via interaction with the pathways that control cell cycle progression. Although we still do not fully understand how initiation of DNA replication occurs in eukaryotes, rapid progress is being made and has been reviewed elsewhere (1, 2). Once initiation occurs, the replication apparatus copies each replicon in a highly efficient process. The fundamental mechanisms that operate at the eukaryotic DNA replication fork are now quite well known and are discussed here. Because the replication of DNA in eukaryotic cells must be coupled to DNA repair and assembly of the DNA into chromatin, the replication fork proteins play prominent roles in maintaining the fidelity of DNA replication, in coordinating replication with cell-cycle progression, and in the inheritance of chromatin complexes.

CELLULAR REPLICATION FORK PROTEINS

Most of what is known about DNA replication in eukaryotes comes from extensive studies performed using cell extracts from mammalian cells that support the complete replication of plasmid DNAs containing the Simian Virus 40 DNA replication origin (SV40 *ori*) (3–6). SV40 DNA replication requires a single virus-encoded protein, the SV40 large tumor antigen (T antigen), which functions both as an initiator protein by binding to the SV40 *ori* and as a DNA helicase at the replication fork; thus the replication is achieved primarily by cellular proteins that also function to duplicate cellular DNA. The purification and reconstitution of DNA replication with purified proteins has yielded great insight into the mechanism of DNA replication as well as other aspects of DNA metabolism such as DNA repair and recombination (7–10).

Investigation of more specific enzymatic reactions and yeast genetic studies have uncovered several proteins thought to be involved directly in DNA synthesis at the replication fork. This review briefly outlines the current understanding of the eukaryotic fork proteins (Table 1) and the reactions in which they

Table 1 Functions of DNA replication fork proteins

Proteins	Functions
RPA	Single-stranded DNA binding; stimulates DNA polymerases; facilitates helicase loading
PCNA	Stimulates DNA polymerases and RFC ATPase
RFC	DNA-dependent ATPase; primer-template DNA binding; stimulates DNA polymerases; PCNA loading
Pol α/primase	RNA-DNA primer synthesis
Pol δ/ε[a]	DNA polymerase; 3'–5' exonuclease
FEN1	Nuclease for removal of RNA primers
RNase HI	Nuclease for removal of RNA primers
DNA ligase I	Ligation of DNA
T antigen[b]	DNA helicase; primosome assembly

[a]A specific function of DNA polymerase ε in replication has not been assigned, although it is known to be essential for S-phase progression in *S. cerevisiae* (196, 197).
[b]T antigen is required for the replication of SV40 DNA. Its functional equivalent in eukaryotic cells has not been identified.

participate. DNA ligases (11, 12) and topoisomerases I and II (13) are also required for replication, but since the function of these enzymes has been reviewed elsewhere, they are not covered extensively here.

DNA Polymerase α/Primase Complex

The DNA polymerase α/primase complex (pol α/primase) is the only enzyme capable of initiating DNA synthesis de novo by first synthesizing an RNA primer and then extending the primer by polymerization to produce a short DNA extension (RNA-DNA primer) (14). Analyses of SV40 *ori*–dependent replication in vivo and in vitro has demonstrated that pol α/primase can synthesize an RNA-DNA primer of approximately 40 nucleotides (nt) in length, including about 10 nt of RNA primer (15–20). The short RNA-DNA then serves as a primer for extension by another polymerase for DNA synthesis on either the leading (continuously synthesized) strand or for each Okazaki fragment on the lagging (discontinuously synthesized) strand (21–26). This process involves a polymerase switch from pol α/primase to either DNA polymerase δ or ε (pol δ and ε) (see below). This switch occurs because, unlike other more complex polymerases, pol α/primase is not capable of processive DNA synthesis and dissociates from the template DNA following primer synthesis (27).

The human cell pol α/primase consists of four subunits (p180, p70, p58 and p48), and similar subunits are found in all eukaryotes examined (for review see 14, 28). cDNAs encoding all four subunits of human, mouse, or yeast (*Saccharomyces cerevisiae* and *Schizosaccharomyces pombe*) pol α/primase have been cloned, and active complexes have been reconstituted from recombinant proteins expressed from baculovirus vectors in infected insect cells (14, 29, 30). The p180 and p48 polypeptides harbor the DNA polymerase and primase catalytic activities, respectively. Extensive mutational analyses of the genes encoding both polymerase and primase catalytic subunits show that they function in DNA replication in vivo but also suggest a regulatory role for the primase subunit (see below; 14, 31–33). The p58 subunit is necessary for the stability and activity of the primase p48 subunit (29, 34, 35). Although no enzymatic activity has been associated with p70 (also known as B subunit or p86 in *S. cerevisiae*) (36, 37), biochemical studies have shown that p70 plays an important role in the assembly of the primosome (see below). In a complementary approach, genetic experiments with temperature-sensitive mutants of *S. cerevisiae* (*pol12-T*) revealed an essential function for the equivalent subunit (p86) in the initiation of DNA replication (37).

The phosphorylation of pol α/primase has been reported in human and *S. cerevisiae* cells (38–40); both human p180 and p70 were phosphorylated predominantly during late G2 and M phases of the cell cycle (38). Yeast p86 (B subunit) is phosphorylated in a manner that depends on the stage of the cell cycle (39), but most phosphorylation occurs late in the cell cycle, suggesting that it might play a role either in coordinating replication and mitosis or in resetting the replication apparatus for the next S phase. Analysis of the initiation of SV40 DNA replication showed that the primase activity during initiation could be suppressed when pol α/primase was phosphorylated by cyclin A-CDK2, but not by cyclin B-CDC2 or cyclin E-CDK2, whereas DNA polymerase or primase activities with synthetic templates were hardly affected by this phosphorylation (41). This observation suggests that the phosphorylation of pol α/primase may play a regulatory role in the initiation of replication, but the in vivo significance of the phosphorylation is unknown.

Replication Protein A (RPA)

Replication protein A (RPA; also reported previously as RFA or HSSB) is a single-stranded DNA-binding protein that exists as a heterotrimeric complex consisting of subunits with apparent masses of approximately 70, 34, and 11 kDa in all eukaryotic cells examined (for review see 42, 43). The trimeric protein was initially identified as a factor essential for SV40 DNA replication in vitro (44–46), but subsequently it was shown to be involved in DNA recombination and repair (43, 47).

RPA promotes extensive unwinding of duplex DNA containing the SV40 *ori* by SV40 T antigen, which in addition to recognizing the *ori* in a sequence-specific manner, is an RPA-stimulated DNA helicase (48–52). RPA also stimulates pol α/primase activity under certain conditions and is required for replication factor C (RFC)– and proliferating cell nuclear antigen (PCNA)–dependent

DNA synthesis by DNA polymerase δ (25, 53–57). In support of these bio-chemical studies, some temperature-sensitive mutations in the gene encoding the large subunit of *S. cerevisiae* RPA showed synthetic lethality, a form of genetic interaction, with mutations in genes encoding pol α, primase, or pol δ (58). Binding assays with purified proteins demonstrated that p70 binds to the primase subunits of pol α/primase and that the RPA heterotrimeric complex, but not p70 alone, binds to SV40 T antigen (59, 60). These interactions are thought to be required for the assembly of the primosome (see below).

cDNAs encoding each of the individual subunits of RPA have been cloned from a variety of species (43, 61). Furthermore, RPA has been produced in bacteria, and the recombinant human trimeric protein can support SV40 DNA replication in vitro (62, 63). Functional RPA has also been produced by infection of recombinant baculoviruses into insect cells (64). Although the human p70 subunit alone can bind single-stranded DNA, it cannot support DNA replication in vitro (55, 65, 66). Mutational studies to probe the structure of p70 have located the DNA-binding region to the N-terminal two-thirds of the subunit; the C-terminal third containing a putative zinc-finger was required for interactions with the other two subunits (65–67).

More recent structural analysis of yeast RPA suggests that it contains a total of four potential single-stranded DNA-binding domains that are distantly related to each other (68). The domains are called SBDs and are made up of about 120 amino acids each; two are in the large subunit and one each is in the middle and the small subunits, although the evidence for the DNA-binding domain in the small subunit is not as convincing as for the large subunit domains. A crystal structure of the two DNA-binding domains derived from the human large subunit (p70) revealed two structurally related subdomains, each corresponding to an SBD (69). It has been suggested that these SBDs, which contain clusters of aromatic amino acids that are similar to structures within the DNA-binding domain of the *Escherichia coli* single-stranded DNA-binding protein (SSB), are responsible for a higher-order assembly of the RPA-DNA complex, where RPA might be wrapped with DNA (68). Indeed DNA-binding studies with human RPA have demonstrated the existence of at least two distinct DNA-binding modes of the protein: one involving 8–10 nucleotides and another involving 30 nucleotides (70, 71). There may well be higher-order interactions between RPA and longer stretches of single-stranded DNA, as occurs for *E. coli* SSB (72).

Both the large (p70) and the middle (p34) subunits of human and yeast RPA are phosphorylated in a cell-cycle-dependent manner (73). Increased levels of similar phosphorylated forms of RPA are also seen in response to DNA damage (74, 75). The phosphorylation of the p34 subunit has been characterized extensively; it is phosphorylated in the S and G2 phases of the cell cycle, and cyclin-dependent and DNA-dependent kinases have been identified as enzymes capable of phosphorylating p34 (76–84). The phosphorylation of RPA is delayed in cells from patients who have ataxia telangiectasia (AT), a cancer-prone disease resulting from loss of the DNA damaging surveillance ATM gene product; and the phosphorylation of RPA is compromised in yeast lacking the *MEC1* gene, a gene with similarities to ATM (74, 85, 86). This is particularly interesting because phosphorylation of RPA seems to decrease the interaction between this protein and the cellular tumor suppresser protein, p53 (87). However, the role of phosphorylation on the function of RPA in either SV40 DNA replication or nucleotide-excision repair is not clear (88–90), and even the link between phosphorylation of RPA and S-phase checkpoint controls has been questioned (91). Thus the functional significance of the RPA phosphorylation remains to be determined.

A little understood aspect of RPA in cellular DNA metabolism is its ability to assemble into discrete foci (pre-replication center) on postmitotic, decondensing chromosomes, which afterward serve as replication foci following the assembly of a nuclear membrane (92). A protein called FFA-1, which is required for the assembly of the RPA foci, has been identified in *Xenopus* egg extracts (93), but formation of the RPA foci required neither subunits of the cellular initiator protein ORC (origin recognition complex), nor the Cdc6 protein (94), both of which are essential for initiation of DNA replication in *Xenopus* extracts and in yeast (2). It is therefore not clear what relationship these replication foci have to ORC-dependent DNA replication or if they form in a normal cell cycle. It is possible that during chromosome decondensation, single-stranded DNA regions created by tortional strain might be bound by RPA, providing local assembly sites for pol α/primase, which binds to RPA. The structure and function of these foci and their role in normal DNA replication in cells remain to be elucidated.

Replication Factor C (RFC)

One of the key proteins involved in loading the replicative polymerases to create the replication fork is replication factor C (RFC), a complex of five subunits (p140, p40, p38, p37, and p36) that is conserved in all eukaryotes (for review see 42). Functional homologs exist in bacteria, some bacteriophages, and Archea (95). The cDNAs encoding the individual subunits of human and the yeast *S. cerevisiae* RFC have been cloned, and all five yeast genes are essential for cell viability (96–104). Sequence comparisons show a high degree of similarity among all five subunits, and based on these similarities and the conserved sequences found among species, short stretches of amino acid sequences called RFC boxes I–VIII have been defined (104). Box I is unique to the larger p140 subunit, is related to sequences in prokaryote DNA ligases, and is distantly related to the BRCT motif present in many proteins that respond to DNA damage in cells (98, 105); boxes III and V are characteristic of sequence motifs present in many ATP- and GTP-binding proteins. The other RFC sequence motifs, particularly RFC box VIII, are found in a number of other replication proteins and proteins of unknown function, in addition to being highly conserved among the RFC subunits (104, 106).

RFC was first identified because it is an essential factor for SV40 DNA replication in vitro (107). It preferentially binds to a primer-template junction created by the annealing of an oligonucleotide to single stranded DNA, or by synthesis of a DNA primer on a single-stranded DNA template. RFC can also bind to a nick in duplex DNA. Binding requires ATP and upon binding to DNA, RFC functions as a DNA-dependent ATPase, an activity stimulated further by PCNA (56, 108–112). One study suggests that ATP hydrolysis is required for the stable loading of PCNA (113), but other studies suggest that this is not the case (56); thus the precise role of ATP in this process remains to be determined. The main role for RFC is to load the trimeric, ring-like structure of PCNA onto DNA at a primer-template junction or to load it onto a nicked site in duplex DNA (56, 108, 109, 114). It has been reported that RFC can load PCNA onto completely duplex DNA, but compared to the interactions with the above-mentioned DNAs, this interaction is very sensitive to physiologic salt concentrations and probably does not represent a reaction that occurs during DNA replication in cells (115). RFC-catalyzed PCNA loading is a prerequisite for assembly of pol δ onto the template DNA to form a processive holoenzyme (25, 108–113, 116), which then functions during synthesis of both the leading and lagging strands at a DNA replication fork (see below).

A functional human RFC complex has been reconstituted using proteins expressed in recombinant baculovirus-infected insect cells (114, 117, 118), and the yeast RFC has been overexpressed in yeast cells (119). Several mutational analyses of human RFC show that distinct regions of the p140 subunit are responsible for DNA and PCNA binding (120, 121). The PCNA-binding domain from the p140 subunit inhibits DNA replication in mammalian cells (120), supporting a role for RFC in DNA replication in vivo. While the N-terminal region that includes RFC box I in p140 has a DNA-binding activity, an RFC complex lacking this region exhibits enhanced activity in a reconstituted SV40 DNA replication reaction as well as enhanced PCNA loading activity (122). This observation suggests a regulatory role for this protein domain that includes the similarity to the BRCT motif. The large subunit of RFC is a target for caspases, the proteases activated during apoptosis or programmed cell death (123, 124), perhaps because it is a significant ATP-regulated enzyme involved in DNA metabolism.

Limited structure-function analysis of the small subunits has been performed, but the C-terminal sequences of each of the small subunits are required for formation of the RFC complex (122, 125). Three small subunits (p40, p37, and p36) form a stable, core complex that has some DNA-dependent ATPase activity, but without the large p140 subunit, this ATPase is no longer stimulated by PCNA (117, 118, 126, 127). The p38 small subunit seems to provide a link between the p140 and the core complex (118, 126).

A cold-sensitive *S. cerevisiae* mutant in the gene encoding the RFC large subunit (*cdc 44*) has been isolated and characterized (100). This mutation has an altered DNA metabolism at the nonpermissive temperature that is consistent with a role for RFC in DNA replication or DNA repair. These defects were suppressed by mutations in the gene encoding PCNA (*pol 30*), supporting the biochemical interaction between these two proteins (128). Interestingly, a mutation in the *RFC5* gene encoding the *S. cerevisiae* p38 subunit homolog causes a defect in a DNA-damage checkpoint signal that transmits to the Rad53 protein and the Tel1 protein, a yeast protein similar to the ataxia telangiectasia gene product ATM (129). This observation suggests that RFC might function in monitoring DNA damage at the replication fork.

Of all the DNA polymerase accessory proteins, the subunits of RFC show both striking sequence and functional similarities to replication fork proteins present in *E. coli* and bacteriophage T4 (γ-complex and gp44/gp62, respectively; 95, 106). These proteins load a ring-like DNA polymerase clamp (e.g. PCNA in eukaryotes and the *E. coli* β-subunit of DNA polymerase III) onto the template DNA and, therefore, are known as clamp-loading proteins. The recent determination of the crystal structure of a member of the γ-complex, the δ' subunit, shows it to have a provocative C-like shape that shows striking structural overlap with the ring-like shape of the *E. coli* β-subunit and PCNA, suggesting how the clamp-loader might open the ring-like clamp and load it onto to DNA (Figure 1; 130). The clamp-loading function of RFC may be one of the key regulated events in DNA metabolism.

Proliferating Cell Nuclear Antigen (PCNA)

Perhaps one of the most intensely studied proteins is proliferating cell nuclear antigen (PCNA), the DNA polymerase clamp. Not only does this protein play a central role in DNA metabolism, but it has become a significant clinical

---→

Figure 1   Structures of a polymerase clamp and a clamp-loading protein subunit. *Top*: stereo view of a ribbon diagram of the structure of the clamp-loader, δ', a part of the *E. coli* DNA polymerase III γ-complex and a functional homolog of the RFC subunits (130). *Bottom*: super-position of the δ' structure and a ribbon diagram of the structure of the *E. coli* DNA polymerase III β-subunit (133), a functional and structural homolog of PCNA. Note the similar shape of the clamp ring surrounding the dimer-dimer interface of the two β-subunits and the clamp-loader C-shape.

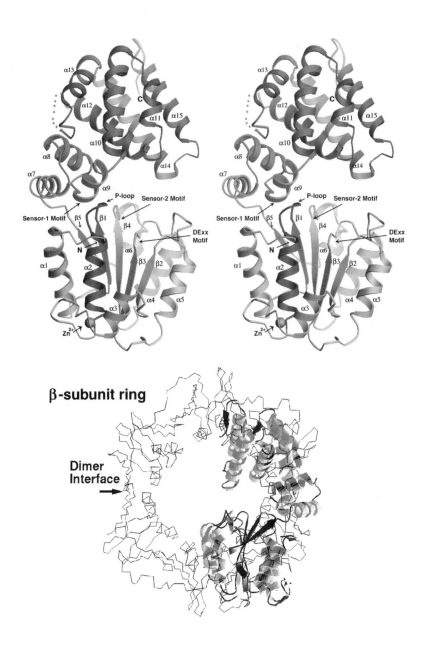

diagnostic marker for proliferating cells. A protein with an apparent mass on polyacrylamide gels of 36 kDa, PCNA forms a homotrimeric complex and functions as a DNA polymerase accessory factor (for reviews see 42, 131, 132). The primary amino acid sequence of PCNA is not highly conserved among species, but yeast and human PCNA nevertheless have an almost identical three-dimensional shape that is also very similar to the structures of the functional homologs of PCNA present in *E. coli* and bacteriophage T4 (i.e. the β-subunit of DNA polymerase III in *E. coli* and the gp45 protein of T4 phage) (95, 132–135). These proteins exist as stable trimers that form a closed ring with a hole in the center that encircles duplex DNA (Figure 1). Each protein monomer of PCNA consists of two structurally similar domains that are linked via an interdomain connecting loop on the surface of the protein (134). The internal surface of the trimer contains six α helices, and each helix is present in a structural repeat in each PCNA monomer. Since RFC loads the PCNA trimer onto DNA, it topologically links the PCNA trimer to the DNA, allowing PCNA to track along the DNA (136). Most probably, RFC also functions to unload the PCNA when DNA synthesis is complete, much like the bacterial and phage T4 counterparts (137–139).

PCNA functions as a processivity factor for pol δ during DNA replication (140–144). Stimulation of pol ε by PCNA has also been detected but only under limited conditions (145–147). PCNA itself does not have DNA-binding activity, but it can be loaded onto the DNA by RFC in an ATP-dependent manner (56, 114). PCNA then associates with pol δ/ε at the primer-template junction (56, 113, 116). Mutational analyses of PCNA have shown that distinct regions of the trimeric ring are required for the stimulation of pol δ or RFC ATPase (148–151); PCNA mutants that alter certain amino acids on the internal surface of the ring failed to stimulate pol δ (149). In contrast, regions on the outer surface, including both the N- and C-termini and the interdomain connecting loop, are necessary for the interaction with pol δ and RFC (149, 151).

PCNA also is capable of binding to other proteins, including the FEN1/Rad27/MF1 nuclease (152, 153) (see below); DNA ligase 1 (154); the p53-inducible, cyclin-dependent kinase (CDK) inhibitor protein p21 (WAF1, CIP1, sdi1) (155, 156); the p53-inducible GADD45 protein (157, 158); the nucleotide excision repair protein XPG (159); DNA-(cytosine-5) methyltransferase (160); the mismatch repair proteins MLH1 and MSH2 (161); and cyclin D (162, 163). These varied interactions with DNA metabolism proteins imply that PCNA is a central factor for the coordination of DNA replication, DNA repair, epigenetic inheritance, and cell-cycle control.

The crystal structure of PCNA bound to a peptide derived from p21 shows that the previously unstructured p21 peptide inserts itself into a cleft in the interdomain connecting loop surface of PCNA (164). This very stable binding causes the inhibition of PCNA stimulation of pol δ activity and competition with the binding of PCNA to DNA ligase, DNA-(cytosine-5) methyltransferase, XPG, and FEN1 (see below). Moreover, a monoclonal antibody that specifically binds to this loop inhibits PCNA-activated DNA synthesis by pol δ (165). Thus the loop serves as an interface for interactions with other cellular proteins.

Immunofluorescent staining of cells with anti-PCNA antibodies has shown that PCNA co-localizes with sites of DNA synthesis in nuclei (replication foci) (166, 167). The PCNA staining pattern varies with the stage in the S phase, corresponding with the region of the genome being replicated. In addition, PCNA becomes resistant to extraction from nuclei with detergents such as Triton X-100, only when cells enter the S phase of the cell cycle (166, 168). PCNA resistance to Triton-extraction also occurs in DNA-damaged cells, even in the G1 phase of the cell cycle, indicating that PCNA localizes to sites of DNA repair (168–172). The variable, detergent-resistant PCNA in nuclei is thought to reflect its topologically closed interaction with DNA at sites of DNA metabolism.

DNA Polymerases δ and ε

Two DNA polymerases function during the S phase of the cell cycle in eukaryotic cells (14). DNA polymerase δ (pol δ) is a heterodimer composed of p125 and p50 subunits (for review see 14), although we consider it likely that the native enzyme in mammalian cells and *S. cerevisiae* contains an additional subunit of approximate mass of 50 kDa. Recent biochemical and genetic evidence shows that the *S. pombe* pol δ contains five subunits, three of which are essential (173–175). p125 is the catalytic subunit for both DNA polymerase activity and a proofreading, $3'$-$5'$ exonuclease activity. cDNAs for each subunit have been cloned (175–183). Moreover, an active human dimeric complex has been reconstituted using proteins expressed in recombinant baculovirus-infected insect cells (184). The N-terminal region of p125 interacts with PCNA (185, 186), and although one report suggests that the polymerase activity of the large subunit alone can be stimulated by PCNA in *S. cerevisiae* (185), a recent study of mammalian pol δ shows that the p50 subunit is required for PCNA stimulation (184). To complicate matters further, a putative PCNA-independent pol δ was also isolated from mouse (187) and *Drosophila* (188). Thus further characterization of the structure of DNA polymerase δ in mammalian cells and *S. cerevisiae* is necessary.

Pol δ is required for DNA synthesis of both the leading and lagging strands during SV40 DNA replication in vitro (22–24, 26, 189). Many studies using a variety of DNA templates support a role for pol δ in RFC- and PCNA-dependent DNA replication (24–26, 56, 109, 113, 146, 184, 189–194). Crosslinking of DNA polymerases to replicating SV40 DNA in vivo also supports the involvement of both pol δ and pol α in SV40 DNA replication (195). Although some reports demonstrate a role for pol ε in cellular DNA replication (113, 146, 196–200), the in vivo crosslinking experiments indicate that pol ε is not essential for SV40 DNA replication (195). This study, however, found that pol ε did crosslink to replicating cell chromosomal DNA, consistent with the essential function of S. cerevisiae pol ε gene (POL2) in cellular DNA replication (196, 197). The precise role of polymerase ε in cellular DNA replication remains to be determined.

FEN1 and RNaseHI

FEN1, a 46-kDa single polypeptide in human and mouse, is a 5′-3′ exo/endonuclease that is required for Okazaki fragment maturation (for reviews see 10, 201). The primary sequence of the protein shows similarity to the repair protein Rad2/XPG and other nucleases (202, 203). FEN1 was identified through purification of the enzyme that specifically cleaves a flap-structure DNA substrate, hence the name flap endonuclease (204). Other researchers independently identified the same protein in different contexts, as follows: MF1 (192), 5′-3′ exonuclease (205), cca/exo (206), DNase IV (207), pol ε-associated nuclease (208), human homolog of S. pombe Rad2 (203), and factor pL (209). The homologs from S. cerevisiae and S. pombe were identified as Rad27 (210, 211) and Rad2 (203), respectively, indicating a role for the protein in DNA repair. FEN1 is required for SV40 DNA replication in vitro (192, 205).

Several biochemical studies have shown that FEN1 functions specifically to remove the RNA primer attached to the 5′-end of each Okazaki fragment (26, 200, 212–215). Extensive studies show that removal of the RNA primer involves other proteins including an RNA-DNA junction endonuclease, PCNA, and Dna2 helicase (10). The mechanism for maturation of Okazaki fragments is described below, but removal of the RNA primer from the 5′-end of the penultimate Okazaki fragment prior to joining to the newly synthesized Okazaki fragment requires strand displacement synthesis to displace the RNA primer, creating a flap structure with a 5′ unpaired RNA-DNA strand.

The enzymatic properties of FEN1 have been examined in some detail. When provided a flap structure containing a 5′-segment of DNA (or RNA) that is not paired to a template DNA, FEN1 efficiently cleaves at the branch point, releasing the unpaired segment (202, 204, 207, 215, 216). However, if the 5′-segment of DNA (or RNA) is completely paired, FEN1 can degrade DNA (or RNA) only exonucleoticaly from the 5′-terminus (192, 200, 202, 204, 214). A substrate that contains a 5′-triphosphorylated ribonucleotide that is annealed to DNA, such as an Okazaki fragment that is completely annealed to a single-stranded DNA, cannot be degraded by FEN1 (200). In such a situation, removal of the RNA requires a ribonuclease in addition to FEN1 (see below).

PCNA and RPA stimulate yeast FEN1 activity under certain conditions (152, 217). These interactions may be important for the maturation of Okazaki fragments during lagging strand replication, as described below. The cyclin-CDK inhibitor p21 disrupts the FEN1-PCNA interaction, suggesting that the step might be regulated (153).

While both *S. cerevisiae RAD27* and *S. pombe rad2+* are not essential for cell viability, both null mutants exhibit elevated chromosome loss rates and increased UV sensitivity (203, 210, 218). Furthermore, a *rad27* null mutant shows temperature-sensitive lethality (210, 211). Interestingly, a novel type of mutation can be generated in cells lacking FEN1 activity. Sequences of 5–108 nucleotide pairs that are flanked by direct repeats of 3–12 nucleotide pairs become duplicated at high frequency in *rad27* null mutants (219). This unique mutation event is thought to result from a defect in lagging-strand DNA synthesis, causing an increased recombination rate due to induced double-strand-break repair of these lesions (220, 221).

RNase H activities in eukaryotic cells have been grouped into two classes, I and II, based on molecular mass of the enzyme and requirements for a cofactor (222). Among them, RNase HI is thought to be involved in the removal of RNA primers during Okazaki fragment synthesis (200, 206, 223–225). While the enzymatic activity of RNase HI has been known for a long time, its precise molecular weight and subunit structure are still unclear. Nevertheless, biochemical characterizations of RNase HI have revealed a unique substrate specificity; it could endonucleolyticaly cleave RNA that is attached to the 5′-end of a DNA strand, such as in an Okazaki fragment, leaving a single ribonucleotide on the 5′-end of the DNA strand (200, 226–228).

DNA Helicases

DNA helicases are enzymes that promote the processive unwinding of duplex DNA, such as occurs at the DNA replication fork to create templates for the polymerases. During SV40 DNA replication, the virus-encoded T antigen functions as the replicative DNA helicase, but for cell chromosome replication, the nature of the replicative helicase remains unclear. Most DNA helicases required for viral or prokaryotic DNA replication form a homomultimeric complex (hexamer in most of the cases) (for review see 229). For example, T antigen functions as a hexamer and assembles at the SV40 *ori* as a do-decamer (two hexamers; 48, 49, 229). As DNA replication proceeds, the two hexamers at the divergent replication forks probably stay connected, forming part of a so-called replisome that the DNA passes through (230). It is highly likely, however, that several distinct helicases would function in cell chromosomal replication and also that the role of an individual helicase might be specialized for certain steps in the replication process. Many eukaryotic cell DNA helicases have been identified, including the helicases associated with pol δ/ε or RFC (see 229), but only a few have been implicated in DNA replication.

DNA2 HELICASE   Yeast Dna2 helicase was identified by screening for replication-defective mutants using a DNA replication assay in permeabilized cells (231, 232). The *DNA2* gene encodes a 172-kDa protein that is essential for cell viability, and the purified protein shows both DNA-dependent ATPase and 3′-to-5′ DNA helicase activities (232, 233). Analysis of ^{3}H-labeled replication intermediates from wild-type and *dna2* temperature-sensitive mutant cells showed that low-molecular-weight intermediates accumulate in the *dna2* mutants but not in wild-type cells, indicating that the defect is at the elongation stage of DNA replication (232).

MOUSE HELICASE B   This mammalian helicase was identified through studies of a temperature-sensitive mutant mouse cell line, tsFT848, which was shown to be defective in DNA replication (234–236). DNA synthesis, but not RNA and protein syntheses, in these mutant cells decreased at the nonpermissive temperature. Comparative analyses of DNA-dependent ATPase activities in fractionated extracts from wild-type and mutant cells showed that one of the major ATPase activities, now designated helicase B, is decreased in the mutant cells. Furthermore, this helicase activity from mutant cells showed heat sensitivity. The DNA chain-elongation rate in the mutant cells, when analyzed by fiber autoradiography, was the same in both wild-type and mutant cells, suggesting that helicase B might be involved in a process that does not determine the elongation rate of the fork.

MINI-CHROMOSOME MAINTENANCE PROTEINS   The mini-chromosome maintenance (MCM) proteins were first identified by yeast genetic studies as proteins required for replication of plasmid DNAs containing cellular origins of DNA replication. Researchers have since determined that these proteins are essential components of the pre-replication complex established prior to the S phase at origins of DNA replication (reviewed in 237, 238). Six MCM proteins have been found in all eukaryotic cells examined to date, and they share a similar amino acid sequence motif called the MCM box, part of which contains a putative ATPase motif (239). Genes encoding MCM proteins have also been found in recent sequences from Archea species, suggesting that they are ancient replication proteins. The MCM protein complex appears to be a hexamer containing equal amounts of each of the six proteins (239a). Two recent observations, when combined, suggest, but do not prove, that the MCM proteins function as a replicative DNA helicase at the cellular replication fork (240, 241). A complex of three MCM proteins (Mcm 4, 6, and 7) is capable of displacing a short oligonucleotide from a larger single-stranded DNA in an ATP-dependent manner, suggesting that it contains DNA helicase activity (241). In addition, some MCM proteins appear to be bound to different regions of a replicon in the yeast genome at different times throughout the S phase of the cell cycle, beginning with the origins of DNA replication where they are assembled in an ORC- and Cdc6-dependent manner prior to S-phase entry (2, 240). This suggests that the MCM proteins might track along the DNA with the DNA replication fork, perhaps acting to unwind the DNA or performing another essential function (240).

Molecular Links Among FEN1, Dna2 Helicase, and DNA Polymerase α/Primase

Accumulating evidence indicates that a multi-enzyme complex exists in cells that contains many of the activities discussed above (193, 242). In addition, biochemical and genetic studies of yeast DNA replication have made connections linking many replication proteins into a multi-protein complex that may function as a so-called replisome. Physical and genetic interactions between the Dna2 helicase and FEN1 have been demonstrated; both proteins co-purified and co-immunoprecipitated, and the overexpression of *FEN1* suppressed the temperature-sensitive growth of a *dna2-1* mutant (243). Conversely, overexpression of *DNA2* suppressed the temperature-sensitive lethality of a *rad27* null mutant (defective in FEN1; 243). This FEN1-Dna2 helicase interaction may play an important role in maturation of the lagging strand.

In an independent study, another allele of *dna2* (*dna2-2*) was isolated in a genetic screen for mutants that show synthetic lethality with the *ctf4-Δ4* mutant (244). Ctf4 protein (Ctf4p), which is identical to Pob1p and Chl15p, was identified as a protein that bound to a pol α catalytic subunit-affinity column (245–247). The *ctf4* null mutant is viable but exhibits elevated chromosome loss, implying a function in some process of DNA metabolism (245, 246).

Cdc68p and Pob3p have also been identified as pol α–binding proteins (244). These proteins seem to compete with Ctf4p for binding to pol α. Biochemical and genetic interactions between pol α and these pol α–binding proteins [Ctf4p, Cdc68p (248–250) and Pob3p] have also been demonstrated (244). Furthermore, the *cdc68-1* mutation was also synthetic lethal with the *dna2-2* mutation (244). These genetic and biochemical data suggest that the synthesis of an RNA-DNA primer to start an Okazaki fragment and maturation of the Okazaki fragment might be coordinately carried out during lagging strand synthesis by a multi-protein complex containing pol α/primase, Ctf4p, Cdc68p, Pob3p, Dna2p helicase, and FEN1. Because FEN1 can bind directly to PCNA, this complex might also contain RFC and pol δ, consistent with biochemical observations (193, 217, 242).

Another allele of RAD27 (*erc 11-2*) was unexpectedly isolated in a genetic screen that sought to identify proteins that interacted with the G1 cyclins Cln1p and Cln2p [*erc* (elevated requirement for *CLN* function) mutations; 251]. The *rad27/erc11-2*, *cln1cln2* mutant strain arrested in the S phase at nonpermissive temperature and gradually lost viability. The temperature-sensitive lethality could be rescued by expression of *CLN1* or *CLN2* but not the other G1 cyclin, *CLN3*. Moreover, overexpression of *DNA2* (referred to as *SEL1* in the original paper) and *CDC9* (DNA ligase) also rescue the temperature-sensitive lethality (251). Although it is unclear how Cln1p/2p can rescue the defect in replication, it is intriguing that both proteins that rescued the defect when overexpressed could interact with PCNA. These studies suggest that the G1 cyclins Cln1p and Cln2p might affect functions at the DNA replication fork.

MECHANISMS OF DNA SYNTHESIS AT A REPLICATION FORK

The above-described studies on the replication proteins in eukaryotic cells have identified their biochemical functions and several specific interactions among these proteins. These interactions underlie the mechanism of DNA synthesis of both the leading and lagging strands at the DNA replication fork. The following sections describe our current understanding of how these proteins cooperate to replicate DNA.

Primosome Assembly

One of the first steps after recognition of the origin of DNA replication and local unwinding of the DNA is to load the pol α/primase complex onto the DNA, a step called primosome assembly. Details about the role of T antigen in origin recognition and local unwinding of the *ori* have been reviewed elsewhere (229, 252). Primosome assembly normally involves a DNA helicase interacting with the pol α/primase, but a cellular helicase that functions in cellular DNA replication in the same way that T antigen functions during SV40 DNA replication has not been identified to date.

T antigen, pol α/primase, and RPA interact with each other and cooperate to initiate DNA synthesis at the SV40 *ori* (16, 27, 36, 60, 254). The protein-protein interactions that have been demonstrated are T antigen–pol α/primase (p70 and/or p180) (36, 255–258), RPA p70–primase (p48 and p58) (59), and RPA–T antigen (59, 60). In addition, the bovine papillomavirus E1 initiator and helicase protein also binds to pol α/primase in a manner analogous to T antigen (259). Primase assays with single-stranded template DNAs have shown that although RPA from either humans or yeast represses primase activity, T antigen can reverse the inhibition only when human RPA, but not *S. cerevisiae* RPA, is coating the DNA (60, 260). In addition, only human pol α/primase, but not calf thymus or mouse pol α/primase, can support primer synthesis in the presence

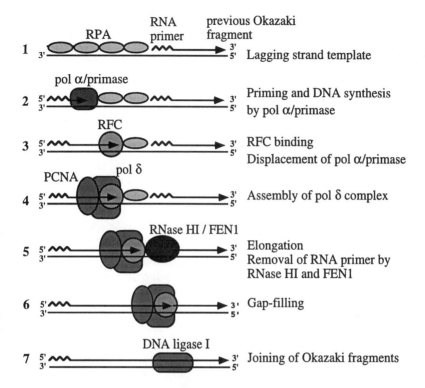

Figure 2 Polymerase switching and maturation of Okazaki fragments on a lagging-strand DNA template. See text for details. Adapted from Reference 313.

of SV40 T antigen and RPA (261), indicating the existence of a species-specific protein-protein interaction during primosome assembly (30, 261, 262). These interactions that promote primosome assembly occur not only during initiation of DNA replication at the *ori* but also for the synthesis of each Okazaki fragment (Figure 2; 10, 26, 56, 200, 212, 260).

Mapping the start sites for RNA-DNA primer synthesis around the SV40 *ori* sequence shows some preference for primer-site selection in vivo and in the crude SV40 replication system (263–265; for review see 20). It has not been determined whether initiation with the purified pol α/primase, RPA, and T antigen occurs at preferred sites for primer synthesis. A number of cellular proteins that associate or cooperate with pol α/primase, such as AAF (266, 267), Ctf4 (Pob1) (245–247), Cdc68 (248–250), and Pob3 (244), might modulate the interaction of the primase with the DNA and affect primer-site selection. In addition, transcription factor activator domains can bind to RPA and stimulate replication in vitro (268, 269), but the precise mechanism of this activation is unclear. In the context of cellular chromatin, these site-specific DNA-binding proteins may aide in recruiting RPA and hence pol α/primase to the DNA.

Polymerase Switching

Biochemical studies using the SV40 DNA replication system reconstituted with purified proteins (21, 24, 26, 54, 190, 191) have shown that two different polymerases, pol α/primase and pol δ, are involved in DNA synthesis and that pol δ is involved in the synthesis of both the leading and lagging strands. The switching from pol α/primase to pol δ occurs during priming of the leading strand (24) and during synthesis of every Okazaki fragment (26). The involvement of two different DNA polymerases in lagging-strand DNA synthesis was suggested by the analysis of SV40 DNA replication in vivo with DNA polymerase inhibitors (15, 17). The mechanisms of initiation of leading strand synthesis and initiation of each Okazaki fragment are apparently very similar: An RNA-DNA primer is produced by pol α/primase, and the 3'-terminus of the initiator DNA (iDNA) is recognized by RFC and PCNA to expel the pol α/primase and load pol δ (Figure 2; 24–26).

As suggested by the model in Figure 2, pol α/primase starts the synthesis of an RNA-DNA primer on an RPA-coated, single-stranded DNA template, perhaps assisted by a putative cellular loading activity (such as T antigen), to yield a short 30-nt primer RNA-DNA (18, 264, 270). Once the RNA-DNA primer is synthesized, RFC binds to the 3'-end of the iDNA, displacing pol α/primase. The turnover of pol α/primase most likely occurs by the inherent nonprocessive nature of the pol α catalytic activity and the tight binding of RFC to the primer-template junction, since both RPA and RFC decrease the length of the iDNA (25). RFC binding triggers the assembly of the primer recognition complex, which is accomplished through the loading of PCNA and subsequent association of PCNA with pol δ. Then the relatively processive pol δ holoenzyme extends the DNA strand (16, 23–25, 191). For the initiation of leading-strand DNA replication, the synthesis by the pol δ/PCNA complex is then processive and continuous, at least for 5–10 kb of DNA. For synthesis on the lagging strand, DNA synthesis of the Okazaki fragment continues until the polymerase encounters the previously synthesized Okazaki fragment. The RNA primer from the preceding Okazaki fragment is then removed by a complex processing reaction described below, and the remaining nick is sealed by DNA ligase I.

Biochemical studies in vitro (described in this review) and studies in vivo (195) show that pol ε is not required for SV40 DNA replication. But given the enzymatic similarities between pol δ and pol ε, such as high processivity, and the observation that polymerase ε is essential for DNA replication (196, 197), pol ε

almost certainly participates in cellular chromosomal replication. A difference between these two systems is the mechanism of initiation, because many cellular proteins are required to achieve what is achieved by T antigen. Therefore, pol ε might be involved in the initiation of DNA replication. Alternatively, each polymerase might be involved specifically in a separate process during DNA replication. A genetic study utilizing *S. cerevisiae* strains that contain mutations in the proofreading exonuclease domains of either pol δ or pol ε suggests that each polymerase is involved in replicating different strands of the DNA (271).

Polymerase ε is also a unique polymerase in that it is involved in cell-cycle checkpoint control and may therefore function at the DNA replication fork to ensure accurate DNA synthesis, perhaps by a postreplicative repair mechanism or another, as yet unrecognized, mechanism (272, 273). It is also possible that pol ε functions in replicating large chromosomes as a specialized enzyme for initiation of DNA replication at sites where DNA synthesis has halted temporarily because the replication fork has to cope with the complex topology of the cell's chromosomes. Clearly, more studies on the role of pol ε are required.

A key protein in the polymerase switching appears to be RFC, and the loading of PCNA by RFC is an essential event for the transition from a priming mode to an extension mode of DNA synthesis. Given the functional similarities between RFC and the *E. coli* pol III γ-complex (95), it is possible that RFC coordinates the synthesis of both the lagging and leading strands. Further investigation of the mechanism for coordinating DNA synthesis of both strands at the eukaryotic cell DNA replication fork is necessary.

Maturation of Okazaki Fragments

In maturation of Okazaki fragment synthesis, the short Okazaki fragments synthesized discontinuously on the lagging strand template are converted into long, ungapped DNA products. This involves several distinct steps including removal of the RNA primer, DNA gap synthesis, and sealing together of the two DNA. Recent studies, and the observation that many of the proteins involved in this process bind to PCNA, suggest that these steps may be regulated coordinately with each other.

The analysis of lagging-strand DNA synthesis using highly purified proteins has revealed the basis of the maturation mechanism. Two different nucleases, RNase HI and FEN1, are involved in the complete removal of the RNA primer (192, 200, 205, 206, 212–216). These nucleases are required for the complete replication of SV40 DNA and for reconstitution of lagging strand synthesis on artificial templates (26, 192, 200). An in vitro assay with a model substrate showed that PCNA binds to FEN1 and stimulates FEN1 activity (152). This observation suggests that the removal of RNA (or RNA-DNA) might be triggered either by the upstream DNA polymerase complex or by the newly synthesized

DNA, creating a duplex DNA region upstream of the RNA at the 5'-end of the Okazaki fragment. Consistent with the latter possibility, an assay for FEN1 using a synthetic oligonucleotide substrate showed that an upstream DNA can influence the cleavage of a downstream flap substrate (204, 213–215).

Okazaki fragment processing and genome integrity during cell chromosome replication might require additional proteins, such as the Dna2 helicase, to complete the same process. Dna2 helicase has been suggested to be involved in removal of RNA primers because of its biochemical and genetic interactions with FEN1/Rad27 nuclease (243; also see above). Dna2 helicase, in conjunction with the polymerase complex that synthesizes the Okazaki fragments, might displace the RNA primer from the template DNA, thereby creating a flap-like substrate for FEN1 endonuclease (see FEN1 and RNase HI section, Figure 3, and Reference 10). A more interesting possibility is that in addition to the RNA at the 5'-end of an Okazaki fragment, the DNA beyond the RNA-DNA

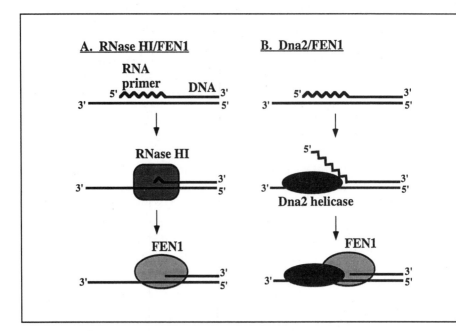

Figure 3 Two mechanisms for the removal of RNA primers. (*A*) RNase HI cleaves the RNA segment attached to the 5'-end of the Okazaki fragment, leaving a single ribonucleotide adjacent to the RNA-DNA junction. FEN1 then removes the remaining ribonucleotide. (*B*) Dna2 helicase displaces the RNA segment (or RNA-DNA). FEN1 then cleaves endonucleolyticaly the branch point, releasing the displaced RNA (or RNA-DNA). Although the Dna2/FEN1-dependent mechanism has not been proved, recent biochemical and genetic studies strongly support this model (see 10).

junction that was synthesized by DNA pol α might also be displaced by Dna2 helicase and cleaved off by FEN1 (10). If this were the case, then the iDNA that was made by pol α would be removed by this Okazaki maturation process, and the gap created would be filled in by either pol δ or pol ε. This would be a significant advantage for cells for maintaining genome integrity because pol α/primase does not have a proofreading activity and thus could not remove any errors that it inserted in the iDNA, whereas if this region of the template DNA were "re-replicated" by pol δ or pol ε, then the proofreading exonucleases from these enzymes would ensure increased accuracy during replication. This mechanism for Okazaki fragment maturation might also prevent inappropriate DNA replication of repeated regions in the genome, such as mini-satellite repeated sequences, a common occurrence in human cancer cells (219, 220).

REPLICATION FORK PROTEINS AND CELL-CYCLE CONTROL

S-Phase Checkpoint Control

When the genome is subjected to excessive DNA damage, or progression of DNA replication forks is blocked, cells arrest progression of the cell cycle at either the G1-S phase transition, the G2-M phase transition, or they slow down S-phase progression. This activates transcription of genes that encode proteins required for the repair of DNA and facilitates the repair process itself (reviewed in 274). The signal transduction mechanisms that detect aberrant replication or DNA damage in the S phase, and then block cell-cycle progression are called S-phase checkpoints. Previous genetic studies in *S. cerevisiae* and *S. pombe* suggest that several replication fork proteins might be involved in these S-phase checkpoints.

In *S. cerevisiae*, mutant cells that have a defect in pol ε (*pol 2*) fail to activate the damage-inducible transcription of certain damage responsive genes in response to DNA damage by methyl methane sulfonate (MMS), or nucleotide-depletion induced by hydroxyurea during the S phase. The mutant cells enter into the M phase without correctly completing DNA replication (272). Thus *POL2* might be involved in an S-phase checkpoint signaling mechanism. The domain in *pol ε* that is responsible for its checkpoint function is separable from its DNA polymerase catalytic domain (272). Recent studies suggest that the Rad53p protein kinase is required for DNA-damage checkpoint signaling by a pol ε- and Rfc5p-dependent mechanism (129, 273, 275).

DPB11, which was isolated as a multicopy suppressor of mutations in genes encoding subunits of *S. cerevisiae* pol ε (*pol2* and *dpb2*), has a checkpoint function (276). A temperature-sensitive mutant in *DPB11* underwent cell division without completion of replication at the nonpermissive temperature, and the mutant cells were sensitive to hydroxyurea, MMS, and UV irradiation. A similar defect was demonstrated in mutants in the cut5$^+$ gene in *S. pombe*, although these mutants were sensitive only to DNA damage and not to nucleotide depletion (277, 278). A recent report showed that the Cut5 protein, which is similar in sequence to Dpb11p from *S. cerevisiae*, binds to the Chk1 protein kinase that is known to control the cyclin-dependent Cdc2 kinase, the principal regulator for M-phase entry (279).

Other replication proteins have been implicated in checkpoint controls. Yeast mutants with defects in genes encoding RPA (*S. pombe* rad11) (280), primase (*S. cerevisiae pri1*) (281), *S. pombe* pol α and pol δ (282–284), and an RFC small subunit (*S. cerevisiae rfc5*) (275, 285) have also been shown to exhibit checkpoint defects. Some of these defects may occur because a replication fork is not established after commitment to cell division, and others may be the result of a real defect in signaling in response to DNA damage or a replication fork block. The primase mutations are particularly interesting because they override a mechanism that slows down S-phase progression in the presence of continuous low doses of a DNA alkylating agent (281). Interestingly, loss of the *MEC1* gene, which is related to the ATM gene that is defective in human ataxia telangiectasia, also causes the same effect (286), suggesting that primase might be part of a signaling mechanism to the Mec1p checkpoint pathway.

PCNA-p21 Interaction

The involvement of replication fork proteins in checkpoint signaling in metazoan species is likely, but additional controls may be imposed on DNA replication in these cells. The identification of proteins that interact specifically with PCNA has led to the emergence of PCNA as a key protein required for the coordinated regulation of replication and other events that take place at the replication fork, such as DNA methylation and DNA repair. In addition, overexpression of PCNA in *S. pombe* results in the delay of entry into the M phase (287), perhaps because PCNA is binding a protein that is essential for this process.

One of the well-characterized PCNA-interacting proteins is the p21 protein, an inhibitor for cyclin-dependent kinase (CDK) (p21 has alternate names of CIP1, WAF1, Sdi1; for review see 288, 289). p21 is induced during mitogenic stimulation of mammalian cells and in response to DNA damage by the tumor suppression protein p53, as well as many other stimuli. The protein binds to and inhibits cyclin-dependent kinase (CDK) activities that are required for G1/S progression, leading to cell-cycle arrest. Thus p21 functions as part of a DNA-damage checkpoint mechanism.

Interest in p21 with respect to the DNA replication fork was triggered by the observation that p21 can form a quaternary complex with a CDK, cyclin, and PCNA in normal cells but not in many transformed cells (162, 290), creating a possible link between cell-cycle progression and DNA replication. p21 inhibits SV40 DNA replication in vitro and in frog cell extracts through its direct binding to PCNA (155, 156, 291, 292). Even though PCNA is essential for nucleotide-excision repair (293, 294), p21 did not inhibit the repair in vitro (295, 296) and in vivo (168). These results suggest that through its binding to both CDK kinase and PCNA, p21 might function to coordinately regulate DNA replication, repair, and cell-cycle progression in response to DNA damage, perhaps inducing a switch from replication to repair.

The biological significance of this interaction is still unclear, even though the p21-PCNA interaction has been characterized extensively, including mapping of the binding domain in p21 (297–303), determination of a crystal structure of PCNA complexed with a p21-derived peptide (164), and investigation of the effect of p21 on DNA synthesis and the loading of PCNA onto DNA (115, 304). A recent observation that p21 modulates the interactions between PCNA and FEN1 endonuclease, DNA (5-cytosine) methyltransferase, and DNA ligase suggests that the levels of p21 in the cell might control a switch from one PCNA-dependent function to another at the DNA replication fork (154, 160, 305). More evidence that the p21-PCNA interaction might be relevant in vivo is provided by the recent observations that the E7 oncoprotein from human papillomavirus type 16 can abrogate a DNA damage–induced cell-cycle arrest by binding to p21 (306, 307). These studies show that the p21-E7 interaction results in both the reversal of p21 inhibition of CDK kinase activity and the p21 inhibition of PCNA function (306).

Regulation of Telomere Length

The appropriate regulation of telomere length and replication is also important for maintaining the integrity of the genome. In addition, the length of telomeres correlates with the replicative potential of cells (for review see 308). Recent studies have indicated that in addition to telomerase, replication fork proteins may directly regulate the length of telomeres. The mutations in the *S. cerevisiae POL1/CDC17* gene encoding polymerase α, as well as the gene encoding the large subunit of RFC (*CDC44* in *S. cerevisiae*), cause elongation of the telomeres (309, 310). Furthermore, pharmacological inhibitors of cellular DNA polymerases disrupt coordinated DNA replication of the G-rich strand that is synthesized by telomerase and the C-rich lagging strand (311). Given that this telomere elongation requires telomerase activity and that both pol α/primase and RFC are involved in lagging strand synthesis, both telomere extension and lagging strand synthesis might be regulated coordinately, in conjunction with the function of *CDC13*, a telomere-sequence-specific, single-stranded DNA-binding protein that is required for telomere maintenance (312). Further biochemical investigation of the mechanism for telomere replication is needed to understand how both strands of the telomere are replicated.

CONCLUDING REMARKS

Modern research on the replication of DNA in eukaryotes can be grouped into three related categories. One active area of research is to characterize the DNA sequences that are important for replicator function and determine where origins of DNA replication occur in chromosomes. Another area is to identify the proteins involved in the mechanism and control of the initiation of DNA replication in eukaryotes. This area is the most active area at the present time, providing interesting links to cell-cycle research, developmental biology, control of gene expression, and chromosome structure and function. A third area is to extend the already remarkable progress in understanding the events at the DNA replication fork, outlined above, to identify other proteins that may cooperate with the known replication fork proteins to ensure accurate and efficient DNA replication. Connections need to be made between the initiation proteins that establish the replication fork and those proteins discussed above that function to replicate DNA after initiation. Proteins that function at the replication fork will likely play significant roles in inheritance of chromosome structures by interacting with other proteins involved in chromatin assembly and epigenetically inherited protein complexes. Clearly much more work needs to be done.

ACKNOWLEDGMENTS

We thank John Kuryan for Figure 1, Dr. Robert Bambara for discussions, and colleagues who provided preprints of research papers. Our work was supported by a grant from the National Cancer Institute (CA13106).

Visit the *Annual Reviews home page* at
http://www.AnnualReviews.org.

Literature Cited

1. Diffley JFX. 1996. *Genes Dev.* 10:2819–30
2. Stillman B. 1996. *Science* 274:1659–64
3. Li JJ, Kelly TJ. 1984. *Proc. Natl. Acad. Sci. USA* 81:6973–77
4. Li JJ, Kelly TJ. 1985. *Mol. Cell. Biol.* 5:1238–46
5. Stillman BW, Gluzman Y. 1985. *Mol. Cell. Biol.* 5:2051–60
6. Wobbe CR, Dean F, Weissbach L, Hurwitz J. 1985. *Proc. Natl. Acad. Sci. USA* 82:5710–14
7. Challberg MD, Kelly TJ. 1989. *Annu. Rev. Biochem.* 58:671–717

8. Stillman B. 1989. *Annu. Rev. Cell. Biol.* 5:197–245
9. Bambara RA, Huang L. 1995. *Prog. Nucleic Acid Res. Mol. Biol.* 51:93–122
10. Bambara RA, Murante RS, Henricksen LA. 1997. *J. Biol. Chem.* 272:4647–50
11. Lindahl T, Barnes DE. 1992. *Annu. Rev. Biochem.* 61:251–81
12. Nash R, Lindahl T. 1996. See Ref. 314, pp. 575–86
13. Andersen AH, Bendixen C, Westergaard O. 1996. See Ref. 314, pp. 587–617
14. Wang TS-F. 1996. See Ref. 314, pp. 461–93
15. Nethanel T, Reisfeld S, Dinter-Gottlieb G, Kaufmann G. 1988. *J. Virol.* 6:2867–73
16. Matsumoto T, Eki T, Hurwitz J. 1990. *Proc. Natl. Acad. Sci. USA* 87:9712–16
17. Nethanel T, Kaufmann G. 1990. *J. Virol.* 64:5912–18
18. Bullock PA, Seo YS, Hurwitz J. 1991. *Mol. Cell. Biol.* 11:2350–61
19. Murakami Y, Eki T, Hurwitz J. 1992. *Proc. Natl. Acad. Sci. USA* 89:952–56
20. Salas M, Miller JT, Leis J, DePamphilis ML. 1996. See Ref. 314, pp. 131–76
21. Prelich G, Stillman B. 1988. *Cell* 53:117–26
22. Lee S-H, Eki T, Hurwitz J. 1989. *Proc. Natl. Acad. Sci. USA* 86:7361–65
23. Weinberg DH, Kelly TJ. 1989. *Proc. Natl. Acad. Sci. USA* 86:9742–46
24. Tsurimoto T, Melendy T, Stillman B. 1990. *Nature* 346:534–39
25. Tsurimoto T, Stillman B. 1991. *J. Biol. Chem.* 266:1961–68
26. Waga S, Stillman B. 1994. *Nature* 369:207–12
27. Murakami Y, Hurwitz J. 1993. *J. Biol. Chem.* 268:11008–17
28. Wang TS-F. 1991. *Annu. Rev. Biochem.* 60:513–52
29. Stadlbauer F, Brueckner A, Rehfuess C, Eckerskorn C, Lottspeich F, et al. 1994. *Eur. J. Biochem.* 222:781–93
30. Brückner A, Stadlbauer F, Guarino LA, Brunahl A, Schneider C, et al. 1995. *Mol. Cell. Biol.* 15:1716–24
31. Longhese MP, Jovine L, Plevani P, Lucchini G. 1993. *Genetics* 133:183–91
32. Copeland WC, Tan X. 1995. *J. Biol. Chem.* 270:3905–13
33. Longhese MP, Fraaschini R, Plevani P, Lucchini G. 1996. *Mol. Cell. Biol.* 16:3235–44
34. Santocanale C, Foiani M, Lucchini G, Plevani P. 1993. *J. Biol. Chem.* 268:1343–48
35. Bakkenist CJ, Cotterill S. 1994. *J. Biol. Chem.* 269:26759–66
36. Collins KL, Russo AAR, Tseng BY, Kelly TJ. 1993. *EMBO J.* 12:4555–66
37. Foiani M, Marini F, Gamba D, Lucchini G, Plevani P. 1994. *Mol. Cell. Biol.* 14:923–33
38. Nasheuer H-P, Moore A, Wahl AF, Wang TS-F. 1991. *J. Biol. Chem.* 266:7893–903
39. Foiani M, Liberi G, Lucchini G, Plevani P. 1995. *Mol. Cell. Biol.* 15:883–91
40. Ferrari M, Lucchini G, Plevani P, Foiani M. 1996. *J. Biol. Chem.* 271:8661–66
41. Voitenleitner C, Fanning E, Nasheuer H-P. 1997. *Oncogene* 14:1611–15
42. Hübscher U, Maga G, Podust VN. 1996. See Ref. 314, pp. 525–43
43. Wold MS. 1997. *Annu. Rev. Biochem.* 66:61–92
44. Wobbe CR, Weissbach L, Borowiec JA, Dean FB, Murakami J, et al. 1987. *Proc. Natl. Acad. Sci. USA* 84:1834–38
45. Fairman MP, Stillman B. 1988. *EMBO J.* 7:1211–18
46. Wold MS, Kelly T. 1988. *Proc. Natl. Acad. Sci. USA* 85:2523–27
47. Sancar A. 1996. *Annu. Rev. Biochem.* 65:43–81
48. Stahl H, Droge P, Knippers R. 1986. *EMBO J.* 5:1939–44
49. Dean FB, Bullock P, Murakami Y, Wobbe CR, Weissbach L, Hurwitz J. 1987. *Proc. Natl. Acad. Sci. USA* 84:16–20
50. Dodson M, Dean FB, Bullock P, Echols H, Hurwitz J. 1987. *Science* 238:964–67
51. Wold MS, Li JJ, Kelly TJ. 1987. *Proc. Natl. Acad. Sci. USA* 84:3643–47
52. Brill SJ, Stillman B. 1989. *Nature* 342:92–95
53. Kenny MK, Lee S-H, Hurwitz J. 1989. *Proc. Natl. Acad. Sci. USA* 86:9757–61
54. Tsurimoto T, Stillman B. 1989. *EMBO J.* 8:3883–89
55. Erdile LF, Heyers W-D, Kolodner R, Kelly TJ. 1991. *J. Biol. Chem.* 266:12090–98
56. Tsurimoto T, Stillman B. 1991. *J. Biol. Chem.* 266:1950–60
57. Braun KA, Lao Y, He Z, Ingles CJ, Wold MS. 1997. *Biochemistry* 36:8443–54
58. Longhese MP, Plevani P, Lucchini G. 1994. *Mol. Cell. Biol.* 14:7884–90
59. Dornreiter I, Erdile LF, Gilbert IU, von Winkler D, Kelly TJ, Fanning E. 1992. *EMBO J.* 11:769–76
60. Melendy T, Stillman B. 1993. *J. Biol. Chem.* 268:3389–95

61. Ishiai M, Sanchez JP, Amin AA, Murakami Y, Hurwitz J. 1996. *J. Biol. Chem.* 271:20868–78
62. Henricksen LA, Umbricht CB, Wold MS. 1994. *J. Biol. Chem.* 269:11121–32
63. He Z, Wong JMS, Maniar HS, Brill SJ, Ingles CJ. 1996. *J. Biol. Chem.* 271:28243–49
64. Stigger E, Dean FB, Hurwitz J, Lee S-H. 1994. *Proc. Natl. Acad. Sci. USA* 91:579–83
65. Gomes XV, Wold MS. 1995. *J. Biol. Chem.* 270:4534–43
66. Kim DK, Stigger E, Lee S-H. 1996. *J. Biol. Chem.* 271:15124–29
67. Lin Y-L, Chen C, Keshav KF, Winchester E, Dutta A. 1996. *J. Biol. Chem.* 271:17190–98
68. Philipova D, Mullen JR, Maniar HS, Lu JA, Gu CY, Brill SJ. 1996. *Genes Dev.* 10:2222–33
69. Bochkarev A, Pfuetzner RA, Edwards AM, Frappier L. 1997. *Nature* 385:176–81
70. Blackwell LJ, Borowiec JA. 1994. *Mol. Cell. Biol.* 14:3993–4001
71. Kim C, Paulus BF, Wold MS. 1994. *Biochemistry* 33:14197–206
72. Mitas M, Chock JY, Christy M. 1997. *Biochem. J.* 324:957–61
73. Din S, Brill S, Fairman MP, Stillman B. 1990. *Genes Dev.* 4:968–77
74. Liu VF, Weaver DT. 1993. *Mol. Cell. Biol.* 13:7222–31
75. Carty MP, Zernik-Kobak M, McGrath S, Dixon K. 1994. *EMBO J.* 13:2114–23
76. Dutta A, Stillman B. 1992. *EMBO J.* 11:2189–99
77. Fotedar R, Roberts JM. 1992. *EMBO J.* 11:2177–87
78. Brush GS, Anderson CW, Kelly TJ. 1994. *Proc. Natl. Acad. Sci. USA* 91:12520–24
79. Pan Z-Q, Amin AA, Gibbs E, Niu H, Hurwitz J. 1994. *Proc. Natl. Acad. Sci. USA* 91:8343–47
80. Boubnov NV, Weaver DT. 1995. *Mol. Cell. Biol.* 15:5700–6
81. Fried LM, Koumenis C, Peterson SR, Green SL, van Zijl P, et al. 1996. *Proc. Natl. Acad. Sci. USA* 93:13825–30
82. Gibbs E, Pan Z-Q, Niu H, Hurwitz J. 1996. *J. Biol. Chem.* 271:22847–54
83. Niu H, Erdjument-Bromage H, Pan Z-Q, Lee S-H, Tempst P, Hurwitz J. 1997. *J. Biol. Chem.* 272:12634–41
84. Zernik-Kobak M, Vasunia K, Connelly M, Anderson CW, Dixon K. 1997. *J. Biol. Chem.* 272:23896–904
85. Brush GS, Morrow DM, Hieter P, Kelly TJ. 1996. *Proc. Natl. Acad. Sci. USA* 93:15075–80
86. Cheng XB, Cheong N, Wang Y, Iliakis G. 1996. *Radiother. Oncol.* 39:43–52
87. Abramova NA, Russell J, Botchan M, Li R. 1997. *Proc. Natl. Acad. Sci. USA* 94:7186–91
88. Henricksen LA, Wold MS. 1994. *J. Biol. Chem.* 269:24203–8
89. Pan Z-Q, Park CH, Amin AA, Hurwitz J, Sancar A. 1995. *Proc. Natl. Acad. Sci. USA* 92:4636–40
90. Henricksen LA, Carter T, Dutta A, Wold MS. 1996. *Nucleic Acids Res.* 24:3107–12
91. Morgan SE, Kastan MB. 1997. *Cancer Res.* 57:3386–89
92. Adachi Y, Laemmli U. 1992. *J. Cell Biol.* 119:1–16
93. Yan H, Newport J. 1995. *Science* 269:1883–85
94. Coleman TR, Carpenter PB, Dunphy WG. 1996. *Cell* 87:53–63
95. Stillman B. 1996. See Ref. 314, pp. 435–60
96. Chen M, Pan Z-Q, Hurwitz J. 1992. *Proc. Natl. Acad. Sci. USA* 89:5211–15
97. Chen M, Pan Z-Q, Hurwitz J. 1992. *Proc. Natl. Acad. Sci. USA* 89:2516–20
98. Bunz F, Kobayashi R, Stillman B. 1993. *Proc. Natl. Acad. Sci. USA* 90:11014–18
99. Burbelo PD, Utani A, Pan Z-Q, Yamada Y. 1993. *Proc. Natl. Acad. Sci. USA* 90:11543–47
100. Howell EA, McAlear MA, Rose D, Holm C. 1994. *Mol. Cell. Biol.* 14:255–67
101. Li XY, Burgers PMJ. 1994. *Proc. Natl. Acad. Sci. USA* 91:868–72
102. Li XY, Burgers PMJ. 1994. *J. Biol. Chem.* 269:21880–84
103. Noskov V, Maki S, Kawasaki Y, Leem S-H, Ono B-I, et al. 1994. *Nucleic Acids Res.* 22:1527–35
104. Cullmann G, Fien K, Kobayashi R, Stillman B. 1995. *Mol. Cell. Biol.* 15:4661–71
105. Bork P, Hofmann K, Bucher P, Neuwald AF, Altschul SF, Koonin EV. 1997. *FASEB J.* 11:68–76
106. O'Donnell M, Onrust R, Dean FB, Chen M, Hurwitz J. 1993. *Nucleic Acids Res.* 21:1–3
107. Tsurimoto T, Stillman B. 1989. *Mol. Cell. Biol.* 9:609–19
108. Tsurimoto T, Stillman B. 1990. *Proc. Natl. Acad. Sci. USA* 87:1023–27

109. Lee S-H, Kwong AD, Pan Z-Q, Hurwitz J. 1991. *J. Biol. Chem.* 266:594–602
110. Yoder BL, Burgers PMJ. 1991. *J. Biol. Chem.* 266:22689–97
111. Fien K, Stillman B. 1992. *Mol. Cell. Biol.* 12:155–63
112. Podust VN, Georgaki A, Strack B, Hübscher U. 1992. *Nucleic Acids Res.* 20:4159–65
113. Burgers PMJ. 1991. *J. Biol. Chem.* 266:22698–706
114. Cai JS, Uhlmann F, Gibbs E, Flores-Rozas H, Lee C-G, et al. 1996. *Proc. Natl. Acad. Sci. USA* 93:12896–901
115. Podust VN, Podust LM, Goubin F, Ducommun B, Hubscher U. 1995. *Biochemistry* 34:8869–75
116. Lee S-H, Hurwitz J. 1990. *Proc. Natl. Acad. Sci. USA* 87:5672–76
117. Podust VN, Fanning E. 1997. *J. Biol. Chem.* 272:6303–10
118. Ellison V, Stillman B. 1998. *J. Biol. Chem.* 273:5979–87
119. Gerik KJ, Gary SL, Burgers PMJ. 1997. *J. Biol. Chem.* 272:1256–62
120. Fotedar R, Mossi R, Fitzgerald P, Rousselle T, Maga G, et al. 1996. *EMBO J.* 15:4423–33
121. Mossi R, Jonsson Z, Allen BL, Hardin SH, Hubscher U. 1997. *J. Biol. Chem.* 272:1769–76
122. Uhlmann F, Cai JS, Gibbs E, O'Donnell M, Hurwitz J. 1997. *J. Biol. Chem.* 272:10058–64
123. Rheaume E, Cohen LY, Uhlmann F, Lazure C, Alam A, et al. 1997. *EMBO J.* 16:6346–54
124. Song QZ, Lu H, Zhang N, Luckow B, Shah G, et al. 1997. *Biochem. Biophys. Res. Commun.* 233:343–48
125. Uhlmann F, Gibbs E, Cai JS, O'Donnell M, Hurwitz J. 1997. *J. Biol. Chem.* 272:10065–71
126. Uhlmann F, Cai JS, Flores-Rozas H, Dean FB, Finkelstein J, et al. 1996. *Proc. Natl. Acad. Sci.* 93:6521–26
127. Cai JS, Gibbs E, Uhlmann F, Phillips B, Yao N, et al. 1997. *J. Biol. Chem.* 272:18974–81
128. McAlear MA, Howell EA, Espenshade KK, Holm C. 1994. *Mol. Cell. Biol.* 14:4390–97
129. Sugimoto K, Ando S, Shimomura T, Matsumoto K. 1997. *Mol. Cell. Biol.* 17:5905–14
130. Guenther B, Onrust R, Sali A, O'Donnell M, Kuriyan J. 1997. *Cell* 91:335–45
131. Jonsson ZO, Hübscher U. 1997. *BioEssays* 19:967–75
132. Kelman Z. 1997. *Oncogene* 14:629–40
133. Kong X-P, Onrust R, O'Donnell M, Kuriyan J. 1992. *Cell* 69:425–37
134. Krishna TSR, Kong X-P, Gary S, Burgers PM, Kuriyan J. 1994. *Cell* 79:1233–43
135. Kelman Z, O'Donnell M. 1995. *Annu. Rev. Biochem.* 64:171–200
136. Tinker RL, Kassavetis GA, Geiduschek EP. 1994. *EMBO J.* 13:5330–37
137. Hacker KJ, Alberts BM. 1994. *J. Biol. Chem.* 269:24209–20
138. Hacker KJ, Alberts BM. 1994. *J. Biol. Chem.* 269:24221–28
139. Stukenberg PT, Turner J, O'Donnell M. 1994. *Cell* 78:877–87
140. Tan CK, Castillo C, So AG, Downey KM. 1986. *J. Biol. Chem.* 261:12310–16
141. Bravo R, Frank R, Blundell PA, MacDonald-Bravo H. 1987. *Nature* 326:515–17
142. Prelich G, Kostura M, Marshak DR, Mathews MB, Stillman B. 1987. *Nature* 326:471–75
143. Prelich G, Tan CK, Kostura M, Mathews MB, So AG, et al. 1987. *Nature* 326:517–20
144. Bauer GA, Burgers PMJ. 1988. *Proc. Natl. Acad. Sci. USA* 85:7506–10
145. Hamatake RK, Hasegawa H, Clark AB, Bebenek K, Kunkel TA, Sugino A. 1990. *J. Biol. Chem.* 265:4072–83
146. Lee S-H, Zhen-Qiang P, Kwong AD, Burgers PMJ, Hurwitz J. 1991. *J. Biol. Chem.* 266:22707–17
147. Chui G, Linn S. 1995. *J. Biol. Chem.* 270:7799–808
148. Ayyagari R, Impellizzeri KJ, Yoder BL, Gary SL, Burgers PMJ. 1995. *Mol. Cell. Biol.* 15:4420–29
149. Fukuda K, Morioka H, Imajou S, Ikeda S, Ohtsuka E, Tsurimoto T. 1995. *J. Biol. Chem.* 270:22527–34
150. Arroyo MP, Downey KM, So AG, Wang TS-F. 1996. *J. Biol. Chem.* 271:15971–80
151. Eissenberg JC, Ayyagari R, Gomes XV, Burgers PM. 1997. *Mol. Cell. Biol.* 17:6367–78
152. Li XY, Li J, Harrington J, Lieber MR, Burgers PMJ. 1995. *J. Biol. Chem.* 270:22109–12
153. Chen JJ, Chen S, Saha P, Dutta A. 1996. *Proc. Natl. Acad. Sci. USA* 93:11597–602
154. Levin DS, Bai W, Yao N, O'Donnell M, Tomkinson AE. 1997. *Proc. Natl. Acad. Sci. USA* 94:12863–68
155. Flores-Rozas H, Kelman Z, Dean FB,

Pan Z-Q, Harper JW, et al. 1994. *Proc. Natl. Acad. Sci. USA* 91:8655–59

156. Waga S, Hannon GJ, Beach D, Stillman B. 1994. *Nature* 369:574–78

157. Smith ML, Chen I-T, Zhan QM, Bae IS, Chen C-Y, et al. 1994. *Science* 266: 1376–80

158. Hall PA, Kearsey JM, Coates PJ, Norman DG, Warbrick E, Cox LS. 1995. *Oncogene* 10:2427–33

159. Gary R, Ludwig DL, Cornelius HL, MacInnes MA, Park MS. 1997. *J. Biol. Chem.* 272:24522–29

160. Chuang LS-H, Ian H-I, Koh T-W, Ng H-H, Xu GL, Li BFL. 1997. *Science* 277: 1996–2000

161. Umar A, Buermeyer AB, Simon JA, Thomas DC, Clark AB, et al. 1996. *Cell* 87:65–73

162. Xiong Y, Zhang H, Beach D. 1992. *Cell* 71:505–14

163. Matsuoka S, Yamaguchi M, Matsukage A. 1994. *J. Biol. Chem.* 269:11030–36

164. Gulbis JM, Kelman Z, Hurwitz J, O'Donnell M, Kuriyan J. 1996. *Cell* 87: 297–306

165. Roos G, Jiang Y, Landberg G, Nielsen NH, Zhang P, Lee MYWT. 1996. *Exp. Cell Res.* 226:208–13

166. Bravo R, MacDonald-Bravo H. 1987. *J. Cell Biol.* 105:1549–54

167. Humbert C, Santisteban MS, Usson Y, Robert-Nicoud M. 1992. *J. Cell Sci.* 103:97–103

168. Li R, Hannon GJ, Beach D, Stillman B. 1996. *Curr. Biol.* 6:189–99

169. Celis JE, Madsen P. 1986. *FEBS Lett.* 209:277–83

170. Toschi L, Bravo R. 1988. *J. Cell Biol.* 107:1623–28

171. Miura M, Domon M, Sasaki T, Kondo S, Takasaki Y. 1992. *Exp. Cell Res.* 201: 541–44

172. Pagano M, Theodoras AM, Tam SW, Draetta GF. 1994. *Genes Dev.* 8:1627–39

173. Hughes DA, MacNeill SA, Fantes PA. 1992. *Mol. Gen. Genet.* 231:401–10

174. MacNeill SA, Moreno S, Reynolds N, Nurse P, Fantes PA. 1996. *EMBO J.* 15: 4613–28

175. Zuo S, Gibbs E, Kelman Z, Wang TS-F, O'Donnell M, et al. 1997. *Proc. Natl. Acad. Sci. USA* 94:11244–49

176. Boulet A, Simon M, Faye G, Bauer GA, Burgers PMJ. 1989. *EMBO J.* 8:1849–54

177. Pignede G, Bouvier D, Recondo A-M, Baldacci G. 1991. *J. Mol. Biol.* 222:209–18

178. Zhang J, Chung DW, Tan C-K, Downey

KM, Davie EW, So AG. 1991. *Biochemistry* 30:11742–50

179. Yang C-L, Chang L-S, Zhang P, Hao H, Zhu L, et al. 1992. *Nucleic Acids Res.* 20:735–45

180. Cullmann G, Hindges R, Berchtold MW, Hübscher U. 1993. *Gene* 134:191–200

181. Park H, Francesconi S, Wang TS-F. 1993. *Mol. Biol. Cell* 4:145–57

182. Sugimoto K, Sakamoto Y, Takahashi O, Matsumoto K. 1995. *Nucleic Acids Res.* 23:3493–500

183. Zhang J, Tan C-K, McMullen B, Downey KM, So AG. 1995. *Genomics* 29:179–86

184. Zhou J-Q, He H, Tan C-K, Downey KM, So AG. 1997. *Nucleic Acids Res.* 25:1094–99

185. Brown WC, Campbell JL. 1993. *J. Biol. Chem.* 268:21706–10

186. Zhang S-J, Zeng X-R, Zhang P, Toomey NL, Chuang R-Y, et al. 1995. *J. Biol. Chem.* 270:7988–92

187. Goulian M, Herrmann SM, Sackett JW, Grimm SL. 1990. *J. Biol. Chem.* 265: 16402–11

188. Chiang C-S, Mitsis PG, Lehman IR. 1993. *Proc. Natl. Acad. Sci. USA* 90: 9105–9

189. Melendy T, Stillman B. 1991. *J. Biol. Chem.* 266:1942–49

190. Weinberg DH, Collins KL, Simancek P, Russo A, Old MS, et al. 1990. *Proc. Natl. Acad. Sci. USA* 87:8692–96

191. Eki T, Matsumoto T, Murakami Y, Hurwitz J. 1992. *J. Biol. Chem.* 267:7284–94

192. Waga S, Bauer G, Stillman B. 1994. *J. Biol. Chem.* 269:10923–34

193. Maga G, Hübscher U. 1996. *Biochemistry* 35:5764–77

194. McConnell M, Miller H, Mozzherin DJ, Quamina A, Tan C-K, et al. 1996. *Biochemistry* 35:8268–74

195. Zlotkin T, Kaufmann G, Jiang Y, Lee MYWT, Uitto L, et al. 1996. *EMBO J.* 15:2298–305

196. Morrison A, Araki H, Clark B, Hamatake RK, Sugino A. 1990. *Cell* 62:1143–51

197. Araki H, Ropp PA, Johnson AL, Johnston LH, Morrison A, Sugino A. 1992. *EMBO J.* 11:733–40

198. Budd ME, Campbell JL. 1993. *Mol. Cell. Biol.* 13:496–505

199. Podust VN, Hübscher U. 1993. *Nucleic Acids Res.* 21:841–46

200. Turchi JJ, Huang L, Murante RS, Kim Y, Bambara RA. 1994. *Proc. Natl. Acad. Sci. USA* 91:9803–7

201. Lieber MR. 1997. *BioEssays* 19:233–40
202. Harrington JJ, Lieber MR. 1994. *Genes Dev.* 8:1344–55
203. Murray JM, Tavassoli M, Al-Harithy R, Sheldrick KS, Lehmann AR, et al. 1994. *Mol. Cell. Biol.* 14:4878–88
204. Harrington JJ, Lieber MR. 1994. *EMBO J.* 13:1235–46
205. Ishimi Y, Claude A, Bullock P, Hurwitz J. 1988. *J. Biol. Chem.* 263:19723–33
206. Goulian M, Richards SH, Heard CJ, Bigsby BM. 1990. *J. Biol. Chem.* 265:18461–71
207. Robins P, Pappin DJC, Wood RD, Lindahl T. 1994. *J. Biol. Chem.* 269:28535–38
208. Siegel G, Turchi J, Myers T, Bambara R. 1992. *Proc. Natl. Acad. Sci. USA* 89:9377–81
209. Kenny MK, Balogh LA, Hurwitz J. 1988. *J. Biol. Chem.* 263:9801–8
210. Reagan MS, Pittenger C, Siede W, Friedberg EC. 1995. *J. Bacteriol.* 177:364–71
211. Sommers CH, Miller EJ, Dujon B, Prakash S, Prakash L. 1995. *J. Biol. Chem.* 270:4193–96
212. Turchi JJ, Bambara RA. 1993. *J. Biol. Chem.* 268:15136–41
213. Murante RS, Huang L, Turchi JJ, Bambara RA. 1994. *J. Biol. Chem.* 269:1191–96
214. Huang L, Rumbaugh JA, Murante RS, Lin RJR, Rust L, Bambara RA. 1996. *Biochemistry* 35:9266–77
215. Murante RS, Rumbaugh JA, Barnes CJ, Norton JR, Bambara RA. 1996. *J. Biol. Chem.* 271:25888–97
216. Murante RS, Rust L, Bambara RA. 1995. *J. Biol. Chem.* 270:30377–83
217. Biswas EE, Zhu FZ, Biswas SB. 1997. *Biochemistry* 36:5955–62
218. Johnson RE, Kovvali GK, Prakash L, Prakash S. 1995. *Science* 269:238–40
219. Tishkoff DX, Filosi N, Gaida GM, Kolodner RD. 1997. *Cell* 88:253–63
220. Gordenin DA, Kunkel TA, Resnick MA. 1997. *Nat. Genet.* 16:116–18
221. Kunkel TA, Resnick MA, Gordenin DA. 1997. *Cell* 88:155–58
222. Cathala G, Rech J, Huet J, Jeanteur P. 1979. *J. Biol. Chem.* 254:7353–61
223. Büsen W. 1980. *J. Biol. Chem.* 255:9434–43
224. DiFrancesco RA, Lehman IR. 1985. *J. Biol. Chem.* 260:14764–70
225. Hagemeier A, Grosse F. 1989. *Eur. J. Biochem.* 185:621–28
226. Eder PS, Walder JA. 1991. *J. Biol. Chem.* 266:6472–79
227. Huang L, Kim Y, Turchi JJ, Bambara RA. 1994. *J. Biol. Chem.* 269:25922–27
228. Rumbaugh JA, Murante RS, Shi S, Bambara RA. 1997. *J. Biol. Chem.* 272:22591–99
229. Borowiec JA. 1996. See Ref. 314, pp. 545–74
230. Wessel R, Schweizer J, Stahl H. 1992. *J. Virol.* 66:804–15
231. Kuo C-L, Huang C-H, Campbell JL. 1983. *Proc. Natl. Acad. Sci. USA* 80:6465–69
232. Budd ME, Campbell JL. 1995. *Proc. Natl. Acad. Sci. USA* 92:7642–46
233. Budd ME, Choe W-C, Campbell JL. 1995. *J. Biol. Chem.* 270:26766–69
234. Matsumoto K, Seki M, Masutani C, Tada S, Enomoto T, Ishimi Y. 1995. *Biochemistry* 34:7913–22
235. Saitoh A, Tada S, Katada T, Enomoto T. 1995. *Nucleic Acids Res.* 23:2014–18
236. Seki M, Kohda T, Yano T, Tada S, Yanagisawa J, et al. 1995. *Mol. Cell. Biol.* 15:165–72
237. Chong JPJ, Thommes P, Blow JJ. 1996. *Trends Biochem. Sci.* 21:102–6
238. Kearsey SE, Maiorano D, Holmes EC, Todorov IT. 1996. *BioEssays* 18:183–90
239. Koonin EV. 1993. *Nucleic Acids Res.* 21:2541–47
239a. Adachi Y, Usukura J, Yanagida M. 1997. *Genes Cells* 2:467–79
240. Aparicio OM, Weinstein DM, Bell SP. 1997. *Cell* 91:59–69
241. Ishimi Y. 1997. *J. Biol. Chem.* 272:24508–13
242. Tom TD, Malkas LH, Hickey RJ. 1996. *J. Cell. Biochem.* 63:259–67
243. Budd ME, Campbell JL. 1997. *Mol. Cell. Biol.* 17:2136–42
244. Wittmeyer J, Formosa T. 1997. *Mol. Cell. Biol.* 17:4178–90
245. Kouprina N, Kroll E, Bannikov V, Bliskovsky V, Gizatullin R, et al. 1992. *Mol. Cell. Biol.* 12:5736–47
246. Miles J, Formosa T. 1992. *Mol. Cell. Biol.* 12:5724–35
247. Miles J, Formosa T. 1992. *Proc. Natl. Acad. Sci. USA* 89:1276–80
248. Prendergast JA, Murray LE, Rowley A, Carruthers DR, Singer RA, Johnston GC. 1990. *Genetics* 124:81–90
249. Malone EA, Clark CD, Chiang A, Winston F. 1991. *Mol. Cell. Biol.* 11:5710–17
250. Rowley A, Singer RA, Johnston GC. 1991. *Mol. Cell. Biol.* 11:5718–26

251. Vallen EA, Cross FR. 1995. *Mol. Cell. Biol.* 15:4291–302
252. Fanning E, Knippers R. 1992. *Annu. Rev. Biochem.* 61:55–85
253. Deleted in proof
254. Murakami Y, Hurwitz J. 1993. *J. Biol. Chem.* 268:11018–27
255. Smale ST, Tjian R. 1986. *Mol. Cell. Biol.* 6:4077–87
256. Gannon JV, Lane DP. 1987. *Nature* 329:456–58
257. Dornreiter I, Höss A, Arthur AK, Fanning E. 1990. *EMBO J.* 9:3329–36
258. Dornreiter I, Copeland WC, Wang TS-F. 1993. *Mol. Cell. Biol.* 13:809–20
259. Park P, Copeland W, Yang L, Wang T, Botchan MR, Mohr IJ. 1994. *Proc. Natl. Acad. Sci. USA* 91:8700–4
260. Collins KL, Kelly TJ. 1991. *Mol. Cell. Biol.* 11:2108–15
261. Schneider C, Weibhart K, Guarino LA, Dornreiter I, Fanning E. 1994. *Mol. Cell. Biol.* 14:3176–85
262. Stadlbauer F, Voitenleitner C, Bruckner A, Fanning E, Nasheuer H-P. 1996. *Mol. Cell. Biol.* 16:94–104
263. Denis D, Bullock PA. 1993. *Mol. Cell. Biol.* 13:2882–90
264. Bullock PA, Tevosian S, Jones C, Denis D. 1994. *Mol. Cell. Biol.* 14:5043–55
265. Bullock PA, Denis D. 1995. *Mol. Cell. Biol.* 15:173–78
266. Goulian M, Heard CJ. 1990. *J. Biol. Chem.* 265:13231–39
267. Goulian M, Heard CJ, Grimm SL. 1990. *J. Biol. Chem.* 265:13221–30
268. He ZG, Brinton BT, Greenblatt J, Hassell JA, Ingles CJ. 1993. *Cell* 73:1223–32
269. Li R, Botchan MR. 1993. *Cell* 73:1207–21
270. Eliasson R, Reichard P. 1978. *J. Biol. Chem.* 253:7469–75
271. Shcherbakova PV, Pavlov YI. 1996. *Genetics* 142:717–26
272. Navas TA, Zhou Z, Elledge SJ. 1995. *Cell* 80:29–39
273. Navas TA, Sanchez Y, Elledge SJ. 1996. *Genes Dev.* 10:2632–43
274. Elledge SJ. 1996. *Science* 274:1664–72
275. Sugimoto K, Shimomura T, Hashimoto K, Araki H, Sugino A, Matsumoto K. 1996. *Proc. Natl. Acad. Sci. USA* 93:7048–52
276. Araki H, Leem S-H, Phongdara A, Sugino A. 1995. *Proc. Natl. Acad. Sci. USA* 92:11791–95
277. Saka Y, Yanagida M. 1993. *Cell* 74:383–93
278. Saka Y, Fantes P, Sutani T, Mclnerny C,

Creanor J, Yanagida M. 1994. *EMBO J.* 13:5319–29
279. Saka Y, Esashi F, Matsusaka T, Mochida S, Yanagida M. 1997. *Genes Dev.* 11:3387–400
280. Parker AE, Clyne RK, Carr AM, Kelly TJ. 1997. *Mol. Cell. Biol.* 17:2381–90
281. Marini F, Pellicioli A, Paciotti V, Lucchini G, Plevani P, et al. 1997. *EMBO J.* 16:639–50
282. D'Urso G, Grallert B, Nurse P. 1995. *J. Cell Sci.* 108:3109–18
283. Francesconi S, De Recondo A-M, Baldacci G. 1995. *Mol. Gen. Genet.* 246:561–69
284. Murakami H, Okayama H. 1995. *Nature* 374:817–19
285. Sugimoto K, Ando S, Shimomura T, Matsumoto K. 1997. *Mol. Cell. Biol.* 17:5905–14
286. Paulovich AG, Hartwell LH. 1995. *Cell* 82:841–47
287. Waseem NH, Labib K, Nurse P, Lane DP. 1992. *EMBO J.* 11:5111–20
288. Sherr CJ, Roberts JM. 1995. *Genes Dev.* 9:1149–63
289. Sherr CJ. 1996. *Science* 274:1672–77
290. Xiong Y, Zhang H, Beach D. 1993. *Genes Dev.* 7:1572–83
291. Strausfeld UP, Howell M, Rempel R, Maller JL, Hunt T, Blow JJ. 1994. *Curr. Biol.* 4:876–83
292. Jackson PK, Chevalier S, Philippe M, Kirschner MW. 1995. *J. Cell Biol.* 130:755–69
293. Nichols AF, Sancar A. 1992. *Nucleic Acids Res.* 20:2441–46
294. Shivji MKK, Kenny MK, Wood RD. 1992. *Cell* 69:367–74
295. Li R, Waga S, Hannon GJ, Beach D, Stillman B. 1994. *Nature* 371:534–37
296. Shivji MKK, Grey SJ, Strausfeld UP, Wood RD, Blow JJ. 1994. *Curr. Biol.* 4:1062–68
297. Chen J, Jackson PK, Kirschner MW, Dutta A. 1995. *Nature* 374:386–88
298. Goubin F, Ducommun B. 1995. *Oncogene* 10:2281–87
299. Luo Y, Hurwitz J, Massague J. 1995. *Nature* 375:159–61
300. Nakanishi M, Robetorye RS, Pereira-Smith OM, Smith JR. 1995. *J. Biol. Chem.* 270:17060–63
301. Warbrick E, Lane DP, Glover DM, Cox LS. 1995. *Curr. Biol.* 5:275–82
302. Chen I-T, Akamatsu M, Smith ML, Lung F-DT, Duba D, et al. 1996. *Oncogene* 12:595–607
303. Chen J, Saha P, Kornbluth S, Dynlacht BD, Dutta A. 1996. *Mol. Cell. Biol.* 16:4673–82

304. Gibbs E, Kelman Z, Gulbis JM, O'Donnell M, Kuriyan J, et al. 1997. *J. Biol. Chem.* 272:2373–81
305. Warbrick E, Lane DP, Glover DM, Cox LS. 1997. *Oncogene* 14:2313–21
306. Funk JO, Waga S, Harry JB, Espling E, Stillman B, Galloway DA. 1997. *Genes Dev.* 11:2090–100
307. Jones DL, Alani RM, Münger K. 1997. *Genes Dev.* 11:2101–11
308. Greider CW, Collins K, Autexier C. 1996. See Ref. 314, pp. 619–38
309. Carson MJ, Hartwell L. 1985. *Cell* 42:249–57
310. Adams AK, Holm C. 1996. *Mol. Cell. Biol.* 16:4614–20
311. Fan X, Price CM. 1997. *Mol. Biol. Cell* 8:2145–55
312. Nugent CI, Hughes TR, Lue NF, Lundblad V. 1996. *Science* 274:249–52
313. Stillman B. 1994. *Cell* 78:725–28
314. DePamphilis ML, ed. 1996. *DNA Replication in Eukaryotic Cells.* Cold Spring Harbor, NY: Cold Spring Harbor Lab

Annu. Rev. Biochem. 1998. 67:753–91

TGF-β SIGNAL TRANSDUCTION

J. Massagué

Cell Biology Program and Howard Hughes Medical Institute, Memorial
Sloan-Kettering Cancer Center, New York, New York 10021;
e-mail: j-massague@ski.mskcc.org

KEY WORDS: TGF-β, receptor serine/threonine kinases, SMADs, growth factors, development,
            cancer

ABSTRACT

The transforming growth factor β (TGF-β) family of growth factors control the
development and homeostasis of most tissues in metazoan organisms. Work over
the past few years has led to the elucidation of a TGF-β signal transduction network. This network involves receptor serine/threonine kinases at the cell surface
and their substrates, the SMAD proteins, which move into the nucleus, where they
activate target gene transcription in association with DNA-binding partners. Distinct repertoires of receptors, SMAD proteins, and DNA-binding partners seemingly underlie, in a cell-specific manner, the multifunctional nature of TGF-β
and related factors. Mutations in these pathways are the cause of various forms
of human cancer and developmental disorders.

CONTENTS

753

0066-4154/98/0701-000753$08.00

INTRODUCTION

The transforming growth factor β (TGF-β) family comprises a large number of structurally related polypeptide growth factors, each capable of regulating a fascinating array of cellular processes including cell proliferation, lineage determination, differentiation, motility, adhesion, and death. Expressed in complex temporal and tissue-specific patterns, TGF-β and related factors play a prominent role in the development, homeostasis, and repair of virtually all tissues in organisms, from fruitfly to human. Collectively, these factors account for a substantial portion of the intercellular signals governing cell fate.

TGF-β and related factors are multifunctional agonists whose effects depend on the state of responsiveness of the target cell as much as on the factors themselves. Given this multifunctional nature, it is not surprising, in retrospect, that the gradual discovery of these factors over the past 15 years has been made through very disparate lines of investigation. For example, the founding member of the family, TGF-β1, was identified as a regulator of mesenchymal growth and, separately, as an antimitogen in epithelial cells (see Table 1 for references). Activins were identified as endocrine regulators of pituitary function and, independently, as inducers of mesoderm in frogs. Bone morphogenetic proteins (BMPs) were identified as bone repair factors and, independently, as dorsalizing agents in *Drosophila*.

A listing of the current members of the TGF-β family and their most representative activities is presented in Table 1 along with citations of articles that review in depth the discovery and biology of these factors. Based on sequence comparisons between the bioactive domains, the TGF-β family can be ordered around a subfamily that includes mammalian BMP2 and BMP4 and their close homologue from *Drosophila*, Dpp. All other known family members progressively diverge from this group, starting with the BMP5 subfamily, followed by the GDF5

Table 1   The transforming growth factor β (TGF-β) family and representative activities[a]

Names [Homologues]	%	Representative activities (References)
BMP2 subfamily		
BMP2 [Dpp[D]]	100	Gastrulation, neurogenesis, chondrogenesis, interdigital
BMP4	92	apoptosis; in frog: mesoderm patterning; in fly: dorsalization, eyes, wings. (1–3)
BMP5 subfamily		
BMP5 [60 A[D]]	61	Along with BMPs 2 and 4, this subfamily participates in the
BMP6/Vgr1	61	development of nearly all organs; many roles
BMP7/OP1	60	in neurogenesis. (1, 2)
BMP8/OP2	55	
GDF5 subfamily		
GDF5/CDMP1	57	Chondrogenesis in developing limbs. (1, 4)
GDF6/CDMP2	54	
GDF7	57	
Vg1 subfamily		
GDF1 [Vg1[X]]	42	Vg1: axial mesoderm induction in frog and fish. (4)
GDF3/Vgr2	53	
BMP3 subfamily		
BMP3/osteogenin	48	Osteogenic differentiation, endochondral bone formation,
GDF10	46	monocyte chemotaxis. (5)
Intermediate members		
Nodal [Xnr 1 to 3[X]]	42	Axial mesoderm induction, left-right asymmetry. (1, 6)
Dorsalin	40	Regulation of cell differentiation within the neural tube. (7)
GDF8	41	Inhibition of skeletal muscle growth. (8)
GDF9	34	
Activin subfamily		
Activin βA	42	Pituitary follicle-stimulating hormone (FSH) production,
Activin βB	42	erythroid cell differentiation; in frog, mesoderm
Activin βC	37	induction. (3, 9, 10)
Activin βE	40	
TGF-β subfamily		
TGF-β1	35	Cell cycle arrest in epithelial and hematopoietic cells, control of
TGF-β2	34	mesenchymal cell proliferation and differentiation, wound
TGF-β3	36	healing, extracellular matrix production, immunosuppression. (11–14)
Distant members		
MIS/AMH	27	Müllerian duct regression. (15, 16)
Inhibin α	22	Inhibition of FSH production and other actions of activin. (9, 10)
GDNF	23	Dopaminergic neuron survival, kidney development. (17)

[a]All members listed have been identified in human and/or mouse. In *brackets*, important homologues from *Drosophila* ([D]) and *Xenopus* ([X]). %, percent of amino acid identity with human bone morphogenetic protein (BMP)2 over the mature polypeptide domain. GDF, growth and differentiation factor. CDMP, cartilage-derived morphogenetic protein. MIS/AMH, Müllerian inhibiting substance/anti-Müllerian hormone. GDNF, glial cell–derived neurotrophic factor.

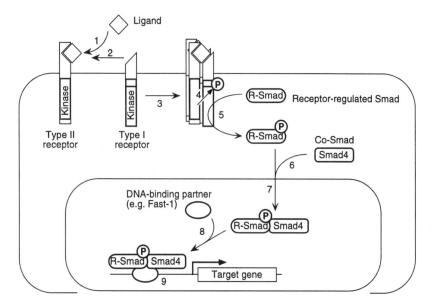

Figure 1   The transforming growth factor β (TGF-β)/SMAD pathway. Binding of a TGF-β family member to its type II receptor (*1*) in concert with a type I receptor (*2*) leads to formation of a receptor complex (*3*) and phosphorylation of the type I receptor (*4*). Thus activated, the type I receptor subsequently phosphorylates a receptor-regulated SMAD (R-Smad) (*5*), allowing this protein to associate with Smad4 (*6*) and move into the nucleus (*7*). In the nucleus, the SMAD complex associates with a DNA-binding partner, such as Fast-1 (*8*), and this complex binds to specific enhancers in targets genes (*9*), activating transcription.

(growth and differentiation factor 5) subfamily, the Vg1 subfamily, the BMP3 subfamily, various intermediate members, the activin subfamily, the TGF-β subfamily, and finally several distantly related members (Table 1) (1–17).

This review is devoted to a major accomplishment in this field over the past few years: the elucidation of a general mechanism by which TGF-β and related factors activate receptors at the cell surface and transduce signals to target genes (Figure 1). Some of these genes encode immediate effectors of ultimate cellular responses, such as cell cycle regulators that mediate antiproliferative responses or extracellular matrix components that determine cell adhesion, positioning, and movement. TGF-β and related factors regulate gene expression by bringing together two types of receptor serine/threonine protein kinases. One of these kinases phosphorylates the other, which in turn phosphorylates SMAD proteins. SMADs are a novel family of signal transducers that move into the nucleus and generate transcriptional complexes of specific DNA-binding ability. This review focuses on the structure and function of the TGF-β receptor family and the SMAD family, their mechanisms of activation and regulation, and their disruption in human disease.

SIGNALING RECEPTORS

TGF-β and related factors signal through a family of transmembrane protein serine/threonine kinases referred to as the TGF-β receptor family. This family came to light with the cloning of an activin receptor (18), now referred to as ActR-II, with properties similar to those of TGF-β receptors identified in ligand cross-linking studies (19) and genetically implicated in TGF-β signal transduction (20). The cloning of ActR-II also revealed a striking similarity between this molecule and Daf-1, a previously identified orphan receptor from *Caenorhabditis elegans* (21). These findings provided the basis and impulse for the rapid identification of many other members of this receptor family.

Extensive evidence has accumulated to indicate that TGF-β family members signal through receptor serine/threonine kinases. One exception is the glial cell–derived neurotrophic factor (GDNF), which signals through the receptor tyrosine kinase Ret (17). GDNF was included in the TGF-β family because it has a set of cysteines that are characteristic of this family (22). However, GDNF is the most divergent family member and shows very little sequence similarity to other members (see Table 1). The next most divergent member, the Müllerian inhibiting substance (MIS; also known as anti-Müllerian hormone, AMH), signals through a TGF-β receptor family member, AMHR (23). GDNF therefore is in a class of its own aligned with the structurally diverse group of factors that signal through receptor tyrosine kinases.

Type I and II Receptor Families

Based on their structural and functional properties, the TGF-β receptor family is divided into two subfamilies: type I receptors and type II receptors (Figure 2). Type I receptors have a higher level of sequence similarity than type II receptors, particularly in the kinase domain. Vertebrate type I receptors form three groups whose members have similar kinase domains and signaling activities. In mammals, one group includes TβR-I, ActR-IB, and ALK7, another includes BMPR-IA and -IB, and the third includes ALK1 and ALK2.

As a result of being simultaneously cloned by different groups, most type I receptors have received different names. One practice has been to use the neutral nomenclature ALK (activin receptor–like kinase) and to adopt a more descriptive name when the physiological ligand becomes known. Thus, the TGF-β type I receptor originally known as ALK5 (24) is now called TβR-I (25). ActR-IB (previously also known as ALK4) (26) is an activin type I receptor (27), and BMPR-IA and -IB (previously known as ALK3 and ALK6, respectively) are

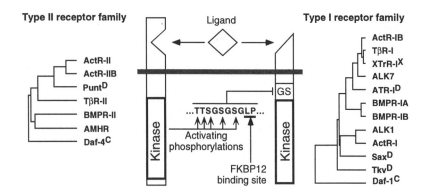

Figure 2 Type I and II TGF-β receptor families. In type I receptors, the protein kinase domain is preceded by the GS domain (GS). The characteristic GS sequence motif of TβR-I is shown, indicating the phosphorylation sites and the FKBP12-binding site. Listed members are from vertebrates unless otherwise indicated: D, *Drosophila*; C, *Caenorhabditis elegans*; X, *Xenopus*. The *dendrograms* indicate the relative level of amino acid sequence similarity in the kinase domain. Over this domain, ActR-II and Daf-4 have 40% sequence identity, and ActR- IB and Tkv have 60% identity.

BMP receptors (28, 29). Mammalian ALK7 (30, 31) and the related receptor XTrR-I from *Xenopus* (32) have no known ligand. ALK1 (also known as TSR-I) binds TGF-β (33) but does so more weakly than TβR-I (34) and is not known to mediate a TGF-β response (33). ALK2 is commonly referred to as ActR-I because it can bind activin and mediate certain activin responses in cultured cells (28, 33). However, the identity of its physiological ligand is a point of debate. ActR-I can also bind BMP2 and 4 (35, 36), and its mouse homologue can bind TGF-β when overexpressed (34, 37). Experiments using *Xenopus* embryo explants have shown that ActR-I/ALK2 mimics the mesoderm ventralizing activity of BMP4 but not the effects of activin or TGF-β, which suggests that ActR-I may function as a BMP receptor in vivo (39). Based on its expression pattern, it has been suggested that ALK2 may also function as an MIS/AMH type I receptor (38).

In vertebrates, the type II receptor subfamily includes TβR-II, BMPR-II, and AMHR, which selectively bind TGF-β (40), BMPs (36, 41, 42), and MIS (23, 43), respectively. ActR-II and -IIB bind activins when expressed alone or in concert with activin type I receptors (18, 44, 45). However, ActR-II and -IIB can bind BMPs 2, 4, and 7 and GDF5 in concert with BMP type I receptors (28, 46, 47).

Members of the TGF-β receptor family in invertebrates include Thick veins (Tkv) and Saxophone (Sax), which act as a Dpp type I receptors in *Drosophila* (48–51). Tkv most closely resembles the mammalian BMPR-I receptors, whereas Sax is somewhat closer to mammalian ALK1 and ALK2. Punt acts as a Dpp type II receptor in concert with Tkv or Sax (52, 53). ATR-I is a *Drosophila* type I receptor closely related to mammalian TβR-I and ActR-IB (54). ATR-I can bind human activin, but its real ligand is unknown. In *C. elegans*, larval development is controlled by Daf-1 (21) and Daf-4 (55), which are thought to be type I and II receptors, respectively, for the BMP-like ligand Daf-7 (56).

Structural Features of the Receptors

THE EXTRACELLULAR DOMAIN   Type I and II receptors are glycoproteins of approximately 55 kDa and 70 kDa, respectively, with core polypeptides of 500 to 570 amino acids including the signal sequence (18, 26, 40, 44, 57). The extracellular region is relatively short (approximately 150 amino acids), N-glycosylated (58, 59), and contains 10 or more cysteines that may determine the general fold of this region. Three of these cysteines form a characteristic cluster near the transmembrane sequence (54). The spacing of other cysteines varies and is more conserved in type I receptors than in type II receptors.

The transmembrane region and the cytoplasmic juxtamembrane region of type I and II receptors have no singular structural features. However, Ser213 in this region of TβR-II is phosphorylated by the receptor kinase in a ligand-independent manner and is required for signaling activity (60). Ser165 in the juxtamembrane region of TβR-I is phosphorylated by TβR-II in a ligand-dependent manner, and this appears to selectively modulate the intensity of different TGF-β responses (61).

THE GS DOMAIN   A unique feature of type I receptors is a highly conserved 30–amino acid region immediately preceding the protein kinase domain (Figure 2). This region is called the GS domain because of a characteristic SGSGSG sequence it contains (62). Ligand-induced phosphorylation of the serines and threonines in the TTSGSGSG sequence of TβR-I by the type II receptor is required for activation of signaling (61–63), and the same happens with the activin type I receptor ActR-IB (64). Immediately following the SGSGSG sequence, all type I receptors have a Leu-Pro motif that serves as a binding site for the immunophilin FKBP12 (65, 66). FKBP12 may act as a negative regulator of the receptor signaling function. The penultimate residue in the GS domain, right at the boundary with the kinase domain, is always a threonine or a glutamine. As shown with TβR-I (63) and several other type I receptors (46, 64, 67–69), mutation of this residue to aspartate or glutamate endows the receptor with elevated kinase activity in vitro and constitutive signaling activity in the cell. Thus, the GS domain is a key regulatory region that may control the catalytic activity of the type I receptor kinase or its interaction with substrates.

THE KINASE DOMAIN   The kinase domain in type I and II receptors conforms to the canonical sequence of a serine/threonine protein kinase domain (18, 24).

Consistent with this, type I receptors have been shown to phosphorylate their substrates—SMAD proteins—on serine residues (68, 70), whereas type II receptors phosphorylate themselves and type I receptors on serine and threonine residues but not tyrosine residues (40, 61–63, 71, 72). Autophosphorylation of TβR-II on tyrosine has been observed in vitro but not in vivo (73).

Conserved residues that in the crystal structure of other protein kinases coordinate ATP phosphate groups are essential for the activity of type I and II receptor kinases. These residues include a universally conserved β3-strand lysine (27, 74) and G217 in the glycine loop of TβR-I (75). The regulatory region known as the T loop in other protein kinases (76) contains two serines in TβR-II whose phosphorylation may enhance or inhibit the signaling activity of the receptor (60). A region of interest in the kinase domain of type I receptor kinases is the L45 loop that links two putative β strands. Replacement of the L45 loop in ActR-I with the L45 loop from TβR-I allows it to mediate TGF-β responses (77). Therefore this region may be involved in substrate recognition.

Type II receptors typically contain a very short C-terminal extension following the kinase domain, whereas type I receptors have essentially no C-terminal extension. Exceptions are the *C. elegans* receptor Daf-4 (55) and an alternative form of human BMPR-II (36, 41, 42, 78) that has long C-terminal extensions of unknown function. The C-terminal extension of TβR-II is phosphorylated (61), but its deletion does not impair signaling (79). This is in contrast to the important role that the C-terminal tail plays in signal transduction by tyrosine kinase receptors (80).

RECEPTOR VARIANTS   Some members of the TGF-β receptor family exist in alternative forms. These forms arise from the presence or absence of the following: a 25–amino acid insert following the signal sequence in TβR-II (81, 82), a 61–amino acid insert in the same position in AMHR-II (23), two alternative N-terminal regions in Tkv (49, 50), two alternative extracellular juxtamembrane regions in ATR-I (54), small inserts in the extracellular and intracellular juxtamembrane regions of ActR-IIB (44), and a long C-terminal extension in BMPR-II (36, 41, 42, 78). The presence of the extracellular insert in ActR-IIB increases the affinity for activin (44). The functional significance of the other receptor variants is unknown.

LIGAND-RECEPTOR INTERACTIONS

The Binding of Ligand to Signaling Receptors

LIGAND STRUCTURE: IMPLICATIONS FOR BINDING   The bioactive forms of TGF-β and related factors are dimers held together by hydrophobic interactions and, in most cases, also by an intersubunit disulfide bond (83). Each monomer contains three disulfide bonds interlocked into a tight structure known as the cystine knot (83). Insights into the possible regions of receptor contact are provided by the crystal structures of TGF-β2 (84, 85) and BMP7/OP-1 (86), the solution structure of TGF-β1 (87), and mutational analysis of TGF-β1 and TGF-β2 (88). The dimeric structure of these ligands suggests that they function by bringing together pairs of type I and II receptors, forming heterotetrameric receptor complexes. The pairing of receptors may be further specified by naturally occurring heterodimeric ligands such as TGF-β1.2 (19), TGF-β2.3 (89), and activin AB (90). The recombinant heterodimer BMP-4/7 is more potent in bioassays than BMP4 or BMP7 homodimers (91).

No species specificity has been described in the ligand-receptor interactions of the TGF-β system. Dpp receptors and Daf-4 can bind human BMPs (49, 50, 55), *dpp* phenotypes in flies can be rescued with a human *BMP4* transgene (92), and recombinant Dpp can induce endochondral bone formation in mammals (93).

TWO MODES OF BINDING   TGF-β and related factors activate signaling by binding to and bringing together pairs of type I and II receptors. Two general modes of binding ligand have been observed (Figure 3). One mode involves direct binding to the type II receptor and subsequent interaction of this complex with the type I receptor, which, in effect, becomes recruited into the complex. This binding mode is characteristic of TGF-β and activin receptors. Type I receptors for these factors can recognize ligand that is bound to the type II

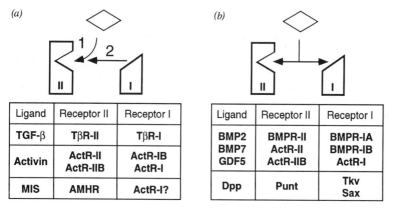

Figure 3   Two modes of ligand binding: (*a*) sequential binding, (*b*) cooperative binding. The ligands that bind according to each mode are listed together with the type I and II receptor combinations that they recognize. TGF, transforming growth factor; BMP, bone morphogenetic protein; GDF, growth and differentiation factor; MIS, Müllerian inhibiting substance.

receptors but not ligand that is free in solution (24, 33, 37, 72). This phenomenon was originally revealed by the receptor phenotype of TGF-β–resistant cell mutants (20, 94). TGF-β1 can bind to TβR-II in cell mutants lacking TβR-I but cannot bind to TβR-I in mutants lacking TβR-II. Restoration of TβR-II ligand-binding function, either by somatic fusion of a TβR-I defective cell with a TβR-II defective cell (95) or by transfection of a TβR-II cDNA (74), restores TGF-β binding to TβR-I. Furthermore, recombinant TβR-II binds TGF-β1 in solution (96–98). Like TβR-II, the type II receptors ActR-II (18), ActR-IIB (44, 45), and AMHR (23) also bind ligand when transfected in the absence of a type I receptor. Indeed, the original cloning of TβR-II (40) and ActR-II (18) was based on the ability of these receptors to bind ligand when overexpressed in COS cells.

The second binding mode is typical of BMP receptors and is cooperative, involving type I and II receptors that bind ligand with high affinity when expressed together but low affinity when expressed separately (36, 41, 42, 47) (Figure 3). Thus, BMPs 2, 4, and 7 and GDF5 bind weakly to the type II receptor BMPR-II expressed alone (36, 41, 42, 47) and to the type I receptors BMPR-IA or BMPR-IB expressed alone (29, 35) or in solution (99). ActR-II and -IIB are bona fide activin receptors that on their own bind BMP poorly if at all. However, ActR-II and -IIB can bind BMPs 2 and 7 in cooperation with BMPR-IA or BMPR-IB (28). This interaction mirrors what is observed with the Dpp receptor system. The Dpp type II receptor Punt, which is more closely related to ActR-II and -IIB than it is to BMPR-II, can recognize human activin (100). However, genetic evidence indicates that Punt acts as a Dpp receptor, and this evidence led to the finding that Punt binds Dpp or BMP poorly on its own but well in the presence of Tkv or Sax (52).

Accessory Receptors: Betaglycan and Endoglin

The original search for cell surface TGF-β–binding proteins using ligand cross-linking methods revealed the existence of binding proteins that were classified, according to their molecular weight, as type I and type II receptors (reviewed above) and type III receptors (19). Type III receptors detected by ligand cross-linking turned out to correspond to either one of two related proteins, betaglycan or endoglin (101–103). The evidence to date suggests that type III receptors do not have an intrinsic signaling function but regulate TGF-β access to the signaling receptors. There is no concrete evidence for type III receptors for other TGF-β family members.

BETAGLYCAN   Betaglycan is a membrane-anchored proteoglycan (58, 104) with an 853–amino acid core protein (101, 102) that carries heparan sulfate and chondroitin sulfate glycosaminoglycan (GAG) chains attached to Ser535 and Ser546 (105, 106). In the cell, betaglycan appears to exist as noncolalent homodimers (107). GAG chains are not required for betaglycan to reach the cell surface or to bind TGF-β, as revealed by studies using cell mutants defective in GAG synthesis (108) and betaglycan mutants defective in GAG attachment sites (105, 106). However, GAG chains of betaglycan can bind fibroblast growth factor (109). The cytoplasmic region of betaglycan is short (43 amino acids) and lacks any discernible signaling motif (101, 102). This region is not required for the TGF-β binding and presentation functions of betaglycan (105), and its function remains unknown. The highest level of sequence similarity between betaglycan and endoglin is found in the cytoplasmic and transmembrane domains (110).

TGF-β binding activity has been demonstrated in separate N-terminal and C-terminal domains of the extracellular region of betaglycan (105, 111, 112). The N-terminal domain has sequence similarity to the corresponding region in endoglin (110). The C-terminal extracellular domain contains the GAG attachment sites (105) and shows sequence similarity to a region of the major urinary protein uromodulin, the pancreatic zymogen granule protein GP-2, and the sperm receptors Zp2 and Zp3 (113). The entire extracellular region of betaglycan may be shed into the medium (114), and it may act as a TGF-β antagonist, inhibiting binding to membrane receptors (105).

Betaglycan binds all three TGF-β forms with high affinity (115, 116) and facilitates TGF-β binding to the type II receptor (102, 117), forming a betaglycan/TGF-β/TβR-II complex in the process (117, 118). The role of betaglycan as a facilitator of TGF-β binding to the signaling receptors is most evident with TGF-β2. Like TGF-β1 and -β3, TGF-β2 signals through TβR-I and TβR-II (74, 95). However, unlike them, TGF-β2 has low intrinsic affinity for TβR-II (116) and is less potent than TGF-β1 in hematopoietic progenitor cells (119), myoblasts (117), and endothelial cells (116) that lack betaglycan. Transfection of betaglycan augments TGF-β2 binding and activity in these cells (117, 120). The ability of betaglycan to equalize the potency of all three TGF-β forms raises the possibility that betaglycan may not only concentrate TGF-β at the cell surface but may also stabilize TGF-βs in a conformation optimal for binding to the signaling receptors.

ENDOGLIN   Endoglin is a cell surface molecule expressed at high levels in endothelial cells and at lower levels in monocytes, erythroid precursors, and other cell types (103, 121). Two splice variants of the cytoplasmic region give rise to human endoglin forms of 625 and 658 amino acids (122), each forming disulfide-linked dimers (103, 122). The sequence similarity between endoglin and betaglycan prompted an analysis of TGF-β binding to endoglin. This revealed that endoglin binds TGF-β1 and -β3, but unlike betaglycan, it does not bind TGF-β2 (110). As with betaglycan, complexes between endoglin and TGF-β receptors have been observed (123).

However, the role of endoglin in TGF-β binding to signaling receptors is unclear. The TGF-β binding activity of endoglin is limited compared to that of betaglycan and is increased by coexpression of TβR-II. In fact, endoglin overexpression can diminish rather than enhance TGF-β responses in monocytes (121). As mentioned below, mutations in *endoglin* and *ALK1* give rise to similar human disorders (124–126). Endoglin and ALK1 therefore might act in the same pathway, with endoglin facilitating ligand binding to ALK1. Given the weak TGF-β–binding activities of both receptors, the common endoglin and ALK1 ligand may not have been identified yet.

Latent Ligands and Soluble Inhibitory Proteins

The activity of TGF-β and related factors is negatively regulated by various soluble proteins that prevent their interaction with membrane receptors (see Figure 6).

THE LATENT TGF-β COMPLEX   Like all other members of its family, TGF-β is synthesized as the C-terminal domain of a precursor form that is cleaved before secretion from the cell (127, 128). However, the TGF-β propeptide, which is referred to as the latency associated peptide (LAP), remains noncovalently bound to TGF-β after secretion, retaining TGF-β in a latent form that cannot bind to betaglycan or the signaling receptors (129). Most cell types secrete TGF-β in this biologically inert form (12). Although LAP may be destroyed in the process of TGF-β activation, recombinant LAP retains TGF-β masking ability, and its injection in mice can inhibit endogenous TGF-β1 action (130).

A third component of the latent TGF-β complex is a large secretory glycoprotein known as latent TGF-β–binding protein (LTBP), which is disulfide-linked to LAP (131). LTBP is not required for the latency of the TGF-β complex but is implicated in the secretion, storage in the extracellular matrix, and eventual activation of this complex (131). LTBP comprises several forms generated from two genes and by alternative splicing: LTBP-1 in short and long forms (132) and LTBP-2 (133). Structurally, LTBPs contain a core of epidermal growth factor (EGF) repeats and eight-cysteine motifs organized in a fashion resembling fibrillin-1 and -2—two microfibrillar proteins whose mutations cause Marfan's syndrome and congenital contractural arachnodactyly, respectively (133). Like fibrillins, LTBP undergoes cross-linking by transglutaminases, forms fibrillar structures, and associates tightly with the extracellular matrix in mesenchymal and endothelial cells (134).

In tissue culture, LTBP associated with the extracellular matrix mediates storage of latent TGF-β and facilitates its activation (134, 135). Latent TGF-β

can be activated in vitro by acid, alkali, heat, limited proteolysis, or incubation by glycosidases (131). In tissue culture, activation of latent TGF-β may involve a combination of steps including the following: LAP proteolysis, binding to the mannose 6-phosphate/type II insulin-like growth factor receptor (Man6P/IGFR-II) via a mannose 6-phosphate group in LAP, cell-cell interactions between endothelial and vascular smooth-muscle cells, and binding to thrombospondin (131, 134–136). However, the physiological activation mechanism or mechanisms remain to be defined.

THE INHIBIN α CHAIN   Inhibin is the name given to heterodimers between the inhibin α chain and an inhibin/activin β chain (137). Inhibin was identified as an inhibitor of follicle-stimulating hormone (FSH) production in pituitary cultures (9). The subsequent identification of activins as β-chain dimers with biological activities opposite those of inhibin led to the idea that inhibins and activins are mutual antagonists (9). Because inhibin can compete for binding to the activin receptors ActR-II and -IIB (18, 44), it might antagonize activin by binding to its receptors without triggering signaling, either by failing to recruit type I receptors or by failing to achieve their activation (138, 139). The inhibin α chain therefore can be regarded as an inhibitor that functions by associating with β chains generating activin receptor antagonists. However, some effects of inhibin could be mediated by as yet unidentified inhibin receptors.

THE ACTIVIN INHIBITOR FOLLISTATIN   Follistatin is a soluble glycoprotein originally identified for its ability to inhibit pituitary FSH production (140) and later found to bind activin (141). Follistatin prevents activin binding to cell surface receptors (142). Paracrine as well as endocrine anti-activin effects of follistatin have been demonstrated in diverse tissues in mammals and *Xenopus* (140, 143–145). Follistatin can also bind to BMP-7, albeit with lower affinity than to activin (28), and may antagonize BMP signaling in vivo (145). Mammalian follistatin exists in forms of 288 and 315 amino acids generated by alternative splicing (146, 147). Follistatin is expressed in diverse mammalian tissues during development and in the adult (148–150) and in the Spemann's organizer in *Xenopus* embryos (145).

THE BMP INHIBITORS NOGGIN AND CHORDIN/SOG   The Spemann's organizer, a signaling center at the dorsal lip of the *Xenopus* gastrula blastopore, secretes BMP antagonists—noggin and chordin—which allow neighboring cells to develop as neural or dorsal mesoderm rather than epidermal or ventral mesoderm tissues (151, 152). Although noggin and chordin are of unrelated primary structure, both bind BMP4 (but not TGF-β or activin), preventing its interaction with cell surface receptors (151, 152). Noggin, a 222–amino acid polypeptide that is secreted as a homodimer, was the first such antagonist to be identified

(153). In the mouse, a noggin homologue is expressed in specific regions of the nervous system (154). Chordin has four cysteine-rich repeats similar to those found in thrombospondin, $\alpha 1$ procollagen, and von Willebrand factor (155). In *Drosophila*, the short gastrulation gene product, Sog, is the structural and functional homologue of chordin (156–158) and prevents Dpp from signaling through its receptors (159). The structural differences between noggin and chordin may result in different abilities to diffuse from their source, interact with extracellular matrix, and/or recognize different members of the large and complex BMP subgroup.

MECHANISM OF RECEPTOR ACTIVATION

Studies on the mechanism of activation of serine/threonine kinase receptors have centered on TGF-β receptors. However, to the extent that these studies have been replicated with activin and BMP receptors, the same basic activation mechanism appears to operate in these receptors as well.

The Basal State

BASAL PHOSPHORYLATION   The TGF-β type I receptor, TβR-I, is not phosphorylated in the basal state (62), but TβR-II, Act-R-II, and ActR-IIB are (40, 62, 64, 71, 160). Their basal phosphorylation is on serine residues and is partially retained in kinase-defective receptor mutants (62, 64). Some of the sites involved are in the C-terminal tail. Their functional significance is unclear: In one study, deletion of this entire region had no detectable effect on receptor signaling (79). Phosphorylation of other sites within TβR-II is dependent, directly or indirectly, on the activity of the receptor kinase (62, 64). In TβR-II, these sites include a serine in the juxtamembrane region and serines in the T-loop region of the kinase domain, and their phosphorylation modulates the signaling activity of TβR-II (60). What regulates the phosphorylation of these sites is not known.

BASAL RECEPTOR OLIGOMERIZATION   The oligomeric state of endogenous TGF-β receptors is not known, but studies with transfected epitope-tagged receptors indicate that TβR-II can form ligand-independent homo-oligomers (107, 161). These complexes are thought to prime the formation of the heteromeric TβR-I/TβR-II receptor complex upon ligand binding.

Type I and II receptors have intrinsic affinity for each other, as manifested by the spontaneous association of TβR-I and TβR-II when overexpressed in insect cells or coincubated in vitro as recombinant proteins (96). In the absence of ligand, TβR-I and TβR-II (162) or ActR-IB and ActR-IIB (64) can form active complexes when overexpressed in mammalian cells. This interaction is mediated, at least in part, by the cytoplasmic regions because these regions interact in a yeast two-hybrid system (36, 78, 96, 160). However, in transfected cells expressing moderate levels of TGF-β receptors (62) or activin receptors (138), the heteromeric receptor complex and, in particular, the phosphorylation and activation of the type I receptor are highly dependent on ligand binding.

FKBP12 BINDING   The cytoplasmic domain of diverse type I receptors interacts with FKBP12 in yeast (36, 163, 164) and mammalian cells (66, 165, 166). FKBP12 is an abundant 12-kDa cytosolic protein with *cis-trans* peptidyl-prolyl isomerase (rotamase) activity (167). FKBP12 binds different proteins, some on its own and some as a target of various natural or synthetic immunosuppressants. On its own, FKBP12 binds to the ryanodine receptor and the inositol 1,4,5-triphosphate receptor, stabilizing the calcium channeling activity of these proteins (168, 169). In complex with the drug FK506, FKBP12 binds calcineurin, inhibiting calcineurin's phosphatase activity and thus its ability to activate the transcription factor NF-AT in the T-cell receptor signal transduction pathway (170). In complex with rapamycin, FKBP12 binds FRAP/RAFT, inhibiting its activity as a kinase in mitogenic signal transduction (171, 172).

FKBP12 binding to TβR-I inhibits TGF-β signaling (66, 166) by inhibiting TβR-I phosphorylation by TβR-II within the oligomeric receptor complex (66). FKBP12-receptor interaction is mediated by the active site of FKBP12 (66, 166) and a conserved Leu-Pro motif adjacent to the phosphorylation sites in the GS domain of the receptor (65, 66) (Figure 2). FKBP12 binds to the TGF-β type I receptor in the basal state and appears to be released upon TGF-β–induced formation of the receptor complex (66, 166). Mutant TβR-I receptors defective in FKBP12 binding have elevated basal signaling activity but normal signaling activity in the presence of ligand (66). Therefore, one function of FKBP12 may be to guard against spurious activation of TGF-β signaling by ligand-independent encounters of type I and II receptors.

OTHER RECEPTOR-BINDING PROTEINS   TRIP-1 was identified as a TβR-II–interacting protein in a yeast two-hybrid screen (173). TRIP-1 contains several WD domains that may mediate protein-protein interactions, but the role of TRIP-1 is unknown. The interaction of TRIP-1 and TβR-II in mammalian cells is independent of ligand, requires the kinase activity of the receptor, and causes TRIP-1 phosphorylation (173).

The TβR-I cytoplasmic domain can interact with the farnesyl transferase-α subunit when both components are overexpressed in yeast or mammalian cells (164, 174, 175). It has been suggested that TGF-β may signal by regulating farnesyl transferase activity (174). However, this notion is controversial because the TGF-β receptor does not associate with the farnesyl transferase holoenzyme

(175). Furthermore, cells do not show a change in farnesyl transferase activity or in the farnesylation pattern of specific proteins in response to TGF-β (175).

The Activated State

RECEPTOR COMPLEX FORMATION   Signals emanate from a TGF-β type I receptor when it is phosphorylated by its activator, the type II receptor. As first shown with TGF-β receptors (74), ligand binding induces the formation of a heteromeric complex of type I and II receptors (24, 25, 27, 33, 36, 41, 62, 64, 74, 138, 176) (Figure 1). Given the dimeric nature of the ligands, each monomer might contact one type I receptor and one type II receptor, thereby generating a heterotetrameric receptor complex. Indeed, that the ligand-induced heteromeric complex contains two or more type I receptor subunits and two or more type II receptor subunits is suggested by analysis of TGF-β receptor complexes on two-dimensional gel electrophoresis (25), coprecipitation of receptors containing distinct epitope tags (75), and genetic complementation between mutant type I receptors (75). The TGF-β receptor complex is extremely stable upon solubilization, resisting dissociation by ionic detergents and chaotropic agents (62). Formation of this complex is required for signaling. Using chimeric receptor constructs containing TβR-I and TβR-II kinase domains in different configurations, signaling is achieved only when type I and II receptor kinase domains are brought together (177–179).

TYPE II RECEPTOR KINASE ACTIVITY   Ligand binding does not increase the overall phosphorylation of the type II receptors TβR-II, ActR-II, or ActR-IIB or their kinase activity in vitro (62, 64, 71, 162). Thus, type II receptors might be constitutively active kinases that require the ligand to interact with the type I receptor as a substrate. One caveat with this notion is that these studies have been done with moderately overexpressed receptors. It remains possible that type II receptors expressed at endogenous levels may undergo a ligand-induced increase in kinase activity. In any case, even when moderately overexpressed, type II receptors require ligand to phosphorylate their substrates, type I receptors.

TRANSPHOSPHORYLATION   Formation of the ligand-induced receptor complex rapidly leads to phosphorylation of the type I receptor (Figure 1), as demonstrated with TGF-β (62, 162) and activin receptors (64, 180). This phosphorylation is catalyzed by the type II receptor, as shown by coexpression of wild-type and kinase-defective type I and II receptors in different combinations (62, 64, 162). TβR-I is phosphorylated by TβR-II at serine and threonine residues in the sequence TTSGSGSGLP of the GS domain (61–63) (Figure 2), and similar sites are phosphorylated in ActR-IB by activin type II receptors (64). In addition to these sites, TβR-II mediates phosphorylation of Ser165 in the juxtamembrane region of TβR-I—a phosphorylation that may positively or negatively affect various TGF-β responses (61). TβR-I can catalyze its own phosphorylation in vitro, but there is no evidence that this occurs in vivo (63, 72, 75).

Signal Flow in the Receptor Complex

The events that transduce TGF-β signals start with type II receptor–mediated activation of the type I receptor. This receptor then phosphorylates and activates SMAD proteins, which carry the signal to the nucleus. This model is based on several lines of evidence. Mammalian cell mutants defective in either TβR-I (94) or TβR-II (20) lack a wide range of TGF-β responses. These responses are recovered in somatic hybrids between these two mutant phenotypes (95) or by transfection of the corresponding wild-type receptor (24, 72, 74). Work in *Drosophila* provides additional genetic evidence that Dpp signaling requires both type I and type II receptors (52, 53). Phosphorylation of serines and threonines in the GS domain of TβR-I is required for signaling (61–63). Alanine or valine mutations of any of these sites in TβR-I does not prevent phosphorylation of the other sites or receptor activation (63). However, mutation of three or more of these sites to alanine, valine, or acidic residues in TβR-I or ActR-IB prevents phosphorylation and signal transduction (63, 64, 180). Signaling is also inhibited when TβR-I phosphorylation is prevented by mutations in TβR-I or TβR-II that impair recognition of TβR-I as a substrate (75, 181), or by FKBP12 binding to the Leu-Pro motif in the GS domain (66).

A role of the type I receptor as the downstream signaling component in the receptor complex was originally inferred from the observation that the kinase activity of TβR-I is required for signal transduction and yet its substrate is neither TβR-I nor TβR-II (62). It was also shown that different type I receptors determine distinct responses to the same agonist (27, 182). Key evidence for a downstream role of the type I receptor was provided by the fact that hyperactive forms of TβR-I (63), ActR-IB (64), BMPR-IA and -IB (46, 68, 69), and Tkv (46, 67), generated by a mutation in the GS domain, have constitutive signaling activity in vivo. Signaling by hyperactive TβR-I also has been demonstrated in TβR-II–defective cells (63). The ability of purified BMP type I receptor to directly phosphorylate the activation sites of Smad1 in vitro (68) provides compelling evidence that in TGF-β receptor complexes, the signal flows from the type II receptor to the type I receptor and on to SMADs.

It is not clear whether activation of the type I receptor is based on an increase in its kinase activity, the appearance of substrate binding sites, or a combination of these two mechanisms. The hyperactive form TβR-I(T204D) has higher autokinase activity in vitro (63), suggesting that receptor activation may involve an increase in intrinsic kinase activity. On the other hand, it has been shown that

TβR-I activation results in Smad2 binding to the receptor complex (70, 183), suggesting that receptor activation may result in the generation of substrate docking sites.

In theory, the type II receptor could also signal independently of the type I receptor by phosphorylating other, as yet unidentified, signal-transducing substrates. However, no TGF-β responses have been described in cells lacking type I receptors. Overexpression of dominant-negative TβR-II receptor constructs can eliminate all TGF-β responses tested (79, 184) or only part of the TGF-β responses tested (185), depending on the assay conditions. Responses requiring a low level of signaling activity may be triggered by a residual level of activity in cells expressing dominant-negative receptors.

SMAD PROTEINS

The proteins of the SMAD family are the first identified substrates of type I receptor kinases and play a central role in the transduction of receptor signals to target genes in the nucleus (Figure 1).

SMADs as Mediators of TGF-β Signaling

The founding member of the SMAD family is the product of the *Drosophila* gene *Mad* (*mothers against dpp*) (186). *Mad* was identified in a genetic screen for mutations that exacerbate the effect of weak *dpp* alleles (187), and its discovery led to the identification of many related genes in nematodes and vertebrates. Three *Mad* homologues were identified in *C. elegans* and called *sma-2*, *-3*, and *-4* because their mutation causes small body size (188). Shortly thereafter, many homologues were described in vertebrates and named SMADs (for SMA/MAD related). *DPC4* (for "deleted in pancreatic carcinoma locus 4"), a gene frequently mutated or deleted in pancreatic cancer (189), also referred to as *Smad4*, was one of the first reported human SMADs. Human, mouse, and/or frog Smads 1–8 were cloned by screening EST (expressed sequence tag) databases or cDNA libraries for Mad homologues (46, 183, 190–198). Smad2 was iedependently identified in a cDNA expression cloning screen for inducers of mesoderm formation in *Xenopus* embryos (199). Smads 6 and 7 were identified as shear stress–induced genes in endothelial cells (200).

Initial evidence that SMADs function downstream of TGF-β receptors was provided by the ability of *Mad* mutations to inhibit signaling by a hyperactive Tkv receptor construct (46, 67). The most compelling evidence came from the observation that in response to TGF-β and related agonists, SMADs are phosphorylated (46, 183, 192–194, 201), accumulate in the nucleus (46, 191, 199), and become transcriptionally active (191). This body of evidence placed SMADs squarely downstream of TGF-β receptors.

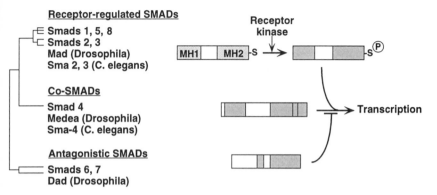

Figure 4 The SMAD family. Listed members are from vertebrates unless otherwise indicated. Vertebrate SMADs are highly conserved between human and *Xenopus*. The *dendrogram* indicates the relative level of amino acid sequence identity between vertebrate SMADs. The highly conserved MH1 and MH2 domains are indicated. Receptor-regulated SMADs are directly phosphorylated by TGF-β family type I receptors, and this phosphorylation allows association with a collaborating SMAD (co-SMAD). Antagonistic SMADs inhibit this SMAD activation process.

SMAD Subfamilies and Their Functions

Based on structural and functional considerations, SMADs fall into three subfamilies (Figure 4): (*a*) SMADs that are direct substrates of TGF-β family receptor kinases, (*b*) SMADs that participate in signaling by associating with these receptor-regulated SMADs, and (*c*) antagonistic SMADs that inhibit the signaling function of the other two groups.

Among the receptor-regulated SMADs, Smad1 and presumably its close homologues Smad5 and Smad8 are substrates for BMPR-I (68) and mediators of BMP signals (46, 190, 191, 202, 203, 203a). Smads 2 and 3 are TβR-I substrates (70, 183) and mediators of TGF-β and activin signals (190, 193, 195, 199, 201, 204). When overexpressed in *Xenopus* early embryos, Smad1 mimics the ability of BMP4 to ventralize mesoderm (190, 191, 202), whereas Smad2 mimics dorsal mesoderm induction and axis formation by activin (190, 199). In mammalian epithelial cells, Smads 2 and 3 mediate growth inhibition and transcriptional activation of TGF-β and activin reporter genes (183, 201). Mad and Sma's 2 and 3 also belong to this subfamily; they act as mediators of Dpp receptor signals (205) and Daf-4 signals (188), respectively.

Signaling by receptor-regulated SMADs requires the participation of a collaborating SMAD. The only known member of this group in vertebrates is Smad4. Smad4 associates with receptor-regulated SMADs when these become phosphorylated by the corresponding receptors (68, 183, 201, 206). Although Smad4 is similar to the receptor-regulated SMADs in overall structure, it normally is not phosphorylated in response to agonists. Smad4 is required for Smad2- or Smad3-dependent growth inhibitory responses in mammalian cells, and a dominant-negative Smad4 construct interferes with Smad1 and Smad2 signaling in frog embryos and mammalian cells (183, 201). Smad4, therefore, participates in TGF-β, activin, and BMP signaling pathways as a shared partner of receptor-regulated SMADs. The *Medea* (206a–c) and *Sma-4* (188) gene products from *Drosophila* and nematode are close homologues of Smad4, and they may fulfill a similar function in these organisms.

Human Smads 6 and 7 and *Drosophila* Dad are a subfamily of structurally divergent SMADs whose only known activity is to inhibit the signaling function of receptor-activated SMADs. Smad6 preferentially inhibits BMP signaling (196, 207), Smad7 can inhibit TGF-β and BMP signaling (197, 208), and Dad inhibits Mad signaling (209). Additional SMADs have been identified in nematode, but their functional properties are complex, as inferred from genetic analysis (210).

Structural Features of SMADs

THE MH1 DOMAIN   SMAD proteins contain highly conserved N-terminal and C-terminal domains (referred to as N and C domains, or MH1 and MH2 domains, respectively) and an intervening linker region that is of variable length and sequence (Figure 5). The MH1 domain has approximately 130 amino acids

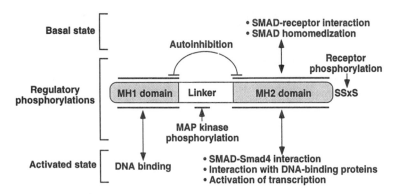

Figure 5   SMAD domains and their functions. In the basal state, SMADs form homo-oligomers and remain in an inactive state through an interaction between the MH1 and MH2 domains. Receptor-regulated SMADs interact with activated type I receptors via the MH2 domain and become activated by receptor-mediated phosphorylation at the C-terminal SS(V/M)S motif. In the activated state, SMADs associate with Smad4 and with DNA-binding proteins via the MH2 domain. The MH1 domain of some SMADs also participates in DNA binding, and the MH2 domain participates in transcriptional activation. MAP kinases phosphorylate some SMADs in the linker region, inhibiting SMAD accumulation in the nucleus.

and is highly conserved in receptor-regulated SMADs and Smad4 but not in inhibitory SMADs. In the basal state, the MH1 domain inhibits the transcriptional (191) and biological (199) activities of the MH2 domain. This inhibitory effect is likely due to the interaction between these two domains. Indeed, the MH1 domains of Smads 2 and 4 can physically interact with the respective MH2 domains, and overexpression of either MH1 domain as a separate protein can prevent the TGF-β–induced association of Smad2 and Smad4 MH2 domains (206).

The MH1 domain does not have a purely inhibitory function because it has DNA-binding activity in the activated state. The DNA-binding activity of the Mad MH1 domain is required for Dpp-induced activation of an enhancer within the *vestigial* wing-patterning gene (211). Likewise, the Smad4 MH1 domain contributes to the DNA-binding activity of a Smad2-Smad4 transcriptional complex (212). The DNA-binding activity of the Mad MH1 domain is inhibited by the presence of the MH2 domain (211), suggesting that the MH1 and MH2 domains may inhibit each other's function in the basal state. The contribution of the MH1 domains to the DNA-binding affinity and specificity of SMAD transcriptional complexes may vary depending on the particular target gene.

THE MH2 DOMAIN   This domain contains receptor phosphorylation sites (in receptor-regulated SMADs) (68, 70), has effector function (191, 199), and is involved in several important protein-protein interactions (Figure 5). The canonical MH2 domain is about 200 amino acids long and contains a characteristic insert in the case of Smad4 and Sma-4 (183). Interactions between MH2 domains support the homo-oligomeric complexes that SMADs from all three subfamilies form in the basal state (201, 206, 207, 213, 214). The MH2 domains also mediate the association of receptor-regulated SMADs with type I receptors (70), with Smad4 upon receptor-mediated phosphorylation (206), and with DNA-binding factors (212, 215) (see below). The Smad2 MH2 domain is biologically active in frog mesoderm induction assays (199), and when fused to the DNA-binding domain of GAL4, the MH2 domains of Smad1 and Smad2 display agonist-independent transcriptional activity (191, 212). Smads 1 and 2 require the presence of the Smad4 MH2 domain to activate transcription (212). In the case of antagonistic SMADs, the MH2 domain is sufficient for their inhibitory effect (200, 207).

The crystal structure of the Smad4 MH2 domain has provided insights into the basis for some of these interactions (214). The Smad4 MH2 domain forms a homotrimer in the crystals, and Smad4 forms a trimer in solution. Each monomer consists of a β-sandwich core flanked by three α-helices in a bundle on one side and several loops and an α-helix on the other side. The trimer interfaces are formed by extensive contacts between the three-helix bundle of one monomer and the loops on the adjacent monomer. Tumor-derived mutations in these interfaces destabilize and inactivate the homotrimer (see below). The trimer has the shape of a disc with the linker region emerging from one face. A loop referred to as the L3 loop protrudes from each monomer on the other face, and an α-helix referred to as helix-2 protrudes from each monomer on the edge of the disc. The L3 loops and the helix-2 may be sites for interaction with other proteins. Indeed, mutations in the L3 loop prevent Smad2 from interacting with the TGF-β receptor (217) and Smad4 from interacting with Smad2 (214). Based on sequence similarities, the overall structure of the MH2 domain is likely to be conserved in the other SMADs. Smads 6 and 7 lack the region corresponding to the third helix of the bundle, so they may form a different type of monomer-monomer interface (207).

THE LINKER REGION   The linker region is highly variable in size and sequence. This region contributes to the formation of SMAD homo-oligomers (206, 213). In receptor-regulated SMADs, the linker region contains MAP-kinase phosphorylation sites (216). As discussed below, phosphorylation of these sites in response to MAP-kinase activation inhibits nuclear translocation of SMADs.

SIGNALING THROUGH SMADs

In the basal state, SMADs exist as homo-oligomers that reside in the cytoplasm (Figures 1 and 6). Upon ligand activation of the receptor complex, the type I receptor kinase phosphorylates specific SMADs, which then form a complex with Smad4 and move into the nucleus. In the nucleus, these complexes, either alone or in association with a DNA-binding subunit, activate target genes by binding to specific promoter elements.

SMADs as Receptor Substrates

PHOSPHORYLATION SITES   SMADs are serine-phosphorylated in response to agonists, as shown with Smad1 in response to BMP2 or 4 (46, 68), Smad2 in response to TGF-β or activin (70, 193, 201), and Smad3 in response to TGF-β (183, 204). Although the kinetics of this phosphorylation are relatively slow ($t_{1/2} \sim 5$ min) when transfected SMADs are used, evidence shows that SMADs are direct substrates of the receptors. Smad1 is phosphorylated by highly purified, bacterially expressed BMPR-I kinase domain (68), Smad2 by immunoprecipitated TGF-β receptor complexes (70), and Smad3 by a TβR-I kinase preparation (183).

In vitro and in vivo, receptor-mediated phosphorylation occurs at serines in the C-terminal motif SS(V/M)S of Smad1 (68) or Smad2 (68, 70). This motif is also present in Smads 3, 5, and 8; *Drosophila* Mad; and *C. elegans* Smas-2 and -3. However, it is not present in the Smad4 subfamily or the inhibitory SMADs. This is consistent with the commonly observed lack of

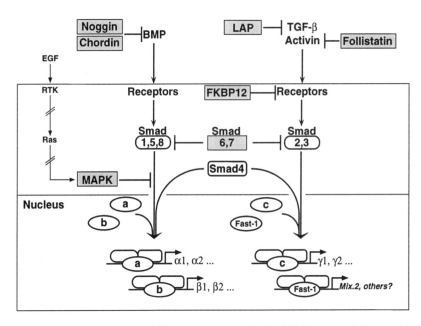

Figure 6   A model for generation of diverse gene responses by the SMAD system and their negative regulation. Smads 1, 5, and 8 are BMP receptor substrates and mediators of BMP gene response, whereas Smads 2 and 3 are substrates and mediators of TGF-β and activin receptors. Hypothetically each SMAD complex associates with different sequence-specific DNA-binding factors, of which Fast-1 is the first known example, and activates a distinct set of target genes. Negative regulation is provided by growth factor–sequestering proteins, FKBP12 binding to type I receptors, antagonistic SMADs, and SMAD phosphorylation by MAP kinases (in *gray boxes*). LAP, latency associated peptide.

agonist-induced phosphorylation in Smads 4, 6, and 7 (183, 196, 197, 201, 207, 208). Mutation of the serines in this sequence inhibits receptor-mediated phosphorylation of Smads 1 and 2 in vivo and in vitro and their association with Smad4 (68), accumulation in the nucleus (68, 70), interaction with DNA-binding proteins (212), and mediation of transcriptional responses (68, 70). Therefore, phosphorylation of this motif is required for SMAD activation.

SMAD-RECEPTOR ASSOCIATION   Smad2 and Smad3 become transiently and selectively associated with the activated TGF-β receptor complex (70, 183, 217). This interaction is required for Smad2 phosphorylation because docking-defective Smad2 mutants are not phosphorylated in response to TGF-β (70, 217). The Smad2 phosphorylation sites themselves along with the adjacent sequence in the 11–amino acid C-tail region are not required for this interaction (217). In fact, phosphorylation of these sites appears to facilitate Smad2 dissociation from the receptor, as either phosphorylation-defective Smad2 mutants or a kinase-defective TGF-β type I receptor mutant enhances SMAD-receptor association (70, 217). The transient nature of the SMAD-receptor interaction is consistent with the role of SMADs as carriers of receptor signals into the nucleus.

Mutational analysis of Smads 1 and 2 has identified their L3 loops as critical determinants of receptor docking interactions (217). The integrity of the L3 loop is necessary for interaction with the receptor and sufficient to dictate the specificity of this interaction. The L3-loop sequence is invariant among TGF-β–activated Smads (Smads 2 and 3) and BMP-activated Smads (Smads 1, 5, 9 and Mad) but differs at two positions between these two groups. Switching these two amino acids switches Smad1 and Smad2 activation by BMP and TGF-β, respectively. However, the isolated L3 loop is not sufficient to fully support this interaction. The SMAD-receptor interaction may require cooperativity provided by the oligomeric state of both the receptors and the SMADs.

Activated SMAD Complexes

Receptor-phosphorylated SMADs associate with Smad4, which functions as a shared partner required for transcriptional activation (Figure 6). Smad1 associates with Smad4 in response to BMPR-I activation (68, 201) and with Smads 2 and 3 in response to TβR-I or ActR-IB activation (201, 204). Smad4 can associate with these SMADs in yeast, which suggests that the interaction is direct (206). Based on structural considerations and the observation that mutations in the Smad4 L3 loop abolish the ability of Smad4 to associate with Smad2, the Smad4 L3 loop appears to mediate the association with receptor-activated SMADs (214). SMAD L3 loops, therefore, are implicated in two distinct types of interactions: (a) interaction with the receptors in the case of receptor-regulated SMADs and (b) interaction with receptor-activated SMADs in the case of Smad4. Functional interactions between receptor-regulated SMADs and a Smad4 family member may also occur in *Drosophila* (187) and *C. elegans* (188).

Nuclear Localization and Its Regulation

Nuclear translocation of receptor-activated SMADs occurs with kinetics that closely follow those of the agonist-induced phosphorylation and association with Smad4. Nuclear translocation of Smads 1 and 2 does not require Smad4, as determined using Smad4-defective cells (212). Smad4 is also translocated into the nucleus in response to TGF-β or BMP (204, 212), and this translocation requires the presence of Smad1 or Smad2 (212). Thus, it appears that receptor-activated SMADs bind Smad4 in the cytoplasm and carry it into the nucleus (212).

As central mediators of TGF-β family signals, SMADs are subject to different types of regulatory mechanisms that integrate and adapt their signaling potential to the status of the cell. One mode of regulation is by phosphorylation of MAP-kinase sites in the linker region, inhibiting the accumulation of SMADs in the nucleus (216) (Figure 6). Agonists that activate Erk MAP kinases, such as epidermal growth factor (EGF) and hepatocyte growth factor, rapidly induce phosphorylation of Smad1 at serines in four PXSP motifs in the linker region. This phosphorylation is catalyzed by Erk MAP kinases and occurs independently of BMP receptor–mediated phosphorylation of Smad1. Erk-mediated phosphorylation inhibits nuclear accumulation of Smad1 without interfering with the association of Smad1 with Smad4. BMP responses that depend on nuclear accumulation of Smad1 are antagonized by activation of the Erk MAP-kinase pathway (216). This mechanism may underlie the ability of EGF to oppose osteogenic differentiation by BMP2 or the ability of fibroblast growth factor (FGF) to oppose the effect of BMP2 during limb bud outgrowth, digit formation, or tooth development (216). Other receptor-regulated SMADs also have potential MAP-kinase phosphorylation sites in their linker region. SMAD regulation by MAP kinases may therefore be a general phenomenon in the regulation of TGF-β signaling.

Transcriptional Complexes

The ability of SMADs to activate transcription was originally detected through the use of GAL4-Smad fusion constructs that activate GAL4 reporter gene (191). GAL4-Smad1 and GAL4-Smad2 constructs activate transcription in response to BMP4 and TGF-β, respectively, and their ability to do so requires Smad4, as determined using Smad4-defective cells (212). The first description of a natural SMAD transcriptional complex was made through studies on the activin response factor (ARF), a DNA-binding complex that forms in *Xenopus* embryo explants in response to activin or an endogenous factor, presumably Vg1 (218). ARF binds to a 50–base pair activin-response element (ARE) in the promoter of the homeobox gene *Mix.2*, an immediate-early activin response gene. The first component of ARF to be identified was the DNA-binding protein Fast-1, based on its ability to interact with a hexanucleotide repeat present in the activin-response element (219). Fast-1 is a novel member of the winged-helix family of putative transcription factors (also known as the HNF-3 family or the forkhead family) (220).

Fast-1 associates with Smad2 and Smad4, forming a ternary complex that binds to the ARE (212, 215) (Figure 6). Because Fast-1 is a nuclear protein (219), it probably binds to incoming Smad2-Smad4 complexes in the nucleus. The interaction involves a region within the C-terminal portion of Fast-1 and the MH2 domain of Smad2 (212, 215). Smad4 is not required for the Smad2-FAST1 interaction but contributes two essential functions to the resulting Smad2/Smad4/FAST-1 complex: Through its MH1 domain, Smad4 promotes binding of the complex to DNA, and through its MH2 domain, Smad4 activates transcription (212).

Other members of the winged-helix family might be DNA-binding partners of SMADs. However, members of structurally unrelated families might play this role as well. For example, the *Drosophila* gene *schnurri*, which encodes a zinc-finger protein with homology to various mammalian transcription factors, is genetically implicated in Dpp signaling (221, 222). Another Dpp-activated gene, *Ubx*, is activated via a cyclic AMP response element (CRE) adjacent to a sequence resembling a Mad-binding site (223). Paradoxically, mutation of this Mad-binding site did not interfere with Dpp activation of *Ubx*. SMADs may interact with certain target enhancers without the involvement of DNA-binding subunits (211, 223a,b), but the biological role of these interactions remains to be ascertained.

Response Elements

Numerous gene responses to TGF-β have been described, but only a fraction of these have the characteristics of an immediate transcriptional response. p15^{Ink4b} and p21^{Cip1} are cyclin-dependent kinase inhibitors whose rapid introduction in response to TGF-β mediates cell cycle arrest (224–227). Clusters of Sp1-like sites near the transcription start site of *p15^{Ink4b}* and *p21^{Cip1}* score as TGF-β–responsive regions in reporter gene assays (228, 229). TGF-β–stimulated expression of interstitial collagens and other extracellular matrix proteins underlies important roles of TGF-β in development and regenerative processes (11–13). The TGF-β–responsive regions of genes encoding such extracellular matrix proteins as collagen α1(I) (230), collagen α2(I) (231, 232), type 1 plasminogen activator inhibitor (PAI-1) (233, 234), elastin (235), and perlecan (236) resemble Sp1 sites or CTF/NF-I sites. However, some of these sequences also resemble the Mad-binding element of *vestigial* (211); thus they might be SMAD-binding sites. TGF-β and related factors can also cause rapid inhibition of gene transcription. Genes affected in this manner include *c-myc* (14) and the Cdk-activating phosphatase *cdc25A* (237); down-regulation of both genes by TGF-β mediates antiproliferative effects. Interestingly, transcriptional activation by TGF-β of PAI-1 (238), retinoic acid receptors (239), collagen α2(I), and other genes (238) appears to require AP-1 activity. Furthermore, a Fos-containing repressor has been implicated in the down-regulation of the secretory protease transin/stromalysin by TGF-β (240). Whether SMADs participate in all or even a majority of TGF-β gene responses is an open question.

Inhibition by Antagonistic SMADs

Vertebrate Smads 6 and 7 and *Drosophila* Dad are inhibitors of signaling by receptor-regulated SMADs (196, 197, 200, 207–209) (Figure 6). When over-expressed, Smad6 can inhibit BMP signaling and, partially, TGF-β signaling

(196), and Smad7 can inhibit TGF-β signaling (197, 208) and BMP signaling (197). At lower concentrations, however, Smad6 is a specific inhibitor of BMP signaling in frog embryos and mammalian cells (207). Dad inhibits Dpp signaling in *Drosophila* wing imaginal discs, and when introduced into frog embryos, Dad exhibits anti-BMP effects (209). Inhibitory SMADs participate in negative feedback loops that may regulate the intensity or duration of TGF-β responses. Thus, Smad7 expression is rapidly elevated in response to TGF-β (197), whereas Dad expression is elevated in response to Dpp (209). The expression of Smads 6 and 7 is elevated by shear stress in vascular endothelial cells (200), a reponse that might be mediated by autocrine TGF-β (241).

Inhibitory SMADs lack a C-terminal SSXS phosphorylation motif, and their N-terminal region has only short segments of MH1 domain homology (196, 197, 207–209). (Smad6 was originally reported as a truncated SMAD structure consisting of the MH2 domain only; see References 200, 242). One mechanism proposed to explain the inhibitory effects of Smads 6 and 7 is based on the observation that each of these SMADs can bind to diverse TGF-β family receptors and interfere with phosphorylation of receptor-regulated SMADs (196, 197, 208). This mechanism could account for the nonselective inhibition of BMP effects and TGF-β effects observed by overexpression of Smads 6 or 7. It is not known whether physiologic levels of inhibitory SMADs can interfere with receptor binding and phosphorylation of receptor-regulated SMADs.

A different mechanism may underlie the selective inhibition of BMP signaling by Smad6 (207). At low levels, Smad6 does not interfere with receptor-mediated phosphorylation of Smad1 but competes with Smad4 for binding to activated Smad1. In a yeast two-hybrid system, the Smad6 MH2 domain interacts with itself and with the Smad1 MH2 domain, but not with the MH2 domains of Smads 2 or 4. Smad6 binding to receptor-phosphorylated Smad1 yields a transcriptionally inert complex. Therefore, Smad6 appears to act as a Smad4 decoy for BMP-activated SMADs.

Other Kinases in TGF-β Signaling

Components of MAP-kinase cascades mediate numerous responses to mitogens, differentiation factors, inducers of apoptosis, radiation, and osmotic stress (243, 244). Several groups investigating whether TGF-β action affects the Erk subfamily of MAP kinases have reported activation (245), inhibition (246, 247), or no change (248) in the activity of these kinases after TGF-β treatment. A novel member of this family, TAK1 (TGF-β–activated kinase 1) was cloned based on its ability to activate a MAP kinase cascade in yeast (249). In mammalian cells, the activity of a transfected TAK1 is rapidly increased in response to TGF-β and BMP4 (249). Overexpression of a kinase-defective TAK1 mutant (249) or a truncated form of the TAK1 activator, TAB1 (250), diminishes the

TGF-β response of a reporter gene construct that contains an AP-1 site, implicating TAK1 in these responses. No effect of TAK1 on other TGF-β responses has been reported. TGF-β activation of the MAP-kinase JNK has been implicated in a similar transcriptional response and tentatively placed downstream of TAK1 (251, 252). However, the JNK-kinase response to TGF-β takes several hours, suggesting that JNK is not a primary transducer of TGF-β signals in these cells.

DISRUPTION OF TGF-β SIGNALING IN HUMAN DISORDERS

Alterations of TGF-β signaling pathways underlie many human disorders. A loss of growth inhibitory responses to TGF-β is often observed in cancer cells (253), and a gain of TGF-β activity is thought to play a central role in fibrotic disorders characterized by excessive accumulation of interstitial matrix material in the lung, kidney, liver, and other organs (254). Abnormal TGF-β activity is also implicated in inflammatory disorders (255–257). The phenotype of mice overexpressing or lacking specific TGF-β family members or their receptors has revealed that these alterations have profound effects on the development or homeostatis of many organs (1, 2, 4). However, direct evidence that disruption of TGF-β signaling is a cause of human disorders is provided by the following cases, in which genes encoding TGF-β family members, their receptors, or SMAD proteins are mutated (Figure 7).

TGF-β Receptor Mutations in Cancer

The effects of TGF-β on target cells include several forms of negative regulation of cell proliferation, such as induction of G1 arrest, promotion of terminal differentiation, or activation of cell death mechanisms (14, 258). Numerous reports have described deficiencies in these types of responses in human tumor–derived cell lines (253). Disruption of TGF-β signaling could therefore predispose or cause cancer.

This prediction was confirmed by the finding that the TGF-β type II receptor is inactivated by mutations in gastrointestinal cancers with microsatellite instability (259, 260). Microsatellite instabilty is common to many sporadic cancers and results from defects in DNA mismatch repair leading to nucleotide additions or deletions in simple repeated sequences—microsatellites—throughout the genome. The human *TβR-II* gene contains one such sequence, a 10-bp polyadenine repeat, starting at nucleotide 709 in the coding region of the extracellular domain. One- or two-base additions or deletions in this repeat occur in most sporadic colon cancers and gastric cancers with microsatellite instability, yielding truncated, inactive TβR-II products (260–262). Mutations in the *TβR-II* polyadenine repeat are also found in colon or gastric tumors from

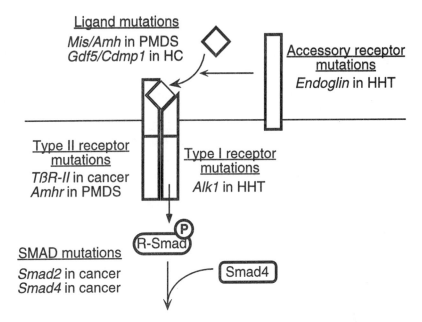

Figure 7   Inactivating mutations in TGF-β signal transduction genes in human disorders. *Mis/Amh* or *Amhr* is mutated in persistent Müllerian duct syndrome (PMDS); *Gdf5/Cdmp1* in hereditary chondrodysplasia (HC); *endoglin* or *Alk1* in hereditary hemorrhagic telangectasia (HHT); *TβR-II* in gastrointestinal cancers with somatic or hereditary microsatellite instability; *Smad2* in colon cancer; and *Smad4* in pancreatic, colon, and other cancers.

individuals with hereditary non-polyposis colon cancer (HNPCC) (263), a familial syndrome characterized by a high incidence of colon, endometrial, and gastric cancers. In most of these cases, both *TβR-II* alleles have mutations in the polyadenine repeat. In some cases, however, the second allele is inactivated by a different mutation, such as (*a*) an addition of a GT dinucleotide to a GT-GTGT sequence in the kinase domain coding region or (*b*) missense mutations, which are also predicted to inactivate this kinase (260, 261, 263). These results indicate that *TβR-II* shares the two-hit inactivation mechanism of other tumor suppressor genes.

Mutations in the *TβR-II* polyadenine repeat are rare in somatic or hereditary cancers of the endometrium, pancreas, liver, and breast (262, 264, 265) or in myelodisplastic syndrome (266) with microsatellite instability. This provides further evidence that mutations in the *TβR-II* polyadenine repeat are not just a random consequence of microsatellite instability but are specifically selected during the progression of colon and gastric cancers. Mutations elsewhere in

TβR-II have been described in T-cell lymphoma, gastric cancers, and head and neck carcinomas (267–269).

SMAD Mutations in Cancer

The TGF-β signaling network is also disrupted in cancer by mutations in *Smad2* and *Smad4/DPC4*. *Smad4/DPC4* was originally identified as a candidate tumor suppressor gene in chromosome 18q21 that was somatically deleted or mutated in half of all human pancreatic carcinomas (189). Biallelic *Smad4/DPC4* inactivation also occurs in a significant proportion of colorectal tumors (270, 271). *Smad4/DPC4* is infrequently mutated in breast (272), ovarian (272), head and neck (273), prostatic (271), esophageal, and gastric cancers (274). In the mouse, *Smad4* inactivation causes intestinal tumors in concert with inactivation of another tumor suppresor gene, *APC* (274a). *Smad2* is also located at 18q21, and it too is the target of inactivating mutations in colon cancer (193, 275, 276). Loss of TGF-β responsiveness in colon cancer therefore may be due to mutations in *TβR-II*, *Smad2*, or *Smad4/DPC4*. Interestingly, the preponderance of *Smad4/DPC4* mutations in pancreatic cancer, together with the low frequency of mutations in *TβR-II* in these tumors (264), raises the possibility that loss of Smad4 function may be selecting for resistance to an endogenous factor other than TGF-β itself.

   Smad2 and *Smad4/DPC4* are inactivated in cancer by missense mutations, nonsense mutations, small deletions, frameshift mutations, or loss of the entire chromosomal region. Most of the missense mutations described fall in the MH2 domain (214), a region that is also the target of mutations in *Mad*, *sma-2*, and *sma-3* inactive alleles (186, 188). The location of these mutations is consistent with the effector role of the MH2 domain in SMAD signaling. Resolution of the crystal structure of the Smad4 MH2 domain has revealed that tumor-derived missense mutations in this domain often affect amino acids that are critical for monomer-monomer interactions within the Smad trimer (214). Such mutations weaken Smad homo-oligomerization and prevent TGF-β–induced Smad2-Smad4 association. Less frequently, tumor-derived mutations destabilize the folding of the MH2 domain (214). Tumor-derived missense mutations have also been identified in the MH1 domains of *Smad2* and *Smad4/DPC4*. These mutations inactivate SMAD function by increasing the affinity of the MH1 domain for the homologous MH2 domain, locking the molecule in an inhibited conformation (206). Several mutations from inactive alleles of *Mad* or *sma* genes map to the region corresponding to the L3 loop and are predicted to interfere with heteromeric Smad interactions (214) or Smad-receptor interactions (217). However, no such mutation has been described in human SMAD genes.

GDF5/CDMP1 Mutations in Hereditary Chondrodysplasia

The phenotypes of mice defective in specific members of the BMP and GDF subfamilies have indicated that despite their similar activities in tissue culture, each of these factors is rate limiting for a distinct subset of developmental processes, including the development of specific skeletal components (1, 2, 4). One example of this is provided by the finding that the *brachypodism* phenotype in mice is due to inactivating mutations in the *Gdf5* gene (277, 278). *Brachypodism* mice have numerous alterations in the length and number of bones in the limbs but retain a normal axial skeleton (277). This finding raised the possibility that the human GDF5 homologue, known as cartilage-derived morphogenetic protein 1 (CDMP1), might likewise be involved in skeletal abnormalities. This possibility was confirmed with the identification of a frameshift mutation in *Cdmp1* in individuals with the recessive chondrodysplasia syndrome, Hunter-Thompson type acromesomelic chondrodysplasia (279). The mutation found in this study is a 22-bp insertion in the mature region of CDMP1 and most likely yields an inactive product. The abnormalities in affected individuals are restricted to the limbs and are most severe in the distal bones, which are short and dislocated (279).

ALK1 and Endoglin Mutations in Hereditary Hemorrhagic Telangectasia

The accessory receptor endoglin (103) and the type I receptor ALK1 (33) are highly expressed in vascular endothelial cells. The genes encoding these products have been identified as the targets of inactivating mutations in human hereditary hemorrhagic telangectasia (124–126). This disorder is characterized by epithelial vascular dysplasia and a high propensity to hemorrhage in the nasal and gastrointestinal mucosa (280). The autosomal dominant nature of this disorder argues that maintenance of appropriate endoglin and ALK1 levels is crucial for vascular homeostasis. The similarity of the phenotypes caused by mutations in either gene suggests that both receptors function in a common pathway controlling the development of the vascular wall. Because endoglin and ALK1 are not effective at binding TGF-β (33, 34, 110), it is possible that these two receptors mediate the action of an as yet unidentified TGF-β family member in the vasculature.

MIS and MIS Receptor Mutations in Persistent Müllerian Duct Syndrome

During the development of the reproductive tract in mammals, the Müllerian duct gives rise to the uterus, fallopian tubes, and upper vagina (15, 16). Regression of the Müllerian duct in males is mediated by MIS/AMH from the

Sertoli cells of the fetal testis acting via its receptor, AMHR, on the mesenchymal cells adjacent to the Müllerian duct epithelium (23, 43, 281). Disruption of this process leads to the appearance of internal pseudohermaphroditism with uterine and oviductal tissues in affected males, a disorder known as persistence of Müllerian duct syndrome (PMDS) (282). PMDS has been shown to result from inactivating mutations in either *Mis/Amh* (283–285) or *Amhr* (285–287). A 27-bp deletion in *Amhr* is a common cause of PMDS (288). The phenotypes of mutations in *Mis/Amh* and *Amhr* are essentially the same, and they are copied in mice defective in the ligand, the receptor, or both (289). These observations suggest that unlike other TGF-β family members, MIS/AMH and its receptor have a highly specific and restricted role during development.

SUMMARY AND PROSPECTS

Recent progress has led to the elucidation of a general TGF-β signaling pathway in which the ligand causes the activation of a heteromeric protein kinase complex that subsequently phosphorylates a subset of SMAD proteins that move into the nucleus, where they activate specific target genes with the agency of DNA-binding partners. The cellular response to a TGF-β factor may be determined not only by the receptors and SMAD isoforms present in the cell but also by the available repertoire of DNA-binding partners. The response is further modulated by regulators of ligand binding, receptor activity, SMAD activation, or nuclear localization. All the central components of these pathways and many of their regulators are novel proteins of previously unknown function.

The combinatorial interactions that configure such TGF-β signaling pathways provide a basis for understanding the multifunctional nature of these factors. In principle, now it should be possible to determine which combination of receptors, SMAD proteins, and DNA-binding partners leads to each particular TGF-β gene response. This signaling process is based on a succession of discrete protein-protein and protein-DNA interactions. The structural elements that mediate each contact can now be investigated to ascertain how signaling specificity is enforced in the pathway. These protein interactions are of limited strength; thus they seem good candidates as drug targets. This prospect is interesting, for either gain or loss of TGF-β signaling processes underlies various developmental disorders, several forms of cancer, and other ailments in humans.

The progress made allows us to explain, in general terms, how a TGF-β signaling pathway works. However, what is described here will likely become, with time, only part of the explanation as the complexity of this pathway is exposed in full. We might yet learn that type II receptors phosphorylate a different set of transducers, or that type I receptors have other substrates besides SMAD proteins, or that SMADs have other functions besides activating transcription. Furthermore, the recent emphasis on the transcriptional effects of TGFs and family members may have sidestepped other important responses to these factors; it is time to investigate these other responses as well. Clearly then, more work and more surprises lie ahead. However, the recent elucidation of the first contiguous TGF-β signaling pathway is a major milestone in this field and provides the framework for future research.

Visit the *Annual Reviews home page* at
http://www.AnnualReviews.org.

Literature Cited

1. Hogan BLM. 1996. *Genes Dev.* 10: 1580–94
2. Mehler MF, Mabie PC, Zhang DM, Kessler JA. 1997. *Trends Neurosci.* 20: 309–17
3. Harland RM. 1994. *Proc. Natl. Acad. Sci. USA* 91:10243–46
4. Kingsley DM. 1994. *Genes Dev.* 10:16–21
5. Cunningham NS, Paralkar V, Reddi AH. 1992. *Proc. Natl. Acad. Sci. USA* 89: 11740–44
6. Beddington R. 1996. *Nature* 381:116–17
7. Basler K, Edlund T, Jessell TM, Yamada T. 1993. *Cell* 73:687–702
8. McPherron AC, Lawler AM, Lee S-J. 1997. *Nature* 387:83–90
9. Vale W, Hsueh A, Rivier C, Yu J. 1990. See Ref. 290, pp. 211–48
10. Gaddy-Kurten D, Tsuchida K, Vale W. 1995. *Recent Prog. Horm. Res.* 50:109–29
11. Massagué J. 1990. *Annu. Rev. Cell. Biol.* 6:597–641
12. Roberts AB, Sporn MB. 1990. See Ref. 290, pp. 419–72
13. Roberts AB, Sporn MB. 1993. *Growth Factors* 8:1–9
14. Alexandrow MG, Moses HL. 1995. *Cancer Res.* 55:1452–57
15. Cate RL, Donahoe PK, MacLaughlin DT. 1990. See Ref. 290, pp. 179–210
16. Josso N, Cate RL, Picard JY, Vigier B, di Clemente N, et al. 1993. *Recent Prog. Horm. Res.* 48:1–49
17. Massagué J. 1996. *Nature* 382:29–30
18. Mathews LS, Vale WW. 1991. *Cell* 65: 973–82
19. Cheifetz S, Weatherbee JA, Tsang ML-S, Anderson JK, Mole JE, et al. 1987. *Cell* 48:409–15
20. Laiho M, Weis FMB, Massagué J. 1990.

J. Biol. Chem. 265:18518–24
21. Georgi LL, Albert PS, Riddle DL. 1990. *Cell* 61:635–45
22. Lin L-FH, Doherty DH, Lile JD, Bektesh S, Collins F. 1993. *Science* 260:1130–32
23. di Clemente N, Wilson C, Faure E, Boussin L, Carmillo P, et al. 1994. *Mol. Endocrinol.* 8:1006–20
24. Franzén P, ten Dijke P, Ichijo H, Yamashita H, Schulz P, et al. 1993. *Cell* 75:681–92
25. Yamashita H, ten Dijke P, Franzén P, Miyazono K, Heldin CH. 1994. *J. Biol. Chem.* 269:20172–78
26. ten Dijke P, Ichijo H, Franzén P, Schulz P, Saras J, et al. 1993. *Oncogene* 8:2879–87
27. Cárcamo J, Weis FMB, Ventura F, Wieser R, Wrana JL, et al. 1994. *Mol. Cell Biol.* 14:3810–21
28. Yamashita H, ten Dijke P, Huylebroeck D, Sampath TK, Andries M, et al. 1995. *J. Cell Biol.* 130:217–26
29. Koenig BB, Cook JS, Wolsing DH, Ting J, Tiesman JP, et al. 1994. *Mol. Cell Biol.* 14:5961–74
30. Tsuchida K, Sawchenko PE, Nishikawa S, Vale WW. 1996. *Mol. Cell. Neurosci.* 7:467–78
31. Rydén M, Imamura T, Jörnvall H, Belluardo N, Neveu I, et al. 1996. *J. Biol. Chem.* 271:30603–9
32. Mahony D, Gurdon JB. 1995. *Proc. Natl. Acad. Sci. USA* 92:6474–78
33. Attisano L, Cárcamo J, Ventura F, Weis FMB, Massagué J, et al. 1993. *Cell* 75: 671–80
34. ten Dijke P, Yamashita H, Ichijo H, Franzén P, Laiho M, et al. 1994. *Science* 264:101–4
35. ten Dijke P, Yamashita H, Sampath TK, Reddi AH, Estevez M, et al. 1994. *J. Biol. Chem.* 269:16985–88

36. Liu F, Ventura F, Doody J, Massagué J. 1995. *Mol. Cell. Biol.* 15:3479–86
37. Ebner R, Chen R-H, Lawler S, Zioncheck T, Derynck R. 1993. *Science* 262: 900–2
38. He WW, Gustafson ML, Hirobe S, Donahoe PK. 1993. *Dev. Dyn.* 196:133–42
39. Armes NA, Smith JC. 1997. *Development* 124:3797–804
40. Lin HY, Wang X-F, Ng-Eaton E, Weinberg RA, Lodish HF. 1992. *Cell* 68:775–85
41. Rosenzweig BL, Imamura T, Okadome T, Cox GN, Yamashita H, et al. 1995. *Proc. Natl. Acad. Sci. USA* 92:7632–36
42. Nohno T, Ishikawa T, Saito T, Hosokawa K, Noji S, et al. 1995. *J. Biol. Chem.* 270:22522–26
43. Baarens WM, van Helmaond MJL, Post M, van der Schoot PJCM, Hoogerbrugge JW, et al. 1994. *Development* 120:189–97
44. Attisano L, Wrana JL, Cheifetz S, Massagué J. 1992. *Cell* 68:97–108
45. Mathews LS, Vale WW, Kintner CR. 1992. *Science* 255:1702–5
46. Hoodless PA, Haerry T, Abdollah S, Stapleton M, O'Connor MB, et al. 1996. *Cell* 85:489–500
47. Nishitoh H, Ichijo H, Kimura M, Matsumoto T, Makishima F, et al. 1996. *J. Biol. Chem.* 271:21345–52
48. Nellen D, Affolter M, Basler K. 1994. *Cell* 78:225–37
49. Penton A, Chen YJ, Staehling-Hampton K, Wrana JL, Attisano L, et al. 1994. *Cell* 78:239–50
50. Brummel TJ, Twombly V, Marques G, Wrana JL, Newfeld SJ, et al. 1994. *Cell* 78:251–61
51. Xie T, Finelli AL, Padgett RW. 1994. *Science* 263:1756–59
52. Letsou A, Arora K, Wrana JL, Simin K, Twombly V, et al. 1995. *Cell* 80:899–908
53. Ruberte E, Marty T, Nellen D, Affolter M, Basler K. 1995. *Cell* 80:889–97
54. Wrana JL, Tran H, Attisano L, Arora K, Childs SR, et al. 1994. *Mol. Cell. Biol.* 14:944–50
55. Estevez M, Attisano L, Wrana JL, Albert PS, Massagué J, et al. 1993. *Nature* 365:644–49
56. Ren PF, Lim C-S, Johnsen R, Albert PS, Pilgrim D, et al. 1996. *Science* 274: 1389–91
57. Ebner R, Chen R-H, Shum L, Lawler S, Zioncheck TF, et al. 1993. *Science* 260:1344–48
58. Cheifetz S, Andres JL, Massagué J. 1988. *J. Biol. Chem.* 263:16984–91
59. Wells RG, Yankelev H, Lin HY, Lodish HF. 1997. *J. Biol. Chem.* 272:11444–51
60. Luo KX, Lodish HF. 1997. *EMBO J.* 16:1970–81
61. Souchelnytskyi S, ten Dijke P, Miyazono K, Heldin CH. 1996. *EMBO J.* 15:6231–40
62. Wrana JL, Attisano L, Wieser R, Ventura F, Massagué J. 1994. *Nature* 370:341–47
63. Wieser R, Wrana JL, Massagué J. 1995. *EMBO J.* 14:2199–208
64. Attisano L, Wrana JL, Montalvo E, Massagué J. 1996. *Mol. Cell. Biol.* 16:1066–73
65. Charng M-J, Kinnunen P, Hawker J, Brand T, Schneider MD. 1996. *J. Biol. Chem.* 271:22941–44
66. Chen YG, Liu F, Massagué J. 1997. *EMBO J.* 16:3866–76
67. Wiersdorff V, Lecuit T, Cohen SM, Mlodzik M. 1996. *Development* 122: 2153–62
68. Kretzschmar M, Liu F, Hata A, Doody J, Massagué J. 1997. *Genes Dev.* 11:984–95
69. Zou HY, Wieser R, Massagué J, Niswander L. 1997. *Genes Dev.* 11:2191–203
70. Macias-Silva M, Abdollah S, Hoodless PA, Pirone R, Attisano L, et al. 1996. *Cell* 87:1215–24
71. Mathews LS, Vale WW. 1993. *J. Biol. Chem.* 268:19013–18
72. Bassing CH, Yingling JM, Howe DJ, Wang TW, He WW, et al. 1994. *Science* 263:87–89
73. Lawler S, Fen XH, Chen R-W, Maruoka EM, Turck CW, et al. 1997. *J. Biol. Chem.* 272:14850–58
74. Wrana JL, Attisano L, Carcamo J, Zentella A, Doody J, et al. 1992. *Cell* 71: 1003–14
75. Weis-Garcia F, Massagué J. 1996. *EMBO J.* 15:276–89
76. Taylor SS, Radzio-Andzelm E. 1994. *Structure* 2:345–55
77. Feng XH, Derynck R. 1997. *EMBO J.* 16:3912–22
78. Kawabata M, Chytil A, Moses HL. 1995. *J. Biol. Chem.* 270:5625–30
79. Wieser R, Attisano L, Wrana JL, Massagué J. 1993. *Mol. Cell Biol.* 13:7239–47
80. Ullrich A, Schlessinger J. 1990. *Cell* 61: 203–12
81. Suzuki A, Shioda N, Maeda T, Tada M, Ueno N. 1994. *FEBS Lett.* 355:19–22
82. Hirai R, Fujita T. 1996. *Exp. Cell Res.* 223:135–41
83. Sun PD, Davies D. 1995. *Annu. Rev. Biophys. Biomol. Struct.* 24:269–91

84. Daopin S, Piez KA, Ogawa Y, Davies DR. 1992. *Science* 257:369–73
85. Schlunegger MP, Grütter M. 1992. *Nature* 358:430–34
86. Griffith DL, Keck PC, Sampath TK, Rueger DC, Carlson WD. 1996. *Proc. Natl. Acad. Sci. USA* 93:878–83
87. Hinck AP, Archer SJ, Quian SW, Roberts AB, Sporn MB, et al. 1996. *Biochemistry* 35:8517–34
88. Qian SW, Burmester JK, Tsang MLS, Weatherbee JA, Hinck AP, et al. 1996. *J. Biol. Chem.* 261:30656–62
89. Ogawa Y, Schmidt DK, Dasch JR, Chang RJ, Glaser CB. 1992. *J. Biol. Chem.* 267:2325–28
90. Ling N, Ying SY, Ueno N, Shimasaki S, Esch F, et al. 1986. *Nature* 321:779–82
91. Aono A, Hazama M, Notoya K, Taketomi S, Yamasaki H, et al. 1995. *Biochem. Biophys. Res. Commun.* 210:670–77
92. Padgett RW, Wozney JM, Gelbart WM. 1993. *Proc. Natl. Acad. Sci. USA* 90:2905–9
93. Sampath TK, Rashka KE, Doctor JS, Tucker RF, Hoffman FM. 1993. *Proc. Natl. Acad. Sci. USA* 90:6004–8
94. Boyd FT, Massagué J. 1989. *J. Biol. Chem.* 264:2272–78
95. Laiho M, Weis FMB, Boyd FT, Ignotz RA, Massagué J. 1991. *J. Biol. Chem.* 266:9108–12
96. Ventura F, Doody J, Liu F, Wrana JL, Massagué J. 1994. *EMBO J.* 13:5581–89
97. Tsang ML, Zhou L, Zheng BL, Wenker J, Fransen G, et al. 1995. *Cytokine* 7:389–97
98. Lin HY, Moustakas A, Knaus P, Wells RG, Henis YI, et al. 1995. *J. Biol. Chem.* 270:2747–54
99. Natsume T, Tomita S, Iemura S, Kinto N, Yamaguchi A, et al. 1997. *J. Biol. Chem.* 272:11535–40
100. Childs SR, Wrana JL, Arora K, Attisano L, O'Connor MB, et al. 1993. *Proc. Natl. Acad. Sci. USA* 90:9475–79
101. López-Casillas F, Cheifetz S, Doody J, Andres JL, Lane WS, et al. 1991. *Cell* 67:785–95
102. Wang X-F, Lin HY, Ng-Eaton E, Downward J, Lodish HF, et al. 1991. *Cell* 67:797–805
103. Gougos A, Letarte M. 1990. *J. Biol. Chem.* 265:8361–64
104. Segarini PR, Seyedin SM. 1988. *J. Biol. Chem.* 263:8366–70
105. López-Casillas F, Payne HM, Andres JL, Massagué J. 1994. *J. Cell Biol.* 124:557–68
106. Zhang L, Esko JD. 1994. *J. Biol. Chem.* 269:19295–99
107. Henis YI, Moustakas A, Lin HY, Lodish HF. 1994. *J. Cell Biol.* 126:139–54
108. Cheifetz S, Massagué J. 1989. *J. Biol. Chem.* 264:12025–28
109. Andres J, DeFalcis D, Noda M, Massagué J. 1992. *J. Biol. Chem.* 267:5927–30
110. Cheifetz S, Bellón T, Calés C, Vera S, Bernabeu C, et al. 1992. *J. Biol. Chem.* 267:19027–30
111. Pepin MC, Beauchemin M, Plamondon J, O'Connor-McCourt MD. 1994. *Proc. Natl. Acad. Sci. USA* 91:6997–7001
112. Kaname S, Ruoslahti E. 1996. *Biochem. J.* 315:815–20
113. Bork P, Sander C. 1992. *FEBS Lett.* 300:237–40
114. Andres JL, Stanley K, Cheifetz S, Massagué J. 1989. *J. Biol. Chem.* 109:3137–45
115. Segarini PR, Rosen DM, Seyedin SM. 1989. *Mol. Endocrinol.* 3:261–72
116. Cheifetz S, Massagué J. 1991. *J. Biol. Chem.* 266:20767–72
117. López-Casillas F, Wrana JL, Massagué J. 1993. *Cell* 73:1435–44
118. Moustakas A, Lin HY, Henis YI, O'Connor-McCourt MD, Lodish HF. 1993. *J. Biol. Chem.* 268:22215–18
119. Ohta M, Greenberger JS, Anklesaria P, Bassols A, Massagué J. 1987. *Nature* 329:539–41
120. Sankar S, Mahooti-Brooks N, Centrella M, McCarthy TL, Madri JA. 1995. *J. Biol. Chem.* 270:13567–72
121. Lastres P, Letamendia A, Zhang HW, Rius C, Almendro N, et al. 1996. *J. Cell Biol.* 133:1109–21
122. Bellón T, Corbí A, Lastres P, Calés C, Cebrián M, et al. 1993. *Eur. J. Immunol.* 23:2340–45
123. Yamashita H, Ichijo H, Grimsby S, Moren A, ten Dijke P, et al. 1994. *J. Biol. Chem.* 269:1995–2001
124. McAllister KA, Grogg KM, Johnson DW, Gallione CJ, Baldwin MA, et al. 1994. *Nat. Genet.* 8:345–51
125. McAllister KA, Baldwin MA, Thukkani AK, Gallione CJ, Berg JN, et al. 1995. *Hum. Mol. Genet.* 4:1983–85
126. Johnson DW, Berg JN, Baldwin MA, Gallione CJ, Marondel I, et al. 1996. *Nat. Genet.* 13:189–95
127. Derynck R, Jarrett JA, Chen EY, Eaton DH, Bell J, et al. 1985. *Nature* 316:701–5
128. Gentry LE, Liobin MN, Purchio AF, Marquardt H. 1988. *Mol. Cell Biol.* 8:4162–68

129. Gentry LE, Webb NR, Lim GJ, Brunner AM, Ranchalis JE, et al. 1987. *Mol. Cell. Biol.* 7:3418–27

130. Böttinger EP, Factor VM, Tsang MLS, Weatherbee JA, Kopp JB, et al. 1996. *Proc. Natl. Acad. Sci. USA* 93:5877–82

131. Miyazono K, Ichijo H, Heldin CH. 1993. *Growth Factors* 8:11–22

132. Olofsson A, Ichijo H, Morén A, ten Dijke P, Miyazono K, et al. 1995. *J. Biol. Chem.* 270:31294–97

133. Morén A, Olofsson A, Stenman G, Sahlin P, Kanzaki T, et al. 1994. *J. Biol. Chem.* 269:32469–78

134. Nunes I, Gleizes PE, Metz CN, Rifkin DB. 1997. *J. Cell Biol.* 136:1151–63

135. Taipale J, Miyazono K, Heldin CH, Keski-Oja J. 1994. *J. Cell Biol.* 124:171–81

136. Schultz-Cherry S, Murphy-Ullrich JE. 1993. *J. Cell Biol.* 122:923–32

137. Forage RG, Ring JM, Brown RW, McInerney BV, Cobon GS, et al. 1986. *Proc. Natl. Acad. Sci. USA* 83:3091–95

138. Lebrun JJ, Vale WW. 1997. *Mol. Cell. Biol.* 17:1682–91

139. Xu JM, McKeehan K, Matsuzaki K, McKeehan WL. 1995. *J. Biol. Chem.* 270:6308–13

140. Ueno N, Ling N, Ying SY, Esch F, Shimasaki S, et al. 1987. *Proc. Natl. Acad. Sci. USA* 84:8282–86

141. Nakamura T, Takio K, Eto Y, Shibai H, Titani K, et al. 1990. *Science* 247:836–38

142. de Winter JP, ten Dijke P, de Vries CJ, van Achterberg TA, Sugino H, et al. 1996. *Mol. Cell. Endocrinol.* 116:105–14

143. Xiao S, Findlay JK. 1991. *Mol. Cell. Endocrinol.* 79:99–107

144. Darland DC, Link BA, Nishi R. 1995. *Neuron* 15:857–66

145. Hemmati-Brivanlou A, Kelly OG, Melton DA. 1994. *Cell* 77:283–95

146. Shimasaki S, Koga M, Esch F, Mercado M, Cooksey K, et al. 1988. *Biochem. Biophys. Res. Commun.* 152:717–23

147. Shimasaki S, Koga M, Esch F, Cooksey K, Mercado M, et al. 1988. *Proc. Natl. Acad. Sci. USA* 85:4218–22

148. Albano RM, Arkell R, Beddington RS, Smith JC. 1994. *Development* 120:803–13

149. Feijen A, Goumans MJ, van den Eijnden-van Raaij AJ. 1994. *Development* 120:3621–37

150. DePaolo LV, Mercado M, Guo YL, Ling N. 1993. *Endocrinology* 132:2221–28

151. Zimmerman LB, De Jesus-Escobar JM, Harland RM. 1996. *Cell* 86:599–606

152. Piccolo S, Sasai Y, Lu B, De Robertis EM. 1996. *Cell* 86:589–98

153. Smith WC, Harland RM. 1992. *Cell* 70:829–40

154. Valenzuela DM, Economides AN, Rojas E, Lamb TM, Nunez L, et al. 1995. *J. Neurosci.* 15:6077–84

155. Sasai Y, Lu B, Steinbeisser H, Geissert D, Gont LK, et al. 1994. *Cell* 79:779–90

156. François V, Solloway M, O'Neill JW, Emery J, Bier E. 1994. *Genes Dev.* 8:2602–16

157. Holley SA, Jackson PD, Sasai Y, Lu B, De Robertis EM, et al. 1995. *Nature* 376:249–53

158. Biehs B, Francois V, Bier E. 1996. *Genes Dev.* 10:2922–34

159. Holley SA, Neul JL, Attisano L, Wrana JL, Sasai Y, et al. 1996. *Cell* 86:607–17

160. Chen R-H, Moses HL, Maruoka EM, Derynck R, Kawabata M. 1995. *J. Biol. Chem.* 270:12235–41

161. Chen R-H, Derynck R. 1994. *J. Biol. Chem.* 269:22868–74

162. Chen F, Weinberg RA. 1995. *Proc. Natl. Acad. Sci. USA* 92:1565–69

163. Wang TW, Donahoe PK, Zervos AS. 1994. *Science* 265:674–76

164. Kawabata M, Imamura T, Miyazono K, Engel ME, Moses HL. 1995. *J. Biol. Chem.* 270:29628–31

165. Okadome T, Oeda E, Saitoh M, Ichijo H, Moses HL, et al. 1996. *J. Biol. Chem.* 271:21687–90

166. Wang TW, Li B-Y, Danielson PD, Shah PC, Rockwell S, et al. 1996. *Cell* 86:435–44

167. Schreiber SL. 1991. *Science* 251:283–87

168. Brillantes A-MB, Ondrias K, Scott A, Kobrinsky E, Ondriasová E, et al. 1994. *Cell* 77:513–23

169. Cameron AM, Steiner JP, Sabatini DM, Kaplin AI, Walensky LD, et al. 1995. *Proc. Natl. Acad. Sci. USA* 92:1784–88

170. Schreiber SL. 1992. *Cell* 70:365–68

171. Sabatini DM, Erdjument-Bromage H, Lui M, Tempst P, Snyder SH. 1994. *Cell* 78:35–43

172. Brown EJ, Albers MW, Shin TB, Ichikawa K, Keith CT, et al. 1994. *Nature* 369:756–58

173. Chen R-H, Miettinen PJ, Maruoka EM, Choy L, Derynck R. 1995. *Nature* 377:548–52

174. Wang TW, Danielson PD, Li BY, Shah PC, Kim SD, et al. 1996. *Science* 271:1120–22

175. Ventura F, Liu F, Doody J, Massagué J. 1996. *J. Biol. Chem.* 271:13931–34
176. Bassing CH, Howe DJ, Segarini PR, Donahoe PK, Wang X-F. 1994. *J. Biol. Chem.* 269:14861–64
177. Okadome T, Yamashita H, Franzén P, Morén A, Heldin C-H, et al. 1994. *J. Biol. Chem.* 269:30753–56
178. Vivien D, Attisano L, Ventura F, Wrana JL, Massagué J. 1995. *J. Biol. Chem.* 270:7134–41
179. Luo KX, Lodish HF. 1996. *EMBO J.* 15:4485–96
180. Willis SA, Zimmerman CM, Li LI, Mathews LS. 1996. *Mol. Endocrinol.* 10:367–79
181. Cárcamo J, Zentella A, Massagué J. 1995. *Mol. Cell. Biol.* 15:1573–81
182. Persson U, Souchelnytskyi S, Franzén P, Miyazono K, ten Dijke P, et al. 1997. *J. Biol. Chem.* 272:21187–94
183. Zhang Y, Feng X-H, Wu R-Y, Derynck R. 1996. *Nature* 383:168–72
184. Brand T, Schneider MD. 1995. *J. Biol. Chem.* 270:8274–84
185. Chen R-H, Ebner R, Derynck R. 1993. *Science* 260:1335–38
186. Sekelsky JJ, Newfeld SJ, Raftery LA, Chartoff EH, Gelbart WM. 1995. *Genetics* 139:1347–58
187. Raftery LA, Twombly V, Wharton K, Gelbart WM. 1995. *Genetics* 139:241–54
188. Savage C, Das P, Finelli AL, Townsend SR, Sun C-Y, et al. 1996. *Proc. Natl. Acad. Sci. USA* 93:790–94
189. Hahn SA, Schutte M, Hoque ATMS, Moskaluk CA, da Costa LT, et al. 1996. *Science* 271:350–53
190. Graff JM, Bansal A, Melton DA. 1996. *Cell* 85:479–87
191. Liu F, Hata A, Baker JC, Doody J, Cárcamo J, et al. 1996. *Nature* 381:620–23
192. Yingling JM, Das P, Savage C, Zhang M, Padgett RW, Wang X-F. 1996. *Proc. Natl. Acad. Sci. USA* 93:8940–44
193. Eppert K, Scherer SW, Ozcelik H, Pirone R, Hoodless P, et al. 1996. *Cell* 86:543–52
194. Lechleider RJ, de Caestecker MP, Dehejia A, Polymeropoulos MH, Roberts AB. 1996. *J. Biol. Chem.* 271:17617–20
195. Chen Y, Lebrun JJ, Vale W. 1996. *Proc. Natl. Acad. Sci. USA* 93:12992–97
196. Imamura T, Takase M, Nishihara A, Oeda E, Hanai J, et al. 1997. *Nature* 389: 622–26
197. Nakao A, Afrakhte M, Morén A, Nakayama T, Christian JL, et al. 1997. *Nature* 389:631–35
198. Watanabe TK, Suzuki M, Omori Y, Hishigaki H, Horie M, et al. 1997. *Genomics* 42:446–51
199. Baker JC, Harland RM. 1996. *Genes Dev.* 10:1880–89
200. Topper JN, Cai J, Qiu Y, Anderson KR, Xu YY, et al. 1997. *Proc. Natl. Acad. Sci. USA* 94:9314–19
201. Lagna G, Hata A, Hemmati-Brivanlou A, Massagué J. 1996. *Nature* 383:832–36
202. Thomsen G. 1996. *Development* 122: 2359–66
203. Suzuki A, Chang CB, Yingling JM, Wang W-F, Hemmati-Brivanlou A. 1997. *Dev. Biol.* 184:402–5
203a. Chen Y, Bhushan A, Vale W. 1997. *Proc. Natl. Acad. Sci. USA* 94:12938–43
204. Nakao A, Imamura T, Souchelnytskyi S, Kawabata M, Ishisaki A, et al. 1997. *EMBO J.* 16:5353–62
205. Newfeld SJ, Chartoff EH, Graff JM, Melton DA, Gelbart WM. 1996. *Development* 122:2099–108
206. Hata A, Lo RS, Wotton D, Lagna G, Massagué J. 1997. *Nature* 388:82–87
206a. Das P, Maduzia LL, Wang H, Finelli AL, Cho SH, et al. 1998. *Development.* In press
206b. Hudson JB, Podos SD, Keith K, Simpson SL, Ferguson EL. 1998. *Development.* In press
206c. Wisotzkey RG, Mehra A, Sutherland DJ, Dobens LL, Liu X, et al. 1998. *Development.* In press
207. Hata A, Lagna G, Massagué J, Hemmati-Brivanlou A. 1998. *Genes Dev.* 12:186–97
208. Hayashi H, Abdollah S, Qiu YB, Cai JX, Xu YY, et al. 1997. *Cell* 89:1165–73
209. Tsuneizumi K, Nakayama T, Kamoshida Y, Kornberg TB, Christian JL, et al. 1997. *Nature* 389:627–31
210. Patterson GI, Koweek A, Wong A, Yanxia L, Ruvkun G. 1997. *Genes Dev.* 11:2679–90
211. Kim J, Johnson K, Chen HJ, Carroll S, Laughon A. 1997. *Nature* 388:304–8
212. Liu F, Pouponnot C, Massagué J. 1997. *Genes Dev.* 11:3157–67
213. Wu R-Y, Zhang Y, Feng X-H, Derynck R. 1997. *Mol. Cell. Biol.* 17:2521–28
214. Shi YG, Hata A, Lo RS, Massagué J, Pavletich NP. 1997. *Nature* 388:87–93
215. Chen X, Weisberg E, Fridmacher V, Watanabe M, Naco G, et al. 1997. *Nature* 389:85–89
216. Kretzschmar M, Doody J, Massagué J. 1997. *Nature* 389:618–22

217. Lo RS, Chen YG, Shi YG, Pavletich N, Massagué J. 1998. *EMBO J.* 17:996–1005
218. Huang H-C, Murtaugh LC, Vize PD, Whitman M. 1995. *EMBO J.* 14:5965–73
219. Chen X, Rubock MJ, Whitman M. 1996. *Nature* 383:691–96
220. Lai E, Clark KL, Burley S, Darnell JE Jr. 1993. *Proc. Natl. Acad. Sci. USA* 90:10421–23
221. Grieder NC, Nellen D, Burke R, Basler K, Affolter M. 1995. *Cell* 81:791–800
222. Arora K, Dai H, Kazuko SG, Jamal J, O'Connor MB, et al. 1995. *Cell* 81:781–90
223. Eresh S, Riese J, Jackson DB, Bohmann D, Bienz M. 1997. *EMBO J.* 16:2014–22
223a. Yingling JM, Datto MB, Wong C, Frederick JP, Liberati NT, Wang XF. 1997. *Mol. Cell. Biol.* 17:7019–28
223b. Zawel L, Dai JL, Buckaults P, Zhou P, Kinzler KW, et al. 1998. *Mol. Cell.* 1:611–17
224. Hannon GJ, Beach D. 1994. *Nature* 371:257–61
225. Datto MB, Li Y, Panus JF, Howe DJ, Xiong Y, Wang X-F. 1995. *Proc. Natl. Acad. Sci. USA* 92:5545–49
226. Reynisdóttir I, Polyak K, Iavarone A, Massagué J. 1995. *Genes Dev.* 9:1831–45
227. Reynisdóttir I, Massagué J. 1997. *Genes Dev.* 11:492–503
228. Datto MB, Li Y, Wang X-F. 1995. *J. Biol. Chem.* 270:28623–28
229. Li JM, Nichols MA, Chandrasekharan S, Xiong Y, Wang X-F. 1995. *J. Biol. Chem.* 270:26750–53
230. Ritzenthaler JD, Goldstein RH, Fine A, Smith BD. 1993. *J. Biol. Chem.* 268:13625–31
231. Rossi P, Karsenty G, Roberts AB, Roche NS, Sporn MB, et al. 1988. *Cell* 5:405–14
232. Inagaki Y, Truter S, Ramirez F. 1994. *J. Biol. Chem.* 269:14828–34
233. Riccio A, Pedone PV, Lund LR, Olesen T, Olsen HS, et al. 1992. *Mol. Cell. Biol.* 12:1846–55
234. Keeton MR, Curriden SA, van Zonneveld A-J, Loskutoff D. 1991. *J. Biol. Chem.* 266:23048–52
235. Marigo V, Volpin D, Vitale G, Bressan GM. 1994. *Biochem. Biophys. Res. Commun.* 199:1049–56
236. Iozzo RV, Pillarisetti J, Sharma B, Murdoch AD, Danielson KG, et al. 1997. *J. Biol. Chem.* 272:5219–28
237. Iavarone A, Massagué J. 1997. *Nature* 387:417–22
238. Chang E, Goldberg H. 1995. *J. Biol. Chem.* 270:4473–77
239. Chen Y, Takeshita A, Ozaki K, Kitano S, Hanazawa S. 1996. *J. Biol. Chem.* 271:31602–6
240. Kerr LD, Miller DB, Matrisian LM. 1990. *Cell* 61:267–78
241. Ohno M, Cooke JP, Dzau VJ, Gibbons GH. 1995. *J. Clin. Invest.* 95:1363–69
242. Riggins GJ, Kinzler KW, Vogelstein B, Thiagalingam S. 1997. *Cancer Res.* 57:2578–80
243. Blenis J. 1993. *Proc. Natl. Acad. Sci. USA* 90:5889–92
244. Davis RJ. 1993. *J. Biol. Chem.* 268:14553–56
245. Hartsough MT, Mulder KM. 1995. *J. Biol. Chem.* 270:7117–24
246. Howe PH, Dobrowolski SF, Reddy KB, Stacey DW. 1993. *J. Biol. Chem.* 268:21448–52
247. Berrou E, Fontenay-Roupie M, Quarck R, McKenzie FR, Levy-Toledano S, et al. 1996. *Biochem. J.* 316:167–73
248. Chatani Y, Tanimura S, Miyoshi N, Hattori A, Sato M, et al. 1995. *J. Biol. Chem.* 270:30686–92
249. Yamaguchi K, Shirakabe K, Shibuya H, Irie K, Oishi I, et al. 1995. *Science* 270:2008–11
250. Shibuya H, Yamaguchi K, Shirakabe K, Tonegawa A, Gotoh Y, et al. 1996. *Science* 272:1179–82
251. Atfi A, Djelloul S, Chastre E, Davis RR, Gespach C. 1997. *J. Biol. Chem.* 272:1429–32
252. Wang W, Zhou G, Hu MCT, Yao Z, Tan TH. 1997. *J. Biol. Chem.* 272:22771–75
253. Fynan TM, Reiss M. 1993. *Crit. Rev. Oncog.* 4:493–540
254. Border WA, Ruoslahti E. 1992. *J. Clin. Invest.* 90:1–7
255. Shull MM, Ormsby I, Kier AB, Pawlowski S, Diebold RJ, et al. 1992. *Nature* 359:693–99
256. Kulkarni AB, Huh C-G, Becker D, Geiser A, Lyght M, et al. 1993. *Proc. Natl. Acad. Sci. USA* 90:770–74
257. Wahl SM. 1992. *J. Clin. Immunol.* 12:61–74
258. Massagué J, Weis-Garcia F. 1996. In *Cancer Surveys Cell Signalling*, ed. T Pawson, P Parker, 27:41–64. London: ICRF
259. Markowitz SD, Roberts AB. 1996. *Cytokine Growth Factor Rev.* 7:93–102
260. Markowitz S, Wang J, Myeroff L, Parsons R, Sun LZ, et al. 1995. *Science* 268:1336–38

261. Parsons R, Myeroff LL, Liu B, Willson JK, Markowitz SD, et al. 1995. *Cancer Res.* 55:5548–50

262. Myeroff LL, Parsons R, Kim S-J, Hedrick L, Cho KR, et al. 1995. *Cancer Res.* 55:5545–47

263. Lu S-L, Zhang W-C, Akiyama Y, Nomizu T, Yuasa Y. 1996. *Cancer Res.* 56:4595–98

264. Vincent F, Hagiwara K, Ke Y, Stoner GD, Demetrick DJ, et al. 1996. *Biochem. Biophys. Res. Commun.* 223:561–64

265. Akiyama Y, Iwanaga R, Saitoh K, Shiba K, Ushio K, et al. 1997. *Gastroenterology* 112:33–39

266. Kaneko H, Horiike S, Taniwaki M, Misawa S. 1996. *Leukemia* 10:1696–99

267. Knaus PI, Lindemann D, DeCoteau JF, Perlman R, Yankelev H, et al. 1996. *Mol. Cell Biol.* 16:3480–89

268. Park KC, Kim SJ, Bang YJ, Park JG, Kim NK, et al. 1994. *Proc. Natl. Acad. Sci. USA* 91:8772–76

269. Garrigue-Antar L, Muñoz-Antonia T, Antonia SJ, Gesmonde J, Vellucci VF, et al. 1995. *Cancer Res.* 55:3982–87

270. Takagi Y, Kohmura H, Futamura M, Kida H, Tanemura H, et al. 1996. *Gastroenterology* 111:1369–72

271. MacGrogan D, Pegram M, Slamon D, Bookstein R. 1997. *Oncogene* 15:1111–14

272. Schutte M, Hruban RH, Hedrick L, Cho KR, Nadasdy GM, et al. 1996. *Cancer Res.* 56:2527–30

273. Kim SK, Fan YH, Papadimitrakopoulou V, Clayman G, Hittelman WN, et al. 1996. *Cancer Res.* 56:2519–21

274. Lei JY, Zou TT, Shi YQ, Zhou XL, Smolinski KN, et al. 1996. *Oncogene* 13:2459–62

274a. Takaku K, Oshima M, Miyoshi H, Matsui M, Seldin MF, Taketo MM, 1998. *Cell* 92:645–56

275. Uchida K, Nagatake M, Osada H, Yatabe Y, Kondo M, et al. 1996. *Cancer Res.* 56:5583–85

276. Riggins GJ, Thiagalingam S, Rozenblum E, Weinstein CL, Kern SE, et al. 1996. *Nat. Genet.* 13:347–49

277. Storm EE, Huynh TV, Copeland NG, Jenkins NA, Kingsley DM, et al. 1994. *Nature* 368:639–43

278. Storm EE, Kingsley DM. 1996. *Development* 122:3969–79

279. Thomas JT, Lin K, Nandedkar M, Camargo M, Cervenka J, et al. 1996. *Nat. Genet.* 12:315–17

280. Marchuk DA. 1997. *Chest* 111:S79–82

281. Teixeira J, He WW, Shah PC, Morikawa N, Lee MM, et al. 1996. *Endocrinology* 137:160–65

282. Guerrier D, Tran D, Vanderwinden JM, Hideux S, Van Outryve L, et al. 1989. *J. Clin. Endocrinol. Metab.* 68:45–52

283. Knebelmann B, Boussin L, Guerrier D, Legeai L, Kahn A, et al. 1991. *Proc. Natl. Acad. Sci. USA* 88:3767–71

284. Carré-Eusèbe D, Imbeaud S, Harbison M, New MI, Josso N, Picard JY. 1992. *Hum. Genet.* 90:389–94

285. Imbeaud S, Carré-Eusèbe D, Rey R, Belville C, Josso N, et al. 1994. *Hum. Mol. Genet.* 3:125–31

286. Imbeaud S, Faure E, Lamarre I, Mattéi M-G, di Clemente N, et al. 1995. *Nat. Genet.* 11:382–88

287. Faure E, Gouedard L, Imbeaud S, Cate R, Picard JY, et al. 1996. *J. Biol. Chem.* 271:30571–75

288. Imbeaud S, Belville C, Messika-Zeitoun L, Rey R, di Clemente N, et al. 1996. *Hum. Mol. Genet.* 5:1269–77

289. Mishina Y, Rey R, Finegold MJ, Matzuk MM, Josso N, et al. 1996. *Genes Dev.* 10:2577–87

290. Sporn MB, Roberts AB, eds. 1990. *Peptide Growth Factors and Their Receptors*, Vol. 95. Berlin: Springer-Verlag

Annu. Rev. Biochem. 1998. 67:793–819

PATHOLOGIC CONFORMATIONS OF PRION PROTEINS

Fred E. Cohen[1] and Stanley B. Prusiner[2]

Departments of [1,2]Biochemistry and Biophysics, [1]Cellular and Molecular
Pharmacology, [1]Medicine, and [2]Neurology, University of California,
San Francisco, California 94143; e-mail: cohen@cgl.ucsf.edu

KEY WORDS: protein folding, CJD, scrapie replication

ABSTRACT

While many aspects of prion disease biology are unorthodox, perhaps the most
fundamental paradox is posed by the coexistence of inherited, sporadic, and infectious forms of these diseases. Sensible molecular mechanisms for prion propagation must explain all three forms of prion diseases in a manner that is compatible
with the formidable array of experimental data derived from histopathological,
biochemical, biophysical, human genetic, and transgenetic studies. In this review, we explore prion disease pathogenesis initially from the perspective of an
autosomal dominant inherited disease. Subsequently, we examine how an intrinsically inherited disease could present in sporadic and infectious forms. Finally,
we explore the phenomenologic constraints on models of prion replication with
a specific emphasis on biophysical studies of prion protein structures.

CONTENTS

0066-4154/98/0701-0793$08.00

THEORY OF PRION DISEASES

The inherited prion diseases include Gerstmann-Sträussler-Scheinker disease (GSS), familial Creutzfeldt-Jakob disease (fCJD), and fatal familial insomnia (FFI). These patients present with characteristic clinical and neuropathologic findings as early as their third or fourth decade of life and their family histories are compatible with an autosomal dominant pattern of inheritance. Molecular genetic studies argue that these diseases are caused by mutations in the prion protein (PrP) gene based on high LOD scores for 5 of the 20 known mutations (1–5). As with many inherited disorders, the pathogenesis of the inherited prion disease is owing to the aberrant behavior of the protein encoded by the mutant PrP gene. The altered physical properties of mutant PrP probably results from a change in the conformation of the mutant protein akin to an allosteric effect; the magnitude of this conformational change can be quite variable. For example, the conformational change in sickle cell hemoglobin or in transthyretin mutants associated with FAP is largely at the level of quaternary structure, while there is evidence that the conformational reorganization of the Alzheimer's βAPP fragment occurs at both the tertiary and quaternary structure level (6–8). Unfortunately, these multimers are long-lived, exhibit pathologic properties, and have a histopathologic record of their existence. Considered in this context it is not surprising that the conformation of the normal cellular isoform of the wild-type (wt) prion protein (PrP^C) is distinct from the disease-causing isoform of the mutant prion protein (PrP$^{Sc/fCJD}$, PrP$^{Sc/FFI}$, PrP$^{Sc/GSS}$) in both conformation and oligomerization states. However, the magnitude of the conformational rearrangement owing to a point mutation is unexpected.

To explain inherited prion diseases, one need only postulate that the protein can exist in two distinct conformations, one that prefers a monomeric state and a second that multimerizes where the wt exhibits a dramatic preference for the monomeric state and the mutant preferentially adopts the multimeric state. The origin of this distinction could be kinetic or thermodynamic. Either the differential stability of the wt and mutant proteins in the monomeric and multimeric states is large, or a kinetic barrier that essentially precludes the conversion of the wt monomer is abrogated by the disease-causing mutations. These two scenarios are contrasted in Figure 1.

Results from a variety of site-directed mutagenesis studies of protein stability suggest that the impact of a single point mutation on the free energy of folding is unlikely to exceed 2–3 kcal (9). If the simple thermodynamic model were operative and the conversion of monomeric PrP^C into multimers were controlled by the differences in the free energies of the ground state, one would expect

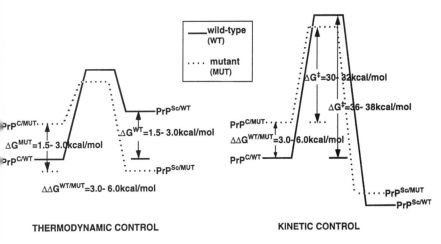

Figure 1 An illustration of the distinction between thermodynamic and kinetic models for the energetics of the conversion of PrP^C carrying the wild-type (WT) and mutant (MUT) sequences into PrP^{Sc}. $\Delta G^{\ddagger}$ is the free energy difference between the PrP^C and PrP^{Sc} states and ΔG is the activation energy barrier separating these two states. $\Delta \Delta G^{WT/MUT}$ is the difference between ΔG^{WT} and ΔG^{MUT}. The free energy diagrams for the wild-type sequences are shown (*solid lines*), as are the mutant sequences (*broken lines*).

that wt PrP^{Sc} production would be approximately 1% as efficient as mutant PrP^{Sc} formation. Given the resistance of the core of wt PrP^{Sc} to proteolytic digestion, PrP^{Sc} would accumulate in the wt setting owing to the difficulties associated with the metabolic clearance of this molecule. This scenario is at odds with neuropathologic and epidemiologic data on the incidence of sporadic CJD (sCJD). Ionizing irradiation experiments have suggested that the minimally infectious PrP^{Sc} particle is a dimer (10). If the energetics of dimer formation is simply the sum of the monomeric terms, then wt PrP^{Sc} formation would be 0.01% as likely as mutant PrP^{Sc} production. This level of infectivity would be detectable using a variety of immunoassays and is in contrast with experimental results. Sensibly, the cooperativity component of dimer formation is unlikely to be substantially different for the wt and mutant forms as disease-causing mutations exist in several distinct regions of the sequence and are distributed throughout the core of the structure (11–13). Thus, the cooperativity component of the free energy difference between the cellular and scrapie isoforms of the wt and mutant proteins ($\Delta \Delta G^{WT/MUT}$) is unlikely to exceed 2.0–3.0 kcal/mol. Under these extreme assumptions, wt PrP^{Sc} in normal cohorts should be 10^{-6} as common as mutant PrP^{Sc} in carriers from affected families. This level of wt PrP^{Sc} would be detected in bioassays of infectivity, a prediction that is at odds with a number of inoculation studies of PrP derived from natural and recombinant sources (14, 15). In contrast, kinetic control over the conversion of PrP^C to PrP^{Sc} provides a simple explanation for the observed clinical and experimental results. PrP^{Sc} would only need to be marginally more stable than PrP^C ($\Delta G = -2.0–3.0$ kcal/mol). To explain the normal absence of PrP^{Sc} the energetic barriers separating the two states would need to be quite large ($\Delta G^{\ddagger} = 36–38$ kcal/mol assuming a transition state argument). In this setting, a small change in the activation barrier (e.g. $\Delta\Delta G^{\ddagger}_{WT/MUT} = 2.0–3.0$ kcal) would result in approximately 100-fold slowing of the rate of conversion. Thus, the 30–40-year prodromal period for inherited prion disease in the mutant setting would become 3000–4000 years with the wt protein if this is the rate-limiting step in disease progression. Cooperative effects could further amplify the distinction between the normal cellular and disease-associated isoforms. For example, if the disease-associated isoform is a dimer, then interactions between the monomeric components could provide additional stability from the thermodynamic perspective. The disease-associated isoform could also impact the kinetic aspects of the conversion process by acting as a template that lowers the activation barrier ($\Delta G^{\ddagger}$) for the conformational change in a manner reminiscent of the way an enzyme's active site orchestrates the positioning of substrates to speed the rate of a reaction. In this setting, the disease-associated isoform would also be disease causing because its presence would dramatically enhance the likelihood of conversion of the normal cellular isoform.

In this context the sporadic occurrence of CJD could arise for two reasons. First, a somatic cell mutation could give rise to a mutant PrP that would prefer the conformation of the disease-causing isoform (16). Initially, propagation of the mutant PrP^{Sc}-like conformer would be limited to the cell in which the somatic mutation had occurred. Either within that cell or surrounding cells, this mutant PrP^{Sc}-like conformer would have to be capable of triggering the conversion of wt PrP^C into PrP^{Sc}. If the particular somatic mutation produces a PrP^{Sc}-like conformer that cannot interact with wt PrP^C, then prion propagation will not occur and disease will not develop. This scenario of sporadic disease caused by a somatic mutation differs from the inherited prion diseases where mutant PrPs with the characteristics of PrP^{Sc} accumulate. Because extracts from the brains of patients carrying the D178N, E200K, or V210L point mutation have transmitted to mice expressing the wt chimeric MHu2M transgene (17) (J Mastrianni & SB Prusiner, manuscript in preparation), it seems likely that any of these point mutations could initiate sporadic prion disease. In contrast, human prions carrying the P102L mutation could not be transmitted to Tg(MHu2M) mice, which argues that this mutation could not initiate sporadic prion disease. Although expression of the P102L mutation in either humans or mice causes neurodegeneration, detection of these prions was greatly facilitated when mice expressing PrPs carrying the same mutation were used as recipients of inocula from ill humans or Tg mice (18–20). A second explanation for sporadic disease is a corollary of the kinetic ideas advanced in the discussion of inherited disease. The kinetic barrier separating the cellular and disease-causing conformers of PrP provides only a stochastic barrier that will be crossed given a sufficient time. While this event will be vanishingly rare in any individual human lifetime, following the logic of the ergodic theorem, the likelihood of a rare event will increase as the size of the population enlarges. In the case of sporadic CJD with an incidence of one per million people, a barrier height of 36–38 kcal/mol would be required following a transition state argument. Once this misfolded conformer is formed, it would be available to act as a template to direct the rapid replication of the disease-causing conformer. If the infectious efficiency of an individual PrP^{Sc} oligomer is small, then this would act as a prefactor in the rate equation that would lower the expected $\Delta G^{\ddagger}$ for a single PrP^{Sc} formation event.

From the perspective of inherited and sporadic neurodegenerative diseases, Alzheimer's and the prion diseases could share similar pathogenic mechanisms. However, a fundamental point of departure arises from the transmissibility of the prion diseases, which has not been demonstrated for Alzheimer's disease (21, 22). Inoculation of tissues from animals suffering from prion disease cause disease in the recipient host. The infectious pathogen can be purified and treated with reagents that modify or hydrolyze polynucleotides without loss of infectivity. Although prions are remarkably resistant to proteolytic degradation, they can be inactivated by prolonged digestion with proteases or by exposure to high concentrations of salts known to denature proteins such as guanidinium thiocyanate (23–25). While these initial biochemical results were viewed with great skepticism, it has become increasingly clear how this protein could replicate. The mechanism of infectious prion disease follows from the second explanation for sporadic disease with the minor modification that the initiation of the process is not truly stochastic at a molecular level but relates to facilitated industrial or ritualistic cannibalism (26, 27). The efficiency of the infection relates to the titer of the inoculum, the mode of entry, and the frequency of exposure; intracerebral inoculation of a large dose of prions most efficiently initiates prion replication (28). In contrast, a single ingestion of foodstuffs containing a small dose of prions is likely to be exceedingly inefficient. The efficiency of this process is also governed by the strength of the interaction of PrP^{C} with PrP^{Sc}. When both isoforms contain the same sequence (homotypic interaction), experimental data confirm that conversion is most likely (29). When the sequences are different, especially in certain regions of the structure, conversion is less likely (30, 31). The inefficiency of heterotypic conversion is commonly referred to as the "species barrier" and has been used to explain why humans have not contracted scrapie from sheep and why nontransgenic mice are largely resistant to human $PrP^{Sc/CJD}$ inocula (19, 32).

A REPLICATION CYCLE FOR PrP^{Sc}

A simple replication cycle for PrP^{Sc} can be constructed. PrP^C exists in equilibrium with a second state, PrP*, that is best viewed as a transient intermediate that participates in PrP^{Sc} formation either through an encounter with PrP^{Sc} or with another PrP* molecule. Under normal circumstances, PrP^C dominates the conformational equilibrium. With infectious diseases, PrP^{Sc} specified here minimally as a PrP^{Sc}/PrP^{Sc} dimer is supplied exogenously. It can bind PrP* to create a heteromultimer that can be converted into a homomultimer of PrP^{Sc} (see Figure 2A). Genetic evidence points to the existence of an auxiliary factor (protein X) in this conversion (19, 33). Protein X preferentially binds PrP^C and is liberated upon the conversion of PrP* to PrP^{Sc}. Protein X can then be recycled and join another heteromultimeric complex (see Figure 2B). The homomultimer can disassociate to form two replication-competent templates creating exponential growth of the PrP^{Sc} concentration. In inherited disease, the concentration of PrP* rises owing to either the destabilizing effect of the mutation on PrP^C or the increased stability of a PrP^C or PrP* dimer (multimer). This increases the likelihood of the presence of a PrP*/PrP* complex that can form PrP^{Sc}/PrP^{Sc} (see Figure 2C) and initiate the replication cycle. Sporadic disease merely requires a rare molecular event, formation of the PrP*/PrP* complex (see Figure 2D), or a somatic cell mutation that follows the mechanism for the initiation of inherited disease. Once formed, the replication cycle is primed for subsequent conversion.

From an analysis of the thermodynamics and kinetics of prion replication and the replication cycle for the inherited, sporadic, and infectious scenarios, several inferences can be made about the biophysical properties of the normal cellular and disease-carrying PrP isoforms.

1. PrP^{Sc} replication requires the presence of the PrP gene in the host cell to direct PrP^C synthesis (34–37). However, while PrP^{Sc} replication requires a PrP gene in the host cell, it does not need to be carried by the infectious pathogen (38).

2. PrP^{Sc} must be more stable than PrP^C, and a sensible origin for this distinction is an extensive network of intermolecular interactions between PrP monomers in a PrP^{Sc} multimer (10). Protease resistance could be a corollary of this increased stability and not necessarily the origin of the increased metabolic stability of PrP^{Sc} (39, 40).

3. A large conformational distinction between PrP^C and PrP^{Sc} would create a substantial kinetic barrier rendering prion disease extremely uncommon in

the wt setting. As a corollary, some region or regions of PrP must exhibit extreme conformational plasticity (41–43).

4. For PrP^{Sc} to provide a useful and efficient template to facilitate PrP^C conversion, the molecular interaction between PrP^{Sc} and PrP^C must be quite specific (29–31). Thus, differences in the sequences of PrP^C and PrP^{Sc} should disrupt or attenuate conversion.

5. When the PrP gene carries amino acid substitutions that destabilize the protein in the PrP^C isoform or stabilize PrP*, the incidence of inherited disease should rise (44, 45). Other amino acid substitutions that do not alter the stability of PrP^C or PrP* significantly will provide sites for polymorphisms between or within species [for review see (46)].

6. If conformational conversion provides the rate-limiting step for prion replication, disease progression should depend on the concentration of PrP (35, 36, 47, 48). As this is most logically a first order process, halving the concentration of PrP should double the time to disease, and doubling the concentration of PrP should halve the time to disease. The standard deviation of measurements of time to disease (t) should be $t^{1/2}$.

7. Small molecules that stabilize the PrP^C isoform could act as therapeutic agents by decreasing the concentration of PrP*, thereby slowing PrP^{Sc} replication (49). Similarly, molecules that abrogate the PrP/protein X interaction or stabilize it sufficiently to prevent protein X recycling could have therapeutic potential (32, 33).

We shall return to these inferences in the context of the experimental results that follow.

EXPERIMENTAL AND COMPUTATIONAL STUDIES OF PrP^C AND PrP^{Sc}

The failure to detect a polynucleotide associated with the infectious prion particle created a "replication" conundrum (50–52). Purification of PrP 27–30 (39, 53), the 27–30-kDa protease resistant core of PrP^{Sc}, and subsequent microsequencing provided sufficient partial sequence information (54) to discover the endogenous PrP gene (38, 55). This gene resides on chromosome 20 in human and on the syntenic chromosome 2 of mouse (56). This is consistent with inference 1 on the requirements of prion replication. Genomic DNA sequencing revealed that the entire PrP coding region was contained in a single open reading frame (57). The sequence revealed many well-known features

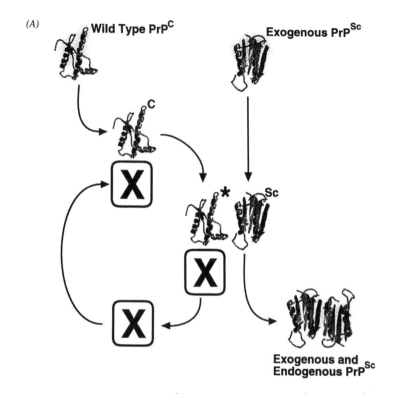

Figure 2   Initiation and replication of PrP^{Sc} synthesis. (*A*) Exogenous PrP^{Sc} initiates PrP^{Sc} synthesis by binding to a PrP$^{C/*}$/X complex. Facilitated by protein X and directed by the PrP^{Sc} template, PrP* changes conformation and forms PrP^{Sc}. PrP^{Sc} no longer binds protein X and so the heteromultimeric complex dissociates yielding recycled protein X and endogenous PrP^{Sc}. (*B*) A replication cycle for PrP^{Sc} synthesis following the creation of endogenous PrP^{Sc}. Again PrP^{Sc} binds to the PrP$^{C/*}$/X complex to form the activated template for conversion of PrP* to PrP^{Sc}. When PrP^{Sc} forms, protein X dissociates and is recycled. The newly generated PrP^{Sc} can then facilitate two replication cycles leading to an exponential rise in PrP^{Sc} formation. (*C*) In inherited disease and perhaps with spontaneous disease in the presence of a somatic cell mutation, mutant PrP^{C} can bind protein X and form the PrP*/X/PrP*/X encounter complex, which can then form PrP$^{Sc/Mut}$ in the absence of a PrP^{Sc} template. Once this event occurs, replication follows the pattern (*B*). (*D*) In spontaneous disease, the rare formation of the PrP*/X/PrP*/X complex could lead to de novo PrP^{Sc} formation. Once this rare event occurs, replication would follow the outline (*B*).

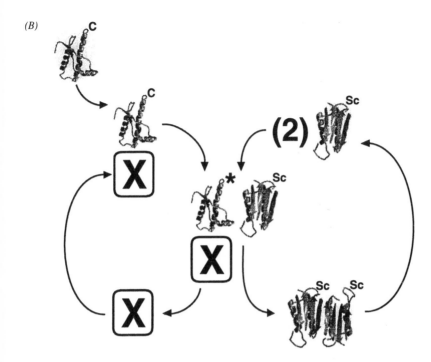

(B)

Figure 2   (Continued)

including a signal sequence and two sites for N-linked glycosylation. In addition, several unusual features were noted including an eight-residue repeating sequence (octarepeat) P-H-G-G-G-W-G-Q that is unlike any known protein structural motif and an alanine-rich region A-G-A-A-A-A-G-A for residues 113–120. Until recently, no extended regions of the PrP sequence were recognizably analogous to non-PrP gene products (58).

While PrP^C and PrP^{Sc} share a common sequence and pattern of posttranslational modification as judged by a variety of biochemical and mass spectroscopic studies, substantial differences at a structural level have been demonstrated between these two conformers. These are summarized in Table 1. Although the magnitude of these conformational distinctions was unexpected by most, it is clear that this feature is an essential aspect of prion biology.

Biochemical Characterization

PrP 27–30 was purified as the protease resistant core of the major, and probably the only, component of the infectious prion particle (39, 53). PrP 27–30 aggregates into rod-shaped polymers that are insoluble in aqueous and organic

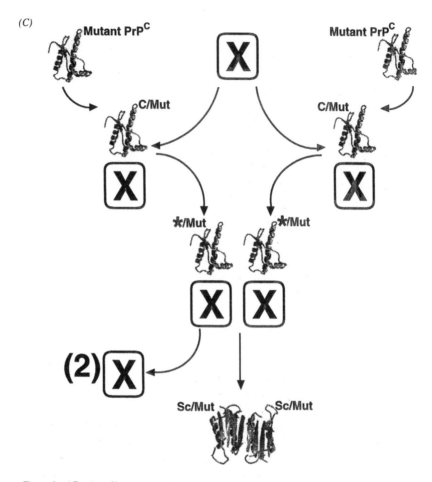

Figure 2   (Continued)

solvents as well as nonionic detergents. These purified prion rods exhibit the tinctorial and ultrastructural properties of amyloid (53). PrP 27–30 is stable at high temperatures for extended periods, consistent with its unusual stability of prion infectivity by comparison with globular proteins from mesophiles. Protein denaturants (e.g. 3 M GdnSCN) that modify the structure of PrP 27–30 also inactivate prion infectivity (25). In contrast, prion infectivity resists inactivation by reagents that disrupt nucleic acid polymers including nucleases, psoralens, and UV irradiation (50, 59–63). Microsequencing of the N-terminus of PrP 27–30 (54) led to the cloning and sequencing of the PrP gene (38, 55).

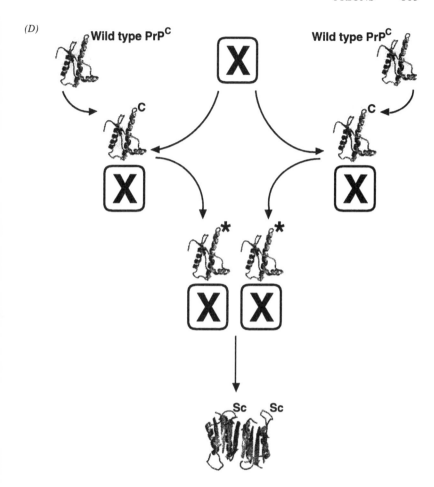

Figure 2   (Continued)

The N-terminus of PrP 27–30 corresponds to approximately codon 90 in the full-length coding sequence (57, 64).

Under physiologic conditions, cells synthesize PrP in the endoplasmic reticulum (ER) with the cleavage of a hydrophobic leader sequence (residues 1–22). Asn-linked carbohydrates that were attached to residues 181 and 197 in the ER are remodeled in the Golgi as PrP transits to the cell surface (65). A glycophosphoinositol (GPI) moiety is attached to residue 231 as 23 C-terminal residues are removed in the ER (66). The biogenesis of PrP is complicated by the presence of a stop transfer signal contained within residues 95–110 (67). While

Table 1 Structural differences between PrP^C and PrP^{Sc}

Property	PrP^C	PrP^{Sc}
Protease resistance	No	Stable core-containing residues 90–231
Disulfide bridge	Yes	Yes
Molecular mass after deglycosylation	16 kDa [rPrP(90–231)]	16 kDa (PrP 27–30)
Glycosylation	2 N-linked sugars	2 N-linked sugars
Glycoforms	Multiple	Multiple
Secondary structure	Dominated by α-helices	Rich in β-structure
Sedimentation rate	Consistent with monomeric species	Multimeric aggregated species
Accessible epitopes[a]	109–112 138–165 225–231	225–231
Free energy of stabilization	$\Delta G = {}^-6 - {}^-8$ kcal/mol	
Predicted secondary structure	α: 109–122; 129–141; 179–191; 200–217	
Experimental data	At pH 4.5[b] α: 144–154; 179–193; 200–217 β: 128–131; 161–164 At pH 5.2[c] α: 144–157; 172–193; 200–227 β: 129–131; 161–163	

[a]90.
[b]12.
[c]13.

the majority of PrP biosynthesis leads to a GPI-anchored form, the channel for PrP export into the ER may disassemble during biogenesis resulting in a transmembrane form of PrP (68, 68a).

PrP secreted to the cell surface under normal conditions is known as PrP^C, a soluble, monomeric protein containing a single disulfide bridge between residues 179 and 214. This protein was initially purified from hamster brain in its glycosylated form (69–71) and subsequently produced by recombinant (r) sources as either the truncated 142-residue molecule SHa rPrP(90–231) corresponding to PrP 27–30 or more recently the nearly full-length molecule SHa rPrP(29–231) (14, 72, 73). The recombinant protein can be oxidized and solubilized in GdnHCl between pH 5 and 8. Dilution of the denaturant and incubation at room temperature for 2–12 h yields a folded material that shares the spectral and immunologic properties of PrP^C purified from SHa brain (14, 40, 43). Thermal and GdnHCl denaturation of rPrP(90–231) indicate a three-state process with a well-defined intermediate (43). The free energy change between the folded and intermediate states was calculated to be 1.9 ± 0.4 kcal/mol. The second transition to the unfolded state was associated with a free energy change of 6.5 ± 1.2 kcal/mol. PrP^C unfolds with a T_m of 50–55°C, substantially lower than PrP^{Sc}. Analytical ultracentrifugation studies reveal that rPrP(90–231) is monomeric with a monomer-dimer equilibrium association constant of 5.4×10^{-5} M^{-1}. rPrP(90–231) refolded under acidic conditions yields a molecule with less helicity as judged by circular dichroism spectroscopy. This form has a greater tendency to aggregate (43).

In contrast, PrP^{Sc}, the full-length infectious conformer of PrP^C, has a tendency to form aggregates but not amyloid fibrils (74). Attempts to identify a covalent distinction between PrP^C and PrP^{Sc} have been unsuccessful (75). In both PrP^C and PrP^{Sc}, the two cysteines form a disulfide bridge that is needed for PrP^{Sc} formation (69, 76). PrP^{Sc} can be denatured with GdnSCN, but refolding conditions that permit recovery of prion infectivity have not been identified (25). In the absence of conditions for reversible unfolding of PrP^{Sc}, it has not been possible to establish the relative stability of PrP^{Sc}, PrP^C, and the unfolded state. However, the difference in T_ms between the molecules suggests that PrP^{Sc} is more stable than PrP^C. This is consistent with inference 2. rPrP(90–231), when refolded at high concentration, forms a β-rich structure that melts at higher temperatures than rPrP(90–231) in the α-helical state. However, the melting curves for β rPrP(90–231) are not reversible, so realistic free energy estimates cannot be derived for this transition (43). Analytic ultracentrifugation reveals a multimer containing 6–7 β-rPrP molecules. While these recombinant molecules folded into a β-rich form share many spectroscopic and biochemical features of PrP^{Sc}, they are not infectious. This limits the conclusions that one can reach from a study of these molecules.

CD and FTIR Spectroscopy

Studies of purified PrP 27–30 showed that it assembled into polymers with the properties of amyloid (53), while earlier investigations had shown that other amyloids have a high β-sheet content (77, 78). In the prion diseases, the amyloid deposits were subsequently shown to contain large amounts of the PrP gene product (79–82). Subsequently, FTIR spectroscopy was used to measure the high β-sheet content of PrP 27–30 (71, 83). FTIR and CD studies showed that PrP^{Sc} contains $\sim$30% α-helical structure and $\sim$40% β-sheet (71, 84). These data are in marked contrast to the structural studies of PrP^C purified from normal brain (71, 85) and rPrP(90–231) (14), which are soluble molecules with substantial α-helical structure (40%) and little β-structure ($\sim$3%) by CD and FTIR spectroscopy. Taken together, these studies confirm inference 3 concerning a large conformational distinction between PrP^C and PrP^{Sc}.

Computational Studies

With experimental evidence of two distinct conformational isoforms in hand, efforts were made to predict the secondary and tertiary structures of PrP^C and PrP^{Sc} (Figure 3, p. C-9, color section at end of volume) (11, 86). A series of secondary structure prediction algorithms were applied to a collection of PrP sequences ranging from species as divergent as chicken and human. No regular secondary structure was identified for the octarepeat region from residues 23–90. Four or five putative structural regions were identified, but there was a relative lack of consistency about the designation of these regions as α-helices or β-strands (11). In an effort to fit the spectroscopic data, four of the regions were identified as structured. We presumed that all these regions adopted α-helical conformation in PrP^C, while two of these regions formed the component strands of a β-sheet in the PrP^{Sc} isoform (86). Combinatorial packing algorithms were applied to obtain plausible models of the tertiary structures of PrP^C and PrP^{Sc}. Each alternative structure was examined for inconsistencies with experimental data from peptide studies, with mutation information derived from patients with inherited prion diseases and with sequence polymorphism data derived from an extensive effort to sequence PrP genes from a variety of species. The result was low-resolution models of the structures of PrP^C and PrP^{Sc} that facilitated the design of a variety of peptides and transgenic constructs.

Antibody Studies

Early studies of distinctions in the antigenic surface of PrP^C and PrP^{Sc} were hampered by the limited antigenicity of PrP^{Sc} (35, 87). Operationally, only two monoclonal antibodies have been developed that recognize PrP: 13A5 and 3F4 (88, 89). While both bind to PrP^C and denatured PrP^{Sc}, neither binds to folded PrP^{Sc} effectively (Figure 4, p. C-9, color section). Recently, we have surveyed a diverse set of Fab fragments of recombinant origin (90). Seven Fabs were identified that bind to distinct linear PrP epitopes: D4, R10, D13, R72, R1, R2 and D2. The last three of these can reliably recognize active PrP^{Sc} as well as PrP^C. Epitope mapping studies demonstrate that 3F4, D4, R10, and D13 bind to an epitope bounded by residues 94–112 (I), while 13A5 and R72 recognize a distinct epitope formed by residues 138–165 (II). The extreme C terminus of the molecule, residues 225–231, contains a binding site for R1, R2, and D2 (III). The large conformational rearrangement between PrP^C and PrP^{Sc} demonstrated spectroscopically is supported by a clearly differentiable antibody binding pattern. From epitope mapping studies, we can now localize the conformationally flexible region to include residues 90–112. This region could extend to residue 138. In contrast, the region between residues 225 and 231 is likely to adopt a similar structure in PrP^C and PrP^{Sc} based on the relatively uniform binding of rFabs R1, R2, and D2 to both isoforms. Whether this conformationally rigid zone extends to include residues 166–224 remains to be determined. This evaluation will require the identification of monoclonal antibodies or Fab fragments that recognize this region. Recent work by Korth et al has identified an IgM that appears to recognize specifically PrP^{Sc} (90a). These results are consistent with the conformational dimorphism of PrP^{C} and PrP^{Sc} and support a model for the structure of PrP^{Sc} similar to that proposed by Huang et al (86).

PrP^{C} BINDS PrP^{Sc} DURING SCRAPIE PRION FORMATION

A cardinal feature of prion transmission studies is the existence of a "species barrier" (91). That is, prions derived from a particular host species are more effective in transmitting disease upon inoculation into an animal of the same species than into one of a more evolutionarily distant species. This concept can be explicitly tested by creating mice expressing a Syrian hamster (SHa) PrP transgene (92). When inoculated with mouse PrP^{Sc} (MoPrP^{Sc}), this transgenic animal produces MoPrP^{Sc}. In contrast, when transgenic (Tg) mice expressing SHaPrP were inoculated with SHa prions containing SHaPrP^{Sc}, the Tg animals produced SHaPrP^{Sc} (29). Thus, the transgene product SHaPrP^{C} must recognize and bind SHaPrP^{Sc} to template a sequence-specific conversion. A similar situation is obtained with the MoPrP^{C}/MoPrP^{Sc} interaction. Thus, homotypic interactions between identical PrP sequences in the PrP^{C} and PrP^{Sc} isoforms are more favorable than the heterotypic alternatives (52). This principle of prion replication was extended using chimeric SHa/MoPrP transgenes in which only the central domain (residues 96–167) rendered the mice susceptible to SHa prions (30, 93). Subsequent biochemical studies demonstrated that PrP^{C} binds PrP^{Sc} respecting sequence-specific preferences (31, 94, 95). These studies provide evidence in support of inference 4 concerning the specificity of the PrP^{C}/PrP^{Sc} heterodimer interface.

PROTEIN X IN PrP^{Sc} FORMATION

Based on the results with chimeric SHa/MoPrP transgenes, mice expressing chimeric Hu/MoPrP transgenes (MHu2M) were created. These Tg mice, which coexpressed MHu2M and MoPrP^{C}, were susceptible to Hu prions from the brains of patients that died of inherited, sporadic, and infectious prion diseases, whereas mice coexpressing Hu and MoPrP^{C} were resistant to Hu prions (19). Only when Tg(HuPrP) mice were crossed onto the Prnp$^{0/0}$ background did they become susceptible to Hu prions. To achieve comparable incubation times, the expression of HuPrP needed to be at least 10-fold higher than that in

Tg(MHu2M)Prnp[0/0] mice. These findings were interpreted in terms of an auxiliary macromolecule that facilitates the conversion of PrP[Sc] formation. Since
this molecule is most likely a protein, it was provisionally designated protein X.
To explain the resistance of Tg mice expressing both MoPrP and HuPrP to Hu
prions, MoPrP[C] must bind to protein X with a higher avidity than does HuPrP[C].
Because the conversion of MHu2MPrP[C] into PrP[Sc] was only weakly inhibited
by MoPrP[C], it was surmised that either the N- or C-terminal domains of PrP but
not the central domain binds to protein X (19). Since the N-terminus of PrP is
required for neither transmission nor propagation of prions (76, 96, 97), it was
then postulated that the binding site for protein X lies at the C-terminus (19).
To test this hypothesis, a chimeric Hu/Mo gene was constructed in which the
N-terminal and central domains were composed of MoPrP and the C-terminal
region of HuPrP. This construct (M3HuPrP) did not support PrP[Sc] formation in
scrapie-infected mouse neuroblastoma (ScN2a) cells, which suggests that its
C-terminus carrying the HuPrP sequence did not bind to Mo protein X. Subsequent studies of the PrP[C]/protein X interaction in ScN2a cells have shown
that MoPrP residues 167, 171, 214, and 218 are essential for protein X binding
(33). Biochemical efforts to purify protein X are under way.

NMR SPECTROSCOPY

Solution phase NMR spectroscopic studies have been completed on three nonglycosylated fragments of PrP derived from synthetic or recombinant sources
and refolded in an effort to replicate the conformation of the native glycosylated membrane-anchored PrP[C]. These fragments are (a) SHaPrP(90–145), a
synthetic peptide that begins with the residues at the N-terminus of PrP 27–30
(54) and ends at a position corresponding to a truncation mutation observed
in a Japanese patient diagnosed with a prion disease (42, 98); (b) MoPrP(121–
231), a recombinant molecule found to form a stable folding domain (99) and
corresponding to most of PrP[C]-II, a normal degradation product of PrP[C] (70,
100–102); and (c) SHaPrP(90–231), a recombinant protein corresponding to
the sequence of PrP 27–30 (14).

In an effort to investigate the structural basis of the conversion of α-helical
structure in PrP[C] into β-sheet structure in PrP[Sc], chemically synthesized peptide fragments corresponding to the SHaPrP(90–145) and subfragment 109–141
were studied. These peptides include regions of the PrP sequence that, from an
evolutionary perspective, are most conserved (58) as well as regions previously
shown to be conformationally plastic (41). A shorter PrP peptide composed of
residues 109–121 corresponding to the first putative α-helix was unexpectedly
found to adopt a β-sheet conformation (103), the structure of which has been
examined by solid state NMR (104). A slightly larger peptide composed of

residues 106–126 has been studied both with respect to its structure and neuro-toxicity using primary cultures of hippocampal neurons (105–107). The longer SHaPrP(90–145) and SHaPrP(109–141) peptides could be induced to form α-helical structures in some organic solvents (e.g. trifluoroethanol) and deter-gents (e.g. sodium dodecyl sulfate) (42). In contrast, acetonitrile or physiologic concentrations of NaCl facilitated the formation of a β-sheet rich structure. Ul-trastructurally, the β-sheet rich form assembled into rod shape polymers. Sta-bilized by the hydrophobic support of detergent micelles, the peptide forms one stable and one more transient helix as judged by NMR spectroscopy. Chemical shift data on PrP(90–145) in 20 mM sodium acetate and 100 mM SDS collected at pH 3.7 and 25°C suggested a stable α-helix in residues 113–122 and a more transient helical structure from 128-139. While it was possible to assign all the proton resonances, only short-range NOEs were identified, suggesting that this PrP fragment has only transient long-range organization, at best. At pH 5.0 and at pH 7.2 to a lesser extent, PrP(90–145) will form a β-sheet rich structure as measured by FTIR.

In efforts to identify a conformationally stable PrP fragment for NMR char-acterization, some investigators chose a MoPrP fragment of 111 residues (99), which is approximately eight residues smaller than PrP^C-II (101). Using this *Escherichia coli*–derived MoPrP(121–231) fragment, a three-dimensional (3-D) structure was obtained using standard heteronuclear spectroscopic meth-ods (12). At pH 4.5 and 20°C in the absence of any buffer, three α-helical and two β-strand regions were identified. As anticipated by the computational and biochemical studies, a disulfide bridge joined the two C-terminal helices. Sev-eral regions of the chain were judged to be too flexible to locate specifically in the three-dimensional structure of this fragment, including residues 121–128, 167–176, and 220–231.

The structure of SHaPrP(90–231), a recombinant protein derived from *E. coli* with a sequence corresponding to PrP 27–30 isolated from scrapie-infected Syrian hamsters (14), has been determined (13). Biochemical and immunologic studies of SHaPrP(90–231), also designated rPrP or rPrP(90–231) (14), suggest a pH labile structure below pH 5.2. As rPrP retains much if not all of its second-ary structure at low pH, it is likely that rPrP adopts an acid-induced "A-state" reminiscent of that observed in apomyoglobin. Fortunately, James et al's studies were carried out at pH 5.2 and 30°C in 20 mM sodium acetate (13).

From the NOE cross-peaks, 2401 experimental distance restraints were used to generate low-resolution structure via standard constrained refinement cal-culations (108). A best-fit superposition of backbone atoms for residues 113–228 of rPrP is shown in Figure 5a (p. C-10, color section) (13). The helices and β-sheet are fairly well defined, and loop regions can also be defined, al-though with less precision. Helix A spans residues 144–157 with the last turn quite distorted, corresponding to helix 144–154 found for MoPrP(121–231). Helix B includes residues 172–193, with the first turn irregular at the present stage of structure refinement. This is about two turns longer than the 179–193 helix found for MoPrP(121–231) (12), which agrees well with predicted helix H3 (179–191) (11). Helix C extends from residue 200 to 227 with the 225–227 turn irregular. This is substantially longer than the helix corresponding to residues 200–217 in MoPrP(121–231) (12). It is notable that predicted helix H4 (residues 202–218) (11) corresponds well with that found in MoPrP(121–231). Two four-residue β-strands (128–131 and 161–164) were identified in the MoPrP(121–231) structure (11). In rPrP, a similar antiparallel β-sheet, as well as S2 spanning residues 161–163 and S1 spanning 129–131 possessing β-sheet characteristics, was found, but the two strands do not manifest standard β-sheet geometry (13). In fact, a β-bridge occurs only between Leu[130] and Tyr[162], and there are extensive cross-strand connectivities of residues in segment 129–134 with proximate residues on the antiparallel segment 159–165.

The loop between S2 and helix B (i.e. residues 165–171) yields resonances clearly exhibiting long-range as well as medium-range restraints in rPrP (13); this contrasts with the absence of resonances for the backbone atoms of residues 167–176 in the shorter MoPrP(121–231) (12). The results of studies on rPrP indicate that the loop is reasonably ordered (13), whereas it was concluded that this region is disordered in MoPrP(121–231) (12).

The NMR results for rPrP(90–231) (13), compared with the structure reported for MoPrP(121–231) (12), support the notion that the core of the PrP^C structure is formed by parts of helices B and C, corresponding largely to the predicted H3 and H4 regions (11), and stabilized by the disulfide, which is essential for α-helical folding (14, 76). As seen in Figure 5 (color section), helices B and C essentially form one side of the protein structure (13). This core is further stabilized by helix A, which lies across helix C with side chains between the two helices interacting (Figure 5b, p. C-11, color section). Strand S2 also lies on this side of the protein and interacts predominantly with helices B and C as well as S1. With or without S2 and S1, we presume this relatively stable folding core is associated with the second unfolding transition.

Under the conditions used for NMR studies, it appears that rPrP(90–231) forms dimers a measurable fraction of the time (13). Recent unfolding experiments on rPrP(90–231) as a function of concentration suggest that unfolding follows a two-state regime at low protein concentrations; thus it is likely that the initial unfolding transition at high protein concentration represents the disruption of a dimer interface (S Marquee, personal communication). Since MoPrP(121–231) exhibits biphasic unfolding at high protein concentration (99), we believe that an essential component of the PrP*:PrP* dimer interface will include residues 90–120. This segment is sequentially and spatially distinct from the protein X binding site and includes some of the species poly-morphisms that are likely contributors to the PrP*:PrP* interface and may be part of the molecular basis of the species barrier (58, 109). Figure 6 (p. C-15, color section) highlights the side chains that are polymorphic in the 90–120 segment. In addition, other polymorphic residues that are spatially adjacent to this subgroup are presented. We hypothesize that this interface is the PrP* portion of the dimerization interface that is essential for PrP^{Sc} replication and that small molecules that bind to this region will disrupt PrP^{Sc} replication. In contrast, polypeptides including antibodies that mimic this interface could act as templates to facilitate PrP^{Sc} replication.

Presence of the additional 31 N-terminal residues of rPrP(90–231), relative to MoPrP(121–231), induces substantial changes in the structure of PrP in-cluding alterations in the C-terminus (13). Helix C is extended by at least nine residues, helix B is up to seven residues longer, and the loop comprising residues 165–171 is sufficiently ordered that many long-range restraints can be observed. A hydrophobic region without regular secondary structure (residues 113–125) predominantly interacts with S1 in the β-sheet (Figure 5c, p. C-12, color section). This may serve to stabilize the observed extension of helix B from 179 in MoPrP(121–231) to 172 in rPrP(90–231). Stability may also be conferred by hydrophobic interactions of Tyr^{128} with Tyr^{163} in the β-sheet, which, in turn, interacts with Val^{176}. The relative stability of the 165–171 loop and the three additional helical turns in helix C are presumably connected to stabilization of the other structural elements.

Within the irregular hydrophobic cluster (13) is a palindromic sequence, $A^{113}GAAAAGA$, that together with surrounding residues is conserved in all species examined to date (58). In humans, the A117V mutation causes GSS (110–112), while an artificial set of mutations consisting of A113V, A115V, and A118V in Tg mice causes spontaneous neurodegeneration and promotes β-sheet formation in recombinant PrP (113). Residues where point mutations lead to human diseases [for review see (46)] are highlighted in Figure 5d (p. C-13, color section). A point mutation in the PrP gene that changes Asp^{178} to Asn^{178} causes FFI if residue 129 is Met (114); the double mutation with Met^{129} changed to Val results in a subtype of CJD instead (115). Residue 178, which is in the extension of helix B seen in rPrP(90–231) (13) but not in MoPrP(121–231) (12), and residue 129 are located opposite one another with strand S2 partially intervening (Figure 5e). If the mutation D178N destabilizes the structure, part of helix B could unravel. This part of helix B is near Arg^{164} in S2, which in turn is adjacent to Met^{129}. Depending upon the identity of residue 129, the structure of the codon 178 mutants may vary in a fashion anticipated by Gambetti and coworkers.

Studies on the transmission of Hu prions to Tg mice suggest that a species-specific non-PrP protein designated protein X as noted above acts to facilitate

PrP^{Sc} formation (19). Information gained from mutagenesis and NMR studies performed in concert suggests the side chains of MoPrP residues Gln^{167}, Gln^{171}, Thr^{214}, and Gln^{218}, which correspond to SHaPrP residues Gln^{168}, Gln^{172}, Thr^{215}, and Gln^{219}, respectively, form the site at which protein X binds to PrP^C (13, 33). While sequentially distant, these residues cluster on one face of the molecule (13). The glycosylation sites, Asn^{181} and Asn^{197}, are evidently not very near this putative binding site (Figure 5e, p. C-14, color section). A comparison of the structure of rPrP(90–231) with MoPrP(121–231) (12) suggests that omission of residues 90–120 destabilizes helix C, resulting in its truncation and consequent disordering of residues 167–176. Perhaps this explains why MoPrP(121–231) is ineligible for conversion into PrP^{Sc} and does not appear to be able to bind to protein X (76) (L Zulianello, K Kaneko, FE Cohen & SB Prusiner, manuscript in preparation). As seen in Figure 5d (p. C-13, color section), Thr^{215} and Gln^{219} lie in register one turn apart on helix C and interact with the loop containing residues 167 and 171. This information also allows us to understand the structural basis of genetic resistance to scrapie. Residue 171 is a Gln in most species. Studies of Suffolk sheep in the United States revealed codon 171 to be polymorphic, encoding either Gln or Arg. All Suffolk sheep with scrapie were found to be Gln/Gln, which indicates that Arg conferred resistance (116). These data suggest that the basic side chain of Arg acts to increase the affinity of PrP^C(R171) for protein X such that PrP^C is not readily released from the complex. Such a scenario where R171 acts through a negative dominant mechanism seems likely, since heterozygous Arg/Gln sheep are also resistant to scrapie. Susceptibility to scrapie in other breeds of sheep is also determined largely by the nature of residue 171 (117–124). Equally important is the observation that ~12% of the Japanese population encode Lys instead of Glu at position 219 (125). No cases of CJD have been found in people with Lys^{219}, which, like Arg, is basic (125a).

Recently, the NMR structures of MoPrP(23–231) and SHaPrP(29–231) have been characterized (125b,c). This work suggests that the N-terminal residues from 23 or 29 to 124 are flexible, under the experimental conditions chosen as judged by the negative NOEs and patterns of chemical shift data. In spite of this flexibility in SHaPrP(29–231), the octarepeat region makes transient interactions with the C-terminal end of helix B, as judged by changes in the C$_\alpha$ chemical shifts for these residues in SHaPrP(29–231) compared with rPrP(90–231). However, it is unlikely that the octarepeat region is truly flexible and disordered under biologically relevant conditions given that full-length, and not truncated, PrP^C is normally purified from animal brain preparations and that the addition of a sixth octarepeat gives rise to an inherited prion disease in humans.

It now seems clear that PrP^C is a metalloprotein and that the highly flexible octarepeat region seen in the NMR experiments may be much more structured in the presence of copper. Each SHaPrP(29–231) molecule has been shown to bind copper preferentially over other divalent cations, following a 2:1 stoichiometry with spectroscopic evidence of tryptophan burial upon metal binding (125d). This work extends earlier work on the binding of copper to synthetic peptides containing the octarepeat sequence (125e–125g). The biological relevance of these findings is supported by the recent findings of alterations of copper levels in Cu/Zn superoxide dismutase activity in PrP-deficient (Prnp$^{0/0}$) mice (125h). The paramagnetic properties of Cu(II) will delay NMR characterization of the complete structure of this metalloprotein.

PRION STRAINS

The existence of prion strains has been considered difficult to reconcile with the protein-only model and the evidence that prion diseases involved protein folding and distinct folded conformations (52, 126–128). However, recent results on the creation of new prion strains and their subsequent biochemical characterization have shown how the strain phenomena is entirely compatible with the protein-only conformational conversion model (17). Prion strains were initially isolated based on different clinical syndromes in goats with scrapie (129); subsequently, strains were isolated in rodents based on different incubation times and neuropathologic profiles (130, 131). New strains have been produced upon passage from one species to another (132) or passage from non-Tg mice to mice expressing a foreign or artificial PrP transgene (133).

Studies of the drowsy (DY) and hyper (HY) prion strains isolated from mink by passage in Syrian hamsters showed that two strains produced PrP^{Sc} molecules with protease-resistant cores (PrP 27–30) of different molecular sizes as judged by gel electrophoresis (134). Since the smaller size of the protease-resistant fragment of the DY strain did not correlate with a particular phenotype, the significance of these findings was unclear (133). In seemingly unrelated studies, brain extracts from patients who died of FFI exhibited a deglycosylated, protease-resistant core of PrP$^{Sc/FFI}$ of 19 kDa, while that found in extracts from patients with familial fCJD(E200K) and sCJD was 21 kDa (135). The smaller size of the PrP$^{Sc/FFI}$ fragment was attributed to the distinct sequence of the PrP$^{Sc/FFI}$ protein. Studies on the transmission of prions from patients with FFI, fCJD(E200K), or sCJD to Tg(MHu2M)Prnp$^{0/0}$ mice have shown that different sequences were not required to maintain the structural distinctions as judged by the molecular sizes of the deglycosylated, protease-resistant core of PrP^{Sc} molecules (17). The Tg(MHu2M)Prnp$^{0/0}$ mice inoculated with extracts from FFI patients developed signs of CNS dysfunction ~200 days later and exhibited a deglycosylated, protease-resistant core of PrP^{Sc} that was 19 kDa in size, whereas mice inoculated with extracts from fCJD(E200K) patients developed signs at ~170 days and exhibited a PrP^{Sc} of 21 kDa. On second passage in Tg(MHu2M)Prnp$^{0/0}$ mice, the animals receiving the mouse FFI extract developed neurologic disease after ~130 days and displayed a 19-kDa PrP^{Sc} molecule, while those injected with the mouse fCJD extract exhibited disease after ~170 days and showed a 21-kDa PrP^{Sc} molecule. The second passage of these distinct prion strains with identical PrP amino acid sequences demonstrates that the size of the protease-resistant core is a stable feature of the strain. Thus, the two prion strains represent distinct PrP^{Sc} structures that can be enciphered on a proteinaceous template that is stable in passaging experiments.

What is the structural basis of these alternative PrP^{Sc} conformers? Work on diphtheria toxin identified distinct crystal forms that displayed different tertiary and quaternary structures for a single polypeptide sequence (136). To describe this observation, the notion of domain swapping was introduced where a region of one monomer displaced the corresponding region in another monomer to create an interlocking molecular handshake (136). This phenomenon has now been observed in a variety of other protein structures with the swapped elements being as small as an isolated α-helix or β-strand and as large as an entire folded domain. We suspect that a similar phenomenon is responsible for prion strains.

Imagine that the protease core of PrP^{Sc} has two subdomains joined by a linking region and that PrP^{Sc} exists as a dimer. Figure 7 (p. C-16, color section) shows a schematic of the fCJD-derived strain that shields a cleavage site near residue 105 but retains the cleavage site near residue 90. In the FFI-derived strain, the smaller subdomain of one monomer would swap with its partner in the dimer in a manner reminiscent of alternative ribonuclease A structures. In this way, the hetero subdomain interface (90–105:110–231) and the homo subdomain interface (90–105:90–105) are maintained, but the proteolytic remnant is quite different. Moreover, the templates provided by each protease-resistant core would present a distinct interface for the conversion of PrP^{C} to PrP^{Sc} during prion replication. Many lines of evidence suggest that the residues between 90 and 120 play a major role in the creation of the PrP^{C}/PrP^{Sc} interface.

Thus, in addition to a structure for PrP^{C} that is distinct from PrP^{Sc}, these results on prion strains suggest that there are multiple approximately isoenergetic PrP^{Sc} conformers. This is an obvious point of departure from earlier work demonstrating that for most proteins, there was a single folded structure that was uniquely encoded in the sequence (137). In an effort to probe the limits of this conformational pluralism for PrP, we have explored this phenomena on a thermodynamically well-understood model of protein structure, the H-P model on a square and cubic lattice (138). In this model two amino acid types are allowed; hydrophobic (H) and hydrophilic (P) residues and the conformations of the chain are limited to those that can be exactly embedded upon the lattice. The energetics of this system are equally discrete, H-H interactions are worth one favorable energy unit, and all other interactions are negligible. While this model lacks many features of real polypeptide chains, for relatively short chains it is possible to enumerate all chain conformations and to identify all sequences with a unique lowest-energy monomeric structure.

For H-P sequences of length 27 on a cubic lattice, one could ask how many sequences that have a unique monomeric structure would prefer a different conformation upon forming a symmetric dimer. With this simplified system, P Harrison et al (personal communication) found that ~5% of all sequences are prion-like in that the lowest-energy structure for the monomer is unlike the most stable dimeric structures. Taking this analogy one step further, it is possible to identify domain swaps between the N- or C-terminal regions of each monomeric member of the dimer that are isoenergetic with the new interleaved dimeric arrangement. This is reminiscent of the experimental work on strains. In this lattice limit, it is easy to see how the strains could act as templates to direct the formation of distinct "scrapie" molecules. It would also follow that proteolysis experiments could reveal different resistant cores. One could also ask how often a mutation to the 27-mer sequence would differentially affect the stability of the dimeric structures. Our preliminary calculations document that this is a relatively common phenomena holding true for 20–30% of the residues especially when residues are mutated from Ps to Hs.

CONCLUSIONS

A wide array of biochemical and biophysical experimental results indicate that a dramatic change in protein conformation is a central feature of the prion diseases. While the transmissible aspect of prion biology has led to a variety of quite disparate hypotheses about the molecular basis of prion diseases, these mechanistic issues are more easily understood when one begins with the tenet that these diseases are fundamentally autosomal-dominant inherited diseases. From a historical perspective, the inherited prion diseases are likely to have been the first to arise, as the stochastic nature of sCJD would seem to require a substantial population of individuals over the age of 60 and cannibalism could only lead to infectious CJD if the victim had preexisting familial or sporadic CJD.

Many autosomal-dominant inherited diseases act by distorting the structure of a protein. The experimental results on the prion diseases are entirely compatible with this notion. In general, the change in protein structure could be under kinetic or thermodynamic control. For the prion diseases, the evidence points toward kinetic control. A special feature of the prion diseases is that the disease-causing isoform, PrP^{Sc}, appears to be capable of acting as a template to lower the kinetic barriers that normally separate PrP^C from PrP^{Sc}. This provides a simple explanation for the inherited prion diseases that can be easily adapted to explain sporadic and infectious prion diseases.

To date, the majority of structural studies have focused on the soluble cellular isoform PrP^C. If we are truly to understand this disease process at a molecular level, we must endeavor to find routes to solubilize the PrP^{Sc} isoform or to identify polypeptides with PrP^{Sc}-like structures (76).

ACKNOWLEDGMENTS

We wish to thank our colleagues for their help with the development of the ideas presented in this manuscript. In particular, we wish to acknowledge S DeArmond, M Scott, A Wallace, K Kaneko, P Harrison, Z Kanyo, D Walther, L Zulianello, T Steitz, D Engelman, S Marqusee, R Fletterick, S Duniach, D Eisenberg, and F Richards. This work was supported by grants from the National Institutes of Health and the Human Frontiers of Science Program. We gratefully acknowledge gifts from the Sherman Fairchild Foundation, the Keck Foundation, the G Harold and Leila Y Mathers Foundation, the Bernard Osher Foundation, the John D French Foundation, and Centeon.

> **Visit the *Annual Reviews* home page at
> http://www.AnnualReviews.org.**

Literature Cited

1. Hsiao K, Baker HF, Crow TJ, Poulter M, Owen F, et al. 1989. *Nature* 338:342–45
2. Dlouhy SR, Hsiao K, Farlow MR, Foroud T, Conneally PM, et al. 1992. *Nat. Genet.* 1:64–67
3. Petersen RB, Tabaton M, Berg L, Schrank B, Torack RM, et al. 1992. *Neurology* 42:1859–63
4. Poulter M, Baker HF, Frith CD, Leach M, Lofthouse R, et al. 1992. *Brain* 115: 675–85
5. Gabizon R, Rosenmann H, Meiner Z, Kahana I, Kahana E, et al. 1993. *Am. J. Hum. Genet.* 33:828–35
6. Kelly JW. 1997. *Structure* 5:595–600
7. Lee JP, Stimson ER, Ghilardi JR, Mantyh PW, Lu YA, et al. 1995. *Biochemistry* 34:5191–200
8. Colon W, Lai Z, McCutchen SL, Miroy GJ, Strang C, Kelly JW. 1996. *Ciba Found. Symp.* 199:228–38
9. Matthews BW. 1996. *FASEB J.* 10:35–41
10. Bellinger-Kawahara CG, Kempner E, Groth DF, Gabizon R, Prusiner SB. 1988. *Virology* 164:537–41
11. Huang Z, Gabriel J-M, Baldwin MA, Fletterick RJ, Prusiner SB, Cohen FE. 1994. *Proc. Natl. Acad. Sci. USA* 91: 7139–43
12. Riek R, Hornemann S, Wider G, Billeter M, Glockshuber R, Wüthrich K. 1996. *Nature* 382:180–82
13. James TL, Liu H, Ulyanov NB, Farr-Jones S, Zhang H, et al. 1997. *Proc. Natl. Acad. Sci. USA* 94:10086–91
14. Mehlhorn I, Groth D, Stöckel J, Moffat B, Reilly D, et al. 1996. *Biochemistry* 35:5528–37
15. Kaneko K, Wille H, Mehlhorn I, Zhang H, Ball H, et al. 1997. *J. Mol. Biol.* 270: 574–86
16. Prusiner SB. 1989. *Annu. Rev. Microbiol.* 43:345–74
17. Telling GC, Parchi P, DeArmond SJ, Cortelli P, Montagna P, et al. 1996. *Science* 274:2079–82
18. Hsiao KK, Groth D, Scott M, Yang S-L, Serban H, et al. 1994. *Proc. Natl. Acad. Sci. USA* 91:9126–30
19. Telling GC, Scott M, Mastrianni J, Gabizon R, Torchia M, et al. 1995. *Cell* 83:79–90
20. Telling GC, Haga T, Torchia M, Trem-

blay P, DeArmond SJ, Prusiner SB. 1996. *Genes Dev.* 10:1736–50

21. Goudsmit J, Morrow CH, Asher DM, Yanagihara RT, Masters CL, et al. 1980. *Neurology* 30:945–50

22. Godec MS, Asher DM, Kozachuk WE, Masters CL, Rubi JU, et al. 1994. *Neurology* 44:1111–15

23. Prusiner SB, McKinley MP, Groth DF, Bowman KA, Mock Nl, et al. 1981. *Proc. Natl. Acad. Sci. USA* 78:6675–79

24. Prusiner SB, Groth DF, McKinley MP, Cochran SP, Bowman KA, Kasper KC. 1981. *Proc. Natl. Acad. Sci. USA* 78: 4606–10

25. Prusiner SB, Groth D, Serban A, Stahl N, Gabizon R. 1993. *Proc. Natl. Acad. Sci. USA* 90:2793–97

26. Wilesmith JW, Ryan JBM, Atkinson MJ. 1991. *Vet. Rec.* 128:199–203

27. Gajdusek DC. 1977. *Science* 197:943–60

28. Prusiner SB, Cochran SP, Alpers MP. 1985. *J. Infect. Dis.* 152:971–78

29. Prusiner SB, Scott M, Foster D, Pan K-M, Groth D, et al. 1990. *Cell* 63:673–86

30. Scott M, Groth D, Foster D, Torchia M, Yang S-L, et al. 1993. *Cell* 73:979–88

31. Kocisko DA, Priola SA, Raymond GJ, Chesebro B, Lansbury PT Jr, Caughey B. 1995. *Proc. Natl. Acad. Sci. USA* 92: 3923–27

32. Prusiner SB. 1997. *Science* 278:245–51

33. Kaneko K, Zulianello L, Scott M, Cooper CM, Wallace AC, et al. 1997. *Proc. Natl. Acad. Sci. USA* 94:10069–74

34. Büeler H, Aguzzi A, Sailer A, Greiner R-A, Autenried P, et al. 1993. *Cell* 73: 1339–47

35. Prusiner SB, Groth D, Serban A, Koehler R, Foster D, et al. 1993. *Proc. Natl. Acad. Sci. USA* 90:10608–12

36. Manson JC, Clarke AR, McBride PA, McConnell I, Hope J. 1994. *Neurodegeneration* 3:331–40

37. Sakaguchi S, Katamine S, Shigematsu K, Nakatani A, Moriuchi R, et al. 1995. *J. Virol.* 69:7586–92

38. Oesch B, Westaway D, Wälchli M, McKinley MP, Kent SBH, et al. 1985. *Cell* 40:735–46

39. Prusiner SB, Bolton DC, Groth DF, Bowman KA, Cochran SP, McKinley MP. 1982. *Biochemistry* 21:6942–50

40. McKinley MP, Bolton DC, Prusiner SB. 1983. *Cell* 35:57–62

41. Nguyen J, Baldwin MA, Cohen FE, Prusiner SB. 1995. *Biochemistry* 34:4186–92

42. Zhang H, Kaneko K, Nguyen JT, Livsh-its TL, Baldwin MA, et al. 1995. *J. Mol. Biol.* 250:514–26

43. Zhang H, Stöckel J, Mehlhorn I, Groth D, Baldwin MA, et al. 1997. *Biochemistry* 36:3543–53

44. Chapman J, Ben-Israel J, Goldhammer Y, Korczyn AD. 1994. *Neurology* 44:1683–86

45. Spudich S, Mastrianni JA, Wrensch M, Gabizon R, Meiner Z, et al. 1995. *Mol. Med.* 1:607–13

46. Prusiner SB, Scott MR. 1997. *Annu. Rev. Genet.* 31:139–75

47. Büeler H, Raeber A, Sailer A, Fischer M, Aguzzi A, Weissmann C. 1994. *Mol. Med.* 1:19–30

48. Carlson GA, Ebeling C, Yang S-L, Telling G, Torchia M, et al. 1994. *Proc. Natl. Acad. Sci. USA* 91:5690–94

49. Cohen FE, Pan K-M, Huang Z, Baldwin M, Fletterick RJ, Prusiner SB. 1994. *Science* 264:530–31

50. Prusiner SB. 1982. *Science* 216:136–44

51. Prusiner SB. 1984. *Sci. Am.* 251:50–59

52. Prusiner SB. 1991. *Science* 252:1515–22

53. Prusiner SB, McKinley MP, Bowman KA, Bolton DC, Bendheim PE, et al. 1983. *Cell* 35:349–58

54. Prusiner SB, Groth DF, Bolton DC, Kent SB, Hood LE. 1984. *Cell* 38:127–34

55. Chesebro B, Race R, Wehrly K, Nishio J, Bloom M, et al. 1985. *Nature* 315:331–33

56. Sparkes RS, Simon M, Cohn VH, Fournier REK, Lem J, et al. 1986. *Proc. Natl. Acad. Sci. USA* 83:7358–62

57. Basler K, Oesch B, Scott M, Westaway D, Wälchli M, et al. 1986. *Cell* 46:417–28

58. Bamborough P, Wille H, Telling GC, Yehiely F, Prusiner SB, Cohen FE. 1996. *Cold Spring Harbor Symp. Quant. Biol.* 61:495–509

59. Alper T, Cramp WA, Haig DA, Clarke MC. 1967. *Nature* 214:764–66

60. Latarjet R, Muel B, Haig DA, Clarke MC, Alper T. 1970. *Nature* 227:1341–43

61. Diener TO, McKinley MP, Prusiner SB. 1982. *Proc. Natl. Acad. Sci. USA* 79: 5220–24

62. Bellinger-Kawahara C, Cleaver JE, Diener TO, Prusiner SB. 1987. *J. Virol.* 61: 159–66

63. Bellinger-Kawahara C, Diener TO, McKinley MP, Groth DF, Smith DR, Prusiner SB. 1987. *Virology* 160:271–74

64. Locht C, Chesebro B, Race R, Keith JM. 1986. *Proc. Natl. Acad. Sci. USA* 83: 6372–76

65. Endo T, Groth D, Prusiner SB, Kobata A. 1989. *Biochemistry* 28:8380–88
66. Stahl N, Baldwin MA, Burlingame AL, Prusiner SB. 1990. *Biochemistry* 29:8879–84
67. Yost CS, Lopez CD, Prusiner SB, Myers RM, Lingappa VR. 1990. *Nature* 343:669–72
68. De Fea KA, Nakahara DH, Calayag MC, Yost CS, Mirels LF, et al. 1994. *J. Biol. Chem.* 269:16810–20
68a. Hegde RS, Mastrianni JA, Scott MR, De-Fea KA, Tremblay P, et al. 1998. *Science* 279:827–34
69. Turk E, Teplow DB, Hood LE, Prusiner SB. 1988. *Eur. J. Biochem.* 176:21–30
70. Pan K-M, Stahl N, Prusiner SB. 1992. *Protein Sci.* 1:1343–52
71. Pan K-M, Baldwin M, Nguyen J, Gasset M, Serban A, et al. 1993. *Proc. Natl. Acad. Sci. USA* 90:10962–66
72. Riek R, Hornemann S, Wider G, Glockshuber R, Wüthrich K. 1997. *FEBS Lett.* 413:282–88
73. Donne DG, Viles JH, Groth D, Mehlhorn I, James TL, et al. 1997. *Proc. Natl. Acad. Sci. USA* 94:13452–57
74. McKinley MP, Meyer R, Kenaga L, Rahbar F, Cotter R, et al. 1991. *J. Virol.* 65:1440–49
75. Stahl N, Baldwin MA, Teplow DB, Hood L, Gibson BW, et al. 1993. *Biochemistry* 32:1991–2002
76. Muramoto T, Scott M, Cohen F, Prusiner SB. 1996. *Proc. Natl. Acad. Sci. USA* 93:15457–62
77. Glenner GG, Eanes ED, Page DL. 1972. *J. Histochem. Cytochem.* 20:821–26
78. Glenner GG. 1980. *N. Engl. J. Med.* 302:1283–92
79. Bendheim PE, Barry RA, DeArmond SJ, Stites DP, Prusiner SB. 1984. *Nature* 310:418–21
80. DeArmond SJ, McKinley MP, Barry RA, Braunfeld MB, McColloch JR, Prusiner SB. 1985. *Cell* 41:221–35
81. Kitamoto T, Tateishi J, Tashima I, Takeshita I, Barry RA, et al. 1986. *Ann. Neurol.* 20:204–8
82. Roberts GW, Lofthouse R, Allsop D, Landon M, Kidd M, et al. 1988. *Neurology* 38:1534–40
83. Caughey BW, Dong A, Bhat KS, Ernst D, Hayes SF, Caughey WS. 1991. *Biochemistry* 30:7672–80
84. Safar J, Roller PP, Gajdusek DC, Gibbs CJ Jr. 1993. *J. Biol. Chem.* 268:20276–84
85. Pergami P, Jaffe H, Safar J. 1996. *Anal. Biochem.* 236:63–73
86. Huang Z, Prusiner SB, Cohen FE. 1996. *Folding Des.* 1:13–19
87. Williamson RA, Peretz D, Smorodinsky N, Bastidas R, Serban H, et al. 1996. *Proc. Natl. Acad. Sci. USA* 93:7279–82
88. Barry RA, Prusiner SB. 1986. *J. Infect. Dis.* 154:518–21
89. Kascsak RJ, Rubenstein R, Merz PA, Tonna-DeMasi M, Fersko R, et al. 1987. *J. Virol.* 61:3688–93
90. Peretz D, Williamson RA, Matsunaga Y, Serban H, Pinilla C, et al. 1997. *J. Mol. Biol.* 273:614–22
90a. Korth C, Stierli B, Streit P, Moser M, Schaller O, et al. 1997. *Nature* 389:74–77
91. Pattison IH. 1965. In *Slow, Latent and Temperate Virus Infections, NINDB Monogr. 2*, ed. DC Gajdusek, CJ Gibbs Jr, MP Alpers, pp. 249–57. Washington, DC: US Gov. Print. Off.
92. Scott M, Foster D, Mirenda C, Serban D, Coufal F, et al. 1989. *Cell* 59:847–57
93. Scott MR, Köhler R, Foster D, Prusiner SB. 1992. *Protein Sci.* 1:986–97
94. Kocisko DA, Come JH, Priola SA, Chesebro B, Raymond GJ, et al. 1994. *Nature* 370:471–74
95. Kaneko K, Peretz D, Pan K-M, Blochberger T, Wille H, et al. 1995. *Proc. Natl. Acad. Sci. USA* 32:11160–64
96. Rogers M, Yehiely F, Scott M, Prusiner SB. 1993. *Proc. Natl. Acad. Sci. USA* 90:3182–86
97. Fischer M, Rülicke T, Raeber A, Sailer A, Moser M, et al. 1996. *EMBO J.* 15:1255–64
98. Kitamoto T, Iizuka R, Tateishi J. 1993. *Biochem. Biophys. Res. Commun.* 192:525–31
99. Hornemann S, Glockshuber R. 1996. *J. Mol. Biol.* 262:614–19
100. Haraguchi T, Fisher S, Olofsson S, Endo T, Groth D, et al. 1989. *Arch. Biochem. Biophys.* 274:1–13
101. Chen SG, Teplow DB, Parchi P, Teller JK, Gambetti P, Autilio-Gambetti L. 1995. *J. Biol. Chem.* 270:19173–80
102. Taraboulos A, Scott M, Semenov A, Avrahami D, Laszlo L, Prusiner SB. 1995. *J. Cell Biol.* 129:121–32
103. Gasset M, Baldwin MA, Lloyd D, Gabriel J-M, Holtzman DM, et al. 1992. *Proc. Natl. Acad. Sci. USA* 89:10940–44
104. Heller J, Kolbert AC, Larsen R, Ernst M, Bekker T, et al. 1996. *Protein Sci.* 5:1655–61
105. Forloni G, Angeretti N, Chiesa R, Monzani E, Salmona M, et al. 1993. *Nature* 362:543–46

106. De Gioia L, Selvaggini C, Ghibaudi E, Diomede L, Bugiani O, et al. 1994. *J. Biol. Chem.* 269:7859–62

107. Brown DR, Schmidt B, Kretzschmar HA. 1996. *Nature* 380:345–47

108. Güntert P, Braun W, Wüthrich K. 1991. *J. Mol. Biol.* 217:517–30

109. Schätzl HM, Da Costa M, Taylor L, Cohen FE, Prusiner SB. 1995. *J. Mol. Biol.* 245:362–74

110. Doh-ura K, Tateishi J, Sasaki H, Kitamoto T, Sakaki Y. 1989. *Biochem. Biophys. Res. Commun.* 163:974–79

111. Hsiao KK, Cass C, Schellenberg GD, Bird T, Devine-Gage E, et al. 1991. *Neurology* 41:681–84

112. Mastrianni JA, Curtis MT, Oberholtzer JC, Da Costa MM, DeArmond S, et al. 1995. *Neurology* 45:2042–50

113. Scott MR, Nguyen O, Stöckel J, Tatzelt J, DeArmond SJ, et al. 1997. *Protein Sci.* 6 (Suppl. 1):84

114. Medori R, Tritschler H-J, LeBlanc A, Villare F, Manetto V, et al. 1992. *N. Engl. J. Med.* 326:444–49

115. Goldfarb LG, Petersen RB, Tabaton M, Brown P, LeBlanc AC, et al. 1992. *Science* 258:806–8

116. Westaway D, Zuliani V, Cooper CM, Da Costa M, Neuman S, et al. 1994. *Genes Dev.* 8:959–69

117. Hunter N, Goldmann W, Benson G, Foster JD, Hope J. 1993. *J. Gen. Virol.* 74: 1025–31

118. Goldmann W, Hunter N, Smith G, Foster J, Hope J. 1994. *J. Gen. Virol.* 75:989–95

119. Belt PBGM, Muileman IH, Schreuder BEC, Ruijter JB, Gielkens ALJ, Smits MA. 1995. *J. Gen. Virol.* 76:509–17

120. Clousard C, Beaudry P, Elsen JM, Milan D, Dussaucy M, et al. 1995. *J. Gen. Virol.* 76:2097–101

121. Ikeda T, Horiuchi M, Ishiguro N, Muramatsu Y, Kai-Uwe GD, Shinagawa M. 1995. *J. Gen. Virol.* 76:2577–81

122. Hunter N, Moore L, Hosie BD, Dingwall WS, Greig A. 1997. *Vet. Rec.* 140:59–63

123. Hunter N, Cairns D, Foster JD, Smith G, Goldmann W, Donnelly K. 1997. *Nature* 386:137

124. O'Rourke KI, Holyoak GR, Clark WW, Mickelson JR, Wang S, et al. 1997. *J. Gen. Virol.* 78:975–78

125. Kitamoto T, Tateishi J. 1994. *Philos. Trans. R. Soc. London Ser. B* 343:391–98

125a. Shibuya S, Higuchi J, Shin R-W, Tateishi J, Kitamoto T. 1998. *Lancet* 351:419

125b. Riek R, Hornemann S, Wider G, Glockshuber R, Wüthrich K. 1997. *FEBS Lett.* 413:282–88

125c. Donne DG, Viles JH, Groth D, Mehlhorn I, James TL, et al. 1997. *Proc. Natl. Acad. Sci. USA* 94:13452–57

125d. Stöckel J, Safar J, Wallace AC, Cohen FE, Prusiner SB. 1998. *Biochemistry.* In press

125e. Hornshaw MP, McDermott JR, Candy JM, Lakey JH. 1995. *Biochem. Biophys. Res. Commun.* 214:993–99

125f. Hornshaw MP, McDermott JR, Candy JM. 1995. *Biochem. Biophys. Res. Commun.* 207:621–29

125g. Miura T, Hori-i A, Takeuchi H. 1996. *FEBS Lett.* 396:248–52

125h. Brown DR, Qin K, Herms JW, Madlung A, Manson J, et al. 1997. *Nature* 390:684–87

126. Bruce ME, Dickinson AG. 1987. *J. Gen. Virol.* 68:79–89

127. Dickinson AG, Outram GW. 1988. In *Novel Infectious Agents and the Central Nervous System. Ciba Found. Symp. 135*, ed. G Bock, J Marsh, pp. 63–83. Chichester, UK: Wiley

128. Ridley RM, Baker HF. 1996. *Neurodegeneration* 5:219–31

129. Pattison IH, Millson GC. 1961. *J. Comp. Pathol.* 71:101–8

130. Fraser H, Dickinson AG. 1973. *J. Comp. Pathol.* 83:29–40

131. Dickinson AG, Fraser HG. 1977. In *Slow Virus Infections of the Central Nervous System*, ed. V ter Meulen, M Katz, pp. 3–14. New York: Springer-Verlag

132. Kimberlin RH, Cole S, Walker CA. 1987. *J. Gen. Virol.* 68:1875–81

133. Scott MR, Groth D, Tatzelt J, Torchia M, Tremblay P, et al. 1997. *J. Virol.* 7:9032–44

134. Bessen RA, Marsh RF. 1994. *J. Virol.* 68:7859–68

135. Monari L, Chen SG, Brown P, Parchi P, Petersen RB, et al. 1994. *Proc. Natl. Acad. Sci. USA* 91:2839–42

136. Bennett MJ, Schlunegger MP, Eisenberg D. 1995. *Protein Sci.* 4:2455–68

137. Anfinsen CB. 1973. *Science* 181:223–30

138. Chan HS, Dill KA. 1996. *Proteins* 24:335–44

139. Pearlman DA, Case DA, Caldwell J, Ross WS, Cheatham TE III, et al. 1995. *AMBER 4.1.* San Francisco: Univ. Calif.

Annu. Rev. Biochem. 1998. 67:821–55

THE AMP-ACTIVATED/SNF1 PROTEIN KINASE SUBFAMILY: Metabolic Sensors of the Eukaryotic Cell?

D. Grahame Hardie,[1] *David Carling,*[2] *and Marian Carlson*[3]
[1]Biochemistry Department, The University, Dundee, DD1 4HN, Scotland, United Kingdom, d.g.hardie@dundee.ac.uk; [2]MRC Molecular Medicine Group, Imperial College School of Medicine, Hammersmith Hospital, London, W12 0NN, United Kingdom, dcarling@rpms.ac.uk; [3]Department of Genetics and Development, College of Physicians and Surgeons, Columbia University, New York, New York 10032; e-mail: mbc1@columbia.edu

KEY WORDS: protein phosphorylation, gene expression, cellular stress, glucose repression, ATP depletion

ABSTRACT

Mammalian AMP-activated protein kinase and yeast SNF1 protein kinase are the central components of kinase cascades that are highly conserved between animals, fungi, and plants. The AMP-activated protein kinase cascade acts as a metabolic sensor or "fuel gauge" that monitors cellular AMP and ATP levels because it is activated by increases in the AMP:ATP ratio. Once activated, the enzyme switches off ATP-consuming anabolic pathways and switches on ATP-producing catabolic pathways, such as fatty acid oxidation. The SNF1 complex in yeast is activated in response to the stress of glucose deprivation. In this case the intracellular signal or signals have not been identified; however, SNF1 activation is associated with depletion of ATP and elevation of AMP. The SNF1 complex acts primarily by inducing expression of genes required for catabolic pathways that generate glucose, probably by triggering phosphorylation of transcription factors. SNF1-related protein kinases in higher plants are likely to be involved in the response of plant cells to environmental and/or nutritional stress.

0066-4154/98/0701-0821$08.00

CONTENTS

INTRODUCTION

Members of the AMP-activated/SNF1-related protein kinase subfamily are central components of highly conserved protein kinase cascades that now appear to be present in most, if not all, eukaryotic cells. Because the downstream targets of the action of these enzymes are many and varied, they have been discovered and rediscovered several times in different guises and by different approaches. Not until the cloning and sequencing of the DNAs encoding these proteins was it realized that all of the ascribed regulatory functions were carried out by members of the same class of protein kinase. Mammalian AMP-activated protein kinase (AMPK) was discovered through biochemical approaches, which has meant that understanding of its function in vivo has lagged behind insight into its regulation and its specificity for protein substrates in cell-free systems. Conversely, the role of the SNF1 system in the yeast Saccharomyces cerevisiae was discovered by genetic approaches, so understanding of its physiological functions in vivo has preceded information about its detailed biochemical properties in vitro. An enhanced level of understanding should be possible by a synthesis of these two approaches, in essence pooling knowledge about the animal and yeast systems. The insights obtained should also provide guidance in investigating the cellular role of SNF1-related kinases in higher plants, where studies are at a much earlier stage. Even though these investigations are ongoing, it is now possible to present some unifying hypotheses for the mechanisms of regulation and the physiological roles of the AMPK/SNF1 subfamily, which is the main objective of this review. Although the hypotheses presented are unlikely to be correct in every detail, their value is in providing a framework for further experimentation.

EARLY HISTORY OF MAMMALIAN AMP-ACTIVATED PROTEIN KINASE

With hindsight it is now possible to date the first experimental observations of AMP-activated protein kinase (AMPK) to two independent studies in 1973 (1, 2), although it was to be 14 years before it was realized that the phenomena studied were related (3) and 16 years before the kinase was given the name by which it is known today (4). In 1973 Beg and coworkers reported that a microsomal preparation of rat liver 3-hydroxy-3-methylglutaryl–coenzyme A reductase (HMG-CoA reductase), a key regulatory enzyme of cholesterol biosynthesis, was inactivated in a time-dependent manner by incubation with MgATP and a cytosolic fraction (1). The same phenomenon was observed in microsomes from human fibroblasts, where evidence was presented that both ADP and ATP were required (5). The apparent requirement for ADP was almost certainly because AMP was being generated from it via the adenylate kinase reaction. It was later shown that the inactivation was associated with phosphorylation of HMG-CoA reductase (6, 7), and the cytosolic factor became known as HMG-CoA reductase kinase. The other event in 1973 was the report by Carlson & Kim that partially purified rat liver acetyl-CoA carboxylase, the key regulatory enzyme of fatty acid biosynthesis, was phosphorylated and inactivated by a protein kinase that contaminated the preparation (2). The protein kinase responsible remained poorly characterized, but in 1980 Yeh et al reported that the inactivation of acetyl-CoA carboxylase was stimulated by 5'-AMP (8).

In 1985, Ferrer et al found that rat liver HMG-CoA reductase kinase was also stimulated by AMP (9). At the time there appears to have been no suggestion that the acetyl-CoA carboxylase kinase studied bore any relationship to the HMG-CoA reductase kinase(s) studied elsewhere. In the mid-1980s, Carling et al (3) partially purified a protein kinase from rat liver, initially called acetyl-CoA carboxylase kinase-3 (ACK3), which phosphorylated and inactivated acetyl-CoA carboxylase. This activity was stimulated by AMP, which suggested that we had purified the activity observed in cruder fractions by Yeh et al (8). However, ACK3 also inactivated HMG-CoA reductase, and both activities were regulated identically by phosphorylation and by AMP (3).

The acetyl-CoA carboxylase and HMG-CoA reductase kinase activities were later shown to copurify from a rat liver extract and corresponded to the only significant HMG-CoA reductase kinase activity detected (10). Because both HMG-CoA reductase and acetyl-CoA carboxylase appeared to be physiological substrates, the enzyme was renamed AMP-activated protein kinase (AMPK) (4).

EARLY HISTORY OF THE IDENTIFICATION OF THE YEAST SNF1 COMPLEX

For the budding yeast *S. cerevisiae*, glucose is the preferred carbon source, and the expression of genes involved in metabolism of alternative carbon sources, in gluconeogenesis, in respiration, and in peroxisome biogenesis, is repressed when glucose is available. Derepression of these genes is a crucial aspect of the response to glucose limitation. The *SNF1* gene was defined by mutations that prevented growth on glycerol [called *CAT1* (11)], ethanol [called *CCR1* (12)], or sucrose [called *SNF1* (13) for sucrose non-fermenting]. *SNF1* function is required for transcription of glucose-repressed genes, sporulation, glycogen storage, thermotolerance, and peroxisome biogenesis (14; see also 15, 16). Genetic evidence suggests that there are functional interactions between the SNF1 and cyclic AMP-dependent protein kinase pathways (16–18). Mutations in another gene, *SNF4 (CAT3)*, were isolated in similar screens and exhibit genetic behavior similar to that of *snf1* mutations (19, 20).

MOLECULAR CHARACTERIZATION

Purification, Characterization, and Cloning of AMP-Activated Protein Kinase Subunits

Although there were earlier reports of the purification of various acetyl-CoA carboxylase and HMG-CoA reductase kinases that may have been related to AMPK (e.g. see References 21–24), the first convincing identification of the catalytic subunit was by Carling et al in 1989 (10). They labeled a preparation purified 700-fold from rat liver with a reactive ATP analog, [^{14}C]fluorosulfonylbenzoyl adenosine (FSBA), and found that a 63-kDa polypeptide was the only species labeled. Two prominent unlabeled polypeptides of 38 and 35 kDa were also present. The kinase was subsequently purified to homogeneity by two different methods: (*a*) chromatography on an ATP-γ-Sepharose affinity column eluting with AMP (25) and (*b*) chromatography on a peptide substrate affinity column (26, 27). In both cases the final preparation contained stoichiometric amounts of three polypeptides of 63, 38, and 35 kDa. Davies et al (25) provided evidence that these proteins represented the components of a heterotrimeric complex,

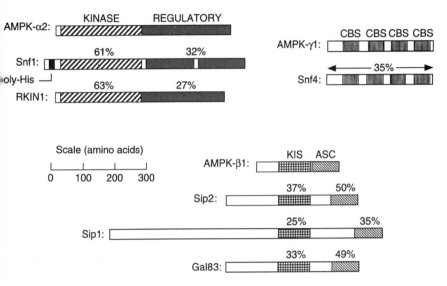

Figure 1 Domain structures of subunits of the AMP-activated protein kinase (AMPK)/SNF1 kinase subfamily. Subunits (*linear bars*) are drawn approximately to scale, with N termini *on the left*. Related domains are represented as *hatched or shaded boxes*, and figures *above these boxes* for the yeast and plant subunits are percent sequence identities with the related domain in the rat subunit shown. Assignment of cystathionine-β-synthase (CBS) domains in the AMPK-γ1/Snf4 sequences as in Reference 36; here only the overall sequence identity of the whole subunits is presented.

which are now referred to as the α (63 kDa), β (38 kDa), and γ (35 kDa) subunits.

Cloning of cDNAs encoding these subunits soon followed, commencing with the catalytic α subunit (28, 29) and then the noncatalytic β and γ subunits (30–32). The most significant, and unexpected, finding to emerge was that all three subunits were closely related to the products of genes previously identified in *S. cerevisiae* (26, 28, 30). The α subunit cDNA encoded a predicted protein of 62 kDa that was 47% identical to the amino acid sequence of the *S. cerevisiae SNF1* gene product, and it contained a typical protein-serine/threonine kinase domain within the N-terminal half. The similarity between AMPK-α and Snf1 extends throughout their lengths, although the C-terminal domains are less closely related than the kinase domains (Figure 1). The AMPK-β cDNA encoded a predicted protein of 30 kDa, significantly smaller than the apparent size (38 kDa) determined by sodium dodecyl sulfate (SDS)–polyacrylamide gel analysis. The reason for this anomaly remains unclear (32). The β sequence contains a predicted myristoylation site at the N terminus, although no covalent modifications have yet been directly demonstrated at the protein level. The AMPK-γ cDNA encoded a 37-kDa protein. The sequences of the β and γ subunits were, like α, clearly related to products of *S. cerevisiae* genes. The β subunit was related to the products of the *SIP1-SIP2-GAL83* gene family (see below), and the γ subunit was 35% identical to the *SNF4* gene product.

The functions of the β and γ subunits remain unclear, except that coexpression of these subunits with the α subunit is essential for recovery of significant kinase activity (33, 34). Indeed, in CCL13 cells (34) there was no detectable expression of any of the subunit proteins unless all three were coexpressed, suggesting that formation of the ternary complex also stabilizes the kinase. Expression and immunoprecipitation in reticulocyte lysates (32) showed that the α and γ subunits each interact directly with the β subunit, but in this system no stable interaction between α and γ was detected. Similar results were obtained when the interactions were studied using the yeast two-hybrid system (32). These findings suggest that formation of the heterotrimeric complex is mediated, at least in part, by the β subunit. This conclusion is consistent with recent two-hybrid analyses of subunit interactions in the yeast SNF1 system (see below).

In 1996, a second isoform of the α subunit was identified and named (somewhat unconventionally) $\alpha1$ (35), making the previously characterized isoform, $\alpha2$. It was originally reported that the $\alpha1$ isoform accounted for 95% of the AMPK activity measurable in rat liver extracts and that the $\alpha2$ isoform was virtually inactive in liver (35). Subsequent studies have challenged this finding, however (34). Immunoprecipitation of AMPK from rat liver extracts using isoform-specific antibodies demonstrates that both $\alpha1$ and $\alpha2$ contribute approximately equally to total activity. Expression studies in mammalian cells also indicate that both isoforms have comparable specific activities (33, 34). A subtle difference between the two isoforms has been found in their preference for particular peptide substrates (34), raising the possibility that there could be some selectivity for downstream targets in vivo. Additional sequences related to the β and γ subunits have been found in the databases of expressed sequence tags. One of our groups has identified and characterized a second isoform of the β subunit ($\beta2$, the original form becoming $\beta1$) (35a) and two additional isoforms ($\gamma2$, $\gamma3$) of the $\gamma1$ subunit (C Thornton, PCF Cheung & DC Carling, unpublished data). The role of each of these different isoforms in mediating the physiological effects of AMPK is unclear. What is certain is that AMPK exists within the cell as a heterogeneous population of complexes composed of different subunit isoforms. A major task that lies ahead is to determine the precise function of each of these complexes in vivo.

Although no obvious domain structure was discerned in the $\gamma1$ subunit when it was first sequenced, it was recently pointed out that it, along with yeast Snf4, comprises four repeats of a protein module known as a cystathionine-β-synthase (CBS) domain (36) (Figure 1). This domain is found in various proteins from archaebacteria to eukaryotes, including the enzyme cystathionine-β-synthase itself. Although the functions of CBS domains are not known, point mutations in the CBS domain of cystathionine-β-synthase lead to homocystinuria, and in this case it appears that the mutations affect regulation of the enzyme (37). The enzyme inosine monophosphate (IMP) dehydrogenase, for which the crystal structure has been determined (38), contains a CBS domain with a compact $\beta\alpha\beta\beta\alpha$ structure. Bateman (36) proposed that pairs of CBS domains might associate to form β-barrel structures.

Genetic and Biochemical Characterization of the Yeast SNF1 Complex

The *SNF1 (CAT1)* (39, 40) and *SNF4 (CAT3)* (41, 42) genes were cloned by complementation of the respective mutants. *SNF1* encodes Snf1, a protein of 633 residues, with an N-terminal protein-serine/threonine kinase domain, followed by a long C-terminal extension that is thought to have regulatory functions. It was subsequently found (28) to be closely related in sequence to AMPK-α (Figure 1). Snf1 has a short-segment N-terminal to the kinase domain that is not conserved in the animal homologue. This region contains a stretch of 13 consecutive histidines that are not required for SNF1 function (43) but fortuitously allow facile purification using Ni^{2+}-chelate affinity chromatography (26, 44, 45). *SNF4* encodes a protein of 322 residues that at the time (41, 42) appeared to be unique but was later shown to be closely related to the mammalian AMPK-γ subunits and to contain four repeats of the CBS motif (Figure 1).

Snf1 in yeast extracts is found in high-molecular-mass complexes (46). In the remainder of this review, the term SNF1 is used to refer to these multimeric complexes, whereas Snf1 is reserved for the catalytic subunit encoded by the *SNF1* gene. SNF1 complexes include the product of the *SNF4* gene (Snf4), as judged both by coimmunoprecipitation (42) and purification via Ni^{2+}-chelate affinity chromatography (26, 44, 45). The Snf1/Snf4 association was used to develop the two-hybrid technique for studying protein-protein interactions (47). Snf4 activates the kinase activity of Snf1 both in vivo (42, 43) and in vitro (48). Genetic approaches subsequently identified three additional components of SNF1 complexes: Sip1, Sip2, and Gal83. The *SIP1* and *SIP2* genes were identified in a two-hybrid screen using Snf1 as "bait," and *SIP1* was also recovered as a multicopy suppressor of *snf4* mutants (49, 50). *GAL83* was originally identified by its effects on glucose repression of *GAL* genes (51, 52). Sip1, Sip2, and Gal83 constitute a family of related proteins that appear to serve as alternative members of SNF1 complexes (50). Each of the three proteins interacts directly and independently with both Snf1 and Snf4, via distinct domains (see Figure 4). A conserved, 80-residue C-terminal region in Sip1, Sip2, and Gal83, designated the *ASC* (association with SNF1 complex) domain (50), binds to Snf4 (45). A conserved internal domain in Sip1, Sip2, and Gal83, designated the *KIS* (kinase interaction sequence) region, mediates interaction with the C-terminal regulatory domain of Snf1 (45). These interactions suggest that one of the functions of each member of the Sip1/Sip2/Gal83 family is to act as a "scaffold," anchoring Snf1 and Snf4 into high-molecular-mass complexes. Neither Sip1, Sip2, nor Gal83 was found in stoichiometric amounts when the active SNF1 complex was purified to homogeneity (26, 30, 44), but this result may be because the purified enzyme is a mixture of different heteromeric complexes. When extracts were prepared from a *sip1sip2gal83* triple mutant, only half of the detectable Snf4 protein copurified with Snf1, in comparison to preparation from wild-type extracts, where copurification of Snf1 and Snf4 was quantitative (45). Although no other genes are closely related to *SIP1, SIP2,* or *GAL83* in the yeast genome, it remains unclear whether other gene products fulfill a similar "scaffold" function to stabilize SNF1 complexes.

Genetic studies have suggested additional roles for Sip1, Sip2, and Gal83 in vivo. Although autophosphorylation activity in an anti-Snf1 immunoprecipitate was greatly reduced in a *gal83*Δ mutant strain (50), the *SIP1, SIP2,* and *GAL83* genes are not essential for many of the functions of the SNF1 pathway in vivo. A triple mutant lacking all three genes is defective in sporulation but does not otherwise display an *snf* phenotype (50, 52). It is possible that these proteins optimize the regulatory response to glucose starvation but are not required for minimally adequate function of the kinase pathway. Increased *SIP1* and *GAL83* gene dosage, and a semidominant *GAL83* allele, have distinct effects on the expression of *SUC* and *GAL* genes (49, 50, 52). These findings suggest that these proteins may serve as "adapters" or "targeting subunits" that direct the SNF1 complex to distinct sets of downstream targets, either by binding to the target proteins or by directing the kinase to specific subcellular locations. The N-terminal domains of Sip1, Sip2, and Gal83 are unrelated in sequence (see Figure 4) and may be involved in this proposed targeting function.

Iin the yeast *Kluyveromyces lactis*, a protein called Fog1, which is closely related to Sip1, Sip2, and Gal83 in their KIS and ASC domains, is essential for expression of glucose-repressed genes, as is Fog2, a Snf1 homologue (53).

SNF1-Related Protein Kinases in Higher Plants

Alderson and coworkers (54) cloned and sequenced a cDNA *(RKIN1)* encoding a Snf1 homologue from the higher plant rye (Figure 1). Transformation of an *snf1* mutant strain of yeast with a low-copy *RKIN1* plasmid restored ability to grow on nonfermentable carbon sources (54), showing that Rkin1 is functionally as well as structurally related to Snf1. Snf1 homologues were subsequently cloned from *Arabidopsis thaliana* (55), barley (56, 57), tobacco (58), and several other plant species (NG Halford, personal communication). As expected, the tobacco DNA *(NPK5)* complemented a yeast *snf1* mutant but not an *snf4* mutant.

While these studies were taking place, MacKintosh et al (59) had detected kinase activities in extracts of several mono- and dicotyledonous plants that phosphorylated the *SAMS* peptide—a peptide based on the major AMPK site in rat acetyl-CoA carboxylase, which is a relatively specific substrate for AMPK (60). This activity was purified from cauliflower inflorescences (59, 61) and termed HMG-CoA reductase kinase-A (HRK-A). Although it was not activated by AMP, in many other respects its biochemical properties were very similar to AMPK, and it could even be activated using mammalian AMP-activated protein kinase kinase (AMPKK) and MgATP (59, 61, 62). The results indicated that HRK-A was a higher-plant homologue of AMPK. Although the plant kinase was not purified to homogeneity, the catalytic subunit was identified using [14C]FSBA labeling as a polypeptide of 58 kDa (61), the mass predicted for higher-plant Snf1 homologues. Antibodies raised against a sequence that is conserved in the plant Snf1 homologues cross-reacted with this polypeptide (63). These results strongly suggest that cauliflower HRK-A is encoded by a homologue of rye *RKIN1* and yeast *SNF1*.

As yet, DNAs encoding the higher-plant homologues of the mammalian or yeast γ (Snf4) subunit or β (Sip1/Sip2/Gal83) subunits have not been cloned. There are strong indications that they must exist, however. Tobacco Npk5 interacts with Snf4 in yeast two-hybrid assays (64), and presumably it must be forming a complex with the yeast accessory subunits when it complements the *snf1* mutation in vivo (58). Although no additional subunits could be identified with certainty when HRK-A was purified from cauliflower, the active kinase had an apparent mass by gel filtration of around 200 kDa, similar to that of mammalian AMPK (61).

REGULATION

AMP-Activated Protein Kinase: Studies of Regulation in Vitro

In 1978, Ingebritsen and colleagues (65) reported that AMPK (then termed HMG-CoA reductase kinase) could be inactivated by treatment with a partially purified protein phosphatase and then reactivated by incubation with MgATP. A factor required for the MgATP-dependent reactivation (an upstream kinase?) could be resolved from AMPK on DEAE-Sepharose (66). Fractions containing

AMPK and the putative upstream kinase were also separated by Beg et al (21). These studies indicated that the system was a protein kinase cascade in which an upstream kinase (now termed AMPKK) phosphorylated and activated the downstream kinase (AMPK). Although the concept of a protein kinase cascade is now very familiar, for 10 years the AMPKK/AMPK couple and the cAMP-dependent protein kinase/phosphorylase kinase pair were the only such cascades known to exist.

Regulation of the system by AMP turned out to be remarkably complex. As well as causing allosteric activation of AMPK (8–10), AMP was absolutely required for phosphorylation of AMPK by AMPKK (67, 68). At first it was not clear whether this effect was due to binding of AMP to the enzyme (AMPKK) or to the substrate (AMPK) or to both. An answer to this question became possible with the finding that the upstream and downstream components would inter-act with the equivalent components in another cascade, namely that involving phosphorylation of calmodulin-dependent kinase I (CaMKI) by calmodulin-dependent kinase I kinase (CaMKK). AMP had no effect on the CaMKK → CaMKI reaction, but activation of CaMKI by AMPKK was stimulated by AMP, as was activation of AMPK by CaMKK. Although these latter two artificial cas-cades probably have no physiological relevance, the results demonstrated (*a*) that AMPKK is an AMP-activated protein kinase and (*b*) that binding of AMP to AMPK promoted its own phosphorylation (presumably because a conforma-tional change makes the phosphorylation site accessible).

Remarkably, AMP has a fourth effect: It inhibits dephosphorylation and inac-tivation of AMPK. The kinase can be inactivated by either protein phosphatase-2A (PP2A) or protein phosphatase-2C (PP2C) in vitro (69), but the dephospho-rylation in intact rat hepatocytes is not sensitive to okadaic acid (67), indicating that PP2C may be the physiologically relevant protein phosphatase. Inactiva-tion of AMPK by PP2C was almost completely blocked by AMP (70). This effect was due to binding of AMP to AMPK rather than to PP2C because the dephosphorylation of other PP2C substrates was not inhibited. AMP there-fore activates the cascade by at least four mechanisms: (*a*) allosteric activation of AMPKK; (*b*) binding to AMPK, making it a better substrate for AMPKK; (*c*) allosteric activation of AMPK; and (*d*) binding to AMPK, making it a worse substrate for PP2C. A simple model to explain these observations, based on the classical Monod/Wyman/Changeux model for allosteric enzymes (71), is pre-sented in Figure 2. AMPK is envisaged as existing in four states, i.e. *R* and *T* conformations, each of which can also exist in phospho and dephospho forms. The dephospho form in the presence or absence or AMP appears to be inactive, but the dephosphorylated *R* state, which is stabilized by AMP binding, is a much better substrate for AMPKK than the *T* state. Phosphorylation of the *R* state

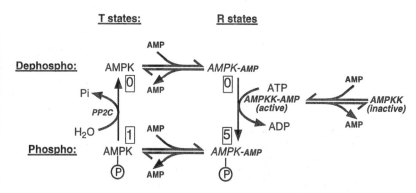

Figure 2 Model for the regulation of AMP-activated protein kinase (AMPK) by AMP and by phosphorylation, based on the classical Monod/Wyman/Changeux model for allosteric enzymes (71). AMPK is proposed to exist in four states, i.e. *R* and *T* states, both of which can exist in phosphorylated and dephosphorylated forms. AMP binding promotes the $T \rightarrow R$ transitions by stabilizing the *R* states. Only the *R* state is a substrate for AMPK kinase (AMPKK), whereas only the *T* state is a substrate for protein phosphatase-2C (PP2C). The figures in *square boxes* indicate the approximate kinase activity of that form, relative to that of the phosphorylated *T* state. Redrawn from Reference 76.

leads to >100-fold activation. Dissociation of AMP from the phosphorylated *R* state leads to partial conversion to the *T* state and a three- to fivefold decrease in activity. More important, in the absence of AMP, the phosphorylation site in this phosphorylated form is exposed to PP2C, leading to complete inactivation. In addition, the upstream kinase is allosterically activated by AMP. There is no evidence that it is regulated by phosphorylation (72).

At least three, if not all, of the mechanisms of activation by AMP are inhibited by high ATP. ATP at 4 mM causes a 10-fold increase in the concentration of AMP required for allosteric activation of AMPK (73) and for inhibition of dephosphorylation (70) (via mechanisms *c* and *d* above). High ATP also inhibits the AMP-dependent reactivation of AMPK by AMPKK (72), although it is not yet clear whether this effect is mediated by binding to AMPK (mechanism *b*), to AMPKK (mechanism *a*), or both. The simplest hypothesis to explain these observations is that AMPK has a single allosteric site that binds both AMP and ATP in a mutually exclusive manner, and that only AMP stabilizes the R conformation. Evidence consistent with this hypothesis includes the effects of the reactive ATP analog, FSBA (24). Incubation of AMPK with FSBA caused irreversible inactivation with a biphasic time course. In the first, rapid phase, FSBA caused inactivation of AMP stimulation, followed by a second slower phase where the basal, AMP-independent activity was lost. This suggests that

FSBA reacts rapidly at the allosteric (AMP/ATP) site and then more slowly at the catalytic (ATP) site. Consistent with that interpretation, both AMP and ATP protected against the first phase of inactivation, whereas only ATP protected against the second (24).

Characterization of AMP-Activated Protein Kinase Kinase

A cDNA encoding the upstream kinase—AMP-activated protein kinase kinase (AMPKK)—is not yet defined in molecular terms, but this enzyme is partially characterized at the protein level. AMPKK has been extensively purified from rat liver using an assay involving its ability to reactivate phosphatase-treated AMPK (72). One intriguing finding was that in the presence of AMPK, AMPKK could be precipitated using an antibody raised against the β subunit of AMPK. This result suggested that the upstream and downstream kinases can form a complex, although they readily dissociate from each other during purification. When AMPKK was coprecipitated with AMPK after incubation with [γ-^{32}P]ATP, a phosphorylated 58-kDa polypeptide was detected. A polypeptide of the same molecular mass was also evident on sodium dodecyl sulfate–polyacrylamide gel electrophoresis (SDS-PAGE) analysis of highly purified AMPKK labeled using either [γ-^{32}P]ATP or [^{14}C]FSBA. These results suggest, but do not prove, that the 58-kDa polypeptide is the catalytic subunit of AMPKK. If this assignment is correct, the molecular masses of the catalytic subunits of AMPKK and AMPK are similar (58 versus 63). The two kinases are also similar in several other physicochemical parameters, i.e. native molecular mass (195,000 Da), Stokes radii, and frictional coefficient (72). Moreover, because both AMPK and AMPKK are activated by AMP (74), it seems likely that they are closely related proteins and that AMPKK is also a heteromeric complex.

The major site on AMPK phosphorylated by AMPKK was found to be Thr-172 (72), located within the so-called activation segment (75), situated between the conserved Asp-Phe-Gly (DFG) and Ala-Pro-Glu (APE) motifs, where many protein kinases require phosphorylation for their activation (Figure 3). Site-directed mutagenesis studies (D Carling, unpublished data) have now confirmed that phosphorylation of T172 is likely to be necessary for activation because a T172A mutant is essentially inactive. It does not appear to be sufficient, however. A T172D mutant, in which the negatively charged aspartic acid might be expected to mimic the introduction of a phosphate group, possesses detectable activity but is still inactivated by protein phosphatase treatment. Therefore at least one other activating phosphorylation site must exist. Because phosphorylation of both residues appears to be essential for activity, it is possible that the second site was missed in the original study (72) because it was not

AMPKα	**DFGL**SNMMSDGEFLR **T**SCGSPNYA**APE**
Snf1	**DFGL**SNIMTDGNFLK **T**SCGSPNYA**APE**
Rkin1	**DFGL**SNVMHDGHFLK **T**SCGSLNYA**APE**
CAMKI	**DFGL**SKMEDPGSVLS **T**ACGTPGYV**APE**
CaMKIV	**DFGL**SKIVEHQVLMK **T**VCGTPGYC**APE**
PKA	**DFG**FAKRVKGRTW**T**LCGTPEYL**APE**
Akt	**DFGL**CKEGIKDGATMK **T**FCGTPEYL**APE**
PKC	**DFG**MCKEHMMDGVTTR **T**FCGTPDYI**APE**
Cdc2	**DFGL**ARAFGIPIRVY **T**HEVVTLWYR**SPE**
MAPK	**DFGL**ARIADPEHDHTGFL **TEY**VATRWYR**APE**
RSK	**DFG**FAKQLRAENGLLM **T**PCYTANFV**APE**
MEK1	**DFG**VSGQLID**S**MAN**S**FVGTRSYM**SPE**
GSK3	**DFG**SAKQLVRGEPNVS **Y**ICSRYYR**APE**

Figure 3 Alignment of the known phosphorylation sites in activation segments (75) of a number of protein kinases. The Asp-Phe-Gly (DFG) and Ala/Ser-Pro-Glu (A/SPE) motifs are in *bold type*, and the sequences are aligned according to the latter. Phosphorylation sites (putative only for Rkin1) are in *bold type and underlined*. AMPKα, rat AMPK-α2; Snf1, *Saccharomyces cerevisiae* Snf1; Rkin1, rye Rkin1; CaMKI, rat calmodulin-dependent protein kinase I; CaMKIV, mouse calmodulin-dependent protein kinase IV; PKA, human cyclic AMP-dependent protein kinase-α catalytic subunit; Akt, human Akt protein kinase; PKC, rat protein kinase C-α; Cdc2, human Cdc2; MAPK, human Erk1 (MAP kinase); RSK, mouse RSK[mo-1]; MEK1, mouse MAPK/Erk kinase-1 (MAP kinase kinase); GSK3, glycogen synthase kinase 3. For further information see References 75, 150, 151.

completely dephosphorylated prior to rephosphorylation with $[\gamma\text{-}^{32}P]$ATP and AMPKK.

It had already been shown that mutation of Thr-210 in Snf1 (analogous to Thr-172 in AMPK-α2, as shown in Figure 3) completely abolished SNF1 function in yeast in vivo (46). It remains to be determined whether phosphorylation of an additional site is necessary for the activity of the yeast enzyme.

AMP-Activated Protein Kinase: Regulation in Intact Cells

Given that the AMPK cascade is activated by elevation of AMP and depletion of ATP, what conditions lead to such changes in vivo? Because of the action of adenylate kinase, AMP and ATP concentrations tend to change in reciprocal directions. Adenylate kinase is a highly active enzyme in apparently all eukaryotic cells. The reaction it catalyzes ($2ADP \leftrightarrow ATP + AMP$) is maintained at close to equilibrium at all times. Because the equilibrium constant ($[ATP][AMP]/[ADP]^2$) is close to one, the AMP:ATP ratio in cells varies approximately as the square of the ADP:ATP ratio; i.e. ($[AMP]/[ATP] \approx ([ADP]/[ATP])^2$). Under optimal conditions, eukaryotic cells typically maintain their ATP:ADP ratio at something on the order of 10:1, so the ATP:AMP ratio is $\sim$100:1. AMP in fully energized cells growing under optimal conditions is therefore extremely low, and AMPK is in the inactive state. But if the cells experience some environmental or nutritional stress such that the ADP:ATP ratio rises, let us say fivefold, the AMP:ATP ratio would rise $\sim$25-fold, which is sufficient to switch on the AMPK cascade. It has been argued that the AMPK system represents a cellular fuel gauge (76) that constantly monitors the energy status of the cell and, if it detects that the energy supply is compromised, AMPK action initiates energy-conserving measures and mobilizes the catabolism of alternative carbon sources if these are available. AMPK therefore protects the cell against environmental and nutritional stresses. An engineering analogy even more apposite than that of a fuel gauge is to the hardware and software systems in laptop computers that monitor the state of the battery charge and, if a problem is detected, request an alternative energy source and activate energy-conserving measures (e.g. dimming the screen).

In yeast, a condition that can dramatically elevate the AMP:ATP ratio is starvation for glucose (see below). In most mammalian cells this situation is much less of a problem because homeostatic mechanisms ensure that even when the animal is starving, blood glucose levels are maintained. Unlike yeast growing in high glucose, most mammalian cells also store glycogen, which acts as a glucose buffer such that short-term glucose deprivation of cells in culture does not usually activate AMPK. However, in mammalian cells, other adverse environmental conditions elevate AMP:ATP and switch on AMPK. Such conditions include treatment of isolated hepatocytes with high fructose (67), with arsenite, or with heat shock (77). High exogenous fructose depletes ATP by trapping phosphate as fructose or triose phosphate esters. A similar mechanism is probably responsible for the depletion of ATP and the inhibition of HMG-CoA reductase and sterol synthesis (indicating activation of AMPK) when hamster fibroblasts are incubated with 2-deoxyglucose (78). Arsenite depletes ATP primarily by inhibiting lipoamide-containing dehydrogenases in the tricarboxylic acid (TCA) cycle. Inhibitors of oxidative phosphorylation such as antimycin A, dinitrophenol, or azide also cause inactivation of acetyl-CoA carboxylase (consistent with activation of AMPK) in Fao hepatoma cells (79). How heat shock causes ATP depletion remains unclear, but almost all of the stresses—e.g. heat shock, hypoxia, arsenite, cadmium ions (77)—that produce the classical cellular stress response (i.e. elevated synthesis of stress proteins) also cause ATP depletion. As long ago as 1964, Ritossa (80) suggested that depletion of ATP might be a trigger for the cellular stress response. Although it remains unclear whether AMPK has any direct role in induction of the stress proteins, we believe that activation of AMPK is nevertheless a central component of the response of cells to stress. Examples of cellular stress events of great medical importance are heart attacks and strokes that are caused by blockage of blood vessels. Kudo et al have shown that ischemia induced by ligation of coronary arteries in perfused rat hearts is associated with elevation of AMP and depletion of ATP and causes phosphorylation and activation of AMPK (81, 82).

All of the AMPK activating events described above can be regarded as pathological rather than physiological. Under normal circumstances, the most important role for AMPK would likely be in cells wherein energy demands fluctuate highly, such as skeletal muscle. Winder & Hardie (83) have shown that AMPK is activated two- to threefold in quadricep muscles within 5 min of commencement of treadmill exercise of rats. This activation can also be demonstrated in muscles stimulated electrically in situ, and in one study, associated changes in the cellular AMP:ATP ratio were also measured (84, 85). The functions of AMPK activation in contracting muscle are considered later.

It seems reasonable to propose that the primary function of AMPK is to act as a fuel gauge that monitors the levels of the key metabolites AMP and ATP and that protects cells during periods of stress when the cellular fuel status is compromised. Earlier suggestions for the physiological role of AMPK were that it was involved in feedback regulation by lipids (86–88) or in regulation of lipid metabolism by hormones (89, 90). The effects reported were generally smaller than the pronounced effects of cellular stress. It is possible that lipid or hormonal signaling pathways may interact with the AMPK cascade, but this role does not seem to be the primary purpose for which the system evolved.

Genetic Studies on Regulation of the SNF1 Complex

In 1986, the protein kinase activity of Snf1 was demonstrated to be essential for its function in gene derepression (39). It is now clear that Snf1 kinase activity is dramatically elevated on removal of glucose from the medium (44, 48; see next section). The localization of the SNF1 complex does not appear to change. In glucose-sufficient and glucose-starved cells, immunofluorescence studies showed that Snf1 and Snf4 remain in both the nucleus and the cytoplasm (39, 42). The known components of the complex also remain associated with Snf1 regardless of glucose availability: Snf4, Sip1, Sip2, and Gal83 coimmunoprecipitate and copurify with Snf1 when extracts are prepared from cells grown in both high and low glucose (26, 30, 42, 45, 46, 49, 50).

A) Inactive conformation (high glucose)

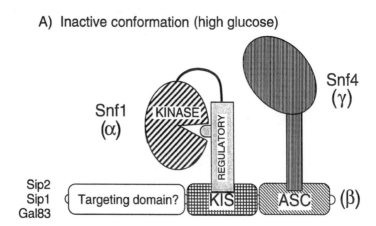

B) Active conformation (low glucose)

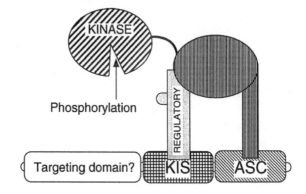

Figure 4 Model for changes in interdomain interactions in the SNF1 complex, based on References 45, 64. In both the inactive and active conformations, the β subunits (Sip1, Sip2, or Gal83) act as a "scaffold" that links Snf1 and Snf4. In the inactive form (repressing conditions, *top*), the kinase domain of Snf1 is inhibited by interaction with the regulatory domain on the same subunit. In response to an unknown signal, this interaction is disrupted (derepressing conditions, *bottom*) and Snf4 interacts with the Snf1 regulatory domain in lieu of the kinase domain. The Snf1 kinase domain is now free to phosphorylate downstream targets.

Although the subunits seem to remain associated, detailed two-hybrid analyses have recently revealed that some of the inter- and intrasubunit interactions are altered by glucose availability (64; see also Figure 4). Snf4 interacts with Snf1 in the two-hybrid system, but only in low-glucose (derepressing) conditions when the kinase would be active. This interaction has been mapped to the C-terminal, regulatory domain of Snf1. Conversely, the regulatory domain of Snf1 interacts with the kinase domain of Snf1 more strongly in high-glucose (repressing) conditions when the kinase is inactive. Snf4 and the Snf1 kinase domains interact with sites on the Snf1 regulatory domain that are distinct but overlapping, and in vitro binding studies and genetic suppressor analyses indicate that both interactions are direct (64). These results are consistent with previous genetic evidence that Snf4 is essential for the function of full-length Snf1 but is not required if a truncated Snf1 is expressed in which the regulatory domain has been deleted (43).

Biochemical Studies on Regulation of the SNF1 Complex

The finding of a close sequence relationship between AMPK-α and Snf1, particularly within the kinase domain, suggested that it might be possible to assay the SNF1 complex using AMPK substrates. This supposition indeed turned out to be correct, and the complex was assayed through use of the *SAMS* peptide. This assay was initially performed using a yeast fraction partially purified by chromatography on DEAE-Sepharose (48), although it was subsequently shown that it was possible to use the assay in a crude extract (44). The method of cell harvesting was found to be critical: The kinase was activated during harvesting if the cells were recovered by a conventional method of centrifugation and resuspension in a homogenization buffer lacking glucose. Rapid filtration and freezing in liquid nitrogen appeared to preserve the in vivo SNF1 activity, and by this method the kinase was found to be activated at least 100-fold within 5 min of glucose removal from the culture medium (44). This situation is reminiscent of the behavior of mammalian AMPK, where rapid freezing of cells is essential to preserve the physiological activation state (91).

The active, phosphorylated form of the SNF1 complex isolated under derepressing conditions could be inactivated by treatment with protein phosphatases and reactivated by adding mammalian AMPKK (44, 48). This result indicated that activation was due to phosphorylation and that the protein kinase cascade is conserved between yeast and mammals. A MgATP-dependent, SNF1-reactivating activity also was observed in partially purified SNF1 and could be resolved from SNF1 on further purification (44). This activity was presumably an upstream protein kinase homologous with AMPKK. Surprisingly, however, a gene encoding a putative upstream kinase has not yet emerged from any of the genetic screens.

The intracellular signals that switch the SNF1 complex on and off remain enigmatic. Snf1 is likely to be the primary target of the high-/low-glucose signal because its overexpression in a *snf1 snf4* double mutant partially restores glucose-repressible gene expression (43). Given that mammalian AMPK is activated by high AMP and low ATP, and given the high degree of conservation between the AMPK and SNF1 complexes, these nucleotides were obvious candidates to be regulators of SNF1. Consistent with this idea, removal of glucose from yeast growing exponentially in a medium containing abundant glucose resulted in large increases in AMP and decreases in ATP, with the AMP:ATP ratio increasing by 200-fold within 5 min (44). This change is perhaps not surprising, given that under growth conditions where high glucose is present, yeast does not store glycogen and uses glucose for ATP production via glycolysis (rather than oxidative phosphorylation). Under a variety of nutritional shifts, such as high to low glucose, low to high glucose, and glucose to sucrose, a reasonable correlation appeared to exist between cellular AMP and ATP levels and SNF1 activity (44). A problem with this hypothesis, however, is the complete failure to find any regulation of SNF1 by these nucleotides in vitro. AMP does not activate homogeneous preparations of SNF1 and does not affect its dephosphorylation by mammalian protein phosphatases. Attempts to demonstrate effects on reactivation by the upstream kinase also have been unsuccessful (44).

Genetic Identification of Elements Upstream of SNF1

Another approach to identify intracellular signals that regulate SNF1 in vivo is to study genes that appear to act upstream of SNF1. Various mutations have been isolated that relieve glucose repression of gene expression and appear to act upstream of Snf1, as judged by genetic interactions with a *snf1* mutation (14, 92). Mutations in three such genes—*HXK2*, *GLC7*, and *REG1*—have been shown to affect the glucose regulation of protein interactions within the SNF1 complex. *HXK2* (*HEX1*) encodes the major hexokinase isoform (PII) that catalyzes the initial step in glucose catabolism. A *hxk2Δ* null mutation affects the two-hybrid interaction of Snf1 and Snf4: Glucose inhibition of this interaction is substantially reduced, indicating a role for hexokinase in regulating SNF1 function (64). *GLC7* is an essential gene that encodes protein phosphatase-1 (PP1) (93), and *REG1* encodes a regulatory/targeting subunit of PP1 that directs its participation in the glucose signaling mechanism (94). Glucose repression is relieved by a specific mutant allele of *GLC7*—*glc7-T152K* (95)—and by *reg1* deletion. These mutations also relieve inhibition by glucose of the Snf1-Snf4 two-hybrid interaction (64). Thus, PP1-Reg1 acts upstream of the kinase to regulate Snf1-Snf4 interactions in response to the glucose signal. Moreover, Reg1 interacts with the Snf1 kinase domain in two-hybrid analysis, which implies that it acts on the SNF1 complex directly rather than through a cascade of regulatory events (Ludin, R Jiang & M Carlson, unpublished data).

Model for Regulation of the SNF1 Complex and Its Relevance for Mammalian AMPK

The results of detailed two-hybrid analyses of the SNF1 complex by Jiang & Carlson (45, 64), combined with recent biochemical studies (44, 48), have suggested a new model for the structure and regulation of the SNF1 complex (see Figure 4). Using the mammalian subunit nomenclature, the β subunit (i.e. Sip1, Sip2, or Gal83) forms a scaffold and the γ subunit (Snf4) attaches to β via the conserved C-terminal *ASC* domain. Snf1 associates with the *KIS* segment of β. Thus the unique, N-terminal domains of the β subunits are left free. A reasonable speculation is that these domains are involved in targeting the complex to specific subcellular locations. Under high-glucose (repressing) growth conditions, the regulatory domain of Snf1 interacts with the kinase domain and blocks its activity (64). When glucose is removed from the medium, the interaction between the regulatory and kinase domains of Snf1 is disrupted. The active conformation is stabilized by binding of Snf4 to the regulatory domain of Snf1. It is not clear whether phosphorylation of Snf1 by an upstream kinase is responsible for disruption of the kinase domain–regulatory domain interaction or whether it is a consequence of it.

This model is likely to be broadly applicable to the mammalian AMPK complex. The regions of the mammalian β subunits that are most similar to the Sip1/Sip2/Gal83 family are the *KIS* and *ASC* domains (Figure 1; see also Reference 45). Woods et al have also shown that mammalian α and γ interact poorly when expressed in reticulocyte lysates, whereas α and β and β and γ interact strongly (32). The reticulocyte lysate contains an ATP-regenerating system, which would maintain AMPK in its inactive state. The yeast model (Figure 4) suggests a lack of direct interaction between α and γ under such conditions.

The events that remain unclear, at least in yeast, are the inputs or signals that cause the initial conformational change. In the mammalian system, the signal would be a rise in AMP coupled with a fall in ATP. Because FSBA blocks the allosteric site (24) and [^{14}C]FSBA reacts exclusively with the α subunit (10), the allosteric site is presumed to be on this subunit, presumably in the C-terminal domain. Binding of AMP at this site may either induce the transition to or stabilize the active conformation. According to the yeast model, the α subunit regulatory domain associates with γ in the active conformation. It does not look likely that the activating signal in the yeast system is AMP. However, large increases in AMP (and decreases in ATP) are present in vivo under conditions where SNF1 is activated. It remains possible that the activating signal in yeast is some other metabolite that varies in concert with AMP.

If the activating signal in yeast is AMP, or some other metabolite whose concentration correlated with it, accommodation of the results with *hxk2* mutants would be straightforward. Because *HXK2* encodes the major isoform of hexokinase, disruption of this gene in cells relying on glycolytic breakdown of glucose would be expected to interfere with ATP production and hence elevate AMP via the adenylate kinase reaction. The simplest explanation of the effects of the *glc7-T152K* and *reg1*Δ mutants is that the Reg1-Glc7 complex is the protein phosphatase that converts SNF1 back to its inactive form. Disruption of this protein phosphatase would cause the SNF1 complex to become constitutively active. Other interpretations are possible, of course. It may be that there are inputs into the SNF1 system other than the putative high-/low-glucose signals, and Reg1-Glc7 may be involved in one of these alternative pathways. Because Snf4 binds to the regulatory domain of Snf1, some of these inputs might be transduced through effects on Snf4.

Regulation of Plant SNF1-Related Kinases

Little is known about regulation of the plant SNF1-related kinases in vitro, and almost nothing is known about their regulation in vivo. Like the yeast system, plant kinases are not allosterically activated by AMP. They are inactivated by protein phosphatases, however, and they can be reactivated by mammalian AMPKK and by a putative upstream kinase in plant extracts that can be removed from the downstream protein kinase on further purification (59). Therefore, plant kinases are probably regulated in a manner similar to that of their animal and yeast counterparts.

DOWNSTREAM TARGETS

The Recognition Motif for Animal, Yeast, and Plant AMPKs

Alignment of six sites phosphorylated by AMPK on protein substrates (96) revealed that the only residues conserved, other than the phosphorylated serine, were hydrophobic residues at $P - 5$ and $P + 4$ (i.e. five residues N-terminal, or four residues C-terminal, to the phosphorylated serine) and a single basic residue at $P - 3$ or $P - 4$. The influence of these residues was addressed through the use of two series of synthetic peptides. One series (96) consisted of variants of the *SAMS* peptide (HMRSAMSGLHLVKRR). The second series consisted of variants of the *AMARA* peptide (AMARAASAAALARRR) (62), a sequence designed with the key determinants at the $P - 5$, $P - 3$, and $P + 4$ positions but with other residues as alanine. (The basic residues at the C terminus allow removal from unreacted ATP via binding to phosphocellulose paper, but they are not essential for phosphorylation.) These studies established the recognition motif *Hyd*-(X, *Bas*)-X-X-Ser/Thr-X-X-X-*Hyd*, where *Hyd* carries a bulky hydrophobic side chain (L, M, I, F, or V), *Bas* carries a basic side chain (R > K > H), and the order of amino acids between parentheses is not critical. Synthetic peptide studies show that AMPK will phosphorylate serine or threonine but not tyrosine, although all of the protein substrates found are phosphorylated on serine. The above recognition motif has recently been confirmed by site-directed mutagenesis of a protein substrate (see section below on HMG-CoA reductase).

Using a limited set of the *AMARA* variant peptides (34), the $\alpha 1$ and $\alpha 2$ isoforms of AMPK appeared to recognize the same motif, except that $\alpha 1$ had a less stringent requirement for the hydrophobic residue at the P + 4 position. Both isoforms prefer a hydrophobic residue at this position, but substitution with glycine causes a larger increase in K_m for the $\alpha 2$ isoform than for the $\alpha 1$ isoform (DG Hardie, unpublished data). Michell and coworkers (27) also have examined the specificity of the $\alpha 1$ isoform using a panel of synthetic peptide substrates. They came to the same conclusions regarding the necessity of a hydrophobic residue at P − 5, but they found that a hydrophobic residue at P + 4 was not essential because the peptide with the lowest K_m of those studied had glutamine at this position. Glutamine was not tested as a replacement for the hydrophobic residues in the other studies (62, 96, 97), and it is possible that this uncharged side chain can be accommodated in the substrate-binding pocket on the kinase that recognizes the P + 4 hydrophobic residue. Alternatively, a hydrophobic side chain at P + 4 may not be essential if a sufficient number of other positive determinants are present.

Most, but not all, of the known protein targets for AMPK have a basic residue at the P − 6 position as well as at the P − 3 or P − 4 position. A histidine is present at this position in the *SAMS* peptide, and a lysine is present in the peptide LKK<u>L</u>T<u>RR</u>P<u>S</u>FSAQ , which was found (27) to be an excellent substrate for AMPK. Although the *AMARA* peptide has alanine at P − 6, this position is also the N-terminal residue of the *AMARA* peptide and the α-amino group may provide the necessary positive charge. A basic residue at the P − 6 position may therefore be an additional recognition selectivity determinant; this proposal needs to be studied systematically.

The availability of these panels of peptides allowed the specificity of yeast SNF1 and the higher-plant SNF1-related kinases to be addressed (61, 62, 96). The recognition motifs for the yeast and plant kinases were very similar to those for their animal homologues, although there were subtle differences. Unlike the animal kinase, neither the yeast nor the plant kinases phosphorylated at significant rates peptides containing threonine in place of serine. The yeast enzyme also had a specific requirement for arginine (not lysine or histidine), which had to be at the P−3 position. The recognition motif for yeast SNF1 (*Hyd*-X-Arg-X-X-Ser-X-X-X-*Hyd*) is therefore more stringent than that of animal or plant kinases.

Targets of Mammalian AMP-Activated Protein Kinase

HMG-COA REDUCTASE   HMG-CoA reductase, which catalyzes the key regulatory step in the biosynthesis of isoprenoids and sterols, was one of the first substrates for AMPK to be identified (1), and it remains perhaps the most conclusively established. AMPK phosphorylates a single site in a catalytically active fragment of rat liver HMG-CoA reductase, corresponding to Ser-871 in the full-length sequence in the enzyme from Chinese hamster, and this site is also phosphorylated in the rat liver enzyme in situ (98). This site was subsequently shown to be the only site phosphorylated on full-length HMG-CoA reductase in intact rat hepatocytes when AMPK was activated in response to treatment of the cells with high fructose (99). HMG-CoA reductase is phosphorylated and inactivated, and sterol synthesis dramatically inhibited, when AMPK is activated in intact rat hepatocytes using either arsenite or heat shock (77) or 5-aminoimidazole-4-carboxamide (AICA) riboside (73). Incubation with AICA riboside is a method for activating AMPK in intact cells that may be more specific than stress agents because it does not usually disturb levels of ATP, ADP, or AMP (73, 100). The nucleoside is taken up by mammalian cells and converted to the monophosphorylated form, ZMP, which mimics the effects of AMP on the AMPK system, not only on allosteric activation but also on phosphorylation (73).

Elegant confirmation of the physiological importance of HMG-CoA reductase phosphorylation was obtained when Sato et al (78) showed that deoxyglucose treatment (which depletes intracellular ATP) caused total inhibition of sterol synthesis in cells expressing wild-type HMG-CoA reductase, but that it had no effect on cells expressing a Ser871Ala mutant. Ching et al (97) have studied the effects of point mutations expressed in bacteria on phosphorylation by AMPK in vitro. Mutations at and around the phosphorylation site were used to confirm, using an enzymically active protein, the recognition motif previously established using synthetic peptides (62, 96). Intriguingly, replacement of the histidine at P − 3 with the more preferred arginine residue increased the rate of phosphorylation, as did replacement of the lysine at P + 1 with alanine. A double mutant (arg at P − 3, ala at P + 1) was phosphorylated fivefold more rapidly than the wild type, although it was still functional as HMG-CoA reductase. It was therefore proposed (97) that unlike acetyl-CoA carboxylase, which has almost the optimal combination of determinants, HMG-CoA reductase may have been selected to be a suboptimal substrate for AMPK. This idea is consistent with findings in isolated rat hepatocytes that a mild heat stress (42°C, which causes only a partial activation of AMPK) inhibits fatty acid but not sterol synthesis, whereas a more severe heat stress (45°C, which causes a much larger activation) inhibits both pathways (77). Conceivably, fatty acid synthesis is a pathway that is completely dispensable in the short term, whereas it may be preferable to maintain at least a low rate of isoprenoid synthesis under moderate stress conditions.

ACETYL-COA CARBOXYLASE AND FATTY ACID SYNTHESIS   Acetyl-CoA carboxylase catalyzes the key regulated step in fatty acid synthesis, and AMP-activated protein kinase was codiscovered independently as a factor that regulated this target (2). AMPK phosphorylates three sites (101, 102) corresponding to Ser-79, Ser-1200, and Ser-1215 on rat acetyl-CoA carboxylase (103), although phosphorylation at Ser-79 seemed to be primarily responsible for enzyme inactivation. This conclusion was later confirmed by expression of a Ser79Ala mutant in COS cells (104). Acetyl-CoA carboxylase is generally thought to exert a major degree of control over fatty acid synthesis. Consistent with this view, activation of AMPK by heat shock or arsenite (77) or by AICA riboside (73, 100) causes almost total inhibition of fatty acid synthesis in isolated rat hepatocytes.

ACETYL-COA CARBOXYLASE AND FATTY ACID OXIDATION   As well as conserving ATP by switching off fatty acid biosynthesis, phosphorylation of acetyl-CoA carboxylase also appears to switch on an ATP-producing pathway, i.e. fatty acid oxidation. The key regulatory step in fatty acid oxidation is the initial uptake of fatty acids into mitochondria catalyzed by carnitine-palmitoyl transferase I (CPT1). The product of acetyl-CoA carboxylase, malonyl-CoA, is an inhibitor of this enzyme (105). Malonyl-CoA is a particularly potent inhibitor in skeletal and cardiac muscle, where a different isoform of CPT1 is expressed (106). These circumstances suggest a mechanism for activating fatty acid oxidation in response to the demand for ATP, which may be particularly important in muscle. Both skeletal and cardiac muscles express novel isoforms of acetyl-CoA carboxylase [probably corresponding to the products of the ACC-β/ACC-2 gene (107, 108)], and the muscle form has been shown to be susceptible to inactivation by AMPK in vitro (109). These tissues do not express fatty acid synthase or carry out de novo fatty acid synthesis, so acetyl-CoA carboxylase is thought to have a purely regulatory role. In resting muscle, or any other cell where the demand for ATP was low, AMPK would be inactive, acetyl-CoA carboxylase would be active, and some acetyl-CoA would be converted to malonyl-CoA, thus inhibiting fatty acid oxidation. On the other hand, in working muscle or any cell type where ATP became depleted, AMPK would be activated, acetyl-CoA carboxylase would be inactivated, malonyl-CoA would be reduced, and fatty acid oxidation would increase to generate more ATP. The evidence for this model is compelling. In skeletal muscle of rats exercising on a treadmill, AMPK is activated, acetyl-CoA carboxylase is inactivated, and malonyl-CoA is significantly depressed (83). The same effects are seen in muscles stimulated electrically, and the activation of AMPK was associated with an increase in AMP and decrease in ATP (84). Another group reported that electrical stimulation of rat gastronemius muscle resulted in phosphorylation and inactivation of acetyl-CoA carboxylase and an activation of AMPK that was confined to the $\alpha 2$ isoform (85). Most convincingly of all, perfusion of rat hind-limb muscle with AICA riboside produces the same effects without disturbing AMP or ATP levels, and these effects are accompanied by a large increase in fatty acid oxidation (110).

In cardiac muscle, AMPK is activated by ischemia and remains high during reperfusion, apparently because of increased phosphorylation. This activation of AMPK is accompanied by phosphorylation and inactivation of acetyl-CoA carboxylase activity and decreased levels of malonyl-CoA, which in turn may be responsible for the highly elevated rates of fatty acid oxidation that occur during subsequent reperfusion (81, 82).

Finally, evidence has recently been obtained that AMPK regulates fatty acid oxidation in tissues other than muscle (111). In isolated rat hepatocytes, AICA riboside profoundly depressed acetyl-CoA carboxylase and malonyl-CoA and stimulated fatty acid oxidation twofold. When a permeabilized cell assay for CPT1 was used, only part of the activation could be explained by the reduction of malonyl-CoA concentration, indicating that additional mechanisms may be in operation.

HORMONE-SENSITIVE LIPASE   Hormone-sensitive lipase (HSL) is a classical target for protein phosphorylation: It is phosphorylated at Ser-563 by cyclic AMP-dependent protein kinase (PKA) (112). This mechanism is responsible for the activation of triglyceride breakdown in adipose tissue in response to catecholamines or glucagon. HSL also breaks down cholesterol–fatty acid esters, and it is responsible for the major neutral cholesterol esterase activity in most tissues (113). AMPK phosphorylates HSL at Ser-565. This phosphorylation does not cause activation, but it completely prevents phosphorylation by PKA at the neighboring activation site, Ser-563. Conditions that activate AMPK are therefore potentially antilipolytic (114). In support of this supposition, incubation of adipocytes with AICA riboside antagonizes the effects on triglyceride breakdown of the β-adrenergic agent, isoproterenol, both decreasing the sensitivity to the agonist and decreasing the maximal extent of lipolysis (73, 115). Because oxidation of fatty acids released by lipolysis can be an important source of ATP, inhibition of lipolysis might first appear not to be consistent with the hypothesis that activation of AMPK is concerned with maintaining adequate levels of ATP. However, fatty acids are usually oxidized in cells other than those in which lipolysis occurs. Thus, the function of AMPK is to conserve ATP in the cell in which it is expressed. If lipolysis occurs at a rate faster than that at which free fatty acids can be processed (e.g. by removal in the circulation), fatty acids simply recycle into triglyceride (116), consuming two acid-anhydride bonds of ATP on the way as they are re-esterified with CoA. Recycling of fatty acids into cholesterol esters has also been demonstrated (117). Phosphorylation of HSL therefore could represent an ATP-conserving mechanism, ensuring that the rate at which triglycerides and cholesterol esters are broken down does not exceed the rate at which free fatty acids can be further processed and preventing excessive rates of recycling.

GLYCOGEN SYNTHASE   AMPK phosphorylates rabbit muscle glycogen synthase, the key regulatory enzyme of glycogen synthesis, at Ser-7 in vitro (118). Phosphorylation at Ser-7 promotes phosphorylation at Ser-10 by casein kinase-1, and together these phosphorylation events produce a large inactivation of the enzyme (119). Because two acid-anhydride bonds of UTP are consumed for every glucose unit transferred from glucose-6-phosphate to glycogen, phosphorylation of glycogen synthase by AMPK could represent an energy-conserving measure. Ser-7 is known to be phosphorylated in vivo, but several protein kinases other than AMPK also phosphorylate this site in vitro, and it remains unclear whether glycogen synthase is a physiological target for AMPK.

RAF-1   The Raf-1 protein kinase is a key intermediate in the pathway from growth-factor receptors to the mitogen-activated protein kinase (MAP kinase) pathway (120). Raf-1 is recruited to the membrane by interaction of its N-terminal regulatory domain with Ras-GTP, and on subsequent activation (by mechanisms that remain unclear), it phosphorylates MAP kinase kinases that in turn phosphorylate and activate MAP kinases. Raf-1 is phosphorylated in intact cells at several sites, including Ser-259 and Ser-621 (121). The sequence around Ser-621 is an excellent fit to the consensus recognition motif for AMPK, and when a kinase from 3T3 cells was found to phosphorylate a peptide corresponding to this sequence, it quickly became apparent that it was AMPK (122). AMPK was also shown to phosphorylate Ser-621 in an N-terminally truncated Raf-1 expressed in bacteria, and in full-length Raf-1 expressed in insect cells, although in the latter case there was an additional site, which appeared to be Ser-259.

The function of Ser-621 phosphorylation has been controversial. Mutation of this site to alanine results in total loss of Raf-1 activity, and it was therefore suggested that phosphorylation at this site is essential for activity (121). It was also claimed that phosphorylation at this site was constitutive, because it was partially phosphorylated under basal conditions and did not change in response to growth factors. More recently, it has been reported that overexpression of the catalytic subunit of cAMP-dependent protein kinase results in phosphorylation at Ser-621 and that this is accompanied by inhibition of Raf-1 (123). This report is consistent with recent findings (DG Hardie, unpublished data) that activation of AMPK in isolated rat hepatocytes using fructose or AICA riboside leads to inhibition of Raf-1 and MAP kinase. Although further study of this system is required, phosphorylation of Raf-1 by AMPK could represent a mechanism for inhibiting the transition of cells from the quiescent to the proliferating state when their energy status is compromised.

Targets of Yeast SNF1

ACETYL-COA CARBOXYLASE   Yeast acetyl-CoA carboxylase is phosphorylated and inactivated by the purified yeast SNF1 complex in vitro (26). Acetyl-CoA carboxylase is also a physiological target for SNF1, because the enzyme is inactivated under conditions that activate SNF1 (starvation for glucose), but this inactivation of acetyl-CoA carboxylase does not occur in a *snf1* Δ mutant strain (48). The effect survives partial purification, which suggests that it is due to phosphorylation. Acetyl-CoA carboxylase is therefore the first target shown to be conserved between mammalian AMPK and yeast SNF1. A puzzling feature, however, is that the Ser-79 site, whose phosphorylation causes inactivation of the mammalian enzyme, is missing in the yeast sequence. Regulation of yeast acetyl-CoA carboxylase must therefore occur at a different site.

MIG1   Most of the other likely downstream targets for SNF1 have been identified by genetic analysis, which does not normally distinguish whether they are direct substrates for the SNF1 complex or whether the effect is transmitted indirectly via a cascade of other proteins. The best characterized of these potential targets is the Mig1 repressor. Mig1 is a Cys_2His_2 zinc-finger protein (124) that binds to the promoters of several glucose-repressible genes, including the *SUC* and *GAL* genes. Mutation of Mig1, or of its binding sites on promoters, partially relieves glucose repression (124–127). Repression of these genes also requires the Ssn6(Cyc8p)-Tup1 corepressor, which is recruited to different promoters by specific DNA-binding proteins and represses transcription of genes regulated by glucose and by other signals (128, 129). An LexA-Mig1 fusion protein represses transcription of reporters containing *lexA* operators, and this repression requires Ssn6-Tup1 and is dependent on glucose (130, 131). These data suggest that Mig1 recruits Ssn6-Tup1 to promoters in response to the glucose signal. Mig1 is differentially phosphorylated in response to glucose, with a greater mobility shift in derepressed cells than in glucose-grown cells (130). Recent work has shown that the subcellular localization of Mig1 is also regulated by glucose, the protein being nuclear when glucose is present and transported to the cytoplasm when cells are limited for glucose (132).

Genetic evidence suggests that the SNF1 complex regulates Mig1. A *mig1* mutation suppresses the *snf1* mutant defects in *SUC2* and *GAL1* expression (126, 127), which indicates that SNF1 functions to inhibit repression by Mig1. SNF1 also inhibits transcriptional activation by a hybrid Mig1-VP16 activator in the absence of glucose (133). In an *snf1* mutant, phosphorylation of Mig1 is reduced to a minimal level in both glucose-repressing and -derepressing conditions (M Treitel & M Carlson, unpublished data). Also, Mig1 is localized in the nucleus in *snf1* mutants, even after a shift to glucose-limiting conditions (132).

Expression of Mig1 as a glutathione-S-transferase (GST) fusion in bacteria revealed that Mig1 is an excellent substrate for purified SNF1 in vitro. Mapping of the sites showed that four sites are present (S Smith, SP Davies, DG Hardie & D Carling, unpublished data). Intriguingly, all of the sites lay within or near the regulatory domain 1, which is required for the response to SNF1 (133). Mutation of these sites suggests that some are phosphorylated in vivo, but none of the mutations reduced phosphorylation of Mig1 as completely as did *snf1* mutations. This finding indicates that the SNF1 complex may also act indirectly to affect function of other Mig1 kinase(s) (M Treitel & M Carlson, unpublished data).

SIP3 AND SIP4   Other potential targets for the SNF1 complex include Sip3 and Sip4, which were identified in two-hybrid screens using Snf1 as bait (49, 134, 135). The interaction of Sip4 with the kinase complex is mediated by Gal83 (Vincent and M Carlson, unpublished data). Sip4 has a Cys_6 zinc cluster motif characteristic of DNA-binding proteins of the Gal4 family, and it is a transcriptional activator. Its expression is repressed by glucose, and its activity is regulated by glucose and depends on Snf1. Moreover, Sip4 is differentially phosphorylated in glucose-repressed and -derepressed cells, and phosphorylation requires Snf1. Thus, Sip4 is most likely a target of Snf1. The promoters that are activated by Sip4 remain to be identified. Genetic evidence also supports the idea of a functional relationship between Sip3 and the SNF1 pathway (134). DNA-bound LexA-Sip3 fusion proteins activate transcription of a reporter gene, which suggests that Sip3 acts downstream of Snf1.

CAT8   Cat8 is a zinc cluster protein that is required for derepression of gluconeogenic genes and for growth on nonfermentable carbon sources (136, 137). Its expression is glucose repressible, and on multicopy expression the gene suppresses the growth defect of an *snf1* mutant on ethanol. A Gal4 DNA-binding domain fused to Cat8 mediates glucose-regulated, Snf1-dependent transcriptional activation. Cat8 is differentially phosphorylated in response to glucose availability, and some of the phosphorylation detected in derepressed cells depends on Snf1. Thus, Cat8 also appears to be a downstream target of SNF1.

Targets of Higher-Plant SNF1-Related Protein Kinases

HMG-COA REDUCTASE AND OTHER METABOLIC ENZYMES   No physiological targets for plant SNF1-related protein kinases have been identified with certainty. However, bacterial expression of the catalytic domain of an HMG-CoA reductase *(HMG1)* from *Arabidopsis thaliana* showed that it was phosphorylated by cauliflower HRK-A at Ser-577 (equivalent to Ser-871 in the Chinese hamster HMG-CoA reductase). This phosphorylation resulted in complete inactivation of the enzyme (138). Because the kinase recognition motif on *HMG1* is conserved on all of the many higher-plant HMG-CoA reductases for which the DNA has been sequenced (Figure 5), HMG-CoA reductase is likely to be a physiological target for the kinase in plants. Plants synthesize a variety of isoprenoids, and overexpression of HMG-CoA reductase in tobacco indicates that it exerts significant control over the whole pathway, even though other controls operate at later branch points (139).

Of the limited number of other metabolic enzymes known to be regulated by phosphorylation in plants, the phosphorylation sites on two enzymes—sucrose phosphate synthase (SPS) (140) and nitrate reductase (NR) (141, 142)—conform to the AMPK/SNF1 recognition motif (Figure 5). SPS catalyzes a key regulatory step in the synthesis of sucrose in photosynthetic tissues (143): The sucrose is transported to nonphotosynthetic tissues (e.g. roots) and zones of active growth (meristems), where it is the major carbon source. NR is involved in the initial conversion of nitrate to nitrite in the cytoplasm: The nitrite is taken up by chloroplasts and assimilated into amino acids and other nitrogen-containing compounds. These two enzymes are therefore involved in major biosynthetic processes of the plant. Spinach SPS is phosphorylated by an uncharacterized protein kinase in vitro at Ser-158 (140), and this phosphorylation causes inactivation in the presence of appropriate concentrations of the substrate, glucose-6-phosphate, and the inhibitor, phosphate. Spinach NR is phosphorylated at Ser-543, which causes total inactivation in the presence of 14-3-3 protein (144, 145). Multiple SPS and NR kinases are present in extracts of spinach leaf (146), and at least two of these kinases appear to be members of the SNF1-related kinase family (147; P Donaghy, C Sugden & DG Hardie, unpublished data).

TRANSCRIPTION   The expression of a number of genes in plants is repressed by high glucose or sucrose in the cell medium (148). Interestingly, one of them is isocitrate lyase (149), an enzyme of the glyoxylate bypass that is also repressed by glucose in *S. cerevisiae*. Plant SNF1-related kinases are therefore likely to be involved in derepression of these genes, just as SNF1 is in yeast. The transcription factors involved in glucose/sucrose repression have not been identified. Transgenic potato plants have been constructed that express in their tubers DNA encoding the potato *SNF1* homologue in antisense orientation. The SAMS peptide kinase activity in the tubers of some lines was reduced by >90%. In these plants the activity of sucrose synthase was greatly reduced and RNA transcripts were completely missing (NG Halford, personal communication). These results suggest that the potato *SNF1* homologue may be involved in derepression of sucrose synthase in the tuber, a storage organ. Despite its name, sucrose synthase is involved in the degradation of sucrose, which is the carbon source supplied to the nonphotosynthetic "sink" tissues (e.g. the tuber) by the

HMG-CoA reductases:

Arabidopsis thaliana	HMKYNR**S**SRDI
Camptotheca acuminata	HMKYNR**S**NKDV
Catharanthus roseus	HMKYNR**S**SKDI
Hevea brasiliensis	HMKYNR**S**SKDM
Nicotiana sylvestris	HMKYNR**S**TKDV
Potato	HMKYNR**S**IKDI
Rice	HMMYNR**S**SKDV
Tomato	HMKYNR**S**TKDV

Sucrose phosphate synthases:

Spinach	RMRRIS**S**VEMM
Potato	KFQRNF**S**DFTL
Rice	RLPRIS**S**VETM
Maize	KFQRNF**S**ELTV
Vicia faba	RLPRIS**S**ADAM

Nitrate reductases:

Spinach	TLKRTA**S**TPFM
Barley	TLKKSV**S**SPFM
Brassica napus	GLKRST**S**TPFM
Chicory	TLKKSV**S**TPFM
Kidney bean	TLKKSV**S**SPFM
Lotus japonicus	ILKKSV**S**SPFM
Maize	ILKKSV**S**SPFM
Petunia	GLKRST**S**TPFM
Rice	TLKKSI**S**TPFM
Soybean	GLKRST**S**TPFM
Arabidopsis thaliana	TLKKSV**S**SPFM
Tobacco	TLKKSI**S**TPFM
Tomato	TLKKSI**S**TPFM
White birch	SLKKSV**S**SPFM
Winter squash	TLKKSV**S**TPFM

Figure 5 Alignment of sequences around the phosphorylation sites on 3-hydroxy-3-methyl-glutaryl–coenzyme A (HMG-CoA) reductases, sucrose phosphate synthases, and nitrate reductases from a number of higher plant species. For each enzyme, the phosphorylation site has been established for the species at the top of the list (138, 140–142). Other sequences are from the databases. Residues that may be important in recognition by the plant SNF1-related kinases are highlighted as follows: phosphorylated serine (*bold, underlined*); basic residues at P − 6, P − 3, P − 4 (*underlined*); hydrophobic residues at P − 5, P + 4 (*bold*).

photosynthetic "source" tissues (e.g. the leaves). Sucrose synthase therefore occupies a position in plants that is analogous to the position of the classical glucose-repressed enzyme, invertase, in yeast.

CONCLUSIONS: UNIFYING HYPOTHESES FOR THE REGULATION AND FUNCTION OF THE AMP-ACTIVATED/SNF1 SUBFAMILY

When the sequence similarities between mammalian AMPK subunits and the proteins of the yeast SNF1 complex were first noticed (26, 28), their physiological roles appeared to be different. We believe that this appearance resulted from the very different approaches through which they were discovered. Both can now be regarded as stress response systems, although the type of stress to which they respond may be different. AMPK is activated by environmental stresses such as heat shock, hypoxia, inhibitors of oxidative phosphorylation, and exercise (in muscle), whereas SNF1 is activated by starvation for glucose. A common feature of these stresses is that they involve depletion of cellular ATP and, via the adenylate kinase reaction, elevation of AMP. In the case of the mammalian AMPK cascade, the rise in AMP and fall in ATP are the actual signals that switch on the system. In yeast, direct regulation by AMP does not appear to occur, but many of the other regulatory features of the system are conserved with AMPK. For example, the activation of Snf1 on glucose removal is due to its phosphorylation, because activation can be reversed by protein phosphatase treatment and can be mimicked by treatment with MgATP and the upstream kinase from the mammalian system (AMPKK) or with a putative upstream kinase from yeast (44). Although the yeast cascade has not been found to be directly regulated by AMP in cell-free systems, large increases in cellular AMP and decreases in ATP occur under conditions where SNF1 is activated (44). Therefore a reasonable working hypothesis is that the signal or signals that activate SNF1 are, if not AMP and ATP, some other metabolite(s) that change in concert with them. Although the metabolites to which the yeast SNF1 kinase and the higher-plant SNF1-related kinases respond have not been found, the high degree of similarity to the mammalian AMP-activated protein kinase justifies the subtitle of this review, i.e. that these systems represent metabolic sensors of the cell.

When considering the functions of these systems in terms of downstream events, the mammalian AMPK system can now be seen both to conserve ATP by switching off anabolic pathways (e.g. fatty acid, cholesterol synthesis) and to promote ATP production by switching on an alternative catabolic pathway (i.e. fatty acid oxidation). Although the latter is achieved by direct phosphorylation of a metabolic enzyme (acetyl-CoA carboxylase) rather than through effects on gene expression, this function is analogous to one of the known functions of the yeast SNF1 system: to derepress genes required for catabolism of alternative carbon sources. Although most of the known downstream targets for SNF1 are transcription factors, SNF1 also directly phosphorylates at least one metabolic enzyme—acetyl-CoA carboxylase (26, 48), one of the classical targets for mammalian AMPK. There is no obvious reason transcription factors should not be important targets for mammalian AMPK; we are actively investigating this area in our laboratories. The genes that may be regulated by AMPK remain a matter for speculation, but a likely place to begin investigations would be with with genes involved in carbon catabolism and maintenance of cellular energy balance.

Studies on the regulation and function of plant SNF1-related kinases are at a much earlier stage. However, if the function of the AMPK/SNF1 subfamily is in the response to cellular stress, members of this subfamily are likely to be playing important roles, because plants are subjected to environmental stress (e.g. lack of water, extremes of temperature) on a regular basis. In plants, reduced carbon is of course normally provided by photosynthesis. An underground tuber in a potato plant receives its carbon fuel in the form of sucrose exported from the leaves. If this amount of carbon fuel were to be reduced (e.g. because of herbivorous animals having grazed the leaves), the tuber would be in a state of starvation, akin to yeast in which glucose had been removed from the medium. It is therefore intriguing that the expression of the sucrose synthase gene (which occupies a metabolic role analogous to yeast invertase) appears to be under the control of an SNF1-related kinase in potato tubers.

To summarize these ideas, we propose the unifying hypothesis that the AMPK, SNF1, and SNF1-related kinase cascades have the overall function of controlling metabolism, gene expression, and perhaps cell proliferation in response to the varying energy status of the cell. These enzymes achieve this regulatory role by directly phosphorylating proteins in the target pathways and indirectly by regulating gene expression. Finally, these cascades appear to be the immediate sensors of metabolic indexes of cellular energy status, such as the levels of AMP and ATP, especially in the case of mammalian AMPK.

ACKNOWLEDGMENTS

Studies in our laboratories were supported by the Wellcome Trust, the Medical Research Council and the Biotechnology and Biological Sciences Research Council (DGH), the Medical Research Council and British Heart Foundation (DC), and National Institutes of Health (NIH) grant GM34095 (MC).

Literature Cited

1. Beg ZH, Allmann DW, Gibson DM. 1973. *Biochem. Biophys. Res. Commun.* 54:1362–69
2. Carlson CA, Kim KH. 1973. *J. Biol. Chem.* 248:378–80
3. Carling D, Zammit VA, Hardie DG. 1987. *FEBS Lett.* 223:217–22
4. Hardie DG, Carling D, Sim ATR. 1989. *Trends Biochem. Sci.* 14:20–23
5. Brown MS, Brunschede GY, Goldstein JL. 1975. *J. Biol. Chem.* 250:2502–9
6. Keith ML, Rodwell VW, Rogers DH, Rudney H. 1979. *Biochem. Biophys. Res. Commun.* 90:969–75
7. Beg ZH, Stonik JA, Brewer HB. 1978. *Proc. Natl. Acad. Sci. USA* 75:3678–82
8. Yeh LA, Lee KH, Kim KH. 1980. *J. Biol. Chem.* 255:2308–14
9. Ferrer A, Caelles C, Massot N, Hegardt FG. 1985. *Biochem. Biophys. Res. Commun.* 132:497–504
10. Carling D, Clarke PR, Zammit VA, Hardie DG. 1989. *Eur. J. Biochem.* 186:129–36
11. Zimmermann FK, Kaufmann I, Rasenberger I, Haussman P. 1977. *Mol. Gen. Genet.* 1951:95–103
12. Ciriacy M. 1977. *Mol. Gen. Genet.* 154:213–20
13. Carlson M, Osmond BC, Botstein D. 1981. *Genetics* 98:25–40
14. Johnston M, Carlson M. 1992. In *The Molecular and Cellular Biology of the Yeast Saccharomyces: Gene Expression*, 2:193–281
15. Simon M, Binder M, Adam G, Hartig A, Ruis H. 1992. *Yeast* 8:303–9
16. Thompson-Jaeger S, Francois J, Gaughran JP, Tatchell K. 1991. *Genetics* 129:697–706
17. Hubbard EJA, Yang XL, Carlson M. 1992. *Genetics* 130:71–80
18. Hardy TA, Huang D, Roach PJ. 1994. *J. Biol. Chem.* 269:27907–13
19. Entian KD, Zimmermann FK. 1982. *J. Bacteriol.* 151:1123–28
20. Neigeborn L, Carlson M. 1984. *Genetics* 108:845–58
21. Beg ZH, Stonik JA, Brewer HB. 1979. *Proc. Natl. Acad. Sci. USA* 76:4375–79
22. Shiao M, Drong RF, Porter JW. 1981. *Biochem. Biophys. Res. Commun.* 98:80–87

23. Lent B, Kim KH. 1982. *J. Biol. Chem.* 257:1897–901
24. Ferrer A, Caelles C, Massot N, Hegardt FG. 1987. *J. Biol. Chem.* 262:13507–12
25. Davies SP, Hawley SA, Woods A, Carling D, Haystead TAJ, Hardie DG. 1994. *Eur. J. Biochem.* 223:351–57
26. Mitchelhill KI, Stapleton D, Gao G, House C, Michell B, et al. 1994. *J. Biol. Chem.* 269:2361–64
27. Michell BJ, Stapleton D, Mitchelhill KI, House CM, Katsis F, et al. 1996. *J. Biol. Chem.* 271:28445–50
28. Carling D, Aguan K, Woods A, Verhoeven AJM, Beri RK, et al. 1994. *J. Biol. Chem.* 269:11442–48
29. Gao G, Widmer J, Stapleton D, Teh T, Cox T, et al. 1995. *Biochim. Biophys. Acta* 1266:73–82
30. Stapleton D, Gao G, Michell BJ, Widmer J, Mitchelhill K, et al. 1994. *J. Biol. Chem.* 269:29343–46
31. Gao G, Fernandez CS, Stapleton D, Auster AS, Widmer J, et al. 1996. *J. Biol. Chem.* 271:8675–81
32. Woods A, Cheung PCF, Smith FC, Davison MD, Scott J, et al. 1996. *J. Biol. Chem.* 271:10282–90
33. Dyck JRB, Gao G, Widmer J, Stapleton D, Fernandez CS, et al. 1996. *J. Biol. Chem.* 271:17798–803
34. Woods A, Salt I, Scott J, Hardie DG, Carling D. 1996. *FEBS Lett.* 397:347–51
35. Stapleton D, Mitchelhill KI, Gao G, Widmer J, Michell BJ, et al. 1996. *J. Biol. Chem.* 271:611–14
35a. Thornton C, Snowden MA, Carling D. 1998. *J. Biol. Chem.* In press
36. Bateman A. 1997. *Trends Biochem. Sci.* 22:12–13
37. Kluijtmans LA, Boers GH, Stevens EM, Renier WO, Kraus JP, et al. 1996. *J. Clin. Invest.* 98:285–89
38. Sintchak MD, Fleming MA, Futer O, Raybuck SA, Chambers SP, et al. 1996. *Cell* 85:921–30
39. Celenza JL, Carlson M. 1986. *Science* 233:1175–80
40. Schuller HJ, Entian KD. 1987. *Mol. Gen. Genet.* 209:366–73
41. Schuller HJ, Entian KD. 1988. *Gene* 67:247–57

42. Celenza JL, Eng FJ, Carlson M. 1989. *Mol. Cell. Biol.* 9:5045–54
43. Celenza JL, Carlson M. 1989. *Mol. Cell. Biol.* 9:5034–44
44. Wilson WA, Hawley SA, Hardie DG. 1996. *Curr. Biol.* 6:1426–34
45. Jiang R, Carlson M. 1997. *Mol. Cell. Biol.* 17:2099–106
46. Estruch F, Treitel MA, Yang XL, Carlson M. 1992. *Genetics* 132:639–50
47. Fields S, Song OK. 1989. *Nature* 340: 245–46
48. Woods A, Munday MR, Scott J, Yang XL, Carlson M, Carling D. 1994. *J. Biol. Chem.* 269:19509–15
49. Yang X, Hubbard EJA, Carlson M. 1992. *Science* 257:680–82
50. Yang X, Jiang R, Carlson M. 1994. *EMBO J.* 13:5878–86
51. Matsumoto K, Toh-e A, Oshima Y. 1981. *Mol. Cell. Biol.* 1:83–93
52. Erickson JR, Johnston M. 1993. *Genetics* 135:655–64
53. Goffrini P, Ficarelli A, Donnini C, Lodi T, Puglisi PP, Ferrero I. 1996. *Curr. Genet.* 29:316–26
54. Alderson A, Sabelli PA, Dickinson JR, Cole D, Richardson M, et al. 1991. *Proc. Natl. Acad. Sci. USA* 88:8602–5
55. LeGuen L, Thomas M, Bianchi M, Halford NG, Kreis M. 1992. *Gene* 120:249–54
56. Hannappel U, Vicente-Carbajosa J, Barker JHA, Shewry PR, Halford NG. 1995. *Plant Mol. Biol.* 27:1235–40
57. Halford NG, Vicente-Carbajosa J, Sabelli PA, Shewry PR, Hannappel U, Kreis M. 1992. *Plant J.* 2:791–97
58. Muranaka T, Banno H, Machida Y. 1994. *Mol. Cell. Biol.* 14:2958–65
59. MacKintosh RW, Davies SP, Clarke PR, Weekes J, Gillespie JG, et al. 1992. *Eur. J. Biochem.* 209:923–31
60. Davies SP, Carling D, Hardie DG. 1989. *Eur. J. Biochem.* 186:123–28
61. Ball KL, Dale S, Weekes J, Hardie DG. 1994. *Eur. J. Biochem.* 219:743–50
62. Dale S, Wilson WA, Edelman AM, Hardie DG. 1995. *FEBS Lett.* 361:191–95
63. Ball KL, Barker J, Halford NG, Hardie DG. 1995. *FEBS Lett.* 377:189–92
64. Jiang R, Carlson M. 1996. *Genes Dev.* 10:3105–15
65. Ingebritsen TS, Lee H, Parker RA, Gibson DM. 1978. *Biochem. Biophys. Res. Commun.* 81:1268–77
66. Ingebritsen TS, Parker RA, Gibson DM. 1981. *J. Biol. Chem.* 256:1138–44
67. Moore F, Weekes J, Hardie DG. 1991. *Eur. J. Biochem.* 199:691–97
68. Weekes J, Hawley SA, Corton J, Shugar D, Hardie DG. 1994. *Eur. J. Biochem.* 219:751–57
69. Clarke PR, Moore F, Hardie DG. 1991. *Adv. Protein Phosphatases* 6:187–209
70. Davies SP, Helps NR, Cohen PTW, Hardie DG. 1995. *FEBS Lett.* 377:421–25
71. Monod J, Wyman J, Changeux JP. 1965. *J. Mol. Biol.* 12:88–118
72. Hawley SA, Davison M, Woods A, Davies SP, Beri RK, et al. 1996. *J. Biol. Chem.* 271:27879–87
73. Corton JM, Gillespie JG, Hawley SA, Hardie DG. 1995. *Eur. J. Biochem.* 229: 558–65
74. Hawley SA, Selbert MA, Goldstein EG, Edelman AM, Carling D, Hardie DG. 1995. *J. Biol. Chem.* 270:27186–91
75. Johnson LN, Noble MEM, Owen DJ. 1996. *Cell* 85:149–58
76. Hardie DG, Carling D. 1997. *Eur. J. Biochem.* 246:259–73
77. Corton JM, Gillespie JG, Hardie DG. 1994. *Curr. Biol.* 4:315–24
78. Sato R, Goldstein JL, Brown MS. 1993. *Proc. Natl. Acad. Sci. USA* 90:9261–65
79. Witters LA, Nordlund AC, Marshall L. 1991. *Biochem. Biophys. Res. Commun.* 181:1486–92
80. Ritossa FM. 1964. *Exp. Cell Res.* 35:601–7
81. Kudo N, Barr AJ, Barr RL, Desai S, Lopaschuk GD. 1995. *J. Biol. Chem.* 270: 17513–20
82. Kudo N, Gillespie JG, Kung L, Witters LA, Schulz R, et al. 1996. *Biochim. Biophys. Acta* 1301:67–75
83. Winder WW, Hardie DG. 1996. *Am. J. Physiol.* 270:E299–304
84. Hutber CA, Hardie DG, Winder WW. 1997. *Am. J. Physiol.* 272:E262–66
85. Vavvas D, Apazidis A, Saha AK, Gamble J, Patel A, et al. 1997. *J. Biol. Chem.* 272:13255–61
86. Arebalo RE, Hardgrave JE, Scallen TJ. 1981. *J. Biol. Chem.* 256:571–74
87. Arebalo RE, Hardgrave JE, Noland BJ, Scallen TJ. 1980. *Proc. Natl. Acad. Sci. USA* 77:6429–33
88. Beg ZH, Reznikov DC, Avigan J. 1986. *Arch. Biochem. Biophys.* 244:310–22
89. Munday MR, Milic MR, Takhar S, Holness MJ, Sugden MC. 1991. *Biochem. J.* 280:733–37
90. Witters LA, Kemp BE. 1992. *J. Biol. Chem.* 267:2864–67
91. Davies SP, Carling D, Munday MR, Hardie DG. 1992. *Eur. J. Biochem.* 203:615–23

92. Neigeborn L, Carlson M. 1987. *Genetics* 115:247–53
93. Stark MJ. 1996. *Yeast* 12:1647–75
94. Tu JL, Carlson M. 1995. *EMBO J.* 14:5939–46
95. Tu JL, Carlson M. 1994. *Mol. Cell. Biol.* 14:6789–96
96. Weekes J, Ball KL, Caudwell FB, Hardie DG. 1993. *FEBS Lett.* 334:335–39
97. Ching YP, Davies SP, Hardie DG. 1996. *Eur. J. Biochem.* 237:800–8
98. Clarke PR, Hardie DG. 1990. *EMBO J.* 9:2439–46
99. Gillespie JG, Hardie DG. 1992. *FEBS Lett.* 306:59–62
100. Henin N, Vincent MF, Gruber HE, Van den Berghe G. 1995. *FASEB J.* 9:541–46
101. Munday MR, Campbell DG, Carling D, Hardie DG. 1988. *Eur. J. Biochem.* 175:331–38
102. Davies SP, Sim ATR, Hardie DG. 1990. *Eur. J. Biochem.* 187:183–90
103. Lopez-Casillas F, Bai DH, Luo XC, Kong IS, Hermodson MA, Kim KH. 1988. *Proc. Natl. Acad. Sci. USA* 85:5784–88
104. Ha J, Daniel S, Broyles SS, Kim KH. 1994. *J. Biol. Chem.* 269:22162–68
105. McGarry JD, Takabayashi Y, Foster DW. 1978. *J. Biol. Chem.* 253:8294–300
106. Weis BC, Cowan AT, Brown N, Foster DW, McGarry JD. 1994. *J. Biol. Chem.* 269:26443–48
107. Abu Elheiga L, Almarza Ortega DB, Baldini A, Wakil SJ. 1997. *J. Biol. Chem.* 272:10669–77
108. Ha J, Lee JK, Kim KS, Witters LA, Kim KH. 1996. *Proc. Natl. Acad. Sci. USA* 93:11466–70
109. Winder WW, Wilson HA, Hardie DG, Rasmussen BB, Hutber CA, et al. 1997. *J. Appl. Physiol.* 82:219–25
110. Merrill GM, Kurth E, Hardie DG, Winder WW. 1997. *Am. J. Physiol.* 36:E1107–12
111. Velasco G, Geelen MJH, Guzman M. 1997. *Arch. Biochem. Biophys.* 337:169–75
112. Garton AJ, Campbell DG, Cohen P, Yeaman SJ. 1988. *FEBS Lett.* 229:68–72
113. Yeaman SJ. 1990. *Biochim. Biophys. Acta* 1052:128–32
114. Garton AJ, Campbell DG, Carling D, Hardie DG, Colbran RJ, Yeaman SJ. 1989. *Eur. J. Biochem.* 179:249–54
115. Sullivan JE, Brocklehurst KJ, Marley AE, Carey F, Carling D, Beri RK. 1994. *FEBS Lett.* 353:33–36
116. Brooks BJ, Arch JR, Newsholme EA. 1983. *Biosci. Rep.* 3:263–67
117. Brown MS, Ho YK, Goldstein JL. 1980. *J. Biol. Chem.* 255:9344–52
118. Carling D, Hardie DG. 1989. *Biochim. Biophys. Acta* 1012:81–86
119. Flotow H, Roach PJ. 1989. *J. Biol. Chem.* 264:9126–28
120. Morrison DK, Cutler RE. 1997. *Curr. Opin. Cell Biol.* 9:174–79
121. Morrison DK, Heidecker G, Rapp UR, Copeland TD. 1993. *J. Biol. Chem.* 268:17309–16
122. Sprenkle AB, Davies SP, Carling D, Hardie DG, Sturgill TW. 1997. *FEBS Lett.* 403:254–58
123. Mischak H, Seitz T, Janosch P, Eulitz M, Steen H, et al. 1996. *Mol. Cell. Biol.* 16:5409–18
124. Nehlin JO, Ronne H. 1990. *EMBO J.* 9:2891–98
125. Nehlin JO, Carlberg M, Ronne H. 1991. *EMBO J.* 10:3373–77
126. Johnston M, Flick JS, Pexton T. 1994. *Mol. Cell. Biol.* 14:3834–41
127. Vallier LG, Carlson M. 1994. *Genetics* 137:49–54
128. Keleher CA, Redd MJ, Schultz J, Carlson M, Johnson AD. 1992. *Cell* 68:709–19
129. Tzamarias D, Struhl K. 1994. *Nature* 369:758–61
130. Treitel MA, Carlson M. 1995. *Proc. Natl. Acad. Sci. USA* 92:3132–36
131. Tzamarias D, Struhl K. 1995. *Genes Dev.* 9:821–31
132. DeVit MJ, Waddle JA, Johnston M. 1997. *Mol. Biol. Cell* 8:1603–18
133. Ostling J, Carlberg M, Ronne H. 1996. *Mol. Cell. Biol.* 16:753–61
134. Lesage P, Yang XL, Carlson M. 1994. *Nucleic Acids Res.* 22:597–603
135. Lesage P, Yang XL, Carlson M. 1996. *Mol. Cell. Biol.* 16:1921–28
136. Hedges D, Proft M, Entian KD. 1995. *Mol. Cell. Biol.* 15:1915–22
137. Randez-Gil F, Bojunga N, Proft M, Entian KD. 1997. *Mol. Cell. Biol.* 17:2502–10
138. Dale S, Arró M, Becerra B, Morrice NG, Boronat A, et al. 1995. *Eur. J. Biochem.* 233:506–13
139. Chappell J, Wolf F, Proulx J, Cuellar R, Saunders C. 1995. *Plant Physiol.* 109:1337
140. McMichael RW Jr, Klein RR, Salvucci ME, Huber SC. 1993. *Arch. Biochem. Biophys.* 307:248–52
141. Douglas P, Morrice N, MacKintosh C. 1995. *FEBS Lett.* 377:113–17
142. Bachmann M, Shiraishi N, Campbell WH, Yoo BC, Harmon AC, Huber SC. 1996. *Plant Cell* 8:505–17
143. Huber SC, Huber JL. 1996. *Annu. Rev. Plant Physiol. Plant Mol. Biol.* 47:431–44

144. Moorhead G, Douglas P, Morrice N, Scarabel M, Aitken A, MacKintosh C. 1996. *Curr. Biol.* 6:1104–13
145. Bachmann M, Huber JL, Liao PC, Gage DA, Huber SC. 1996. *FEBS Lett.* 387:127–31
146. McMichael RW Jr, Bachmann M, Huber SC. 1995. *Plant Physiol.* 108:1077–82
147. Douglas P, Pigaglio E, MacKintosh C. 1997. *Biochem. J.* In press
148. Jang JC, Sheen J. 1994. *Plant Cell* 6:1665–79
149. Grahame IA, Denby KJ, Leaver CJ. 1994. *Plant Cell* 6:761–72
150. Hardie DG, Hanks SK. 1995. *The Protein Kinase Factsbook: Protein-Serine Kinases.* London: Academic. 418 pp.
151. Alessi DR, Andjelkovic M, Caudwell B, Cron P, Morrice N, et al. 1996. *EMBO J.* 15:6541–51

AUTHOR INDEX

857

347, 358, 366, 367, 369, 373,
380
Seraphin B, 157, 174, 177, 279
Serban A, 797–99, 802, 804–6
Serban D, 807
Serban H, 796, 806
Sergiev PV, 127
Serr M, 358
Serrano R, 39, 40
Serunian LA, 502, 503
Serve H, 461
Serwer P, 557
Sessa WC, 202, 207, 217, 218
Seto E, 564, 566
Seufert W, 430, 431
Seufferlein T, 28
Severns CW, 237
Severs NJ, 212
Sevinsky JR, 202
Seyedin SM, 762, 763
Shaeffer JR, 445
Shafer HM, xvii
Shah G, 727
Shah PC, 767, 783
Shalom A, 19
Shalon D, 45
Shamu CE, 39
Shan S-O, 698, 700
Shank DD, 110
Shapiro A, 454, 471
Shareck F, 158
Sharif WD, 519, 522, 526, 527,
529, 533
Sharma B, 618, 620, 778
Sharma K, 285–88, 290
Sharma OK, 256
Sharma R, 660
Sharnick SV, 268
Sharova N, 11
Sharp PA, 5, 10, 102, 105, 106,
120, 126, 281
Shaul PW, 202, 203, 207, 217
Shaw AS, 216, 230
Shaw GM, 16, 661–63
Shaw L, 52
Shaw R, 254
Shaw SK, 412
Shayman JA, 37, 39
Shcherbakova PV, 739
Shea RG, 104
Sheaff RJ, 449–51
Shear JB, 526
Sheardown SA, 337
Shearer M, 235, 240
Shearn A, 467
Sheehan JK, 641
Sheehan KCF, 233, 250
Sheen J, 849
Shelby RD, 533
Sheldrick KS, 732, 733
Shelton-Inloes BB, 411

Shematek EM, 321, 324
Shen C-P, 388
Shen J-C, 186
Shen R, 252
Shen Y, 245
Shen ZY, 431
Sheng N, 16
Shenk TE, 245, 282
Shennan DB, 56
Shenoy-Scaria AM, 206, 207, 216
Shenvi AB, 701
Shepard SRP, 62
Shepherd PR, 486
Shepherd S, 35
Sher A, 249
Sher DA, 252
Sherman F, 439
Sherman MY, 465
Sherr CJ, 438, 449, 450, 742
Sheshberadaran H, 150
Shetty K, 109
Sheu YJ, 327
Shevenko A, 455
Shewring L, 623
Shewry PR, 829
Shi S, 733
Shi Y, 280
Shi Y-B, 573
Shi YG, 773–76, 782
Shi YQ, 782
Shiao M, 824
Shiba K, 781
Shibai H, 765
Shibasaki F, 283, 297
Shibasaki Y, 496, 497
Shibuya H, 779
Shibuya S, 812
Shichi H, 655, 678
Shigematsu K, 798
Shigeyoshi Y, 136, 138, 139, 141
Shih M, 675, 678, 680
Shih S, 460
Shillingford JM, 56
Shilowski D, 171
Shimamoto T, 270
Shimasaki S, 761, 765
Shimizu A, 404
Shimizu F, 671
Shimomura O, 510, 511, 515
Shimomura T, 728, 741, 742
Shimonaka M, 629
Shimura Y, 154, 156, 157, 281,
288
Shin R-W, 812
Shin SJ, 52, 62
Shin TB, 496, 767
Shin TH, 436, 446
Shinagawa M, 812
Shinjo K, 315
Shinomura T, 624, 627, 632, 638
Shinowara NL, 202, 207

Shinozaki K, 175
Shinzato T, 621
Shioda N, 760
Shiozaki K, 678, 680
Shiraishi N, 848, 849
Shirakabe K, 779
Shires A, 559, 560
Shivji MKK, 743
Shkedy D, 437
Shoelson SE, 488, 489
Shore D, 559
Shorr R, 655
Shortle D, 62
Showman RM, 298
Shrivastava A, 644
Shteinberg M, 447
Shu HH, 158, 162
Shuai K, 232, 269, 298
Shub DA, 73, 75, 76, 89, 91
Shugar D, 65, 830
Shui Z, 674
Shulga N, 267, 280
Shull MM, 780
Shulman LM, 240
Shum L, 759
Shworak NW, 610
Shyng SL, 208
Sibley D, 655, 657, 659, 674
Sicheri F, 15, 490
Sick C, 243
Siderovski DP, 245, 256, 686
Sidrauski C, 243
Siebenlist U, 298, 454
Siebler J, 244
Siede W, 732, 733
Siedlecki CA, 410
Siegel G, 732
Siegel MS, 534
Siegel MW, 206
Siegel NR, 430
Siegel RW, 159, 171, 172
Siegel V, 385
Siemering KR, 512, 529, 532
SIGLER PB, 581–608; 107, 182,
191, 593, 594, 598, 661, 666
Sigurdsson ST, 125
Siguret V, 418
Sikorski RS, 436
Sil A, 341, 354, 362, 388, 572
Silbert JE, 611
Silengo L, 532
Silhol M, 244
Silva CM, 522, 524, 535
Silvennoinen O, 252
Silver J, 14
Silver PA, 269, 276–79, 287, 290,
291, 299, 431, 522, 523, 534
Silver R, 139, 149
Silverman N, 561, 562
SILVERMAN RH, 227–64; 243,
244

SUBJECT INDEX

A

Abasic nucleosides
 modified oligonucleotides and,
  116, 118–19
Abdominal cell fate specification
 RNA localization in
  development and, 385–86
Acceptor stems
 ribonuclease P and, 171
Accessory proteins
 Golgi and endoplasmic
  reticulum transporters, 59
 HIV-1 life cycle and, 1, 13–19
 TGF-β signal transduction and,
  762–64
Acetylation
 nucleosome structure and
  transcriptional regulation,
  545–73
Acetylcholine
 matrix proteoglycans and,
  622–23
Acetyl-CoA carboxylase
 AMP-activated/SNF1 protein
  kinase subfamily and,
  843–44, 846
Acetylgalactosamine
 Golgi and endoplasmic
  reticulum transporters,
  52–53
Acetylglucosamine
 Golgi and endoplasmic
  reticulum transporters,
  51–54, 56–57, 59–66
N-Acetylglucosaminyltransferase
 Golgi and endoplasmic
  reticulum transporters, 55
Acetyltransferases
 histone
  nucleosome structure and
   transcriptional regulation,
   559–63
Acidification
 S. cerevisiae sphingolipids and,
  39–40
Acquired immunodeficiency
 syndrome (AIDS)
  HIV-1 life cycle and, 1–22
Actin
 small G proteins in yeast
  growth and, 307, 315–18
Activation cascade
 interferons and, 232, 238–39
Activation energy

enzymatic transition states and,
  693
Active indicator
 green fluorescent protein and,
  534–39
Activin
 TGF-β signal transduction and,
  760–62, 765–66, 767–69
Acylated proteins
 caveolae and, 202
ACT1 gene
 Golgi and endoplasmic
  reticulum transporters, 62
Adaptins
 HIV-1 life cycle and, 14
Adaptors
 AMP-activated/SNF1 protein
  kinase subfamily and, 828
 nucleocytoplasmic transport
  and, 282–88, 294–96
 ubiquitin system and, 437, 467
Additive function
 of localization elements
  RNA localization in
   development and, 371
Adenosine deaminase
 enzymatic transition states and,
  700, 707–9
Adenosine monophosphate (AMP)
 AMP-activated/SNF1 protein
  kinase subfamily and,
  821–51
Adenosine triphosphate (ATP)
 AMP-activated/SNF1 protein
  kinase subfamily and,
  821
 GroEL-mediated protein
  folding and, 581–82,
  584–85, 588–96, 598–600,
  602–3, 605–6
 nucleocytoplasmic transport
  and, 267
 nucleosome structure and
  transcriptional regulation,
  545, 567–73
 ribonucleotide reductases and,
  74, 76–78
 TGF-β signal transduction and,
  760
 transporters and, 49–66
 ubiquitin system and, 426–28,
  430–31, 444, 465, 470–71
Adenosylcobalamin
 ribonucleotide reductases and,
  71–76, 81–83, 94

S-Adenosyl-L-homocysteine
 base flipping and, 182
S-Adenosyl-L-methionine
 base flipping and, 182
S-Adenosylmethionine synthetase
 enzymatic transition states and,
  706–7
Adenoviruses
 interferons and, 238, 241, 244
ADP-ribosylating toxins
 enzymatic transition states and,
  712–16
Aequora victoria
 green fluorescent protein and,
  509–41
Aerobes
 ribonucleotide reductases and,
  71, 73–74, 76, 91
A-form geometry
 ribonuclease P and, 163
African swine fever virus
 ribonucleotide reductases and,
  92–93
Aggrecan
 matrix proteoglycans and,
  625–28
Agonists
 G protein-coupled receptor
  kinases and, 653, 659–61
Agrin
 matrix proteoglycans and, 611,
  613, 615, 621–23
AHR nuclear transporter protein
 circadian behavioral rhythms in
  Drosophila and, 138–39
Alanine
 AMP-activated/SNF1 protein
  kinase subfamily and,
  832–33, 841–42, 846
 green fluorescent protein and,
  515, 522–23
 GroEL-mediated protein
  folding and, 587, 604
 HIV-1 life cycle and, 12–13
 prion diseases and, 801
 TGF-β signal transduction and,
  769
 ubiquitin system and, 471
 von Willebrand factor and, 407
Alcaligenes eutrophus
 ribonucleotide reductases and,
  90, 92–93
Algae
 circadian behavioral rhythms in
  Drosophila and, 139

920

CUMULATIVE INDEXES

CONTRIBUTING AUTHORS, VOLUMES 63–67

CHAPTER TITLES, VOLUMES 63-67

RNA

Enzymes and Binding Proteins

Methodology

Ribozymes and Molecular Evolution

Splicing, Posttranscriptional Processing and Modification

Structure

Transcription and Gene Regulation

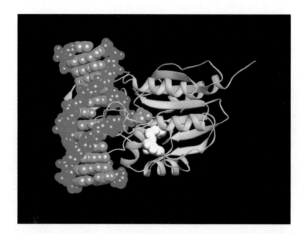

Figure 1   M.*Hha*I complexed to its substrate DNA (pdb code 1mht) (9). DNA bases are shown in *blue*, sugar-phosphate backbone in *red*, and protein in *gold*. AdoHcy is shown in *white*.

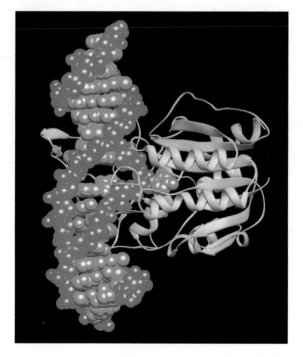

Figure 2   M.*Hae*III with DNA bound (pdb code 1dct) (11). DNA bases are shown in *blue*, sugar-phosphate backbone in *red*, and protein in *gold*.

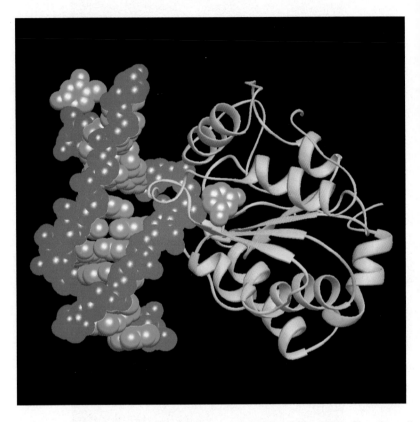

Figure 3   Human uracil DNA glysosylase complexed to DNA with a flipped uracil (coordinates provided by C Mol and J Tainer) (14). The N-C1' glycosylic bond of the flipped nucleotide has been cleaved, and the free uracil is bound in the specificity pocket. DNA bases are shown in *blue*, sugar-phosphate backbone in *red*, and protein in *gold*.

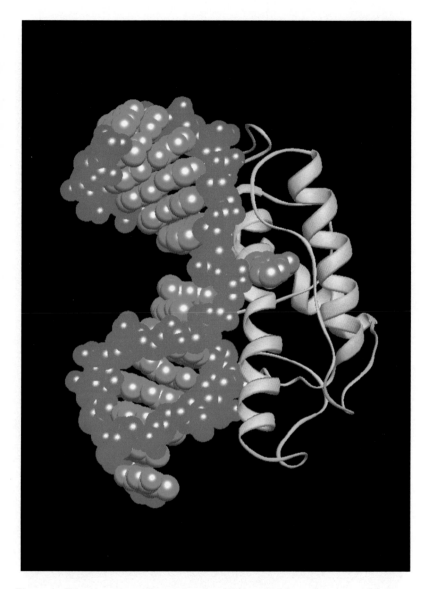

Figure 4   T4 endonuclease V complexed to DNA containing a thymine cyclobutane dimer (pdb code 1vas) (12). The adenine opposite the 5'-thymine residue of the pyrimidine dimer is flipped. DNA bases are shown in *blue*, sugar-phosphate backbone in *red*, and protein in *gold*.

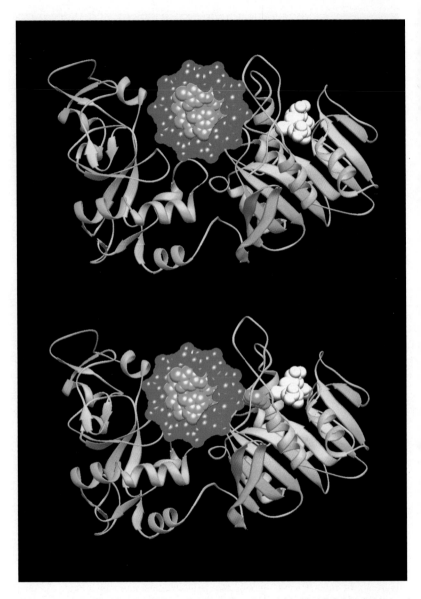

Figure 5   M.*Taq*I-DNA docking model (coordinates provided by G Schluckebier and W Saenger) (48) with (*top*) B-DNA and (*bottom*) DNA containing a flipped adenine. DNA bases are shown in *blue*, sugar-phosphate backbone in *red*, and protein in *gold*. AdoMet is shown in *white*.

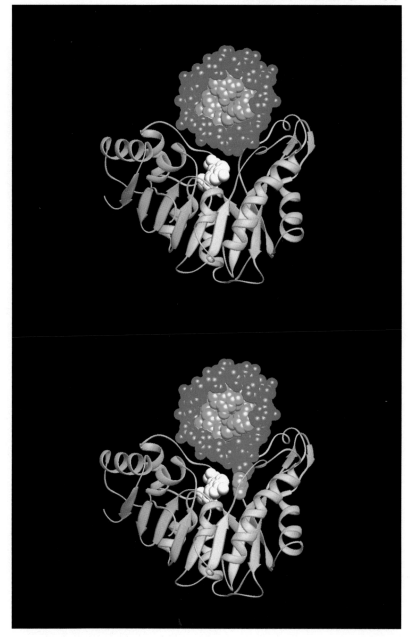

Figure 6   M.*Pvu*II-DNA docking model (40) with (*top*) B-DNA and (*bottom*) DNA with a flipped cytosine. DNA bases are shown in *blue*, sugar-phosphate backbone in *red*, and protein in *gold*. AdoMet is shown in *white*.

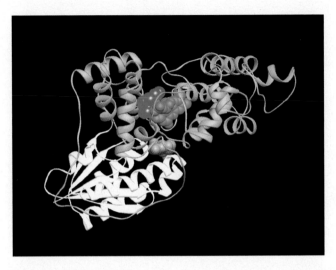

Figure 7 Ribbon diagram of *Escherichia coli* DNA photolyase (pdb code 1dnp) (41). The α/β domain is shown in *white*, methenyl-tetrahydrofolylpolyglutamate (MTHF) in *light gray*, and the helical domain in *gold*. The flavin mononucleotide group of flavin adenine dinucleotide (FAD) is shown in *green*, and the AMP unit is in *red*.

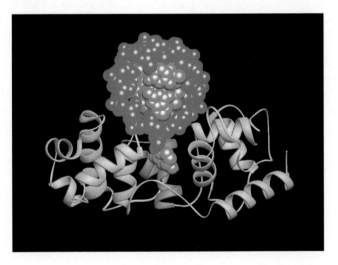

Figure 8   Escherichia coli endonuclease III–DNA docking model with a flipped nucleotide (coordinates provided by M Thayer and J Tainer) (42). DNA bases are shown in *blue*, sugar-phosphate backbone in *red*, and protein in *gold*. The HhH motif is shown in *green*.

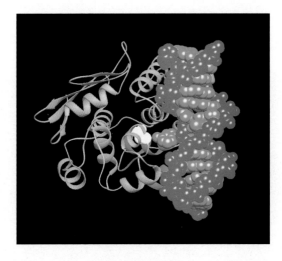

Figure 9 *Escherichia coli* 3-methyladenine DNA glycosylase II–DNA docking model with a flipped nucleotide (coordinates provided by Y Yamagata) (43). DNA bases are shown in *blue*, sugar-phosphate backbone in *red*, and protein in *gold*. The HhH motif is shown in *green*, and the catalytically essential Asp is in *white*.

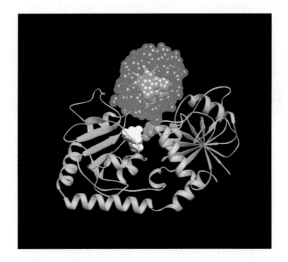

Figure 10 T4 β-glucosyltransferase (pdb code 2bgu)–DNA docking model with a flipped nucleotide (coordinates provided by P Freemont) (46). DNA bases are shown in *blue*, sugar-phosphate backbone in *red*, and protein in *gold*. Uridine diphosphoglucose is shown in *white*.

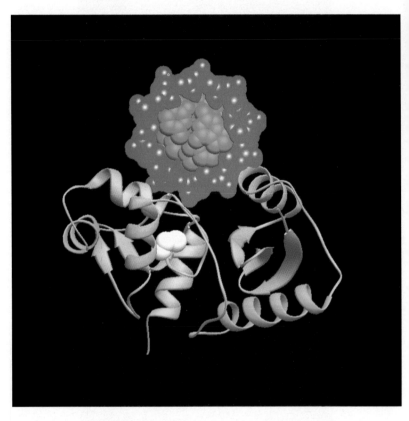

Figure 11   Ribbon diagram of ada *O*6-methylguanine DNA methyltransferase (pdb code 1sfe) docked with a B-DNA. DNA bases are shown in *blue*, sugar-phosphate backbone in *red*, and protein in *gold*. The side chain of the catalytic cysteine is shown in *white* with a *red border*.

Figure 3 Ribbon diagram of a predicted structure for PrP^Sc. The two C-terminal helices correspond to helices B and C in PrP^C. The *red* and *green* strands correspond to the region between residues 90 and 145 that are predicted to adopt a β structure in PrP^Sc. Side chains of some residues implicated in the species barrier are also included and noted to cluster on one face of the putative β-sheet.

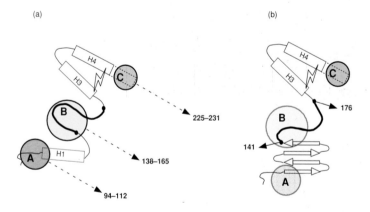

Figure 4 Schematic diagram of the secondary structure of PrP^C (residues 90–231) and a model of PrP^Sc. The distinct epitopes A, B, and C have been identified, and the boundaries of these linear epitopes mapped. Antibodies 3F4, D4, R10, and D13 recognize epitope A in PrP^C, but this epitope is cryptic in PrP^Sc. Antibodies 13A5 and R72 recognize epitope B in PrP^C and, to a lesser extent, epitope B in PrP^Sc. Finally, antibodies R1, R2, and D2 recognize epitope C in PrP^C and PrP^Sc.

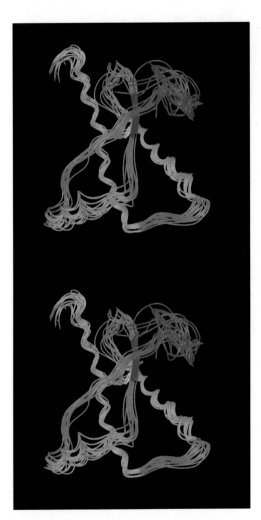

Figure 5A NMR structure of SHa rPrP (90–231). Comparison of the 15 best-scoring structures of rPrP shown with a best-fit superposition of backbone atoms for residues 113–227 (stereoview). In all figures except Figure 5B, the color scheme is disulfide between Cys[179] and Cys[214], *yellow*; sites of glycosidation in PrP[C], i.e. Asn[181] and Asn[197], *gold*; hydrophobic cluster composed of residues 113–126, *red*; helices, *pink*; loops, *gray*; residues 129–134, *green*, encompassing strand S1 and residues 159–165, *blue*, encompassing strand S2; the *arrows* span residues 129–131 and 161–163, as these show a closer resemblance to β-sheet. The structures were generated with the program DIANA (108), followed by energy minimization with AMBER 4.1 (139). Structure generation parameters are as follows: 2401 distance restraints [intraresidue, 858; sequential ($i \rightarrow i+1$), 753; ($i \rightarrow i+2$), 195; ($i \rightarrow i+3$), 233; ($i \rightarrow i+4$), 109; and ($i \rightarrow i + \geq 5$), 253 for amino acid]; hydrogen bond restraints, 44; distance restraint violations >0.5Å per structure, 30; AMBER energy, −1443±111 kcal/mol. Precision of structures: atomic RMSD for all backbone heavy atoms of residues 128–227, <1.9Å. The distance restraint violations and precision in some molecular moieties reflect the conformational heterogeneity of rPrP.

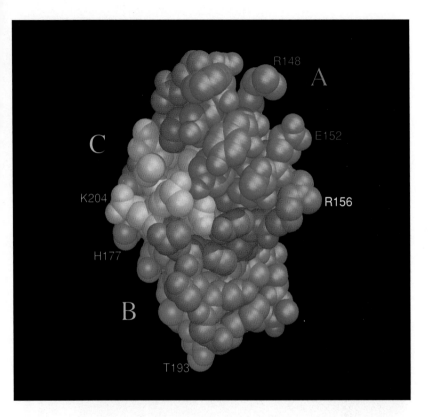

Figure 5B   van der Waals surface of rPrP turned approximately 180° from Figure 5*A*, illustrating the interaction of helix A with helix C. Helices A, B, and C are colored *magenta*, *cyan*, and *gold* respectively.

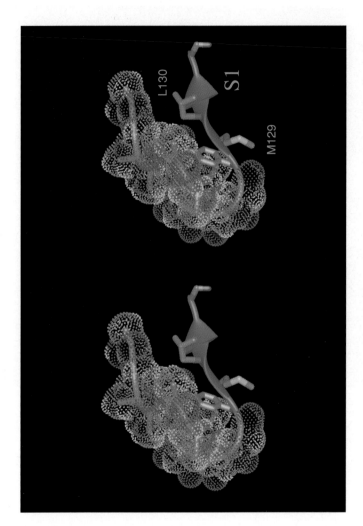

Figure 5C   Residues 113–132 illustrating (stereoview) in one representative structure the interaction of the hydrophobic cluster, with van der Waals rendering of atoms in residues 113–127, with the first β-strand. For color legend, see Figure 5A.

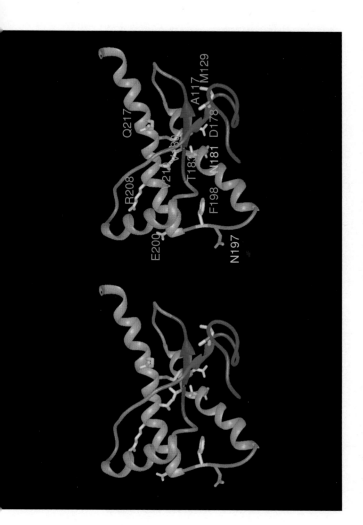

Figure 5D   Stereoview, highlighting in *white* the residues corresponding to point mutations that lead to human prion diseases. For color legend, see Figure 5A.

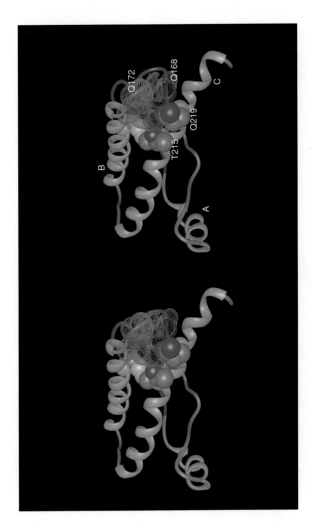

Figure 5E   Stereoview, using ribbonjr, illustrating the proximity of helix C to the 165–171 loop and the end of helix B, where residues Gln[168] and Gln[172] are depicted with a low-density van der Waals rendering and helix C residues Thr[215] and Gln[219] are depicted with a high-density van der Waals rendering.   Illustrations were generated with Midasplus.   For color legend, see Figure 5A.

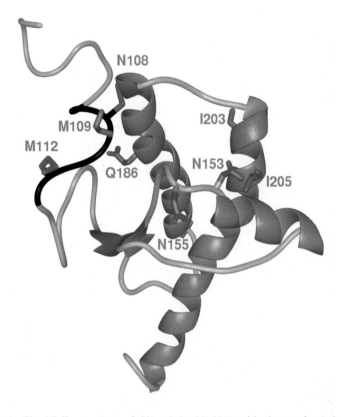

Figure 6 The NMR structure of SHa rPrP (90–231) with the conformation of residues 107–112 constrained to follow the structure of this peptide as seen in an X-ray structure of the peptide bond to an Fab fragment of the 3F4 antibody (Z Kanyo, personal communication). Residues with substantial polymorphisms across known sequences (109) that are not part of the protein X binding interface are shown. These residues form a likely site for the PrP^C:PrP^{Sc} interaction.

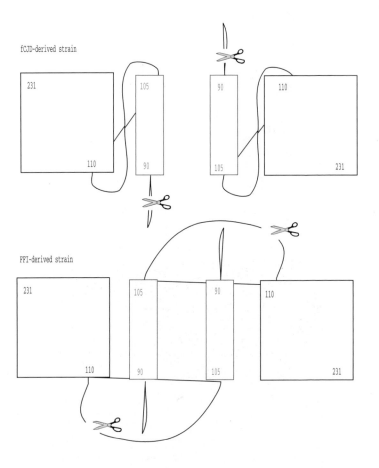

Figure 7   A cartoon of two strains of PrP^Sc formed by domain swapping. The tw
schematic structures are approximately isoenergetic but are expected to have distinc
proteolytic cleavage patterns.

ANNUAL REVIEW OF:

	INDIVIDUALS U.S.	INDIVIDUALS Other countries	INSTITUTIONS U.S.	INSTITUTIONS Other countries
ANTHROPOLOGY				
Vol. 27 (avail. Oct. 1998)	$55	$60	$110	$120
Vol. 26 (1997)	$55	$60	$110	$120
ASTRONOMY & ASTROPHYSICS				
Vol. 36 (avail. Sept. 1998)	$70	$75	$140	$150
Vol. 35 (1997)	$70	$75	$140	$150
BIOCHEMISTRY				
Vol. 67 (avail. July 1998)	$68	$74	$136	$148
Vol. 66 (1997)	$68	$74	$136	$148
BIOPHYSICS & BIOMOLECULAR STRUCTURE				
Vol. 27 (avail. June 1998)	$70	$75	$140	$150
Vol. 26 (1997)	$70	$75	$140	$150
CELL & DEVELOPMENTAL BIOLOGY				
Vol. 14 (avail. Nov. 1998)	$64	$69	$128	$138
Vol. 13 (1997)	$64	$69	$128	$138
COMPUTER SCIENCE (suspended)				
Call Customer Service or see our Web site for pricing.				
EARTH & PLANETARY SCIENCES				
Vol. 26 (avail. May 1998)	$70	$75	$140	$150
Vol. 25 (1997)	$70	$75	$140	$150
ECOLOGY & SYSTEMATICS				
Vol. 29 (avail. Nov. 1998)	$60	$65	$120	$130
Vol. 28 (1997)	$60	$65	$120	$130
ENERGY & THE ENVIRONMENT				
Vol. 23 (avail. Oct. 1998)	$76	$81	$152	$162
Vol. 22 (1997)	$76	$81	$152	$162
ENTOMOLOGY				
Vol. 43 (avail. Jan. 1998)	$60	$65	$120	$130
Vol. 42 (1997)	$60	$65	$120	$130
FLUID MECHANICS				
Vol. 30 (avail. Jan. 1998)	$60	$65	$120	$130
Vol. 29 (1997)	$60	$65	$120	$130
GENETICS				
Vol. 32 (avail. Dec. 1998)	$60	$65	$120	$130
Vol. 31 (1997)	$60	$65	$120	$130
IMMUNOLOGY				
Vol. 16 (avail. April 1998)	$64	$69	$128	$138
Vol. 15 (1997)	$64	$69	$128	$138
MATERIALS SCIENCE				
Vol. 28 (avail. Aug. 1998)	$80	$85	$160	$170
Vol. 27 (1997)	$80	$85	$160	$170
MEDICINE				
Vol. 49 (avail. Feb. 1998)	$60	$65	$120	$130
Vol. 48 (1997)	$60	$65	$120	$130
MICROBIOLOGY				
Vol. 52 (avail. Oct. 1998)	$60	$65	$120	$130
Vol. 51 (1997)	$60	$65	$120	$130
NEUROSCIENCE				
Vol. 21 (avail. March 1998)	$60	$65	$120	$130
Vol. 20 (1997)	$60	$65	$120	$130
NUCLEAR & PARTICLE SCIENCE				
Vol. 48 (avail. Dec. 1998)	$70	$75	$140	$150
Vol. 47 (1997)	$70	$75	$140	$150
NUTRITION				
Vol. 18 (avail. July 1998)	$60	$65	$120	$130
Vol. 17 (1997)	$60	$65	$120	$130
PHARMACOLOGY & TOXICOLOGY				
Vol. 38 (avail. April 1998)	$60	$65	$120	$130
Vol. 37 (1997)	$60	$65	$120	$130
PHYSICAL CHEMISTRY				
Vol. 49 (avail. Oct. 1998)	$64	$69	$128	$138
Vol. 48 (1997)	$64	$69	$128	$138
PHYSIOLOGY				
Vol. 60 (avail. March 1998)	$62	$67	$124	$134
Vol. 59 (1997)	$62	$67	$124	$134
PHYTOPATHOLOGY				
Vol. 36 (avail. Sept. 1998)	$62	$67	$124	$134
Vol. 35 (1997)	$62	$67	$124	$134
Vol. 34 (1996)	$54	$59	$54	$59
Vol. 33 (1995) and 10 Year CD-ROM Archive (volumes 24-33)	$49	$54	$49	$54
10 Year CD-ROM Archive only	$40	$45	$40	$45
PLANT PHYSIOLOGY & PLANT MOLECULAR BIOLOGY				
Vol. 49 (avail. June 1998)	$60	$65	$120	$130
Vol. 48 (1997)	$60	$65	$120	$130
POLITICAL SCIENCE *New Series!*				
Vol. 1 (avail. June 1998)	$60	$65	$120	$130
PSYCHOLOGY				
Vol. 49 (avail. Feb. 1998)	$55	$60	$110	$120
Vol. 48 (1997)	$55	$60	$110	$120
PUBLIC HEALTH				
Vol. 19 (avail. May 1998)	$64	$69	$128	$138
Vol. 18 (1997)	$64	$69	$128	$138
SOCIOLOGY				
Vol. 24 (avail. Aug. 1998)	$60	$65	$120	$130
Vol. 23 (1997)	$60	$65	$120	$130

Also Available From Annual Reviews:

	U.S.	Other	U.S.	Other
The Excitement & Fascination Of Science				
Vol. 4 (1995)	$50	$55	$50	$55
Vol. 3 (1990) 2-part set, sold as set only	$90	$95	$90	$95
Vol. 2 (1978)	$25	$29	$25	$29
Vol. 1 (1965)	$25	$29	$25	$29
Intelligence and Affectivity				
by Jean Piaget (1981)	$8	$9	$8	$9
Paperback Collections				
The Cytoskeleton			$21	$21
Genetic Flow			$21	$21
AIDS	$15	$18	$15	$18
Origins of Planets and Life	$15	$20	$15	$20
Hydrologic Processes from Catchment to Continental Scales	$15	$20	$15	$20

BACK VOLUMES ARE AVAILABLE
Visit www.AnnualReviews.org for a list and prices

Annual Reviews

BB98

A nonprofit scientific publisher
4139 El Camino Way • P.O. Box 10139
Palo Alto, CA 94303-0139 USA

STEP 1 : ENTER YOUR NAME & ADDRESS

NAME

ADDRESS

CITY _____ STATE/PROVINCE _____ COUNTRY _____ POSTAL CODE

TODAY'S DATE _____ DAYTIME PHONE

E-MAIL ADDRESS _____ FAX NUMBER

Phone 800-523-8635 (U.S. or Canada)
Orders 650-493-4400 ext. 1 (worldwide)
8 a.m. to 4 p.m. Pacific Time, Monday–Friday

FAX 650-424-0910
Orders 24 hours a day

Mention
priority code
BB98
when placing
phone orders

STEP 4 : CHOOSE YOUR PAYMENT METHOD

☐ Check or Money Order Enclosed (US funds, made payable to "Annual Reviews")

☐ Bill Credit Card ☐ AmEx ☐ MasterCard ☐ VISA

Account No. _____

Signature _____

STEP 2 : ENTER YOUR ORDER

QTY	ANNUAL REVIEW OF:	VOL.	Place on Standing Order? SAVE 10% NOW WITH PAYMENT	PRICE	TOTAL
		#	☐ Yes, save 10% ☐ No	$	$
		#	☐ Yes, save 10% ☐ No	$	$
		#	☐ Yes, save 10% ☐ No	$	$
		#	☐ Yes, save 10% ☐ No	$	$
		#	☐ Yes, save 10% ☐ No	$	$

30% STUDENT/RECENT GRADUATE DISCOUNT (past 3 years) Not for standing orders. Include proof of status.

CALIFORNIA CUSTOMERS: Add applicable California sales tax for your location. $

CANADIAN CUSTOMERS: Add 7% GST (Registration # 121449029 RT). $

STEP 3 : CALCULATE YOUR SHIPPING & HANDLING

HANDLING CHARGE (Add $3 per volume, up to $9 max. per location). **Applies to all orders.** $

SHIPPING OPTIONS:
(No UPS to P.O. boxes)

U.S. Mail 4th Class Book Rate (surface). Standard option. FREE. $ **N/C**

UPS Ground Service ($3/ volume. 48 contiguous U.S. states.) $

Please note expedited
shipping preference:
☐ UPS Next Day Air ☐ UPS Second Day Air ☐ US Airmail
☐ UPS Worldwide Express ☐ UPS Worldwide Expedited

Note option at left. We will calculate
amount and add to your total

TOTAL $

Abstracts and content lists available on the World Wide Web at
www.AnnualReviews.org. **E-mail orders: service@annurev.org**

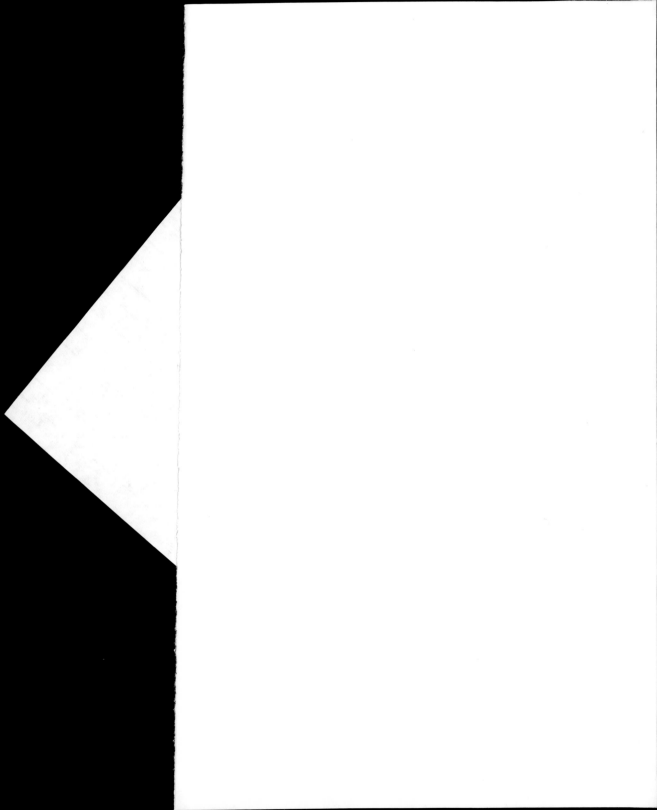

DATE DUE

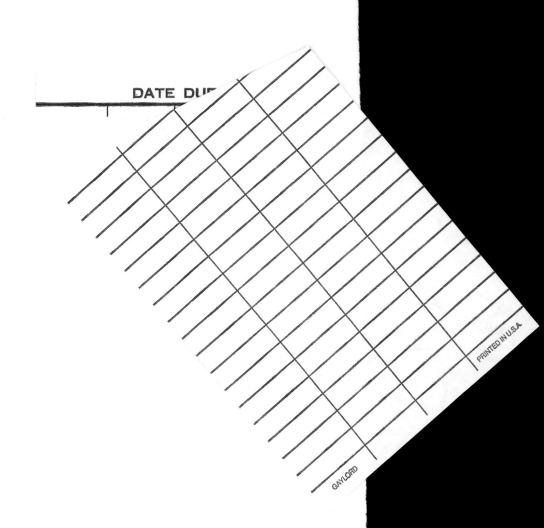

GAYLORD

PRINTED IN U.S.A.